Targeting Cellular Signalling Pathways in Lung Diseases

Kamal Dua • Raimar Löbenberg •
Ângela Cristina Malheiros Luzo •
Shakti Shukla • Saurabh Satija
Editors

Targeting Cellular Signalling Pathways in Lung Diseases

Springer

Editors
Kamal Dua
Graduate School of Health, Discipline of Pharmacy
University of Technology Sydney
Sydney, NSW, Australia

Raimar Löbenberg
Drug Development and Innovation Centre
University of Alberta
Edmonton, AB, Canada

Ângela Cristina Malheiros Luzo
Haematology and Hemotherapy Center
State University of Campinas
Campinas, São Paulo, Brazil

Shakti Shukla
Graduate School of Health, Discipline of Pharmacy
University of Technology Sydney
Sydney, NSW, Australia

Saurabh Satija
School of Pharmaceutical Sciences &
Division of Research & Development
Lovely Professional University
Phagwara, Punjab, India

ISBN 978-981-33-6826-2 ISBN 978-981-33-6827-9 (eBook)
https://doi.org/10.1007/978-981-33-6827-9

This Springer imprint is published by the registered company Springer Nature Singapore Pte Ltd.
The registered company address is: 152 Beach Road, #21-01/04 Gateway East, Singapore 189721, Singapore

Preface

Chronic lung diseases are among the leading causes of mortality and morbidity globally and are significantly associated with the economic burden as well as drastically reduced quality of life and productivity. The currently available therapies are mostly effective in managing the aggravated symptoms associated with the lung diseases, especially the non-communicable diseases such as asthma and chronic obstructive pulmonary disease (COPD). This warrants consistent rigorous research into discovering novel therapeutic targets that could be then harnessed pharmacologically to improve the quality of life of patients. This would require understanding the underlying central/key molecular mechanisms that are generally referred as "disease drivers." Various molecules that drive the disease are key immune cells, associated cytokines or chemokines, disease-associated genes, microRNAs, signalling pathways, pathogen/damage associate molecules sensing receptors and disease-associated oxidative stress molecules/pathways. Another key aspect of management of lung diseases involves lack of effective drug delivery strategies that could improve the bioavailability and subsequent efficacy of the existing/upcoming therapies.

The primary objective of this book, *Targeting Cellular Signalling Pathways in Lung Diseases*, is to comprehensively provide an overview of various cellular and molecular mechanisms involved in chronic lung diseases and then to suggest the currently available or upcoming therapeutic targets that could be efficacious in managing these diseases. The book adequately highlights a myriad of drug delivery approaches that could be developed and tested before implementing in real-life scenarios for improving the drug bioavailability and efficacy in lungs and/or target cells/tissues/pathways. Collectively, the book could be immensely informative for readers who are interested in recent developments in lung diseases in terms of novel mechanisms that could be targeted therapeutically, as well as the innovative drug delivery systems that may become mainstream in coming years for effectively treating the lung regions/cells/pathways that demonstrate associated pathology.

The book rightly begins with an elaborated introduction to the lung diseases and then expands to the comprehensive chapters covering the crucial pathological mechanisms in non-communicable respiratory diseases, i.e., asthma, chronic obstructive pulmonary disease, lung cancer, and idiopathic pulmonary fibrosis. Thereafter, the book contains comprehensive chapters on microbial infections of the lung, such as tuberculosis, Haemophilus influenzae, and viral infections (including the novel Coronavirus Disease 2019). This section is followed by chapters comprising discussions around the master molecular regulators of lung diseases, including, phosphoinositide 3-kinase, the cholinergic system, Nrf2-Keap1 signalling pathway, Toll-like receptor (TLR). There is a dedicated subsection on the importance of plant-based or natural compounds that could be potential in the management of lung diseases. The final section of book exclusively summarizes the prospects and challenges in developing and validating various drug delivery systems for respiratory diseases. The chapters in the book have extensive visual illustrations that make it easier for readers to understand the complex disease mechanisms.

We believe that this book will be a valuable resource for academicians, researchers, and industry engaged in the field of respiratory biology. It is also a valuable resource for translational researchers, graduates, and postgraduates (Master's, PhD, and Post-Doctoral Researchers) of various disciplines including Pharmaceutical Sciences, Biotechnology, Immunology, and Medical and Health Sciences.

The editorial team has extensive research experience in the field of respiratory diseases, respiratory infections, lung inflammation, and drug delivery systems for respiratory system. The editors of this book would like to express their sincere gratitude to all the authors for their time and for their valuable contributions in the production of this book.

Sydney, NSW, Australia — Kamal Dua
Edmonton, AB, Canada — Raimar Löbenberg
Campinas, São Paulo, Brazil — Ângela Cristina Malheiros Luzo
Sydney, NSW, Australia — Shakti Shukla
Phagwara, Punjab, India — Saurabh Satija

Acknowledgment

The publication of this book was finalized during the coronavirus (COVID-19) pandemic. We would like to dedicate this book to all those who were affected by the pandemic and, in particular, to our health workforce around the world for their dedication and care during this difficult time.

Contents

About the Editors

Kamal Dua is a Senior Lecturer in the Discipline of Pharmacy at the Graduate School of Health, University of Technology Sydney (UTS), Australia. He has research experience of over 12 years in the field of drug delivery systems targeting inflammatory diseases. Dr. Dua is also a Node Leader of Drug Delivery Research in the Centre for Inflammation at Centenary Institute/UTS, where the targets identified from the research projects are pursued to develop novel formulations as the first step towards translation into clinics. Dr. Dua researches in two complementary areas; drug delivery and immunology, specifically addressing how these disciplines can advance one another helping the community to live longer and healthier. This is evidenced by his extensive publication record in reputed journals. Dr. Dua's research interests focus on harnessing the pharmaceutical potential of modulating critical regulators such as interleukins and microRNAs and developing new and effective drug delivery formulations for the management of chronic airway diseases. He has published more than 80 research articles in peer-reviewed international journals and authored or co-authored four books. He is an active member of many national and international professional societies.

Raimar Löbenberg holds a BS in Pharmacy from the Johannes Gutenberg-University in Mainz, Germany, and received his PhD in Pharmaceutics from the Johann Wolfgang Goethe-University. His research interests are in Biopharmaceutics and inhalable nanoparticles to treat lung diseases. He is a founder and Director of the Drug Development and Innovation Centre at the University of Alberta. He was president of the Canadian Society for Pharmaceutical Sciences 2014–2015 and a member of the United States Pharmacopeia Dietary Supplement Expert Committee. He is a Vice-chair of the Specialty Committee of Traditional Chinese Medicine in Pharmaceutics of the World Foundation of Chinese Medicine Science. He is a member of the Health Canada Scientific Advisory Committee on Pharmaceutical Sciences and Clinical Pharmacology.

Ângela Cristina Malheiros Luzo is a full Professor from the Surgical Science Post Graduation Discipline at the Faculty of Medical Science Surgery Department, UNICAMP. She is an active researcher at the Biology Institute of the State University of Campinas (UNICAMP). She served as a member of the liver transplantation team and was responsible for the installation of the Umbilical Cord Blood Bank (BSCUP) at UNICAMP. She obtained her medical degree from Faculty of Medical Sciences of Jundiaí and holds Master's and PhD degrees in the field of biology and cell therapy of umbilical cord blood stem cells from the Clinical Medicine Department of the Faculty of Medical Science, UNICAMP. She has three licensed patents related to fibrin glue that allows the attachment of mesenchymal stromal stem cells in suture filaments providing a complete regeneration of enterocutaneous fistula, tracheal injury, and correction of urethral stenosis in animal models (rat, pig, and rabbit).

Shakti Shukla is a trained microbiologist and completed his PhD in Medical Studies from the University of Tasmania. Dr. Shukla has been actively involved in various aspects of chronic respiratory diseases, including the pathophysiology and immunology of respiratory diseases especially cigarette smoking-related chronic obstructive pulmonary disease (COPD), lung cancer, and asthma. Dr. Shukla's primary research interest involves the crucial role of microorganisms, in particular bacteria, in the development, progression, and exacerbations of COPD/asthma. His more recent research focuses on understanding the role of gastrointestinal microbiomes in the progression of COPD and whether the gut microbes could be utilized as a potential treatment for COPD. Dr. Shukla has published more than 65 publications in the last 6 years and has won Young Investigator Award from Thoracic Society of Australia and New Zealand (Tasmania Branch) for his research.

Saurabh Satija is currently working as Assistant Dean at School of Pharmaceutical Sciences and Division of Student Welfare, Lovely Professional University (LPU), India, and also an academic with Discipline of Pharmacy, Graduate School of Health, University of Technology, Sydney, Australia. He has extensive experience in the field of natural products research, drug development, analytical method development, and nanotechnology-based novel drug delivery systems. He has received various national and international awards, scholarships, and research grants. He also carries an impressive bibliography of scientific papers published in journals of international repute. He has more than 100 research publications and two patents published to his credit.

Introduction to Lung Diseases

1

Shivraj Popat Jadhav, Himmat Singh, Salman Hussain, Ritu Gilhotra, Anurag Mishra, Parteek Prasher, Anand Krishnan, and Gaurav Gupta

Abstract

Lungs come in contact with external environmental factors such as smoke, pollens, microbes, dust, environmental chemicals, and various other pollutants continuously. Exposure of lungs to these factors increases the risk of lungs to develop various diseases and disorders. In the United States, lung diseases are the third top reason of deaths. In infants, the majority of deaths are due to lung diseases and other breathing problems. Lung diseases may affect the whole respiratory system including the trachea, bronchi, bronchioles, alveoli, pleurae, pleural cavity, and nerves and muscles responsible for respiration or part of the respiratory system. Respiratory diseases make gas exchange difficult in air-breathing animals. This can lead to mild and self-limiting to life-threatening conditions. Lung diseases include asthma, chronic obstructive pulmonary disease (COPD), idiopathic pulmonary fibrosis (IPF), lung cancer, and various bacterial as well as viral infections like influenza, pneumococcal, respiratory syncytial virus, etc. These conditions if are not treated well within time can lead to serious

S. P. Jadhav
Divine College of Pharmacy, Nashik, Maharashtra, India

H. Singh · R. Gilhotra · A. Mishra · G. Gupta (✉)
School of Pharmacy, Suresh Gyan Vihar University, Jaipur, India

S. Hussain
Department of Pharmaceutical Medicine, School of Pharmaceutical Education and Research, Jamia Hamdard, New Delhi, India

P. Prasher
Department of Chemistry, University of Petroleum & Energy Studies, Dehradun, India

A. Krishnan
Department of Chemical Pathology, School of Pathology, Faculty of Health Sciences and National Health Laboratory Service, University of the Free State, Bloemfontein, South Africa

K. Dua et al. (eds.), *Targeting Cellular Signalling Pathways in Lung Diseases*,
https://doi.org/10.1007/978-981-33-6827-9_1

health complications and may often produce life-threatening situations. The quality and productivity of the life of the patient are severely hampered because of respiratory diseases due to the related symptoms like breathlessness and chronic cough. This chapter introduces different kinds of respiratory disorders along with their etiology, epidemiology, pathophysiology, sign and symptoms, and treatment.

Keywords

Respiratory disorders · Respiratory infections · Etiology · Epidemiology · Lung

Lungs do continuous work of gaseous exchange to fulfill the body's need for oxygen. For this purpose, oxygen from the external environment is inhaled and carbon dioxide is exhaled continuously. Along with oxygen, various other environmental factors are also inhaled inside the lungs such as smoke, pollen grains, microorganisms, dust particles, chemicals, and various other pollutants which at all are not necessary for the human body. Exposure to these unnecessary factors may lead to several lung diseases and disorders such as asthma, chronic obstructive lung disease (COPD), lung cancer, and many more. Breathlessness, chronic cough, and hypoxia are some of the complications associated with lung diseases which severely hamper the quality and productivity of a patient's life [1].

Patients suffering from lung diseases such as asthma or COPD are affected in multiple ways in their day-to-day life. While the effect of lung diseases is different for each patient, the main effects include difficulty in breathing, reduced day-to-day activities, lowered self-esteem, lack of energy, disturbed relations, nervousness, and depression [2, 3]. About 40% of patients with COPD suffer from depression [4] and about 34% of patients suffer from anxiety [2]. There is a high incidence of anxiety and depression in patients suffering from asthma. The major five lung diseases across the world include chronic obstructive pulmonary disorder (COPD), asthma, acute respiratory infections, tuberculosis, and lung cancer [5]. This chapter will address epidemiology and causes, etiology, pathophysiology, and treatment in short of each disease.

1.1 Chronic Obstructive Pulmonary Disorder (COPD)

COPD is a reversible disease of the lungs and is one of the major causes of mortality and morbidity throughout the world. In the United States, after heart disease, cancer, and cerebrovascular diseases, COPD is the fourth leading cause of death [6, 7]. As projected, COPD may become the third leading cause of death worldwide [5]. COPD is a common term generally used for various medical conditions that are responsible for impairment of airflow into the lungs which are not fully reversible [8]. Hence, COPD could be better seen as a clinical syndrome identified as chronic respiratory symptoms, pulmonary abnormalities like airway obstruction, emphysema, or airflow limitation [9]. Patients suffering from COPD will be at a higher risk of death due to the development of coexisting conditions than patients without COPD [10, 11].

1.1.1 Definition

Several different definitions of COPD are present as per the different medical societies. Some of them are as follows:

The American Thoracic Society (ATS) defined COPD as "a disease state characterized by the presence of airflow limitation due to chronic bronchitis or emphysema" [12]. According to the European Respiratory Society (ERS), COPD is "reduced maximum expiratory flow and slow forced emptying of the lungs, which is slowly progressive and mostly irreversible to present medical treatment" [13]. The Global Initiative for Chronic Obstructive Lung Disease (GOLD) defined COPD as "a disease state characterized by airflow limitation that is not fully reversible. The airflow limitation is usually both progressive and associated with an abnormal inflammatory response of the lungs to noxious particles or gases" [14].

1.1.2 Epidemiology

Global estimation of COPD is difficult as different methods are used for calculation of prevalence [15]. About 174 million cases are present globally according to the Global Burden of Disease Study 2015 [16]. According to population-based health administrative data, there is a risk of 28% of getting COPD by the age of 80 in Canada [17]. There is a maximum increase in COPD cases between 1990 and 2010 of about 119% in the Eastern Mediterranean and 102% in African regions [18]. COPD is underdiagnosed in low-income countries, in younger patients, in smokers, and in lower education groups [19]. COPD is more common in men than in women, but due to increase in smoking among women in high-income countries and exposure to different pollutants like biomass fuel in low-income countries, the risk of COPD in women also increases. Low- and middle-income countries contributed about 90% of deaths worldwide [20].

1.1.3 Causes

Globally, tobacco smoking remains one of the main causes of COPD [21]. The World Health Organization evaluates that about 73% of COPD mortality in high-income countries and about 40% mortality of COPD in low-income countries are due to smoking [22]. Genetic factors may also be responsible for COPD as indicated by the study. The deficiency of enzyme serine protease $\alpha 1$ antitrypsin is related to COPD in 1–3% of patients. The genetic risk of COPD increases if the patient is a smoker or is exposed to other pollutants [23]. Exposure to various chemicals, pollutants, dust, vapors, and fumes at workplaces is also responsible for COPD.

The incidence of COPD related to the workplace is more in countries where laws are not stringent regarding exposure to pollutants [24]. Exposure to indoor air pollutants such as biomass fuel like coal, dung, straw, wood, and crop residues may also increase the risk of COPD. According to WHO, 35% of cases of COPD from low- and middle-income countries is because of exposure to biomasses [22]. Other causes for COPD are related to age, various infections, asthma, and socioeconomic factors (Fig. 1.1).

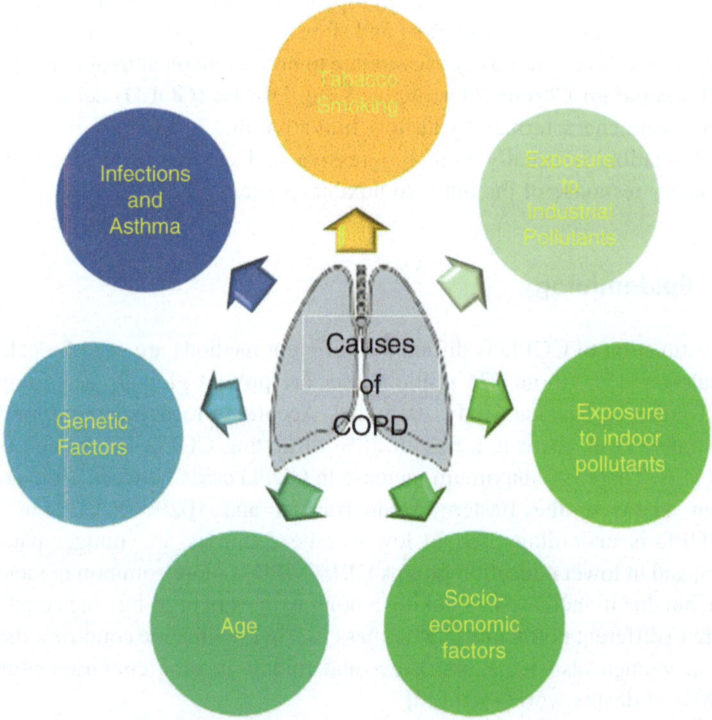

Fig. 1.1 Causes of COPD

1.1.4 Symptoms (Fig. 1.2)

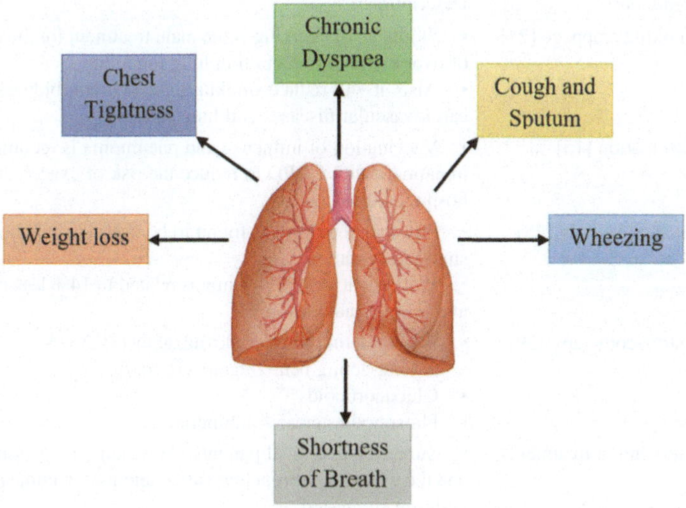

Fig. 1.2 Symptoms of COPD

1.1.5 Pathophysiology

Tobacco smoking is the major cause behind COPD. Harmful agents present in tobacco smoke cause injury to the airway epithelium tissue and start the main process which leads to airway inflammation and obstruction. Other changes include increased oxidative stress and protease-antiprotease imbalance. In COPD small airways and lung, parenchyma tissues are more affected, and the destruction of the proximal airway epithelium is less. Walls of the small conducting airways are thickened and obstruction worsens because of mucus exudates. Oxidative stress is increased because of bronchial inflammation involving phagocytes like neutrophils and macrophages. Squamous cell metaplasia and goblet cell hyperplasia are induced because of smoking. There is an increase in smooth muscle mass [25]. Other major changes observed are mucus hypersecretion, ciliary dysfunction, hyperinflation, abnormalities in gaseous exchange, and other systemic effects [26].

1.1.6 Treatment

The damage done to the lungs in COPD is irreversible, but by the use of several medications and self-discipline, COPD can be lowered. The following table summarizes several approaches (Table 1.1):

Table 1.1 Various approaches for the treatment of COPD

Sr. no.	Treatment	Description
1	Smoking stoppage [27]	• Stoppage of smoking is the main treatment for the cessation of over the time reduction in lung functions • Also, it will reduce smoking-related comorbidities like cardiovascular diseases and lung cancer
2	Vaccination [15]	• Vaccination of influenza and pneumonia is recommended in patients with COPD to reduce the risk of exacerbations and hospital admission
3	Physical activity [28]	• Physical activity was found to be useful as much as smoking cessation • Walking at least for 15 min is related to 14% less morbidity and hospital admission
4	Pharmacotherapy [29]	• A long-acting muscarinic antagonist (LAMA) • A long-acting beta-2 agonist (LABA) • Glucocorticoids • Phosphodiesterase-4 inhibitors
5	Intervention treatments	• Surgeries in critical patients like volume reduction surgery and the use of endobronchial valves and lung volume reduction coils are suggested
6	Oxygen and ventilation support [30]	• Patients with severe resting hypoxia are recommended oxygen support

1.2 Asthma

Asthma is among the most common chronic types of disease affecting 334 million people throughout the world [31]. It is a more common chronic type of disease in childhood [32]. Asthma is considered as a disorder localized to the lungs, but new studies indicate a component of systemic airway disease which involves the entire respiratory tract and often associated with other disorders like allergic rhinitis [33]. Asthma control remains difficult although there is an increase in diagnosis and management over the past decade. Control of asthma in a substandard way may lead to unnecessary morbidity and limitation so of day-to-day activity of the patient and overall degradation of life of the patient [31].

1.2.1 Definition

Asthma is defined as a chronic inflammatory disease of airways. Inflammation is associated with hyperresponsiveness of airways due to specific triggers like viruses, allergens, or exercise which leads to repetitive episodes of chest tightness, wheezing, breathlessness, and coughing. These episodes often cause airflow obstruction and may go away spontaneously or with appropriate treatment [33].

1.2.2 Epidemiology

The global occurrence of physician-diagnosed asthma in adults is about 4.3%. This occurrence has a large variation between country to country and age groups. The occurrence of asthma is high in developed countries but low in developing countries [34]. Epidemiological data may be underestimated in resource-poor countries where patients do not get basic medications for asthma. The occurrence of asthma is higher in boys as compared to girls, but the interesting fact is that the chances of a woman getting asthma are more than 20% as compared to men which may indicate a switch in prevalence during puberty [35]. Reemission of asthma is more in boys and patients who are sensitized to furred animals. However, certain factors like obesity and tobacco smoking may lead to the development of asthma [36]. Most countries suffering from asthma are low- and middle-income countries as diagnosis and treatment are poor which may lead to significant mortality. However, from the past 25 years, there is a significant reduction in overall mortality because of increased use of inhaled corticosteroids.

1.2.3 Causes (Fig. 1.3)

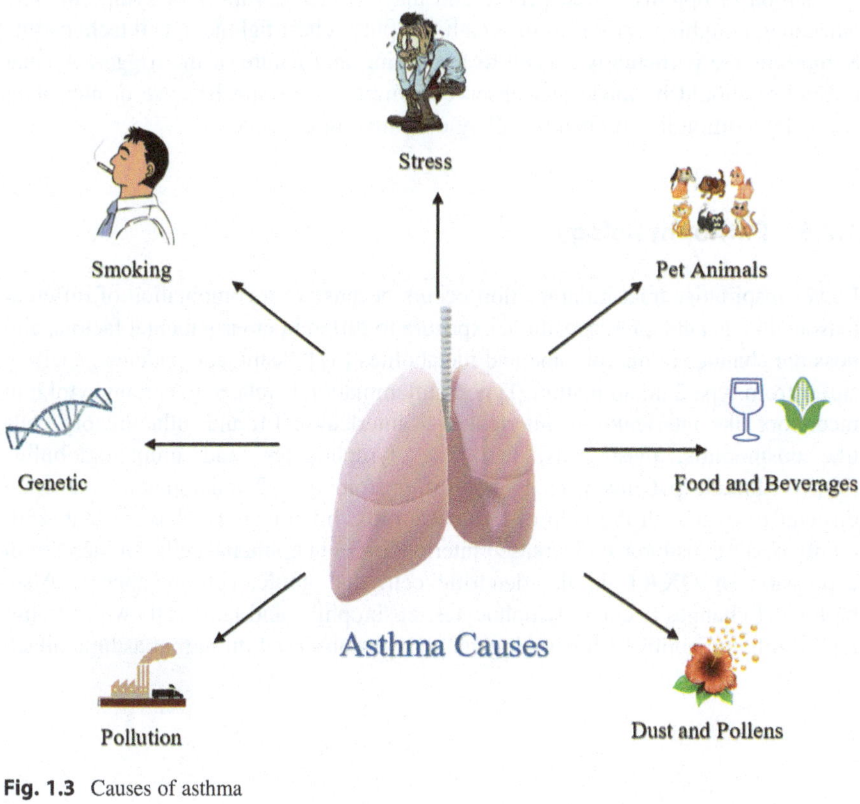

Fig. 1.3 Causes of asthma

1.2.4 Symptoms (Fig. 1.4)

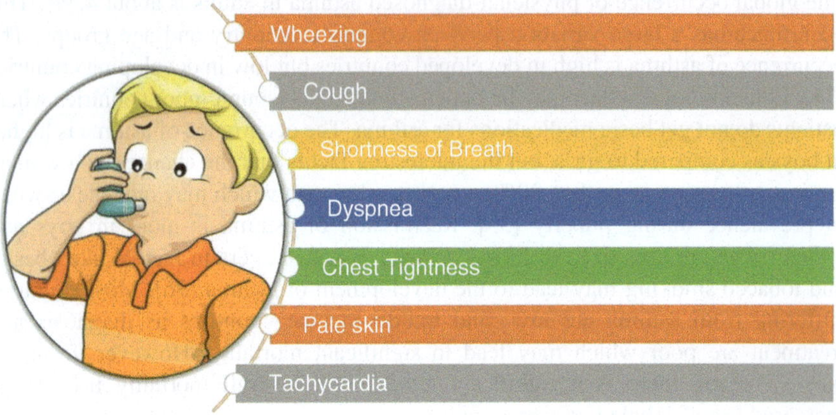

Fig. 1.4 Symptoms of asthma

Asthma symptoms are unspecific and may include a variety of symptoms like wheezing, coughing, shortness of breath, dyspnea, chest tightness, and tachycardia. Symptoms are particularly related to the timing and nature of the trigger. A clear difference should be made that observed symptoms are not because of any other clinical condition like bronchitis, allergic rhinitis, or eczema.

1.2.5 Pathophysiology

Lower respiratory tract inflammation occurs because of a combination of different reasons like genetic predisposition, exposure to different environmental factors, and possible changes in microbiome and metabolites [37]. Asthmatic patients generally suffer from type 2 inflammation. Type 2 inflammation is related to certain cytokinin mediators like interleukin-4, interleukin-5, interleukin-14, and inflammatory cells like eosinophils, mast cells, basophils, lymphocytes, and immunoglobulins [38]. Asthmatic patients who do not suffer from type 2 inflammation are often difficult to treat with the help of corticosteroids. Inflammatory changes that commonly occur in asthma are increased interleukin-33 in epithelial cells, an increase in expression of OX40L in the dendritic cell, and goblet cell metaplasia. Also, biological changes occur in lymphocytes, eosinophils, and mast cells which cause type 2 inflammation. Histopathological changes observed during an asthma attack include hypertrophy of smooth muscles, hyperplasia of goblet cells, fibrosis of subepithelial tissues, and deposition of collagen, increased number of blood vessels in the submucosa, and edema of the submucosa [39].

1.2.6 Types of Asthma [40]

It is important to diagnose the type of asthma for proper treatment.

(a) **Allergic asthma**
Episodes of allergic asthma may arise if asthmatic patients come in contact with common allergens like dust, pollutants, pollen grains, or animal hairs. When these allergens enter into the respiratory system, a series of events start that lead to an asthma attack. Keeping away from these trigger factors is the best practice to avoid asthma episodes.

(b) **Exercise-induced asthma**
Exercise or physical exertion may lead to the induction of exercise asthma. Airway narrowing is maximum after 5–20 min of starting exercise. Such patients may be prescribed asthma inhalers before the start of the exercise.

(c) **Occupational asthma**
This type of asthma may trigger from the workplace environment. The patient may suffer from asthma symptoms like difficulty in breathing because of various trigger factors present at the job. Some typical jobs which may cause occupational asthma are hairdressers, painters, and woodworkers.

(d) **Nocturnal asthma**
Patients having asthma may experience episodes of asthma during nighttime. This is because of the circadian rhythm. This type of asthma attack can be dangerous and may be related to changes in the body such as hormonal secretion, cooling of airways, and sleeping position. Treatment of such type of asthma often includes medications before going to bed.

1.2.7 Treatment

Treatment of asthmatic patients includes both preventive and curative measures. Initial treatment at the start of asthma symptoms is the best way of asthma management [41]. Patients with a high risk of death require special attention in terms of monitoring, self-care, and education. The following points should be considered while battling with asthma [42]:

- Patients and family members should be made aware of the severity of asthma.
- A written home plan for tackling early symptoms of asthma.
- Making patients aware of necessary skills for recognizing early symptoms of asthma.
- Making patients aware of the importance of communication with a physician about any serious complication of asthma.

Pharmacological management of asthma includes the use of various chemical agents like inhaled β2-agonists like salbutamol, anticholinergics like ipratropium bromide, methylxanthines such as theophylline, corticosteroids like methylprednisolone, epinephrine, and terbutaline. Other treatments may include the use of antibiotics in case of infection. Mechanical ventilation along with sedation can be used for patients with status asthmaticus. For sedation purposes, midazolam, morphine sulfate, ketamine, and propofol are prescribed [43].

1.3 Lung Cancer

Lung cancer is the most fatal malignancy in the modern world and responsible for over 40,000 deaths in the United Kingdom per year [44]. Smoking tobacco remains one of the leading causes of lung cancer. From studies, it is indicated that only 10% of smokers develop lung cancer; hence, other factors such as genetic and dietary risk factors are also responsible. As compared to other types of cancers like breast cancer, prostate cancer, and cervical cancer, lung cancer is diagnosed at an early metastatic stage. The maximum incidence of lung cancer occurs between 75 and 80 years of age in the United Kingdom. Over 50% population of 500,000 diagnosed with lung cancer is above 70 years and above [44]. The high hope of lung cancer cure is when it is detected in the early stages. Radiation therapy along with chemotherapy plays an important role in the treatment of lung cancer [45]. Lung cancer is of two types: non-small cell carcinoma (NSCC) and small cell lung carcinoma (SCLC). NSCC accounts for 80% of cases, while SCLC accounts for the remaining 20% [46].

1.3.1 Classification

WHO classifies lung cancer as follows [46] (Fig. 1.5):

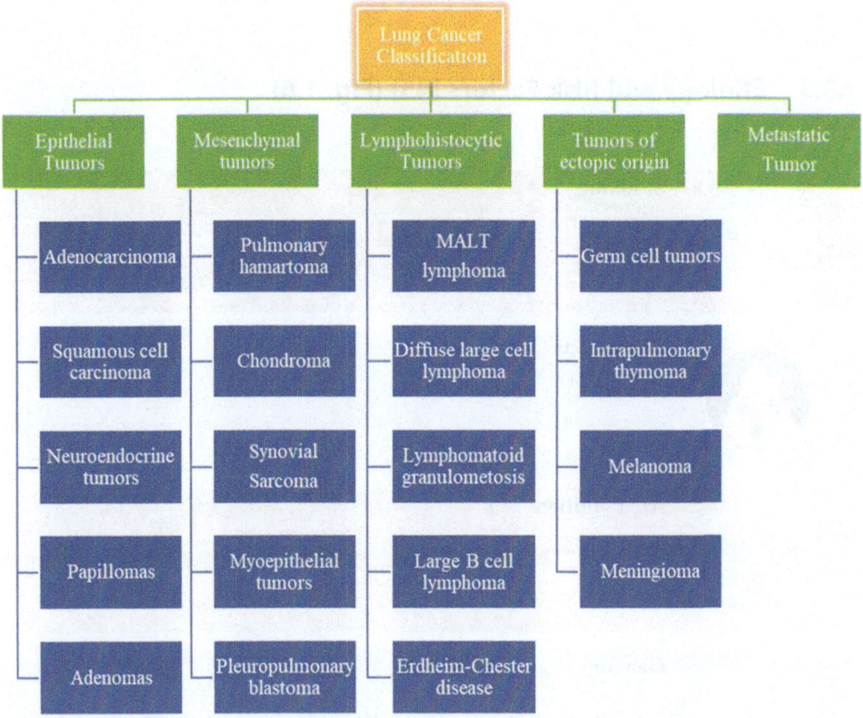

Fig. 1.5 Classification of lung cancer

1.3.2 Epidemiology

Six million new cases of lung cancer were diagnosed in the year 2008 which is about 12.7% of the world's total cancer incidence [47]. In the United States, lung cancer is accountable for 160,340 deaths in the year 2012, which is about 28% of all cancer deaths. There are marked regional variations in the global geographic distribution of lung cancer. Lung cancer is the commonest type of cancer among men throughout the world and the number four cancer type in a woman. The highest rate of occurrence in men is from North America, East Asia, and central-eastern and southern Europe. In underdeveloped countries like West Asia, South Africa, and the Caribbean, the rate of incidence of lung cancer is high [47]. Data of Surveillance, Epidemiology, and End Results (SEER) from 2004 to 2008 indicated the common age for diagnosis of cancer is 71 years, while no cancer is detected in men below 20 years of age [48]. Around 0.2% of lung cancer incidents are from age 20 to 34, 1.5% between age 35 and 44, 8.8% from age 45 to 54, 21% between age 55 and 64, and a maximum 31% between age 65 and 74. A huge advancement in diagnostic, surgical, and curative methods fails to increase the overall survival rate in the last 5 years which is 15.6% in the United States, and globally, it is worse in developing countries which is around 8.9% [49].

1.3.3 Etiology and Risk Factors [45] (Fig. 1.6)

Fig. 1.6 Causes of lung cancer

1.3.4 Sign and Symptoms

No specific signs and symptoms are there which can conclude that a person is suffering from lung cancer [50]. Lung cancer may have similar signs as that of a smoker person or person suffering from other disorders like upper respiratory tract infection. Signs and symptoms are related to the size and location of the tumor.

Common symptoms seen in a lung cancer patient are cough, wheezing, dyspnea, and hemoptysis. Symptoms may worsen as the tumor increases in size and starts to spread. Blood-filled sputum may be observed as the condition becomes worse. Additional symptoms may include edema in the face, neck, and shoulders. Shoulder and arm pain are common [45].

1.3.5 Diagnosis

Complete medical, as well as personal history, along with regular physical examination can help to diagnose lung cancer in an early stage. Smokers, who are at high risk of lung cancer, suffer from cough and produce sputum regularly [50]. Any change in the amount of sputum is important along with symptoms like shortness of breath, chest pain, and blood in sputum or frequent respiratory infection. Chest radiography is one of the first tests which can be performed for a possible diagnosis. Bronchoscopy is another technique that uses a brush or forceps to obtain a sample from lesions. Drawback related to bronchoscopy is that it can only detect lesions present in the airways but fails to detect lesions deep within the bronchial tree. Fine-needle biopsy of the lesion is another technique that can produce 95% positive results. Mediastinoscopy and thoracotomy are other diagnostic procedures for the detection of lung cancer.

1.3.6 Staging of Lung Cancer

Two main systems are designed for an indication of lung cancer severity. One of the systems is the "TNM" system which denotes lung cancer depending upon T, primary tumor, N, the extent of involvement of the lymph node; and M, distant metastases. Each letter is designated with a particular number which denotes the advancement of tumors. Based on TNM, characteristic numbers are designated from 0 to 4. NSCC can be staged with the help of the TNM system [51, 52]. Veterans Affairs lung cancer classification system is used for the staging of SCLC. The disease is classified into two stages in this system, limited and extensive [53].

1.3.7 Treatment [54]

Depending upon the stage and type of lung cancer, there are different treatments. Treatment differs in small cell lung cancer and non-small cell lung cancer. The most common type of lung cancer is non-small cell lung cancer and adenocarcinoma is the commonest subtype. Treatment of this type of lung cancer can be done with surgery, chemotherapy, radiation therapy, or targeted therapy. The type of treatment will depend upon the type and stage of cancer and if the patient has any other health conditions. The following table gives an idea about treatments used for non-small cell lung cancer depending upon stages (Table 1.2).

Table 1.2 A treatment option for non-small cell lung cancer

Stage	Treatment
I	– The first treatment is the removal of the lesion by surgery – If surgery is not possible, then radiation
II	– Surgery – Radiation if individual can't undergo surgery along with adjuvant chemotherapy
IIIA	– Surgery, chemotherapy, and radiation
IIIB	– Surgery, chemotherapy, and radiation
IV	– Chemotherapy or targeted therapy and for symptomatic relief palliative radiation treatment can be used

Table 1.3 Different types of lung surgeries

Name of surgery	Description
Wedge resection	Removal of an infected small piece of the lung
Segmentectomy	Removal of a larger part of the lung but not the whole lung
Lobectomy	Removal of the complete lobe of the lung
Pneumonectomy	The entire right or left lung is removed

Surgery involves the removal of the lung in part or full. Depending upon it, lung surgery is of different types (Table 1.3).

For the treatment of small cell lung cancer, if diagnosed in an early stage, surgery can be an option. But in the case of advanced stages, combination of chemotherapy and radiation is used. For prevention of spread to the brain, prophylactic radiation to the brain is used. There are several side effects of radiation and chemotherapy like nausea, vomiting, alopecia, fatigue, increased infection risk, anemia, kidney damage, and nerve damage.

1.4 Acute Respiratory Tract Infections

Acute respiratory tract infections are classified depending upon the area of infection, i.e., upper respiratory tract infection (URI) and lower respiratory tract infection (LRI). Different organs like nostrils, larynx, sinuses, and the middle ear are the parts of the upper respiratory tract. The lower respiratory tract consists of the trachea, bronchi, bronchioles, and alveoli. Acute respiratory tract infections are the cause of illness and mortality in children all over the world [55, 56]. But the majority of cases are found in high- and middle-income countries. As a lack of specific etiological factors, severity of LRIs has become more worse in developing countries for children below 5 years of age. Around 10.8 million children die annually [57]. According to a study, 1.9 million children died because of ARIs. Among them, 70% were from Africa and Southeast Asia [58]. According to the WHO, two million children are susceptible to die because of pneumonia below the age of 5 [59].

1.4.1 Upper Respiratory Tract Infections

These infections are the most common and include common cold (also termed as the common cold), ear infection, sinusitis, pharyngitis, epiglottitis, and laryngitis. Among these ear infections, pharyngitis may cause more complications like deafness and rheumatic fever. The main cause of URIs is viruses. Twenty-five to thirty percent of URIs are because of rhinovirus; 25–35% of UTIs are because of influenza, parainfluenza, metapneumovirus, and adenovirus; coronavirus accounts for 10%; and the remaining is because of anonymous reasons [60].

1.4.1.1 Acute Pharyngitis

This term is used to indicate common cold and other disorders that occurred due to viral rhinitis. It is among the most common type of infection in childhood. Children of age younger than 5 years may suffer five to eight recurrent occurrences per year. Various viruses such as rhinovirus, coronavirus, influenza virus, adenovirus, and coxsackievirus are responsible for pharyngitis [61]. Inflammation of the nasal mucosa may lead to obstruction of Eustachian tubes and can cause secondary bacterial infection. More complications can lead to acute childhood asthma [62].

Symptoms

- Runny nose
- Cough and fatigue
- Pharyngitis
- Fever and chills
- Diarrhea
- Vomiting
- Abdominal pain and headache

Mode of Transmission

- Viruses may get transmitted in the form of droplets by coughing or sneezing or by a handshake.

Communicability

- Infection is communicable in closed and semi-closed areas like school and daycare centers.

Incubation period

- Varies from 2 to 10 days

Treatment

- Rest, hydration, and air humidification
- Nasal hygiene and decongestion
- Antipyretics and pain killer
- Oral antitussive and antihistaminic drugs
- Antimicrobial agents

1.4.1.2 Acute Sinusitis

It is a bacterial infection of paranasal sinuses which can remain up to 30 days. After 30 days, symptoms resolve completely on their own. In acute sinusitis, maxillary and ethmoid sinuses are more infected. Bacteria such as *S. pneumoniae* and *H. influenzae* are responsible for acute sinusitis. Some viruses may also be associated with sinusitis [63]. If an infection is not treated well, then possible complications may include meningitis, thrombosis of the cavernous sinus, and brain abscess.

Sign and symptoms

- Cough in the daytime and worsen at night
- Fever
- Eyelid edema, headache, discomfort, and pain at the infection site or teeth [61]

Diagnosis

- Clinical history with a physical examination
- Hemogram
- Nasal swab study
- X-ray [64]
- Compound tomography
- Nasal endoscopy

Treatment

- Rest and air humidification
- Painkiller and antipyretic
- Topical or systemic decongestants
- Antimicrobial agents
- Surgical treatment

1.4.1.3 Acute Viral Laryngitis

This infection consists of inflammation of a specific portion of the larynx due to respiratory viral infection. Airway obstruction may be caused due to congestion and edema. Acute viral laryngitis generally occurs in infants and preschool children with a peak at age of 2. Causative agents involved in infection are parainfluenza I and II and syncytial virus along with the influenza virus and adenovirus in some cases [65].

Sign and Symptoms

- Nasal congestion, dry cough, and fever of low grade
- Restlessness, intensified stridor, intense dyspnea, and cyanosis
- Seizure, apnea, and death in severe cases

Treatment

- Hydration and air humidification
- Inhaled corticosteroids

1.4.2 Lower Respiratory Tract Infection

Pneumonia and bronchitis are the two major infections of LRI. Respiratory syncytial viruses (RSV) are the most common cause of LRI. These viruses are highly seasonal and contagious. The global burden of disease study indicated that infection of the lower respiratory tract is the third most common cause of mortality after heart disease and cerebrovascular diseases [66].

1.4.2.1 Pneumonia

Worldwide, *Streptococcus pneumoniae* is the main causative pathogen behind community-acquired pneumonia (CAP) [67]. Bacteria as well as viruses are responsible for pneumonia. *Haemophilus influenzae*, generally type B (Hib), along with *Staphylococcus aureus* is also responsible for pneumonia infection. Other pathogens responsible for pneumonia are *Mycoplasma pneumoniae* and *Chlamydia pneumoniae*. Forty to fifty percent of cases of infection in infants and children are due to viruses such as the measles virus, parainfluenza virus, and adenovirus [68]. A report from Jain et al. in 2015 indicated that there was an increase in cases of pneumonia as age is increased [69]. According to a study, around 24.8 episodes per 10,000 adults were reported with the highest incidence around age 65–69. The study also indicated that approximately one in three adults will die after hospitalization within a year. According to a statistical health report, influenza and pneumonia are responsible as the sixth leading cause of death in South Africa [70].

Classification of Pneumonia

1. By location:
   - Community-acquired
   - Hospital-acquired
2. By cause:
   - Dust pneumonia
   - Chemical pneumonia
   - Chemical pneumonia
   - Aspiration pneumonia

Sign and Symptoms of pneumonia

- Fever and chills
- Shortness of breath
- Chest pain during cough
- Nausea, vomiting, and diarrhea
- Cough with phlegm

Treatment

- Antibiotics
- Antivirals for viral pneumonia
- Antifungals for fungal pneumonia
- Oxygen therapy if the blood oxygen level is low

1.4.2.2 Influenza

Influenza is caused by a variety of viruses. Generally, influenza is responsible for causing upper respiratory tract infection. But it is more and more being recognized for lower respiratory tract infection. It is the second most common cause of hospitalization of children with acute respiratory tract infection [71]. The influenza virus is of three types, mainly, A, B, and C. Among these types, influenza A and influenza B are responsible for causing human infection. Influenza A is more dangerous and often responsible for global pandemics especially in developing countries [72]. Chances of emergence of new strains of influenza A virus are maximum as occurred in the 1950s and 1960s. Apart from this, influenza virus is capable of transferring from animals, birds, to humans with improved genetic structure because of gene shifting ability. Influenza A virus is responsible for 90% of flu cases. These viruses are capable of transferring from human to human through various activities like coughing, sneezing, and talking. Influenza is highly seasonal especially in autumn and winter [73].

Sign and Symptoms of Influenza
- Fever, chills, and sweat
- Muscle pain and fatigue
- Headache
- Dry continuous cough
- Difficulty in breathing
- Sore throat
- Stuffed or runny nose
- Eye pain, vomiting, and diarrhea
- Chest pain

Treatment
- Antiviral drugs like oseltamivir, zanamivir, and peramivir.
- Vaccination is highly recommended before starting of flu season.

1.5 Idiopathic Pulmonary Fibrosis (IPF)

Interstitial lung diseases (ILDs) are a diverse family of parenchyma lung disease in which lung inflammation and fibrosis occur. Idiopathic pulmonary fibrosis is the most destructive among interstitial lung diseases. It is characterized by growing fibrosis, cellular proliferation, interstitial inflammation of the alveolar wall with irreversible unstoppable destruction of lung function, and respiratory failure leading to high mortality [74]. Tissue damage of alveoli will be heal but if the process is repeated the healing process stops which causes the thickening of the walls of alveoli, due to which oxygen transfer from alveoli into capillaries becomes difficult. For optimized treatment, accurate diagnosis is necessary. Progression of IPF in some patients is unnoticeable, while in some, it progresses very fast [75, 76]. Adults aged 18 to 64 years may experience annual cases of 6 per 100,000 persons, while for adults 65 years and above, incidence increases up to 94 cases per 100,000 persons [77, 78]. Increasing age, male sex, tobacco smoking, environmental exposures, genetic factors, and microbial infection are some of the risk factors for IPF [79]. If a person is a tobacco smoker, then there are 60% more chances of him getting IPF [80].

1.5.1 Epidemiology

IPF prevails throughout the world. IPF seems to be higher in North America and Europe as compared to South America and South Asia. In the United States, IPF is reported in 10–60 cases among 100,000 individuals. The rate of hospitalization is increasing which suggests an increase in the burden of disease [81]. Data can be variable as results will depend upon the data collection method. The regional variation between the numbers of cases can be observed depending upon the method of data collection, exposure to different caustic agents, and occupational risk factors. The mortality rate in IPF is high. It may be due to increased diagnosis [82]. It is well-known that IPF is a heterogeneous disease with a changing disease course.

1.5.2 Causes

Pulmonary fibrosis causes thickening of air sacs which prevents oxygen transfer. This damage can be caused by several factors like:

- Occupational and environmental factor
  - Asbestos
  - Silica
  - Dust
  - Long-term exposure to animals and livestock
  - Metal dust
- Radiation
- Pharmacological agents like chemotherapeutic agents, CVS medications, anti-inflammatory drugs, and antibiotics
- Also, some medical conditions like GERD, rheumatoid arthritis, pneumonia, and polymyositis
- Genetic factors
- Tobacco smoking

1.5.3 Pathophysiology

Traditionally, IPF was considered to be an inflammatory disease of the lungs which leads to fibrosis of lungs. IPF is now considered as a result of genetic as well as environmental risk factors that repetitively cause tissue damage causing thickening of walls of alveolar sacs. In IPF, the aging of the alveolar epithelium tissue occurs due to repetitive injury. These etiological factors start as unusual fibroblast outgrowth, myofibroblasts cause the production of the matrix, and due to matrix accumulation, the lung interstitium tissue becomes remolded [83]. Gas exchange between alveoli and nearby capillaries is lost permanently.

1.5.4 Symptoms

- Dry cough
- Tiredness
- Shortness of breath and rapid and shallow breathing
- Fatigue
- Swelling of fingers called as clubbing
- Muscle and joint pain
- Weight loss

1.5.5 Diagnosis

- Checking of patient medical history and physical examination
- Various diagnostic tests like chest X-ray, high-resolution chest computerized tomography scan, and lung biopsy
- Pulmonary function test, blood test, and genetic testing

1.5.6 Treatment

Presently, there is no absolute cure for IPF. Depending upon the patient condition, physicians may attempt one or more of the following treatments:

- Medicines
  - Kinase inhibitors: nintedanib and pirfenidone
  - Antacids to relieve GERD
- Oxygen therapy
- Ventilation therapy
- Surgery like a lung transplant

Apart from the abovementioned diseases, there are numerous diseases which can affect the lungs or respiratory systems like sarcoidosis, tuberculosis, silicosis, cystic fibrosis, bronchitis, pleural effusion, SARS, and COVID-19.

References

1. Anand N, Kollef MH (2009) The alphabet soup of pneumonia: CAP. Semin Respir Crit Care Med 30(1):3–9
2. Chetta A, Marangio E (2005) Psychological implications of respiratory health and disease. Respiration 72:210–215
3. Cleland JA, Lee AJ, Hall S (2007) Associations of depression and anxiety with gender, age, health-related quality of life and symptoms in primary care COPD patients. Fam Pract (Oxford Univ) 24(May):217–223
4. van Manen JG (2002) PB. Risk of depression in patients with chronic obstructive pulmonary disease and its determinants. Thorax 57:412–416
5. Schraufnagel D (2013) The world respiratory diseases report. Int J Tuberc Lung Dis 17(12). https://doi.org/10.5588/ijtld.13.0743
6. Tu ZY, Ning FH, Yu WJ (2014) Study about recognition of digital meter dial reading based on SVM. Appl Mech Mater 615:194–197. https://doi.org/10.4028/www.scientific.net/AMM.615.194
7. National Heart, Lung, and Blood Institute (2003) Data fact sheet: chronic obstructive pulmonary disease. National Institutes of Health Publication 03–5229. US Department of Health and Human Services, Bethesda, MD
8. Rabe KF, Watz H (2017) Chronic obstructive pulmonary disease. Lancet 389 (10082):1931–1940. https://doi.org/10.1016/S0140-6736(17)31222-9
9. Celli BR, Agustí A (2018) COPD: time to improve its taxonomy? ERJ Open Res 2:1–8. https://doi.org/10.1183/23120541.00132-2017
10. Mannino DM, Thorn D, Swensen A, Holguin F (2008) Prevalence and outcomes of diabetes, hypertension and cardiovascular disease in COPD. Eur Respir J 32(4):962–969. https://doi.org/10.1183/09031936.00012408
11. Divo MJ, Celli BR, Poblador-Plou B, Calderón-Larrañaga A, de-Torres JP, Gimeno-Feliu LA, Bertó J, Zulueta JJ, Casanova C, Pinto-Plata VM, Cabrera-Lopez C, Polverino F, Píréz JC, Prados-Torres A, Marin JM (2018) Chronic Obstructive Pulmonary Disease (COPD) as a disease of early aging: evidence from the EpiChron Cohort. PLoS One 13(2):1–16. https://doi.org/10.1371/journal.pone.0193143
12. COPD EOF, Obstruction I (1995) Definitions, epidemiology, pathophysiology, diagnosis, and staging. Am J Respir Crit Care Med 152(subset 10):S77–S121
13. Siafakas NM, Vermeire P, Pride NB, Paoletti P, Gibson J, Howard P, Yernault JC, Decramer M, Higenbottam T, Postma DS, Rees J (1995) Optimal assessment and management of chronic obstructive pulmonary disease (COPD). Eur Respir J 8(8):1398–1420. https://doi.org/10.1183/09031936.95.08081398
14. Buist AS, Anzueto A, Calverley P, DeGuia TS, Fukuch Y (2006) GOLD—Global Initiative for Chronic Obstructive Lung Disease. p 12. https://www.who.int/respiratory/copd/GOLD_WR_06.pdf. Assessed 16 Sept 2020
15. GOLD (2017) Global strategy for the diagnosis, management and prevention of COPD. https://goldcopd.org/wp-content/uploads/2017/02/wms-GOLD-2017-FINAL.pdf. Assessed 16 Sept 2020
16. Incidence I, Collaborators P (2016) (2017). Global, regional, and national incidence, prevalence, and years lived with disability for 310 diseases and injuries, 1990 – 2015: a systematic analysis for the Global Burden of Disease Study. Lancet 388:1545–1602. https://doi.org/10.1016/S0140-6736(16)31678-6
17. Gershon AS, Warner L, Cascagnette P, Victor JC, To T (2011) Lifetime risk of developing chronic obstructive pulmonary disease: a longitudinal population study. Lancet 378 (9795):991–996. https://doi.org/10.1016/S0140-6736(11)60990-2
18. Adeloye D, Chua S, Lee C et al (2015) Global and regional estimates of COPD prevalence: systematic review and meta-analysis. J Glob Health 5(2):020415. https://doi.org/10.7189/jogh.05-020415

19. Lamprecht B, Soriano JB, Studnicka M et al (2015) Determinants of underdiagnosis of COPD in national and international surveys. Chest 148(4):971–985. https://doi.org/10.1378/chest.14-2535
20. World Health Organization. Chronic respiratory diseases. Burden of COPD. http://www.who.int/respiratory/copd/burden/en/. Accessed 16 Sept 2020
21. Mannino DM, Buist AS (2007) Global burden of COPD: risk factors, prevalence, and future trends. Lancet 370(9589):765–773. https://doi.org/10.1016/S0140-6736(07)61380-4
22. Lopez AD, Mathers CD, Ezzati M, Jamison DT, Murray CJL (2006) Global burden of disease and risk factors. The World Bank, Washington
23. Cox DW (2013) Alpha1-antitrypsin deficiency. Brenner's Encyclopedia Genet 365(1):85–87. https://doi.org/10.1016/B978-0-12-374984-0.00040-1
24. Trupin L, Earnest G, San Pedro M, Balmes JR, Eisner MD, Yelin E, Katz PP, Blanc PD (2003) The occupational burden of chronic obstructive pulmonary disease. Eur Respir J 22 (3):462–469. https://doi.org/10.1183/09031936.03.00094203
25. Bourdin A, Burgel PR, Chanez P, Garcia G, Perez T, Roche N (2009) Recent advances in COPD: pathophysiology, respiratory physiology and clinical aspects, including comorbidities. Eur Respir Rev 18(114):198–212. https://doi.org/10.1183/09059180.00005509
26. Matsuba K, Wright JL, Wiggs BR, Pare PD, Hogg JC (1989) The changes in airways structure associated with reduced forced expiratory volume in one second. Eur Respir J 2(9):834–839
27. Anthonisen NR (2005) The effects of a smoking cessation intervention on 14.5-year mortality. Ann Intern Med 143(8):615. https://doi.org/10.7326/0003-4819-143-8-200510180-00019
28. Wen CP, Wai JPM, Tsai MK, Yang YC, Cheng TYD, Lee MC, Chan HT, Tsao CK, Tsai SP, Wu X (2011) Minimum amount of physical activity for reduced mortality and extended life expectancy: a prospective cohort study. Lancet 378(9798):1244–1253. https://doi.org/10.1016/S0140-6736(11)60749-6
29. Celli BR, Wedzicha JA (2019) Update on clinical aspects of chronic obstructive pulmonary disease. N Engl J Med 381(13):1257–1266. https://doi.org/10.1056/NEJMra1900500
30. Ekström M, Ph D (2016) Clinical usefulness of long-term oxygen therapy in adults. N Engl J Med 375(17):1683–1684. https://doi.org/10.1056/NEJMe1611742
31. Vos T, Flaxman AD, Naghavi M, Lozano R, Michaud C, Ezzati M, Shibuya K, Salomon JA, Abdalla S, Aboyans V, Abraham J, Ackerman I, Aggarwal R, Ahn SY, Ali MK, Almazroa MA, Alvarado M, Anderson HR, Anderson LM, Murray CJL (2012) Years lived with disability (YLDs) for 1160 sequelae of 289 diseases and injuries 1990-2010: a systematic analysis for the Global Burden of Disease Study 2010. Lancet 380(9859):2163–2196. https://doi.org/10.1016/S0140-6736(12)61729-2
32. Akinbami LJ, Moorman JE, Garbe PL, Sondik EJ (2009) Status of childhood asthma in the United States, 1980-2007. Pediatrics 123(SUPPL. 3). https://doi.org/10.1542/peds.2008-2233C
33. Bourdin A, Gras D, Vachier I, Chanez P (2009) Upper airway · 1: allergic rhinitis and asthma: united disease through epithelial cells. Thorax 64(11):999–1004. https://doi.org/10.1136/thx.2008.112862
34. To, T, Stanojevic S, Moores G, Gershon AS, Bateman ED, Cruz AA, Boulet LP (2012) Global asthma prevalence in adults: findings from the cross-sectional world health survey. BMC Public Health 12(1):204. https://doi.org/10.1186/1471-2458-12-204
35. Leynaert B, Sunyer J, Garcia-Esteban R, Svanes C, Jarvis D, Cerveri I, Dratva J, Gislason T, Heinrich J, Janson C, Kuenzli N, De Marco R, Omenaas E, Raherison C, Real FG, Wjst M, Zemp E, Zureik M, Burney PGJ, Neukirch F (2012) Gender differences in prevalence, diagnosis and incidence of allergic and non-allergic asthma: a population-based cohort. Thorax 67 (7):625–631. https://doi.org/10.1136/thoraxjnl-2011-201249
36. Raghavan D, Jain R (2016) Increasing awareness of sex differences in airway diseases. Respirology 21(3):449–459. https://doi.org/10.1111/resp.12702
37. Bunyavanich S, Schadt EE (2015) Systems biology of asthma and allergic diseases: a multiscale approach. J Allergy Clin Immunol 135(1):31–42. https://doi.org/10.1016/j.jaci.2014.10.015

38. Fahy JV (2015) Type 2 inflammation in asthma-present in most, absent in many. Nat Rev Immunol 15(1):57–65. https://doi.org/10.1038/nri3786
39. Mims JW (2015) Asthma: definitions and pathophysiology. Int Forum Allergy Rhinol 5(April): S2–S6. https://doi.org/10.1002/alr.21609
40. Brown J (2013) Types of asthma. Asthma Clinical Research Network (ACRN), pp 1–2. https://www.cochranelibrary.com/central/doi/10.1002/central/CN-01507572/full
41. Petty TL (1989) Treat status asthmaticus three days before it occurs. J Intensive Care Med 4 (4):135–136. https://doi.org/10.1177/088506668900400401
42. National Asthma and Prevention Program (2007) Expert panel report 3: guidelines for the diagnosis and management of asthma full report 2007. https://www.nhlbi.nih.gov/sites/default/files/media/docs/EPR-3_Asthma_Full_Report_2007.pdf. Accessed 17 Sept 2020
43. Spyros P, Anastasia K, Katerina M, Charis R (2002) Clinical review: severe asthma. Crit Care 6:30–40. https://link.springer.com/content/pdf/10.1186/cc1451.pdf. Accessed 17 Sept 2020
44. Sethi T (2002) Lung cancer: introduction. Thorax 57:992–993
45. Robert C (2002) Lung cancer: a review. Clinical reviews 59:611–642
46. Travis WD, Brambilla E, Nicholson AG, Yatabe Y, Austin JHM, Beasley MB, Chirieac LR, Dacic S, Duhig E, Flieder DB, Geisinger K, Hirsch FR, Ishikawa Y, Kerr KM, Noguchi M, Pelosi G, Powell CA, Tsao MS, Wistuba I (2015) The 2015 World Health Organization Classification of lung tumors: impact of genetic, clinical and radiologic advances since the 2004 classification. J Thorac Oncol 10(9):1243–1260. https://doi.org/10.1097/JTO.0000000000000630
47. Ferlay J, Shin HR, Bray F, Forman D, Mathers C, Parkin DM (2010) Estimates of worldwide burden of cancer in 2008: GLOBOCAN 2008. Int J Cancer 127(12):2893–2917. https://doi.org/10.1002/ijc.25516
48. Howlader N, Noone AM, Krapcho M et al (eds) (2010) SEER cancer statistics review, 1975–2008. National Cancer Institute, Bethesda, MD. http://seer.cancer.gov/csr/1975_2008/
49. Dela Cruz CS, Tanoue LT, Matthay RA (2011) Lung cancer: epidemiology, etiology, and prevention. Clin Chest Med 32(4):605–644. https://doi.org/10.1016/j.ccm.2011.09.001
50. Ginsberg RJ, Vokes EE, Rosenzweig K (2001) Non-small cell lung cancer. In: DeVita VT Jr, Hellman S, Rosenberg SA (eds) Cancer principles and practice of oncology, 6th edn. Lippincott Williams & Wilkins, Philadelphia, pp 925–983
51. Fleming ID, Cooper JS, Henson DE et al (eds) (1998) AJCC cancer staging handbook. Lippincott-Raven, Philadelphia, pp 117–127
52. Mountain CF (1997) Revisions in the international system for staging lung cancer. Chest 111 (6):1710–1717. https://doi.org/10.1378/chest.111.6.1710
53. Osterlind K, Ihde DC, Ettinger DS, Gralla RJ, Karrer K, Krauss S, Maurer LH, Rørth M, Sörenson S, Vincent R (1983) Staging and prognostic factors in small cell carcinoma of the lung. Cancer Treat Rep 67(1):3–9
54. American thoracic society patient information series: treatment of lung cancer (2014). Am J Respir Crit Care Med 189:P1–P3. https://www.thoracic.org/patients/patient-resources/resources/lung-cancer-treatment.pdf. Accessed 18 Sept 2020
55. Kamath KR, Feldman RA, Sundar PSSRRAO, Webb JKG (1969) Infection and disease in a group of South Indian families: general morbidity patterns in families and family members. Am J Epidemiol 89(4):375–383. https://doi.org/10.1093/oxfordjournals.aje.a120951
56. Monto AS, Ullman BM (1974) Acute respiratory illness in an American community. The Tecumseh study. JAMA 227(2):164–169
57. Black RE, Morris SS, Bryce J (2003) Where and why are 10 million children dying every year? Lancet 361(9376):2226–2234
58. Williams BG, Gouws E, Boschi-Pinto C, Bryce J, Dye C (2002) Estimates of world-wide distribution of child deaths from acute respiratory infections. Lancet Infect Dis 2(1):25–32. https://doi.org/10.1016/S1473-3099(01)00170-0
59. Bryce J, Boschi-Pinto C, Shibuya K, Black RE (2005) WHO estimates of the causes of death in children. Lancet 365(9465):1147–1152. https://doi.org/10.1016/S0140-6736(05)71877-8

60. Denny FW (1995) The clinical impact of human respiratory virus infections. Am J Respir Crit Care Med 152(4 II). https://doi.org/10.1164/ajrccm/152.4_pt_2.s4
61. Herendeen NE, Szilagy PG (2000) Infections of the upper respiratory tract. In: Behrman RE, Kliegman RM, Jenson HB (eds) Nelson textbook of pediatrics, 16th edn. W. B. Saunders Company, Philadelphia, pp 1261–1266
62. Freymuth F, Vabret A, Brouard J, Toutain F, Verdon R, Petitjean J, Gouarin S, Duhamel JF, Guillois B (1999) Detection of viral, Chlamydia pneumoniae and Mycoplasma pneumoniae infections in exacerbations of asthma in children. J Clin Virol 13(3):131–139. https://doi.org/10.1016/S1386-6532(99)00030-X
63. Pöyry A, Starck M, Savolainen S, Pöyry T, Suomalainen I, Hyypï T, Carpen O, Vaheri A (2001) Rhinovirus RNA in the maxillary sinus epithelium of adult patients with acute sinusitis. Clin Infect Dis 33(6):909–911. https://doi.org/10.1086/322678
64. Conrad DA, Jenson HB (2002) Management of acute bacterial rhinosinusitis. Curr Opin Pediatr 14(1):86–90. https://doi.org/10.1097/00008480-200202000-00015
65. Orenstein DM (2000) Acute inflammatory upper airway obstruction. In: Behrman RE, Kliegman RM, Jenson HB (eds) Nelson textbook of pediatrics, 16th edn. W. B. Saunders Company, Philadelphia, pp 1274–1279
66. Wang H, Naghavi M, Allen C, Barber RM, Carter A, Casey DC, Charlson FJ, Chen AZ, Coates MM, Coggeshall M, Dandona L, Dicker DJ, Erskine HE, Haagsma JA, Fitzmaurice C, Foreman K, Forouzanfar MH, Fraser MS, Fullman N et al (2016) Global, regional, and national life expectancy, all-cause mortality, and cause-specific mortality for 249 causes of death, 1980–2015: a systematic analysis for the Global Burden of Disease Study 2015. Lancet 388 (10053):1459–1544. https://doi.org/10.1016/S0140-6736(16)31012-1
67. Blasi F, Mantero M, Santus P, Tarsia P (2012) Understanding the burden of pneumococcal disease in adults. Clin Microbiol Infect 18(SUPPL. 5):7–14. https://doi.org/10.1111/j.1469-0691.2012.03937.x
68. Hortal M, Mogdasy C, Russi JC, Deleon C, Suarez A (1990) Microbial agents associated with pneumonia in children from Uruguay. Rev Infect Dis 12(April 1988):S915–S922. https://doi.org/10.1093/clinids/12.Supplement_8.S915
69. Jain S, Self WH, Wunderink RG, Fakhran S, Balk R, Bramley AM, Reed C, Grijalva CG, Anderson EJ, Courtney DM, Chappell JD, Qi C, Hart EM, Carroll F, Trabue C, Donnelly HK, Williams DJ, Zhu Y, Arnold SR, Finelli L (2015) Community-acquired pneumonia requiring hospitalization among U.S. adults. N Engl J Med 373(5):415–427. https://doi.org/10.1056/NEJMoa1500245
70. Boyles TH, Brink A, Calligaro GL, Cohen C, Dheda K, Maartens G, Richards GA, Smit R v Z, Smith C, Wasserman S, Whitelaw AC, Feldman C (2017) South African guideline for the management of community acquired pneumonia in adults. J Thorac Dis 9(6):1469–1502. https://doi.org/10.21037/jtd.2017.05.31
71. Neuzil KM, Zhu Y, Griffin MR, Edwards KM, Thompson JM, Tollefson SJ, Wright PF (2002) Burden of interpandemic influenza in children younger than 5 years: a 25-year prospective study. J Infect Dis 185(2):147–152. https://doi.org/10.1086/338363
72. Peiris JSM, Yu WC, Leung CW, Cheung CY, Ng WF, Nicholls JM, Ng TK, Chan KH, Lai ST, Lim WL, Yuen KY, Guan Y (2004) Re-emergence of fatal human influenza A subtype H5N1 disease. Lancet 363(9409):617–619. https://doi.org/10.1016/S0140-6736(04)15595-5
73. Alguacil-Ramos AM, Portero-Alonso A, Pastor-Villalba E, Muelas-Tirado J, Díez-Domingo J, Sanchis-Ferrer A, Lluch-Rodrigo JA (2019) Rapid assessment of enhanced safety surveillance for influenza vaccine. Public Health 168:137–141. https://doi.org/10.1016/j.puhe.2018.12.013
74. American Thoracic Society (Ats), & European Espiratory Society (Ers) (2000) American thoracic society idiopathic pulmonary fibrosis: diagnosis and treatment. Am J Respir Crit Care Med 161:646–664. https://doi.org/10.1164/ajrccm.161.2.ats3-00
75. Buendía-Roldán I, Mejía M, Navarro C, Selman M (2017) Idiopathic pulmonary fibrosis: clinical behavior and aging associated comorbidities. Respir Med 129:46–52. https://doi.org/10.1016/j.rmed.2017.06.001

76. Brownell R, Moua T, Henry TS, Elicker BM, White D, Vittinghoff E, Jones KD, Urisman A, Aravena C, Johannson KA, Golden JA, King TE, Wolters PJ, Collard HR, Ley B (2017) The use of pretest probability increases the value of high-resolution CT in diagnosing usual interstitial pneumonia. Thorax 72(5):424–429. https://doi.org/10.1136/thoraxjnl-2016-209671
77. Raghu G, Chen SY, Hou Q, Yeh WS, Collard HR (2016) Incidence and prevalence of idiopathic pulmonary fibrosis in US adults 18-64 years old. Eur Respir J 48(1):179–186. https://doi.org/10.1183/13993003.01653-2015
78. Raghu G, Chen SY, Yeh WS, Maroni B, Li Q, Lee YC, Collard HR (2014) Idiopathic pulmonary fibrosis in US Medicare beneficiaries aged 65 years and older: incidence, prevalence, and survival, 2001-11. Lancet Respir Med 2(7):566–572. https://doi.org/10.1016/S2213-2600(14)70101-8z
79. Raghu G, Collard HR, Egan JJ, Martinez FJ, Behr J, Brown KK, Colby TV, Cordier JF, Flaherty KR, Lasky JA, Lynch DA, Ryu JH, Swigris JJ, Wells AU, Ancochea J, Bouros D, Carvalho C, Costabel U, Ebina M, Schünemann HJ (2011) An official ATS/ERS/JRS/ALAT statement: idiopathic pulmonary fibrosis: evidence-based guidelines for diagnosis and management. Am J Respir Crit Care Med 183(6):788–824. https://doi.org/10.1164/rccm.2009-040GL
80. Baumgartner KB, Samet JM, Stidley CA, Colby TV, Waldron JA (1997) Cigarette smoking: a risk factor for idiopathic pulmonary fibrosis. Pneumologie 51(8):868. https://doi.org/10.1164/ajrccm.155.1.9001319
81. Hutchinson J, Fogarty A, Hubbard R, McKeever T (2015) Global incidence and mortality of idiopathic pulmonary fibrosis: a systematic review. Eur Respir J 46(3):795–806. https://doi.org/10.1183/09031936.00185114
82. Navaratnam V, Fleming KM, West J, Smith CJP, Jenkins RG, Fogarty A, Hubbard RB (2011) The rising incidence of idiopathic pulmonary fibrosis in the UK. Thorax 66(6):462–467. https://doi.org/10.1136/thx.2010.148031
83. Richeldi L, Collard HR, Jones MG (2017) Idiopathic pulmonary fibrosis. Lancet 389 (10082):1941–1952. https://doi.org/10.1016/S0140-6736(17)30866-8

Targeting Molecular and Cellular Mechanisms in Asthma

2

Archita Ray, Sabita Singh, Joytri Dutta, and Ulaganathan Mabalirajan

Abstract

Many a times, a very thin line exists between physiology and pathology. Airway obstruction seen in the asthmatic subjects is exemplary of such phenomena. Physiologically, airway obstruction is needed to guard the master regulator of lung homeostasis, i.e. the alveolar epithelial cells, against the massive entry of exogenous irritants like noxious particles, allergens and microbes into the airway. Surprisingly, genetic predisposition with prolonged exposure to these exogenous irritants sets off the pathological clock causing mild-moderate- severe asthma. With time, numerous innate and adaptive immune cell types along with cytokines have been pinpointed which participates in the pathogenesis of asthma. Though Th2 cytokines form the foundation of the majority of the asthmatic situation, researchers started dividing the patients as Th2-high and Th2-low asthma. The recent concept of classifying asthma based on its endotypes is a more granular approach to avert the progression of asthma rather than the phenotypic classification. The identification of novel biomarkers based on the -omics technology will leverage the molecular data and provide a precision-based care. Here, in this chapter, we have described about the various immune cells responsible for the heterogeneity of the disease such that it can be targeted for substantial improvisation of asthma, along with current lacuna in the understanding of asthma pathogenesis.

Archita Ray, Sabita Singh, Joytri Dutta contributed equally to this work

A. Ray · S. Singh · J. Dutta · U. Mabalirajan (✉)
Molecular Pathobiology of Respiratory Diseases, Cell Biology and Physiology Division, CSIR-Indian Institute of Chemical Biology, Kolkata, India

Academy of Scientific and Innovative Research (AcSIR), Ghaziabad, India

K. Dua et al. (eds.), *Targeting Cellular Signalling Pathways in Lung Diseases*, https://doi.org/10.1007/978-981-33-6827-9_2

Keywords

Asthma · Airway inflammation · Bronchial epithelium · Biomarkers · Endotypes · Personalised approach

2.1 Introduction

2.1.1 Rationale to the Study

Asthma causes a significant socioeconomic burden as well as family burden. The major symptoms are wheezing (vibrations in the small airways that are almost closed off due to obstruction), chest tightening, shortness of breath, cough with or without sputum production and night waking due to combination of the above symptoms. Asthma is an important global public health problem as 339 million people are affected worldwide as of the 2016 report from World Health Organization (WHO) [1]. It is one of the very few diseases that predominantly affects the people in developed nations. This is in accordance with the hygiene hypothesis proposed by Strachan in 1989 [2]. However, a dramatic increase in the prevalence, morbidity and mortality due to asthma has been noted in developing countries according to a recent report from the Global Initiative for Asthma [3] in spite of advanced understandings and therapeutic strategies.

2.1.2 Asthma: A Global Health Burden

Despite emergence of newer therapeutic strategies, the prevalence of asthma keeps on increasing [4]. It is the leading cause of hospitalisation among children. Asthma had secured the 16th and 28th position worldwide in causing years lived with disability and adjusted life years, respectively [2]. Its incidence varies from country to country. It has been estimated that 34% of the man days in India are lost due to airway-related disorders, of which asthma is the major cause [5]. It is also important to note that higher mortality is observed in patients having features of refractory asthma. This is due to limited therapeutic options for different pathophysiological features such as greater involvement of neutrophils and increased airway remodelling.

2.1.3 Asthma: GINA Definition

Asthma can be defined as 'a heterogeneous disease, usually characterised by chronic airway inflammation. It is defined by the history of respiratory symptoms such as wheeze, shortness of breath, chest tightness and cough that vary over time and in intensity, together with variable expiratory airflow limitation'.

Since Hippocrates time, the concept of asthma had changed from psychological disease to smooth muscle disease to inflammatory disease to current airway remodelling. Parallel to change in the concept, the therapeutic strategy has also changed from mind calming to bronchodilators to anti-inflammatory drugs to thermoplasty. Though all these available traditional therapy are mostly non-specific approach, the knowledge we obtained in the last two decades converts this mere non-specific approach to combined non-specific coupled with pathway-specific approach. In this chapter, we are going to witness the same.

2.2 Traditional Understanding and Parallel Non-specific Therapy

2.2.1 Major Types of Asthma

There are two major types of asthma: atopic and non-atopic. Asthma which gets triggered by allergens is called extrinsic or atopic asthma [6]. Atopic asthma is characterised by heightened expression of cytokines like interleukin-4 (IL-4), interleukin-5 (IL-5) by T helper 2 (Th2) cell, interleukin-2 (IL-2) and granulocyte-macrophage colony-stimulating factor (GM-CSF) in bronchoalveolar lavage (BAL) fluid [7]. These atopic asthma patients have increased levels of allergen-specific immunoglobulin E (IgE) in serum and skin prick positivity to susceptible allergens [8]. Intrinsic, non-atopic, asthma that sets off by running, exercise, etc. generally appears at a later life and is non-allergic in nature. Occupational asthma, one subtype of non-atopic asthma, is triggered by the exposure to irritants found in workplace.

2.2.2 Allergens Are Major Inducers for Atopic Asthma

The inducers for the asthma symptoms vary depending on the type of asthma. The inducers for atopic asthma are various types of allergens like different types of common pollens, cockroach allergen, house dust mite, danders of cat and dog, etc. Importantly, these allergens sensitise only allergy-prone individuals who are having genetic tendency. After sensitisation, when they are exposed to same allergens again, it leads to bronchoconstriction. In general, other environmental factors like air pollutants may not instigate asthma development in such atopic individuals, but they can aid in promoting the asthma initiation. For example, by damaging airway epithelial barrier, cigarette smoke is seen to increase the accessibility of allergens. As a result, air pollution could potentiate the allergen uptake following which they get processed by antigen-presenting cells. On the other hand, in non-atopic asthma, both indoor and outdoor pollutants could initiate the asthma development and this leads to cause occupational asthma. Thus, children residing in the urban areas have an increased tendency to develop asthma than non-urban children [9].

2.2.3 Pathophysiology

In Greek language, the word asthma means panting and being out of breath. Initially, asthma was misunderstood to have a psychosomatic history. But actually, this is not completely a misunderstanding because neurogenic inflammation does happen in asthma pathogenesis. Asthma is a complex disease commonly identified as hyperactivity in the bronchial airways. Such hyperresponsiveness in the bronchial region increases before the onset of the allergic trigger and reverts only after treatment. There is an increase in the infiltration of inflammatory cells like eosinophils. This eosinophil-dominant airway inflammation is associated with goblet cell metaplasia with overproduction of mucus in the airways and hypertrophy and hyperplasia of airway smooth muscles [10]. Airway hyperresponsiveness occurs due to the increased sensitivity of the sensory nerves to allergen. In case of serious asthmatic conditions, mucus plugs are formed which are made up of plasma proteins and mucus glycoprotein [11]. Most of these asthma features are believed to be mediated by Th2 response. Repeated airway inflammation also causes irreversible structural changes in the airways. All these features, termed 'airway remodelling', is characterised by airway wall thickening due to deposition of extracellular matrix proteins underneath the airway epithelial layer. Such increase in the wall tissue of the airways further causes more narrowing of the walls of the airways.

2.2.4 Symptomatic Therapy

Bronchodilators and corticosteroids are the two major types of anti-asthma drugs that are in use for a long time. Both of these drugs fall under the category of symptomatic treatment. There are other drugs like antihistamine agents and leukotriene modifiers that are considered as additional supplementary therapy. Though leukotriene modifiers showed promising results at the initial times of launching, all asthmatics did not show good improvement in lung function.

Bronchodilators are group of drugs which function by relaxing the tightened airway muscles. This facilitates intake of air and improves breathing. They symptomatically act in managing asthma [12]. Depending on their nature, bronchodilators can be subdivided into two categories. These are beta-2-agonist drug and anticholinergic drug. Based on the time of action, beta-2-agonist drugs can be further classified as long-acting beta-2-agonist (LABA) and short-acting beta-2-agonist (SABA) [13]. Ultra-LABA is also used [12]. Tiotropium and theophylline are long-acting *muscarinic* antagonists (LAMA). Salmeterol and formoterol are long-acting beta-2- agonists that are administered by inhalation. Theophylline, whose chemical name is dimethylxanthine, has been used for asthma therapeutics for a long time. It works by inhibiting phosphodiesterase 3 (PDE3). Activation of PDE4 and histone deacetylase-2 is also inhibited by theophylline. However, long-term theophylline use can pose some harmful effects. Low dose is recently recommended to avoid therapeutic resistance [14].

Inhaled corticosteroids (ICS) are common drugs to manage acute asthma [15]. Combination therapy of both beta-2-agonists and ICS is the frequently used strategy to control moderate to severe asthmatic conditions. LABA in combination with ICS proves to be an effective treatment for asthma [12]. Corticosteroids inhibit the inflammatory activities through transactivation and/or transrepression. The former leads to the transcription of numerous anti-inflammatory mediators, whereas the latter suppresses the transcription of pro-inflammatory mediators. However, 5–10% asthmatic patients do not respond to corticosteroids, and this condition is called steroid-insensitive condition. This has been discussed in a separate chapter in this book.

Anti-leukotriene drugs work by inhibiting the leukotriene receptors. Though lot of expectations were there on leukotriene modifiers when these were launched, the clinical trial findings were disappointing. However, leukotriene receptor antagonists have been shown to be beneficial in certain asthmatic conditions like aspirin-sensitive asthma, obese asthma, senile asthma and smoking-associated asthma [16]. The effects of these leukotriene modifiers on airway remodelling and identification of biomarkers of good responders are yet to be investigated [16].

The above-mentioned drugs are non-specific symptomatic therapies available in asthma, except leukotriene modifiers. Though this non-specific therapy can be successful in controlling the disease for a short term, it seems that this may not be sufficient in a considerable number of patients when they have to control the disease for a long time. With the emergence of multiple endotypes in asthma including neutrophilic asthma, steroid-resistant asthma, obese asthma and so on, the asthma field needs to be revisited. This would provide a better understanding of the disease pathogenesis which would in turn aid in formulating more effective therapeutic strategies. In this chapter, we have attempted to achieve the same. In this journey, one can realise how non-specific therapy can be complemented with specific, personalised strategy in asthma management.

2.3 Current Understanding and Steps Towards Pathway-Specific Therapy (Fig. 2.1)

2.3.1 Cells Involved in Asthma Pathogenesis and Respective Targets

2.3.1.1 Airway Epithelium

The respiratory organ is lined by epithelial cells. Being a frontline defence barrier, the airway epithelium acts as a protective interface between the lung and environment [17]. Earlier, it was believed that airway epithelium is a mere victim of most of the recruited immune cells. However, recent reports indicate that airway epithelium can govern airway inflammation. For example, airway epithelium secretes Th2-polarising cytokines on being exposed to mild irritants like allergen but secretes interleukin-17 (IL-17) when they are exposed to severe irritants like cigarette smoke [17]. The entire epithelial layer present in the airways provides two major barriers.

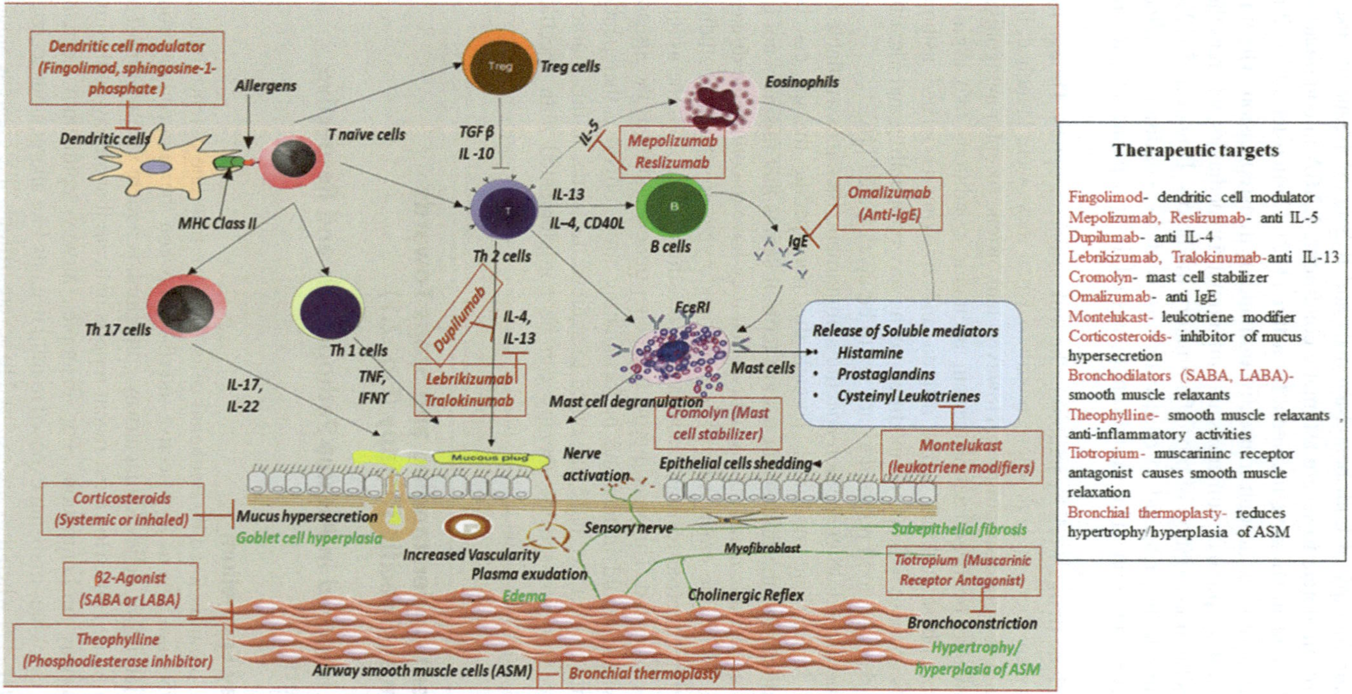

Fig. 2.1 Current understanding and steps towards pathway- or molecule-specific therapy in asthma. Upon allergen exposure, the allergens are sampled by dendritic cells that are located in between the two subsequent airway epithelia in the mucosal layer, processed and presented to naïve T cells for the conversion of Th2 cells. These Th2 cells release various Th2 cytokines like IL-4, interleukin-13 (IL-13) and IL-5. This leads to various asthma features like IgE class switching in B lymphocytes followed by release of IgE that further binds to IgE receptors present in basophils and mast cells to release various bronchoconstrictors, goblet cell metaplasia, sub-epithelial fibrosis and recruitment of eosinophils. In contrast to traditional symptomatic therapy in asthma, current understanding in asthma

pathogenesis emphasised the presence of various endotypes in asthma patients and also pave the way for pathway-oriented therapy or cytokine-oriented biologic to control asthma symptoms. However, the combination of both symptomatic therapy and cytokine-targeted therapy are complimentary. It is important to note that cytokine-targeted therapy may not be universal, but it depends on molecular involvement of individual patients. This laid a foundation of personalised approach in current therapeutic strategy. IL, interleukin; CD, cluster of differentiation; IgE, immunoglobulin E

They act as physical barrier by forming tightly connected airway epithelial layer that does not allow foreign particles like allergens to enter. Airway epithelia also function as a chemical barrier. The epithelial layer secretes a number of innate immune cytokines that can further act as a signal to initiate the acquired immune response [17]. This will then bring the right type of immune cells to the site of insult that can lead to resolution. The protective mechanism also includes the mucus layer which has two layers: gel and sol. These layers are responsible for entrapping foreign particles. Such particles will be sent upwards in the airways by ciliary beating to be removed from the airway [17]. In addition, the airways also have other physiological mechanisms like sneezing and coughing. They are primarily responsible for elimination of foreign particles from the airway with the goal of protecting the alveoli, the functional unit of lungs. Indeed, the partial closure of the airways, bronchoconstriction, is meant for the prevention of massive entry of allergens and other toxic particles from the environment into the airways. Thus, the airway epithelial layer acts as a security force to protect the alveoli.

2.3.1.2 Dendritic Cells

Dendritic cells recognise foreign unknown agents which have a role in promoting asthma. When the airway lumen is exposed to any antigen or allergen, they are sampled by these dendritic cells. Dendritic cells insert their teeth-like projections between two subsequent bronchial epithelia. Upon identifying allergens, dendritic cell has the job to initiate allergic responses. They process the antigens and then present them to the naïve T helper (Th0) cell. They also decide whether the response would be vigorous or can be tolerated [18]. Studies have shown that dendritic cells have a part in the differentiation of the CD4+ T cells. In response to aeroallergen, dendritic cells dominate eosinophilic inflammation by regulating the activation of lung Th2 cells. They may also have a role in tuning the immune balance between T regulatory cell (Treg) and Th2 cells [19].

Dendritic Cell Modulator as a Target

It is known that successful migration of lymphocytes from the lymph node to the site of inflammation is crucial for the development of active immune response. This migration is facilitated by a number of events, and one such event is the gradient of sphingosine-1-phosphate (S1P) [20, 21]. In general, the levels of S1P are always high in both lymph nodes and blood compared to the tissues. Upon its receptor interaction, S1P can inhibit the exit of lymphocytes from the lymph node. Applying this strategy, fingolimod (FTY720), a synthetic sphingolipid, was developed. Upon phosphorylation, fingolimod phosphate acts as an analogue of S1P. Fingolimod prevents the movement of DCs present in the lung to mediastinal lymph nodes and can thus function as an immunosuppressant in asthma [20, 21]. However, it could cause lymphopenia that has important adverse effects. Studies have reported that inhalation of FTY720 ameliorates the asthma features without causing lymphopenia [20]. Though fingolimod seemed to be successful in murine asthma, its effects in asthmatic patients are controversial. Thus, more investigation is needed to see whether it provides beneficial effects in asthma.

2.3.1.3 Lymphocytes

Lymphocytes have a crucial role in asthma pathogenesis and both B and T cells are under its umbrella. Naïve T cells endorse T cell receptor (TCR) which, upon recognising any antigen, differentiates and proliferates into effector T cells like T helper 1 (Th1), Th2, T helper 17 (Th17) and Treg cells. Th2 lymphocytes lead to the commencement and maintenance of allergic inflammation. Such phenomena occur in asthma and it is dependent on various environmental factors like allergen exposure. CD25+ Treg, a subclass of T lymphocytes, also has a part in asthma progression. Treg cells play an integral role in managing the allergy coupled asthmatic response through immunotherapy [22]. They subdue the cytokine production and proliferating ability of CD4+ CD25- T cells. Treg cells also produce interleukin-10 (IL-10) that have a role in safeguarding atopic sensitivity in humans [22].

2.3.1.4 T Helper Type 2 (Th2) Cells

They are dominantly involved in atopic asthmatic condition [23]. However, its involvement is also reported in non-atopic asthma. GATA-3, the transcription factor for Th2 cell subtype, gets overexpressed in asthma. In contrast, the transcription factor for Th1 cell is T-bet whose expression gets reduced in asthma. The cytokines released by Th2 cell like IL-13 have the capacity to induce airway hyperresponsiveness (AHR), and subsequently inflammation in asthma, independently without IgE and eosinophils. Allergic asthma which gets instigated by the Th2 cells is called eosinophilic asthma.

Th2 cytokines like IL-4, IL-5 and IL-13, which are secreted by these Th2 cells, are responsible for most of the asthma features. IL-4 is also released from mast cells upon initial allergen exposure, and this event is crucial for the conversion of naïve T cells to Th2 cells. IL-4 secreted by Th2 cells leads to a number of asthma features. Since both IL-4 and IL-13 share a common receptor, IL-13Rα1/IL-4Rα complex, most of the events, like signal transducer and activator of transcription 6 (STAT-6) phosphorylation and IgE class switching, are common in IL-4 and IL-13 signalling. In addition, IL-4 also leads to IL-13-independent effects, thanks to IL-4-specific receptor subunit that is not shared with IL-13 [24].

IL-13 is a key Th2 cytokine that regulates major airway inflammatory events in asthma. Genetic polymorphisms in IL-13, its receptor components, and its downstream molecules such as STAT-6 have been shown to be associated with asthma. Therapies have been developed which will target IL-13/IL-4/STAT-6 pathway and control this heterogeneous airway disease amount [25]. Mere administration of recombinant IL-13 seems to be sufficient to induce most of the asthma features in naïve mice even without allergen exposure. This indicates the dominant nature of IL-13 in asthma pathogenesis. In addition to its participation in allergic airway inflammation, IL-13 signalling also leads to airway remodelling changes like goblet cell metaplasia and sub-epithelial fibrosis.

Th2 Cytokines as Target

In order to block the interaction between IL-13 and its receptors, development of some monoclonal antibodies, that are presently having asthma based clinical trials, took place. Anrukinzumab monoclonal antibody (mAb) is of human origin that works by blocking IL-13 and inhibiting the activation of IL-13 receptor subunits IL-13Rα1 and IL-13Rα2 [26]. Lebrikizumab, which is also used in the treatment of atopic dermatitis, is a human immunoglobulin G4 (IgG4) mAb that has high affinity for IL-13, binds to IL-13 and inhibits signalling pathway that occurs through the IL-4Rα/IL-13Rα1 heterodimeric complex. Humanised IgG4 mAb, tralokinumab, that specifically targets and neutralises IL-13, is used for treating severe asthma. It affects the mRNA level of IL-13 regulated dipeptide peptidase-4 (DPP-4) in serum [27].

Inhibition of IL-4 can be done either by direct blocking of the IL-4 cytokine or by inhibiting IL-4/IL-13 receptor. One such drug is dupilumab. Being a mAb, dupilumab blocks IL-4Rα/IL-13Rα1 receptor complex and thus prevents interaction between IL-4 with IL-4R and IL-13 with IL-13Rα1. Another such IL-4Rα/IL-13Rα1 antagonist is pitrakinra [28]. AMG17 is another humanised monoclonal antibody which binds to IL-4Rα and inhibits both IL-4 and IL-13 pathways. However, it is not effective for all groups of asthmatic patients [29]. Antagonists can also be designed in the laboratory. By joining crystallisable IL-13Rα2 fragment with general antagonist mu11B11 through SCORPIO method, a research group developed dual IL-4/IL-13 antagonist that has a role in inhibiting both the cytokine and lung inflammation in OVA-challenged mice [30]. IL-13 also shares signalling pathway with thymic stromal lymphopoietin (TSLP). A research group came up with bispecific nature anti-IL-13/anti-TSLP antibodies for therapeutic benefit in Th2 asthma. These antibodies are called Zweimab and Doppelmab which are monovalent bispecific and bivalent bispecific, respectively. Both of them are very potent and target both IL-13 and TSLP [31]. It is important to note that various IL-13 and IL-4 targeting therapeutic molecules cannot be useful in all asthmatic patients. But it would be useful in a subset of patients who are having eosinophilic asthma phenotype along with increased levels of periostin [32].

2.3.1.5 B Lymphocytes and IgE

Allergen-induced T cells induce and activate B lymphocytes that further release IgE after class switching. Indeed, atopicity is determined mainly by the existence of higher IgE levels. The receptor proteins involved in such IgE signalling pathway are high-affinity Fc receptor for IgE (FcεRI) and CD23 which has low affinity. Apart from basophils and mast cells that constitutively express FcεRI protein, dendritic cells also express FcεRI. Expression of FcεRI protein is also found in the airway smooth muscle cells and airway epithelial cells of asthmatic individuals [33]. Apart from IgE, low-affinity CD23 receptor has other ligands like CD21 and integrin. CD23 causes transfer of IgE to the mucosal tissue. Concentration of IgE is maintained in tissue as masts cells express FcεRI in high amount and the rate of dissociation of IgE from FcεRI receptor protein is low [34].

Importantly, anti-IgE therapy reduces asthma features significantly. Omalizumab, a humanised monoclonal anti-IgE antibody, selectively binds to IgE and prevents the binding of IgE to its receptor [35]. This diminishes both early and late allergic responses. IgE-stimulated allergic reaction response is also inhibited. It provides a long-term effect and reduces the use of inhaled corticosteroids. Administration of omalizumab reduces the expression of FcεRI in basophils. IgE can also bind to its low-affinity receptor found on B cells. Membrane-bound IgE on the B lymphocytes, IgE+ B cell, has a critical role in controlling synthesis of IgE. Currently under clinical trial, quilizumab is an immunoglobulin G1 (IgG1) mAb that acts on the IgE+ B cells and further regulates the generation of IgE. It thus reduces the quantity of serum IgE in allergic asthma [36]. However, IgE therapy may not be effective for treating non-allergic asthma. Even in allergic asthma, it is being prescribed mostly to patients who are having severe refractory asthma as it is very expensive. More importantly, IgE mAb therapy would be useful in a subset of patients who are having very high IgE levels [32].

2.3.1.6 Mast Cells

They belong to the haematopoietic lineage and are located near the airway epithelium in the mucosal region. In airway mucosa, they are present at the junction point where the foreign antigen enters the host. When activated, degranulation of mast cells happens, following which, the progenitors of mast cells are recruited to inflammatory sites [37]. There is a significant difference in the mast cells found near airway smooth muscle between asthmatic patients and normal individuals. In asthma, airway smooth muscle cross talks with infiltrating mast cells [38]. Airway smooth muscle uses chemokine (C-X-C motif) ligand 10 (CXCL10) or chemokine (C-X-C motif) ligand 3 (CXCR3) axis for the recruitment of mast cells. Degranulation of the infiltrated mast cells is positively correlated with airway obstruction by mucus inside the airway lumen. Inhibiting the secretion by mast cells can be a therapeutic treatment for asthmatic patient [39].

Type 1 hypersensitivity-mediated mast cell degranulation in asthma is controlled by mast cell stabilising agents. Cromolyn sodium, a chemical derived from khellin, obtained from the medicinal plant *Ammi visnaga*, is a stabilising compound that blocks the release of inflammation-mediating agents from the mast cells [40]. Atopic asthmatic individuals who undergo pre-treatment with cromolyn sodium show decrease in asthma features [41]. Cromolyn, which is given in aerosolised form via nebuliser, acts on both early and late phases of asthma. Nedocromil sodium, an agonist of G protein-coupled receptor 35 (GPCR35), is another mast cell stabilising compound [42] that works by acting against the flux of chloride ion in the mast cells. Though it is not used commonly in asthmatic patients, it is the main therapy in case of other allergies like ocular allergy [42].

2.3.2 Th2-High Eosinophilic Asthma Versus Th2-Low Neutrophilic Asthma

Depending on the nature of infiltrated cells inside the airways, there can be two major types of asthma. One is the allergic mild to moderate **eosinophilic asthma** which is regulated by Th2 cells. It has an early onset and occurs in children at a young age. The other more severe type is the neutrophilic asthma which is controlled by Th1 and Th17 cells. They are mediated by two different pathways and they regulate different cytokines and co-stimulatory molecules [43]. It is also called Th2-high asthma and Th2-low asthma. The person having **neutrophilic Th2-low asthma** will not respond when they are given the same drug which is used for treating eosinophilic inflammation. Therefore, the approach for treating asthma is gradually changing from one size for all to personalised therapeutic. Hence, it is very important to study the pathobiology in asthma.

The most commonly defined phenotype of asthma is the IgE-regulated eosinophilic asthma.

2.3.2.1 Eosinophilic Asthma/Th2-High Asthma

Airway eosinophilia accompanied with shredding of the bronchial epithelium is the hallmark of asthma. The major basic protein (MBP) found inside the eosinophil granules is toxic to the respiratory epithelial barrier. It results in desquamation followed by destruction of the ciliated cells. Heightened MBP levels in the sputum are an important marker of asthma [44]. Eosinophilia can be identified by measuring the eosinophil count in the sputum. Reduction in eosinophilia also indicates decrease in exacerbations that happen in asthma [45]. Reduced apoptosis of eosinophils is noted in asthma. GM-CSF production is primarily the reason behind this. In vitro studies have showed that β_2-agonist has a role in lengthening the lifespan of eosinophils and inhibits apoptosis [46]. Hence, it is concluded that in asthma, eosinophilia occurs both in blood and in tissue. Eosinophil maturation that happens in the bone marrow, recruitment of eosinophils by chemokine receptor 3 agonists, eosinophil transition and their survival are the steps that occur in asthma. Therapeutic strategies aimed at inhibiting eosinophil maturation and trafficking are some strategies to combat asthma [47].

Though it was believed that Th2 response is entirely due to Th2, acquired immune cells, the similar response can also be generated through type 2 innate lymphoid cell (ILC2) that secretes Th2 cytokines. However, type 2 response in asthma via ILC2 cells is carried out by interleukin-33 (IL-33) and interleukin-25 (IL-25) cytokines and TSLP. These cytokines, airway epithelial signature proteins, stimulate the ILC2 cells residing in the pulmonary region to produce more IL-5 and IL-13. IL-5 has the major role for mediating the maturation, growth, migration and viability of eosinophils. Cytokine profiles can cluster population in various subsections as they possess diverse patterns in inflammation. The subgroups can be either mixed granulocytic 'IL-5-high and IL-17F-high' or eosinophilic 'IL-4- or IL-13-high' [48].

This Th2-high asthma accounts for about 50% of asthma which can be of mild to moderate severity. They are receptive to corticosteroids. Overall, this Th2-high asthma encompasses the following features: hypersensitiveness to aeroallergen by IgE, epithelial barrier lining the airways get activated and effector cells like mast cells and basophils come into action which in turn lead to airway remodelling. The second type is Th2-low asthma that shows neutrophilic inflammation. They are non-reactive to glucocorticoids and also not receptive to Th2-high treatment [49]. Th2-low asthma is regulated by Th17 cells. This subgroup of T cell causes emission of cytokine IL-17 and interleukin-22 (IL-22) in the bronchial region. It leads to AHR and hypersecretion of mucus followed by obstruction in the airways. Degree of IL-17 is correlated with the seriousness of disease [50]. These cytokines in Th2-low asthma elevates the growth of airway smooth muscle cell and propels deposition of collagen in the airways. Interleukin-8 (IL-8) guides the recruitment of neutrophils. Transient receptor potential vanilloid 1 (TRPV1) channel expressed in the bronchial region is activated, and this incites the production of more neutrophils [51].

IL-13 signature genes like periostin, chloride channel accessory 1 and serpin β2 are considered to be biomarker in Th2-high asthma. Th2-high asthma shows a unique gene phenotype in airway epithelial cells (AECs). Th2 asthmatic patients have high level of exhaled nitric oxide fraction (F_{eNO}). They also exhibit increased level of eosinophils in blood and sputum. Analysis has showed that there is not much difference in the IL-4, IL-5 and IL-13 levels in both Th2-high and Th2-low asthma. Mucosal C-C motif chemokine ligand 26 (CCL26) gene expression was higher in Th2 asthma. Statistical studies showed that still better biomarkers are required for easy detection at bedside level [52].

Interleukin-5, a central Th2 cytokine, helps egress of immature eosinophils from the bone marrow, activates them and recruits them into the airways, increases their survival and also causes the eosinophil degranulation upon allergen induction. IL-5 is another therapeutic target in asthma. Mepolizumab is an anti-IL-5 monoclonal antibody which is of human origin. Mepolizumab administration is safe and causes a decline in the number of circulating eosinophils in tissues and peripheral blood [53]. Mepolizumab got a FDA approval for treating eosinophilic phenotype asthma. Reslizumab is another FDA-approved IL-5 antagonist, monoclonal anti-IL-5 antibody given in severe asthma [54]. Compared to placebo, IL-5 neutralising reslizumab causes reduction in the eosinophilic counts and also improves the airway obstruction happening in eosinophilic asthma [55]. Another drug undergoing clinical trial is benralizumab. Being a humanised mAb, benralizumab acts on the α-subunit of IL-R receptor IL-5αR. Upon administration, benralizumab also causes a decrease in eosinophilic amount in sputum and peripheral blood. A decrease in asthma exacerbations is also observed [56].

2.3.2.2 Neutrophilic Asthma

In case of severe asthma, the Th1 and Th17 pathways also get activated. This then brings about neutrophilic inflammatory asthmatic response [43]. Upon facing insults from different allergens, various chemotactic factors are released. They recruit the inflammatory cells to the site of injury in the lung. The different types of structural cells in the lung like epithelial cell, endothelial cell and fibroblast also release some mediators that promote inflammation. In asthma, both branches of immune response get involved. B cell and T lymphocytes provide the adaptive immunity. Innate immunity is provided by eosinophils, neutrophils, mast cells, macrophages and dendritic cells [11]. Acute asthma often occurs by viruses. In respiratory virus-induced asthma, increased neutrophilic inflammation is observed. There are incidences of heightened neutrophilic degranulation in sputum of asthmatic patients.

Increase in cell lysis was also observed [57]. Interleukin-8 (IL-8) has a role in the neutrophil infiltration and is upregulated in severe asthma. Upon IL-8 stimulation, neutrophils cause transmembrane migration of eosinophils. This in turn leads to neutrophil accumulation in the airways [58]. Detailed analysis of patient's sputum has showed that there are some sub-phenotypes in asthma. Multivariate analysis of clusters was performed. In mild to moderate sub-phenotype of asthma, there is more eosinophil. However, in case of moderate to severe asthma, there is more predominance of neutrophils [59]. Monocyte-derived immune cells, macrophages, have an unclear role in asthma. They can secrete both pro-inflammatory and anti-inflammatory cytokines. However, it is predicted that the anti-inflammatory cytokine is reduced in asthma [11]. There is another study which showed that alveolar macrophages (AM) are the source of pro-inflammatory cytokine Th17. In OVA-induced allergic mice model, the IL-17 positive cells were primarily $CD11b^{+}F4/80^{+}$ macrophages. Neutralising IL-17 or depletion of AM diminishes the effect of OVA-induced asthma inflammation and reduces the infiltration of inflammatory cells [60]. Thus, AM has a variety of roles in the context of immune regulation in asthma. It is like a double-edged player, where, on the one hand, it can promote inflammation and on the other hand its anti-inflammatory role can maintain lung homeostasis. Further studies are required to figure out macrophage's function in different phenotypes of asthma [61].

2.3.3 Asthma: A Complex Syndrome with Multiple Endotypes

Depending on the types of various inflammatory cells in asthmatic sputum, asthma is differentiated into different phenotypes. In addition to eosinophilic and neutrophilic asthma phenotypes, the third phenotype is mixed granulocytic asthma and the last one is steroid-insensitive paucigranulocytic asthma (PGA) [62]. But why do we need such phenotypic and endotypic classifications? In non-Th2 asthma, Th2 cell-targeted asthma therapeutics may not be beneficial [63]. Thus, phenotyping asthma helps in planning better treatment procedure. For doing it, understanding the molecular and cellular mechanism behind this disease is of utmost importance. Thus, a better understanding of these asthma phenotypes will aid in the development of specific biomarkers for each of the phenotypes [64], and phenotype-specific molecules or mechanisms can be targeted for treating asthma.

There are some comorbidities that can further heighten the asthma triggers. Additional disease pathologies found in the upper airway tract are chronic rhinosinusitis, polyposis in nasal areas, allergic rhinitis and so on. These comorbidities in the upper airway tract augment complications and take part in worsening asthma [65]. Hence, it becomes very important to deeply understand the complex pathophysiology of this heterogeneous airway inflammatory disease. Owing to various risk factors, sometimes asthma therapy becomes complicated. It becomes refractory to the standard treatments. One such refractory condition is called steroid-resistant asthma, and this is elaborately discussed in the chapter entitled 'Targeting Cellular and Molecular Mechanisms in Steroid-Resistant Asthma' in this book.

Among various comorbid conditions, **obesity** is found to be an important condition in recent times, thanks to westernisation in food and lifestyle [66]. In mice model, it is noted that the obese mice have airway hyperresponsiveness at an innate level. Such mice also show a heightened allergic response to several asthma-triggering agents [67]. There are many emerging studies which showed that there is some association between asthma and obesity. During obesity, a systemic pro-inflammatory response gets triggered in the adipose tissue. The lung volume capacity is also decreased in obese people with reduction in the peripheral airway diameter, which can further contribute to airway hyperresponsiveness. Alteration in the smooth muscle function was also observed [68]. Another study proposed that the oxidative stress that happens during obesity has a role in augmenting the airway inflammation. These also reduce the efficacy of the inhaled corticosteroids, which are commonly given drugs in asthma [69]. Asthma has also become partly a lifestyle disease.

Western lifestyle also introduced a hygiene hypothesis that has a role in an individual's susceptibility to asthma. Evidences have shown that children who were given antibiotics at a sensitive childhood age have a genetic predisposition to develop asthma [70].

2.4 Nitro-Oxidative Stress in Asthma Pathogenesis

2.4.1 Oxidative Stress

The superoxide (O_2^-), hydroxyl radical and hypochlorous acid that are reactive oxygen species (ROS) are constantly formed in the lung for physiological reasons to guard against a number of environmental irritants including microbes. However, in asthmatic condition, there is a release of more ROS into the airways by various recruited inflammatory cells like Th2 cells, eosinophils, etc. The dominant role of ROS in causing various asthma features has been demonstrated in earlier literature [71, 72]. On the other hand, the lung also produces antioxidants. Once the balance between oxidants and antioxidants is disturbed, radicals can both damage the cell and initiate, amplify the inflammatory response. In the cell, these radicals damage DNA, lipids, proteins and carbohydrates. The lipid peroxidation product, 8-isoprostane, is a known bronchoconstrictor [71]. Similarly, oxidative DNA damage product, 8-hydroxydeoxyguanosine, also leads to bronchoconstriction. Excessive oxidative free radicals can also induce apoptosis of airway epithelium. This could further induce the airway remodelling by activating epithelial mesenchymal trophic unit (EMTU) to cause further airway hyperresponsiveness. In addition, oxidative free radicals combine with nitric oxide (NO) free radicals to form peroxynitrite, which is a powerful bronchoconstrictor. Overall, oxidative stress seems to play a detrimental role in asthma pathogenesis. Despite being beneficial in animal models of asthma, exogenous antioxidants have not shown any drastic improvement in asthma. Such discrepancy could either be due to species difference or it could be also due to physiological benefits of ROS. In any event, using antioxidants in lung diseases becomes controversial as it may also lead to development of lung cancer.

2.4.2 Mitochondrial Dysfunction in Asthma

On the one hand, oxidative stress has a major role in causing airway epithelial injury, and on the other hand, mitochondria are the major sources of reactive oxygen species. So theoretically, one can expect dysfunctional mitochondria in asthma pathogenesis. Our lab and others have demonstrated such mitochondrial dysfunction in bronchial epithelium of asthmatic subjects [73, 74]. Further, we have demonstrated how 15-lipoxygenase (15-LOX) leads to mitochondrial dysfunction. Though 5-LOX has been studied in detail in asthma pathogenesis, detailed studies of 15-LOX revealed more interesting observation in asthma pathogenesis [75, 76]. Further, 15-LOX inhibitors like esculetin and baicalein have attenuated asthma features [74]. Also, we have further demonstrated that mesenchymal stem cell can replace the dysfunctional mitochondria present in airway epithelia which can in turn cause attenuation of asthmatic features [77]. All these indicate that improvement of mitochondrial function in airway epithelia could be an attractive therapeutic strategy. This also emphasises the importance of airway epithelia such that epithelial protective agents could be an effective therapeutic strategy in the future. Traditionally, airway epithelium was considered as a victim of immune cells. In contrast to earlier belief, a number of recent literatures indicate that a healthy airway epithelium is crucial to have a healthy lung.

2.4.3 Nitric Oxide (NO)

Nitric oxide is produced from L-arginine, catalysed by constitutive nitric oxide synthases (cNOS), namely, endothelial NOS (eNOS) and neuronal NOS (nNOS), and their isoform, inducible nitric oxide synthase (iNOS). Under normal physiological conditions, NO produced from cNOS functions as a bronchodilator and has an anti-inflammatory action in the airways through the cyclic guanosine monophosphate (cGMP) pathway [78]. NO produced by iNOS reacts with superoxide and leads to peroxynitrite radical formation which has negative effects like inflammation and bronchoconstriction in the airways [78]. L-arginine is also utilised as a substrate by arginase, the last hepatic urea cycle enzyme, which catalyses the conversion of L-arginine to L-ornithine and urea. It also occurs in non-hepatic tissues like the airways and has two isoforms – arginase1 and arginase 2. cNOS, iNOS and arginase vie for their common substrate L-arginine. A rise in iNOS expression in asthmatic condition limits the bioavailability of L-arginine for cNOS. In asthma, there is also an upregulation of arginase by transforming growth factor β (TGFβ) and Th2 cytokines like IL-4 and IL-13, reducing the substrate availability of cNOS [79]. This competition for substrate causes depletion of L-arginine in the asthmatic airways. Previous work from our lab has shown how L-arginine supplementation at high doses can alleviate allergic airway inflammation [78]. The harmful form of NO, made from arginine utilisation by iNOS, causes L-arginine depletion and promotes peroxynitrite production. Peroxynitrite causes mitochondrial dysfunction by haem degradation in cytochrome c oxidase, a crucial mitochondrial electron transport chain enzyme. Our lab has demonstrated that L-arginine administration is able to lower peroxynitrite generation and rescue the impaired mitochondrial function in allergic airway inflammation [80].

Asymmetric dimethylarginine (ADMA), an endogenous NOS antagonist, is increased in asthmatic condition. It is a product of protein methylation formed during post-translational modification catalysed by protein methyltransferases. ADMA is a naturally occurring uncoupler for the all NOS isoforms, halting NO production by the NOS, which promotes superoxide production. Besides, an increased level of ADMA is also found to upregulate arginase activity [81]. Rise in arginase activity boosts L-ornithine production that functions as a precursor for polyamines and L-proline. Polyamines are involved in cell proliferation, whereas L-proline promotes collagen production and fibrosis [79]. Indeed, ADMA-induced arginase upregulation is found to contribute to collagen deposition and airway hyperresponsiveness in murine lungs [81].

NO, derived from iNOS activity, is highly prevalent in asthmatics. However, considerable amount of variation has been found in the FeNO in asthma patients, with some patients having FeNO levels within normal range. Nevertheless, a major population of asthmatics with high FeNO levels represent a distinct phenotype defined by their higher arginine metabolism and a more severe form of the disease. The higher arginine metabolism is attributed to the increased levels of iNOS and arginase present in their airways [82]. This highlights the clinical relevance of understanding the role of arginine biology in asthma.

2.5 Airway Remodelling in Asthma: The Knowledge that Will Shape the Future Treatment

Now, it is very clear that the airway epithelium works beyond mere mechanical barrier in the airways. It indeed maintains healthy lung homeostasis by protecting against obnoxious particles entering the airways [83]. The injury to the airway epithelium is traditionally believed as a converging event of various inflammatory pathways. In contrast to this view, recent evidences indicate that epithelial injury and its subsequent airway remodelling changes can be parallel set of events happening along with airway inflammation. This is the possible reason why anti-inflammatory agents could not reduce the initiation of irreversible airway remodelling changes. The challenges faced by the epithelium every minute are commendable, and its structural alteration leads to 'airway remodelling' [83].

Defining 'airway remodelling' is very challenging. It is a collective term to indicate the overall changes happening in the airway wall. Almost every cell type in the airways increases in number and size that further causes obstruction in the airway diameter. Airway narrowing, which occurs in early stages of asthma, seems to be physiological and temporary, with smooth muscle contraction and mucosal oedema. However, thanks to the airway remodelling changes, airway narrowing that occurs in late stages of asthma seems to be a combination of physiological and anatomical changes. As a result, even without allergen exposure, at a later stage, asthmatic patients constantly have airway narrowing. As a consequence, there occurs a continuous difficulty in breathing in later stage compared to earlier stages. Airway remodelling is thus a characteristic hallmark of asthma that has a notable functional implication in asthma pathogenesis.

Epithelial alterations in asthmatic airway remodelling are characterised by features like epithelial cell shedding, goblet cell hyperplasia and metaplasia, loss of ciliated cells [84] and compensatory hypertrophy and hyperplasia of epithelial cells. Clinically, the extent of epithelial injury is correlated to airway hyperresponsiveness [83]. **Sub-epithelial fibrosis**, another important feature of airway remodelling, is due to the dense deposition of a variety of extracellular matrix (ECM) proteins like proteoglycan, tenascin and collagens, predominantly in the lamina reticularis [85]. The deposition of ECM proteins in sub-epithelial regions is facilitated by the transformation of fibroblasts to myofibroblasts. Myofibroblasts are a kind of hybrid cells that have the functions of both muscle and fibroblasts. These hybrid cells are majorly responsible for the overall thickening of airway wall. The imbalance between proteases and antiproteases may favour a profibrotic balance. A number of matrix metalloproteinases (MMPs) like MMP-2, MMP-3, MMP-9 and MMP-12 are seen to have a role in asthmatic airway remodelling [84]. **Hypertrophy and hyperplasia of airway smooth muscle cells (ASM) are other key features in airway remodelling. This leads to** prominence of smooth muscle mass that has been correlated to clinical severity and duration of asthma. Hypertrophy and hyperplasia seem to comprehend both the small and large airways. Distinctly, various molecular mechanisms, involving pro-inflammatory cytokines, chemokines and ECM proteins, can be attributed to causing hyperplasia or hypertrophy [86]. The **goblet cell and mucous gland hyperplasia leads to** bronchial mucous plugging [83, 84] in the airways of chronic asthma. This causes enhanced production of sputum, airway narrowing and congestion from sputum secretion followed by increase in the thickness of airway walls [86]. The interstices and folds of the airway surface are filled with liquid and mucus. These features could be conducive, leading to the narrowing of the airways by amplifying the surface tension at the air-liquid interface. **Angiogenesis, a part vascular remodelling,** includes increased vascularity, vasodilatation, angiogenesis and leakage in microvascular region [84]. Various cytokines, mediators released from mast cells like histamine, leukotriene B_4 and prostaglandins, hypoxia-inducible factor 1 (HIF-1) and vascular endothelial growth factor (VEGF) [84] are important factors contributing to angiogenesis.

Even though it was believed that inflammation-induced airway remodelling occurs via epithelial injury, the incidences of airway remodelling changes even in young children indicate the possibility of inflammation and its subsequent repair-induced airway remodelling. This further reemphasises the need to explore the possible genes involved in airway remodelling happening in asthma pathogenesis [83, 84, 87]. More research is required to explore the possible surrogate markers of future airway remodelling that could help us to segregate and treat high-risk populations at early stages.

2.6 Non-pharmacological Intervention for Asthma: Bronchial Thermoplasty

Medication is important in the treatment of asthma, to prevent and control asthma attacks. However, many people like to do something more than just taking medications. Such non-pharmacologic approaches for asthma like focused breathing techniques, pulmonary rehabilitation, avoiding triggers and bronchial thermoplasty [88] have a significant, yet underappreciated, role. In asthmatic patients who are unresponsive to the medical therapies and have evidence of airway remodelling, bronchial thermoplasty is carried out. Those patients not only exhibited long-term improvement in the quality of life but also showed reduced severity of exacerbations and health-care utilisation.

Asthma is characterised by the sub-epithelial deposition of collagen and remodelling of the bronchial architecture. In late asthma, permanent airway narrowing and airflow obstruction in asthmatic patients further leads to increase in the airway smooth muscle mass. This causes a gradual decline in lung functions [89]. Hence, a new concept of bronchial thermoplasty (BT) has been developed [88–90] that intents to decrease the airway smooth muscle mass and sub-epithelial fibrosis. The objective of bronchial thermoplasty is to modify the airway remodelling and diminish the bronchial constriction, thus ameliorating asthma symptoms. Thus, thermoplasty is a new therapeutic option that has to be used along with other pharmacological options in the treatment of uncontrolled severe asthma.

When thermal energy is applied to airway wall in a controlled manner, it heats the tissue that causes a reduction in the airway smooth muscle mass. A thermal energy of 65 °C temperature is delivered through bronchial thermoplasty system [90]. It comprises a sterile, single-use catheter and a radiofrequency electrical generator. BT targets for treating the distal intra-parenchymal airways up to the main stem bronchi, down to airways 3 mm in diameter. BT promotes a marked improvement in the quality of life, along with a decrease in steroid use and reduced number of patients visiting the emergency room with exacerbations. The improvement monitored is not short term; rather, it seems to be committed for a long-term positive impact. A large number of controlled human trials have shown FEV_1 improvements along with improvements in the lung function [91]. However, thermoplasty-treated patients have shown adverse effects in the form of worsening and infection of the upper respiratory tract occurring during the first week of the thermoplasty session.

Although BT seems to be tedious, it is an attractive strategy that specifically targets ASM and improves the standard of asthma-related life. Therefore, future studies are called for in order to explore the possible mechanisms behind the improvements. There is also a pressing need to know the asthmatics' phenotypes that get the maximal benefits from BT as no uniform improvement has been seen among asthmatics.

2.7 Tailoring the Patient Needs Via 'Personalised Approach' in Asthma

Now that we are living in an era of 'magic bullets', a personalised approach to asthma will undoubtedly be a fairy tale for asthmatic patients who are striving to cope up with the difficulties. Personalised medicine is an emerging approach (opposite of the approach 'one size fits all') to treat and prevent disease by assessing the individual's variability in genes, environment and psychosocial characteristics.

Now we have moved from traditional gross phenotypic classification of asthma to refined molecular classification called endotypic classification [92]. After correlating with clinical and cellular phenotypes, various endotypes have been demonstrated in asthmatic patients in the form of clusters. Such endotypic classification is gradually being refined with aid from other technological advancements like analysis of genes and proteins through genomic, microbiolomic and metabolonomic strategies [92]. As a result, we can pinpoint the dominant biochemical pathways that are responsible for the appearances of symptom in a given patient. Thus, this approach can be called 'treating the right patient with the right drug at the right time' [93]. Biomarkers are surrogate measures that refer several indicators. They can be measured and evaluated to assess the normalcy of pathological state or any physiological state. Endotypes are thus significantly different forms of classification from phenotypes and distinctly describe the entities of the disease with a defining aetiology and/or a pathophysiological mechanism.

Asthma is considered as an umbrella condition. It comprises numerous distinct diseases (endotypes) caused by distinct pathological mechanisms. Hence, understanding the underlying heterogeneity is very vital. The observable characteristics (phenotype) in asthma are the various physiological, clinical, morphological and biochemical characteristics based upon which medications are provided [51]. All the asthma patients manifest with the similar symptoms of wheezing or squealing, chest tightness and shortness of breath and fatigue and are accompanied by variable airflow obstruction. Despite this similarity in symptoms, there is a huge variation in response to common therapy. One such example of a treatment is omalizumab that works only for IgE-dominant asthmatic patients [94]. The mechanism of action for omalizumab is focused predominantly on reducing the circulatory IgE. This further inhibits its binding to FcεRI receptors of mast cells and basophils. Sometimes, omalizumab treatment results in exacerbations of symptoms. Similar to IgE, there are many other biomarkers like sputum eosinophils, sputum neutrophils, blood eosinophil count and fractional exhaled nitric oxide. These biomarkers could help respiratory clinicians to subcategorise asthmatic patients like eosinophil-dominant airway inflammation and neutrophil-dominant airway inflammation. Such sub-classification can in turn help clinicians to choose corticosteroid for eosinophilic-dominant patients. Similarly, some biomarkers like periostin has been used as a promising surrogate marker of severe asthma [51]. Hence, it has become very crucial to look for more generic medicines instead of population-based medications. Thus, with the available huge number of biologic agents, a befitting endotypic classification with well-characterised biomarkers is essential to correlate the molecular data to phenotypes and provide tailored medicines for asthma patients.

2.8 Conclusions and Future Perspectives

Asthma is not a single entity; instead, it's a heterogeneous, complex disorder with overlapping syndromes which makes it difficult to treat. The traditional approach of 'blockbuster drugs' to treat asthma is losing their edge as the drugs available in the markets are unable to alter the course of the disease. Thus, there is a pressing need to adopt the 'personalised approach' based on the endotype-driven classification of asthma. In this report, we have elucidated the exigency to understand the underlying disease mechanisms in different individuals with different endotypic asthma. We have also presented the idea of various cellular and molecular drivers of allergic airway inflammation that can be targeted to design novel therapies in the near future. The concept of asthma has continuously evolved, from being presumed to be an intrinsic smooth muscle abnormality of the airways to the notion accrediting the 'inflammation theory' for the asthmatic diathesis. Recently, the researchers are giving weightage to the concept of 'airway remodelling' as the primary event responsible for effectuating asthma. We have also highlighted the important facet of airway remodelling. A milieu of factors is involved in mediating the airway structural alterations including cytokines (IL-5, IL-13 and TGF-β), growth factors (VEGF) and MMPs (MMP-9, MMP-12). Targeting these genes will undoubtedly present us with a curative pharmacological intervention in the immediate future. However, caution is required in targeting MMPs as it could create havoc in lung homeostasis as they are required to act against various microbes that are entering into the airways from the environment.

Steroid-resistant asthma handicaps the situation of treating patients with corticosteroids and has emerged as a global problem for the clinicians. With the various inconclusive underlying mechanisms, like neutrophilia, obesity, bacterial or viral infection, etc., that contribute to the steroid refractory condition, it has become obligatory to understand the mechanisms intricately in order to identify novel therapeutic treatments. We have discussed about steroid-resistant asthma elaborately in the chapter entitled 'Targeting Cellular and Molecular Mechanisms in Steroid-Resistant Asthma'. The expenses of asthma due to frequent hospitalisations and the intangible expense in terms of reduced standard of living have increased due to poor management and knowledge of the exact underlying mechanism. Future research should focus on personalisation of anti-asthmatic therapeutic strategy to satiate the individual needs based on the distinct disease endotypes.

Acknowledgements This work was supported by the projects MLP137 (MISSION LUNG) at CSIR-Indian Institute Chemical Biology, Council of Scientific and Industrial Research, Govt. of India. SS, JD, AR and AJ acknowledge the Academy of Scientific and Innovative Research (AcSIR), Ghaziabad, India, for their PhD registrations.

References

1. World Health Organisation (2020) Asthma. https://www.who.int/news-room/q-a-detail/asthma Accessed 10 Sep 2020
2. Dharmage SC, Perret J, Custovic A (2019) Epidemiology of asthma in children and adults. Front Pediatr 7:246
3. Masoli M, Fabian D, Holt S et al (2004) The global burden of asthma: executive summary of the GINA Dissemination Committee report. Allergy 59:469–478
4. Lugogo NL, Kraft M (2006) Epidemiology of asthma. Clin Chest Med 27:1–15
5. Smith KR (2000) Inaugural article: national burden of disease in India from indoor air pollution. Proc Natl Acad Sci U S A 97:13286–13293
6. Romanet-Manent S, Charpin D, Magnan A, Lanteaume A, Vervloet D, EGEA Cooperative Group (2002) Allergic vs nonallergic asthma: what makes the difference? Allergy 57:607–613
7. Robinson DS, Hamid Q, Ying S et al (1999) Predominant TH2-like bronchoalveolar T-lymphocyte population in atopic asthma. N Engl J Med 326:298–304
8. Pearce N, Pekkanen J, Beasley R (1999) How much asthma is really attributable to atopy? Thorax 54:268–272
9. Toskala E, Kennedy DW (2015) Asthma risk factors. Int Forum Allergy Rhinol 5:S11–S16
10. Boushey HA, Fahy JV (1995) Basic mechanisms of asthma. Environ Health Perspect 103:229–233
11. Barnes PJ (2017) Cellular and molecular mechanisms of asthma and COPD. Clin Sci 131:1541–1558
12. Cazzola M, Segreti A, Matera MG (2010) Novel bronchodilators in asthma. Curr Opin Pulm Med 16:6–12
13. Cazzola M, Rogliani P, Calzetta L, Matera MG (2019) Bronchodilators in subjects with asthma-related comorbidities. Respir Med 151:43–48
14. Barnes PJ (2013) Theophylline. Am J Respir Crit Care 188:901–906
15. Alangari AA (2014) Corticosteroids in the treatment of acute asthma. Ann Thorac Med 9:187–192
16. Trinh HKT, Lee SH, Cao TBT, Park HS (2019) Asthma pharmacotherapy: an update on leukotriene treatments. Expert Rev Respir Med 13:1169–1178
17. Frey A, Lunding LP, Ehlers JC et al (2020) More than just a barrier: the immune functions of the airway epithelium in asthma pathogenesis. Front Immunol 11:761
18. Gill MA (2012) The role of dendritic cells in asthma. J Allergy Clin Immunol 129:889–901
19. Kuipers H, Lambrecht BN (2004) The interplay of dendritic cells, Th2 cells and regulatory T cells in asthma. Curr Opin Immunol 16:702–708
20. Idzko M, Hammad H, van Nimwegen M et al (2006) Local application of FTY720 to the lung abrogates experimental asthma by altering dendritic cell function. J Clin Invest 116:2935–2944
21. Marciniak A, Camp SM, Garcia JGN, Polt R (2018) An update on sphingosine-1-phosphate receptor 1 modulators. Bioorg Med Chem Lett 28:3585–3591
22. Zhao ST, Wang CZ (2018) Regulatory T cells and asthma. J Zhejiang Univ Sci B 19:663–673
23. Nelson RK, Bush A, Stokes J et al (2020) Eosinophilic asthma. J Allergy Clin Immunol Pract 8:465–473
24. Moran A, Pavord ID (2020) Anti-IL-4/IL-13 for the treatment of asthma: the story so far. Expert Opin Biol Ther 20:283–294
25. Ingram JL, Kraft M (2012) IL-13 in asthma and allergic disease: asthma phenotypes and targeted therapies. J Allergy Clin Immunol 130:829–844

26. Majumdar S, Ghosh A, Saha S (2018) Modulating interleukins and their receptors interactions with small chemicals using in silico approach for asthma. Curr Top Med Chem 18:1123–1134
27. Ridolo E, Pucciarini F, Nizi MC et al (2020) Mabs for treating asthma: omalizumab, mepolizumab, reslizumab, benralizumab, dupilumab. Hum Vaccin Immunother:1–8
28. Bagnasco D, Ferrando M, Varricchi G, Passalacqua G, Canonica GW (2016) A critical evaluation of anti-IL-13 and anti-IL-4 strategies in severe asthma. Int Arch Allergy Immunol 170:122–131
29. Corren J, Busse W, Meltzer EO et al (2010) A randomized, controlled, phase 2 study of AMG 317, an IL-4Ralpha antagonist, in patients with asthma. Am J Respir Crit Care Med 181:788–796
30. Kasaian MT, Marquette K, Fish S et al (2013) An IL-4/IL-13 dual antagonist reduces lung inflammation, airway hyperresponsiveness, and IgE production in mice. Am J Respir Cell Mol Biol 49:37–46
31. Venkataramani S, Low S, Weigle B et al (2018) Design and characterization of Zweimab and Doppelmab, high affinity dual antagonistic anti-TSLP/IL13 bispecific antibodies. Biochem Biophys Res Commun 504:19–24
32. Hambly N, Nair P (2014) Monoclonal antibodies for the treatment of refractory asthma. Curr Opin Pulm Med 20:87–94
33. Pelaia G, Canonica GW, Matucci A, Paolini R, Triggiani M, Paggiaro P (2017) Targeted therapy in severe asthma today: focus on immunoglobulin E. Drug Des Devel Ther 11:1979–1987
34. Gould HJ, Sutton BJ (2008) IgE in allergy and asthma today. Nat Rev Immunol 8:205–217
35. Busse WW, Humbert M, Haselkorn T et al (2020) Effect of omalizumab on lung function and eosinophil levels in adolescents with moderate-to-severe allergic asthma. Ann Allergy Asthma Immunol 124(2):190–196
36. Hu J, Chen J, Ye L, Cai Z, Sun J, Ji K (2018) Anti-IgE therapy for IgE-mediated allergic diseases: from neutralizing IgE antibodies to eliminating IgE^+ B cells. Clin Transl Allergy 8:27
37. Krystel-Whittemore M, Dileepan KN, Wood JG (2016) Mast cell: a multi-functional master cell. Front Immunol 6:620
38. Brightling CE, Bradding P, Symon FA et al (2002) Mast-cell infiltration of airway smooth muscle in asthma. N Engl J Med 346:1699–1705
39. Bradding P, Walls AF, Holgate ST (2006) The role of the mast cell in the pathophysiology of asthma. J Allergy Clin Immunol 117:1277–1284
40. Zhang T, Finn DF, Barlow JW, Walsh JJ (2016) Mast cell stabilisers. Eur J Pharmacol 778:158–168
41. Asfour MH, Kassem AA, Salama A, Abd El-Alim SH (2020) Hydrophobic ion pair loaded self-emulsifying drug delivery system (SEDDS): a novel oral drug delivery approach of cromolyn sodium for management of bronchial asthma. Int J Pharm 585:119494
42. Sinniah A, Yazid S, Bena S et al (2019) Endogenous Annexin-A1 negatively regulates mast cell-mediated allergic reactions. Front Pharmacol 10:1313
43. Pelaia G, Vatrella A, Busceti MT et al (2015) Cellular mechanisms underlying eosinophilic and neutrophilic airway inflammation in asthma. Mediat Inflamm 2015:879783
44. Frigas E, Gleich GJ (1986) The eosinophil and the pathophysiology of asthma. J Allergy Clin Immunol 77:527–537
45. Green RH, Brightling CE, McKenna S et al (2002) Asthma exacerbations and sputum eosinophil counts: a randomised controlled trial. Lancet 360:1715–1721
46. Kankaanranta H, Lindsay MA, Giembycz MA et al (2000) Delayed eosinophil apoptosis in asthma. J Allergy Clin Immunol 106:77–83
47. Rosenberg HF, Phipps S, Foster PS (2007) Eosinophil trafficking in allergy and asthma. J Allergy Clin Immunol 119:1303–1310
48. Sven S, Scheers H, Marijsse G et al (2015) Th2-high asthma: a heterogeneous asthma population? Clin Transl Allergy 5:O1

49. Fahy JV (2015) Type 2 inflammation in asthma—present in most, absent in many. Nat Rev Immunol 15:57–65
50. Robinson D, Humbert M, Buhl R et al (2017) Revisiting type 2-high and type 2-low airway inflammation in asthma: current knowledge and therapeutic implications. Clin Exp Allergy 47:161–175
51. Kuruvilla ME, Lee FEH, Lee GB (2019) Understanding asthma phenotypes, endotypes, and mechanisms of disease. Clin Rev Allergy Immunol 56:219–233
52. Pavlidis S, Takahashi K, Kwong FNK et al (2019) "T2-high" in severe asthma related to blood eosinophil, exhaled nitric oxide and serum periostin. Eur Respir J 53:1800938
53. Garrett JK, Jameson SC, Thomson B et al (2014) Anti-interleukin-5 (mepolizumab) therapy for hypereosinophilic syndromes. J Allergy Clin Immunol 113:115–119
54. Maselli DJ, Velez MI, Rogers L (2016) Reslizumab in the management of poorly controlled asthma: the data so far. J Asthma Allergy 9:155–162
55. Castro M, Mathur S, Hargreave F et al (2011) Reslizumab for poorly controlled, eosinophilic asthma: a randomized, placebo-controlled study. Am J Respir Crit Care Med 184:1125–1132
56. Tabatabaian F, Ledford DK, Casale TB (2017) Biologic and new therapies in asthma. Immunol Allergy Clin N Am 37:329–343
57. Wark PAB, Johnston SL, Moric I et al (2002) Neutrophil degranulation and cell lysis is associated with clinical severity in virus-induced asthma. Eur Respir J 19:68–75
58. Nakagome K, Matsushita S, Nagata M (2012) Neutrophilic inflammation in severe asthma. Int Arch Allergy Immunol 158:96–102
59. Moore WC, Hastie AT, Li X et al (2014) Sputum neutrophil counts are associated with more severe asthma phenotypes using cluster analysis. J Allergy Clin Immunol 133:1557–1563
60. Song C, Luo L, Lei Z et al (2008) IL-17-producing alveolar macrophages mediate allergic lung inflammation related to asthma. J Immunol 181:6117–6124
61. Balhara J, Gounni AS (2012) The alveolar macrophages in asthma: a double-edged sword. Mucosal Immunol 5:605–609
62. Tliba O, Panettieri RA Jr (2019) Paucigranulocytic asthma: uncoupling of airway obstruction from inflammation. J Allergy Clin Immunol 143:1287–1294
63. Ray A, Oriss TB, Wenzel SE (2015) Emerging molecular phenotypes of asthma. Am J Physiol Lung Cell Mol Physiol 308:L130–L140
64. Wenzel SE (2006) Asthma: defining of the persistent adult phenotypes. Lancet 368:804–813
65. Licari A, Brambilla I, De Filippo M et al (2017) The role of upper airway pathology as a co-morbidity in severe asthma. Expert Rev Respir Med 11:855–865
66. Agrawal A, Mabalirajan U, Ahmad T, Ghosh B (2011) Emerging interface between metabolic syndrome and asthmaAm J Respir. Cell Mol Biol 44:270–275
67. Shore SA, Johnston RA (2006) Obesity and asthma. Pharmacol Ther 110:83–102
68. Beuther DA, Weiss ST, Sutherland ER (2006) Obesity and asthma. Am J Respir Crit Care 174:112–119
69. Holguinn F, Fitzpatrick A (2010) Obesity, asthma, and oxidative stress. J Appl Physiol 108:754–759
70. Droste JHJ, Wieringa MH, Weyler JJ et al (2000) Does the use of antibiotics in early childhood increase the risk of asthma and allergic disease? Clin Exp Allergy 30:1548–1553
71. Barnes PJ, Chung KF, Page CP (1998) Inflammatory mediators of asthma: an update. Pharmacol Rev 50:515–596
72. Nadeem A, Chhabra SK, Masood A et al (2003) Increased oxidative stress and altered levels of antioxidants in asthma. J Allergy Clin Immunol 111:72–78
73. Mabalirajan U, Dinda AK, Kumar S et al (2008) Mitochondrial structural changes and dysfunction are associated with experimental allergic asthma. J Immunol 181:3540–3548
74. Mabalirajan U, Dinda AK, Sharma SK et al (2009) Esculetin restores mitochondrial dysfunction and reduces allergic asthma features in experimental murine model. J Immunol 183:2059–2067
75. Mabalirajan U, Rehman R, Ahmad T et al (2013a) 12/15-lipoxygenase expressed in non-epithelial cells causes airway epithelial injury in asthma. Sci Rep 3:1540

76. Mabalirajan U, Rehman R, Ahmad T et al (2013b) Linoleic acid metabolite drives severe asthma by causing airway epithelial injury. Sci Rep 3:1349
77. Ahmad T, Mukherjee S, Pattnaik B et al (2014) Miro1 regulates intercellular mitochondrial transport & enhances mesenchymal stem cell rescue efficacy. EMBO J 33:994–1010
78. Mabalirajan U, Ahmad T, Leishangthem GD et al (2010a) Beneficial effects of high dose of L-arginine on airway hyperresponsiveness and airway inflammation in a murine model of asthma. J Allergy Clin Immunol 125:626–635
79. Meurs H, Zaagsma J, Maarsingh H et al (2019) Recent patents in allergy/immunology: use of arginase inhibitors in the treatment of asthma and allergic rhinitis. Allergy 74:1206–1208
80. Mabalirajan U, Ahmad T, Leishangthem GD et al (2010b) L-arginine reduces mitochondrial dysfunction and airway injury in murine allergic airway inflammation. Int Immunopharmacol 10:1514–1519
81. Tajti G, Papp C, Kardos L et al (2018) Positive correlation of airway resistance and serum asymmetric dimethylarginine (ADMA) in bronchial asthma patients lacking evidence for systemic inflammation. Allergy Asthma Clin Immunol 14:2
82. Xu W, Comhair S, Janocha AJ et al (2017) Arginine metabolic endotypes related to asthma severity. PLoS One 12:e0183066
83. Fehrenbach H, Wagner C, Wegmann M (2017) Airway remodeling in asthma: what really matters. Cell Tissue Res 367:551–569
84. Bergeron C, Al-Ramli W, Hamid Q (2009) Remodeling in asthma. Proc Am Thorac Soc 6:301–305
85. Hough KP, Curtiss ML, Blain TJ et al (2020) Airway Remodeling in asthma. Front Med 7:191
86. Elias JA (2000) Airway remodeling in asthma.Unanswered questions. Am J Respir Crit Care 161:S168–S171
87. Ergan-Arsava B, Karakaya G, Firat P et al (2011) Smooth muscle hyperplasia in an asthmatic patient: do we know it all? Respiration 81:152–156
88. Hall C, Nici L, Sood S et al (2017) Nonpharmacologic therapy for severe persistent asthma. J Allergy Clin Immunol Pract 5:928–935
89. Singh SK, Tiwari KK (2017) Bronchial thermoplasty: a non-pharmacological approach. Clin Respir J 11:13–20
90. Wahidi MM, Kraft M (2012) Bronchial thermoplasty for severe asthma. Am J Respir Crit Care 185:709–714
91. Laxmanan B, Hogarth DK (2015) Bronchial thermoplasty in asthma: current perspectives. J Asthma Allergy 8:39–49
92. Svenningsen S, Nair P (2017) Asthma endotypes and an overview of targeted therapy for asthma. Front Med 4:158
93. Weiss ST (2012) New approaches to personalized medicine for asthma: where are we? J Allergy Clin Immunol 129:327–334
94. Godar M, Blanchetot C, de Haard H et al (2018) Personalized medicine with biologics for severe type 2 asthma: current status and future prospects. MAbs 10:34–45

Various Cellular and Molecular Axis Involved in the Pathogenesis of Asthma

3

Gurjit Kaur Bhatti, Amit Khurana, Debapriya Garabadu, Prashant Gupta, Snehal Sainath Jawalekar, Jasvinder Singh Bhatti, and Umashanker Navik

Abstract

Asthma is a chronic inflammatory disease described by impaired lung function, airway hyperresponsiveness, episodic wheezing, and dyspnea. Asthma prevalence has risen drastically, and it is estimated that more than 339 million individuals worldwide had asthma with marked heterogeneity in pathophysiology

G. K. Bhatti
Department of Medical lab Technology, University Institute of Applied Health Sciences, Chandigarh University, Mohali, Punjab, India

A. Khurana
Centre for Biomedical Engineering (CBME), Indian Institute of Technology (IIT), Delhi, India

Department of Veterinary Pharmacology and Toxicology, College of Veterinary Science (CVSc), PVNRTVU, Hyderabad, Telangana, India

D. Garabadu
Department of Pharmacology, Central University of Punjab, Bathinda, India
e-mail: debapriya.garabadu@cup.edu.in

P. Gupta
Department of Pharmacology, Shree S K Patel College of Pharmaceutical Education and Research, Mehsana, Gujrat, India

S. S. Jawalekar
Department of Biotechnology, National Institute of Pharmaceutical Education and Research, S.A.S. Nagar, Punjab, India

J. S. Bhatti
Department of Human Genetics and Molecular Medicine, School of Health Sciences, Central University of Punjab, Bathinda, India

U. Navik (✉)
Department of Pharmacology, School of basic and applied Sciences, Central University of Punjab, Bathinda, India

K. Dua et al. (eds.), *Targeting Cellular Signalling Pathways in Lung Diseases*,
https://doi.org/10.1007/978-981-33-6827-9_3

and etiology. Several factors involved in the progression and development of asthma include allergens, pollutants, obesity, viruses, antigens, and many more, eliciting strong inflammatory and immune responses, causing airflow obstruction, and tightening of respiratory smooth muscle causing the characteristic asthma symptoms. Multiple complex molecular pathways are involved in asthma pathophysiologies such as immunoglobulin E, cytokines, nitric oxide, dendritic cells, leukotrienes, oxidative stress, and inflammatory infiltrate of mast cells, neutrophils, eosinophils, lymphocytes, innate immunity, and many more. The current chapter focuses on illustrating the various molecular pathways that contribute to asthma development and its progression.

Keywords

Asthma · Airway hyperresponsiveness · Cytokines · Inflammation · Immunity

3.1 Introduction

Asthma is a clinical syndrome characterized as chronic inflammation of airways, bronchial hyperresponsiveness with reversible airway obstruction, wheezing, and increased mucus production, ultimately leading to breathing problems, particularly at night or in the early morning [1]. Asthma is associated with impaired lung function parameters such as either diminished peak expiratory flow or forced expiratory volume in 1 second [2]. Asthma is characterized by multiple reasons including genetic and environmental factors. Asthma of childhood and young adults is associated with atopic asthma (triggered by an allergen) leading to modulation of the Ig (immunoglobulin) E-dependent pathway. However, asthma that emerges later in life is non-atopic asthma (non-allergic) and most commonly reported in women and smokers [3, 4]. Asthma in the adult is often non-allergic to the normal level of IgE, and usually, treatment will not work. Patients are generally intolerant to nonsteroidal anti-inflammatory drugs leading to diminished prostaglandin E_2, thereby presenting inflammatory conditions, rhinosinusitis, and nasal polyps [5, 6]. Atopic asthma is the most common form of endotype of disease representing over 60% of cases. In contrast, phenotype non-atopic eosinophilic represents about 25–30% of the cases and 5–10% of patients having severe refractory asthma. Clinical evidence showed that the pathology of asthma involved 55% of eosinophilic forms, 20% of neutrophilic forms, 18% of paucigranulocytic forms, and 6% of mixed forms [7].

In the nineteenth century, Henry Hyde Salter has contributed to understanding asthma pathology. In 1908, the first asthma patient died and presented eosinophilic infiltration in the airways [8, 9]. However, until now, the inflammation was not believed to be involved in asthma pathology. Later in the 1980s, the role of inflammation has been reported using fiberoptic bronchoscopy, and subsequent involvement of other mediators including Th (T helper) 2 cells, Th2 cytokines, and CD (cluster of differentiation) 4 Th subsets has been reported in the airways of asthmatics [8]. Besides, several molecular mechanisms are also involved in the pathogenesis of asthma, including mast cells, eosinophil, neutrophils, macrophages, cytokines, autacoids, nitric oxide, epithelial cells, and many more. Therefore, this chapter aims to concentrate primarily on the various cellular and molecular axes involved in the pathophysiology of asthma.

3.2 Epidemiology and Risk Factors

Worldwide, asthma is ranked 16th among the foremost causes of years of life lived with disability and 28th among the leading causes of the burden of disease, as measured by disability-adjusted life years [10]. According to the report of the World Health Organization (WHO), there are more than 339 million people in the world had asthma, and it is a common disease among children. Asthma-related deaths was most commonly observed in low- and lower-middle-income countries, but the increase in occurrence seems to have plateaued in some developed countries, while asthma prevalence is higher among high-income countries. Globally, there were 417,918 deaths reported due to asthma [11]. Out of the total cases reported, 25 million of asthmatics were Americans, of which 6.4 million are children. Childhood asthma is more common among boys, while adult asthma is commonly observed among women indicating that sex hormones play a crucial role in asthma pathology [10, 12]. Since 1960, there was a sharp rise in the prevalence of asthma in developed countries, and the cause of the epidemic was unclear. There are several hypotheses stated to describe the outbreak; for example, in the late 1980s, it was believed that exposure to allergens like house dust mite, fungi, cat, furniture plush, carpets, and tighter insulation of modern house might increase the risk of allergies and asthma. Further, in 1989, "hygiene hypothesis" was proposed by Strachan and explains that the reduced exposure to unhygienic conditions in the early stages of life made the individual more prone to allergies [13]. Later, supporting this hypothesis, in 2003, Rook et al. suggested that reduced exposure to nonpathogenic microbes and commensal organisms also enhances asthma and allergic disease prevalence. Thus, a new hypothesis, named "microbial diversity," has emerged indicating microbes of the gut, mucosa, and respiratory tract are the crucial factors in grooming and regulating the immune system [10, 14]. The incidence of asthma prevalence is rising among patients with other immune-mediated diseases, including multiple sclerosis, inflammatory bowel disorder, and type 1 diabetes mellitus [15].

The risk factor for asthma is multifactorial including a combination of genetic predisposition, age, and immunological and host factors (Fig. 3.1). Factors such as environmental exposure to allergens, tobacco smoke, chemical irritants in the workplace (like varnishes, paint, isocyanate, and adhesives), pollution, cold air, and extreme emotional arousal irritate the airways, thereby leading to asthma attacks. Furthermore, drugs such as beta-blockers, aspirin, and other nonsteroidal anti-inflammatory drugs also provoke asthma attacks [16–20]. Besides, asthma is well correlated with infectious respiratory pathogens such as human respiratory syncytial virus, coronaviruses, influenza viruses, and rhinoviruses as well as atypical bacterial

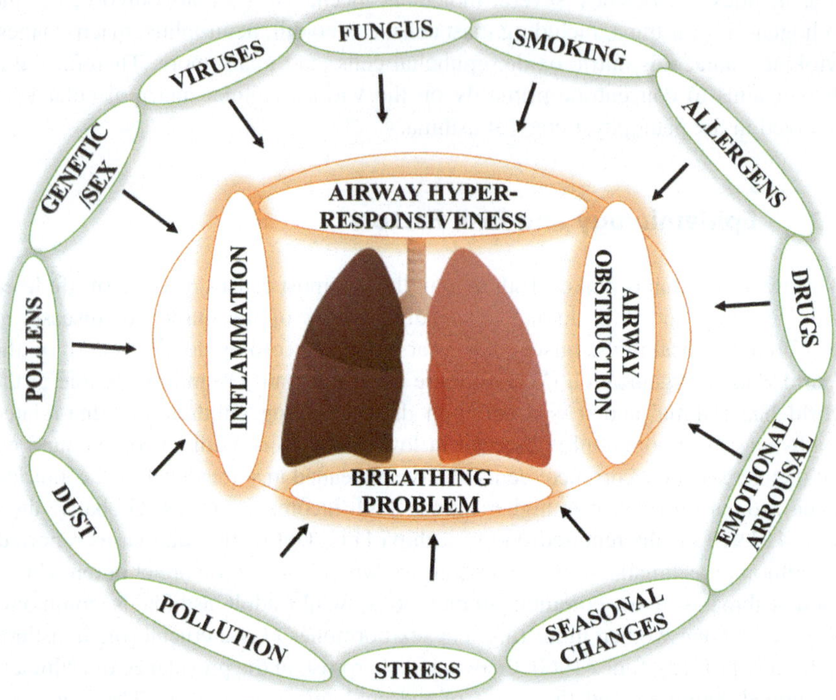

Fig. 3.1 Risk factors for asthma

infections (*Mycoplasma pneumoniae* and *Chlamydia pneumoniae*) and fungi [21–23].

3.3 Granulocytes: As Predictors of Asthma

Immune cells like basophils, eosinophils, and neutrophils contain specialized granules in their cytoplasm and are categorized as granulocytes. Other types of immune cells that contain such granules are termed as mast cells, but as they primarily reside in tissues, their classification as granulocytes is seldom stereotypical. The majority of granulocytes contain antimicrobial peptides like cathelicidins and inflammatory mediators packed in their cytoplasmic granules. Basophils, eosinophils, neutrophils, and mast cells are essential for preserving normal lung function as they alone, or in combination, not only evade latent and budding infectious agents but also stimulate lung repair responses. Granulocytes are thus vital for maintaining lung homeostasis. Conversely, overstimulation, or recruitment, of granulocytes can induce or exacerbate lung injury, i.e., bronchoconstriction that is responsible for the augmentation of asthma [24, 25]. Asthma is a phenotypically heterogeneous long-lasting ailment of the airway and often follows either neutrophilic or eosinophilic inflammatory patterns [24, 26].

3.3.1 Basophils and Allergic Inflammation

Recently a growing body of evidence cemented that basophils, even though being a minor population of granulocytes, portray a significant role in allergic inflammation. Discovered in 1879 by Paul Ehrlich, basophils are now known to be involved in IgE-independent and IgE-dependent allergic inflammation, consequently due to their mobilization at the location of inflammation, releasing several chemokines, cytokines, and proteases. Another major focus is their capability to release IL (interleukin)-4 and IL-13 after being stimulated and known to exacerbate allergic inflammation in combination with serine proteases and other cells like macrophages, fibroblasts, endothelial cells, and some innate lymphoid cells. Moreover, reports have advocated that basophils are involved in Th2 differentiation and amplification of humoral memory responses [27–30].

3.3.1.1 Stimuli for Basophil Activation

Basophils are activated by a plethora of pathways arbitrated by antibodies, cytokines, serine proteases, complement component C5a, or antigens directly. In the case of antibody-mediated basophil activation, the most well-known activation modality is via IgE and FcεEIα cross-linking, which results in the release of leukotrienes (LT) (LTC_4, LTD_4, and LTE_4) and histamines responsible for the contraction of airway smooth muscles. Besides, IgE-, IgD-, and IgG1-dependent antigen complexes also mediate basophils, activation [30, 31]. From the perspective of cytokine-mediated activation of basophils, the most prominent mediator is IL-3. Several reports point toward the induction of basophil production from bone marrow precursors by IL-3. Also, the survival of basophils is enhanced by IL-3 wherein it activates nuclear factor kappa-light-chain-enhancer of activated B cells (NF-κB) and upregulates serine/threonine kinase Pim1 (proto-oncogene serine/threonine-protein kinase) [32, 33]. Further, IL-3 and IgE in liaison increase the production of IL-13 and IL-4 in IgE-facilitated activation [30]. Another medium of basophil activation is via serine proteases like Derp1 (*Dermatophagoides pteronyssinus*), a house dust mite protein that was found to stimulate IL-4, IL-5, and IL-13 production in human basophil cell lines [34, 35]. *Necator americanus*, *Nippostrongylus brasiliensis*, *Strongyloides venezuelensis*, schistotome-derived IPSE-alpha-1, HIV-derived GP120 (glycoprotein 120), and TLR (Toll-like receptor) ligands are some antigenic bodies thought to activate basophils in IgE-dependent or IgE-independent manner [28, 30, 36, 37]. Another example of IgE-independent basophil activation is of primary epithelial cell-derived cytokine thymic stromal lymphopoietin (TSLP) responsible for IL-4 release from basophils [30].

3.3.1.2 Inflammatory Mediators of Basophils

In addition to IL-4, other primary inflammatory mediators were released by basophils including LTC_4, platelet-activating factor (PAF), other IL (3, 6, 8, and 13), vascular endothelial growth factor (VEGF), heparin, histamine, and macrophage inflammatory protein 1-α (MIP-1α) [38]. Several of these basophil products are major inflammatory mediators or else are involved in chemotaxis or help in the propagation of inflammation. Further, it has also been discovered that basophils can also produce cyclooxygenase and 5-lipoxygenase metabolites after stimulation and thereby act as a central propeller of inflammation [28, 39].

3.3.1.3 Basophil Adhesion Molecules

Adhesion of basophils to the endothelium is a landmark event before it migrates to the site of allergy or any inflammatory stimulant. IL-3, GM-CSF (granulocyte-macrophage colony-stimulating factor), IL-5, and IL-33 are responsible for adhesion of basophils before their transendothelial migration upregulation of surface β2-integrin (CD11b/CD18) expression [40]. A study by Lim et al. reported that under physiological shear flow conditions after IL-3 treatment of endothelial cells, basophil adhesion was facilitated through the P-selectin and β1 integrins (CD49d and CD49e/CD29) [41].

3.3.2 Neutrophil Granule Contents

Neutrophil granules are classified into five different types: (A) azurophilic granules that are mainly antimicrobial; (B) specific; (C) gelatinase granules, which are vital for migration through tissues and possess antimicrobial functions; (D) ficolin-1-rich granules, which are required in neutrophil locomotion, firm adhesion, and transendothelial migration; and (E) secretory vesicles, which are specialized for neutrophil recruitment and are enriched in proteins required for extravasation from the circulation [42–44]. The granule proteins present in neutrophils are described in Table 3.1.

3.3.2.1 Mechanisms in Neutrophil Activation

After acute infection or trauma, neutrophils effectively participate in the activation of innate immunity, either in a resting state or activated state depending upon their cell membrane and secretory activity. Conscription of circulating or resting neutrophils by innate immune system occurs precipitously via two major ways: (A) adhesion of neutrophils to the vascular endothelium via cell adhesion molecules (CAMs) upon leukocytes and on endothelial cells in postcapillary venules and (B) neutrophil transmigration via the endothelium and its ECM (extracellular matrix) by virtue of chemotactic factors. In order to escape from blood vessels, resting neutrophils must adhere to activated vascular endothelial cells [45]. This occurs in response to several innate stimuli like a physical wound and LPS (lipopolysaccharide), thrombin, leukotriene, or histamine release (Fig. 3.2). The process is initiated by neutrophil rolling wherein a reversible loose bond is established between nonactivated endothelial cells and neutrophils. The molecules involved in this process belong to three

Table 3.1 Contents of neutrophil granule

Granule type	Primary (azurophilic)	Secondary (specific)	Tertiary (gelatinase)	Ficolin-1-rich	Secretary vesicles
Explicit proteins	Lysozyme Elastase Myeloperoxidase	Collagenase Lysozyme Lactoferrin	Gelatinase Lysozyme	Ficolin-1	CD35/CR1, CD16/FCγRIII
Matrix proteins	Cathepsin-G Proteinase-3 Cathepsin-D α1-Antitrypsin Defensins Azurocidin	β2-Macroglobulin Lactoferrin hCAP-18 NADPH oxidase Heparinase	β2-Macroglobulin MMP25 Gelatinase Lysozyme Acetyltransferase	Human serum albumin, actin	Plasma proteins
Membrane proteins	Presenilin1 V type H^+-ATPase CD63 CD68	Fibronectin-R G protein$_\alpha$-subunit Cytochrome-B_{588} TNF-R Mac-1 CD15 CD66 CD67 fMLP-R	Mac-1 V type H^+-ATPase Cytochrome-B_{588} fMLP-R Leukolysin	CR1 Vanin-2 LFA-1	Mac-1 CD10 CD13 CD14 Alkaline phosphatase fMLP-R Leukolysin

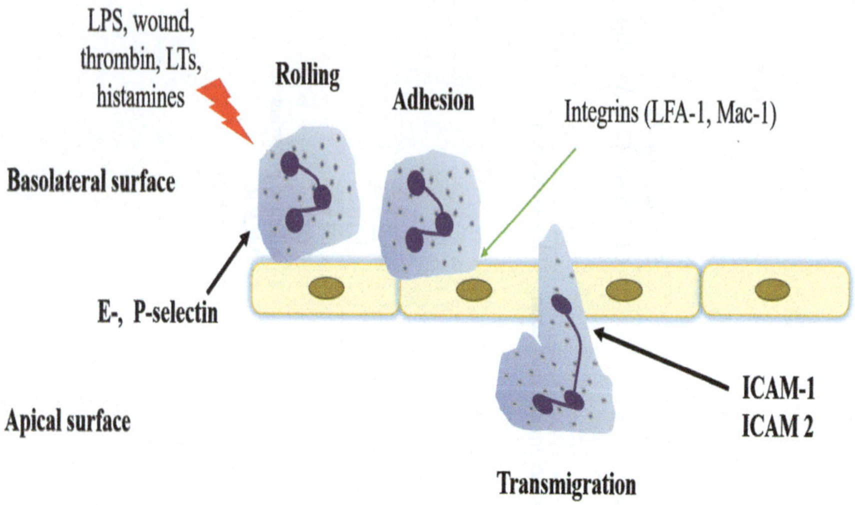

Fig. 3.2 Mechanisms involved in neutrophil activation. It occurs in response to several innate stimuli like physical wound and LPS, thrombin, leukotriene, or histamine release. The process is initiated by neutrophil rolling wherein a reversible loose bond is established between nonactivated endothelial cells and neutrophils. Consequently, neutrophils adhered to endothelial cells, and this phase is reliant on integrin (LFA-1 and Mac-1) stimulation, which is facilitated by GPCRs interacting with chemokines presented on the endothelium. After that, neutrophils crawl beside the endothelium, through ICAM-1 and ICAM-2 interactions with Mac-1 and LFA-1, till they reach their site of transendothelial migration. Abbreviations: GPCR, G protein-coupled receptor; ICAM, intracellular adhesion molecule; Mac-1, macrophage-1 antigen; LFA-1, lymphocyte function-associated antigen-1; LPS, lipopolysaccharide

different families of CAM, namely, (a) selectin (E-, L-, P-selectin), (b) integrins (LFA-1, Mac-1), and (c) immunoglobulin superfamily molecules (ICAM-1 and ICAM-2) [46].

Post-immune stimulation, nonactivated neutrophils express L-selectin, which helps in binding weakly to endothelial cells, and then endothelial cells get activated and bind to neutrophils using E- and P-selectin initiating adhesion. Following these, various factors like cytokines, PAF, and endotoxins induce β-integrin upregulation in neutrophils which leads to the shedding of selectins. Subsequently, neutrophils, by virtue of immunoglobulin superfamily molecules, bind firmly to endothelial cells [47]. The next stage after adhesion is transmigration carried out by short-range signaling molecules called chemotactic factors, which attach themselves to neutrophil membrane receptors. Some noteworthy chemotactic factors include fibrin split products, C5a like complement products, substance P, bacteria-derived fMLP, leukotrienes, and IL-8. The most efficient triggers of neutrophil activation are bacteria and their endotoxins like LPS. Alternate innate or adaptive immune mechanisms like complement pathway and other chemical mediators like PAF and leukotrienes also carry out neutrophil activation. Finally, post-activation, instead of recirculating like other lymphocytes, neutrophils get perished [48, 49].

3.3.2.2 Neutrophil in Airway Inflammation and its Signal Transduction Pathways

Post-activation of lung neutrophils, granule proteins are released such as myeloperoxidase and human neutrophil elastase, both involved in bronchial inflammation in carrying out structural changes leading to emphysema and peribronchiolar fibrosis. Generally, neutrophils are recruited to the airways via IL-1B, IL-8, TNF-α (tumor necrosis factor- α), and LTB_4 [50].

A varied range of endogenous and exogenous agonists are involved in neutrophil signaling pathways like leukotrienes, IL-8, C5a, formylated peptides, LPS, immunoglobulins, C3b, GM-CSF, tumor necrosis factor, IL-1, fibrinogen, ICAM-1, etc. These agonists mediate their signaling via explicit receptors on plasma membranes like GPCRs, Fc receptors, integrin and selectin ligand receptors, cytokine receptors, and some innate immune receptors [51, 52]. There are multiple pathways of neutrophil signal transduction, such as G protein-coupled receptor (GPCR), calcium, protein kinase C (PK C), phosphatidylinositol 3-kinase (PI3K), and tyrosine kinase. A varied number of neutrophil agonists interact with GPCRs that share the physiognomies signal transduction via interaction with heterotrimeric GTP binding proteins (G proteins). These G proteins are involved in GPCR activation in neutrophils, but additional studies need to be carried out to understand their role completely. Some studies have pointed out the major role of Gi family proteins based on chemoattractants' sensitivity to pertussis toxin [53]. Further, a transient increase in cytoplasmic free calcium occurs during neutrophil activation, which has a varied role in controlling neutrophil phases and its properties, including but not limited to shape changes, adhesion, degranulation, secretion, motility, and NADPH oxidase activation [54]. Moreover, PKCα, PKCβ, and PKCδ are expressed in neutrophils. In neutrophils, phorbol esters (PKC stimulants) effectively trigger, among other things, an oxidative burst, inducing the degranulation of some distinct granules like the specific granules [55].

PI3K enzymes are responsible for phosphorylation of phosphatidylinositols (and PtdIns (4,5,) P_2 in particular) in the third position. This is important for the functional responsiveness of neutrophils. The main PI3Ks involved in the early events of neutrophil activation are of class IA and IB PI3K. These enzymes are involved in the regulation of adhesion of neutrophils, chemotaxis, phagocytosis, and antibacterial activity through the regulation of the assembly and activation of the NADPH complex [56].

Tyrosine kinases are involved majorly in the phosphorylation of tyrosine residues, which are related to neutrophil activation. For example, dimerization of opsonin receptors of neutrophils, namely, CD32a and CD16b, leads to recruitment and further activation of Src, Syk, and Tec which are the three prominent families of tyrosine kinases. Enhanced levels of tyrosine phosphorylation upon engagement of CD32a or CD16b have been observed [51, 52].

3.3.2.3 Neutrophil Apoptosis: The Resolution of Inflammation

Apoptosis occurs due to the activation of intrinsic, extrinsic, or endoplasmic reticulum stress pathways. Generation of reactive oxygen species is likely responsible for the activation of the intrinsic pathway and loss in mitochondrial transmembrane potential, thereby releasing cytochrome c, endonuclease G, and apoptosis-inducing factor (AIF). The released cytochrome c induces oligomerization of apoptotic protease activating factor 1 (Apaf-1) that stimulates caspase-9, followed by caspase-3 [57]. The inhibitor of apoptosis protein (IAP) family members can increase cell survival via inhibiting both caspase-9 and caspase-3. The problem is that neutrophils express such IAPs at very low levels and so their effect on neutrophils is still uncertain. Extrinsic pathway causes cell death after ligation of death receptors present on cell surfaces like TRAIL receptors or Fas or TNF-α forming death-inducing signaling complex (DISC). The downstream signaling consists of adaptor proteins like the Fas-associated death domain (FADD), which results in cleavage of caspase-8. The extrinsic pathway is not fundamentally involved in neutrophil apoptosis. The Fas-facilitated signaling supersedes the anti-apoptotic effects of the GM-CSF by binding SHP-1 (Src homology 2 domain-containing protein tyrosine phosphatase 1). The resistance to Fas-facilitated apoptosis contributes to the deferred neutrophil apoptosis [58]. TNF-α effect on neutrophils is intricate as it can generate signals for neutrophil survival that was eliminated by TNF-α.

Further, TNF-α exerts rather complex actions on neutrophils and could generate survival signals in neutrophils that were not initially killed by this cytokine. Another mechanism regulating caspase-8 activity is cathepsin D translocation from azurophilic granules to the cytosol in a caspase-independent and ROS-dependent manner. Any type of genetic or pharmacologic inhibition of cathepsin D may result in a delay of neutrophil apoptosis. Calpains (noncaspase cysteine proteases) and calpastatin (calpain-1 inhibitor) are also present in neutrophils. The expression of calpastatin is decreased in neutrophils undergoing apoptosis. This results in an increased activity of calpain-1, which then cleaves and activates Bax, a proapoptotic factor. The role of autophagy in fundamental neutrophil apoptosis is still unclear, but some studies have pointed out that autophagy like cell death is observed in the case of autoantibody response [59].

3.3.3 Eosinophilic Asthma

The innate immune system consists of some cytotoxic granules containing effector cells known as eosinophils which post degranulation release toxic proteins like eosinophil cationic protein (ECP), eosinophil-derived neurotoxin (EDN), eosinophil peroxidase, IL-5, major basic protein (MBP), and cysteinyl leukotrienes (cysLTs). During asthma tissue, eosinophilia is also considered along with blood eosinophilia to track disease pathology. Accumulation of eosinophils at the site of allergic inflammation is one of the factors contributing to bronchial asthma. This might play a role in the remodeling of the airway through TGF-β (transforming growth factor-beta) and cysLTs. Furthermore, they induce airway hyperresponsiveness with the production of MBP. The transendothelial migration of eosinophils is induced by LTD_4 which subsequently leads to the release of specific granule proteins via cysLT1 receptor and β2-integrin [60]. The development and conservation of inflammation mediated by eosinophils in the airways occurred by cysLTs in combination with the Th2 network. Vascular cell adhesion molecule-1 (VCAM-1)/α4 integrin is involved in eosinophil adhesion to endothelial cells leading to their recruitment. IL-4 and IL-13 upregulate VCAM-1 in endothelial cells responsible for the induction of eosinophil activation. Also, in asthmatic patients, the expression of eotaxin (eosinophil-specific chemoattractant) and its receptor CCR3 (C-C chemokine receptor type 3) is increased [61].

3.3.3.1 Eosinophil-Derived Cytokines and Associated Bronchial Hyperresponsiveness

A proliferation-inducing ligand (APRIL); GM-CSF; interleukin-1α, interleukin-1β, interleukin-2, interleukin-3, interleukin-4, interleukin-5, interleukin-6, interleukin-10, interleukin-11, interleukin-12, interleukin-13, interleukin-16, interleukin-17, and interleukin-25; TNF-α; and interferon γ are the cytokines derived from eosinophils [62]. Eosinophils appear to be using a specific tubulovesicular system for the transportation of some of these cytokines via crystalloid granule to the cell membrane. This is known as piecemeal degranulation, and it allows shuttling of granule contents to the cell surface via swiftly mobilizable secretory vesicles that bud from the surface of crystalloid granule [63].

Chronic airway inflammation in asthma causes precipitous contraction of airway smooth muscles leading to the development of bronchial hyperresponsiveness. Eosinophil chemotactic factor secreted by T-lymphocyte propels eosinophil infiltration [64]. The cytotoxic eosinophilic proteins like MBP, ECP, and cysLTs mediate damage in the bronchial epithelium leading to bronchial hyperresponsiveness in asthma patients [65]. Bronchial hyperresponsiveness is caused by a decrease in the function of neuronal M2 muscarinic receptors on the parasympathetic nerves of the lungs. This occurs in response to the release of MBP from granules of eosinophils [66].

3.4 Fibroblasts and Myofibroblasts

The spindle-shaped cells associated with ECM are termed as fibroblasts. They reside in the majority of tissue and organ systems of the body. In an adult lung, they reside in the adventitia of vascular structures and airways. A distinct feature of fibroblasts includes vimentin expression in the absence of α-smooth muscle actin and desmin. Fibroblasts embryologically are of mesenchymal origin with a range of phenotypic entities, including noncontractile fibroblasts, contractile myofibroblasts, and other intermediates like protomyofibroblast. Myofibroblasts not only have the properties of fibroblasts but also contain α-smooth muscle actin with stress fibers. However, they still lack smooth muscle markers like desmin and smooth muscle myosin, which distinguishes them from smooth muscle cells. Myofibroblasts arise from the transdifferentiation of smooth muscle cells and fibroblasts [67, 68].

3.4.1 Extracellular Matrix Production in the Airways

ECM is produced and maintained by fibroblasts and airway smooth muscle cells in the respiratory tract. Additionally, fibroblasts also synthesize and secrete ECM components like collagen, fibronectin, proteoglycans, laminin, and tenascin. ECM is implicated in the airway remodeling occurring in asthma and is regarded as the bridge between airways and lung parenchyma. It is responsible for the preservation of pulmonary structure and function, promoting the dissemination and adhesion of inflammatory cells, elasticity, and fluid balance, and acts as a repository of inflammatory mediators. Some mediators like TGF-β induce fibroblasts to increase the production of ECM components. Airway smooth muscle cells also produce a variety of ECM proteins in the lungs. In some respiratory disorders, excess deposition of ECM proteins occurs as a result of profibrotic factors like TGF-β1 (released by eosinophils), which mediates Wnt and Smad signaling pathway [67, 69, 70].

3.4.2 Fibroblast to Myofibroblast Transition

In patients with moderate to severe asthma, airway remodeling is a frequently observed phenomenon followed by a narrowing of the bronchial tree. In asthmatic patients, upon histological examination, several epithelial abnormalities were observed, followed by subepithelial fibrosis, smooth muscle cell proliferation, angiogenesis, and ECM accumulation. This suggests the role of fibroblasts and other mesenchymal subtypes. An increase in airway smooth muscle mass occurs in the bronchi of the asthmatic patient due to cell hypertrophy and proliferation accompanied by an influx of blood-derived mesenchymal progenitors and transdifferentiation of fibroblasts into a contractile phenotype [71]. Thus, fibroblast to myofibroblast transition (FMT) may take place, which increases ECM production and causes hyperplasia of smooth muscle in asthmatic bronchi. This phenotypic switch can occur in response to several stimuli like growth factors, pro-inflammatory cytokines, mechanical tension, and epithelial-mesenchymal interactions [72].

3.5 Pathological Signaling Pathways Involved in the Activation of Lung-Resident Macrophages and Dendritic Cells (DCs)

3.5.1 Macrophage Activation Pathways

3.5.1.1 IL-4/IL-13 Signaling Pathway

IL-4 and IL-13 have unique physiological roles based on their immune response phase. First is the afferent immunity, which is the induction part of physiological immune response wherein T and B cells mature after the initial encounter with antigens and respond to the local immune system-mediated signals. Second is the efferent immunity where effects are driven by the mature T and B cells. IL-4 plays a critical role in afferent immunity. On the other hand, IL-13 elicits vital functions during the phase of effector immunity [73].

The primary features of IL-4 and IL-13 in asthma phenotypes relate mainly to afferent immunity. Some of the critical presentation of asthma, like airway obstruction and excess mucus production, are directly related to the airways and are mediated through the effector immunity [73, 74]. Both IL-4 and IL-13 have crucial pathological roles in the effector immune response of asthma; however, IL-13 is the major effector molecule. The vital role of IL-13 in murine lungs and the deviation of action of IL-4 and IL-13 have been shown through different approaches. Antigen induction in IL-4 knockout mice exhibits severe defects in Th2 cell generation and directs important asthma phenotype [75, 76]. Further, deficiency of IL-4Rα in mice leads to even more detrimental effects in asthma phenotype, and the response is shared owing to the response of IL-4 and IL-13 receptors [77, 78]. This unique asthma phenotype is the same as the simultaneous deficiency of IL-4 and IL-13 in mice. This phenotype persists irrespective of Th cell developmental defects and is the result of knockout of IL-4Rα [79]. Also, T cells deficient in IL-4 can produce all other Th2 cytokines, which can induce asthma-related phenotype, but the airway eosinophilia is absent. These observations appreciate the role of cytokines other than IL-4, likely IL-13 and others, in airway obstruction and are mainly driven by mature Th2 immunity. The vital role of IL-13 was proved in studies wherein IL-13 was specifically inhibited by the synthetic antagonist, soluble IL-13 receptor α2 (sIL-13Rα2). It neutralizes the IL-13 but does not affect IL-4 [80]. The administration of sIL-13Rα2 or polyclonal antiserum against IL-13 in inflammation significantly decreased the severity of the asthma phenotype. However, the treatment does not completely reverse the disease. On the other side, inhibition of IL-4 during airway inflammation does not have any major effects except the minor reduction in the recruitment of airway eosinophils [81, 82]. In conclusion, these findings show that during mature Th cell immunity, both IL-4 and IL-13 play a significant role in asthma phenotypes. However, IL-13 outweighs IL-4 in terms of critical pathological role [83].

3.5.1.2 TNF-α Signaling Pathway

TNF-α is an ubiquitous cytokine which is responsible for a wide array of normal physiological functions, like the stimulation of growth, inhibition of growth, inflammation, blood vessel formation, cytotoxicity, and immunomodulation [84]. TNF-α plays critical roles in a variety of inflammatory conditions like pancreatitis, psoriasis, fibrosis, and diabetes [85–88]. It is mainly produced by activated macrophages. Besides, some other immune cells like mast cells, lymphocytes, and natural killer cells produce it. It is also secreted by stromal cells like fibroblasts, endothelial cells, and microglial cells. Its responses are mediated and regulated by the TNF receptors 1 (TNF-R1) and 2 (TNF-R2). The TNF receptors are membrane-bound glycoproteins, which specifically binds with TNF. However, the two TNF receptors are different in terms of expression, affinity, tail structure, and activation of downstream pathways [89, 90]. On the other side, the cytoplasmic regions of TNF receptors are linked to a plethora of intracellular pathways and are unrelated to each other.

The role of TNF-α is widely known in many of the lung diseases. It plays a critical role in asthma, chronic obstructive pulmonary disease (COPD), bronchitis, acute respiratory distress syndrome, the pathogenesis of SARS-CoV-2 viral infection, and acute lung injury [90, 91]. It is expressed in the airways of asthma patients. It has a crucial role in enhancing severe inflammation via activation of AP-1, NF-κB, and other important cellular transcription factors. The sputum and the bronchoalveolar lavage fluid (BALF) of asthma subjects tend to have increased levels of TNF [92]. Further, TNF overexpression in mast cells, lung macrophages, and bronchial epithelial cells is well reported. TNF-α is responsible for the increased expression of a plethora of epithelial cell genes, like chemokines, cytokines, mucins, adhesion molecules, extracellular matrix (ECM) glycoproteins, etc. It stimulates the activated eosinophils for adhesion to the respiratory epithelial cells and increases the chemotaxis of neutrophils [93]. TNF-α is also responsible for transient bronchial hyperresponsiveness (BHR) as it decreases the expression of M2 muscarinic receptors. Further, it has been shown that the polymorphism of the TNF-α gene is associated with asthma, its severity, and BHR [94].

3.5.1.3 TGF-β Signaling Pathway

TGF-β is a multifunctional profibrotic cytokine and an important growth factor. It is extensively recognized as a master player that significantly drives the pathogenesis of asthma [95]. It presents a superfamily of growth factors which include TGF-β1, TGF-β2, TGF-β3, activins, inhibins, bone morphogenetic proteins (BMP), and growth differentiation factors (GDFs). It is involved in embryogenesis, modulation of the immune system, adhesion, migration, and cellular differentiation. Further, it participates in ECM deposition and angiogenesis [96–100]. Its effects are mediated by its receptors, including TGF-β R1, R2, and R3, which regulate the function of serine-threonine kinases. TGF-β R1 is also known as activin receptor-like kinases (ALK) with seven subtype receptors ALK 1–7, where ALK 1 and 5 are mainly implicated in the pathogenesis of asthma. TGF-β R2 consists of five receptor subtypes, and TGF-β R3, also recognized as beta glycans, inhibits TGF-β1 degradation that results in augmented cellular response [101–103].

Despite the versatile role of TGF-β involvement in various diseases, its role in the pathogenesis of wound healing and asthma is fundamental [104, 105]. TGF-β acts via a stringently synchronized process, which involves reactive oxygen species (ROS), proteases, and integrins. The integrin proteins engage a key responsibility in instigating latent TGF-β [106]. Integrins belong to a transmembrane family that connects intracellular and extracellular environments and regulates ECM synthesis and cytoskeleton of the cells. Integrins exist as two subtypes, namely, α and β, of which α subtype of integrins forms dimerization with TGF-β, which is responsible for the activation of latent TGF-β, thus triggering the asthmatic events [107, 108]. Besides integrins, thrombospondin-1 and proteases are also found to be potent activators of TGF-β. However, the activity of TGF-β1 in healthy tissues is harmonized by an endogenous pro-peptide, namely, latency-associated peptide (LAP), that inhibits the association of TGF-β1 with its target receptors [109].

3.5.2 Reactive Oxygen and Nitrogen Species

The imbalance in the redox system, which is defined as an impairment of oxidant and antioxidant production contributing to cellular and tissue dysfunction, is involved in inflammatory responses [110, 111]. Shreds of evidence suggest the potential role of oxidative stress (OS) in asthma pathogenesis [112, 113]. The reactive oxygen species (ROS) are the key players in the OS-mediated damage to the lungs during asthma. As the lung is extremely sensitive to OS compared to other tissues, several exogenous oxidants and other chemical exposures trigger oxidants' production, which promotes the inflammatory cells to generate free radicals. Medications and exposure to allergens are well-known causes of asthma [112].

Malondialdehyde (MDA) is a critical pathological OS biomarker. MDA relates to the extent of pulmonary lipid peroxidation. Further, myeloperoxidase (MPO) is an important mediator responsible for the overproduction of ROS. MPO concentration is directly linked to the degree of cellular damage [114]. Isoprostanes (IsoPs) are prostaglandin isomers which serve as important biomarkers of OS and lipid peroxidation [115]. Earlier, the levels of hydrogen peroxide (H_2O_2) and nitric oxide were also examined in the breath condensate of asthma patients [116]. H_2O_2 is generated from the reduction of O_2 which is catalyzed by superoxide dismutase (SOD) enzyme. H_2O_2 can internalize through the cellular membrane and elicits some of its actions as a signaling molecule. During OS in the lungs, H_2O_2 can react with metal-complexed proteins, like ceruloplasmin (Cu^{2+}) and ferritin (Fe^{2+}), and generates extremely reactive hydroxyl radical. Further, in the Haber-Weiss reaction, superoxide anion reacts with H_2O_2 further worsening ROS burden. In addition, reaction of H_2O_2 with halide ions led to production of highly reactive compounds which trigger a variety of ROS mediated damage, e.g., hypochlorous acid, which is produced under the influence of MPO enzyme [117].

Nitrosative stress is another important component of redox imbalance. In physiological conditions, endothelial NO synthase (eNOS) generates nitric oxide (NO), which acts as a secondary messenger. NO affects the blood vessel smooth muscles and acts as blood flow and pressure regulator. The cytokines IL-1β, IL-6, TNF-α, and interferon-γ activate macrophages via sustained induction of inducible NOS (iNOS) and trigger the release of massive amounts of NO [118]. This activates a cascade of events leading to the stimulation of xanthine oxidase and NADPH oxidase. These in turn stimulate the mast cells to secrete activators of selectin, e.g., histamine and thrombin, which promote the first contact of white blood cells with endothelial cells and aid in cell rolling [119]. These events stimulate the secretion of adhesion-promoting molecules like LTB_4, PAF, and C5 and thus help in the adhesion of leukocytes to endothelial cells and further migration to the inflammatory site [120].

The superoxide radicals can directly elicit their effects or couple with NO, synthesized from iNOS during inflammation, resulting in reactive nitrogen species (RNS). Several antioxidants, including superoxide dismutase (SOD), glutathione, etc., protect the lungs from oxidative insult [121–124]. These antioxidant enzymes present in airway epithelial cells, macrophages, and extracellular spaces of the lung are modulated by nuclear factor erythroid-related factor 2 (Nrf2) as evident from a study, wherein deficiency of Nrf2 augmented asthma [125]. A study examined the expression of neutrophils, and the lung tissue demonstrated that OS decreased the intracellular concentrations of reduced glutathione (GSH). OS either directly or indirectly is significantly implicated in the pathogenesis of asthma [126]. Besides, various inflammatory cytokines and growth factors, including angiotensin II, PDGF, and TGF-β, were found as pro-oxidative and drive inflammatory responses [127]. Further, a growing evidence suggested the enhanced differentiation of pathological cells mediated by TGF-β via NADPH oxidase and NADPH oxidase (Nox) 4. It is understood from numerous studies that the stimulus of OS promotes a wide spectrum of asthmatic events.

3.5.3 Antigen Presentation by Dendritic Cells and their Role in the Pathology of Asthma

Dendritic cells (DCs) are critical immune cells, and their immature form is distributed ubiquitously in the lungs [128]. DCs are crucial in the control of host immune response to inhaled antigens. The basement of the respiratory epithelium harbors a network of airway DCs [129]. It has been shown that antigens which cannot pass the epithelial tight junction (TJ) are taken up by the airway DCs. These airway DCs, within 12 hours, take it to the T-cell area of mediastinal lymph nodes (MLNs) [130]. The DCs which encounter antigen in the periphery are immature and do not have the ability to induce T cells [131]. However, once the DCs carrying antigen reach MLN T-cell area, a cascade of events triggers an array of co-stimulatory factors and an increased concentration of MHC class II molecules capable of stimulating growth and proliferation of naive T cells [132, 133]. It has been proved that airway adoptive transfer of T-cell receptor (TCR) transgenic T cells to lipopolysaccharide (LPS) free antigens induces significantly high T-cell proliferation. The proliferation is mainly seen in lymph nodes and MLNs, and not in the lungs directly [134, 135]. This results in tolerance, which is probably due to the fact that nontoxic antigens or self-antigens dead cells to not activate the network of airway DC completely. It has been proven that antigen presentation by airway DCs which present inducible co-stimulatory molecule ligand (ICOSL) and release IL-10 stimulates the generation of T-regulatory (Treg) cells. The Treg cells have the potential to inhibit consequent inflammatory responses [136].

It has been reported that rodents with experimental asthma exhibit an 80-fold rise in the number of myeloid DCs in the airway mucosa as well as in the bronchoalveolar lavage (BAL) fluid [137, 138]. Interestingly, the DCs in naive animals were immature, whereas the airway DCs of ovalbumin-challenged mice exhibited mature phenotype. This shows that the interaction of DCs with primed T cells takes place in the airways [139]. The airway DCs form clubs in the airway mucosa with primed T cells, resulting in local maturation of DCs. The conditions in which DCs locally interact with primed T cells in lung airways are pathologically very much possible as airway DCs produce chemokines (CCL17) which selectively promote chemotaxis of CCR4 expressing Th2 memory cells to lungs [140].

3.6 Bronchial Epithelial Cells

The type of pulmonary injury and the resultant immunological response govern the cellular involvement and the pathological phenotypes cells in the normal and asthmatic pulmonary epithelium [128, 141]. Inflammatory cytokines promote goblet cell metaplasia and excessive mucus production and recognized by upregulation of mucus-related gene expression in epithelial cells, like MUC5AC or MUC5B. There are two cell lineages that originate from the epithelial cells which are incredibly relevant for mucus production. The ciliated cells and the secretory cells mainly the goblet cells and the club cells [142]. It has been shown by detailed single-cell analysis of inflammatory and structural cells that the asthmatic and healthy phenotypes have specific disease-related proinflammatory genes in the epithelial cells. This indicates the basis of distinct epithelial cellular states. The goblet cell hyperplasia is responsible for the mucous cell hyperplasia in asthma patients and is mainly driven by the mucous-producing phenotype of ciliated cells. The mucous-secreting ciliated cells found in asthma conditions are detected by simultaneous co-expression of ciliary and mucous genes and other critical asthma pathogenesis-involved genes generating a transition ciliary cell state characterized by mucous metaplasia. The production of mucous ciliated cells is critically based on the IL-4/IL-13 signaling pathway. Hence, this signaling pathway is associated with increased mucus production and inflammation in asthma patients [143]. The ovalbumin challenge in mice not only aggravates mucus production but simultaneously stimulates the metaplasia of club cells and transforms them into mucus-secreting cells [144]. The ethmoid sinus epithelium basal cells in patients with chronic rhinosinusitis with polyposis show effects of IL-4/IL-13 recognized by Wnt pathway activation, which are absent in non-polyposis patients. This indicates the role of inflammatory memory as a result of chronic type 2 cytokine exposure. This may be the responsible factor for epithelial cell differentiation leading to increased severity of disease [145]. However, in the case of asthma patients, a pro-inflammatory cellular role of epithelial cells in result to chronic cytokine exposure has not been described yet.

3.6.1 Mechanisms of Endothelial Cell Activation

3.6.1.1 Cytokine Activation

The type 2 cytokines like IL-4- and IL-13-mediated signaling is associated with mucous metaplasia. Further, it has been implicated in MUC5AC overexpression in the ciliated cells [146, 147]. Further, IL-4 and IFN-γ (type 1 cytokine) elicit opposite genetic patterns in epithelial cells of the bronchi, indicating that they have a type 1/type 2 driven opposite impact on epithelial gene regulation [148]. The biological response of type 1 cytokines in asthma indicates opposing effects as compared to overexpression of mucous-related genes by type 2 cytokines. The neutrophil-derived inflammation in asthma is associated with the activation of the inflammasome and IL-1β-/IL-17-mediated recruitment of neutrophils [149, 150]. The comparative analysis between neutrophilic and eosinophilic murine asthma models showed distinct and differential expression of tight junction proteins. Further, overexpression of mucin genes, like MUC5AC and MUC5B, has been mainly seen in neutrophil-driven inflammation. Moreover, Gob5 is a highly inducible gene linked with excessive mucus production in asthma patients [151]. This gene also was found to be overexpressed in neutrophil-mediated asthmatic inflammation. The bronchial epithelial cell stimulation with inflammasome-associated IL-1β and IL-17 cytokines induces enhanced mucus synthesis and disturbance of epithelial cell barrier function [152]. Further, the activation of heat shock protein 90 (HSP90) is an essential condition that involved in the hyperplasia of goblet cells mediated IL-13 and IL-17 [153]. IL-1β and IL-17 stimulate MUC5AC and MUC5B genes by Janus kinase (JAK2)-mediated autocrine/paracrine loop signaling and NF-κB pathway [154, 155]. On the other side, type 1 cytokine IFN-γ inhibits MUC5AC transcription via suppression of MUC5AC expression through JAK/STAT signaling pathway [156]. It was reported in the ovalbumin-induced murine model of asthma that the involvement of activated Th1 and Th2 cells causes Th2 function inhibition through IFN-γ. Thus, IFN-γ may be responsible for eliciting a regulatory role in the pathogenesis of asthma [157].

3.6.1.2 Role of Epithelial Cells in Type 2-Driven and Non-type 2 Allergic Asthma

Allergic asthma driven by type 2 mechanisms is the most common and the most studied type of asthma (Fig. 3.3). For this reason, our current understanding of the role of epithelial cells in asthma pathogenesis is obtained from the investigations of type 2 allergic asthma. Earlier, it was believed that allergic asthma is initiated by abnormal stimulation of type 2 immune response upon allergen challenge. Consequently, it causes local changes in airways, like alteration of the epithelium, e.g., hyperplasia of the goblet cells, and disturbed integrity of barrier and tissue remodeling through the epithelial-to-mesenchymal transition (EMT). This theory was countered by a hypothesis stating that asthma is mainly an epithelial disease wherein a dysfunctional barrier leads to immune activation, e.g., by viruses, entry of pollutants, or allergens [158]. However, it is believed that both mechanisms are involved in asthma pathogenesis, i.e., the epithelial changes and the overactivation

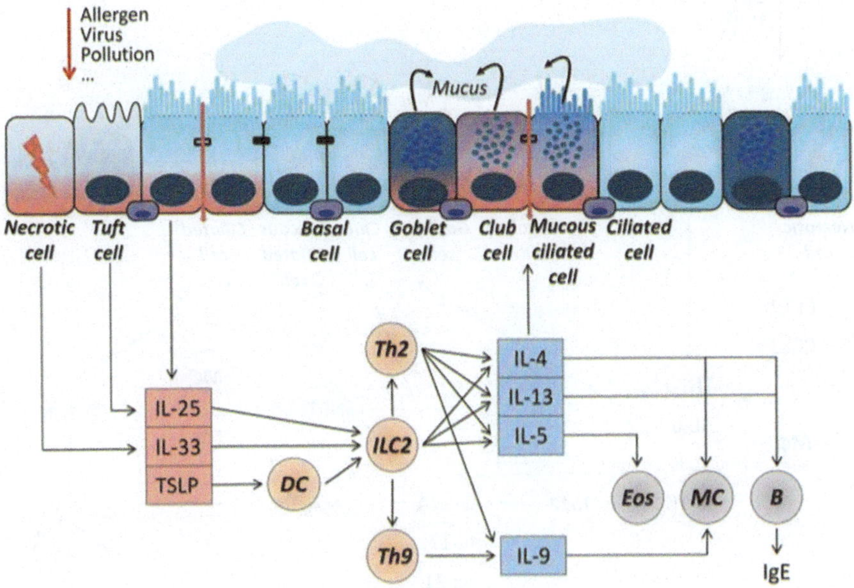

Fig. 3.3 Involvement of epithelial cells in pathomechanisms leading to type 2-driven asthma phenotypes. Abbreviations: IL, interleukin; TSLP, thymic stromal lymphopoietin; IgE, immunoglobulin E; DC, dendritic cells, Th, T helper cells; ILC, innate lymphoid cell; Eos, eosinophils; MC mast cells. The figure is reproduced with permission from reference [142]

of type 2 inflammatory pathway. These develop simultaneously and are interdependent [159]. Further, the attenuation in epithelial barrier features and multiple airway epithelial cell types is integral to the initial innate immune response which drives the increased type 2 adaptive immune reaction. This response is mediated by the secretion of many epithelial cell-derived cytokines formed upon induction due to a variety of environmental insults. Along with the chemokines and inflammatory cytokines, alarmin molecules such as IL-25 and IL-33 are released [142]. The airway epithelial cells elicit the expression of IL-25 physiologically, but the levels remarkably increase upon stimulation with allergens containing protease activity like dust mite. It was shown that the apically expressed protease-activated receptor 2 (PAR2) mediates the signaling mechanisms causing expression of epithelial IL-25 [160]. Further, tuft cells are the major IL-25-secreting cells in the airways indicating its crucial role in the regulation of type 2 immune mechanisms [161].

On the other side, the pathogenesis of non-type 2 asthma is poorly understood. In general, the non-type 2 asthmatic subjects undergo severe disease state and respond poorly to corticosteroid treatment [162]. The inducers of neutrophil-mediated airway inflammation include chronic obstructive pulmonary disease (COPD), smoking, viral infection, and obesity-related inflammation. Sustained secretion of mediators like TNF-α, IL-6, IFN-γ, and IL-17 in fat tissue leads to a state of chronic low-grade inflammation. This process stimulates the development and pathological features of

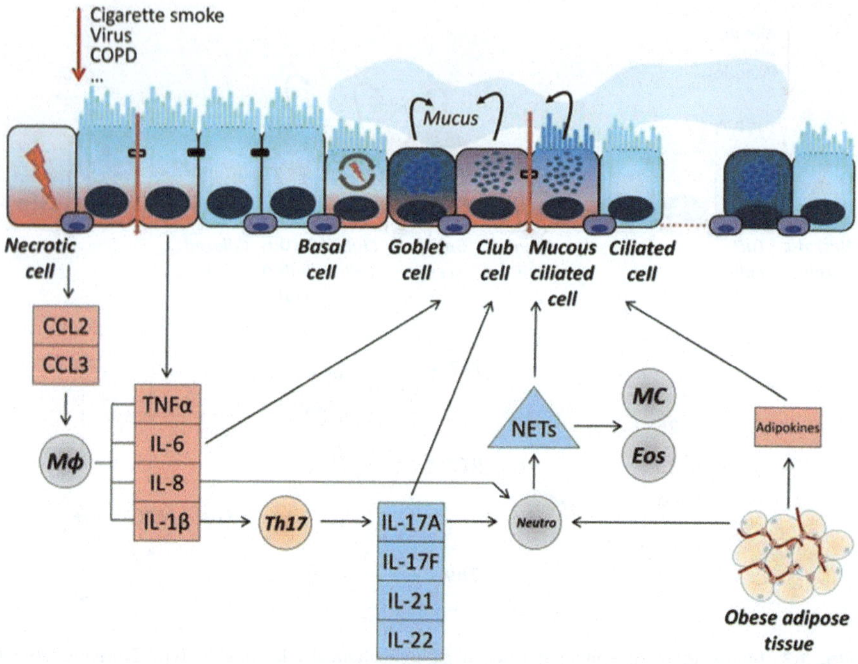

Fig. 3.4 Involvement of epithelial cells in pathomechanisms leading to non-type 2-driven asthma phenotypes. Abbreviations: COPD, chronic obstructive pulmonary disease; CCL, CeC type chemokine ligand; TNF, tumor necrosis factor; IL, interleukin; MΦ, macrophages; DC, dendritic cells, Th, T helper cells; Neutro, neutrophils; Eos, eosinophils; MC mast cells; NETs, neutrophil extracellular traps. The figure is reproduced with permission from reference [142]

non-type 2 asthma in obese people [163]. The age when disease initiates plays a critical role in asthma development, and obese people have an extremely increased bronchial hyperresponsiveness. Due to this, obese people tend to have type 2-associated phenotypes of asthma [164]. In a large number of patients, asthma due to neutrophil inflammation is mediated by Th17 signaling, enhanced IL-8 synthesis, and increased innate immune system stimulation [162]. Figure 3.4 shows the pathomechanism of non-type 2 asthma.

3.7 Airway Epithelial and Smooth Muscle Cells

Asthma is a disorder of the respiratory system where dynamic structural alterations are observed in all the layers of the airways of bronchi [165]. Airway remodeling is a collective event of recurring tissue injuries followed by cellular reactions that leads to abnormal repair in the respiratory tract [166]. The normal growth or repair process in lungs, aging, and transient response to bronchial injury with the re-establishment of normal anatomical structures constitute the phenomenon of airway remodeling.

Airway remodeling is considered as a consequence of airway inflammation due to either allergen exposure in sensitized patients or other environmental hazards. On the other hand, a response to chronic injury or inflammation results in persistent aberrant airway structure and function. Moreover, mechanical stress or early life events also contribute to the pathophysiology of airway remodeling. The changes in epithelium, sub-epithelium, and airway smooth muscle thickening and bronchial neoangiogenesis comprise airway remodeling in asthma [167]. Furthermore, epithelial damage and angiogenesis are major etiological factors of asthma in both children and adults. It is well-known that airway inflammation and bronchial remodeling are major pathological manifestations of chronic asthma. The reciprocal intercellular communications between airway inflammation and remodeling involve several cytokines and growth factors. The pathobiology of airway remodeling in asthma includes shedding of the epithelium, hyperplasia of goblet cells, fibrosis of the sub-epithelium, abnormal deposition of the extracellular matrix and increased thickness of the smooth muscle layer, and neoangiogenesis of bronchial vasculature [168].

3.7.1 Physiological Barrier

The airway epithelium is the first barrier to environmental agents such as pollution, viral infection, and allergens. The cells in the airway epithelium exhibit the properties of mucociliary cells. The pathogens are trapped in the mucus of inflammatory airways and subsequently removed by the ciliary movement of the epithelial cells, which is known as the mucociliary escalator phenomenon. Moreover, the inflammatory cells are mobilized by chemokines and cytokines to remove the microbes [169, 170]. Epithelial cells also regulate tissue homeostasis through modulating endogenous chemicals, including growth factors, antioxidants, chemokines, cytokines, and lipid mediators [171]. Additionally, the epithelium can also reduce inflammation in the airways to maintain homeostasis. On the contrary, the exaggerated protective mechanism triggers the onset of chronic inflammation of airways in asthma.

The ciliated and goblet cells within the columnar epithelial layer undergo changes in structure and function during asthma, and these are fragile and susceptible to injury. Intrinsically, they are susceptible to destructive damage and apoptosis [172, 173]. Additionally, the infiltrating cells such as eosinophils and mast cells in asthmatic airways also produce TGF-β that can also induce bronchial epithelial apoptosis [174]. One of the clinical manifestations of remodeling of airway epithelium is hyperplasia or metaplasia of goblet cells that is present in between columnar ciliated cells. These cells are mostly responsible for increased secretion of mucus [175, 176]. Indeed, a damaged epithelium promotes the proliferation of smooth muscles, increases angiogenesis, increases collagen deposition, and thickens the reticular basement membrane. These events collectively may involve airway remodeling.

Epithelial cells form sheets with adjacent cells giving rise to a typical "cobblestone" pattern [177, 178]. The tight and adherens junctions comprise the array of cell-cell junctions in the epithelium closest to the apical junction of airway lumen [179]. The tight junction (TJ) is selective and permeable to ions and solutes through the paracellular pathway [180]. The permeability in TJ can be regulated by TNF-α, IFN-γ, interleukins, and growth factors [181]. Gap junctions in the epithelium also provide a link between the cytoplasms of adjacent cells such as desmosomes and hemidesmosomes. Thus, additional junctions can be considered as an important factor in the pathophysiology of asthma [182].

The increased proliferation and cytokine production from bronchial smooth muscle cells are considered as other characteristic features of severe asthma [183]. It is well established that airway smooth muscle hypertrophy is the consequence of mechanical stretch that is the result of persistent bronchoconstriction [184]. Several growth factors including insulin-like and platelet-derived growth factors exert proliferative action in the airway smooth muscles [185, 186].

3.7.2 Airway Epithelial Cell-Derived Mediators

3.7.2.1 Nitric Oxide (NO)

It is a potent bronchodilator of endogenous as well as an exogenous origin [187]. In airways, the major source of NO is airway epithelium [188], nonadrenergic/noncholinergic innervations [189], and airway smooth muscles [190]. Nitric oxide synthase (NOS) catalyzes the conversion of L-arginine to NO. The different types of NOS such as neuronal, endothelial, and inducible are expressed within the respiratory tract [191, 192]. The inducible NOS is mostly expressed in epithelial and inflammatory cells of the airway and is induced after exposure to pro-inflammatory cytokines [193]. NO produced by iNOS contributes maximally to asthmatic inflammation, epithelial injury, and clinical exacerbations in asthma; hence, the inhibition of NOS can be used to manage the respiratory illness [194, 195]. The measurement of NO flux is considered as a diagnostic marker for airway diseases [196]. An increased level of NO is reported in exhaled air and BALF in patients suffering from asthma [197].

3.7.2.2 Endothelin

Endothelin (ET), in particular ET-1, is reported to exhibit bronchoconstriction, airway inflammation, and structural remodeling of the mammalian airways. ETs exert their effect through ETA and ETB receptors [198, 199]. ET-1 exerts pro-inflammatory activity through mediators such as IL-6, IL-8, and GM-CSF [200]. It is reported that the level of ET-1 is significantly elevated in asthmatic cats, and therefore, ET-1 could be considered as one of the diagnostic markers for asthma [201]. The primary source of ET-1, the most studied of the three ET isoforms, is cells of epithelial, endothelial, smooth muscle, and inflammatory cells of airways [202]. Clinical studies reveal that both ET-1 and ET-3, as well as their precursors, are present in the airways [203]. ET-1 increases mucus secretion, contracts bronchial smooth muscle, enhances pro-inflammatory mediator release, and stimulates collagen production by pulmonary fibroblasts [202]. ET-1 contributes to airway inflammation through the upregulation of VCAM-1 and promotes migration and adhesion of leukocytes in the respiratory system of the mammalians. ET antagonist, BQ-123, attenuates the recruitment of inflammatory cells in the airways and total cell counts assessed in lung lavage. Thus, it has been reported that antagonizing ET is an effective strategy against inflammation and remodeling [204].

3.7.2.3 Arachidonic Acid Metabolites

Arachidonic acid metabolites play an important role in the pathogenesis of airway disorders. The enzyme cyclooxygenases (COX) catalyzes the conversion of arachidonic acid into prostaglandins (PGs), thromboxanes (TXs), and other eicosanoids [205, 206]. Each metabolite is involved in the process of inflammation [207]. The level of prostanoids in bronchoalveolar lavage fluid increases in asthma. Reports suggest increased activity of both COX-1 and COX-2 in the asthmatic airways. The isoprostanes activate the thromboxane receptor and exert similar biological activity to that of thromboxane A2 [208]. It has been reported that prostaglandin D_2 (PGD_2) is released in large amounts in asthma patients [209]. Further, a major culprit for asthma are leukotrienes, and LTB_4, a potent lipid mediator of inflammation, mediates multiple biological functions through LTB4 receptor 1 (BLT1) and LTB_4 receptor 2 (BLT2). The LTB_4/BLT1 pathway is well established in the pathophysiology of asthma [210, 211]. The LTB_4 receptor antagonists in asthma fail to exhibit therapeutic benefit from asthma and COPD. However, selective BLT1 antagonists exhibit potential therapeutic effects in asthma and other related airway disorders [212]. Further, drugs like aspirin precipitate the asthma in patients through the modulation of lipoxygenase leading to overproduction of inflammatory leukotrienes [213].

3.7.2.4 Inflammatory Cytokines

The biomarkers, type 2 cytokine IL-13 like periostin in serum and blood eosinophilia, are also the clinical manifestations of asthma [214]. The inflammatory cytokines can induce inflammation through the generation of allergy-specific immunoglobulin E. Thus, they promote recruitment, growth, and differentiation of eosinophils [215, 216]. The elevated interferon (IFN)-γ is related to antigen-induced airway hyperresponsiveness and eosinophilic infiltration [217].

3.7.2.5 Cell Adhesion Molecules

The increased level of adhesion molecules on the surfaces of respiratory epithelial cells is an important cause in the pathogenesis of asthma [218]. Surface adhesion molecules such as intercellular adhesion molecule-1 (ICAM-1) and vascular cell adhesion molecule-1 (VCAM-1) play a significant role in the pathophysiology of asthma [219]. Tumor necrosis factor-α (TNF-α) exerts pro-inflammatory and immunomodulatory properties in the airways. TNF-α increases the expression of ICAM-1 and VCAM-1 on epithelial cells of the airways [220]. Therefore, strategies can be implicated to suppress TNF-α-induced biological functions [221]. ALCAM has pro-inflammatory functions in a model of delayed-type hypersensitivity. The ALCAM levels increase in the bronchoalveolar lavage fluid (BALF) and serum and decreases in the allergic mice lungs. Further, ALCAM deficiency induced a significant amelioration in airway dysfunction in the experimental condition [222]. ICAM-1 causes intercellular adhesion by binding with macrophage adhesion ligand 1, leukocyte function-associated antigen 1, fibrinogen, and human rhinovirus [223, 224].

3.7.2.6 Platelet-Activating Factor (PAF)

PAF is one of the important mediators of inflammatory reactions in asthma. Preclinical studies suggest that PAF can directly exert bronchial obstruction and increase bronchial tree hyperreactivity in the mammals [225, 226]. PAF antagonists such as Y-24180, GB, and others can relieve the symptoms of asthma and improve the respiratory function in airway inflammation [227]. PAF exerts its action through phospholipase A2 and phospholipase C. PAF stimulation can activate MAPKs, including Erk1/2, p38, and JNK in different cell types [228, 229]. PAF stimulation can also increase the level of IL-6, IL-1β, and TNF-α and thus augments the airway inflammation in asthma [230].

3.7.2.7 Tachykinin

Tachykinins exert a wide range of physiological actions in the mammalian system. The neurokinins act mainly through neurokinin receptors (NK1, NK2, and NK3). These are potent mediators of the physiology of airways [231]. Substance P is the predominant tachykinin neuropeptide within the mammalian airways that is released from the nonadrenergic-noncholinergic system by inflammatory, thermal, chemical, or mechanical stimuli. The NK1 receptor is predominantly located in the bronchial vessels, epithelial cells, submucosal glands and vascular endothelium [232]. However, the NK2 receptor is located in airway smooth muscle and NK3 receptor is also involved in the regulation of airway functions in healthy and diseased conditions [233]. Selective tachykinin receptor antagonists attenuate the late allergic and airway hyperresponsiveness and thus ameliorate eosinophilic infiltration and vascular permeability in the guinea pigs [234]. Substance P induces bronchoconstriction and airway hyperresponsiveness in the mammalian system and also binds to NK2 receptors in the mammalian system [235, 236].

3.7.2.8 Histamine

Histamine is predominantly produced from mast cells through the stimulation of a vast array of receptors on their surface such as FcεR1 and FcγRI and receptors for nerve growth factor (NGF), vasoactive intestinal peptide (MrgX2), complement components (C3aR and C5aR), substance P, adenosine phosphate, etc. [237, 238]. Basophils and other immune cells can also produce histamine; however, the higher concentration of histamine is found in the bronchial tissues, skin, and intestinal mucosa [239]. Histamine exerts bronchoconstriction of smooth muscle through H1 receptor-mediated mechanism in the mammalian respiratory system [240]. Histamine also increases the secretion of mucus and glycoprotein from the human airways. This action is inhibited by the H2 receptor antagonist cimetidine which is not done by H1 receptor antagonists. Further, histamine facilitates the chloride ion transport of the airway epithelial cells to promote water transfer in the airways [241]. The H4R stimulation enhances the migration of eosinophils and the recruitment of mast cells which subsequently leads to the amplification of chronic inflammation and immune responses in the airways [242]. Similarly, H4R stimulation promotes T-cell differentiation and dendritic cell activation [243], and its inhibition of H4R exhibits anti-inflammatory activity in a murine asthma model [244].

3.7.2.9 Adenosine

Adenosine exhibits immunomodulatory activities, an endogenous purine nucleoside. The increased level of adenosine is observed in lung inflammation with patients of bronchial asthma [245, 246]. The role of adenosine and its receptors in the activation of monocytes and monocyte-derived mature myeloid cells (macrophages and inflammatory cells) is well established in the animal models of lung inflammation [247, 248]. A1 receptors are upregulated in asthmatic airways due to chronic inflammation and can mediate hyperresponsiveness in the mammalian bronchi [249]. It is suggested that the stimulation of A2A receptors on lymphoid cells leads to inhibition of the inflammatory response in the mammalian system. Therefore, a selective agonist to A2AR could exhibit anti-inflammatory potential in airway inflammations [250]. In addition, relatively high density of functionally active A3Rs is expressed in human eosinophils [251]. Transcript levels of the A3Rs are elevated in the lungs of asthma or COPD patients and inhibit the eosinophil chemotaxis when stimulated in the airways [252]. Furthermore, selective A3R agonist, IBMECA, inhibits pro-inflammatory functions of human eosinophils in respiratory dysfunctions [253]. Adenosine 5′-monophosphate (AMP), an indirect bronchostimulant, stimulates specific mast cell surface adenosine A2b receptors and contracts the airway smooth muscle. Thus, bronchial response to adenosine in humans is an indirect mechanism that involves mast cell activation probably through A2B receptor activation and contributes in either acute or chronic symptoms of asthma [254, 255].

3.8 Role of Innate Immunity in Asthma

Lungs, being the primary organs of the respiratory system in humans, are continuously exposed to a diverse array of microbes and microbial products, in addition to small particulate matter such as smoke, pollen, and aerosols, on a daily basis. This constant exposure necessitates the provision of an inherent defense system to cope with the unknown harmful threats. In fact, the respiratory airways are strategically protected with well-regulated components of the cellular and humoral innate defense system. This immune system is active and immediately responsive to distinguish and combat potentially harmful foreign elements, determining the outcome of host-pathogen interactions within the respiratory airways. The primary cells that orchestrate this dynamic response are the T cells. Studies have shown that the airways of asthmatic patients are endowed with T cells [256]. Moreover, asthma, recognized to be a heterogeneous disease, is classically labeled as a T helper 2 cell-driven inflammatory disease [257]. This topic reviews the role of different immune cells with their contribution to the immunology of asthma that has led to the concept that asthma is more than just a Th2-type disease.

3.8.1 CD4+ and CD8+ T Cells in Asthma

CD4+ T cells remain the central players in the pathophysiology of asthma as they are the cells that initiate and direct myriad inflammatory cascades in respiratory airways [258]. CD4+ T cells (named as they express CD4 on their surface) basically assist other immune components, including maturation of B cells, activation of cytotoxic T cells, and production of several cytokines. Therefore, these cells are also termed as T helper cells (Th cells). CD4+ helper T cells further differentiate into one of the several Th subtypes such as Th1, Th2, Th17, Th9, and regulatory T cells, according to their specific functions. The long-established Th1-Th2 prototype is considered as the backbone for helper T cell-mediated inflammatory pathways. Earlier, Th1 helper cells were anticipated to play an inhibitory role in asthma. Interleukin-12 (IL-12) and interferon-γ (IFN-γ) are the specific cytokines that set up the downstream signaling cascade for the development of Th1 helper cells. However, when based on this earlier hypothesis, IFN-γ (Th1-associated cytokine) was administered to patients with asthma; no improvement in the associated symptoms was observed [258]. In fact, later studies suggested that Th1 helper cells do not possess any inhibitory role and might even exacerbate lung inflammatory responses [259].

Looking into the profile of Th2 helper cells, it is widely accepted that Th2 cell activation initiates as well as perpetuates the asthma pathogenesis. Th2 cells induce the proliferation of basophils and mast cells (interleukin-3, interleukin-4) and stimulate the differentiation and proliferation of eosinophils (interleukin-5). Th2 cells also signal B cells (via IL-4, IL-5, and IL-13) to produce IgE antibodies, mediating type1 hypersensitivity reactions. Thereby, the salient features of Th2-type immune response in asthma include eosinophilic inflammation, edema, and increased mucus production, which subsequently results in airway obstruction. A distinct class of helper T cells, Th 17 cells, produce interleukin-17 (IL-17) which plays a pleiotropic role in asthmatic airways. IL-17 is known to induce neutrophilic airway inflammation, by stimulating the expression of mucin MUC5B implicated in airway remodeling. Moreover, IL-17 has been observed to be significantly elevated in sputum and bronchial biopsies obtained from asthmatic patients [260]. Th 9 cells are another subset of T cells mainly secreting interleukin-9 (IL-9, the signature cytokine), but are also known to produce interleukin-10 (IL-10) and interleukin-21 (IL-21) cytokines. Studies have demonstrated that IL-9 is sufficient in itself to initiate the bronchial hyperresponsiveness by significantly enhancing Th2 cytokine secretions [261, 262].

With regard to these CD4+ T cell subsets, it appears that regulation by T-regulatory (Treg) cells is the shared controlling factor, particularly in allergic asthma, as Treg cells are able to suppress the effector cells. This Treg-mediated negative regulation is carried out through the transforming growth factor-β (TGF-β) and interleukin-10 (IL-10). Any defect in Treg cell functioning is being viewed as a sufficient requirement for the pathogenesis of asthma [263, 264] .

Most of the research has been focused upon CD4+ T cell subsets in elucidating the role of immune cells in the asthmatic inflammatory response. In contrast, CD8+ T cells have not gained much attention in this context. CD8+ T cells (named as they express CD8 marker on their surface) are also known as cytotoxic T cells or killer T cells as they can directly kill the infected cells and, therefore, are more important in virus-infected pathological states. However, CD8+ T cells are known to induce interleukin-12 (IL-12) and interleukin-18 (IL-18) and regulate IgE production. They also play a moderating role in the polarization of Th2 cells during allergic sensitization. However, the profound role of CD8+ T cells in allergic or inflammatory asthma demands further exploratory studies.

3.8.2 Cellular Receptors of Innate Immunity in Asthma

The cellular receptors that distinguish self from non-self in the development of allergic immune response in asthma through the interaction with microbes or microbial products play a crucial role in determining the disease manifestation. Toll-like receptors (TLRs) are the chief members among these receptors. TLRs belong to a class of pattern recognition receptors (PRRs), usually expressed on the surface of immune cells that recognize specific structurally conserved molecules derived from pathogens. The respiratory tract, being continuously exposed to the environment, is particularly much susceptible to foreign pathogens. Therefore, as the first line of defense, the TLRs are expressed on the airway epithelial cells (AECs) to recognize the foreign and microbial components. Among the ten active TLRs (TLR1–TLR10) that have been explored in humans so far, particularly TLR2 and TLR4 are accountable for the inflammatory response in pulmonary disorders. TLR2 and TLR4 recognize the lipopolysaccharides and lipoproteins, respectively. These receptors are known to induce the transcription of MAPK and NF-κB by guiding their migration into the nucleus and then further triggering the release of IL-1β and IL-8, initiating the cascade for neutrophilic asthma. Another Toll-like receptor, TLR7, which is mainly involved in autoimmune disorders, is also present intracellularly in AECs as well as on the surface of innate immune cells (such as dendritic cells and natural killer cells). TLR7 recognizes the nucleic acid component of respiratory viruses (mainly ssRNA) and induces the activation of IFN-γ via Th1 response in immune cells. Besides, it also reduces the Th2 cytokines such as IL-4 and IL-5 and thereby prevents the local eosinophilic inflammation. Altogether, these various TLRs present on the AECs pose as the early responders of innate immunity that hold the asthmatic inflammatory response in check.

Furthermore, the AECs drive the airway hyperresponsiveness and associated inflammation through transmitting the signals to various innate immune cells such as macrophages, NKT cells, and innate lymphoid cells (ILCs). ILCs are the innate counterpart of T cells that present the immune responses via signaling molecules. Based on the signaling molecules produced and the difference in their differentiation pathways, ILCs have been categorized into three groups: ILC1s (produce IFN-γ), ILC2s (secrete Th2 cytokines), and ILC3s (produce IL-17 and/or IL-22). Several studies have linked ILC2 cells directly or indirectly to allergic inflammation [265, 266]. These studies reveal that early exposure to allergens can activate ILC2 cells that triggers the release of IL-5 and IL-13 inducing eosinophilic inflammation and airway hypersensitivity, respectively, in the lungs. Moreover, during the effector phase, ILC2 cells express the major histocompatibility complex (MHC) class II and subsequently activate CD4+ T cells. However, this MHC II expression by ILC2 cells is quite low in lung tissues, which questions this mechanism to be relevant to asthma.

3.8.3 Innate-like T Cells and Asthma

Natural killer T (NKT) cells are the subset of T cells that express a variety of natural killer cell markers, constituting only 0.1% of the total peripheral blood T cells. Unlike conventional T cells, these NKT cells express invariant T cell receptors (TCRs) presented as CD1d and MHC-related molecule (MR1). The classical NKT cells carry an invariant TCR α-chain paired with β-chains. For instance, human invariant NKT (iNKT) cells carry a Vα24-Jα18 chain paired with the Vβ11 chain. These iNKT cells can be detected and activated by the environment microbial components (mainly glycolipids) through their invariant TCR. The activated iNKT cells in the lungs stimulate airway hyperresponsiveness and neutrophilic inflammation by producing IL-13 and IL-17, respectively. They are also involved in the exacerbation of allergic airway inflammation. Surprisingly, a study exploring the role of these iNKT cells in asthma found that the number of iNKT cells detected in the bronchoalveolar lavage fluid from asthmatic patients was almost similar to the numbers observed in those of healthy individuals [267]. The study demonstrated that in the asthmatic patients, iNKT cells produced the cytokines IL-4 and IL-13 but not IFN-γ. However, iNKT cells from healthy individuals were producing all three cytokines. Therefore, it's the question of research if the allergic asthma is causing iNKT activation or the activated iNKT cells are causing the development of asthma.

Gamma delta (γδ) T cells are another type of unconventional T cell and, similar to NKT cells, represent only a small subset of peripheral blood T cells. The heterodimeric T cell receptors expressed on these cells are composed of γ- and δ-chains, as indicated by their name. Their prime function is stress surveillance in response to their environmental signaling, as reflected during wound healing. Unlike their low prevalence in peripheral blood cells, the predominant sequestration (8–20% of pulmonary lymphocytes) in mucosal and airway epithelial tissues makes the γδ cells as "sentries patrolling the border." These cells are reportedly elevated in asthmatic patients [268] and investigated as the active negative regulators of inflammation. Moreover, they have been recognized to mediate the resolution of acute airway hyperresponsiveness and airway inflammation via the production of IL-17 [269]. Furthermore, γδ T cells are also involved in inflammation-associated airway remodeling. These multiple protective facets of γδ cells clearly express the obligation of these cells in conserving normal airway homeostasis.

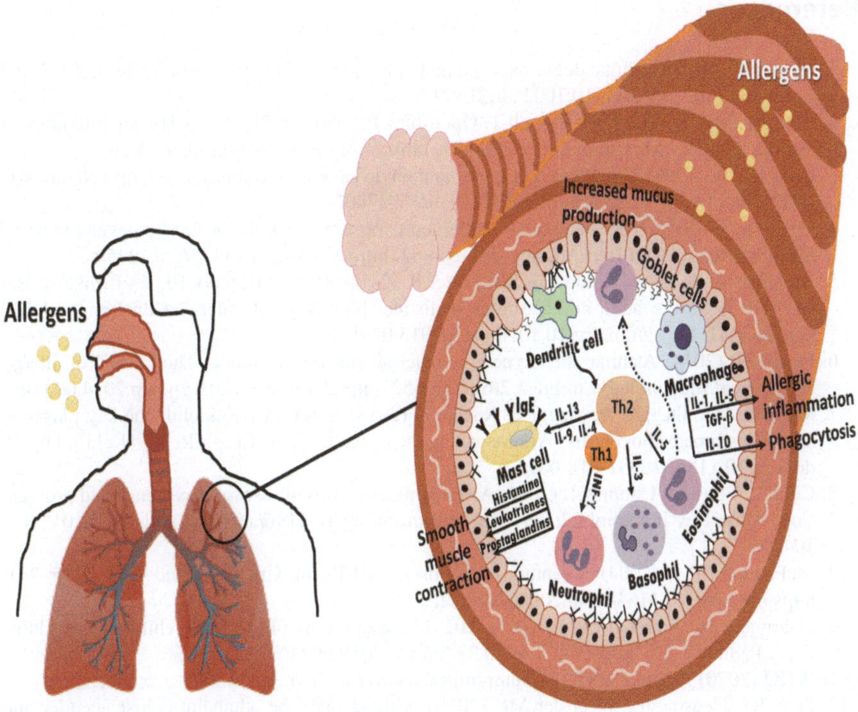

Fig. 3.5 Cellular and molecular pathways in asthma pathophysiology. Abbreviations: *IL* interleukin, *TGF-β* transforming growth factor-β, *Th* T helper, *IFN-γ* interferon gamma

3.9 Conclusion

The increasing prevalence of asthma cases and the diverse pattern of its clinical features within a population have increased the researchers' interest in the mechanistic insights of asthma. Various cellular and molecular pathways discussed in this chapter are briefly summarized in the figure that direct new interventions targeting asthma (Fig. 3.5). The available therapeutic agents for the treatment of asthma are mainly suppressing the associated symptoms. With the recognition and understanding of the underlying molecular targets in the pathophysiology of asthma, we can reach the stage where it would be possible to cure the disease rather than just suppressing the symptoms.

Conflict of Interest The authors declare that they have no known competing financial interests or personal relationships that could have appeared to influence the work reported in this paper.

Funding Source None.

References

1. Mims JW (2015) Asthma: definitions and pathophysiology. Int Forum Allergy Rhinol 5(Suppl 1):S2–S6. https://doi.org/10.1002/alr.21609
2. Edwards MR, Bartlett NW, Hussell T, Openshaw P, Johnston SL (2012) The microbiology of asthma. Nat Rev Microbiol 10(7):459–471. https://doi.org/10.1038/nrmicro2801
3. Cookson W (1999) The alliance of genes and environment in asthma and allergy. Nature 402 (6760 Suppl):B5-11. https://doi.org/10.1038/35037002
4. Custovic A (2015) To what extent is allergen exposure a risk factor for the development of allergic disease? Clin Exp Allergy 45(1):54–62. https://doi.org/10.1111/cea.12450
5. Amelink M, de Groot JC, de Nijs SB, Lutter R, Zwinderman AH, Sterk PJ, ten Brinke A, Bel EH (2013) Severe adult-onset asthma: a distinct phenotype. J Allergy Clin Immunol 132 (2):336–341. https://doi.org/10.1016/j.jaci.2013.04.052
6. Peters SP (2014) Asthma phenotypes: nonallergic (intrinsic) asthma. The journal of allergy and clinical immunology. In Pract 2(6):650–652. https://doi.org/10.1016/j.jaip.2014.09.006
7. Matucci A, Vultaggio A, Maggi E, Kasujee I (2018) Is IgE or eosinophils the key player in allergic asthma pathogenesis? Are we asking the right question? Respir Res 19(1):113. https://doi.org/10.1186/s12931-018-0813-0
8. Cohn L, Elias JA, Chupp GL (2004) Asthma: mechanisms of disease persistence and progression. Annu Rev Immunol 22:789–815. https://doi.org/10.1146/annurev.immunol.22.012703.104716
9. McFadden ER Jr (2004) A century of asthma. Am J Respir Crit Care Med 170(3):215–221. https://doi.org/10.1164/rccm.200402-185OE
10. Dharmage SC, Perret JL, Custovic A (2019) Epidemiology of asthma in children and adults. Front Pediatr 7:246. https://doi.org/10.3389/fped.2019.00246
11. WHO (2020) Asthma. WHO [online] https://www.who.int/news-room/q-a-detail/asthma
12. Zein JG, Denson JL, Wechsler ME (2019) Asthma over the adult life course: gender and hormonal influences. Clin Chest Med 40(1):149–161. https://doi.org/10.1016/j.ccm.2018.10.009
13. Strachan DP (1989) Hay fever, hygiene, and household size. BMJ (Clin Res ed) 299 (6710):1259–1260. https://doi.org/10.1136/bmj.299.6710.1259
14. Rook GA, Martinelli R, Brunet LR (2003) Innate immune responses to mycobacteria and the downregulation of atopic responses. Curr Opin Allergy Clin Immunol 3(5):337–342. https://doi.org/10.1097/00130832-200310000-00003
15. Holgate ST, Wenzel S, Postma DS, Weiss ST, Renz H, Sly PD (2015) Asthma. Nat Rev Dis Primers 1(1):15025. https://doi.org/10.1038/nrdp.2015.25
16. Stern J, Pier J, Litonjua AA (2020) Asthma epidemiology and risk factors. Semin Immunopathol 42(1):5–15. https://doi.org/10.1007/s00281-020-00785-1
17. Stocks SJ, Jones K, Piney M, Agius RM (2015) Isocyanate exposure and asthma in the UK vehicle repair industry. Occup Med (Oxford, England) 65(9):713–718. https://doi.org/10.1093/occmed/kqv108
18. Subbarao P, Mandhane PJ, Sears MR (2009) Asthma: epidemiology, etiology and risk factors. CMAJ 181(9):E181–E190. https://doi.org/10.1503/cmaj.080612
19. Törmänen S, Lauhkonen E, Riikonen R, Koponen P, Huhtala H, Helminen M, Korppi M, Nuolivirta K (2018) Risk factors for asthma after infant bronchiolitis. Allergy 73(4):916–922. https://doi.org/10.1111/all.13347
20. Williams WR (2018) Asthma exacerbation by aspirin and chemical additives: use of a nucleotide template model to investigate potential mechanisms. Gen Physiol Biophys 37 (4):461–468. https://doi.org/10.4149/gpb_2018003
21. Darveaux JI, Lemanske RF Jr (2014) Infection-related asthma. The journal of allergy and clinical immunology. In Pract 2(6):658–663. https://doi.org/10.1016/j.jaip.2014.09.011
22. Fireman P (2002) Virus-provoked rhinitis in patients who have allergies. Allergy Asthma Proc 23(2):99–102

23. Veerapandian R, Snyder JD, Samarasinghe AE (2018) Influenza in asthmatics: for better or for worse? Front Immunol 9:1843. https://doi.org/10.3389/fimmu.2018.01843
24. Pelaia G, Vatrella A, Busceti MT, Gallelli L, Calabrese C, Terracciano R, Maselli R (2015) Cellular mechanisms underlying eosinophilic and neutrophilic airway inflammation in asthma. Mediat Inflamm 2015:879783. https://doi.org/10.1155/2015/879783
25. Tavares LP, Peh HY, Tan WSD, Pahima H, Maffia P, Tiligada E, Levi-Schaffer F (2020) Granulocyte-targeted therapies for airway diseases. Pharmacol Res 157:104881. https://doi.org/10.1016/j.phrs.2020.104881
26. Brightling CE (2014) Destination airway: tracking granulocytes in asthma. EBioMedicine 1 (2–3):105–106. https://doi.org/10.1016/j.ebiom.2014.11.002
27. Kay AB (2016) Paul Ehrlich and the early history of granulocytes. Microbiol Spectr 4:4. https://doi.org/10.1128/microbiolspec.MCHD-0032-2016
28. Miyake K, Karasuyama H (2017) Emerging roles of basophils in allergic inflammation. Allergol Int 66(3):382–391. https://doi.org/10.1016/j.alit.2017.04.007
29. Raap U, Sumbayev VV, Gibbs BF (2015) The role of basophils in allergic inflammation. Allergo J 24(5):28–33. https://doi.org/10.1007/s15007-015-0883-y
30. Siracusa MC, Kim BS, Spergel JM, Artis D (2013) Basophils and allergic inflammation. J Allergy Clin Immunol 132(4):789–801.; quiz 788. https://doi.org/10.1016/j.jaci.2013.07.046
31. Siracusa MC, Comeau MR, Artis D (2011) New insights into basophil biology: initiators, regulators, and effectors of type 2 inflammation. Ann N Y Acad Sci 1217:166–177. https://doi.org/10.1111/j.1749-6632.2010.05918.x
32. Nakashima C, Otsuka A, Kabashima K (2018) Recent advancement in the mechanism of basophil activation. J Dermatol Sci 91(1):3–8. https://doi.org/10.1016/j.jdermsci.2018.03.007
33. Schroeder JT, Chichester KL, Bieneman AP (2009) Human basophils secrete IL-3: evidence of autocrine priming for phenotypic and functional responses in allergic disease. J Immunol 182(4):2432–2438. https://doi.org/10.4049/jimmunol.0801782
34. Phillips C, Coward WR, Pritchard DI, Hewitt CR (2003) Basophils express a type 2 cytokine profile on exposure to proteases from helminths and house dust mites. J Leukoc Biol 73 (1):165–171. https://doi.org/10.1189/jlb.0702356
35. Reithofer M, Jahn-Schmid B (2017) Allergens with protease activity from house dust mites. Int J Mol Sci 18:7. https://doi.org/10.3390/ijms18071368
36. Bieneman AP, Chichester KL, Chen YH, Schroeder JT (2005) Toll-like receptor 2 ligands activate human basophils for both IgE-dependent and IgE-independent secretion. J Allergy Clin Immunol 115(2):295–301. https://doi.org/10.1016/j.jaci.2004.10.018
37. Patella V, Florio G, Petraroli A, Marone G (2000) HIV-1 gp120 induces IL-4 and IL-13 release from human fc epsilon RI+ cells through interaction with the VH3 region of IgE. J Immunol 164(2):589–595. https://doi.org/10.4049/jimmunol.164.2.589
38. Bosch X, Ramos-Casals M (2020) Granulocytes: neutrophils, basophils, eosinophils. In: The autoimmune diseases. Elsevier, pp 243–262. https://doi.org/10.1016/B978-0-12-812102-3.00013-0
39. Lie WJ, Homburg CH, Kuijpers TW, Knol EF, Mul FP, Roos D, Tool AT (2003) Regulation and kinetics of platelet-activating factor and leukotriene C4 synthesis by activated human basophils. Clin Exp Allergy 33(8):1125–1134. https://doi.org/10.1046/j.1365-2222.2003.01726.x
40. Yamaguchi M, Koketsu R, Suzukawa M, Kawakami A, Iikura M (2009) Human basophils and cytokines/chemokines. Allergol Int 58(1):1–10. https://doi.org/10.2332/allergolint.08-RAI-0056
41. Lim LH, Burdick MM, Hudson SA, Mustafa FB, Konstantopoulos K, Bochner BS (2006) Stimulation of human endothelium with IL-3 induces selective basophil accumulation in vitro. J Immunol 176(9):5346–5353. https://doi.org/10.4049/jimmunol.176.9.5346
42. Borregaard N, Sørensen OE, Theilgaard-Mönch K (2007) Neutrophil granules: a library of innate immunity proteins. Trends Immunol 28(8):340–345. https://doi.org/10.1016/j.it.2007.06.002

43. Jasper AE, McIver WJ, Sapey E, Walton GM (2019) Understanding the role of neutrophils in chronic inflammatory airway disease. F1000Res 8. https://doi.org/10.12688/f1000research.18411.1
44. Lawrence SM, Corriden R, Nizet V (2018) The ontogeny of a neutrophil: mechanisms of granulopoiesis and homeostasis. Microbiol Mol Biol Rev: MMBR 82:1. https://doi.org/10.1128/mmbr.00057-17
45. Maas SL, Soehnlein O, Viola JR (2018) Organ-specific mechanisms of transendothelial neutrophil migration in the lung, liver, kidney, and aorta. Front Immunol 9:1–24. https://doi.org/10.3389/fimmu.2018.02739
46. Gane J, Stockley R (2012) Mechanisms of neutrophil transmigration across the vascular endothelium in COPD. Thorax 67:553–561. https://doi.org/10.1136/thoraxjnl-2011-200088
47. Rossaint J, Zarbock A (2013) Tissue-specific neutrophil recruitment into the lung, liver, and kidney. J Innate Immun 5:348–357. https://doi.org/10.1159/000345943
48. Rosales C, Lowell CA, Schnoor M, Uribe-Querol E (2017) Neutrophils: their role in innate and adaptive immunity 2017. J Immunol Res 2017:9748345. https://doi.org/10.1155/2017/9748345
49. Selders GS, Fetz AE, Radic MZ, Bowlin GL (2017) An overview of the role of neutrophils in innate immunity, inflammation and host-biomaterial integration. Regener Biomater 4 (1):55–68. https://doi.org/10.1093/rb/rbw041
50. Gernez Y, Tirouvanziam R, Chanez P (2010) Neutrophils in chronic inflammatory airway diseases: can we target them and how? Eur Respir J 35:467–469. https://doi.org/10.1183/09031936.00186109
51. Naccache PH (2013) Signalling in neutrophils: a retro look. ISRN Physiol 2013:1–13. https://doi.org/10.1155/2013/986320
52. Naccache PH (2014) Cell signalling in neutrophils. Encycl Inflammatory Dis. https://doi.org/10.1007/978-3-0348-0620-6_93-1
53. Futosi K, Fodor S, Mócsai A (2013) Neutrophil cell surface receptors and their intracellular signal transduction pathways. Int Immunopharmacol 17:638–650. https://doi.org/10.1016/j.intimp.2013.06.034
54. Han H, Stessin A, Roberts J, Hess K, Gautam N, Kamenetsky M, Lou O, Hyde E, Nathan N, Muller WA, Buck J, Levin LR, Nathan C (2005) Calcium-sensing soluble adenylyl cyclase mediates TNF signal transduction in human neutrophils. J Exp Med 202:353–361. https://doi.org/10.1084/jem.20050778
55. Bertram A, Ley K (2011) Protein kinase C isoforms in neutrophil adhesion and activation. Arch Immunol Ther Exp 59:79–87. https://doi.org/10.1007/s00005-011-0112-7
56. Luo HR, Mondal S (2015) Molecular control of PtdIns(3,4,5)P3 signaling in neutrophils. EMBO Rep 16:149–163. https://doi.org/10.15252/embr.201439466
57. Fox S, Leitch AE, Duffin R, Haslett C, Rossi AG (2010) Neutrophil apoptosis: relevance to the innate immune response and inflammatory disease. J Innate Immun 2:216–227. https://doi.org/10.1159/000284367
58. Kebir DE, Filep JG (2013) Modulation of neutrophil apoptosis and the resolution of inflammation through β2 integrins. Front Immunol 4:1–15. https://doi.org/10.3389/fimmu.2013.00060
59. El Kebir D, Filep JG (2010) Role of neutrophil apoptosis in the resolution of inflammation. TheScientificWorldJournal 10:1731–1748. https://doi.org/10.1100/tsw.2010.169
60. Carr TF, Zeki AA, Kraft M (2018) Eosinophilic and noneosinophilic asthma. Am J Respir Crit Care Med 197:22–37. https://doi.org/10.1164/rccm.201611-2232PP
61. Boonpiyathad T, Sözener ZC, Satitsuksanoa P, Akdis CA (2019) Immunologic mechanisms in asthma. Seminars in Immunology, vol 46. Elsevier. doi:https://doi.org/10.1016/j.smim.2019.101333
62. Spencer LA, Bonjour K, Melo RCN, Weller PF (2014) Eosinophil secretion of granule-derived cytokines. Front Immunol 5:1–9. https://doi.org/10.3389/fimmu.2014.00496

63. Davoine F, Lacy P (2014) Eosinophil cytokines, chemokines, and growth factors: emerging roles in immunity. Front Immunol 5. https://doi.org/10.3389/fimmu.2014.00570
64. Iwashita H, Morita S, Sagiya Y, Nakanishi A (2006) Role of eosinophil chemotactic factor by T lymphocytes on airway hyperresponsiveness in a murine model of allergic asthma. Am J Respir Cell Mol Biol 35(1):103–109. https://doi.org/10.1165/rcmb.2005-0134OC
65. Saeki M, Nishimura T, Kitamura N, Hiroi T, Mori A, Kaminuma O (2019) Potential mechanisms of T cell-mediated and eosinophil-independent bronchial hyperresponsiveness. Int J Mol Sci 20:12. https://doi.org/10.3390/ijms20122980
66. Kay AB (2005) The role of eosinophils in the pathogenesis of asthma. Trends Mol Med 11 (4):148–152. https://doi.org/10.1016/j.molmed.2005.02.002
67. McAnulty RJ (2007) Fibroblasts and myofibroblasts: their source, function and role in disease. Int J Biochem Cell Biol 39(4):666–671. https://doi.org/10.1016/j.biocel.2006.11.005
68. Phan SH (2008) Biology of fibroblasts and myofibroblasts. Proc Am Thorac Soc 5 (3):334–337. https://doi.org/10.1513/pats.200708-146DR
69. Brightling CE, Gupta S, Gonem S, Siddiqui S (2012) Lung damage and airway remodelling in severe asthma. Clin Exp Allergy 42(5):638–649. https://doi.org/10.1111/j.1365-2222.2011.03917.x
70. Ito JT, Lourenço JD, Righetti RF, Tibério I, Prado CM, Lopes F (2019) Extracellular matrix component remodeling in respiratory diseases: what has been found in clinical and experimental studies? Cell 8:4. https://doi.org/10.3390/cells8040342
71. Kollmannsberger P, Bidan CM, Dunlop JWC, Fratzl P, Vogel V (2018) Tensile forces drive a reversible fibroblast-to-myofibroblast transition during tissue growth in engineered clefts. Sci Adv 4(1):eaao4881. https://doi.org/10.1126/sciadv.aao4881
72. Wójcik-Pszczoła K, Jakieła B, Plutecka H, Koczurkiewicz P, Madeja Z, Michalik M, Sanak M (2018) Connective tissue growth factor regulates transition of primary bronchial fibroblasts to myofibroblasts in asthmatic subjects. Cytokine 102:187–190. https://doi.org/10.1016/j.cyto.2017.09.002
73. Bagnasco D, Ferrando M, Varricchi G, Passalacqua G, Canonica GW (2016) A critical evaluation of anti-IL-13 and anti-IL-4 strategies in severe asthma. Int Arch Allergy Immunol 170(2):122–131
74. Caminati M, Le Pham D, Bagnasco D, Canonica GW (2018) Type 2 immunity in asthma. World Allergy Organ J 11(1):13
75. Corry DB, Folkesson HG, Warnock ML, Erle DJ, Matthay MA, Wiener-Kronish JP, Locksley RM (1996) Interleukin 4, but not interleukin 5 or eosinophils, is required in a murine model of acute airway hyperreactivity. J Exp Med 183(1):109–117
76. Watanabe A, Mishima H, Kotsimbos TC, Hojo M, Renzi PM, Martin JG, Hamid QA (1997) Adoptively transferred late allergic airway responses are associated with Th2-type cytokines in the rat. Am J Respir Cell Mol Biol 16(1):69–74
77. Cohn L, Tepper JS, Bottomly K (1998) Cutting edge: IL-4-independent induction of airway hyperresponsiveness by Th2, but not Th1, cells. J Immunol 161(8):3813–3816
78. Grünig G, Warnock M, Wakil AE, Venkayya R, Brombacher F, Rennick DM, Sheppard D, Mohrs M, Donaldson DD, Locksley RM (1998) Requirement for IL-13 independently of IL-4 in experimental asthma. Science 282(5397):2261–2263
79. Walter DM, McIntire JJ, Berry G, McKenzie AN, Donaldson DD, DeKruyff RH, Umetsu DT (2001) Critical role for IL-13 in the development of allergen-induced airway hyperreactivity. J Immunol 167(8):4668–4675
80. Marone G, Granata F, Pucino V, Pecoraro A, Heffler E, Loffredo S, Scadding GW, Varricchi G (2019) The intriguing role of interleukin 13 in the pathophysiology of asthma. Front Pharmacol 10
81. Bagnasco D, Ferrando M, Varricchi G, Puggioni F, Passalacqua G, Canonica GW (2017) Anti-interleukin 5 (IL-5) and IL-5Ra biological drugs: efficacy, safety, and future perspectives in severe eosinophilic asthma. Front Med 4:135

82. Haspeslagh E, Debeuf N, Hammad H, Lambrecht BN (2017) Murine models of allergic asthma. In: Inflammation. Springer, pp 121–136
83. Karo-Atar D, Bitton A, Benhar I, Munitz A (2018) Therapeutic targeting of the interleukin-4/interleukin-13 signaling pathway: in allergy and beyond. BioDrugs 32(3):201–220
84. Lykouras D, Sampsonas F, Kaparianos A, Karkoulias K, Spiropoulos K (2008) Role and pharmacogenomics of TNF-alpha in asthma. Mini Rev Med Chem 8(9):934–942
85. Cuchacovich R, Espinoza CG, Virk Z, Espinoza LR (2008) Biologic therapy (TNF-α antagonists)-induced psoriasis: a cytokine imbalance between TNF-α and IFN-α? JCR. J Clin Rheumatol 14(6):353–356
86. Karkale S, Khurana A, Saifi MA, Godugu C, Talla V (2018) Oropharyngeal administration of silica in Swiss mice: a robust and reproducible model of occupational pulmonary fibrosis. Pulm Pharmacol Ther 51:32–40
87. Khurana A, Tekula S, Godugu C (2018) Nanoceria suppresses multiple low doses of streptozotocin-induced type 1 diabetes by inhibition of Nrf2/NF-κB pathway and reduction of apoptosis. Nanomedicine 13(15):1905–1922
88. Pasari LP, Khurana A, Anchi P, Saifi MA, Annaldas S, Godugu C (2019) Visnagin attenuates acute pancreatitis via Nrf2/NFκB pathway and abrogates associated multiple organ dysfunction. Biomed Pharmacother 112:108629
89. Liu Z-g (2005) Molecular mechanism of TNF signaling and beyond. Cell Res 15(1):24–27
90. Mukhopadhyay S, Hoidal JR, Mukherjee TK (2006) Role of TNFα in pulmonary pathophysiology. Respir Res 7(1):125
91. Del Valle DM, Kim-Schulze S, Huang H-H, Beckmann ND, Nirenberg S, Wang B, Lavin Y, Swartz TH, Madduri D, Stock A (2020) An inflammatory cytokine signature predicts COVID-19 severity and survival. Nat Med:1–8
92. Charrad R, Berraïes A, Hamdi B, Ammar J, Hamzaoui K, Hamzaoui A (2016) Anti-inflammatory activity of IL-37 in asthmatic children: correlation with inflammatory cytokines TNF-α, IL-β, IL-6 and IL-17A. Immunobiology 221(2):182–187
93. Lilly CM, Nakamura H, Kesselman H, Nagler-Anderson C, Asano K, Garcia-Zepeda EA, Rothenberg ME, Drazen JM, Luster AD (1997) Expression of eotaxin by human lung epithelial cells: induction by cytokines and inhibition by glucocorticoids. J Clin Invest 99(7):1767–1773
94. Mei D, Tan WSD, Wong WSF (2019) Pharmacological strategies to regain steroid sensitivity in severe asthma and COPD. Curr Opin Pharmacol 46:73–81
95. Wortley MA, Bonvini SJ (2019) Transforming Growth Factor-β1: A Novel Cause of Resistance to Bronchodilators in Asthma? American Thoracic Society
96. Ferrari G, Cook BD, Terushkin V, Pintucci G, Mignatti P (2009) Transforming growth factor-beta 1 (TGF-β1) induces angiogenesis through vascular endothelial growth factor (VEGF)-mediated apoptosis. J Cell Physiol 219(2):449–458
97. Lafontaine L, Chaudhry P, Lafleur M-J, Van Themsche C, Soares MJ, Asselin E (2011) Transforming growth factor Beta regulates proliferation and invasion of rat placental cell lines. Biol Reprod 84(3):553–559
98. Roberts AB, McCune BK, Sporn MB (1992) TGF-β: regulation of extracellular matrix. Kidney Int 41(3):557–559
99. Sanjabi S, Oh SA, Li MO (2017) Regulation of the immune response by TGF-β: from conception to autoimmunity and infection. Cold Spring Harb Perspect Biol 9(6):a022236
100. Wu MY, Hill CS (2009) TGF-β superfamily signaling in embryonic development and homeostasis. Dev Cell 16(3):329–343
101. Bartram U, Speer CP (2004) The role of transforming growth factor β in lung development and disease. Chest 125(2):754–765
102. Massagué J (2012) TGFβ signalling in context. Nat Rev Mol Cell Biol 13(10):616
103. Zarzynska JM (2014) Two faces of TGF-beta1 in breast cancer. Mediat Inflamm 2014

104. Lakos G, Takagawa S, Chen S-J, Ferreira AM, Han G, Masuda K, Wang X-J, DiPietro LA, Varga J (2004) Targeted disruption of TGF-β/Smad3 signaling modulates skin fibrosis in a mouse model of scleroderma. Am J Pathol 165(1):203–217
105. Sayed N, Khurana A, Saifi MA, Singh M, Godugu C (2019) Withaferin a reverses bile duct ligation-induced liver fibrosis by modulating extracellular matrix deposition: role of LOXL2/Snail1, vimentin, and NFκB signaling. Biofactors 45(6):959–974
106. Boo S, Dagnino L (2013) Integrins as modulators of transforming growth factor beta signaling in dermal fibroblasts during skin regeneration after injury. Adv Wound Care 2(5):238–246
107. Johansson MW, Mosher DF (2013) Integrin activation states and eosinophil recruitment in asthma. Front Pharmacol 4:33
108. Sidhu SS, Yuan S, Innes AL, Kerr S, Woodruff PG, Hou L, Muller SJ, Fahy JV (2010) Roles of epithelial cell-derived periostin in TGF-β activation, collagen production, and collagen gel elasticity in asthma. Proc Natl Acad Sci 107(32):14170–14175
109. Khalil N, Parekh T, O'connor R, Antman N, Kepron W, Yehaulaeshet T, Xu Y, Gold L (2001) Regulation of the effects of TGF-β1 by activation of latent TGF-β1 and differential expression of TGF-β receptors (TβR-I and TβR-II) in idiopathic pulmonary fibrosis. Thorax 56 (12):907–915
110. Babalola O, Mamalis A, Lev-Tov H, Jagdeo J (2014) NADPH oxidase enzymes in skin fibrosis: molecular targets and therapeutic agents. Arch Dermatol Res 306(4):313–330
111. Herrick A, Cerinic MM (2001) The emerging problem of oxidative stress and the role of antioxidants in systemic sclerosis. Clin Exp Rheumatol 19(1):4–8
112. Kinnula VL, Fattman CL, Tan RJ, Oury TD (2005) Oxidative stress in pulmonary fibrosis: a possible role for redox modulatory therapy. Am J Respir Crit Care Med 172(4):417–422
113. Kliment CR, Oury TD (2010) Oxidative stress, extracellular matrix targets, and idiopathic pulmonary fibrosis. Free Radic Biol Med 49(5):707–717
114. Niki E, Yoshida Y, Saito Y, Noguchi N (2005) Lipid peroxidation: mechanisms, inhibition, and biological effects. Biochem Biophys Res Commun 338(1):668–676
115. Peel AM, Crossman-Barnes C-J, Tang J, Fowler SJ, Davies GA, Wilson AM, Loke YK (2017) Biomarkers in adult asthma: a systematic review of 8-isoprostane in exhaled breath condensate. J Breath Res 11(1):016011
116. Emelyanov A, Fedoseev G, Abulimity A, Rudinski K, Fedoulov A, Karabanov A, Barnes PJ (2001) Elevated concentrations of exhaled hydrogen peroxide in asthmatic patients. Chest 120 (4):1136–1139
117. Ekmekci O, Donma O, Sardoğan E, Yildirim N, Uysal O, Demirel H, Demir T (2004) Iron, nitric oxide, and myeloperoxidase in asthmatic patients. Biochem Mosc 69(4):462–467
118. Batra J, Chatterjee R, Ghosh B (2007) Inducible nitric oxide synthase (iNOS): role in asthma pathogenesis. Indian journal of biochemistry & biophysics 44 (5):303–309
119. Dunford PJ, Holgate ST (2010) The role of histamine in asthma. In: Histamine in inflammation. Springer, pp 53–66
120. Kleniewska P, Pawliczak R (2017) The participation of oxidative stress in the pathogenesis of bronchial asthma. Biomed Pharmacother 94:100–108
121. Fattman CL, Schaefer LM, Oury TD (2003) Extracellular superoxide dismutase in biology and medicine. Free Radic Biol Med 35(3):236–256
122. Kinnula VL, Crapo JD (2003) Superoxide dismutases in the lung and human lung diseases. Am J Respir Crit Care Med 167(12):1600–1619
123. Kinnula VL, Crapo JD, Raivio KO (1995) Generation and disposal of reactive oxygen metabolites in the lung. Lab Invest 73(1):3–19
124. Powis G, Mustacich D, Coon A (2000) The role of the redox protein thioredoxin in cell growth and cancer. Free Radic Biol Med 29(3–4):312–322
125. Rangasamy T, Guo J, Mitzner WA, Roman J, Singh A, Fryer AD, Yamamoto M, Kensler TW, Tuder RM, Georas SN (2005) Disruption of Nrf2 enhances susceptibility to severe airway inflammation and asthma in mice. J Exp Med 202(1):47–59

126. Sahiner UM, Birben E, Erzurum S, Sackesen C, Kalayci Ö (2018) Oxidative stress in asthma: part of the puzzle. Pediatr Allergy Immunol 29(8):789–800
127. Gabrielli A, Svegliati S, Moroncini G, Amico D (2012) Suppl 1: new insights into the role of oxidative stress in scleroderma fibrosis. Open Rheumatol J 6:87
128. Hammad H, Lambrecht BN (2008) Dendritic cells and epithelial cells: linking innate and adaptive immunity in asthma. Nat Rev Immunol 8(3):193–204
129. Schon-Hegrad MA, Oliver J, McMenamin PG, Holt PG (1991) Studies on the density, distribution, and surface phenotype of intraepithelial class II major histocompatibility complex antigen (Ia)-bearing dendritic cells (DC) in the conducting airways. J Exp Med 173 (6):1345–1356
130. Vermaelen KY, Carro-Muino I, Lambrecht BN, Pauwels RA (2001) Specific migratory dendritic cells rapidly transport antigen from the airways to the thoracic lymph nodes. J Exp Med 193(1):51–60
131. Tan JK, O'Neill HC (2005) Maturation requirements for dendritic cells in T cell stimulation leading to tolerance versus immunity. J Leukoc Biol 78(2):319–324
132. Akbari O, Freeman GJ, Meyer EH, Greenfield EA, Chang TT, Sharpe AH, Berry G, DeKruyff RH, Umetsu DT (2002) Antigen-specific regulatory T cells develop via the ICOS–ICOS-ligand pathway and inhibit allergen-induced airway hyperreactivity. Nat Med 8(9):1024–1032
133. Ortiz-Stern A, Kanda A, Mionnet C, Cazareth J, Lazzari A, Fleury S, Dombrowicz D, Glaichenhaus N, Julia V (2011) Langerin+ dendritic cells are responsible for LPS-induced reactivation of allergen-specific Th2 responses in postasthmatic mice. Mucosal Immunol 4 (3):343–353
134. Brimnes MK, Bonifaz L, Steinman RM, Moran TM (2003) Influenza virus–induced dendritic cell maturation is associated with the induction of strong T cell immunity to a coadministered, normally nonimmunogenic protein. J Exp Med 198(1):133–144
135. Lambrecht BN, Pauwels RA, Groth BFS (2000) Induction of rapid T cell activation, division, and recirculation by intratracheal injection of dendritic cells in a TCR transgenic model. J Immunol 164(6):2937–2946
136. Akbari O, DeKruyff RH, Umetsu DT (2001) Pulmonary dendritic cells producing IL-10 mediate tolerance induced by respiratory exposure to antigen. Nat Immunol 2(8):725–731
137. El-Gammal A, Oliveria J-P, Howie K, Watson R, Mitchell P, Chen R, Baatjes A, Smith S, Al-Sajee D, Hawke TJ (2016) Allergen-induced changes in bone marrow and airway dendritic cells in subjects with asthma. Am J Respir Crit Care Med 194(2):169–177
138. Lambrecht BN, Carro-Muino I, Vermaelen K, Pauwels RA (1999) Allergen-induced changes in bone-marrow progenitor and airway dendritic cells in sensitized rats. Am J Respir Cell Mol Biol 20(6):1165–1174
139. Vermaelen K, Pauwels R (2003) Accelerated airway dendritic cell maturation, trafficking, and elimination in a mouse model of asthma. Am J Respir Cell Mol Biol 29(3):405–409
140. Vermaelen KY, Cataldo D, Tournoy K, Maes T, Dhulst A, Louis R, Foidart J-M, Noël A, Pauwels R (2003) Matrix metalloproteinase-9-mediated dendritic cell recruitment into the airways is a critical step in a mouse model of asthma. J Immunol 171(2):1016–1022
141. Tata PR, Rajagopal J (2017) Plasticity in the lung: making and breaking cell identity. Development 144(5):755–766
142. Potaczek DP, Miethe S, Schindler V, Alhamdan F, Garn H (2020) Role of airway epithelial cells in the development of different asthma phenotypes. Cell Signal 69:109523
143. Braga FAV, Kar G, Berg M, Carpaij OA, Polanski K, Simon LM, Brouwer S, Gomes T, Hesse L, Jiang J (2019) A cellular census of human lungs identifies novel cell states in health and in asthma. Nat Med 25(7):1153–1163
144. Leishangthem GD, Mabalirajan U, Singh VP, Agrawal A, Ghosh B, Dinda AK (2013) Ultrastructural changes of airway in murine models of allergy and diet-induced metabolic syndrome. Int Scholarly Res Not 2013

145. Ordovas-Montanes J, Dwyer DF, Nyquist SK, Buchheit KM, Vukovic M, Deb C, Wadsworth MH, Hughes TK, Kazer SW, Yoshimoto E (2018) Allergic inflammatory memory in human respiratory epithelial progenitor cells. Nature 560(7720):649–654
146. Turner J, Roger J, Fitau J, Combe D, Giddings J, Heeke GV, Jones CE (2011) Goblet cells are derived from a FOXJ1-expressing progenitor in a human airway epithelium. Am J Respir Cell Mol Biol 44(3):276–284
147. Tyner JW, Kim EY, Ide K, Pelletier MR, Roswit WT, Morton JD, Battaile JT, Patel AC, Patterson GA, Castro M (2006) Blocking airway mucous cell metaplasia by inhibiting EGFR antiapoptosis and IL-13 transdifferentiation signals. J Clin Invest 116(2):309–321
148. Zissler U, Chaker A, Effner R, Ulrich M, Guerth F, Piontek G, Dietz K, Regn M, Knapp B, Theis F (2016) Interleukin-4 and interferon-γ orchestrate an epithelial polarization in the airways. Mucosal Immunol 9(4):917–926
149. Fogli LK, Sundrud MS, Goel S, Bajwa S, Jensen K, Derudder E, Sun A, Coffre M, Uyttenhove C, Van Snick J (2013) T cell–derived IL-17 mediates epithelial changes in the airway and drives pulmonary neutrophilia. J Immunol 191(6):3100–3111
150. Simpson JL, Phipps S, Baines KJ, Oreo KM, Gunawardhana L, Gibson PG (2014) Elevated expression of the NLRP3 inflammasome in neutrophilic asthma. Eur Respir J 43 (4):1067–1076
151. Nakanishi A, Morita S, Iwashita H, Sagiya Y, Ashida Y, Shirafuji H, Fujisawa Y, Nishimura O, Fujino M (2001) Role of gob-5 in mucus overproduction and airway hyperresponsiveness in asthma. Proc Natl Acad Sci 98(9):5175–5180
152. Tan HTT, Hagner S, Ruchti F, Radzikowska U, Tan G, Altunbulakli C, Eljaszewicz A, Moniuszko M, Akdis M, Akdis CA (2019) Tight junction, mucin, and inflammasome-related molecules are differentially expressed in eosinophilic, mixed, and neutrophilic experimental asthma in mice. Allergy 74(2):294–307
153. Pezzulo AA, Tudas RA, Stewart CG, Buonfiglio LGV, Lindsay BD, Taft PJ, Gansemer ND, Zabner J (2019) HSP90 inhibitor geldanamycin reverts IL-13–and IL-17–induced airway goblet cell metaplasia. J Clin Invest 129(2):744–758
154. Chen Y, Thai P, Zhao Y-H, Ho Y-S, DeSouza MM, Wu R (2003) Stimulation of airway mucin gene expression by interleukin (IL)-17 through IL-6 paracrine/autocrine loop. J Biol Chem 278(19):17036–17043
155. Fujisawa T, Chang MM-J, Velichko S, Thai P, Hung L-Y, Huang F, Phuong N, Chen Y, Wu R (2011) NF-κB mediates IL-1β–and IL-17A–induced MUC5B expression in airway epithelial cells. Am J Respir Cell Mol Biol 45(2):246–252
156. Oyanagi T, Takizawa T, Aizawa A, Solongo O, Yagi H, Nishida Y, Koyama H, Saitoh A, Arakawa H (2017) Suppression of MUC5AC expression in human bronchial epithelial cells by interferon-γ. Allergol Int 66(1):75–82
157. Mitchell C, Provost K, Niu N, Homer R, Cohn L (2011) IFN-γ acts on the airway epithelium to inhibit local and systemic pathology in allergic airway disease. J Immunol 187(7):3815–3820
158. Holgate ST, Davies DE (2009) Rethinking the pathogenesis of asthma. Immunity 31 (3):362–367
159. Holgate ST (2011) The sentinel role of the airway epithelium in asthma pathogenesis. Immunol Rev 242(1):205–219
160. Kouzaki H, Tojima I, Kita H, Shimizu T (2013) Transcription of interleukin-25 and extracellular release of the protein is regulated by allergen proteases in airway epithelial cells. Am J Respir Cell Mol Biol 49(5):741–750
161. O'Leary CE, Schneider C, Locksley RM (2019) Tuft cells—systemically dispersed sensory epithelia integrating immune and neural circuitry. Annu Rev Immunol 37:47–72
162. Desai M, Oppenheimer J (2016) Elucidating asthma phenotypes and endotypes: progress towards personalized medicine. Ann Allergy Asthma Immunol 116(5):394–401
163. Miethe S, Guarino M, Alhamdan F, Simon H-U, Renz H, Dufour J-F, Potaczek DP, Garn H (2018) Effects of obesity on asthma: immunometabolic links. Pol Arch Internal Med 128 (7–8):469–477

164. Shahana S, Björnsson E, Lúdvíksdóttir D, Janson C, Nettelbladt O, Venge P, Roomans GM (2005) Ultrastructure of bronchial biopsies from patients with allergic and non-allergic asthma. Respir Med 99(4):429–443
165. Turato G, Barbato A, Baraldo S, Zanin ME, Bazzan E, Lokar-Oliani K, Calabrese F, Panizzolo C, Snijders D, Maestrelli P, Zuin R, Fabbri LM, Saetta M (2008) Nonatopic children with multitrigger wheezing have airway pathology comparable to atopic asthma. Am J Respir Crit Care Med 178(5):476–482. https://doi.org/10.1164/rccm.200712-1818OC
166. Trejo Bittar HE, Yousem SA, Wenzel SE (2015) Pathobiology of severe asthma. Annu Rev Pathol 10:511–545. https://doi.org/10.1146/annurev-pathol-012414-040343
167. Salter B, Pray C, Radford K, Martin JG, Nair P (2017) Regulation of human airway smooth muscle cell migration and relevance to asthma. Respir Res 18(1):156. https://doi.org/10.1186/s12931-017-0640-8
168. Pascual RM, Peters SP (2005) Airway remodeling contributes to the progressive loss of lung function in asthma: an overview. J Allergy Clin Immunol 116(3):477–486.; quiz 487. https://doi.org/10.1016/j.jaci.2005.07.011
169. Kim KC (2012) Role of epithelial mucins during airway infection. Pulm Pharmacol Ther 25 (6):415–419. https://doi.org/10.1016/j.pupt.2011.12.003
170. Whitsett JA, Alenghat T (2015) Respiratory epithelial cells orchestrate pulmonary innate immunity. Nat Immunol 16(1):27–35. https://doi.org/10.1038/ni.3045
171. Lambrecht BN, Hammad H (2013) Asthma: the importance of dysregulated barrier immunity. Eur J Immunol 43(12):3125–3137. https://doi.org/10.1002/eji.201343730
172. Cohen L, E X, Tarsi J, Ramkumar T, Horiuchi TK, Cochran R, DeMartino S, Schechtman KB, Hussain I, Holtzman MJ, Castro M (2007) Epithelial cell proliferation contributes to airway remodeling in severe asthma. Am J Respir Crit Care Med 176(2):138–145. https://doi.org/10.1164/rccm.200607-1062OC
173. Druilhe A, Wallaert B, Tsicopoulos A, Lapa e Silva JR, Tillie-Leblond I, Tonnel AB, Pretolani M (1998) Apoptosis, proliferation, and expression of Bcl-2, Fas, and Fas ligand in bronchial biopsies from asthmatics. Am J Respir Cell Mol Biol 19(5):747–757. https://doi.org/10.1165/ajrcmb.19.5.3166
174. Makinde T, Murphy RF, Agrawal DK (2007) The regulatory role of TGF-beta in airway remodeling in asthma. Immunol Cell Biol 85(5):348–356. https://doi.org/10.1038/sj.icb.7100044
175. Fahy JV (2002) Goblet cell and mucin gene abnormalities in asthma. Chest 122 (6 Suppl):320s–326s. https://doi.org/10.1378/chest.122.6_suppl.320s
176. Rubin BK, Priftis KN, Schmidt HJ, Henke MO (2014) Secretory hyperresponsiveness and pulmonary mucus hypersecretion. Chest 146(2):496–507. https://doi.org/10.1378/chest.13-2609
177. Boublil L, Assémat E, Borot MC, Boland S, Martinon L, Sciare J, Baeza-Squiban A (2013) Development of a repeated exposure protocol of human bronchial epithelium in vitro to study the long-term effects of atmospheric particles. Toxicol in vitro 27(2):533–542. https://doi.org/10.1016/j.tiv.2012.11.008
178. Cozens AL, Yezzi MJ, Kunzelmann K, Ohrui T, Chin L, Eng K, Finkbeiner WE, Widdicombe JH, Gruenert DC (1994) CFTR expression and chloride secretion in polarized immortal human bronchial epithelial cells. Am J Respir Cell Mol Biol 10(1):38–47. https://doi.org/10.1165/ajrcmb.10.1.7507342
179. Bruewer M, Luegering A, Kucharzik T, Parkos CA, Madara JL, Hopkins AM, Nusrat A (2003) Proinflammatory cytokines disrupt epithelial barrier function by apoptosis-independent mechanisms. J Immunol 171(11):6164–6172. https://doi.org/10.4049/jimmunol.171.11.6164
180. Sawada N, Murata M, Kikuchi K, Osanai M, Tobioka H, Kojima T, Chiba H (2003) Tight junctions and human diseases. Med Electron Microsc 36(3):147–156. https://doi.org/10.1007/s00795-003-0219-y
181. Steed E, Balda MS, Matter K (2010) Dynamics and functions of tight junctions. Trends Cell Biol 20(3):142–149. https://doi.org/10.1016/j.tcb.2009.12.002

182. Green KJ, Jones JC (1996) Desmosomes and hemidesmosomes: structure and function of molecular components. FASEB J 10(8):871–881. https://doi.org/10.1096/fasebj.10.8.8666164
183. Girodet PO, Ozier A, Bara I, Tunon de Lara JM, Marthan R, Berger P (2011) Airway remodeling in asthma: new mechanisms and potential for pharmacological intervention. Pharmacol Ther 130(3):325–337. https://doi.org/10.1016/j.pharmthera.2011.02.001
184. Tschumperlin DJ, Drazen JM (2001) Mechanical stimuli to airway remodeling. Am J Respir Crit Care Med 164(10 Pt 2):S90–S94. https://doi.org/10.1164/ajrccm.164.supplement_2.2106060
185. Kaminska M, Foley S, Maghni K, Storness-Bliss C, Coxson H, Ghezzo H, Lemière C, Olivenstein R, Ernst P, Hamid Q, Martin J (2009) Airway remodeling in subjects with severe asthma with or without chronic persistent airflow obstruction. J Allergy Clin Immunol 124 (1):45–51.e41-44. https://doi.org/10.1016/j.jaci.2009.03.049
186. Ramos-Barbón D, Fraga-Iriso R, Brienza NS, Montero-Martínez C, Verea-Hernando H, Olivenstein R, Lemiere C, Ernst P, Hamid QA, Martin JG (2010) T cells localize with proliferating smooth muscle alpha-actin+ cell compartments in asthma. Am J Respir Crit Care Med 182(3):317–324. https://doi.org/10.1164/rccm.200905-0745OC
187. Folkerts G, Nijkamp FP (2006) Nitric oxide in asthma therapy. Curr Pharm Des 12 (25):3221–3232. https://doi.org/10.2174/138161206778194141
188. Bove PF, van der Vliet A (2006) Nitric oxide and reactive nitrogen species in airway epithelial signaling and inflammation. Free Radic Biol Med 41(4):515–527. https://doi.org/10.1016/j.freeradbiomed.2006.05.011
189. Belvisi M, Ward J, Mitchell J, Barnes P (1995) Nitric oxide as a neurotransmitter in human airways. Archives Internationales de Pharmacodynamie et de Therapie 329(1):97–110
190. Nijkamp FP, Folkerts G (1995) Nitric oxide and bronchial hyperresponsiveness. Arch Int Pharmacodyn Ther 329(1):81–96
191. Ricciardolo FL, Sterk PJ, Gaston B, Folkerts G (2004) Nitric oxide in health and disease of the respiratory system. Physiol Rev 84(3):731–765. https://doi.org/10.1152/physrev.00034.2003
192. Shaul PW (2002) Regulation of endothelial nitric oxide synthase: location, location, location. Annu Rev Physiol 64:749–774. https://doi.org/10.1146/annurev.physiol.64.081501.155952
193. Robbins RA, Barnes PJ, Springall DR, Warren JB, Kwon OJ, Buttery LD, Wilson AJ, Geller DA, Polak JM (1994) Expression of inducible nitric oxide in human lung epithelial cells. Biochem Biophys Res Commun 203(1):209–218. https://doi.org/10.1006/bbrc.1994.2169
194. Kharitonov SA, O'Connor BJ, Evans DJ, Barnes PJ (1995) Allergen-induced late asthmatic reactions are associated with elevation of exhaled nitric oxide. Am J Respir Crit Care Med 151 (6):1894–1899
195. Prado CM, Leick-Maldonado EA, Yano L, Leme AS, Capelozzi VL, Martins MA, Tibério IF (2006) Effects of nitric oxide synthases in chronic allergic airway inflammation and remodeling. Am J Respir Cell Mol Biol 35(4):457–465. https://doi.org/10.1165/rcmb.2005-0391OC
196. Pavord ID, Shaw D (2008) The use of exhaled nitric oxide in the management of asthma. J Asthma 45(7):523–531. https://doi.org/10.1080/02770900801978557
197. Andreadis AA, Hazen SL, Comhair SA, Erzurum SC (2003) Oxidative and nitrosative events in asthma. Free Radic Biol Med 35(3):213–225. https://doi.org/10.1016/s0891-5849(03)00278-8
198. Pollock DM, Keith TL, Highsmith RF (1995) Endothelin receptors and calcium signaling 1. FASEB J 9(12):1196–1204
199. Rubanyi G (1994) Endothelins: molecular biology, biochemistry, pharmacology, physiology, and pathophysiology. Pharmacol Rev 46:325–415
200. Kılkıl G, Muz MH, Deveci F, Turgut T, İlhan F, Özercan İ (2007) Effect of bosentan on the production of proinflammatory cytokines in a rat model of emphysema. Exp Mol Med 39 (5):614–620

201. Sharp CR, Lee-Fowler TM, Reinero CR (2013) Endothelin-1 concentrations in bronchoalveolar lavage fluid of cats with experimentally induced asthma. J Vet Intern Med 27(4):982–984. https://doi.org/10.1111/jvim.12119
202. Webb DJ, Monge JC, Rabelink TJ, Yanagisawa M (1998) Endothelin: new discoveries and rapid progress in the clinic. Trends Pharmacol Sci 19(1):5–8. https://doi.org/10.1016/s0165-6147(97)01144-9
203. Boscoe MJ, Goodwin AT, Amrani M, Yacoub MH (2000) Endothelins and the lung. Int J Biochem Cell Biol 32(1):41–62. https://doi.org/10.1016/s1357-2725(99)00115-6
204. Labram B, Namvar S, Hussell T, Herrick SE (2019) Endothelin-1 mediates aspergillus fumigatus-induced airway inflammation and remodelling. Clin Exp Allergy 49(6):861–873. https://doi.org/10.1111/cea.13367
205. Mitchell JA, Warner TD (1999) Cyclo-oxygenase-2: pharmacology, physiology, biochemistry and relevance to NSAID therapy. Br J Pharmacol 128(6):1121–1132. https://doi.org/10.1038/sj.bjp.0702897
206. Montuschi P, Macagno F, Parente P, Valente S, Lauriola L, Ciappi G, Kharitonov SA, Barnes PJ, Ciabattoni G (2005) Effects of cyclo-oxygenase inhibition on exhaled eicosanoids in patients with COPD. Thorax 60(10):827–833. https://doi.org/10.1136/thx.2004.035592
207. Ramalho TC, Rocha M, da Cunha EF, Freitas MP (2009) The search for new COX-2 inhibitors: a review of 2002–2008 patents. Expert Opin Ther Pat 19(9):1193–1228. https://doi.org/10.1517/13543770903059125
208. Shiraki A, Kume H, Oguma T, Makino Y, Ito S, Shimokata K, Honjo H, Kamiya K (2009) Role of Ca2+ mobilization and Ca2+ sensitization in 8-iso-PGF2α-induced contraction in airway smooth muscle. Clin Exp Allergy 39(2):236–245
209. Robinson C, Hardy CC, Holgate ST (1985) Pulmonary synthesis, release, and metabolism of prostaglandins. J Allergy Clin Immunol 76(2 Pt 2):265–271. https://doi.org/10.1016/0091-6749(85)90640-2
210. Kostikas K, Gaga M, Papatheodorou G, Karamanis T, Orphanidou D, Loukides S (2005) Leukotriene B4 in exhaled breath condensate and sputum supernatant in patients with COPD and asthma. Chest 127(5):1553–1559. https://doi.org/10.1378/chest.127.5.1553
211. Wenzel SE, Trudeau JB, Kaminsky DA, Cohn J, Martin RJ, Westcott JY (1995) Effect of 5-lipoxygenase inhibition on bronchoconstriction and airway inflammation in nocturnal asthma. Am J Respir Crit Care Med 152(3):897–905. https://doi.org/10.1164/ajrccm.152.3.7663802
212. Watanabe M, Machida K, Inoue H (2014) A turn on and a turn off: BLT1 and BLT2 mechanisms in the lung. Expert Rev Respir Med 8(4):381–383. https://doi.org/10.1586/17476348.2014.908715
213. Babu KS, Salvi SS (2000) Aspirin and asthma. Chest 118(5):1470–1476. https://doi.org/10.1378/chest.118.5.1470
214. Eum S-Y, Haile S, Lefort J, Huerre M, Vargaftig BB (1995) Eosinophil recruitment into the respiratory epithelium following antigenic challenge in hyper-IgE mice is accompanied by interleukin 5-dependent bronchial hyperresponsiveness. Proc Natl Acad Sci 92(26):12290–12294
215. Factor P (2003) Gene therapy for asthma. Mol Ther 7(2):148–152. https://doi.org/10.1016/s1525-0016(03)00003-0
216. Hansen G, Berry G, DeKruyff RH, Umetsu DT (1999) Allergen-specific Th1 cells fail to counterbalance Th2 cell-induced airway hyperreactivity but cause severe airway inflammation. J Clin Invest 103(2):175–183. https://doi.org/10.1172/jci5155
217. Proud D, Leigh R (2011b) Epithelial cells and airway diseases. Immunol Rev 242(1):186–204
218. Proud D, Leigh R (2011a) Epithelial cells and airway diseases. Immunol Rev 242(1):186–204. https://doi.org/10.1111/j.1600-065X.2011.01033.x
219. Cook-Mills JM, Deem TL (2005) Active participation of endothelial cells in inflammation. J Leukoc Biol 77(4):487–495. https://doi.org/10.1189/jlb.0904554

220. Silvestri M, Bontempelli M, Giacomelli M, Malerba M, Rossi GA, Di Stefano A, Rossi A, Ricciardolo FL (2006) High serum levels of tumour necrosis factor-alpha and interleukin-8 in severe asthma: markers of systemic inflammation? Clin Exp Allergy 36(11):1373–1381. https://doi.org/10.1111/j.1365-2222.2006.02502.x
221. Santos RF, Oliveira L, Carmo AM (2016) Tuning T cell activation: the function of CD6 at the immunological synapse and in T cell responses. Curr Drug Targets 17(6):630–639. https://doi.org/10.2174/1389450116666150531152439
222. Semitekolou M, Xanthou G (2018) Activated leukocyte cell adhesion molecule: a novel regulator of allergic inflammation in the airways. Am J Respir Crit Care Med 197(8):973–975. https://doi.org/10.1164/rccm.201801-0196ED
223. Greve JM, Davis G, Meyer AM, Forte CP, Yost SC, Marlor CW, Kamarck ME, McClelland A (1989) The major human rhinovirus receptor is ICAM-1. Cell 56(5):839–847. https://doi.org/10.1016/0092-8674(89)90688-0
224. Staunton DE, Merluzzi VJ, Rothlein R, Barton R, Marlin SD, Springer TA (1989) A cell adhesion molecule, ICAM-1, is the major surface receptor for rhinoviruses. Cell 56(5):849–853. https://doi.org/10.1016/0092-8674(89)90689-2
225. Cuss FM, Dixon CM, Barnes PJ (1986) Effects of inhaled platelet activating factor on pulmonary function and bronchial responsiveness in man. Lancet 2(8500):189–192. https://doi.org/10.1016/s0140-6736(86)92489-x
226. Pretolani M, Vargaftig BB (1993) From lung hypersensitivity to bronchial hyperreactivity. What can we learn from studies on animal models? Biochem Pharmacol 45(4):791–800. https://doi.org/10.1016/0006-2952(93)90161-o
227. Chu X, Ci X, He J, Wei M, Yang X, Cao Q, Li H, Guan S, Deng Y, Pang D, Deng X (2011) A novel anti-inflammatory role for ginkgolide B in asthma via inhibition of the ERK/MAPK signaling pathway. Molecules 16(9):7634–7648. https://doi.org/10.3390/molecules16097634
228. Maruoka S, Hashimoto S, Gon Y, Takeshita I, Horie T (2000) PAF-induced RANTES production by human airway smooth muscle cells requires both p38 MAP kinase and Erk. Am J Respir Crit Care Med 161(3 Pt 1):922–929. https://doi.org/10.1164/ajrccm.161.3.9906059
229. Miike S, Kurasawa K, Saito Y, Iwamoto I (2000) Platelet-activating factor activates mitogen-activated protein kinases through the activation of phosphatidylinositol 3-kinase and tyrosine kinase in human eosinophils. J Leukoc Biol 67(1):117–126. https://doi.org/10.1002/jlb.67.1.117
230. Guo X, Zheng M, Pan R, Zang B, Gao J, Ma H, Jin M (2019) Hydroxysafflor yellow a (HSYA) targets the platelet-activating factor (PAF) receptor and inhibits human bronchial smooth muscle activation induced by PAF. Food Funct 10(8):4661–4673. https://doi.org/10.1039/c9fo00896a
231. Groneberg DA, Harrison S, Dinh QT, Geppetti P, Fischer A (2006) Tachykinins in the respiratory tract. Curr Drug Targets 7(8):1005–1010. https://doi.org/10.2174/138945006778019318
232. O'Connor TM, O'Connell J, O'Brien DI, Goode T, Bredin CP, Shanahan F (2004) The role of substance P in inflammatory disease. J Cell Physiol 201(2):167–180. https://doi.org/10.1002/jcp.20061
233. Canning BJ (2006) Neurokinin3 receptor regulation of the airways. Vasc Pharmacol 45(4):227–234. https://doi.org/10.1016/j.vph.2005.08.031
234. Schuiling M, Zuidhof AB, Zaagsma J, Meurs H (1999) Involvement of tachykinin NK1 receptor in the development of allergen-induced airway hyperreactivity and airway inflammation in conscious, unrestrained Guinea pigs. Am J Respir Crit Care Med 159(2):423–430. https://doi.org/10.1164/ajrccm.159.2.9804125
235. Boichot E, Biyah K, Germain N, Emonds-Alt X, Lagente V, Advenier C (1996) Involvement of tachykinin NK1 and NK2 receptors in substance P-induced microvascular leakage hypersensitivity and airway hyperresponsiveness in Guinea-pigs. Eur Respir J 9(7):1445–1450. https://doi.org/10.1183/09031936.96.09071445

236. Boot JD, de Haas S, Tarasevych S, Roy C, Wang L, Amin D, Cohen J, Sterk PJ, Miller B, Paccaly A, Burggraaf J, Cohen AF, Diamant Z (2007) Effect of an NK1/NK2 receptor antagonist on airway responses and inflammation to allergen in asthma. Am J Respir Crit Care Med 175(5):450–457. https://doi.org/10.1164/rccm.200608-1186OC
237. Saito H (2014) Mast cell research. Chem Immunol Allergy 100:165–171. https://doi.org/10.1159/000358733
238. Tatemoto K, Nozaki Y, Tsuda R, Konno S, Tomura K, Furuno M, Ogasawara H, Edamura K, Takagi H, Iwamura H, Noguchi M, Naito T (2006) Immunoglobulin E-independent activation of mast cell is mediated by Mrg receptors. Biochem Biophys Res Commun 349(4):1322–1328. https://doi.org/10.1016/j.bbrc.2006.08.177
239. Saluja R, Ketelaar ME, Hawro T, Church MK, Maurer M, Nawijn MC (2015) The role of the IL-33/IL-1RL1 axis in mast cell and basophil activation in allergic disorders. Mol Immunol 63(1):80–85. https://doi.org/10.1016/j.molimm.2014.06.018
240. Curry JJ (1946) The effect of antihistamine substances and other drugs on histamine bronchoconstriction in asthmatic subjects. J Clin Invest 25(6):792–799. https://doi.org/10.1172/jci101765
241. Nadel JA, Davis B, Phipps RJ (1979) Control of mucus secretion and ion transport in airways. Annu Rev Physiol 41:369–381. https://doi.org/10.1146/annurev.ph.41.030179.002101
242. Kay LJ, Suvarna SK, Peachell PT (2018) Histamine H(4) receptor mediates chemotaxis of human lung mast cells. Eur J Pharmacol 837:38–44. https://doi.org/10.1016/j.ejphar.2018.08.028
243. Hofstra CL, Desai PJ, Thurmond RL, Fung-Leung WP (2003) Histamine H4 receptor mediates chemotaxis and calcium mobilization of mast cells. J Pharmacol Exp Ther 305(3):1212–1221. https://doi.org/10.1124/jpet.102.046581
244. Dunford PJ, O'Donnell N, Riley JP, Williams KN, Karlsson L, Thurmond RL (2006) The histamine H4 receptor mediates allergic airway inflammation by regulating the activation of CD4+ T cells. J Immunol 176(11):7062–7070. https://doi.org/10.4049/jimmunol.176.11.7062
245. Driver AG, Kukoly CA, Ali S, Mustafa SJ (1993) Adenosine in bronchoalveolar lavage fluid in asthma. Am Rev Respir Dis 148(1):91–97. https://doi.org/10.1164/ajrccm/148.1.91
246. Huszár E, Vass G, Vizi E, Csoma Z, Barát E, Molnár Világos G, Herjavecz I, Horváth I (2002) Adenosine in exhaled breath condensate in healthy volunteers and in patients with asthma. Eur Respir J 20(6):1393–1398. https://doi.org/10.1183/09031936.02.00005002
247. Belikoff BG, Vaickus LJ, Sitkovsky M, Remick DG (2012) A2B adenosine receptor expression by myeloid cells is proinflammatory in murine allergic-airway inflammation. J Immunol 189(7):3707–3713. https://doi.org/10.4049/jimmunol.1201207
248. Young HW, Sun CX, Evans CM, Dickey BF, Blackburn MR (2006) A3 adenosine receptor signaling contributes to airway mucin secretion after allergen challenge. Am J Respir Cell Mol Biol 35(5):549–558. https://doi.org/10.1165/rcmb.2006-0060OC
249. Brown RA, Clarke GW, Ledbetter CL, Hurle MJ, Denyer JC, Simcock DE, Coote JE, Savage TJ, Murdoch RD, Page CP, Spina D, O'Connor BJ (2008) Elevated expression of adenosine A1 receptor in bronchial biopsy specimens from asthmatic subjects. Eur Respir J 31(2):311–319. https://doi.org/10.1183/09031936.00003707
250. Feoktistov I, Garland EM, Goldstein AE, Zeng D, Belardinelli L, Wells JN, Biaggioni I (2001) Inhibition of human mast cell activation with the novel selective adenosine a(2B) receptor antagonist 3-isobutyl-8-pyrrolidinoxanthine (IPDX)(2). Biochem Pharmacol 62(9):1163–1173. https://doi.org/10.1016/s0006-2952(01)00765-1
251. Kohno Y, Ji X, Mawhorter SD, Koshiba M, Jacobson KA (1996) Activation of A3 adenosine receptors on human eosinophils elevates intracellular calcium. Blood 88(9):3569–3574
252. Knight D, Zheng X, Rocchini C, Jacobson M, Bai T, Walker B (1997) Adenosine A3 receptor stimulation inhibits migration of human eosinophils. J Leukoc Biol 62(4):465–468. https://doi.org/10.1002/jlb.62.4.465

253. Caruso M, Alamo A, Crisafulli E, Raciti C, Fisichella A, Polosa R (2013) Adenosine signaling pathways as potential therapeutic targets in respiratory disease. Expert Opin Ther Targets 17 (7):761–772. https://doi.org/10.1517/14728222.2013.795220
254. Feoktistov I, Biaggioni I (1995) Adenosine A2b receptors evoke interleukin-8 secretion in human mast cells. An enprofylline-sensitive mechanism with implications for asthma. J Clin Invest 96(4):1979–1986. https://doi.org/10.1172/jci118245
255. Forsythe P, Ennis M (1999) Adenosine, mast cells and asthma. Inflamm Res 48(6):301–307. https://doi.org/10.1007/s000110050464
256. Holt PG, Strickland DH, Wikström ME, Jahnsen FL (2008) Regulation of immunological homeostasis in the respiratory tract. Nat Rev Immunol 8(2):142–152. https://doi.org/10.1038/nri2236
257. Coverstone AM, Seibold MA, Peters MC (2020) Diagnosis and management of T2-high asthma. Journal Allergy Clin Immunol Pract 8(2):442–450. https://doi.org/10.1016/j.jaip.2019.11.020
258. Muehling LM, Lawrence MG, Woodfolk JA (2017) Pathogenic CD4(+) T cells in patients with asthma. J Allergy Clin Immunol 140(6):1523–1540. https://doi.org/10.1016/j.jaci.2017.02.025
259. Herrick CA, Bottomly K (2003) To respond or not to respond: T cells in allergic asthma. Nat Rev Immunol 3(5):405–412. https://doi.org/10.1038/nri1084
260. Molet S, Hamid Q, Davoine F, Nutku E, Taha R, Pagé N, Olivenstein R, Elias J, Chakir J (2001) IL-17 is increased in asthmatic airways and induces human bronchial fibroblasts to produce cytokines. J Allergy Clin Immunol 108(3):430–438. https://doi.org/10.1067/mai.2001.117929
261. Temann UA, Laouar Y, Eynon EE, Homer R, Flavell RA (2007) IL9 leads to airway inflammation by inducing IL13 expression in airway epithelial cells. Int Immunol 19 (1):1–10. https://doi.org/10.1093/intimm/dxl117
262. Temann UA, Ray P, Flavell RA (2002) Pulmonary overexpression of IL-9 induces Th2 cytokine expression, leading to immune pathology. J Clin Invest 109(1):29–39. https://doi.org/10.1172/jci13696
263. Baatjes AJ, Smith SG, Watson R, Howie K, Murphy D, Larché M, Denburg JA, Inman MD, O'Byrne PM (2015) T regulatory cell phenotypes in peripheral blood and bronchoalveolar lavage from non-asthmatic and asthmatic subjects. Clin Exp Allergy 45(11):1654–1662. https://doi.org/10.1111/cea.12594
264. Joetham A, Schedel M, O'Connor BP, Kim S, Takeda K, Abbott J, Gelfand EW (2017) Inducible and naturally occurring regulatory T cells enhance lung allergic responses through divergent transcriptional pathways. J Allergy Clin Immunol 139(4):1331–1342. https://doi.org/10.1016/j.jaci.2016.06.051
265. Bartemes KR, Iijima K, Kobayashi T, Kephart GM, McKenzie AN, Kita H (2012) IL-33-responsive lineage- CD25+ CD44(hi) lymphoid cells mediate innate type 2 immunity and allergic inflammation in the lungs. J Immunol 188(3):1503–1513. https://doi.org/10.4049/jimmunol.1102832
266. Gorski SA, Hahn YS, Braciale TJ (2013) Group 2 innate lymphoid cell production of IL-5 is regulated by NKT cells during influenza virus infection. PLoS Pathog 9(9):e1003615. https://doi.org/10.1371/journal.ppat.1003615
267. Akbari O, Faul JL, Hoyte EG, Berry GJ, Wahlström J, Kronenberg M, DeKruyff RH, Umetsu DT (2006) CD4+ invariant T-cell-receptor+ natural killer T cells in bronchial asthma. N Engl J Med 354(11):1117–1129. https://doi.org/10.1056/NEJMoa053614
268. Hamzaoui A, Kahan A, Ayed K, Hamzaoui K (2002) T cells expressing the gammadelta receptor are essential for Th2-mediated inflammation in patients with acute exacerbation of asthma. Mediat Inflamm 11(2):113–119. https://doi.org/10.1080/09629350220131971
269. Murdoch JR, Lloyd CM (2010) Resolution of allergic airway inflammation and airway hyperreactivity is mediated by IL-17-producing {gamma}{delta}T cells. Am J Respir Crit Care Med 182(4):464–476. https://doi.org/10.1164/rccm.200911-1775OC

4 Targeting Molecular and Cellular Mechanisms in Steroid-Resistant Asthma

Joytri Dutta, Sabita Singh, Archita Ray, and
Ulaganathan Mabalirajan

Abstract

The 'era of corticosteroids' brought about a revolutionary change in the course of asthma treatment in the mid-twentieth century. A few decades later, a lack of response to corticosteroids was recognised in some asthma patients, thereby revealing a distinct endotype of the disease called 'steroid-resistant asthma'. This finding set in motion a plethora of research to elucidate its inducers and unravel the underlying mechanisms driving this variant. Surprisingly, apart from possible genetic factors and infection, lifestyle-related conditions like stress and obesity were found to invoke this disease. Alteration in the pathways mediating anti-inflammatory effect of glucocorticoids, defective autophagy, certain immune mediators and micro RNAs has been found to slacken the steroid response in asthma. This chapter focuses on the fundamental cell types and key mechanistic pathways involved in steroid-resistant asthma and also discusses the treatment strategies that have been adopted to combat the disease in the recent past.

Keywords

Steroid resistant asthma · Glucocorticoid receptor · Airway neutrophilia

Joytri Dutta, Sabita Singh and Archita Ray contributed equally to this work

J. Dutta · S. Singh · A. Ray · U. Mabalirajan (✉)
Molecular Pathobiology of Respiratory Diseases, Cell Biology and Physiology Division, CSIR-Indian Institute of Chemical Biology, Kolkata, India

Academy of Scientific and Innovative Research (AcSIR), Ghaziabad, India

K. Dua et al. (eds.), *Targeting Cellular Signalling Pathways in Lung Diseases*,
https://doi.org/10.1007/978-981-33-6827-9_4

4.1 Introduction

The year 1935 marked the advent of the 'wonder drug' corticosteroid, when it was first isolated from bovine adrenal glands by Edward Calvin Kendall and found to be effective in enhancing muscle strength in adrenalectomised dogs or rats. But it was not until 1946, when Kendall's Compound E (cortisol) among all the isolated steroids was first chemically synthesised by an industrial chemist Louis Sarett, that it was available for clinical testing. In 1948, Rheumatologist Philip Hench tested it on a woman with rheumatoid arthritis which yielded spectacular results within just 3 days. The 1950 Nobel Prize awardee Dr. Kendall and Dr. Hench shared their honour with Dr. Tadeus Reichstein in the field of physiology and medicine for the discovery of cortisone [1]. Several decades after this path-breaking discovery, corticosteroids still continue to be a blockbuster drug, noted for its powerful anti-inflammatory and immunosuppressant properties, albeit with quite a few twists and turns on its journey.

Synthetic glucocorticoids (or corticosteroids or simply steroids) are biologically active synthetic derivatives of cortisol [2] which are widely used in the treatment of different inflammatory and autoimmune diseases like RA (rheumatoid arthritis), nephrotic syndrome (NS), sarcoidosis, multiple sclerosis and systemic lupus erythematosus [3]. All the while cortisol was being glorified as an elixir drug, its efficacy was found to be limited by excessive salt and water retention as it also has mineralocorticoid activity [1]. Eventually, modern glucocorticoids like prednisone, prednisolone, fluticasone, budesonide, dexamethasone, etc. were developed by modification of the cortisol structure to enhance the anti-inflammatory action and reduce mineralocorticoid receptor binding of steroids [4].

Owing to its high efficacy and cost-effectiveness, about 1% of adults in the UK use oral glucocorticoids at any particular time, which generates a market size of 10 billion dollars yearly [3]. The clinicians prescribe glucocorticoids (GCs) for numerous diseases as it has broad-spectrum biological activities [1, 2, 5, 6]. However, 20–30% of patients belonging to otherwise steroid-responsive diseases develop steroid insensitivity that precludes them to use GCs [7–12]. The management of steroid insensitive/steroid-resistant patients is really a big challenge for clinicians who use either very high-dose GCs or secondary immunosuppressants such as calcineurin inhibitors, methotrexate, etc. Both of these strategies lead to worser side effects including early death [13–16]. Thus, understanding mechanisms to improve steroid sensitivity would help us to use GCs in an effective and safer way.

The diseases like RA, inflammatory bowel disease, NS, ulcerative colitis and Crohn's disease can be quite effectively treated with GCs, yet there are some patients showing poor responsiveness to them [17]. As in case of NS, after the diagnosis, NS patients will be prescribed prednisolone (2 mg/kg/day or 60 mg/m^2/day) in single or divided doses for 4 weeks. If there is an absence of proteinuria for three subsequent days, it is considered as remission or steroid sensitive NS. Steroid-resistant NS is defined if there is no such remission in spite of the abovementioned treatment [18–21].

Systemic corticosteroids started being used for asthma therapy soon after its arrival in the market, in the 1950s, and there was a steady rise in its popularity in the 1960s. Meanwhile, when the side effects of systemic corticosteroids were becoming evident, the first inhaled corticosteroid beclomethasone dipropionate was introduced in the early 1970s, to ensure direct delivery into the lung tissue and minimise systemic side effects [5, 22]. Yet, soon enough, steroid resistance emerged as a new menace in the course of asthma treatment. Steroid responsiveness in asthma represents a spectrum which ranges from steroid insensitivity to complete resistance. In steroid insensitive asthma, there is a relative resistance to corticosteroid treatment in some asthma patients, requiring higher doses of inhaled corticosteroids (ICS), maybe even oral corticosteroids. However, absolute steroid-resistant asthma is a rare case, where some asthmatics show complete resistance to corticosteroid therapy, even high oral doses [7, 23]. **Steroid-resistant asthma** is defined as <15% FEV_1 (forced expiratory volume in 1 s) improvement after subsequent oral administration of prednisolone (40 mg/day) for 14 days or high inhaled corticosteroid dose [23, 24].

About 5–10% of the asthmatics show diminished response to glucocorticoids which accounts for around 50% of the economic burden of asthma [24]. Steroid resistance in asthma was first noted in 1968, in six patients who showed no clinical response or decline in blood eosinophil count to systemic glucocorticoid treatment even in high doses. Later, steroid-resistant asthma was identified in a larger group of patients. These patients neither had cortisol deficiency nor had familial glucocorticoid resistance with sex hormone abnormalities. Diminished gastrointestinal absorption or other pharmacokinetic mechanism was also ruled out as a reason for inefficient steroid response. Moreover, patients with glucocorticoid-resistant asthma are found to have normal levels of plasma cortisol and adrenal suppression upon administration of glucocorticoids exogenously, and due to the use of corticosteroids, they tend to manifest the conventional Cushingoid side effects. These evidences indicate that the lack of responsiveness to steroids is probably due to their impaired anti-inflammatory effect in these patients and not their endocrine or metabolic function [17]. This calls for a detailed understanding of the cellular and molecular mechanisms accountable for the ineffectiveness of the steroids to exert their anti-inflammatory effects that culminates into steroid-resistant asthma.

4.2 Inducers of Steroid Resistance in Asthma

The SARP III cohort study on severe asthma showed that there is a distinction in the context of corticosteroid response between children and adults. This indicates that ageing might have a role in steroid responsiveness in asthma [25]. This link has been mechanistically explored by Dietz et al., who showed that there are alterations in remodelling factors with age and they contribute to steroid-resistant airway remodelling in asthma. These factors are, namely, transglutaminase 2 (TGM2), group X secretory phospholipase A2 (sPLA2-X), Wnt 5a and leukotrienes (LTs) [26].

Stressful experience can lead to steroid resistance in asthma. In both the conditions of acute and chronic stress, patients with childhood asthma have reported 5.5-fold decrease in their glucocorticoid receptor (GR) mRNA, such that it affects the sensitivity to anti-inflammatory action of glucocorticoids [27]. Another study showed that psychosocial stress in asthmatic mice can trigger corticosterone release, activate the hypothalamic-pituitary-adrenal (HPA) axis and also reduce the level of glucocorticoid receptor (GR) expression, which in part contributes to the steroid insensitivity [28, 29].

Steroid insensitivity has been reported in obese children as well as obese adults with asthma [30, 31]. The previous study in our lab has demonstrated that the steroid resistance is due to an altered nitric oxide metabolism in obese and house dust mite-induced asthmatic mice [32]. Another study from our lab showed that a metabolite of dietary lipid, 13-S-hydroxyoctadecadienoic acid (HODE), induces steroid insensitivity in asthmatic mice by activation of NF-κB and reduction of glucocorticoid receptor α (GRα). This highlights that how food choices such as omega-6 fatty acids like linoleic acid-rich foods giving rise to the lipid metabolite HODE can not only predispose individuals to the risk of obesity but also steroid insensitive asthma [33].

Genetic susceptibility is a vital contributing factor in invoking steroid refractory asthma. Mutation in the glucocorticoid receptor (*GR*) gene might result in a protein malfunction, thus engendering steroid resistance. The D641V variant of the *GR* was found to give rise to steroid-resistant asthma in the Chinese Han population [34]. McGeachie et al. identified Family With Sequence Similarity 129 Member A gene (*FAM129A*) as a modulator of steroid responsiveness in asthma [35]. However, mechanistic studies are necessary to provide a clear picture on how this gene regulates steroid response. The gene cysteine-rich secretory protein LCCL domain-containing 2 is comprehended to be linked with lung development and endotoxin response. It was identified in GWAS studies to have SNPs with a moderate association with steroid resistance and bronchodilator response in asthma. Later, this association was validated by functional studies in airway smooth muscle cells [36]. Fc fragment of IgE (*FCER2* gene) encodes CD23, a low-affinity IgE receptor. Koster et al. have highlighted that the T2206C variant of the FCER2 gene was notably related with a severe outcome in asthmatic children having a diminished response to ICS [37]. Glucocorticoid-induced transcript 1 (*GLCCI1*) is a gene induced by glucocorticoids that may act as an early marker in glucocorticoid-mediated apoptosis. Two studies have identified the functional variant rs37973 of the *GLCCI1* gene that reduces the expression of GLCCI1, to be associated with a marked decline in response to ICS in the asthmatics of white non-Hispanic and Han Chinese population, respectively [38, 39]. Recently, it was shown that *GLCCI1* knockout in asthmatic mice hinders the activation of GR and consequently the downstream mitogen-activated protein kinase phosphatase-1 pathway, leading to the phosphorylation of p38 MAPK and subsequent production of pro-inflammatory cytokines. However, the mechanism by which GLCCI1 influences GR activation is not yet clear [40].

Respiratory infections are also significant contributors to the development of steroid insensitivity in asthma. The effects of bacteria like *Chlamydia* and

Haemophilus influenzae and viruses like influenza virus and respiratory syncytial virus (RSV) have been investigated in this regard. It was found that infection caused by these agents was able to hamper the steroid sensitivity in asthmatic mice by inducing the micro RNA, miR-21. miR-21 is capable of supressing histone deacetylase 2 (HDAC2), a key molecule controlling glucocorticoid response, via PI3K/AKT pathway [41]. Another study by the same group reported that infection by *Chlamydia* and *Haemophilus* activates the NLRP3 inflammasome, triggering an IL-1β response that drives the steroid-resistant neutrophilic inflammation in asthmatic condition. However, the exact mechanism by which IL-1β brings about this steroid-resistant neutrophilic inflammation remains to be determined [42]. Similarly, LPS, a bacterial component, was shown to drive asthma exacerbation through a steroid-resistant IL-13 pathway [43]. Another group showed that LPS and fungal β-glucan can cumulatively cause steroid refractoriness in asthma via Dectin-1 and TLR4 [44]. Apart from transient bacterial infection, the effect of a persistent low-dose bacterial colonisation has also been explored in this context. It was found that a long-term low-dose *H. influenzae* infection induced IL-17-mediated steroid resistance. The authors reasoned that this insensitivity to glucocorticoids might be attributed to the role of IL-17 in diminishing the expression of GRα (receptor for corticosteroid binding) and heightening the expression of GRβ (inhibitor of GRα) [45].

Therefore, apart from the fact that family history of asthma is common in asthmatics with poor response to steroids [46], unhealthy diet, stressful life experiences, respiratory infections and progressive age are the principal factors that predispose individuals to steroid insensitive asthma.

4.3 Shedding Light on the Cellular Mechanisms Underlying the Glucocorticoid Receptor Signalling Pathway-Mediated Steroid Resistance

4.3.1 Neutrophils

Out of the different types of white blood cells that serve as the armamentarium of our body system, neutrophils are rendered less sensitive to glucocorticoids [47]. The poor response of neutrophils to glucocorticoids can be attributed to the fact that the cascade of events involved in neutrophil activation, including adhesion, chemotaxis, phagocytosis, degranulation and the release of the arachidonic acid metabolite is not effectively inhibited by glucocorticoids [48]. Asthmatic patients characterised by neutrophilic inflammation and obesity-induced asthma have been categorised as the non-Th2/type 2 asthma by Wenzel [49, 50]. Accumulating evidence has shown an increased percentage of neutrophils in the sputum, BAL (bronchoalveolar lavage) fluid and peripheral blood samples from patients suffering from severe asthma and acute asthma exacerbation. Although there's no compelling evidence that acknowledges how the neutrophils induce a steroid-resistant condition, the increased levels of various cytokines are believed to be involved in driving this condition

[49]. Reports from mice and humans have revealed elevated levels of transcripts for IL-17A produced from Th7 in the sputum of asthmatic patients [51]. IL-1β and IL-23 released by the APCs (antigen-presenting cells) are crucial in differentiating the naïve CD4+ T cells into Th17 cell lineages. The pro-inflammatory cytokine IL-17 appears to participate in ushering a condition of neutrophil-predominant phenotype leading to steroid-refractory asthma [51]. The chemokine CXCL8 (IL-8) which is a potent chemoattractant engaged in recruiting neutrophils is also found to be increased in the sputum of asthmatic lungs. The increased release of IL-17 and IL-8 was correlated with the severity of asthma [52]. Adoptive transfer of Th17 cells compared to Th2 cells exhibited severe airway neutrophilia and airway hyperresponsiveness and was way more resistant to corticosteroid treatment as observed in a preclinical model [53]. Additionally, TNF-α level was also found to be significantly enhanced in asthma patients who showed an inadequate response to glucocorticoid medications [52]. In a murine model demonstrating neutrophilic airway inflammation, TNF-α was highlighted to be associated with manifesting a steroid-resistant asthmatic condition [54].

NETosis: In 2004, first emerged the concept of specific cell death of neutrophils or 'NETosis' as reported by Brinkmann and his colleagues. Neutrophil extracellular traps or NET formations are a new mechanism employed by the innate immune system in response to infections and injuries [49, 55]. Following injuries, the neutrophils are activated and infiltrate in the area of injured tissues or organs, cumulating to form NETs. Various stimuli including reactive oxygen species (ROS) generation can activate the peptidyl arginine deiminase 4 (PAD4) [56]. Activated PAD4 citrullinates the histones, which is crucial for the formation of NETs. This causes decondensation of the chromatin material and finally the ousting of the chromosomal DNA [56, 57]. The surge of NETs can turn out to be a double-edged sword wherein the constituents of the NETs like the proteases and others can damage the epithelial and endothelial cells, hence exacerbating the inflammation-induced tissue or organ damage.

Patients with allergic asthma have been reported to have increased levels of NET formation and autophagy in their BAL fluids, sputum granulocytes and peripheral blood cells [58]. NETosis has been highlighted in contributing to lung pathology by stimulating airway obstruction, damaging the alveolar-capillary network and disrupting the host cellular matrix. In an asthmatic PAD4-deficit mice model, the allergen-induced airway neutrophilia was found to be reduced along with a reduction in NETosis [59]. Recently, the formation of neutrophil extracellular traps (NETs) in asthma has been implicated in rendering steroid resistance conditions via the IL-17-mediated pathway. Krishnamoorthy and his colleagues have demonstrated that the NETs did not trigger the neutrophilic airway inflammation, but the neutrophilic cytoplasts were responsible for inducing Th17 cell-mediated inflammation in the murine model [59]. Patients with severe asthma indicated high levels of both the neutrophilic cytoplast and NETs which showed a positive correlation with the IL-17 levels present in the lungs. Hence, it is critical to ascertain whether NETs and neutrophilic cytoplasts are responsible for the manifestation of the steroid-refractory asthma or not.

4.3.2 Factors Inducing Airway Neutrophilia

Leukotriene B4: Another factor that possibly contributes to the booming of steroid-refractory condition in asthma is the pro-inflammatory lipid mediator leukotriene B4 (LTB4) [60]. It is an oxidative product of the arachidonic acid metabolism pathway. Various respiratory diseases including COPD (chronic obstructive pulmonary disease) and allergic rhinitis are linked with remarkably elevated levels of LTB4 [60]. Glucocorticoids which are the cornerstone therapy against various inflammatory diseases seem to behave paradoxically with neutrophils by increasing their longevity. Similarly, the LTB4 has been approved of having an anti-apoptotic effect on the neutrophils [61]. Studies have reported that upon administration of corticosteroid, CD8+ T cells expressing BLT1 (the high-affinity receptor of LTB4) are comparatively more resistant than CD4+ T cells [60, 62]. The CD8+ T cells were found to be activated with enhanced functionality in the presence of corticosteroids by upregulating the steady-state expression of the BLT1 by increasing the expression of the IL-2 receptor [62, 63]. The upregulated BLT1 in turn intensifies the airway inflammation and hyperresponsiveness in an allergen-induced asthmatic murine model. Reports from asthmatic patients have shown increased LTB4 levels in serum, nasal and BAL fluid samples when medicated with corticosteroids [63].

Concomitant respiratory infections: Microbial virulence are critical in the induction and development of steroid-resistant asthma [64]. Several studies have substantiated that patients with *Chlamydia pneumoniae* infection tend to have severe asthma featuring frequent exacerbations along with airway neutrophilia which makes them resistant to steroid treatments [64]. Clinical studies have shown that infection with another respiratory pathogen, *Haemophilus influenzae*, is commonly associated with the manifestation of IL-17-driven neutrophilia in the airways of the asthmatics [65]. Enhanced neutrophils correlate with the severity of the disease, aggravated symptoms and decline in lung function. Viral respiratory infections also contribute to deteriorating the standard of life of asthmatic patients. Infections caused due to organisms like influenza, rhinovirus and respiratory syncytial virus (RSV) [66] further contribute to health hazards as they are responsible for severe airway inflammation and wheezing and thus insinuating asthma exacerbations. The virus-induced infections increase the severity of the diseases that causes a significant reduction in FEV1 and elevated levels of infiltration of mixed immune cells, i.e. both neutrophil and eosinophil in the airways. This makes the patients incompetent to respond to steroid medications [51]. Thus, uncovering the mechanisms underlying the respiratory microbiome load contributing to asthma severity can be interrogated in the future to underpin the steroid-refractory pathogenesis.

4.3.3 Eosinophils

Eosinophilic asthma is elucidated as the sub-phenotypes with eosinophils as the major inflammatory cells infiltrating in the airways effectuating the pathobiology of both childhood-onset asthma and adult asthma. Eosinophils co-migrating with other inflammatory cells release an array of cytotoxic products that range from eosinophils-derived neurotoxins, cationic proteins and peroxidases that contributes to the features of severe asthma [67]. Although asthmatic patients with eosinophilic inflammation respond fairly when treated with corticosteroids, recently, a large section of asthmatic patients with eosinophilic airway inflammation have been distinguished and categorised as 'refractory eosinophilic asthma' [68]. This subgroup of asthmatic patients is associated with poor prognosis and is resistant to any sort of treatment either inhaled or oral corticosteroids. Reports have proclaimed that the clinical features of the patients suffering from late-onset eosinophilic asthma were characterised by features of low FEV1, chronic rhinosinusitis with nasal polyposis and systemic inflammation in the entire respiratory tract, i.e. from paranasal sinuses up to the distal airways [69]. Even high doses of ICS (inhaled corticosteroids) were found not to be effective against these features of eosinophilic asthma, presumably, because it's not feasible for the ICS or topical corticosteroids to reach the distant airways [69]. The pathophysiological mechanism of eosinophilic asthma orchestrates the involvement of both the adaptive immunity (Th2 cells) and the innate immunity. Recent studies have underscored the participation of IL-33 in inducing robust type 2 immune responses and consequently an eosinophilic inflammation which is steroid-resistant [70]. Enhanced secretion of IL-33 has been implicated in the poor control of asthma and the steroid-refractory phenotype of asthma. Due to the lack of signal sequence on IL-33, they are released from the epithelial cells instead of being secreted when hurled upon with injuries or stress conditions [71]. IL-33 is responsible for the release of IL-5-producing memory-type Th2 cells, which induces eosinophilia in the lungs. IL-33 also induces the production of IL-13 that is accountable for smooth muscle cell hyperplasia and mucus production [72]. IL-33 is demonstrated to confer steroid-refractory features upon memory ST2 positive CD4+ T cells which plays a decisive role in the eosinophilic phenotype of lung inflammation [71, 73]. Elevated levels of IL-33 are contemplated as a ramification of severe asthma, chronic airway inflammation and bronchial defacement [74]. Nevertheless, subsequent studies are crucial to demonstrate the exact role of IL-33 in mediating steroid-resistant eosinophilic asthma to pave the way for more effectual therapeutics.

4.3.4 Group 2 Innate Lymphoid Cells (ILC-2)

ILC-2, a subset of innate lymphoid cells, is involved in initiating and fine-tuning of immune responses by dint of chemical mediators and cytokines. Recent studies have started unveiling the importance of ILC-2 in mediating steroid resistance in allergic airway inflammation (AAI) [75]. Upon stimulation with allergens, viruses or fungi, bronchial epithelial cells release endogenous mediators (alarmins) like IL-25, IL-33 and TSLP (thymic stromal lymphopoietin) which initiate the activation of ILC-2 s [76]. ILC-2 s sequentially initiate a type 2 immune response, releasing elevated amounts of cytokines like IL-5 and IL-13 that provoke severe eosinophilic inflammation and airway hyperreactivity, respectively [76]. A growing body of evidence has implicated the increased frequency of ILC-2 production in bronchoalveolar lavage as well as the sputum of allergic asthma patients. Kabata et al. have recently highlighted that in a murine model, ILC-2 s develop steroid resistance which is TSLP (thymic stromal lymphopoietin) dependent [77]. Accumulating evidence has shown the involvement of TSLP being biased towards the Th2 (T helper) type immune response, a hallmark of the allergic inflammation. Liu and his co-workers have established the prominence of the TSLP-STAT5 (signal transducer and activator of transcription 5) axis in steroid resistance in a murine model of asthma [78]. They have identified IL-7 as the possible modifier of corticosteroid sensitivity as they cause induction of the STAT5 pathway which upregulates the expression of Bcl-2 and Bcl-xL (anti-apoptotic proteins), which insulates the cells from apoptosis [78]. Upregulation of IL-7Rα (interleukin-7 receptor subunit alpha) was found to be crucial in developing steroid resistance in ILC-2 s of human, as only IL-7Rα agonists and no other ILC-2-stimulating cytokines were capable of rendering steroid resistance condition. Out of the three cytokines (IL-25, IL-33 and TSLP) secreted from the epithelia, only TSLP which is a dexamethasone (a corticosteroid medication)-regulated IL-7Rα ligand induces steroid-refractory condition in ILC-2 s [78].

4.3.5 T Helper Type 9 Cells (Th9 Cells)

Identified a decade ago, the Th9 cells are a new subset of CD4+ T helper cells that secretes IL-9 characteristically. Upon priming with TGF-β (transforming growth factor-beta) and IL-4, naïve CD4+ T cells are differentiated to produce Th9 cells [79]. IL-9 released from these cells kindles the various asthmatic features like airway inflammation with excessive infiltration of eosinophils, goblet cell hyperplasia, bronchial hyperresponsiveness, mucous secretion, etc. [80]. An increased level of IL-9 in blood and sputum samples of patients with bronchial asthma demonstrates the clinical significance of Th9 cells. Researchers have shown the involvement of IL-9 in nurturing the survivability of ILC-2 and that ILC-2 might be responsible for driving the steroid resistance of Th9 cell-mediated airway inflammation [80]. Kara et al. reported a probable mechanism by which the Th9 cells migrate to cause steroid resistance [81]. They demonstrated that Th9 cells exhibited expression of chemokine receptors like CCR3, CCR6 and CXCR3, and on the other hand, Th2 cells express CCR3 and CCR4 [81]. Treatment with corticosteroid has been elucidated to upregulate the expression of IP-10 (interferon gamma-induced protein 10), a ligand for CXCR3, in the airway mucosal region of asthmatics [80]. Therefore, dexamethasone is believed to facilitate the infiltration of Th9 cells by the means of upregulated IP-10. However, subsequent researches are necessitated to elucidate the exact mechanism recruited for Th9-mediated steroid resistance in the light of chemokine-related responses.

4.3.6 $CD8^+$ T Cells

In the murine models of asthma, along with ILC-2 and $CD4^+$ Th2 cells, another subset of the T cells, i.e. $CD8^+$ T cells, are also proposed to drive the asthma pathogenesis and mediating a steroid-refractory condition [82]. $CD8^+$ T cells display plasticity and undergo transdifferentiation, switching their functional activities. Interferon-gamma (IFN-γ)-producing $CD8^+$ T cells sustain through various phases of transcription and translation when subjected to IL-4 in vitro that resulted in epigenetic poising [83]. This transforms the IFN-γ-producing $CD8^+$ T cells into $CD8^+$ T cells that are competent enough to generate Th2 cytokines like IL-5 and IL-13 [83]. These cytokines elicit the recruitment and survival of eosinophils in the airways leading to inflammation. Studies have revealed that activated $CD8^+$ T cells are comparatively less susceptible to corticosteroid than $CD4^+$ T cells. A possible explanation could be the lowered expression of the ATF2 (activating transcription factor-2), a DNA-binding protein that is capable of specifically acetylating the histones H2B and H4 in in vitro conditions [83, 84]. ATF2 is necessary for the corticosteroid-stimulated transactivation. A diminished corticosteroid-stimulated transactivation is demonstrated by these $CD8^+$ T cells, in addition to reduced induction of IL-10. Many questions remain undetermined; many mechanistic studies are needed to demonstrate how these cells engage in driving the Th2 cytokines leading to steroid-resistant eosinophilic asthma.

4.4 Possible Molecular Mechanisms behind Steroid-Resistant (SR) Asthma (Fig. 4.1)

Though asthma cannot be fully cured, it can be regulated through various medications and lifestyle changes. Severe asthmatic exacerbations positively react to steroids and provide quick relief. Generally, asthmatic patients respond to steroid therapy. However, a small group of people do not exhibit the normal response to steroid treatment. This can be due to various reasons like incorrect disease diagnosis, inadequate quantity of steroid reaching airway mucosal barrier or unidentifiable factors. Other grounds could be inappropriate use of drug and unregulated disease management [85]. Further detailed studies have to be carried out. Glucocorticoids are the major therapeutic drug group in asthma. It is seen that this group of people unresponsive to steroid treatment possess alterations in the GR signalling. The glucocorticoid drugs cannot bind to the GR properly. Such GR binding abnormality makes them unresponsive to the therapeutic drugs. After oral steroid treatment, steroid sensitive individuals exhibit increase in FEV1 (forced expiratory volume 1) by about 30 per cent. However, such increase was not observed in steroid-resistant individuals. They exhibit only about 15 per cent improvement in FEV1. These steroid-resistant individuals also show less number of GR in the PBMC (peripheral blood mononuclear cells). Even after incubation with cytokines like IL-2 and IL-4, the number didn't improve [86].

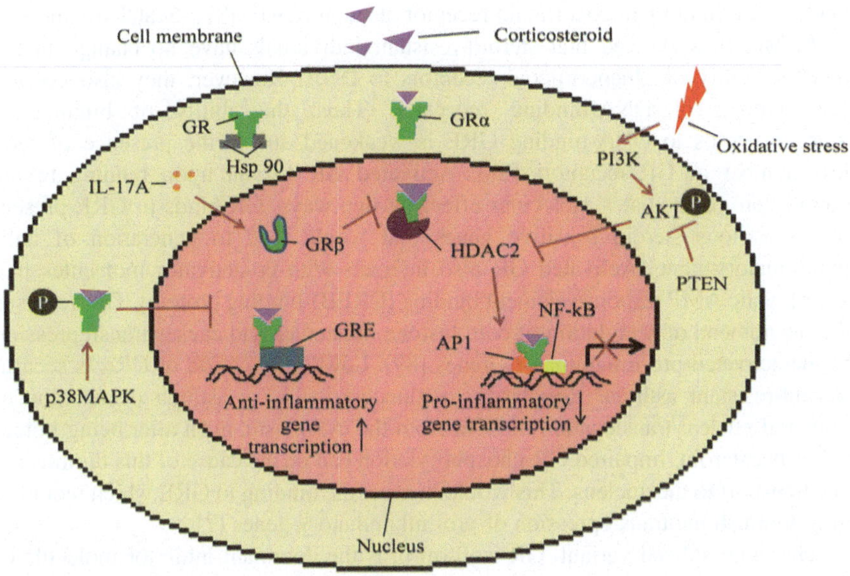

Fig. 4.1 Schematic diagram showing molecular mechanisms of steroid-resistant asthma. Steroid resistance in asthma involves diverse signalling pathways. Before being activated, glucocorticoid receptor (GR) remains in the cytoplasm by complexing with Hsp90. Upon binding with corticosteroids, GR undergoes conformational change in such a way that it dissociates from Hsp90 and translocates to the nucleus. In the nucleus, GR binds with glucocorticoid receptor element (GRE) and causes transactivation of anti-inflammatory genes. By tethering with other transcription factors, GR also causes transrepression of pro-inflammatory genes. Such transrepression occurs through activation of HDAC2 molecules. There are two isoforms of GR: GRα and GRβ. While GRα is the main isoform involved in the corticosteroid-induced effects, GRβ acts against GRα and its expression is increased in steroid-resistant asthma. Decoy receptor GRβ competes with GRα in binding with DNA. Cytokines secreted by Th17 cells can promote the upregulation of GRβ. On facing oxidative stress, PI3K gets phosphorylated and further causes phosphorylation of AKT kinase that leads to phosphorylation and inactivation of HDAC2. p38MAPK also causes phosphorylation in serine 226 residue in GRα. This will hinder the binding of GRα to ligand and hinder the translocation of GRα into the nucleus

4.4.1 Bioavailability of Glucocorticoid Receptor (GR) and the Underlying Molecular Mechanism

Glucocorticoid or glucocorticosteroid, which is the main therapy for asthma, can be inhaled or orally administered. Upon inhalation, ICS gets diffused into the tissues of the airways and activation of GR occurs. Alternative splicing of the GR mRNA causes generation of two isoforms of GR. GRα isoform is found in the cytoplasm and GR β resides in the nucleus. GRα is the only form of GR which binds to corticosteroids. GRα mediates anti-inflammatory effect by ligand binding to the transcription factor. GR binds to ligand and undergoes dissociation from Hsp90 kD protein complexes, and following conformational change, they migrate to the nucleus and bind to glucocorticoid receptor element (GRE) [87]. Scatchard analysis of PBMCs has showed that steroid-resistant individuals have no change in the binding ability of glucocorticoid receptors to DNA. However, they also possess less number of DNA-binding receptors. Thus, the ability of binding of glucocorticoids to DNA-binding GRE is weakened due to the presence of such less number of GR receptors [88]. Activated GR variant upon binding to the corticosteroids mediates molecular effect by two ways. GR binds to GRE present in the various steroid sensitive genes that would lead to generation of anti-inflammatory gene. Activated GR also interacts with co-activator molecules like CBP [cyclic AMP response element binding (CREB)-binding protein]. This reverses the phenomena of acetylation of core histone molecules and causes transrepression of the activated pro-inflammatory genes [89]. Lower expression of GRα is seen in steroid-resistant asthma. Resistance to glucocorticoids in asthma occurs due to impaired nuclear translocation of GRα from the cytoplasm, even after being bound by corticosteroid. Impaired GR phosphorylation can be the cause of this diminished GR transport to the nucleus. This would hamper the binding to GRE which would in turn diminish the transrepression of pro-inflammatory genes [7].

The other spliced variant, GR isoform β, is the dominant inhibitor molecule of GRα which does not bind to corticosteroids but interacts with DNA. Expression level of GR isoform β is seen to increase under steroid insensitive asthmatic conditions. Upon stimulation with pro-inflammatory cytokines, GRβ competes with GRα for binding with GRE. As GRβ does not bind to ligand, it is often regarded as the decoy receptor [87].

There occurs abnormal tethering of GR with other transcription factors. GR interacts with transcription factors like activated protein-1 (AP-1) and NF-kB. AP-1 forms a complex with activated GR and blocks GR from binding with DNA. This would reduce the activated GR found in the nucleus for binding to GRE and thus lead to steroid resistance phenomena [90]. Further ligand binding analysis has shown that in steroid sensitive individuals, there is decreased interaction between AP-1 and GR. However, interactions with other transcription factors remain unhindered. An increase in the AP-1 protein is also observed in such steroid-resistant individuals. All these suggest that in steroid resistance, either increased level of AP-1 or altered binding capacity accounts for such impaired binding between GR and AP-1 [91]. Studies by Serra et al. have showed that repeated allergen exposure in A/J mice changes the nature of their responsiveness to steroid and they become steroid-resistant. With increase in the number of exposures, loss in glucocorticoid receptor bioavailability takes place [92]. Thus, glucocorticoid receptors play a pivotal role in modulating steroid responsiveness in asthma.

4.4.2 IL-17A High and IFN-γ High Phenotypes

Steroid-resistant asthmatic airway inflammation is mediated via Th17 subtype T cells. These Th17 cells release IL-17, IL-17F and IL-22 cytokines that further stimulate several pro-inflammatory cytokines and growth mediators. Studies were done in the PBMCs where researchers analysed the level of Th17 transcription factor RORγt. It is seen that compared to healthy individuals, heightened expression of retinoic acid-related orphan receptor γt (RORγt) is seen in the PBMCs of severe allergic asthmatics. With increase in the disease progression, IL-17 and IL-22 levels increase both in PBMC and in plasma. IL-23 which is a Th17 cell maintenance cytokine also increases simultaneously and showed positive correlation with IL-17 level [93]. These Th17 cytokines cause upregulation of the innate isoform glucocorticoid receptor GRβ. There also occurs a subsequent decrease in the expression pattern of active glucocorticoid receptor GRα which is mediated by IL-2 or IL-4. This leads to development of insensitivity upon treatment with the steroid, dexamethasone. Thus, Th17 cytokines found in the airways play a part in causing steroid insensitivity in PBMCs in severe grade asthma patients [94]. IL-17 also had other varied functions which changes the profile in bronchial epithelial cell. IL-17 has a major role in neutrophilic inflammation that happens in severe asthma. IL-17 causes generation of chemokines like CXCL-8 from the bronchial epithelial cell and thus coordinates the recruitment of granulocytes like neutrophils. Indirectly, IL-17 also stimulates the production of cytokines like IL-6, chemokines and CSF from epithelial cells and IL-17 secreting ILC3 cells [95]. Another group of researchers found that although IL-17A-stimulated IL-8 secretion is sensitive to glucocorticoid (GC) treatment, IL-17A pre-treatment done on TNF-α-induced IL-8 cytokines makes them insensitive to GCs. Cytokine IL-17A not only works by regulating downstream signalling components p38 and PI3K but also affects the HDAC pathway. IL-17A cytokine causes a decrease in the activity of HDAC gene. Overexpression of HDAC2 gene causes a reversion in the IL-17A-induced steroid insensitivity effect. However, it is seen that IL-17A does not affect or hinder the binding affinity of the GC receptor. Thus, by modulating GC sensitivity, IL-17A can be a potent therapeutic target in steroid insensitive asthma [96].

Steroid-resistant asthmatic individuals show irresponsiveness to the oral drug prednisone. Despite drug administration, mRNA of BAL fluid shows no alteration in the IL-4 and IL-5 level. However, the level of pleiotropy characterised by the IFN-γ cytokine, which is released by Th1 cells, increases. These indicate the possibility of an imbalance between Th1 and Th2 cytokines in steroid-resistant asthma [97]. It is seen that in addition to Th17 cytokines, cytokines secreted by Th1 cells also have a role in mediating the steroid resistance in asthma. TNF-α and IFN-γ are released by Th1 subtype of T cell and have a role in causing corticosteroid resistance in airway smooth muscle. Here, IFN-γ augments other inflammatory pathways and amplifies the effect of pro-inflammatory cytokine, TNF-α. Studies were carried out to see the effect of TNF-α or IFN-γ on human airway smooth muscle cell in vitro. It was observed that compared to each of their individual effect, when administered in combination, both TNF-α and IFN-γ pose greater curbing effect on the chemokine inhibiting activity of steroid fluticasone propionate. This highlights the significance of the interaction between TNF-α and IFN-γ in steroid-resistant asthma [98]. Th1 cells produce IFN-γ and its level gets heightened in severe asthma. Mast cells secrete chemokine CXCL10 that acts downstream of IFN-γ, which also contributes to steroid unresponsiveness occurring in Th1 high asthma. The mRNA level of

CXCL10 is found to be increased in those severe asthma patients having poor disease prognosis. Thus, it can be concluded that such Th1 signalling IFN-γ and CXCL10 axis that occurs via the CXCR3 receptor have a role in steroid resistance [99]. Another cytokine secreted by macrophages that can also trigger the release of IFN-γ by Th1 cell is IL-27. The effect of IL-27 and IFN-γ occurs via MYD88 pathway which inhibits the translocation of the glucocorticosteroid receptors. Thus, such host defence mechanisms encompassing IFN-γ and Th17 cells occur in steroid-resistant non-allergic asthma [100, 101]. Studies on PBMCs by Chambers et al. have shown supporting evidences in this regard. Compared to steroid sensitive individuals, steroid-resistant asthmatic individuals exhibit higher level of IL-17A and IFN-γ. IL-17A was observed in sputum and BAL fluid and correlates to AHR. With dexamethasone treatment, IFN-γ level seems to slightly decrease in steroid-resistant individuals, whereas upon dexamethasone treatment, both IFN-γ and IL-17A levels decrease in steroid sensitive and steroid-resistant individuals. It is also seen that the increased levels of IFN-γ and IL-17A in steroid-resistant individuals are non-overlapping and do not get affected by each other [102]. Elevated expression of IL-17A can serve as a risk factor identifier and also a biomarker in steroid-resistant asthma. In addition, IFN high phenotype is also observed.

4.4.3 PI3K/HDAC Signalling Pathway

Phosphatidylinositol 3-kinase (PI3K) signalling pathway plays a deciding role in orchestrating the steroid resistance in asthmatic airways. The lipid kinase PI3K has several isoforms which are PI3Kα, PI3Kß, PI3Kγ and PI3Kδ. The isoform phosphoinositide 3-kinase-δ (PI3K-δ) has an important role in steroid-resistant asthma. PI3K-δ phosphorylates the downstream serine threonine kinase, AKT [103]. When oxidative stress happens in asthma, PI3K-δ is phosphorylated. HDAC2 gene undergoes phosphorylation and is inactivated. This occurs via phosphorylated AKT [7]. Besides engaging with GR, glucocorticosteroid given to asthma patients also recruits histone deacetylase (HDAC) molecules for transrepression. Experiments have shown that there is a decrease in the HDAC2 level in BAL fluid of steroid-resistant patients but not in steroid sensitive asthmatic individuals. Further study by this group shows that GRβ suppresses HDAC2 expression level and thus regulates steroid insensitive asthma [100, 101].

Generally, upon stimulation, transcription factor like NF-kB which is found in alveolar macrophages activates histone acetylation. This in turn causes acetylation of lysine residue which is followed by activation of the pro-inflammatory genes. In steroid sensitive asthmatic individuals, corticosteroids promote the GR migration into the nucleus. Then, there occurs interaction between GRs and histone 4 which then leads to acetylation of K5 and K16 lysine residues. In some steroid-resistant asthmatic patients, despite proper nuclear localisation of GR, there occurs impaired acetylation of K5 residue. This hinders the GR-bound steroid-mediated effect leading to corticosteroid resistance. Corticosteroids cause recruitment of HDAC-2, which aids in reversing the effect. This also stops activation of inflammatory genes by NF-kB. However, oxidative stress and peroxynitrite formation has the potential to hinder HDAC-2 activity. In such case, not only the activation of inflammatory genes by NF-kB is increased, but also the anti-inflammatory response mediated by HDAC2 is reduced [23].

HDAC2 is also expressed in lower amount in the peripheral lung region and alveolar macrophages of severe and steroid insensitive asthmatics. Here, PI3K-δ plays a major role. Thus, by inhibiting PI3K-δ, there can be selective upregulation and activation of HDAC2 [87]. Being a tumour suppressor, PTEN (phosphatase and tensin homologue) is the naturally occuring endogenous PI3K inhibitor found in the cell. Increased expression of PTEN causes diminished PI3K activity and heightened nuclear expression of HDAC2. Pan-PI3K inhibitor LY294002 exhibits similar effects on severe steroid-resistant acute airway disease (SSRAAD) [104]. PI3K inhibitor, CL27c, is another pan prodrug which also has an inhibiting effect on asthma. It is seen that in allergic asthma which exhibits corticosteroid resistance, administration of both CL27c and dexamethasone diminishes the inflammation in the lung and restores the lost lung function. While single administration of drug dexamethasone did not show change, single administration of CL27c remarkably reduced the score of lung damage. Such pan-PI3K inhibitors are more preferred over direct PI3K inhibitor and mTOR inhibitor. As pan-PI3K inhibitors act upstream of mTOR pathway, additional off-target side effects and lung toxicity are reduced [105].

4.4.4 p38 MAPK

The p38 mitogen-associated protein kinase (p38 MAPK), a serine threonine kinase, exerts its effect on transcription factors like NF-kB and AP-1. Severe asthmatic people who are unresponsive to corticosteroids show an increased expression of p38 MAPK [106]. Presence of p38 MAPK is observed both in epithelial cells in the alveolar region and in alveolar macrophages. An association is seen between severity of asthma and the presence of p-p38 in the airways. Western blot analysis has shown the presence of activated phospho-p38 MAPK and stress activated protein kinase-1. Compared to steroid sensitive patients, phosphorylated form of p38 is found in increased amount in the CD14+ monocytes of SR patients. Downstream of p38 MAPK, there is presence of phosphorylated mitogen- and stress-activated protein kinase 1 (MSK1). MSK1 kinase can interact with other pro-inflammatory transcription factors that cause more synthesis of inflammatory genes. This would in turn lead to steroid resistance. Phosphorylated form of MSK1 is also found in the PBMC lysates of steroid-resistant individuals. Hence, both MSK1 and p38 phosphorylation can serve as biomarkers in SR asthmatic patients [107].

Studies by Bhavsar et al. have showed that p38 MAPK has the potential to cause corticosteroid insensitivity in severe asthma. PBMCs and alveolar macrophages were isolated from severe and non-severe asthma patients who were then subjected to LPS both in the presence and absence of p38 MAPK inhibitor SD282 and drug dexamethasone. Compared to solo dexamethasone use, heightened inhibition on the release of pro-inflammatory cytokines in alveolar macrophages was observed when SD282 and dexamethasone were administered together. Hence, it was noted that use of p38 MAPK inhibitors can have a positive effect in reversing steroid resistance [106].

P38 MAPK stimulates phosphorylation of the glucocorticoid receptor. GR phosphorylation imparts a negative effect and curbs the ability of dexamethasone. Studies in healthy individual's PBMCs have shown that IL-2 and IL-4 incubation alters the GR's ligand binding activity inside the nucleus as indicated by Kd value. This will in turn reduce the binding capacity of GR to drug dexamethasone. Additionally, it is observed that incubation with p38 MAPK inhibitor SB203580 lessens the IL-2- and IL-4-stimulated reduction of glucocorticosteroid receptor's binding affinity. Thus, IL-2/IL-4 controls p38 MAPK-mediated GR phosphorylation and the presence of SB203580 can inhibit this phenomenon [108]. The region of phosphorylation in the GR is serine 226. There are some phosphatases that dephosphorylates the action of kinase p38 MAPK. Phosphatase MKP-1, the endogenously found p38MAPK inhibitor, gets activated by corticosteroids. It is seen that SR asthmatic individuals have diminished expression of MKP-1 and increased level of p38 MAPK in alveolar macrophages. The other phosphatase whose expression gets reduced in asthma is serine/threonine phosphatase protein phosphatase 2A (PP2A) [7]. Thus, selective p38 MAPK inhibitors might be propitious in treating steroid-refractory conditions of asthma.

4.4.5 NLPR3 Inflammasome

Multiprotein nucleotide-binding domain and leucine-rich repeat containing (NLR) protein inflammasome works on complex signalling pathways. It mediates the release of several cytokines that exert pro-inflammatory effects. NLR family pyrin domain-containing 3 (NLPR3) inflammasome encompasses various apoptosis-related subunits. The pyrin domain recruits procaspase 1 and activates caspase-1. Ideally, inflammasome requires two major events like assembly and activation before exerting its effector function. A major adaptor protein component involved here is apoptosis-associated speck-like protein containing a CARD (ASC). Several stimuli like danger associated molecular pattern, infection or allergen plays a role in activating the inflammasome. After being activated, NLPR3 interacts with ASC. Then, NLPR3 inflammasome intervenes in cleaving procaspase 1 into activated caspase-1. This caspase-1 in turn causes cleavage of pro-IL-1ß into active IL-1ß. Along with IL-1ß, pro-IL-18 is also released [109].

In neutrophilic asthma, the level of NLPR3 and caspase activity increases. The protein level of IL-1ß also increases in the sputum of neutrophilic asthmatic patients. A correlation could be seen between neutrophil chemoattractants IL-8 and IL-1ß [110]. IL-1ß-dependent inflammatory response happens in steroid-resistant asthma. Another group of researchers also showed increased level of IL-1ß and NLPR3 mRNA in sputum of patients who displayed clinical parameters of severe steroid-resistant asthma. A correlation is noted between increased IL-1ß and NLPR3 with more neutrophil occurrence and heightened airway obstruction. These in turn are associated with fall in steroid responsiveness and incidence of neutrophil-mediated airway inflammation. Treatment of IL-1ß intranasally to naive mice generated hallmark features like steroid-resistant *Chlamydia* infection. When caspase-1 is inhibited by pan-caspase inhibitor z-VAD-fmk (ZVAD), the IL-1ß level and *Chlamydia* infection-induced severe steroid-resistant allergic airway disease (SSRAAD) characteristics get reduced. In SSRAAD, which is induced by *Chlamydia* infection, it is seen that on administration of NLPR3-specific inhibitor MCC950, both IL-1ß and asthmatic features get diminished. Thus, inflammasome-driven caspase-1/IL-1ß signalling pathway has a potential role in promoting AHR in steroid-resistant asthma [41, 42]. A recent study by this group has shown that inflammasome has a role in obesity-driven steroid-resistant asthma. In high-fat diet-induced obesity-related asthma, it is seen that NLPR3 level is more in the airways, and there occurs activation of caspase-1. However, along with MCC950, use of anti-IL-5 and anti-IL-13 neutralising antibodies is also seen to reduce NLPR3-mediated inflammatory AHR responses [111]. Studies by Wood et al. also highlighted the presence of inflammasome in the obese asthma. It is seen that compared to non-obese asthmatic adults, obese asthmatic individuals displayed upregulation of the NLPR3 and nucleotide oligomerisation domain 1 NOD1 gene expression. They also have more IL-1ß in the sputum. Subsequently, the non-obese asthmatic individuals were given food filled with saturated fatty acid FA. An increase in the neutrophils, TLR4 and NLPR3 were found in them after 4 h of eating [112]. Thus, NLPR3 inflammasome does have an important role in regulating the IL-1ß pathway in steroid-resistant asthma.

4.4.6 Lack of Autophagy

Autophagy is involved in asthma and has a role in regulating airway remodelling that occurs in asthma. In human asthma, autophagy-related gene 5 (Atg5) is prone to undergo single nuclear polymorphism. There is an association seen between increase in Atg5 gene polymorphism and reduction in the value of FEV1%. Atg5 gene polymorphism and variation in the gene promoter region are seen to have an association with childhood asthma [113]. In Atg5 knockdown mice, it is seen that the AHR occurring is Th17 dependent. There is neutrophil accumulation in BAL fluid, and even though eosinophil count is a bit reduced, steroid resistance is seen. HDM was administered a day after neutralising anti-IL-17A antibody in both Atg5 knockout mice and wild-type mice. It was seen that AHR and neutrophil count was reduced in Atg5 knockout mice; however, no such change was observed in wild-type mice after anti-IL-17A antibody treatment. On the contrary, steroid treatment altered inflammatory effects on steroid-resistant asthma, and no significant improvement was seen in Atg5 knockout mice. Further studies showed that in asthma, autophagy is an important phenomenon in immune cells. CD11c-specific ATG5−/− cells in mice can develop severe grade neutrophilic asthma. Impairing autophagy in Cd11c-positive cells can also lead to increased neutrophilia in BAL fluid. It is also seen that in ATG5−/− in bone marrow dendritic cells can lead to increased secretion of IL-17 and IL-23 after HDM or LPS administration. All these indicate polarisation into Th17 subset of T cell. Thus, it is seen that Atg5-mediated autophagy is important, and its absence can lead to the development of severe grade steroid-resistant neutrophilic asthma [114].

4.4.7 Other Key Mechanisms

IL-33 cytokine released by epithelial cell is the ligand for Th2-linked ST2 receptor. It is seen that in allergic asthma, ILC2 cell that is activated by IL-33 becomes resistant to glucocorticoid treatment [49]. IL-33 can stimulate AHR upon allergen exposure, and IL-33/ST2 signalling pathway has the potential to activate eosinophils. IL-33 is also identified as an important mediator in severe therapy-resistant asthma. In endobronchial specimens obtained from asthmatic children, increased expressions of IL-33 are associated with thickening of the basement membrane. In the neonatal mice who were suffering from allergic airway disorder, blocking of IL-33 is also seen to partially inhibit the AHR. IL-33, which acts on various cell types having ST2 surface receptor, thus plays the role of a mediator in steroid-resistant asthma [115].

MicroRNAs that are small RNAs of non-coding nature have major interfering role in the inflammatory response that happens in asthma. Studies show that the miRNA-21 level is enhanced in steroid-resistant individual than in steroid responsive people. The miRNA-21 level also correlated with FEV1 [116]. It is seen that in *Chlamydia*-induced severe, steroid-insensitive allergic airway disease (SSIAAD), the level of miRNA-21 is elevated. miRNA-21 targets PTEN, and the level of PTEN is less in SSIAAD. However, with the use of miRNA-21-specific antagomir, improvement could be seen. This in turn highlights the role of miRNA-21 in promoting steroid resistance [41, 42].

Thus, steroid-resistant asthma is a multidimensional approach with complex mechanism. More clinical-based studies should be carried out to identify and treat such patients at the hospital bedside. Asthma itself poses several therapeutic challenges. Often, there are incidences like communication gap, poor medical devices used in the treatment and medicine unavailability. Proper identification of the grade of severity in asthma is of utmost importance [117]. Even there are situations when patients underestimate their own symptoms. Such delayed diagnosis and non-follow up of the medications is a major problem in asthma [118]. Asthma management requires perseverance and effective medications from both the doctor and patient's side.

4.5 Tailoring Steroid-Refractory Asthma: The Possible Diagnosis and Therapeutics

Although the proportion of patients with steroid-refractory asthma is low, they are the major contributors when it comes to the frequent hospitalisations, visit to the emergency ward and enhanced health-care-related costs. Clinicians face gruelling challenges while treating this subgroup of patients. The paucity of knowledge underlying the steroid resistance mechanism and the limited availability of medications for diagnosing these patients need every effort that will outweigh the risks associated with steroid resistance. In this section, we will talk about the appropriate alternative therapies to diagnose steroid-refractory conditions in asthmatic patients.

4.5.1 Immunosuppressive Drugs

Immunosuppressive drugs mediate its action by inhibiting the production of various inflammatory mediators coming from mast cells and basophils. Several immunosuppressive drugs have been screened to achieve significant improvement in the case of steroid-resistant asthma. Various immunosuppressants that can be used to treat severe asthma and the steroid-resistant conditions include **cyclosporine A (CsA)**, **mycophenolate mofetil** (MMF), **tacrolimus**, **azathioprine**, **auranofin** [119], etc. Here, we have discussed a few of them that can be our drugs of choice in the future.

Cyclosporine A (CsA)—It is a metabolite ensued from the fungus *Tolypocladium inflatum* that was identified having immunosuppressive properties in the 1970s by Borel and Stahelin. They selectively inhibit the activation of T lymphocytes along with eosinophils and mast cells [120] and have been rigorously used for preventing allograft rejection and other inflammatory ailments like RA (rheumatoid arthritis), nephrotic syndrome, asthma, etc. Steroid-resistant asthmatic patients when administered with a daily dose of 5 mg/kg of cyclosporine A have shown improvements in FEV1 and peak expiratory flow rate (PEFR) [120, 121]. The results of human trials obtained provide a gleam of hope such that it can be used judiciously in severe asthmatic condition and steroid-refractory conditions.

Mycophenolate Mofetil (MMF)—This is another immunosuppressant that can be used in treating asthma patients with steroid resistance. In vitro studies have elucidated the immunosuppressive properties of MMF, wherein MMF suppresses T and B lymphocyte proliferation and enhances the apoptosis of T cells [122]. The T cells are an important component of steroid-resistant asthma pathogenesis, and the use of MMF would be a great choice in averting the severity of the disease. MMF has been demonstrated in alleviating the severity of asthma by decreasing the production inflammatory genes like TNF-α and IL-1 [123], and by preventing the transition of fibroblast to myofibroblast, thus reducing airway remodelling [123, 124]. However, sufficient trials are not available to confirm its efficacy.

Tacrolimus—It is an immunosuppressive agent that works similar to cyclosporine A [125]. Reports have suggested that tacrolimus suppresses the activation of T lymphocytes along with the release of pro-inflammatory cytokines [126] like IL-4. Taniguchi et al. have demonstrated that thwarting of T cell activation subdues the manifestation of the airway inflammation in addition to improvement in AHR and reduced mucus secretion [125]. However, more studies are needed to validate the use of tacrolimus, such that in the future, this drug could be considered while treating steroid-resistant asthma.

4.5.2 Biologic Therapies

As the researchers are moving from the idea of 'one-size-fits-all', they have recently developed 'biological therapies' for treating steroid-resistant asthma which provides targeted therapy to this population of patients. These sorts of medications are derived from the cells of living organisms which are meant to target the inflammatory pathways that are involved in the pathogenesis of steroid-refractory asthma [127]. Currently, the FDA has recommended five biologics for treating asthma: **omalizumab**, **reslizumab**, **mepolizumab**, **benralizumab** and **dupilumab** [128].

Omalizumab—It is a recombinant, monoclonal antibody that downregulates the IgE receptors on mast cells. B cells are responsible for producing IgE antibodies upon activation by allergens [128, 129]. IgE is averted from attaching to the FcεRI (high-affinity IgE receptor), found on mast cells and basophils when treated with omalizumab, hence suppressing the progression of allergic response. The use of omalizumab has shown favourable outcomes including improvement in lung functions and reduced exacerbations and hospitalisations in asthmatic patients [130].

Mepolizumab, **Reslizumab** and **Benralizumab**—These are anti-IL-5 biologics that target the eosinophilic pathway [130] and hence suppress the eosinophilic airway inflammation. Trials have indicated positive response in terms of increase in FEV1, ameliorated lung function and reduced eosinophilic count when medicated with these therapies [131].

Dupilumab—This is an anti-IL-4R (interleukin-4 receptor) biologic therapy that targets the receptor for two crucial cytokines, IL-4 and IL-13 [132, 133] linked with the pathogenesis of asthma. These cytokines stimulate the IgE production and infiltration of inflammatory cells in the airways leading to mucus production and goblet cell hyperplasia. Clinical response of dupilumab has shown improvement in lung function and AHR along with a reduction in asthma exacerbations [130, 133].

However, further efforts are needed to understand the pathogenesis of asthma and develop new biologic agents that instead of targeting the downstream Th2 inflammatory pathways could also target the inflammatory upstream modulators which include IL-25, IL-33, TSLP, etc.

4.5.3 Highly Potent Glucocorticoids

Inhaled glucocorticoids (GCs) form the mainstay therapy of asthma. However, treatment with inhaled GCs faces obstacles when it comes to treating steroid-resistant asthmatic patients, and excessive use leads to adverse side effects. Hence, researchers have developed GCs with utmost potency which has increased binding affinity to the receptors and has an immediate course of action which suppresses the inflammation response in conditions of severe steroid-refractory asthma [134, 135]. Recently, an extremely potent glucocorticoid, i.e. VSG158, has been developed [135] that displayed excellent pharmacokinetic properties with appreciative pulmonary circulation distribution. This compound was found to be very effective in alleviating asthma symptoms and reversing the steroid-resistant conditions in the murine model of asthma [135]. They exhibited a striking ability to repress the airway hyperresponsiveness and accumulation of inflammatory cells like neutrophils and eosinophils. It was found to be ten times more efficient than the former most fruitful clinical glucocorticoid, fluticasone furoate (FF), for treating steroid-resistant asthma. It elicited negligible side effects, and thus it is a strong candidate for future treatment of severe asthma which is unresponsive to glucocorticoid treatments. Earlier, another potent GCs was discovered, VSGC12 [134]. It had demonstrated maximal anti-inflammatory properties and effectively subdued the lung inflammation at low dosage compared to FF which makes it a potent drug for clinical investigations against severe asthma.

4.5.4 Vitamin D

Vitamin D (calciferol), a fat-soluble nutrient, has been reported to be crucial in regulating various immune responses and respiratory infections [136]. It improves steroid sensitivity by participating in various cellular and molecular pathways like HDAC2, Il-17, IL-33, TSLP, etc. Studies have demonstrated that vitamin D behaves synergistically by augmenting the effects of corticosteroids or reversing the steroid-refractory conditions when administered along with corticosteroids [137]. In 2016, researchers have elucidated that in in vitro and in vivo models of asthma, vitamin D supresses the manifestation of steroid resistance [137].

Vitamin D had displayed to upgrade SR asthma by reducing neutrophil-mediated airway inflammation. The production of TNF-α, a key cytokine involved in the accumulation of neutrophils [54], is arrested when vitamin D is used. Besides, vitamin D downregulates the TNF-α expression by obstructing the MAPK (mitogen-activated protein kinase) pathway [138].

Th17 cells are implicated in neutrophil chemotaxis in airways leading to steroid resistance in asthma. Nanzer et al. had shown the inhibitory role of vitamin D on the development and function of Th17 cells [139], thus increasing the sensitivity of the SR asthma. Studies have shown that histone deacetylase 2 (HDAC2) deacetylates the glucocorticoid receptor (GR) [16]. Reduced expression of HDAC2 correlates with the manifestation of SR conditions [140]. The use of vitamin D had exhibited increased histone acetylation and enabled gene transcription for robust GR activity.

Another cellular pathway involved in the GC resistance is the IL-33/NH (natural helper cells) axis. NH cells are induced by the TSLP protein during airway inflammation resulting in glucocorticoid resistance by phosphorylating STAT5 [77]. Vitamin D had shown to inhibit the expression of IL-33 by increasing the expression of ST2, hence preventing airway neutrophilia that leads to steroid resistance in asthma [141].

4.5.5 Sulforaphane

Sulforaphane, a dietary isothiocyanate compound, is present in cruciferous vegetables. It has gained noticeable recognition due to its inherent antioxidative and anti-inflammatory properties [142]. The deficit of antioxidants is related to deterioration in lung function and alleviated airway inflammation in asthma conditions. Disrupted oxidant-antioxidant balances are also inflicted in mediating SR asthma [143, 144]. The Nrf2 (nuclear factor erythroid-related factor 2) signalling pathway is crucial in triggering the first line of cellular homeostasis by regulating the synthesis of various cytoprotective enzymes, and this anti-oxidant and anti-inflammatory gene suppresses the cytokine-mediated airway inflammation [142]. Nrf2 knockout murine model has elicited a role of impaired Nrf2 in prompting severe eosinophilic inflammation and the release of pro-inflammatory Th2 cytokines [28, 29, 142, 143]. The perturbed Nrf2 pathway is associated with steroid-refractory state due to magnified oxidative stress conditions. Hence, this proposes that the upregulated Nrf2 pathway could be a promising therapeutic target for treating SR asthma [143]. Studies have highlighted the appreciative role of sulforaphane in activating the Nrf2 pathway which leads to the upregulation of the antioxidant enzymes. In a mixed-granulocyte murine model of asthma, treatment with sulforaphane had shown a significant reduction in neutrophil counts and dampened MPO (myeloperoxidase) activity through activation of Nrf2 [143, 145]. The combined therapy of corticosteroids with sulforaphane has proven to be more beneficial compared to only corticosteroid treatments as the combined treatment impedes the Th17/Th2 immune responses in asthma [143]. Clinical studies have shown a significant reduction in hyperresponsiveness, an increase in the luminal area of the airways and improvement in lung function in asthmatic individuals. These observations raise the prospect of using sulforaphane as an adjuvant to curb airway neutrophilia and combat SR asthma.

4.6 Conclusions and Future Outlook

The use of a couple of puffs of steroids acts as a magic potion for asthmatic patients with mild to chronic symptoms. However, a subset of patients exists that struggles with the calming effect of these anti-inflammatory steroids. In 1981, Carmichael and his co-workers identified these patients who responded poorly to systemic corticosteroids and labelled them as 'steroid-resistant' patients. They demonstrate a therapeutic dilemma to the clinicians and consume a pronounced share of medical resources. Hence, monitoring the inflammation effectively and treating the SR patients timely is of paramount significance. In this chapter, we have discussed the probable aetiologies underlying this condition and the pharmacological interventions that have proved promising in the preliminary studies and need further validation. Alternative therapies such as immune-modulators, high-potent GCs and biologics have shown notable success in relieving asthmatic symptoms and reversing the SR asthma state. Mastering the underlying cellular and molecular mechanism will benefit us with more therapies in the long run.

As researchers, we need to continue our venture to unravel the complexity of this disease. Evaluating endotypic responders based on the underlying cellular and molecular players and use of novel biomarkers will provide the needful means for pulling the strings to control the severity of the disease. Moreover, stratifying patients based upon endotypes and bringing forth a 'personalised approach' will always be a cherry on the cake.

Acknowledgements This work was supported by the projects MLP137 (MISSION LUNG) at CSIR-Indian Institute Chemical Biology, Council of Scientific and Industrial Research, Govt. of India. JD, SS and AR acknowledge the Academy of Scientific and Innovative Research (AcSIR), Ghaziabad, India, for their PhD registrations.

References

1. Hillier SG (2007) Diamonds are forever: the cortisone legacy. J Endocrinol 195:1–6
2. Shaikh S, Verma H, Yadav N et al (2012) Applications of steroid in clinical practice: a review. Int Sch Res Notices 2012:1–11
3. Cain DW, Cidlowski JA (2015) Specificity and sensitivity of glucocorticoid signaling in health and disease. Best Pract Res Clin Endocrinol Metab 29:545–556
4. Adcock IM, Mumby S (2017) Glucocorticoids. Handb Exp Pharmacol 237:171–196
5. Rhen T, Cidlowski JA (2005) Antiinflammatory action of glucocorticoids—new mechanisms for old drugs. N Engl J Med 353:1711–1723
6. Strehl C, Buttgereit F (2013) Optimized glucocorticoid therapy: teaching old drugs new tricks. Mol Cell Endocrinol 380:32–40
7. Barnes PJ (2013) Corticosteroid resistance in patients with asthma and chronic obstructive pulmonary disease. J Allergy Clin Immunol 131:636–645
8. Chikanza IC, Kozaci DL (2004) Corticosteroid resistance in rheumatoid arthritis: molecular and cellular perspectives. Rheumatology (Oxford) 43:1337–1345
9. Creed TJ, Probert CS (2007) Review article: steroid resistance in inflammatory bowel disease - mechanisms and therapeutic strategies. Aliment Pharmacol Ther 25:111–122
10. Creed TJ, Lee RW, Newcomb PV et al (2009) The effects of cytokines on suppression of lymphocyte proliferation by dexamethasone. J Immunol 183:164–171
11. Lombel RM, Gipson DS, Hodson EM (2013) Treatment of steroid-sensitive nephrotic syndrome: new guidelines from KDIGO. Pediatr Nephrol 28:415–426
12. Sinha A, Sharma A, Mehta A (2013) Calcineurin inhibitor induced nephrotoxicity in steroid resistant nephrotic syndrome. Indian J Nephrol 23:41–46
13. Goleva E, Hauk PJ, Boguniewicz J et al (2007) Airway remodeling and lack of bronchodilator response in steroid-resistant asthma. J Allergy Clin Immunol 120:1065–1072
14. Mekahli D, Liutkus A, Ranchin B et al (2009) Long-term outcome of idiopathic steroid-resistant nephrotic syndrome: a multicenter study. Pediatr Nephrol 24:1525–1532
15. Peters SP, Ferguson G, Deniz Y et al (2006) Uncontrolled asthma: a review of the prevalence, disease burden and options for treatment. Respir Med 100:1139–1151
16. Trevor JL, Deshane JS (2014) Refractory asthma: mechanisms, targets, and therapy. Allergy 69:817–827
17. Barnes PJ, Adcock IM (2009) Glucocorticoid resistance in inflammatory diseases. Lancet 373:1905–1917
18. Bagga A, Hari P, Moudgil A et al (2003) Mycophenolate mofetil and prednisolone therapy in children with steroid-dependent nephrotic syndrome. Am J Kidney Dis 42:1114–1120

19. Indian Pediatric Nephrology Group, Indian Academy of Pediatrics, Bagga A, Ali U, Banerjee S et al (2008) Management of steroid sensitive nephrotic syndrome: revised guidelines. Indian Pediatr 45:203–214
20. Indian Society of Pediatric Nephrology, Gulati A, Bagga A, Gulati S et al (2009) Management of steroid resistant nephrotic syndrome. Indian Pediatr 46:35–47
21. Sabry A, El-Husseini A, El-Dahshan K et al (2009) Single-center experience with cyclosporine for treatment of idiopathic minimal change nephrotic syndrome in children. Iran J Kidney Dis 3:127–135
22. Tanaka A (2015) Past, present and future therapeutics of asthma: a review. J Gen Fam Med 16:158–169
23. Barnes PJ (2004) Corticosteroid resistance in airway disease. Proc Am Thorac Soc 1:264–268
24. Ito K, Chung KF, Adcock IM (2006) Update on glucocorticoid action and resistance. J Allergy Clin Immunol 117:522–543
25. Phipatanakul W, Mauger DT, Sorkness RL et al (2017) Effects of age and disease severity on systemic corticosteroid responses in asthma. Am J Respir Crit Care Med 195:1439–1448
26. Dietz K, de Los Reyes Jiménez M, Gollwitzer ES et al (2017) Age dictates a steroid-resistant cascade of Wnt5a, transglutaminase 2, and leukotrienes in inflamed airways. J Allergy Clin Immunol 139:1343–1354.e6
27. Miller GE, Chen E (2006) Life stress and diminished expression of genes encoding glucocorticoid receptor and beta2-adrenergic receptor in children with asthma. Proc Natl Acad Sci U S A 103:5496–5501
28. Li B, Duan XH, Wu JF et al (2013a) Impact of psychosocial stress on airway inflammation and its mechanism in a murine model of allergic asthma. Chin Med J 126:325–334
29. Li N, Wang M, Barajas B et al (2013b) Nrf2 deficiency in dendritic cells enhances the adjuvant effect of ambient ultrafine particles on allergic sensitization. J Innate Immun 5:543–554
30. Forno E, Lescher R, Strunk R et al (2011) Decreased response to inhaled steroids in overweight and obese asthmatic children. J Allergy Clin Immunol 127:741–749
31. Sutherland ER, Goleva E, Strand M et al (2008) Body mass and glucocorticoid response in asthma. Am J Respir Crit Care Med 178:682–687
32. Singh VP, Mabalirajan U, Pratap K et al (2018) House dust mite allergen causes certain features of steroid resistant asthma in high fat fed obese mice. Int Immunopharmacol 55:20–27
33. Panda L, Gheware A, Rehman R et al (2017) Linoleic acid metabolite leads to steroid resistant asthma features partially through NF-κB. Sci Rep 7:9565
34. Zhao F, Zhou G, Ouyang H et al (2015) Association of the glucocorticoid receptor D641V variant with steroid-resistant asthma: a case-control study. Pharmacogenet Genomics 25:289–295
35. McGeachie MJ, Clemmer GL, Hayete B et al (2018) Systems biology and in vitro validation identifies family with sequence similarity 129 member a (FAM129A) as an asthma steroid response modulator. J Allergy Clin Immunol 142:1479–1488.e12
36. Himes BE, Jiang X, Wagner P et al (2014) RNA-Seq transcriptome profiling identifies CRISPLD2 as a glucocorticoid responsive gene that modulates cytokine function in airway smooth muscle cells. PLoS One 9:e99625
37. Koster ES, Maitland-van der Zee AH, Tavendale R et al (2011) FCER2 T2206C variant associated with chronic symptoms and exacerbations in steroid-treated asthmatic children. Allergy 66:1546–1552
38. Tantisira KG, Lasky-Su J, Harada M et al (2011) Genomewide association between GLCCI1 and response to glucocorticoid therapy in asthma. N Engl J Med 365:1173–1183
39. Xu Y, Wu H, Wu X et al (2017) GLCCI1 rs37973: a potential genetic predictor of therapeutic response to inhaled corticosteroids in Chinese asthma patients. Medicine 96:e9442
40. Hu CP, Xun QF, Li XZ et al (2018) Effects of glucocorticoid-induced transcript 1 gene deficiency on glucocorticoid activation in asthmatic mice. Chin Med J 131:2817–2826

41. Kim RY, Horvat JC, Pinkerton JW et al (2017a) MicroRNA-21 drives severe, steroid-insensitive experimental asthma by amplifying phosphoinositide 3-kinase-mediated suppression of histone deacetylase 2. J Allergy Clin Immunol 139:519–532
42. Kim RY, Pinkerton JW, Essilfie AT et al (2017b) Role for NLRP3 Inflammasome-mediated, IL-1β-dependent responses in severe, steroid-resistant asthma. Am J Respir Crit Care Med 196:283–297
43. Hadjigol S, Netto KG, Maltby S et al (2020) Lipopolysaccharide induces steroid-resistant exacerbations in a mouse model of allergic airway disease collectively through IL-13 and pulmonary macrophage activation. Clin Exp Allergy 50:82–94
44. Hadebe S, Kirstein F, Fierens K et al (2015) Microbial ligand costimulation drives neutrophilic steroid-refractory asthma. PLoS One 10:e0134219
45. Yang X, Wang Y, Zhao S et al (2018) Long-term exposure to low-dose *Haemophilus influenzae* during allergic airway disease drives a steroid-resistant neutrophilic inflammation and promotes airway remodeling. Oncotarget 9:24898–24913
46. Carmichael J, Paterson IC, Diaz P et al (1981) Corticosteroid resistance in chronic asthma. Br Med J 282:1419–1422
47. Schleimer RP (2004) Glucocorticoids suppress inflammation but spare innate immune responses in airway epithelium. Proc Am Thorac Soc 1:222–230
48. Schleimer RP, Freeland HS, Peters SP et al (1989) An assessment of the effects of glucocorticoids on degranulation, chemotaxis, binding to vascular endothelium and formation of leukotriene B4 by purified human neutrophils. J Pharmacol Exp Ther 250:598–605
49. Nabe T (2020) Steroid-resistant asthma and neutrophils. Biol Pharm Bull 43:31–35
50. Wenzel SE (2012) Asthma phenotypes: the evolution from clinical to molecular approaches. Nat Med 18:716–725
51. Ray A, Kolls JK (2017) Neutrophilic inflammation in asthma and association with disease severity. Trends Immunol 38:942–954
52. Bullens DMA, Truyen E, Coteur L et al (2006) IL-17 mRNA in sputum of asthmatic patients: linking T cell driven inflammation and granulocytic influx? Respir Res 7:135
53. McKinley L, Alcorn JF, Peterson A et al (2008) T H 17 cells mediate steroid-resistant airway inflammation and airway Hyperresponsiveness in mice. J Immunol 181:4089–4097
54. Dejager L, Dendoncker K, Eggermont M et al (2015) Neutralizing TNFα restores glucocorticoid sensitivity in a mouse model of neutrophilic airway inflammation. Mucosal Immunol 8:1212–1225
55. Vargas A, Boivin R, Cano P et al (2017) Neutrophil extracellular traps are downregulated by glucocorticosteroids in lungs in an equine model of asthma. Respir Res 18:1–11
56. Remijsen Q, Kuijpers TW, Wirawan E et al (2011) Dying for a cause: NETosis, mechanisms behind an antimicrobial cell death modality. Cell Death Differ 18:581–588
57. Papayannopoulos V, Metzler KD, Hakkim A et al (2010) Neutrophil elastase and myeloperoxidase regulate the formation of neutrophil extracellular traps. J Cell Biol 191:677–691
58. Dworski R, Simon HU, Hoskins A et al (2011) Eosinophil and neutrophil extracellular DNA traps in human allergic asthmatic airways. J Allergy Clin Immunol 127:1260–1266
59. Krishnamoorthy N, Douda DN, Brüggemann TR et al (2018) Neutrophil cytoplasts induce TH17 differentiation and skew inflammation toward neutrophilia in severe asthma. Sci Immunol 3:eaao4747
60. Ohnishi H, Miyahara N, Gelfand EW (2008) The role of leukotriene B4 in allergic diseases. Allergol Int 57:291–298
61. Pétrin D, Turcotte S, Gilbert AK et al (2006) The anti-apoptotic effect of leukotriene B4 in neutrophils: a role for phosphatidylinositol 3-kinase, extracellular signal-regulated kinase and Mcl-1. Cell Signal 18:479–487
62. Stankova J, Turcotte S, Harris J et al (2002) Modulation of leukotriene B 4 Receptor-1 expression by dexamethasone: potential mechanism for enhanced neutrophil survival. J Immunol 168:3570–3576

63. Gelfand EW (2017) Importance of the leukotriene B4-BLT1 and LTB4-BLT2 pathways in asthma. Semin Immunol 33:44–51
64. Webley WC, Hahn DL (2017) Infection-mediated asthma: etiology, mechanisms and treatment options, with focus on chlamydia pneumoniae and macrolides. Respir Res 18:1–12
65. Essilfie AT, Simpson JL, Horvat JC et al (2011) Haemophilus influenzae infection drives IL-17-mediated neutrophilic allergic airways disease. PLoS Pathog 7:1002244
66. Sekhri K, Bhasin D (2015) Asthma exacerbations: understanding role of viral respiratory tract infections and possible treatment strategies. Med J Dr DY Patil Univ 8:285
67. Pelaia G, Vatrella A, Busceti MT et al (2015) Cellular mechanisms underlying eosinophilic and neutrophilic airway inflammation in asthma. Mediat Inflamm 2015:879783
68. Brinke AT, Zwinderman AH, Sterk PJ et al (2004) "Refractory" eosinophilic airway inflammation in severe asthma: effect of parenteral corticosteroids. Am J Respir Crit Care Med 170:601–605
69. De Groot JC, Ten BA, Bel EHD (2015) Management of the patient with eosinophilic asthma: a new era begins. ERJ Open Res 1:00024-2015
70. Castanhinha S, Sherburn R, Walker S et al (2015) Pediatric severe asthma with fungal sensitization is mediated by steroid-resistant IL-33. J Allergy Clin Immunol 136:312–322
71. Hirahara K, Mato N, Hagiwara K et al (2018) The pathogenicity of IL-33 on steroid-resistant eosinophilic inflammation via the activation of memory-type ST2 + CD4 + T cells. J Leukoc Biol 104:895–901
72. Fulgheri G, Malinowski B (2011) The role of IL-33 in the inflammation process of asthma and atherosclerosis. EJIFCC 22:79–91
73. Mato N, Hirahara K, Ichikawa T et al (2017) Memory-type ST2+CD4+ T cells participate in the steroid-resistant pathology of eosinophilic pneumonia. Sci Rep 7:1–12
74. Bahrami Mahneh S, Movahedi M, Aryan Z et al (2016) Serum IL-33 is elevated in children with asthma and is associated with disease severity. Int Arch Allergy Immunol 168:193–196
75. Chang Y-J, DeKruyff RH, Umetsu DT (2013) The role of type 2 innate lymphoid cells in asthma. J Leukoc Biol 94:933–940
76. Gauvreau GM, Sehmi R, Ambrose CS et al (2020) Thymic stromal lymphopoietin: its role and potential as a therapeutic target in asthma. Expert Opin Ther Targets 24:777–792
77. Kabata H, Moro K, Fukunaga K et al (2013) Thymic stromal lymphopoietin induces corticosteroid resistance in natural helper cells during airway inflammation. Nat Commun 4:1–10
78. Liu S, Verma M, Michalec L et al (2018) Steroid resistance of airway type 2 innate lymphoid cells from patients with severe asthma: the role of thymic stromal lymphopoietin. J Allergy Clin Immunol 141:257–268
79. Kaplan MH (2013) Th9 cells: differentiation and disease. Immunol Rev 252:104–115
80. Saeki M, Kaminuma O, Nishimura T et al (2017) Th9 cells induce steroid-resistant bronchial hyperresponsiveness in mice. Allergol Int 66:S35–S40
81. Kara EE, Comerford I, Bastow CR et al (2013) Distinct chemokine receptor axes regulate Th9 cell trafficking to allergic and autoimmune inflammatory sites. J Immunol 191:1110–1117
82. Gelfand EW, Hinks TSC (2019) Is there a role for type 2 CD8+ T cells in patients with steroid-resistant asthma? J Allergy Clin Immunol 144:648–650
83. Hinks TSC, Hoyle RD, Gelfand EW (2019) CD8+ Tc2 cells: underappreciated contributors to severe asthma. Eur Respir Rev 28:190092
84. Li NB, Leung DYM, Strand MJ et al (2007) ATF2 impairs glucocorticoid receptor-mediated transactivation in human CD8+ T cells. Blood 110:1570–1577
85. Woolcock AJ (1993) Steroid resistant asthma: what is the clinical definition? Eur Respir J 6:743–747
86. Sher ER, Leung DY, Surs W et al (1994) Steroid-resistant asthma. Cellular mechanisms contributing to inadequate response to glucocorticoid therapy. J Clin Invest 93:33–39
87. Panda L, Mabalirajan U (2018) Recent updates on corticosteroid resistance in asthma. EMJ 3:49–57

88. Adcock IM, Lane SJ, Brown CR et al (1995a) Abnormal glucocorticoid receptor-activator protein 1 interaction in steroid-resistant asthma. J Exp Med 182:1951–1958
89. Barnes PJ (2010) Inhaled corticosteroids. Pharmaceuticals 3:514–540
90. Adcock IM (1996) Steroid resistance in asthma: molecular mechanisms. Discussion. Am J Respir Crit Care Med 154:S58–S61
91. Adcock IM, Lane SJ, Brown CR et al (1995b) Differences in binding of glucocorticoid receptor to DNA in steroid-resistant asthma. J Immunol 154:3500–3505
92. Serra MF, Cotias AC, Pão CR et al (2018) Repeated allergen exposure in a/J mice causes steroid-insensitive asthma via a defect in glucocorticoid receptor bioavailability. J Immunol 201:851–860
93. Zhao Y, Yang J, Gao YD et al (2010) Th17 immunity in patients with allergic asthma. Int Arch Allergy Immunol 151:297–307
94. Vazquez-Tello A, Halwani R, Hamid Q et al (2013) Glucocorticoid receptor-beta up-regulation and steroid resistance induction by IL-17 and IL-23 cytokine stimulation in peripheral mononuclear cells. J Clin Immunol 33:466–478
95. Chesné J, Braza F, Mahay G et al (2014) IL-17 in severe asthma. Where do we stand? Am J Respir Crit Care Med 190:1094–1101
96. Zijlstra GJ, Ten Hacken NHT, Hoffmann RF et al (2012) Interleukin-17A induces glucocorticoid insensitivity in human bronchial epithelial cells. Eur Respir J 39:439–445
97. Leung DY, Martin RJ, Szefler SJ et al (1995) Dysregulation of interleukin 4, interleukin 5, and interferon gamma gene expression in steroid-resistant asthma. J Exp Med 181:33–40
98. Britt RD Jr, Thompson MA, Sasse S et al (2019) Th1 cytokines TNF-α and IFN-γ promote corticosteroid resistance in developing human airway smooth muscle. Am J Physiol Lung Cell Mol Physiol 316:L71–L81
99. Gauthier M, Chakraborty K, Oriss TB et al (2017) Severe asthma in humans and mouse model suggests a CXCL10 signature underlies corticosteroid-resistant Th1 bias. JCI Insight 2:e94580
100. Li JJ, Wang W, Baines KJ et al (2010a) IL-27/IFN-γ induce MyD88-dependent steroid-resistant airway hyperresponsiveness by inhibiting glucocorticoid signaling in macrophages. J Immunol 185:4401–4409
101. Li LB, Leung DY, Martin RJ et al (2010b) Inhibition of histone deacetylase 2 expression by elevated glucocorticoid receptor β in steroid-resistant asthma. Am J Respir Crit Care Med 182:877–883
102. Chambers ES, Nanzer AM, Pfeffer PE et al (2015) Distinct endotypes of steroid-resistant asthma characterized by IL-17Ahigh and IFN-γ high immunophenotypes: potential benefits of calcitriol. J Allergy Clin Immunol 136:628–637
103. Jeong JS, Kim JS, Kim SR et al (2019) Defining bronchial asthma with phosphoinositide 3-kinase delta activation: towards endotype-driven management. Int J Mol Sci 20:3525
104. Wadhwa R, Dua K, Adcock IM et al (2019) Cellular mechanisms underlying steroid-resistant asthma. Eur Respir Rev 28:190096
105. Campa CC, Silva RL, Margaria JP et al (2018) Inhalation of the prodrug PI3K inhibitor CL27c improves lung function in asthma and fibrosis. Nat Commun 9:1–16
106. Bhavsar P, Khorasani N, Hew M et al (2010) Effect of p38 MAPK inhibition on corticosteroid suppression of cytokine release in severe asthma. Eur Respir J 35:750–756
107. Li LB, Leung DY, Goleva E (2015) Activated p38 MAPK in peripheral blood monocytes of steroid resistant asthmatics. PLoS One 10:e0141909
108. Irusen E, Matthews JG, Takahashi A et al (2002) p38 mitogen-activated protein kinase–induced glucocorticoid receptor phosphorylation reduces its activity: role in steroid-insensitive asthma. J Allergy Clin Immunol 109:649–657
109. Hansbro PM, Kim RY, Starkey MR et al (2017) Mechanisms and treatments for severe, steroid-resistant allergic airway disease and asthma. Immunol Rev 278:41–62
110. Simpson JL, Phipps S, Baines KJ et al (2014) Elevated expression of the NLRP3 inflammasome in neutrophilic asthma. Eur Respir J 43:1067–1076

111. Pinkerton J, Kim R, Brown A et al (2019) IL-5/IL-13 drive NLRP3 inflammasome-mediated, steroid-resistant AHR in a model of obesity-associated asthma. Eur Respir J 54:PA3345
112. Wood LG, Li Q, Scott HA et al (2019) Saturated fatty acids, obesity, and the nucleotide oligomerization domain–like receptor protein 3 (NLRP3) inflammasome in asthmatic patients. J Allergy Clin Immunol 143:305–315
113. Zeki AA, Yeganeh B, Kenyon NJ et al (2016) Autophagy in airway diseases: a new frontier in human asthma? Allergy 71:5–14
114. Suzuki Y, Maazi H, Sankaranarayanan I et al (2016) Lack of autophagy induces steroid-resistant airway inflammation. J Allergy Clin Immunol 137:382–1389
115. Saglani S, Lui S, Ullmann N et al (2013) IL-33 promotes airway remodeling in pediatric patients with severe steroid-resistant asthma. J Allergy Clin Immunol 132:676–685
116. Elbehidy RM, Youssef DM, Amal S et al (2016) MicroRNA–21 as a novel biomarker in diagnosis and response to therapy in asthmatic children. Mol Immunol 71:107–114
117. Onyedum C, Ukwaja K, Desalu O et al (2013) Challenges in the management of bronchial asthma among adults in Nigeria: a systematic review. Ann Med Health Sci Res 3:324–329
118. Barnes P, Virchow JC, Sanchis J et al (2005) Asthma management: important issues. Eur Respir Rev 14:147–151
119. Corrigan CJ, Bungre JK, Assoufi B et al (1996) Glucocorticoid resistant asthma: T-lymphocyte steroid metabolism and sensitivity to glucocorticoids and immunosuppressive agents. Eur Respir J 9:2077–2086
120. Coren ME, Rosenthal M, Bush A (1997) The use of cyclosporin in corticosteroid dependent asthma. Arch Dis Child 77:522–523
121. Lock SH, Kay AB, Barnes NC (1996) Double-blind, placebo-controlled study of cyclosporin a as a corticosteroid-sparing agent in corticosteroid-dependent asthma. Am J Respir Crit Care Med 153:509–514
122. Backer V, Hjardem E, Karlsmark T (2009) Treatment with mycophenolate Mofetil of steroid-dependent asthma—one case of severe asthma. J Allergy 2009:1–3
123. Allison AC, Eugui EM (2005) Mechanisms of action of mycophenolate mofetil in preventing acute and chronic allograft rejection. Transplantation 80:S181–S190
124. Allison AC, Eugui EM (2000) Mycophenolate mofetil and its mechanisms of action. Immunopharmacology 47:85–118
125. Taniguchi H, Tokui K, Iwata Y et al (2011) A case of severe bronchial asthma controlled with tacrolimus. J Allergy 2011:1–4
126. Sano T, Nakamura Y, Matsunaga Y et al (1995) FK506 and cyclosporin a inhibit granulocyte/macrophage colony-stimulating factor production by mononuclear cells in asthma. Eur Respir J 8:1473–1478
127. Godar M, Blanchetot C, de Haard H et al (2018) Personalized medicine with biologics for severe type 2 asthma: current status and future prospects. MAbs 10:34–45
128. McGregor MC, Krings JG, Nair P et al (2019) Role of biologics in asthma. Am J Respir Crit Care Med 199:433–445
129. Manka LA, Wechsler ME (2018) Selecting the right biologic for your patients with severe asthma. Ann. Allergy. Asthma Immunol 121:406–413
130. Rogliani P, Calzetta L, Matera MG et al (2020) Severe asthma and biological therapy: when, which, and for whom. Pulm Ther 6:47–66
131. Humbert M (2009) Biologics in severe difficult-to-treat asthma: find the right niche. Eur Respir Rev 18:51–53
132. Kariyawasam HH, James LK, Gane SB (2020) Dupilumab: clinical efficacy of blocking IL-4/IL-13 signalling in chronic rhinosinusitis with nasal polyps. Drug Des Devel Ther 14:1757–1769
133. Vatrella A, Fabozzi I, Calabrese C et al (2014) Dupilumab: a novel treatment for asthma. J Asthma Allergy 7:123–130
134. He Y, Shi J, Yi W et al (2015) Discovery of a highly potent glucocorticoid for asthma treatment. Cell Discov 1:1–13

135. He Y, Shi J, Nguyen QT et al (2019) Development of highly potent glucocorticoids for steroid-resistant severe asthma. Proc Natl Acad Sci U S A 116:6932–6937
136. Prietl B, Treiber G, Pieber TR et al (2013) Vitamin D and immune function. Nutrients 5:2502–2521
137. Litonjua AA (2013) Vitamin D and corticosteroids in asthma: synergy, interaction and potential therapeutic effects. Expert Rev Respir Med 7:101–104
138. Kerley CP, Elnazir B, Faul J et al (2015) Vitamin D as an adjunctive therapy in asthma. Part 1: a review of potential mechanisms. Pulm Pharmacol Ther 32:60–74
139. Nanzer AM, Chambers ES, Ryanna K et al (2013) Enhanced production of IL-17A in patients with severe asthma is inhibited by 1α,25-dihydroxyvitamin D3 in a glucocorticoid-independent fashion. J Allergy Clin Immunol 132:297–304
140. Vargas JE, Porto BN, Puga R et al (2016) Identifying a biomarker network for corticosteroid resistance in asthma from bronchoalveolar lavage samples. Mol Biol Rep 43:697–710
141. Saglani S (2017) Childhood severe asthma: new insights on remodelling and biomarkers. Paediatr. Res Rev 24:11–13
142. Brown RH, Reynolds C, Brooker A et al (2015) Sulforaphane improves the bronchoprotective response in asthmatics through Nrf2-mediated gene pathways. Respir Res 16:106
143. Al-Harbi NO, Nadeem A, Ahmad SF et al (2019) Sulforaphane treatment reverses corticosteroid resistance in a mixed granulocytic mouse model of asthma by upregulation of antioxidants and attenuation of Th17 immune responses in the airways. Eur J Pharmacol 855:276–284
144. Nadeem A, Masood A, Siddiqui N (2008) Therapeutic advances in respiratory disease oxidant-antioxidant imbalance in asthma: scientific evidence, epidemiological data and possible therapeutic options. Ther Adv Respir Dis 2:215–235
145. Sakurai H, Morishima Y, Ishii Y et al (2018) Sulforaphane ameliorates steroid insensitivity through an Nrf2-dependent pathway in cigarette smoke-exposed asthmatic mice. Free Radic Biol Med 129:473–485

Targeting Molecular and Cellular Mechanisms in Chronic Obstructive Pulmonary Disease

5

Shvetank Bhatt, Jovita Kanoujia, A. N. Nagappa, and
K. Sreedhara R. Pai

Abstract

Chronic obstructive pulmonary disease (COPD), also known as chronic obstructive airway disease (COAD) or chronic obstructive lung disease (COLD), is the leading cause of disability worldwide. It is an increasing worldwide health issue and becomes the third most important reason of mortality and the fifth commonest cause of ill health and compromised quality of life in the world by 2020. Cigarette smoking is one of the prominent causes for pathological development of COPD. In addition, alpha1-antitrypsin (AAT) deficiency is an inherited condition, which is associated with COPD. The two main peculiar features of the disease are chronic bronchitis (contraction of airways/bronchi due to inflammation) and emphysema (damage of alveolar wall). COPD is recognised by increased number of cytotoxic T-lymphocytes, neutrophils (NPHs) and alveolar macrophages (MPs) and number of inflammatory mediators like growth factors, cytokines and chemokines. In addition, reactive oxygen species (ROS) and imbalance between oxidant and antioxidant mechanisms are also involved in pathophysiological progression of inflammatory COPD. The pulmonary inflammation may also responsible for growth, progression and development of lung cancer. Plasma levels of elastolytic enzymes such as serine proteases, cathepsins and matrix metalloproteinase (MMP) are highly increased in COPD. The pulmonary inflammation leads to development of systemic inflammation and other comorbid disorders. The disease is progressive and inflammation is predominant in

S. Bhatt (✉) · J. Kanoujia · A. N. Nagappa
Amity Institute of Pharmacy, Amity University Madhya Pradesh (AUMP), Gwalior, India
e-mail: sbhatt@gwa.amity.edu

K. S. R. Pai
Manipal College of Pharmaceutical Sciences (MCOPS), Manipal Academy of Higher Education (MAHE), Manipal, Karnataka, India

K. Dua et al. (eds.), *Targeting Cellular Signalling Pathways in Lung Diseases*,
https://doi.org/10.1007/978-981-33-6827-9_5

comparison to asthma which seems to be resistant towards corticosteroid treatment. Specific treatment options that are working against the remodelling and inflammation need to be developed for the treatment of COPD. Hence, the present book chapter will discuss about the strategies for targeting COPD at cellular and molecular levels including their signalling pathways.

Keywords

Cellular signalling · Emphysema · Macrophages · Matrix metalloproteinase · Reactive oxygen species

5.1 Introduction

COPD is a progressive disorder of the lung that exhibits emphysema and chronic bronchitis as peculiar features. Coughs, excessive mucus production, chest tightness and difficulty in breathing with shortness of breath are the most common manifestations of the disease. COPD patient is not able to sleep properly due to difficulty in breathing. It is a progressive condition as disease get worsens over a period of time. Smoking cigarette is the primary cause of the disease [1]. Around 75% of COPD patients have a habit of smoking [2, 3]. However, 25% of the patient never smoked directly but may have affected with the passive smoking or chronic exposure respiratory system irritants such as chemical fumes, air pollution, dusts, etc. which may contribute to COPD. In addition, one important genetic cause of COPD is deficiency of AAT enzyme [4, 5]. Alveoli of the normal lung vs alveoli of the patient affected with COPD have been shown in Fig. 5.1.

In COPD, the elasticity of alveoli is lost and walls between the air sacs are destroyed that is known as 'emphysema'. The airways lose their peculiar spherical shape and become floppy. Emphysema causes overall reduction in effective surface

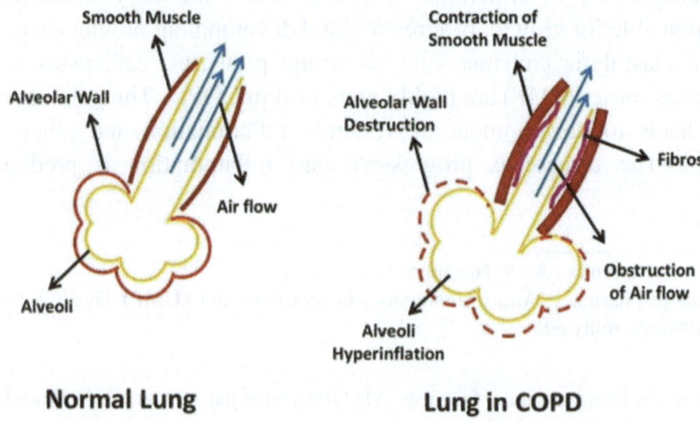

Fig. 5.1 Alveoli of the normal lung vs the COPD-affected lung

Table 5.1 Comparison between asthma and COPD on the basis of clinical characteristics [10]

S. N.	Clinical characteristics	Asthma	COPD
1.	Onset of disease	Common in children and young population	Usually >35 years old
2.	Parenchyma involvement	Inflammation involvement in all airways is common but does not have association with the lung parenchyma	Bronchioles and parenchyma are involved in COPD
3.	Cell involvement	Eosinophil (Eos) and CD4-T-lymphocytes are involved	MPs, NPHs various inflammatory mediators and CD8 (cytotoxic) T-lymphocytes
4.	Smoking history	Reason for the disease in some patients	Common in 90% of sufferers
5.	Chronic cough and sputum production with difficulty in sleeping at night	Rare	Common

area of alveoli and leads to difficulty in holding gases, resulting in difficulty in breathing [6]. Moreover, it also exhibits the feature or symptom of inflammation in airways that is known as 'bronchitis'. Thick mucus is produced in airways and it becomes clogged. The comprehensive burden of disease study reports a prevalence of 251 million cases of COPD worldwide in 2016. The data suggests an estimated 3.17 million mortalities were caused by this disease in 2015. The death ratio is higher in low- and middle-income countries [7]. COPD is the third leading cause of death worldwide, and the global burden of the disease can be increased further. The previous data demonstarted that, the main reason of COPD death was found to be cigarette smoke or second-hand smoke [3]. In addition, long-term cases of asthma also have chances to get converted into COPD. Avoidance or early cessation of smoking could prevent progression of COPD; however, the disease is not curable, but the treatment can reduce the symptoms, leading to an improvement in the quality of life as well as reducing the death risk [8, 9]. The differences between asthma and COPD are mentioned in Table 5.1.

Diagnosis of COPD is done on the basis of an obstructive pre-bronchodilator spirometry (FEV1/FVC < 0.70) in agreement with the Global Initiative for Chronic Obstructive Lung Disease (GOLD) guidelines. This approach is identified to be easy to operate; however, it must be appreciated that these merits should be well adjusted against the demerits of possible false diagnoses [11,12].

The management of COPD depends on individual's condition, evaluation/assessment of symptoms of the patient and forthcoming risk of exacerbations. The patients who smoke are stringently advised and supported to quit. The management of the disease seeks reduction in symptoms as well as future risk of exacerbations. The treatment approach includes both non-pharmacological and pharmacological interventions.

5.2 Pathophysiology of COPD

COPD is associated with poorly reversible obstruction of airflow and highly exaggerated inflammatory responses inside the lungs and respiratory tract. The inflammatory responses have participation of innate and adaptive immune components to chronic exposure to noxious particulate matter and poisonous gases, specifically cigarette smoke [13]. The extent of limitation of airflow is determined by the degree of inflammation. It is seen that all cigarette smokers have certain types of lung inflammation, but individuals who develop COPD have unusual response to inhaling toxic substances [14]. Around 20% of all cigarette smokers have shown that abnormal reaction. This exaggerated response may lead to mucous hyper-secretion (chronic bronchitis), tissue destruction (emphysema) and deterioration of normal physiological repair and defence mechanisms causing inflammation of the small airway and fibrosis (bronchiolitis). Moreover, increased resistance to airflow is also a peculiar feature of COPD [15].

In particular, COPD is characterised by increased numbers of cells such as NPHs, MPs and T-lymphocytes (CD8) in the lungs. These inflammatory cells are responsible for proliferation of various inflammatory cytokines and mediators that have predominant role in inflammation and obstruction of airways [16]. The inflammatory pattern of COPD is predominantly different from asthma. The inflammatory mediators that are involved in asthma are as follows:

5.2.1 Inflammatory Mediators and Cells

The obstruction of airway in COPD is due to various types of cytokines and mediators. The important ones are as follows: (a) epithelial cells (EP), MPs and NPHs which are involved in the production of leukotriene B4 and T-cell chemoattractant [17]; (b) chemotactic factors which are produced by MPs and EP have mediated CXC chemokines interleukin-8 (IL-8) and oncogene α-related pro-inflammatory response and growth [18]; (c) other important pro-inflammatory cytokines, namely, tumour necrosis factor-α (TNF-α) and interleukin-1β (IL-1β) and interleukin-6 (IL-6) [19]; and (d) transforming growth factor-β (TGF-β) which is involved in immunosuppression and causes fibrosis of the airways [20]. The mechanism of production of fibrosis is either direct or involving other cytokines, connective tissues or growth factors.

RANTES and eotaxin chemokine expression may underlie the airway-associated eosinophilia in some COPD patients [21]. ROS can also enhance the expression of many inflammatory mediators like TNF-α and IL-1β [22]. Moreover, extracellular matrix glycoproteins like tenascin are produced by bronchial EP, and MMP-9 is produced by MPs under the influence of inflammatory mediators [23]. Increased TGF-β and epidermal growth factor (EGF) expression takes place in the submucosal cell and epithelium in chronic bronchitis. Stimulation of TGF-β and EGF receptor results in expression of mucin gene [24]. The involvement of various cells is discussed in detail in this section.

5.2.1.1 Neutrophils (NPHs)

Human patients and murine cell culture with AAT deficiency to develop COPD have shown the involvement of NPHs-derived proteases and NPHs elastase as key mediators of tissue damage. The elevated levels of NPHs have been found in sputum and BAL fluid of COPD patients [25]. However, they are raised comparatively less in the airway and lung parenchyma. Smoking leads to accumulation of NPHs in the lungs. The serine proteases released by NPHs such as elastase, cathepsin G and MMP-3, MMP-8, MMP-9 and MMP-12 are involved in great extent in the destruction of alveoli. In addition, these serine proteases also stimulate the cough or mucus production [26]. Recruitment of NPHs in the parenchyma and airways consists of adhesion of endothelial cells, and E-selectin is upregulated on the endothelial cells in the airway of COPD patients [27]. Several chemotactic signals are involved in NPHs signalling or recruitment such as LTB4, IL-8 and related CXC chemokines [28]. These well-recognised cytokines are involved in modulating NPHs functions as well as survival. Moreover, it is evident that hypoxia influences NPHs function [29]. Involvement of various cytokines, proteases and chemokines via involvement of MP and fibroblast signalling on hypersecretion of mucus, emphysema and fibrosis has been shown in Fig. 5.2.

5.2.1.2 Macrophages (MPs)

MPs are important cells that have a pivotal role in the pathophysiology of COPD. Predominantly increased levels of MPs have been observed in airways, parenchyma of lungs, BAL fluid and sputum of COPD patients. Cigarette smoke triggers MPs and associated inflammatory pathways [30]. Alveolar MPs also secrete elastolytic enzymes such as MMP-2, MMP-9, and MMP-12; cathepsins K, L and S; and NPHs elastase taken up from NPHs [28, 31]. MPs present in alveoli of COPD patients release proteins linked with inflammation and marked loss of elasticity in chronic smokers. MMP-9 is an important elastolytic enzyme released from MPs [32]. MPs of MMP-12−/− mice have a significantly reduced capacity to degenerate extracellular matrix components. In addition, MMP-12−/− MPs are specifically not able penetrate reconstituted basal membranes both in vitro and in vivo [33, 34]. COPD patients and smokers have shown significantly amplified expression of the anti-apoptotic protein Bcl-XL and p21CIP/WAF1 in the cytoplasm that leads to prolong survival of MPs [13]. This suggests that MPs may have a prolonged life span in smokers and COPD patients.

5.2.1.3 T-Lymphocyte

CD8+ T-cells' amount in lungs of the COPD patients can be directly linked with the degree of limitation of airways. Increased number of T-lymphocyte is seen in the lung parenchyma and peripheral and central airways of COPD patients [14]. CD8+ T-cells increased more than CD4+ T-cells [35]. A correlation is observed between the amount of destruction in alveoli and airway obstruction and the number of T-cells. In addition, CXCR3 activating chemokines are also involved in the activation of the T-lymphocytes in airways of COPD patient [36]. Through release of

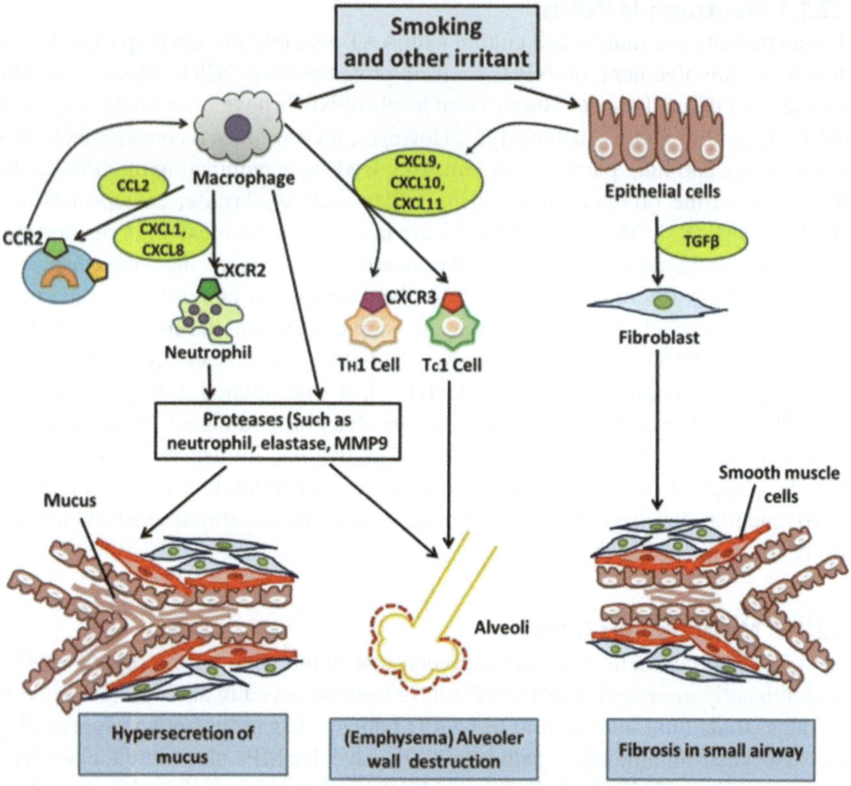

Fig. 5.2 Involvement of cytokines, proteases and chemokines on COPD progression

mediators like perforins, granzyme-B and TNF-α, CD8+ T-cells have the capacity to cause cytolysis and apoptosis of alveolar EP [37].

Eosinophils (Eos)

The significance of Eos in asthma is well established; however, their role in the pathological progression of COPD is uncertain [38]. Some reports showed increased numbers of Eos in lavage and airways of COPD patients, whereas others have not reported its increased numbers in biopsies of the airway, BAL or induced sputum [39]. During acute exacerbations of chronic bronchitis in bronchial biopsies and BAL fluid, increased levels of Eos have been observed [40]. The patient subset with COPD has shown the involvement of Eos directed airway inflammation. In addition, these patients showed good response to corticosteroids. Moreover, sputum Eos are increased in COPD patients, and use of corticosteroids reduced the observed numbers of Eos count. However, the prevalence of Eos in COPD patient is unknown [28, 40].

5.2.1.4 Dendritic Cells (DCs)

DCs offer immune surveillance of the small airways, important in health but significantly participate in COPD progression [41]. DCs/antigen presenting cells play a predominant role in the starting of innate as well as adaptive immune response [42]. The lungs and airways contain rich numbers of DCs to initiate its signalling on the entry of foreign particle or antigen. The activation of DCs leads to initiation of a variety of other inflammatory and immune cells, including MPs, NPHs and T- and B-lymphocytes [43]. The glycoprotein obtained from cigarette smoking of tobacco exhibits strong immunostimulatory actions. In the lungs of the mice, increased number of DCs was seen after cigarette smoke [44]. In addition, lung DCs can participate substantially in the initial development of tertiary lymphoid tissues, including the largest lymphoid follicle [41].

5.2.1.5 Epithelial Cells (EP)

EP of the airway and alveoli is activated by smoking and produces mediators of inflammation such as TNF, IL-1, GM-CSF and IL-8 [45, 46]. It is also an important source of TGF which is important for the induction of local fibrosis [47]. In addition, cigarette smoke also leads to proliferation of EP, and this may contribute also to proliferation of basal cell, resulting in carcinoma of bronchi and squamous metaplasia [48].

5.2.2 Protease and Antiprotease Imbalance

The disproportion of protease and antiprotease also results in worsening the condition of COPD patients. The reason of the imbalance is increased in the production of proteases and inactivation or less production of the protective antiproteases. Smoking is the prime cause of this imbalance with enhanced levels of oxidative stress (OS) markers and inflammatory markers. The increased OS stimulates several inflammatory cells to release a protease combination and inactivation of antiproteases [49]. The examples of various main proteases are (1) NPHs derived, serine proteases such as elastase, protease 3 and cathepsin G; (2) MPs derived, cysteine proteases and cathepsins E, A, L and S; and (3) MMP, MMP-8, MMP-9 and MMP-12 [50]. On the other hand, the important antiproteases that are involved in the pathological progression of COPD are AAT, secretory leucoprotease inhibitor and tissue inhibitors of MMP [49].

5.2.3 Oxidative Stress (OS)

Increased OS is one of the important pathological features involved in the progression of COPD [51]. An imbalance in oxidant and antioxidant mechanism is observed in COPD. The ROS and reactive nitrogen species (RNS) are involved in oxidative damage of macromolecules such as DNA, lipids, carbohydrates and proteins [52]. Effects of ROS on progression of OS have been shown in Fig. 5.3. They

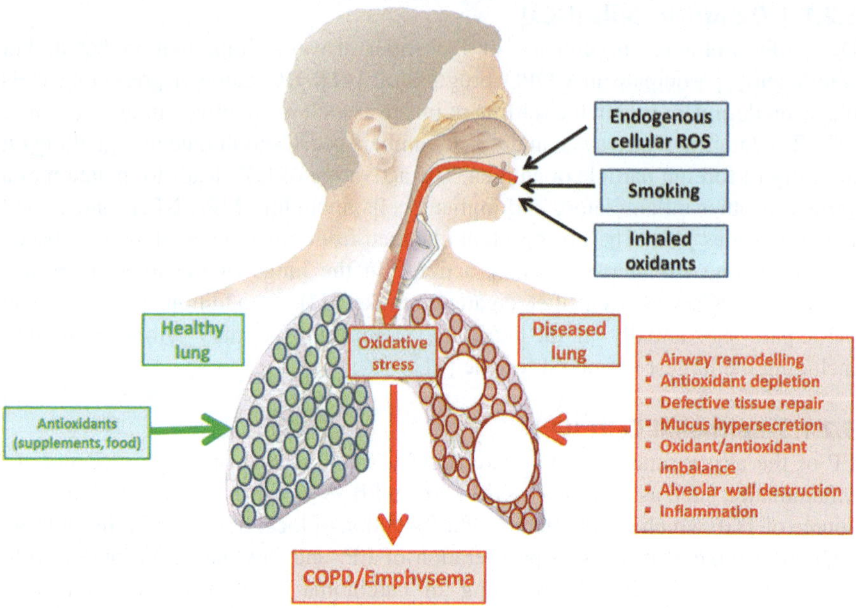

Fig. 5.3 Effect of ROS on progression of COPD

further cause activation of resident cells of the lung specifically EP and alveolar MPs, to generate chemotactic molecules that recruit other inflammatory cells such as monocytes, NPHs and lymphocytes into the lungs [53]. Collectively, these events lead to involvement of various inflammatory markers accompanied by chronic OS which finally leads to imbalance between the protease and antiprotease balance, abnormality in tissue repair processes, accelerated apoptosis and increased autophagy in lung cells. These all are associated with pathological progression of COPD [54].

Cigarette smoke, ROS and nitrogen species released from inflammatory cells are the main source of oxidants [55]. ROS induces the peroxidation of lipids and leads to the formation of malondialdehyde (MDA). This MDA is responsible for inducing pulmonary inflammation. Increased levels of MDA have been observed in the plasma of patients affected with COPD [56]. The increased MDA and associated inflammation also targets signalling transduction cascade molecules like small G proteins and various pro-inflammatory transcription factors (e.g. NF-κB), to induce and sustain a pro-inflammatory state that stimulates the formation of cell-derived ROS [57]. The main source of ROS production in inflammatory cells is NADPH oxidase [22]. In addition, xanthine and haem peroxidase enzyme system is also upregulated in patients affected with COPD [57]. In a similar way, inducible nitric oxide synthase (iNOS) is formed or generated in the RNS pathway, which in the presence of superoxide anion forms a more dangerous and comparatively powerful peroxynitrite anion. Upregulation of this iNOS pathway is considered as the main culprit that triggers emphysema and pulmonary hypertension which can be linked with cigarette smoking [58]. Plasma levels of antioxidant enzymes like catalase, superoxide dismutase (SOD) and glutathione peroxidase (GPx) are highly reduced in the patients affected with the disorder [59]. The increased OS results in inactivation of antiproteases or overproduction of mucus. It can also increase inflammation by enhancing the activation of transcription factor (such as nuclear factor κB) and hence gene expression of pro-inflammatory mediators. This finally leads to activation of caspase pathway and responsible for cell death. Moreover, cigarette smoke also compromises levels of reduced glutathione (GSH) [60].

The above-mentioned dysfunctions lead to functional abnormalities in mucus hypersecretion and ciliary dysfunction, obstruction of airflow and pulmonary hypertension including other systemic effects.

5.3 Mediators Involved in COPD

5.3.1 Lipid Mediators

5.3.1.1 Prostaglandin

The higher level COX-2 enzyme expression is observed in alveolar MPs of COPD patients. COX-2 enzyme is having a substantial role in inflammation process and involved in the production of prostaglandins by arachidonic acid pathways. Excessive formation of prostanoids and its signalling might be responsible for the increased rate of infection that may lead to exacerbations of COPD and reduction of alveolar repair that leads to emphysema [28, 61].

The increased levels of prostaglandin E2 have been demonstrated in exhaled breath of COPD patients. Prostaglandin E2 is a bronchodilator and has an anti-inflammatory action [62]. It is also involved in the enhancement of anti-inflammatory actions of PDE4 inhibitors. The other prostaglandin increased in the breath is PGF2α. It has a bronchoconstrictor action and is also involved in hypersecretion of cough [63].

5.3.1.2 Thromboxane

Concentration of thromboxane (TX) B2 is also found to be increased in the COPD patient's breath [64]. The metabolite of TXB2, 11-dehydro-thromboxane, is found in higher amounts in urine, and low-dose aspirin was linked with more than 90% inhibition of excretion of thromboxane metabolite, demonstrating that it originated from platelets [65].

5.3.1.3 Leukotrienes (LT)

It is an important mediator for both asthma and COPD progression. Leukotriene B4 (LTB4) is found to be increased in the serum and breath of COPD patients like other autacoids [66]. LTB4 is a powerful chemoattractant of NPHs. The cells involved in the production of LT are NPHs and MPs [28]. Their levels are further found to be increased in the plasma of the patients affected with COPD. There is an observed selective increase of LTB4 in patients with COPD which may be resistant to inhaled corticosteroid therapy. The meta-analysis study conducted with leukotriene receptor antagonists (LTRA) suggests neither acute nor chronic exposure of LTRA can improve the function of lung decline in COPD [67].

5.3.1.4 Platelet-Activating Factor (PAF)

The role of PAF in COPD patient is uncertain though it is a potent chemoattractant and activator of NPHs and alveolar MPs [68].

5.3.2 Peptide Mediators

5.3.2.1 Endothelins (ET)

These are the potent vasoconstrictor peptide autacoids [69]. Endothelin-1 (ET-1) levels are increased in the lungs and airways of COPD patients. It has role in asthma and may be involved in the pathogenesis of COPD. This potent vasoconstrictor substance is involved in pulmonary endothelial cell and vascular smooth muscle hyperplasia. ET-1 is also associated with vascular remodelling and linked to hypoxic pulmonary hypertension. Sputum levels of ET-1 are increased in COPD patients and may involve in inflammatory changes of airways during exacerbations [70].

5.3.2.2 Bradykinin (BK)

BK is an effective and potent vasodilator peptide. It is studied well in asthma; however, its role in COPD is not clear. BK is a vasoactive pro-inflammatory peptide intermediating acute responses in asthma [71]. Airways of COPD patients exhibited enhanced expression of B1 and B2-BK receptors. BK is a potent bronchoconstrictor at small airways. It is also involved in more protruding mucus production and remodelling of cells [72].

5.3.2.3 Tachykinins (TK)

Substance P is the important member of TK family involved in the pathogenesis of COPD via induced sputum production and chronic bronchitis [73]. The receptor NK (neurokinin)-1 and NK-2 for substance P are found in airways, submucosal glands and blood vessels [28]. Moreover, NK-2 receptors are also found in the inflammatory cells [74]. In addition, substance P also contributed in the secretion from airways of human in in vitro studies [75]. The TK antagonists have predominant potential against COPD and worked by reduction of mucus production [76].

5.3.2.4 Chemokines

Chemokines are small molecules that belong to the cytokine superfamily. They produce their effect through interaction with GPCR and mediate the cellular process like chemotaxis. Four subgroups of chemokines (CC, CXC, XC and CX3C) have been identified based on the positions of sequentially conserved residues and their quaternary structures [77]. Several chemokines have been identified which have an important role in the initiation of inflammatory and immune response as well as trafficking of inflammatory and immune cells to the target organs [78]. So, specific chemokine receptors develop a great interest in the development of treatment strategies for COPD, and targeting chemokine receptors could be important as involved in long-term inflammation and remodelling of bronchi [79].

5.3.2.5 Interleukin-8 (IL-8)

The CXC chemokine IL-8 (CXCL8) is a strong chemoattractant of NPHs and implicated in the progression of COPD [80]. In the sputum and BAL fluid of COPD patients, levels of IL-8 are increased significantly and also co-relate with the levels of NPHs [81]. The IL-8 levels are increased in emphysema cases due to the deficiency of AAT enzyme [82]. In addition, IL-8 levels are also found to be increased during exacerbations of COPD. The airway EP releases IL-8 under the influence of TNF-α, il-1β, LPS certain viruses, ROS and cigarette smoke [83]. In addition, NF-KB also has stimulatory effect on the production of IL-8. It works through CXCR-1 and CXCR-2 receptors. Neutralisation of IL-8 with a blocking antibody reduces the NPHs chemotactic activity in sputum of patients with COPD [28].

5.4 Potential Targets and Therapies for COPD

The potential targets/therapies which have shown some marked response in COPD patients have been discussed in this section of the chapter.

5.4.1 PDE (Phosphodiesterase) Inhibitors

The PDE enzyme is involved in various cellular processes such as release of mediators of inflammation and relaxation of smooth muscles by hydrolysing the cAMP and cGMP [84]. PDE3 is involved in the bronchoconstriction and inhibition of this enzyme leading to bronchodilation. PDE4 enzyme is present in most inflammatory cell types and is considered as one of the important targets for emerging COPD therapies [85, 86]. The PDE4 inhibitor roflumilast (ROF) is already approved by the US Food and Drug Administration (FDA) as a potential treatment option for COPD. Despite its relatively poor tolerability, ROF was approved by the FDA in 2011. The main drawback of ROF is its gastrointestinal adverse events [87]. Low-dose combinations of PDE4 and PI3Kδ inhibitors predominantly reduced the apoptosis of EP of lungs induced by the cigarette smoke, NPHs elastase production MP secretion of TNF-α, phosphorylated protein kinase B and MMP-9 in in vitro studies [88]. PDE4 inhibitors' inhaled AWD-12-281 (Elbion/GSK) showed preclinical efficacy in a variety of species. Development of this compound has been discontinued in 2006 due to poor efficacy [89]. Tofimilast (Pfizer) also failed to demonstrate its efficacy and was discontinued. UK-500,001 (Pfizer) was also discontinued because it failed to show efficacy in COPD patients. However, some companies are still working on targeting PDE enzyme such as AstraZeneca, Glide Pharma, Eisai, etc. [89].

5.4.2 Hypersecretion of Mucus

The hypersecretion of mucus can be modulated by blocking the overproduction/secretion of it. Various signalling mechanisms are involved in the secretion of mucus such as by-products of bacteria, cytokines, cholinergic drugs, elastases, MMP and activation of epidermal growth factor receptor (EGFR) [90]. Cigarette smoke induces EGFR- and hypoxia-inducible factor-1 (HIF-1)-mediated signalling and thus can induce hyperplasia of goblet cells involved in mucin production [91]. NPHs recruited to the epithelium of airways also potentiate the production of mucus via NPHs proteases and OS. N-Acetylcysteine and carbocysteine, EGFR tyrosine kinase receptors, muscarinic receptor antagonist and corticosteroids are some examples which have been studied for their effect on mucus [92]. Gefitinib, an EGFR inhibitor, has been demonstrated to inhibit spontaneous and cigarette smoke-induced MUC5AC production in cultured airway EP. However, the potential of gefitinib on secretion of mucus in clinics is yet to be established. In a similar way, N-acetylcysteine or carbocysteine showed reduced MUC5AC production in cell culture, ex vivo and preclinical studies of COPD [90].

5.4.3 Bronchodilators

β_2-Adrenoreceptor agonists (β_2-AA) and muscarinic receptor antagonists are used as treatments for COPD. β_2-AA mediate bronchodilation through stimulation of bronchi. Muscarinic receptors (M_2 and M_3) affect the tone of bronchi and mucus secretion. M_2 receptors indirectly affect contraction of airway smooth muscle, while M_3 receptors predominantly lead to constriction of airway smooth muscles and mucus secretion [93]. Selective M_3-muscarinic receptor inhibition leads to bronchodilation as well as reduction of mucus production. The short-acting β_2-AA (SABA), namely, levalbuterol and salbutamol, and short-acting muscarinic antagonist (SAMA) such as ipratropium bromide are used. The combination of SABA and SAMA is also used such as fenoterol/ipratropium and salbutamol/ipratropium [94]. On the other hand, the long-acting β_2-AA (LABA), namely, arformoterol, formoterol, indacaterol and salmeterol, and long-acting muscarinic antagonist (LAMA) such as aclidinium bromide, glycopyrronium bromide and tiotropium are used. The combination of LABA and LAMA is also used such as formoterol/aclidinium, formoterol/glycopyrronium, indacaterol/glycopyrronium, vilanterol/umeclidinium and olodaterol/tiotropium. In patients who have further exacerbations, LABA in combination with inhaled corticosteroids (ICS), e.g. formoterol/beclomethasone, formoterol/budesonide, etc., are used [95]. Treatment approaches with LAMA, LABA and ICS have been shown in Fig. 5.4.

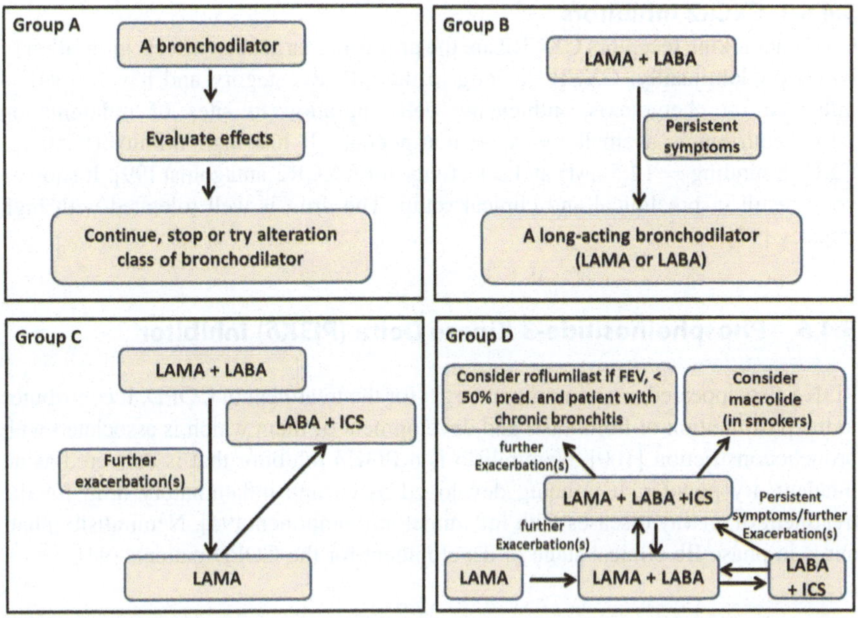

Fig. 5.4 Treatment approach of inhaled corticosteroids, beta agonists and muscarinic antagonists

5.4.4 Glucocorticoids (Inhaled Corticosteroids)

Inhaled corticosteroids are well-known for their anti-inflammatory activity. These inhaled corticosteroids are mainly working through glucocorticoid receptor (GR) [96]. This receptor is an example of a nuclear receptor which is expressed intracellularly. Patients with peripheral Eos counts >300 cells/μL may be best treated with LABA + ICS as the preferred choice for preventions of exacerbations, on the basis of clinical evidence. Once the COPD is controlled, ICS can be removed from the combination. However, clinical evidence suggests the limited beneficial effects of the ICS therapy in COPD patients. Despite this, 40–50% of patients with COPD receive ICS therapy [94].

5.4.5 Cystic Fibrosis Transmembrane Conductance Regulator (CFTR)

Chronic bronchitis, asthma and cystic fibrosis (CF) have many common characteristics. Moreover, a COPD model is developed by gene mutations in the CFTR. Some recent reports also suggested that cigarette smoke leads to dysfunction of CFTR. The smokers exhibit reduced CFTR functions [97]. The dysfunction of CFTR is also closely interlinked with chronic bronchitis and is closely related with reduced forced expiratory volume (FEV1) in the lungs of patients with COPD [98].

5.4.5.1 CXCR2 Inhibitors

CXC chemokine receptors CXCR2 are the promising target that affects infiltration of NPHs predominantly. CXCR2 belong to the GPCR category and have a marked influence on chemotaxis, influencing cell migration to sites of inflammation [79]. Danirixin is a small molecule non-peptide. It has high affinity (IC50 for CXCL8 binding = 12.5 nM) and selectivity for CXCR2 antagonist [99]. It showed good result in preclinical and clinical setup. The drug is well tolerated with high efficacy [94].

5.4.6 Phosphoinositide-3 Kinase Delta (PI3Kδ) Inhibitor

PI3Kδ also appeared as an important target for the treatment of COPD. It contributes to the proliferation of fibroblasts and development of them which is associated with bronchoconstriction [100]. Nemiralisib is a PI3Kδ inhibitor that is delivered as an inhalant dry powder. It is being developed as an anti-inflammatory drug for the treatment of airway diseases with inflammatory component [94]. Nemiralisib inhalant is in phase IIb clinical stage of development for the COPD patients [94].

5.4.7 Targeting Inflammatory Markers

Inflammation is a peculiar feature of COPD; hence, various inflammatory markers can be targeted by anti-inflammatory drugs. Markers such as interleukin (IL)-1β, IL-13, IL-17A, IKK2, iNOS, β2-integrin, TNF-α, p38, secretory phospholipase A2 and adenosine receptor 2a are some important examples of this category [28, 94]. IL-5 receptor antagonists are considered as one of the important investigational targets/therapies which efficiently reduce inflammation mediated through Eos and lower asthma exacerbations [101]. However, mepolizumab and benralizumab failed to show their effectiveness in COPD patients [102]. Both drugs were unable to show their potential in phase III trials [102, 103]. Recent studies also have showed the importance of IL-17 and IL-22 in the progression of COPD. Increased IL-17 secretion leads to recruitment of NPHs that is responsible for chronic inflammation and emphysema. Moreover, pathogen-linked infections and disease exacerbations are mediated by defective IL-22 response [94]. Altered formation of these markers results in destructive functions of antigen-presenting cells and complex network of immune system. Targeting these inflammatory mediators may be serving as a useful approach for the treatment for COPD.

5.4.8 Other COPD Treatments

Tools helpful in smoking cessation are also considered as important treatment options for COPD. In this connection, varenicline is a potential treatment which is helpful to leave smoking habits. Nicotine lozenges, patches and non-combustible cigarettes, like E-cigarettes, are also considered as beneficial options to reduce the exposure to poisonous and toxic chemicals found in cigarette smoke [94].

5.5 Conclusion

COPD is a progressive disease which is characterised by emphysema and chronic bronchitis. The cellular and molecular level mechanisms underlying COPD are not understood well. Cigarette smoke is the one of the main causes for stimulation of various inflammatory markers. As compared to asthma, it is complicated to hamper the progression of the disease. The disease has complex pathophysiology characterised by the involvement of various mediators such as inflammatory and lipid mediators.

There is no effective treatment available for complete reversal of the disease. There are recently no treatment approaches that markedly reduce disease progression or mortality and prevent COPD exacerbations. However, drugs like long-acting beta agonist and steroids as well as anticholinergic drugs can reduce the progress of the disease to some extent. The newer agents and targets such as PDE inhibitors, CFTR and CXCR2 inhibitors have shown some promising results in some preclinical and clinical studies.

References

1. Devine JF (2008) Chronic obstructive pulmonary disease: an overview. Am Health Drug Benefits 1(7):34–42
2. https://www.who.int/respiratory/copd/burden/en/ Date accessed: 15.09.20
3. Centers for Disease Control and Prevention (US), National Center for Chronic Disease Prevention and Health Promotion (US), Office on Smoking and Health (US) (2010) How tobacco smoke causes disease: the biology and behavioral basis for smoking-attributable disease: a report of the surgeon general. Centers for Disease Control and Prevention (US), Atlanta, GA. 7, Pulmonary Diseases. Available from: https://www.ncbi.nlm.nih.gov/books/NBK53021/ Date accessed: 15.09.20
4. Hersh CP, Campbell EJ, Scott LR, Raby BA (2019) Alpha-1 antitrypsin deficiency as an incidental finding in clinical genetic testing. Am J Respir Crit Care Med 199(2):246–248
5. Tasch JJ, McLaughlan AT, Nasir AA (2018) A novel approach to screening for Alpha-1 antitrypsin deficiency: inpatient testing at a teaching institution. Chronic Obstr Pulm Dis 5 (2):106–110
6. Sarkar M, Niranjan N, Banyal PK (2017) Mechanisms of hypoxemia. Lung India 34(1):47–60. [published correction appears in Lung India]
7. https://www.nhp.gov.in/world-copd-day-2018_pg Date accessed 14.09.20
8. Chee A, Sin DD (2008) Treatment of mild chronic obstructive pulmonary disease. Int J Chron Obstruct Pulmon Dis 3(4):563–573
9. https://www.who.int/news-room/fact-sheets/detail/chronic-obstructive-pulmonary-disease-(copd) date accessed 14.09.20
10. Buist AS (2003) Similarities and differences between asthma and chronic obstructive pulmonary disease: treatment and early outcomes. Eur Respir J Suppl 39:30s–35s

11. D'Urzo AD, Tamari I, Bouchard J, Jhirad R, Jugovic P (2011) New spirometry interpretation algorithm: primary care respiratory Alliance of Canada approach. Can Fam Physician 57 (10):1148–1152
12. https://www.who.int/respiratory/copd/GOLD_WR_06.pdf Date accessed 14.09.20
13. Akata K, Van Eeden SF (2020) Lung macrophage functional properties in chronic obstructive pulmonary disease. Int J Mol Sci 21(3):853
14. MacNee (2005) Pathogenesis of chronic obstructive pulmonary disease. Proc Am Thorac Soc 2(4):258–291
15. Papandrinopoulou D, Tzouda V, Tsoukalas G (2012) Lung compliance and chronic obstructive pulmonary disease. Pulm Med 2012:542769
16. Barnes PJ, Cosio MG (2004) Characterization of T lymphocytes in chronic obstructive pulmonary disease. PLoS Med 1(1):e20
17. Metzemaekers M, Gouwy M, Proost P (2020) Neutrophil chemoattractant receptors in health and disease: double-edged swords. Cell Mol Immunol 17(5):433–450
18. Pease JE, Sabroe I (2002) The role of interleukin-8 and its receptors in inflammatory lung disease: implications for therapy. Am J Respir Med 1(1):19–25
19. Mukhopadhyay S, Hoidal JR, Mukherjee TK (2006) Role of TNFalpha in pulmonary pathophysiology. Respir Res 7(1):125
20. Ojiaku CA, Yoo EJ, Panettieri RA Jr (2017) Transforming growth factor β1 function in airway remodeling and hyperresponsiveness. The missing link? Am J Respir Cell Mol Biol 56 (4):432–442
21. Chung KF (2005) Inflammatory mediators in chronic obstructive pulmonary disease. Curr Drug Targets Inflamm Allergy 4(6):619–625
22. Mittal M, Siddiqui MR, Tran K, Reddy SP, Malik AB (2014) Reactive oxygen species in inflammation and tissue injury. Antioxid Redox Signal 20(7):1126–1167
23. Chung KF (2001) Cytokines in chronic obstructive pulmonary disease. Eur Respir J Suppl 34:50s–59s
24. Shu DY, Hutcheon AEK, Zieske JD, Guo X (2019) Epidermal growth factor stimulates transforming growth factor-beta receptor type II expression in corneal epithelial cells. Sci Rep 9(1):8079
25. Lacoste JY, Bousquet J, Chanez P (1993) Eosinophilic and neutrophilic inflammation in asthma, chronic bronchitis and chronic obstructive pulmonary disease. J Allergy Clin Immunol 92:537–548
26. Pandey KC, De S, Mishra PK (2017) Role of proteases in chronic obstructive pulmonary disease. Front Pharmacol 8:512
27. Rovina N, Koutsoukou A, Koulouris NG (2013) Inflammation and immune response in COPD: where do we stand? Mediat Inflamm 2013:413735
28. Barnes PJ (2004) Mediators of chronic obstructive pulmonary disease. Pharmacol Rev 56 (4):515–548
29. Hoenderdos K, Condliffe A (2013) The neutrophil in chronic obstructive pulmonary disease. Am J Respir Cell Mol Biol 48(5):531–539
30. O'Donnell R, Breen D, Wilson S, Djukanovic R (2006) Inflammatory cells in the airways in COPD. Thorax 61(5):448–454
31. Punturieri A, Filippov S, Allen E, Caras I, Murray R, Reddy V, Weiss SJ (2000) Regulation of elastinolytic cysteine proteinase activity in normal and cathepsin K-deficient human macrophages. J Exp Med 192:789–800
32. Lim S, Roche N, Oliver BG, Mattos W, Barnes PJ, Fan CK (2000b) Balance of matrix metalloprotease-9 and tissue inhibitor of metalloprotease-1 from alveolar macrophages in cigarette smokers. Regulation by interleukin-10. Am J Respir Crit Care Med 162:1355–1360
33. Madala SK, Pesce JT, Ramalingam TR (2010) Matrix metalloproteinase 12-deficiency augments extracellular matrix degrading metalloproteinases and attenuates IL-13-dependent fibrosis. J Immunol 184(7):3955–3963

34. Demedts IK, Morel-Montero A, Lebecque S, Pacheco Y, Cataldo D, Joos GF, Pauwels RA, Brusselle GG (2006) Elevated MMP-12 protein levels in induced sputum from patients with COPD. Thorax 61:196–201
35. Chen L, Chen G, Zhang MQ (2016) Imbalance between subsets of CD8(+) peripheral blood T cells in patients with chronic obstructive pulmonary disease. PeerJ:e2301
36. Groom JR, Luster AD (2011) CXCR3 in T cell function. Exp Cell Res 317(5):620–631
37. Freeman CM, Han MK, Martinez FJ (2010) Cytotoxic potential of lung CD8(+) T cells increases with chronic obstructive pulmonary disease severity and with in vitro stimulation by IL-18 or IL-15. J Immunol 184(11):6504–6513
38. George L, Brightling CE (2016) Eosinophilic airway inflammation: role in asthma and chronic obstructive pulmonary disease. Ther Adv Chronic Dis 7(1):34–51
39. Turato G, Zuin R, Saetta M (2001) Pathogenesis and pathology of COPD. Respiration 68:117–128
40. Tashkin DP, Wechsler ME (2018) Role of eosinophils in airway inflammation of chronic obstructive pulmonary disease. Int J Chron Obstruct Pulmon Dis 13:335–349
41. Freeman CM, Curtis JL (2017) Lung dendritic cells: shaping immune responses throughout chronic obstructive pulmonary disease progression. Am J Respir Cell Mol Biol 56(2):152–159
42. Ito T, Connett JM, Kunkel SL, Matsukawa A (2013) The linkage of innate and adaptive immune response during granulomatous development. Front Immunol 4:10
43. Chaplin DD (2010) Overview of the immune response. J Allergy Clin Immunol 125(2 Suppl 2):S3–S23
44. D'Hulst A, Vermeulen KY, Pauwels RA (2002) Cigarette smoke exposure causes increase in pulmonary dendritic cells. Am J Respir Crit Care Med 164:A604
45. Thorley AJ, Tetley TD (2007) Pulmonary epithelium, cigarette smoke, and chronic obstructive pulmonary disease. Int J Chron Obstruct Pulmon Dis 2(4):409–428
46. Hellermann GR, Nagy SB, Kong X, Lockey RF, Mohapatra SS (2002) Mechanism of cigarette smoke condensate-induced acute inflammatory response in human bronchial epithelial cells. Respir Res 3:22–28
47. Takizawa H, Tanaka M, Takami K, Ohtoshi T, Ito K, Satoh M, Okada Y, Yamasawa F, Nakahara K, Umeda A (2001) Increased expression of transforming growth factor-1 in small airway epithelium from tobacco smokers and patients with chronic obstructive pulmonary disease (COPD). Am J Respir Crit Care Med 163:1476–1483
48. Franklin WA, Veve R, Hirsch FR, Helfrich BA, Bunn PA Jr (2002) Epidermal growth factor receptor family in lung cancer and premalignancy. Semin Oncol 29:3–14
49. MacNee W (2006) Pathology, pathogenesis, and pathophysiology. BMJ 332 (7551):1202–1204
50. Korkmaz B, Horwitz MS, Jenne DE, Gauthier F (2010) Neutrophil elastase, proteinase 3, and cathepsin G as therapeutic targets in human diseases. Pharmacol Rev 62(4):726–759
51. McGuinness AJ, Sapey E (2017) Oxidative stress in COPD: sources, markers, and potential mechanisms. J Clin Med 6(2):21
52. Nita M, Grzybowski A (2016) The role of the reactive oxygen species and oxidative stress in the pathomechanism of the age-related ocular diseases and other pathologies of the anterior and posterior eye segments in adults. Oxidative Med Cell Longev 2016:3164734
53. Moldoveanu B, Otmishi P, Jani P, Walker J, Sarmiento X, Guardiola J, Saad M, Yu J (2009) Inflammatory mechanisms in the lung. J Inflamm Res 2:1–11
54. Bagdonas E, Raudoniute J, Bruzauskaite I, Aldonyte R (2015) Novel aspects of pathogenesis and regeneration mechanisms in COPD. Int J Chron Obstruct Pulmon Dis 10:995–1013
55. Foronjy R, D'Armiento J (2006) The effect of cigarette smoke-derived oxidants on the inflammatory response of the lung. Clin Appl Immunol Rev 6(1):53–72
56. Wofniak A, Górecki D, Szpinda Micha B, Mila-Kierzenkowska C, Wofniak B (2013, 2013) Oxidant-antioxidant balance in the blood of patients with chronic obstructive pulmonary disease after smoking cessation. Oxidative Med Cell Longev:1–9

57. van Eeden SF, Sin DD (2013) Oxidative stress in chronic obstructive pulmonary disease: a lung and systemic process. Can Respir J 20(1):27–29
58. Pacher P, Beckman JS, Liaudet L (2007) Nitric oxide and peroxynitrite in health and disease. Physiol Rev 87(1):315–424
59. Rahman I, MacNee W (2012) Antioxidant pharmacological therapies for COPD. Curr Opin Pharmacol 12(3):256–265
60. Ma Q (2013) Role of nrf2 in oxidative stress and toxicity. Annu Rev Pharmacol Toxicol 53:401–426
61. Radi ZA, Meyerholz DK, Ackermann MR (2010) Pulmonary cyclooxygenase-1 (COX-1) and COX-2 cellular expression and distribution after respiratory syncytial virus and parainfluenza virus infection. Viral Immunol 23(1):43–48
62. Machado-Carvalho L, Roca-Ferrer J, Picado C (2014) Prostaglandin E2 receptors in asthma and in chronic rhinosinusitis/nasal polyps with and without aspirin hypersensitivity. Respir Res 15(1):100
63. Zuo H, Cattani-Cavalieri I, Musheshe N, Nikolaev VO, Schmidt M (2019) Phosphodiesterases as therapeutic targets for respiratory diseases. Pharmacol Ther 197:225–242
64. Montuschi P, Macagno F, Parente P, Valente S, Lauriola L, Ciappi G, Kharitonov SA, Barnes PJ, Ciabattoni G (2005) Effects of cyclo-oxygenase inhibition on exhaled eicosanoids in patients with COPD. Thorax 60(10):827–833
65. Lopez LR, Guyer KE, Torre IG, Pitts KR, Matsuura E, Ames PR (2014) Platelet thromboxane (11-dehydro-thromboxane B2) and aspirin response in patients with diabetes and coronary artery disease. World J Diabetes 5(2):115–127
66. Seggev JS, Thornton WH Jr, Edes TE (1991) Serum leukotriene B4 levels in patients with obstructive pulmonary disease. Chest 99(2):289–291
67. Liu L, Wang JL, Xu XY, Feng M, Hou Y, Chen L (2018) Leukotriene receptor antagonists do not improve lung function decline in COPD: a meta-analysis. Eur Rev Med Pharmacol Sci 22 (3):829–834
68. Shindo K, Koide K, Fukumura M (1998) Platelet-activating factor increases leukotriene B4 release in stimulated alveolar macrophages from asthmatic patients. Eur Respir J 11:1098–1104
69. Davenport AP, Hyndman KA, Dhaun N, Southan C, Kohan DE, Pollock JS, Pollock DM, Webb DJ, Maguire JJ (2016) Endothelin. Pharmacol Rev 68(2):357–418
70. Bacakoglu F, Atasever A, Ozhan MH, Gurgun C, Ozkilic H, Guzelant A (2003) Plasma and bronchoalveolar lavage fluid levels of endothelin-1 in patients with chronic obstructive pulmonary disease and pulmonary hypertension. Respiration 70:594–599
71. Ricciardolo FL, Sabatini F, Sorbello V, Benedetto S, Defilippi I, Petecchia L, Usai C, Gnemmi I, Balbi B, De Rose V, Ten Hacken NH, Postma DS, Timens W, Di Stefano A (2013) Expression of vascular remodelling markers in relation to bradykinin receptors in asthma and COPD. Thorax 68(9):803–811
72. Barnes PJ (1992) Effect of bradykinin on airway function. Agents Actions Suppl 38 (Pt 3):432–438
73. De Swert KO, Bracke KR, Demoor T, Brusselle GG, Joos GF (2009) Role of the tachykinin NK1 receptor in a murine model of cigarette smoke-induced pulmonary inflammation. Respir Res 10(1):37
74. Renzi D, Pellegrini B, Tonelli F, Surrenti C, Calabrò A (2000) Substance P (neurokinin-1) and neurokinin A (neurokinin-2) receptor gene and protein expression in the healthy and inflamed human intestine. Am J Pathol 157(5):1511–1522
75. Rogers DF, Aursudkij B, Barnes PJ (1989) Effects of tachykinins on mucus secretion on human bronchi in vitro. Eur J Pharmacol 174:283–286
76. Joos GF, Pauwels RA (2001) Tachykinin receptor antagonists: potential in airways diseases. Curr Opin Pharmacol 1:235–241
77. Zlotnik A, Yoshie O, Nomiyama H (2006) The chemokine and chemokine receptor superfamilies and their molecular evolution. Genome Biol 7:243

78. Sokol CL, Luster AD (2015) The chemokine system in innate immunity. Cold Spring Harb Perspect Biol 7(5):a016303
79. Henrot P, Prevel R, Berger P, Dupin I (2019) Chemokines in COPD: from implication to therapeutic use. Int J Mol Sci 20(11):2785
80. Ha H, Debnath B, Neamati N (2017) Role of the CXCL8-CXCR1/2 Axis in cancer and inflammatory diseases. Theranostics 7(6):1543–1588
81. Feng E, Wan R, Yang S, Yan Z, Wang S, He W, Zhang Y, Yin H, Chen Z, Liu R (2013) Expression levels of induced sputum IL-8 and IL-10 and drug intervention effects in patients with acute exacerbated COPD complicated with chronic cor pulmonale at high altitude. Exp Ther Med 6(3):747–752
82. Dunlea DM, Fee LT, McEnery T, McElvaney NG, Reeves EP (2018) The impact of alpha-1 antitrypsin augmentation therapy on neutrophil-driven respiratory disease in deficient individuals. J Inflamm Res 11:123–134
83. Strzelak A, Ratajczak A, Adamiec A, Feleszko W (2018) Tobacco smoke induces and alters immune responses in the lung triggering inflammation, allergy, asthma and other lung diseases: a mechanistic review. Int J Environ Res Public Health 15(5):1033
84. Keravis T, Lugnier C (2012) Cyclic nucleotide phosphodiesterase (PDE) isozymes as targets of the intracellular signalling network: benefits of PDE inhibitors in various diseases and perspectives for future therapeutic developments. Br J Pharmacol 165(5):1288–1305
85. Banner KH, Press NJ (2009) Dual PDE3/4 inhibitors as therapeutic agents for chronic obstructive pulmonary disease. Br J Pharmacol 157(6):892–906
86. Li H, Zuo J, Tang W (2018) Phosphodiesterase-4 inhibitors for the treatment of inflammatory diseases. Front Pharmacol 9:1048
87. Cazzola M, Calzetta L, Rogliani P, Matera MG (2016) The discovery of roflumilast for the treatment of chronic obstructive pulmonary disease. Expert Opin Drug Discov 11:733–744
88. Lakshmi SP, Reddy AT, Reddy RC (2017) Emerging pharmaceutical therapies for COPD. Int J Chron Obstruct Pulmon Dis 12:2141–2156
89. Phillips JE (2020) Inhaled phosphodiesterase 4 (PDE4) inhibitors for inflammatory respiratory diseases. Front Pharmacol 11:259
90. Martin C, Frija-Masson J, Burgel PR (2014) Targeting mucus hypersecretion: new therapeutic opportunities for COPD? Drugs 74(10):1073–1089
91. Polosukhin VV, Cates JM, Lawson WE, Milstone AP, Matafonov AG, Massion PP, Lee JW, Randell SH, Blackwell TS (2011) Hypoxia-inducible factor-1 signalling promotes goblet cell hyperplasia in airway epithelium. J Pathol 224(2):203–211
92. Zhang T, Zhou X (2014) Clinical application of expectorant therapy in chronic inflammatory airway diseases (review). Exp Ther Med 7(4):763–767
93. Buels KS, Fryer AD (2012) Muscarinic receptor antagonists: effects on pulmonary function. Handb Exp Pharmacol 208:317–341
94. Haarst AV, McGarvey L, Paglialunga S (2019) Review of drug development guidance to treat chronic obstructive pulmonary disease: US and EU. Perspectives 106:1222–1235
95. Malerba M, Foci V, Patrucco F, Pochetti P, Nardin M, Pelaia C, Radaeli A (2019) Single Inhaler LABA/LAMA for COPD. Front Pharmacol 10:390
96. Barnes PJ (2010) Inhaled corticosteroids. Pharmaceuticals (Basel) 3(3):514–540
97. Solomon GM, Fu L, Rowe SM, Collawn JF (2017) The therapeutic potential of CFTR modulators for COPD and other airway diseases. Curr Opin Pharmacol 34:132–139
98. Raju SV, Solomon GM, Dransfield MT, Rowe SM (2016) Acquired cystic fibrosis transmembrane conductance regulator dysfunction in chronic bronchitis and other diseases of mucus clearance. Clin Chest Med 37(1):147–158
99. Miller BE, Mistry S, Smart K (2015) The pharmacokinetics and pharmacodynamics of danirixin (GSK1325756)—a selective CXCR2 antagonist – in healthy adult subjects. BMC Pharmacol Toxicol 16:18

100. Wilson R, Cahn A, Deans A (2013) Safety, tolerability and pharmacokinetics (PK) of single and repeat nebulised doses of a novel phosphoinositide 3-kinase δ inhibitor (PI3Kδ), GSK2269557, administered to healthy male subjects in a phase I study. Eur Respir J 42:729
101. Brussino L, Heffler E, Bucca C, Nicola S, Rolla G (2018) Eosinophils target therapy for severe asthma: critical points. Biomed Res Int 2018:7582057
102. Mkorombindo T, Dransfield MT (2019) Mepolizumab in the treatment of eosinophilic chronic obstructive pulmonary disease. Int J Chron Obstruct Pulmon Dis 14:1779–1787
103. https://www.fda.gov/media/114447/download Date accessed 14.09.20

Probing the Cellular and Molecular Mechanisms Underlying in the Pathogenesis of Chronic Obstructive Pulmonary Disease

6

C. Sarath Chandran, Alan Raj, and T. K. Shahin Muhammed

Abstract

Chronic obstructive pulmonary disease (COPD) is a chronic inflammatory lung disease which is characterised by the blockade and destruction of small airways, emphysema, structural remodelling, etc. The COPD Foundation classifies COPD into three subtypes, including chronic bronchitis, emphysema and small airway disease. Males are reported to be more susceptible to COPD than females. There are various inflammatory reactions that occur in COPD patients, which may persist even after cessation of smoking. This may be due to the production of memory T cells and bacterial/viral colonisation. The inflammatory reactions were triggered by the presence of various inflammatory cells, which included epithelial cells, macrophages, dendritic cells, etc. The inflammatory mediators such as lipids, free radicals, cytokine and growth factors have a role in the pathogenesis of COPD. The comprehensive knowledge on the cellular and molecular mechanisms involved in triggering COPD helps to develop novel treatment strategies.

Keywords

COPD · Cigarette · Smoke · Emphysema · Inflammatory · Cells · Mediators

6.1 Introduction

Chronic obstructive pulmonary disease (COPD) is a chronic inflammatory lung disease which is characterised by the blockade and destruction of small airways, emphysema, structural remodelling, etc. The elasticity of the lung tissue was

C. Sarath Chandran (✉) · A. Raj · T. K. Shahin Muhammed
College of Pharmaceutical Sciences, Government Medical College, Kannur, Kerala, India

K. Dua et al. (eds.), *Targeting Cellular Signalling Pathways in Lung Diseases*,
https://doi.org/10.1007/978-981-33-6827-9_6

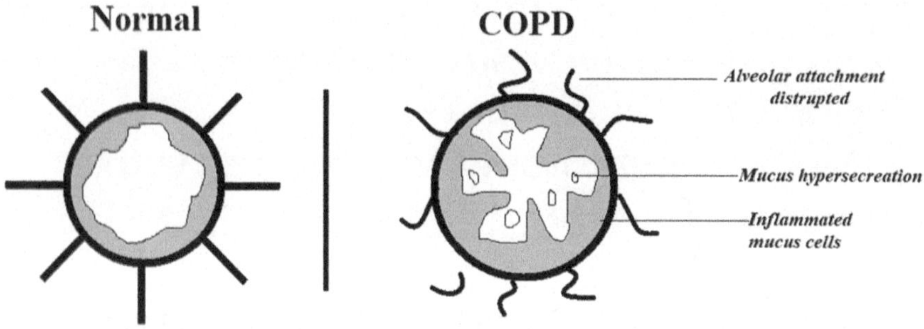

Fig. 6.1 Airway epithelium in normal and COPD conditions

decreased because of the alveolar wall destruction. The patients may surrender to death due to the slow destruction of lung tissue, with systemic infections like lung cancer and cardiovascular disease in COPD patients [1]. Males are reported to be more susceptible to COPD than females. COPD is mainly seen in cigarette smokers and is caused by inhaling smoke from burning carbon compounds and other irritants. It was found that only 25% of the patients developed COPD due to cigarette smoking. There may be some other predisposing factors that trigger chronic lung infections [2]. Exposure to passive smoking may also lead to COPD [3]. There are various inflammatory reactions that occur in COPD patients, which may persist even after cessation of smoking. This may be due to the production of memory T cells and bacterial/viral colonisation. The airways of non-COPD individuals were expanded with the help of alveolar attachments during breathing process, which allows alveolar emptying. But in case of COPD patients, alveolar attachment is altered due to emphysema. The poor mucociliary clearance may lead to thick mucus-filled airway (Fig. 6.1). The inflammatory reactions were triggered by the presence of various inflammatory cells, which included epithelial cells, macrophages, dendritic cells, etc. [4]. The comprehensive knowledge on the cellular and molecular mechanisms involved in triggering COPD helps to develop novel treatment strategies.

6.2 Subtypes of COPD

The COPD Foundation classifies COPD into three subtypes, including chronic bronchitis, emphysema and small airway disease.

6.2.1 Chronic Bronchitis

Chronic bronchitis is a disease condition in which the bronchial tubules were subjected to inflammations or lesions due to the activation of innate and adaptive immune system followed by the production of sputum and severe cough [5, 6]. Severe cough may be due to the hyper-secretion of mucus in the airway tract [7]. The high amount of mucus will be deposited in the small airways due to the disruption in the epithelial barrier cells and reduction in mucociliary clearance [8]. The airway will become thick due to the over-deposition of connective tissue [9].

Various predisposing factors including smoking, chemical irritants and smog may cause chronic bronchitis. The microorganisms like influenza A and B and *Staphylococcus* and *Streptococcus* may induce bronchitis. Patients with a history of asthma and cystic fibrosis are more susceptible to this condition.

6.2.2 Emphysema

It is a disease condition in which the elastic recoil force of the lung is reduced, followed by shortness of breath [10, 11]. It is mainly caused by the inhalation of chemicals, irritant gases, etc. The hereditary autoimmune disease, deficiency in alpha-1-antitrypsin (AAT), may lead to emphysema [12–14]. The elastic nature of the lung wall is maintained by the balance of protease-anti-protease activity, which is required for the proper lung function. Any imbalance in this activity will lead to emphysema [15]. Respiratory bronchiole, alveolar sac and alveoli were also destroyed in emphysema. Due to these problems, the lung may not exchange gases efficiently [16].

6.2.3 Small Airway Disease

Small airway disease is seen in the terminal part of lungs, i.e. it did not include the alveoli. The internal diameter of the small airway is less than 2 mm [17, 18]. It was reported that the structure of small airways was altered in smokers without COPD [19–21]. But in case of COPD patients, airways will be filled with thick mucus [22, 23].

6.3 Inflammatory Cells Involved in COPD

There are various cellular processes that occur in COPD patients (Fig. 6.2). They are innate immunity and adaptive immunity. The innate immunity includes neutrophils, macrophages, mast cells, eosinophils, natural killer cells, gamma and delta T cells and dendritic cells, and the adaptive immunity includes T and B lymphocytes. There are a few other inflammatory cells included, i.e. airway and alveolar epithelial cells, fibroblasts, endothelial cells, cellular activation, extra-pulmonary effects, somatic mutation, etc. Cigarette smoke/other chemical irritants stimulate alveolar macrophages in the respiratory tract, which release massive amount of interleukins and leukotriene like IL-8 and LTB_4, respectively. These chemotactic factors activate the protease enzymes that break the connective tissues present in the lung parenchyma, resulting in emphysema and chronic bronchitis. These protease enzymes are blocked by protease enzyme inhibitors present in our body like AAT, tissue inhibitor

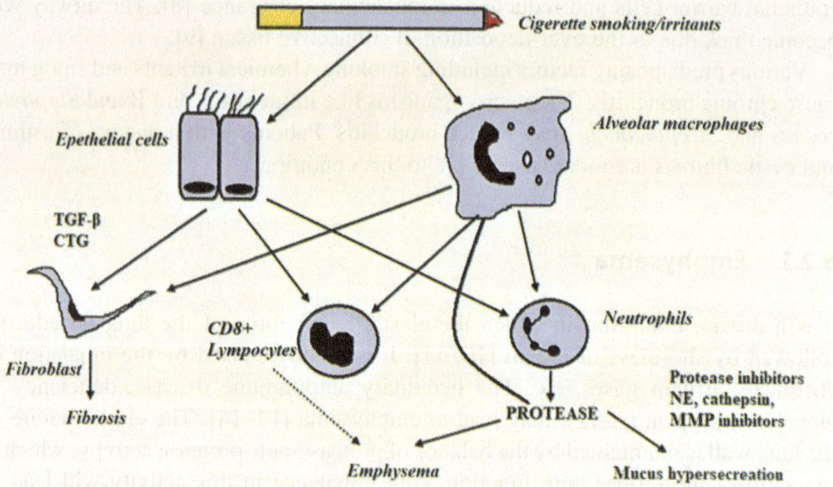

Fig. 6.2 Inflammatory cells involved in pathogenesis of COPD

of metalloproteinase (TIMP), etc. But deficiency of these enzymes may lead to lung tissue destruction. Fibrosis occurs due to the activation of fibroblast cells with the help of growth factors released from macrophages and epithelial cells.

6.3.1 Epithelial Cells

The lung epithelial cells have a crucial role in COPD pathogenesis. When compared to healthy population, numerous epithelial goblet cells were found in the lungs of COPD patients. The goblet cells of COPD patients became hyperplastic and produced inflammatory reactions. The stimulation of goblet cells resulted in the enhanced mucin production with a thick mucus as well as enhanced mucin gene synthesis. This impaired function made the lung more susceptible to viral/bacterial infections [24, 25]. The stimulated epithelial cells produced not only the mucin but also some inflammatory mediators like tumour necrosis factor (TNF)-α, interleukin (IL)-1, IL-8 and granulocyte-macrophage colony-stimulating factor. The cellular apoptosis that occurred in human adult lungs may be an important component in reducing the inflammation. But in COPD patients, the alveolar epithelial apoptosis is high in comparison with healthy population. The habit of smoking may drastically affect the epithelial lung repair in COPD patients.

Immunoglobulin A (IgA), transported by epithelial cells, had a vital role in adaptive immunity of a healthy individual. In cigarette smokers, there may be an imbalance between innate and adaptive immunity, thereby the airway epithelium may be prone to inflammation and infection. Squamous metaplasia was observed in COPD patients, which may be due to the high proliferation of epithelial cells

[26]. The epithelial repair was not proper because of the overexpression of epithelial growth factor (EGF) receptors [27].

6.3.2 Macrophages

The alveolar macrophages have a crucial role in triggering COPD. The increased number of macrophages (about 10–25-fold) was seen in the lung parenchyma, bronchoalveolar lavage fluid (BALF) and sputum of COPD patients (Fig. 6.3). The habit of cigarette smoking and presence of other irritants may trigger the production of macrophages and increase the inflammatory mediators. The main inflammatory mediator produced was protease enzyme, which included matrix metalloproteinase (MMP)-2, MMP-9 and cathepsins K, L and S. The overproduction of these inflammatory mediators and protease enzyme may destroy the lung parenchyma [28, 29]. In COPD patients with smoking history, the alveolar macrophages secrete more inflammatory mediators which have high elastolytic activity than in normal smokers, and it will increase upon further exposure to cigarette smoking [30–32]. MMP-9 shows the highest elastolytic activity among other inflammatory mediators secreted by the alveolar macrophage. The up-regulation of majority of the inflammatory mediators secreted from the alveolar macrophage is controlled by

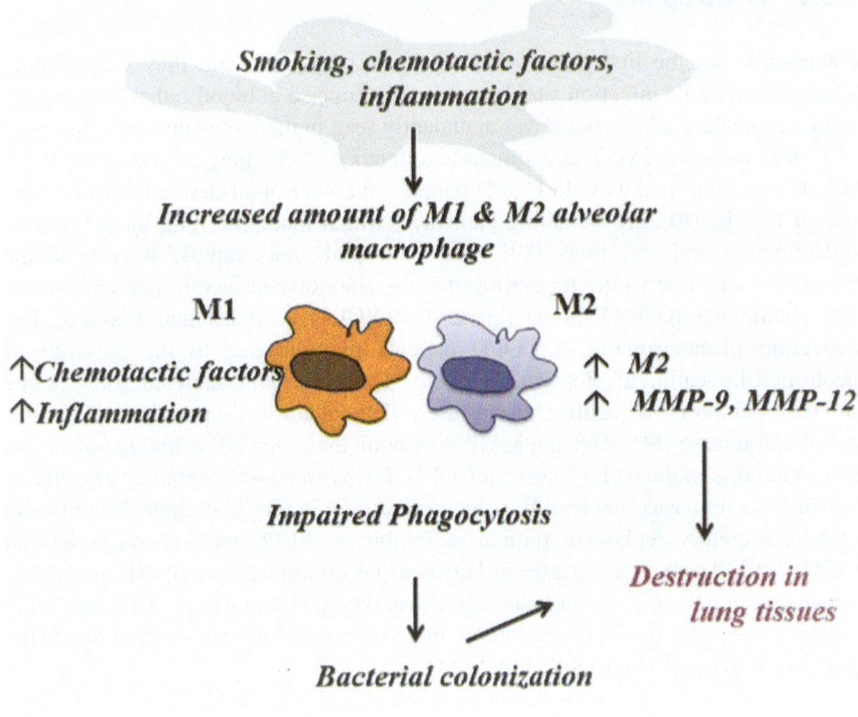

Fig. 6.3 Role of macrophage in lung tissue destruction

the nuclear factor kappa B (NF-κB). This is particularly stimulated by the help of alveolar macrophage at worse COPD condition of patients [33, 34]. It is difficult to find out the survival time of macrophages in COPD patients directly. An alternative method is to determine the overexpression of antiapoptotic protein B-cell lymphoma-extra-large (Bcl-X_L) and cyclin-dependent kinase inhibitor 1 ($p21^{CIP/WAF1}$) in the cytoplasm of COPD patients [35]. Host defence is one of the main activities of alveolar macrophage by phagocytic action. It engulfs bacteria and other foreign substances. But any impairment in this function led to the formation of bacterial colonisation in lung tissues of COPD patients. Phosphatidylserine is overexpressed on the cell surface of apoptotic cells, which is recognised by macrophage by special receptors, which is capable of binding with phosphatidylserine [36]. Alveolar macrophage cannot identify the apoptotic neutrophils because of the cleavage that occurred in phosphatidylserine receptor by neutrophil elastase, resulting in the presence of high concentration of apoptotic neutrophils in the airway tract [37]. The distinct phenotypes of macrophages were identified as M1 and M2. M1 macrophage produced pro-inflammatory cytokines, high level of reactive oxygen species and oxygen intermediates. M2 macrophage secreted anti-inflammatory cytokines which included interleukin-10, TGF-β, CCL-8, etc. M1 macrophage is secreted more in COPD patients which led to destruction of the lung tissue.

6.3.3 Neutrophils

Neutrophils are the first line of defence against infection and they were rapidly accumulated at the infection site. It was mainly concise in blood rather than in lung tissues of healthy adults, but it was abundantly seen in the lumen and bronchial wall of COPD patients [38]. The antimicrobial defence of the lung is due to the IL22/IL22R signalling pathway. In COPD patients, the neutrophil-derived protease may impair the IL22/IL22R signalling pathways, which made the lung more prone to inflammation and infections [39]. The neutrophils may rapidly migrate to the infectious cells due to the triggering of some chemotactic factors like leukotriene B4, chemokine (C-X-C motif) ligand 1 (CXCL1), CXCL5 and CXCL8. The activation of neutrophils in COPD patients may be due to the presence of neutrophil-lipocalin, myeloperoxidase, etc. The neutrophil secreted some serine protease composed of neutrophil elastase (NE), cathepsin G, proteinase-3, matrix metalloproteinase (MMP)-8 and MMP-9. Among these, the NE is highly potent and has a vital role in the pathogenesis of COPD. It was an elastin degrading enzyme, in which its action was blocked by some protease inhibitors like alpha-1-antitrypsin (AAT), secretory leukocyte proteinase inhibitors (SLPI) and α2-macroglobulin (α2M) [40]. There is a balance in between the concentrations of NE and AAT, which was impaired in the presence of reactive oxygen intermediate (ROI) in COPD patients. The AAT-deficient patients are more susceptible to lung diseases due to the protease activity of neutrophil elastase [41].

As we know, the sputum is rich in neutrophils, which protect it from infection, but in cigarette smokers, it was considerably lesser than non-smokers. Cigarette smoking may induce neutrophil cell death via mitochondrial dysfunction, damage-associated molecular pattern (DAMP) release, etc. DAMP is a set of cytotoxic endogenous molecules which was abundantly released from dying cells. It may trigger the activation of innate immune system by binding with pattern recognition receptors (PRRs), which included toll-like receptors (TLR)2, TLR4 and TLR9 [42]. The serum concentration of DAMP gene expression is high. The activation of PRR receptor TLR2/TLR4 may enhance the migration of neutrophil into the lung, which may lead to elevated airway inflammation [43]. It was found that cigarette smoking extract (CSE) induced the release of one chemotactic factor CXCL8 by TLR9 activation. CSE promoted the degranulation of secondary neutrophils. These two processes contributed in the massive accumulation of neutrophils in the airway of COPD patients which led to inflammation [44].

6.3.4 Eosinophil

Eosinophil plays a vital role in triggering inflammation in COPD pathogenesis. The concentration of eosinophil is high in sputum and the airway walls [45]. The presence of interleukin-8 has a chemotactic property on eosinophil. It was reported that elevated level of eosinophil count was seen in sputum, but low in airway biopsy in COPD patients with chronic bronchitis syndrome compared to those who are asymptomatic [46–48].

Generally, in normal condition, eosinophil is concentrated in general circulation, but upon some stimuli like pro-inflammatory mediators (IL3, IL5, granulocyte-macrophage colony-stimulating factor). It activates and accumulates at the site of infection. Pro-inflammatory mediator IL5 is the predominant one, which plays a crucial role in triggering inflammation [49, 50]. Eosinophil migration is controlled by some chemotactic factors which include chemokine (C-C motif) ligand (CCL)5, CCL7, CCL11 (exotoxin), CCL-B (MCP-4), etc. Eosinophil had the ability to synthesis and secrete some pro-inflammatory cytokines and growth factors which include IL-2, IL-3, IL-4, IL-5, IL-10, IL-13, etc. and tumour necrosis factor and transforming growth factor (TGF) α/β which trigger immune response [51].

6.3.5 Dendritic Cells

Dendritic cells form a bridge in between the innate immunity and adaptive immunity. It recognises various danger signals like foreign bodies, infection and cell damage and initiates the adaptive immunity [52, 53]. Dendritic cells are mainly located in the airways and lungs, so it can easily recognise the foreign body [54]. Dendritic cells are antigen-presenting cells which activate the T cell. The T cell plays a vital role in the initiation of specific antigen response [55]. The activation of immune system by tobacco smoking is not clear; some glycoproteins present in tobacco was isolated which potentially activates the immune system [56]. An experiment in rat showed that upon exposure to cigarette smoke, there is an increase in the number of dendritic cells in rat lungs, and a similar condition is observed in the walls of COPD patients with smoking [57–59]. A similar experiment was done by Bruggemann et al. on mice; the animal was exposed to cigarette smoke for 30 min twice a day for 12 days. The data showed that high amount of dendritic cells is present in the airway as well as in the lung parenchyma [60].

Dendritic cell is subdivided into two, i.e. mature dendritic cells and immature dendritic cells. The immature dendritic cells are latter transferred into mature dendritic cells upon migration from peripheral to lymphoid tissues [61]. Pulmonary immature dendritic cells are more seen in COPD patients rather than the normal ones. Initially, the amount of mature dendritic cells is low in non-COPD patients with active smoking, but its level increases as smoking ceases [62]. The activated dendritic cells trigger adaptive immune response which includes T-helper cells, cluster of differentiation 4^+ ($CD4^+$), cluster of differentiation 8^+ ($CD8^+$), T cell, B cell, etc. responses. The overexpression of CD80 and CD86 and the release of cytokine interferon (IFN)-α were seen in the dendritic cells of COPD patients [63].

6.3.6 T-Lymphocytes

T cells are also called T-lymphocytes. They are produced in the thymus gland. In the thymus gland, they are matured and differentiated into T-helper cells ($CD4^+$ cells) and cytotoxic T cells/memory T cells ($CD8^+$ cells) [64]. The lung T cells were activated in the presence of biologic mediators. Upon antigenic activation, $CD4^+$ cells are activated and release cytokine. Studies show that $CD4^+$ cells have less contribution in developing COPD. The autoimmune response in COPD patients was triggered by $CD4^+$ T lymphocytes. The increased production of $CD4^+$ cells indicates chronic antigenic exposure [65]. The $CD4^+$ T-helper cells were subdivided into two, i.e. type 1 T-helper cells (Th1) and type 2 T-helper cells (Th2). The interferon and interleukins like IL-2 and TNF-γ, and TNF-α was produced by Th1, while the IL-4, IL-5, IL-9 and IL-13 were produced by Th2 cells [66]. The reports shows that the presence of Th1 and Th2 was higher in COPD patients than the healthy ones [67].

$CD8^+$ cells play a vital role in the pathogenesis in COPD. It is found in the airway epithelium and airway lumen of COPD patients. Previous reports revealed that $CD8^+$ lymphocytes secrete cytokines including increased production of IFN-γ, interferon-inducible protein-10 (IP-10) and monokine. These cytokines will cause tissue destruction by inducing matrix metalloproteinase production. $CD8^+$ lymphocytes cause cell death because of secretion of cytotoxic agents such as granzyme, perforin and FaS (a type 2 membrane protein under tumour necrosis factor family) [68, 69]. IP-10 will induce the production of matrix metalloproteinase-12, which causes degradation of elastin. The degraded elastin fragments will serve as chemotactic agents, leading to macrophage-mediated lung destruction [70].

Hodge et al. conducted a study on 48 COPD patients and found out the potential of $CD8/CD8^{null}$ cells in triggering COPD. In this study, the production of pro-inflammatory cytokine, granzyme, perforin and co-stimulatory molecules was measured in smoke-induced murine models. IFN-γ, OX-40, 4-IBB, CTLA4, granzyme and perforin were found to be in increased production than CD8/CD28^{null} cell stimulation. Overall, these produced cytokines by the stimulation of CD8/CD28^{null} cells play a vital role in autoimmune response or inflammatory response in COPD patients [71].

6.4 Inflammatory Mediators Involved in COPD

Not only the inflammatory cells like macrophages, T and B lymphocytes, etc. but also some inflammatory mediators have a role in the pathogenesis of COPD. It includes lipids, free radicals, cytokine and growth factors [72]. The mediators are secreted from the inflammatory cells and interact with the lung tissues.

6.4.1 Lipid Mediators

Prostaglandin E_2 is a lipid-derived substance from arachidonic acid that is generated by the action of cyclooxygenase (COX), which plays a vital role in the pathogenesis of COPD. It acts as a pro-inflammatory mediator. The level of PG E_2 is higher in COPD patients, which is due to the presence/stimulation of inflammatory cytokines [73]. COX-2 and PG E_2 were overexpressed in COPD patients which suggested the possibility of airway inflammation [74]. PG E_2 potentially stimulates the hyper-secretion of mucus in the airway tract and induces coughing in COPD cases [75].

6.4.2 Pro-inflammatory Cytokines

Pro-inflammatory cytokines like TNF-α, IL-1β, and IL-6 are at high concentrations in COPD and they induce inflammations. TNF-α, a pro-inflammatory cytokine which is activated by macrophage, causes damage to the lung tissues [76]. Nuclear factor kappa B (NF-κB) plays a great role in triggering inflammation, especially in COPD patients. And TNF-α is a potential activator of NF-κB.

IL-1B is also a pro-inflammatory cytokine which is activated by macrophage for the secretion of inflammatory cytokine in COPD patients. The level of IL-6 is high in sputum of patients. Certain studies showed that the level of C-reactive protein (CRP) is high, which can be effectively used as a biomarker in the detection of COPD [77]. IL-6 is a potential inducer of CRP. IL-32 is correlated with CD8$^+$ cells, tumour necrosis factor-α, etc., suggesting that IL-32 cytokine contributes in the pathogenesis of COPD [78].

Thymic stromal lymphopoietin (TSLP) from IL-7 cytokine family is overexpressed in the airway epithelium of COPD patients [79]. It has a major role in the stimulation of dendritic cells for the massive release of chemokine (C-C motif) ligand (CCL) 17 and CCL22, which attract CD4$^+$ Th-2 cells towards the airways [80]. It also triggers the release of some chemotactic factors like CXCL-10, and this attracts Th-1 and cytotoxic T (Tc)-type 1 cells to the lung tissues. TSLP is overexpressed in the smooth muscles of COPD patients [81].

6.4.3 Inflammasomes

The lung is continuously exposed to various foreign bodies like smoke, irritants, viruses, bacteria, etc., and the adaptive as well as innate immune system plays a key role in the protection of the lung from these foreign bodies. They protect the lungs by activation of special receptors like toll-like receptors (TLRs), nucleotide-binding oligomerisation domain-like receptors (NLRs), etc., which release massive amount of inflammatory mediators [82]. Interleukin-1 (IL-1) family which includes IL-1β and IL-18 produces high proteolytic activity. This is controlled by a large macromolecular signalling complex called inflammasomes [83–86]. Various inflammasomes are present, in that nucleotide-binding oligomerisation domain receptor protein 3 (NLRP3) is mainly involved in the triggering of various inflammatory reactions [87, 88]. Continuous activation of NLRP3 by irritants may lead to the progression of chronic pulmonary disease like COPD, asthma, etc. The higher concentration of inflammasomes was reported in lung tissues of COPD patients [89]. The effective blocking of the activation of NLRP3 is a better way to reduce the disease pathogenesis in COPD [90].

6.4.4 Role of Reactive Oxygen Species (ROS) and Oxidative Stress in COPD

Reactive oxygen species (ROS) plays a pivotal role in the pathogenesis of COPD. Some endogenous and exogenous factors like cigarette smoke, irritants, and ROS, respectively (Fig. 6.4), will cause an impairment in oxidant/antioxidant balance (oxidative stress), leading to damage of lipids, proteins and DNA and finally to cell death [91–93]. The lipids react with ROS and undergo lipid peroxidation followed by formation of malondialdehyde. This compound forms cross-linkage with the cellular proteins and inactivates them, followed by the stimulation of pulmonary inflammation, and finally leads to destruction of alveolar walls [94, 95]. 4-Hydroxynonenal is also an end product formed after lipid peroxidation, which has cytotoxic activity towards lung tissues [96]. The oxidative stress inhibits the anti-protease activity of AAT, and secretory leukocyte protease inhibitor's activity, thereby an imbalance between the protease/anti-protease activity occurs followed by destruction of lung tissues [97–100]. Upon oxidative stress, carbonylated proteins were produced, which leads to the formation of autoantibodies which trigger the inflammatory reactions [101].

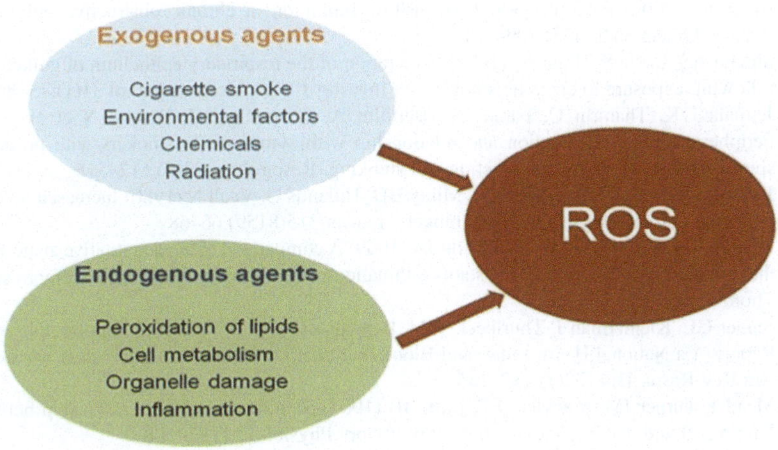

Fig. 6.4 Factors affecting ROS production

6.5 Conclusion

COPD is a chronic lung disorder which is characterised by emphysema, destruction of air walls, etc. The factors such as cigarette smoking, inhalation of chemicals, environmental factors, etc. may trigger the pathogenesis of COPD. Many inflammatory mediators were involved in the pathogenesis of COPD. They are activated and released from inflammatory cells like alveolar macrophages and neutrophils and also from some structural cells like epithelial cells, etc. These mediators not only alter the functions of lungs by tissue damage but also produce some systemic effects like muscle wasting and cachexia. A deep understanding of the cellular and molecular mechanisms involved in the pathogenesis of COPD is essential in the development of therapeutic strategy as well as drug delivery systems to effectively manage this condition.

References

1. Chung KF, Adcock IM (2008) Multifaceted mechanisms in COPD: inflammation, immunity, and tissue repair and destruction. Eur Respir J 31(6):1334–1356
2. Barnes PJ (2014) Cellular and molecular mechanisms of chronic obstructive pulmonary disease. Clin Chest Med 35(1):71–86
3. Kalyan M, Kanitkar S, Gupta A, Sharma D, Singh S (2016) Effect of passive smoking as a risk factor for chronic obstructive pulmonary disease in normal healthy women. Int J Res Med Sci 4(11):4970–4973
4. Rufino R, Silva JRL (2006) Cellular and biochemical bases of chronic obstructive pulmonary disease. J Bras Pneumol 32(3). http://www.scielo.br/scielo.php?script=sci_arttext&pid=S1806-37132006000900011&lng=en&nrm=iso&tlng=en

5. Hogg JC (2004) Pathophysiology of airflow limitation in chronic obstructive pulmonary disease. Lancet 364(9435):709–721
6. Simani AS, Inoue S, Hogg JC (1974) Penetration of the respiratory epithelium of guinea pigs following exposure to cigarette smoke. Lab Investig J Tech Methods Pathol 31(1):75–81
7. Jetmalani K, Thamrin C, Farah CS, Bertolin A, Chapman DG, Berend N et al (2018) Peripheral airway dysfunction and relationship with symptoms in smokers with preserved spirometry: small airway dysfunction and smoking. Respirology 23(5):512–518
8. Jones JG, Lawler P, Crawley JCW, Minty BD, Hulands G, Veall N (1980) Increased alveolar epithelial permeability in cigarette smokers. Lancet 315(8159):66–68
9. Dunnill MS, Massarella GR, Anderson JA (1969) A comparison of the quantitative anatomy of the bronchi in normal subjects, in status asthmaticus, in chronic bronchitis, and in emphysema. Thorax 24(2):176–179
10. Snider GL, Kleinerman J, Thurlbeck WM, Bengali ZH (1985) The definition of emphysema. Report of a National Heart, Lung, and Blood Institute, Division of Lung Diseases workshop. Am Rev Respir Dis 132(1):182–185
11. Mead J, Turner JM, Macklem PT, Little JB (1967) Significance of the relationship between lung recoil and maximum expiratory flow. J Appl Physiol 22(1):95–108
12. Kim WD, Eidelman DH, Izquierdo JL, Ghezzo H, Saetta MP, Cosio MG (1991) Centrilobular and panlobular emphysema in smokers: two distinct morphologic and functional entities. Am Rev Respir Dis 144(6):1385–1390
13. Heard BE (1959) Further observations on the pathology of pulmonary emphysema in chronic bronchitis. Thorax 14(1):58–70
14. Thurlbeck WM (1963) The incidence of pulmonary emphysema, with observations on the relative incidence and spatial distribution of various types of emphysema. Am Rev Respir Dis 87:206–215
15. Tuder RM, McGrath S, Neptune E (2003) The pathobiological mechanisms of emphysema models: what do they have in common? Pulm Pharmacol Ther 16(2):67–78
16. Sharafkhaneh A, Hanania NA, Kim V (2008) Pathogenesis of emphysema: from the bench to the bedside. Proc Am Thorac Soc 5(4):475–477
17. Burgel P-R, de Blic J, Chanez P, Delacourt C, Devillier P, Didier A et al (2009) Update on the roles of distal airways in asthma. Eur Respir Rev 18(112):80–95
18. Van Brabandt H, Cauberghs M, Verbeken E, Moerman P, Lauweryns JM, Van de Woestijne KP (1983) Partitioning of pulmonary impedance in excised human and canine lungs. J Appl Physiol 55(6):1733–1742
19. Reid LM (1955) Correlation of certain bronchographic abnormalities seen in chronic bronchitis with the pathological changes. Thorax 10(3):199–204
20. Matsuba K (1989) The changes in airways structure associated with reduced forced expiratory volume in one second. Eur Respir J 2(9):834–839
21. Bosken CH, Wiggs BR, Paré PD, Hogg JC (1990) Small airway dimensions in smokers with obstruction to airflow. Am Rev Respir Dis 142(3):563–570
22. Hogg JC, Buzatu L, Coxson HO (2004) The nature of small-airway obstruction in chronic obstructive pulmonary disease. N Engl J Med 350(26):2645–2653
23. Ryu JH, Myers JL, Swensen SJ (2003) Bronchiolar disorders. Am J Respir Crit Care Med 168 (11):1277–1292
24. Molfino NA (2004) Genetics of COPD. Chest 125(5):1929–1940
25. Nadel J (2001) The role of epidermal growth factor in mucus production. Curr Opin Pharmacol 1(3):254–258
26. Demoly P, Simony-Lafontaine J, Chanez P, Pujol JL, Lequeux N, Michel FB et al (1994) Cell proliferation in the bronchial mucosa of asthmatics and chronic bronchitis. Am J Respir Crit Care Med 150(1):214–217
27. de Boer WI, Hau CM, van Schadewijk A, Stolk J, van Krieken JHJM, Hiemstra PS (2006) Expression of epidermal growth factors and their receptors in the bronchial epithelium of subjects with chronic obstructive pulmonary disease. Am J Clin Pathol 125(2):184–192

28. Barnes PJ, Shapiro SD, Pauwels RA (2003) Chronic obstructive pulmonary disease: molecular and cellular mechanisms. Eur Respir J 22(4):672–688
29. Vlahos R (2014) Role of alveolar macrophages in chronic obstructive pulmonary disease. Front Immunol 5:435. http://journal.frontiersin.org/article/10.3389/fimmu.2014.00435/abstract
30. Russell REK, Thorley A, Culpitt SV, Dodd S, Donnelly LE, Demattos C et al (2002) Alveolar macrophage-mediated elastolysis: roles of matrix metalloproteinases, cysteine, and serine proteases. Am J Physiol-Lung Cell Mol Physiol 283(4):L867–L873
31. Lim S, Roche N, Oliver BG, Mattos W, Barnes PJ, Fan Chung K (2000) Balance of matrix metalloprotease-9 and tissue inhibitor of metalloprotease-1 from alveolar macrophages in cigarette smokers: regulation by interleukin-10. Am J Respir Crit Care Med 162(4):1355–1360
32. Russell REK, Culpitt SV, DeMatos C, Donnelly L, Smith M, Wiggins J et al (2002) Release and activity of matrix metalloproteinase-9 and tissue inhibitor of metalloproteinase-1 by alveolar macrophages from patients with chronic obstructive pulmonary disease. Am J Respir Cell Mol Biol 26(5):602–609
33. MacNee W (2005) Pathogenesis of chronic obstructive pulmonary disease. Proc Am Thorac Soc 2(4):258–266
34. Di Stefano A, Capelli A, Lusuardi M, Balbo P, Vecchio C, Maestrelli P et al (1998) Severity of airflow limitation is associated with severity of airway inflammation in smokers. Am J Respir Crit Care Med 158(4):1277–1285
35. Tomita K, Caramori G, Lim S, Ito K, Hanazawa T, Oates T et al (2002) Increased p21CIP1/WAF1 and B cell lymphoma leukemia-xL expression and reduced apoptosis in alveolar macrophages from smokers. Am J Respir Crit Care Med 166(5):724–731
36. Fadok VA, Bratton DL, Rose DM, Pearson A, Ezekewitz RAB, Henson PM (2000) A receptor for phosphatidylserine-specific clearance of apoptotic cells. Nature 405:6
37. Vandivier RW, Fadok VA, Hoffmann PR, Bratton DL, Penvari C, Brown KK et al (2002) Elastase-mediated phosphatidylserine receptor cleavage impairs apoptotic cell clearance in cystic fibrosis and bronchiectasis. J Clin Invest 109(5):661–670
38. Hoenderdos K, Condliffe A (2013) The neutrophil in chronic obstructive pulmonary disease. Too little, too late or too much, too soon? Am J Respir Cell Mol Biol 48(5):531–539
39. Guillon A, Jouan Y, Brea D, Gueugnon F, Dalloneau E, Baranek T et al (2015) Neutrophil proteases alter the interleukin-22-receptor-dependent lung antimicrobial defence. Eur Respir J 46(3):771–782
40. Jasper AE, McIver WJ, Sapey E, Walton GM (2019) Understanding the role of neutrophils in chronic inflammatory airway disease. F1000Research 8:557
41. Hiemstra PS (2013) Altered macrophage function in chronic obstructive pulmonary disease. Ann Am Thorac Soc 10(Supplement):S180–S185
42. Roh JS, Sohn DH (2018) Damage-associated molecular patterns in inflammatory diseases. Immune Network 18(4):e27
43. Sabroe I, Dower SK, Whyte MKB (2005) The role of toll-like receptors in the regulation of neutrophil migration, activation, and apoptosis. Clin Infect Dis 41(Supplement_7):S421–S426
44. Liu J, Pang Z, Wang G, Guan X, Fang K, Wang Z et al (2017) Advanced role of neutrophils in common respiratory diseases. J Immunol Res 2017:1–21
45. Turato G, Zuin R, Saetta M (2001) Pathogenesis and pathology of COPD. Respiration 68 (2):117–128
46. Snoeck-Stroband JB, Lapperre TS, Gosman MME, Boezen HM, Timens W, ten Hacken NHT et al (2008) Chronic bronchitis sub-phenotype within COPD: inflammation in sputum and biopsies. Eur Respir J 31(1):70–77
47. Saetta M, Mariani M, Panina-Bordignon P, Turato G, Buonsanti C, Baraldo S et al (2002) Increased expression of the chemokine receptor CXCR3 and its ligand CXCL10 in peripheral airways of smokers with chronic obstructive pulmonary disease. Am J Respir Crit Care Med 165(10):1404–1409

48. Saetta M, Turato G, Maestrelli P, Mapp CE, Fabbri LM (2001) Cellular and structural bases of chronic obstructive pulmonary disease. Am J Respir Crit Care Med 163(6):1304–1309
49. Molfino NA, Gossage K, Kolbeck R (2012) Molecular and clinical rationale for therapeutic targeting of interleukin-5 and its receptor. Clin Exp Allergy 42(5):712–737
50. Barnes PJ (2017) Cellular and molecular mechanisms of asthma and COPD. Clin Sci 131 (13):1541–1558
51. George L, Brightling CE (2016) Eosinophilic airway inflammation: role in asthma and chronic obstructive pulmonary disease. Ther Adv Chronic Dis 7(1):34–51
52. Cook PC, MacDonald AS (2016) Dendritic cells in lung immunopathology. Semin Immunopathol 38(4):449–460
53. Banchereau J, Briere F, Caux C, Davoust J, Lebecque S, Liu Y-J et al (2000) Immunobiology of dendritic cells. Annu Rev Immunol 18(1):767–811
54. Holt PG, Stumbles PA (2000) Regulation of immunologic homeostasis in peripheral tissues by dendritic cells: the respiratory tract as a paradigm. J Allergy Clin Immunol 105(3):421–429
55. Zhang M, Yin T, Lu Y, Feng H (2016) The application of cytidyl guanosyl oligodeoxynucleotide can affect the antitumor immune response induced by a combined protocol of cryoablation and dendritic cells in lewis lung cancer model. Med Sci Monit 22:1309–1317
56. Francus T, Francus Y, Weksler ME, Becker CG (1987) Effects of tobacco glycoprotein (TGP) on the immune system. Cell Immunol 105(1):1–8
57. Zeid NA, Konrad Muller H (1995) Tobacco smoke induced lung granulomas and tumors: association with pulmonary langerhans cells. Pathology (Phila) 27(3):247–254
58. Casolaro MA, Bernaudin J-F, Saltini C, Ferrans VJ, Crystal RG (1988) Accumulation of langerhans' cells on the epithelial surface of the lower respiratory tract in normal subjects in association with cigarette smoking. Am Rev Respir Dis 137(2):406–411
59. Soler P, Moreau A, Basset F, Hance AJ (1989) Cigarette smoking-induced changes in the number and differentiated state of pulmonary dendritic cells/langerhans cells. Am Rev Respir Dis 139(5):1112–1117
60. Brüggemann TR, Fernandes P, de Mendonça Oliveira L, Sato MN, de Arruda Martins M, Arantes-Costa FM (2017) Cigarette smoke increases CD8α+ dendritic cells in an ovalbumin-induced airway inflammation. Front Immunol 8:718
61. Zheng X, Zhang L, Chen J, Gu Y, Xu J, Ouyang Y (2018) Dendritic cells and Th17/Treg ratio play critical roles in pathogenic process of chronic obstructive pulmonary disease. Biomed Pharmacother 108:1141–1151
62. Tsoumakidou M, Bouloukaki I, Koutala H, Kouvidi K, Mitrouska I, Zakynthinos S et al (2009) Decreased sputum mature dendritic cells in healthy smokers and patients with chronic obstructive pulmonary disease. Int Arch Allergy Immunol 150(4):389–397
63. Nurwidya F, Damayanti T, Yunus F (2016) The role of innate and adaptive immune cells in the immunopathogenesis of chronic obstructive pulmonary disease. Tuberc Respir Dis 79(1):5
64. Bhat TA, Panzica L, Kalathil SG, Thanavala Y (2015) Immune dysfunction in patients with chronic obstructive pulmonary disease. Ann Am Thorac Soc 12:7
65. Lambers C, Hacker S, Posch M, Hoetzenecker K, Pollreisz A, Lichtenauer M et al (2009) T cell senescence and contraction of T cell repertoire diversity in patients with chronic obstructive pulmonary disease. Clin Exp Immunol 155(3):466–475
66. Baraldo S, Oliani KL, Turato G, Zuin R, Saetta M (2007) The role of lymphocytes in the pathogenesis of asthma and COPD. Curr Med Chem 14(21):2250–2256
67. Wei B, Sheng Li C (2018) Changes in Th1/Th2-producing cytokines during acute exacerbation chronic obstructive pulmonary disease. J Int Med Res 46(9):3890–3902
68. Duncan S (2008) Role of T-lymphocytes and pro-inflammatory mediators in the pathogenesis of chronic obstructive pulmonary disease. Int J Chron Obstruct Pulmon Dis 3:531–541
69. Kojima H, Shinohara N, Hanaoka S, Someya-Shirota Y, Takagaki Y, Ohno H et al (1994) Two distinct pathways of specific killing revealed by perforin mutant cytotoxic T lymphocytes. Immunity 1(5):357–364

70. Maeno T, Houghton AM, Quintero PA, Grumelli S, Owen CA, Shapiro SD (2007) CD8+ T cells are required for inflammation and destruction in cigarette smoke-induced emphysema in mice. J Immunol 178(12):8090–8096
71. Hodge G, Mukaro V, Reynolds PN, Hodge S (2011) Role of increased CD8/CD28null T cells and alternative co-stimulatory molecules in chronic obstructive pulmonary disease: increased CD8/CD28null T cells in COPD. Clin Exp Immunol 166(1):94–102
72. Barnes PJ (2004) Mediators of chronic obstructive pulmonary disease. Pharmacol Rev 56 (4):515–548
73. Dagouassat M, Gagliolo J-M, Chrusciel S, Bourin M-C, Duprez C, Caramelle P et al (2013) The cyclooxygenase-2–prostaglandin E2 pathway maintains senescence of chronic obstructive pulmonary disease fibroblasts. Am J Respir Crit Care Med 187(7):703–714
74. Chen Y, Chen P, Hanaoka M, Droma Y, Kubo K (2008) Enhanced levels of prostaglandin E2 and matrix metalloproteinase-2 correlate with the severity of airflow limitation in stable COPD. Respirology 13(7):1014–1021
75. Borchers MT, Carty MP, Leikauf GD (1999) Regulation of human airway mucins by acrolein and inflammatory mediators. Am J Physiol-Lung Cell Mol Physiol 276(4):L549–L555
76. Tanni SE, Pelegrino NR, Angeleli AY, Correa C, Godoy I (2010) Smoking status and tumor necrosis factor-alpha mediated systemic inflammation in COPD patients. J Inflamm 7(1):29
77. Heidari B (2012) The importance of C-reactive protein and other inflammatory markers in patients with chronic obstructive pulmonary disease. Casp J Intern Med 3(2):428–435
78. Calabrese F, Baraldo S, Bazzan E, Lunardi F, Rea F, Maestrelli P et al (2008) IL-32, a novel proinflammatory cytokine in chronic obstructive pulmonary disease. Am J Respir Crit Care Med 178(9):894–901
79. Ying S, O'Connor B, Ratoff J, Meng Q, Fang C, Cousins D et al (2008) Expression and cellular provenance of thymic stromal lymphopoietin and chemokines in patients with severe asthma and chronic obstructive pulmonary disease. J Immunol 181(4):2790–2798
80. Liu Y-J, Soumelis V, Watanabe N, Ito T, Wang Y-H, de Waal Malefyt R et al (2007) TSLP: an epithelial cell cytokine that regulates T cell differentiation by conditioning dendritic cell maturation. Annu Rev Immunol 25(1):193–219
81. Zhang K, Shan L, Rahman MS, Unruh H, Halayko AJ, Gounni AS (2007) Constitutive and inducible thymic stromal lymphopoietin expression in human airway smooth muscle cells: role in chronic obstructive pulmonary disease. Am J Physiol-Lung Cell Mol Physiol 293(2):L375–L382
82. Toews GB (2001) Cytokines and the lung. Eur Respir J 18(1):3–17
83. Wilson SP, Cassel SL (2010) Inflammasome mediated autoinflammatory disorders. Postgrad Med 122(5):125–133
84. Guo H, Callaway JB, JP-Y T (2015) Inflammasomes: mechanism of action, role in disease, and therapeutics. Nat Med 21(7):677–687
85. Lamkanfi M, Dixit VM (2012) Inflammasomes and their roles in health and disease. Annu Rev Cell Dev Biol 28(1):137–161
86. De Nardo D, De Nardo CM, Latz E (2014) New insights into mechanisms controlling the NLRP3 inflammasome and its role in lung disease. Am J Pathol 184(1):42–54
87. Jo E-K, Kim JK, Shin D-M, Sasakawa C (2016) Molecular mechanisms regulating NLRP3 inflammasome activation. Cell Mol Immunol 13(2):148–159
88. Eltom S, Stevenson CS, Rastrick J, Dale N, Raemdonck K, Wong S et al (2011) P2X7 receptor and caspase 1 activation are central to airway inflammation observed after exposure to tobacco smoke. PLoS One 6(9):e24097
89. Hosseinian N, Cho Y, Lockey RF, Kolliputi N (2015) The role of the NLRP3 inflammasome in pulmonary diseases. Ther Adv Respir Dis 9(4):188–197
90. Yang W, Ni H, Wang H, Gu H (2015) NLRP3 inflammasome is essential for the development of chronic obstructive pulmonary disease. Int J Clin Exp Pathol 8(10):13209–13216
91. MacNee W (2001) Oxidative stress and lung inflammation in airways disease. Eur J Pharmacol 429(1–3):195–207

92. Rahman I, MacNee W (1996) Role of oxidants/antioxidants in smoking-induced lung diseases. Free Radic Biol Med 21(5):669–681
93. Boukhenouna S, Wilson MA, Bahmed K, Kosmider B (2018) Reactive oxygen species in chronic obstructive pulmonary disease. Oxidative Med Cell Longev 2018:1–9
94. Rahman I (2006) Oxidative stress and redox regulation of lung inflammation in COPD. Eur Respir J 28(1):219–242
95. Siu GM, Draper HH (1982) Metabolism of malondialdehyde in vivo and in vitro. Lipids 17 (5):349–355
96. Breitzig M, Bhimineni C, Lockey R, Kolliputi N (2016) 4-Hydroxy-2-nonenal: a critical target in oxidative stress? Am J Physiol-Cell Physiol 311(4):C537–C543
97. Van Eeden SF, Sin DD (2013) Oxidative stress in chronic obstructive pulmonary disease: a lung and systemic process. Can Respir J 20(1):27–29
98. Hirata H, Takahashi A, Kobayashi S, Yonehara S, Sawai H, Okazaki T et al (1998) Caspases are activated in a branched protease cascade and control distinct downstream processes in Fas-induced apoptosis. J Exp Med 187(4):587–600
99. Perlmutter DH (1996) Alpha-1-antitrypsin deficiency: biochemistry and clinical manifestations. Ann Med 28(5):385–394
100. Pandey KC, De S, Mishra PK (2017) Role of proteases in chronic obstructive pulmonary disease. Front Pharmacol 8:512
101. Kirkham PA, Caramori G, Casolari P, Papi AA, Edwards M, Shamji B et al (2011) Oxidative stress–induced antibodies to carbonyl-modified protein correlate with severity of chronic obstructive pulmonary disease. Am J Respir Crit Care Med 184(7):796–802

Chronic Obstructive Pulmonary Disease: Molecular Basis of Pathogenesis and Targeted Therapeutic Approaches

7

Sushweta Mahalanobish, Sayanta Dutta, and Parames C. Sil

Abstract

Chronic obstructive pulmonary disease (COPD) is a heterogeneous inflammatory disorder that is strongly associated with cigarette smoking. Its prominent symptoms include chronic bronchitis, dyspnea, and emphysema. Goblet cell-mediated mucus hypersecretion and chronic inflammation in the respiratory tract lead to the congestion and narrowing of small airways and alveolar wall destruction (emphysema). Increased number of alveolar macrophages, cytotoxic T lymphocyte, and neutrophils along with upregulated multiple inflammatory mediators (cytokines, chemokines, growth factors, etc.) cause the exaggerated inflammatory reaction. Sustained oxidative stress additionally amplifies this inflammatory cascade. Moreover, increased elastolysis by several elastolytic enzymes, including serine, proteases, cathepsins, and matrix metalloproteinases, causes the alveolar wall destruction. Besides this, various genetic and epigenetic factors are also responsible for the early onset of COPD. This inflammatory disorder, in marked contrast to asthma, is resistant to corticosteroid therapy. Different physiological activities along with bronchodilator therapy, antioxidant intake, and oxygen supplementation can effectively subdue the disease outcome and improve the quality of life of COPD patients.

Keywords

Chronic obstructive pulmonary disease · Cigarette smoke · Lung · Inflammation · Oxidative stress

S. Mahalanobish · S. Dutta · P. C. Sil (✉)
Division of Molecular Medicine, Bose Institute, Kolkata, West Bengal, India
e-mail: parames@jcbose.ac.in

K. Dua et al. (eds.), *Targeting Cellular Signalling Pathways in Lung Diseases*,
https://doi.org/10.1007/978-981-33-6827-9_7

Abbreviations

BTVA	Bronchoscopic thermal vapour ablation
CHARGE	Cohorts for Heart and Aging Research in Genomic Epidemiology
DEFB1	Defensin beta 1
ENA-78	Epithelial neutrophil activating protein of 78 kDa
G-CSF	Granulocyte colony-stimulating factor
GM-CSF	Granulocyte-macrophage colony-stimulating factor
GRO-α	Growth-related oncoprotein
HAT	Histone acetylate
HDAC	Histone deacetylase
IL	Interleukin
IP-10	Interferon gamma-induced protein-10
I-TAC	Interferon-inducible T-cell alpha-chemoattractant
LTB-4	Leukotriene B4
LVR	Lung volume reduction
MCP-1	Monocyte chemoattractant protein-1
MIG	Monokine induced by gamma interferon
NE	Neutrophil elastase
PCNA	Proliferative cell nuclear antigen
PS	Phosphatidylserine
SOD	Superoxide dismutase
TLD	Targeted lung denervation
TNF	Tumour necrosis factor

7.1 Introduction

Chronic obstructive pulmonary disease (COPD) is a major health-care burden all over the world which causes the third worldwide mortality. Persistent inflammation in the airways and the destruction of lung parenchyma lead to the onset of chronic bronchitis and emphysema in COPD patients [1]. Shortness of breath, whizzing, and continuous production of cough along with sputum are the clinical symptoms of COPD patients. In recent years, advancement in lung imaging technique explored the detailed view of airway and lung parenchyma that help to design novel endoscopic interventions for emphysema management. No specific drug treatments can target emphysema except α1-antitrypsin treatment in α1-antitrypsin deficiency associated emphysema [2]. Based upon a large clinical trial studies, dual bronchodilator therapy has been used extensively than others anti-inflammatory corticosteroids therapy to prevent the airway suffering [3, 4]. In this chapter, we aimed to discuss the molecular basis of COPD pathogenesis and various aspects of therapeutic approaches to inhibit the disease progression.

7.2 Epidemiology

COPD is responsible for worldwide morbidity and mortality. Tobacco smoking is the main stimulating factor of COPD development. The estimation of COPD prevalence is quite difficult due to the adaption of different approaches (like spirometer-confirmed airflow survey) to calculate the disease incidence. In 2010, the prevalence of COPD was 384 million globally based on the spirometric analysis of the epidemiological cohort [5]. In 2015, the prevalence of COPD was almost about 174 million according to the Global Burden of Disease Study [6]. In Canada, the risk for COPD development is near about 28% [7]. These data indicate that the burden of COPD occurrence is similar to the risk of diabetes in high-income countries. Although, the prevalence of COPD is more common in men than women, the increase in popularity of tobacco smoking among women in well-developed countries also might lead a similar prevalence of COPD development in the future. It has been found that one-third of the COPD affected patients died due to cardiovascular diseases [8, 9]. The rate of global mortality in COPD is high due to an increased number of ageing populations. Majority of deaths in COPD occurs mainly in low-income and middle-income countries. Additionally, smoke inhalation from biomass fuel, second-hand smoke, and ambient particulate matter are also associated with COPD development in low-income countries [10–12]. It has been estimated that long-term exposure to indoor smoke from biomass fuels accounts for 35% of COPD cases in the middle- and low-income countries [13].

7.3 Aetiology of Disease Progression

Numerous discrete and overlapping phenotypes comprise the symptoms of COPD. Not all the cigarette smoker develops the airway obstruction which indicated there are other susceptibility factors like genetic, epigenetic, and environmental issues that also trigger the onset of COPD. A detailed investigation for a better understanding of complex cellular mechanisms and molecular pathways involved in COPD progression is needed for the early detection of this disease. Figure 7.1 shows various factors responsible for COPD progression, and their roles have been discussed below:

7.3.1 Inflammation

COPD is a heterogeneous inflammatory disorder where numerous inflammatory cells and various inflammatory mediators are involved to switch on the inflammatory cascade. Although the presence of various inflammatory cells in COPD has been recognized, the exact relationship between these cell types, their sequential appearance, and persistence are basically undetermined. After analysing the cellular profile of alveoli and small airway, it has been found that neutrophils, B and T lymphocytes, and macrophages are mainly responsible for COPD progression [14].

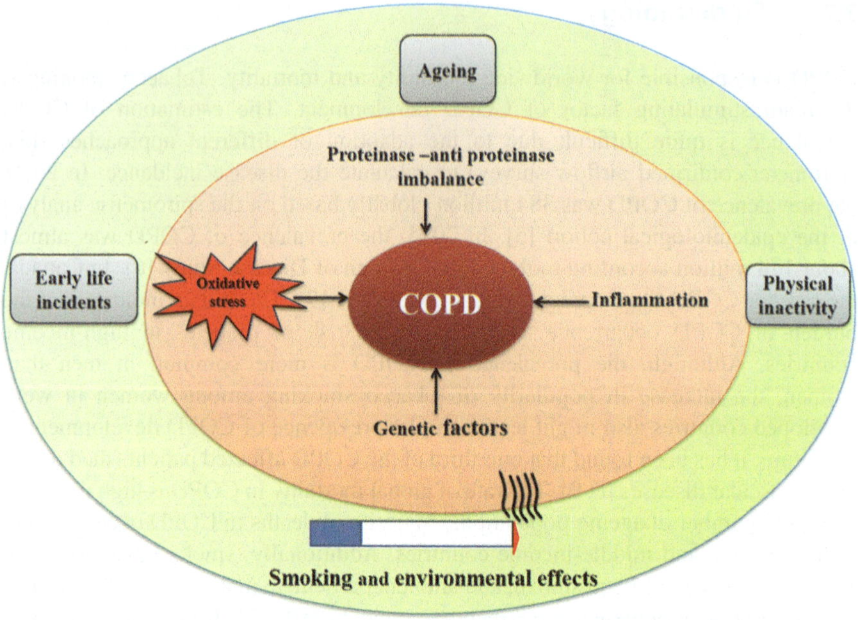

Fig. 7.1 The role of various factors responsible for COPD progression

7.3.1.1 Neutrophils

In COPD patients, though there is an increased percentage of neutrophils in the sputum and BAL fluid [15, 16], its abundance in lung parenchyma or airways is still relatively small [17]. The secretory product of neutrophils is serine proteases in nature that include cathepsin G, neutrophil elastase (NE), matrix metalloproteinase (MMP)-8, and MMP-9, all of which play a significant role in alveolar destruction. Besides their matrix degradation property, these proteases are also potent mucus stimulators for submucosal glands and goblet cells in the airway tract [18, 19]. Circulating neutrophils adhere to the endothelial wall of blood vessels with the help of E-selectins, and there is upregulated E-selectin expression in endothelial cells of airways in COPD patients [20]. From blood vessels, neutrophils transmigrate to the respiratory tract by the activity of various chemoattractants like interleukin (IL)-8, leukotriene B4 (LTB-4) growth-related oncoprotein (GRO-α), epithelial neutrophil activating protein of 78 kDa (ENA-78), etc. The process of adhesion and transmigration of neutrophil differs between systemic and pulmonary circulation. Inside the respiratory tract, cytokines, such as granulocyte-macrophage colony-stimulating factor (GM-CSF) and granulocyte colony-stimulating factor (G-CSF), help to neutrophil survival. Due to significant difference in timing of neutrophil transition in different area of lungs, a different form of emphysema like centriacinar, panacinar, and paraseptal has been developed. The process of survival and apoptosis of neutrophils in the airways of COPD patients is not studied in detail. It has been found that smoking directly stimulates the generation of granulocyte and its subsequent release from bone marrow [21, 22]. The rate of lung function (ratio of forced expiratory volume and forced vital capacity) decline is directly proportional to increased neutrophil number in the lung parenchyma [15, 23]. The increased neutrophil accumulation leads to respiratory outburst which induces tissue damage and inflammation [24]. GM-CSF assist prolong survival of neutrophil, but the culture of neutrophils from sputum samples is very difficult. Although in some diseases like cystic fibrosis, bronchiectasis, etc. chronic neutrophilia is developed in airway, there is an absence of neutrophil elastolytic activity. This indicates the involvement of other factors in emphysema generation. Therefore, airway neutrophilia is more related to mucus hypersecretion in bronchitis other than alveolar destruction.

7.3.1.2 Macrophages

Macrophages account for most of the pathophysiological conditions associated with COPD [25, 26]. Patients with COPD show a marked increase in macrophages number in the airways, lung parenchyma, BAL fluid, and sputum of patients. There is a 25-fold increase in total macrophages in alveoli in patients with emphysema compared to normal smokers [14]. Moreover, in the site alveolar destruction, a huge number of macrophage have been identified in emphysema patients. Therefore, there is a strong correlation in macrophage number in airway and COPD severity [23]. The increased number of macrophages is due to monocyte infiltration from circulation in response to monocyte selective cytokines. The level of monocyte chemoattractant protein-1 (MCP-1) is upregulated in BAL fluid and sputum of COPD patients along with increased number of macrophages [27]. Cigarette smoke-induced activated macrophages release various inflammatory mediators like tumour necrosis factor (TNF)-α, IL-8, MCP-1, LTB-4, etc. Moreover, alveolar macrophage-mediated secretion of various elastolytic enzymes like MMP-2, MMP-9, MMP-12, cathepsins K, and S play an important role of elastolytic activity for COPD patients [28, 29]. CXC chemokines also attract monocytes via CXC chemokine receptor 2 (CXCR2), and the level of CXC chemokine GRO-α is noticeably upregulated in the BAL fluid and sputum of COPD patients [30]. Macrophages also release various chemokines like interferon c-inducible protein (IP-10), interferon-inducible T-cell alpha-chemoattractant (I-TAC), monokine induced by gamma interferon (MIG), all of which are potent chemoattractant for $CD8^+$ T_C 1 cells by interacting with CXCR3 receptor expressed on $CD8^+$ T_C 1 cells [31]. In COPD, the increased number of macrophage is due to increased monocytes recruitment from the circulation and their increased proliferation and long-term sustention inside the lung microenvironment. It has been found that there is upregulation of proliferative cell nuclear antigen (PCNA), a cell proliferation marker, in the lung tissue [32]. Moreover, there is an increased expression of anti-apoptotic protein BCL-X_L and p21CIP/WAF1 in the macrophages of smokers [32]. All of these results indicated the prolonged survival of macrophages in COPD patients.

The phagocytic activity of macrophages is responsible for providing the host defence mechanism. The presence of specific receptors in the macrophage cell

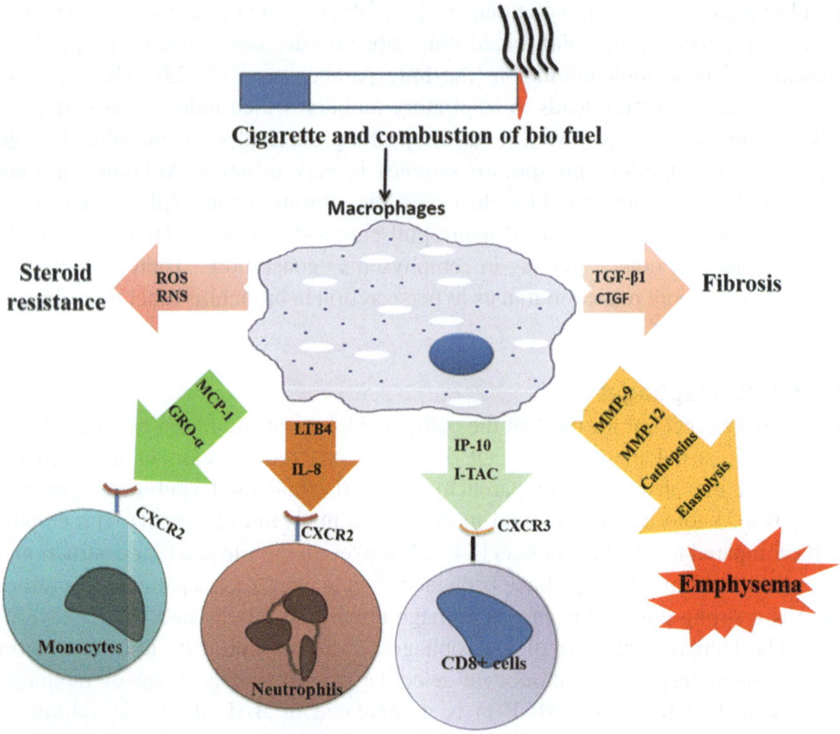

Fig. 7.2 The role of macrophages in COPD pathogenesis

surface helps to recognize the phosphatidylserine (PS) moiety present on the outer leaflet of apoptotic cells and subsequent phagocytosis to remove these damaged cells [33]. It has been found that in COPD patients, there is phagocytic impairment that results the bacterial upsurge in the respiratory tract. This is due to the increased neutrophils elastolytic activity which cleaves the PS receptor of macrophage and therefore impairs the macrophage ability to recognize and engulf the apoptotic neutrophil and resulting inflammation in the respiratory tract [34]. Figure 7.2 represents the role of macrophages in COPD pathogenesis.

7.3.1.3 T Lymphocytes

The central and peripheral airways of COPD patients show an increased number of total T lymphocytes. Mainly, increased number of CD8$^+$ T lymphocytes compared to CD4$^+$ T lymphocytes have been found in the airways of COPD patients [14, 17, 35–37]. The exact mechanism by which CD8$^+$ T lymphocytes and to a lesser extent CD4$^+$ T lymphocytes is upregulated in the airways and lung parenchyma in COPD is still not clear. The homing of T cells to lung required initial activation, adhesion, and chemoattractant-mediated migration. In the peripheral airways, T cells exhibit upregulated expression of CXCR3 which is recognized by chemoattractant like IP-10, MIG, I-TAC, etc. It has been found that in bronchiolar epithelial cells, the increased IP-10 expression caused CXCR3 receptor expressing $CD8^+$ T lymphocytes accumulation [31]. Dendritic cell migration from airways to regional lymph nodes stimulates the proliferation of $CD8^+$ and $CD4^+$ T lymphocytes. The number of $CD8^+$ T lymphocytes are often upregulated in airway infection due to the accumulation of bacterial pathogens in the lower respiratory tract [38]. Additionally, neutrophil and macrophage-mediated breakdown of lung tissue or cigarette smoke-induced alveolar destruction perhaps produces peptides which are recognized by T cells as an antigenic molecules and initiate T cell-mediated inflammation. Therefore, T cell-mediated lung injury due to an antigenic stimulus originating in the lung indicates COPD could be an autoimmune disease induced by smoking [39].

7.3.1.4 Eosinophils

In COPD, the exact role of eosinophil is undefined. Some study reports show that there is increased eosinophil in airways and BAL fluid of COPD patients, wherein in another studies, there was no significant increase of eosinophil numbers in sputum and BAL fluid [40]. However, the increase of eosinophil in COPD indicates a coexistence of asthma [41]. During acute exacerbations of chronic bronchitis, eosinophil upregulation is reported in bronchial biopsies [42]. Interestingly, one study showed that instead in the absence of eosinophil, the level of eosinophil basic proteins was upregulated in sputum of COPD. Perhaps, the neutrophil elastase activity caused the degranulation of eosinophil, and therefore these degranulated eosinophil are no longer detectable under microscope [43, 44].

7.3.1.5 Dendritic Cells

The role of dendritic cells to elicit the innate and adaptive immune response in our body is indispensable [45]. The presence of a rich network of dendritic cells in airways and lungs assists to detect inhaled foreign substances [46]. The activation of various inflammatory cells like macrophages, neutrophils, and B and T lymphocytes is dependent upon dendritic cells [47]. Therefore, the dendritic cell has a critical role to evoke inflammatory response in the lung after cigarette smoke exposure in COPD. It has been found that in the alveolar walls and the airways of smokers, a number of dendritic cells were highly upregulated [48]. Dendritic cell granulomata in the lung causes significant destruction of lung parenchyma (which is similar to emphysema) is known as pulmonary histiocytosis [49]. This disease ensues almost entirely in smokers. The role of dendritic cells in regulating other effector cells in COPD deserves the advance research. Figure 7.3 represents the role of dendritic cells to stimulate the proliferation of other inflammatory cells in COPD pathogenesis.

7.3.1.6 Epithelial Cells

Airway and alveolar epithelial cells are the major sources of proteases and inflammatory cytokines production. Upon cigarette smoke exposure, epithelial cells lead to the production of various inflammatory mediators like TNF-α, IL-1β, IL-8, GM-CSF, etc. [50–52]. In the small airways, epithelial cell-mediated TGF-β secretion causes local fibrosis in the small airway [53]. Interestingly, airway epithelial cells play an important role in airway defence mechanism. Goblet cell-mediated

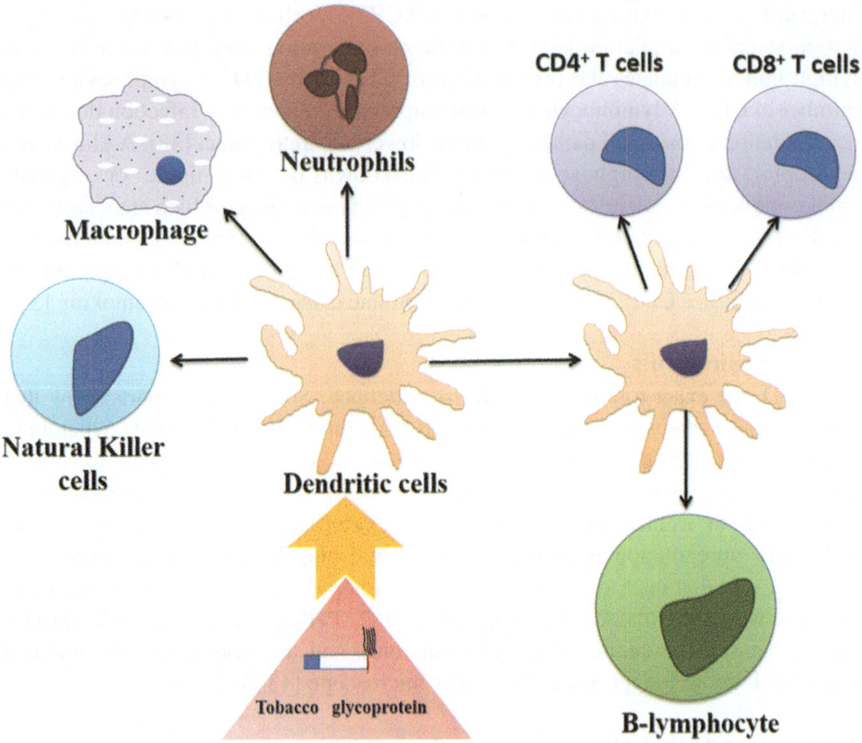

Fig. 7.3 The role of dendritic cells to stimulate the proliferation of other inflammatory cells in COPD pathogenesis

mucus secretion helps to trap the bacteria and inhaled particulate matters [54]. Besides goblet cells, epithelial cells produce various antimicrobial agents like defensins, cationic peptides as a part of innate defence mechanism [55]. Furthermore, epithelial cell-mediated immunoglobulin A transport arouses the adaptive immunity in the lung [56]. Exposure of cigarette smoke and other noxious substances disrupts the immune response of airway epithelial cells and increases the susceptibility of infection in the airways. In chronic bronchitis, increased proliferation of epithelial cells results squamous metaplasia which is correlated with PCNA upregulation in airway epithelial cells of smokers [57]. Moreover, the increased expression of epithelial growth factor receptors in airway epithelial cells suggests that various growth factors stimulate basal cell proliferation to promote squamous cell metaplasia and resulting bronchial carcinoma [58].

7.3.2 Oxidative Stress

Oxidative stress has a significant role in COPD pathogenesis [59–61]. Excessive production of reactive oxygen species (ROS) causes the damage of lipids, proteins, and DNA by impairing antioxidant defence mechanisms [62–64]. In the airway of COPD patients, various inflammatory cells like macrophages, neutrophils, and airway epithelial cells generate a huge amounts of ROS [60]. NADPH oxidase present inside the neutrophils produces the superoxide anions which is subsequently converted into hydrogen peroxide (H_2O_2) by superoxide dismutase, and then the catalase is converted H_2O_2 into water. In the presence of free iron, H_2O_2 interacts with free iron to form highly reactive hydroxyl radical (OH). Oxidative stress-mediated arachidonic acid oxidation generates isoprostanes which lead to bronchoconstriction and plasma exudation [65–67].

Antioxidants system efficiently counteracts with oxidants produced in the respiratory tract [40]. Major antioxidant enzymes in our lung environment include superoxide dismutase (SOD), catalase, and glutathione. It has been found that the concentration of reduced glutathione is increased several folds after cigarette smoking [68]. Moreover, epithelial cell-mediated glutathione peroxidase secretion is upregulated after cigarette smoke exposure to mitigate the ROS as well as RNS production [69].

In the airways, cigarette smoke-induced ROS production causes the activation of NF-κB molecule which in turn switches on several inflammatory genes and intensified inflammatory responses [70, 71]. In airway epithelial cells, oxidative stress upregulated histone acetyltransferase (HAT) activity that unwinds the chromatin structure and turns on the transcription of inflammatory gene [72, 73]. Activation of NF-κB and AP-1 causes the upregulation of IL-8 and other chemokines to set up neutrophilic inflammation. Moreover, oxidative stress impedes the activity of anti-proteases like α-1 antitrypsin, secretory protease leukocyte inhibitor, etc. to speed up the process of elastin breakdown of lung parenchyma [74]. Additionally, oxidative stress also triggers the apoptosis of alveolar epithelial and endothelial cells. Specially, stress-induced apoptosis of type 1 pneumocytes causes the emphysema development. The survival of alveolar endothelial cells is regulated by VEGF. Inhibition of VEGF receptors results apoptosis of alveolar septal cells and resulting emphysema [75].

The presence of oxidants in cigarette smoke and inhaled particulates also induces the mitogen-activated protein (MAP) kinase pathway. H_2O_2 triggers the activation of P-38 MAP kinase pathways which in turn upregulate the expression of inflammatory genes and stimulate the spreading of macrophages [76]. Besides cigarette smoke, these activated macrophages and neutrophils also act as a potent source of ROS [77]. ROS directly damage the DNA by inhibiting the activity of polyadenosine diphosphate ribose. Various markers of oxidative stress, such as 8-isoprostane and H_2O_2 have been detected in the exhaled breath condensate of COPD patients [78]. Immunocytochemistry study on COPD lung showed increase expression of 4-hydoxy-2-nonenal (product of lipid peroxidation) which forms an adduct with amino acid residues of protein [79, 80]. Moreover, oxidative stress negatively regulates the translocation of glucocorticoid receptors from cytoplasm to nucleus. In COPD patients, the marked reduction of histone deacetylase (HDAC2) expression of alveolar macrophage results reduced corticosteroid response as corticosteroid recruit HDAC2 for deacetylating hyperacetylated promoter region of inflammatory gene [81–83]. This limits the corticosteroids therapy in COPD as it does not render anti-inflammatory protection against COPD [81, 84–86].

7.3.3 Genetic Predisposition

Although the risk of COPD development is associated with smoking, other factors also stimulate COPD progression. Based on the pedigree analysis and twin study report, it has been found that at least 30% of COPD risk is associated with heritability [87–89]. First-degree relatives of COPD patients are more susceptible to develop COPD in comparison to non-smokers in the general population which indicates genetic determinants have a role to enhance the risk of COPD development. Different Mendelian syndromes like α1-antitrypsin deficiency and cutis laxa are associated with emphysema progression. In α1-antitrypsin deficiency, mutation of the Z allele in *SERPINA1* gene results a very little production of α1-antitrypsin. Genotype with PIZZ has only 10% of α1-antitrypsin level in serum which is highly insufficient to prevent neutrophil elastase destructive activity upon lung parenchyma. Mutation of S allele in SERPINA1 gene produces the enzyme moderately low level [90]. The PIZZ α1-antitrypsin deficiency is responsible for approximately 1% of COPD cases. Although the exact role of PISZ genotype in COPD is still controversial, several recent studies showed PISZ heterozygote individuals are under the risk of COPD if they smoke. Additionally, in cutis laxa, elastin (*ELN*) 29 gene mutation hampers the production of elastin which is a major constituent of lung extracellular matrix. Due to the disrupted formation and assembly of elastic fibre, lung flexibility to expand and contract is lost resulting emphysema [91]. Whole genome sequencing approach can identify the rare genetic determinants of COPD and comprehensive genomic analysis of chain smokers without COPD help to identify the possible genetic determinants of COPD resistance. In general population, genome-wide studies have been performed in the Cohorts for Heart and Aging Research in Genomic Epidemiology (CHARGE) and Spirometa consortia to explored the genetic loci associated with FEV_1 and its ratio to FVC [92, 93]. It has been found that eight loci (*ADAM19*, *AGER-PPT2*, *GPR126*, *HHIP*, *PTCH1*, *PID1*, *FAM13A*, and *HTR4*) are related with FEV_1/FVC, and one locus (*INTS12-GSTCD-NPNT*) is related with FEV_1. Moreover genome-wide association studies (GWAS) revealed the susceptible loci associated with COPD-related traits. For example, COPD-susceptible locus like 15q25 locus is strongly linked with lung cancer and nicotine addiction. Recently whole-exome sequence analysis revealed that a coiled coil domain containing 38 (*CCDC38*) variant, which is associated with ciliary function, provides COPD resistance. The presence of cilia in lung epithelial cells helps to sweep out inhaled particulates; therefore the ciliary abnormalities contribute the COPD development.

7.3.4 Epigenetics

Imbalance in HDAC/HAT activity leads to anomalies in histone acetylation and resulting change in gene expression [94]. In the case of histone methylation, the functional outcome of gene expression is dependent upon the target residues [95, 96]. In COPD, the presence of oxidants in cigarette smoke alters the activity of both HATs and HDACs to stimulate NF-κB-dependent inflammatory gene expression [82, 97]. In vitro exposure of human airway epithelial cells causes remarkable changes in histone acetylation and methylation and thereby modifies the cellular phenotype [98]. Short-term exposure of cigarette smoke in primary human airway epithelial cells causes phosphorylation and acetylation of NF-κB p65 and histone residue (H3S10-P, H3K9-Ac, and H4K12-Ac) and results in inflammation in the airway [99]. On the other hand, HDAC, suppresses the gene expression by recruiting corepressor proteins to switch off gene transcription [100]. It is the prerequisite molecule for corticosteroids to turn off numerous inflammatory genes. Therefore, decreased HDAC2 expression in alveolar macrophages of COPD patients leads to reduced glucocorticoid sensitivity. This reducing activity of HDAC2 is strongly correlated with upregulated inflammatory genes expression in alveolar macrophages [101, 102]. Moreover, the reduced level of total HDAC in peripheral lung and alveolar macrophages of COPD patients is associated with increased CXCL8 gene expression by increased histone acetylation of NF-κB binding site on the CXCL8 promoter region [103, 104]. Patients with advanced stage of COPD (GOLD stage 4), the HDAC2 expression is below 5% compared to healthy one. It has been found that the expression of HDACs 2, 3, 5, and 8 was decreased and HDACs 4 and 6 increased in a selective manner [95]. The level of defensin beta 1 (DEFB1) mRNA was upregulated in the lung of COPD patients which has strong correlation with HDAC5 mRNA expression. Chromatin immunoprecipitation revealed that upregulated DEFB1 expression was related with H3K4 methylation [105]. Furthermore, a reduced level of mRNA suggested that there may be transcriptional downregulation of HDAC2 gene or fragility of its mRNA. A recent GWAS on COPD showed that approximately 27,000 CpGs were examined for methylation [106]. It has been found that hypomethylation of CpGs occurs near to SERPINA1 gene (coding alpha1-antitrypsin) in COPD patient. Inclusively, 349 CpGs in 330 genes showed altered methylation along with disease progression. Genome-wide study to detect epigenetic change in COPD patients should be carried out [107].

7.3.5 Activity of Proteases

The balance between proteases and anti-protease is very much crucial for maintaining normal lung homeostasis. When this homeostasis is disrupted, anti-proteases are insufficient to neutralize the proteases mediated breakdown of connective tissue component. Elastin, the major connective tissue of lung parenchyma, is the main target of protease enzymes as its degradation causes loss of elasticity in lung parenchyma and resulting emphysema. Once emphysema has been developed, elastin cannot be regenerated in an active form. Excessive excretion of desmosine, involved in elastin cross-linking, has been reported to COPD patients due to elastin degradation [180]. Although, initially neutrophil elastase gained attention for emphysema, now it has been found other proteases also degradation [108].

7.3.5.1 Neutrophil Wlastase

Neutrophil elastase (NE) is a serine protease which is inhibited by the action of antitrypsin in the lung parenchyma. Tracheal instillation of NE stimulates neutrophil infiltration and induces emphysema progression in animal models [109]. NE induces the secretion of various chemokines like IL-8 and CCL-20 in airway epithelial cells [110]. It is a potent mucus secretor, and it stimulates MUC5AC expression in airway epithelial cell line [111]. NE causes phosphatidylserine receptor inactivation of macrophage and inhibits phagocytosis of apoptotic cells [89]. Although, several NE inhibitors are showed promising effect in guinea pig model of emphysema, but their effect on human emphysema is not sufficient. MR889, a neutrophil inhibitor administration, shows reduced urinary desmosine concentration in short-time COPD patients but no effect for long-term COPD smokers [112]. However, flurithromycin, erythromycin, and macrolide antibiotics inhibit NE activity and reduce mucus hypersecretion [113]. Besides elastase, neutrophils also contain two other serine proteases cathepsin G and proteinase 3 which have similar functional properties like neutrophil elastase [18, 19].

7.3.5.2 Cysteine Proteases

Cathepsins and lysosomal cysteine proteases, are involved in COPD progression [114, 115]. Interferon-γ induced the cathepsin S expression in different cell types, including smooth muscle cells. Overexpression of interferon-γ causes emphysema development by upregulating the cathepsins S, L, B, D, and H expression [116]. The use of cathepsin inhibitors reduced the emphysema progression that ensured its elastolytic potential [117]. Patients with emphysema show upregulated expression of cathepsin L in BAL fluid [118].

7.3.5.3 Matrix Metalloproteinases

Increasing evidences suggest that MMPs play a significant role in COPD patients [119]. The expression of MMP-1 and MMP-9 is upregulated in the BAL fluid of COPD patients [120–122]. In emphysema patients, type 2 pneumocytes showed increased expression of MMP-1 [123]. In alveolar macrophages, MMP-9 expression is increased at a significant level that also accounts for its elastolytic activity for emphysema progression in COPD patients [28]. Chronic exposure of cigarette smoke in MMP-12 null mice (macrophage metalloelastase) did not show emphysema progression [124]. When IL-13-mediated emphysema induction was done in these MMP-12 null mice, there was a marked reduction of IFN-γ and monocyte recruitment in lung tissue [116, 117]. This is due to MMP-mediated production of chemotactic peptides which stimulate the recruitment of macrophages in airways and lung parenchyma. MMPs cause activation of TGF-β from its latent form. Mice with

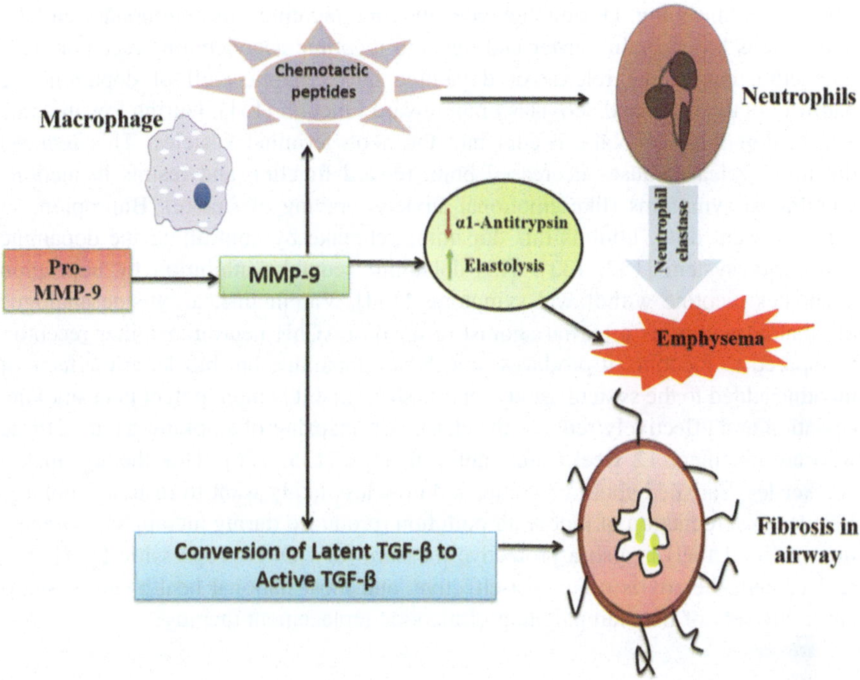

Fig. 7.4 Relationship in between elastolysis and TGF-β1-mediated fibrosis under MMP-9 induction

integrin avb6 knockdown (Itgb6-null mice) were unable to activate TGF-β and developed emphysema [125]. This indicates that under normal conditions, TGF-β downregulates MMP-12 expression, and in its absence, excessive MMP-12 leads to emphysema development. Cigarette smoke exposure in MMP-9 null mice still causes emphysema without development of small airway fibrosis [126]. Therefore, MMP-9-induced proteolytic activation of TGF-β1 links MMP-9-mediated elastolysis and TGF-β-mediated fibrosis progression [127, 128]. Figure 7.4 depicts a link in between elastolysis and TGF-β1-mediated fibrosis under MMP-9 induction.

7.4 Current Treatments and Their Drawbacks

7.4.1 Quit Smoking

In COPD patients, progressive decline of lung function and smoke-induced comorbidities (cardiovascular disease, lung cancer) have been found over time [129–131]. Quitting smoking is one of the major steps to prevent COPD development as 40% of COPD patients are chain smokers in nature. It has been found that nicotine has a very strong psycho-additive effect, so quitting smoking requires extensive counselling. During cigarette smoking, nicotine from pulmonary circulation crosses blood-brain barrier and binds to nicotinic acetylcholine receptor. This interaction causes the release of dopamine into synaptic cleft of dopaminergic pathway in the brain and activates brain reward function [131]. During low nicotine intake, dopamine reuptake occurs into the axon terminal vesicles. This reduced dopamine release causes decreased brain reward function and results in nicotine withdrawal symptoms (like emotional anxiety, craving of smoke). Bupropion, an antidepressant drug, inhibits this dopamine reuptake by controlling the dopamine transporter system [132, 133]. This dopamine reuptake inhibition by bupropion attenuates nicotine withdrawal symptoms [134]. Varenicline, a cytisine analogue of alkaloid nature, is a partial agonist of $\alpha 4$ $\beta 2$ nAChR neurotransmitter receptor. Compared to nicotine, it produces less (50%) dopamine but blocks any effects of nicotine added to the system. Study reports show that it is more potent than smoking cessation and effectively reduces the chance of relapsing of smoking habit of those who are abstinent 12 weeks after initial therapy [135, 136]. This therapy makes smoker less satisfied about cigarette, and they less likely want to resume smoking. Besides this, reduction of indoor air pollution (produced during incomplete combustion of fossil fuel) is also a protective measure for COPD progression [137]. The replacement therapy is quite cost-effective, and most national health-care systems are indifferent of free campaigning of nicotine replacement therapy.

7.4.2 Vaccination

Though, there is no vaccine designed for COPD specifically, vaccination against influenza virus and streptococcus pneumonia provides a secondary protection by minimizing the effect of COPD exacerbations due to respiratory infections. Influenza and pneumonia attack impair the lung functions, and people with COPD tend to develop pneumonia and other respiratory diseases due to compromised immune system. PCV13 and PPSV23 are widely used as pneumococcal vaccines which can be used to COPD patients to minimize the chance of exacerbation of COPD [138, 139].

7.4.3 Physical Activity

In daily life, physical workout is effective for a better quality of lifestyle. As comorbidity like diabetes, hypertension, and cardiovascular disease is associated with COPD, exercise can help to reduce the rate of morbidity and mortality of COPD patient. People who are physical inactive during early course of COPD are more susceptible of hospitalization and mortality [140, 141]. Physical exercise also helps to cope up with nicotine withdrawal symptoms like anxiety, depression, and insomnia. Walking for 15 min per day reduces 14% risk of all kinds of mortality, and that for 600 steps per day decreases the chance of hospitalization of COPD patients

[142, 143]. Therefore these data indicate little exercise can be clinically significant for COPD patient.

7.4.4 Pharmacological Treatment

Muscarinic receptor antagonist is an anticholinergic agent that blocks the activity of the muscarinic acetylcholine receptor. Inhalation of long-acting muscarinic antagonists (LAMAs) reduces air trapping and improves the rate of airflow in COPD patients [144]. Long-acting β2 agonists (LABAs), another long-acting bronchodilator, are used from several years that also improve the lung function and reduce exacerbations rate [145]. But the polymorphism of β2-adrenergic receptor causes LABA less effective over LAMA tiotropium [146–148]. Moreover, though inhalation of LABA is well tolerated, occasionally adverse effects like tremors and palpitations occur in COPD patients. Therefore, LAMA monotherapy is always preferred over LABA monotherapy in patients with COPD exacerbations [149]. Dual bronchodilation has advanced effect over monotherapy, but it is difficult to detect these advanced benefit based on patient report outcome. In one study, it has been reported that during the treatment of LAMA-LABA (indacaterol-glycopyrronium) dual therapy, significantly less exacerbation occurs than only LAMA monotherapy (glycopyrronium) [150]. Furthermore, the addition of LAMA and LABA along with inhaled corticosteroids provides a better outcome than LAMA corticosteroid alone [151]. Future research should be carried out to elucidate the dual effect of inhaled corticosteroids along with bronchodilator therapy for COPD management [152].

Though bronchodilator therapy (LAMA, LABA, or their combinational use) is generally safe, patients with cardiovascular problems are usually excluded from these treatments [153]. If a patient has a history of pneumonia and body mass index is less than 25 kg/m^2, inhaled corticosteroids treatment enhances the risk of pneumonia [154–156]. COPD patients with low eosinophil counts in blood also develop the chance of pneumonia attack when fluticasone (an inhaled corticosteroid) is used with LABA treatment [157]. This indicates that high counts of blood eosinophil can outweigh the risk of pneumonia during their corticosteroid therapy.

Though inhaled therapy shows optimized results, patients with severe COPD still have exacerbations. Anti-inflammatory drugs like phosphodiesterase-4 (PDE4) inhibitors show efficacy and acceptable tolerability COPD patients [129, 158]. Roflumilast, a second-generation PDE4 inhibitor, inhibits chemotaxis and cytokine production in COPD [159]. It reduces neutrophil and eosinophil counts in the sputum of COPD patients [160, 161]. In particular, in hospitalized patients due to COPD exacerbation, roflumilast intake with long-acting bronchodilators improves the lung function. Side effects of roflumilast include diarrhoea, weight loss, nausea, and headache and therefore need to be considered carefully. Macrolide therapy is another option especially for ex-smoker [162, 163]. Macrolides decrease mucus secretion and suppress cytokines, adhesion molecules, and chemokine production to inhibit neutrophils and macrophages in airway [164]. However, macrolide-related side effects limit its application as bacterial resistance developed after 1 year treatment period [165].

Over the past decade, considerable research has been carried out for the discovery of new therapeutic agents to mitigate COPD-related inflammation [166]. Targeting a single inflammatory pathway is not enough to block COPD-mediated inflammation [167, 168]. When CXCR2 antagonist is used to targeting the neutrophilic airway inflammation, it improves lung function and lessen COPD exacerbations, but severe side effects like neutropenia might exclude its further implication [169].

7.4.5 Emerging Antioxidants for an Alternative Therapeutic Approach

Various studies reveal the clinical effectiveness of antioxidant agents for COPD treatment [170]. A certain class of antioxidants such as thiol derivatives (e.g. N-acetyl cysteine [NAC], carbocysteine), Nrf2 activating agents (e.g. sulforaphane), antioxidant vitamins (e.g. vitamin C and E), and plant-derived phenolic compounds (e.g. curcumin and resveratrol) show promising effects to mitigate COPD-induced pathophysiological condition. Carbocysteine-mediated therapy on COPD patients for 1 year period significantly reduced the exacerbation rate of COPD [171]. The use of NAC also lessens the exacerbation frequency and improves the forced expiratory flow 25–75% by reducing small airway obstruction [172]. The antioxidant and anti-inflammatory properties of NAC effectively restores the natural innate immunity which is diminished due to cigarette smoking [173]. Therefore, thiol antioxidants can be considered as a COPD therapeutic "toolbox", to attenuate exacerbation frequency, which can slow down the progression of the disease.

Nrf2 has a significant inhibitory effect on ROS production, which is especially generated by cigarette smoking. Nrf2 acts as a transcription factor that binds to the antioxidant response element of stress response genes and thereby controls the activity of antioxidant and phase II cytoprotective enzyme. Sulforaphane, an ingredient of broccoli sprouts, is a potent Nrf2 activator, which has been used for clinical trials in COPD patients. The treatment of sulforaphane in alveolar macrophages of COPD patients helps to restore the glucocorticoid sensitivity in a glutathione-dependent mechanism to increase HDAC2 activity [174]. Similar to sulforaphane, application of curcumin in a monocytic cell line effectively restored the HDAC2 expression by mitigating cigarette smoke extract or peroxide-mediated steroid resistance [175]. Oral intake of curcumin can significantly reduce the number of neutrophils and macrophages as well as emphysemal break out in COPD mice model [176]. Resveratrol, another polyphenol present in red wine, also reduced bacterial endotoxin-induced cytokine production in alveolar macrophages of COPD patients [177].

Intake of bioflavonoids like catechins, flavonols, flavones, etc. in diet is positively related with increased FEV1/FVC ratio and inversely related with cough production [177]. Increased uptake of vitamin E in European patient inverse the trend of 20-year

COPD mortality rate with a 24% lower COPD mortality risk [178]. Consumption of food and vegetables over 3-year period increases the survival rate of COPD patients by increasing FEV1 percentage and decreasing mean annual exacerbation rate. Supplementation of vitamins at specific doses is helpful for COPD management. However, it has been found that high dose of a particular nutrient tempers respiratory pathology [179]. Intake of some vitamins (like vitamin A) at higher dose can enhance the chance of lung cancer development in susceptible individuals [180–182]. Supplementation of vitamin E or C up to 12 weeks in isolated white blood cells from stable COPD patients developed peroxide-induced DNA breakages resistance [183]. In another study, intravenous infusion of ascorbate in COPD patients improved the skeletal muscle fatigue resistance [184]. The effect of vitamin D for COPD management showed varying results. Vitamin D is an essential immune system regulator where reduced level of vitamin D causes the onset of various chronic diseases (autoimmune disease, infections, etc.) [185]. It has been found that when the level of 25-hydroxyvitamin D (25-OH vitamin D) in serum of COPD patients is below 20 ng/mL, patients showed an increased risk of exacerbations and declined lung function. Study reports showed that COPD patients generally developed vitamin D deficiency where vitamin D supplementation showed beneficial effects [186].

7.4.6 Interventional Treatments

Endoscopic and surgical lung volume reduction (LVR) is beneficial for patients with COPD and severe emphysema. LVR surgery effectively reduces hyperinflation and thereby improves lung performance. COPD patients suffering upper lobe emphysema, after LVR surgery improved their exercise ability and quality of life [187]. However, increased risk of morbidity and mortality is associated with LVR surgery. Endoscopic lung volume reduction (ELVR) has been used as an alternative, minimally invasive therapeutic approach with minor risk aptitude [1]. Several endoscopic processes like intrabronchial valves, coil implants, and thermal vapour ablation process provide satisfactory performance by improving respiratory mechanics. An implication of these techniques helps to determine the characteristics of emphysema (homogenous vs. heterogeneous, intact lobular fissure or collateral ventilation [188, 189]. Endoscopic valve therapy is used for lobar atelectasis of the emphysematous lung, and it provides the survival benefit [190, 191]. In bronchoscopic thermal vapour ablation (BTVA) technique, instillation of water vapour causes a local inflammatory reaction and results in scar in the lung and fibrosis after some weeks [192, 193]. Besides ELVR techniques, targeted lung denervation (TLD) removes the parasympathetic pulmonary nerves to induce anticholinergic action in COPD. The removal of parasympathetic innervations causes decreased acetylcholine activity and reduced airflow obstruction and mucus production in COPD patients [194].

7.4.7 Oxygen and Ventilator Support

Long-term oxygen supplement does not provide benefits for the patient with moderate resting or exercise-induced survival rate of COPD patients suffering in severe hypoxemia. However, oxygen therapy for long duration can be effective for the patients with moderate hypoxemia ($PaO_2 \leq 60$ mm Hg) or severe resting hypoxemia ($PaO_2 \leq 55$ mm Hg) and the symptoms of pulmonary hypertension, polycythaemia, or heart failure. Non-invasive positive airway pressure ventilation in patient with stable hypercapnia does not always give positive outcome. Non-invasive positive airway pressure ventilation with constant hypercapnia and high inspiratory pressure can cut minimum 20% partial pressure of CO_2 in arterial blood ($PaCO_2$) and improve the survival percentage [195].

7.4.8 Comorbidity Treatment

Smoking of tobacco intensifies the risk of lung cancer, coronary heart disease, and heart failure. Moreover, pulmonary artery disease, obesity, metabolic syndrome, diabetes, and systemic venous thromboembolism—all of these—are responsible for comorbidities of COPD patients. Chronic systemic inflammation is the common ground of all these extra-pulmonary manifestations. Therefore, the treatment of comorbidities is essential to manage the patient's survival. The treatment effect of some drugs that have been used in COPD is beyond the lung. Fluticasone furoate-vilanterol (inhaled corticosteroid and LABA) combinational therapy gives promising effect against hyperinflation associated under-filling of the heart in COPD patients [196]. Although, use of roflumilast can treat major cardiovascular problem in a pooled analysis [197], but no randomized controlled study has been conducted to measure the beneficial effect of roflumilast in cardiovascular problems associated with COPD. Patient with COPD and diabetes showed improved outcome after roflumilast treatment compared to patients with diabetes but without COPD [198]. Figure 7.5 represents the role of various therapeutic approaches against COPD.

7.5 Conclusion

COPD is a significant health concern all over the world for its irreversible nature. Rapid demographic growths in high-income countries and a significant upsurge of non-communicable diseases in low-income countries will escalate this health issue. Further research should be carried out in the field of genetic and molecular biology of this disease to innovate drug development.

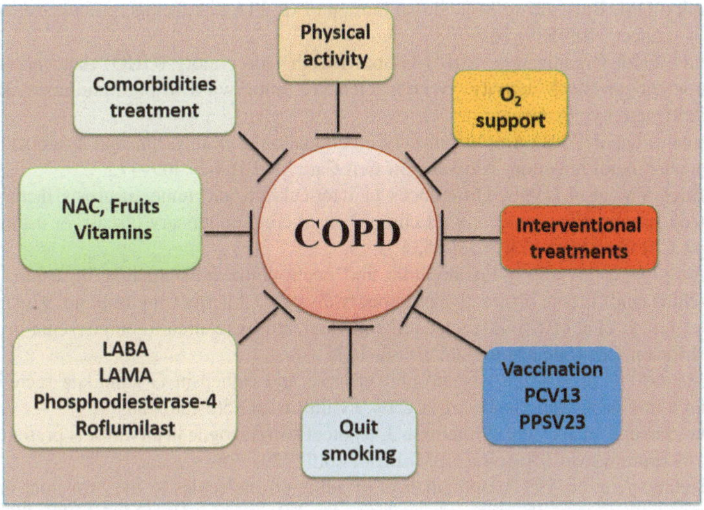

Fig. 7.5 The role of various therapeutic approaches against COPD

References

1. Rabe KF, Watz H (2017) Chronic obstructive pulmonary disease. Lancet 389 (10082):1931–1940
2. Chapman KR et al (2015) Intravenous augmentation treatment and lung density in severe α1 antitrypsin deficiency (RAPID): a randomised, double-blind, placebo-controlled trial. Lancet 386(9991):360–368
3. Wedzicha JA et al (2016) Indacaterol–glycopyrronium versus salmeterol–fluticasone for COPD. N Engl J Med 374(23):2222–2234
4. Pascoe SJ et al (2016) A phase III randomised controlled trial of single-dose triple therapy in COPD: the IMPACT protocol. Eur Respir J 48(2):320–330
5. Vos T et al (2016) Global, regional, and national incidence, prevalence, and years lived with disability for 310 diseases and injuries, 1990–2015: a systematic analysis for the Global Burden of Disease Study 2015. Lancet 388(10053):1545–1602
6. Adeloye D et al (2015) Global and regional estimates of COPD prevalence: systematic review and meta-analysis. J Glob Health 5(2)
7. Viegi G et al (2001) Epidemiology of chronic obstructive pulmonary disease (COPD). Respiration 68(1):4–19
8. Anthonisen NR et al (2005) The effects of a smoking cessation intervention on 14.5-year mortality: a randomized clinical trial. Ann Intern Med 142(4):233–239
9. Calverley PM et al (2007) Salmeterol and fluticasone propionate and survival in chronic obstructive pulmonary disease. N Engl J Med 356(8):775–789
10. Forouzanfar MH et al (2016) Global, regional, and national comparative risk assessment of 79 behavioural, environmental and occupational, and metabolic risks or clusters of risks, 1990–2015: a systematic analysis for the Global Burden of Disease Study 2015. Lancet 388 (10053):1659–1724
11. Salvi SS, Barnes PJ (2009) Chronic obstructive pulmonary disease in non-smokers. Lancet 374(9691):733–743

12. Mannino DM, Buist AS (2007) Global burden of COPD: risk factors, prevalence, and future trends. Lancet 370(9589):765–773
13. World Health Organization (2016) Global Health Observatory (GHO) data: prevalence of insufficient physical activity. WHO, Geneva. http://www.who.int/gho/ncd/risk_factors/physical_activity_text/en
14. Retamales I et al (2001) Amplification of inflammation in emphysema and its association with latent adenoviral infection. Am J Respir Crit Care Med 164(3):469–473
15. Keatings VM et al (1996) Differences in interleukin-8 and tumor necrosis factor-alpha in induced sputum from patients with chronic obstructive pulmonary disease or asthma. Am J Respir Crit Care Med 153(2):530–534
16. Lacoste J-Y et al (1993) Eosinophilic and neutrophilic inflammation in asthma, chronic bronchitis, and chronic obstructive pulmonary disease. J Allergy Clin Immunol 92(4):537–548
17. Finkelstein R et al (1995) Alveolar inflammation and its relation to emphysema in smokers. Am J Respir Crit Care Med 152(5):1666–1672
18. Sommerhoff C et al (1990) Neutrophil elastase and cathepsin G stimulate secretion from cultured bovine airway gland serous cells. J Clin Invest 85(3):682–689
19. Witko-Sarsat V et al (1999) Proteinase 3, a potent secretagogue in airways, is present in cystic fibrosis sputum. Am J Respir Cell Mol Biol 20(4):729–736
20. Di Stefano A et al (1994) Upregulation of adhesion molecules in the bronchial mucosa of subjects with chronic obstructive bronchitis. Am J Respir Crit Care Med 149(3):803–810
21. Terashima T et al (1997) Phagocytosis of small carbon particles (PM10) by alveolar macrophages stimulates the release of polymorphonuclear leukocytes from bone marrow. Am J Respir Crit Care Med 155(4):1441–1447
22. Stănescu D et al (1996) Airways obstruction, chronic expectoration, and rapid decline of FEV1 in smokers are associated with increased levels of sputum neutrophils. Thorax 51(3):267–271
23. Di Stefano A et al (1998) Severity of airflow limitation is associated with severity of airway inflammation in smokers. Am J Respir Crit Care Med 158(4):1277–1285
24. Richards GA et al (1989) Spirometric abnormalities in young smokers correlate with increased chemiluminescence responses of activated blood phagocytes. Am Rev Respir Dis 139 (1):181–187
25. Shapiro SD (1999) The macrophage in chronic obstructive pulmonary disease. Am J Respir Crit Care Med 160(supplement_1):S29–S32
26. Barnes PJ (2002) Current and future therapies for airway mucus hypersecretion. In Novartis Found Symp. Wiley Online Library
27. de Boer WI et al (2000) Monocyte chemoattractant protein 1, interleukin 8, and chronic airways inflammation in COPD. J Pathol 190(5):619–626
28. Russell RE et al (2002) Alveolar macrophage-mediated elastolysis: roles of matrix metalloproteinases, cysteine, and serine proteases. Am J Phys Lung Cell Mol Phys 283(4): L867–L873
29. Punturieri A et al (2000) Regulation of elastinolytic cysteine proteinase activity in normal and cathepsin K–deficient human macrophages. J Exp Med 192(6):789–800
30. Traves S et al (2002) Increased levels of the chemokines GROα and MCP-1 in sputum samples from patients with COPD. Thorax 57(7):590–595
31. Saetta M et al (2002) Increased expression of the chemokine receptor CXCR3 and its ligand CXCL10 in peripheral airways of smokers with chronic obstructive pulmonary disease. Am J Respir Crit Care Med 165(10):1404–1409
32. Tomita K et al (2002) Increased p21CIP1/WAF1 and B cell lymphoma leukemia-xL expression and reduced apoptosis in alveolar macrophages from smokers. Am J Respir Crit Care Med 166(5):724–731
33. Fadok VA et al (2000) A receptor for phosphatidylserine-specific clearance of apoptotic cells. Nature 405(6782):85–90
34. Vandivier RW et al (2002) Elastase-mediated phosphatidylserine receptor cleavage impairs apoptotic cell clearance in cystic fibrosis and bronchiectasis. J Clin Invest 109(5):661–670

35. Majo J, Ghezzo H, Cosio M (2001) Lymphocyte population and apoptosis in the lungs of smokers and their relation to emphysema. Eur Respir J 17(5):946–953
36. Saetta M et al (1999) CD8+ ve cells in the lungs of smokers with chronic obstructive pulmonary disease. Am J Respir Crit Care Med 160(2):711–717
37. O'Shaughnessy TC et al (1997) Inflammation in bronchial biopsies of subjects with chronic bronchitis: inverse relationship of CD8+ T lymphocytes with FEV1. Am J Respir Crit Care Med 155(3):852–857
38. Hill AT et al (2000) Association between airway bacterial load and markers of airway inflammation in patients with stable chronic bronchitis. Am J Med 109(4):288–295
39. Cosio MG, Majo J, Cosio MG (2002) Inflammation of the airways and lung parenchyma in COPD: role of T cells. Chest 121(5):160S–165S
40. Turato G, Zuin R, Saetta M (2001) Pathogenesis and pathology of COPD. Respiration 68 (2):117–128
41. Papi A et al (2000) Partial reversibility of airflow limitation and increased exhaled NO and sputum eosinophilia in chronic obstructive pulmonary disease. Am J Respir Crit Care Med 162 (5):1773–1777
42. Saetta M et al (1994) Airway eosinophilia in chronic bronchitis during exacerbations. Am J Respir Crit Care Med 150(6):1646–1652
43. Keatings VM, Barnes PJ (1997) Granulocyte activation markers in induced sputum: comparison between chronic obstructive pulmonary disease, asthma, and normal subjects. Am J Respir Crit Care Med 155(2):449–453
44. Liu H et al (1999) Neutrophil elastase and elastase-rich cystic fibrosis sputum degranulate human eosinophils in vitro. Am J Phys Lung Cell Mol Phys 276(1):L28–L34
45. Banchereau J et al (2000) Immunobiology of dendritic cells. Annu Rev Immunol 18 (1):767–811
46. Holt PG, Stumbles PA (2000) Regulation of immunologic homeostasis in peripheral tissues by dendritic cells: the respiratory tract as a paradigm. J Allergy Clin Immunol 105(3):421–429
47. Huang Q et al (2001) The plasticity of dendritic cell responses to pathogens and their components. Science 294(5543):870–875
48. Soler P et al (1989) Cigarette smoking-induced changes in the number and differentiated state of pulmonary dendritic cells/Langerhans cells. Am J Respir Crit Care Med 139(5):1112–1117
49. D'Hulst A, Vermeulen K, Pauwels R (2002) Cigarette smoke exposure causes increase in pulmonary dendritic cells. Am J Respir Crit Care Med 164:A604
50. Mio T et al (1997) Cigarette smoke induces interleukin-8 release from human bronchial epithelial cells. Am J Respir Crit Care Med 155(5):1770–1776
51. Hellermann GR et al (2002) Mechanism of cigarette smoke condensate-induced acute inflammatory response in human bronchial epithelial cells. Respir Res 3(1):1–8
52. Floreani AA et al (2003) Smoke and C5a induce airway epithelial intercellular adhesion molecule-1 and cell adhesion. Am J Respir Cell Mol Biol 29(4):472–482
53. Takizawa H et al (2001) Increased expression of transforming growth factor-β 1 in small airway epithelium from tobacco smokers and patients with chronic obstructive pulmonary disease (COPD). Am J Respir Crit Care Med 163(6):1476–1483
54. Adler KB, Li Y (2001) Airway epithelium and mucus: intracellular signaling pathways for gene expression and secretion. Am J Respir Cell Mol Biol 25(4):397–400
55. Aarbiou J, Rabe KF, Hiemstra PS (2002) Role of defensins in inflammatory lung disease. Ann Med 34(2):96–101
56. Pilette C et al (2001) Lung mucosal immunity: immunoglobulin—a revisited. Eur Respir J 18 (3):571–588
57. Demoly P et al (1994) Cell proliferation in the bronchial mucosa of asthmatics and chronic bronchitics. Am J Respir Crit Care Med 150(1):214–217
58. Franklin WA et al (2002) Epidermal growth factor receptor family in lung cancer and premalignancy. In: Seminars in oncology. Elsevier

59. Repine JE et al (1997) Oxidative stress in chronic obstructive pulmonary disease. Am J Respir Crit Care Med 156(2):341–357
60. MacNee W (2001) Oxidative stress and lung inflammation in airways disease. Eur J Pharmacol 429(1–3):195–207
61. Henricks PA, Nijkamp FP (2001) Reactive oxygen species as mediators in asthma. Pulm Pharmacol Ther 14(6):409–421
62. Das J et al (2009) Taurine protects rat testes against NaAsO2-induced oxidative stress and apoptosis via mitochondrial dependent and independent pathways. Toxicol Lett 187 (3):201–210
63. Manna P, Sinha M, Sil PC (2008) Arsenic-induced oxidative myocardial injury: protective role of arjunolic acid. Arch Toxicol 82(3):137–149
64. Manna P, Sinha M, Sil PC (2009) Taurine plays a beneficial role against cadmium-induced oxidative renal dysfunction. Amino Acids 36(3):417
65. Kawikova I et al (1996) 8-Epi-PGF2 alpha, a novel noncyclooxygenase-derived prostaglandin, constricts airways in vitro. American journal of respiratory and critical care medicine. 153 (2):590–596
66. Janssen L (2001) Isoprostanes: an overview and putative roles in pulmonary pathophysiology. Am J Phys Lung Cell Mol Phys 280(6):L1067–L1082
67. Van Der Vliet A et al (1999) Reactive nitrogen species and tyrosine nitration in the respiratory tract: epiphenomena or a pathobiologic mechanism of disease? Am J Respir Crit Care Med 160 (1):1–9
68. Cantin AM et al (1990) Antioxidant macromolecules in the epithelial lining fluid of the normal human lower respiratory tract. J Clin Invest 86(3):962–971
69. Comhair SA, Erzurum SC (2002) Antioxidant responses to oxidant-mediated lung diseases. Am J Phys Lung Cell Mol Phys 283(2):L246–L255
70. Adcock IM et al (1994) Oxidative stress induces NFκB DNA binding and inducible NOS mRNA in human epithelial cells. Biochem Biophys Res Commun 199(3):1518–1524
71. Rusznak C et al (1996) Ozone-induced mediator release from human bronchial epithelial cells in vitro and the influence of nedocromil sodium. Eur Respir J 9(11):2298–2305
72. Tomita K, Barnes P, Adcock I (2003) The effect of oxidative stress on histone acetylation and IL-8 release. Biochem Biophys Res Commun 301(2):572–577
73. Rahman I (2003) Oxidative stress, chromatin remodeling and gene transcription in inflammation and chronic lung diseases. J Biochem Mol Biol 36(1):95–109
74. Taggart C et al (2000) Oxidation of either methionine 351 or methionine 358 in α1-antitrypsin causes loss of anti-neutrophil elastase activity. J Biol Chem 275(35):27258–27265
75. Kasahara Y et al (2000) Inhibition of VEGF receptors causes lung cell apoptosis and emphysema. J Clin Invest 106(11):1311–1319
76. Ogura M, Kitamura M (1998) Oxidant stress incites spreading of macrophages via extracellular signal-regulated kinases and p38 mitogen-activated protein kinase. J Immunol 161 (7):3569–3574
77. Pryor WA, Stone K (1993) Oxidants in cigarette smoke radicals, hydrogen peroxide, peroxynitrate, and peroxynitrite A. Ann N Y Acad Sci 686(1):12–27
78. Praticò D et al (1998) Chronic obstructive pulmonary disease is associated with an increase in urinary levels of isoprostane F2 α-III, an index of oxidant stress. Am J Respir Crit Care Med 158(6):1709–1714
79. Rahman I et al (1996) Systemic oxidative stress in asthma, COPD, and smokers. Am J Respir Crit Care Med 154(4):1055–1060
80. Rahman I et al (2002) 4-Hydroxy-2-nonenal, a specific lipid peroxidation product, is elevated in lungs of patients with chronic obstructive pulmonary disease. Am J Respir Crit Care Med 166(4):490–495
81. Group, L.H.S.R (2000) Effect of inhaled triamcinolone on the decline in pulmonary function in chronic obstructive pulmonary disease. N Engl J Med 343(26):1902–1909

82. Ito K et al (2001) Cigarette smoking reduces histone deacetylase 2 expression, enhances cytokine expression, and inhibits glucocorticoid actions in alveolar macrophages. FASEB J 15 (6):1110–1112
83. Ito K, Barnes PJ, Adcock IM (2000) Glucocorticoid receptor recruitment of histone deacetylase 2 inhibits interleukin-1β-induced histone H4 acetylation on lysines 8 and 12. Mol Cell Biol 20(18):6891–6903
84. Vestbo J et al (1999) Long-term effect of inhaled budesonide in mild and moderate chronic obstructive pulmonary disease: a randomised controlled trial. Lancet 353(9167):1819–1823
85. Pauwels RA et al (1999) Long-term treatment with inhaled budesonide in persons with mild chronic obstructive pulmonary disease who continue smoking. N Engl J Med 340 (25):1948–1953
86. Burge PS et al (2000) Randomised, double blind, placebo controlled study of fluticasone propionate in patients with moderate to severe chronic obstructive pulmonary disease: the ISOLDE trial. BMJ 320(7245):1297–1303
87. Ingebrigtsen T et al (2010) Genetic influences on chronic obstructive pulmonary disease–a twin study. Respir Med 104(12):1890–1895
88. McCloskey SC et al (2001) Siblings of patients with severe chronic obstructive pulmonary disease have a significant risk of airflow obstruction. Am J Respir Crit Care Med 164 (8):1419–1424
89. Silverman EK et al (1998) Genetic epidemiology of severe, early-onset chronic obstructive pulmonary disease: risk to relatives for airflow obstruction and chronic bronchitis. Am J Respir Crit Care Med 157(6):1770–1778
90. Zorzetto M et al (2008) SERPINA1 gene variants in individuals from the general population with reduced α1-antitrypsin concentrations. Clin Chem 54(8):1331–1338
91. Kelleher CM et al (2005) A functional mutation in the terminal exon of elastin in severe, early-onset chronic obstructive pulmonary disease. Am J Respir Cell Mol Biol 33(4):355–362
92. Repapi E et al (2010) Genome-wide association study identifies five loci associated with lung function. Nat Genet 42(1):36–44
93. Artigas MS et al (2011) Genome-wide association and large-scale follow up identifies 16 new loci influencing lung function. Nat Genet 43(11):1082–1090
94. Khan SN, Khan AU (2010) Role of histone acetylation in cell physiology and diseases: an update. Clin Chim Acta 411(19–20):1401–1411
95. Greer EL, Shi Y (2012) Histone methylation: a dynamic mark in health, disease and inheritance. Nat Rev Genet 13(5):343–357
96. Rice JC, Allis CD (2001) Histone methylation versus histone acetylation: new insights into epigenetic regulation. Curr Opin Cell Biol 13(3):263–273
97. Yang S-R et al (2006) Cigarette smoke induces proinflammatory cytokine release by activation of NF-κB and posttranslational modifications of histone deacetylase in macrophages. Am J Phys Lung Cell Mol Phys 291(1):L46–L57
98. Alashkar Alhamwe B et al (2020) Epigenetic Regulation of Airway Epithelium Immune Functions in Asthma. Front Immunol 11:1747
99. Sundar IK et al (2012) Mitogen- and stress-activated kinase 1 (MSK1) regulates cigarette smoke-induced histone modifications on NF-κB-dependent genes. PLoS One 7(2):e31378
100. Barnes PJ (2005) Targeting histone deacetylase 2 in chronic obstructive pulmonary disease treatment. Expert Opin Ther Targets 9(6):1111–1121
101. Garten RJ et al (2009) Antigenic and genetic characteristics of swine-origin 2009 A (H1N1) influenza viruses circulating in humans. Science 325(5937):197–201
102. Ito T et al (2001) A comprehensive two-hybrid analysis to explore the yeast protein interactome. Proc Natl Acad Sci 98(8):4569–4574
103. Szulakowski P et al (2006) The effect of smoking on the transcriptional regulation of lung inflammation in patients with chronic obstructive pulmonary disease. Am J Respir Crit Care Med 174(1):41–50

104. Ito K et al (2005) Decreased histone deacetylase activity in chronic obstructive pulmonary disease. N Engl J Med 352(19):1967–1976
105. Andresen E et al (2011) Increased expression of beta-defensin 1 (DEFB1) in chronic obstructive pulmonary disease. PLoS One 6(7):e21898
106. Morrow JD et al (2016) DNA methylation profiling in human lung tissue identifies genes associated with COPD. Epigenetics 11(10):730–739
107. Yuan C et al (2017) Genetic polymorphism and chronic obstructive pulmonary disease. Int J Chron Obstruct Pulmon Dis 12:1385–1393
108. Stockley RA (2001) Proteases and antiproteases. In: Novartis Found Symp. Wiley Online Library
109. Senior RM et al (1977) The induction of pulmonary emphysema with human leukocyte elastase. Am Rev Respir Dis 126(3):469–475
110. Voynow JA et al (1999) Neutrophil elastase increases MUC5AC mRNA and protein expression in respiratory epithelial cells. Am J Phys Lung Cell Mol Phys 276(5):L835–L843
111. Fischer BM, Voynow JA (2002) Neutrophil elastase induces MUC 5AC gene expression in airway epithelium via a pathway involving reactive oxygen species. Am J Respir Cell Mol Biol 26(4):447–452
112. Luisetti M et al (1996) MR889, a neutrophil elastase inhibitor, in patients with chronic obstructive pulmonary disease: a double-blind, randomized, placebo-controlled clinical trial. Eur Respir J 9(7):1482–1486
113. Goswami SK, Kivity S, Marom Z (1990) Erythromycin inhibits respiratory glycoconjugate secretion from human airways in vitro. Am Rev Respir Dis 141(72–78):1
114. Turk B, Turk V, Turk B, Turk D (2001) Lysosomal cysteine proteases: facts and opportunities. EMBO J 20(17):4629–4633
115. Chapman HA, Riese RJ, Shi G-P (1997) Emerging roles for cysteine proteases in human biology. Annu Rev Physiol 59(1):63–88
116. Wang Z et al (2000) Interferon γ induction of pulmonary emphysema in the adult murine lung. J Exp Med 192(11):1587–1600
117. Zheng T et al (2000) Inducible targeting of IL-13 to the adult lung causes matrix metalloproteinase–and cathepsin-dependent emphysema. J Clin Invest 106(9):1081–1093
118. Takeyabu K et al (1998) Cysteine proteinases and cystatin C in bronchoalveolar lavage fluid from subjects with subclinical emphysema. Eur Respir J 12(5):1033–1039
119. Shapiro SD, Senior RM (1999) Matrix metalloproteinases: matrix degradation and more. Am J Respir Cell Mol Biol 20(6):1100–1102
120. Finlay GA et al (1997) Matrix metalloproteinase expression and production by alveolar macrophages in emphysema. Am J Respir Crit Care Med 156(1):240–247
121. Betsuyaku T et al (1999) Neutrophil granule proteins in bronchoalveolar lavage fluid from subjects with subclinical emphysema. Am J Respir Crit Care Med 159(6):1985–1991
122. Culpitt SV et al (1999) Effect of high dose inhaled steroid on cells, cytokines, and proteases in induced sputum in chronic obstructive pulmonary disease. Am J Respir Crit Care Med 160 (5):1635–1639
123. Imai K et al (2001) Human collagenase (matrix metalloproteinase-1) expression in the lungs of patients with emphysema. Am J Respir Crit Care Med 163(3):786–791
124. Hautamaki RD et al (1997) Requirement for macrophage elastase for cigarette smoke-induced emphysema in mice. Science 277(5334):2002–2004
125. Morris DG et al (2003) Loss of integrin αvβ6-mediated TGF-β activation causes Mmp12-dependent emphysema. Nature 422(6928):169–173
126. Lanone S et al (2002) Overlapping and enzyme-specific contributions of matrix metalloproteinases-9 and-12 in IL-13–induced inflammation and remodeling. J Clin Invest 110(4):463–474
127. Yu Q, Stamenkovic I (2000) Cell surface-localized matrix metalloproteinase-9 proteolytically activates TGF-β and promotes tumor invasion and angiogenesis. Genes Dev 14(2):163–176

128. Dallas SL et al (2002) Proteolysis of latent-TGFβ binding protein-1 by osteoclasts-a cellular mechanism for release of TGFβ from bone matrix. J Biol Chem 277(24):21352–21360
129. Anthonisen NR et al (2005) The effects of a smoking cessation intervention on 14.5-year mortality: a randomized clinical trial. Ann Intern Med 142(4):233–239
130. Scanlon PD et al (2000) Smoking cessation and lung function in mild-to-moderate chronic obstructive pulmonary disease. The Lung Health Study. Am J Respir Crit Care Med 161(2 Pt 1):381–390
131. Rooney CP et al (2001) Anti-proteinase 3 antibody activation of neutrophils can be inhibited by alpha1-antitrypsin. Am J Respir Cell Mol Biol 24(6):747–754
132. Wilkes S (2008) The use of bupropion SR in cigarette smoking cessation. Int J Chron Obstruct Pulmon Dis 3(1):45–53
133. Nomikos GG et al (1992) Effects of chronic bupropion on interstitial concentrations of dopamine in rat nucleus accumbens and striatum. Neuropsychopharmacology 7(1):7–14
134. Benowitz NL (2009) Pharmacology of nicotine: addiction, smoking-induced disease, and therapeutics. Annu Rev Pharmacol Toxicol 49:57–71
135. Spruit MA et al (2015) Pulmonary rehabilitation and physical activity in patients with chronic obstructive pulmonary disease. Am J Respir Crit Care Med 192(8):924–933
136. Beeh KM et al (2014) Aclidinium improves exercise endurance, dyspnea, lung hyperinflation, and physical activity in patients with COPD: a randomized, placebo-controlled, crossover trial. BMC Pulm Med 14(1):209
137. Romieu I et al (2009) Improved biomass stove intervention in rural Mexico: impact on the respiratory health of women. Am J Respir Crit Care Med 180(7):649–656
138. Vila-Corcoles A et al (2012) Clinical effectiveness of 23-valent pneumococcal polysaccharide vaccine against pneumonia in patients with chronic pulmonary diseases: a matched case-control study. Hum Vaccin Immunother 8(5):639–644
139. Stanbrook MB (2017) In COPD, injectable polyvalent pneumococcal vaccines reduce risk for community-acquired pneumonia. Ann Intern Med 166(10):JC52–JC52
140. Garcia-Aymerich J et al (2006) Regular physical activity reduces hospital admission and mortality in chronic obstructive pulmonary disease: a population based cohort study. Thorax 61(9):772–778
141. Waschki B et al (2011) Physical activity is the strongest predictor of all-cause mortality in patients with COPD: a prospective cohort study. Chest 140(2):331–342
142. Rochester CL et al (2015) An official American Thoracic Society/European Respiratory Society policy statement: enhancing implementation, use, and delivery of pulmonary rehabilitation. Am J Respir Crit Care Med 192(11):1373–1386
143. Demeyer H et al (2016) The minimal important difference in physical activity in patients with COPD. PLoS One 11(4):e0154587
144. Malerba M et al (2019) Single Inhaler LABA/LAMA for COPD. Front Pharmacol 10:390
145. Rossi A, Khirani S, Cazzola M (2008) Long-acting beta2-agonists (LABA) in chronic obstructive pulmonary disease: efficacy and safety. Int J Chron Obstruct Pulmon Dis 3 (4):521–529
146. Vogelmeier C et al (2011) Tiotropium versus salmeterol for the prevention of exacerbations of COPD. N Engl J Med 364(12):1093–1103
147. Decramer ML et al (2013) Once-daily indacaterol versus tiotropium for patients with severe chronic obstructive pulmonary disease (INVIGORATE): a randomised, blinded, parallel-group study. Lancet Respir Med 1(7):524–533
148. Rabe KF et al (2014) Effect of ADRB2 polymorphisms on the efficacy of salmeterol and tiotropium in preventing COPD exacerbations: a prespecified substudy of the POET-COPD trial. Lancet Respir Med 2(1):44–53
149. Rodrigo GJ et al (2017) LABA/LAMA combinations versus LAMA monotherapy or LABA/ICS in COPD: a systematic review and meta-analysis. Int J Chron Obstruct Pulmon Dis 12:907–922

150. Wedzicha JA et al (2013) Analysis of chronic obstructive pulmonary disease exacerbations with the dual bronchodilator QVA149 compared with glycopyrronium and tiotropium (SPARK): a randomised, double-blind, parallel-group study. Lancet Respir Med 1(3):199–209
151. Singh D et al (2016) Single inhaler triple therapy versus inhaled corticosteroid plus long-acting β2-agonist therapy for chronic obstructive pulmonary disease (TRILOGY): a double-blind, parallel group, randomised controlled trial. Lancet 388(10048):963–973
152. Pascoe SJ et al (2016) A phase III randomised controlled trial of single-dose triple therapy in COPD: the IMPACT protocol. Eur Respir J 48(2):320–330
153. Lahousse L et al (2016) Cardiac effects of current treatments of chronic obstructive pulmonary disease. Lancet Respir Med 4(2):149–164
154. Vestbo J et al (2016) Fluticasone furoate and vilanterol and survival in chronic obstructive pulmonary disease with heightened cardiovascular risk (SUMMIT): a double-blind randomised controlled trial. Lancet 387(10030):1817–1826
155. Finney L et al (2014) Inhaled corticosteroids and pneumonia in chronic obstructive pulmonary disease. Lancet Respir Med 2(11):919–932
156. Crim C et al (2009) Pneumonia risk in COPD patients receiving inhaled corticosteroids alone or in combination: TORCH study results. Eur Respir J 34(3):641–647
157. Pavord ID et al (2016) Blood eosinophil count and pneumonia risk in patients with chronic obstructive pulmonary disease: a patient-level meta-analysis. Lancet Respir Med 4 (9):731–741
158. (2016) Global, regional, and national life expectancy, all-cause mortality, and cause-specific mortality for 249 causes of death, 1980–2015: a systematic analysis for the Global Burden of Disease Study 2015. Lancet 388(10053):1459–1544
159. Kawamatawong T (2017) Roles of roflumilast, a selective phosphodiesterase 4 inhibitor, in airway diseases. J Thorac Dis 9(4):1144–1154
160. Sun XJ, He ZY (2019) Macrolides for treatment of chronic obstructive pulmonary disease. Chin Med J 132(11):1261–1263
161. Calverley PM et al (2007) Salmeterol and fluticasone propionate and survival in chronic obstructive pulmonary disease. N Engl J Med 356(8):775–789
162. Albert RK et al (2011) Azithromycin for prevention of exacerbations of COPD. N Engl J Med 365(8):689–698
163. Han MK et al (2014) Predictors of chronic obstructive pulmonary disease exacerbation reduction in response to daily azithromycin therapy. Am J Respir Crit Care Med 189 (12):1503–1508
164. Kanoh S, Rubin BK (2010) Mechanisms of action and clinical application of macrolides as immunomodulatory medications. Clin Microbiol Rev 23(3):590–615
165. Vogelmeier CF et al (2017) Global strategy for the diagnosis, management, and prevention of chronic obstructive lung disease 2017 report. GOLD executive summary. Am J Respir Crit Care Med 195(5):557–582
166. Barnes PJ, Hansel TT (2004) Prospects for new drugs for chronic obstructive pulmonary disease. Lancet 364(9438):985–996
167. Watz H et al (2014) Efficacy and safety of the p38 MAPK inhibitor losmapimod for patients with chronic obstructive pulmonary disease: a randomised, double-blind, placebo-controlled trial. Lancet Respir Med 2(1):63–72
168. Brightling CE et al (2014) Benralizumab for chronic obstructive pulmonary disease and sputum eosinophilia: a randomised, double-blind, placebo-controlled, phase 2a study. Lancet Respir Med 2(11):891–901
169. Rennard SI et al (2015) CXCR2 antagonist MK-7123. A phase 2 proof-of-concept trial for chronic obstructive pulmonary disease. Am J Respir Crit Care Med 191(9):1001–1011
170. Mahalanobish S et al (2020) Melatonin induced suppression of ER stress and mitochondrial dysfunction inhibited NLRP3 inflammasome activation in COPD mice. Food Chem Toxicol 144:111588

171. Zheng JP et al (2008) Effect of carbocisteine on acute exacerbation of chronic obstructive pulmonary disease (PEACE Study): a randomised placebo-controlled study. Lancet 371 (9629):2013–2018
172. Tse HN et al (2013) High-dose N-acetylcysteine in stable COPD: the 1-year, double-blind, randomized, placebo-controlled HIACE study. Chest 144(1):106–118
173. Wu W et al (2011) Cigarette smoke extract suppresses the RIG-I-initiated innate immune response to influenza virus in the human lung. Am J Physiol Lung Cell Mol Physiol 300(6): L821–L830
174. Malhotra D et al (2011) Denitrosylation of HDAC2 by targeting Nrf2 restores glucocorticosteroid sensitivity in macrophages from COPD patients. J Clin Invest 121 (11):4289–4302
175. Meja KK et al (2008) Curcumin restores corticosteroid function in monocytes exposed to oxidants by maintaining HDAC2. Am J Respir Cell Mol Biol 39(3):312–323
176. Suzuki M et al (2009) Curcumin attenuates elastase- and cigarette smoke-induced pulmonary emphysema in mice. Am J Physiol Lung Cell Mol Physiol 296(4):L614–L623
177. Knobloch J et al (2011) Resveratrol impairs the release of steroid-resistant cytokines from bacterial endotoxin-exposed alveolar macrophages in chronic obstructive pulmonary disease. Basic Clin Pharmacol Toxicol 109(2):138–143
178. Walda IC et al (2002) Diet and 20-year chronic obstructive pulmonary disease mortality in middle-aged men from three European countries. Eur J Clin Nutr 56(7):638–643
179. Schols A (2013) Nutrition as a metabolic modulator in COPD. Chest 144(4):1340–1345
180. Omenn GS et al (1996) Effects of a combination of beta carotene and vitamin A on lung cancer and cardiovascular disease. N Engl J Med 334(18):1150–1155
181. Omenn GS et al (1996) Risk factors for lung cancer and for intervention effects in CARET, the Beta-Carotene and Retinol Efficacy Trial. J Natl Cancer Inst 88(21):1550–1559
182. Goodman GE et al (2004) The Beta-Carotene and Retinol Efficacy Trial: incidence of lung cancer and cardiovascular disease mortality during 6-year follow-up after stopping beta-carotene and retinol supplements. J Natl Cancer Inst 96(23):1743–1750
183. Wu TC et al (2007) Vitamin E and vitamin C supplementation in patients with chronic obstructive pulmonary disease. Int J Vitam Nutr Res 77(4):272–279
184. Rossman MJ et al (2013) Ascorbate infusion increases skeletal muscle fatigue resistance in patients with chronic obstructive pulmonary disease. Am J Physiol Regul Integr Comp Physiol 305(10):R1163–R1170
185. Lehouck A et al (2012) High doses of vitamin D to reduce exacerbations in chronic obstructive pulmonary disease: a randomized trial. Ann Intern Med 156(2):105–114
186. Tsiligianni IG, van der Molen T (2010) A systematic review of the role of vitamin insufficiencies and supplementation in COPD. Respir Res 11(1):171
187. Darwiche K, Aigner C (2018) Clinical management of lung volume reduction in end stage emphysema patients. J Thorac Dis 10(Suppl 23):S2732–s2737
188. Shah PL et al (2017) Lung volume reduction for emphysema. Lancet Respir Med 5 (2):147–156
189. Herth FJ et al (2016) Endoscopic lung volume reduction: an expert panel recommendation. Respiration 91(3):241–250
190. Chapman KR et al (2015) Intravenous augmentation treatment and lung density in severe α1 antitrypsin deficiency (RAPID): a randomised, double-blind, placebo-controlled trial. Lancet 386(9991):360–368
191. Wedzicha JA et al (2016) Indacaterol-glycopyrronium versus salmeterol-fluticasone for COPD. N Engl J Med 374(23):2222–2234
192. (2016) Global, regional, and national incidence, prevalence, and years lived with disability for 310 diseases and injuries, 1990–2015: a systematic analysis for the Global Burden of Disease Study 2015. Lancet 388(10053):1545–1602
193. Adeloye D et al (2015) Global and regional estimates of COPD prevalence: systematic review and meta-analysis. J Glob Health 5(2):020415

194. Lamprecht B et al (2015) Determinants of underdiagnosis of COPD in national and international surveys. Chest 148(4):971–985
195. Köhnlein T et al (2014) Non-invasive positive pressure ventilation for the treatment of severe stable chronic obstructive pulmonary disease: a prospective, multicentre, randomised, controlled clinical trial. Lancet Respir Med 2(9):698–705
196. Stone IS et al (2016) Lung deflation and cardiovascular structure and function in chronic obstructive pulmonary disease. A randomized controlled trial. Am J Respir Crit Care Med 193 (7):717–726
197. White WB et al (2013) Cardiovascular safety in patients receiving roflumilast for the treatment of COPD. Chest 144(3):758–765
198. Wouters EF et al (2012) Effect of the phosphodiesterase 4 inhibitor roflumilast on glucose metabolism in patients with treatment-naive, newly diagnosed type 2 diabetes mellitus. J Clin Endocrinol Metab 97(9):E1720–E1725

Exploring the 'Dormancy Activation Switch' in the Tumour Microenvironment for Metastatic Lung Cancer: The Possible Role of MicroRNA

8

Amnani Aminuddin, Siti Fathiah Masre, Sin-Yeang Teow, and Pei Yuen Ng

Abstract

Lung cancer produces the highest mortality rate among cancers, mainly due to its late diagnosis accompanied by single or multiple organ metastasis to the lymph node, pleura, liver, brain and/or bones. In this chapter, we will review the major cellular signalling pathways that promote the proliferation and metastasis of lung cancer cells to each of these metastatic sites, hopefully to shine the light on the possible therapeutic targets that could be developed against these pathways for better therapeutic outcome. Cellular pathways including MAPK, CXCR4/CXCL12, ALK, Rho/ROCK, PI3K and RANK/RANKL signalling will be elaborated in relation to their roles in lymph node, pleura, liver, brain and bone metastasis of lung cancer, respectively. The blooming understanding of microRNA (miRNA) in lung cancer in these recent years has shown that it could play bi-faceted roles in modulating oncogenic pathways. Thus, we also briefly described the candidates of miRNA that may either promote or inhibit the key pathways in lung cancer with various metastatic sites.

Both Siti Fathiah Masre and Amnani Aminuddin are first authors of the book chapter.

A. Aminuddin · P. Y. Ng (✉)
Drug and Herbal Research Centre, Faculty of Pharmacy, Universiti Kebangsaan Malaysia, Kuala Lumpur, Malaysia
e-mail: pyng@ukm.edu.my

S. F. Masre
Biomedical Science Programme, Centre for Toxicology and Health Risk Studies, Faculty of Health Sciences, Universiti Kebangsaan Malaysia, Kuala Lumpur, Malaysia

S.-Y. Teow
School of Medical and Life Sciences, Sunway University, Jalan Universiti, Subang Jaya, Selangor Darul Ehsan, Malaysia

K. Dua et al. (eds.), *Targeting Cellular Signalling Pathways in Lung Diseases*,
https://doi.org/10.1007/978-981-33-6827-9_8

Keywords

Metastatic lung cancer · Tumour microenvironment · Cellular signalling · MicroRNA · Cancer dormancy

8.1 Lung Cancer

Lung cancer is the leading cause of cancer death globally, with higher cancer-related deaths even after breast, colon and prostate cancer incidences combined [1]. Most cases of lung cancer among the US population are often diagnosed at a late stage, up to 22% regional spread to lymph nodes and 57% cancer that has metastasised to distant sites, which leads to poor prognosis and low 5-year patient survival of 31.7% and 5.8%, respectively (https://seer.cancer.gov/statfacts/html/lungb.html).

There are two major subtypes of lung cancer which are non-small cell lung cancer (NSCLC) and small cell lung cancer (SCLC). The NSCLC is the dominant subtype that accounts for 95% of all cases of lung cancer, which can be further subdivided into three main subtypes based on their histomorphology characteristics, namely, lung adenocarcinoma (AD), lung squamous cell carcinoma (SCC) and large cell carcinoma (LCC) (Fig. 8.1) [2]. Based on SEER statistics, the 5-year survival rate for NSCLC is 22% (AD, 27.1%; SCC, 21.1%; LCC, 17.8%), with an even lower rate of 6.6% for SCLC despite numerous preclinical and clinical studies (https://seer.cancer.gov/explorer/application.html).

Based on all lung cancer cases from MSK-IMPACT clinical sequencing initiative [3] (data accessed from cBio Cancer Genomics Portal database in September 2020), major differences in the genomic profiles between NSCLC and SCLC are frequently observed among the patients (Fig. 8.1). The genetic mutations in *RB1*, *TP53*, *NOTCH*, *KMT2D*, *CREBBBP* and *ATRX* are most likely to be present in the SCLC than in the NSCLC cases. Moreover, the genomic structural variation, such as the changes in the gene copy number within a chromosome, is also common in SCLC compared to NSCLC cases. Several significant copy number variations (CNVs) in SCLC cases involve the homozygous deletions of *RB1*, *TGFBR2*, *TNFRSF14* and *FLT4* genes and the amplifications of *MYCL* and *MPL* genes. Nevertheless, approximately 7.6% of the NSCLC cases had an amplification of the *EGFR* gene, while it was totally absent in SCLC cases. These genomic differences may explain their variability in molecular and metastatic capacities.

Although most of lung cancer cases are the NSCLC types, the percentage of metastatic lung cancer is slightly higher in SCLC than in NSCLC with 60.9% and 40.3% of their total cases (Fig. 8.2). Although different subtypes of lung cancer may have their preferred metastatic location [4–7], the most common metastatic sites for SCLC include lymph node followed by the liver, mediastinum of thoracic cavity and brain (52, 16, 10 and 8% of total metastatic SCLC cases, respectively). Meanwhile, the major metastatic site for NSCLC is the lymph node with 34.8% of the total

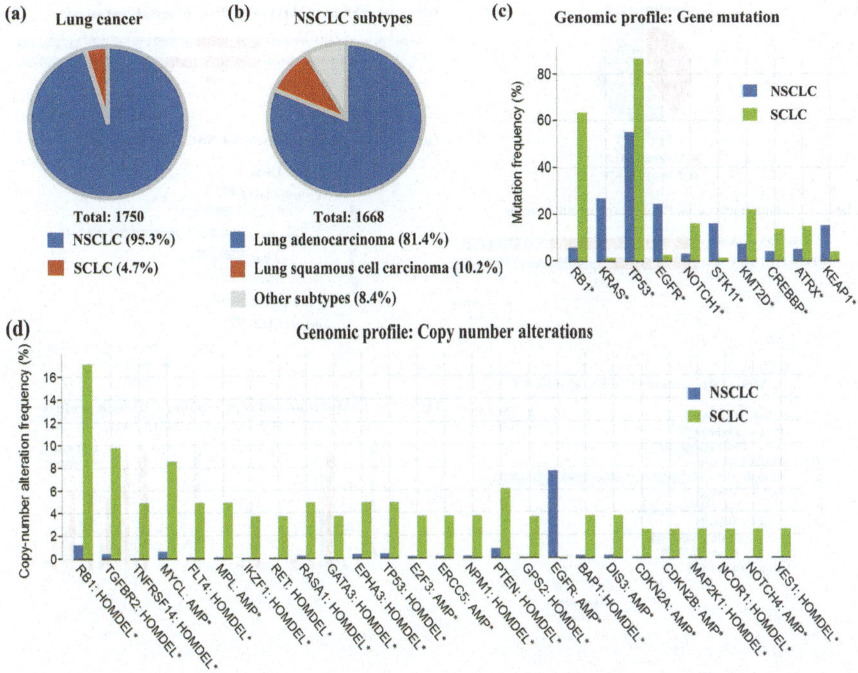

Fig. 8.1 General summary of lung cancer cases based on large-scale cancer genomic studies from cBioPortal databases. (**a**) The total clinical samples were from 1750 patients with SCLC and NSCLC. (**b**) NSCLC can be divided into various subtypes including its major subtypes, namely, lung adenocarcinoma and lung squamous cell carcinoma. Significant genomic profile differences, particularly (**c**) the gene mutation and (**d**) the copy number alterations profiles, between NSCLC and SCLC. *The difference between groups is considered significant when q-value derived from Benjamini-Hochberg procedure is less than 0.05. Abbreviation: *NSCLC* non-small cell lung cancer, *SCLC* small cell lung cancer

metastatic NSCLC cases, followed by the liver (12.4%), pleura (11.3%), brain (9.2%) and bone (7.4%) (Fig. 8.2) [8, 9].

Additionally, the metastatic cases of NSCLC is more frequent in AD compared to SCC subtypes (41.2 and 26.5% of their total cases, respectively). Successively after lymph node (AD, 35.1%; SCC, 31.1%), the major metastatic sites for AD are the liver (12.3%), pleura (12.2%), brain (9.5%) and bone (7.2%), while for SCC are the bone (15.6%), pleura (8.9%), brain (6.7%) and adrenal gland (4.4%) (Fig. 8.2). In-depth genomic data analysis demonstrated that CNVs are the common events in metastatic AD. They include the homozygous deletions of *CDKN2A* and *CDKN2B* and the amplifications of *FOXA1*, *NKX2-1*, *MYC* and *RECQL4* (Fig. 8.2). Nevertheless, no variation between genomic profiles of primary and metastatic SCLC and NSCLC SCC can be observed as mutations such as *EGFR* and *KRAS* could be present in both the primary and metastatic tumour [10].

Tobacco smoking is best recognised as one of the risk factors for lung cancer. In 2019, the WHO European Region estimated that the attributable proportion for the

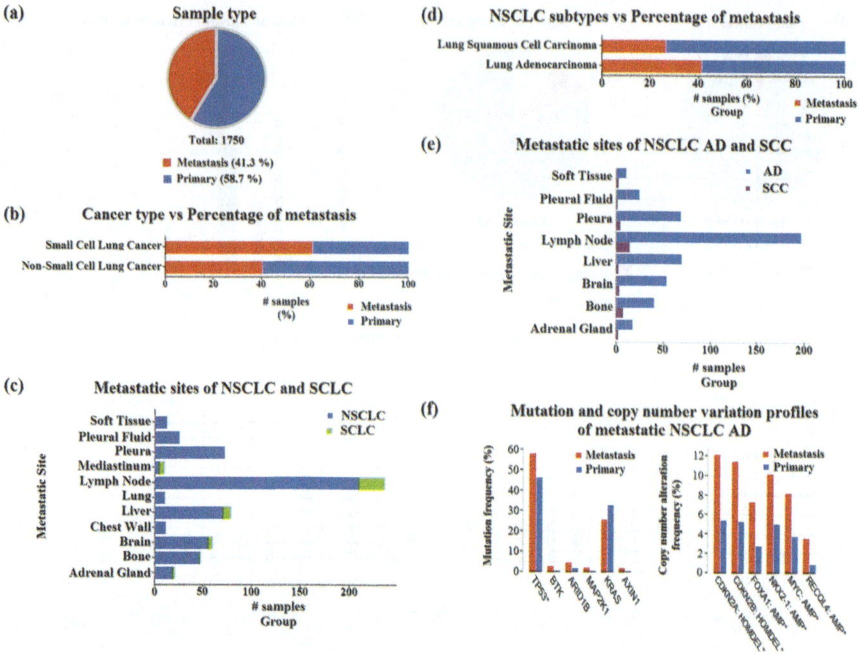

Fig. 8.2 Summary of metastatic lung cancer cases based on large-scale cancer genomic studies from cBioPortal databases. (**a**) Percentage of the primary and metastatic lung cancer cases among total 1750 patients. (**b**) Percentage of primary and metastatic lung cancer cases in two types of lung cancer, NSCLC and SCLC. (**c**) The most frequent metastatic sites of NSCLC and SCLC. (**d**) Percentage of primary and metastatic lung cancer cases in two major subtypes of NSCLC, AD and SCC. (**e**) The most frequent metastatic sites of NSCLC AD and SCC. (**f**) The differences in (I) mutational and (II) copy number variation profiles between primary and metastatic NSCLC AD cases. *The difference between groups is considered significant when q-value derived from Benjamini-Hochberg procedure is less than 0.05. Abbreviation: *NSCLC* non-small cell lung cancer, *SCLC* small cell lung cancer, *AD* adenocarcinoma, *SCC* squamous cell carcinoma.

trachea, bronchus and lung cancer due to smoking among European population was 92% [11]. The risk of lung cancer varies among smokers due to several factors, including age, cigarette count, duration of smoking and family history [12, 13]. Apart from smoking, there are many other environmental factors that may lead to lung cancer, including exposure to asbestos, radiation and pollution [14, 15]. In order to identify those with higher risk of lung cancer, potential application of novel biomarkers that underlies the risk and prognosis from either the host, the tumour microenvironment (TME), or the tumour itself. Assessing microRNA (miRNA), a small non-coding RNA that plays a role in gene regulation, is one of the various approaches used to classify biomarkers in relation to lung cancer. Identification of miRNAs from asymptomatic high-risk persons showed 80% accuracy in the early stage identification of the NSCLC case [16]. Also, miRNA expression can accurately identify particular subtypes of lung cancer in differential diagnosis [17], and miRNA was able to predict recurrence in stage 1 NSCLC patients [18]. Moreover, several reviews have highlighted the potential of targeting cancer-specific miRNAs in treating lung cancer [19, 20]. It is clear that the main cause of death in most cases of cancer, including lung cancer, is the spreading of cancer cells by metastasis process [21]. It is therefore important to understand the biology of metastasis and the possible contributors to metastasis in lung cancer in order to develop a research strategy to reduce the number of cancer deaths.

8.2 Metastatic Niche in Different Sites

Initiation of the metastasis process may begin with the local invasion of cancer cells, followed by intravasation into blood and/or lymphatic circulations, which may spread and colonise to distant sites [22, 23]. Cancer cells are documented to have the potential to acquire self-renewal characteristics through the epithelial-mesenchymal transition (EMT) that can also contribute to the metastasis process [24]. Different metastatic sites have unique mechanisms that enable the lung cancer cells to survive and proliferate in the new distant organ. In turn, the switch on of mesenchymal-epithelial transition of (MET) lung cancer cells requires adaptation to the new site, utilising the available 'resources' there. Interestingly, lung cancer patients with and without metastasis may exhibit different miRNA profiles. The miRNA profile could be also different in lung cancer at different metastatic sites. They can either act as tumour suppressor or tumour promoter depending of the types of miRNAs and the tumour microenvironment. In the following sections, we will discuss the molecular characteristics and their major signalling pathways of the most common metastatic organs for lung cancer based on their sequential order of incidence rate, namely, lymph node, pleura, liver, brain and bone metastasis (Fig. 8.3), and their respective miRNA expression profiles (Table 8.1).

8.3 Lymph Node Metastasis

The main metastasis site of all types of lung cancer is the lymph nodes, or more commonly known as stage 3 or N2 in the cancer diagnosis. As it is a temporary 'storeroom' and 'gate' before further invasion and metastasis to further sites, lymph node metastasis is an ideal treatment point to prevent further worsening of the cancer situation, whereby a combination of surgery, radiation and/or chemotherapy is commonly applied. For early-stage lung cancer patients, lymphatic invasion is one of the important prognostic factors. However, the conventional scanning such as chest CT has low sensitivity. More sensitive molecular biomarker examination using patients' blood or tumour tissues may be superior as compared to the conventional methods. Several miRNAs have been found in promoting lung cancer which are positively correlated with lymph node metastasis such as miR-21 [25], miR-19a and miR-19b [26], miR-494 [27], miR-210 [28], miR-421 [29], and miR-411 [30]. On the other hand, miRNAs such as miR-138 [31], miR-133a-3p [32], miR-452-5p [33],

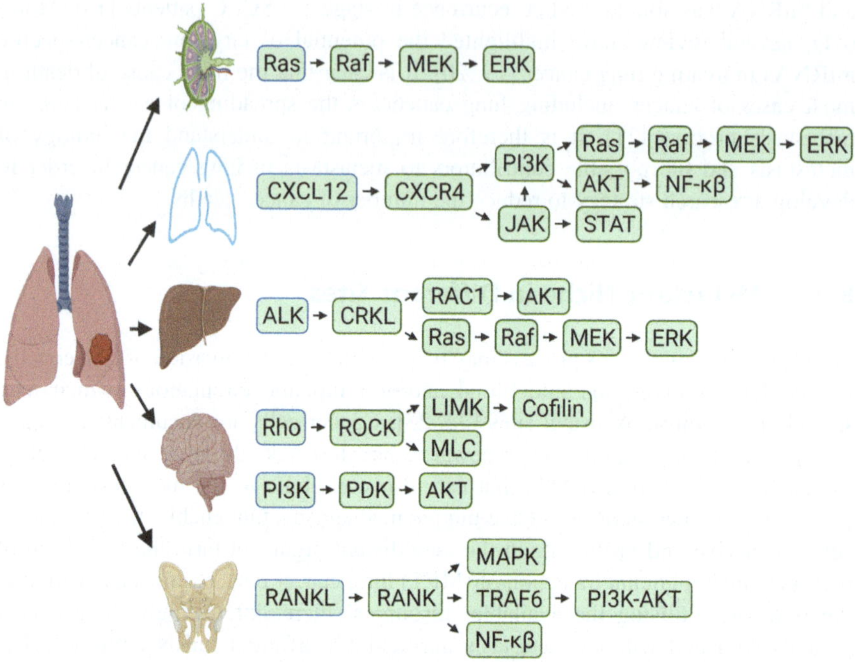

Fig. 8.3 An overview of major signalling pathways of the common metastatic organs for lung cancer, namely, the lymph node, pleura, liver, brain and bone. Abbreviation: *JAK* Janus kinase, *LIMK* Lim kinase, *PDK* phosphoinositide-dependent kinase, *TRAF6* tumour necrosis factor receptor-associated factor 6. The figure was created with BioRender.com

miR-145-5p [34] and miR-101-3p [35] were found to inhibit oncogenic processes in lung cancer with lymph node metastasis.

The cancer cells adapt to survive longer in the lymph nodes as, theoretically, they act as a natural barrier to cancer cell metastasis as there resides numerous dendritic cells, macrophages and T and B lymphocytes that perform immune surveillance on clearing cells that express foreign antigens [60]. Due to the high genomic instability of cancer cells due to their rapid turnover, cancer cells often express neoantigens that can trigger immunological NK and T cells that, theoretically, should eliminate the cancer cells [61]. However due to the heterogenicity nature of cancer, surviving immune-silent cancer cells could escape to develop its immunosuppressive TME, allowing it to continue to proliferate and metastasise [62].

8.3.1 MAPK Signalling in Lymph Node Metastasis of Lung Cancer

Formation of lymphatic vessels by the lung cancer cells is essential to aid its migration and metastatic spread into the lymph nodes. Comparison of 17 lung cancer

Table 8.1 Role of miRNAs in lung cancer in different metastatic sites

Metastatic sites	miRNAs	Effects	Related target	References
Lymph node	miR-21	Tumourigenesis	PTEN	Liu et al. [25]
	miR-19	Tumourigenesis	–	Wu et al. [26]
	miR-494	Tumourigenesis	–	Zhang et al. [27]
	miR-210	Tumourigenesis	–	Osugi et al. [28]
	miR-421	Tumourigenesis	–	Li et al. [29]
	miR-411	Tumourigenesis	–	Wang et al. [30]
	miR-138	Suppression	PDK1, Sirt1, YAP1	Han et al. [31]
	miR-133a-3p	Suppression	–	Yang et al. [32]
	miR-452-5p	Suppression	SMAD-4, SMAD-2, $p27^{Kip1}$, YWHAE, YWHAB	Gan et al. [33]
	miR-145-5p	Suppression	AEG-1/MTDH RIOK2, NOB1 N-cadherin	Gan et al. [34]
	miR-101-3p	Suppression	SOX9	Lu et al. [35]
	miR-200c	Tumourigenesis	–	Liu et al. [25]
		Suppression	USP25	Li et al. [36]
	miR-125a-5p	Tumourigenesis	NEDD9	Hou et al. [37]
	miR-125a-3p	Suppression	IGF2, CCL4	Yu et al. [38]
	miR-130	Tumourigenesis	–	Wang et al. [39]
		Suppression	PTEN	Ye et al. [40]
Pleura	miR-1-3p	Tumourigenesis	–	Roman-Canal et al. [41]
	miR-144-5p	Tumourigenesis	–	Roman-Canal et al. [41]
	miR-150-5p	Tumourigenesis	–	Roman-Canal et al. [41]
	miR-182	Tumourigenesis	–	Tamiya et al. [42]
	miR-210	Tumourigenesis	–	Tamiya et al. [42]

(continued)

Table 8.1 (continued)

Metastatic sites	miRNAs	Effects	Related target	References
Brain	miR-21	Tumourigenesis	SPRY2, TIMP3, PTEN, CDKN1A, SERPINB5	Singh et al. [43]
	miR-330-3p	Tumourigenesis	GRIA3	Wei et al. [44]
	miR-423-5p	Tumourigenesis	MTSS1	Sun et al. [45]
	miR-490-3p	Tumourigenesis	PCBP1	Li et al. [46]
	miR-328	Tumourigenesis	PRKCA	Arora et al. [47]
	miR-378	Tumourigenesis	MMP-2, MMP-9, VEGF	Chen et al. [48]
	miR-95-3p	Suppression	Cyclin D1	Hwang et al. [49]
	miR-145-5p	Suppression	TPD52	Zhao et al. [50]
	miR-375	Suppression	VEGF, MMP-9	Chen et al. [51]
	miR-590	Suppression	ADAM9	Wang et al. [52]
	miR-4317	Suppression	FGF9, CCND2	He et al. [53]
Bone	miR-21	Tumourigenesis	COX19	Guo et al. [54]
	Hsv-miR-H9-5p	Tumourigenesis	SOCS2	Wang et al. [55]
	miR-33a	Suppression	PTHrP	Kuo et al. [56]
	miR-139-5p	Suppression	Notch1	Xu et al. [57]
	miR-192	Suppression	ICAM-1, PTPRJ	Valencia et al. [58]
	miR-203	Suppression	TGF-beta/SMAD2	Wei et al. [59]

cell lines with differential lymphangiogenesis capacity has revealed that vascular endothelial growth factor C (VEGF-C) gene copy number, expression and activity clearly drive the formation of lymphangiogenic tumours [63]. Although VEGF-C could phosphorylate multiple downstream signalling, namely, AKT, ERK and p38 MAPK, binding of cell adhesion molecule CD146 and its various truncated forms with VEGF-C revealed that the lymphangiogenesis process of VEGF-C is dependent on the phosphorylation of ERK and p38 MAPK. Furthermore, other activators of MAPK can also present in abundance in the lymph nodes to promote lung cancer cell proliferation and lymph node formation, including Rho guanine nucleotide exchange factor ARHGEF19, IL6, IGFBP7, ERBB2 and Crabp2 (cellular retinoic acid-binding protein 2) [64–68].

In general, MAPK signalling pathway is a three-tier activation of MAPKKK-MAPKK-MAPK. Although MAPK signalling consists of four sub-pathways, ERK1/2 and p38 MAPK are consisted found to be elevated and activated in lymph node tissues with metastasised lung cancer of patient samples and in vitro models, but not for JNK or ERK5 signalling [69]. Knockout of MAPK or one of their upstream activators, such as GLK/MAP 4K3, consistently reduced the metastatic capacity of lung cancer cells [70].

Furthermore KRAS, an upstream component of ERK1/2, is a common driver mutation and upregulation in primary and metastatic lung cancer (Fig. 8.2) [71]. Ras activation leads to the subsequent phosphorylation of Raf-MEK-ERK1/2 signalling. *KRAS* mutation that induces elevated activation of the ERK signalling has been associated with increased pro-cancerous gene expression. Specifically on metastatic-related genes, upregulation of Krt18, Krt8, MMP14, vimentin, N-cadherin and Wnt with downregulation of genes known to inhibit invasion such as Reck, Gsn and Cav1 have been reported [72, 73]. It is also worth to note that different *KRAS* mutations can confer variations in their downstream gene profile, leading to different migration and metastatic capacities as observed for $KRAS^{G12V}$, $KRAS^{G12D}$ and $KRAS^{G12C}$ [73].

8.4 Pleura Metastasis

Pleura is one of the common metastatic sites for lung cancer, particularly lung adenocarcinoma [74]. This is mainly attributed to its anatomic proximity from the primary tumour site. In brief, pleura consists of two-layered serous membranes which fold back onto themselves, thus forming a pleural cavity that contains a small amount of viscous lubricant known as pleural fluid. Visceral pleura is the inner membrane covering the outer surface of the lung. Meanwhile, parietal pleura is the outer membrane that attached to the thoracic wall [75].

Pleura metastasis can be characterised by the involvement of visceral pleural invasion (VPI) and malignant pleural effusion (MPE). MPE is the most frequent pathological complication associated with the metastatic lung cancer. It is characterised by the abnormal build-up of pleural fluid in the pleural cavity due to aberrant cycle of fluid secretion and absorption, accompanied by the presence of metastatic cancer cells, serous proteins, and lymphoid and myeloid immune cells [76–78]. Aside from the close proximity of the site from lung, another important factor that promotes the development of MPE includes the paracrine signalling between lung tumour and mesothelial cells of pleura, which mainly influenced by the distinctive molecular profiles of the cells. In particular, the abundance secretion of specific cytokines or chemokines by the pleura cells creates the molecular gradient and thus directs recruitment or chemotaxis of tumour cells expressing the corresponding chemokine receptors into the pleural space. CXC chemokine receptor type 4 (CXCR4)/stromal cell-derived factor 1 (CXCL12) axis has been reported as the most common regulatory pathway associated with pleura metastasis [79].

8.4.1 CXCR4/CXCL12 Signalling and Pleura Metastasis in Lung Cancer

Briefly, CXCR4 is a G protein-coupled receptor that interacts with its specific ligand, CXCL12 chemokine. A meta-analysis study among lung cancer patients demonstrated that a high level of CXCR4 expression in lung cancer cells was significantly associated with metastasis and poor prognosis [80]. On the other hand, CXCL12 expression was found to be highly expressed in the pleura mesothelium [81]. The binding of CXCL12 ligand to seven-transmembrane domain of CXCR4 receptor causes the activation of the coupled heterotrimeric G protein, which made up of alpha (Gα), beta (Gβ) and gamma (Gγ) subunits, through guanosine diphosphate (GDP) to guanosine triphosphate (GTP) exchange on Gα. This, in turn, causes G protein structural change and the dissociation of Gα and Gβ/γ heterodimer complex from the membrane-bound receptor, both of which further lead to the simultaneous activation of various downstream signalling pathways associated with cell proliferation, invasion and migration, including mitogen-activated protein kinase (MAPK), nuclear factor (NF)-κβ and signal transducer and activator of transcription (STAT) pathways, in phosphatidylinositol 3 kinase (PI3K)-dependent manner [82].

In general, the dissociated, activated Gβ/γ complex is the critical mediator for the activation of PI3K. The activated PI3K can further stimulate NF-κβ signalling through Akt activation and MAPK signalling through multiple activation of downstream target proteins, including the important intermediates tyrosine-protein kinase c-Src, Ras and Raf proteins [82]. These signalling can further promote matrix metallopeptidases (MMPs) activity to promote cellular migration. In a NSCLC study, the induction of CXCR4/CXCL12 signalling promote cell migration into pleural space, and therefore causing MPE, through the activation of p44/42 MAPK signalling in PI3K-dependent manner [81]. Furthermore, the activated p44/42 MAPK signalling following the CXCR4/CXCL12 activation can activate c-Myc, a critical proto-oncogene transcription factor, to further promote the upregulation of CXCR4 expression [83].

Interestingly, other ligands that are also available in abundance in the pleura effusion, such as EGF and LPA, could also increase the expression and concentration of the CXCR4 and/or CXCL12 to modulate the CXCR4/CXCL12 signalling axis. EGFR, a tyrosine kinase receptor, via Src could also interact and activate the CXCR4/CXCL12 axis to phosphorylate the downstream PI3K/Akt and p44/42 MAPK signalling pathways [84]. Mutation and overexpression of *EGFR* have been identified in lung cancer cells [85], and its exon 19 deletions and exon 21 L858R common mutations in NSCLC have been highly associated with the greater incidence of pleura metastasis [86]. In vitro validation of NSCLC cell lines reported that the activation of EGFR by EGF upregulated the CXCR4 expression on the cancer cells via PI3K/Akt/mammalian target of rapamycin (mTOR) pathway and hypoxia-inducible factor (HIF) 1α-dependent transcription of CXCR4 successively [87]. In addition, EGFR signalling can also induce the activation of various other downstream targets, such as cytoskeletal proteins (actin, tensin and talin) and Ras homologous (Rho) family members (Rho, Rac, and Cdc42), to mediate cancer cell motility and invasion [88].

Besides that, it has been reported that the expression of both CXCR4 and CXCL12 can be regulated by lysophosphatidic acid (LPA) [89]. In brief, LPA is synthesised via the hydrolysis of lysophosphatidylcholine (LPC) into LPA by autotaxin (ATX). ATX has been reported to be highly expressed in the pleura mesothelium, specifically in the visceral pleura compared to its parietal counterpart [90]. In addition to the regulation of CXCR4/CXCL12 axis expression, ATX can directly involve in the pleura metastasis of lung cancer through ATX/LPA signalling pathway [91]. Recent in vitro study on lung cancer cell line demonstrated that the expression of LPA receptor, particularly subtype 2 (LPA2), was highly abundant in migratory cells derived from A549 lung cancer cells [92]. The synthesised LPA by ATX binds to specific G protein-coupled LPA receptors in the cancer cells, thus activating the downstream signalling pathways that modulate the cellular metastatic capacity [93].

Several miRNAs such as miR-1-3p, miR-144-5p and miR-150-5p have been identified as potential diagnostic biomarker in pleural fluids of lung cancer patients when compared to the samples collected from non-cancer controls [41]. However, the role of these miRNAs in lung cancer with pleural metastasis remains unexplored. Similarly, another study reported the high level of miR-182 and miR-210 in MPE related to lung cancer, highlighting their diagnostic potentials [42].

8.5 Liver Metastasis

One of the most common metastatic organs from lung cancer is the liver. Nevertheless, single distant metastasis in the liver was less frequent compared to the bone or brain in lung cancer patients, as multiple organ metastasis have already occurred upon the diagnosis of liver metastasis [94–96]. Among different subtypes of lung cancer, SCLC is reported to have the highest frequency to develop liver metastasis as more than 61% of liver metastasis occurred in patients of this lung cancer subtype [94, 97, 98]. Furthermore, the median survival time for lung cancer patients that have liver metastasis was just about 3 months, as compared to brain or bone metastasis which was up to 7 months [94]. Thus, liver metastasis has much higher risk or mortality which was shown to be 1.5-fold higher than bone or brain metastasis in lung cancer patients [6, 99]. Poor prognosis outcome is also reported in liver metastasis lung cancer patients after chemotherapy, and the exact underlying molecular mechanisms remains to be investigated [100].

8.5.1 ALK and Liver Metastasis in Lung Cancer

Due to the low incidence of de novo liver metastasis, the studies on its exclusive cellular mechanism and miRNA are very limited. Two independent NSCLC cohort studies have shown that co-expression of *KRAS* or *EGFR* mutations with anaplastic lymphoma kinase rearrangements (*ALK*+) led to higher incidences of liver metastasis [101, 102], suggesting that the direct or indirect signalling interaction between ALK with MAPK or EGFR signalling could drive the liver metastatic process.

Generally, ALK is a transmembrane receptor tyrosine kinase that are phosphorylated upon the binding of its ligands, such as pleiotrophin, midkine or heparin, to activate subsequent signal transductions involving the kinases AKT and ERK 1/2 [103, 104]. Echinoderm microtubule-associated protein-like 4 (EML4)-ALK fusion proteins, which resulted from *EML-ALK* genes rearrangement, are commonly detected in adenocarcinoma subtype [105], and they play a critical role in metastasis of NSCLC via directing EMT [106]. The EML-ALK+ NSCLC cells demonstrated an aberrant, ligand-independent activation of the ALK that further activated its critical downstream target, Crk-like protein (CRKL). In turn, CRKL stimulates downstream pathways that modulate cellular motility and migration such as Ras-Raf-MAPK/ERK and Rac1 signalling [107, 108].

Additionally, it has been described that the presence of *KRAS* mutation in ALK+ NSCLC cells is responsible for the enhanced metastasis capability and primary resistance to ALK tyrosine kinase inhibitor, thus complicating the management of NSCLC patients [101, 109]. Besides *ALK*/*KRAS* mutation, *EGFR* mutation has also been reported in ALK+ NSCLC cells [109–111]. Importantly, the ALK targeted-therapy showed a limited activity on co-altered *ALK*/*EGFR* NSCLC cells due to EGFR bypass pathway activation, similarly inducing the downstream signalling pathways of ALK [112, 113]. Additionally, elevated expression of upstream activator of EGFR, namely, adaptor protein CIP4 (Cdc42-interacting protein 4) in the lung cancer tissues, may further aid EGFR trafficking and signalling via coordinating actin polymerisation to promote migration and metastasis to the liver [114–116].

8.6 Brain Metastasis

SCLC has reported the highest percentage for the incidence of brain metastasis by subtype, followed by lung AD, LCC and lung SCC, respectively [117, 118]. Increasing cases of lung cancer brain metastasis over the years could be attributed to several factors, such as a high number of late-stage lung cancer patients, lung cancer patients with a longer survival period and developments and wider accessibility of neuroimaging technology [119, 120].

Upon transmigration of cancer cells into the brain, they are capable of evading pro-apoptotic cytokine by releasing serpins that allowed them to proliferate [121]. Subsequently, PI3K and MAPK signalling pathways can be activated as cancer cells trigger astrocytes to release endothelin-1 (ET-1), leading to cancer cell survival in the brain [122]. Cancer cells could also express photocadherin-7 (PCDH7) to form a network of carcinoma-astrocyte gap junctions which enable cancer cells to transfer cyclic GMP-AMP (cGAMP) to astrocytes [123]. Subsequently, astrocytes may activate the STING pathway to release inflammatory cytokines including interferon-α (IFNα) and tumour necrosis factor (TNF), leading to proliferation and chemoresistance of metastatic cells in the brain.

8.6.1 Rho/ROCK Signalling and Brain Metastasis in Lung Cancer

Rho-associated protein kinase (ROCK), a downstream effector of Rho GTPase family, is a serine-threonine kinase that plays a variety of pharmacological roles such as cytoskeletal reorganisation, cell proliferation, migration and adhesion [124–126]. ROCK signalling pathway is also known to regulate cellular permeability and cell junctions in endothelial cells [127]. In the brain, there is a special networking of endothelial cells called blood-brain barrier (BBB) that guard selective molecules to pass through and enter the brain [128]. Therefore, the specific underlying mechanisms on how cancer cells manage to invade and metastasise through the BBB remain to be elucidated.

As ROCK pathway mediates actomyosin contraction and cytoskeletal reorganisation to increase cell motility, therefore, during malignancy, ROCK pathway is upregulated to disrupt the tight junction in BBB to facilitate metastatic cancer cells to migrate across the BBB [129, 130]. The disruption of tight junction was due to increment of endothelial cofilin and myosin light chain (MLC) activities with ROCK activation as its upstream regulator [130, 131]. Their findings were in line with previous data that reported that ROCK signalling mediates the crosstalk of lung cancer cells to human brain microvascular endothelial cells (HBMEC), an in vitro version of BBB [132, 133]. The tight junction disassembly was prevented upon inhibition of ROCK in NCI-H209 SCLC cells. Also, incubation of HBMEC with NCI-H209 SCLC cells for 2 h has resulted in the upregulation of endothelial ROCK activity. In addition, ROCK could also mediate placental growth factor (PLGF)-induced disruption of tight junction in the brain endothelial cells as no changes to tight junction occurred upon inhibition of ROCK activity [134]. Since it is a big challenge to singly predict which lung cancer patient that may develop brain metastasis, therefore, ROCK signalling pathway may be used as one of the valuable biomarkers to predict the occurrence of brain metastasis in lung cancer patients.

8.6.2 PI3K/AKT Signalling and Brain Metastasis in Lung Cancer

Previous next-generation sequencing study revealed that PI3K/AKT signalling pathway is involved in the brain metastasis of NSCLC cases [135]. The PI3K signalling plays various functional roles such as cell migration, apoptosis, proliferation and cell cycle regulation [136–138]. It is reported that AKT phosphorylation (pAKT), the downstream component of PI3K signalling, is capable to induce activation of other signalling pathways which may promote tumour formation [139, 140], epithelial mesenchymal transmission (EMT) event [141] and influence malignancy progression leading to poor prognosis [142]. Hence, the detailed mechanism on how PI3K-AKT signalling impacts the brain metastasis in lung cancer patients needs to be fully investigated.

Increased level of pAKT expression has been associated with the high occurrence of brain metastasis in NSCLC patients as compared to patients with little to no expression of pAKT through the Kaplan-Meier and Cox proportional hazard analysis [143]. This elevated AKT activity could be attributed by the single nucleotide polymorphisms (SNPs) in the *AKT1* gene at rs2498804 and rs2494732 and *PIK3CA* gene at rs2699887 [143, 144]. Hence, these SNPs could act as biomarkers on the prognosis of brain metastasis risk among NSCLC patients.

Similarly, miRNAs can act as both promoter and suppressor in NSCLC with brain metastasis. MiRNAs such as miR-21 [43], miR-330-3p [44], miR-423-5p [45] and miR-490-3p [46] were reported to promote tumour growth and progression in NSCLC. Some miRNAs were shown to enhance the growth of both lung and brain tumours such as miR-328 [47] and miR-378 [48]. On the other hand, miR-95-3p [49], miR-145 [50], miR375 [51], miR-590 [52] and miR-4317 [53] could inhibit various tumourigenic processes such as cell proliferation, migration, invasion and colony formation.

8.7 Bone Metastasis

Bone is one of the most favoured metastatic sites for lung cancer, in particular NSCLC of the AD subtype with up to 40% of NSCLC patients having bone metastasis [145, 146]. Previous studies found that the mean interval between diagnosis of NSCLC and development of bone metastasis was 8 months and the mean interval between diagnosis of lung cancer and bone metastasis diagnosis was 4 months [145, 147]. Metastasis is not common in all bone types; pelvic, rib and vertebrae are the more common metastasis sites while the bones in the hands or feet rarely occurred [148]. This could be due to several factors such as higher bone turnover rate, trabecular bone and the presence of vascularisation [149].

In general, bone remodelling, which consists of two primary processes, including bone formation and bone resorption by osteoblasts and osteoclasts, respectively, serves as the basis of bone turnover [150]. Osteocytes are another type of bone cells that are the master regulator of bone remodelling as they control the activities of osteoblasts and osteoclasts [151]. The imbalance activities between bone-forming osteoblasts and bone-resorbing osteoclasts result in the dysregulation of bone homeostasis that influence bone turnover rate, thus promoting tumour growth.

Bone metastasis can be either osteolytic (bone destruction), osteoblastic (abnormal bone formation) or mix of osteolytic and osteoblastic metastases [152]. In brief, the distinct molecular profiles of lung cancer cells, characterised by elevated expression of cytokines and growth factors that act as pro-osteoblastic and pro-osteolytic factors, can promote the development of bone metastasis [153]. In the bone, the continual bone turnover activity offers a favourable bone microenvironment with an abundant of growth factors and cytokines, for the localisation and survival of the primary tumour.

Major regulatory pathways involved in osteoblastic metastasis are Wnt [154], endothelin-1 (ET-1) [155] and bone morphogenetic protein (BMP) pathways. Proteins secreted by the metastasised tumour cells, such as Wnt, ET-1 and BMPs, including BMP-2, BMP-6 and BMP-7, promote the activation and proliferation of osteoblast or inhibition of osteoclast activity, thus maintaining a predominant bone-forming state [153]. Importantly, the secreted growth factors by the activated osteoblasts during the bone-forming activity, such as connective tissue growth factor (CTGF) [156], transforming growth factor β (TGF-β) [157], interleukin-6 (IL-6) [158] and BMPs [159], are used by the tumour cells to support for their growth and survival in the bone, thus establishing a continuous cycle of tumour cells and bone cells interaction in the bone microenvironment for tumour pathogenesis.

Apart from osteoblast and osteoclast involvement, osteocytes can also contribute to bone metastasis. Increased intraosseous pressure by bone metastasis has been mimicked in the in vitro setting through the induction of hydrostatic pressure to osteocytes. As a result, the viability of lung cancer cell lines was promoted together with increased motility and invasion of cancer cells via the expression of chemokine ligand 5 (CCL5) and matrix metalloproteases (MMPs) [160].

In lung cancer with bone metastasis, miRNAs such as miR-21 [54] and Hsv-miR-H9-5p [55] were reported to promote the osteoclastogenesis and tumourigenesis. In contrast, miR-33a [56], miR-139-5p [57], miR-192 [58] and miR-203 [59] were found to act as a suppressor in NSCLC with bone metastasis by inhibiting cell proliferation, migration and angiogenesis.

8.7.1 RANK/RANKL Signalling and Bone Metastasis in Lung Cancer

Importantly, it has been reported that the bone metastasis from lung cancer is predominantly osteolytic metastasis [161, 162]. Osteolytic metastasis is mainly regulated by the receptor activator of nuclear factor-κB (RANK)/RANK ligand (RANKL) signalling pathway [163, 164]. During the development of osteolytic metastasis, tumour cells release various pro-osteoclastogenic factors, such as parathyroid hormone-related peptide (PTHrP) [165], interleukin-11 (IL-11) [166], macrophage colony-stimulating protein (M-CSP) [149] and intercellular adhesion molecule 1 (ICAM-1) [167]. These proteins in turn promote the upregulation of RANKL expression on osteoblasts. In brief, RANKL is a transmembrane molecule expressed by osteoblasts. It acts as a ligand for RANK, a membrane-bound tumour necrosis factor (TNF) receptor on osteoclasts. The binding of RANKL to RANK activates the intracellular signalling cascades of precursor osteoclasts, such as nuclear factor-κB, p38 mitogen-activated protein kinase, cellular Src kinases and Jun amino-terminal kinases, thus promoting osteoclast differentiation and maturation [153]. The increase in formation and activation of osteoclasts further causes increased osteolysis or bone resorption.

Osteolysis causes further release of numerous growth factors stored in the bone matrix, such as TGF-β, BMPs and insulin-like growth factors (IGFs), for vicious cycle of tumour growth in the bone, thus advancing the bone lesions [162]. The interaction of RANK and RANKL can be inhibited by osteoprotegerin (OPG), which acts as a 'decoy' receptor for RANKL and therefore an inhibitor of osteolytic activity. Thus, the upregulation of RANKL expression is commonly observed in the lung cancer patients with bone metastasis [168, 169]. Among the lung cancer-related miRNAs, miR-33a has been reported to specifically reduce the effect of RANKL on osteoblasts in lung cancer cell lines [56].

8.8 Controversial Role miRNA in Lung Cancer Metastasis

In addition, there are several miRNAs which are controversial and are believed to play multifaceted role in lung cancer with lymph node metastasis as summarised in Table 8.1. For example, miR-200c was shown to be significantly associated with lymph node metastasis and poor clinical outcome [25], while it was reported to inhibit multiple tumourigenic events in another study [36]. Similarly, two mature miRNAs, miR-125a-3p and miR-125a-5p (3′ and 5′ ends of pre-miR-125a), were reported to associate with lymph node metastasis [37] and inhibit the cell proliferation and invasion [38] in lung cancer, respectively. MiR-130 is another controversial miRNA, and it was shown to be downregulated in NSCLC tumour and cell lines, and inversely correlated with lymph node metastasis [40]. In another study, high miR-130 level was observed in NSCLC tissues and was strongly associated with lymph node metastasis and poor prognosis [39].

8.9 Concluding Remarks

The important pathways in different lung cancer metastasis sites have been comprehensively covered in this chapter as summarised graphically in Fig. 8.3. More molecular studies are still required especially on the pleura and liver metastasis of lung cancer to improve the disease outcome of these patients. We also proposed that application of miRNA may potentially serve as a diagnostic and/or prognostic screening for lung cancer with different metastatic sites due to the differential miRNA profiling. As miRNAs play dual roles in either promoting or suppressing the oncogenic pathways, they also carry a huge potential in the exploration of therapeutic targets. This review however could not cover those recently discovered novel targets and pathways with unclear pathophysiological actions. For example, transcriptomic analysis of TCGA lung cancer data samples have identified *RN7SL494P* gene to be a promising lymph node metastatic marker, but its gene function and downstream linkage to the proposed JAK-STAT signalling or other pathways would need to be experimentally verified in a lung cancer setup [170]. All these exciting molecular discoveries via big data could be the key to open the door for better therapeutic prevention and treatment of lung cancer metastasis in the near future.

References

1. Siegel RL, Miller KD, Jemal A (2018) Cancer statistics, 2018. CA Cancer J Clin 68:7–30. https://doi.org/10.3322/caac.21442
2. Petersen I (2011) The morphological and molecular diagnosis of lung cancer. Dtsch Aerzteblatt Online 108:525–531. https://doi.org/10.3238/arztebl.2011.0525
3. Rutering J, Ilmer M, Recio A et al (2016) Mutational Landscape of Metastatic Cancer Revealed from Prospective Clinical Sequencing of 10,000 Patients. Nat Med 5:1–8. https://doi.org/10.4172/2157-7633.1000305.Improved
4. Popper HH (2016) Progression and metastasis of lung cancer. Cancer Metastasis Rev 35:75–91. https://doi.org/10.1007/s10555-016-9618-0
5. Shin DY, Na II, Kim CH et al (2014) EGFR mutation and brain metastasis in pulmonary adenocarcinomas. J Thorac Oncol 9:195–199. https://doi.org/10.1097/JTO.0000000000000069
6. Tamura T, Koichi K, Kensuke N et al (2015) Specific organ metastases and survival in metastatic non-small-cell lung cancer. Mol Clin Oncol 3:217–221. https://doi.org/10.3892/mco.2014.410
7. Wilbertz T, Wagner P, Petersen K et al (2011) SOX2 gene amplification and protein overexpression are associated with better outcome in squamous cell lung cancer. Mod Pathol 24:944–953. https://doi.org/10.1038/modpathol.2011.49
8. Cerami E, Gao J, Dogrusoz U et al (2012) The cBio Cancer Genomics Portal: an open platform for exploring multidimensional cancer genomics data. Cancer Discov 2:401–404. https://doi.org/10.1158/2159-8290.CD-12-0095
9. Gao J, Aksoy BA, Dogrusoz U et al (2013) Integrative analysis of complex cancer genomics and clinical profiles using the cBioPortal. Sci Signal 6(269):p11. https://doi.org/10.1126/scisignal.2004088
10. Wang S, Wang Z (2015) Meta-analysis of epidermal growth factor receptor and KRAS gene status between primary and corresponding metastatic tumours of non-small cell lung cancer. Clin Oncol 27:30–39. https://doi.org/10.1016/j.clon.2014.09.014
11. World Health Organization (2019) European tobacco use: trends report 2019
12. Hasegawa Y (2014) Lung cancer: progress in diagnosis and treatments. Topics: I. Epidemiology and pathogenesis; 2. The etiology of lung cancer. Nihon Naika Gakkai Zasshi 103:1261–1266. https://doi.org/10.2169/naika.103.1261
13. Thun MJ, Carter BD, Feskanich D et al (2013) 50-Year trends in smoking-related mortality in the United States. N Engl J Med 368:351–364. https://doi.org/10.1056/NEJMsa1211127
14. Nasim F, Sabath BF, Eapen GA (2019) Lung cancer. Med Clin North Am 103:463–473. https://doi.org/10.1016/j.mcna.2018.12.006
15. O'Malley M, King AN, Conte M et al (2014) Effects of cigarette smoking on metabolism and effectiveness of systemic therapy for lung cancer. J Thorac Oncol 9:917–926. https://doi.org/10.1097/JTO.0000000000000191
16. Bianchi F, Nicassio F, Marzi M et al (2011) A serum circulating miRNA diagnostic test to identify asymptomatic high-risk individuals with early stage lung cancer. EMBO Mol Med 3:495–503. https://doi.org/10.1002/emmm.201100154
17. Huang W, Hu J, Yang DW et al (2012) Two microRNA panels to discriminate three subtypes of lung carcinoma in bronchial brushing specimens. Am J Respir Crit Care Med 186:1160–1167. https://doi.org/10.1164/rccm.201203-0534OC
18. Lu Y, Govindan R, Wang L et al (2012) MicroRNA profiling and prediction of recurrence/relapse-free survival in stage I lung cancer. Carcinogenesis 33:1046–1054. https://doi.org/10.1093/carcin/bgs100
19. Wu KL, Tsai YM, Lien CT et al (2019b) The roles of microRNA in lung cancer. Int J Mol Sci 20(7):1611
20. Wu SG, Chang TH, Liu YN, Shih JY (2019c) MicroRNA in lung cancer metastasis. Cancers (Basel) 11:265

21. Dillekås H, Rogers MS, Straume O (2019) Are 90% of deaths from cancer caused by metastases? Cancer Med 8:5574–5576. https://doi.org/10.1002/cam4.2474
22. Hanahan D, Weinberg RA (2011) Hallmarks of cancer: the next generation. Cell 144:646–674. https://doi.org/10.1016/j.cell.2011.02.013
23. Paduch R (2016) The role of lymphangiogenesis and angiogenesis in tumor metastasis. Cell Oncol 39:397–410. https://doi.org/10.1007/s13402-016-0281-9
24. Mani SA, Guo W, Liao MJ et al (2008) The Epithelial-Mesenchymal Transition Generates Cells with Properties of Stem Cells. Cell 133:704–715. https://doi.org/10.1016/j.cell.2008.03.027
25. Liu XG, Zhu WY, Huang YY et al (2012) High expression of serum miR-21 and tumor miR-200c associated with poor prognosis in patients with lung cancer. Med Oncol 29:618–626. https://doi.org/10.1007/s12032-011-9923-y
26. Wu C, Cao Y, He Z et al (2014) Serum levels of miR-19b and miR-146a as prognostic biomarkers for non-small cell lung cancer. Tohoku J Exp Med 232:85–95. https://doi.org/10.1620/tjem.232.85
27. Zhang J, Wang T, Zhang Y et al (2018) Upregulation of serum miR-494 predicts poor prognosis in non-small cell lung cancer patients. Cancer Biomark 21:763–768. https://doi.org/10.3233/CBM-170337
28. Osugi J, Kimura Y, Owada Y et al (2015) Prognostic impact of hypoxia-inducible miRNA-210 in patients with lung adenocarcinoma. J Oncol 2015:1–8. https://doi.org/10.1155/2015/316745
29. Li Y, Cui X, Li Y et al (2018a) Upregulated expression of miR-421 is associated with poor prognosis in non-small-cell lung cancer. Cancer Manag Res 10:2627–2633. https://doi.org/10.2147/CMAR.S167432
30. Wang SY, Li Y, Jiang YS, Li RZ (2017a) Investigation of serum miR-411 as a diagnosis and prognosis biomarker for non-small cell lung cancer. Eur Rev Med Pharmacol Sci 21:4092–4097
31. Han L, Zhang G, Zhang N et al (2014) Prognostic potential of microRNA-138 and its target mRNA PDK1 in sera for patients with non-small cell lung cancer. Med Oncol 31:1–7. https://doi.org/10.1007/s12032-014-0129-y
32. Yang ZQ, Wu CA, Cheng YX (2018) Prognostic Value of microRNA-133a expression and its clinicopathologic significance in non-small cell lung cancer: a comprehensive study based on meta-analysis and the tcga database. Oncol Res Treat 41:762–768. https://doi.org/10.1159/000492343
33. Gan X, Luo J, Tang R et al (2017b) Clinical value of miR-452-5p expression in lung adenocarcinoma: a retrospective quantitative real-time polymerase chain reaction study and verification based on The Cancer Genome Atlas and Gene Expression Omnibus databases. Tumor Biol 39:101042831770575. https://doi.org/10.1177/1010428317705755
34. Gan TQ, Xie ZC, Tang RX et al (2017a) Clinical value of miR-145-5p in NSCLC and potential molecular mechanism exploration: a retrospective study based on GEO, qRT-PCR, and TCGA data. Tumor Biol 39:1–23. https://doi.org/10.1177/1010428317691683
35. Lu HM, Yi WW, Ma YS et al (2018) Prognostic implications of decreased microRNA-101-3p expression in patients with non-small cell lung cancer. Oncol Lett 16:7048–7056. https://doi.org/10.3892/ol.2018.9559
36. Li J, Tan Q, Yan M et al (2014) MiRNA-200c inhibits invasion and metastasis of human non-small cell lung cancer by directly targeting ubiquitin specific peptidase 25. Mol Cancer 13:166. https://doi.org/10.1186/1476-4598-13-166
37. Hou L, Luo P, Ma Y et al (2017) MicroRNA-125a-3p downregulation correlates with tumorigenesis and poor prognosis in patients with non-small cell lung cancer. Oncol Lett 14:4441–4448. https://doi.org/10.3892/ol.2017.6809
38. Yu X, Wei F, Yu J et al (2015) Matrix metalloproteinase 13: a potential intermediate between low expression of microRNA-125b and increasing metastatic potential of non-small cell lung cancer. Cancer Genet 208:76–84. https://doi.org/10.1016/j.cancergen.2015.01.006

39. Wang XC, Tian LL, Wu HL et al (2010) Expression of miRNA-130a in nonsmall cell lung cancer. Am J Med Sci 340:385–388. https://doi.org/10.1097/MAJ.0b013e3181e892a0
40. Ye L, Wang Y, Nie L et al (2017) MiR-130 exerts tumor suppressive function on the tumorigenesis of human non-small cell lung cancer by targeting PTEN. Am J Transl Res 9:1856–1865
41. Roman-Canal B, Moiola CP, Gatius S et al (2019) EV-associated miRNAs from pleural lavage as potential diagnostic biomarkers in lung cancer. Sci Rep 9:15057. https://doi.org/10.1038/s41598-019-51578-y
42. Tamiya H, Mitani A, Saito A et al (2018) Exosomal MicroRNA expression profiling in patients with lung adenocarcinoma-associated malignant pleural effusion. Anticancer Res 38:6707–6714. https://doi.org/10.21873/anticanres.13039
43. Singh M, Garg N, Venugopal C et al (2015) STAT3 pathway regulates lung-derived brain metastasis initiating cell capacity through miR-21 activation. Oncotarget 6:27461–27477. https://doi.org/10.18632/oncotarget.4742
44. Wei C-H, Wu G, Cai Q et al (2017a) MicroRNA-330-3p promotes cell invasion and metastasis in non-small cell lung cancer through GRIA3 by activating MAPK/ERK signaling pathway. J Hematol Oncol 10:125. https://doi.org/10.1186/s13045-017-0493-0
45. Sun G, Ding X, Bi N et al (2018) MiR-423-5p in brain metastasis: potential role in diagnostics and molecular biology. Cell Death Dis 9:936. https://doi.org/10.1038/s41419-018-0955-5
46. Li J, Feng Q, Wei X, Yu Y (2016) MicroRNA-490 regulates lung cancer metastasis by targeting poly r(C)-binding protein 1. Tumor Biol 37:15221–15228. https://doi.org/10.1007/s13277-016-5347-9
47. Arora S, Ranade AR, Tran NL et al (2011) MicroRNA-328 is associated with (non-small) cell lung cancer (NSCLC) brain metastasis and mediates NSCLC migration. Int J Cancer 129:2621–2631. https://doi.org/10.1002/ijc.25939
48. Chen LT, Xu SD, Xu H et al (2012) MicroRNA-378 is associated with non-small cell lung cancer brain metastasis by promoting cell migration, invasion and tumor angiogenesis. Med Oncol 29:1673–1680. https://doi.org/10.1007/s12032-011-0083-x
49. Hwang SJ, Lee HW, Kim HR et al (2015) Overexpression of microRNA-95-3p suppresses brain metastasis of lung adenocarcinoma through downregulation of cyclin D1. Oncotarget 6:20434–20448. https://doi.org/10.18632/oncotarget.3886
50. Zhao C, Xu Y, Zhang Y et al (2013) Downregulation of miR-145 contributes to lung adenocarcinoma cell growth to form brain metastases. Oncol Rep 30:2027–2034. https://doi.org/10.3892/or.2013.2728
51. Chen LJ, Li XY, Zhao YQ et al (2017) Down-regulated microRNA-375 expression as a predictive biomarker in non-small cell lung cancer brain metastasis and its prognostic significance. Pathol Res Pract 213:882–888. https://doi.org/10.1016/j.prp.2017.06.012
52. Wang FF, Wang S, Xue WH, Cheng JL (2016) microRNA-590 suppresses the tumorigenesis and invasiveness of non-small cell lung cancer cells by targeting ADAM9. Mol Cell Biochem 423:29–37. https://doi.org/10.1007/s11010-016-2822-y
53. He X, Chen S, Yang Z et al (2018) miR-4317 suppresses non-small cell lung cancer (NSCLC) by targeting fibroblast growth factor 9 (FGF9) and cyclin D2 (CCND2). J Exp Clin Cancer Res 37:230. https://doi.org/10.1186/s13046-018-0882-4
54. Guo Q, Zhang H, Zhang L et al (2015) MicroRNA-21 regulates non-small cell lung cancer cell proliferation by affecting cell apoptosis via COX-19. Int J Clin Exp Med 8:8835–8841
55. Wang X, Liu S, Zhou Z et al (2017b) A herpes simplex virus type 2-encoded microRNA promotes tumor cell metastasis by targeting suppressor of cytokine signaling 2 in lung cancer. Tumor Biol 39(5):1010428317701633. https://doi.org/10.1177/1010428317701633
56. Kuo PL, Liao SH, Hung JY et al (2013) MicroRNA-33a functions as a bone metastasis suppressor in lung cancer by targeting parathyroid hormone related protein. Biochim Biophys Acta Gen Subj 1830:3756–3766. https://doi.org/10.1016/j.bbagen.2013.02.022

57. Xu S, Yang F, Liu R et al (2018) Serum microRNA-139-5p is downregulated in lung cancer patients with lytic bone metastasis. Oncol Rep 39:2376–2384. https://doi.org/10.3892/or.2018.6316
58. Valencia K, Luis-Ravelo D, Bovy N et al (2014) MiRNA cargo within exosome-like vesicle transfer influences metastatic bone colonization. Mol Oncol 8:689–703. https://doi.org/10.1016/j.molonc.2014.01.012
59. Wei Y, Li D, Wang D et al (2017b) Evaluation of microRNA-203 in bone metastasis of patients with non-small cell lung cancer through TGF-β/SMAD2 expression. Oncol Rep. https://doi.org/10.3892/or.2017.5987
60. Hu X, Luo J (2018) Heterogeneity of tumor lymphangiogenesis: progress and prospects. Cancer Sci 109:3005–3012. https://doi.org/10.1111/cas.13738
61. Aguadé-Gorgorió G, Solé R (2019) Genetic instability as a driver for immune surveillance. J Immunother Cancer 7:345. https://doi.org/10.1186/s40425-019-0795-6
62. Mittal D, Gubin MM, Schreiber RD, Smyth MJ (2014) New insights into cancer immunoediting and its three component phases-elimination, equilibrium and escape. Curr Opin Immunol 27:16–25. https://doi.org/10.1016/j.coi.2014.01.004
63. Regan E, Sibley RC, Cenik BK et al (2016) Identification of gene expression differences between lymphangiogenic and non-lymphangiogenic non-small cell lung cancer cell lines. PLoS One 11(3):e0150963. https://doi.org/10.1371/journal.pone.0150963
64. Huang Q, Zhang Z, Liao Y et al (2018) 17β-estradiol upregulates IL6 expression through the ERβ pathway to promote lung adenocarcinoma progression. J Exp Clin Cancer Res 37:133. https://doi.org/10.1186/s13046-018-0804-5
65. Li Y, Ye Z, Chen S et al (2018b) ARHGEF19 interacts with BRAF to activate MAPK signaling during the tumorigenesis of non-small cell lung cancer. Int J Cancer 142:1379–1391. https://doi.org/10.1002/ijc.31169
66. Wu JI, Lin YP, Tseng CW et al (2019a) Crabp2 promotes metastasis of lung cancer cells via HuR and integrin β1/FAK/ERK signaling. Sci Rep 9:845. https://doi.org/10.1038/s41598-018-37443-4
67. Zhang N, Zhang SW (2019) Identification of differentially expressed genes between primary lung cancer and lymph node metastasis via bioinformatic analysis. Oncol Lett 18:3754–3768. https://doi.org/10.3892/ol.2019.10723
68. Zhao W, Wang J, Zhu B et al (2016) IGFBP7 functions as a potential lymphangiogenesis inducer in non-small cell lung carcinoma. Oncol Rep 35:1483–1492. https://doi.org/10.3892/or.2015.4516
69. Zhang ZF, Ma JQ, Zhang L (2005) Expression and clinical significance of MAPK in non-small cell lung cancer. Natl Med J China 85:339–342
70. Chuang HC, Chang CC, Teng CF et al (2019) MAP 4K3/GLk promotes lung cancer metastasis by phosphorylating and activating IQGAP1. Cancer Res 79:4978–4993. https://doi.org/10.1158/0008-5472.CAN-19-1402
71. Guo Y, Pan W, Liu S et al (2020) ERK/MAPK signalling pathway and tumorigenesis (Review). Exp Ther Med 19:1997–2007. https://doi.org/10.3892/etm.2020.8454
72. Hung P-S, Huang M-H, Kuo Y-Y, Yang JC-H (2020) The Inhibition of Wnt Restrain KRASG12V-Driven Metastasis in Non-Small-Cell Lung Cancer. Cancers (Basel) 12:837. https://doi.org/10.3390/cancers12040837
73. Wakamatsu N, Collins JB, Parker JS et al (2008) Gene expression studies demonstrate that the K-ras/Erk MAP kinase signal transduction pathway and other novel pathways contribute to the pathogenesis of cumene-induced lung tumors. Toxicol Pathol 36:743–752. https://doi.org/10.1177/0192623308320801
74. Du Rand I, Maskell N (2010) British Thoracic Society pleural disease guideline 2010. Br Thorac Soc 65:ii1–ii3. https://doi.org/10.1136/thx.2010.137042
75. Charalampidis C, Youroukou A, Lazaridis G et al (2015) Pleura space anatomy. J Thorac Dis 7:S27–S32. https://doi.org/10.3978/j.issn.2072-1439.2015.01.48

76. Donnenberg AD, Luketich JD, Dhupar R, Donnenberg VS (2019) Treatment of malignant pleural effusions: the case for localized immunotherapy. J Immunother Cancer 7:5–9. https://doi.org/10.1186/s40425-019-0590-4
77. Psallidas I, Kalomenidis I, Porcel JM et al (2016) Malignant pleural effusion: from bench to bedside. Eur Respir Rev 25:189–198. https://doi.org/10.1183/16000617.0019-2016
78. Semaan R, Feller-Kopman D, Slatore C et al (2016) Malignant pleural effusions. Am J Respir Crit Care Med 194:P11–P12. https://doi.org/10.1164/rccm.1946P11
79. Li T, Li H, Wang Y et al (2011) The expression of CXCR4, CXCL12 and CXCR7 in malignant pleural mesothelioma. J Pathol 223:519–530. https://doi.org/10.1002/path.2829
80. Liang JX, Gao W, Liang Y, Zhou XM (2015) Chemokine receptor CXCR4 expression and lung cancer prognosis: a meta-analysis. Int J Clin Exp Med 8:5163–5174
81. Oonakahara K, Matsuyama W, Higashimoto I et al (2004) Stromal-derived factor-1α/CXCL12-CXCR 4 axis is involved in the dissemination of NSCLC cells into pleural space. Am J Respir Cell Mol Biol 30:671–677. https://doi.org/10.1165/rcmb.2003-0340OC
82. Cavallaro S (2013) CXCR4/CXCL12 in non-small-cell lung cancer metastasis to the brain. Int J Mol Sci 14:1713–1727. https://doi.org/10.3390/ijms14011713
83. Guo F, Wang Y, Liu J et al (2016) CXCL12/CXCR4: a symbiotic bridge linking cancer cells and their stromal neighbors in oncogenic communication networks. Oncogene 35:816–826. https://doi.org/10.1038/onc.2015.139
84. Porcile C, Bajetto A, Barbieri F et al (2005) Stromal cell-derived factor-1α (SDF-1α/CXCL12) stimulates ovarian cancer cell growth through the EGF receptor transactivation. Exp Cell Res 308:241–253. https://doi.org/10.1016/j.yexcr.2005.04.024
85. Sigismund S, Avanzato D, Lanzetti L (2018) Emerging functions of the EGFR in cancer. Mol Oncol 12:3–20. https://doi.org/10.1002/1878-0261.12155
86. Zou JY, Bella AE, Chen ZG et al (2014) Frequency of EGFR mutations in lung adenocarcinoma with malignant pleural effusion: implication of cancer biological behaviour regulated by EGFR mutation. J Int Med Res 42:1110–1117. https://doi.org/10.1177/0300060514539273
87. Phillips RJ, Mestas J, Gharaee-Kermani M et al (2005) Epidermal growth factor and hypoxia-induced expression of CXC chemokine receptor 4 on non-small cell lung cancer cells is regulated by the phosphatidylinositol 3-kinase/PTEN/AKT/mammalian target of rapamycin signaling pathway and activation of hypoxia ind. J Biol Chem 280:22473–22481. https://doi.org/10.1074/jbc.M500963200
88. Maulik G, Kijima T, Salgia R (2003) Role of receptor tyrosine kinases in lung cancer. In: Lung cancer. Humana Press, Totowa, NJ, pp 113–126
89. Wang H, Liu W, Wei D et al (2014) Effect of the LPA-mediated CXCL12-CXCR4 axis in the tumor proliferation, migration and invasion of ovarian cancer cell lines. Oncol Lett 7:1581–1585. https://doi.org/10.3892/ol.2014.1926
90. Shelton EL, Galindo CL, Williams CH et al (2013) Autotaxin signaling governs phenotypic heterogeneity in visceral and parietal mesothelia. PLoS One 8:e69712. https://doi.org/10.1371/journal.pone.0069712
91. Chu X, Wei X, Lu S, He P (2015) Autotaxin-LPA receptor axis in the pathogenesis of lung diseases. Int J Clin Exp Med 8:17117–17122
92. Ueda N, Minami K, Ishimoto K, Tsujiuchi T (2020) Effects of lysophosphatidic acid (LPA) receptor-2 (LPA2) and LPA3 on the regulation of chemoresistance to anticancer drug in lung cancer cells. Cell Signal 69:109551. https://doi.org/10.1016/j.cellsig.2020.109551
93. Houben AJS, Moolenaar WH (2011) Autotaxin and LPA receptor signaling in cancer. Cancer Metastasis Rev 30:557–565. https://doi.org/10.1007/s10555-011-9319-7
94. Ren Y, Dai C, Zheng H et al (2016) Prognostic effect of liver metastasis in lung cancer patients with distant metastasis. Oncotarget 7:53245–53253. https://doi.org/10.18632/oncotarget.10644
95. Shiroyama T, Suzuki H, Tamiya M et al (2018) Clinical characteristics of liver metastasis in nivolumabtreated patients with non-small cell lung cancer. Anticancer Res 38:4723–4729. https://doi.org/10.21873/anticanres.12779

96. Xu Z, Yang Q, Chen X et al (2019) Clinical associations and prognostic value of site-specific metastases in non-small cell lung cancer: a population-based study. Oncol Lett 17:5590–5600. https://doi.org/10.3892/ol.2019.10225
97. Kagohashi K, Satoh H, Ishikawa H et al (2003) Liver metastasis at the time of initial diagnosis of lung cancer. Med Oncol 20:25–28. https://doi.org/10.1385/MO:20:1:25
98. Mishima S, Nozaki Y, Mikami S et al (2015) Diffuse liver metastasis of small-cell lung cancer presenting as acute liver failure and diagnosed by transjugular liver biopsy: a rare case in whom nodular lesions were detected by enhanced CT examination. Case Rep Gastroenterol 9:81–87. https://doi.org/10.1159/000381140
99. Riihimäki M, Hemminki A, Fallah M et al (2014) Metastatic sites and survival in lung cancer. Lung Cancer 86:78–84. https://doi.org/10.1016/j.lungcan.2014.07.020
100. Nakazawa K, Kurishima K, Tamura T et al (2012) Specific organ metastases and survival in small cell lung cancer. Oncol Lett 4:617–620. https://doi.org/10.3892/ol.2012.792
101. Doebele RC, Lu X, Sumey C et al (2012) Oncogene status predicts patterns of metastatic spread in treatment-naive nonsmall cell lung cancer. Cancer 118:4502–4511. https://doi.org/10.1002/cncr.27409
102. Kuijpers CCHJ, Hendriks LEL, Derks JL et al (2018) Association of molecular status and metastatic organs at diagnosis in patients with stage IV non-squamous non-small cell lung cancer. Lung Cancer 121:76–81. https://doi.org/10.1016/j.lungcan.2018.05.006
103. Golding B, Luu A, Jones R, Viloria-Petit AM (2018) The function and therapeutic targeting of anaplastic lymphoma kinase (ALK) in non-small cell lung cancer (NSCLC). Mol Cancer 17:1–15. https://doi.org/10.1186/s12943-018-0810-4
104. Murray PB, Lax I, Reshetnyak A et al (2015) Heparin is an activating ligand of the orphan receptor tyrosine kinase ALK. Sci Signal 8:1–8. https://doi.org/10.1126/scisignal.2005916
105. Sakagami H, Hashimoto KEN, Suzuki F et al (2008) Tumor-specificity and type of cell death induced by vitamin K2 derivatives and prenylalcohols. Anticancer Res 158:151–158
106. Voena C, Varesio LM, Zhang L et al (2016) Oncogenic ALK regulates EMT in non-small cell lung carcinoma through repression of the epithelial splicing regulatory protein 1. Oncotarget 7:33316–33330. https://doi.org/10.18632/oncotarget.8955
107. An R, Wang Y, Voeller D et al (2016) CRKL mediates EML4-ALK signaling and is a potential therapeutic target for ALK-rearranged lung adenocarcinoma. Oncotarget 7:29199–29210. https://doi.org/10.18632/oncotarget.8638
108. Cooke M, Baker MJ, Kazanietz MG (2020) Rac-GEF/Rac Signaling and Metastatic Dissemination in Lung Cancer. Front Cell Dev Biol 8:1–7. https://doi.org/10.3389/fcell.2020.00118
109. Schmid S, Gautschi O, Rothschild S et al (2017) Clinical outcome of ALK-Positive Non–Small Cell Lung Cancer (NSCLC) patients with de novo EGFR or KRAS co-mutations receiving tyrosine kinase inhibitors (TKIs). J Thorac Oncol 12:681–688. https://doi.org/10.1016/j.jtho.2016.12.003
110. Marino FZ, Ronchi A, Accardo M, Franco R (2017) Concomitant ALK/KRAS and ALK/EGFR mutations in non small cell lung cancer: different profile of response to target therapies. Transl Cancer Res 6:S457–S460. https://doi.org/10.21037/tcr.2017.03.77
111. Sweis RF, Thomas S, Bank B et al (2016) Concurrent EGFR mutation and ALK translocation in non-small cell lung cancer. Cureus 8. https://doi.org/10.7759/cureus.513
112. Gonzales CB, De La Chapa JJ, Saikumar P et al (2016) Co-targeting ALK and EGFR parallel signaling in oral squamous cell carcinoma. Oral Oncol 59:12–19. https://doi.org/10.1016/j.oraloncology.2016.05.007
113. Miyawaki M, Yasuda H, Tani T et al (2017) Overcoming EGFR bypass signal-induced acquired resistance to ALK tyrosine kinase inhibitors in ALK-translocated lung cancer. Mol Cancer Res 15:106–114. https://doi.org/10.1158/1541-7786.MCR-16-0211
114. Meng DF, Xie P, Peng LX et al (2017) CDC42-interacting protein 4 promotes metastasis of nasopharyngeal carcinoma by mediating invadopodia formation and activating EGFR signaling. J Exp Clin Cancer Res 36:21. https://doi.org/10.1186/s13046-016-0483-z

115. Tonucci FM, Almada E, Borini-Etichetti C et al (2019) Identification of a CIP4 PKA phosphorylation site involved in the regulation of cancer cell invasiveness and metastasis. Cancer Lett 461:65–77. https://doi.org/10.1016/j.canlet.2019.07.006
116. Truesdell P, Ahn J, Chander H et al (2015) CIP4 promotes lung adenocarcinoma metastasis and is associated with poor prognosis. Oncogene 34:3527–3535. https://doi.org/10.1038/onc.2014.280
117. Gaspar LE (2004) Brain metastases in lung cancer. Expert Rev Anticancer Ther 4:259–270. https://doi.org/10.1586/14737140.4.2.259
118. Mujoomdar A, Austin JHM, Malhotra R et al (2007) Clinical predictors of metastatic disease to the brain from non-small cell lung carcinoma: primary tumor size, cell type, and lymph node metastases. Radiology 242:882–888. https://doi.org/10.1148/radiol.2423051707
119. Barnholtz-Sloan JS, Sloan AE, Davis FG et al (2004) Incidence proportions of brain metastases in patients diagnosed (1973 to 2001) in the Metropolitan Detroit Cancer Surveillance System. J Clin Oncol 22:2865–2872. https://doi.org/10.1200/JCO.2004.12.149
120. Davis FG, Dolecek TA, McCarthy BJ, Villano JL (2012) Toward determining the lifetime occurrence of metastatic brain tumors estimated from 2007 United States cancer incidence data. Neuro-Oncology 14:1171–1177. https://doi.org/10.1093/neuonc/nos152
121. Valiente M, Obenauf AC, Jin X et al (2014) Serpins promote cancer cell survival and vascular Co-option in brain metastasis. Cell 156:1002–1016. https://doi.org/10.1016/j.cell.2014.01.040
122. Kim SW, Choi HJ, Lee HJ et al (2014) Role of the endothelin axis in astrocyte- and endothelial cell-mediated chemoprotection of cancer cells. Neuro-Oncology 16:1585–1598. https://doi.org/10.1093/neuonc/nou128
123. Chen Q, Boire A, Jin X et al (2016) Carcinoma-astrocyte gap junctions promote brain metastasis by cGAMP transfer. Nature 533:493–498. https://doi.org/10.1038/nature18268
124. Ellenbroek SIJ, Collard JG (2007) Rho GTPases: functions and association with cancer. Clin Exp Metastasis 24:657–672. https://doi.org/10.1007/s10585-007-9119-1
125. Masre SF, Rath N, Olson MF, Greenhalgh DA (2017) ROCK2/rasHa co-operation induces malignant conversion via p53 loss, elevated NF-κB and tenascin C-associated rigidity, but p21 inhibits ROCK2/NF-κB-mediated progression. Oncogene 36:2529–2542. https://doi.org/10.1038/onc.2016.402
126. Masre SF, Rath N, Olson MF, Greenhalgh DA (2020) Epidermal ROCK2 induces AKT1/GSK3β/β-catenin, NFκB and dermal tenascin C; but enhanced differentiation and p53/p21 inhibit papilloma. Carcinogenesis. https://doi.org/10.1093/carcin/bgz205
127. Wojciak-Stothard B, Ridley AJ (2002) Rho GTPases and the regulation of endothelial permeability. Vasc Pharmacol 39:187–199. https://doi.org/10.1016/S1537-1891(03)00008-9
128. Abbott NJ, Patabendige AAK, Dolman DEM et al (2010) Structure and function of the blood–brain barrier. Neurobiol Dis 37:13–25. https://doi.org/10.1016/j.nbd.2009.07.030
129. Hanibuchi M, Kim SJ, Fidler IJ, Nishioka Y (2014) The molecular biology of lung cancer brain metastasis: an overview of current comprehensions and future perspectives. J Med Investig 61:241–253
130. Li B, Zhao WD, Tan ZM et al (2006) Involvement of Rho/ROCK signalling in small cell lung cancer migration through human brain microvascular endothelial cells. FEBS Lett 580:4252–4260. https://doi.org/10.1016/j.febslet.2006.06.056
131. Maekawa M, Ishizaki T, Boku S et al (1999) Signaling from Rho to the actin cytoskeleton through protein kinases ROCK and LIM-kinase. Science 285:895–898. https://doi.org/10.1126/science.285.5429.895
132. Chang YC, Stins MF, McCaffery MJ et al (2004) Cryptococcal yeast cells invade the central nervous system via transcellular penetration of the blood-brain barrier. Infect Immun 72:4985–4995. https://doi.org/10.1128/IAI.72.9.4985-4995.2004
133. Lee TH, Avraham HK, Jiang S, Avraham S (2003) Vascular endothelial growth factor modulates the transendothelial migration of MDA-MB-231 breast cancer cells through regulation of brain microvascular endothelial cell permeability. J Biol Chem 278:5277–5284. https://doi.org/10.1074/jbc.M210063200

134. Li B, Wang C, Zhang Y et al (2013a) Elevated PLGF contributes to small-cell lung cancer brain metastasis. Oncogene 32:2952–2962. https://doi.org/10.1038/onc.2012.313
135. Hauck CR, Thibodeau BJ, Ahmed S et al (2017) Targeted DNA sequencing of non–small cell lung cancer identifies mutations associated with brain metastases. Int J Radiat Oncol 99:S199–S200. https://doi.org/10.1016/j.ijrobp.2017.06.495
136. Cantley LC (2002) The phosphoinositide 3-kinase pathway. Science 296:1655–1657
137. Engelman JA, Luo J, Cantley LC (2006) The evolution of phosphatidylinositol 3-kinases as regulators of growth and metabolism. Nat Rev Genet 7:606–619. https://doi.org/10.1038/nrg1879
138. Vanhaesebroeck B, Leevers SJ, Panayotou G, Waterfield MD (1997) Phosphoinositide 3-kinases: a conserved family of signal transducers. Trends Biochem Sci 22:267–272. https://doi.org/10.1016/S0968-0004(97)01061-X
139. Gray JW (2016) PI3 kinase pathway mutations in human cancers. JAMA Oncol 2:1543–1544. https://doi.org/10.1001/jamaoncol.2016.0875
140. Steelman LS, Stadelman KM, Chappell WH et al (2008) Akt as a therapeutic target in cancer. Expert Opin Ther Targets 12:1139–1165. https://doi.org/10.1517/14728222.12.9.1139
141. Tian X, Zhou D, Chen L et al (2018) Polo-like kinase 4 mediates epithelial–mesenchymal transition in neuroblastoma via PI3K/Akt signaling pathway. Cell Death Dis 9:54. https://doi.org/10.1038/s41419-017-0088-2
142. David O, Jett J, LeBeau H et al (2004) Phospho-Akt overexpression in non-small cell lung cancer confers significant stage-independent survival disadvantage. Clin Cancer Res 10:6865–6871. https://doi.org/10.1158/1078-0432.CCR-04-0174
143. Jin Y, Yuan Y, Yi M et al (2019) Phosphorylated-Akt overexpression is associated with a higher risk of brain metastasis in patients with non-small cell lung cancer. Biochem Biophys Rep 18:100625. https://doi.org/10.1016/j.bbrep.2019.100625
144. Li Q, Yang J, Yu Q et al (2013b) Associations between single-nucleotide polymorphisms in the PI3K-PTEN-AKT-mTOR pathway and increased risk of brain metastasis in patients with non-small cell lung cancer. Clin Cancer Res 19:6252–6260. https://doi.org/10.1158/1078-0432.CCR-13-1093
145. da Silva GT, Bergmann A, Thuler LCS (2019) Incidence and risk factors for bone metastasis in Non-Small Cell Lung Cancer. Asian Pacific J Cancer Prev 20:45–51. https://doi.org/10.31557/APJCP.2019.20.1.45
146. Kuchuk M, Kuchuk I, Sabri E et al (2015) The incidence and clinical impact of bone metastases in non-small cell lung cancer. Lung Cancer 89:197–202. https://doi.org/10.1016/j.lungcan.2015.04.007
147. Decroisette C, Monnet I, Berard H et al (2011) Epidemiology and treatment costs of bone metastases from lung cancer: a French prospective, observational, multicenter study (GFPC 0601). J Thorac Oncol 6:576–582. https://doi.org/10.1097/JTO.0b013e318206a1e3
148. Kakhki VRD, Anvari K, Sadeghi R et al (2013) Pattern and distribution of bone metastases in common malignant tumors. Nucl Med Rev 16:66–69. https://doi.org/10.5603/NMR.2013.0037
149. Andrade K, Fornetti J, Zhao L et al (2017) RON kinase: a target for treatment of cancer-induced bone destruction and osteoporosis. Sci Transl Med 9:eaai9338. https://doi.org/10.1126/scitranslmed.aai9338
150. Fazzalari NL (2008) Bone remodeling: a review of the bone microenvironment perspective for fragility fracture (osteoporosis) of the hip. Semin Cell Dev Biol 19:467–472. https://doi.org/10.1016/j.semcdb.2008.08.003
151. Bellido T (2014) Osteocyte-driven bone remodeling. Calcif Tissue Int 94:25–34. https://doi.org/10.1007/s00223-013-9774-y
152. Roodman GD (2004) Mechanisms of Bone Metastasis. N Engl J Med 350:1655–1664. https://doi.org/10.1056/NEJMra030831
153. Virk MS, Lieberman JR (2007) Tumor metastasis to bone. Arthritis Res Ther 9:1–10. https://doi.org/10.1186/ar2169

154. Stern PH, Tatrai A, Semler DE et al (1995) Endothelin receptors, second messengers, and actions in bone. J Nutr 125:2028S–2032S. https://doi.org/10.1093/jn/125.suppl_7.2028S
155. Hall CL, Kang S, MacDougald OA, Keller ET (2006) Role of WNTS in prostate cancer bone metastases. J Cell Biochem 97:661–672. https://doi.org/10.1002/jcb.20735
156. Khosla S (2001) Minireview: the OPG/RANKL/RANK system. Endocrinology 142:5050–5055. https://doi.org/10.1210/endo.142.12.8536
157. Kang Y, Siegel PM, Shu W et al (2003) A multigenic program mediating breast cancer metastasis to bone. Cancer Cell 3:537–549. https://doi.org/10.1016/S1535-6108(03)00132-6
158. Shariat SF, Andrews B, Kattan MW et al (2001) Plasma levels of interleukin-6 and its soluble receptor are associated with prostate cancer progression and metastasis. Urology 58:1008–1015. https://doi.org/10.1016/S0090-4295(01)01405-4
159. Bobinac D, Marić I, Zoričić S et al (2005) Expression of bone morphogenetic proteins in human metastatic prostate and breast cancer. Croat Med J 46:389–396
160. Sottnik JL, Dai J, Zhang H et al (2015) Tumor-induced pressure in the bone microenvironment causes osteocytes to promote the growth of prostate cancer bone metastases. Cancer Res 75:2151–2158. https://doi.org/10.1158/0008-5472.CAN-14-2493
161. Krzeszinski JY, Wan Y (2015) New therapeutic targets for cancer bone metastasis. Trends Pharmacol Sci 36:360–373. https://doi.org/10.1016/j.tips.2015.04.006
162. Luo Q, Xu Z, Wang L et al (2016) Progress in the research on the mechanism of bone metastasis in lung cancer. Mol Clin Oncol 5:227–235. https://doi.org/10.3892/mco.2016.917
163. D'Antonio C, Passaro A, Gori B et al (2014) Bone and brain metastasis in lung cancer: recent advances in therapeutic strategies. Ther Adv Med Oncol 6:101–114. https://doi.org/10.1177/1758834014521110
164. Tobeiha M, Moghadasian MH, Amin N, Jafarnejad S (2020) RANKL/RANK/OPG pathway: a mechanism involved in exercise-induced bone remodeling. Biomed Res Int 2020:1–11. https://doi.org/10.1155/2020/6910312
165. Miki T, Yano S, Hanibuchi M et al (2004) Parathyroid hormone-related protein (PTHRP) is responsible for production of bone metastasis, but not visceral metastasis, by human small cell lung cancer SBC-5 cells in natural killer cell-depleted scid mice. Int J Cancer 108:511–515. https://doi.org/10.1002/ijc.11586
166. Roato I (2014) Bone metastases: when and how lung cancer interacts with bone. World J Clin Oncol 5:149–155
167. Ell B, Mercatali L, Ibrahim T et al (2013) Tumor-induced osteoclast miRNA changes as regulators and biomarkers of osteolytic bone metastasis. Cancer Cell 24:542–556. https://doi.org/10.1016/j.ccr.2013.09.008
168. Karapanagiotou EM, Terpos E, Dilana KD et al (2010) Serum bone turnover markers may be involved in the metastatic potential of lung cancer patients. Med Oncol 27:332–338. https://doi.org/10.1007/s12032-009-9214-z
169. Peng X, Guo W, Ren T et al (2013) Differential expression of the RANKL/RANK/OPG system is associated with bone metastasis in human non-small cell lung cancer. PLoS One 8: e58361. https://doi.org/10.1371/journal.pone.0058361
170. Zhu X, Luo H, Xu Y (2019) Transcriptome analysis reveals an important candidate gene involved in both nodal metastasis and prognosis in lung adenocarcinoma. Cell Biosci 9:82. https://doi.org/10.1186/s13578-019-0356-1

Therapeutic Strategies Targeting Signaling Pathways in Lung Cancer

9

Gurjit Kaur Bhatti, Paras Pahwa, Anshika Gupta, Umashanker Navik, and Jasvinder Singh Bhatti

Abstract

Recent knowledge of the role of signaling pathways and their underlying mechanisms in the pathogenesis of several diseases may lead to the development of therapeutic strategies. In the recent time, several drug molecules have been developed which target the cell signaling pathways and may be used in combination with other standard therapies for the synergistic effects in reducing the lung cancer pathophysiology across the world. Further, some of predictive biomarkers have been identified. The current chapter deals with the involvement of signaling pathways in the development of lung cancer and further new therapeutic approaches that intend to pave the way for the development of improved clinical treatment of lung cancer.

Keywords

Lung cancer · Signalling pathways · Drug molecules · Vaccine

G. K. Bhatti
Department of Medical Lab Technology, University Institute of Applied Health Sciences, Chandigarh University, Mohali, Punjab, India

P. Pahwa · A. Gupta
Department of Biotechnology, Sri Guru Gobind Singh College, Chandigarh, India

U. Navik
Department of Pharmacology, Central University of Punjab, Bathinda, Punjab, India

J. S. Bhatti (✉)
Department of Human Genetics and Molecular Medicine, School of Health Sciences, Central University of Punjab, Bathinda, Punjab, India
e-mail: jasvinder.bhatti@cup.edu.in

K. Dua et al. (eds.), *Targeting Cellular Signalling Pathways in Lung Diseases*,
https://doi.org/10.1007/978-981-33-6827-9_9

9.1 Introduction

Lung cancer is considered as extremely rare disease in early twentieth century [1], but now lung cancer is one of the most commonly diagnosed cancer especially in a developed country. The major risk factor for the development of lung cancer is linked with smoking [2–4]; however, there are cases reported that even nonsmokers are also diagnosed with the lung cancer thus indicating there are other unknown risks factors that are also involved with dissimilar molecular markers and posing to complex etiology of cancer [5, 6]. Worldwide around 1.8 million people are annually diagnosed with lung cancer, and of them, 1.6 million demise has been reported [7]. There is increased incidence of mortality rate among patients with lung cancer which is related to a widespread advanced diagnosis that impedes the restorative treatment and a low 5-year survival rate, expected at 15% of all phases combined [8, 9]. Smoking leads to 80–90% of lung cancer cases thus constitute the major risk factor for the disease development and progression. According to the World Health Organization (WHO), lung cancer has the following two wide-ranging histological subtypes: first is non-small cell lung cancer (NSCLC), and second is small-cell lung cancer (SCLC), which accounts for 85% and 15% of cancer, respectively [10]. Lung adenocarcinoma (LUAD), lung squamous carcinoma (LUSC), and broad subtypes of cell carcinoma are included in the category of NSCLC [10].

The signaling pathways with a critical function in cell development, proliferation, angiogenesis, apoptosis, and metastasis are dysregulated by genomic mutations of cancer cells [11]. The molecular mechanisms that are involved in lung cancer growth and progression are complex. Genetic and epigenetic modifications arise after exposure to chronic inflammation and cigarette smoking, including activation of various oncogenes and inactivation of the tumor suppressor gene due to DNA damage caused by oxidative stress. Such changes contribute to the multistep development and progression of lung cancer through apoptosis and cell proliferation. Changes in gene expressions and molecular abnormalities involved in signaling pathways have been documented in a number of studies, despite the fact that by changing chromatin structures and the exact articulation of genes, epigenetic adjustments in the cellular breakdown in the lungs emphatically lead to cell transformation, e.g., modification of histone and chromatin protein and micro-RNA and DNA methylation, which are all liable for silencing genes that stifle the tumor while improving oncogene expression. Both genetic and epigenetic pathways associated with lung tumorigenesis vary among smokers and nonsmokers and go about as devices for cancer analysis, forecast, and clinical development and focused on treatments [12].

The firm drop in both occurrence and demise rate has resulted due to the introduction of programs like "tobacco control," "low-dose spiral computed tomography screening," and developments of multidisciplinary therapies. However, 52–58%of patients with lung cancer have the advanced-stage disease, and even through surgery, the immense number of these patients does not recover. Likewise, due to the elevated relapse rate and early formation of micro metastasis, the prognosis remains low even in locally advanced diseases [13]. Unveiling of mutations involved in cancer and the development of various molecules that target a crucial enzymes such as tyrosine kinase are one of the most significant therapeutic developments in the treatment of lung cancer in the last decade [14]. In 2009, the US Food and Drug Administration (FDA) approved the first drug Erlotinib that targets selectively epidermal growth factor receptor (EGFR) which is involved in cell proliferation and development of cancer [15]. Afterward, crizotinib was developed which is used for NSCLC with anaplastic lymphoma kinase (ALK) through the inhibition of MET (mesenchymal-to-epithelial transition/hepatocyte growth factor receptor) having around 70% of response rate, but drug resistance is the common problem with all the treatment strategies. Therefore, new generation TKIs have been developed to overcome the drug resistance [16]. Further many novel targets which have been discovered including EGFR, BRAF, ALK, MET, ROS1, RET, and many more are under investigation. Next-generation sequencing and cell-free DNA technologies, in addition to these drugs, have provided quick and suitable methods for checking for gene mutations and designing targeted therapies [17]. In addition, personalized medicine has become part of everyday practice, and tailoring care is becoming a reality for individual patients.

9.2 Epidemiology of Lung Cancers

Overall, the cancer is the second leading causes of death globally [18]. Figure 9.1 shows the different types of lung cancers which are associated with higher mobidity and mortality accross the world. A dropping degree of economic growth indicates no disparities in deaths from cancer in men but a higher rate of death in women of lung cancer compared to developing countries with developed countries.

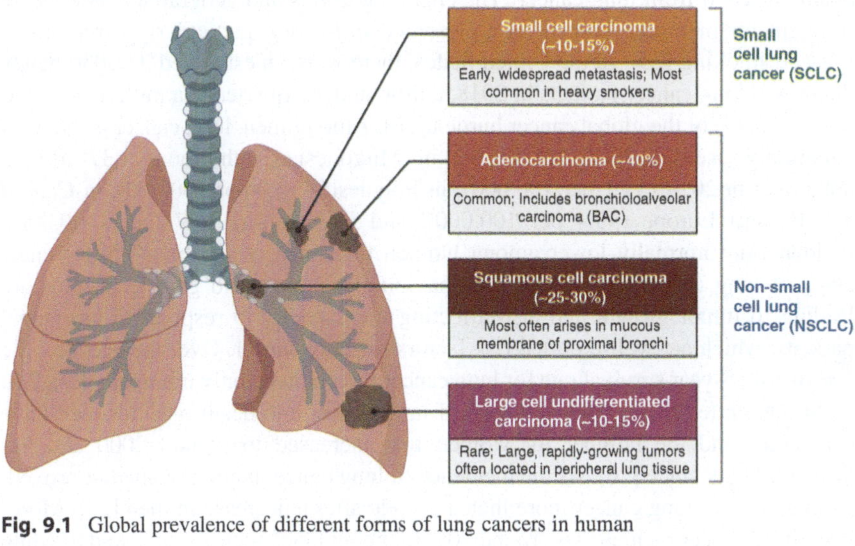

Fig. 9.1 Global prevalence of different forms of lung cancers in human

Females suffering from lung cancer were lagged behind those caused by breast cancer in developed countries [19]. The prevalence and mortality of lung cancer were closely related to the trend of cigarette smoking. As smoking rates usually peak in men first, followed by female lung cancer occurrence and mortality, they increased in the decades before the start of comprehensive tobacco control programs (20–22). In developed countries, these patterns have arisen earlier than in developing countries. The prevalence of lung cancer and death rates in the United States and the United Kingdom, in fact, has declined since the 1990s. Emerging countries (BRICS), in comparison, also have high cigarette smoking rates in both men and women—like Brazil, Russia, India, China, and South Africa (BRICS). They have a lower cancer rate than developing countries but a rising mortality burden. Coherent access to healthcare leading to retarded diagnosis and treatment, environmental infection, and sociocultural challenges are the reasons for this trend [20]. Lung carcinoma has evolved in the last century from a mysterious dark disease to the world's most frequent cancer and cause of death due to cancer. In the late 1840s, only 22 lung cancer cases were identified [21, 22]. In a study conducted by Siegel and colleagues in 2010 in reviewed the cancer data and estimated that around 239,320 were new cases and 161,250 deaths were reported with lung cancer in the United States [23]. The statistics reflects 2007 data and thus possibly underestimates the present burden of pulmonary cancer. Lung cancer has been the world's most prevalent incidence and mortality cancer since 1985. Globally, lung cancer is the highest diagnosis contributor (1,350,000 new cases and 12.4% of the total new cases) and cancer death (1,180,000 deaths, with 17.6% of the total deaths). The United States' 5-year survival rate for lung cancer is 15.6%, although the survival rate has risen in the last few decades. In 2012, 1.8 million new cases worldwide were detected in the most recent global statistical study, with 1.6 million demises in the same year [24]. This is growing from 1.6 million new diagnoses and 1.4 million deaths in 2008 from lung cancer. The changing trends and geographic patterns are different for men and women, as represented in history, particularly variations in tobacco smoking [25]. In the United States, there were an estimated 234,030 newly diagnosed lung cancer patients in 2018, a little under a quarter of a million accounting for 12.1% of the global cancer burden [26]. Among men, lung cancer is the most commonly used diagnosis of cancer in Micronesia, with about 1.37 million diagnoses in 2018 (54.1 per 100,000), in Polynesia (52.0 per 100,000), in Central and Eastern Europe (49.4 per 100,000), and in East Asia (47.2 per 100,000). Incidents are normally lower among women than men, with about 725,000 new cases of lung cancer in 2018 [27]. Behavioural, climate, and genetic hazards are leading to tumor growth and also affecting the capacity to respond to individual patients, which are well-reported risk factors for lung cancer. Over the decades, the low overall 5-year survival rate for lung cancer has changed only minimally [26, 28].

In the United States, the incidence of lung cancer and death rates for men have decreased, and the number for women first increased to around 2000 and has decreased ever since [26]. As the incidence of lung cancer among women improved, women died of lung cancer more than a decade after a decrease in men [29]. Moreover, lung cancer incidence is around 20% higher in black men as compared to white men. On the opposite, Asian Americans, Pacific Islanders, and Hispanic women have the lowest incidence and mortality. In patients younger than 40 years, the prevalence of lung cancer remains low. It then starts to grow steadily, with peaks between 65 and 84 years old. The median age of diagnosis of lung cancer in the United States is 71 years and 90% of the patients are diagnosed after 55 years of age [29]. Present trends in smoking are a big indicator of potential trends in the incidence of lung cancer. However, 19% of cases are never-smoking women and 9% are men in the United States with a rising incidence in young women [30, 31].

9.2.1 Mortality

Global and regional trends of lung cancer-related deaths closely match those attributed to poor survival and high mortality rates. Since the early 1950s, lung cancer was the main cause of demise in men, and in 1987 it was the major reason of deaths in women in the United States [32]. Globally around 1.38 million deaths have been reported in 2008 (18.2% of the total cases). While lung cancer mortality in the United States has increased since the 1950s, recent statistics recorded an annual 1.9% decrease in men's mortality from 1993 to 2005 and 0.9% decrease in women in the same period [2]. The number of deaths of both men and women from lung cancer is higher than any other cancer. An estimated 160,340 deaths in the United States were predicted to occur in 2012, nearly 28% of all deaths from cancer. The rate of mortality among men started to decline in 1991 and decreased by 2.6% per year from 2004 to 2008; the most recent study indicates that mortal death rates were 61.9 per 100,000 for men. Improvements in the death rate for females from lung cancer were back in the background but began to decline in 2003 and by 0.9% a year between 2004 and 2008. The most recent death rate for women was 38.5 per 100,000 for lung cancer. The mortality rate in males and females decreased by around 4% per year from 2012 to 2016. An estimated 142,670 deaths, or approximately 23.5% of all cancer-related deaths, were also projected to occur in 2019. The prevalence of lung cancer, including the highest levels found in the south, is similar in geography.

9.2.2 Survival

While survival advances in recent years have dramatically improved for most other forms of cancer in the United States, the survival of patients diagnosed with lung cancer has improved slightly. This lack of progress largely is due to the late-stage illness that is diagnosed with most patients with unstable survival rates. For both non-small cell lung cancer (NSCLC) and small-cell lung cancer (SCLC) combined, the 5-year relative survival rate of all lung cancers is 19%, and NSCLC 5-year survival (23%) is greater than SCLC (6%). The next wave of lung cancer is the same despite the high mortality rates and the poor survival outcomes of targeted medications, and the development of inhibitors of the immune control point shows long-term survival in patient subgroups. As such, these treatments can play an important part in improving lung cancer patient results that contribute to curable early diagnosis of lung cancer and persistent and management-consistent disease in advanced and metastatic patients.

9.3 Lung Cancer

Lung cancer is the foremost reason of deaths among all different cancer types [19], and it is categorized to NSCLC and SLC which accounts for 85% and 15% of cancer, respectively. NSLC histology presents key features of squamous and adeno cell carcinoma, along with molecular stratification through the specific DNA mutations [33]. The identification of NSCLC at early stages and then the surgical resection offer a favorable prognosis with increased survival rates (5 years) of 70–90% for small localized tumors (stage I) [34–36]. Despite advancement in the diagnostic tools and treatments, most of the patients (around 75%) were diagnosed at advanced stage of disease (stage III/IV) thereby leading to poor survival of the patients [37]. In 2014, it was confirmed by the UK Office for National Statistics that patients have only a 1-year survival rate of only 15–19%, which were diagnosed with distant metastatic disease (stage IV), compared with 81–85% for stage [38].

Cigarette smoking is a closely linked, violent neuroendocrine cancer of SCLC. Patients usually have brief symptoms, and most (60–65%) have metastatic disease. SCLC is a heterogeneous disease of highly chemical sensitive clones. This is why a high percentage of patients respond to first-line chemotherapy but quickly succumb to the disease. Generally, SCLC has a small and comprehensive division into two levels. Normal restricted stage illness treatment encompasses the four-cycle mixture of cisplatin and etoposide chemotherapy, early initiated chest radiation, and treatment of prophylactic crane radiation in the subset of successful response patients. In TNM stages I and II, surgery may play a role. The first-line treatment norm in the United States is again used in extensive disorders, platinum agents, and etoposide in combination. But thoracic radiation therapy is primarily used in patients with essential local control and unclear PCI gain. Despite this, the prognosis appears to be low, and new treatments are needed in order to improve survival [39]. The statistics of cancer has shown, however, that the incidence of SCLC has declined over the last two decades in the United States, along with the decline in cigarette smoking prevalence among the population. In the year 1980, the average incidence rate of lung cancer was at 11/100,000 people/year, which was then declined to 8–9/100,000 people/year by the year 2002. The incidence of SCLC also decreased by 17–20% in the late 1980s as compared to total cases of lung cancer.

SCLC is characterized by a short doubling time, high growth fraction, and widely distributed metastasis in comparison with non-small cell lung cancers (NSCLC). Therefore, at diagnosis (defined as the cancer that cannot be able to include in one single radiation field due to spread of cancer beyond the ipsilateral lung and regional lymph nodes), 60–70% of patients are suffering from an extensive stage disease (ES). Recently, the national lung screening test demonstrated the vigorous characteristic of SCLC and its ability to achieve characteristic early hematogenous expansion [40]. This study demonstrated the importance of the early detection of lung cancer for high-risk patients with low-dose computed tomography (CT). Early detection did not decrease the number of patients diagnosed with ES disease and had no apparent effect on survival for patients with SCLC in patients who had SCLC detected (in comparison to patients with NSCLC). Of the 125 patients detected with SCLC via this early detection program, 86% were diagnosed with advanced disease. Presently there are no effective diagnostic tools for the early detection of SCLC based on these findings (insert reference).

9.4 Signaling Pathways Involved in Lung Cancers

In terms of signaling mechanisms, rather than relying on human causes, genetic abnormalities associated with the risk of lung cancer can be seen. Oncogenes contribute to malignancy, and the escape from apoptosis are the most stimulating signaling pathways. The abnormal role of the tumor cells, dubbed "oncogenetic addiction," is due to mutated oncogenic proteins [41].

9.4.1 Epidermal Growth Factor Receptor (EGFR)

Approximately 90 recognized tyrosine kinase receptors (RTKs) form a series of 58 ligand-mediated TK cell surface factor growth receptors [42]. Whereas RTK is strictly regulated in normal rest cells, mutations or deregulated expression can lead to their role as powerful oncogenes. The prototypical members of four RTKs include EGFR (ERBB1, HER1), ERBB2 (HER2, Neu), ERBB3 (HER3), and ERBB4 (HER4) [43]. The receptor ERBB comprises of a ligand-binding extracellular domain, transmembrane segment and intracellular TK domain, and a C-terminal segment for regulatory applications.TK (59–81% identity) is the largest sequence homologies among the four genes. The different members are activated by multiple ligands, including the EGFRs, the growth factor-α transition, and amphiregulin. Ligand-binding permits the development and the subsequent activation of TK and transphosphorylation of homo- or heterodimer complexes. In turn, it creates docking sites for a range of cytoplasmic signals and leads to activation of Ras, phosphatidylinositol-3-kinase (PI3K) pathway. Multiple tumor types, including NSCLCs, have been observed to deregulate EGFR. In NSCLCs [44] of squamous cells and ADC subtypes, Hirsch et al. reported that EGFR protein was overexpressed (62%) and is often related to adverse prognosis [45]. The RTK cell surface receptor superfamily acts as a mediator of cell signaling through extracellular factors of growth. ERBB members of the RTK family gained substantial attention because of their close interaction with malignant proliferation [46]. The protein kinase and PI3K/Akt pathways are fundamental signaling networks connecting EGFR to cell expansion and endurance. In NSCLC (and infrequently in different tumors), EGFR flagging pathway genes were found to have transformed. EGFR and KRAS transformations in ~10 30% of NSCLCs were accounted for relying upon the topographical position. EGFR changes are connected to ADC histology, Eastern Asian identity, never-smoking, and female sexual orientation [47]. KRAS transformations are regularly focused on ADC histology; however, some cases are unique in relation to EGFR changes, since they are uncommon in East Asia and happen in guys and smokers all the more frequently. Other pathway EGFR genes, for example, HER2 (~2%), HER4 (~2%), BRAF (~2%) [22%), and PIK3CA (~4%), were less habitually recognized in somatic mutations. The changes center around basic areas of the TK domain related to downstream signaling (exons 18–21) and include deletion, inclusions, and point-initiating mutation in an assortment of structures. However, ~85% of mutations are accounted for by two main types: exon 19 removal and L858R mutation in exon 21 [48]. Figure 9.2 shows the involvement of EGFR signaling pathway in cancer.

EGFR became one of the first rationally selected molecules for the targeted therapy as the result of repeated deregulation by EGFR pathway genes in NSCLC. Although initial approaches have been focused on monoclonal antibodies blocking the interaction between ligands and receptors, recent approaches have been based on small molecular reversible TK inhibitors (TKIs). For biochemical responses triggered by this receptor, the TK activity of the EGFR is essential. In the treatment of advanced or chronic NSCLC, two TKIs and erlotinib have been widely utilized [49].

EGFR mutations have subsequently been found to predict the TKI response in the same patient with the TK domain. According to the meta-analysis of 1170 patients, >70% of the EGFR mutations in NSCLCs reacted to TKIs, while 10% of the EGFR mutations were non-EGFR tumors. Not all active mutations are, however, associated with the TKI reaction and secondary resistance to a point mutation T790 M [50]. Furthermore, insertion mutation is also related to primary resistance in exon 20. Further studies have shown that after TKI therapy, factors other than EGFR mutations can be used to assess responsiveness and survival. In a major, unselected study with adequate controls, the increase in the EGFR gene copy number was associated with substantial improvement in TKI sensitivity and survival. Moreover, other EGFR family members may include substantial factors of TKI sensitivity, HER2, and EGFR3. The clinical discovery of inherent resistance to TKIs by somatic mutations of KRAS is another complication [51].

Though mutations are aimed at the histology of ADC, protein upregulation in squamous carcinomas is normal. Autocrine loops resulting from the formation of ligands by tumor cells and ligand release proteins from the surface of the cells also constitute additional mechanisms of upregulation. Different methods have shown that mutations are an early characteristic of the multistage pathogenic phase with a small field effect on tumors, while an increased copy count is a relatively late tumor phenotype or metastasis occurrence [52].

9.4.2 Mutations in Other EGFR Signaling Pathway Genes

An enormous number of genes interaction and sub-pathways comprise the complex EGFR signaling mechanism. One of the well-documented and often activated

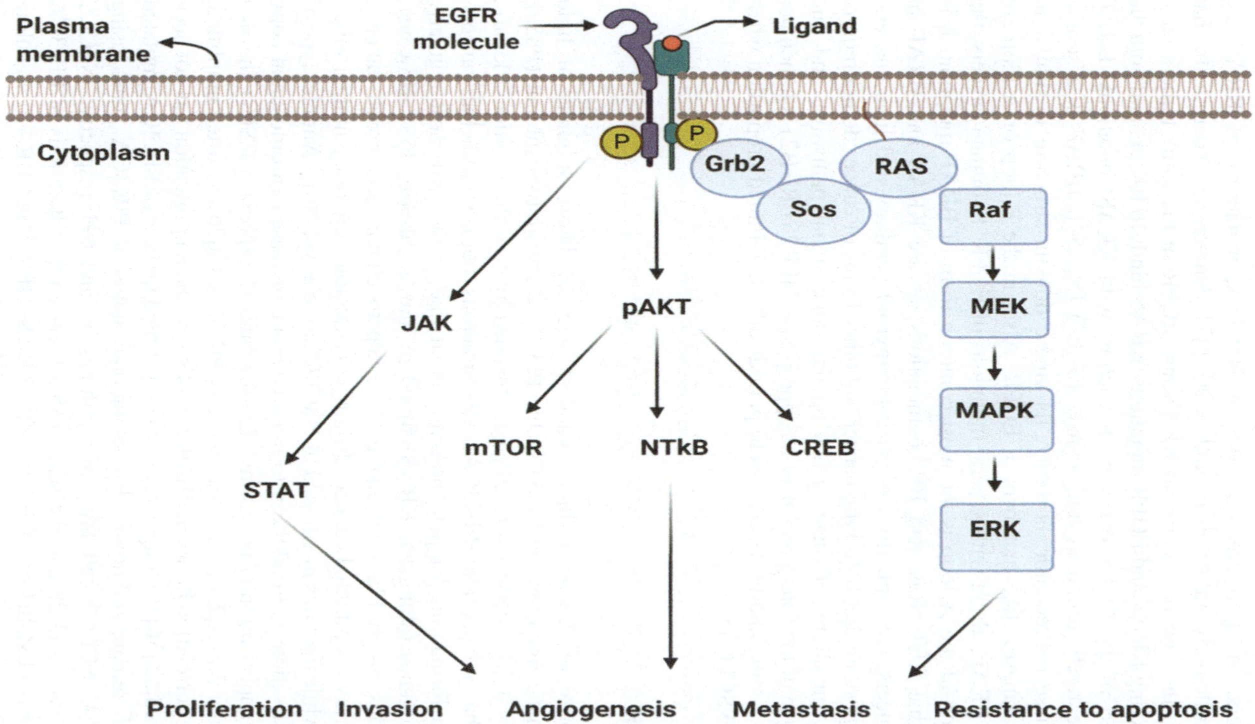

Fig. 9.2 EGFR signaling pathway in cancer. EGFR is activated either as a homo- or heterodimer resulting in regulation of multiple pathways. In particular the RAS/RAF/MAPK, AKT, and JAK/STAT pathways downstream of EGFR play integral roles in cell migration, proliferation, and survival, respectively. Anti-EGFR antibodies are targeted to the external ligand-binding domains, while the small molecule inhibitors or tyrosine kinase inhibitors (TKIs) target its cytoplasmic kinase domains

oncogenes, the downstream KRAS that encodes the small, guanosine 5′ triphosphate-binding protein, is triggered by missense mutations in several kinds of human cancers that make it recognize the most common oncogenes activated in human cancers. In approximately 20% of NSCLCs, mutations of KRAS are identified, particularly in smokers and ADC [47]. Numerous experiments that examine both the mutation status of KRAS and EGFR in the same tumors have shown conflicting KRAS and EGFR mutations. KRAS binds to BRAF, and both the genes are part of the EGFR cascade. In contrast with KRAS mutations, BRAF mutations are rarely found in lung cancer (0–3%) [53, 54]. BRAF is a serine/threonine kinase nonreceptor; however, its kinase domain structure is similar to other protein kinases, like members of EGFR. Also, in the P-loop or A-loop are mutations of BRAF, along with several mutations of EGFR. In human cancers, the V600 mutation of the A-loop is the most common form of BRAF mutation. It is noteworthy that RET, Ras, and BRAF mutations, as are KRAS and BRAF in colorectal cancers, are mutually exclusive for thyroid papillary and lung cancers. These findings show that both lung cancer and other forms of cancer do not require simultaneous mutations of many genes on the same signal pathways, and one mutation in one of the four genes may be appropriate. HER2 (ERBB2) is a member of the EGFR family, and mutations occur periodically in a limited proportion when amplified in NSCLC (59).

9.4.3 PI3K

PI3Ks comprise of the heterodimer, catalyst, and regulatory subunits of lipid kinases. Somatic mutation are found in the PI3 K administrative subunit (P85a), fundamentally in its synergist subunit areas, encoded by phosphoinositide-3-kinase, catalytic alpha polypeptide (PIK3CA). In numerous human epithelial cancers, PIK3CA transformations happen, bringing about one of the two most regularly changed oncogenes (alongside KRAS) found in human diseases [55]. However, with high rates of glioblastoma and gastric, hepatocellular, and breast cancers, individual forms of epithelial cancers display considerable variability in their mutational rates, while the rates mentioned in NSCLC are relatively low. Another type of oncogenic activation is, in addition to mutations, the increased chromosomal copy number (via amplification or polysomy). In lung cancers (especially CSCs), the area of chromosome 3q (3q25–27), where there is PIK3CA (3q26), is often amplified. However, the correlation between PIK3CA mutations and amplification has not been thoroughly studied [56]. In lung cancers, the functional impacts of PIK3CA, mutated or intensified, remain unknown. The mutational status of PIK3CA genes using 86 NSCLC, 43 SCLC3 cell lines extrapulmonary, and 691 respective NSCLC tumors in exons 9 and 20 and the relationship between the alterations of PIK3CA and the mutational status of EGFR (EGFR, KRAS, HER2, and BRAF) signaling tract genes. The expression and activity of PIK3CA were also determined, and results were related to cell growth effects. 4.7% of NSCLC cell lines and 1.6% of tumors of all major histological types were identified with mutations. In SCC

(33.1%) the gains in PIK3CA copy numbers were more common compared to adenocarcinoma (6.2%) or the SCLC (4.7%), respectively. Thus, deregulation of the PI3 K pathway has been recognized as one of the only known molecular modifications in SCCs rather than adenocarcinomas. PIK3CA mutations or gains are present and functional in a subgroup of lung cancers [57]. Figure 9.3 shows the cascade of P13K, RAS/MAPK, and JAK/STAT signaling pathways involved in cancer development and as a therapeutic target in the development of new drug molecules.

9.4.4 p53 Gene

A 53-kDa nuclear protein encodes the p53 gene, located in chromosome region 17p13.1. The protein serves as a transcription factor for DNA injury response, inhibits the development of cells via the cell cycle in the late G1 phase, and activates apoptosis. DNA damage is the main phosphorylation signal for p53, which is catalyzed by the kinase encoded by the ataxia-telangiectasia gene ATM. P53 For various genes, including p21/WAF1/CIP1, MDM2, GADD45, BAX, and cyclin G, p53 acts as a DNA-binding transcription factor for many genes. These genes' activation contributes to apoptosis, cell cycle arrest, and repair of DNA [58]. p53 gene mutations constitute some of the most common cancer-related genetic changes and cause loss of tumor suppressor function and loss of apoptosis induction capacity. Benzo(a)pyrene-induced damage from cigarette smoking appears to be related to this form of mutation. A substantially higher risk compared to nonsmokers for p53 mutations was observed in smokers. Alcohol consumption, however, appears to have an effect on the incidence of p53 mutations found more commonly in alcohol drinkers who smoke than in nondrinkers who smoke or in nonsmokers. Missense mutations in p53 contribute to increased protein half-life, which leads to higher levels of p53, an immunohistochemistry easily detectable protein [59]. In 40–70% of SCLCs and 40–60% of NSCLCs, abnormal p53 expression via immunostaining was reported. There have been several reports that p53 abnormalities have been correlated with the forecasts of NSCLC patients, but the findings have been divisive in whether p53 abnormalities affect the prognosis. In patients having adenocarcinomas, the negative prognostic effect of alterations was very considerable, while it was not the same for patients having squamous cell carcinomas [60]. The incidences of alterations in p53 were smaller than in protein studies in adenocarcinomas and were substantially lower than those in the case of p53 overexpression and mutations in adenocarcinomas. As a new approach to treatment, p53 was initiated into clinical studies with direct tumor retroviral and adenoviral gene therapy, which initially showed positive antitumor reactions. Recent research analyzed the more benefits of adenoviral p53 gene therapy that has been injecting directly into tumors in NSCLC first-line patients [61]. There were, however, no variations in the response rate or survival rate between the p53 treatment group and the chemotherapy treatment group alone. The systemic performance of liposome delivery of p53 has recently been demonstrated in lung cancer and must be further

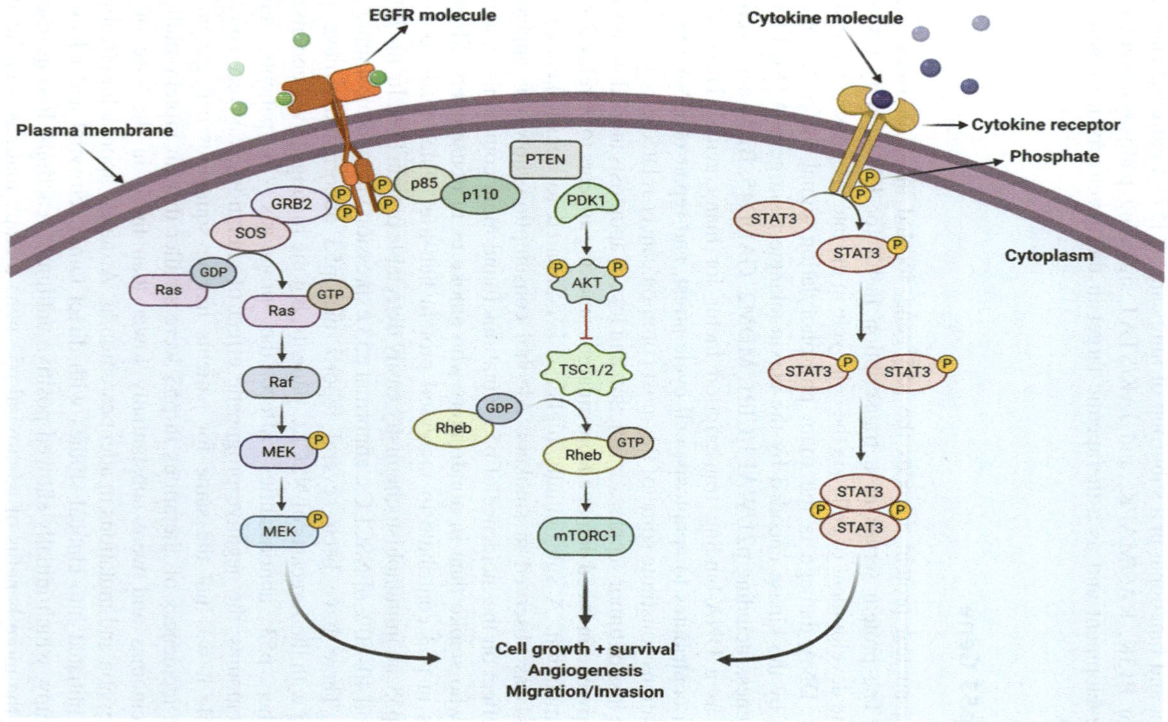

Fig. 9.3 Cascade of P13K, RAS/MAPK, JAK/STAT signaling pathway

examined for treating primary and disseminated lung cancer. Vaccine experiments were also performed with mutant p53 peptides [62].

9.5 Therapeutic Strategies

In the last few years, the 5-year survival has been meager, despite the several improvements in the treatment of advanced lung cancer. Also, the development of therapeutic vaccines in lung cancer has been very unsatisfactory, primarily due to the lack of effective predictive biomarkers. The patients could eventually achieve a success story if the patient population is better able to respond to therapy. A revolution in the treatment of lung cancer has come into force with the development of targeted agents. The following are some of the therapeutic approaches that are used against lung cancer pathologies.

9.5.1 Immunotherapy

The immune system has a major role to play in tumor progression management. When tumor-related antigens are detected, it begins to respond appropriately to cells' removal [63, 64]. However, the cancer cells can resist the sensation of the immune system and hinder the antitumor's effects and result in the continuing growth of and possible spread of cancer in the body. In terms of its interaction with the host immune system and form of response, immunotherapy has been defined as active or passive in nature. The active immune response includes humoral and/or cell-mediated immunity via recombinant cytokines, biochemotherapy, cancer vaccines, and monoclonal immunomodulatory antibodies (70, 71). Passive immune response, however, does not require the immune system to be activated; temporary antitumor activity is characterized by the use of preformed, target-specific, monoclonal antigens that bind to tumor-associated antigens and activate the immune system for the clearance of cancer cells. Examples include goal-specific monoclonal antibodies, oncolytic viruses, and adoptive T-cell therapy (T-cell) therapy [65, 66].

9.5.2 Anti PD-1 and anti-PD-L1

The PD-1, a T-cell surface receptor, is a part of B7-CD28 expressed on T cells, B cell, natural killer cell (NK), activated monocytes, and dendritic cells [67]. PD-1's function in normal human physiology is to restrict the autoimmunity of cells, including the tumor-infiltrating lymphocytes, as an immune control point expressed on the surface [68]. The death receptor ligand (PD-L1/B7-H1) and 2 (PD-L2/B7-DC) [69] are programmed to be used by two ligands. Clinical research has included multiple agents targeting PD-1 pathways such as "nivolumab" (BMS-936558, completely human IgG4 anti-PD-1), "lambrolizumab" (MK-3475, humanized IgG4 anti-PD-1), "MEDI473" (anti-PD-L1), "BMS-936559" (formerly

MDX-1105, completely man-made anti-PD-L1 IgG4), and "MPDL-3280" (anti-PD-L1). In separate phase I studies of nivolumab, lambrolizumab, and MPDL-3280, early effects on NSCLC have been very promising. PD-L1 is uncertain about its prognostic function. Different studies have shown that better prognostic, worse prognostic, or no prognostic values are correlated with the expression of PD-L1. The current application of immunohistochemistry (IHC) methods for measuring PD-L1 concentrations in tissues may provide an explanation for these discordant findings. Another alternative is that PD-L1 may actually differ among different cohorts for lung cancer.

9.5.3 Nivolumab

In patients with solid refractory tumors, nivolumab was tested in a major phase 1 dose-escalation trial. Two hundred ninety-six extremely pre-treated patients (with advance NSCLC, melanoma, renal cell carcinoma, castration-resistant prostate cancer, and advanced colorectal cancer) received doses intravenous nivolumab of 0.1–10 mg/kg every 2 weeks [70]. Responses were evaluated after each 8-week cycle of treating using RECIST, but care for clinically stable patients should continue past apparent initial disease progression before progression is confirmed. Until disease progression or maximum response, patients received up to 12 8-week cycles. Nivolumab toxicity, including fatigue and diarrhea, was experienced by 14% of patients with grade 3 or 4 toxicity. It should be noted that three patients had fatal pneumonitis during this early study. As a result of this complication, the early use of immunosuppressors is now part of treatment plans, and this seems to minimize toxicity.

Encouraging signs of efficacy were observed in renal cell carcinoma, melanoma, and, most shockingly, NSCLC, where about 16% of cases had an objective response, and 33% were six-month free of tumor progression [70]. In 9 out of 48 patients (18.8%) with squamous NSCLC and 11 out of 73 patients (15.1%) with non-squamous NSCLC, responses were noted, indicating activity in both histologic subtypes [70]. The overall response ratio of 129 patients was 17.2% (16.7% squamous, 17.6% squamous), with a median response period of more than 18 months [71]. Drug-related adverse effects (any grade) occurred in 71% of NSCLC patients, with 14% reporting drug-related adverse effects in grades 3–4. In 6% of the patients, drug-related pneumonitis occurred; 2% of these were grades 3–4, and two deaths from pneumonitis occurred in NSCLC patients. Present proposed treatment algorithms for patients with suspected anti-PD-1-related grade 2 pneumonitis include medication discontinuation, prompt systemic steroid therapy, and empirical antibiotic consideration due to the difficulties involved in differentiating infective versus drug-induced pneumonitis [72]. Grade 3–4 pneumonitis includes immediate pharmaceutical discontinuation, early administration of a high dose of steroids, and adjunctive immunosuppressive treatment such as infliximab, mycophenolate mofetil, or cyclophosphamide [72]. Furthermore, to prevent recurrence of the adverse event, patients may require prolonged tapering of immunosuppression.

Compared with second-line single-agent chemotherapy for advanced squamous and non-squamous NSCLC, nivolumab has now entered phase III clinical investigation as a single agent. A possible marker of response to anti-PD-1 is the expression of PD-L1 through immunohistochemical tumor cell analysis. Recently published results using the absence of expression of PD-L1 using these assays do not preclude the possibility of immunomodulator gain [73, 74].

9.6 Epidermal Growth Factor (EGFR) as a Therapeutic Target

The *EGFR* mutations and anaplastic lymphoma kinase (*ALK*) rearrangements are targeted for the treatment of NSCLCs. Development of some small and highly effective molecules called tyrosine kinase inhibitors (TKIs), which are administered orally, has made it possible to replace the standard chemotherapy. The first-line treatment drugs (TKIs) that were approved for use were erlotinib, afatinib, and gefitinib, and these are inhibitors of EGFR mutations [75, 76]. Monoclonal antibodies, such as cetuximab, are also used for the same [77]. Several clinical trials, including domestic studies, have been performed to check the efficacy of TKIs and compare them with the platinum-based chemotherapies. Second-generation inhibitor, such as afatinib, is an irreversible binder to EGFR mutant. Some studies have also been conducted to compare the efficacies of first-generation and second-generation inhibitors. In a study by Fujiwara et al. [78], there was not much difference found in the failure time and mPFS between both kinds of drugs (first and second-generation), while in another study [79], it was found that the mPFS of afatinib was longer than that of gefitinib but similar to erlotinib. Osimertinib is a third-generation inhibitor that has been shown to have more impact on *EGFR* T790M-positive disease than platinum-pemetrexed, as salvage treatment. Studies also showed "osimertinib" as more effective than erlotinib or gefitinib for the first-line treatment.

ALK inhibitors like crizotinib may substantially extend the survival of human beings having ALK gene mutations as compared to chemotherapy due to longer duration, faster onset time, and fewer side effects [80]. In a 2017 study, it was shown that alectinib showed better results than crizotinib as the first-line treatment in possessing better ability to cross brain-barrier and can be used rather than waiting until the disease progresses to the CNS.

9.6.1 PK13-Akt Pathway

The phosphatidylinositol-3 kinases, PI3Ks, are a group of proteins that are initiated by the signaling of receptor tyrosine kinase, and the signaling pathway of PI3K-Akt is basic for various tumors to create and survive. PI3K is categorized by their capacity to phosphorylate inositol ring 3′-OH group in inositol phospholipids to produce phosphatidylinositol-3,4,5-trisphosphate (PI-3,4,5-P(3)) [81]. This pathway is known to play crucial role in the development of various cancer types and can be exploited as a major therapeutic strategy in the development of new drugs [76, 81–84]. Through tyrosine kinase receptors and G protein-coupled receptors, serine/threonine kinase Akt, one of PI3 K's most significant downstream targets, can be initiated [82]. Akt generates oncogenic signals and mediates several cell reactions including cell development, motility, and separation to survival [84]. The PI3K-Akt pathway is engaged with the guideline of an assortment of cell pathways, including the hindrance of supportive of apoptotic proteins, for example, Bad, Bix, and Bid dependent on phosphorylation and the actuation of against apoptotic proteins [85]. PI3K inhibitors increment affectability in NSCLC cell lines to apoptosis-advancing therapeutics and predict in vitro development and formation of the colony. There are many PI3K inhibitor which can be used in specific combinations with other cancer drugs to control the development of lung cancer. Downstream, the serine/threonine kinase mTOR intervenes in the PI3K-Akt pathway, which directs fundamental cell function, for example, cell development and multiplication. The PI3K/Akt pathway is involved in the functioning of several new drugs; thus it may help us in better understanding of the new therapeutic drug targets in future.

9.6.2 Vaccine Therapy

Vaccine therapy is not a very promising way to deal with lung cancer because the results are not very encouraging despite the fact that clinical trials have played an important role in bringing light on the fact that what challenges are needed to be overcome to successfully employ tumor vaccine therapy. At this point, there are multiple vaccinations present which are approved by the FDA and clinically trialled, among which most of them are utilized by the NSCLC and a very small portion being developed for SCLC. Moreover, anticancer vaccines are designed to elicit antigen-specific immune responses in lung cancer. Tumor/whole-cell vaccines consisting of either autologous, allogeneic tumor cells, on the other hand, expose the immune system to a number of tumor antigens that are mostly unknown [86, 87].

Melanoma-associated antigen-A3 (MAGE-A3) is an antigen that is expressed in 35% cases of NSCLCs, with higher levels of expression associated with more advanced disease and poor prognosis [88, 89]. In the phase II clinical trial, 182 resected early-stage NSCLC patients were injected by recombinant MAGE-A3 protein to check its efficacy as a therapeutic vaccine [90]. Patients were vaccinated every 3 weeks for five cycles with MAGE-A3 protein of placebo, followed by eight vaccinations every 3 months. In contrast to placebo, no statistically significant period changes in vaccine therapy have been seen or overall disease-free survival or survival. The MAGRIT trial was a phase III clinical study for the MAGE-A3 tumor expression in resected NSCLC patients [91]. The trial failed to achieve its key endpoint, although the vaccine was well tolerated.

TG4010 is a mucin-1 (MUC1) antigen-targeted vaccine consisting of a recombinant vaccine virus (modified Ankara virus or MVA) encoding human MUC1 and IL-2 (MVA-MUC1-IL-2). The median PFS of the TG4010 group in phase IIb of TG4010 and first-line chemotherapy for advanced NSCLC (TIME) was 5.9 months versus 5.1 months in the placebo group. When the baseline values of CD16, CD56, and CD56 are taken into account, PFS was substantially improved by the addition of TG4010 to chemotherapy with CD69 triple-positive activated lymphocytes (TrPAL value) Q3 (third quartile of TrPAL distribution), while no gain was observed in patients with TrPAL value > Q3. In patients with non-squamous histology and TrPAL value per Q3, the highest benefit was noted [92].

"Tecemotide" (L-BLP25) is a liposome-based vaccine which is derived from the tandem repeat region of MUC1, a peptide overexpressed in NSCLC. Several preclinical studies found that MUC1-directed immunotherapy successfully induced a cellular immune reaction characterized by T-cell proliferation and production of IFN-c in a mouse model of NSCLC [93]. The association between overall 1-year survival and endogenous MUC1 antibodies has also been found in NSCLC patients [94]. 1513 unresectable NSCLC patients who achieved either stable disease or objective reactions with either complete or sequential chemoradiation were enrolled in the START trial. In patients with 8 weeks of treatments per week and then per 6 weeks before improvement, a tecemotide or placebo ratio of 2:1 was assigned. Although the study has failed to reach its endpoint of better overall survival, the overall survival rate of placebo was increased compared to the patient subgroup review of patients with concurrent chemoradiation [95].

9.7 Conclusions and Future Challenges

Although novel therapeutic drug targets continue to emerge, however, the best current clinical response rate is still not satisfactory. The emergence of novel molecular mechanisms that are involved in progression of lung cancer and present treatments may provide an opportunity to put together the recent available information from multiple therapeutic approaches for a better understanding of the disease and future treatment opportunities. A lot of information about the identity of different types of lung tumors is constantly being collected; however, novel therapeutics and improved functional studies are needed to meet the pace of dataset generation and all the aspects of tumor heterogeneity. The trans-omics strategies needed to be integrated to establish investigate the new approaches to make an impact on this devastating disease.

Conflict of Interest None declared.

References

1. Proctor RN (2001) Tobacco and the global lung cancer epidemic. Nat Rev Cancer 1:82–86
2. Jemal A, Thun MJ, Ries LA, Howe HL, Weir HK, Center MM, Ward E, Wu XC, Eheman C, Anderson R, Ajani UA, Kohler B, Edwards BK (2008) Annual report to the nation on the status of cancer, 1975-2005, featuring trends in lung cancer, tobacco use, and tobacco control. J Natl Cancer Inst 100:1672–1694
3. Siegel R, Naishadham D, Jemal A (2013) Cancer statistics, 2013. CA Cancer J Clin 63:11–30

4. Weiss W (1997) Cigarette smoking and lung cancer trends. A light at the end of the tunnel? Chest 111:1414–1416
5. Bryant A, Cerfolio RJ (2007) Differences in epidemiology, histology, and survival between cigarette smokers and never-smokers who develop non-small cell lung cancer. Chest 132:185–192
6. Dutu T, Michiels S, Fouret P, Penault-Llorca F, Validire P, Benhamou S, Taranchon E, Morat L, Grunenwald D, Le Chevalier T, Sabatier L, Soria JC (2005) Differential expression of biomarkers in lung adenocarcinoma: a comparative study between smokers and never-smokers. Ann Oncol 16:1906–1914
7. Ferlay J, Soerjomataram I, Dikshit R, Eser S, Mathers C, Rebelo M, Parkin DM, Forman D, Bray F (2015) Cancer incidence and mortality worldwide: sources, methods and major patterns in GLOBOCAN 2012. Int J Cancer 136:E359–E386
8. Gatta G, Mallone S, van der Zwan JM, Trama A, Siesling S, Capocaccia R, Hackl M, Van Eycken E, Henau K, Hedelin G, Velten M, Launoy G, Guizard AV, Bouvier AM, Maynadié M, Woronoff AS, Buemi A, Colonna M, Ganry O, Grosclaude P, Holleczek B, Ziegler H, Tryggvadottir L, Bellù F, Ferretti S, Serraino D, Dal Maso L, Bidoli E, Birri S, Zucchetto A, Zainer L, Vercelli M, Orengo MA, Casella C, Quaglia A, Federico M, Rashid I, Cirilli C, Fusco M, Traina A, Michiara M, De Lisi V, Bozzani F, Giacomin A, Tumino R, La Rosa MG, Spata E, Signora A, Mangone L, Falcini F, Giorgetti S, Ravaioli A, Senatore G, Iannelli A, Budroni M, Piffer S, Franchini S, Crocetti E, Caldarella A, Intrieri T, La Rosa F, Stracci F, Cassetti T, Contiero P, Tagliabue G, Zambon P, Guzzinati S, Berrino F, Baili P, Bella F, Ciampichini R, Gatta G, Margutti C, Micheli A, Minicozzi P, Sant M, Trama A, Caldora M, Capocaccia R, Carrani E, De Angelis R, Francisci S, Grande E, Inghelmann R, Lenz H, Martina L, Roazzi P, Santaquilani M, Simonetti A, Tavilla A, Verdecchia A, Langmark F, Rachtan J, Mężyk R, Góźdź S, Siudowska U, Zwierko M, Bielska-Lasota M, Safaei Diba C, Primic-Zakelj M, Mateos A, Izarzugaza I, Torrella Ramos A, Zurriaga O, Marcos-Gragera R, Vilardell ML, Izquierdo A, Ardanaz E, Moreno-Iribas C, Galceran J, Klint Å, Talbäck M, Jundt G, Usel M, Frick H, Ess SM, Bordoni A, Konzelmann I, Dehler S, Siesling S, Visser O, Otter R, Coebergh JWW, Greenberg DC, Wilkinson J, Roche M, Verne J, Meechan D, Poole J, Lawrence G, Gavin A, Brewster DH, Black RJ, Steward JA (2013) Cancer prevalence estimates in Europe at the beginning of 2000. Ann Oncol 24:1660–1666
9. Siegel RL, Miller KD, Jemal A (2017) Cancer Statistics, 2017. CA Cancer J Clin 67:7–30
10. Marx A, Chan JK, Coindre JM, Detterbeck F, Girard N, Harris NL, Jaffe ES, Kurrer MO, Marom EM, Moreira AL, Mukai K, Orazi A, Strobel P (2015) The 2015 World Health Organization classification of tumors of the thymus: continuity and changes. J Thorac Oncol 10:1383–1395
11. Cho WC (2012) Targeting the signaling pathways in cancer therapy. Expert Opin Ther Targets 16:1–3
12. Brambilla E, Gazdar A (2009) Pathogenesis of lung cancer signalling pathways: roadmap for therapies. Eur Respir J 33:1485–1497
13. Goldstraw P, Crowley J, Chansky K, Giroux DJ, Groome PA, Rami-Porta R, Postmus PE, Rusch V, Sobin L (2007) The IASLC lung cancer staging project: proposals for the revision of the TNM stage groupings in the forthcoming (seventh) edition of the TNM classification of malignant tumours. J Thorac Oncol 2:706–714
14. Smith AD, Roda D, Yap TA (2014) Strategies for modern biomarker and drug development in oncology. J Hematol Oncol 7:70
15. Zhou C, Wu YL, Chen G, Feng J, Liu XQ, Wang C, Zhang S, Wang J, Zhou S, Ren S, Lu S, Zhang L, Hu C, Hu C, Luo Y, Chen L, Ye M, Huang J, Zhi X, Zhang Y, Xiu Q, Ma J, Zhang L, You C (2015) Final overall survival results from a randomised, phase III study of erlotinib versus chemotherapy as first-line treatment of EGFR mutation-positive advanced non-small-cell lung cancer (OPTIMAL, CTONG-0802). Ann Oncol 26:1877–1883

16. Xing P, Wang S, Hao X, Zhang T, Li J (2016) Clinical data from the real world: efficacy of Crizotinib in Chinese patients with advanced ALK-rearranged non-small cell lung cancer and brain metastases. Oncotarget 7:84666–84674
17. Sun W, Yuan X, Tian Y, Wu H, Xu H, Hu G, Wu K (2015) Non-invasive approaches to monitor EGFR-TKI treatment in non-small-cell lung cancer. J Hematol Oncol 8:95
18. Fitzmaurice C, Dicker D, Pain A, Hamavid H, Moradi-Lakeh M, MacIntyre MF, Allen C, Hansen G, Woodbrook R, Wolfe C, Hamadeh RR, Moore A, Werdecker A, Gessner BD, Te Ao B, McMahon B, Karimkhani C, Yu C, Cooke GS, Schwebel DC, Carpenter DO, Pereira DM, Nash D, Kazi DS, De Leo D, Plass D, Ukwaja KN, Thurston GD, Yun Jin K, Simard EP, Mills E, Park EK, Catalá-López F, deVeber G, Gotay C, Khan G, Hosgood HD 3rd, Santos IS, Leasher JL, Singh J, Leigh J, Jonas JB, Sanabria J, Beardsley J, Jacobsen KH, Takahashi K, Franklin RC, Ronfani L, Montico M, Naldi L, Tonelli M, Geleijnse J, Petzold M, Shrime MG, Younis M, Yonemoto N, Breitborde N, Yip P, Pourmalek F, Lotufo PA, Esteghamati A, Hankey GJ, Ali R, Lunevicius R, Malekzadeh R, Dellavalle R, Weintraub R, Lucas R, Hay R, Rojas-Rueda D, Westerman R, Sepanlou SG, Nolte S, Patten S, Weichenthal S, Abera SF, Fereshtehnejad SM, Shiue I, Driscoll T, Vasankari T, Alsharif U, Rahimi-Movaghar-V, Vlassov VV, Marcenes WS, Mekonnen W, Melaku YA, Yano Y, Artaman A, Campos I, MacLachlan J, Mueller U, Kim D, Trillini M, Eshrati B, Williams HC, Shibuya K, Dandona R, Murthy K, Cowie B, Amare AT, Antonio CA, Castañeda-Orjuela C, van Gool CH, Violante F, Oh IH, Deribe K, Soreide K, Knibbs L, Kereselidze M, Green M, Cardenas R, Roy N, Tillmann T, Li Y, Krueger H, Monasta L, Dey S, Sheikhbahaei S, Hafezi-Nejad N, Kumar GA, Sreeramareddy CT, Dandona L, Wang H, Vollset SE, Mokdad A, Salomon JA, Lozano R, Vos T, Forouzanfar M, Lopez A, Murray C, Naghavi M (2015) The global burden of cancer 2013. JAMA Oncol 1:505–527
19. Torre LA, Bray F, Siegel RL, Ferlay J, Lortet-Tieulent J, Jemal A (2015) Global cancer statistics, 2012. CA Cancer J Clin 65:87–108
20. Goss PE, Strasser-Weippl K, Lee-Bychkovsky BL, Fan L, Li J, Chavarri-Guerra Y, Liedke PE, Pramesh CS, Badovinac-Crnjevic T, Sheikine Y, Chen Z, Qiao YL, Shao Z, Wu YL, Fan D, Chow LW, Wang J, Zhang Q, Yu S, Shen G, He J, Purushotham A, Sullivan R, Badwe R, Banavali SD, Nair R, Kumar L, Parikh P, Subramanian S, Chaturvedi P, Iyer S, Shastri SS, Digumarti R, Soto-Perez-de-Celis E, Adilbay D, Semiglazov V, Orlov S, Kaidarova D, Tsimafeyeu I, Tatishchev S, Danishevskiy KD, Hurlbert M, Vail C, St Louis J, Chan A (2014) Challenges to effective cancer control in China, India, and Russia. Lancet Oncol 15:489–538
21. Debakey M (1999) Carcinoma of the lung and tobacco smoking: a historical perspective. Ochsner J 1:106–108
22. Hasse C (1846) Cancerous tumors in the respiratory organs. An anatomical description of the diseases of the organs of the circulation and respiration. Sydeham Society, England, London, pp 370–375
23. Jemal A, Bray F, Center MM, Ferlay J, Ward E, Forman D (2011) Global cancer statistics. CA Cancer J Clin 61:69–90
24. Arnold M, Soerjomataram I, Ferlay J, Forman D (2015) Global incidence of oesophageal cancer by histological subtype in 2012. Gut 64:381–387
25. Ferlay J, Colombet M, Soerjomataram I, Dyba T, Randi G, Bettio M, Gavin A, Visser O, Bray F (2018) Cancer incidence and mortality patterns in Europe: estimates for 40 countries and 25 major cancers in 2018. Eur J Cancer 103:356–387
26. Siegel RL, Miller KD, Jemal A (2018) Cancer statistics, 2018. CA Cancer J Clin 68:7–30
27. Schabath MB, Cote ML (2019) Cancer progress and priorities: lung cancer. Cancer Epidemiol Biomark Prev 28:1563–1579
28. Dela Cruz CS, Tanoue LT, Matthay RA (2011) Lung cancer: epidemiology, etiology, and prevention. Clin Chest Med 32:605–644

29. Kohler BA, Ward E, McCarthy BJ, Schymura MJ, Ries LA, Eheman C, Jemal A, Anderson RN, Ajani UA, Edwards BK (2011) Annual report to the nation on the status of cancer, 1975-2007, featuring tumors of the brain and other nervous system. J Natl Cancer Inst 103:714–736
30. Jemal A, Miller KD, Ma J, Siegel RL, Fedewa SA, Islami F, Devesa SS, Thun MJ (2018) Higher Lung Cancer Incidence in Young Women Than Young Men in the United States. N Engl J Med 378:1999–2009
31. Wakelee HA, Chang ET, Gomez SL, Keegan TH, Feskanich D, Clarke CA, Holmberg L, Yong LC, Kolonel LN, Gould MK, West DW (2007) Lung cancer incidence in never smokers. J Clin Oncol 25:472–478
32. Forman D, de Martel C, Lacey CJ, Soerjomataram I, Lortet-Tieulent J, Bruni L, Vignat J, Ferlay J, Bray F, Plummer M (2012) Global burden of human papillomavirus and related diseases. Vaccine 30:F12–F23
33. Greenman C, Stephens P, Smith R, Dalgliesh GL, Hunter C, Bignell G, Davies H, Teague J, Butler A, Stevens C, Edkins S, O'Meara S, Vastrik I, Schmidt EE, Avis T, Barthorpe S, Bhamra G, Buck G, Choudhury B, Clements J, Cole J, Dicks E, Forbes S, Gray K, Halliday K, Harrison R, Hills K, Hinton J, Jenkinson A, Jones D, Menzies A, Mironenko T, Perry J, Raine K, Richardson D, Shepherd R, Small A, Tofts C, Varian J, Webb T, West S, Widaa S, Yates A, Cahill DP, Louis DN, Goldstraw P, Nicholson AG, Brasseur F, Looijenga L, Weber BL, Chiew YE, DeFazio A, Greaves MF, Green AR, Campbell P, Birney E, Easton DF, Chenevix-Trench G, Tan MH, Khoo SK, Teh BT, Yuen ST, Leung SY, Wooster R, Futreal PA, Stratton MR (2007) Patterns of somatic mutation in human cancer genomes. Nature 446:153–158
34. Goldstraw P, Chansky K, Crowley J, Rami-Porta R, Asamura H, Eberhardt WE, Nicholson AG, Groome P, Mitchell A, Bolejack V (2016) The IASLC lung cancer staging project: proposals for revision of the TNM stage groupings in the forthcoming (eighth) edition of the TNM classification for lung cancer. J Thorac Oncol 11:39–51
35. Nesbitt JC, Putnam JB Jr, Walsh GL, Roth JA, Mountain CF (1995) Survival in early-stage non-small cell lung cancer. Ann Thorac Surg 60:466–472
36. Shah R, Sabanathan S, Richardson J, Mearns AJ, Goulden C (1996) Results of surgical treatment of stage I and II lung cancer. J Cardiovasc Surg 37:169–172
37. Walters S, Maringe C, Coleman MP, Peake MD, Butler J, Young N, Bergström S, Hanna L, Jakobsen E, Kölbeck K, Sundstrøm S, Engholm G, Gavin A, Gjerstorff ML, Hatcher J, Johannesen TB, Linklater KM, McGahan CE, Steward J, Tracey E, Turner D, Richards MA, Rachet B (2013) Lung cancer survival and stage at diagnosis in Australia, Canada, Denmark, Norway, Sweden and the UK: a population-based study, 2004-2007. Thorax 68:551–564
38. Broggio JBN (2016) Cancer survival by stage at diagnosis for England. Office for National Statistics, Newport, UK
39. Bernhardt EB, Jalal SI (2016) Small Cell Lung Cancer. Cancer Treat Res 170:301–322
40. Blandin Knight S, Crosbie PA, Balata H, Chudziak J, Hussell T, Dive C (2017) Progress and prospects of early detection in lung cancer. Open Biol 7:170070
41. Weinstein IB, Joe A (2008) Oncogene addiction. Cancer Res 68:3077–3080; discussion 3080
42. Robinson DR, Wu YM, Lin SF (2000) The protein tyrosine kinase family of the human genome. Oncogene 19:5548–5557
43. Sharma SV, Bell DW, Settleman J, Haber DA (2007) Epidermal growth factor receptor mutations in lung cancer. Nat Rev Cancer 7:169–181
44. Rowinsky EK (2004) The erbB family: targets for therapeutic development against cancer and therapeutic strategies using monoclonal antibodies and tyrosine kinase inhibitors. Annu Rev Med 55:433–457
45. Hirsch FR, Varella-Garcia M, Bunn PA Jr, Di Maria MV, Veve R, Bremmes RM, Barón AE, Zeng C, Franklin WA (2003) Epidermal growth factor receptor in non-small-cell lung carcinomas: correlation between gene copy number and protein expression and impact on prognosis. J Clin Oncol 21:3798–3807

46. Nicholson RI, Gee JM, Harper ME (2001) EGFR and cancer prognosis. Eur J Cancer 37(Suppl 4):S9–S15
47. Shigematsu H, Gazdar AF (2006) Somatic mutations of epidermal growth factor receptor signaling pathway in lung cancers. Int J Cancer 118:257–262
48. Gazdar AF, Shigematsu H, Herz J, Minna JD (2004) Mutations and addiction to EGFR: the Achilles 'heal' of lung cancers? Trends Mol Med 10:481–486
49. Pao W, Wang TY, Riely GJ, Miller VA, Pan Q, Ladanyi M, Zakowski MF, Heelan RT, Kris MG, Varmus HE (2005) KRAS mutations and primary resistance of lung adenocarcinomas to gefitinib or erlotinib. PLoS Med 2:e17
50. Zhou BB, Peyton M, He B, Liu C, Girard L, Caudler E, Lo Y, Baribaud F, Mikami I, Reguart N, Yang G, Li Y, Yao W, Vaddi K, Gazdar AF, Friedman SM, Jablons DM, Newton RC, Fridman JS, Minna JD, Scherle PA (2006) Targeting ADAM-mediated ligand cleavage to inhibit HER3 and EGFR pathways in non-small cell lung cancer. Cancer Cell 10:39–50
51. Zhang J, Iwanaga K, Choi KC, Wislez M, Raso MG, Wei W, Wistuba II, Kurie JM (2008) Intratumoral epiregulin is a marker of advanced disease in non-small cell lung cancer patients and confers invasive properties on EGFR-mutant cells. Cancer Prev Res 1:201–207
52. Yonesaka K, Zejnullahu K, Lindeman N, Homes AJ, Jackman DM, Zhao F, Rogers AM, Johnson BE, Jänne PA (2008) Autocrine production of amphiregulin predicts sensitivity to both gefitinib and cetuximab in EGFR wild-type cancers. Clin Cancer Res 14:6963–6973
53. Ji H, Wang Z, Perera SA, Li D, Liang MC, Zaghlul S, McNamara K, Chen L, Albert M, Sun Y, Al-Hashem R, Chirieac LR, Padera R, Bronson RT, Thomas RK, Garraway LA, Jänne PA, Johnson BE, Chin L, Wong KK (2007) Mutations in BRAF and KRAS converge on activation of the mitogen-activated protein kinase pathway in lung cancer mouse models. Cancer Res 67:4933–4939
54. Zhang W, Stabile LP, Keohavong P, Romkes M, Grandis JR, Traynor AM, Siegfried JM (2006) Mutation and polymorphism in the EGFR-TK domain associated with lung cancer. J Thorac Oncol 1:635–647
55. Samuels Y, Diaz LA Jr, Schmidt-Kittler O, Cummins JM, Delong L, Cheong I, Rago C, Huso DL, Lengauer C, Kinzler KW, Vogelstein B, Velculescu VE (2005) Mutant PIK3CA promotes cell growth and invasion of human cancer cells. Cancer Cell 7:561–573
56. Garnis C, Lockwood WW, Vucic E, Ge Y, Girard L, Minna JD, Gazdar AF, Lam S, MacAulay C, Lam WL (2006) High resolution analysis of non-small cell lung cancer cell lines by whole genome tiling path array CGH. Int J Cancer 118:1556–1564
57. Meyerson M (2007) Cancer: broken genes in solid tumours. Nature 448:545–546
58. Olivier M, Petitjean A, Marcel V, Pétré A, Mounawar M, Plymoth A, de Fromentel CC, Hainaut P (2009) Recent advances in p53 research: an interdisciplinary perspective. Cancer Gene Ther 16:1–12
59. Brambilla E, Gazzeri S, Lantuejoul S, Coll JL, Moro D, Negoescu A, Brambilla C (1998) p53 mutant immunophenotype and deregulation of p53 transcription pathway (Bcl2, Bax, and Waf1) in precursor bronchial lesions of lung cancer. Clin Cancer Res 4:1609–1618
60. Jeanmart M, Lantuejoul S, Fievet F, Moro D, Sturm N, Brambilla C, Brambilla E (2003) Value of immunohistochemical markers in preinvasive bronchial lesions in risk assessment of lung cancer. Clin Cancer Res 9:2195–2203
61. Eymin B, Gazzeri S, Brambilla C, Brambilla E (2002) Mdm2 overexpression and p14(ARF) inactivation are two mutually exclusive events in primary human lung tumors. Oncogene 21:2750–2761
62. Gazzeri S, Della Valle V, Chaussade L, Brambilla C, Larsen CJ, Brambilla E (1998) The human p19ARF protein encoded by the beta transcript of the p16INK4a gene is frequently lost in small cell lung cancer. Cancer Res 58:3926–3931
63. Kwak JJ, Tirumani SH, Van den Abbeele AD, Koo PJ, Jacene HA (2015) Cancer immunotherapy: imaging assessment of novel treatment response patterns and immune-related adverse events. Radiographics 35:424–437
64. Schuster M, Nechansky A, Kircheis R (2006) Cancer immunotherapy. Biotechnol J 1:138–147

65. Hu Z, Ott PA, Wu CJ (2018) Towards personalized, tumour-specific, therapeutic vaccines for cancer. Nat Rev Immunol 18:168–182
66. Rasmi RR, Sakthivel KM, Guruvayoorappan C (2020) NF-kappaB inhibitors in treatment and prevention of lung cancer. Biomed Pharmacother 130:110569
67. Keir ME, Butte MJ, Freeman GJ, Sharpe AH (2008) PD-1 and its ligands in tolerance and immunity. Annu Rev Immunol 26:677–704
68. Topalian SL, Drake CG, Pardoll DM (2012a) Targeting the PD-1/B7-H1(PD-L1) pathway to activate anti-tumor immunity. Curr Opin Immunol 24:207–212
69. Dong H, Strome SE, Salomao DR, Tamura H, Hirano F, Flies DB, Roche PC, Lu J, Zhu G, Tamada K, Lennon VA, Celis E, Chen L (2002) Tumor-associated B7-H1 promotes T-cell apoptosis: a potential mechanism of immune evasion. Nat Med 8:793–800
70. Topalian SL, Hodi FS, Brahmer JR, Gettinger SN, Smith DC, McDermott DF, Powderly JD, Carvajal RD, Sosman JA, Atkins MB, Leming PD, Spigel DR, Antonia SJ, Horn L, Drake CG, Pardoll DM, Chen L, Sharfman WH, Anders RA, Taube JM, McMiller TL, Xu H, Korman AJ, Jure-Kunkel M, Agrawal S, McDonald D, Kollia GD, Gupta A, Wigginton JM, Sznol M (2012b) Safety, activity, and immune correlates of anti-PD-1 antibody in cancer. N Engl J Med 366:2443–2454
71. Rajan A, Gulley JL (2014) Nivolumab (anti-PD-1, BMS-936558, ONO-4538) in patients with advanced non-small cell lung cancer. Transl Lung Cancer Res 3:403–405
72. Topalian SL, Sznol M, Brahmer JR, McDermott DF, Smith DC, Gettinger SN, Taube JM, Drake CG, Pardoll DM, Powderly JD, Carvajal RD, Sosman JA, Atkins MB, Antonia SJ, Spigel DR, Lawrence DP, Kollia G, Gupta AK, Wigginton JM, Hodi FS (2013) Nivolumab (anti-PD-1; BMS-936558; ONO-4538) in patients with advanced solid tumors: survival and long-term safety in a phase I trial. J Clin Oncol 31:3002–3002
73. Grosso J, Horak CE, Inzunza D, Cardona DM, Simon JS, Gupta AK, Sankar V, Park J-S, Kollia G, Taube JM, Anders R, Jure-Kunkel M, Jim Novotny J, Taylor CR, Zhang X, Phillips T, Simmons P, Cogswell J (2013) Association of tumor PD-L1 expression and immune biomarkers with clinical activity in patients (pts) with advanced solid tumors treated with nivolumab (anti-PD-1; BMS-936558; ONO-4538). J Clin Oncol 31:3016–3016
74. Spigel DR, Gettinger SN, Horn L, Herbst RS, Gandhi L, Gordon MS, Cruz C, Conkling P, Cassier PA, Antonia SJ, Burris HA, Fine GD, Mokatrin A, Kowanetz M, Shen X, Chen DS, Soria J-C (2013) Clinical activity, safety, and biomarkers of MPDL3280A, an engineered PD-L1 antibody in patients with locally advanced or metastatic non-small cell lung cancer (NSCLC). J Clin Oncol 31:8008–8008
75. Metro G (2018) EGFR targeted therapy for lung cancer: are we almost there? Transl Lung Cancer Res 7:S142–S145
76. Ray MR, Jablons D, He B (2010) Lung cancer therapeutics that target signaling pathways: an update. Expert Rev Respir Med 4:631–645
77. Heist RS, Christiani D (2009) EGFR-targeted therapies in lung cancer: predictors of response and toxicity. Pharmacogenomics 10:59–68
78. Fujiwara A, Yoshida M, Fujimoto H, Nakahara H, Ito K, Nishihama K, Yasuma T, Hataji O, Taguchi O, D'Alessandro-Gabazza CN, Gabazza EC, Kobayashi T (2018) A retrospective comparison of the clinical efficacy of gefitinib, erlotinib, and afatinib in Japanese patients with non-small cell lung cancer. Oncol Res 26:1031–1036
79. Tu CY, Chen CM, Liao WC, Wu BR, Chen CY, Chen WC, Hsia TC, Cheng WC, Chen CH (2018) Comparison of the effects of the three major tyrosine kinase inhibitors as first-line therapy for non-small-cell lung cancer harboring epidermal growth factor receptor mutations. Oncotarget 9:24237–24247
80. Hu Z, Li M, Chen Z, Zhan C, Lin Z, Wang Q (2019) Advances in clinical trials of targeted therapy and immunotherapy of lung cancer in 2018. Transl Lung Cancer Res 8:1091–1106
81. Fresno Vara JA, Casado E, de Castro J, Cejas P, Belda-Iniesta C, Gonzalez-Baron M (2004) PI3K/Akt signalling pathway and cancer. Cancer Treat Rev 30:193–204

82. Andrew S (1999) PIK3CA: determining its role in cellular proliferation and ovarian cancer. Clin Genet 56:190–191
83. Lu Y, Wang H, Mills GB (2003) Targeting PI3K-AKT pathway for cancer therapy. Rev Clin Exp Hematol 7:205–228
84. Osaki M, Oshimura M, Ito H (2004) PI3K-Akt pathway: its functions and alterations in human cancer. Apoptosis 9:667–676
85. Kandasamy K, Srivastava RK (2002) Role of the phosphatidylinositol 3′-kinase/PTEN/Akt kinase pathway in tumor necrosis factor-related apoptosis-inducing ligand-induced apoptosis in non-small cell lung cancer cells. Cancer Res 62:4929–4937
86. Brahmer JR (2013) Harnessing the immune system for the treatment of non-small-cell lung cancer. J Clin Oncol 31:1021–1028
87. Holt GE, Podack ER, Raez LE (2011) Immunotherapy as a strategy for the treatment of non-small-cell lung cancer. Therapy 8:43–54
88. Gure AO, Chua R, Williamson B, Gonen M, Ferrera CA, Gnjatic S, Ritter G, Simpson AJG, Chen Y-T, Old LJ, Altorki NK (2005) Cancer-testis genes are coordinately expressed and are markers of poor outcome in non-small cell lung cancer. Clin Cancer Res 11:8055–8062
89. Sienel W, Varwerk C, Linder A, Kaiser D, Teschner M, Delire M, Stamatis G, Passlick B (2004) Melanoma associated antigen (MAGE)-A3 expression in Stages I and II non-small cell lung cancer: results of a multi-center study. Eur J Cardiothorac Surg 25:131–134
90. Vansteenkiste J, Zielinski M, Linder A, Dahabreh J, Gonzalez EE, Malinowski W, Lopez-Brea M, Vanakesa T, Jassem J, Kalofonos H, Perdeus J, Bonnet R, Basko J, Janilionis R, Passlick B, Treasure T, Gillet M, Lehmann FF, Brichard VG (2013) Adjuvant MAGE-A3 immunotherapy in resected non-small-cell lung cancer: phase II randomized study results. J Clin Oncol 31:2396–2403
91. Tyagi P, Mirakhur B (2009) MAGRIT: the largest-ever phase III lung cancer trial aims to establish a novel tumor-specific approach to therapy. Clin Lung Cancer 10:371–374
92. Quoix E, Lena H, Losonczy G, Forget F, Chouaid C, Papai Z, Gervais R, Ottensmeier C, Szczesna A, Kazarnowicz A, Beck JT, Westeel V, Felip E, Debieuvre D, Madroszyk A, Adam J, Lacoste G, Tavernaro A, Bastien B, Halluard C, Palanché T, Limacher JM (2016) TG4010 immunotherapy and first-line chemotherapy for advanced non-small-cell lung cancer (TIME): results from the phase 2b part of a randomised, double-blind, placebo-controlled, phase 2b/3 trial. Lancet Oncol 17:212–223
93. Xia W, Wang J, Xu Y, Jiang F, Xu L (2014) L-BLP25 as a peptide vaccine therapy in non-small cell lung cancer: a review. J Thorac Dis 6:1513–1520
94. Hirasawa Y, Kohno N, Yokoyama A, Kondo K, Hiwada K, Miyake M (2000) Natural autoantibody to MUC1 is a prognostic indicator for non-small cell lung cancer. Am J Respir Crit Care Med 161:589–594
95. Butts C, Socinski MA, Mitchell PL, Thatcher N, Havel L, Krzakowski M, Nawrocki S, Ciuleanu TE, Bosquée L, Trigo JM, Spira A, Tremblay L, Nyman J, Ramlau R, Wickart-Johansson G, Ellis P, Gladkov O, Pereira JR, Eberhardt WE, Helwig C, Schröder A, Shepherd FA (2014) Tecemotide (L-BLP25) versus placebo after chemoradiotherapy for stage III non-small-cell lung cancer (START): a randomised, double-blind, phase 3 trial. Lancet Oncol 15:59–68

Modulation of Signaling Pathways by Immunotherapeutics in Lung Cancer

10

Paramita Mandal, Anindita Goswami, Sarmistha Adhikari, and Subham Sarkar

Abstract

Lung cancer is the leading cause of mortality for both males and females among all types of cancer. Histologically lung cancer is classified into non-small cell lung cancer (NSCLC) and small cell lung cancer (SCLC). Conventional surgery and chemotherapy are often associated with severe toxicity and multiple drug resistance among lung cancer patients. Recent advancement of immunotherapy can elicit immune-mediated destruction of tumor cells. Wide variations of immunotherapeutic approaches were undertaken to stimulate immune response against lung cancer cells including immunomodulators, therapeutic vaccines, and monoclonal antibodies, and those were directed towards checkpoint proteins in cancer cells. Checkpoint inhibitors targeting cytotoxic T lymphocyte-associated antigen 4 (CTLA-4) and the programmed death-1 (PD-1) pathway were employed as immunotherapy with the achievement of progression free survival and associated with minimal toxicity among lung cancer patients. Various categories of therapeutic vaccines were employed in lung cancer, and those were associated with improved survival rate and quality of life. Recently, a combination of radiotherapy and immunotherapy was employed in lung cancer patients and found to be more effective with no significant generation of toxicity.

Keywords

Lung cancer · Non-small cell lung cancer (NSCLC) · Small-cell lung cancer (SCLC) · Immune checkpoints · Vaccine

P. Mandal (✉) · A. Goswami · S. Adhikari · S. Sarkar
Biomedical Genetics Laboratory, Department of Zoology, The University of Burdwan, Burdwan, West Bengal, India

K. Dua et al. (eds.), *Targeting Cellular Signalling Pathways in Lung Diseases*,
https://doi.org/10.1007/978-981-33-6827-9_10

10.1 Introduction

Lung cancer is the leading cause of cancer deaths in the world for both men and women. According to GLOBOCAN, 2018, the prevalence of lung cancer was highest (11.6%) among all the cancers for both sexes in the world. Worldwide data revealed that the number of deaths due to lung cancer was 1,76,11,007 in 2018 among both the sexes for all ages which was highest among all the cancers (https://gco.iarc.fr/today/data/factsheets/cancers/15-Lung-fact-sheet.pdf). The 5-year survival statistics for lung cancer has remained unchanged since the 1970s compared to other most prevalent types of cancers including prostate, breast, and colorectal cancer [1]. Based on histological diagnosis, there are two types: small-cell lung cancer (SCLC) and non-small cell lung cancer (NSCLC). Nearly 85% of lung cancer cases are classified as NSCLC, and majority of the patients were diagnosed at advanced stage of cancer [2]. NSCLCs are again classified into three subtypes including adenocarcinoma, squamous cell carcinoma, and large-cell carcinoma. Most NSCLC were diagnosed as locally advanced or metastatic disease with stage IIIB and stage IV based on the International Staging System and was associated with poor prognosis [3]. The median survival of NSCLC was 6 months, and 5-year survival was 2% [4]. 40% 5-year survival rate was attained after complete surgical resection, and it is the best option for treatment. But unfortunately, 75% of the lung cancer patients are diagnosed with advanced or metastatic disease which is unsuitable for surgery [5]. Apart from surgical resection, platinum-based chemotherapy was employed in NSCLC patients, but such treatment was often associated with severe toxicity and multiple drug resistance [6]. Therefore, searching for alternative and complementary treatments is an essential concern to reduce the side effects of chemotherapy.

Molecular basis of carcinogenesis and improved understanding about lung cancer biology can help to overcome the therapeutic plateau for the treatment of advanced stages of NSCLC. Currently, research efforts were focused on developing novel agents that target signaling pathways relevant for lung cancer. Molecular-targeted therapies, such as erlotinib and crizotinib for a limited group of NSCLC patients harboring specific EGFR mutations, had improved median overall survival [7]. But, popular EGFR or HER-2 therapies may not be sufficient to address this treatment regimen [8]. Unfortunately, for a large group of NSCLC and SCLC, molecular alterations are not available which could be associated with targeted therapies.

Recently, a broad class of therapy termed as immunotherapeutics has been developed which can elicit immune-mediated destruction of tumor cells. Previous studies have found that lung cancer cells can evade the immune system by the loss of major histocompatibility complex expression, downregulation of molecules relevant for inhibition of T-cell activation, and secretion of cytokines related to immunosuppression. Another important part of the immune system is to use immune checkpoint proteins in cells, which need to be activated or inhibited to start an immune response. Cancer cells mostly use these checkpoints to escape the attack by the immune system. Different immunotherapeutic approaches were undertaken to stimulate immune response against various types of cancer including immunomodulators, therapeutic vaccines, and monoclonal antibodies, and those were directed toward checkpoint proteins in cancer cells. Checkpoint inhibitors targeting cytotoxic T lymphocyte-associated antigen 4 (CTLA-4) and the programmed death-1 (PD-1) pathway were recently employed as immunotherapy with achievement of prolong clinical responses with minimal toxicity among lung cancer patients. This article summarizes the types of immunotherapeutic agents for NSCLC and SCLC.

10.2 Immune Checkpoint Blockade in NSCLC

Like the conventional way of dealing with any shrewd element in our body, cancer-specific neo-antigens derived from tumor cells are too phagocytosed and processed by antigen-presenting cells (APCs) such as dendritic cells (DCs). Next, DCs flag the processed antigens onto major histocompatibility complex (MHC) class I molecules, resulting in the activation of T lymphocytes. Then the activated CTLs, after receiving the "tip" from APCs, hurl through blood vessels and infiltrate the deadly tumor. Finally, CTLs identify cancer cells through the interaction between the T-cell receptor (TCR) and the cancer-specific antigen and kill them by immune attack. Immune checkpoints are regulators of activation of immune reaction that have a key role in maintenance of immune homeostasis and autoimmunity prevention. They are very necessary to maintain self-tolerance and protect the host from tissue damage.

In cancer, immune checkpoint mechanisms are often activated to subdue antitumorigenic effects. This has led to the development of many checkpoint inhibitors that are currently under clinical trials, while some of those have already been approved to treat a number of cancers [9]. Here, we briefly describe the mechanism of action of some major immune checkpoints and their inhibition strategy by immune checkpoint inhibitors (ICIs).

10.2.1 CTLA-4

To clamp its inhibitory effects on T cells, cytotoxic T lymphocyte-associated protein 4 (CTLA-4) utilizes both signaling and non-signaling mechanisms. CTLA-4 blockade induces immune responses that are dependent on CD4+ T cells and also inhibits the immunosuppressive function of regulatory T cells (Tregs), as these cells express this immune checkpoint protein constitutively [10]. Two monoclonal antibodies targeting CTLA-4 are ipilimumab and tremelimumab which had been designed to promote antitumor immune response by T-cell activation and proliferation. Subsequently, ipilimumab in the case of advanced NSCLC had been tested both as monotherapy and in combination with standard chemotherapy [11]. Both the cases showed significantly prolonged tumor response and overall survival (OS) with a manageable safety profile [12, 13]. Similarly, tremelimumab had shown a much higher progression free survival (PFS) of 10.3 months than carboplatin or platinum-based chemotherapy (PFS = 6 months) [9].

10.2.2 PD-1/PD-1L

In contrast to CTLA-4, Programmed cell death protein 1 (PD-1) mostly regulates activity of effector T cell within both tissues and tumors as opposed to regulating T-cell activation in lymphoid organs [10]. While CTLA-4 mainly targets naïve T cells, PD-1 is expressed on mature T cells in peripheral tissues and the tumor microenvironment. It is also expressed on non-T-cell subsets like B cells, professional APCs, and natural killer (NK) cells. Like CTLA-4, PD-1 is highly expressed on Tregs and enhances proliferation and suppressive activity when bound to ligand [10]. PD-1 has two ligands, PD-L1 (also known as B7-H1 and CD274) and PD-L2 (also known as B7-DC and CD273). PD-L1 interacts with the CTLA-4 and CD28 ligand, CD80, to block T-cell proliferation by sending an inhibitory signal through CD80 [14]. The functions of PD-1 and PD-L1 on tumor cells are illustrated in Fig. 10.1.

Four ICIs, namely, nivolumab and pembrolizumab (anti-PD-1 antibodies); atezolizumab and avelumab (anti-PD-L1 antibodies) were approved for the treatment of advanced NSCLC patients in Japan, the USA and the European Union (EU). Single-agent treatment with docetaxel (DOC) in NSCLC patients associated with significant survival benefits in two landmark phase III trials (TAX317 and TAX320) [15, 16]. Several randomized controlled trials (RCTs) were conducted to evaluate the efficacy of ICIs (nivolumab, pembrolizumab, atezolizumab) in comparison with DOC [17].

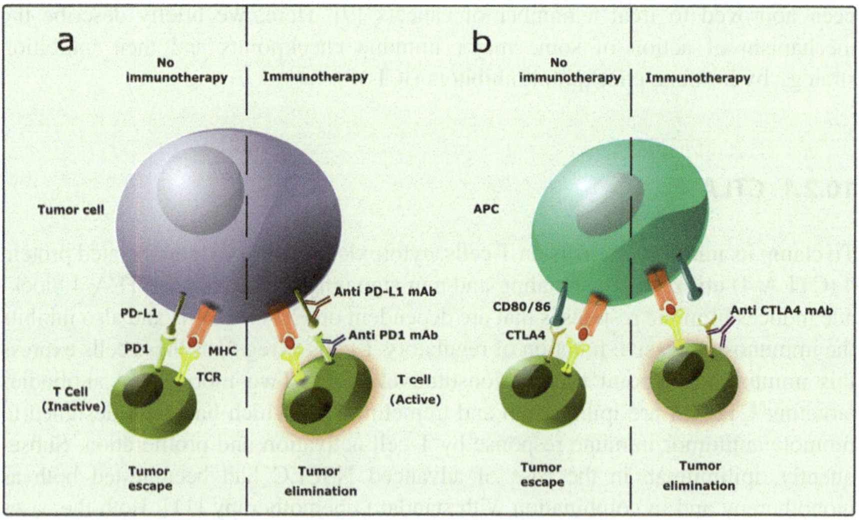

Fig. 10.1 (**a**) The interaction of PD-1 receptor on T cells with PD-L1 on tumor cells promotes T-cell anergy and tumor escape. (**b**) Interactions between tumor cells and activated T cells. CTL-4: T lymphocyte-associated antigen 4; PD-1: programmed cell death-1; APC: antigen-presenting cells; MHC: major histocompatibility complex; TCR: T-cell receptor

National Comprehensive Cancer Network (NCCN), the International Association for the Study of Lung Cancer (IASLC), College of American Pathologists (CAP), and the Association for Molecular Pathology (AMP) shared guidelines, and they recommend that the PD-L1 expression should be tested in patients [18, 19] As a result, phase III study on Keynote 024 was done in 305 patients with previously untreated advanced NSCLC, harboring above 50% PD-L1 expression on biopsy. They were randomized to either pembrolizumab- or platinum-based chemotherapy. The result showed that pembrolizumab significantly longer PFS and OS than other chemotherapy. Pembrolizumab is now the first-line treatment for advanced NSCLC in patients with PD-L1 expression more than 50%. It can also be used in the second-line treatment for advanced or unresectable NSCLC cases where the situation has progressed despite platinum-based chemotherapy [20, 21].

One of the promising approaches to improve the efficacy of these therapies is combinatorial therapy. These include more than 250 combining therapeutic agents targeting CTLA-4 and PD-1. Combinations can be achieved with novel agents targeting other costimulatory molecules (e.g., TIM3, LAG3). Recent studies indicate that PD-1 inhibits CD28 signaling in addition to proximal signaling elements of the TCR complex [22]. CTLA-4 CD28+ primarily attenuates activation of T cells by limiting costimulation. These findings indicate that dual blockade of CTLA-4 and PD-1 therapy is adequate to elicit unique cellular responses compared with either monotherapy [23].

10.2.3 TIM3

It is a glycoprotein with extracellular immunoglobulin and mucin domains. T-cell immunoglobulin mucin 3 (TIM3) is expressed on a number of cells like activated T cells as well as tissues such as the liver, kidney, spleen, intestine, thymus, lung, muscle, and brain [24]. TIM3 plays an important part in CD8+ T-cell exhaustion, regulates NK cells and Tregs. Additionally, in CD8+ T cells co-expressing TIM3 and PD-1 leads to greater deterioration in cell cycle progression and cytokine production compared to CD8+ T cells expressing PD-1 alone [25]. A study in NSCLC tissues showed that more than 60% of Tregs was TIM3-positive and was associated with nodal metastasis and advanced-stage disease [26]. In addition, TIM3 blockade improved the cytotoxicity of NK cells, and it is being considered as a promising target for immunotherapy in lung cancer [27, 28].

Two anti-TIM3 antibodies (TSR-022 and MBG453) are under the clinical trials (NCT02817633 and NCT02608268). In combination with anti-PD-1 antibodies as well as independently, both of the drugs showed affirmative response (Table 10.1).

10.2.4 LAG3

Lymphocyte activation gene 3 (LAG3) is a member of the Ig superfamily that plays an important part in lymphocyte homeostasis. LAG3 (CD223) is a CD4 homolog

Table 10.1 Ongoing studies of selected ICIs in lung cancer NSCLC

Checkpoint proteins	Blockers with trade names	Class
CTLA-4	Ipilimumab (Yervoy®) Tremelimumab	IgG1 mAb IgG2 mAb
PD-1	Pembrolizumab (Keytruda®) Nivolumab (Opdivo®)	IgG4κ mAb IgG4 mAb
PD-L1	Atezolizumab (Tecentriq®) Avelumab (Bevancio®) Durvalumab (Imfinzi®)	IgG1 mAb IgG1 mAb IgG1κ mAb
TIM3	TSR-022 (Cobolimab®) MBG453 (Sabatolimab®)	IgG4 mAb IgG4 mAb
LAG3	BMS-986016 (Relatlimab®)	mAb
KIR	BMS-986015 (Lirilumab®)	IgG4 mAb

(Triebel et al. 1990). Similar to CD4, MHC II is the only known ligand for LAG3. LAG3 is also critical for dampening T-cell immune responses. LAG3 specifically blocks CD8+ effector T-cell functions and enhances the suppressive activity of Tregs [29, 30].

Anti-LAG3 antibodies have proven to be effective in activating T effectors and immune response. As PD-1 and LAG3 are very often co-expressed on anergic T cells, dual blockade of these receptors has proven to reverse the anergy in a chronic infection scenario [31]. LAG3 knockout mice show less severe autoimmune effects than CTLA-4 and PD-1 knockout animals [32]. Burova et al. [33] showed that the treatment with a combination of antihuman PD-1 and antihuman LAG3 in mice enhanced antitumor efficacy. This confirms that LAG3 may play a subtle role in regulating T-cell function [34]. A preclinical study revealed that blocking LAG3 significantly increased CD4+ T-cell proliferation in a manner that relied on the presence of IL-2 and signal transducer and activator of transcription (STAT)-5. It also inhibits Treg induction [35].

In humans, LAG3 has been observed in surgically removed NSCLC tumors and was associated with PD-L1 expression. Moreover, a positive correlation of LAG3 expression with early recurrence and poor prognosis was also found [36]. Till now, an anti-LAG3 monoclonal antibody (BMS-986016) was put under scrutiny in two clinical trials in advanced lung cancer (Phase I NCT02966548 and phase I/IIa NCT01968109) alone or in combination with an anti-PD-1 antibody [11].

10.2.5 KIRs

Killer cell inhibitory receptors (KIRs) are a large clan of glycoprotein receptors that primarily bind MHC I molecules and inhibit NK cells [37]. In addition to NK cells, these receptors are also expressed specially on CD8+ T cells and APCs [38]. KIRs contain 2–3 Ig ectodomains (2D and 3D) and varying lengths of cytoplasmic tails: long (L) and short (S).

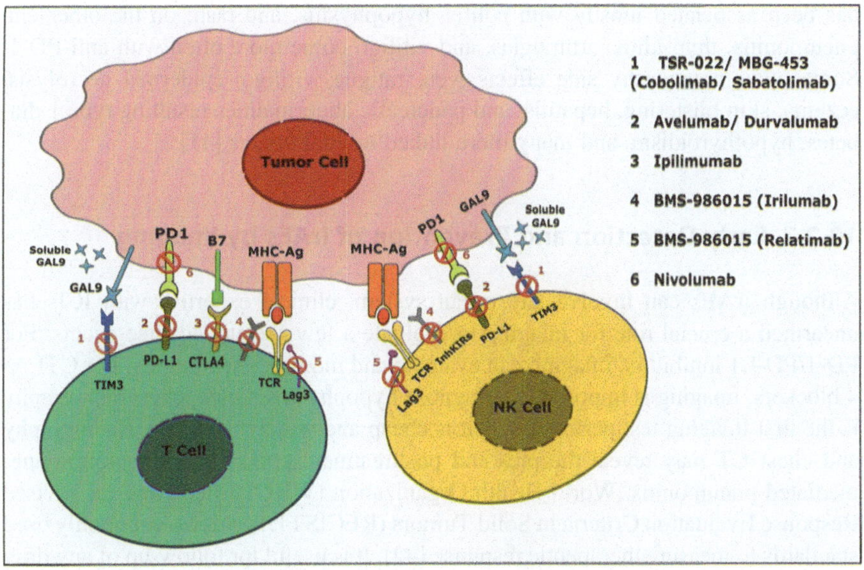

Fig. 10.2 Major immune checkpoints in lung cancer and their blockades. MHC: major histocompatibility complex; APC: antigen-presenting cell; TCR: T-cell receptor; CTLA-4: cytotoxic T lymphocyte antigen 4; PD-1: programmed cell death-1; PD-L1: PD ligand 1; TIM3: T-cell immunoglobulin-3; GAL9: galectin-9; LAG3: lymphocyte activation gene 3; KIR: killer cell immunoglobulin-like receptor; KIRs: killer cell inhibitory receptors

Interestingly, tumor cells expressing HLA-I in lesser quantities blinds the T cells. As they cannot recognize tumor antigens without HLA-TCR complex, immunological escape is promoted. At the same time inhibitory KIRs on T and NK cells exerts lower tumor-specific cytotoxic activity. A study on 62 NSCLC patients [39] had shown that patients with negative KIR expression on tumor cells had a significantly longer overall survival (OS) than patients with positive KIR on tumor cells. The most advanced anti-KIR monoclonal antibody (lirilumab) recognizes KIR-2DL-1, KIR-2DL-2, KIR-2DL-3 and can hinder the interaction with its HLA-I ligands [40]. Another trial (NCT01714739) with a monoclonal antibody (BMS-986015) is on and is in its last phase. A phase I study of lirilumab and ipilimumab combination in advanced NSCLC has been completed and efficacy was established (Fig. 10.2) [11].

10.2.6 Immune-Related Adverse Effects (irAEs)

Immune checkpoint inhibitors sometimes come with adverse side effects. Collectively they are called "immune-related adverse events" (irEAs) which appear depending on the inhibiting agent, type of malignancy, and individual's susceptibility. It had been reported to have serious irEAs as high as 27%, 16%, and 55% with the use of anti CTLA-4 and anti-PD-1. For instance, the use of anti CTLA-4 therapy has been associated mostly with colitis, hypophysitis, and rash; on the other end pneumonitis, thyroiditis, arthralgias, and vitiligo come more often with anti-PD-1. Some other noteworthy side effects were fatigue, vitiligo, epidermal necrolysis, eczema, skin blistering, hepatitis, and pancreatic abnormalities resulting type 1 diabetes, hypothyroidism, and many more linked to vital organs [41].

10.2.7 Early Detection and Prevention of irAEs by Imaging

Although irAEs can involve any organ system, clinical expertise with ICIs has unearthed a crucial role for imaging to evaluate a few specific adverse events. For PD-1/PD-L1 inhibitors, imaging can evaluate and monitor pneumonitis. For CTLA-4 blockers, imaging is important to diagnose hypophysitis. Often, chest radiography is the first imaging test performed as it is cheap and widely available. Radiography and chest CT may reveal the pre- and posttreatment condition like autoimmune-mediated pneumonitis. World Health Organization (WHO) criteria and the revised Response Evaluation Criteria in Solid Tumors (RECIST1.1) are two commonly used standards to measure therapeutic response [42]. It is useful for follow-up of any drug failure and required change in therapy. As the drugs differ from conventional cytotoxic chemotherapy in both treatment response and related adverse events, radiologists must be knowledgeable regarding the tumor response pattern and common adverse events associated with this novel therapeutic approach [43].

10.2.8 ICIs on Elderly People Having NSCLC

Unfortunately, the older segment of the population remains underrepresented in NSCLC clinical trials, and most of the evidence comes from selected study populations. Aging comes with decline and impairment of immune function (immunosenescence). Different data had shown that expression of PD-1 and CTLA-4 hampered with age. Furthermore, it is well-known that modulation of different checkpoint molecules due to exhaustion of T effector cells and a decline in Treg cells are observed in senescence. These data suggest that elderly patients may have a peculiar response to immune checkpoint-targeting therapies [44, 45]. The approved drugs are advised to be administered as consolidation therapy after chemo-RT in unresectable stage III NSCLC expressing PD-L1 at levels 1%. Data regarding elderly population of these molecules was limited as most studies have involved a small number of elderly patients. Immunosenescence may play a central role in irAEs as well.

10.2.9 Influence of Prognostic Factors

The assessment of requirement of any specific ICIs can be confirmed by looking at the associated biomarker expression status of a patient. For instance, tumor mutational burden (TMB) can be a potential biomarker as high TMBs, treated with pembrolizumab, showed higher objective responses (ORs) and PFS, than those with low TMBs in NSCLC [46] (CHECKMATE-227 clinical trial). This approach showed a potential clinical benefit for patients treated with anti-PD-1/PD-L1 drugs [47, 48]. CTLA-4 seems to be associated with FoxP3 upregulation which in turn increases TGF-β production and again promotes FoxP3 expression [49]. This mechanism is also implicated in the conversion of naive T cells to Treg cells, thus inducing immune suppression [50]. Some highly or poorly proliferative tumors are resistant to ICIs. So assessing the expressivity of proliferation-related genes by RNA-seq data in biopsies stands out as a promising strategy for improving clinical decision making [51].

10.2.10 Influence of Probiotics on ICIs in NSCLC

Bifidobacterium promotes antitumor immunity and enhances the efficacy of ICI [52]. A study done over a period of 4 years and 5 months among 118 Japanese people with NSCLC revealed the potential positive influence of probiotic *Clostridium butyricum* therapy (CBT) on ICI efficacy. In NSCLC, *C. butyricum* reduced systemic Th17 cells in mice which were associated with invasion and migration of cancer cells and cancer stem cell-like properties by STAT3 signaling [53]. Antitumor immunity promoting *Bifidobacterium* facilitated efficacy of ICIs significantly if used in combination with probiotic CBT and clindamycin [54, 55]. It was also suggested that combination therapy of probiotic CBT and antibiotics could potentially improve responses to immune checkpoint blockade (ICB) therapy [56]. These findings provide a rationale for combining immunotherapies with probiotics that manipulates commensal microbiota which may improve the efficacy of ICB in lung cancer patients [57].

10.3 Vaccination Strategies in NSCLC

Application of vaccines in cancer is an obvious extension of their utility, but the journey toward the ultimate goal is not so easy [58]. Nowadays, various evidences have been reported for the development of different types of vaccines against NSCLC tumors to improve quality of life.

10.3.1 Types of Vaccines and Their Mechanisms of Action

The concept of vaccine against cancer originated from the identification of aberrant proteins, expressed by tumor cells, which are also called tumor-associated antigens (TAAs) and can be classified into fetal antigens (normally absent in healthy people) and overexpressed normal proteins. Several vaccination strategies have been examined for the treatment of NSCLC, which are as follows:

10.3.1.1 Whole-Cell Vaccines

Belagenpumatucel-L is a whole-cell vaccine, which was prepared by using four different cell lines (one squamous cell carcinoma, one large-cell carcinoma, two adenocarcinoma) with a plasmid containing transforming growth factor β2 (TGF-β2) antisense transgene. TGF-β suppresses the activity of natural killer cells, cytotoxic T lymphocytes (CTLs), and dendritic cells and effectively shuts down the antitumor immune function. It also has the capacity to convert the immature T cells to immunosuppressive regulatory T cells by inducing the transcription factor forkhead box P3 (FOXP3). High TGF-β2 levels are related with poor prognosis in NSCLC. Belagenpumatucel-L helps in depletion of the amount of TGF-β2 in the tumor microenvironment and thereby immunity against the tumor is restored. Various studies demonstrated that the phase II clinical trials showed the safety and efficacy of belagenpumatucel-L in patients with NSCLC. Phase III trials demonstrated an overall survival benefit in several subgroup of patients, but further studies are required in order to select patients who will be the main beneficiaries of this vaccine [59–65].

10.3.1.2 Protein- and Peptide-Based Vaccines

Along with whole tumor cell vaccines, the use of proteins or peptides is one of the strategies for the treatment of cancer, but it has several limitations. It requires co administration of adjuvant to stimulate the immune system because cancer antigen shows low immunogenicity which exerts low response rate [66–69]. Besides this the absence of proteins exclusively expressed in cancer cells increased the risk of triggering autoimmune responses. Antigen proteins present as complex glycosylation patterns, which are difficult to purify.

Despite this, several protein vaccines had been developed for the effective treatment against NSCLC. The melanoma-associated antigen A3 (MAGE-A3) is a specific tumor antigen which was exclusively expressed in various types of tumor cells but not in normal cells. MAGE-A3 protein is expressed in several types of cancer including 35% of NSCLC. The rate of expression of this protein is related with disease progression and also associated with poor prognosis. Among all types of NSCLC, MAGE-A3 is found mostly in squamous cell carcinoma. The MAGE-A3 vaccine consists of a fusion protein of MAGE-A3 and *Haemophilus influenzae* protein D combined with an immunoadjuvant. Administration of this vaccine had shown positive results, although not clinically effective [64, 69–71].

Another therapeutic vaccine TG4010 based on a viral vector Ankara virus is an attenuated vaccinia virus which was genetically modified to encode the mucinous glycoprotein-1 protein (MUC-1) and interleukin 2(IL-2). MUC-1 is a transmembrane protein normally found on the apical surface of cells. MUC-1 is overexpressed in many types of malignancies and under glycosylated or aberrantly glycosylated which can uncover epitopes on the peptide core that may act as a tumor-associated antigen. MUC-1 is abnormally expressed in half of all NSCLC and inhibits physiologic T-cell proliferation. Due to immune suppression, high levels of serum MUC-1 are associated with poor survival. Its glycosylation pattern on the abnormally expressed MUC-1 in NSCLC makes it an attractive target for immunotherapy. But vaccine based on entire MUC-1 protein did not exert notable positive results in several clinical trials [63, 64, 69, 72–74]. Another vaccine BLP25 liposome vaccine (L-BLP25) also designed to target MUC-1, and it can induce cellular immune response, which may lead to immune rejections of tumor tissues that express MUC-1 antigen. But there were no notable results as such [63, 75, 76].

The CIMAvax-EGF is a conjugated anticancer vaccine of epidermal growth factor (EGF) and P64 protein derived from the meningitis B bacteria and Montanide ISA 51 as adjuvant that was developed entirely in Cuba. EGFR overexpression can trigger malignant transformation of normal cells, metastasis, angiogenesis, cell proliferation, anti-apoptotic signals and invasiveness. This is overexpressed in squamous malignancies including NSCLC. Targeted therapy with EGFR gene mutation is commonly associated with treatment among NSCLC. This vaccine showed immune responses against a molecular driver of cancer cell EGF and block cancer cell proliferation. After phase II trial, it was found to be safe and immunogenic in advanced NSCLC patients. After phase III trial, this vaccine shows overall survival, safety, and immunogenicity. Its administration was approved in Cuba, Venezuela, and Peru for stage IIIB and IV NSCLC patients after a first line of chemotherapy, but some studies revealed that CIMAvax-EGF along with PD-1 inhibitors nivolumab gives better outcomes for patients with advanced NSCLC [63, 77–83].

10.3.1.3 mRNA Vaccines

At first, research on vaccines was focused mainly on DNA rather than RNA because of the power of stability. Gradual appearance of drug delivery nanosystem and safety of mRNA in terms of mutagenicity and ease of internalization favor mRNA vaccines [69, 84, 85]. RNActive® CV9201 is one of the promising vaccines available in this category, which consists of mixture of five NSCLC-associated antigens that activate the immune response after the extraction of dendritic cells and subsequent delivery to the patient or after the direct administration of mRNA [69, 84]. After phase Ib clinical trial, this vaccine is associated with better immune response in over 65% of patients, but the strength of response varies between patients [69, 86].

Most vaccines were reported to be safe and well tolerated for humans but with poor benefit of response rate, overall survival, and progression free survival. Currently several trials are now ongoing with various categories of vaccine. For NSCLC patients, new designs are now being explored such as vaccine plus chemotherapy, vaccine with immune checkpoint inhibitors, vaccine plus chemotherapy, and immune checkpoint inhibitors.

10.4 Immunotherapy in SCLC

SCLC is a form of neuroendocrine tumor which is traditionally classified into limited stage SCLC (LS-SCLC) and extensive stage SCLC (ES-SCLC) [87]. SCLC is commonly known for its aggressive behavior, rapid doubling time, growth, and distant sites [88, 89]. Besides these, alterations in chromosome 3p and mutations in

RB1, TP53, RASSFI, MYC, FGFR1, and PTEN were reported to be associated with tumor progression [90, 91]. The following immunotherapeutic strategies were employed in SCLC patients.

10.4.1 Immune Checkpoint Inhibitors in SCLC

10.4.1.1 CTLA-4

A single-arm, non-randomized phase II trial was performed in United Kingdom back in 2011–2014, and the efficacy of ipilimumab was being evaluated [92], and ipilimumab with standard chemotherapy was found to be beneficial in case of ES-SCLC patients.

10.4.1.2 PD-L1/PD-1

Back in the period 2013–2015, a ½ phase multicenter trial was conducted among ES-SCLC patients where the patients are treated with nivolumab plus ipilimumab [93]. Nivolumab monotherapy and nivolumab plus ipilimumab were clinically effective, and an acceptable safety profile was identified among ES-SCLC patients [94]. Data on distribution of PD-L1 expression in SCLC across various stages are very limited in patients. But PD-L1 expression seems to lower in advanced stage than in early stage among SCLC patients [95, 96] and also than in NSCLC [97, 98].

10.4.1.3 CD 47

CD 47 is integrin-associated protein (IAP) which is also a multifunctional counter-receptor for signal regulatory protein alpha ligand (SIRPα) and also associated with cellular responses, proliferation, and cellular migration [99–101]. It is also known as a key anti-phagocytic molecule that renders tumor cell resistance to host-immune surveillance [99, 100]. CD47 is a cell surface molecule that is present in higher amounts on SCLC cell surfaces [88]. In the case of SCLC, blockade of CD47/SIRPα axis by monoclonal antibody was used as a potential immunotherapeutic strategy [102].

10.4.2 Ongoing Clinical Trials

10.4.2.1 IMpower 133

IMpower 133 was under first trial with positive results in first-line treatment of ES-SCLC. This was a phase III, double-blinded, randomized, placebo-controlled trial where addition of atezolizumab's efficacy and safety had been evaluated in ES-SCLC patients who did not receive conventional chemotherapy [103]. Atezolizumab is a humanized monoclonal antibody of IgG isotype against PD-L1 protein [103]. Results showed that atezolizumab along with carboplatin and etoposide provided a significant improvement in OS and PFS among ES-SCLC which set the new standard of care in this particular scenario. Keynote 604 trial was much similar with IMpower 133, and it focused mainly on the OS and PFS of patients with ES-SCLC [104].

10.4.2.2 CheckMate 331 and CheckMate 451

CheckMate 331 was under phase III randomized trial where the use of nivolumab along with chemo-topotecan in the case of relapsed SCLC was assessed [105]. CheckMate 451 was under another trial to assess the OS and PFS of patients with nivolumab as monotherapy or with addition of ipilimumab as first-line maintenance therpy [106].

10.4.2.3 Keynote 028 and Keynote 158

Keynote 028 was under a Ib phase trial to evaluate potentiality of pembrolizumab in ES-SCLC patients. Pembrolizumab is an IgG4 isotype antibody targeting PD-1 receptor of lymphocytes and enhancing immune response to cancer evasion. Result indicated the promising antitumor activity in ES-SCLC patients [107]. Keynote 158 was also under phase II trial to prove beneficial efficiency of pembrolizumab, and the outcome of this trial was safe and satisfactory [108]. Pembrolizumab was the first FDA-approved immunotherapeutic agent with a recommended dosage of 200 mg and applied intravenously as monotherapy [109].

Though various targets were identified (PD-1/PD-L1, CTLA-4, CD47), and various trials are ongoing, but the endpoint of all these are not satisfactory up to the mark, and they remain efficient in the early stages of the disease. For those limitations, further study should be focused on different biomarkers, and innovative trials with combination of immunotherapy and radiotherapy may be explored as a new strategy.

10.5 Combination of Radiotherapy and Immunotherapy in Lung Cancer

Radiotherapy is known as an effective treatment for cancers including lung cancer. Apart from only radiation, chemotherapy combined with radiation showed significant results in some cases of lung cancer. It was also reported that by Yuan and colleagues that combination of radiotherapy and immunotherapy potentially overcome resistance to immune checkpoint blockade [110, 111]. Apart from the tumor size regression, data also suggested that combined therapy also doesn't have any significant additive toxicity in the case of lung cancer [110, 112, 113].

10.5.1 Combined Therapy Trials

Following are the list of ongoing combined therapy clinical trials:

10.5.1.1 Keynote 001

This trial involved 97 patients in which 24 patients received thoracic radiotherapy. The approximate median interval between radiotherapy and immunotherapy was nearly 1 year. PD-1 inhibitor pembrolizumab was used to block systemic immune responses. In this trial, significantly longer PFS and OS were observed. But a higher frequency of pulmonary toxicities was also observed (63% of all patients). A subgroup treated with extracranial radiation therapy showed remarkable results and registered as an effective strategy [114, 115].

10.5.1.2 Pacific Trial

The trial was conducted with 713 patients which were analyzed after sequential immunotherapeutic treatment. Platinum-based chemotherapy was given to those patients 1–42 days prior to initiation of immunotherapy. IgG1-PD-L1 inhibitors with durvalumab were administered in patients with stage III NSCLC which were not progressing after the chemoradiation [116]. As a result, a higher frequency of low-grade pulmonary toxicity was also observed in patients who received chemoradiation earlier. Based on the data, the US FDA approved durvalumab on 31 July 2017 as a therapeutic agent for NSCLC patients [115–117]. Both Keynote 001 and pacific trials showed excellent results and significantly increased median progression free survival.

10.5.1.3 Bevacizumab

Vascular growth factor plays an essential role in tumor angiogenesis in NSCLC [118–121]. The most studied inhibitor for VEGF is bevacizumab which showed increased PFS and OS in patients with non-squamous NSCLC [118, 119, 121] when combined with standard cytotoxic chemotherapy. But this combination also showed a high incidence of tracheoesophageal fistula formation [116, 117]. There are so many ongoing trials which evaluate bevacizumab with other platinum-based combined therapy and other targeted agents like erlotinib and ramucirumab [118, 119, 121].

10.5.1.4 Caspian

This is an ongoing phase III trial which investigates the effect of durvalumab with standard radiotherapy in 795 patients and associated with significantly high PFS and OS [122]. Durvalumab is an FDA-approved IgG1κ monoclonal antibody against PD-1/PD-L1 [122].

10.5.1.5 Meru

This is a phase III trial with the use of Rova-T in 740 patients with four cycles of first-line platinum doublet chemotherapy. Rovalpituzumab tesirine (Rova-T) is an experimental antibody drug conjugate targeting protein DLL3 in tumor cells [123, 124].

10.6 Conclusion

Large numbers of clinical trials are ongoing on immunotherapeutic agents in lung cancer. Promising immunotherapeutic agents like checkpoint inhibitors are associated with greater survival rate among lung cancer patients. Antigen-specific therapeutic vaccines were also found to be effective to augment the immune response. Combined therapy with radiation and immunotherapy has become the most useful tool till today in the lung cancer scenario. Increasing knowledge in molecular signaling pathways, evaluation of potential inhibitors, and more clinical trials can overcome the drug resistance and toxicities in normal cells in future.

Acknowledgments The authors acknowledge Department of Zoology, The University of Burdwan.

Conflicts of Interest There are none.

References

1. National Cancer Institute (2008) Non-small cell lung and bronchus cancer (invasive) survival rates, by race, sex, diagnosis year, state and age. In: SEER Cancer Statistics Review 1975–2004. National Cancer Institute, Bethesda, MD
2. Alberg AJ, Ford JG, Samet JM (2007) Epidemiology of lung cancer: ACCP evidence-based clinical practice guidelines (2nd edition). Chest 132:29s–55s
3. Mountain CF (1997) Revisions in the international system for staging lung cancer. Chest 111:1710–1717
4. Shepherd FA, Crowley J, Van Houtte P, Postmus PE, Carney D, Chansky K et al (2007) The International Association for the Study of Lung Cancer lung cancer staging project: proposals regarding the clinical staging of small cell lung cancer in the forthcoming (seventh) edition of the tumor, node, metastasis classification for lung cancer. J Thorac Oncol 2:1067–1077
5. Greenlee R, Murray T, Bloden S, Wingo PA (2000) Cancer statistics. CA Cancer J Clin 50:7–33
6. Bunn PJ, Kelly K (2000) New combinations in the treatment of lung cancer: a time for optimism. Chest 117:138–143
7. Barlesi F (2010) Targeted therapies in non-small-cell lung cancer (NSCLC): how to proceed to aim at the good target? Eur J Cardio Thorac 38:37–38
8. Kim EK, Kim KA, Lee CY, Shim HS (2017) The frequency and clinical impact of HER2 alterations in lung adenocarcinoma. PLoS One 12(2):e0171280
9. Yoneda K, Imanishi N, Ichiki Y, Tanaka F (2018) Immune checkpoint inhibitors (ICIs) in non-small cell lung Cancer (NSCLC). J UOEH 40(2):173–189
10. Pardoll DM (2012) The blockade of immune checkpoints in cancer immunotherapy. Nat Rev Cancer 12:252–264
11. Dal Bello MG, Alama A, Coco S, Vanni I, Grossi F (2017) Understanding the checkpoint blockade in lung cancer immunotherapy. Drug Discov Today 22(8):1266–1273
12. Hodi FS et al (2010) Improved survival with ipilimumab in patients with metastatic melanoma. N Engl J Med 363:711–723
13. Robert C et al (2011) Ipilimumab plus decarbazine for previously untreated metastatic melanoma. N Engl J Med 364:2517–2526
14. Butte MJ et al (2007) PD-L1 interacts specifically with B7-1 to inhibit T cell proliferation. Immunity 27:111–122

15. Shepherd FA, Dancey J, Ramlau R et al (2000) Prospective randomized trial of docetaxel versus best supportive care in patients with non–small-cell lung cancer patients previously treated with platinum-based chemotherapy. J Clin Oncol 18:2095–2103
16. Fossella FV, DeVore R, Kerr RN et al (2000) Randomized phase III trial of docetaxel versus vinorelbine or ifosfamide in patients with advanced non–small-cell lung cancer previously treated with platinum-containing chemotherapy. J Clin Oncol 18:2354–2362
17. Bagley SJ, Bauml JM, Langer CJ (2015) PD-1/PD-L1 immune checkpoint blockade in non-small cell lung cancer. Clin Adv Hematol Oncol 13(10):676–683
18. Herbst RS, Soria JC, al KM (2014) Predictive correlates of response to the anti-PD-L1 antibody MPDL3280A in cancer patients. Nature 515(7528):563–567
19. Spigel D, Gettinger SN, Horn L et al (2013) Clinical activity, safety, and biomarkers of MPDL3280A, an engineered PD-L1 antibody in patients with locally advanced or metastatic non-small cell lung cancer (NSCLC) [ASCO abstract 8008]. J Clin Oncol 31(15):8008
20. Garon EB, Rizvi NA, Hui R et al (2015) Pembrolizumab for the treatment of non-small-cell lung cancer. N Engl J Med 372:2018–2028
21. Reck M, Rodríguez-Abreu D, Robinson AG et al (2016) Pembrolizumab versus chemotherapy for PDL1positive non-small-cell lung cancer. N Engl J Med 375:1823–1833
22. Hui E, Cheung J, Zhu J et al (2017) T cell costimulatory receptor CD28 is a primary target for PD-1-mediated inhibition. Science 355(6332):1428–1433
23. Wei SC, Anang NAS, Sharma R et al (2019) Combination anti-CTLA-4 plus anti-PD-1 checkpoint blockade utilizes cellular mechanisms partially distinct from monotherapies. Proc Natl Acad Sci U S A 116(45):22699–22709
24. Wada J, Kanwar YS (1997) Identification and characterization of galectin-9, a novel β-galactoside-binding mammalian lectin. J Biol Chem 272:6078–6086
25. Sundar R et al (2014) Immunotherapy in the treatment of non-small cell lung cancer. Lung Cancer 85:101–109
26. Gao X et al (2012) TIM-3 expression characterizes regulatory T cells in tumor tissues and is associated with lung cancer progression. PLoS One 7:e30676
27. Anderson AC (2014) Tim-3: an emerging target in the cancer immunotherapy landscape. Cancer Immunol Res 2:393–398
28. Zhu C et al (2005) The Tim-3 ligand galectin-9 negatively regulates T helper type 1 immunity. Nat Immunol 6:1245–1252
29. Grosso JF et al (2007) LAG-3 regulates CD8+ T cell accumulation and effector function in murine self- and tumor-tolerance systems. J Clin Invest 117:3383–3392
30. Huang CT, Workman CG, Flies D, Pan X, Marson AL, Zhou G, Hipkiss EL, Ravi S, Kowalski J, Levitsky HI, Powell JD, Pardoll DM, Drake CG, Vijnali DA (2004) Role of LAG-3 in regulatory T cells. Immunity 21(4):503–513
31. Blackburn SD et al (2009) Coregulation of CD8+ T cell exhaustion during chronic viral infection by multiple inhibitory receptors. Nat Immunol 10:29–37
32. Miyazaki T et al (1996) LAG-3 is not responsible for selecting T helper cells in CD4-deficient mice. Int Immunol 8:725–729
33. Burova E et al (2016) 31st annual meeting and associated programs of the Society for Immunotherapy of Cancer (SITC 2016): part two. J Immunother Cancer 4(Suppl. 1):73
34. Nirschl CJ, Drake CG (2013) Molecular pathways: co-expression of immune checkpoint molecules: Signalling pathways and implications for cancer immunotherapy. Clin Cancer Res 19:4917–4924
35. Durham NM et al (2014) Lymphocyte activation gene 3 (LAG-3) modulates the ability of CD4 T-cells to be suppressed in vivo. PLoS One 9:e109080
36. He Y et al (2017) LAG-3 protein expression in non-small cell lung cancer and its relationship with PD-1/PD-L1 and tumor-infiltrating lymphocytes. J Thorax Oncologia 12:814–823
37. Lanier LL (2008) Up on the tightrope: natural killer cell activation and inhibition. Nat Immunol 9(5):495–502

38. Mingari MC et al (2005) Human cytolytic T lymphocytes expressing HLA class I inhibitory receptors. Curr Opin Immunol 17:312–319
39. He Y et al (2016) KIR 2 D (L1, L3, L4, S4) and KIR 3 DL1 protein expression in non-small cell lung cancer. Oncotarget 7:82104–82111
40. Kohrt HE et al (2014) Anti-KIR antibody enhancement of anti-lymphoma activity of natural killer cells as monotherapy and in combination with anti-CD20 antibodies. Blood 123:678–686
41. Marin-Acevedo JA, Chirila RM, Dronca RS (2019) Immune checkpoint inhibitor toxicities. Mayo Clin Proc 94(7):1321–1329
42. Kwak JJ, Tirumani SH, Van den Abbeele AD, Koo PJ, Jacene HA (2015) Cancer immunotherapy: imaging assessment of novel treatment response patterns and immune-related adverse events. Radiographics 35:424–437
43. Wang GX, Guo LQ, Gainor JF, Fintelmann FJ (2017) Immune checkpoint inhibitors in lung Cancer: imaging considerations. AJR Am J Roentgenol 209(3):567–575
44. Barber DL, Wherry EJ, Masopust D, Zhu B, Allison JP, Sharpe AH, Freeman GJ, Ahmed R (2006) Restoring function in exhausted CD8 T cells during chronic viral infection. Nature 439:682–687
45. Canaday DH, Parker KE, Aung H, Chen HE, Nunez-Medina D, Burant CJ (2013) Age-dependent changes in the expression of regulatory cell surface ligands in activated human T-cells. BMC Immunol 14:45
46. Rizvi NA, Hellmann MD, Snyder A, Kvistborg P et al (2015) Mutational landscape determines sensitivity to PD-1 blockade in non-small cell lung cancer. Science 348:124–128
47. Gandara DR, Paul SM, Kowanetz M, Schleifman E, Zou W et al (2018) Blood-based tumor mutational burden as a predictor of clinical benefit in non-small-cell lung cancer patients treated with atezolizumab. Nat Med 24:1441–1448
48. Khagi Y, Goodman AM, Daniels GA, Patel SP, Sacco AG et al (2017) Hypermutated circulating tumor DNA: correlation with response to checkpoint inhibitor-based immunotherapy. Clin Cancer Res 23:5729–5736
49. Zheng SG, Wang JH, Stohl W et al (2006) TGF-beta requires CTLA-4 early after T cell activation to induce FoxP3 and generate adaptive CD4+CD25+ regulatory cells. J Immunol 176:3321–3329
50. Bianco A, Perrotta F, Barra G, Malapelle U, Rocco D, De Palma R (2019) Prognostic factors and biomarkers of responses to immune checkpoint inhibitors in lung cancer. Int J Mol Sci 20 (19):4931
51. Pabla S, Conroy JM, Nesline MK et al (2019) Proliferative potential and resistance to immune checkpoint blockade in lung cancer patients. J Immunother Cancer 7(1):27
52. Matson V, Fessler J, Bao R, Chongsuwat T, Zha Y, Alegre ML et al (2018) The commensal microbiome is associated with anti-PD-1 efficacy in metastatic melanoma patients. Science 359:104–108
53. Wang R, Yang L, Zhang C, Wang R, Zhang Z, He Q et al (2018) Th17 cell derived IL-17A promoted tumor progression via STAT3/NF-kB/Notch1 signaling in non-small cell lung cancer. Onco Targets Ther 7:e1461303
54. Hagihara M, Yamashita R, Matsumoto A, Mori T, Inagaki T, Nonogaki T et al (2019) The impact of probiotic Clostridium butyricum MIYAIRI 588 on murine gut metabolic alterations. J Infect Chemother 25:571–577
55. Hagihara M, Yamashita R, Matsumoto A, Mori T, Kuroki Y, Kudo H et al (2018) The impact of Clostridium butyricum MIYAIRI 588 on the murine gut microbiome and colonic tissue. Anaerobe 54:8–18
56. Helmink BA, Khan MAW, Hermann A, Gopalakrishnan V, Wargo JA (2019) The microbiome, cancer, and cancer therapy. Nat Med 25:377–388
57. Tomita Y, Ikeda T, Sakata S et al (2020) Association of probiotic Clostridium butyricum therapy with survival and response to immune checkpoint blockade in patients with lung cancer. Cancer Immunol Res 8:1236–1242. https://doi.org/10.1158/2326-6066.cir-20-0051

58. Hollingsworth RE, Jansen K (2019) Turning the corner on therapeutic cancer vaccines. NJP Vaccines 4(7):1–10
59. Giaccone G, Bazhenova LA, Nemunaitis J et al (2015) A phase III study of Belagenpumatucel-L, an allogenic tumor cell vaccine, as maintenance therapy for non-small cell lung cancer. Eur J Cancer 51(16):2321–2329
60. Kong F, Jirtle RL, Huang DH, Clough RW, Anscher MS (1999) Plasma transforming growth factor-beta1 level before radiotherapy correlates with long term outcome of patients with lung carcinoma. Cancer 86(9):1712–1719
61. Nemunaitis J, Dillman RO, Schwarzenberger PO et al (2006) Phase II study of belagenpumatucel-L, a transforming growth factor beta-2 antisense gene-modified allogenic tumor cell vaccine in non-small-cell lung cancer. J Clin Oncol 24(29):4721–4730
62. Nemunaitis J, Nemunaitis M, Senzer N et al (2009) Phase II trial of Belagenpumatucel-L, a TGF-beta2 antisense gene modified allogenic tumor vaccine in advanced non-small cell lung cancer (NSCLC) patients. Cancer Gene Ther 16(8):620–624
63. Olivers H, Caglevic C, Passiglia F, Taverna S, Smits E, Rolfo C (2018) Vaccine and immune cell therapy in non-small cell lung cancer. J Thorac Dis 10(suppl 13):S1602–S1614
64. Socola F, Scherfenberg N, Raez LE (2013) Therapeutic vaccines in non-small cell lung cancer. Immuno Target Ther 2:115–124
65. Suzuki H, Owada Y, Watanabe Y et al (2014) Recent advances in immunotherapy for non-small-cell lung cancer. Hum Vaccines Immunother 10(2):352–357
66. Banchereau J, Palucka K (2018) Cancer vaccines on the move. Nat Rev Clin Oncol 15:9–10
67. Cortes-jofre M, Uranga R, Torres Pombert A, Arango Prado MDC, Caballero Aguirrechu I, Pacheco C, Ortiz Reyes RM, Chuecas F, Mas Bermejo PI (2019) Therapeutic vaccines for advanced non-small cell lung cancer. Cochrane Database Syst Rev 2019:1–13
68. Cuppens K, Vansteenkiste J (2014) Vaccination therapy for non-small-cell lung cancer. Curr Opin Oncol 26(2):165–170
69. Fernandez CG, Fornaguera C, Borros S (2020) Nanomedicine in non-small cell lung cancer: from conventional treatments of immunotherapy. Cancer 12(1609):1–26
70. Sienel W, Varwerk C, Linder A et al (2004) Melanoma associated antigen (MAGE)-A3 expression in stages I and II non-small cell lung cancer: results of a multi-center study. Eur J Cardio Thorac Surg 25(1):131–134
71. Vansteenkiste J, Zielinski M, Linder A, Dahabreh J, Gonzalez EE, Malinowski w L-BM, Vanakesa T, Jassem J, Kalofonos H et al (2013) Adjuvant MAGE-A3 immunotherapy in resected non-small-cell lung cancer: phase II randomized study results. J Clin Oncol 31 (19):2396–2403
72. Limacher JM, Quoix E (2012) TG4010: a therapeutic vaccine against MUC1 expressing tumors. Onco Targets Ther 1(5):791–792
73. Quoix E, Ramlau R, Westeel V, Papai Z, Madroszyk A, Riviere A, Koralewski P, Breton JL, Stoelben E, Braun D et al (2011) Therapeutic vaccination with TG4010 and first-line chemotherapy in advanced non-small-cell lung cancer: a controlled phase 2b trial. Lancet Oncol 12 (12):1125–1133
74. Raina D, Kosugi M, Ahmad R et al (2011) Dependence on the MUC1-C oncoprotein in non-small cell lung cancer cells. Mol Cancer Ther 10(5):806–816
75. Ho SB, Niehans GA, Lyftogt C et al (1993) Heterogenicity of mucin gene expression in normal and neoplastic tissues. Cancer Res 53:641–651
76. Palmer M, Parker J, Modi S et al (2001) Phase I study of the BLP25 (MUC1) peptide liposomal vaccine for active specific immunotherapy in stage IIIB/IV non-small-cell lung cancer. Clin Lung Cancer 3:49–57
77. Hirsch FR, Varella-García M, Cappuzzo F (2009) Predictive value of EGFR and HER2 overexpression in advanced non-small-cell lung cancer. Oncogene 28:S32–S37
78. Luo SY, Lam DC (2013) Oncogenic driver mutations in lung. Transl Respir Med 1:6
79. Mendelsohn J, Baselga J (2006) Epidermal growth factor receptor targeting in cancer. Semin Oncol 33(4):369–385

80. Rodriguez PC, Rodriguez G, Gonzalez G, Lage A (2010) Clinical development and perspectives of CIMAvax EGF, Cuban vaccine for non-small-cell lung cancer therapy. MEDICC Rev 12(1):17–23
81. Rotow J, Bivona TG (2017) Understanding and targeting resistance mechanisms in NSCLC. Nat Rev Cancer 17(11):637–658
82. Saavedra D, Crombet T (2017) CIMAvax-EGF: a new therapeutic vaccine for advanced non-small cell lung cancer patients. Front Immunol 8:269
83. Toffoli G, De Mattia E, Cecchin E, Biason P, Masier S, Corona G (2007) Pharmacology of epidermal growth factor inhibitors. Int J Biol Markers 22(1):S24–S39
84. Fiedler K, Lazzaro S, Lutz J, Rauch S, Heidenreich R (2016) MRNA cancer vaccines. Recent Results Cancer Res 209:61–84
85. Midoux P, Pichon C (2014) Lipid based MRNA vaccine delivery system. Expert Rev Vaccines 14(2):221–234
86. Sebastian M, Von Boehmer L, Zippelius A, Mayer F, Reck M, Atanackovic D, Thomas M, Schneller F, Stoehlmacher J, Goekkurt E et al (2011) Messenger RNA vaccination in NSCLC: findings from a phase I/IIa clinical trial. J Clin Oncol 2584s:29
87. Huang J, Kondo H, Marom EM et al (2016) The IASLC lung cancer staging project: proposals for coding T categories for subsolid nodules and assessment of tumor size in part-solid tumors in the forthcoming eighth edition of the TNM classification of lung cancer. J Thorac Oncol 11 (8):1204–1223
88. Calles A, Aguado G, Sandoval C, Alvarez R (2019) The role of immunotherapy in small cell lung cancer. Clin Transl Oncol 21(8):961–976
89. Winston WT, Maghfoor I (2019) Small Cell Lung Cancer. https://emedicine.medscape.com/article/280104-overview#a5
90. D'Angelo SP, Pietanza MC (2010) +e molecular pathogenesis of small cell lung cancer. Can Biol Ther 10(1):1–10
91. Kahnert K, Kauffmann-Guerrero D, Huber RM (2016) SCLC–state of the art and what does the future have in store? Clin Lung Cancer 17(5):325–333
92. Arriola E, Wheater M, Galea I et al (2016) Outcome and biomarker analysis from a multicenter phase 2 study of ipilimumab in combination with carboplatin and etoposide as first-line therapy for extensive-stage SCLC. J Thor Oncol 11(9):1511–1521
93. Antonia SJ, Lopez-Martin JA, Bendell J, Ott PA, Taylor M, Eder JP et al (2016) Nivolumab alone and nivolumab plus ipilimumab in recurrent small-cell lung cancer (CheckMate 032): a multicentre, open-label, phase 1/2 trial. Lancet Oncol 17(7):883–895
94. Tsiouprou I, Zaharias A, Spyratos D (2019) The role of immunotherapy in extensive stage small-cell lung cancer: a review of the literature. Can Respir J 2019:6860432
95. Bonanno L, Pavan A, Dieci MV, Di Liso E, Schiavon M, Comacchio G et al (2018) The role of immune microenvironment in small-cell lung cancer: distribution of PD-L1 expression and prognostic role of FOXP3-positive tumour infiltrating lymphocytes. Eur J Cancer 101:191–200
96. Ishii H, Azuma K, Kawahara A, Yamada K, Imamura Y, Tokito T et al (2015) Significance of programmed cell death-ligand 1 expression and its association with survival in patients with small cell lung cancer. J Thorac Oncol 10(3):426–430
97. Ancevski HK, Socinski MA, Villaruz LC (2018) PD-L1 testing in guiding patient selection for PD-1/PD-L1 inhibitor therapy in lung cancer. Mol Diagn Ther 22(1):1–10
98. He J, Hu Y, Hu M, Li B (2015) Development of PD-1/PD-L1 pathway in tumor immune microenvironment and treatment for non-small cell lung cancer. Sci Rep 5:13110
99. Piccione EC, Juarez S, Tseng S, Liu J, Stafford M, Narayanan C et al (2016) SIRPa antibody fusion proteins selectively bind and eliminate dual antigen expressing tumor cells. Clin Cancer Res 22:5109–5119
100. Vonderheide RH (2015) CD47 blockade as another immune checkpoint therapy for cancer. Nat Med 21:1122–1123

101. Zhang X, Fan J, Wang S et al (2017) Targeting CD47 and autophagy elicited enhanced antitumor effects in non-small cell lung cancer. Cancer Immunol Res 5(5):363–375
102. Weiskopf K, Jahchan NS, Schnorr PJ, Cristea S, Ring AM, Maute RL et al (2016) CD47-blocking immunotherapies stimulate macrophage-mediated destruction of small-cell lung cancer. J Clin Invest 126:2610–2620
103. Horn L, Mansfield AS, Szczesna A et al (2018) First-line atezolizumab plus chemotherapy in extensive-stage small-cell lung cancer. N Engl J Med 379(23):2220–2229
104. Rudin C, Shen L, Pietanza MC (2017) P2.04-007 KEYNOTE-604: phase 3 randomized, double-blind trial of pembrolizumab/placebo plus etoposide/platinum for extensive stage-SCLC. J Thorac Oncol 12(11):S2400
105. Horn L, Reck M, Gettinger SN et al (2016) CheckMate 331: an open-label, randomized phase III trial of nivolumab versus chemotherapy in patients (pts) with relapsed small cell lung cancer (SCLC) after first-line platinum-based chemotherapy (PT-DC). J Clin Oncol 34(15): TPS8578
106. Ready N, Owonikoko TK, Postmus PE et al (2017) CheckMate 451: a randomized, double-blind, phase III trial of nivolumab (nivo), nivo plus ipilimumab (ipi), or placebo as maintenance therapy in patients (pts) with extensive-stage disease small cell lung cancer (ED-SCLC) after first-line platinum-based doublet chemotherapy (PT-DC). J Clin Oncol 34(15):TPS8579
107. Ott PA, Elez E, Hiret S et al (2017) Pembrolizumab in patients with extensive-stage small-cell lung cancer: results from the phase ib KEYNOTE-028 study. J Clin Oncol 35(34):3823–3829
108. Chung HC, Lopez-Martin JA, Kao SCH et al (2018) Phase 2 study of pembrolizumab in advanced small-cell lung cancer (SCLC): KEYNOTE-158. J Clin Oncol 36(15):8506
109. FDA, FDA Approves Pembrolizumab for Metastatic Small Cell Lung Cancer (2019). https://www.fda.gov/drugs/resourcesinformation-approved-drugs/fda-approvespembrolizumabmetastatic-small-cell-lung-cancer
110. Ko EC, Raben D, Formenti SC (2018) The integration of radiotherapy with immunotherapy for the treatment of non-small cell lung cancer. Clin Cancer Res 24(23):5792–5806
111. Yuan Z, Fromm A, Ahmed KA, Grass GD, Yang GQ, Oliver DE et al (2017) Radiotherapy rescue of a nivolumab-refractory immune response in a patient with PD-L1 negative metastatic squamous cell carcinoma of the lung. J Thorac Oncol 12:e135–e136
112. Butts C, Socinski MA, Mitchell PL, Thatcher N, Havel L, Krzakowski M et al (2014) Tecemotide (L-BLP25) versus placebo after chemoradiotherapy for stage III non-small cell lung cancer (START): a randomised, double-blind, phase 3 trial. Lancet Oncol 15:59–68
113. Mitchell P, Thatcher N, Socinski MA, Wasilewska-Tesluk E, Horwood K, Szczesna A et al (2015) Tecemotide in unresectable stage III non-small-cell lung cancer in the phase III START study: updated overall survival and biomarker analyses. Ann Oncol 26:1134–1142
114. Shaverdian N, Lisberg AE, Bornazyan K, Veruttipong D, Goldman JW, Formenti SC, Garon EB, Lee P (2017) Previous radiotherapy and the clinical activity and toxicity of pembrolizumab in the treatment of non-small-cell lung cancer: a secondary analysis of the KEYNOTE-001 phase 1 trial. Lancet Oncol 18:895–903
115. Wirsdörfer F, de Leve S, Jendrossek V (2018) Combining radiotherapy and immunotherapy in lung cancer: can we expect limitations due to altered normal tissue toxicity? Int J Mol Sci 20 (1):24
116. Lazzari C, Karachaliou N, Bulotta A et al (2018) Combination of immunotherapy with chemotherapy and radiotherapy in lung cancer: is this the beginning of the end for cancer? Ther Adv Med Oncol 10:1758835918762094
117. Antonia SJ, Villegas A, Daniel D, Vicente D, Murakami S, Hui R, Yokoi T, Chiappori A, Lee KH, deWit M et al (2017) Durvalumab after chemoradiotherapy in stage III non-small-cell lung cancer. N Engl J Med 377:1919–1929
118. Reck M, von Pawel J, Zatloukal P et al (2010) Overall survival with cisplatin-gemcitabine and bevacizumab or placebo as first-line therapy for nonsquamous non-small-cell lung cancer: results from a randomised phase III trial (AVAiL). Ann Oncol 21:1804–1809

119. Sandler A, Gray R, Perry MC et al (2006) Paclitaxel-carboplatin alone or with bevacizumab for non-small-cell lung cancer. N Engl J Med 355:2542–2550
120. Simone CB 2nd, Burri SH, Heinzerling JH (2015) Novel radiotherapy approaches for lung cancer: combining radiation therapy with targeted and immunotherapies. Transl Lung Cancer Res 4(5):545–552
121. Soria JC, Mauguen A, Reck M et al (2013) Systematic review and meta-analysis of randomised, phase II/III trials adding bevacizumab to platinum-based chemotherapy as first-line treatment in patients with advanced non-small-cell lung cancer. Ann Oncol 24:20–30
122. Paz-Ares L, Jiang H, Huang Y, Dennis P (2017) P2.04-002 CASPIAN: phase 3 study of first-line durvalumab ± tremelimumab + platinum-based chemotherapy vs chemotherapy alone in ED-SCLC. J Thorac Oncol 12(11):S2398
123. ClinicalTrials.gov: a study of rovalpituzumab tesirine (SC16LD6.5) in the frontline treatment of patients with deltalike protein 3-expressing extensive stage small cell lung cancer. https://clinicaltrials.gov/ct2/show/NCT02819999?term=NCT02819999&rank=1
124. Lashari BH, Vallatharasu Y, Kolandra L, Hamid M, Uprety D (2018) Rovalpituzumabtesirine: a novel DLL3-targeting antibody-drug conjugate. Drugs R D 18(4):255–258

Underpinning the Cellular and Molecular Mechanisms with Nanotheranostics for Lung Cancer

11

Asiya Mahtab, Monika Yadav, Karishma Niveria, and Anita Kamra Verma

Abstract

Out of all the cancers, lung cancer is the most devastating causing mortality worldwide out of which ~90% of deaths are from non-small-cell lung cancer (NSCLC). Early diagnosis and ineffective traditional therapies lead to poor prognosis and percent survival in patients suffering from lung cancer. Redox signalling enacts a crucial role in controlling numerous disease biology and cellular signalling pathways. Lately, nanomedicine (application of nanotechnology in biology and medicine) has been revealed to normalize the growth of cancer. With the dawn of robust proteomics and sequencing techniques, immunohistochemistry and identification of novel conclusive biomarkers coupled with improved understanding of the molecular mechanisms regarding cancer are quintessential for targeting redox biology. Herein, a detailed overview of the recent advances in therapeutics includes nano-strategies over conventional therapeutics for targeting redox biology thereby affecting the various cell death mechanisms.

Keywords

Lung cancer · Nanotheranostics · Redox biology · Cellular and molecular mechanism

A. Mahtab · M. Yadav · K. Niveria · A. K. Verma (✉)
Nanobiotech Lab, Department of Zoology, Kirori Mal College, University of Delhi, New Delhi, India
e-mail: akverma@kmc.du.ac.in

K. Dua et al. (eds.), *Targeting Cellular Signalling Pathways in Lung Diseases*,
https://doi.org/10.1007/978-981-33-6827-9_11

11.1 Introduction

Cancer is portrayed as an aggressive and one of the deadliest diseases, prophesied to be the prime cause of mortality and the major hindrance to increase life expectancy in the next few decades. In 2018, International Agency for Research on Cancer released a status report approximating 18.1 million new cancer cases and 9.6 million cases of cancer-related deaths worldwide [1]. The recent futuristic trends predict the uninterrupted increase in new cases in the next 10 years. To overcome the maladies of the traditional cancer treatments including surgery, radiotherapy and chemotherapy, there is a need to have a paradigm change in the therapeutic regimen to substantially enhance the efficacy in the clinic. Alternative approaches for diagnostics by cancer imaging with simultaneous therapy are not only safe but economical too [2].

11.2 Types of Lung Cancer

Lung cancer can be widely categorized into two main types depending upon the histologic representation: (1) small-cell lung cancer (SCLC) and (2) non-small-cell lung cancer (NSCLC). Among them, NSCLC is the most common type that accounts for about 80–85% of lung cancer cases, while the remaining 15% has small-cell lung cancer. However, SCLC is more destructive, and if left untreated it reduces the patient survival with mean survival rate of 4 months. The cause of life-threatening lethality of SCLC is its fast metabolism, early metastasis and fast proliferation rate. SCLC is decorated with neurofilaments and neurosecretory vesicles owing to its origin from neuroendocrine tumours. Adenocarcinoma is the most communal type among the NSCLC and originates in the mucus-producing gland cells in the airway lining. In addition to this, it has been observed that several other aspects counting genetic modification (ROS1 genes, MET, ALK), familial predilection to lung cancer and *Helicobacter pylori* infection create a small class of lung cancer initiators [3]. Another one that roots in the flat cells covering the surface of airways and develops adjacent to centre of lung is known as squamous cell carcinoma. However, currently none of the lung cancer types are druggable.

11.3 Waging a War on Cancer by Redox

Almost five decades ago, a war was declared by the National Cancer Act to fight cancer, cure still eludes and the intensity worldwide is unstoppable. Despite the advancements in overcoming this dreadful disease, the treatments are limited by the stringent success rates. Chemotherapy proves an ultimate pharmacological barrier to be conquered that can pave the way for future drug diagnostics, delivery and discovery. Essentially the twentieth-century biological revolution has paved the way to comprehend cancer by understanding the pathophysiology, the diagnosis, treatment and prevention; still, the complete holistic cure remains a serious challenge and eliminating the patient suffering a far-reaching, unmet goal. Further, owing to improved life expectancy ensuing a demographic transferral in the direction of senior population globally, coupled with age-related rise in cancer occurrence, this disease definitely embodies a foremost health dare of our period [4].

Therapeutic revolutions have been accomplished in numerous areas counting urogenital oncology aiming testicular cancer and paediatric oncology aiming childhood leukaemia where chemotherapeutic intermediation can nowadays result into remarkable survival rates [4]. Contrariwise, slight advancement has been crafted in the chemotherapeutic treatment of another crucial cancers comprising glioblastoma, pancreatic cancer and metastatic melanoma where currently no effective chemotherapeutic treatments are offered and survival of patient is calculated in months after diagnosis [5].

11.4 Reactive Oxygen Species (ROS)

ROS is the main arbitrators of redox deregulation and cellular oxidative stress implicated in cancer instigation and advancement, and in recent time it has been arisen as an encouraging aim for discovery of anticancer drugs; widespread research has recognized a relevant contribution of redox modulations in cancer advancement, mainly for oncological signs with reduced possibilities for successful treatment, that includes pancreatic carcinoma and metastatic melanoma [6, 7]. Nowadays, this is extensively acknowledged that constitutively enhanced oxidative stress levels in cells coupled with need of anti-apoptotic and mitogenic ROS-signalling in tumours signify specific susceptibility that allows selective targeting by pro- and antioxidants that may act directly or indirectly and modulators of redox balance that together will be discussed as redox chemotherapeutics. This chapter will impart an update on target recognition, drug discovery and mechanisms of action of investigational redox chemotherapeutics with prominence on mediators that have illustrated effectiveness in innovative preclinical animal models and nowadays progressed into human clinical trials.

11.5 Reactive Oxygen Species (RNS) and Reactive Oxygen Species (ROS)

11.5.1 ROS as Signalling Molecules

ROS are defined as chemically reactive molecules containing oxygen that comprise of superoxide anion ($O2^-$), hydrogen peroxide (H_2O_2) and hydroxyl radical (OH•). These radicals are identified as toxic molecules and deterrents for diseases like cancer [8]. Nowadays, ROS can perform as signalling particles to normalize various biological and physiological reactions such as proliferation, growth factors signalling and acclimatization to hypoxia is extremely appreciated [9, 10].

11.5.1.1 H_2O_2-Mediated Oxidation of Cysteine

H_2O_2 is a distinct biomarker mediating its signalling through cysteine oxidation within proteins to alter their functions [11–13]. H_2O_2 further stimulates P13K/Akt survival pathway by impeding the activity of phosphatases (such as PP2A, PTP1B, PTEN), and reversibly oxidizes the crucial cysteine thiol groups (84, 78). Besides, during hypoxia H_2O_2 mediates the oncogenic steadiness of HIF-1α protein through prolyl hydroxylase domain protein-2 (PHD2) oxidation [14]. H_2O_2 first reversibly oxidizes the thiolate anion (S^-) within target proteins into sulfenic form (SO^-), that can consequently create reversible sulfenic-amide (S-N) or disulphide (S-S) bonds, resulting in a modified protein conformation, leading to altered activity and may shield the target protein from irreversible oxidation of cysteine. The thioredoxin (TRX) and glutaredoxin again reform the reduced S− by their enzymatic activity. At high level of H2O2, the additional H_2O_2 molecules further oxidize the SO^- intermediate (hyper-oxidation) to produce the sulfonic ($SO3^-$) and sulfinic ($SO2^-$) acid. The $SO3^-$ formed here usually signifies an irretrievable oxidative alteration [12].

11.5.1.2 H_2O_2-Mediated Signal Transduction

Despite of widespread multi-compartmental cellular antioxidant system, how H_2O_2 oxidizes the target proteins so unambiguously and proficiently is still uncertain. The mechanisms of facilitated oxidation are alluring, as the manifestation of extremely reactive antioxidant system makes the direct oxidation of aimed proteins by H_2O_2 dubious. The first mechanism to stimulate the oxidation of aimed proteins is known as *redox relay* that exploits the capability of scavenging enzymes to distinguish H_2O_2 signal [15]. It is a two-step mechanism where H_2O_2 oxidation is first received and then transferred to the aimed proteins by scavengers enzymes (for instance, GPX, PRX) and has been described in mammalian cells as well (PRX2-STAT3) [16]. In recent times, proton spread has been illustrated to upsurge the responsiveness of GAPDH to H_2O_2 [17]. The second mechanism is 'flood-gate model' where phosphorylation and hyper-oxidation lead to inactivation of scavenging enzymes that further the accumulation of H_2O_2 for oxidation of aimed protein [18]. The stimulation of growth factor receptor causes a surge in production of NOX-derived H_2O_2 and Src kinase activation ensuing PRX1 inactivation through phosphorylation [18]. Similarly, in mitochondria, ROS stimulates inter-membranous Lyn (Src family) and Syk kinase pathway to regulate cellular routes including cell cycle regulation, metabolism and transcription [19]. Moreover, possibly oxidized scavengers can be recycled to their reduced form and the signal transferred to the target protein via thiol comprising molecules such as reduced TRX or GSH [20].

11.5.2 ROS Promotes Pro-tumourigenic Signalling

Literature survey suggests the pro-tumourigenic function of ROS by stimulating MAPK/ERK and PI3K/Akt/mTOR signalling cascades. Oncogenic Ras triggers ROS generation from NOX4 to boost multiplication [14]. In a majority of cancers, ROS causes oxidation and inactivation of phosphatases (PTP1B, PTEN) thus ensuing hyper-activation of PI3K/Akt/mTOR survival pathway [21–23]. Moreover, ROS production further intensifies the mutation of Akt thereby encouraging tumour proliferation and survival [24]. ROS can further stimulate MAPK/ERK pro-proliferative signalling and activation of growth factor receptor by oxidizing and inactivating MAPK phosphatases [25]. In lung cancer, mitochondrial ROS (mROS) through the MAPK/ERK pathway controls Kras-prompted anchorage-independent growth of cancer cells. Besides, reduced levels of ROS interruption of the mitochondrial respiratory chain leads to abridged tumourigenesis [26]. The transcription factors such as NRF2 (master regulator) and NF-κB that boost the expression of antioxidant system to dodge the ROS-mediated tumour cell death and to avert the build-up of ROS by intensifying flux via pentose phosphate pathway (PPP) to produce NADPH can also be activated by ROS to trigger cancer cell survival [27]. Furthermore, ROS supports cancer metastasis, angiogenesis and inter-related processes that can lead to poor prognosis. With the proliferation of tumour, the vascular supply becomes inadequate, the regions within the tumour become hypoxic and glucose levels drop thereby further enhancing the ROS levels. To sustain proliferation and survival under extreme conditions, the tumour cells undertake metabolic adjustments like AMPK activation, HIF stabilization and one-carbon metabolism pathway instead of PPP (as glucose level drops) to promote NADPH generation and conserve redox equilibrium [28, 29].

11.5.3 ROS Promote Anti-tumourigenic Signalling

As the elevated ROS levels cross threshold, cell cycle ceases leading to senescence and cell death [30]. ROS-mediated cell death may happen via ASK1/p38 and ASK1/JNK signalling pathway activation too [31]. The inactivated ASK1 interconnects with reduced thioredoxin (TRX), though H_2O_2 promoted TRX oxidation resulting in ASK1 disconnection and stimulation, with continuous stimulation of MAP kinase MKK3/MKK6/p38 and MKK4/MKK7/JNK pathways prompting the inhibition of anti-apoptotic factors [32, 33]. Various studies indicated that deactivating mutations in JNK and p38 pathway suggest that these signal transduction pathways stimulate tumour cell death [34]. The initiation of JNK and p38 signalling pathways by ROS can also help cells to obstruct cell cycle and prevent cell division [35]. Therefore, distressing the ROS modifying pathways can be a possible therapeutic strategy for quashing metastasis and cell multiplication [36, 37]. When the cancer cells lose their cell-cell adhesion and are cut off from extracellular matrix to cross the basement membrane to finally come in contact with the oxidizing environment of blood, the ROS level further elevates [38]. Cancer cells then promote their antioxidant system by one-carbon metabolism pathways and by enhanced production of NADPH in mitochondria through isocitrate dehydrogenase-1 (IDH-1)-mediated reductive carboxylation, to back anchorage liberated metastasis and growth [22]. The level of ROS beneath the threshold for cell death is essential for tumour cells to advance into metastasis and invasion.

11.6 ROS as Targets for Effective Cancer Therapy

Any possible reason that causes depletion of ROS scavengers or elevation in production of ROS places the cell in a condition of oxidative stress. Pathophysiology of cancer involves oxidative stress, moreover increased ROS level as a result of aerobic glycolysis trailed by oxidation of pyruvate in mitochondria (Warburg effect), induction of oxidizing enzymes or growth factor-dependent pathways, enhanced oncogene and receptor activity prompt genetic uncertainty [39]. Additionally, high intracellular ROS levels can impair DNA, proteins and lipids, and this capability has been exploited to be the therapeutic window for anticancer therapy.

11.6.1 Cellular Pathways

11.6.1.1 Apoptosis (Type 1 Programmed Cell Death)

Apoptosis is a tightly regulated caspase-mediated programmed cell death that can be induced by various intrinsic, and extrinsic stresses like DNA damage, ROS and RNS have relevance in cancer therapy [40]. Cells undertaking apoptosis exhibit several biochemical and morphological attributes like membrane blebbing, cell shrinkage, nuclear fragmentation and condensation of chromatin leading to formation of apoptotic bodies that are scavenged by macrophages thus minimizing destruction to neighbouring cells and tissue [41]. Apoptosis can be triggered by ROS through pathways such as endoplasmic reticulum stress-stimulated pathway, mitochondria-dependent pathway or death receptor-dependent apoptotic pathway. ROS can induce all forms of DNA damage such as DNA cross-linking, strand breakage, proteins and base modifications that are related with cancer beginning and expansion [42]. Indeed, mitochondrial ROS forms the bulk of the intracellular ROS generated that entails opening of mitochondrial permeability transition pore (PTP) by disrupting the membrane of mitochondria, consequently hampering the mitochondrial electron transfer chain, prompting cytochrome-c (Cyt-c) release. In cytosol, Cyt-c together with procaspase-9 and apoptotic peptidase-activating factor-1 (Apaf-1) forms 'apoptosomes' resulting in caspase-9 activation that further causes stimulation of effector caspases (caspase-3), which causes breakage of cellular proteins and eventually apoptotic cell death [43]. H_2O_2 may trigger either caspase-independent or caspase-dependent apoptosis. Most of the caspases are inclined to oxidation (H_2O_2) at their catalytic site, the cysteine residue, while caspase-3, procaspase-3 and procaspase-9 are sensitive to S-glutathione. Undeniably, one of the most vital, potent and direct inducers of apoptosis among the ROS groups is H_2O_2 [44].

11.6.1.2 Necrosis

Necrosis is a disrupted cell death, a bioenergetic devastation that takes place in the absence of nuclear condensation, unlike apoptosis. The morphological features of cell undergoing necrosis include distension of cellular organelles, random DNA degradation, loss of plasma membrane integrity and release of molecules like LDH and HMGB1 in the microenvironment generating an immune response

[42, 45]. Necrosis can be triggered by broad range of stresses or stimuli such as infection, trauma, calcium and ROS (mitochondrial or non-mitochondrial) [46, 47]. As, necrotic cell death occurs at low ATP level unlike apoptosis and autophagy, it is usually known as 'passive distortion' of cell [48]. Necrotic cell death mostly occurs in core of solid tumours as they develop hypoxic condition, are devoid of nutrient supply and start downregulating programmed cell death by itself. Due to development of necrotic core inside the tumour, necrosis is considered as having tumour-promoting functions due to release of danger-associated molecular patterns (DAMP), for instance, HMGB1 in the surrounding space thus stimulating immune response and promoting tumour advancement. Contrarily, apoptosis can act as tumour-suppressor, while autophagy is known to have both tumour-suppressor as well as tumour-promoting potential depending upon the cellular situation [45, 48, 49].

11.6.1.3 Necroptosis (Type 3 Programmed Cell Death)

Necroptosis is a greatly controlled mechanism of cell death that can be activated by death receptor (TNF) signalling, interferon receptors (IFN), toll-like receptors (TLR), ROS and pathogens [50]. Necroptosis is limited by mixed lineage kinase domain-like protein (MLKL) and receptor interacting protein kinases (RIPK1, RIKP2). Unlike apoptosis, necroptosis generates an inflammatory response due to release of cellular content into the surrounding tissue [51]. With the onset of stress stimulus (ROS), RIPK1 gets activated, and when the level of both RIPK1 and MLKL is adequately elevated and caspase-8 is defective or absent, necroptosis is induced by the phosphorylation of RIPK3 that further recruits and phosphorylates MLKL, followed by the formation of a core complex termed 'necrosomes' (RIPK1, RIPK2 and MLKL). The necrosome then translocates to the plasma membrane causing rapid disruption of plasma membrane that initiates inflammation by releasing cytokines [41, 52]. Necroptosis and ROS through metabolic signalling can form a positive feedback loop. Both mitochondrial (oxidative phosphorylation) and extra-mitochondrial ROS (NOX1) facilitate RIPK1 autophosphorylation at S161 that leads to RIPK3 recruitment and induction of necroptosis. RIPK1 can stimulate ROS generation by elevating ATP and blocking adenine nucleotide translocate (ANT), whereas RIPK3 stimulates ROS generation by modulating enzymes such as pyruvate dehydrogenase (PDH) activation by glycogen phosphorylase that stimulate oxidative respiration and by necrosome formation, glutamate ammonia ligase upregulation along with glutamate dehydrogenase-1 to yield ROS [53, 54].

11.6.1.4 Autophagy (Type 2 Programmed Cell Death)

Biologically, autophagy is a self-degradative process that can be induced by several metabolic insults including elevated levels of ROS, hypoxia, drug treatment, organelle damage and ER stress. The role of ROS in induction of autophagy to kill cancer cells is well documented [55, 56]. However, despite being recognized as cell survival practice, autophagy can act via tumour-suppressor mechanism causing cell demise in caspase-independent manner. This specifies the possibility of ROS being the signalling target for survival associated to autophagy [57]. Currently, the possibilities of

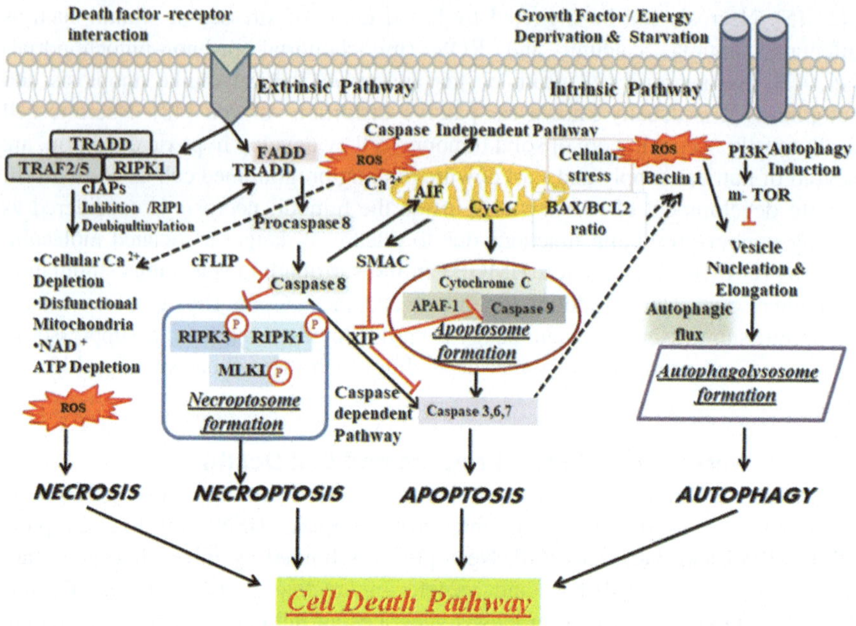

Fig. 11.1 Cellular pathways regulating cell death

application of ROS-mediated autophagy in therapeutic approaches for cancer treatment are being explored [58]. A complete set of autophagy-associated genes (Atg) and mammalian target of rapamycin (mTOR) kinase signalling pathway regulates the process of autophagy. H_2O_2-induced oxidation of autophagy-associated 4A cysteine peptidase (ATG4), crucial for the de-lipidation of the ATG8 protein causes stimulation of autophagy. The inactivation of ATG4 by H_2O_2-induced oxidation consequently directs the elevation in production of LC3-associated autophagosomes [58]. In contrast, ROS-dependent autophagy is controlled by AMP-activated protein kinase (AMPK) pathway also. The AMPK activation by ataxia-telangiectasia mutated (ATM)-serine/threonine kinase-mediated oxidation inhibits mTORC1; a crucial negative regulator of autophagy thus ensues induction of autophagy. Moreover, activation of AMPK pathway can be regulated by oxidative stress through AMPKK phosphorylation (AMPK kinase) that surges H_2O_2 production thereby triggering apoptosis [58]. Furthermore, the expression of autophagy-related genes (p62/SQSTM1 or ATG6/BECLIN1) can be modulated by transcription factors like NF-κB that leads to ROS-dependent autophagy in cancer [59, 60]. Thus, it can be hypothesized that increased levels of ROS, and its modulation stimulates autophagy in tumour cells (Fig. 11.1).

11.6.2 Molecular Pathway

Anomalous intracellular signalling is a prominent contributor to lung cancer, causing mutation and activation of oncogenes, coupled with genetic enhancement of critical proteins altering signal transduction pathways. Recent advancements in assessing the molecular mechanisms of lung cancer help conceptualize specific turnover conditions in signalling networks that support tumour oxidative microenvironment by stimulating cell proliferation, hijacking the death signals and their ability to repair damages.

11.6.2.1 MAPK

Apparently, three well-known subfamilies of mitogen-activated protein kinases (MAPKs), namely, extracellular signal regulated kinases (ERKMAPK, Ras/Raf1/MEK/ERK), c-Jun N-terminal kinases (JNK) and p38MAPK are critical. Constitutive expression of these activated signal transduction pathway is quintessential hallmark of cancer progression, genetic mutation or epigenetic modification that leads to the initiation and metastatic spread of lung cancers. MAPK pathway regulates the physiological mechanisms of cell death and survival, potentially cross-talks with multiple effector molecules of signal transduction pathway. MAPK is conventionally regulated by the stimulation of a series of three conserved protein kinases via phosphorelay cascade [61]. Reportedly, mutations in upstream activators such as Ras are sensitive for ROS generation and intertwined with redox signalling. Recently reports suggest that 90% mutations in lung cancer are caused by Kras that activates the RAS/RAF/MEK/MAPK pathway and downstream activator-MYC (cell proliferation regulator) [62]. Overexpression of EGFR receptor (phosphorylation and tyrosine kinase) is a critical metabolic regulator for cancerous cells as enhanced intracellular activity results in aberrant cell multiplication, angiogenesis, invasion and metastasis in NSCLC, having a frequency of 40–89% [63]. Occurrence of BRAF mutation by atypical signalling of EGFR/ALK rearrangement has appeared in 2–4% in NSCLC [64]. Basically, effector molecules impacting MAPK pathway recruited in lung tumourigenesis include p38 MAPK, ERK1/2 and MMP-12(matrix metalloproteinases) – a downstream substrate of MAPK, along with other biomarkers such as COX-2 and p53.

11.6.2.2 FOXO Transcription Factors

The Forkhead box-O (FOXO) is a wide group of transcription factor that regulates decision of cell fate. FOXO is well characterized as a tumour-suppressor, by its ability to control cell proliferation, angiogenesis, cell death, cell migration, metastasis and ROS scavengers [65]. Antagonist mechanisms regulating normal cell activity include (1) insulin growth factor signalling via PI3K/Akt pathway inhibitory activity of FOXO protein. Since PI3K/Akt pathway is in an inactive state (via PTEN or PI3K inhibitors), it results in nuclear localization of FOXO, thereby promoting cell death. Also, (2) JNK-mediated pathway is activated by ROS generation that further mediates the nuclear translocation of FOXO, leading to the progression of transcription and tumour-suppressing activity. (3) Proteosomal degradation of FOXO by poly-ubiquitylation regulator (14–3-3 protein) [66]. Aberrant activity of FOXO proteins (loss of PTEN or by activation of RTK) in an apoptosis pathway clearly gives an edge to lung cancer cells to survive despite elevated oxidative stress. Earlier reports suggest the constitutive overexpression of Akt in NSCLC cells, making it the most prevalent indicator in NSCLC irrespective of the independent of tumour phase. In murine lung cancer model, it is reported that inhibiting PI3K pathway triggers FOXO and redirects the expression of $p27^{KIP1}$ [67]. Emerging reviews suggest the atypical activity of FOXO1, FOXO3, FOXO4 and FOXO6 that are responsible for the migration, invasion and metastasis of lung cancer. In context to tumour micro-environment, upregulation of tumour-associated fibroblasts to promote tumourigenesis and migration in lung cancer have been observed [68].

11.6.2.3 Keap1-Nrf2 System

The Keap1-Nrf2 system is a prominent defensive driver against ROS (electrophile). Nuclear factor erythroid 2-related factor-2 (Nrf2) is a dominant transcription effector providing the maintenance of cytoprotective genes. While its negative regulator Kelch-like erythroid cell-derived protein with CNC homology (ECH)-associated protein-1 (Keap1) produces ubiquitin E3 ligase with CUL3, the Keap1-CUL3 complex ubiquitinates leads to degradation of NRF2 under normal condition [69]. If subjected to the positive outcomes from NRF2 function, it often can hijack by cancerous cells. The somatic germline mutations that involved in Keap-Nrf2 pathway such as KEAP1, CUL3 and NRF2 genes rapidly occur in the lung cancer. Thereby constitutive activation of NRF2 is a prevalent factor observed in NSCLC and strictly associated with tumour development [70]. The emergences of recently studied data from NSCLC tumour samples have also reported that NRF2 and KEAP1 mutations are mutually exclusive but related with different physiologies. Recent advancement in treating cancer displaying aberrant activation of NRF2 and, consequently, in clinical medicine NRF2 inhibitors re-emerged as a promising therapeutic approach for lung cancer (Fig. 11.2) [71].

11.7 Exploiting ROS for Lung Cancer Nanotherapeutics

Exploiting nanotechnology for cancer nanotherapeutics necessitates the urgency to overcome the limitations of conventional therapeutic strategies against lung cancers. Nanoscale therapeutic agents include liposomes, polymeric, metal and bio-based nanoparticles [72]. Furthermore, NPs do exhibit multifunctional properties such as imaging, diagnostics, therapeutics and sensing that may help choose benefits to exploit nanomaterials for multifaceted biomedical applications in lung cancer theragnostic. A schematic representation depicts the approaches (Fig. 11.3).

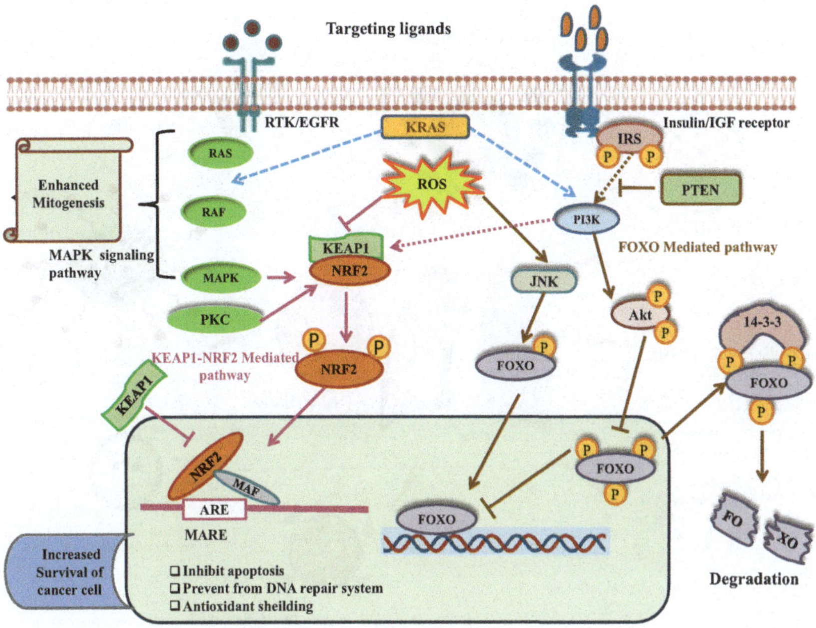

Fig. 11.2 Aberrant signalling pathway in lung cancer

11.7.1 Nanotheranostics for Lung Cancer

11.7.1.1 Lipid-Based Nanoparticles (Liposomes)

Liposomes are artificially generated bilayer vesicles that can be surface engineered by attaching polyethylene glycol (PEG). Lin et al. reported a twin ligand (anti-carbonic anhydrase IX (anti-CA IX) antibody and CPP33) altered liposomes encapsulated with triptolide designed especially for pulmonary delivery of triptolide. The bio-toxicity was evaluated by apoptosis. Penetration of tumour and progress inhibition efficacy of liposome were then established by means of 3D tumour spheroids [73]. Similarly, Makale and co-workers in 2010 reported Dox-encapsulated PEGylated liposome having a dextran core, comprising FeO for MRI contrast to established a vigorous imaging capability of NPs in murine model (Lewis) of lung cancer [74].

11.7.1.2 Polymeric Nanoparticles

For the development of a stable nanotheranostic drug delivery system, polymeric nanoparticles offer a stage for insertion of therapeutically potential drug/gene/proteins, an imaging compound and a suitable targeting moiety. Chitosan, alginic acid, gelatin, poly(ε-caprolactone) (PCL), poly(lactide-co-glycolide) (PLGA) and polylactic acid (PLA) are the most regularly used polymeric systems in the lung cancer therapeutics. PLGA has been verified as a prospective carrier molecule for

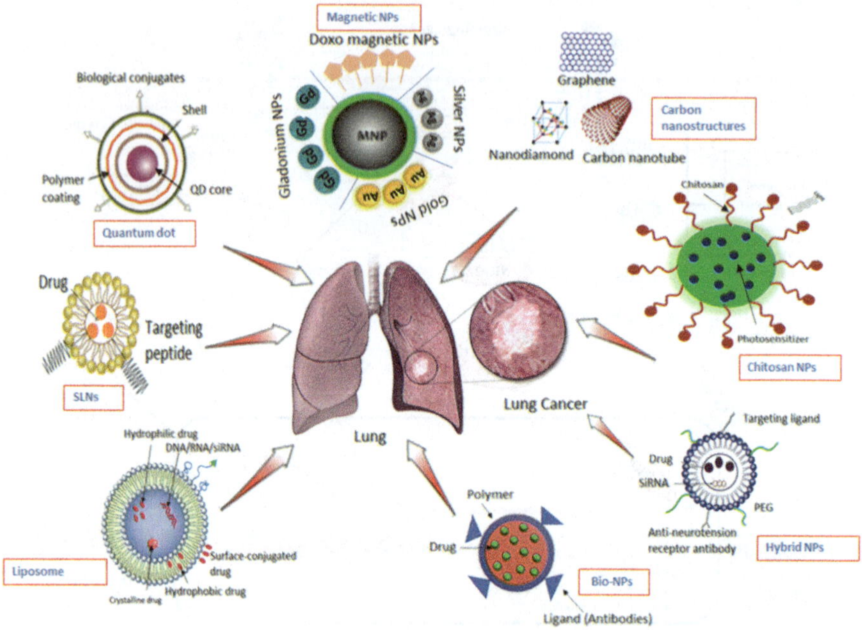

Fig. 11.3 Diagrammatic representation of nanotheranostics for lung cancer therapy

lung cancer treatment, as PLGA microspheres encapsulated with endostatin were prepared by Wu and Wang [75]. Simultaneous delivery of cytotoxic and antiangiogenic drugs exploits their synergistic effect for effective NSCLC treatment. Sengupta et al. [76] reported PLGA NPs coated with bi-phospholipid consisting of doxorubicin-PLGA conjugation and combretastatin mixed with phospholipid forming the outer lipid bilayer. Nguyen et al. further reported the effective transfection in human lung epithelial (H1299 luc) cell lines with tertiary-amine-engineered PVA attached over PLGA for siRNA delivery [77].

The cationic nature of chitosan facilitates binding and transport of nucleic acids in lung cancer cells. Okamoto et al. reported the efficacy of low molecular weight chitosan to deliver pCMV-Luc gene into lung cancer cells and then delver in a mice model via nasal administration [78]. Chitosan-dextran-based delivery system fabricated by Ventura et al. has proved its efficacy in delivering gemcitabine to NSCLC cells [79].

11.7.1.3 Metal Nanoparticles

In the existing world, metal nanoparticles frequently contaminate our water, food, medicine and cosmetics, because they are extensively utilized in a diversity of daily appliances. These metal NPs also have use in lung cancer therapy and diagnosis such as gold nanoparticles (AuNPs) that have been used for the treatment of lung cancer and diagnosis-theranostics. A photothermal therapeutic hollow Au/Ag NPs with a dendritic morphological structure has been used in treatment of A549 cells.

Biomolecular assay for detection of biosensors and biomarkers for tumour identification can be effectively achieved by silica nanoparticles. This strategy may be exploited for biomedical applications, like drug/DNA delivery substances in cancer therapy due to its biocompatibility and fast renal clearance [3].

11.7.1.4 Bio-nanoparticles

Although metal-based nanotherapeutics have matured, their used in in vivo application for lung cancer treatment is severely limited due to its toxicity. This has led to a paradigm shift for exploiting bio-nanotechnology-based therapeutic systems, in which a pre-existing biological constituent is combined to the therapeutic NPs [3].

Apoferritin-loaded NPs enter the target cancer cell either by receptor-mediated endocytosis [80], clathrin-mediated endocytosis or macropinocytosis process [81]. Further, Li et al. created a ferritin-based multifaceted nanostructure that was utilized for analysis of human lung adenocarcinoma A549 cells through fluorescence and MR imaging [82].

11.7.1.5 Viral Nanoparticles

Viral NPs (VNPs) have developed as promising candidates for biomedical applications due to their biocompatible nature, extensive range of shapes and sizes and easy modification of surface by various functional entities and ability to transfect cells [83].

Intrinsic and acquired drug resistance develops in lung cancer for the existing small molecule-based anticancer agents. One such effort was made by Beljanski and Hiscott, where they have worked on chemotherapeutic drug conventionally through genetically altered oncolytic viruses (OVs) for the lung cancer treatment [97]. Table 11.1 represents various nanocarriers for the therapeutic delivery to lung cancer.

11.7.2 Modulation of Redox Homeostasis by Nanoparticles

A precarious equilibrium maintained by ROS generation and elimination is termed as redox homeostasis [98]. ROS are generated in cells through numerous oxidases and may work as secondary messengers that control diverse signal transduction pathways. As per ROS rheostat theory, ROS controls cell fate in a dose-dependent manner. Essentially, ease of exploiting altered redox status in cancer cells has encouraged researchers to devise novel strategies to achieve effective cancer treatment [99]. NPs effectively modulate redox homeostasis either by reducing scavenging pathways or by triggering ROS generation. The physicochemical properties and the use of transition metals in composition of NPs may help generate ROS [98]. Modifications in NPs structural and physicochemical properties can alter biological activities conferring NP-associated toxicities. Oxidative stress in cells could be made by engineered NP primarily by acellular features such as surface of particle, size, composition and presence of metals or by cellular factors like interaction with NP-cell membrane interaction, cytosolic detoxifying enzymes, mitochondrial respiration, etc. leading to ROS-mediated damage [100].

Table 11.1 Nanotheranostics for the delivery to lung cancer

Carrier	Therapeutic agent	Delivery	Model system under study	Main findings	Reference
Chitosan NPs	Insulin	Microencapsulated insulin loaded CS NP; rat model	Intratracheal administration Sprague-Dawley	Improved absorption for systemic action	[84]
	Palmitic acid modified exendin 4 (PalEx4)	Nebulization of NPs in rat model	Sprague-Dawley	Enhanced cellular uptake a delayed release. 3.1-fold higher hypoglycaemic effectiveness than that of unmodified drug	[85]
	Paclitaxel	Chitosan-modified PLGA NPs	Male CDF1 mice and male SD rats	Temporary formation of aggregates in the blood stream followed by improved trapping in the lung capillaries. Electrical interaction-mediated higher uptake across the endothelial cells of the lung tumour capillary	[86]
PLGA NPs	Oridonin	(PLGA) porous microspheres	Male Wistar rats	The oridonin-loaded EPMs are promising dry powder inhalers for the local therapy of primary lung cancer	[87]
	V1Jns DNA plasmid	PLGA-PEI NPs	Calu-3 cell line	Cell toxicity can be reduced by using lower PEI-DNA ratio that will bear the same characteristics and loading efficiency as the higher PEI-DNA ratios	[88]
	Celecoxib	PLGA NPs	A549 cancer cell line	PLGA NPs incorporating Cxb showed more cytotoxicity against A549 tumour cells only at higher concentration than Cxb itself	[89]
PEG NPs	NF-κB decoy	PEG-PLGA	Explanted lungs from patients with PAH and rat models	NF-κB plays a primary role in the pathogenesis of PAH and, thus, represents a new target for therapeutic intervention in PAH	[90]
	Poly-L-lysine (PLL) modified PEG NPs	Genomic DNA of *E. coli*	C57BL/6 mice dosed with compacted DNA nanoparticles	Stable compacted DNA nanoparticles transfer exogenous genes to airway	[91]

				epithelium and show promise for lung gene therapy	
	PEGylated gelatin nanoparticle	pCMV β-gal	Lewis lung carcinoma (LLC) bearing female C57BL/6 J mice		[92]
Magnetic NPs	Iron oxide NPs	Doxorubicin, cy 5.5	Tumour-bearing mice	Dox@TCL-SPION showed exceptional antitumour effects without any systemic toxicity	[93]
	PTX-loaded core-shell magnetic electrosprayed NPs	Paclitaxel	A549 cells	PTX-loaded nanoparticles and CAP synergistically inhibited the growth of A549 cells more effectively	[94]
Carbon nanotube	Carbon nanotube gold hybrid	Doxorubicin-HCl	A549 cell line	In vitro cellular activity proved that nanostructures can proficiently administer doxorubicin inside the cells	[95]
	Amino-functionalized carbon nanotube	siRNA	Human lung xenograft model	Carbon nanotube facilitated delivery of siRNA by intratumoural route showed effective and statistically substantial suppression of tumour volume	[96]

Transition metals such as lanthanide cerium have redox stimulating characteristics, due to shift in the 4+ and 3+ oxidation states [101]. Large use of cerium in nanomedicine is in the form of cerium oxide NPs that display redox-modulatory activities which imitates both superoxide dismutase and catalase activities [102]. Carbon NPs/carbon nanodots, fullerenes and CNTs also produce ROS via NP-associated redox-active functional groups [103].

Modification of NPs can be done to control release of drug in response to the exogenous and endogenous stimuli exploiting the distinctive modifications of tumour metabolism [104]. Owing to the exceptional physicochemical properties and ease of surface modification, carbon-based nanostructured materials have been widely explored for cancer theranostics.

11.8 Emerging Anticancer Redox Nanotherapeutics Targeting the Pathways

NPs provide safe and successful vehicles for anticancer agents and utilize various modes of therapeutics including chemotherapy, immunotherapy, gene-mediated therapy, receptor-mediated therapy, photodynamic therapy, etc. Herein we discuss redox nanotherapeutic strategies to target the different set of pathways to combat lung cancer.

11.8.1 Receptor-Mediated Therapy

Targeting the EGFR signalling in lung cancer using iron oxide NPs effectively reduces the tumour burden and is a promising strategy for imaging and targeting in in vitro and in vivo models. Additionally, multifunctional IONPs combat Kras-mutated lung cancer by receptor-mediated pathway [105, 106]. Delivery of hexadentate-polyD,L-lactic acid-co-glycolic acid polymer functionalized with PD98059 (MAPK inhibitor) in lung cancer model may hijack the MAPK signalling pathway, which inhibits ERK phosphorylation, stimulates apoptosis and reduces cell proliferation [107] (3). Dual-drug loaded multifunctional liposomes decorated by octreotide (synthetic peptide that mimic somatostatin) designed to target NSCLC has shown enhanced accumulation by overexpressed somatostatin receptor in the tumour niche in in vivo model. Liposomal delivery resulted in reduced metastasis and inhibited vasculogenic channel by downregulating PI3K, MMP-9, PI3K, VE-cadherin and activated caspase-3 [108].

11.8.2 Gene-Mediated Therapy

It has been reported that magnetic iron oxide nanoparticles act as an ideal carrier of PTEN gene expression vector in cisplatin-resistant A549/CDDP cells leading to inhibition of cell proliferation via apoptosis [109]. Silica-polymer nanocomposites carrying p53 gene have significantly reduced growth both in in vitro and in vivo lung cancer model as demonstrated by Wu et al. [110] Carbon dots have been used as multimodal nanoagents for bioimaging and mediating siRNA delivery (cyclin B1 and EGFR) to lung tumours. Enhanced efficient tumour response by gene silencing of the target lung cancer cells via receptor-mediated endocytosis has been suggested [108]. Although chitosan has been extensively studied in siRNA delivery to H1299 lung cancer cell lines, its efficacy is limited in in vivo systems due to non-specific binding. Therefore, to overcome this Varkouhi et al. [111] developed siRNA/thiolated chitosan complex exhibiting enhanced gene silencing activity in H1299 cells with improved targeted delivery in vivo [3]. Currently, NLC nanoparticles conjugated with LHRH (luteinizing hormone-releasing hormone) decapeptide and co-encapsulated siRNA and pactitaxel (LHRH-NLC-siRNAs-TAX system) have effectively targeted four variants of EGFR-TK receptor and induced cell death in lung cancer [112].

Polyethylenimine-(5) myristic acid/poly(ethylene glycol)-oleic acid/cholesterol (PEI600-MA5/PEG-OA/Cho) nanoparticles have been reportedly utilized to deliver pro-angiogenic (FOXM1/FOXF1) cDNA expression vector via pulmonary circulation to enhance angiogenesis and stimulate lung regeneration [113].

11.8.3 Stimuli Responsive Chemotherapy

Currently, pH/redox based dual responsive nanosystems like EMCI (ICG-encapsulated mesoporous silica nanoparticles gated along with ZnO QDs and capped via erlotinib-modified chitosan) were designed to specific bioimaging and synergistic therapy for EGFR driven lung cancers. Due to the dual responsive element, pH/redox in tumour microenvironment disaggregates the nanocomplex and releases ICG and Er under NIR irradiation thereby activating fluorescence emission of ICG to that efficiently differentiates Er-resistant or Er-sensitive NSCLC exhibiting synergistic responses to ROS with upregulated apoptosis [114]. EGCG (epigallocatechin 3-gallate) co-encapsulating magnetite nanoparticles (MNPs) in BSA that trigger apoptosis via ROS generation causing loss of mitochondrial membrane potential in A549 cells have been reported. Mechanistic studies revealed increased expression of Nrf2 and Keap1 by EGCG that regulates apoptosis in lung cancer [115]. Similar studies were demonstrated by doxorubicin-loaded polyvinylpyrrolidone-conjugated gold nanoparticles [116] and by green synthesis of silver nanoparticles (AgNPs) when used against lung cancer both in in vitro and in vivo model by stimulated caspase-3 and decreased bcl-2 causing apoptosis, via inhibition of NF-κB pathway [117].

11.8.4 Naturally Derived Product Therapy

An exhaustive list of natural phenolic compounds incorporated in polymeric nanocarrier delivery system in lung cancer, including curcumin, naringenin, resveratrol, etc., has been reported with anticancer property mediated by regulating innumerable pathways such as protein kinase suppression and inhibition of transcription factors, namely, NF-κB, EGFR, AP-1 and STATs, culminating in cell death by different mechanisms. Muller et al. reported co-delivery of curcumin and doxorubicin in poly (butyl cyanoacrylate) nanovector against breast cancer cells that resulted in reversion of multidrug resistance and enhanced apoptosis by doxorubicin [118].

11.9 Conclusion

Despite the emerging variety of pulmonary and therapeutic delivery approaches, lung cancer still stays an important cause of death. The foremost downside of existing lung cancer treatment processes in practice is the dearth of strategies for early diagnosis coupled with ineffective drug delivery and poor response. This calls for a concerted effort in evolving nanotherapeutic approach to holistically manage lung cancer, underpinning the critical redox homeostasis and resetting redox balance to alter tumour metabolism. The multifaceted association amid cancer and ROS levels is fundamentally constructed on precise calibration among production of ROS and their scavenging. The instigation and advancement of cancer influence slender boost in levels of ROS. Therefore, the dual role of ROS needs to be entirely understood, and its redox state in malignancies is still mysterious in spite of widespread research. Future medicines will be based on integration of redox-active nanoparticles coupled with cellular metabolic will revolutionize cancer therapy. Additionally, nano-toxicological issues should be resolved to clinch the lung cancer management by nanotheranostics.

References

1. Nagai H, Kim YH (2017) Cancer prevention from the perspective of global cancer burden patterns. J Thorac Dis 9:448–451. https://doi.org/10.21037/jtd.2017.02.75
2. Arruebo M, Vilaboa N, Sáez-Gutierrez B et al (2011) Assessment of the evolution of Cancer treatment therapies. Cancers 3:3279–3330. https://doi.org/10.3390/cancers3033279
3. Sukumar UK, Bhushan B, Dubey P et al (2013) Emerging applications of nanoparticles for lung cancer diagnosis and therapy. Int Nano Lett 3:45. https://doi.org/10.1186/2228-5326-3-45
4. Wondrak GT (2009) Redox-directed Cancer therapeutics: molecular mechanisms and opportunities. Antioxid Redox Signal 11:3013–3069. https://doi.org/10.1089/ars.2009.2541
5. Meyskens FL, Farmer PJ, Yang S, Anton-Culver H (2007) New perspectives on melanoma pathogenesis and chemoprevention. Cancer Prev 174:191–195
6. Cabello CM, Bair WB, Wondrak GT (2007) Experimental therapeutics: targeting the redox Achilles heel of cancer. Curr Opin Investig Drugs 12:1022–1037
7. Fruehauf JP, Meyskens FL (2007) Reactive oxygen species: a breath of life or death? Clin Cancer Res 13:789–794. https://doi.org/10.1158/1078-0432.CCR-06-2082
8. Collin F (2019) Chemical basis of reactive oxygen species reactivity and involvement in neurodegenerative diseases. IJMS 20:2407. https://doi.org/10.3390/ijms20102407

9. Holmström KM, Finkel T (2014) Cellular mechanisms and physiological consequences of redox-dependent signalling. Nat Rev Mol Cell Biol 15:411–421
10. Schieber M, Chandel NS (2014) ROS function in redox signaling and oxidative stress. Curr Biol 24:R453–R462. https://doi.org/10.1016/j.cub.2014.03.034
11. Finkel T (2011) Signal transduction by reactive oxygen species. J Cell Biol 194:7–15. https://doi.org/10.1083/jcb.201102095
12. Reczek CR, Chandel NS (2015) ROS-dependent signal transduction. Curr Opin Cell Biol 33:8–13. https://doi.org/10.1016/j.ceb.2014.09.010
13. D'Autréaux B, Toledano MB (2007) ROS as signalling molecules: mechanisms that generate specificity in ROS homeostasis. Nat Rev Mol Cell Biol 8:813–824. https://doi.org/10.1038/nrm2256
14. Lee G, Won H-S, Lee Y-M et al (2016) Oxidative dimerization of PHD2 is responsible for its inactivation and contributes to metabolic reprogramming via HIF-1α activation. Sci Rep 6:18928. https://doi.org/10.1038/srep18928
15. Winterbourn CC (2013) The biological chemistry of hydrogen peroxide. In: Methods in enzymology. Elsevier, pp 3–25
16. Sobotta MC, Liou W, Stöcker S et al (2015) Peroxiredoxin-2 and STAT3 form a redox relay for H2O2 signaling. Nat Chem Biol 11:64–70. https://doi.org/10.1038/nchembio.1695
17. Peralta D, Bronowska AK, Morgan B et al (2015) A proton relay enhances H2O2 sensitivity of GAPDH to facilitate metabolic adaptation. Nat Chem Biol 11:156–163. https://doi.org/10.1038/nchembio.1720
18. Woo HA, Yim SH, Shin DH et al (2010) Inactivation of Peroxiredoxin I by phosphorylation allows localized H2O2 accumulation for cell signaling. Cell 140:517–528. https://doi.org/10.1016/j.cell.2010.01.009
19. Patterson HC, Gerbeth C, Thiru P et al (2015) A respiratory chain controlled signal transduction cascade in the mitochondrial intermembrane space mediates hydrogen peroxide signaling. PNAS 112:5679–5688
20. Shao D, Oka S, Liu T et al (2014) A redox-dependent mechanism for regulation of AMPK activation by Thioredoxin1 during energy starvation. Cell Metab 19:232–245. https://doi.org/10.1016/j.cmet.2013.12.013
21. Zhang Q, Claret FX (2012) Phosphatases: the new brakes for Cancer development? Enzyme Res 2012:1–11. https://doi.org/10.1155/2012/659649
22. Reczek CR, Chandel NS (2017) The two faces of reactive oxygen species in Cancer. Annu Rev Cancer Biol 1:79–98. https://doi.org/10.1146/annurev-cancerbio-041916-065808
23. Cantley LC (2002) The phosphoinositide 3-kinase pathway. Science 296:1655–1657. https://doi.org/10.1126/science.296.5573.1655
24. Los M, Maddika S, Erb B, Schulze-Osthoff K (2009) Switching Akt: from survival signaling to deadly response. BioEssays 31:492–495. https://doi.org/10.1002/bies.200900005
25. Son Y, Cheong Y-K, Kim N-H et al (2011) Mitogen-activated protein kinases and reactive oxygen species: how can ROS activate MAPK pathways? J Signal Transduct 2011:1–6. https://doi.org/10.1155/2011/792639
26. Weinberg F, Hamanaka R, Wheaton WW et al (2010) Mitochondrial metabolism and ROS generation are essential for Kras-mediated tumorigenicity. Proc Natl Acad Sci 107:8788–8793. https://doi.org/10.1073/pnas.1003428107
27. Morgan MJ, Liu Z (2011) Crosstalk of reactive oxygen species and NF-κB signaling. Cell Res 21:103–115. https://doi.org/10.1038/cr.2010.178
28. Ye J, Fan J, Venneti S et al (2014) Serine catabolism regulates mitochondrial redox control during hypoxia. Cancer Discov 4:1406–1417. https://doi.org/10.1158/2159-8290.CD-14-0250
29. Jeon S-M, Chandel NS, Hay N (2012) AMPK regulates NADPH homeostasis to promote tumour cell survival during energy stress. Nature 485:661–665. https://doi.org/10.1038/nature11066

30. Moon D-O, Kim M-O, Choi YH et al (2010) Butein induces G2/M phase arrest and apoptosis in human hepatoma cancer cells through ROS generation. Cancer Lett 288:204–213. https://doi.org/10.1016/j.canlet.2009.07.002
31. Papaconstantinou J (2019) The role of signaling pathways of inflammation and oxidative stress in development of senescence and aging phenotypes in cardiovascular disease. Cell 8:1383. https://doi.org/10.3390/cells8111383
32. Tobiume K, Matsuzawa A, Takahashi T et al (2001) ASK1 is required for sustained activations of JNK/p38 MAP kinases and apoptosis. EMBO Rep 2:222–228. https://doi.org/10.1093/embo-reports/kve046
33. Zhang J, Wang X, Vikash V et al (2016) ROS and ROS-mediated cellular signaling. Oxidative Med Cell Longev 2016:1–18. https://doi.org/10.1155/2016/4350965
34. Han J, Sun P (2007) The pathways to tumor suppression via route p38. Trends Biochem Sci 32:364–371. https://doi.org/10.1016/j.tibs.2007.06.007
35. Thornton TM, Rincon M (2009) Non-classical P38 map kinase functions: cell cycle checkpoints and survival. Int J Biol Sci 5:44–52. https://doi.org/10.7150/ijbs.5.44
36. Wu X, Zhu Y, Yan H et al (2010) Isothiocyanates induce oxidative stress and suppress the metastasis potential of human non-small cell lung cancer cells. BMC Cancer 10:269. https://doi.org/10.1186/1471-2407-10-269
37. Adhikary A, Mohanty S, Lahiry L et al (2010) Theaflavins retard human breast cancer cell migration by inhibiting NF-κB via p53-ROS cross-talk. FEBS Lett 584:7–14. https://doi.org/10.1016/j.febslet.2009.10.081
38. Chiang AC (2008) Molecular basis of metastasis. N Engl J Med 359:2814–2823
39. Perillo B, Di Donato M, Pezone A et al (2020) ROS in cancer therapy: the bright side of the moon. Exp Mol Med 52:192–203. https://doi.org/10.1038/s12276-020-0384-2
40. Ikwegbue P, Masamba P, Oyinloye B, Kappo A (2017) Roles of heat shock proteins in apoptosis, oxidative stress, human inflammatory diseases, and cancer. Pharmaceuticals 11:2. https://doi.org/10.3390/ph11010002
41. Redza-Dutordoir M, Averill-Bates DA (2016) Activation of apoptosis signalling pathways by reactive oxygen species. Biochim Biophys Acta Mol Cell Res 1863:2977–2992. https://doi.org/10.1016/j.bbamcr.2016.09.012
42. Zong W-X (2006) Necrotic death as a cell fate. Genes Dev 20:1–15. https://doi.org/10.1101/gad.1376506
43. Chen Q, Kang J, Fu C (2018) The independence of and associations among apoptosis, autophagy, and necrosis. Signal Transduct Target Ther 3:18. https://doi.org/10.1038/s41392-018-0018-5
44. Xiang J, Wan C, Guo R, Guo D (2016) Is hydrogen peroxide a suitable apoptosis inducer for all cell types? Biomed Res Int 2016:1–6. https://doi.org/10.1155/2016/7343965
45. Lee SY, Ju MK, Jeon HM et al (2018) Regulation of tumor progression by programmed necrosis. Oxidative Med Cell Longev 2018:1–28. https://doi.org/10.1155/2018/3537471
46. Zong W-X (2004) Alkylating DNA damage stimulates a regulated form of necrotic cell death. Genes Dev 18:1272–1282. https://doi.org/10.1101/gad.1199904
47. Ricci MS, Zong W (2006) Chemotherapeutic approaches for targeting cell death pathways. Oncologist 11:342–357. https://doi.org/10.1634/theoncologist.11-4-342
48. Miyoshi N, Watanabe E, Osawa T et al (2008) ATP depletion alters the mode of cell death induced by benzyl isothiocyanate. Biochim Biophys Acta (BBA) - Mol Basis Dis 1782:566–573. https://doi.org/10.1016/j.bbadis.2008.07.002
49. White E (2015) The role for autophagy in cancer. J Clin Invest 125:42–46. https://doi.org/10.1172/JCI73941
50. Vanden Berghe T, Kaiser WJ, Bertrand MJ, Vandenabeele P (2015) Molecular crosstalk between apoptosis, necroptosis, and survival signaling. Mol Cell Oncol 2:e975093. https://doi.org/10.4161/23723556.2014.975093

51. Qiu X, Zhang Y, Han J (2018) RIP3 is an upregulator of aerobic metabolism and the enhanced respiration by necrosomal RIP3 feeds back on necrosome to promote necroptosis. Cell Death Differ. https://doi.org/10.1038/s41418-018-0075-x
52. Hanson B (2016) Necroptosis: a new way of dying? Cancer Biol Ther 17:899–910. https://doi.org/10.1080/15384047.2016.1210732
53. Wang T, Liu L, Chen X et al (2018) MYCN drives glutaminolysis in neuroblastoma and confers sensitivity to an ROS augmenting agent. Cell Death Dis 9:220. https://doi.org/10.1038/s41419-018-0295-5
54. Hsu S-K, Chang W-T, Lin I-L et al (2020) The role of necroptosis in ROS-mediated Cancer therapies and its promising applications. Cancers 12:2185. https://doi.org/10.3390/cancers12082185
55. Scherz-Shouval R, Elazar Z (2011) Regulation of autophagy by ROS: physiology and pathology. Trends Biochem Sci 36:30–38. https://doi.org/10.1016/j.tibs.2010.07.007
56. Ferro F, Servais S, Besson P et al (2020) Autophagy and mitophagy in cancer metabolic remodelling. Semin Cell Dev Biol 98:129–138. https://doi.org/10.1016/j.semcdb.2019.05.029
57. Ghavami S, Asoodeh A, Klonisch T et al (2008) Brevinin-2R1 semi-selectively kills cancer cells by a distinct mechanism, which involves the lysosomal-mitochondrial death pathway. J Cell Mol Med 12:1005–1022. https://doi.org/10.1111/j.1582-4934.2008.00129.x
58. Poillet-Perez L, Despouy G, Delage-Mourroux R, Boyer-Guittaut M (2015) Interplay between ROS and autophagy in cancer cells, from tumor initiation to cancer therapy. Redox Biol 4:184–192. https://doi.org/10.1016/j.redox.2014.12.003
59. Boyer-Guittaut M, Poillet L, Liang Q et al (2014) The role of GABARAPL1/GEC1 in autophagic flux and mitochondrial quality control in MDA-MB-436 breast cancer cells. Autophagy 10:986–1003. https://doi.org/10.4161/auto.28390
60. Lin Y, Jiang M, Chen W et al (2019) Cancer and ER stress: mutual crosstalk between autophagy, oxidative stress and inflammatory response. Biomed Pharmacother 118:109249. https://doi.org/10.1016/j.biopha.2019.109249
61. Rahman SMT, Zhou W, Deiters A, Haugh JM (2020) Optical control of MAP kinase kinase 6 (MKK6) reveals that it has divergent roles in pro-apoptotic and anti-proliferative signaling. J Biol Chem 295:8494–8504. https://doi.org/10.1074/jbc.RA119.012079
62. Pradhan R, Singhvi G, Dubey SK et al (2019) MAPK pathway: a potential target for the treatment of non-small-cell lung carcinoma. Future Med Chem 11:793–795. https://doi.org/10.4155/fmc-2018-0468
63. Shen T, Guo Q (2020) EGFR signaling pathway occupies an important position in cancer-related downstream signaling pathways of Pyk2. Cell Biol Int 44:2–13. https://doi.org/10.1002/cbin.11209
64. Mu Y, Yang K, Hao X et al (2020) Clinical characteristics and treatment outcomes of 65 patients with BRAF-mutated non-small cell lung Cancer. Front Oncol 10:603. https://doi.org/10.3389/fonc.2020.00603
65. Farhan M, Wang H, Gaur U et al (2017) FOXO signaling pathways as therapeutic targets in Cancer. Int J Biol Sci 13:815–827. https://doi.org/10.7150/ijbs.20052
66. Hornsveld M, Dansen TB, Derksen PW, Burgering BMT (2018) Re-evaluating the role of FOXOs in cancer. Semin Cancer Biol 50:90–100. https://doi.org/10.1016/j.semcancer.2017.11.017
67. Pérez-Ramírez C, Cañadas-Garre M, Molina MÁ et al (2015) PTEN and PI3K/AKT in non-small-cell lung cancer. Pharmacogenomics 16:1843–1862. https://doi.org/10.2217/pgs.15.122
68. Jiramongkol Y, Lam EW-F (2020) FOXO transcription factor family in cancer and metastasis. Cancer Metastasis Rev 39:681–709. https://doi.org/10.1007/s10555-020-09883-w
69. Tong Y-H, Zhang B, Fan Y, Lin N-M (2015) Keap1–Nrf2 pathway: a promising target towards lung cancer prevention and therapeutics. Chron Dis Transl Med 1:175–186. https://doi.org/10.1016/j.cdtm.2015.09.002
70. Barrera-Rodríguez R (2018) Importance of the Keap1-Nrf2 pathway in NSCLC: is it a possible biomarker? (Review). Biom Rep. https://doi.org/10.3892/br.2018.1143

71. Saigusa D, Motoike IN, Saito S et al (2020) Impacts of NRF2 activation in non–small-cell lung cancer cell lines on extracellular metabolites. Cancer Sci 111:667–678. https://doi.org/10.1111/cas.14278
72. Mukherjee A, Paul M, Mukherjee S (2019) Recent Progress in the Theranostics application of nanomedicine in lung Cancer. Cancers 11:597. https://doi.org/10.3390/cancers11050597
73. Lin C, Zhang X, Chen H et al (2018) Dual-ligand modified liposomes provide effective local targeted delivery of lung-cancer drug by antibody and tumor lineage-homing cell-penetrating peptide. Drug Deliv 25:256–266. https://doi.org/10.1080/10717544.2018.1425777
74. Erten A, Wrasidlo W, Scadeng M et al (2010) Magnetic resonance and fluorescence imaging of doxorubicin-loaded nanoparticles using a novel in vivo model. Nanomedicine 6:797–807. https://doi.org/10.1016/j.nano.2010.06.005
75. Wu XS, Wang N (2001) Synthesis, characterization, biodegradation, and drug delivery application of biodegradable lactic/glycolic acid polymers. Part II: biodegradation. J Biomater Sci Polym Ed 12:21–34. https://doi.org/10.1163/156856201744425
76. Sengupta S et al (2005) Temporal targeting of tumour cells and neovasculature with a nanoscale delivery system. Nature 436:568–572. https://doi.org/10.1038/nature03794568
77. Nguyen J, Steele TWJ, Merkel O et al (2008) Fast degrading polyesters as siRNA nano-carriers for pulmonary gene therapy. J Control Release 132:243–251. https://doi.org/10.1016/j.jconrel.2008.06.010
78. Okamoto H, Nishida S, Todo H et al (2003) Pulmonary gene delivery by chitosan–pDNA complex powder prepared by a supercritical carbon dioxide process. J Pharm Sci 92:371–380. https://doi.org/10.1002/jps.10285
79. Ventura CA, Cannavà C, Stancanelli R et al (2011) Gemcitabine-loaded chitosan microspheres. Characterization and biological in vitro evaluation. Biomed Microdevices 13:799–807. https://doi.org/10.1007/s10544-011-9550-6
80. Liu X, Wei W, Wang C et al (2011) Apoferritin-camouflaged Pt nanoparticles: surface effects on cellular uptake and cytotoxicity. J Mater Chem 21:7105. https://doi.org/10.1039/c1jm10575b
81. Liu X, Wei W, Yuan Q et al (2012) Apoferritin–CeO_2 nano-truffle that has excellent artificial redox enzyme activity. Chem Commun 48:3155–3157. https://doi.org/10.1039/C1CC15815E
82. Li K, Zhang Z-P, Luo M et al (2012) Multifunctional ferritin cage nanostructures for fluorescence and MR imaging of tumor cells. Nanoscale 4:188–193. https://doi.org/10.1039/C1NR11132A
83. Mao C, Liu A, Cao B (2009) Virus-based chemical and biological sensing. Angew Chem Int Ed 48:6790–6810. https://doi.org/10.1002/anie.200900231
84. Al-Qadi S, Grenha A, Carrión-Recio D et al (2012) Microencapsulated chitosan nanoparticles for pulmonary protein delivery: in vivo evaluation of insulin-loaded formulations. J Control Release 157:383–390. https://doi.org/10.1016/j.jconrel.2011.08.008
85. Islam N, Ferro V (2016) Recent advances in chitosan-based nanoparticulate pulmonary drug delivery. Nanoscale 8:14341–14358. https://doi.org/10.1039/C6NR03256G
86. Yang R, Yang S-G, Shim W-S et al (2009) Lung-specific delivery of paclitaxel by chitosan-modified PLGA nanoparticles via transient formation of microaggregates. J Pharm Sci 98:970–984. https://doi.org/10.1002/jps.21487
87. Zhu L, Li M, Liu X, Jin Y (2017) Drug-loaded PLGA electrospraying porous microspheres for the local therapy of primary lung cancer via pulmonary delivery. ACS Omega 2:2273–2279. https://doi.org/10.1021/acsomega.7b00456
88. Bivas-Benita M, Romeijn S, Junginger HE, Borchard G (2004) PLGA–PEI nanoparticles for gene delivery to pulmonary epithelium. Eur J Pharm Biopharm 58:1–6. https://doi.org/10.1016/j.ejpb.2004.03.008
89. Emami J, Pourmashhadi A, Sadeghi H et al (2015) Formulation and optimization of celecoxib-loaded PLGA nanoparticles by the Taguchi design and their *in vitro* cytotoxicity for lung cancer therapy. Pharm Dev Technol 20:791–800. https://doi.org/10.3109/10837450.2014.920360

90. Kimura S, Egashira K, Chen L et al (2009) Nanoparticle-mediated delivery of nuclear factor κB decoy into lungs ameliorates monocrotaline-induced pulmonary arterial hypertension. Hypertension 53:877–883. https://doi.org/10.1161/HYPERTENSIONAHA.108.121418
91. Ziady A-G, Gedeon CR, Miller T et al (2003) Transfection of airway epithelium by stable PEGylated poly-l-lysine DNA nanoparticles in vivo. Mol Ther 8:936–947. https://doi.org/10.1016/j.ymthe.2003.07.007
92. Kaul G, Amiji M (2005) Tumor-targeted gene delivery using poly(ethylene glycol)-modified gelatin nanoparticles: in vitro and in vivo studies. Pharm Res 22:951–961. https://doi.org/10.1007/s11095-005-4590-3
93. Yu MK, Jeong YY, Park J et al (2008) Drug-loaded superparamagnetic iron oxide nanoparticles for combined cancer imaging and therapy in vivo. Angew Chem Int Ed 47:5362–5365. https://doi.org/10.1002/anie.200800857
94. Yu H, Wang Y, Wang S et al (2018) Paclitaxel-loaded core–shell magnetic nanoparticles and cold atmospheric plasma inhibit non-small cell lung cancer growth. ACS Appl Mater Interfaces 10:43462–43471. https://doi.org/10.1021/acsami.8b16487
95. Minati L, Antonini V, Dalla Serra M, Speranza G (2012) Multifunctional branched gold–carbon nanotube hybrid for cell imaging and drug delivery. Langmuir 28:15900–15906. https://doi.org/10.1021/la303298u
96. Podesta JE, Al-Jamal KT, Herrero MA et al (2009) Antitumor activity and prolonged survival by carbon-nanotube-mediated therapeutic siRNA silencing in a human lung xenograft model. Small NA-NA. https://doi.org/10.1002/smll.200801572
97. Beljanski V, Hiscott J (2012) The use of oncolytic viruses to overcome lung cancer drug resistance. Curr Opin Virol 2:629–635. https://doi.org/10.1016/j.coviro.2012.07.006
98. Ciccarese F, Raimondi V, Sharova E et al (2020) Nanoparticles as tools to target redox homeostasis in cancer cells. Antioxidants 9:211. https://doi.org/10.3390/antiox9030211
99. Marengo B, Nitti M, Furfaro AL et al (2016) Redox homeostasis and cellular antioxidant systems: crucial players in cancer growth and therapy. Oxidative Med Cell Longev 2016:1–16. https://doi.org/10.1155/2016/6235641
100. Manke A, Wang L, Rojanasakul Y (2013) Mechanisms of nanoparticle-induced oxidative stress and toxicity. Biomed Res Int 2013:1–15. https://doi.org/10.1155/2013/942916
101. Xu C, Qu X (2014) Cerium oxide nanoparticle: a remarkably versatile rare earth nanomaterial for biological applications. NPG Asia Mater 6:e90–e90. https://doi.org/10.1038/am.2013.88
102. Alili L, Sack M, von Montfort C et al (2013) Downregulation of tumor growth and invasion by redox-active nanoparticles. Antioxid Redox Signal 19:765–778. https://doi.org/10.1089/ars.2012.4831
103. Baldrighi M, Trusel M, Tonini R, Giordani S (2016) Carbon nanomaterials interfacing with neurons: an in vivo perspective. Front Neurosci 10. https://doi.org/10.3389/fnins.2016.00250
104. Tang Q, Yu B, Gao L et al (2018) Stimuli responsive nanoparticles for controlled anti-cancer drug release. Curr Med Chem 25:1837–1866
105. Sadhukha T, Wiedmann TS, Panyam J (2013) Inhalable magnetic nanoparticles for targeted hyperthermia in lung cancer therapy. Biomaterials 34:5163–5171. https://doi.org/10.1016/j.biomaterials.2013.03.061
106. Kallu J, Banerjee T, Sulthana S et al (2019) Nanomedicine-assisted combination therapy of NSCLC: new platinum-based anticancer drug synergizes the therapeutic efficacy of Ganetespib. Nano 3:120–134. https://doi.org/10.7150/ntno.28468
107. Basu S, Harfouche R, Soni S et al (2009) Nanoparticle-mediated targeting of MAPK signaling predisposes tumor to chemotherapy. Proc Natl Acad Sci 106:7957–7961. https://doi.org/10.1073/pnas.0902857106
108. Wu Y-F, Wu H-C, Kuan C-H et al (2016) Multi-functionalized carbon dots as theranostic nanoagent for gene delivery in lung cancer therapy. Sci Rep 6:21170. https://doi.org/10.1038/srep21170

109. Min L, He L, Chen Q et al (2012) Magnetic iron oxide nanoparticles carrying PTEN gene to reverse cisplatin-resistance of A549/CDDP cell lines. J Cent S Univ Technol 19:331–339. https://doi.org/10.1007/s11771-012-1009-2
110. Wu H, Zhao Y, Mu X et al (2015) A silica–polymer composite nano system for tumor-targeted imaging and p53 gene therapy of lung cancer. J Biomater Sci Polym Ed 26:384–400. https://doi.org/10.1080/09205063.2015.1012035
111. Varkouhi AK et al (2010) Gene silencing activity of siRNA polyplexes based on thiolated N,N,N-trimethylated chitosan. Bioconjug Chem 21:2339–2346. https://doi.org/10.1021/bc1003789
112. Garbuzenko OB, Kuzmov A, Taratula O et al (2019) Strategy to enhance lung cancer treatment by five essential elements: inhalation delivery, nanotechnology, tumor-receptor targeting, chemo- and gene therapy. Theranostics 9:8362–8376. https://doi.org/10.7150/thno.39816
113. Bolte C, Ustiyan V, Ren X et al (2020) Nanoparticle delivery of proangiogenic transcription factors into the neonatal circulation inhibits alveolar simplification caused by Hyperoxia. Am J Respir Crit Care Med 202:100–111. https://doi.org/10.1164/rccm.201906-1232OC
114. Zhang Y, Zhang L, Lin X et al (2019) Dual-responsive nanosystem for precise molecular subtyping and resistant reversal of EGFR targeted therapy. Chem Eng J 372:483–495. https://doi.org/10.1016/j.cej.2019.04.140
115. Velavan B, Divya T, Sureshkumar A, Sudhandiran G (2018) Nano-chemotherapeutic efficacy of (−) -epigallocatechin 3-gallate mediating apoptosis in A549 cells: involvement of reactive oxygen species mediated Nrf2/Keap1signaling. Biochem Biophys Res Commun 503:1723–1731. https://doi.org/10.1016/j.bbrc.2018.07.105
116. Ramalingam V, Varunkumar K, Ravikumar V, Rajaram R (2018) Target delivery of doxorubicin tethered with PVP stabilized gold nanoparticles for effective treatment of lung cancer. Sci Rep 8:3815. https://doi.org/10.1038/s41598-018-22172-5
117. He Y, Du Z, Ma S et al (2016) Effects of green-synthesized silver nanoparticles on lung cancer cells in vitro and grown as xenograft tumors in vivo. Int J Nanomed 11:1879–1887. https://doi.org/10.2147/IJN.S103695
118. Muller J, Huaux F, Moreau N et al (2005) Respiratory toxicity of multi-wall carbon nanotubes. Toxicol Appl Pharmacol 207:221–231. https://doi.org/10.1016/j.taap.2005.01.008

12 Targeting Molecular and Cellular Mechanisms in Idiopathic Pulmonary Fibrosis

Banudevi Sivanantham and Vijayageetha Bose

Abstract

Idiopathic pulmonary fibrosis (IPF) is a persistent and an aggravating interstitial alveolar disease, marked by thickening and scarring of the lung parenchyma with unknown etiology leading to early death among the elderly population worldwide. The vast knowledge on the understanding of the pathogenesis, diagnosis, and therapeutic management has been raised significantly in recent years to delay the progression of IPF. Currently, nintedanib and pirfenidone are two medications used to treat IPF which significantly restores the alveolar epithelial functions; however, it is associated with a few demerits. Thus, new approaches are needed to overcome hurdles raised from IPF pathogenesis. Up-to-date approaches in pulmonary rehabilitation, non-pharmacological strategies, lung transplantation, and comorbidity management are mainly involved in subsiding the symptoms and thereby attempt to improve patient's health outcomes. Hence clinical trial studies are in search of novel molecular and cellular targets to overcome IPF. This chapter highlights the understanding of up-to-date molecular and cellular targets and potential approaches which will create an avenue to design and develop novel therapeutics to defend against this complex and injurious disease.

B. Sivanantham (✉)
Department of Bioengineering, School of Chemical & Biotechnology, SASTRA Deemed-to-be University, Thirumalaisamudram, Thanjavur, Tamilnadu, India
e-mail: banudevi@scbt.sastra.edu

V. Bose
Department of Biotechnology, Rajalakshmi Engineering College (Autonomous), Affiliated to Anna University, Thandalam, Chennai, Tamilnadu, India
e-mail: vijayageetha.b@rajalakshmi.edu.in

K. Dua et al. (eds.), *Targeting Cellular Signalling Pathways in Lung Diseases*, https://doi.org/10.1007/978-981-33-6827-9_12

Keywords

Coagulation cascade · Fibrogenesis · Idiopathic pulmonary fibrosis · Interstitial lung diseases · Myofibroblast · Usual interstitial pneumonia

12.1 Introduction

Interstitial lung diseases (ILDs) exemplify the heterogeneous characteristics of pulmonary parenchymal diseases emanating from various systemic diseases, idiopathic conditions, and different environmental insults. Among ILDs, idiopathic pulmonary fibrosis (IPF) is the most endemic and dreadful form of lung fibrosis with unknown etiology and has life sustenance of 2–6 years after diagnosis in a patient lacking lung transplantation [1]. IPF is well-characterized by myofibroblast proliferation, excessive extracellular matrix (ECM) deposition, increased collagen synthesis, and aberrant wound healing process resulting in irreversible devastation of lung parenchymal morphology evidenced by severe scarring that ultimately ends in organ impairment and death due to respiratory failure. The term IPF is closely analogous to the usual interstitial pneumonia (UIP), which is detailed by the occurrence of temporal heterogeneity, honey-combing structure, and finally development of subpleural fibrosis. Further, the evolution of IPF pathobiology continues with supporting manifestation like repetitive lung injuries particularly to the alveolar epithelial cells (AECs) which guide the morphological interaction with fibroblast resulting in the accumulation of ECM thereby destructing the gas exchanging regions [2]. Smoking, viral infections such as SARS-COVID-19, aerosolized environmental factors, occupational toxins exposure, traction injuries to the peripheral lung tissue, chemotherapeutic drugs, and radiotherapy result in the development of chronic alveolar epithelium damage [1, 3]. It also occurs in patients with chronic host graft disease who has undergone bone marrow transplantation and inflammatory diseases such as rheumatoid arthritis and scleroderma [3]. Although its causative factor remains elusive, recent advancement led us to understand the cellular and molecular mechanisms implicated in the initiation and progression of IPF [4].

The morbidity and mortality rate of IPF is very high, and there has been a rapid breakthrough in overcoming this prevalent disease including diagnosis, development of approachable therapies, and genetic predisposition identification. Currently, two FDA-approved therapeutics such as nintedanib and pirfenidone are widely accepted medications to delay the IPF progression [5, 6]. However, these two drugs do not revamp the normal pulmonary activities, and neither of the drugs has been examined in patients synchronized with IPF and lung cancer (LC) [7], which indicates that 22% of patients bearing IPF is prone to develop LC which corresponds to five-fold higher probability than the normal population [8]. Similarly, LC risk is surpassing 20 times higher in the patient who has undergone lung transplantation to overcome IPF pathology [9, 10]. This distressing information warrants efforts to identify cellular and molecular targets to overcome this endemic disease and to improve the patient's survival. This chapter explores the understanding of emerging cellular and novel molecular targets to handle the hurdles caused by IPF and further accelerate a novel drug design and develop valuable therapeutics to prevent IPF progression.

12.1.1 IPF Subclassification

The relevance of discrete pathological, radiological, and clinical characteristics is essential to categorize IPF and also required for further diagnosis and treatment. Clinical classification of IPF is based on the patient's well-defined cellular or molecular phenotypes anticipating differences in progression rate. In this line, IPF patients are subcategorized based on their histopathological features as (a) *rapid progression rate* if IPF patients have a greater number of fibroblast foci [11]; (b) *Slow progression rate* if IPF patients have greater number of lung mast cells [12]. Based on the genetic changes, it is further subcategorized as (a) patients with polymorphic for mucin (MUC) promoter SNP (single nucleotide polymorphism), i.e., TT or GT MUC5B genotype experiences slower progression rate; (b) patients possessing GG MUC5B type have rapid progression of IPF; and (c) increased risk for mortality is experienced in patients with a minor genetic change in allele rs5743890 (an SNP located 3 kb upstream of the start of transcription from the MUC5B gene) in TOLLIP (Toll-interacting protein) gene [13]. Similarly, circulating molecular markers also help to sub-phenotype IPF patients. The elevated level of molecular markers such as matrix metalloproteinases-7 (MMP-7), C-C motif chemokine ligand (CCL), and serum KL (mucin-like high-molecular-weight glycoprotein) are prone to increase the risk of IPF progression and thus render poor survival [14, 15]. Thus, this clinical relevance of IPF patients may be essential for diagnosis and provide a clue for targeted therapy against IPF. However, these subtype categorization needs further scientific evidence to pinpoint the exact disease type, different stages, and expected response toward therapy.

12.2 Pathogenesis of IPF

Current IPF pathogenesis presumes as an aberrant wound healing activity concerning repetitive injurious factors resulting in alveolar epithelial cell apoptosis, fibroblast activation, and ECM accumulation. An overview of IPF pathogenesis has been represented in Fig. 12.1.

The pathological feature of IPF lung displays typical fibrosis, distributed along the inferior segments of the alveolar lobes with subpleural accentuation. The histopathological alterations in IPF acute exacerbation comprise the lung tissue manifesting the changes of diffuse alveolar scarring including type 2 pneumocyte hyperplasia, the formation of hyaline membrane, and edematous alveolar septa [16].

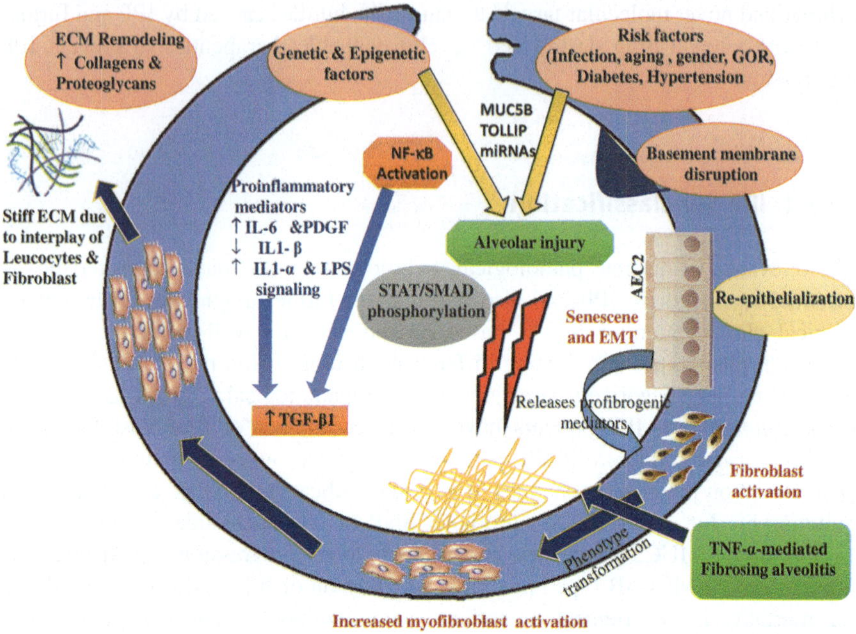

Fig. 12.1 An overview of pathogenesis of IPF. Recurrent epithelial alveolar cell injury in genetically susceptible individuals causes senescence of epithelial cells and epithelial-mesenchymal transition (EMT), releasing profibrogenic mediators induces fibrocyte/fibroblast migration and differentiation into profibrotic macrophages/myofibroblasts, resulting in aberrant matrix deposition with destructing lung structure. AEC2, alveolar epithelial cells type 2; IL-1 and IL-6, interleukin-1and interlukin-6; STAT, signal transducers of activation; SMAD, signal transducer; LPS, lipopolysaccharide

12.2.1 Modifications of Genetic and Epigenetic Factors in IPF Development

IPF, a late-age-of-onset lung disease causes extensive lung function impairment along with hypoxemia and dyspnea, ultimately ending in respiratory failure and alveolar cell death. IPF prevalence is projected to be 63/1, 00,000 population globally. The prevalence and mortality of pulmonary fibrosis are increasing as the population ages [17]. Several genetic loci are prone to initiate and progress this devastating disease. A wide array of rare genetic variants and a common MUC5B SNP (single nucleotide polymorphism) have been well-correlated with IPF occurrence [13]. Experimentally in bleomycin-induced pulmonary fibrosis, it was revealed that the mucociliary dysfunction might play a causative role in overexpressing MUC5B in distal airspaces that can serve as a potential therapeutic target in IPF patients [13]. Besides, the dysfunction of telomere maintenance has been associated with the pathogenesis of IPF. An earlier study showed that shortening of telomeres critically affects alveolar epithelial cells 2 (AEC2 or AT2) in fibrotic regions, implying telomere length (TL) as a cause of fibrogenesis. Furthermore, short lung TL is associated with decreased survival, and the genetic mutations in telomere-responsive genes have been considered as a causative factor for familial IPF.

Nevertheless, it has been clear that telomere shortening and telomere syndromes are linked with different types of lung diseases [18]. The expression of REVERBα, an orphan nuclear receptor regulates circadian rhythm, and metabolism was found to be increased in human IPF lung tissue. Targeting REVERBα by pharmacological agents ceases not only the activation of myofibroblast resides in IPF fibroblasts but also decreased collagen secretion in organotypic cultures from IPF patients, implies that REVERBα could be a possible therapeutic approach to manage IPF complications [19]. New insight into the IPF mechanisms related to the mutation in surfactant protein A1 (*SFTPA1)* promotes necroptosis of AEC2 through c-jun N-terminal kinase (JNK)-mediated upregulation of RIPK3 (receptor-interacting serine-threonine kinase 3), highlighting the necroptosis pathway as a possible therapeutic target to overcome the pathogenesis of IPF [20].

IPF signifies the alveolar epithelium injury followed by the vast release of fibroproliferative and pro-inflammatory mediators that delays the response associated with normal tissue repair. However, these repair processes are never sorted out and intensifying fibrosis for unknown reasons [21]. Several line of studies evidence that epigenetic factors impart to the poorly regulated gene expression in IPF lung, and the risk factors such as cigarette smoking, gender, age, and genetic variants influence epigenetic markers that predispose to IPF. It is also reported that epigenetic mechanisms strongly affect the expression and binding of cytokine and transcription factors, respectively, that regulate T-helper cells-1 (Th1), Th2, T-regulator cells (Treg), and Th17 cells lineages in IPF conditions [22].

Methylation and acetylation are the two important modifications of histone tails, and manage gene expression by regulating DNA availability to the enzyme, RNA polymerase II, and nuclear transcription factors. Histone deacetylases (HDACs), histone acetyltransferases (HATs), and bromodomain (Brd) proteins play a crucial role in channeling of epigenetic memory across cell divisions and gene regulation. Similarly, histone demethylases (HDMs) and histone methyltransferases (HMTs) enumerated modifications at residues specific sites regulating chromatin availability [23, 24]. Genome-wide profiles of DNA methylation to focus specific genetic loci demonstrated extensive changes in DNA methylation in IPF alveolar tissue and its considerable effect on gene expression involved in epigenetic mechanisms. Dysregulation of noncoding RNA, i.e., microRNA (miR), has also been considered as an epigenetic factor, and about 90% of the human genes is silenced by miRNAs through target mRNA degradation or protein synthesis inhibition. miRNAs such as miR-21, miR-26a, miR-155, miR9-5p, miR-31 displays a significant role in IPF pathogenesis and progression [25]. Studies have reported that significant changes in regulatory miRNAs levels were observed in IPF patients compared with normal subjects [26]. Differentially methylated sites of genetic loci associated with IPF pathogenesis strongly suggest that genetic and epigenetic factors proceed in concert to dysregulate genes involved in the progression of IPF [27].

12.2.2 Role of Pro-inflammatory Mediators in IPF

The imbalance between pro-inflammatory and anti-inflammatory mechanisms that favors the release of pro-inflammatory mediators resulting in chronic diffuse lung disease [28]. Even cytokine balance such as tumor necrosis factor (TNF), interleukin-1 (IL-1) contributes to lung injury at the initial stages during lung inflammation (Fig. 12.1). Potential inhibitors against IL-1 and TNF suggest that an imbalance in T-helper cells (Th1), Th2, and cytokine profiles seems to be the potential criteria in chronic lung disorders and fibrotic effects [29]. In hyperoxia-mediated lung injury, increased insulin-like growth factor (IGF) signaling pathway performs a critical role in alveolar epithelial cell proliferation and differentiation during tissue remodeling [29]. Angiogenesis, a process of formation of new blood vessels was found to increase significantly in chronic inflammation because of imbalance in angiogenic mediators during chronic lung inflammation. Thus dysregulated angiogenesis seems to be a prime cellular process in the aid of fibroplasia and ECM deposition that happens during IPF [30]. These studies strongly indicate that drugs targeting inflammatory signaling pathways that are responsible for the development of chronic fibroproliferative lung diseases are geared in the near future to overcome IPF pathogenesis.

12.2.3 Recruitment of Leukocytes and Fibroblast by Chemokine

Chemokines and their receptors play a crucial role in cell recruitment during granulomatous inflammatory reactions. Particularly, interferon-γ (IFN-γ) regulates chemokines and its receptors expression as well as cell types that infiltrate the lungs in vivo [31]. These regulatory mediators, i.e., chemokines, its receptors, and transcription factors in a particular inflammatory condition possess an ability to direct the Th17 cell response to prefer profibrotic or anti-fibrotic function [32].

IL-1α and IL-1β are considered as proximal profibrotic mediators of severe lung inflammation-promoting fibrosis via activation of fibroblasts to synthesize cytokines including additional IL-1, IL-6, CXC, and CC chemokines. An increase in pro-inflammatory cytokines such as IL-6 and TNF, PDGF, profibrotic cytokine, and TGF-β1 levels suggests that IL-1β has a pivotal role in lung injury at the initial stage resulting in self-perpetuating fibrotic response [33]. Similar to IL-1 and lipopolysaccharide (LPS), TNF-α signaling happens through the activation of nuclear factor kappa B (NF-κB), which promotes gene transcription that mediates innate immune responses [34]. TNF overexpression has a significant chance in directing pulmonary fibrosis through induction of severe lung inflammation and speckled interstitial fibrosis with increased TGF-β expression and increased fibroblast differentiation into myofibroblasts [35]. The prolonged expression of TNF may be considered as an essential factor in fibrosing alveolitis pathogenesis, and curtailing the cytokine functioning might be beneficial in reviving the lung architecture and function of IPF patients.

12.3 Comorbidities Associated with IPF

The conditions like hypertension, depression, cardiovascular disease, and diabetes mellitus influence the patients with IPF and have the greatest impact on mortality [36, 37]. Pulmonary hypertension is a frequent complication that occurs during progression and in advanced stages of IPF. An array of vascular comorbidities has been furnished in a number of studies right from arrhythmias or atrial fibrillation, systemic arterial hypertension, ischemic heart disease, cerebral ischemia, heart failure, and strokes.

Recently, ample of studies are intended to understand the association between IPF and sleep apnea, frequently reveal a moderate-to-severe condition of obstructive sleep apnea (OSA) in IPF patients. Next in the line is telomerase mutations in TERT or TERC and shortening of telomere during aging are closely associated with IPF pathogenesis [38, 39]. The comorbidities during IPF share the same probability of risk factors of pulmonary cancer and chronic obstructive pulmonary disease (COPD), e.g., cigarette smoking. The majority of IPF patients are ex-smokers or smokers, and not surprisingly the IPF patients may have symptoms related to COPD or emphysema.

Gastroesophageal reflux (GOR) diseases are frequent comorbidities; both symptomatic and asymptomatic forms were observed in IPF patients. Microaspiration procedures also cause repetitive lung injury that could result in disease pathogenesis [40]. Those IPF patients often display symptoms like depression and anxiety, and these clinical symptoms have not been appropriately recognized.

It is not feasible to explore the etiology of progressive lung fibrosis without considering the importance of the virus. Initial analysis from COVID-19 patients reported a high rate of abnormalities in lung performance concerning gas transfer impairment of about 47% which may reflect partially in cardiopathy, a key emerging feature of COVID-19. Whereas in SARS patients, it is about 16% abnormalities in gas transfer, and even abnormal chest X-ray was observed in 30% patients and 62% patients who have CT evidence of lung fibrosis soon after discharge [41, 42]. The significance of other respiratory virus-inducing pulmonary fibrosis is less understood and generally limited to acute exacerbations of IPF and in vivo models [3].

12.4 Cellular and Molecular Mechanism Underlying IPF Progression

12.4.1 Cellular Players

Aberrant fibrogenesis occurring in IPF is a dynamic cellular process entailing complicated interaction between fibroblasts, immune cells specifically T cells and macrophages, endothelial and epithelial cells [43, 44].

12.4.1.1 Fibroblast Activation via Mesenchymal Phenotype Transformation

Aberrant myofibroblast population is responsible for the fibrotic scar formation and further destruction of lung architecture in IPF [11]. The activated myofibroblast contributes to increased synthesis of collagen, a component of extracellular matrix (ECM), and secretes an increased amount of glycoprotein and proteoglycans. Further activation of α-smooth muscle actin (SMA), contractile protein transforms the resident fibroblast into myofibroblast in case of disease progression [45]. Increased expression of transforming growth factor β (TGF-β) is the central mediator in the development of lung fibrosis that induces epithelial to mesenchymal transition (EMT), which is part of an unending form of the wound healing process in which alveolar epithelial cells type 2 (AT2) present them as a source for the aberrant myofibroblast population that results in lung deterioration if the primary inflammatory insults causing wound healing is not attenuated [45]. Further, increased TGF-β and endothelin-1 causes the circulating fibrocytes, mesenchymal progenitor cells originating from bone marrow home and extravasate into the region of lung injury and differentiate into myofibroblast-like phenotype [45]. Myofibroblast activation is also routed by pulmonary endothelial cells to mesenchymal transition (EnMT) that bestows to lung fibrosis and hypertension and further it is associated with IPF development [46] indicating poor prognosis (Fig. 12.2). Finally, pleural mesothelial cell or lung pericytes serve as a remarkable source for increased myofibroblast population in IPF [46].

12.4.1.2 Cellular Mechanism Pertaining to Various Targets for Therapeutic Intervention

Cellular Senescence and Deregulated Metabolism in IPF

Cellular senescence particularly alveolar epithelial cell (AEC) is the prime factor for the development of progressive lung fibrosis [47]. The cellular events that result in AEC senescence are AEC mitochondrial dysfunction, telomere shortening, and endoplasmic reticulum (ER) stress [47]. AECs and fibroblasts are the two typical cell types prone to cellular senescence following repetitive lung injury. Senolytic drugs such as dasatinib and quercetin are employed exclusively to remove senescent cells through increased apoptosis and have been evidenced to eliminate profibrotic senescent fibroblast and AECs in vitro and ex vivo, respectively [48–50]. Further, a combination approach of dasatinib and quercetin displayed positive effects for improvement of alveolar function in bleomycin-induced pulmonary fibrosis; however, fibrotic effects were not altered significantly [49]. Metformin is also used to reduce pulmonary fibrosis in vivo [51, 52]. Clarification on senolytic drugs is essential whether these drugs selectively stamp out pathogenic cell types, not the cells needed for regeneration in IPF resolution. Therefore nowadays research is more focused on molecular signaling pathways involved in these cellular events that may be preferred as a target for therapeutic intervention to surpass the fibrotic effects caused by IPF.

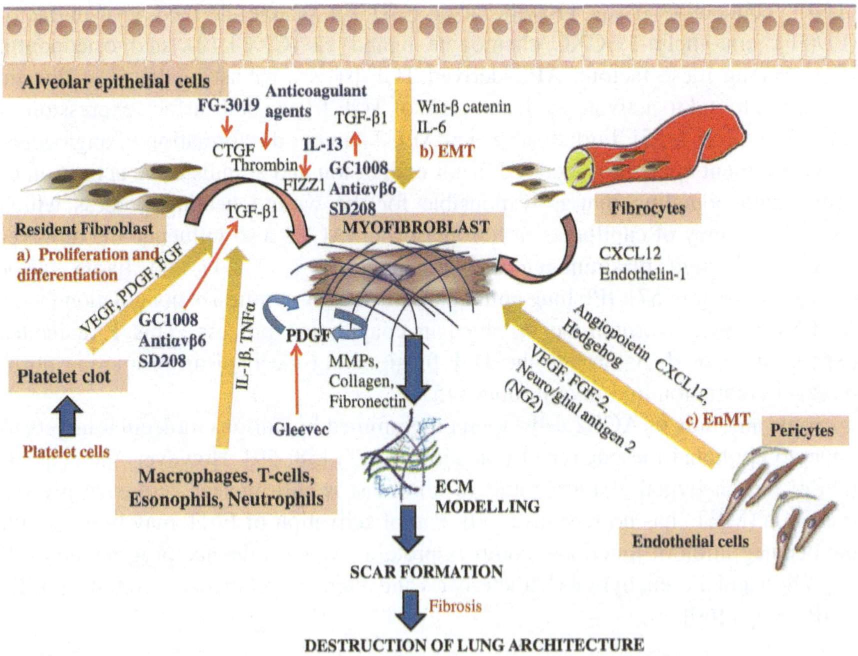

Fig. 12.2 Cellular targets of IPF. Schematic representation of the different types of cells involved into myofibroblast formation, a vital effectors cell type in IPF progression. (a) Involvement of resident fibroblast proliferation and differentiation into myofibroblast and other cell types such as leukocytes and platelets also contributes to myofibroblast formation. (b) Alveolar epithelial cells type 2 involved in epithelial-mesenchymal transition (EMT). (c) Endothelial cells participate in endothelial-mesenchymal transition (EnMT). Red arrow represents IPF cellular targets therapeutic management employed in clinics

Activation of Alveolar Epithelial Type 2 Cells (AEC2 or AT2)

Several lines of research studies have evidenced that alveolar cell dysfunction associated with senescence, mutations, and stress following repetitive lung injury and remodeling process has a pronounced effect in both sporadic and familial IPFs [1]. IPF epithelial cells possess a unique and aberrant phenotype that is well-characterized by different expression markers of type 1 and type 2 AEC cells (AEC1, AEC2) and airway cells. Numbers of canonical signaling pathways are activated in IPF epithelial cells such as phosphoinositide (PI3K)/protein kinase B (AKT), TGF-β1, p53, HIPPO-YAP, and WNT signaling [53]. This activated epithelial cell generates all types of essential mediators prone to enhance the mesenchymal cell migration of different origins like fibrocytes and resident fibroblasts and induces myofibroblast differentiation. The differentiated myofibroblast in turn promotes the extensive accumulation of ECM proteins specifically collagen and fibronectin, thereby demolishing the lung structure and function (Fig. 12.2).

An array of growth factors and other potential mediators that promote profibrotic effects are expressed in ACE2 cell type including TGF-β1, tumor necrosis factor

(TNF), connective tissue growth factor (CTGF), platelet-derived growth factor (PDGF), endothelin-1, CXC chemokine ligand 12 (CXCL12), and osteopontin [54]. Among these factors, AEC-derived TGF-β1 is a robust mediator of fibrotic tissue and found to activate the latent form of TGF-β1 via cell surface expression of integrin αvβ6 [54, 55]. Further activated AEC2 hamper the formation of angiogenesis via pigment epithelium-derived factor expression and stimulate the activation of coagulation signaling that is responsible for the wound healing process which indicates truancy of capillaries in fibroblast foci. ACE2 also influence the turnover of fibrin through plasminogen activator inhibitor-1 (PAI1) and tissue factor (TF) secretion [56, 57]. IPF lung epithelial cells locally synthesize coagulation factor X (FX), a crucial proteinase involved in coagulation process. This FXa incites expression of α-SMA through the TGF-β1 signaling mechanism, promoting fibroblast differentiation into myofibroblast [57].

Some amounts of ACE2 cells are reprogrammed by various molecular targets to undergo epithelial-mesenchymal transition (EMT) [58, 59]. However, the acquisition of mesenchymal characteristics to fibroblast is still unclear, and even reverse transition (MET) has no response [60]. Partial activation of EMT may bestow with the disintegration of junctional complex proteins which influence programmed cell migration and mesenchymal characteristics and alters the epithelial secretome profile in IPF lungs [60].

Modified AEC2-Fibroblast Signaling Mechanisms in IPF

Interaction of AEC2-fibroblast interactions are tightly coordinated cellular processes in pulmonary tissue development and perform repair process of the damaged lung following repetitive injury [61–64]. In the IPF lung, the cross talk of epithelial-mesenchymal (EM) is distinctly disorganized such that ACE2 cells obtain profibrotic phenotype characteristics that aberrantly secrete profibrotic mediators that enhance and activate fibroblast [65]. HIPPO-YAP signaling induces activation of target genes such as CTGF and may partially involve in the reactivation of development signaling in IPF thereby contributes to disease progression [66]. Hedgehog signaling, a key player in epithelial-mesenchymal interactions during lung fibrosis [62, 67, 68], regulates EM unit by sustaining a quiescent state of fibroblast at homeostasis and during resolution following lung injury in vivo [62, 69]. IPF lung expresses decreased levels of prostaglandin E2 (PGE2) and seems to be a crucial factor that suppresses proliferation and activation of fibroblast [70]. Bone morphogenetic factor (BMP) is also found to drive an inhibitory effect on fibroblast [71]. Thus, AEC2 promotes not only lung fibrosis and its progression via the release of profibrotic mediators but the loss of negative reinforcement factors such as BMP and PGE2 can contribute to fibroproliferation.

Thus ACE2 cells act as a driver of IPF disease progression as supported by several lines of evidence that highlights their loss and dysfunction to exhibit a vital role in fibroproliferation which can be considered as a significant cellular target for IPF therapeutic management. Cellular heterogeneity of differential ACE2 responses to different factors such as genetic and epigenetic, environmental insults, age-related changes that contribute to repetitive injury needs greater research attention to overcome IPF pathogenesis.

12.4.2 Molecular Players as Therapeutic Targets and Their Mechanism in IPF

Multiple lines of evidence highlight the significance of an array of molecular targets involved in IPF therapy. Some of the essential molecular players which end up in increased myofibroblast formation are detailed in Table 12.1. A few novel unarticulated molecular targets and their inhibitory mechanism have been discussed for the learner's understanding (Fig. 12.3).

12.4.2.1 TGF-β-Integrin αvβ6

TGF-β is the central mediator in lung tissue fibrosis, and integrin αvβ6 seems to be an essential accelerator of TGF-β signaling in the alveolar tissue development [89]. The exact mechanism of integrin-directed TGF-β signaling is still indecisive; however, the interplay of integrin αvβ6 with RGF motif of LAP (latency-associated protein) of TGF-β complex is essential for tissue fibrosis [90]. The curbing of TGF-β activation induced by integrin αvβ6 could be an essential therapeutic target for IPF management [89]. Studies have shown that ACE 2 exhibits a very high expression of integrin αvβ6 following lung injury and inflammation [89]. Anti-αvβ6 monoclonal antibody has been employed to act against bleomycin-induced lung fibrosis and the antibody effectively induced partial inhibition of TGF-β thereby decreased the collagen level at low doses, whereas at high doses it expressed a significant change in lung fibrotic condition. Further, an increased impediment of integrin αvβ6 upregulated matrix metalloproteinase-12 (MMP-12) expression in macrophages and emphysema mediator in vivo [91]. These studies evidence that the partial inhibition of integrin αvβ6 halting TGF-β activation without augmenting inflammation could be the possible mechanism to be targeted to curb the IPF pathogenesis.

12.4.2.2 Matrix Metalloproteinase-19 (MMP-19)

Increased accumulation of ECM proteins is one of the important characteristics of IPF. Generally, MMPs are involved in the regulation of ECM proteins; however, studies have shown that MMPs contribute to IPF disease progression [92]. Most of the MMPs exert anti-fibrotic effects, whereas MMP-19 associated with ECM degradation during the tissue modeling process [93]. MMP-19 cleaves several components of ECM including fibronectin, laminin-5γ2, aggrecan, collagen type 4, and cartilage oligomeric matrix protein [94] and also hydrolyzes IGF binding protein-3 (IGFBP-3) and tenascin C which are the essential molecules for bleomycin-mediated pulmonary fibrosis in vivo [95].

MMP-19 overexpression has been evidenced in IPF patients and in vivo plays a key role in the wound healing process [96, 97]. Increased expression of MMP-19 promotes prostaglandin endoperoxide synthase 2 (PTGS2) which in turn results in PGE2 and PGD2 synthesis. PEG2 exerts an anti-fibrotic effect through inhibition of

Table 12.1 List of molecular targets and their inhibitory mechanism implicated in IPF

Molecular targets	Cellular process involved	Clinical trials and drugs against targets	Inhibitory mechanisms implicated in IFP	Reference
RhoA/ROCK	Wound healing	Pirfenidone or nintedanib (NCT02688647) – phase 2 trial	LPA-LPA1, S1P-S1P1, thrombin-proteinase-activated receptor, TGF-β-RTK	[72]; [73]; [74]
Lysyl oxidase-like 2 (LOXL2)	Formation and organization of collagen fibers	RAINER (NCT017691960); ATLAS (NCT01759511) --phase 2 trial; Simtuzumab – phase 2 trial (NCT01766817)	TGF-β1 and its signaling, ET-1, CXCL12, LOXL2 reduction in vivo	[75]
PI3K/AKT/ mTOR	Cell metabolism, proliferation, differentiation, survival, and ER-related stress	Omipalisib (NCT01725139) – phase 1B trial	p-S6K, YAP signaling, CTGF, AXL, AJUBA IL-3, PDGF-A and TGF-β1 levels in vivo	[76]; [77]
Connective tissue growth factor (CTGF)	Tissue remodeling (ECM deposition, Myofibroblast activation, cell adhesion and invasion)	Pamrevlumab (NCT01262001) – phase 2A trial and phase 2B trial (NCT01890265)	Reduced CTGF activity, infiltration of mast cells and macrophages in vivo	[78]; [79]
Forkhead box FOXM1	Lung embryonic development and adult tissue homeostasis		Attenuates fibroproliferation and TGF-β1-driven Col1 expression in vivo	[80]
Transforming growth factor (TGF-β)	Profibrotic factor	GSK3008348 – phase I trial (NCT02612051); BG00011/STX-100 (NCT01371305)	Inhibition of TGF-β1-integrin αvβ6 pathway	[81]; [82]
Lysophosphatidic acid receptor-1	Wound healing	BMS-986020, LPA1 receptor antagonist – phase 2 trial (NCT01766817)	Inhibition of LPA/LPA1 receptor signaling	[83]

(continued)

Table 12.1 (continued)

Molecular targets	Cellular process involved	Clinical trials and drugs against targets	Inhibitory mechanisms implicated in IFP	Reference
Interleukin-13	Fibrogenesis, lung remodeling and repair	Lebrikizumab (NCT01872689), Tralokinumab (NCT01629667)	Results awaited	[84]; [85]; [86]
TGF-β/PI-3K/AKT	Pulmonary fibrosis retardation	Tubastatin	HDAC6-independent mechanism	[87]
FoxO3	Profibrotic factor	UCN-01, a staurosporine derivative	Reverting profibrotic signaling	[88]

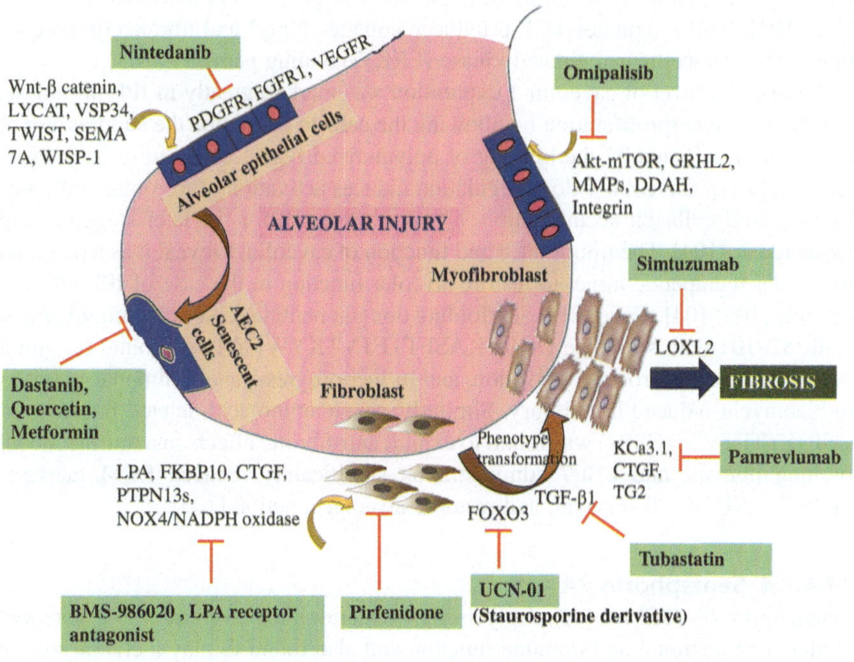

Fig. 12.3 Molecular targets and their cell types in IPF. Schematic represents targeting of cells expressing these molecular targets or molecular pathway by respective pharmacological agents to rectify the pathogenesis of IPF. LPA, lipopolysaccharides; LYCAAT, lysocardiolipin acetyltransferase; WISP-1, WNT-1 inducible signaling pathway protein-1; PTPN13s, protein tyrosine phosphates non-receptor type 13; TG2, transglutaminase 2; VSP34, vesicle transporting protein-34; TWIST, basic helix-loop-helix transcription factor; SEMA7A, semaphorin 7A; MMP, matrix metalloproteinase; DDAH, dimethylarginine dimethylaminohydrolase; GRL2, leukocyte glycoprotein antibody 2; KCa3.1, calcium-dependent potassium channel, LOX2, lysyl oxidase 2

fibroblast proliferation and upregulates the apoptotic factors, whereas PTGS2 revamped cell migration and wound healing [98, 99]. Hence the studies have strongly correlated the effect of MMP-19 as a potential target mechanism favoring IPF clinical management and employed for reversal of alveolar epithelial function subjected to repetitive lung injury. Hence substantial studies are vindicated before it is considered as a possible therapeutic target for clinical use.

12.4.2.3 Caveolin-1

Caveolin-1 is a principal component of caveolae that performs vesicular trafficking and compartmentalization of specific signaling cascades [100]. Caveolin-1 is highly expressed in different cell types such as epithelial cells (AEC2), endothelial, adipocytes, and fibroblast [101]. Activation of caveolin-1 inhibits activation of EGFR and PDGFR in turn inhibits mitogen-activated protein kinase (MAPK) and AKT signaling pathway resulting in decreased cell growth and increased apoptosis [102, 103]. It also regulates TGF-β-induced collagen type 1 and fibronectin production in the extracellular-regulated kinase (ERK) signaling pathway [103].

Suppressed level of caveolin-1 expression was marked greatly in IPF fibroblasts which drive hyperproliferation by allowing the cells to overcome the suppression of proliferation initiated by the binding of polymerized type 1 collagen to its integrin receptor [104]. Caveolin-1 downregulation induces activation of fibroblast followed by increased collagen accumulation, a feature essential for the later stage of lung tissue repair [104]. The importance and function of caveolin-1 reveal it as a potential molecular therapeutic target to restore alveolar function in the case of IPF affected patients [103, 104]. Caveolin-1 scaffolding domain peptide (CSF) containing amino acids 82–101 with sequence DGIWKASFTTFTVTKYWFYR was found to inhibit AEC2 apoptosis, fibroblast activation, and further progression of pulmonary fibrosis in bleomycin-induced lung injury. Similarly, seven amino acid-deleted fragment of CSP (FTTFTVT) (CSP7) was found to exhibit anti-fibrotic effects in a murine model of lung fibrosis, and CSP7 administration significantly reduced ECM markers, increased AEC2 cell survival, and restored alveolar function [105].

12.4.2.4 Semaphorin 7A

Semaphorin 7A (SEMA 7A), a glycosylphosphatidylinositol protein is a potential mediator of neuronal and immune function and also found to play a crucial role in tissue remodeling [106]. SEMA 7A was observed to be strongly involved in inducing TGF-β1-mediated lung fibrosis, increased myofibroblast activation, alveolar remodeling, and cell death. Further SEMA 7A was found to upregulate TGF-β1-mediated PI3K/AKT signaling pathway in lung fibrosis displaying it as a beneficial therapeutic target to inhibit the progression of pulmonary fibrosis [107]. No such studies have been reported so far in clinical trials. However, few studies have evidenced the status of SEMA 7A expression on T_{reg} cells which was significantly increased in the circulation of IPF patients and found to induce TGF-β1-induced lung fibrosis and remodeling in murine model confirming its role in IPF pathogenesis [108].

12.4.2.5 Calcium-Activated Potassium Channel (KCa3.1)

KCa3.1 actively regulates various cellular functions concerning lung fibrosis such as the proliferation of myofibroblast, wound healing process, secretion of collagen, and contraction. Ion channels maintain a negative membrane potential to regulate calcium signaling during cell activation [109]. Thus KCa3.1 channel blockers are involved in cellular events contributing to the progression of IPF [109]. Human lung myofibroblast has pronounced expression of $K_{Ca}3.1$ in IPF patients which is co-localized with an area of α-SMA. TGF-β upregulation in IPF-derived myofibroblast shows a larger ion current and also upregulated mRNA levels of $K_{Ca}3.1$ [109].

$K_{Ca}3.1$ effectively modulates TGF-β1-dependent profibrotic signaling in human lung myofibroblast [110], and IPF cells overexpresses α-SMA, Smad2/3 expression, and stress fiber formation. The Smad signaling is regulated by TGF-β, which is partially controlled by $K_{Ca}3.1$ channels, and its expression can also be inhibited by $K_{Ca}3.1$ blockers. The channel blockers depolarize the cell membrane, upregulating intracellular Ca^{2+} levels, and suppressing extracellular Ca^{2+} level resulting in its nuclear translocation inhibition [110]. Blocker can also decrease α-SMA and stress fiber formation; however, suppression of the activity of $K_{Ca}3.1$blockers causes dedifferentiation of IPF cells to a more quiescent fibroblast phenotype [111]. Thus targeting the profibrotic effects of myofibroblast via ion channel could be requisite for positive and beneficial therapeutic approach to restore fibrotic matrix conditions in IPF patients.

12.4.2.6 Micro RNA-26a (miR-26a) and Micro-RNA let-7d (miR let-7d)

Approximately 10% of microRNAs such as miR-21, miR let-7d, and miR-26a play a vital role in IPF pathogenesis and its progression [112, 113]. There is a significant downregulation of miRNA let-7d that has been reported in IPF scientific studies [114], which are implicated by increased expression of TGF-β resulting in high-mobility group proteins (HMGA2), a protein involved in the modification of DNA structure [115]. Further, both miR-26a and miR let-7d inhibition causes changes in EMT and results in increased alveolar cell thickening and deposition of collagen [112, 115] indicating its profibrotic effects and importance as an effective target for management of IPF demerits.

12.4.2.7 Transglutaminase 2 (TG2)

TG2 resides in different tissues and cell types including fibroblast and greatly promotes lung tissue fibrosis through diverse cellular mechanisms. TG2 catalyzes posttranslational interactions between proteins [116] which can cross-link fibronectin and collagen making it difficult to break them down [117]. TG2 serve as a G protein to enroll its function in cell survival and cell cycle and stimulate fibroblast formation and maturation [118]. TG2 inhibitors have evidenced that its downregulation significantly reduced pulmonary fibrosis in bleomycin-induced lung fibrosis via decreased level of collagen deposition and fibronectin [117]. TG2 levels were found to be overexpressed in IPF patients, and it is partly regulated by TGF-β which in turn upregulated membrane-associated and extracellular TG2 level, not mRNA levels. Further TG2 does not halt the transformation of fibroblast into myofibroblast but resulted in matrix functional defects, cell organization, migration, and contraction. Thus the partial cessation of TG2 expression is a prime therapeutic strategy to reduce the progression of IPF [117].

Anti-fibrotic effects of small electrophilic compounds such as 15-deoxy-$\delta^{12,14}$-prostaglandin J_2 (15d-PGJ_2) and 2-cyano-3,12-dioxolane-1,9-dien28-oic acid (CDDO) has been reported to halt the progression of lung fibrosis through the inhibition of TG2. These compounds effectively inhibit differentiation of myofibroblast, collagen, and fibronectin formation through downregulation of TG2. Also, it was observed that TG2 levels in IPF fibroblast is not as low as healthy fibroblast and it can be regulated by ERK signaling pathway [117]. Thus ERK inhibitors, CDDO, and 15d-PGJ_2 can inhibit the ERK signaling and reduces lung fibrosis in vivo indicating the significance of these inhibitors along with TG2 serve as a potential targets for the cessation of IPF progression.

Other potential targets involving cellular and molecular pathways that exhibit an essential role in IPF pathogenesis are summarized in Table.12.1.

12.5 Current Clinical Implications of IPF

The most widely employed therapeutic strategy to manage IPF is corticosteroids or related compounds possessing immunosuppressant or anti-inflammatory activities. However, these approaches end in vain and failed to show any improvement in lung function of IPF patients.

The existing therapy for IPF using pirfenidone and nintedanib revealed mild to moderate retrieval of lung functioning [119] because of its usage in late-stage patients. Though these two drugs displayed a substantial reduction in IPF development, early-stage disease diagnosis is essential for proper drug evaluation (Fig. 12.3). Pirfenidone exhibits antioxidant, anti-inflammatory, and anti-fibrotic effects [120, 121] which has been employed in about 555 IPF patients in four phase 3 randomized controlled clinical trials. Pirfenidone doses were found to be safe and well-tolerated by patients, and it significantly decreases the progression of IPF after 52 weeks of its administration which in turn contributed to increased survival and decreased death risk by 48% at 1 year [122].

Nintedanib functions by targeting the three tyrosine kinases receptor such as PDGF, vascular endothelial growth factor (VEGF), and fibroblast growth factor (FGF) [123, 124]. Nintedanib acts by inhibiting the proliferation of fibroblast, fibroblast transformation into myofibroblast, and lung inflammation in vivo [124]. Nintedanib specifically targets the growth factors and anti-fibrotic signaling which effectively slows the functional decline and further IPF progression. The positive results of pirfenidone and nintedanib are not favoring IPF progression; however, reversal of IPF effects requires a drug combination approach to target multifaceted signaling pathways responsible for IPF development [125]. To date, no single drug has been ratified which establishes a potential ability to reverse lung tissue scarring in IPF. Hence it is crucial to identify, design, and develop novel therapeutics toward various cellular and molecular targets of IPF.

Contemporary medical therapies are not fully operative at limiting impermanence in patients with IPF, and novel therapies are instantly needed. The prime clinical management of IPF is multifaceted and not only necessitates anti-fibrotic treatment but also other relevant vaccinations, assessment of nutritional status, oxygen supplementation, psychological support, and patient's understanding of disease progression. Pulmonary rehabilitation, symptom remedy, palliative care, and comorbidity management indicates future research domains of clinical intervention [126].

12.6 Impediments in IPF Basic Research

Several in vitro *and* in vivo *model systems are* required for the conception of pathological mechanisms that occurs in IPF; nevertheless, these systems have certain constraints. For instance, primary culturing of human alveolar epithelial cells seems to be a great challenge for in vitro testing. Further, in vivo model such as bleomycin-induced lung fibrosis fails to culture human primary AECs, which is another greater challenge for in vitro testing. Furthermore, different animal models including bleomycin-induced pulmonary fibrosis fail to mimic the clinical features of IPF as experienced in patients. Despite these pitfalls, the thrust of basic science research on IPF disease understanding continues and contributes to unknot the mystery by the development of newer in vitro and in vivo model systems such as precision ex vivo lung models, genetically engineered mice models, induced pluripotent stem cells generated AECs, and organ-on-chip technology to replace lung transplantation [127–129]. Thus these novel model systems give positive vibes for a complete understanding of progressive pulmonary fibrosis and aid in developing potential therapies in near future to combat IPF.

12.7 Conclusion

Idiopathic pulmonary fibrosis is considered as one of the common and highly lethal lung diseases initially predicted as a chronic inflammatory disease. Thus anti-inflammatory and immunosuppressant drugs were prescribed to treat IPF. Recent advancement in IPF pathogenesis has identified TGF-β signaling involvement in disease progression which is prone to induce fibrotic responses in lung tissue. Pirfenidone and nintedanib are two currently employed FDA-approved medications to reverse lung functions via targeting growth factor signaling and subsiding disease symptoms; however, they fail to improve the patient's survival. This chapter highlights a few crucial cellular and molecular signaling targets that are strongly implicated in the development and progression of IPF. Further, understanding the interaction of these molecular targets with several cellular events such as injury to epithelial and endothelial cells, hypercoagulation condition, activation of fibroblast, differentiation into myofibroblast, EMT, recruitment of fibrocyte, deposition of

ECM, angiogenesis, and irregular repair mechanisms during IPF progression will delineate the complex spatial and temporal relationship between these molecular pathways. Individual therapeutic strategies are no more effective against IPF treatment; rather combinational approach of drugs is essential to restore the lung function or reverse the fibrotic effects. Emerging advancement in gene-editing technology such as CRISPR/CAS9-based system could edit the genes responsible for IPF initiation and progression, and also cell-based technology might be a beneficial approach to replace or repair the damaged lung cell architecture. Hence identification and diagnosis of marker genes involved in the destruction of lung cells may be used as a biomarker for personalized IPF therapy. Thus this devastating and progressive disease demands supercilious treatment over current existing medications, and the input of substantial effort in tracking novel therapeutics may provide a new avenue for the betterment of IPF patients in near future.

Acknowledgments The work was supported in part by TRR fund, SASTRA Deemed-to-be University, Thanjavur, Tamilnadu, India.

References

1. Selman M, King TE, Pardo A (2001) Idiopathic pulmonary fibrosis: prevailing and evolving hypotheses about its pathogenesis and implications for therapy. Ann Intern Med 134:136–151
2. Wolters PJ, Blackwell TS, Eickelberg O, Loyd JE, Kaminski N et al (2018) Time for a change: Is idiopathic pulmonary fibrosis still idiopathic and only fibrotic? Lancet Respir Med 6:154–160
3. Jenskin G (2020) Demystifying pulmonary fibrosis. Am J Physiol Lung Cell Mol Physiol 319: L554–L559
4. Raghu G, Collard HR, Egan JJ, Martinez FJ, Behr J et al (2011) An official ATS/ERS/JRS/ ALAT statement: idiopathic pulmonary fibrosis: evidence-based guidelines for diagnosis and management. Am J Respir Crit Care Med 183:788–824
5. Iwata T, Yoshida S, Nagato K, Nakajima T, Suzuki H et al (2015) Experience with perioperative pirfenidone for lung cancer surgery in patients with idiopathic pulmonary fibrosis. Surg Today 45:1263–1270
6. Iwata T, Yoshida S, Fujiwara T, Wada H, Nakajima T (2016) Effect of perioperative Pirfenidone treatment in lung cancer patients with idiopathic pulmonary fibrosis. Ann Thorac Surg 102:1905–1910
7. Karampitsakos T, Tzilas V, Tringidou R, Steiropoulos P, Aidinis V et al (2017) Lung cancer in patients with idiopathic pulmonary fibrosis. Pulm Pharmacol Ther 45:1–10
8. Goto T (2018) Measuring surgery outcomes of lung Cancer patients with concomitant pulmonary fibrosis: a review of the literature. Cancers 10:223
9. Daniels CE, Jett JR (2005) Does interstitial lung disease predispose to lung cancer? Curr Opin Pulm Med 11:431–437
10. Hendriks LE, Drent M, van Haren EH, Verschakelen JA, Verleden GM (2012) Lung cancer in idiopathic pulmonary fibrosis patients diagnosed during or after lung transplantation. Respir Med Case Rep 5:37–39
11. King TE Jr, Schwarz MI, Brown K, Tooze JA, Colby TV et al (2001) Idiopathic pulmonary fibrosis: relationship between histopathologic features and mortality. Am J Respir Crit Care Med 164:1025–1032
12. Cha SI, Chang CS, Kim EK, Lee JW, Matthay MA et al (2012) Lung mast cell density defines a subpopulation of patients with idiopathic pulmonary fibrosis. Histopathology 61:98–106

13. Noth I, Zhang Y, Ma SF, Flores C, Barber M et al (2013) Genetic variants associated with idiopathic pulmonary fibrosis susceptibility and mortality: a genome-wide association study. Lancet Resp Med 1:309–317
14. Prasse A, Probst C, Bargagli E, Zissel G, Toews GB et al (2009) Serum CC-chemokine ligand 18 concentration predicts outcome in idiopathic pulmonary fibrosis. Am J Respir Crit Care Med 179:717–723
15. Rosas IO, Richards TJ, Konishi K, Zhang Y, Gibson K et al (2008) MMP1 and MMP7 as potential peripheral blood biomarkers in idiopathic pulmonary fibrosis. PLoS Med 5:e93
16. Wolters PJ, Collard HR, Jones KD (2014) Pathogenesis of idiopathic pulmonary fibrosis. Annu Rev Pathol 9:157–179
17. Ley B, Collard HR (2013) Epidemiology of idiopathic pulmonary fibrosis. Clin Epidemiol 5:483–492
18. Hoffman TW, van Moorsel CHM, Borie R, Crestani B (2018) Pulmonary phenotypes associated with genetic variation in telomere-related genes. Curr Opin Pulm Med 24:269–280
19. Cunningham PS, Meijer P, Nazgiewicz A, Anderson SG, Borthwick LA et al (2020) The circadian clock protein REVERBα inhibits pulmonary fibrosis development. Proc Natl Acad Sci U S A 117:1139–1147
20. Takezaki A, Tsukumo S, Setoguchi Y, Ledford JG, Goto H et al (2019) A homozygous SFTPA1 mutation drives necroptosis of type II alveolar epithelial cells in patients with idiopathic pulmonary fibrosis. J Exp Med 216:2724–2735
21. Meltzer EB, Noble P (2008) Idiopathic pulmonary fibrosis. Orphanet J Rare Dis 3:3–8
22. Cuddapah S, Barski A, Zhao K (2010) Epigenomics of T cell activation, differentiation, and memory. Curr Opin Immunol 22:341–347
23. Arrowsmith CH, Bountra C, Fish PV, Lee K, Schapira M (2012) Epigenetic protein families: a new frontier for drug discovery. Nat Rev Drug Discov 11:384–400
24. Rivera CM, Ren B (2013) Mapping human epigenomes. Cell 155:39–55
25. Li H, Zhao X, Shan H, Liang H (2016) MicroRNAs in idiopathic pulmonary fibrosis: involvement in pathogenesis and potential use in diagnosis and therapeutics. Acta Pharm Sin B 6:531–539
26. Kidd C, Hayden BY (2015) The psychology and neuroscience of curiosity Celeste. Neuron 88:449–460
27. Helling BA, Yang IV (2015) Epigenetics in lung fibrosis: from pathobiology to treatment perspective. Curr Opin Pulm Med 21:454–462
28. Keane MP, Strieter RM (2002) The importance of balanced pro-inflammatory and anti-inflammatory mechanisms in diffuse lung disease. Respir Res 3:5
29. Telugu AN, Chen H, Weng T, Bhaskaran M, Jin N et al (2006) Expression profile of IGF system during lung injury and recovery in rats exposed to hyperoxia: a possible role of IGF-1 in alveolar epithelial cell proliferation and differentiation. Cell Biochem 97:984–998
30. Camelo A, Dunmore R, Sleeman MA, Clarke DL (2013) The epithelium in idiopathic pulmonary fibrosis: breaking the barrier. Front Pharmacol 4:173
31. Souto JT, Aliberti JC, Campanelli AP, Livonesi MC, Maffei CML et al (2003) Chemokine production and leukocyte recruitment to the lungs of *Paracoccidioides brasiliensis*-infected mice is modulated by interferon-γ. Am J Pathol 163:583–590
32. Linthout SV, Miteva K, Tschöpe C (2014) Crosstalk between fibroblasts and inflammatory cells. Cardiovasc Res 102:258–269
33. Kolb M, Margetts PJ, Anthony DC, Pitossi F, Gauldie J (2001) Transient expression of IL-1beta induces acute lung injury and chronic repair leading to pulmonary fibrosis. J Clin Invest 107:1529–1536
34. Genersch E, Hayess K, Neuenfeld Y, Haller H (2000) Sustained ERK phosphorylation is necessary but not sufficient for MMP-9 regulation in endothelial cells: involvement of Ras-dependent and -independent pathways. J Cell Sci 113:4319–4330
35. Sime PJ, Marr RA, Gauldie D, Xing Z, Hewlett BR (1998) Transfer of tumor necrosis factor-alpha to rat lung induces severe pulmonary inflammation and patchy interstitial fibrogenesis

with induction of transforming growth factorbeta1 and myofibroblasts. Am J Pathol 153:825–832
36. Hyldgaard C, Hilberg O, Bendstrup E (2014) How does comorbidity influence survival in idiopathic pulmonary fibrosis? Respir Med 108:647–653
37. Kreuter M, Ehlers-Tenenbaum S, Palmowski K, Bruhwyler J, Oltmanns U et al (2016) Impact of comorbidities on mortality in patients with idiopathic pulmonary fibrosis. PLoS One 11:151425
38. Tsakiri KD, Cronkhite JT, Kuan PJ, Xing C, Raghu G et al (2007) Adult-onset pulmonary fibrosis caused by mutations in telomerase. Proc Natl Acad Sci U S A 104:7552–7557
39. Vancheri C, Failla M, Crimi N, Raghu G (2010) Idiopathic pulmonary fibrosis: a disease with similarities and links to cancer biology. Eur Respir J 35:496–504
40. Raghu G (2003) The role of gastroesophageal reflux in idiopathic pulmonary fibrosis. Am J Med 115:60–64
41. Antonio GE, Wong KT, Hui DS, Wu A, Lee N et al (2003) Thin-section CT in patients with severe acute respiratory syndrome following hospital discharge: preliminary experience. Radiology 228:810–815
42. Hui DS, Joynt GM, Wong KT, Gomersall CD, Li TS et al (2005) Impact of severe acute respiratory syndrome (SARS) on pulmonary function, functional capacity and quality of life in a cohort of survivors. Thorax 60:401–409
43. Barkauskas CE, Noble PW (2014) Cellular mechanisms of tissue fibrosis. New insights into the cellular mechanisms of pulmonary fibrosis. Am J Physiol Cell Physiol 306:C987–C996
44. Wynn TA (2011) Integrating mechanisms of pulmonary fibrosis. J Exp Med 208:1339–1350
45. Kropski JA, Blackwell TS (2019) Progress in understanding and treating idiopathic pulmonary fibrosis. Annu Rev Med 70:211–224
46. Yanagihara T, Sato S, Upagupta C, Kolb M (2019) What have we learned from basic science studies on idiopathic pulmonary fibrosis? Eur Respir Rev 28:190029
47. Selman M, Pardo A (2014) Revealing the pathogenic and aging-related mechanisms of the enigmatic idiopathic pulmonary fibrosis. An integral model. Am J Respir Crit Care Med 189:1161–1172
48. Lehmann M, Korfei M, Mutze K, Klee S, Skronska-Wasek W et al (2017) Senolytic drugs target alveolar epithelial cell function and attenuate experimental lung fibrosis ex vivo. Eur Respir J 50:1602367
49. Schafer MJ, White TA, Iijima K, Haak AJ, Ligresti G et al (2017) Cellular senescence mediates fibrotic pulmonary disease. Nat Commun 8:14532
50. Zhu Y, Tchkonia T, Pirtskhalava T, Gower AC, Ding H et al (2015) The Achilles' heel of senescent cells: from transcriptome to senolytic drugs. Aging Cell 14:644–658
51. Rangarajan S, Bone NB, Zmijewska AA, Jiang S, Won-park D et al (2018) Metformin reverses established lung fibrosis in a bleomycin model. Nat Med 24:1121–1127
52. Sato N, Takasaka N, Yoshida M, Tsubouchi K, Minagawa S et al (2016) Metformin attenuates lung fibrosis development via NOX4 suppression. Respir Res 17:107
53. Xu Y, Mizuno T, Sridharan A, Du Y, Guo M et al (2016) Single-cell RNA sequencing identifies diverse roles of epithelial cells in idiopathic pulmonary fibrosis. JCI Insight 1:e90558
54. Horan GS, Wood S, Ona V, Li DJ, Lukashev ME et al (2008) Partial inhibition of integrin alpha (v) beta6 prevents pulmonary fibrosis without exacerbating inflammation. Am J Respir Crit Care Med 177:56–65
55. Munger JS, Huang X, Kawakatsu H, Griffiths MJ, Dalton SL et al (1999) The integrin alpha v beta 6 binds and activates latent TGF beta 1: a mechanism for regulating pulmonary inflammation and fibrosis. Cell 96:319–328
56. Cosgrove GP, Brown KK, Schiemann WP, Serls AE, Parr JE et al (2004) Pigment epithelium-derived factor in idiopathic pulmonary fibrosis: a role in aberrant angiogenesis. Am J Respir Crit Care Med 170:242–251

57. Kotani I, Sato A, Hayakawa H, Urano T, Takado Y et al (1995) Increased procoagulant and antifibrinolytic activities in the lungs with idiopathic pulmonary fibrosis. Thromb Res 77:493–504
58. Kim KK, Kugler MC, Wolters PJ, Robillard L, Galvez MG et al (2006) Alveolar epithelial cell mesenchymal transition develops in vivo during pulmonary fibrosis and is regulated by the extracellular matrix. Proc Natl Acad Sci U S A 103:13180–13185
59. Willis BC, Liebler JM, Luby-Phelps K, Nicholson AG, Crandall ED et al (2005) Induction of epithelial-mesenchymal transition in alveolar epithelial cells by transforming growth factor beta1: potential role in idiopathic pulmonary fibrosis. Am J Pathol 166:1321–1332
60. Nieto MA (2013) Epithelial plasticity: a common theme in embryonic and cancer cells. Science 342:1234850
61. Nabhan AN, Brownfield DG, Harbury PB, Krasnow MA, Desai TJ (2018) Single-cell Wnt signaling niches maintain stemness of alveolar type 2 cells. Science 359:1118–1123
62. Peng T, Frank DB, Kadzik RS, Morley MP, Rathi KS et al (2015) Hedgehog actively maintains adult lung quiescence and regulates repair and regeneration. Nature 526:578–582
63. Tsao PN, Matsuoka C, Wei SC, Sato A, Sato S, Hasegawa K et al (2016) Epithelial notch signaling regulates lung alveolar morphogenesis and airway epithelial integrity. Proc Natl Acad Sci U S A 113:8242–8247
64. Xu K, Nieuwenhuis E, Cohen BL, Wang W, Canty AG et al (2010) Lunatic fringe-mediated notch signaling is required for lung alveogenesis. Am J Physiol Lung Cell Mol Physiol 298: L45–L56
65. Selman M, Pardo A (2019) The leading role of epithelial cells in the pathogenesis of idiopathic pulmonary fibrosis. Cell Signal 66:109482
66. Bueno M, Zank D, Buendia-Roldan I, Fiedler K, Mays BG et al (2019) PINK1 attenuates mtDNA release in alveolar epithelial cells and TLR9 mediated profibrotic responses. PLoS One 14:e0218003
67. Raredon MSB, Adams TS, Suhail Y, Schupp JC, Poli S et al (2019) Single-cell connectomic analysis of adult mammalian lungs. Sci Adv 5:3851
68. Wang C, Cassandras M, Peng T (2019) The role of hedgehog signaling in adult lung regeneration and maintenance. J Dev Biol 7:14
69. Hu B, Liu J, Wu Z, Liu T, Ullenbruch MR et al (2015) Reemergence of hedgehog mediates epithelial-mesenchymal crosstalk in pulmonary fibrosis. Am J Respir Cell Mol Biol 52:418–428
70. Moore BB, Peters-Golden M, Christensen PJ, Lama V, Kuziel WA et al (2003) Alveolar epithelial cell inhibition of fibroblast proliferation is regulated by MCP-1/CCR2 and mediated by PGE2. Am J Physiol Lung Cell Mol Physiol 284:L342–L349
71. Tan Q, Ma XY, Liu W, tal MJA e (2019) Nascent lung organoids reveal epithelium- and Bone morphogenetic protein-mediated suppression of fibroblast activation. Am J Respir Cell Mol Biol 61:607–619
72. Jiang C, Huang H, Liu J, Wang Y, Lu Z et al (2012) Fasudil, a rho-kinase inhibitor, attenuates bleomycin-induced pulmonary fibrosis in mice. Int J Mol Sci 13:8293–8307
73. Bei Y, Hua-Huy T, Duong-Quy S, Nguyen V, Chen W et al (2013) Long-term treatment with fasudil improves bleomycin-induced pulmonary fibrosis and pulmonary hypertension via inhibition of Smad2/3 phosphorylation. Pulm Pharmacol Ther 26:635–643
74. Knipe RS, Probst CK, Lagares D, Franklin A, Spinney J et al (2018) The rho kinase isoforms ROCK1 and ROCK2 each contribute to the development of experimental pulmonary fibrosis. Am J Respir Cell Mol Biol 58:471–481
75. Barry-Hamilton V, Spangler R, Marshall D, McCauley S, Rodiguez HM et al (2010) Allosteric inhibition of lysyl oxidase-like-2 impedes the development of a pathologic microenvironment. Nat Med 16:1009–1017
76. Gokey JJ, Sridharan A, Xu Y, Green J, Carraro G et al (2018) Active epithelial Hippo signaling in idiopathic pulmonary fibrosis. JCI Insight 3:e98738

77. Axell-House VYDB, Zhang Z, Burdick MD, Strieter RM, Mehrad B (2018) Fibrocytes in pulmonary fibrosis: double-blind placebo-controlled crossover pilot study of Sirolimus in idiopathic pulmonary fibrosis. Am J Respir Crit Care Med 197:A4353
78. Peter E, Huber SB, Peter P, Alexandra T, Ute W et al (2010) Reversal of established fibrosis by treatment with the anti-CTGF monoclonal antibody FG-3019 in a murine model of radiation-induced pulmonary fibrosis. Am J Respir Crit Care Med 181:A1054
79. Raghu G, Scholand MB, de Andrade J, Lancaster L, Mageto Y et al (2016) FG-3019 anti-connective tissue growth factor monoclonal antibody: results of an open-label clinical trial in idiopathic pulmonary fibrosis. Eur Respir J 47:1481–1491
80. Melboucy-Belkhir S, Pradere P, Tadbiri S, Habib S, Baccrat A et al (2014) Forkhead box F1 represses cell growth and inhibits COL1 and ARPC2 expression in lung fibroblasts in vitro. Am J Physiol Lung Cell Mol Physiol 307:L838–L847
81. Shi M, Shu J, Wang R, Chen X, Mi L et al (2011) Latent TGF-beta structure and activation. Nature 474:343–349
82. Henderson NC, Arnold TD, Katamura Y, Giacomini MM, Rodriguez JD et al (2013) Targeting of alpha integrin identifies a core molecular pathway that regulates fibrosis in several organs. Nat Med 19:1617–1624
83. Tager AM, LaCamera P, Shea BS, Campanella GS, Selman M et al (2008) The lysophosphatidic acid receptor LPA1 links pulmonary fibrosis to lung injury by mediating fibroblast recruitment and vascular leak. Nat Med 14:45–54
84. Zhu Z, Homer RJ, Wang Z, Chen Q, Geba GP (1999) Pulmonary expression of interleukin-13 causes inflammation, mucus hypersecretion, subepithelial fibrosis, physiologic abnormalities, and eotaxin production. J Clin Invest 103:779–788
85. Belperio JA, Dy M, Burdick MD, Xue YY, Li K et al (2002) Interaction of IL-13 and C10 in the pathogenesis of bleomycin-induced pulmonary fibrosis. Am J Respir Cell Mol Biol 27:419–427
86. Murray LA, Zhang H, Oak SR, Coelho AL, Herath A et al (2014) Targeting interleukin-13 with tralokinumab attenuates lung fibrosis and epithelial damage in a humanized SCID idiopathic pulmonary fibrosis model. Am J Respir Cell Mol Biol 50:985–994
87. Saito S, Zhuang Y, Shan B, Danchuk S et al (2017) Tubastatin ameliorates pulmonary fibrosis by targeting the TGFβ-PI3K-Akt pathway. PLoS One 12:e0186615
88. Sellares J, Veraldi KL, Thiel KJ et al (2019) Intracellular heat shock protein 70 deficiency in pulmonary fibrosis. Am J Respir Cell Mol Biol 60:629–636
89. Agarwal SK (2014) Integrins and cadherins as therapeutic targets in fibrosis. Front Pharmacol 5:1–7
90. Annes JP, Chen Y, Munger JS, Rifkin DB (2004) Integrin alphaVbeta6-mediated activation of latent TGF-beta requires the latent TGF-beta binding protein-1. J Cell Biol 165:723–734
91. Morris DG, Huang X, Kaminski N, Wang Y, Shapiro SD et al (2003) Loss of integrin alpha (v) beta6-mediated TGF-beta activation causes Mmp12-dependent emphysema. Nature 422:169–173
92. Craig VJ, Zhang L, Hagood JS, Owen CA (2015) Matrix metalloproteinases as therapeutic targets for idiopathic pulmonary fibrosis. Am J Respir Cell Mol Biol 53:585–600
93. Kolb C, Mauch S, Peter HH, Krawinkel U, Sedlacek R (1997) The matrix metalloproteinase RASI-1 is expressed in synovial blood vessels of a rheumatoid arthritis patient. Immunol Lett 57:83–88
94. Sadowski T, Dietrich S, Koschinsky F, Ludwig A, Proksch E et al (2005) Matrix metalloproteinase 19 processes the laminin 5 gamma 2 chain and induces epithelial cell migration. Cell Mol Life Sci 62:870–880
95. Carey WA, Taylor GD, Dean WB, Bristow JD (2010) Tenascin-C deficiency attenuates TGF-s mediated fibrosis following murine lung injury. Am J Physiol Lung Cell Mol Physiol 299: L785–L793

96. Jara P, Calyeca J, Romero Y, Placido L, Yu G et al (2015) Matrix metalloproteinase (MMP)-19-deficient fibroblasts display a profibrotic phenotype. Am J Physiol Lung Cell Mol Physiol 308:L511–L522
97. Yu G, Kovkarova-Naumovski E, Jara P, Parwani A, Kass D et al (2012) Matrix metalloproteinase-19 is a key regulator of lung fibrosis in mice and humans. Am J Respir Crit Care Med 186:752–762
98. Smith WL, Langenbach R (2001) Why there are two cyclooxygenase isozymes. J Clin Invest 107:1491–1495
99. White ES, Atrasz RG, Dickie EG, Aronoff DM, Stambolic V, Mak TW (2005) Prostaglandin E (2) inhibits fibroblast migration by E-prostanoid 2 receptor-mediated increase in PTEN activity. Am J Respir Cell Mol Biol 32:135–141
100. Anderson RG (1998) The caveolae membrane system. Annu Rev Biochem 67:199–225
101. Okamoto T, Schlegel A, Scherer PE, Lisanti MP (1998) Caveolins, a family of scaffolding proteins for organizing "preassembled signaling complexes" at the plasma membrane. J Biol Chem 273:5419–5422
102. Labrecque L, Royal I, Surprenant DS, Patterson C, Gingras D, Beliveau R (2003) Regulation of vascular endothelial growth factor receptor-2 activity by caveolin-1 and plasma membrane cholesterol. Mol Biol Cell 14:334–347
103. Wang XM, Zhang Y, Kim HP, Zhou Z, Feghali-Bostwick CA et al (2006) Caveolin-1: a critical regulator of lung fibrosis in idiopathic pulmonary fibrosis. J Exp Med 203:2895–2906
104. Shivshankar P, Brampton C, Miyasato S, Kasper M, Thannickal VJ et al (2012) Caveolin-1 deficiency protects from pulmonary fibrosis by modulating epithelial cell senescence in mice. Am J Respir Cell Mol Biol 47:28–36
105. Marudamuthu AS, Bhandary YP, Fan L, Radhakrishnan V, MacKenzie B et al (2019) Caveolin-1–derived peptide limits development of pulmonary fibrosis. Caveolin-1–derived peptide limits development of pulmonary fibrosis. Sci Transl Med 11:2848
106. Pasterkamp RJ, Peschon JJ, Spriggs MK, Kolodkin AL (2003) Semaphorin 7A promotes axonoutgrowth through integrins and MAPKs. Nature 424:398–405
107. Kang HR, Lee CG, Homer RJ, Elias JA (2007) Semaphorin 7A plays a critical role in TGF-beta1-induced pulmonary fibrosis. J Exp Med 204:1083–1093
108. Reilkoff RA, Peng H, Murray LA, Peng X, Russell T et al (2013) Semaphorin 7a+ regulatory T cells are associated with progressive idiopathic pulmonary fibrosis and are implicated in transforming growth factor-beta1-induced pulmonary fibrosis. Am J Respir Crit Care Med 187:180–188
109. Roach KM, Duffy SM, Coward W, Feghali-Bostwick C, Wulff H (2013) The K^+channel KCa3.1 as a novel target for idiopathic pulmonary fibrosis. PLoS One 8:e85244
110. Roach KM, Wulff H, Feghali-Bostwick C, Amrani Y, Bradding P (2014) Increased constitutive alphaSMA and Smad2/3 expression in idiopathic pulmonary fibrosis myofibroblasts is KCa3.1- dependent. Respir Res 15:155
111. Roach KM, Feghali-Bostwick C, Wulff H, Amrani Y, Bradding P (2015) Human lung myofibroblast TGFbeta1-dependent Smad2/3 signalling is ca(2+)-dependent and regulated by KCa3.1 K(+) channels. Fibrogenesis Tissue Repair 8:5
112. Liang H, Xu C, Pan Z, Zhang Y, Xu Z et al (2014) The antifibrotic effects and mechanisms of microRNA-26a action in idiopathic pulmonary fibrosis. Mol Ther J Am Soc Gene Ther 22:1122–1133
113. Pandit KV, Milosevic J (2015) MicroRNA regulatory networks in idiopathic pulmonary fibrosis. Biochem Cell Biol 93:129–137
114. Pandit KV, Milosevic J, Kaminski N (2011) MicroRNAs in idiopathic pulmonary fibrosis. Transl Res J Clin Med 157:191–199
115. Pandit KV, Corcoran D, Yousef H, Yarlagadda M, Tzouvelekis A et al (2010) Inhibition and role of let-7d in idiopathic pulmonary fibrosis. Am J Respir Crit Care Med 182:220–229
116. Lorand L, Graham RM (2003) Transglutaminases: crosslinking enzymes with pleiotropic functions. Nat Rev Mol Cell Biol 4:140–156

117. Olsen KC, Sapinoro RE, Kottmann RM, Kulkarni AA, Iismaa SE et al (2011) Transglutaminase 2 and its role in pulmonary fibrosis. Am J Respir Crit Care Med 184:699–707
118. Akimov SS, Belkin AM (2001) Cell-surface transglutaminase promotes fibronectin assembly via interaction with the gelatin-binding domain of fibronectin: a role in TGF beta-dependent matrix deposition. J Cell Sci 114:2989–3000
119. Spagnolo P, Wells AU, Collard HR (2015) Pharmacological treatment of idiopathic pulmonary fibrosis: an update. Drug Discov Today 20:514–524
120. du Bois RM (2010) Strategies for treating idiopathic pulmonary fibrosis. Nat Rev Drug Discov 9:129–140
121. Taniguchi H, Ebina M, Kondoh Y, Ogura T, Azuma A et al (2010) Pirfenidone in idiopathic pulmonary fibrosis. Eur Respir J 35:821–829
122. King TE Jr, Bradford WZ, Castro-Bernardini S, Fagan EA, Glaspole I et al (2014) A phase 3 trial of pirfenidone in patients with idiopathic pulmonary fibrosis. N Engl J Med 370:2083–2092
123. Hilberg F, Roth GJ, Krssak M, Kautschitsch S, Sommergruber W et al (2008) BIBF 1120: triple angiokinase inhibitor with sustained receptor blockade and good antitumor efficacy. Cancer Res 68:4774–4782
124. Richeldi L, du Bois RM, Raghu G, Azuma A, Brown KK et al (2014) Efficacy and safety of nintedanib in idiopathic pulmonary fibrosis. N Engl J Med 370:2071–2082
125. Spagnolo P, Rossi G, Cavazza A (2014) Pathogenesis of idiopathic pulmonary fibrosis and its clinical implications. Expert Rev Clin Immunol 10:1005–1017
126. Caminati A, Lonati C, Cassandro R, Elia D, Pelosi G, Torre O, Zompatori M, Uslenghi E, Harari S (2019) Comorbidities in idiopathic pulmonary fibrosis: an underestimated issue. Eur Respir Rev 28:190044
127. Alsafadi HN, Staab-Weijnitz CA, Lehmann M, Lindner M, Peschel B et al (2017) An *ex vivo* model to induce early fibrosis-like changes in human precision-cut lung slices. Am J Physiol Lung Cell Mol Physiol 312:L896–L902
128. Konar D, Devarasetty M, Yildiz DV, Atala A, Murphy SV (2016) Lung-on-a-chip technologies for disease modeling and drug development. Biomed Eng Comput Biol 7:17–27
129. Tamò L, Hibaoui Y, Kallol S, Alves MP, Albrecht C et al (2018) Generation of an alveolar epithelial type II cell line from induced pluripotent stem cells. Am J Physiol Lung Cell Mol Physiol 315:L921–L932

A Refined Approach to Target the Molecular and Cellular Mechanisms in Pulmonary Fibrosis

13

Sabita Singh, Joytri Dutta, Archita Ray, Ashish Jaiswal, and Ulaganathan Mabalirajan

Abstract

The COVID-19 scenario has heated up the entire scientific community including those dealing with pulmonary fibrosis. The fibrotic sequelae of SARS-CoV-2 have given value to the anti-fibrotic therapies that are being evaluated to prevent the severity of the pandemic. However, our understanding of the precise mechanism that drives fibrosis and knowledge about effective management of pulmonary fibrosis are still in the state of darkness. A landscape of pulmonary fibrosis (PF) and dismal prognosis continues to mar the progression of society over the past decades. It was in 2014 that two "umbrella" therapies were approved by the FDA for IPF management, nintedanib and pirfenidone, post which there are no significant additions in this field. An interplay between genetic and environmental factors leads to cause microinjuries to the alveolar epithelium. The maladaptive repair process over time contributes to the fibroblast proliferation and epithelium-mesenchymal crosstalk leading to the pathogenesis of PF. Although there are several hurdles to combat this deadly disease, our next step is to develop efficacious treatment regimens that can ameliorate survivability and functional quality of life. This chapter presents recent updates in PF pathogenesis and possible novel therapeutic strategies.

The authors declare no competing financial interests.

Sabita Singh, Joytri Dutta and Archita Ray contributed equally with all other contributors.

S. Singh · J. Dutta · A. Ray · A. Jaiswal · U. Mabalirajan (✉)
Molecular Pathobiology of Respiratory Diseases, Cell Biology and Physiology Division, CSIR-Indian Institute of Chemical Biology, Kolkata, India

Academy of Scientific and Innovative Research (AcSIR), Ghaziabad, India

K. Dua et al. (eds.), *Targeting Cellular Signalling Pathways in Lung Diseases*,
https://doi.org/10.1007/978-981-33-6827-9_13

Keywords

Pulmonary fibrosis · Alveolar epithelial injury · Fibroblast · Epithelium-mesenchymal crosstalk

13.1 Introduction

13.1.1 Pulmonary Fibrosis: A Global Challenge

Pulmonary fibrosis (PF) is a chronic restrictive lung disease that is highly heterogeneous [1, 2] and refractory to any sort of treatment. The literal meaning of pulmonary fibrosis is lung scarring [2]. It is pathologically characterized by the progressive and irreversible deposition of excessive extracellular matrix (ECM) proteins such as elastin, fibronectin, hyaluronan, and collagen in lung interstitial regions [1] and remodelling of the lung architecture leading to thickening of the alveolar and peribronchial walls [3] and additionally characterized by noticeable clinical, physiological, and radiographic findings. Lung fibrosis develops due to various factors like exposure to various noxious chemicals, smoke, and also certain viral infections [1, 2]. Cancer patients who are undergoing radiotherapy and taking chemotherapy can also develop lung fibrosis. In addition, various toxins present in the air could cause microinjuries in gradual manner to the alveolar epithelium to initiate fibrogenesis in lungs [1].

September is the Global Pulmonary Fibrosis Awareness Month, and the awareness to combat year-on-year increase in mortality rate due to pulmonary fibrosis is the need of the hour. A much-needed awareness among patients with pulmonary fibrosis can avert its progression. However, part of the trouble comes while raising awareness of pulmonary fibrosis because of its complexity of the disease. Indeed, the term pulmonary fibrosis itself is an umbrella term as it comprises numerous similar lung diseases, and these diseases are together also called as interstitial lung diseases [4]. Due to the similarity of various diseases, it is very difficult to calculate the incidence of lung fibrosis.

13.1.2 Historical Perspective of Pulmonary Fibrosis

It is well known that dust-rich occupation is strongly related to incidence of numerous lung diseases. Even Hippocrates observed the relation between respiratory problems and metal mining in his times [5, 6]. Later in the 1500s, the relation between the exposure to the dusty environment and the development of severe lung diseases has been reported [5]. The current illustration of pulmonary fibrosis crossed the mind of Hamman and Rich in the early twentieth century [4, 6] in which they found interstitial fibrosis in the lungs of few patients. In contrast to their observations of fast progression in the development of lung fibrosis, many patients have gradual progression. Due to the complexity and multifactorial nature of lung fibrosis, various clinicians and scientists have coined different terminology for the lung fibrosis [5].

13.1.3 Types of Pulmonary Fibrosis Based on the Known Cause

When a clinician sees a patient with the features of lung fibrosis, he cannot immediately identify the cause of lung fibrosis unless if there is any history of exposure to fibrogenic agents. If he can rule out all possible known reasons of lung fibrosis, it would be termed as "idiopathic" [7]. These are the five main known etiologies of pulmonary fibrosis:

(a) Iatrogenic: Various medications are implicated in causing PF. Drugs like chemotherapeutics [1], antiarrhythmic (amiodarone), and certain anti-inflammatory drugs (methotrexate) cause inflammation, injury, and scarring in the lungs.
(b) Radiation-induced: Exposure to radiations can cause lung fibrosis [8].
(c) Environmental: Pulmonary fibrosis is also caused due to environmental cues like exposures to mold spores, agricultural/farming, bacteria, animal droppings (especially from caged/indoor animals), or other known triggers [1, 8].
(d) Autoimmune: With an autoimmune disease, a person's own immune system attacks the lungs, inflicting inflammation and scarring that can impair lung function and breathing. Lung fibrosis is one of the pathological features of few systemic autoimmune diseases like Sjogren's syndrome [1, 9].
(e) Occupational: Various work environments which are related to exposure to fibrogenic materials devastate the lung tissue, like silica, beryllium, etc.

13.1.4 Idiopathic Pulmonary Fibrosis (IPF): The Expert's Conundrum

A historical change has been swayed in the definition of idiopathic pulmonary fibrosis (IPF) since Hamman and Rich recounted the pathogenesis of pulmonary fibrosis. **IPF,** which is a devastating and a progressive form of interstitial lung disease with no known etiology, has now been taken up seriously by the researchers. Earlier it was believed that inflammation caused by exogenous irritants could drive the progression of lung fibrosis. However, this concept has changed as many started believing that the entire pathophysiology roots from alveolar epithelial dysfunction after lung epithelia are subjected to repetitive microinjuries followed by anomalous crosstalk between epithelial and mesenchymal cells, creating a disturbance in the balance of pro-fibrotic and anti-fibrotic mediators [10, 11].

Phenotypically, IPF is characterized as a chronic condition wherein the lung tissue, primarily the interstitium, and the space or tissue around the alveoli of the lungs become increasingly scarred and thick, leading to chronic cough, breathlessness, and, ultimately, respiratory failure and death [10, 12]. Reports suggest that the death in IPF patients is solely due to the disease, followed by rapid deterioration of lung health and manifestation of lung cancer and cardiac disorders.

Based on intense research using multiple strategies and models, relatively successful drugs like pirfenidone and nintedanib were discovered to reduce the development of lung fibrosis [1]. However, these drugs could not reverse the already established fibrogenesis, and as a result, these drugs cannot restore the lung function completely. Thus, intrinsic insight into the roles of the various mediators contributing in this condition is obligatory for developing considerable number of therapeutic interventions.

13.1.5 The Nexus Between Pulmonary Fibrosis and COVID-19

In December 2019, news and reports surfaced out from Wuhan, China, of the outbreak of a severe acute respiratory syndrome coronavirus 2 (SARS-CoV-2) [13]. As of September 1, 2020, nearly 26.6 million people tested positive, and around 875, 000 people have died from coronavirus disease 2019 (COVID-19) worldwide. Patients with confirmed SARS-CoV-2 complain of diverse range of symptoms, from mere sneezing to severe respiratory distress [14]. Reports insinuate that all COVID-19-related serious consequences accentuate pneumonia. Interestingly, both COVID-19 and lung fibrosis are having common risk factors like old age, male gender, and disease comorbid conditions like diabetes and hypertension [13, 14].

Pulmonary fibrosis and COVID-19 have a stark resemblance, and at first sight, it won't seem evident [15]. It solely depends upon us, what learning we can take from these two dreadful conditions in order to benefit patients who are having permanent scars in their lungs. COVID-19 and pulmonary fibrosis are grievous diseases, and both embark with a lung injury [15]. It is proposed that pulmonary fibrosis is an consequence of acute lung injury caused by viruses. The mechanisms through which SARS-CoV-2 hurls damage to the lung are only partially known [16], but probable contributors include a "cytokine storm" triggered by the viral antigen upon entry into the system resulting in a hyperactive inflammatory reaction, drug-induced pulmonary toxicity, and intubation or invasive ventilation upon hypoxemia, and high airway pressure induces acute lung injury [16].

13.2 Genetic Factors Involved in the Development of Pulmonary Fibrosis

The heritable form of pulmonary fibrosis is called familial pulmonary fibrosis (FPF). It is characterized by a history of pulmonary fibrosis in two or more members of a family. It is an autosomal dominant disease with low penetrance. Around 10% of the IPF cases are FPF [17].

13.2.1 MUC5B Gene Polymorphism

A gene polymorphism for *MUC5B*, a gene encoding a mucin glycoprotein, has been extensively linked to both FPF and sporadic IPF. Homozygosity for the mutant allele (T) in single-nucleotide polymorphism of the MUC5B gene on chromosome 11p15.5 was found to increase the risk of having sporadic or familial PF by 20-fold [18]. This association has also been found in non-idiopathic fibrotic lung conditions like rheumatoid arthritis-associated interstitial lung disease (RA-ILD) and chronic hypersensitivity pneumonitis but not in some other forms of lung fibrosis like scleroderma-associated interstitial lung disease, sarcoidosis, and asbestosis [19]. This gain-of-function variant is associated with a heightened expression of the gene in the bronchoalveolar region. But, how the *MUC5B* variant increases susceptibility to PF is poorly understood. Overexpression of this gene impairs mucociliary clearance and enhances retention of inhaled particles and lung injury which might culminate into fibrosis in the long run [20]. The overexpression was found to ameliorate lung fibrosis but did not induce spontaneous fibrosis [19]. Quite contrary to expectations, only a small fraction of carriers of this risk allele was actually found to develop the disease indicating that genetic susceptibility alone is not enough to trigger the pathogenesis of PF [18].

13.2.2 Surfactant Protein Mutations

Surfactant protein C (SP-C), a constituent of surfactant, is solely synthesized by the alveolar type 2 epithelial cells (AEC2). More than 60 mutations of the surfactant protein C gene (*SPTPC*) have been linked to the pathogenesis of PF till date which amply highlights the importance of studying its mechanistic involvement in the disease [21]. The mutant protein has been found to accumulate in the endoplasmic reticulum (ER) causing ER stress and activating unfolded protein response (UPR). Prolonged or severe UPR activation may cause AEC2 cell apoptosis [18]. Recently, Nureki et al. demonstrated in a knock-in murine model of *SPTPC* mutation that this mutation indeed causes spontaneous PF in vivo. This shows that ER stress induced by the mutant *SPTPC* in AEC2 cells is involved in development of fibrosis in the lungs [22]. Surfactant protein A (SP-A), a surfactant protein secreted by the airway cells, is formed by the SPA1 and SPA2 proteins encoded by the genes *SFTPA1* and *SFTPA2*, respectively [21]. Mutations in both *SFTPA1* and *SFTPA2* have been reported to have association with the pathogenesis of PF [21, 23]. SFTPA2 mutant protein was found to be retained in the ER and hence not secreted, subsequently activates UPR [21]. Takezaki et al. have recently shown that a homozygous *SFTPA1* mutation induces AEC2 cell necroptosis and initiates spontaneous fibrosis in a knock-in murine model of the mutation [23].

13.2.3 Mutations in Other Genes

Rare mutations in telomere-related genes like telomerase reverse transcriptase (*TERT*), dyskerin pseudouridine synthase 1 (*DKC1*), and regulator of telomere elongation helicase 1 (*RTEL1*) have been linked to FPF pathogenesis [17]. Short telomeres have been detected in PF patients in comparison to age-matched controls, irrespective of whether they carry telomerase gene loss-of-function mutations [18]. Gable et al. showed that zinc finger CCHC-type domain containing 8 protein (ZCCHC8) is necessary for telomerase RNA component (TERC) maturation, and its heterozygous loss-of-function mutation leads to short telomere length and FPF [24].

Polymorphism in and around Toll-interacting protein (*TOLLIP*) gene, a regulator of Toll-like receptors, has been implicated in IPF pathogenesis [25]. The protective role of TOLLIP in PF is further supported by another study that showed TOLLIP to have an inhibitory effect in the pro-fibrotic transforming growth factor (TGFβ) signalling in lung epithelial cells [26].

Genetic variations in desmoplakin (*DSP*), involved in cell adhesion, and A-kinase-anchoring protein (*AKAP13*), predispose an individual to develop IPF. The expression of these proteins has been found to be elevated in IPF lungs [25]. In recent times, AKAP13 was shown to activate TGFβ (a key mediator of PF) in lung epithelial cells via Rho-αvβ6 pathway, thus providing a functional basis to its link with IPF pathogenesis [27].

13.3 Various Inducers for the Development of Pulmonary Fibrosis

PF can occur in response to a persisting lung condition like chronic inflammation or due to an unknown etiology, known as IPF. It can also occur due to certain fibrogenic agents like ambient particulate matter and soluble chemicals or secondary to some existing conditions like rheumatoid arthritis, sarcoidosis, etc.

There is a controversy regarding wrongful labelling some patients as having IPF that might have occurred because of environmental or occupational exposure. Although this can be due to improper diagnosis, it can be also attributed to insufficient association between the intensity of exposure of the etiological factor and severity of the disease. Therefore, occupational or environmental exposures may partially prompt the development of IPF. Some of these factors can be exemplified as organic dust from activities like agriculture, livestock, and farming or exposure to wood dust, mineral dust, asbestos, metal, and ambient particulate matter [28].

13.3.1 Silica

Silica or silicon dioxide (SiO_2) exists in a number of crystalline forms or polymorphs. Inhalation of fine crystalline silica of respirable conformation (<10 μm size) predisposes to the risk of developing a lung condition called silicosis.

In silicosis, the fine silica particles are deposited in the macrophages. Thus, an inflammatory response ensues that incites the fibroblasts to multiply and produce collagen. The silica is enwrapped by collagen that leads to fibrosis and formation of nodular lesions in the lungs, which are typical of the disease [29].

13.3.2 Asbestos

Asbestos are crystalline mineral fibers that find commercial use due to their durable and heat-resistant nature. There are quite a few types, the chrysotile or "curly" fibers and the straight rod-like crocidolite fibers, which are the most common. Prolonged heavy exposure to asbestos gives rise to a diffuse PF called asbestosis. After inhalation, asbestos is taken up by the alveolar epithelial cells (AECs) and alveolar macrophages that lead to oxidative stress in the lungs. Both reactive oxygen species (ROS) and reactive nitrogen species (RNS) are generated on the surface of the fibers through cell-free mechanism. These oxidants damage the cellular macromolecules, especially DNA, triggering the apoptosis of AECs. Epithelial injury is known to provoke repair responses by releasing pro-fibrotic cytokines, leading to fibrosis [30]. Furthermore, macrophages on phagocytosis release pro-fibrotic mediators that trigger deposition of connective tissue by fibroblasts. Since asbestos fibers are naturally resistant to digestion, the macrophages perish, releasing more such cytokines and enticing more macrophages and fibroblasts to generate fibrous tissue [31].

13.3.3 Iatrogenic

Bleomycin-induced pulmonary fibrosis mice model is a well-known animal model for studying PF. Bleomycin is a glycopeptide antibiotic and anticancer drug that has been first isolated in 1966 from *Streptomyces verticillus*. For almost 50 years, bleomycin has been used in parallel to chemotherapy in squamous cell carcinomas, germ cell cancers, and malignant lymphomas which has proved to be highly effective due to its low immunosuppression and myelosuppression. It also showed potent effects against other types of cancers like cervical cancer, ovarian cancer, melanoma, and sarcoma. However, its high therapeutic efficiency is severely restricted by its potential to cause PF [32].

13.3.4 Paraquat

Paraquat (PQ) is the most common herbicide used worldwide. It accumulates in the lung epithelial cells after exposure and causes significant damage. As a result of this, repair responses are triggered, leading to the development of PF [33]. In one study, collagen deposition was found to occur after 2 h of PQ administration. A separate study has shown epithelial to mesenchymal transition (EMT) to be a central event in the development and progression of PF. PQ being extremely toxic to human beings has a high mortality rate of 60–87.5%, PF being detected in most of the survivors [34].

13.3.5 Polyhexamethylene Guanidine (PHMG)

The PHMG is a biocide and disinfectant which was initially thought to be safe and nontoxic. Later, when it was used as a humidifier disinfectant, it was shown to be associated with PF [35]. A study by Kim et al. showed that PHMG can trigger ROS generation, causing airway barrier injury, followed by fibrotic repair response with increased synthesis of collagen and fibronectin and increased ECM deposition [36].

13.3.6 Carbon Nanotubes (CNTs)

The CNTs are 1-atom-wide graphene sheets with tube-like structures. They are extensively used for commercial and industrial purposes owing to some of their useful properties like remarkable strength and high thermal conductivity. Individuals involved in the manufacture of CNT-containing products and the consumers of such items are extensively exposed to this material. CNTs act like airborne fibers due to their nanoscale size and shape that is a like a fiber, thus increasing the risk of being inhaled. The inhaled fibers can accumulate in the airways and the alveoli. They are capable of inducing oxidative stress and pass through the plasma membrane, leading to epithelial cell lesions. Disruption of the epithelial barrier in the airways and type 2 alveolar epithelial cells leads to the secretion of alarmins that elicit type 1 or type 2 immune responses. Activated T helper 2 (Th2) cells and M2 macrophage induce the release of pro-fibrotic factors to promote tissue repair and lung fibrosis in type 2 immune response [37].

13.3.7 Lung Fibrosis as a Part of Other Diseases

Sarcoidosis is an inflammatory disease induced by a yet to be identified antigen, characterized by the formation of granuloma, most frequently in the lungs. Due to its unidentified etiology, specific treatment is still unavailable. About 20% of the patients having sarcoidosis are found to develop PF, which transforms this benign disease into a fatal one. In such cases, fibrosis starts at the fringes of the sarcoid granulomas. Progressive collagen deposition over time gives rise to mature fibrosis that damages the lung parenchyma, ushering in the "end-stage" sarcoidosis [38].

Rheumatoid arthritis (RA) is an inflammatory condition of autoimmune nature, affecting the joints. RA patients have sometimes been found to have interstitial lung disease, named rheumatoid arthritis-associated interstitial lung disease (RA-ILD). Clinically evident RA-ILD has been found in almost 10% of the RA patients [39]. Although the exact mechanism is not yet clear, two theories have been put forward based on available evidence regarding the pathology of this disease. First, external insults like cigarette smoke cause citrullination of proteins in the lungs, which stimulates the synthesis of anti-citrullinated protein antibodies (ACPA). This triggers breakdown of self-tolerance in RA. The second theory propounds that in local or systemic inflammation of chronic nature (as in RA), circulating inflammatory cytokines [interferon γ (IFNγ), tumor necrosis factor α (TNFα)] act on the lungs inducing an increase in expression of adhesion molecules on the pulmonary vascular endothelial cells. These signals attract the circulating monocytes/macrophages to the lungs and also activate lung-resident macrophages and other immune cells. On exposure to external factors like cigarette smoke, antigen presentation (of citrullinated proteins) occurs, triggering the activation of secondary or adaptive immune response and the resultant production of ACPA. The second bout of inflammation also stimulates the release of pro-fibrotic TGFβ and subsequent fibroblast activation and collagen deposition, leading to lung fibrosis [40].

13.4 Lung Injury: A Predominant Inducer for Pulmonary Fibrosis

Injury to the lung tissue can occur as a result of various insults, such as mechanical stretch, exposure to high doses of radiation, microbial infection, or gastroesophageal reflux.

13.4.1 Mechanical Stretch

Mechanical ventilation (MV) is an essential supportive therapy for patients having acute respiratory distress syndrome (ARDS), but it is capable of causing lung injury, named as ventilator-induced lung injury (VILI). VILI is caused either by recurrent opening or closure of the alveoli, i.e., the lungs can become derecruited with each expiration and recruited once again in the subsequent inspiration (atelectrauma), MV-associated inflammation (biotrauma), or overdistention of the lungs (volutrauma) [41]. Mechanical stretches to the alveolar epithelial cells lead to disruption of the tight junction and cellular attachments. Biotrauma heightens the production of ECM, triggering the repair response. The loss of alveolar epithelial integrity and overexuberant repair response induces EMT. There is also an increase in circulating fibrocytes and fibroblast proliferation. These events collectively contribute to the development of PF [42].

13.4.2 Exposure to High Doses of Radiation

Treatment by radiation of malignancies present in the lung, breast, and esophagus increases the vulnerability of patients to radiation-induced lung injury (RILI). The sensitivity of the normal lung parenchyma is the deciding factor for the dose of radiation in the malignancies of the chest. When the radiation penetrates the soft tissue, the localized aggregation of energy packets induces the breakdown of the strong chemical bonds of mainly water molecules, generating free radicals. These free radicals damage the cellular macromolecules, primarily DNA, and also lipids and proteins. This DNA damage sets in motion the downstream processes, which culminate into cell death in both normal and cancer cells. To increase the therapeutic efficacy and reduce toxicity of radiation, the total dose of radiation has to be broken down into small daily doses. This ensures enough time for recovery of the normal epithelial cells in comparison to malignant cells, owing to their differential DNA repair capacity and radiation sensitivity. Apart from the cytotoxic effects, strong inflammatory responses are also incited which eventually culminates into PF, the final phase of RILI. In this phase, pro-fibrotic mediators are released, and these lead to deposition of collagen by fibroblasts in the alveolar spaces and resultant reduction in lung volume [43].

13.4.3 Lung Fibrosis as a Post-SARS Sequel

Evident from the clinical studies and radiography, many persons who were infected by severe acute respiratory syndrome (SARS) in 2003 epidemic were found to have PF, as fibrosis of varying degrees was noticed on autopsy of the deceased patients. Although such fibrotic changes have also been observed in other respiratory viral infections, it was found to be more frequent in post-SARS coronavirus (SARS-CoV) infection. Increased levels of TGFβ, a pro-fibrotic mediator, have been found in the mouse models of SARS-CoV infection [44]. The SARS-CoV strain has genetic similarity with SARS-CoV-2, the cause of the 2020 pandemic that is wreaking havoc worldwide. Preliminary analysis of the COVID-19 patients on hospital discharge indicates that greater than one-third of these recovered individuals have fibrotic abnormalities [14].

13.4.4 Link with Gastroesophageal Reflux

Gastroesophageal reflux is the motion of gastric fluid in a retrograde direction toward the esophagus [45]. In states of waned consciousness as when a patient is under the influence of general anesthesia or conditions like cerebral vascular ischemia, trauma, and metabolic encephalopathies, the gastric refluxate may aspirate into the lower airways. The aspirate material can be composed of acid, bile, food particulates, pepsin, and microbes [46]. In order to reach the airways, the aspirate has to evade a lot of barriers such as esophageal peristalsis, cough reflex, swallow reflex, and mucociliary barrier [45]. The gastric contents are capable of inflicting injury to the lung epithelium and consequently trigger sterile inflammation [46]. Chronic microaspiration causes recurrent injury which eventually over a period of time results in lung fibrosis [45].

13.5 Alveolar Epithelial Dysfunction as a Converging Point for the Development of Pulmonary Fibrosis

The distal lung is formed of two alveolar epithelial cell types, namely, type 1 and type 2. Type 1 alveolar epithelial cells (AEC1) are situated near the endothelial cells and form a surface for gaseous exchange. AEC2 function as stem cells by differentiating into AEC1 cells for repair and renewal of the alveolar epithelium. Besides, they are also responsible for surfactant production [47]. Haschek and Witschi proposed a theory, in the 1990s, challenging the pre-existing notion that lung fibrosis is an inflammatory disease. After that, there was a gradual shift in the concept when it was realized that although inflammation contributes to the disease pathogenesis, it is not the central event that drives PF. AEC2 dysfunction was eventually recognized to have a key role in development of PF [21].

Multiple factors like genetic predisposition, ROS, hypoxia, and infection trigger ER stress in the AECs. There is a buildup of unfolded proteins inside the ER which activates the unfolded protein response. A downstream process ensues with the increase in chaperone proteins in order to aid protein folding. The AEC2 cells in the fibrotic lungs are unable to deal with ER stress owing to defective autophagy, which normally helps to remove the expanded ER [47]. There is also downregulation of lipid synthesis enzymes like stearoyl-CoA desaturase (SCD1), which is needed for reducing ER stress in the AECs [48]. Persistent ER stress leads to the activation of the ER stress-associated transcription factor, C/EBP homologous protein (CHOP), triggering apoptosis of the AEC2 cells [47]. Hypoxic microenvironment in the alveoli has been shown to stabilize hypoxia-inducible factor 1α (HIF1α) that triggers ER stress and CHOP-mediated transcription of apoptotic genes like B cell lymphoma 2 (*Bcl2*), Bcl2-like protein 11 (*Bim*), and cation transport regulator-like protein 1 (*Chac1*) [49]. Kamp et al. showed that oxidative stress induced by asbestos fibers induces ER stress followed by mitochondria-mediated apoptosis via ER Ca^{2+} release [50]. AEC2 cells, being metabolically active, possess a large number of mitochondria [51]. Mitochondrial fusion proteins mitofusin 1 and mitofusin 2 were shown to be involved in the production of surfactant lipids in AEC2 cells, and their absence triggers spontaneous lung fibrosis. This shows that damaged mitochondria affect the AEC2 surfactant production, thus compromising the integrity of epithelial barrier eventually leading to fibrosis [52].

Instead of apoptosis, AECs can also undergo premature senescence, a permanent form of cell cycle arrest, in response to injury [47]. Senescence in AECs is regulated by the activation of PTEN/NF-kB pathway that drives collagen deposition by fibroblasts and, consequently, fibrosis [53]. Moreover, plasminogen activator inhibitor-1 (PAI-1) has also been shown to induce AEC2 senescence by activating the cell cycle repressor p53-p21-Rb pathway in fibrotic lungs [54]. AEC2 senescence in lung fibrosis has been also attributed to dysfunctional telomere [51]. Senescence limits the capacity of transdifferentiation of AEC2 cells into AEC1 cells as observed in PF. Activation of developmental pathways like Notch and Hedgehog significantly affects AEC2 differentiation in fibrotic lungs [51]. Homeobox only protein X (HOPX), a protein involved in the development of distal lung, was found to promote AEC2 to AEC1 differentiation in early stages of alveolar injury but decreases with progression of IPF, leading to the loss of reparative capacity of AEC2 cells [55]. Recently, Wu et al. showed that failure to differentiate into AEC1 generates mechanical tension in AEC2 cells which activates the pro-fibrotic TGFβ pathway [56]. In previous studies, it has been shown that mechanical stress in AECs induces epithelial cell-restricted integrin αvβ6 to bring about conformational changes in the inactive precursor latent complex of TGFβ in the ECM, revealing the active TGFβ that can bind to its receptor on neighboring cells and exert its fibrotic effects [57]. The impact of this mechanical tension progress from the periphery to the center which provides a possible explanation for the periphery to center progression observed in IPF [56].

AEC2 cells have also been observed to go through EMT and drive fibrosis through epithelial-fibroblast crosstalk [58].

Thus, external insults and genetic factors set in motion a series of events in the AECs such as ER stress, impaired autophagy, intrinsic apoptosis, and cellular senescence, triggering an aberrant repair response in the lungs which culminates into fibrosis.

13.6 Role of Inflammatory Cells and Mediators in Pulmonary Fibrosis

In general, the role of innate immune cells like macrophage and neutrophil in the development of lung fibrosis is well known. However, the role of inflammation or immunological involvement in IPF is controversial as a number of studies have ruled out the involvement of immunological mechanisms in the IPF development [59]. Based on the earlier observation like the existence of more neutrophils in lung interstitium of IPF patients, it was hypothesized that IPF could be a chronic inflammatory disease. Later, a number of studies could not verify the involvement of inflammation. Initially, it was believed that the infiltration of inflammatory cells is the primary reason for the development of fibrosis. Currently, this concept has been modified in such a way that inflammation may have a role only after the establishment of lung fibrosis [59]. The immunopathogenesis could involve both innate and adaptive immune cells. The neutrophils, macrophages, fibrocytes, monocytes, dendritic cell, mast cells, and type 2 innate lymphoid cells are the innate immune cells, whereas almost all kinds of helper T cells (Th1, Th2, Th17), T_{reg}, and B cells are adaptive immune cells [59].

Macrophages can be of two types: (a) M1 if the macrophages are induced by TNFα or IFNγ and (b) M2 if the macrophages are induced by cytokines like IL-4, IL-13, IL-10, or TGFβ1. While M1 macrophages are formed in stages of acute inflammation, M2 macrophages are formed when the inflammation is getting resolved. During the resolution of inflammation, fibrosis occurs. In IPF, the M2 macrophages are activated, and these macrophages secrete TGFβ1 [59, 60]. Though neutrophils are recruited at the peak stage of inflammation, its role in IPF is beyond the inflammation as it helps in tissue remodelling through their enzymes like neutrophil elastase [59, 60]. Though the role of extracellular neutrophil traps has been shown to be present in a number of fibrotic conditions, its involvement in IPF is not clear yet. Fibrocytes, derived from blood monocytes, are present in the circulation. But when there is an injury, these cells are recruited to the site of injury. When they get activated, these cells further activate the fibroblast to induce the deposition of various ECM proteins. While Th2 cytokines like IL-4 and IL-13 can participate in the development of lung fibrosis by activating M2 macrophages, type 2 ILCs (innate lymphoid cells) also secrete most of these Th2 cytokines along with amphiregulin. IL-13 secreted by type 2 ILCs participate in lung interstitium remodelling along with the deposition of ECM proteins [59]. Similarly, Th2 cells of adaptive immune response also secrete cytokines like IL-4 and IL-13. Both of these cytokines are capable of activating myofibroblasts so that they secrete more ECM protein to cause lung fibrosis. On the other hand, IFNγ secreted by the Th1 cells are known to attenuate the features of lung fibrosis [59]. While B cells can produce autoantibodies against epithelial antigens to initiate an autoimmune response, regulatory T cells have been shown to reduce autoimmune response along with its anti-fibrotic role in lung fibrosis.

13.7 Role of Oxidative Stress in Pulmonary Fibrosis

Though the involvement of immunopathogenesis or inflammation in IPF is controversial, the role of oxidative stress in lung fibrosis especially in IPF is well demonstrated [61, 62]. In any event, its causative role in IPF is not proven yet. As IPF affects mostly elder people, it is also important to consider whether aging acts as a factor to induce oxidative stress, because dysfunction of mitochondria, reduction in proteostasis, and other oxidative stress-related mechanisms of aging could be involved in the development of IPF [62]. Oxidative stress participates in a number of ways in the pathogenesis of IPF, starting from the predisposition stage to the progression stage.

Both structural cells like alveolar epithelial cells, fibroblasts, vascular endothelial cells, and immune cells like macrophages and neutrophils can generate ROS and RNS. On the other hand, most of these cells also generate various antioxidant enzymes. However, the antioxidant mechanisms are reduced with aging. Each and every abovementioned cell has a unique way of participation in promoting lung fibrosis. For example, oxidative stress helps in the polarization of macrophage to promote lung fibrosis. While ROS can be pro-inflammatory and thus polarize macrophages toward M1, ROS released by dysfunctional mitochondria and ER stress could polarize the macrophage into pro-fibrotic M2 macrophage through TGFβ, platelet-derived growth factor (PDGF), and tissue inhibitor of metalloproteinase [62]. Oxidative stress-induced cellular senescence observed in structural cells is linked to the development of lung fibrosis. Whether the cellular senescence is involved with pro-fibrosis or anti-fibrosis depends on the organ and type of senescent cells. In case of lung fibrosis, senescence of alveolar epithelial cells and myofibroblasts has been well demonstrated. The senescent lung epithelia may reduce the rate of re-epithelialization to produce pro-fibrotic cytokines. Senolytics, a new variant of drugs that selectively targets the senescent cells, have been shown to reduce the deposition of ECM proteins [62].

Lung epithelial cell apoptosis can happen due to mitochondrial dysfunction or ER stress due to oxidative stress. The alveolar epithelial apoptosis leads to aberrant activation of fibroblasts and consequently ushering in lung fibrosis [61, 62]. In contrast, the apoptosis of myofibroblasts is reduced in lung fibrotic conditions. This could lead to the accumulation of more number of myofibroblasts in the lung interstitium, and these activated myofibroblasts secrete a variety of ECM proteins to cause dense lung fibrosis. Not only activated myofibroblasts participate in the deposition of ECM proteins, but also senescent fibroblasts can also promote lung fibrosis through attaining senescence-associated secretory phenotype that leads to express a number of pro-fibrotic genes including collagen, fibronectin, etc. [62]. While ROS released by various cells lead to fibrosis through various mechanisms, oxidation of even ECM proteins could lead to worsening of lung fibrosis by hardening of ECM along with resistance to proteolysis [62].

Thus, oxidative stress serves as a pro-fibrotic mechanism in a peculiar and a context-dependent manner. The role of antioxidants in reducing lung fibrosis seems to be controversial due to the additional anti-fibrotic role of ROS. However, finding effective and site-targeted antioxidants may be beneficial in reducing lung fibrosis.

13.8 Role of MMPs in the Development of Pulmonary Fibrosis

As mentioned above, alveolar epithelial cells are injured to initiate the development of pulmonary fibrosis. These alveolar epithelial cells secrete multiple mediators that promote fibroblast generation and myofibroblast accumulation which in turn alters the normal lung architecture. Zinc-dependent matrix metalloproteases (MMPs) belong to a larger family known as M10A metallopeptidase family. They degrade the ECM and can also cause casting away of cell membrane proteins. MMPs lead to activation of other mediators like cytokines, growth factors, chemokines, etc. In IPF, imperfect regulation of MMPs has a crucial role. MMP family consists of many genes. On one hand, some MMPs have a protective role against fibrosis. On the other hand, MMPs can promote and also contribute to disease progression [63]. MMPs cause parenchymal remodelling by accumulation of ECM constituents in lung interstitium which further aggravates the progression of the disease. Anti-protease treatment is thus thought to be beneficial to stabilize the balance between excessive synthesis and degradation of ECM [64].

Bleomycin-induced fibrosis rat model had shown initial elevation of MMP-9 level followed by reduction. Similarly, MMP-2 level increases and decreases more rapidly. In BAL (bronchoalveolar lavage) fluid, MMP-2 and MMP-9 activity is seen to be heightened in the early phase of pulmonary fibrosis; however, the level then gradually decreases. The level of change in MMP-9 was more prominent when compared to MMP-2. This may be due to the reason that MMP-9 is secreted by neutrophils. Alveolar macrophages and neutrophils release these MMPs which degrades the basement membrane. The breached basement membrane then allows the infiltration of inflammatory cell, and here the role of MMP-9 seems to be more. MMP-2 is hypothesized to act more in the later fibrosis phase and has a role in the generation of the fibrotic foci [65]. Tissue inhibitor of metalloprotease (TIMP) family proteins blocks the action of MMPs. Myofibroblast residing in the intra-alveolar fibrosis showed predominant reaction toward TIMP-2. Thus, TIMP-2 might have a role in irreversible lung structure remodelling observed in IPF. Matrix metalloprotease-1 (MMP-1) level is also found to be overexpressed in IPF. There is a slight paradox regarding MMP-1. In lung obtained from IPF, MMP-1 level is found to be localized in the AEC and not in fibroblasts [66]. Along with MMP-1, the level of MMP-7 (matrilysin) is seen to be increased in pulmonary fibrosis. The level of MMP-7 gets increased in lung and BAL fluid in comparison to control subjects. The surge in MMP-7 and MMP-1 level is correlated with the increase in the severity of disease. These can be the rationale for taking MMP-7 as a biomarker in asymptomatic patients suffering from early interstitial lung disease. Enhanced level of MMP-7, found in peripheral blood in IPF, shows correlation with pulmonary function test. Thus it can be a potential biomarker in IPF pathobiology [67]. MMP-3 has a unique role in promoting IPF. There is heightened expression of MMP-3 mRNA in the IPF lung. Experiments carried out in vitro on epithelial cells have shown that MMP-3 mediates EMT. MMP-3 works via inducing the ß-catenin/WNT pathway and causes the cleavage of epithelial cell marker E-cadherin. Upon MMP-3 treatment in lung epithelial cells, there is an increase in the expression level of mesenchymal marker, vimentin [68]. Another novel biomarker in fibrotic lung disease is MMP-10. In comparison with control and COPD patients, MMP-10 level is higher in sera of IPF patients. Immunohistochemistry studies have shown that apart from alveolar epithelial cells, macrophages and peribronchial epithelial cells also express MMP-10 [69]. MMP-8 (collagenase-2) is another pro-fibrotic MMP that promotes PF. In IPF, the MMP-8 level gets increased in BAL fluid. Studies revealed that MMP-8 decreases the expression of macrophage inflammatory protein-1 α (MIP-1α) and IFNγ-inducible protein 10 (IP-10) in the lung. Fibrocytes have been shown to express MMP-8. In PF, migration of fibrocytes gets increased. Fibrocytes incubated with MMP-8 inhibitor showed a decrease in their migratory activity. This accentuates the fact that MMP-8 heightens PF by regulating the migratory capacity of fibrocytes [70].

Few MMPs exert anti-fibrotic effects. MMP-13 (collagenase-3) is a primary enzyme that causes degradation of cartilages [70]. Compared to donor lungs, an increase in the levels of MMP-13 is observed in lungs of IPF patients. In bleomycin-exposed MMP-13 knockout mice, a severe increase in symptoms of fibrosis along with early onset of inflammation is observed [71]. Another anti-fibrotic regulator in IPF is MMP-19. When MMP-19 knockdown mice were exposed to bleomycin, heightened PF characteristics were observed. Another study showed that fibroblasts obtained from MMP-19-deficient mice showed an increase in fibrillar collagen proteins, collagens present in the basement membrane along with an elevation in collagen production. The growth pace of the fibroblast was also found to be increased. Along with these, production of myosin genes gets increased, and an increase in myofibroblast is also observed. MMP-19 thus has a controlling effect on IPF by regulating fibroblast growth and its differentiation into myofibroblast [72].

MMP has a regulatory effect on fibrosis. With MMP expression, various pathways also get affected. Thus, MMP can be a potent therapeutic target in PF. But we need to be cautious for the following reasons: (a) dynamicity of MMP in different stages of PF and (b) MMP inhibition that could cause havoc in lung homeostasis as a number of MMPs are required to fight against various infections by recruiting various immune cells and to activate key cytokines and chemokines that are involved in immune cell recruitment.

13.9 Disturbance in Balance of Pro-fibrotic and Anti-fibrotic Molecules

Fibrosis is an irreversible degenerative biological phenomenon which encompasses numerous other smaller events. Upon sensing the tissue degradation, scar formation takes place. This happens through excessive production of the ECM proteins and deposition of connective tissue like collagen. The cellular environment also undergoes certain alterations to decide whether the fibrotic pathway will be activated or not. There are certain pro-fibrotic agents that promote fibrosis. They are type 2 CD4 T lymphocyte cells and CD40 ligand-receptor interaction. There are also other mediators like cytokines and growth factors that accelerate the fibrotic process. Anti-fibrotic agent like interferon γ also exists [73]. CD40 receptors are expressed on the surface of fibroblasts. Immune cells like mast cells and T cells express CD40 ligand. In PF, CD40 ligand-receptor interaction stimulates the production of pro-inflammatory mediators and cell adhesion molecules like cyclooxygenase 2 (COX2) which then increases the production of ECM proteins. Prostaglandin 2 (PGE2) synthesis happens in the human fibroblasts which then disturbs the type 1/type 2 cytokine balance. Type 2 cytokine productions are also increased. On the other hand, generation of IL-12 and IFNγ which belong to type 1 cytokine family are inhibited [74]. T lymphocytes are primary inflammatory cells in fibrosis. Compared to wild type, bleomycin-administered T cell-deficient athymic mice showed less inflammation and fibroblast deposition on ECM. Immunosuppressive T regulatory (T_{reg}) cells exert pro-fibrotic effect via pro-fibrotic cytokines like TGFβ and PDGF [6]. There is a balance transpiring between pro-fibrotic and anti-fibrotic cytokine mediators. They are like the two sides of a seesaw. The pro-fibrotic cytokines are TGFβ, TNFα, vasoconstrictor molecule endothelin-1 (ET-1), and interleukins. TGFβ family assists in the production of fibroblast collagen gene. TNFα which gets upregulated in PF promotes the replication of fibroblasts and escalates the synthesis of collagen in the lung. ET-1 found in the lungs of fibrotic patients escalates proliferation of fibroblasts, promotes chemotaxis, and increases pro-collagen synthesis [75]. Interleukin-9 (IL-9) is another anti-inflammatory cytokine that has a role in attenuating lung fibrosis. B lymphocytes have a role in regulating the protective role of IL-9 [6]. Coagulation-regulating proteinases like thrombin and factor VIIa employ pro-fibrotic effect in the remodelled lung. These act by regulating the proteinase-activated receptors 1–4 (PAR-1–PAR-4). PAR-1 which imparts its effect by factor Xa and thrombin is found to have a prominent role in IPF. Experimentally it is seen that inhibiting the pro-inflammatory mediator thrombin prevents fibrosis in the IPF setting. PAR-1-deficient mice are also seen to be insulated from bleomycin-mediated injury [76]. In normal lung, there is a balance between pro-fibrotic collagen-synthesizing cytokines and anti-fibrotic collagen-inhibiting cytokines. Such balance existing between the positive and negative collagen-producing cytokines is found to be hampered in fibrotic lung. The synthesis of collagen-positive cytokines is higher in the fibrotic lungs. Pro-fibrotic cytokines work in a redundant manner. These cytokines act on diverse cell phenotypes and affect various mediator responses. Apart from affecting collagen synthesis, cytokines regulate recruitment of adhesion molecules, impact leukocytes, and also affect the inflammatory pattern in PF. These pro-fibrotic cytokines do have a role in accelerating the growth of fibroblast cells at injury sites. Hence, this cytokine balance is very crucial in the pathology of PF. The cytokine-dependent therapeutic strategy is another approach targeting PF. Further studies should aim at finding more anti-fibrotic agents to maintain the positive-negative collagen balance in PF [75].

13.10 Therapeutic Targets in Pulmonary Fibrosis

While numerous clinical trials aimed at developing therapies for pulmonary fibrosis are in the pipeline, pirfenidone and nintedanib have been shown to be relatively successful [77] and are available in the market. Our understanding of the complex pathobiology of pulmonary fibrosis has refined gradually over time, and this has modified the perspective to treatment. Nevertheless, most of these trials conducted have been gloomy, probably due to the milieu of mediators, complexity of the disease, and signalling pathways involved in the fibrotic process of IPF. Thus, a deeper investigation of the cellular and molecular mechanisms involved in the pathobiology is necessary in order to harness the knowledge and utilize them in developing new therapeutic targets and drugs. Below we discuss the evolving therapeutic targets which hold a promising future in combating the development and progression of pulmonary fibrosis.

13.10.1 Leukotriene Receptor Antagonists

Leukotrienes, derived from the oxidative metabolism of arachidonic acid, are inflammatory mediators and pro-fibrotic metabolites. In 1996, it was first demonstrated that leukotrienes were found to be elevated in the lungs of IPF patients. Thus, leukotrienes may be targeted for developing therapies for PF. Tipelukast (MN-001), an antagonist of the leukotriene receptor, is currently being explored [78]' and it is undergoing a phase 2 trial. Initial data obtained from this study suggests that it is generally safe and well tolerated [78, 79].

13.10.2 Targeting B Lymphocyte

Various abnormalities in B cell functions including the presence of increased autoantibodies have been demonstrated in IPF patients indicating the importance of B lymphocyte in pathogenesis of IPF [80]. And, now that there is evidence showing autoimmunity is involved in the genesis of fibrotic condition, drugs targeting autoimmune processes will be an effective therapy. Hence, the CD20 surface molecule of B lymphocytes is being targeted to rescue the lung damage inflicted by the autoantibodies generated during IPF. Rituximab, an antibody targeting the CD20 surface molecule of B lymphocytes, is currently being assessed in IPF patients [78, 79, 81]. Results of these trials are still pending; however, positive results of these investigations will substantiate the efficacy and pharmacokinetics of rituximab treatment in IPF patients.

13.10.3 Protein Kinase Inhibitors

The pro-fibrogenic role of several tyrosine kinases including PDGF, vascular endothelial growth factor (VEGF), and epidermal growth factor (EGF) has been implicated in lung fibrosis [82]. So, more exploration has been done to identify novel drugs using various relevant kinases as targets. Two such kinases were Rho-associated coiled-coil-containing protein kinase (ROCK2) and c-Jun N-terminal kinases (JNK). So KD025 and CC-90001 were discovered as inhibitors of ROCK2 and JNK, respectively. Both of these inhibitors have reached phase 2 trial [79, 83].

13.10.4 Phosphoinositide 3-Kinase/Protein Kinase B/Mammalian Target of Rapamycin (PI3K/Akt/mTOR) Pathway Inhibitors

A variety of vital cellular functions starting from cell differentiation to proliferation are tightly regulated by this PI3K/Akt/mTOR pathway [84]. The conversion of fibroblasts to myofibroblasts is associated with the increase in the expression of various isoforms of PI3K. Thus, the inhibition of PI3K might halt the fibrosing processes. This hypothesis was proven by the demonstration of the effects of omipalisib, which inhibit PI3K as observed in the phase 1 study [78, 83] Similarly, sirolimus, a mTOR inhibitor, had reached the phase 2 trial with anti-fibrotic properties. On the one hand, PI3K/AKT is the upstream regulator of ER stress, and on the other hand, ER stress-associated proteins were found to be increased in pulmonary fibrosis [84]. These evidences indicate that either ER stress inhibitors or a PI3K inhibitor could halt lung fibrosis [84].

13.10.5 Anti-Integrin Antibodies

Integrins are cell adhesion molecules that facilitate the adhesion and movement of cells on ECM. Beyond cell adhesion and migration, integrins were also shown to activate crucial kinases like epidermal growth factor receptor (EGFR) [78]. Reports have demonstrated that integrin family members are the key mediators of tissue fibrosis. The activation of TGFβ needs the help of αvβ6 integrin that is present exclusively in the lung epithelium. Thus, one can expect the inhibition of TGFβ activation along with a reduction in the features of lung fibrosis by inhibiting αvβ6 integrin [78, 79]. In a mice model of lung fibrosis, αvβ6 integrin antibody-mediated partial neutralization has been shown to reduce the features of lung fibrosis. In this context, the humanized monoclonal antibody against the αvβ6 integrin (BG00011) trial (phase 2) is ongoing after a pilot study that was performed to determine the efficacy of the antibody on TGFβ signalling inhibition. The outcomes of the study are still awaited [79].

13.10.6 Anti-Connective Tissue Growth Factor

Connective tissue growth factor (CTGF), a key glycoprotein, is involved in the process of fibrotic conditions along with TGFβ [78, 79]. Reports have suggested that CTGF due to its capability of producing collagen and fibronectin deposition in wound healing is also involved in the production of the extracellular matrix. Excess expression of CTGF upregulates several growth factors such as TGFβ [79]. It activates myofibroblasts which are responsible for fibrosis and tissue remodelling, consequently leading to various pathological conditions. Researchers have recently taken upon targeting CTGF which seems to be a promising therapeutic approach to combat the development of fibrosis. Pamrevlumab is a human monoclonal antibody against CTGF [85]. Preclinical studies and open-label phase 2 trials indicate the effectiveness of pamrevlumab. A randomized phase 3 clinical trial (NCT03955146) is set to start for pamrevlumab. More recently, PBI-4050 was developed and has demonstrated anti-fibrotic activities by reducing levels of CTGF [78, 86]. It has undergone a phase 2 evaluation, open-label, and it seems it does not have any safety issues.

13.10.7 Inhibitors of Autotaxin-Lysophosphatidic Acid

Autotaxin (ATX) is an extracellular enzyme that produces lysophosphatidic acid (LPA), a lipid molecule responsible for releasing pro-inflammatory mediators and recruitment of fibroblast and epithelial apoptosis [87]. The autotaxin-lysophosphatidic acid (ATX-LPA) pathway has been demonstrated to be involved in various pathological and fibroproliferative disorders, including pulmonary fibrosis [88]. The increased levels of autotaxins have been found in IPF patients indicating that the ATX-LPA pathway could yield successful anti-fibrotic inhibitors

[83]. Cuozzo and his group have screened a number of inhibitors of autotaxin and obtained a potent inhibitor, X-165 [89]. The FDA has cleared the way for X-165 and has moved into the first clinical test as a therapy candidate.

13.10.8 Pentraxin-2 (PTX-2) Analogues

The PTX-2 is a naturally occurring plasma protein that controls various facets of the innate immune system as it has the ability to regulate the conversion of blood monocytes into pro-fibrotic macrophages and is also involved in the process of wound healing [78]. Studies have shown reduced levels of PTX-2 in IPF conditions correlating with the severity of the disease. Administration of PTX-2 blocks the bleomycin-induced fibrotic damage in mice [78, 90]. Preclinical data showed a reduction in fibrotic conditions upon using PRM-151. Based on the phase 1 trial, PRM-151 was seen to significantly increase the blood levels of PTX-2 [78, 79, 90]. A very recent double-blind, phase 2 study (NCT02550873) of PRM-151 has demonstrated a significantly better exercise capacity compared to the placebo as well as a slower decline in lung function. Further studies are needed to evaluate its safety and efficacy.

13.10.9 Targeting the Respiratory Microbiota

The doctrine of "lung sterility" which persisted in the medical literature for decades has long been debunked. Researches and evidence now indicate how the lung and microbiota interact and exist. Since the Nobel Prize discovery by Marshall et al. for identifying the bacterium *H. pylori* and its causative role in gastric inflammatory disease [91], the perspective toward the involvement of microbes in chronic conditions has changed. The exploitation of the flora has now become a pragmatic therapeutic and prophylactic approach for many infectious and inflammatory diseases [91, 92]. Various reports have demonstrated that the severity of IPF is strongly correlated with the presence of certain microbiome in the lungs [92].

13.11 Pirfenidone and Nintedanib: Success in Pulmonary Fibrosis

Though the immunological role in the development of IPF is controversial, it was believed that immunosuppressive therapy was the primary option for IPF therapy earlier [93]. Indeed, corticosteroids and azathioprine were used for the treatment of IPF. Later N-acetyl cysteine was also added in this combination [94]. Until the 2012 PANTHER study, this triple therapy was the preferable mode of therapy in IPF. However, the PANTHER study indicated the higher mortality due to this triple therapy [93]. Then disease-modifying drugs including pirfenidone and nintedanib were used after a number of clinical trials. Though the exact mechanism of pirfenidone is not known, reduction in the proliferation of fibroblasts along with the reduction in the secretion of pro-fibrotic growth factors like TGFβ and PDGF has been demonstrated [93]. Initially, it was approved in Japan followed by many other countries. The anti-fibrotic effects of nintedanib were evident after the 3rd phase of INPULSIS 1 and 2 trials. It is a tyrosine kinase inhibitor as it inhibits tyrosine kinases like PDGF, fibroblast growth factor (FGF), and VEGF [93]. Thus, the possible relative success story behind these two drugs might be attributable to their multiple targets and actions. As one of the long-term complications of COVID-19 is lung fibrosis, many believe that these two drugs could stop the fibrotic complications post-COVID-19 [95].

13.12 Conclusion

The pathogenesis of pulmonary fibrosis seems to be a sort of a labyrinth and is multifactorial. The development of lung fibrosis is similar to wound repair or healing. In spite of so many pathways involved, researchers have succeeded in developing only two drugs, pirfenidone and nintedanib, which are currently approved by the FDA and available in the market to avert the progression of the disease. Immunosuppressants and glucocorticoid medications are given to the IPF patients that have poor efficiency and various toxic effects. In this review, we have highlighted some of the regulatory mechanisms that are linked to the development of lung fibrosis. With this in mind, an intricate understanding of these pathways and the identification of various biomarkers are needed for designing more number of efficacious therapies for this fatal lung disease.

13.13 Future Perspectives

In contemplation of the experience from the positive and negative outcomes of the studies, inhibition of the single pathway of IPF might only have a modest effect on the fibrotic condition. The accomplishment of the drugs such as pirfenidone and nintedanib that inhibit multiple pathways of IPF has motivated the researchers and clinicians to design and assess the anti-multifactorial pathogenesis drugs with novel molecular scaffolds to attenuate the progression of IPF. Clinical investigations with anti-fibrotic therapeutics targeting IL-13 and TGFβ, chemokine receptor antagonists, and inhibitors of angiogenesis are required. An integrative effort by researchers and clinicians across various disciplines could make this a climbable mountain (Fig. 13.1).

Acknowledgments This work was supported by the project MLP137 (MISSION LUNG) at CSIR-Indian Institute Chemical Biology, Council of Scientific and Industrial Research, Govt. of India. SS, JD, AR, and AJ acknowledge the Academy of Scientific and Innovative Research (AcSIR), Ghaziabad, India, for their Ph.D. registrations.

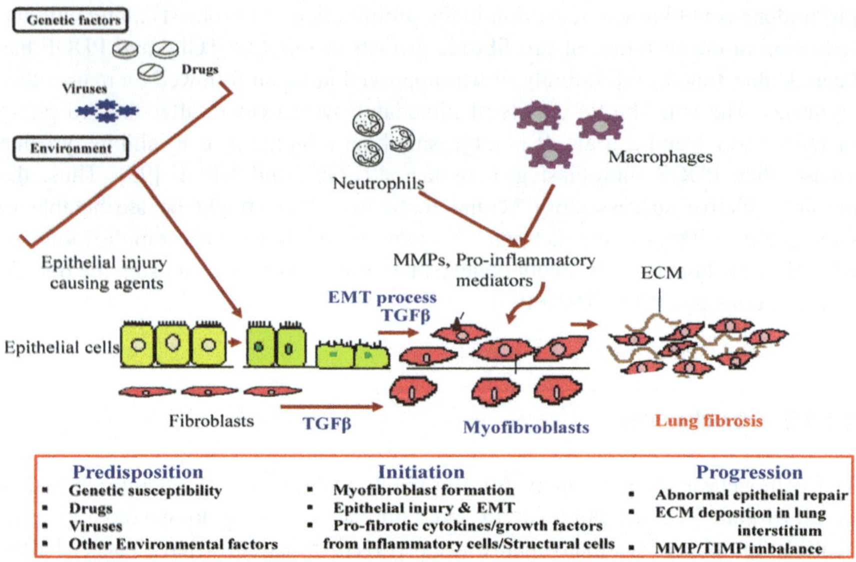

Fig. 13.1 Schematic diagram to show the pathogenesis of pulmonary fibrosis. The development of IPF has three major stages: (a) predisposition, (b) initiation, and (c) progression. A single or milieu of factors like exogenous environmental cues, drugs, certain viruses, and genetic susceptibility predisposes the individuals to cause alveolar epithelial injury with or without inflammation. The alveolar epithelial injury initiates a multitude of pathways that leads to loss of epithelial cells, conversion of fibroblast to myofibroblasts, and EMT. The imbalance between pro- and anti-fibrotic mediators, MMP/TIMP imbalance, also evokes an abnormal epithelial repair along with excess deposition of ECM constituents leading to the development of lung fibrosis

References

1. Kolahian S, Fernandez IE, Eickelberg O et al (2016) Immune mechanisms in pulmonary fibrosis. Am J Respir Cell Mol Biol 55:309–322
2. Wynn TA (2011) Integrating mechanisms of pulmonary fibrosis. J Exp Med 208:1339–1350
3. Crouch E (1990) Pathobiology of pulmonary fibrosis. Am J Physiol Lung Cell Mol Physiol 259: L159–L184
4. Ryu JH, Colby TV, Hartman TE (1998) Idiopathic pulmonary fibrosis: current concepts. Mayo Clin Proc 73:1085–1101
5. Orr CR, Jacobs WF (1926) Pulmonary fibrosis. Radiology 7:318–325
6. Todd NW, Luzina IG, Atamas SP (2012) Molecular and cellular mechanisms of pulmonary fibrosis. Fibrogenes. Tissue Repair 5:1–24
7. Wuyts WA, Agostini C, Antoniou KM et al (2013) The pathogenesis of pulmonary fibrosis: a moving target. Eur Respir J 41:1207–1218
8. Wilson MS, Wynn TA (2009) Pulmonary fibrosis: pathogenesis, etiology and regulation. Mucosal Immunol 2:103–121
9. Thannickal VJ, Toews GB, White ES et al (2004) Mechanisms of pulmonary fibrosis. Annu Rev Med 55:395–417
10. Kinoshita T, Goto T (2019) Molecular mechanisms of pulmonary fibrogenesis and its progression to lung cancer: a review. Int J Mol Sci 20:1461

11. Wells AU, Brown KK, Flaherty KR et al (2018) What's in a name? That which we call IPF, by any other name would act the same. Eur Respir J 51:1800692
12. Nalysnyk L, Cid-Ruzafa J, Rotella P et al (2012) Incidence and prevalence of idiopathic pulmonary fibrosis: review of the literature. Eur Respir Rev 21:355–361
13. Zhu N, Zhang D, Wang W et al (2020) A novel coronavirus from patients with pneumonia in China, 2019. N Engl J Med 382:727–733
14. Vasarmidi E, Tsitoura E, Spandidos DA et al (2020) Pulmonary fibrosis in the aftermath of the COVID-19 era (review). Exp Ther Med 20:2557–2560
15. Jenkins G (2020) Demystifying pulmonary fibrosis. Am J Physiol Cell Mol Physiol 319:L554–L559
16. Spagnolo P, Balestro E, Aliberti S et al (2020) Pulmonary fibrosis secondary to COVID-19: a call to arms? Lancet Respir Med 8:750–752
17. Borie R, Crestani B (2019) Familial pulmonary fibrosis: a world without frontiers. J Bras Pneumol 45:e20190303
18. Spagnolo P, Cottin V (2017) Genetics of idiopathic pulmonary fibrosis: from mechanistic pathways to personalised medicine. J Med Genet 54:73–83
19. Yanagihara T, Sato S, Upagupta C et al (2019) What have we learned from basic science studies on idiopathic pulmonary fibrosis? Eur Respir Rev 28:190029
20. Schwartz DA (2018) Idiopathic pulmonary fibrosis is a genetic disease involving mucus and the peripheral airways. Ann Am Thorac Soc 15:S192–S197
21. Mulugeta S, Nureki SI, Beers MF (2015) Lost after translation: insights from pulmonary surfactant for understanding the role of alveolar epithelial dysfunction and cellular quality control in fibrotic lung disease. Am J Physiol Lung Cell Mol Physiol 309:L507–L525
22. Nureki SI, Tomer Y, Venosa A et al (2018) Expression of mutant Sftpc in murine alveolar epithelia drives spontaneous lung fibrosis. J Clin Invest 128:4008–4024
23. Takezaki A, Tsukumo SI, Setoguchi Y et al (2019) A homozygous SFTPA1 mutation drives necroptosis of type II alveolar epithelial cells in patients with idiopathic pulmonary fibrosis. J Exp Med 216:2724–2735
24. Gable DL, Gaysinskaya V, Atik CC et al (2019) ZCCHC8, the nuclear exosome targeting component, is mutated in familial pulmonary fibrosis and is required for telomerase RNA maturation. Genes Dev 33:1381–1396
25. Barros A, Oldham J, Noth I (2019) Genetics of idiopathic pulmonary fibrosis. Am J Med Sci 357:379–383
26. Li X, Kim SE, Chen TY et al (2020) Toll interacting protein protects bronchial epithelial cells from bleomycin-induced apoptosis. FASEB J 34:9884–9898
27. Organ L, Porte J, John A et al (2019) Investigating the role of AKAP13 on epithelial cell TGFß activation. Eur Respir J 54:PA2429
28. Trethewey SP, Walters GI (2018) The role of occupational and environmental exposures in the pathogenesis of idiopathic pulmonary fibrosis: a narrative literature review. Medicina 54:108
29. Pollard KM (2016) Silica, silicosis, and autoimmunity. Front Immunol 7:97
30. Robledo R, Mossman B (1999) Cellular and molecular mechanisms of asbestos-induced fibrosis. J Cell Physiol 180:158–166
31. Matsuzaki H, Maeda M, Lee S et al (2012) Asbestos-induced cellular and molecular alteration of immunocompetent cells and their relationship with chronic inflammation and carcinogenesis. J Biomed Biotechnol 2012:492608
32. Wang K, Zhang T, Lei Y et al (2018) Identification of ANXA2 (annexin A2) as a specific bleomycin target to induce pulmonary fibrosis by impeding TFEB-mediated autophagic flux. Autophagy 14:269–282
33. Dong J, Yu X, Porter DW et al (2016) Common and distinct mechanisms of induced pulmonary fibrosis by particulate and soluble chemical fibrogenic agents. Arch Toxicol 90:385–402
34. Zhu Y, Tan J, Xie H et al (2016) HIF-1α regulates EMT via the snail and β-catenin pathways in paraquat poisoning-induced early pulmonary fibrosis. J Cell Mol Med 20:688–697

35. Seo C, Kim SH, Lee HS et al (2019) Metabolomic study on bleomycin and polyhexamethylene guanidine phosphate-induced pulmonary fibrosis mice models. Metabolomics 15:111
36. Kim HR, Lee K, Park CW et al (2016) Polyhexamethylene guanidine phosphate aerosol particles induce pulmonary inflammatory and fibrotic responses. Arch Toxicol 90:617–632
37. Dong J, Ma Q (2018) Type 2 immune mechanisms in carbon nanotube-induced lung fibrosis. Front Immunol 9:1120
38. Bonham CA, Strek ME, Patterson KC (2016) From granuloma to fibrosis: sarcoidosis associated pulmonary fibrosis. Curr Opin Pulm Med 22:484–491
39. Paulin F, Doyle TJ, Fletcher EA et al (2015) Rheumatoid arthritis-associated interstitial lung disease and idiopathic pulmonary fibrosis: shared mechanistic and phenotypic traits suggest overlapping disease mechanisms. Rev Investig Clin 67:280–286
40. Wu EK, Ambrosini RD, Kottmann RM et al (2019) Reinterpreting evidence of rheumatoid arthritis-associated interstitial lung disease to understand etiology. Curr Rheumatol Rev 15:277–289
41. Chen L, Xia HF, Shang Y et al (2018) Molecular mechanisms of ventilator-induced lung injury. Chin Med J 131:1225–1231
42. Cabrera-Benitez NE, Laffey JG, Parotto M et al (2014) Mechanical ventilation-associated lung fibrosis in acute respiratory distress syndrome: a significant contributor to poor outcome. Anesthesiology 121:189–198
43. Bledsoe TJ, Nath SK, Decker RH (2017) Radiation pneumonitis. Clin Chest Med 38:201–208
44. Venkataraman T, Frieman MB (2017) The role of epidermal growth factor receptor (EGFR) signaling in SARS coronavirus-induced pulmonary fibrosis. Antivir Res 143:142–150
45. Lee AS, Ryu JH (2018) Aspiration pneumonia and related syndromes. Mayo Clin Proc 93:752–762
46. Mizushina Y, Karasawa T, Aizawa K et al (2019) Inflammasome-independent and atypical processing of IL-1β contributes to acid aspiration-induced acute lung injury. J Immunol 203:236–246
47. Winters NI, Burman A, Kropski JA et al (2019) Epithelial injury and dysfunction in the pathogenesis of idiopathic pulmonary fibrosis. Am J Med Sci 357:374–378
48. Kamp DW (2018) Mitigating lung fibrosis by targeting dysfunctional alveolar epithelial cell lipid metabolism. Am J Respir Cell Mol Biol 59:139–140
49. Delbrel E, Soumare A, Naguez A et al (2018) HIF-1α triggers ER stress and CHOP-mediated apoptosis in alveolar epithelial cells, a key event in pulmonary fibrosis. Sci Rep 8:1–14
50. Kamp DW, Liu G, Cheresh P et al (2013) Asbestos-induced alveolar epithelial cell apoptosis: the role of endoplasmic reticulum stress response. Am J Respir Cell Mol Biol 49:892–901
51. Parimon T, Yao C, Stripp BR et al (2020) Alveolar epithelial type II cells as drivers of lung fibrosis in idiopathic pulmonary fibrosis. Int J Mol Sci 21:2269
52. Chung KP, Hsu CL, Fan LC et al (2019) Mitofusins regulate lipid metabolism to mediate the development of lung fibrosis. Nat Commun 10:1–17
53. Tian Y, Li H, Qiu T et al (2018) Loss of PTEN induces lung fibrosis via alveolar epithelial cell senescence depending on NF-κB activation. Aging Cell 18:e12858
54. Jiang C, Liu G, Luckhardt T et al (2017) Serpine 1 induces alveolar type II cell senescence through activating p53-p21-Rb pathway in fibrotic lung disease. Aging Cell 16:1114–1124
55. Ota C, Ng-Blichfeldt JP, Korfei M et al (2018) Dynamic expression of HOPX in alveolar epithelial cells reflects injury and repair during the progression of pulmonary fibrosis. Sci Rep 8:1–10
56. Wu H, Yu Y, Huang H et al (2020) Progressive pulmonary fibrosis is caused by elevated mechanical tension on alveolar stem cells. Cell 180:107–121
57. Camelo A, Dunmore R, Sleeman MA et al (2014) The epithelium in idiopathic pulmonary fibrosis: Breaking the barrier. Front Pharmacol 4:173
58. Hill C, Jones M, Davies D et al (2019) Epithelial-mesenchymal transition contributes to pulmonary fibrosis via aberrant epithelial/fibroblastic cross-talk. J Lung Heal Dis 3:31–35

59. Desai O, Winkler J, Minasyan M, Herzog EL (2018) The role of immune and inflammatory cells in idiopathic pulmonary fibrosis. Front Med (Lausanne) 5:43
60. Heukels P, Moor CC, von der Thüsen JH et al (2019) Inflammation and immunity in IPF pathogenesis and treatment. Respir Med 147:79–91
61. Hosseinzadeh A, Javad-Moosavi SA, Reiter RJ et al (2018) Oxidative/nitrosative stress, autophagy and apoptosis as therapeutic targets of melatonin in idiopathic pulmonary fibrosis. Expert Opin Ther Targets 22:1049–1061
62. Otoupalova E, Smith S, Cheng G, Thannickal VJ (2020) Oxidative stress in pulmonary fibrosis. Compr Physiol 10:509–547
63. Pardo A, Cabrera S, Maldonado M et al (2016) Role of matrix metalloproteinases in the pathogenesis of idiopathic pulmonary fibrosis. Respir Res 17:23
64. Corbel M, Belleguic C, Boichot E et al (2002) Involvement of gelatinases (MMP-2 and MMP-9) in the development of airway inflammation and pulmonary fibrosis. Cell Biol Toxicol 18:51–61
65. Kim JY, Choeng HC, Ahn C et al (2009) Early and late changes of MMP-2 and MMP-9 in bleomycin-induced pulmonary fibrosis. Yonsei Med J 50:68–77
66. Checa M, Ruiz V, Montaño M et al (2008) MMP-1 polymorphisms and the risk of idiopathic pulmonary fibrosis. Hum Genet 124:465–472
67. Rosas IO, Richards TJ, Konishi K et al (2008) MMP1 and MMP7 as potential peripheral blood biomarkers in idiopathic pulmonary fibrosis. PLoS Med 5:0623–0633
68. Yamashita CM, Dolgonos L, Zemans RL et al (2011) Matrix metalloproteinase 3 is a mediator of pulmonary fibrosis. Am J Pathol 179:1733–1745
69. Sokai A, Handa T, Tanizawa K et al (2015) Matrix metalloproteinase-10: a novel biomarker for idiopathic pulmonary fibrosis. Respir Res 16:1–8
70. Craig VJ, Zhang L, Hagood JS et al (2015) Matrix metalloproteinases as therapeutic targets for idiopathic pulmonary fibrosis. Am J Respir Cell Mol Biol 53:585–600
71. Nkyimbeng T, Ruppert C, Shiomi T et al (2013) Pivotal role of matrix metalloproteinase 13 in extracellular matrix turnover in idiopathic pulmonary fibrosis. PLoS One 8:e73279
72. Jara P, Calyeca J, Romero Y et al (2015) Matrix metalloproteinase (MMP)-19-deficient fibroblasts display a profibrotic phenotype. Am J Physiol Lung Cell Mol Physiol 308:L511–L522
73. Molina V, Blank M, Shoenfeld Y (2002) Fibrotic diseases. Harefuah 141:973–978
74. Sime PJ, O'reilly KMA (2001) Fibrosis of the lung and other tissues: new concepts in pathogenesis and treatment. Clin Immunol 99:308–319
75. Coker RK, Laurent GJ (1998) Pulmonary fibrosis: cytokines in the balance. Eur Respir J 11:1218–1221
76. Laurent GJ, McAnulty RJ, Hill M et al (2008) Escape from the matrix: multiple mechanisms for fibroblast activation in pulmonary fibrosis. Proc Am Thorac Soc 5:311–315
77. Vancheri C, Kreuter M, Richeldi L et al (2018) Nintedanib with add-on pirfenidone in idiopathic pulmonary fibrosis: results of the INJOURNEY trial. Am J Respir Crit Care Med 197:356–363
78. Calvello M, Flore MC, Richeldi L (2019) Novel drug targets in idiopathic pulmonary fibrosis. Expert Opin Orphan Drugs 7:125–146
79. Sato S, Yanagihara T, Kolb MRJ (2019) Therapeutic targets and early stage clinical trials for pulmonary fibrosis. Expert Opin Investig Drugs 28:19–28
80. Xue J, Kass DJ, Bon J et al (2013) Plasma B lymphocyte stimulator and B cell differentiation in idiopathic pulmonary fibrosis patients. J Immunol 191:2089–2095
81. Keir GJ, Maher TM, Ming D et al (2014) Rituximab in severe, treatment-refractory interstitial lung disease. Respirology 19:353–359
82. Grimminger F, Günther A, Vancheri C (2015) The role of tyrosine kinases in the pathogenesis of idiopathic pulmonary fibrosis. Eur Respir J 45:1426–1433
83. Torrisi SE, Kahn N, Vancheri C et al (2020) Evolution and treatment of idiopathic pulmonary fibrosis. Press Medicale 49:104025

84. Hsu HS, Liu CC, Lin JH et al (2017) Involvement of ER stress, PI3K/AKT activation, and lung fibroblast proliferation in bleomycin-induced pulmonary fibrosis. Sci Rep 7:1–11
85. Raghu G, Scholand MB, De Andrade J et al (2016b) FG-3019 anti-connective tissue growth factor monoclonal antibody: results of an open-label clinical trial in idiopathic pulmonary fibrosis. Eur Respir J 47:1481–1491
86. Khalil N, Manganas H, Ryerson CJ et al (2019) Phase 2 clinical trial of PBI-4050 in patients with idiopathic pulmonary fibrosis. Eur Respir J 53:1800663
87. Swaney JS, Chapman C, Correa LD et al (2010) A novel, orally active LPA 1 receptor antagonist inhibits lung fibrosis in the mouse bleomycin model. Br J Pharmacol 160:1699–1713
88. Funke M, Zhao Z, Xu Y et al (2012) The lysophosphatidic acid receptor LPA 1 promotes epithelial cell apoptosis after lung injury. Am J Respir Cell Mol Biol 46:355–364
89. Cuozzo JW, Clark MA, Keefe AD et al (2020) Novel Autotaxin inhibitor for the treatment of idiopathic pulmonary fibrosis: a clinical candidate discovered using DNA-encoded chemistry. J Med Chem 63:7840–7856
90. Raghu G, van den Blink B, Hamblin MJ et al (2019) Long-term treatment with recombinant human pentraxin 2 protein in patients with idiopathic pulmonary fibrosis: an open-label extension study. Lancet Respir Med 7:657–664
91. Varone F, Gibiino G, Gasbarrini A et al (2019) Evaluation of the lung microbiome as a therapeutic target in the management of idiopathic pulmonary fibrosis: role of antioxidant/antibiotic combination therapy. Eur Rev Med Pharmacol Sci 23:6379–6386
92. Salisbury ML, Han MLK, Dickson RP et al (2017) Microbiome in interstitial lung disease: from pathogenesis to treatment target. Curr Opin Pulm Med 23:404–410
93. Hayton C, Chaudhuri N (2017) Current treatments in the management of idiopathic pulmonary fibrosis: pirfenidone and nintedanib. Clin Med Insights Ther 9:1–4
94. Idiopathic Pulmonary Fibrosis Clinical Research Network, Raghu G, Anstrom KJ, King TE Jr, Lasky JA, Martinez FJ (2012) Prednisone, azathioprine, and N-acetylcysteine for pulmonary fibrosis. N Engl J Med 366:1968–1977
95. George PM, Wells AU, Jenkins RG (2020) Pulmonary fibrosis and COVID-19: the potential role for antifibrotic therapy. Lancet Respir Med 8:807–815

Targeting Molecular and Cellular Mechanisms in Tuberculosis

14

Lubhan Singh, Kamal Dua, Sokindra Kumar, Deepak Kumar, and Sagarika Majhi

Abstract

The major global problems are the prevalence rate of *Mycobacterium tuberculosis* (Mtb) and processes of resistance against continuing therapy. The shortage of possible drug candidates and consumer recognition along with unhygienic procedures are the key explanations for MDR, TDR and XDR Mtb strains in rapid emergence. Mtb's powerful molecular structure and drug resistance pathways, demands expertise to develop new anti-tuberculosis therapies. Eventually, the synthesis of modern genomic knowledge of drug resistance mechanisms in Mtb will offer a new path for combinatorial drug development and provide considerable support for highly successful anti-tubercular drugs.

After a time of relative lack of interest in cell envelope targeting inhibitors, the immediate necessity of new therapeutic approaches has motivated renewed work towards a deeper understanding of the cell envelope, its biogenesis and function during Mtb's stages of proliferation, survival and reactivation. As a result, new appealing goals for drugs were identified and followed, with varying effects, in

L. Singh · S. Kumar
Kharvel Subharti College of Pharmacy, Swami Vivekanand Subharti University, Meerut, India

K. Dua
Discipline of Pharmacy, Graduate School of Health, University of Technology Sydney, Sydney, NSW, Australia

Priority Research Centre for Healthy Lungs, University of Newcastle & Hunter Medical Research Institute, New Lambton Heights, Newcastle, NSW, Australia

D. Kumar
Faculty of Pharmaceutical Sciences, Shoolini University, Bajhol, Himachal Pradesh, India

S. Majhi (✉)
I. T. S College of Pharmacy, Muradnagar, Ghaziabad, India

K. Dua et al. (eds.), *Targeting Cellular Signalling Pathways in Lung Diseases*,
https://doi.org/10.1007/978-981-33-6827-9_14

the form of target-based scanning and other target-based strategies. The challenge of detecting compounds whose inhibitory action against filtered targets translated into entire Mtb cell activity prompted several field investigators to go back to cell-based screens. This strategy leads to several new groups of inhibitors being identified, and some of those are now in preclinical research.

Keywords

Mycobacterium tuberculosis (Mtb) · Drug resistance · Cell envelope · Bacilli · Tuberculosis (TB)

14.1 Introduction

The Great White Plague also was earlier defined as tuberculosis. *Mycobacterium tuberculosis* (Mtb), a slow-growing imperative aerobic, discretionary intracellular parasite, is still the leading cause of death. It has an envelope of complex and immunomodulatory cells. The cell envelope differs structurally and functionally over the lifetime of the cell and throughout the cycle of infection. This variation helps the bacterium to control the host immune system, to resist antibiotic therapy and to respond to the complex host environment. Only one vaccine generated several years earlier, the live-attenuated *Bacillus* Calmette-Guerin strain, provides very limited defence against Mtb [1]. Moreover, the new therapy is lengthy, allowing several, potentially dangerous antibiotics to be administered over a span of months [2]. Thus, the processes by which Mtb interferes with the host immune response remain essential both to improve vaccine approaches and to investigate therapies that facilitate sterilizing host immunity as a supplement to antibiotics.

Mycobacteria are small rod-shaped (0.2–0.5 μm by 2–4 μm), aerobic, non-spore-forming, non-motile, facultative, curved intracellular rods, and their cell walls contain mycolic acid-rich, long-chain glycolipids and phospholipoglycans (mycocides) which shield mycobacteria against cell lysosomal attack as well as preserve red basic fuchsin dye during acid rinsing (acid-fast stain). They can be classified into three main groups as shown in Fig. 14.1.

Its treatment depends on its forms, viz. multidrug-resistant (MDR) TB, extensively drug-resistant (XDR) TB and total drug-resistant (TDR) TB. Since XDR and TDR strains cannot be effectively treated with antibiotics currently available, several more compounds need to meet the 'pipeline' drug production to effectively tackle tuberculosis (TB) problem. New anti-TB compounds can fix problems with recent therapy. The ideal anti-TB drug has to exhibit high efficacy, especially towards drug-resistant strains, and it has an appropriate safety profile. Drugs will also be involved in counteracting latent and replicating types of *M. tuberculosis* and restricted drug/drug interactions, with antiretroviral agents in particular. Tuberculosis can be subdivided into different categories based on signs, their symptoms and diagnostic tests as illustrated in Fig. 14.2.

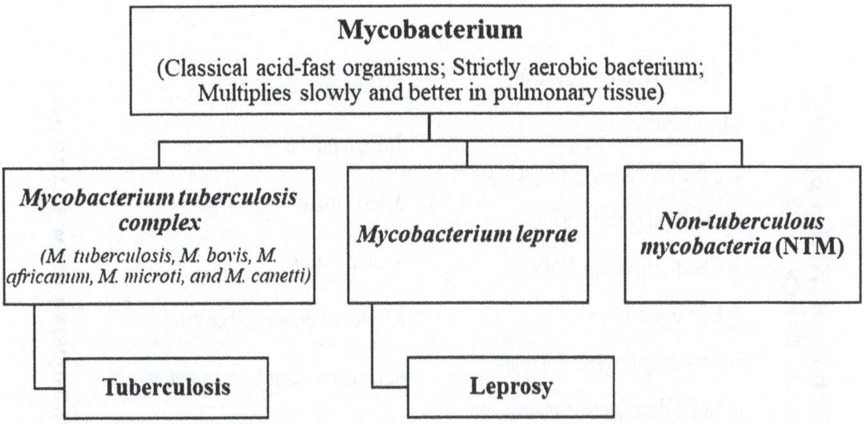

Fig. 14.1 Classification of mycobacterium

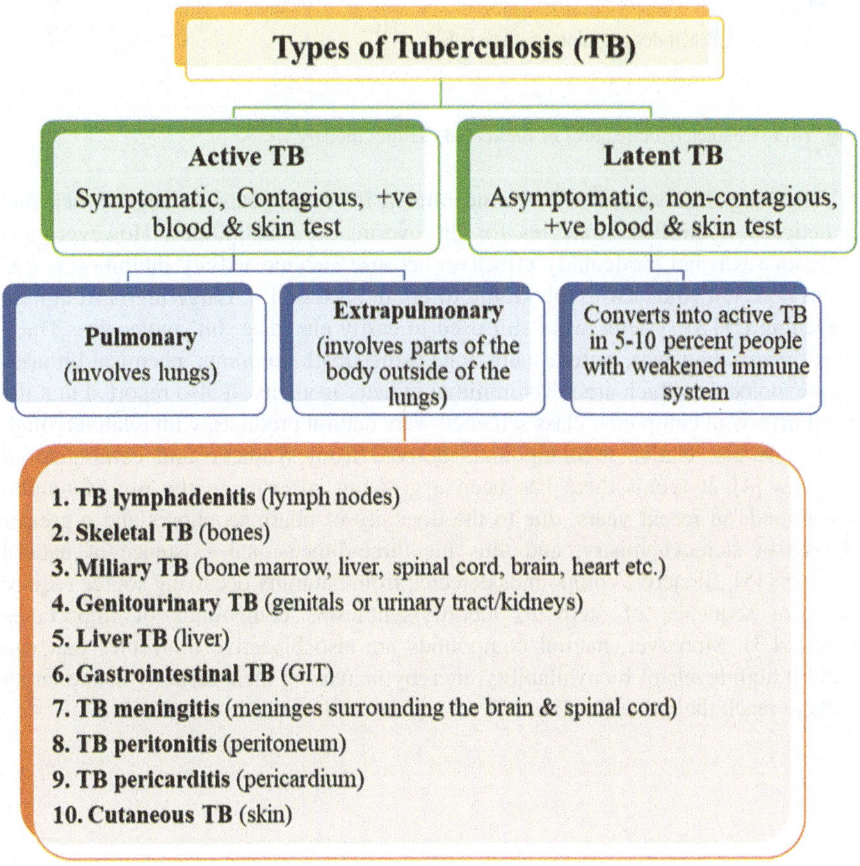

Fig. 14.2 Classification of types of tuberculosis (TB)

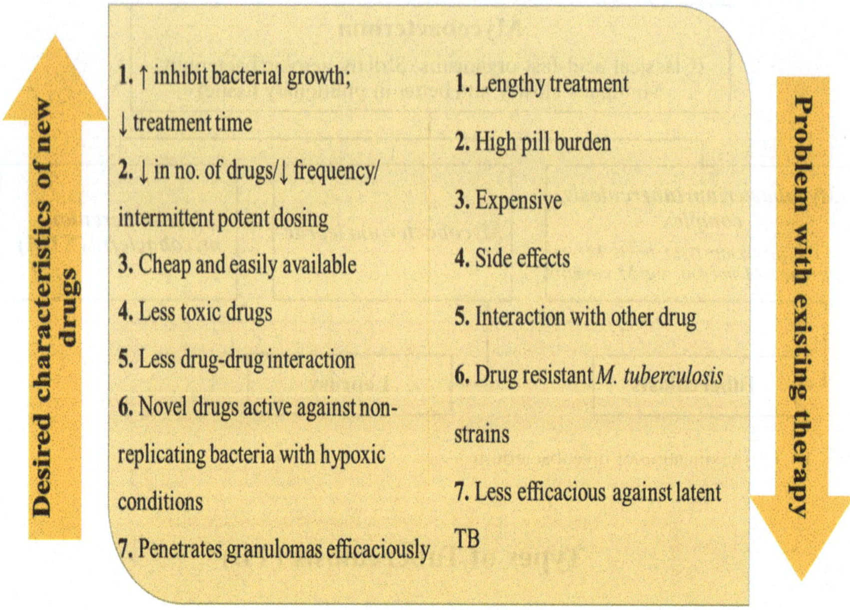

Fig. 14.3 Characteristic features of future and existing therapy

The drug discovery area has concentrated in recent years on target-based and genetically controlled strategies for discovering new antibiotics. However, this technique was not particularly effective, because enzyme activity inhibition sometimes does not equate with the killing of entire bacteria [3]. Large high-throughput screening (HTS) systems were still used to easily elucidate 'hit' molecules. These experiments, however, were usually done using small 'corporate' chemical libraries of the molecule, which are fairly minimal in heterogenicity. It also reported that the most divergent compound class screened were natural products, with relatively high impact levels relative to compounds derived from synthetic and combinatorial libraries [4]. It seems there has been a growing curiosity in the use of natural compounds in recent years, due to the diversity of pharmacophores and a greater degree of stereochemistry, and thus the three-dimensional existence of natural products [5]. Bioactive compounds detected from naturally occurring source require a given sequence of steps to identify/synthesize compounds of importance (Fig. 14.3). Moreover, natural compounds are also bioactive molecules that can exhibit high levels of bioavailability, thereby increasing their capacity inside target cells to reach their site of action.

14.2 Aetiology and Symptoms/Clinical Features

A strong immune system battles tuberculosis bacterium. But if you have reduced body weight and are undernourished and have HIV/AIDS, diabetes, serious kidney disease and neck and head cancers, you may not be able to ward off aggressive TB disease. Tuberculosis risk increases in medications for cancer such as chemotherapy; immunosuppressive agents; drugs for treating rheumatoid arthritis, Crohn's disease and psoriasis; etc. Babies and young children are even more likely to have it because their immune systems aren't fully developed. We may also have a greater risk of having TB if:

- You are a smoker/puffer.
- A relative, co-worker or member of his family has active TB.
- You live in an area where TB is popular or have travelled there.
- For TB patients, you are working or staying in a TB hospital/nursing home or healthcare provider.

Infections of TB often involve the lungs. Less often, TB infections arise in locations far outside lungs, such as in the small glands that make up the immune system. This includes the lymph nodes, bones, joints, gastrointestinal tract, urinary bladder, reproductive system and nervous system. TB that affects other areas of the body is more frequent in those with a compromised immune system. Symptoms can include those provided in Fig. 14.4.

14.3 Transmission Cycle [6]

M. tuberculosis is transferred from human to human and is generally transmitted by the air. Patients with pulmonary or laryngeal tuberculosis (TB) that cough up bacilli are the cause of infection. The patient develops tiny infectious droplets when coughing, talking or sneezing. These droplet nuclei have a diameter of around 1–5 μm. Based on the environment, the droplet nuclei can remain suspended in the air over several hours. Contamination may occur during inhalation of these air droplets. Sunlight, UV radiation and ventilation are efficient at reducing droplet capacity to penetrate the lung. The other transmission modes are considerably less general. Cutaneous or phlegm inoculation happens rarely, although these cases do arise in laboratory personnel. A patient's contagiousness is correlated to the volume of bacilli in his sputum. Patients with smear-positive test are the most infectious. Less resilient are those of smear-negative/culture-positive effects. Patients whose community and sputum smear microscopy are both unfavourable are normally not infectious.

Patients diagnosed with *M. tuberculosis* but have no active disease cannot spread TB. Extrapulmonary (EP) variants of TB are infectious even under rare circumstances. Children are usually much less infectious than adults. This may be attributed to poor cough dynamics, decreased development of sputum and lower

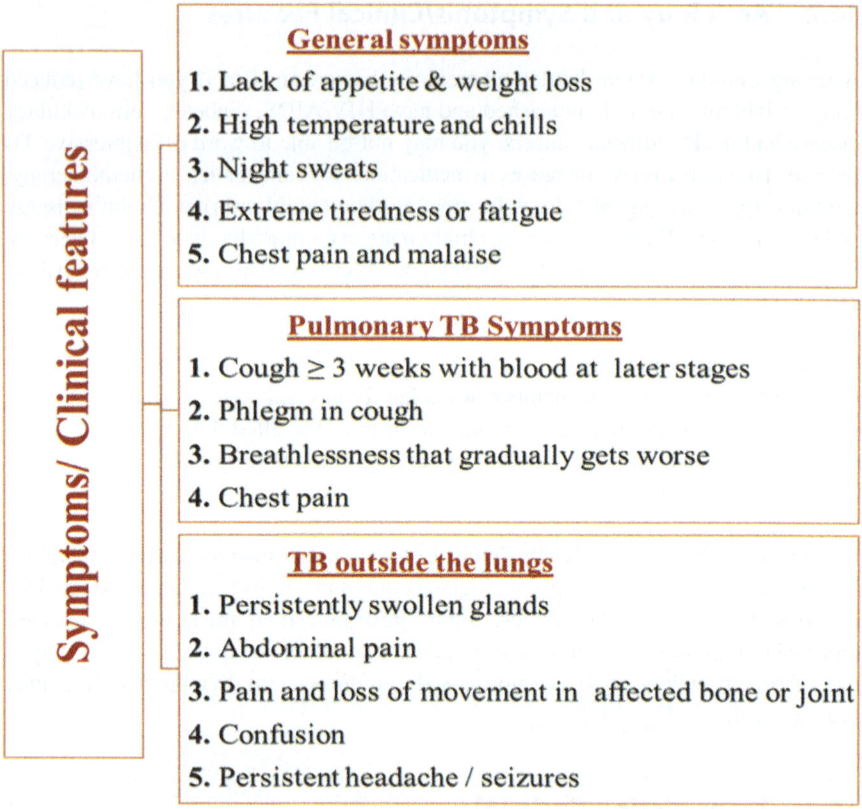

Fig. 14.4 Clinical symptoms of tuberculosis

bacillary load. Not everybody exposed to infectious TB gets infected with *M. tuberculosis*. The probability that TB will be transmitted depends on three factors:

(a) Contagiousness of the source (the greatest factor)
(b) Environment where the exposure occurred
(c) Duration of exposure

There are many diagnostic tests available for tuberculosis with their own advantages and disadvantages. Some of them are listed in Table 14.1.

Table 14.1 Diagnostic tests of tuberculosis [7, 8]

Sr. no.	Test/marker	Characteristic features
1	Symptom screen	Individuals can go with diagnosis if they have one or more of the symptoms described here: Existing coughing (for any duration and with blood at later stages), sweating at night, weight loss and fever
2	Chest radiograph	An X-ray of the chest is often used to identify anomalies. Lesions can occur in the lungs and may vary in size, shape and density. However, a chest X-ray could be used to eliminate the possibility of pulmonary infection in an individual who has had a positive reaction to a blood test for TST or TB and therefore has no disease symptoms
3	Smear microscopy	Analysis of sputum under a microscope to recognize mycobacterium is relatively quick, whereas precision is subject to debate
4	Mycobacterial culture	Identification of *Mycobacterium tuberculosis* growth over an incubation time of 6 weeks; tests of the culture are correct but sluggish
5	Nucleic acid amplification testing (NAAT) (a) *GeneXpert MTB/RIF.* (b) *Line probe assay (LPA)/ GenoType MTBDR.* (c) *Loop-mediated isothermal amplification (LAMP).*	It is a genotypic or molecular testing: (a) A molecular diagnostic procedure to identify TB and rifampicin resistance (MDR-TB surrogate marker) in just 2 hours; quicker and more flexible than conventional smear microscopy; somewhat costly and requires continuous electrical supply (b) Identifies mutations in the DNA of the TB bacterium using a colouring agent that shows gene mutations linked to drug resistance. This can diagnose TB and resistance in 1 day; with sufficient equipment and qualified staff (c) LAMP test is a manual NAAT. With LAMP, DNA magnification and identification of a gene can be done in a single step. The analysis somehow doesn't require advanced equipment. It produces a result which can be observed under ultraviolet light with the naked human eye
6	Tuberculin skin test (TST)	Tuberculin skin test (TST) and chest radiography are two widely used TB diagnostic examinations. They need refrigeration and an injection of a TB protein derivative to be administered under the skin. Chest X-rays in patients with positive TSTs can be used to rule out or validate active pulmonary tuberculosis. TSTs cannot differentiate the active from the latent TB
7	MGIT (mycobacteria growth indicator tube)	The MGIT method uses liquid culture to check whether the presence of different TB drugs can allow TB bacteria to develop. If TB grows, then perhaps the bacterium becomes tolerant to the test drug. MGIT tests take some days but they are much more accessible than traditional solid culture. It must be accommodated in a laboratory facility with appropriate facilities and qualified staff

(continued)

Table 14.1 (continued)

Sr. no.	Test/marker	Characteristic features
8	Interferon-gamma release assays (IGRAs)	Many TB studies that cannot discriminate between latent and active TB are interferon-gamma release assays (IGRAs). People who have been vaccinated with BCG before will go for this test. In TB patients, white blood cells, expresses interferon-gamma when combined with protein derivatives of bacterium
9	Genedrive	A recent NAAT has been designed for the identification of TB and rifampin-resistant TB; however, the responsiveness (zero percent in smear negative) and accuracy (45.6 percent in smear positive) are lower than in smear microscopy. Thus, Genedrive generates high false-negative results and provides little benefit over smear microscopy

14.4 Mechanisms of Drug Resistance

Drug-resistant TB (DR-TB) is a rising concern globally, with no nation or territory exempted. Multidrug-resistant TB (MDR-TB) is classified as at least isoniazid- and rifampicin-resistant TB. Extensively drug-resistant TB (XDR-TB) is classified as isoniazid- and rifampin-resistant TB, with resistance to any fluoroquinolone and at least one second-line injectable drugs. Various reasons behind anti-tubercular therapy are as follows: *impermeable cell wall*, *slow metabolism mechanism* and *possession of numerous efflux pumps* [9–11]. Some major genes responsible for drug resistance in tuberculosis are shown in Fig. 14.5.

14.5 Treatment of Tuberculosis

The primary drug therapy for tuberculosis includes the following drugs shown in Fig. 14.6.

Apart from the drugs shown in Fig. 14.6, there are many drugs along with their molecular targets that have been identified as shown in Table 14.2.

14.6 Conclusion

The present treatment methods have little effect on management of TB; therefore, there is an immediate need for new therapies. Few strains of *M. tuberculosis* are resistant to all currently available antibiotics used in the therapy of TB, which illustrates the need for new medications with new mechanism of action. The high degree of target-based molecule invention and high-throughput screening utilizing

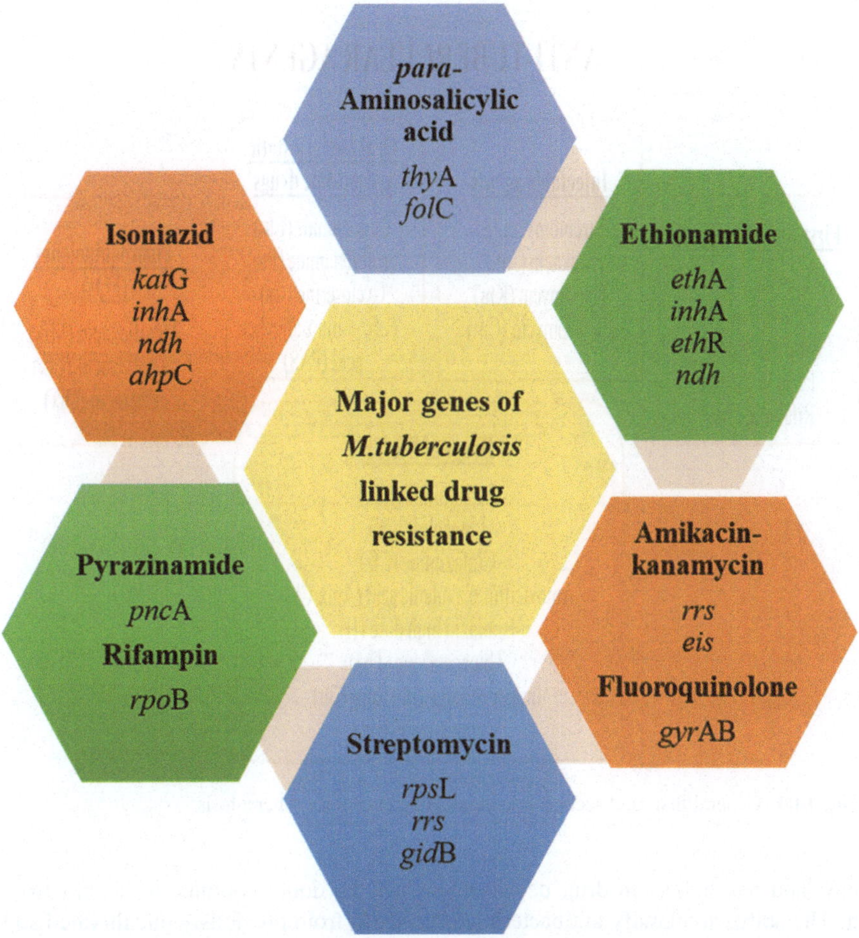

Fig. 14.5 Some major genes of *M. tuberculosis* linked to drug resistance

synthetic compound repositories have revived interest in natural ingredients as a source of various antibacterial action bioactive compounds. A large number of natural substances show promising results as clinically potentially anti-mycobacterial agents. Natural compounds are rich and underused source of unique chemical scaffolds. But purifying active ingredients from natural source is cumbersome, laborious as well as time-consuming technique. Natural substance screens focus on bioassay-guided derivatization and involve advanced structure determination, as compared to screening of chemical compound libraries, which start with pure compounds with known structure. Another problem lies with intellectual property rights, which are much more difficult to implement and secure when the origin of a novel drug begins to grow freely in the environment. Even after these concerns, natural products continue to stay as the most productive source of drug leads to this

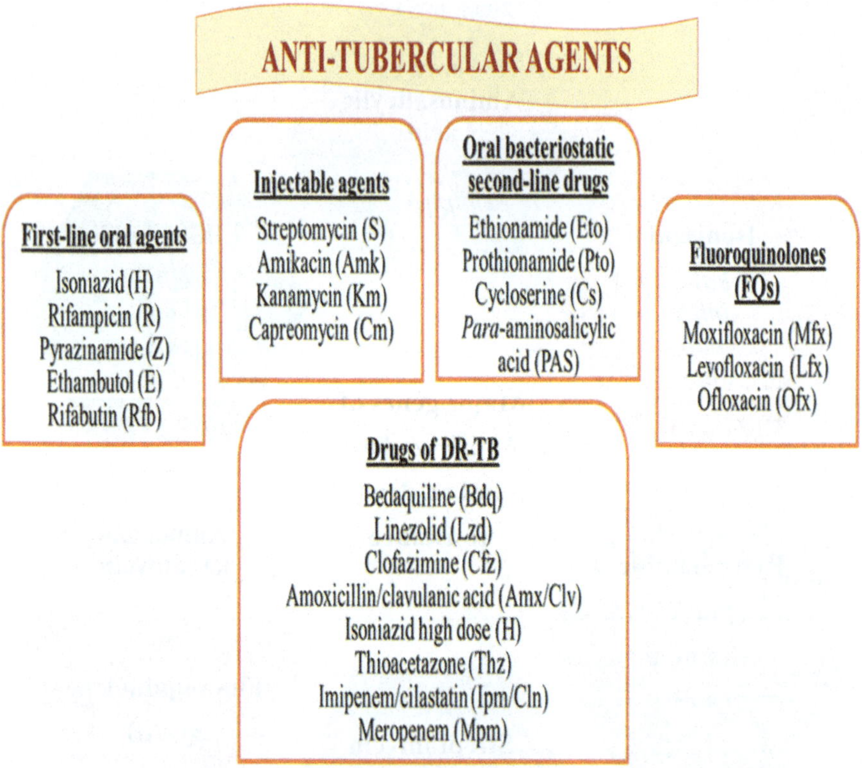

Fig. 14.6 General first- and second-line therapeutic agents for tuberculosis

day, and researchers in drug development will no doubt continue to benefit from it. The ability to classify antibacterial compounds from previously uncultivated soil bacteria species, the metagenomic methods to investigate microbial diversity and the successful clinical use of marine natural products provide hope that nature can provide the required starting points for managing the TB epidemic.

In *M. tuberculosis*, drug resistance is not the result of a single homogeneous genetic entity. Perhaps that is because of repeated mutations in multiple genes that encode for antibiotic resistance. The slow metabolism during an extended dormant period often greatly increases its drug resistance; the waxy impermeable cell wall with the help of multiple efflux pumps is necessary for withstanding the potency of antibiotics. Having sufficient information on drug resistance's molecular pathways in *M. tuberculosis* can be of assistance in identifying potential drug development goals. The number of new lead chemical compounds is also too limited to fulfil therapeutic needs. Whole-cell screens are expected to control future screening initiatives and to recognize more inhibitors that target processes linked to the cell envelope. Regardless of the inhibitor screening method, continued experiments to better understand Mtb's basic physiology and to establish screening conditions that

Table 14.2 Drugs/molecular targets against tuberculosis disease

Sr. no.	Category	Drug/molecule	Molecular targets	Mechanism	Features	References
1	Isonicotinic acid	Isoniazid	InhA, a NADH-dependent enoyl-acyl carrier protein reductase	Inhibition of synthesis of mycolic acid and cell wall	Approved first-line drug	[12]
2	Rifamycin	Rifampicin Rifapentine	RNA polymerase, β-subunit	Inhibition of RNA synthesis which catalyses transcription of DNA to RNA	Approved first-line drug	[13, 14]
3	Ethylenediamine	Ethambutol	Arabinosyl transferase	Inhibition of arabinogalactan synthesis	Approved first-line drug	[15, 16]
4	Pyrazine	Pyrazinamide	Pyrazinamidase	Ribosomal protein 1; 30S ribosomal subunit in cytoplasm	Approved first-line drug	[17, 18]
5	Aminoglycoside	Streptomycin	30S ribosomal protein; 7-methyl-guanosine methyltransferase	Inhibition of arabinogalactan and protein synthesis	Approved first-line drug	[19, 20]
6	Anti-leprotic	Clofazimine	Type II NADH dehydrogenase (NDH-2)	Inhibits the point of entry of electrons into the respiratory chain	Repurposed for MDR-TB	[21, 22]
7	Neuroleptic (phenothiazines)	Thioridazine	Type II NADH dehydrogenase (NDH-2)	Inhibits the point of entry of electrons into the respiratory chain	Repurposed for MDR-TB	[23–25]
8	Imidazopyridines	Q203	Cytochrome bc1 complex	Strongly suppresses ATP synthesis by mycobacterial membrane vesicles	Phase 1 phenotypic screening using infected macrophages	[26]
9	Diarylquinolines	BDQ	ATP synthase	Drops cellular ATP level	Phenotypic screening for MDR-TB	[27]

(continued)

Table 14.2 (continued)

Sr. no.	Category	Drug/molecule	Molecular targets	Mechanism	Features	References
10	Amides of squaric acids	Squaramides	ATP synthase	Drops cellular ATP level	Preclinical screening for ATP synthesis	[28]
11	Antimicrobial agent	Pyrazinamide	Proton-motive force	Downregulates biosynthetic pathways	Approved murine model for DS[a] and MDR-TB	[29, 30]
12	1,2-ethylene diamine	SQ109	Proton-motive force	Targets MmpL3 (essential gene and function as a transporter of mycolic acids)	Phenotypic phase 2 screening for DS and MDR-TB	[31, 32]
13	Mycobactin/ carboxymycobactin	–	MmpL4	Implicated in transport of virulence factors; required for Mtb growth in early stages of infection	Cell-based screening	[33]
14	Adamantyl-urea inhibitor	AU1235	MmpL3	Involved in transport of mycolic acids; implicated in heme transport; essential gene	Cell-based screening	[33]
15	Phthiocerol dimycocerosate (PDIM)	–	MmpL7	Implicated in transport of virulence factors; required for Mtb growth in early stages of infection; transport of PDIM	Cell-based screening	[33]
16	Sulfolipid	–	MmpL8	Transport of sulfolipids; required for survival during chronic stages of infection	Cell-based screening	[33]
17	Long-chain TAG/mycolate wax esters	–	MmpL11	Transport of heme; required for survival during chronic stages of infection	Cell-based screening	[33]
18	Carbapenem-β-lactamase inhibitor combination	Meropenem/ clavulanate	L,D-Transpeptidase; penicillin-binding proteins	β-Lactam inhibiting PG cross-linking; β-lactamase inhibitor	Repurposing of existing antimicrobials	[34]

19	Translocase I inhibitors	SQ641 (lead capuramycin analogue) (Sequella)	Phospho-N-acetylmuramyl-pentapeptide translocase	Inhibits translocase I	Cell- and target-based	[34]
20	DprE1 inhibitor	Benzothiazinone BTZ043	DprE1	Inhibits the epimerization of decaprenylphosphoryl ribose to decaprenylphosphoryl arabinose (arabinose donor in the building of arabinan domains of arabinogalactan and lipoarabinomannan)	Cell-based screening methods	[34]
21	Putative cell wall inhibitor	Dipiperidine SQ609	Unknown	Inhibits the synthesis of major cell envelope constituents	Cell-based and target-based phenotypic screen	[34]
22	EthR inhibitors	BDM31343 and analogues	EthR	–	Target- and structure-based lead optimization	[34]
23	Protein kinase inhibitors	Several chemical scaffolds	PknA, PknB, PknG	Inhibits the activity of multiple enzymes and transporters	Target-based hit to lead development	[34]
24	InhA inhibitors	Novel structural classes	InhA	Increases bactericidal activity	Target-based lead optimization	[34]
25	MEP pathway inhibitors	Fosmidomycin analogues	Dxr	Competitive inhibitor of enzyme of the pathway, 1-deoxy-d-xylulose-5-phosphate reducto-isomerase (DXR)	Target-based hit to lead development	[34]
26	Caprazamycin derivative	CPZEN-45	WecA	Inhibits transferase involved in arabinogalactan biosynthesis	In vitro against *M. tuberculosis*	[35]

(continued)

Table 14.2 (continued)

Sr. no.	Category	Drug/molecule	Molecular targets	Mechanism	Features	References
27	Mycins	Thiolactomycin	FAS-II	Blocks mycolic acid synthesis; inhibits bacterial fatty acid biosynthesis	–	[36]
28		Lipiarmycin A3	–	RNA polymerase inhibitors	Effective against *M. tuberculosis* H37Rv in vitro	[37]
29	Antimicrobial peptides	Teixobactin Sansanmycins	Peptidoglycan	Inhibits cell wall biosynthesis	Target-based lead optimization	[38]
30		Cyclomarin A Lassomycin Ecumicin	Caseinolytic protease C1 (ClpC1)	Increases the ATPase activity and inhibits protein degradation; alters cellular homeostasis	Early stages of lead optimization	[39]
31		Griselimycin	DnaN	Inhibits interaction of DnaN with DNA polymerase involved in DNA repair	Early stages of lead optimization	[40]
32	Anti-mycobacterial peptides	Lariatins	–	–	Early stages of lead optimization	[41]
33	Aminolipopeptides	Trichoderins	–	Inhibits ATP synthesis	Early stages of lead optimization	[42]

accurately represent the physical environments experienced by TB bacillus in humans will be a key to their success.

References

1. Mangtani P, Abubakar I, Ariti C, Beynon R, Pimpin L, Fine PEM et al (2014) Protection by BCG vaccine against tuberculosis: a systematic review of randomized controlled trials. Clin Infect Dis 58:470–480
2. Gillespie SH, Crook AM, McHugh TD, Mendel CM, Meredith SK, Murray SR et al (2014) REMox TB consortium. Four-month moxifloxacin-based regimens for drug-sensitive tuberculosis. N Engl J Med 371:1577–1587
3. Payne DJ, Gwynn MN, Holmes DJ, Pompliano DL (2007) Drugs for bad bugs: confronting the challenges of antibacterial discovery. Nat Rev Drug Discov 6:29–40
4. Sukuru SC, Jenkins JL, Beckwith RE, Scheiber J, Bender A, Mikhailov D et al (2009) Plate-based diversity selection based on empirical HTS data to enhance the number of hits and their chemical diversity. J Biomol Screen 14:690–699
5. Harvey AL, Edrada-Ebel R, Quinn RJ (2015) The re-emergence of natural products for drug discovery in the genomics era. Nat Rev Drug Discov 14:111–129
6. https://www.nap.edu/read/9144/chapter/4 (Last assessed on 08 September 2020)
7. http://www.tbonline.info/diagnostics/ (Last assessed on 08 September 2020)
8. https://www.treatmentactiongroup.org/wp-content/uploads/2017/04/TB-Diagnostics-Guide.pdf (Last assessed on 08 September 2020)
9. Goswam A, Chakraborty U, Bhattacharya B, Pal NK (2016) Association of generation time with anti-tubercular drug(s) resistance pattern of *Mycobacterium tuberculosis* isolates among treatment failure pulmonary tuberculosis patients. Asian J Pharm Clin Res 9(1):258–261
10. Nguyen L (2016) Antibiotic resistance mechanisms in *M. tuberculosis*: an update. Arch Toxicol 90(7):1585–1604
11. Somoskovi A, Parsons LM, Salfinger M (2001) The molecular basis of resistance to isoniazid, rifampin, and pyrazinamide in *Mycobacterium tuberculosis*. Respir Res 2(3):164–168
12. Timmins GS, Deretic V (2006) Mechanisms of action of isoniazid. Mol Microbiol 62:1220–1227
13. Sensi P (1983) History of the development of rifampin. Rev Infect Dis 5:S402–S406
14. Campbell EA, Korzheva N, Mustaev A et al (2001) Structural mechanism for rifampicin inhibition of bacterial RNA polymerase. Cell 104:901–912
15. Pourakbari B, Mamishi S, Mohammadzadeh M, Mahmoudi S (2016) First-line anti-tubercular drug resistance of *Mycobacterium tuberculosis* in IRAN: a systematic review. Front Microbiol 28:1139
16. Forbes M, Kuck NA, Peets EA (1962) Mode of action of ethambutol. J Bacteriol 84:1099–1103
17. Konno K, Feldmann FM, McDermott W (1967) Pyrazinamide susceptibility and amidase activity of tubercle bacilli. Am Rev Respir Dis 95:461–469
18. Zhang S, Che J, Shi W et al (2013) Mutations in panD encoding aspartate decarboxylase are associated with pyrazinamide resistance in *Mycobacterium tuberculosis*. Emerg Microbes Infect e34:2
19. Honore N, Cole ST (1994) Streptomycin resistance in mycobacteria. Antimicrob Agent Chemother 38:238–242
20. Dookie N, Rambaran S, Padayatchi N et al (2018) Evolution of drug resistance in *Mycobacterium tuberculosis*: a review on the molecular determinants of resistance and implications for personalized care. J Antimicrob Chemother 73:1138–1151
21. Yano T, Kassovska-Bratinova S, Teh JS, Winkler J, Sullivan K, Isaacs A, Schechter NM, Rubin H (2011) Reduction of clofazimine by mycobacterial type 2 NADH: quinone oxidoreductase: a

pathway for the generation of bactericidal levels of reactive oxygen species. J Biol Chem 286:10276–10287
22. Grosset JH, Tyagi S, Almeida DV, Converse PJ, Li SY, Ammerman NC et al (2013) Assessment of clofazimine activity in a second-line regimen for tuberculosis in mice. Am J Respir Crit Care Med 188:608–612
23. Boshoff HI, Myers TG, Copp BR, McNeil MR, Wilson MA, Barry CE III (2004) The transcriptional responses of *Mycobacterium tuberculosis* to inhibitors of metabolism: novel insights into drug mechanisms of action. J Biol Chem 279:40174–40184
24. Weinstein EA, Yano T, Li LS, Avarbock D, Avarbock A, Helm D, McColm AA, Duncan K, Lonsdale JT, Rubin H (2005) Inhibitors of type II NADH: menaquinone oxidoreductase represent a class of antitubercular drugs. Proc Natl Acad Sci U S A 102:4548–4553
25. Kristiansen JE, Dastidar SG, Palchoudhuri S, Roy DS, Das S, Hendricks O, Christensen JB (2015) Phenothiazines as a solution for multidrug resistant tuberculosis: from the origin to present. Int Microbiol 18:1–12
26. Pethe K, Bifani P, Jang J, Kang S, Park S, Ahn S et al (2013) Discovery of Q203, a potent clinical candidate for the treatment of tuberculosis. Nat Med 19:1157–1160
27. Koul A, Dendouga N, Vergauwen K, Molenberghs B, Vranckx L, Willebrords R et al (2007) Diarylquinolines target subunit c of mycobacterial ATP synthase. Nat Chem Biol 3:323–324
28. Tantry SJ, Markad SD, Shinde V, Bhat J, Balakrishnan G, Gupta AK et al (2017) Discovery of imidazo[1,2-α]pyridine ethers and squaramides as selective and potent inhibitors of mycobacterial adenosine triphosphate (ATP) synthesis. J Med Chem 60:1379–1399
29. Zhang Y, Wade MM, Scorpio A, Zhang H, Sun Z (2003) Mode of action of pyrazinamide: disruption of *Mycobacterium tuberculosis* membrane transport and energetics by pyrazinoic acid. J Antimicrob Chemother 52:790–795
30. Zhang Y, Shi W, Zhang W, Mitchison D (2014) Mechanisms of pyrazinamide action and resistance. Microbiol Spectr 2:MGM2-0023-2013
31. Li K, Schurig-Briccio LA, Feng X, Upadhyay A, Pujari V, Lechartier B et al (2014) Multi-target drug discovery for tuberculosis and other infectious diseases. J Med Chem 57:3126–3139
32. Reddy VM, Einck L, Andries K, Nacy CA (2010) *In vitro* interactions between new antitubercular drug candidates SQ109 and TMC207. Antimicrob Agents Chemother 54:2840–2846
33. Geetha VR (2014) MmpL3 a potential new target for development of novel anti-tuberculosis drugs. Expert Opin Ther Targets 18(3):247–256
34. Mary J, Michael RM, Patrick JB (2013) Progress in targeting cell envelope biogenesis in *Mycobacterium tuberculosis*. Future Microbiol 8(7):855–875
35. Ishizaki Y, Hayashi C, Inoue K, Igarashi M, Takahashi Y, Pujari V et al (2013) Inhibition of the first step in synthesis of the mycobacterial cell wall core, catalyzed by the GlcNAc-1-phosphate transferase WecA, by the novel caprazamycin derivative CPZEN-45. J Biol Chem 288:30309–30319
36. Hayashi T, Yamamoto O, Sasaki H, Kawaguchi A, Okazaki H (1983) Mechanism of action of the antibiotic thiolactomycin inhibition of fatty acid synthesis of Escherichia coli. Biochem Biophys Res Commun 115:1108–1113
37. Sergio S, Pirali G, White R, Parenti F (1975) Lipiarmycin, a new antibiotic from Actinoplanes III. Mechanism of action. J Antibiot (Tokyo) 28:543–549
38. Xie Y, Chen R, Si S, Sun C, Xu H (2007) A new Nucleosidyl-peptide antibiotic, Sansanmycin. J Antibiot 60:158–161
39. Lee H, Suh JW (2016) Anti-tuberculosis lead molecules from natural products targeting Mycobacterium tuberculosis ClpC1. J Ind Microbiol Biotechnol 43:205–212
40. Kling A, Lukat P, Almeida DV, Bauer A, Fontaine E, Sordello S et al (2015) Targeting DnaN for tuberculosis therapy using novel griselimycins. Science 348:1106–1112
41. Iwatsuki M, Uchida R, Takakusagi Y, Matsumoto A, Jiang C-L, Takahashi Y et al (2007) Lariatins, novel anti-mycobacterial peptides with a lasso structure, produced by *Rhodococcus jostii* K01-B0171. J Antibiot 60:357–363

42. Pruksakorn P, Arai M, Liu L, Moodley P, Jacobs WR Jr, Kobayashi M (2011) Action mechanism of trichoderin a, an anti-dormant mycobacterial aminolipopeptide from marine sponge-derived Trichoderma sp. Biol Pharm Bull 34:1287–1290

Cellular and Molecular Mechanisms of Repurposed Antidiabetic Drug as an Adjunctive Treatment for Tuberculosis

15

Rupesh K. Gautam, Ritu Mishra, Kanika Sharma, and Manju Sharma

Abstract

Apart from the accessibility of antitubercular therapy (ATT), tuberculosis (TB) emerged to be a chief cause of mortality around the world. The present ATT had a disadvantage of lengthy period that creates a problem of noncompliance in patients and growth of resistance. The greater price and delayed temperament of TB medication development coupled with low advantages lead to repurposing of complementary drugs which may contribute as an innovative pharmaceutical approach. Metformin has currently engrossed a major consideration as a host-directed adjunctive therapy (HDT) and has several complementary roles on cellular and molecular metabolism, immunity of host, and transcription of genes engaged in innate host responses to *M. tuberculosis.* It has an inhibitory effect on mitochondrial complex I and has been found to increase AMP/ATP ratio, with the help of a series of several pathways, and causes bacterial killing. This chapter would discuss in detail about the cellular and molecular mode of action of metformin including its impact on T helper cell 1 (TH1) along with trends which metformin demonstrates in reference to CD4+ and CD8+ cells. The necessity for adjunctive host-targeted therapy and the synergistic role of metformin with other antitubercular medications have been thoroughly debated. Novel

R. K. Gautam (✉)
Department of Pharmacology, MM School of Pharmacy, Maharishi Markandeshwar University, Sadopur-Ambala, Haryana, India

R. Mishra · M. Sharma
Department of Pharmacology, School of Pharmaceutical Education and Research, Jamia Hamdard, New Delhi, India

K. Sharma
Department of Pharmacy, Ram-Eesh Institute of Vocational and Technical Education, Greater Noida, Uttar Pradesh, India

K. Dua et al. (eds.), *Targeting Cellular Signalling Pathways in Lung Diseases*,
https://doi.org/10.1007/978-981-33-6827-9_15

strategy to fight drug-resistant TB in concurrence with future perspectives has been discussed in this chapter.

Keywords

Tuberculosis · Antitubercular therapy · Drug repurposing · Metformin · Cellular and molecular mechanism

15.1 Introduction

Tuberculosis (TB) is a communicable ailment and a global health epidemic with a considerable morbidity and mortality [1]. Approximately ten million individuals develop TB as per 2018 global TB report. The appearance of resistance has been found to pose a problem for current anti-TB treatment and creates a serious concern to community health around the world. The WHO has reported 484,000 fresh cases of rifampicin resistance, the most powerful first-line antitubercular medication [2]. The continuous increase of multidrug-resistant tuberculosis (MDR-TB) creates a foremost challenge to TB management all over the world. Management of such resistant strains is difficult and needs a use of expensive, lesser potent medications which contributes to increased death [3]. For the past few years, treatment of TB had been using several potent medications for treating infection. TB should primarily be managed for 6 months. The existing anti-TB medications have various drawbacks such as adverse events and interactions with concomitant medications [4]. In addition, high price and lengthy characteristics of TB medication screening coupled with less profit have hindered novel anti-TB medication manufacture, and therefore repositioning of current medications as anti-TB drug can be useful.

Individuals with weaker immune function like diabetes mellitus (DM) have a greater chance of progression from latent TB to active tubercular infection. The interrelation between TB and DM is well recognized. DM is related with decreased innate and adaptive immune responses that are requisite to fight *Mycobacterium tuberculosis* (*Mtb*) intracellular proliferation [5]. Host cell identification in diabetics is reduced resulting in reduced immune response, making diabetics more vulnerable to infections. There are evidences which show that DM is responsible for causing TB and may influence ailment presentation and clinical outcomes and decreases TB control [6–8]. Moreover, TB induces glucose resistance and exacerbates glycemic levels in persons with diabetes [9–12].

15.2 Relationship of Tuberculosis and Diabetes

15.2.1 Diabetes Causes TB

Diabetes is responsible for lower respiratory tract infections like TB. TB is related with several immunosuppressive conditions like HIV, and diabetes serves to be a crucial factor for tubercular infections [13]. A research by Stevenson et al. confirmed that diabetes raises the likelihood of TB from 1.5 to 7.8 times [14], whereas Jeon and Murray meta-analysis established that the probabilities for TB in peoples with diabetes were 3.11 [15, 16]. An American study once again stated that the ratio of odds for multidrug-resistant (MDR) TB coupled with peoples of diabetes is 2.1 [17]. While T2DM is more common, the chances of TB in type 1 diabetes are three to five times greater because of comparatively weaker regulation, less body weight, and younger age of influenced people. TB coupled with DM can also exhibit greater rates of hemoptysis, fever, and unusual presentations in contrast to nondiabetics with TB [18]. In 2007, Alisjahbana et al. stated that peoples with DM and TB had chronic symptoms, without any serious ailment, and higher percentage of positive sputum for AFB after 2 months of microscopic assessment compared to their nondiabetic counterparts. Moreover, after calibrating for chest X-ray anomalies and mycobacterial load prior to management, this was no longer statistically considerable. Other findings also showed a pattern toward enhanced time for conversion of sputum [19–22], while others reported no association in diabetes and rate of conversion of sputum at the termination of intensive phase [21–24]. In persons with TB, DM raises the chances for failure of treatment, mortality, and relapse. A research by Baker et al. investigated individuals with DM had a risk ratio (RR) for results of mortality of 1.69, whereas RR of mortality in TB management in 23 research findings was 1.89 [25]. DM is also related with raised risk of relapse (RR, 3.89) [25]. Furthermore, there was no confirmation for an enhanced chance of TB in peoples with DM. The Brazilian research performed via compilation of Brazilian national dataset from 2001 to 2011 to evaluate sociodemographic and clinical parameters that could impact various TB outcomes, and it had been investigated that the increase of MDR-TB was more linked with relapse. Diabetes may influence pharmacokinetics of anti-TB medications, mainly rifampicin, by minimizing their plasma levels. Consequently, the treatment schedule for TB for both diabetics and nondiabetics generally remains unchanged [13].

15.2.2 TB as a Diabetes Risk Factor

While a two-way correlation in DM and TB has been documented, there are few devoted findings to establish whether TB enhances the chances of diabetes [26–29]. TB might result in impaired glucose tolerance (IGT) [29, 30] and DM. In general, IGT becomes normal after TB had been treated effectively but continues to be the main reason for causing type 2 diabetes mellitus [31]. It is hard to mark TB as the causative factor for hyperglycemia or DM with formerly undiagnosed patients of TB [32–34]. In a research of TB patients without DM, Basoglu et al. [27, 29] observed glucose intolerance in 10.4% and DM in 8.6%, and compared to control groups of community-acquired pneumonia, 17.4% had DM, and nobody had glucose intolerance. There was no considerable disparity in both the groups ($P > 0.05$). The values of the oral glucose tolerance test (OGTT) changed to usual readings in TB and pneumonia groups post treatment.

15.2.3 The Fundamental Pathophysiological Role

The escalated chance of TB in persons of DM had several factors, and there have been many proposed mechanisms. There is a reduction in cellular immunity caused by decreased T lymphocyte values, minimal neutrophil count, and their functions [34, 35]. Individuals with diabetes demonstrate decreased T helper 1 (TH1) cytokine responses and tumor necrosis factor (TNF-alpha and TNF-beta), interleukin-1, and interleukin-6 generation in contrast to their nondiabetic patients. Moreover macrophage dysfunction has been found in diabetes that causes lesser generation of reactive oxygen species and phagocytic and chemotactic activity [36]. In patients of diabetes, chemotaxis of monocytes was also observed, an imperfection that does not get better with insulin. Hyperglycemia is also believed to damage the capacity of respiratory burst in eliminating pathogens [37–39]. While such putative mechanisms are reasonable, it is crucial that more research should be performed to authenticate them. The stress reaction to infection also causes dysglycemia, a condition mediated by the role of interleukin-1 (IL-1), interleukin-6 (IL-6), and TNF-alpha (Fig. 15.1). This temporal association has been seen in some studies in which 42.6% of patients of TB were observed to have IGT or DM with a considerable decrease in the rates after management. With some progressive diabetes, a greater percentage of patients sustained to have glucose intolerance. Even though it is assumed that the greater irregular test findings acquired a month, subsequent to the start of treatment, may represent exact glucose intolerance instead of stress reaction to infection [40]. Later TB can develop into TB pancreatitis and pancreatic endocrine hypofunction that contributes to IGT or DM and deteriorates its regulation. TB pancreatitis is apparent after the individual progresses to DM. It is believed that malnutrition had been anticipated as a cause for dysglycemia and infections and also body mass index was not related with IGT or DM.

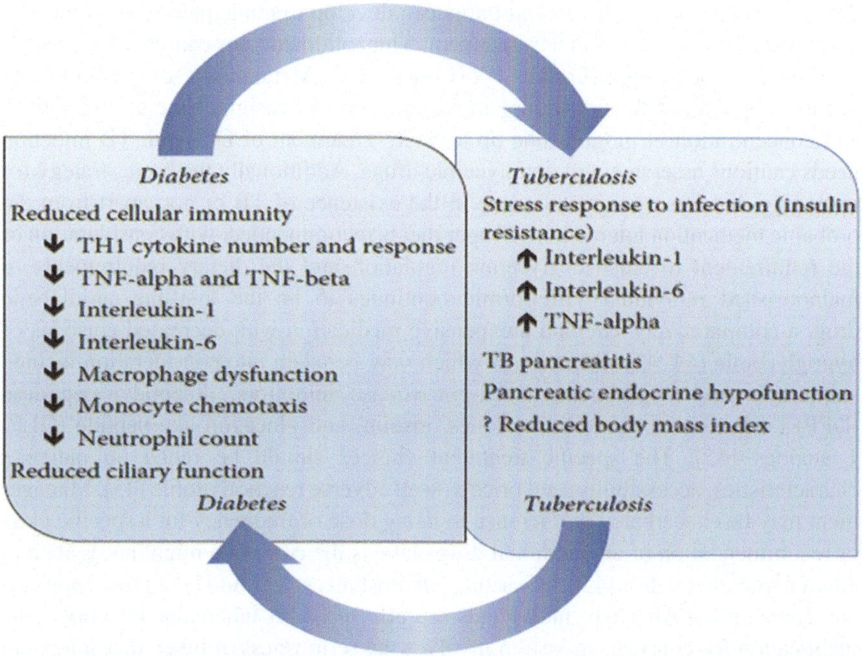

Fig. 15.1 Pathophysiological mechanism involved in TB-diabetes association [35–39]. *TH1* T helper 1, *TNF* tumor necrosis factor, *TB* tuberculosis

15.2.4 Treatment of Comorbid Diabetes and TB

Apart from the fact that diabetes may cause more serious ailments and mortality, the medications and length of anti-TB medications in peoples with TB and without TB are almost similar. Conventionally, most programs typically manage TB for 6 months, consisting of 2 months intensive phase that includes drugs like isoniazid, rifampicin, pyrazinamide, and ethambutol and 4 months continuation phase that includes rifampicin and isoniazid. There is a putative pharmacokinetic and pharmacodynamic interaction among anti-TB medications and antidiabetic drugs. Rifampicin is essential as an anti-TB drug and enhances metabolism of sulphonylureas and biguanides, by enzyme induction minimizing their plasma concentration and hence contributing to hyperglycemia. It also augments intestinal glucose absorption among nondiabetics. Isoniazid opposes the effect of sulphonylureas and deteriorates glycemic control. While isoniazid reduces metabolism of antiglycemic drugs and enhances their plasma levels like cytochrome P2C9 (CYP2C9) engaged in sulphonylurea metabolism whereas the inducing action of rifampicin is assumed to be greater than the inhibitory action [41, 42]. It can also hinder release of insulin in nondiabetics which contributes to hyperglycemia. Isoniazid and rifampicin are not affecting the insulin collapse significantly as insulin degradation occurred by hydrolysis of disulfide bonds by insulin-degrading enzyme in the liver [43]. In addition, dipeptidyl peptidase (DPP) IV inhibitors can develop immune paresis and probably deteriorate clinical results in TB treatment. Thiazolidinediones can be substrates for rifampicin-induced cytochrome P450 enzymes. Metabolism of rosiglitazone occurred by CYP2C8, and rifampicin reduces level of rosiglitazone up to 54–65% and concentration of pioglitazone up to 54%. Treatment of DM with TB infection needs cautious assessment of antiglycemic drugs. Additionally the basic strategy for managing diabetes remains the same in the existence of TB or not, apart from the probable medication interactions. Proper diet is recommended, with consideration of the requirement to balance glycemic regulation and the dietary requirements of malnourished individuals. Metformin continues to be the first-line antidiabetic drug, a comparatively safe and inexpensive medication with decreased episodes of hypoglycemia [44, 45]. Other drugs which may be taken into consideration include sulphonylureas, meglitinides, alpha-glucosidase inhibitors, dipeptidyl peptidase (DPP) IV inhibitors, thiazolidinediones, insulin, and glucagon-like peptide (GLP) 1 analogs [45]. The specific treatment choices should be relied on patient's characteristics, accessibility, and price as well adverse reaction profile [45]. Management may be exacerbated in reference to rising dose or frequency for a specific class or recommendation of an additional drug class as the condition might necessitate to attain glycemic control [45]. In various circumstances, insulin is the most appropriate agent in T2DM where individuals are affected with tubercular infection. The justification for considering insulin involves the seriousness of tubercular infection, loss of body tissues, requirement for raised anabolism, pancreatic hypofunction, interaction among antidiabetic drugs and some antitubercular drug, and likelihood of liver disease that may prohibit use of oral drugs. For mentioned factors, administration of antidiabetic agents for individuals with existing DM can be changed to insulin in TB once diagnosed, or if still administering insulin, changes should be done to exacerbate glycemic regulation. If the glucotoxicity is controlled and infection is reduced, the requisite of insulin may fall. However, once the appetite enhances and food consumption improves, requirement can rise again. Insulin preferences must be relied on safety, efficacy, price, and patient requirements. It has to be considered that if the infection is minimized, antidiabetic agents could be cautiously recommended, and regular glucose monitoring is vital for optimum control. This can contribute to identification of plausible adverse events like hypoglycemia from few antidiabetic drugs like sulphonylurea and insulin. However, in developing nations, patients need to give price for glucometers, and no settlement is provided. This hinders the capability to accomplish goals and recognize and validate hypoglycemia. In addition, patient education is crucial in identifying the characteristics of disease (TB and diabetes), treatment period, adverse effects of medications, and disease complications and promotion of healthy lifestyle [46–48].

15.2.5 Pharmacological Considerations in the Co-treatment of Diabetes Mellitus and Tuberculosis

Infection is considered to hinder diabetic control and there is no exclusion for TB. While TB may lead to glucose intolerance and makes individuals more vulnerable to DM, the medications prescribed to cure TB may deteriorate glycemic management in diabetics. In the co-management of TB and DM, like peripheral neuropathy produced by the management of TB with isoniazid, intersecting of toxicities should also be considered. Due to the chances of peripheral neuropathy, isoniazid and pyridoxine must be administered throughout tuberculosis management in diabetics. Additionally management with rifampicin may cause hyperglycemia through interactions with hypoglycemic medications [49, 50].

Rifampicin is a possible inducer of a host of metabolizing enzymes and comprises phase II enzymes and cytochrome P450 system enzymes [51], and enzyme induction can cause metabolism of medications prescribed with rifampicin and minimized management results. Sulfonylureas are frequently prescribed antidiabetic medications for individuals of non-insulin dependent diabetes mellitus (NIDDM). Glyburide and glipizide are two substrates of cytochrome P450 isoenzyme 2C9 (CYP2C9), and their pharmacokinetic analysis indicates that serum levels of both medications are reduced by 39% and 22% while prescribed with rifampicin. Pharmacodynamic readings suggest that hypoglycemic action of glyburide is minimized when recommended with rifampicin.

Thiazolidinediones are mainly utilized as substrates for cytochrome P450 enzymes [52–55]. Another antidiabetic medication like repaglinide experiences a reduction in an area under curve by 31–57% when prescribed with rifampicin, even though its glucose-controlling action was decreased in a research finding and remains unaffected in a different study [56, 57]. In insulin dependent diabetes mellitus (IDDM) patients, insulin need may enhance when rifampicin is used [56]. It is already established that rifampicin causes early-phase hyperglycemia with related hyperinsulinemia in individuals without DM [58, 59]. A direct and indirect role of rifampicin on glycemic managements recommends cautious observation with proper dose modifications of antidiabetic drugs which is crucial in diabetics along with tuberculosis. It is already established that tuberculosis medications influence diabetes management. Diabetes may also modify the pharmacokinetics of antitubercular medications. In a study carried out in Indonesia, it was suggested that patients of TB and DM had rifampicin serum concentration of 53% lesser than in TB patients with no DM, and there should be an indirect association among fasting glucose and rifampicin concentrations [60]. The lower levels of antitubercular medications have been related to TB management failure. Diabetes may lead to alterations in oral absorption, reducing drug protein binding, renal insufficiency, or fatty liver with reduced clearance of drugs [61]. Its role on antitubercular medication levels had not been officially observed in poor response cases to TB management in patients of DM with TB [62].

15.3 The Requirement for Adjunctive Host-Targeted Therapy

Even though safe and efficient treatment regimens are accessible for the treatment of TB with greater than 95% treatment success, the lengthy period of such treatment regimens has created a problem for successful TB management [63]. The present antitubercular therapy (ATT) is related to drug toxicity and may cause multidrug-resistant TB. The shortcomings of the present ATT led to a vital requirement for potent anti-TB drugs, treatment regimens, and newer drugs for curing TB, and hence there is a crucial requirement to think alternate adjunctive therapy that may boost the host immunity for the complete eradication of TB bacilli. The greater cost and extended characteristics of TB medication development combined with less productivity contributes to repositioning and revival of medications like metformin which may contribute to a new pharmaceutical strategy to manage particular diseases like TB [64, 65]. A competent and effective immune system is vital to curb and minimize the increase of the Mtb [64]. A host-targeted adjunct therapeutic intervention improves defensive host immune reaction and decreases chances of growth of resistance. Presently, among the medications which are recommended in the treatment of DM, metformin has engrossed notable concern as a host-directed adjunctive therapy (HDT). Its functions comprise increasing effectors of macrophage, decreasing inflammation, and defending the lung from injury. Metformin is an AMPK regulator which minimizes the intracellular load of *Mtb*, minimizes ailment immunopathology, and raises the potency of anti-TB medications [66]. The anticipated mechanisms of metformin successful management results in TB involve a rise in mitochondrial reactive oxygen species (mROS) and increased killing of *Mtb*. Metformin therapy increases the number of CD8 and CD4 T cell, which fight against bacterial infection, and help patient for managing Mtb infection. The defensive role of metformin in TB patients may be because of its efficiency to produce immunity against TB infection [67].

15.4 Repurposing of Adjuvant Antitubercular Drug: Novel Approach to Fight Drug-Resistant TB

The poor clinical results, slow improvement in production, and screening of novel ATT have improved the evolution of various therapies for possible utilization as adjuvant drugs for treating TB [68]. Moreover, new therapeutic inventions are needed to cure complicated TB cases. One new approach known as adjunctive host-directed therapies (HDTs) includes targeting host factors directly instead of parts of the pathogens. HDTs decrease bacterial production and inflammatory reactions via boosting host immunity. HDTs have effective treatment results with decreased morbidity, mortality, tissue injury, and TB therapy period. A two-way relationship in TB and DM proves to be harmful. Due to weaker immune status, a diabetic is more prone to catching the disease. In addition diabetes reasons for deterioration of symptoms and worsens signs of disease in peoples having tuberculosis-diabetes comorbidity. TB symptom aggravates in peoples that have

Table 15.1 Effect of metformin in the control of tuberculosis: drug repositioning

Adjuvant drug with antitubercular properties	Current utilization	Molecular mechanism as an antitubercular agent	Preclinical and clinical evidences	References
Metformin	T2DM	Increases CD8 T cell responses and also improves mitochondrial reactive oxygen species-mediated intracellular *Mtb* killing	The in vitro studies carried on Mtb-infected hmDM cells proves that the percentage of Mtb survival was lower in MET-treated group when compared to untreated group (Control, CG: 299 and Case, SG: 152) which comprises 299 controls, 128 diabetics not on MET, and 171 diabetics on MET. Among the 152 cases, 111 were MET non-users, and the remaining 38 were on MET. MET use at any time was associated with reduced risk of TB	[72–80]

weak glycemic status [9–11]. Recommendation of metformin for curing diabetes in TB-DM sufferers may decrease the inflammatory reactions and also reduces harshness of tuberculosis. Metformin treatment added to normal tuberculosis management routine is linked to favorable results on therapeutic results in active TB. Moreover, MET management showed beneficial impact on latent TB in diabetics. This could be hypothesized that treatment results of tuberculosis and diabetes can be enhanced by metformin because of its efficiency in decreasing microbial burden [69, 70]. Metformin is presently recommended as a first-line drug in management of NIDDM. Because of oblique operationalization of AMP-activated protein kinase (AMPK) pathways, it functions as an insulin sensitizer. Targeting of AMPK contributes to its potent antitubercular effect (Table 15.1).

It also prevents hepatic gluconeogenesis and synthesis of fatty acid and augments glucose uptake and fatty acid oxidation (FAO). As it increases mitochondrial reactive oxygen species (ROS) and plays a crucial role in reducing pulmonary bacterial burden in mice via increased mitochondrial turnover, MET could be used as an efficient adjuvant anti-TB drug [71, 72]. MET has been reported to enhance the potency of present anti-TB drugs, and it was established in both acute and chronic nonclinical models of TB [72]. In a mouse model of TB, MET therapy was documented to potentiate the efficacy of the first-line anti-TB drugs like isoniazid and ethionamide. The potency of MET was also investigated in model of *Mtb* infection. When compared with untreated or isoniazid-treated mice, mice treated with MET along with standard ATT showed reduced bacillary load in lungs. MET is therefore a possible adjuvant drug along with existing anti-TB drugs that can increase *Mtb* clearance. The novel trend and pharmaceutical strategy for treating specific disease such as TB with drugs that are already FDA approved for other diseases is the repositioning of drugs. In the age of antibiotic resistance and global emergency, many molecules or medicines can be repurposed as novel anti-TB drug that were previously used to cure other infectious diseases or TB [73, 74]. Few of the medications which are already approved like statins, metformin, ibuprofen, aspirin, valproic acid, adalimumab, bevacizumab, zileuton, and vitamin D3 have been investigated to interfere biologically relevant cellular checkpoints and can thus be repurposed to achieve better clinical results in TB patients. Host-directed therapies (HDTs) have a useful effect by strengthening the host immune defenses, targeting the inflammatory pathway, and interfering with the host mechanisms used by the pathogens to survive in host tissues. HDTs will be helpful in TB management by reducing the treatment period, decreasing the number of antibiotics, and enhancing the efficiency of antitubercular medications. Recent researches proved that several medications including etanercept, statins, vitamin D, cyclooxygenase inhibitors, carbamazepine, bevacizumab (vascular endothelial growth factor inhibitor), metformin, and various other medications may be used as HDTs in the management of TB. These drugs may change the host immune response in various ways [75]:

- By damaging granuloma: they enhance medication penetration into the cell.
- By inducing autophagy of infected cells: intracellular bacteria will be damaged.
- By anti-inflammatory response.
- By improving cell-mediated immune response.

15.5 Synergistic Effects of Repurposed Adjuvant Drug with Other Antitubercular Drugs

Mtb-infected mice were administered with MET alone or MET in combination with either isoniazid (INH) or ethionamide (ETH), starting on day 7 or 42 after infection. In five different acute model experiments, mice treated with MET (500 mg/kg) alone had reduced bacillary load in both the lung and spleen. This dose is equivalent to a MET dose of 2430 mg/day for a 60-kg human and lower than the maximum daily dose for MET therapy in diabetic patients (3000 mg/day). Furthermore, MET administration enhanced the efficacy of the conventional first-line anti-TB drug INH, as demonstrated by decreased bacillary load in the lungs of mice co-treated with INH + MET when compared with mice that received INH alone. We next evaluated the efficacy of MET adjunctive therapy when used in combination with the second-line anti-TB drug ETH [76].

To significantly improve current TB therapy, anti-Mtb drugs should also promote resolution of tissue pathology in addition to accelerating bacillary clearance [4]. Lungs and spleens from Mtb-infected mice treated with MET were smaller than those of untreated mice at 35 days after infection. As expected, mice treated with the conventional anti-TB drug INH or ETH exhibited a clear reduction in organ size and tissue lesions, and combination therapy of MET with INH or ETH further reduced tissue pathology. Histopathologic evaluation of the infected lungs of untreated control mice revealed diffuse coalescent lung lesions with large numbers of infiltrating macrophages and lymphocytes and scattered intracellular acid-fast bacilli (AFB). In contrast, MET treatment was associated with reduced numbers of AFB and increased lymphocyte infiltration of the infected tissues which has previously been linked with improved Mtb control in mice [77, 78]. Mice treated with INH alone had few residual lesions in the lungs with areas of only increased cellularity. No granulomas were observed in the lungs of mice treated with INH + MET, with some areas of the lungs appearing completely normal.

The T helper cell immune responses play a crucial role in controlling *Mtb* infection. It was observed that mice treated with MET showed a larger number of CD4 and CD8 T cells when compared with untreated mice. In *Mtb*-infected mice, MET treatment was associated with an increased percentage and number of mycobacteria-specific IFN-gamma-secreting CD8 T cells compared with untreated control animals. MET restricts mycobacterial growth by inducing mitochondrial ROS production, reduces TB-induced tissue pathology, and enhances immune response.

Amit Singhal et al. [72] performed an in vitro study on *Mtb*-infected human monocyte-derived macrophage (hmDM) cells and THP-1 cells and reported that MET restricts mycobacterial growth by inducing mitochondrial ROS production. MET administration enhances the efficacy of the conventional first-line anti-TB drug like isoniazid as demonstrated by a decreased bacillary load in the lungs of mice co-treated with isoniazid + MET when compared with isoniazid alone. MET therapy reduces TB-induced tissue pathology and enhances the host immune system for faster and complete elimination of the TB bacilli. CD4 cells are glycoprotein found on the surface of immune cells such as T helper cells, monocytes, macrophages, and dendritic cells. Similarly, CD8 T cells are generated in the thymus often called cytotoxic T lymphocytes, and they are very important for immune defense against intracellular pathogens including viruses and bacteria. Studies have validated that MET treatment enhances CD8 and CD4 T cells which contributes to *Mtb* infection control and boosts immune responses of the patients. Srujitha Marupuru et al. [79] observed that the protective effect of MET against TB was found to be 3.9-fold in DM patients (OR 1/4 0.256, 0.16–0.40) and there has been an improved sputum culture conversion rate in patients with cavitary pulmonary TB who have higher bacterial loads. Therefore, it can be concluded that MET could be an effective adjunctive antitubercular agent to improve sputum culture conversion. However, these observations need to be proved in well-designed randomized trials [76–78].

15.6 Cellular and Molecular Mechanism

Metformin has recently attracted significant attention as a host-directed adjunctive therapy (HDT) as it has an inhibitory effect on mitochondrial complex I, inhibition of which has been found to increase the AMP/ATP ratio [72].

The altered cellular energy status induces activation of AMPK, a serine/threonine kinase, and acts as an energy sensor. Activation of AMPK by metformin stimulates endothelial nitric oxide synthase activity which leads to bacterial killing. Metformin also acts through AMPK-independent mechanisms. It promotes phagocytosis, phagolysosome fusion, and bactericidal capacity autophagy in macrophages. Macrophages exposed to metformin had higher bactericidal capacity attributed to increased mitochondrial reactive oxidative species (ROS) production required for bacterial killing (Fig. 15.2). Metformin has been reported to enhance the efficacy of the conventional first-line anti-TB drug like isoniazid by decreasing bacillary load in the lungs. It enhances the host immune system for faster and complete elimination of the TB bacilli [79]. Our findings, in conjunction with previous efficacy results, are critical for different stakeholders and policymakers as metformin can be used as an adjuvant antitubercular drug particularly in those who had comorbidity of TB and DM [80]. The protective role of metformin is mediated by increased host cell production of mROS and increased acidification of mycobacterial phagosome. Indeed, mROS produced upon mitochondrial recruitment to phagosomes is instrumental in killing of intracellular bacteria by macrophages. The anti-inflammatory effect is mediated by activation of AMPK, a negative regulator of inflammation [81, 82].

Interleukin-17 (IL-17) is a pro-inflammatory cytokine mainly produced by Th17 lymphocytes, which play an immunoregulatory role by producing a unique spectrum of pro-inflammatory cytokines. These cytokines could induce activation and recruitment of neutrophils, macrophages, and Th1 lymphocytes into the site of infection, which contributes to delimitation of the damaged area in lung tissue, as well as inhibition of Mtb growth, and hence IL-17 is associated in the protection of Mtb infection [83].

15.7 Conclusion and Future Perspective

Despite the availability of potent antitubercular drugs, TB still remains one of the world's leading causes of death. The current antitubercular therapy (ATT) suffers from a drawback of longer duration and adverse effects that imposes a major challenge of patient noncompliance and resistance development. Identification of new adjunctive therapies that can improve the treatment outcomes of TB patients should be considered as a priority, and repurposing of the adjuvant drugs with antitubercular effects should be fast tracked. Such therapeutic approaches and pharmaceutical interventions should potentially (1) enhance the efficacy of antibiotic-based treatment, (2) shorten treatment duration, (3) reduce the

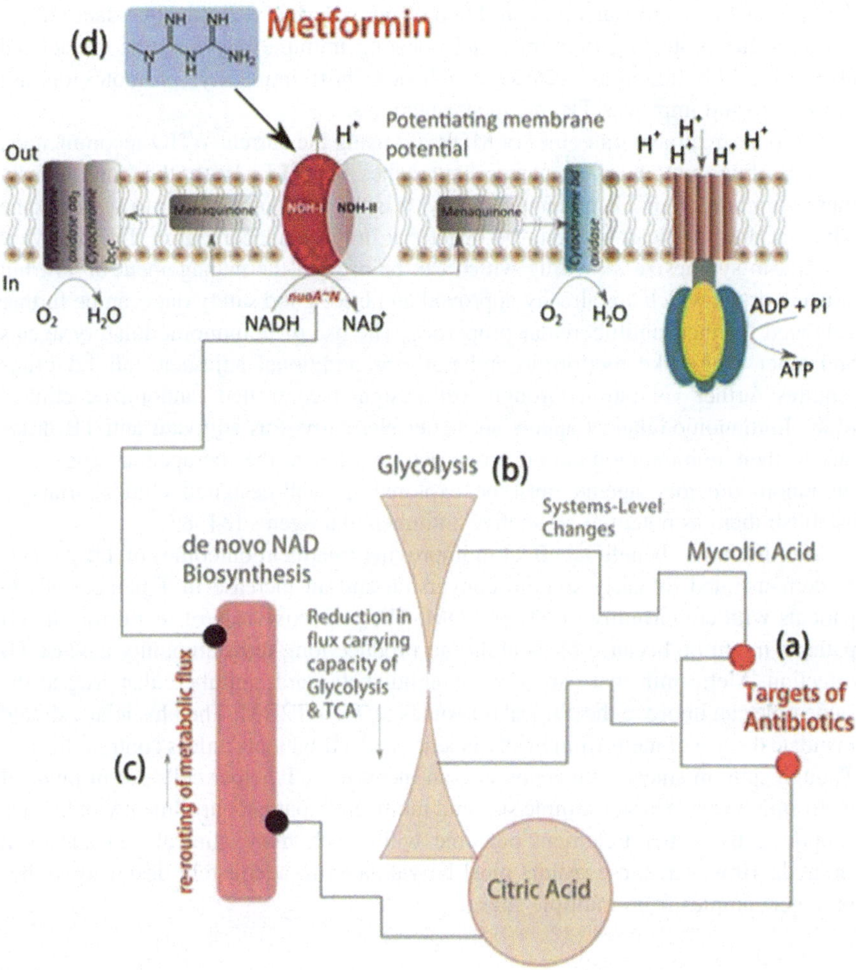

Fig. 15.2 Metformin, an FDA-approved drug for type 2 diabetes as a potential combination therapy for TB with existing antibiotics. (**a**) Antibiotic targeting mycolic acid biosynthesis; (**b**) systems-level changes resulting in the reduction of flux-carrying capacity of glycolysis and citric acid cycle; (**c**) resulting re-routing of metabolic fluxes through de novo NAD biosynthesis pathway and electron transport through NDH-I; (**d**) possibility of targeting NDH-I with metformin [82]

immunopathology associated with TB, and (4) augment protective host immune responses, thus enhancing bacterial clearance from the system.

In the last few years, there is an increasing interest in the use of metformin in improving TB treatment outcomes, as it is a widely prescribed drug in T2DM. Despite the impressive efficacy of metformin in the management of TB, it remains to be clarified how the protective immunity and immune responses play a significant role in Mtb infection control. Hence, we sought to evaluate the effect of metformin therapy on treatment outcomes in TB patients who had T2DM. Our data clearly suggests that protecting immunity and boosting immune responses had beneficial effects on Mtb infection because it enhances host immunity, promotes disease resolution, and improves TB treatment outcomes.

The poor treatment outcomes of MDR-TB using the current WHO-recommended anti-TB drug now warrant newer approaches to ATT. With the continuously increasing burden of TB patients globally, it is the need of the hour to screen some adjuvant drug that can prove to be beneficial in the management of TB. It is very difficult to synthesize new drug which can be used in the management of TB, but existing drugs which are already approved and have good safety data can be further validated for their antitubercular properties. The use of immunomodulatory agents and other drugs like metformin and statin as additional adjuvant anti-TB drugs requires further validation through well-designed controlled randomized clinical trials. Immunomodulatory agents are better alternatives as adjuvant anti-TB drugs due to their immunomodulation properties, and hence the therapeutic efficacy of immunomodulatory agents must be explored in well-designed clinical trials to establish them as potential adjunctive antitubercular agents [84–87].

Metformin has beneficial effect in improving treatment outcomes of TB patients as demonstrated by early sputum conversion and an increase in T helper cells in patients with comorbidity of TB and DM. The protective role of metformin in TB patients might be because of its ability to produce long-term immunity against TB infection. Metformin that can serve as antidiabetic and antitubercular drug at the same time can improve the clinical outcomes of TB + T2DM. The physicians should mandate the use of metformin in all diabetics and TB patients unless contraindicated. Though sputum smear conversion and an increase in T helper cells are in favor of metformin users, a larger sample size and multicenter patient enrollment would have supported the better treatment outcome with much more clinical and statistical strength. However, these results must be validated in adequately designed studies in larger samples from multiple sites.

References

1. Zaman K (2010) Tuberculosis: a global health problem. J Health Popul Nutr 28(2):111. https://doi.org/10.3329/jhpn.v28i2.4879
2. World Health Organization (2018) Global tuberculosis report 2018. WHO United Nations. https://www.who.int/tb/global-report-2019
3. Uhlin M, Andersson J, Zumla A, Maeurer M (2012) Adjunct immunotherapies for tuberculosis. J Infect Dis 205(suppl_2):S325–S334. https://doi.org/10.1093/infdis/jis197
4. Schnippel K, Rosen S, Shearer K, Martinson N, Long L, Sanne I, Variava E (2013) Costs of inpatient treatment for multi-drug-resistant tuberculosis in South Africa. Tropical Med Int Health 1:109–116. https://doi.org/10.1111/tmi.12018
5. Jeon CY, Murray MB (2008) Diabetes mellitus increases the risk of active tuberculosis: a systematic review of 13 observational studies. PLoS Med 5:e152. https://doi.org/10.1371/journal.pmed.0050152

6. Siddiqui AN, Hussain S, Siddiqui N, Khayyam KU, Tabrez S, Sharma M (2018) Detrimental association between diabetes and tuberculosis: an unresolved double trouble. Diabetes Metab Synd Clin Res Rev 12(6):1101–1107. https://doi.org/10.1016/j.dsx.2018.05.009
7. Yu X, Li L, Xia L, Feng X, Chen F, Cao S, Wei X (2019) Impact of metformin on the risk and treatment outcomes of tuberculosis in diabetics: a systematic review. BMC Infect Dis 19 (1):859. https://doi.org/10.1186/s12879-019-4548-4
8. Degner NR, Wang JY, Golub JE, Karakousis PC (2018) Metformin use reverses the increased mortality associated with diabetes mellitus during tuberculosis treatment. Clin Infect Dis 66 (2):198–205. https://doi.org/10.1093/cid/cix819
9. Gomez DI, Twahirwa M, Schlesinger LS, Restrepo BI (2013) Reduced Mycobacterium tuberculosis association with monocytes from diabetes patients that have poor glucose control. Tuberculosis (Edinb) 93:192–197. https://doi.org/10.1016/j.tube.2012.10.003
10. Kumar Nathella P, Babu S (2017) Influence of diabetes mellitus on immunity to human tuberculosis. Immunology 152:13–24. https://doi.org/10.1111/imm.12762
11. Baker MA, Harries AD, Jeon CY et al (2011) The impact of diabetes on tuberculosis treatment outcomes: a systematic review. BMC Med 9:81. https://doi.org/10.1186/1741-7015-9-81
12. Alisjahbana B, Sahiratmadja E, Nelwan EJ et al (2007) The effect of type 2 diabetes mellitus on the presentation and treatment response of pulmonary tuberculosis. Clin Infect Dis 45:428–435. https://doi.org/10.1086/519841
13. Niazi K, Kalra S (2012) Diabetes and tuberculosis: a review of the role of optimal glycemic control. J Diabetes Metab Disord 11(1):28
14. Stevenson R, Critchley JA, Forouhi NG et al (2007) Diabetes and the risk of tuberculosis: a neglected threat to public health? Chronic Illn 3(3):228–245
15. Jeon Y, Murray MB (2008) Diabetes mellitus increases the risk of active tuberculosis: a systematic review of 13 observational studies. PLoS Med 5(7):0050152
16. Alisjahbana B, Sahiratmadja E, Nelwan EJ et al (2007) The effect of type 2 diabetes mellitus on the presentation and treatment response of pulmonary tuberculosis. Clin Infect Dis 45 (4):428–435
17. Feleke Y, Abdulkadir J, Aderaye G (1999) Prevalence and clinical features of tuberculosis in Ethiopian diabetic patients. East Afr Med J 76(7):361–364
18. Swai BM, Mclarty DG, Mugusi F (1990) Tuberculosis in diabetic patients in tanzania. Trop Dr 20(4):147–150
19. Restrepo I, Fisher-Hoch SP, Crespo JG et al (2007) Type 2 diabetes and tuberculosis in a dynamic bi-national border population. Epidemiol Infect 135(3):483–491
20. Shaikh MA, Singla R, Khan NB, Sharif NS, Saigh MO (2003) Does diabetes alter the radiological presentation of pulmonary tuberculosis. Saudi Med J 24(3):278–281
21. Wang J-Y, Lee L-N, Hsueh P-R (2005) Factors changing the manifestation of pulmonary tuberculosis. Int J Tuberc Lung Dis 9(7):777–783
22. Mugusi F, Swai ABM, Alberti KGMM, McLarty DG (1990) Increased prevalence of diabetes mellitus in patients with pulmonary tuberculosis in Tanzania. Tuberculosis 71(4):271–276
23. Heysell SK, Moore JL, Keller SJ, Houpt ER (2010) Therapeutic drug monitoring for slow response to tuberculosis treatment in a state control program, Virginia, USA. Emerg Infect Dis 16(10):1546–1553
24. Chang J-T, Dou H-Y, Yen C-L et al (2011) Effect of type 2 diabetes mellitus on the clinical severity and treatment outcome in patients with pulmonary tuberculosis: a potential role in the emergence of multidrug-resistance. J Formos Med Assoc 110(6):372–381
25. Baker MA, Harries AD, Jeon CY et al (2011) The impact of diabetes on tuberculosis treatment outcomes: a systematic review. BMC Med 9:81
26. Reis-Santos B, Gomes T, Locatelli R et al (2014) Treatment outcomes in tuberculosis patients with diabetes: a polytomous analysis using Brazilian surveillance system. PLoS One 9(7): e100082

27. Young F, Wotton CJ, Critchley JA, Unwin NC, Goldacre MJ (2010) Increased risk of tuberculosis disease in people with diabetes mellitus: record-linkage study in a UK population. J Epidemiol Community Health 2010:114595
28. Oluboyo PO, Erasmus RT (1990) The significance of glucose intolerance in pulmonary tuberculosis. Tuberculosis 71(2):135–138
29. Basoglu OK, Bacakoglu F, Cok G, Saymer A, Ates M (1999) The oral glucose tolerance test in patients with respiratory infections. Monaldi Arch Chest Dis 54:307–310
30. Jeon CY, Harries AD, Baker MA et al (2010) Bi-directional screening for tuberculosis and diabetes: a systematic review. Tropical Med Int Health 15(11):1300–1314
31. Guptan A, Shah A (2000) Tuberculosis and diabetes: an appraisal. Ind J Tub 47(3):2–8
32. Bansal V, Asmar NE, Selman WR, Arafah BM (2015) Pitfalls in the diagnosis and management of Cushing's syndrome. Neurosurg Focus 38(2):E4
33. Atkin SL, Masson EA, Bodmer CW, Walker BA, White MC (1993) Increased Insulin Requirement in a Patient with Type 1 Diabetes on Rifampicin. Diabet Med 10(4):392–392
34. Lebovitz HE (1990) Oral hypoglycemic agents. In: Rifkin H, Porte D Jr (eds) Ellenberg and Rifkins's Diabetes Mellitus, 4th edn. Elsevier, New York, NY
35. Geerlings SE, Hoepelman AI (1999) Immune dysfunction in patients with diabetes mellitus (DM). FEMS Immunol Med Microbiol 26(3–4):259–265
36. Tsukaguchi K, Yoneda T, Yoshikawa M et al (1992) Case study of interleukin-1β, tumor necrosis factor ALPHA, and interleukin-6 production peripheral blood monocytes patients diagnosed with diabetes mellitus complicated by pulmonary tuberculosis. Kekkaku (Tuberculosis) 67(12):755–760
37. Sidibé EH (2007) Diabéte et tuberculosepulmonaire: aspects épidémiologiques, physiopathologiques et symptomatologiques. Cahiers d'´Etudes et de Recherches Francophones/Santé 17(1):29–32
38. Moutschen MP, Scheen AJ, Lefebvre PJ (1992) Impaired immune responses in diabetes mellitus: analysis of the factors and mechanisms involved. Relevance to the increased susceptibility of diabetic patients to specific infections. Diabetes Metab 18(3):187–201
39. Yorke E, Atiase Y, Akpalu J, Sarfo-Kantanka O, Boima V, Dey ID (2017) The bidirectional relationship between tuberculosis and diabetes. Tuberculosis research and treatment
40. Zack MB, Fulkerson LL, Stein E (1973) Glucose intolerance in pulmonary tuberculosis. Am Rev Respir Dis 108(5):1164–1169
41. Aguirre F, Brown A, Cho NH, Dahlquist G, Dodd S, Dunning T, Hirst M, Hwang C, Magliano D, Patterson C, Scott C, Shaw J, Soltesz G, Usher-Smith J, Whiting D (2013) IDF diabetes atlas: sixth edition, 6th edn. International Diabetes Federation
42. Venkatesan K (1992) Pharmacokinetic drug interactions with rifampicin. Clin Pharmacokinet 22(1):47–65
43. Duckworth WC, Bennett RG, Hamel FG (1998) Insulin degradation: progress and potential. Endocr Rev 19(5):608–624
44. Madsbad S (2009) Treatment of type 2 diabetes with incretin-based therapies. Lancet 373 (9662):438–439
45. American Diabetes A (2016) Standards of Medical Care in Diabetes—2016 Abridged for Primary Care Providers. Clin Diabetes 34(1):3–21
46. Beran D, Ewen M, Laing R (2016) Constraints and challenges in access to insulin: a global perspective. Lancet Diabetes Endocrinol 4(3):275–285
47. Riza L, Pearson F, Ugarte-Gil C et al (2014) Clinical management of concurrent diabetes and tuberculosis and the implications for patient services. Lancet Diabetes Endocrinol 2(9):740–753
48. World Health Organisation (2011) Collaborative framework for care and control of tuberculosis and diabetes
49. Niemi M, Backman JT, Neuvonen M, Neuvonen PJ, Kivisto KT (2001) Effects of rifampin on the pharmacokinetics and pharmacodynamics of glyburide and glipizide. Clin Pharmacol Ther 69:400–406

50. Atkin S, Masson E, Bodmer C, Walker B, White M (1992) Increased insulin requirement in a patient with type 1 diabetes on rifampicin. Diab Med 10:202
51. Niemi M, Backman JT, Fromm MF, Neuvonen PJ, Kivisto KT (2003) Pharmacokinetic interactions with rifampicin: clinical relevance. Clin Pharmacokinet 42:819–850
52. Niemi M, Backman JT, Neuvonen PJ (2004) Effects of trimethoprim and rifampin on the pharmacokinetics of the cytochrome P450 2C8 substrate rosiglitazone. Clin Pharmacol Ther 76:239–249
53. Park JY, Kim KA, Kang MH, Kim SL, Shin JG (2004) Effect of rifampin on the pharmacokinetics of rosiglitazone in healthy subjects. Clin Pharmacol Ther 75:157–162
54. Jaakkola T, Backman JT, Neuvonen M, Laitila J, Neuvonen PJ (2006) Effect of rifampicin on the pharmacokinetics of pioglitazone. Br J Clin Pharmacol 61:70–78
55. Niemi M, Backman JT, Neuvonen M, Neuvonen PJ (2003) Effect of rifampicin on the pharmacokinetics and pharmacodynamics of nateglinide in healthy subjects. Br J Clin Pharmacol 56:427–432
56. Hatorp V, Hansen KT, Thomsen MS (2003) Influence of drugs interacting with CYP3A4 on the pharmacokinetics, pharmacodynamics, and safety of the prandial glucose regulator repaglinide. J Clin Pharmacol 43:649–660
57. Niemi M, Backman JT, Neuvonen M, Neuvonen PJ, Kivisto KT (2000) Rifampin decreases the plasma concentrations and effects of repaglinide. Clin Pharmacol Ther 68:495–500
58. Takasu N, Yamada T, Miura H et al (1982) Rifampicin-induced early phase hyperglycemia in humans. Am Rev Respir Dis 125:23–27
59. Waterhouse M, Wilson C, White VL, Chowdhury TA (2005) Resolution of insulin-requiring diabetes after cessation of chemotherapy for tuberculosis. J R Soc Med 98:270–271
60. Nijland HM, Ruslami R, Stalenhoef JE et al (2006) Exposure to rifampicin is strongly reduced in patients with tuberculosis and type 2 diabetes. Clin Infect Dis 43:848–854
61. Gwilt PR, Nahhas RR, Tracewell WG (1991) The effects of diabetes mellitus on pharmacokinetics and pharmacodynamics in humans. Clin Pharmacokinet 20:477–490
62. Peloquin CA (2002) Therapeutic drug monitoring in the treatment of tuberculosis. Drugs 62:2169–2183
63. Churchyard GJ, Kaplan G, Fallows D, Wallis RS, Onyebujoh P, Rook GA (2009) Advances in immunotherapy for tuberculosis treatment. Clin Chest Med 30:769–782. https://doi.org/10.1016/j.ccm.2009.08.009
64. Hawn TR, Matheson AI, Maley SN, Vandal O (2013) Host directed therapeutics for tuberculosis: can we harness the host? Microbiol Mol Biol 77:608–627. https://doi.org/10.1128/MMBR.00032-13
65. Zumla A, Chakaya J, Hoelscher M, Toumi FN, Rustomjee R, Vilaplana C, Yeboah-Manu D, Rasolofo V, Munderi P, Singh N, Aklillu E (2015) Towards host-directed therapies for tuberculosis. Nat Rev Drug Discov 14(8). https://doi.org/10.1038/nrd4696.511.14:511-512
66. Kim YR, Yang CS (2018) Host-directed therapeutics as a novel approach for tuberculosis treatment. J Microb Biotechnol 27(9):1549–1558. https://doi.org/10.3389/fcimb.2018.00275
67. Zumla AI, Gillespie SH, Hoelscher M, Philips PP, Cole ST, Abubakar I, McHugh TD, Schito M, Maeurer M, Nunn AJ (2014) New antituberculosis drugs, regimens, and adjunct therapies: needs, advances, and future prospects. Lancet Infect Dis 14(4):327–340. https://doi.org/10.1016/S1473-3099(13)70328-1
68. Sharma D, Dhuriya YK, Deo N, Bisht D (2017) Repurposing and revival of the drugs: a new approach to combat the drug resistant tuberculosis. Front Microbiol 11:2452. https://doi.org/10.3389/fmicb.2017.02452.8
69. Padmapriyadarsini C, Bhavani PK, Natrajan M, Ponnuraja C, Kumar H, Gomathy SN, Guleria R, Jawahar SM, Singh M, Balganesh T, Swaminathan S (2019) Evaluation of metformin in combination with rifampicin containing antituberculosis therapy in patients with new, smear-positive pulmonary tuberculosis (METRIF): study protocol for a randomised clinical trial. BMJ Open 9(3):e024363

70. Yang CS, Kim JJ, Lee HM, Jin HS, Lee SH, Park JH et al (2014) The AMPK-PPARGC1A pathway is required for antimicrobial host defense through activation of autophagy. Autophagy 20(5):785–802. https://doi.org/10.4161/auto.28072.10
71. Adil M, Khan RA, Kalam A, Venkata SK, Kandhare AD, Ghosh P, Sharma M (2017) Effect of antidiabetic drugs on bone metabolism: evidence from preclinical and clinical studies. Pharmacol Rep 69(6):1328–1340. https://doi.org/10.1016/j.pharep.2017.05.008
72. Singhal A, Jie L, Kumar P, Hong GS, Leow MKS, Paleja B, Tsenova L, Kurepina N, Chen J, Zolezzi F, Kreiswirth B (2014) Metformin as adjunct antituberculosis therapy. Sci Transl Med 6 (263):59. https://doi.org/10.1126/scitranslmed.3009885
73. Eikawa S, Nishida M, Mizukami S, Yamazaki C, Nakayama E, Udono H (2015) Immune-mediated antitumor effect by type 2 diabetes drug metformin. Proc Natl Acad Sci 10 (6):1809–1814. https://doi.org/10.1073/pnas.1417636112
74. Siddiqui AN, Hussain S, Siddiqui N, Khayyam KU, Tabrez S, Sharma M (2018) Detrimental association between diabetes and tuberculosis: an unresolved double trouble. Diabetes Metab Syndr 12(6):1101–1107. https://doi.org/10.1016/j.dsx.2018.05.009
75. Mishra R, Krishan S, Siddiqui AN, Kapur P, Khayyam KU, Sharma M (2020) Potential role of adjuvant drugs on efficacy of first line oral antitubercular therapy: drug repurposing. Tuberculosis 11:101902. https://doi.org/10.1016/j.tube.2020.101902
76. Koo MS, Manca C, Yang G, O'Brien P, Sung N, Tsenova L, Subbian S, Fallows D, Muller G, Ehrt S, Kaplan G (2011) Phosphodiesterase 4 inhibition reduces innate immunity and improves isoniazid clearance of Mycobacterium tuberculosis in the lungs of infected mice. PLoS One 6: e17091
77. Lyadova V, Eruslanov EB, Khaidukov SV, Yeremeev VV, Majorov KB, Pichugin AV, Nikonenko BV, Kondratieva TK, Apt AS (2000) Comparative analysis of T lymphocytes recovered from the lungs of mice genetically susceptible, resistant, and hyperresistant to Mycobacterium tuberculosis-triggered disease. J Immunol 165:5921–5931
78. Chackerian A, Behar SM (2003) Susceptibility to Mycobacterium tuberculosis: lessons from inbred strains of mice. Tuberculosis 83:279–285
79. Marupuru S, Senapati P, Pathadka S, Miraj SS, Unnikrishnan MK (2017) Protective effect of metformin against tuberculosis infections in diabetic patients: an observational study of south Indian tertiary healthcare facility. Braz J Infect Dis 21(3):312–316. https://doi.org/10.1016/j.bjid.2017.01.001
80. Lee YJ, Han SK, Park JH, Lee JK, Kim DK, Chung HS, Heo EY (2018) The effect of metformin on culture conversion in tuberculosis patients with diabetes mellitus. Korean J Intern Med 33 (5):933. https://doi.org/10.3904/kjim.2017.249
81. Yew WW, Chang KC, Chan DP, Zhang Y (2019) Metformin as a host-directed therapeutic in tuberculosis: is there a promise? Tuberculosis 115:76–80. https://doi.org/10.1016/j.tube.2019.02.004
82. Vashisht R, Brahmachari SK (2015) Metformin as a potential combination therapy with existing front-line antibiotics for tuberculosis. J Transl Med 13:83. https://doi.org/10.1186/s12967-015-0443-y
83. Lachmandas E, Eckold C, Böhme J, Koeken VA, Marzuki MB, Blok B, Arts RJ, Chen J, Teng KW, Ratter J, Smolders EJ (2019) Metformin alters human host responses to Mycobacterium tuberculosis in healthy subjects. J Infect Dis 220(1):139–150. https://doi.org/10.1093/infdis/jiz064
84. Dooley KE, Chaisson RE (2009) Tuberculosis and diabetes mellitus: convergence of two epidemics. Lancet Infect Dis 9(12):737–746
85. Dooley KE, Tang T, Golub JE, Dorman SE, Cronin W (2009) Impact of diabetes mellitus on treatment outcomes of patients with active tuberculosis. Am J Trop Med Hygiene 80 (4):634–639
86. Singla R, Khan N, Al-Sharif N, Al-Sayegh MO, Shaikh MA, Osman MM (2006) Influence of diabetes on manifestations and treatment outcome of pulmonary TB patients. Int J Tuberc Lung Dis 10(1):74–79
87. Kameda K, Kawabata S, Masuda N (1990) Follow-up study of short course chemotherapy for pulmonary tuberculosis complicated with diabetes mellitus. Kekkaku 65(12):791–803

Targeting Host and Bacterial Signaling Pathways in Tuberculosis: An Effective Strategy for the Development of Novel Anti-tubercular Therapies

16

Samreen Fatima, Bhavya Bhardwaj, and Ved Prakash Dwivedi

Abstract

Tuberculosis (TB) is a pandemic disease caused by an obligate intracellular pathogen *Mycobacterium tuberculosis (M.tb)*. The current TB therapy, Directly Observed Treatment Short-Course (DOTS), consists of the prolonged use of four antibiotics (Rifampicin, Isoniazid, Pyrazinamide, and Ethambutol) that must be administered alone or in combination for at least 6 months to patients affected by drug-sensitive pulmonary TB. Although this therapy is efficient in eliminating *M. tb*, it has numerous side effects such as liver toxicity, poor compliance, and development of multidrug-resistant strains. Therefore, there is an urgent need for the development of new drug targets for the effective management of the disease and to also ensure the prevention of reinfection and reactivation of the disease. Here, in this chapter, we have discussed the presently available drugs and their mechanism of action for the treatment of TB as well as various other drugs, which have been repurposed and deployed for the treatment against drug-sensitive and drug-resistant TB.

Keywords

Intra-macrophage survival · Drug target · Signaling mechanisms · Target identification · Virulence factors

S. Fatima · B. Bhardwaj · V. P. Dwivedi (✉)
Immunobiology Group, International Centre for Genetic Engineering and Biotechnology, New Delhi, India
e-mail: ved@icgeb.res.in

K. Dua et al. (eds.), *Targeting Cellular Signalling Pathways in Lung Diseases*,
https://doi.org/10.1007/978-981-33-6827-9_16

16.1 Introduction

Mycobacterium tuberculosis (*M.tb*) is an intracellular obligate pathogen which is the main cause of tuberculosis (TB) infection in humans. According to the WHO report, despite control in incidence of new cases and mortality, TB still remains a major reason of morbidity and mortality worldwide. The majority of the global bulk of the global burden of new infection and tuberculosis death is borne by developing countries with six countries, India, Indonesia, China, Nigeria, Pakistan, and South Africa, accounting for 60% of TB death in 2015 [1]. *M.tb* enters the human system by inhalation in the form of aerosol droplets and reaches the alveoli of the lungs. Within the lungs, the alveolar macrophages internalize the bacteria and prevent them from spreading to the other organs. However, from the lungs, infected macrophages can spread to other organs in the form of secondary infections [2]. The current TB therapy, Directly Observed Treatment Short-Course (DOTS), consists of prolonged use of four antibiotics (rifampicin, isoniazid, pyrazinamide, and ethambutol) that must be administered alone or in combination for at least 6 months to patients affected by drug-sensitive pulmonary TB. Although this therapy is efficient at eliminating *M.tb*, it has numerous side effects including liver toxicity, poor compliance, and development of multidrug-resistant strains. Moreover, the global situation has become precarious due to inefficacy and high variability of bacille Calmette-Guerin (BCG) vaccine in adults, HIV-TB co-infection, emergence of drug-resistant forms of the disease, and other additional factors such as immunodeficiencies and diabetes that increase the risk of developing TB, creating a lifelong reservoir of awaiting disease and infection, which may become a pandemic [3]. The emergence of resistance to existing drugs has further aggravated the situation. *M.tb* increases the duration of their survival and viability by releasing certain virulence factors and by modulating host signaling pathways [4]. Therefore, there is an urgent need for the development of new drug targets for the effective management of the disease and to also ensure prevention of reinfection and reactivation of the disease. There has not been much success in the drug development programs against TB due to myriad reasons. However, the status of search for new anti-TB drugs has improved a lot over the last 10 years, as evident by the approval of two drugs, bedaquiline and delamanid, for treatment of multidrug-resistant (MDR) and extremely drug-resistant (XDR) TB [5]. This shows that development of novel drugs is a challenging task. Therefore, providing novel chemical inhibitors of targets and promoting the development of the candidate drugs that are already present in trials of drug discovery pipeline are therefore of outmost importance in order to shorten anti-TB treatment. The propensity of *M.tb* to survive within host macrophages and evade the host immune system represents a key challenge for the eradication of the bacteria from the infected individuals. Understanding the signaling pathways behind virulence, infection, and host evasion mechanisms of *M.tb* is also critical for the identification of new drug targets and the development of new drugs.

Therefore, in this chapter, we will discuss both the host and bacterial factors and signaling mechanisms employed in survival, persistence, and host evasion of the bacteria by avoiding degradation within macrophages. Specifically, we will highlight those host and bacterial factors which can be manipulated by drugs or compounds and therefore be used as efficient drug targets in the future to control TB pathogenesis.

***M. tuberculosis* pathogenesis**: *M.tb* pathogenesis in most of the infected humans manifests itself in the form of asymptomatic disease wherein bacterial replication is kept at bay with no evidence of infection [6]. The only indication of infection is the presence of immune system against the bacterial antigens. This condition is known as latent infection, and approximately one-third of the world's population is affected by it. Of the total infected population, in around 5–10% of infected individuals, the person is affected by the active form of the disease, which is clinically transmissible [7]. Although latent infection is characterized by a lifelong standoff of the bacteria in most of the individuals, it is a huge reservoir of an upcoming pandemic. These infected individuals cause an estimated 1.5 million deaths each year, which is the highest from an infectious agent [7]. After getting inhaled through respiration, *M. tuberculosis* reaches the alveoli of the lungs and has their first encounter with the alveolar macrophages. Other cells such as neutrophils and dendritic cells (DCs) also phagocytose bacteria and play a vital role in control of the infection [8]. The receptors on these phagocytic cells such as C-type lectin receptors, scavenger receptors, and complement receptors interact with the mycobacteria and internalize them. The bacteria display dynamic responses in the host, depending on the immune status of the host to exhibit different outcomes such as survival, replication, and transmission [9]. Upon exposure to certain cytokines such as interferon-gamma (IFN-γ), tumor necrosis factor-alpha (TNF-α), and vitamin D, the macrophages can get activated to kill *M. tuberculosis* [10]. IFN-γ activates the antimicrobial functions of the macrophages. Pretreatment of macrophages with IFN-γ before *M. tuberculosis* infection leads to the clearance of significantly high number of infecting bacteria by production of nitric oxide synthase 2 (NOS-2) and reactive nitrogen intermediates (RNI) [10]. IFN-γ also eliminates the bacteria by enhancing phagosome maturation and lysosomal degradation of *M. tuberculosis* [11]. However, bacterial factors which contribute to the virulence of *M. tuberculosis*, such as secretory proteins and biologically active lipids, pave way for the long-lasting survival of the bacteria by modulating intracellular trafficking, granuloma formation, and perturbations in the cell death pathways such as apoptosis. Granuloma is the hallmark of TB infection (Fig. 16.1). Granulomas are populated by persistent *M. tuberculosis* as well as immune cells (macrophages, epithelioid macrophages,

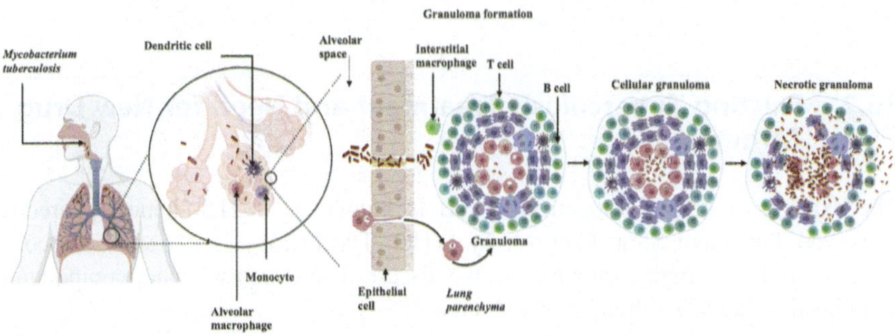

Fig. 16.1 Pathogenesis of tuberculosis and granuloma formation

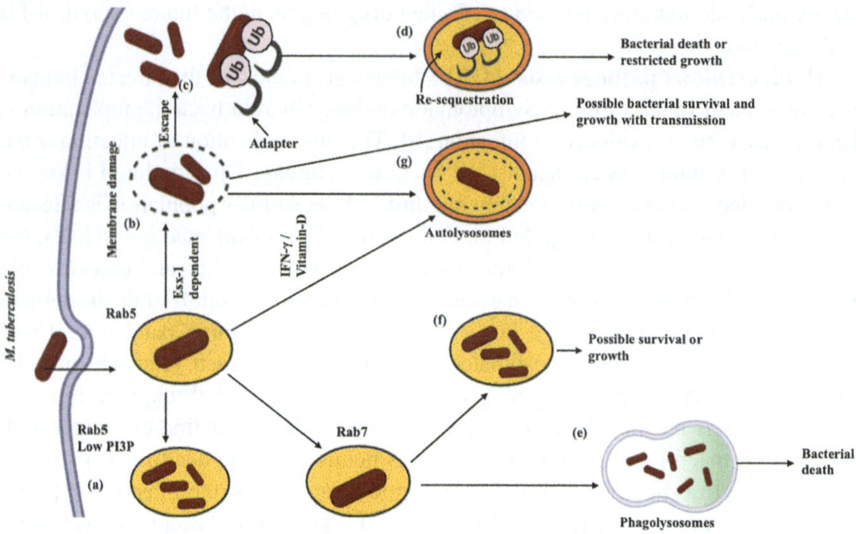

Fig. 16.2 Fate of *M. tuberculosis* and elimination after entering the host macrophages

foamy macrophages, and multinucleated giant cells) and act as a standoff between the host and pathogen [12]. Evidence proves that granulomas benefit both the host and the pathogen. At the stage of granuloma formation, the fate of the infection is determined if it will get arrested or progress to active infection based on the power of the host cellular immune response. This stage of infection also known as latency or dormant TB can persist in an asymptomatic and non-transmissible state throughout the life of the infected person. In most of the infected individuals, with strong cell-mediated immunity, the infection is prevented permanently at this stage with the subsequent healing of the granuloma. However, in individuals with compromised cell-mediated immunity, the person's immunity cannot arrest the infection at the initial stage, and the bacteria in the granuloma replicate uncontrollably and spread within the lungs and other tissues via the lymphatic system (active pulmonary TB and miliary or extrapulmonary TB, respectively) [13]. Elimination of actively replicating bacteria requires antibiotic treatment or DOTS therapy [14]. The fate of *M. tuberculosis* after entry in the host and its elimination are depicted in Fig. 16.2.

16.2 Existing Tuberculosis Treatment and Need for New Drug Targets

The existing strategy for controlling TB is known as DOTS therapy (Directly Observed Treatment Short-Course), by WHO. The therapy is known to be cost-effective and has a high curing rate for the disease. There are five major components associated with DOTS therapy, which are:

- Commitment by the government
- Sputum smear testing for TB detection
- 6–9 months of standard treatment with at least 2 months under the supervision of health officials
- Constant drug supply
- Recording and reporting the treatment results in a standardized manner for assessment

16.3 Drugs Involved In DOTS Therapy

The most important drugs of the TB treatment regime are isoniazid, rifampicin, and pyrazinamide. Their mode of action is discussed as [15]:

16.3.1 Isoniazid

INH (isoniazid) is a prodrug used along with rifampicin, pyrazinamide, and streptomycin or ethambutol for active tuberculosis but is used by itself for latent TB. Initial activation of INH is done by bacterial catalase-peroxidase (KatG), which catalyzes isonicotinic acyl radical formation. This product then forms nicotinoyl-nicotinamide adenine dinucleotide (NAD) adduct by coupling with nicotinamide adenine dinucleotide hydrogen (NADH). The complex then blocks natural enoyl-AcpM substrate and fatty acid synthase II (FAS II) action by binding tightly enoyl-acyl carrier protein reductase, Inh A. This in turn inhibits the synthesis of mycolic acids, which are an essential component of *M.tb* cell wall.

16.3.2 Rifampicin

Rifampicin (rifampin) is known to hinder the synthesis of DNA-dependent RNA by discouraging bacterial DNA-dependent RNA polymerase. The drug binds to beta subunit of RNA polymerase and prevents RNA synthesis, which in turn inhibits the synthesis of host bacterial proteins. It is a bacteriostatic drug against the actively growing TB bacteria, which inhibits the formation of cellular wall of mycobacteria. Mycolic acid forms a complex with arabinose residues of arabinogalactan, which is responsible for the disruption of arabinogalactan synthesis and finally inhibition of another complex formation that leads to high permeability in cell wall.

16.3.3 Pyrazinamide

Pyrazinamide is used along with INH and rifampicin to treat active TB. It diffuses inside granuloma of TB bacteria and gets converted into active pyrazinoic acid by the action of pyrazinamidase. The active form of pyrazinamide gets converted into

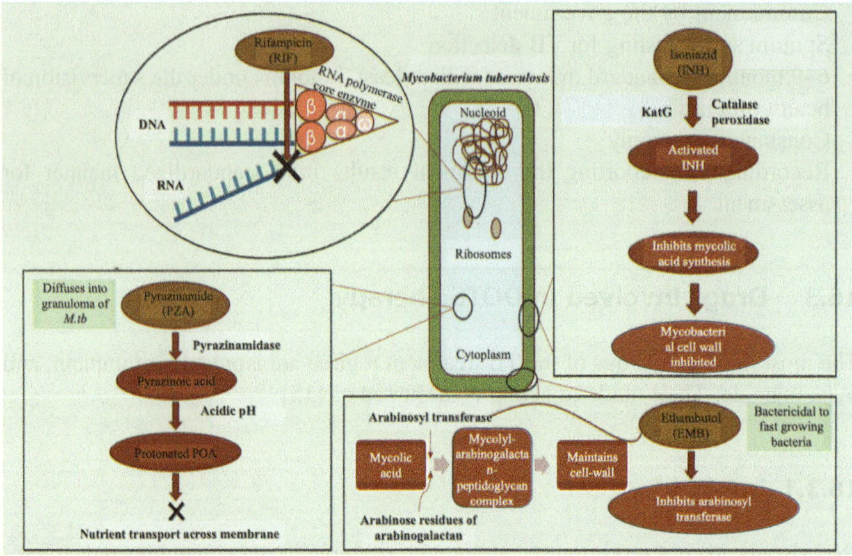

Fig. 16.3 Primary mechanism of action of drugs involved in DOTS therapy

protonated conjugate acid, under pH level of 5–6. This gets accumulated inside *M.tb* and leads to inhibition of nutrient transport across membrane.

The primary mode of action of these drugs is shown with the help of Fig. 16.3.

However, there are multiple drawbacks of the DOTS therapy. It is lengthy which often leads to noncompliance with the treatment, consists of multiple antibiotics, and causes severe hepatotoxicity [3]. Since the treatment is for a long duration, patients often discontinue the treatment, once they become sputum negative, which paves way for drug resistance and the problem of persistence.

16.4 Elimination of *M. tuberculosis* by Targeting of Bacterial Signaling Pathways

M. tuberculosis employs a wide arsenal of strategies to avoid killing by the host immune system. Some of these strategies include releasing protein and lipid kinases and phosphatases. The kinases and phosphatases of *M. tuberculosis* regulate intracellular signal transduction as well as interfere with the host immune machinery. Several of the mycobacterial components are currently under investigation as novel targets for anti-TB therapy.

16.5 PE_PGRS in TB Pathogenesis

Any interaction of the microbe with the host cells requires calcium-dependent adhesion. This initial interaction via adhesion between host cells and *M. tuberculosis* occurs via PE_PGRS proteins that leads to a decrease in the intracellular Ca2+ concentration, further leading to inhibition of phagolysosomal maturation which helps in survival of *M. tuberculosis* [16]. Phagosomal maturation requires activation of calmodulin-dependent protein kinases, which is dependent on Ca2+ levels in the cell. *M. tuberculosis* uses different survival strategies by blocking phagosomal maturation. A decrease in intracellular Ca2+ levels is a way of decreasing cytosolic Ca2+ levels in the infected cells to create a niche for the bacteria [17]. These proteins contain multiple calcium-binding motifs. These repeated motifs or sequences form β-sheets or parallel helix structures. These motifs interact with intracellular calcium and mediate bacterial survival within the host cells [17, 18]. These proteins use proline-glutamic acid (PE) domain for their localization on the bacterial cells. These proteins can be targeted along with the genes controlling their expression as drug target against TB.

16.5.1 Tyrosine Phosphatases

Tyrosine phosphatases such as PtpA and PtpB are utilized by the bacteria for their successful survival in the host. These phosphatases are secreted by the mycobacterium in the cytosol of the macrophages where it prevents phagolysosomal degradation of the bacteria and promotes survival and long-term persistence [19]. Vacuolar proteins such as V-ATPases, which regulate membrane fusion and are responsible for vesicle trafficking, are substrates of the tyrosine phosphatases [20]. PtpA interacts with V-ATPases and interferes with acidification of the lysosomes [54]. PtpB allows the bacteria to interfere with calcium signaling of the host by interfering with phosphoinositide metabolism [21]. Therefore, PtpA and PtpB both are promising targets for discovery of novel drug. Many inhibitors of PtpA and PtpB are under research for their anti-mycobacterial activities [22]. Certain indole group derivatives, isoxazoles and thiazolidinones, inhibit PtpB and have shown promising results [23, 24]. We need a better approach in selection of inhibitors, which may be used as drug targets of the future.

16.5.2 Tyrosine Kinase

Tyrosine phosphorylation was the first Ser/Thr/Tyr phosphorylation to be detected in *M. tuberculosis* [25]. Mycobacterial tyrosine kinase A (PtkA) lacks the walker A and walker B domain characteristic of tyrosine kinases and therefore belongs to a different set of protein kinases [26]. PtkA lies immediately upstream to PtpA in the same operon, pointing at the functional similarity between these two, but the exact role of PtkA has not been studied very well till date. This kinase is capable of autophosphorylation [27]. However, the exact role of PtpA phosphorylation by PtkA in vivo still needs to be determined. Thus, this kinase also aids in intracellular shielding of the bacteria and therefore may be an important kinase in terms of drug development.

16.6 Serine/Threonine Protein Phosphatase

PstP, a serine/threonine phosphatase, gets activated upon phosphorylation by serine/threonine kinases. It is the known substrate for serine/threonine kinases pknA and pknB [28]. Together, these kinases and phosphatases are present on the same operon. Remarkably, this phosphatase is reported to dephosphorylate the kinases [29]. Both the kinases and phosphatases together contribute in regulating the fatty acid synthase complex, which is responsible for mycolic acid synthesis [30]. Targeting PstP may reduce the level of mycolic acids and compromise cell wall biosynthesis in *M. tuberculosis*. Therefore, inhibitors against this phosphatase may reduce the bacterial burden in the host.

16.7 Serine/Threonine Protein Kinases (STPKs)

M. tuberculosis STPKs contribute to the effective functioning of various processes such as cell division, bacterial virulence, regulation of molecular machinery, metabolism, and stress. These kinases and phosphatases orchestrate the phosphorylation- and dephosphorylation-dependent response of the bacteria and also help in protecting the bacteria from the host machinery. The mycobacterial kinases have been organized into five clades depending on the structural and sequence similarities: PknA, PknB, and PknL, ABL group (Clade I); PknH, PknE, and PknD, HED group (Clade II); PknF, PknI, and PknJ, FIJ group (Clade III); PknK (Clade IV); and PknG (Clade V) [31].

The most studied and significantly important serine-threonine protein kinases in the context of bacterial survival are PknA and PknB [32]. They have been reported to possess other functions such as in cell division and maintenance of bacterial morphology [33, 34]. Many proteins and enzymes of the mycobacterial cell wall biosynthesis pathway and cell division are substrates of PknA and PknB such as InhA, mtFabH, FipA, and FtsZ [35–39]. These kinases also phosphorylate and activate PstP as already discussed. Mitoxantrone, an anthraquinone derivative and an ATP-competitive inhibitor, is an inhibitor of PknB. Treatment of bacteria with mitoxantrone prevents the mycobacterial growth. Structure analysis confirms that mitoxantrone occupies the adenine-binding pocket of PknB, giving insight into the structure of PknB. This gives understanding of the design of inhibitors, which mimic mitoxantrone in structure and are higher in affinity and potency [40].

PknD has been reported to contribute in osmosensory signaling in the mycobacteria [41]. It has been reported to control the transcriptional machinery of *M. tuberculosis* by stimulating phosphorylation of sigma factor regulators [42]. More studies have reported that PknD acts as a potential phosphate receptor used for mycobacterial survival under low phosphate environments [43]. Researchers have also studied that the sensor domain of PknD assists in mycobacterial adhesion and invasion into the brain endothelium and therefore has a role in development of brain TB affecting the central nervous system [44]. This was further confirmed by using inhibiting PknD that reduced the invasion of *M. tuberculosis* in the brain tissues [44].

Other kinases, such as PknE, have been suggested to regulate apoptosis in *M. tuberculosis*-infected macrophages by suppressing the process and thus promoting persistence in the host [45]. Therefore, it is another interesting candidate as a target for future drugs.

PknG is important for growth under nutrient-deprived conditions such as in host, subverting host-induced stresses, maintaining redox homeostasis and cellular metabolism. It is required for mycobacterial survival in macrophages, by inhibiting phagolysosomal fusion [46]. Indeed, gene deletion or chemical inhibition of PknG resulted in killing of the bacteria in acidified lysosomes [46]. The role of PknG in survival of pathogenic bacteria can be seen by the fact that it is expressed to a very minimal level in nonpathogenic bacteria such as *Mycobacterium smegmatis*, which makes their lysosomal degradation very easy [46, 47].

PknG is different from the other members of mycobacterial protein kinase family in having a long N-terminal extension in front of the catalytic domain and in lacking transmembrane domain. It has three functionally altered domains: the N-terminal rubredoxin domain containing CXXCG motifs, the kinase domain, and the C-terminal domain, which is responsible for the dimerization of the kinase [48, 49]. The CXXCG motifs at the N-terminal are crucial for the catalytic activity of PknG, and the C-terminal tetratricopeptide repeat (TPR) domain is required to prevent lysosomal delivery of mycobacteria [50]. PknG does not require autophosphorylation to get activated. However, autophosphorylation prevents delivery of endosomal cargo to lysosomes helping in evasion of the host defense system [50, 51]. A small inhibitory compound, a tetrahydrobenzothiophene termed AX20017, was used to study the structure of PknG that binds to the catalytic domain. This compound inhibits the function of PknG very selectively [48]. The structure and affinity of AX20017 have been used by the researchers to develop more potent inhibitors of this kinase. A plant secondary metabolite withanolide derived from *Withania somnifera* has been found to be a better inhibitor of PknG compared with AX20017 [52]. More inhibitors and structural analogues are under research to target this functionally important kinase.

Like PknA and PknB, PknH phosphorylates InhA and therefore is vital for cell wall biosynthesis [78]. Its function has been inhibited by the use of O6-cyclohexylmethylguanine, an ATP-competitive cyclin-dependent kinase 1 [53].

PknK has function in regulation of the cell wall composition and cell size [54]. PknK is differentially expressed during the early phase of the infection and at later phase. It has high expression during the establishment of an infection, and later PknK limits mycobacterial growth within macrophages. To further confirm the role of PknK, we need more research on it.

Host tyrosine kinases are also involved in the entry and survival of mycobacteria within the antigen-presenting cells. Inhibition of tyrosine kinases by imatinib (Gleevec®), a tyrosine kinase inhibitor, has been reported to strongly reduce intracellular survival of *M. tuberculosis*, including that of drug-resistant strains. Imatinib has been used along with the drugs of DOTS therapy and has been found to be quite effective [55].

Inhibition of one particular kinase or phosphatase may block many signaling pathways as they regulate many pathways together. Therefore, combining two or more inhibitors to target many regulatory kinases and phosphatases might be an intelligent strategy to increase overall efficacy.

16.8 Lipid Phosphatase

Lipid phosphatases also contribute to the intracellular survival of *M. tuberculosis* in the host cells by interfering with phagolysosomal fusion [56]. In the secreted form, SapM, a lipid phosphatase, prevents the formation of phagolysosome and thus helps in the bacterial survival in the host [57]. SapM (secretion of an acid phosphatase) has been found to possess PI3P phosphatase activity and is responsible for PI3P removal, a membrane trafficking regulatory lipid, which is necessary for phagosomes to acquire lysosomal constituents [58]. Therefore, lipid phosphatases can also be used as drug targets in the future.

16.9 By Targeting Bacterial Protein Secretion System and Proteosomal System

Many mycobacterial survival factors in the form of secreted proteins are released in the cytosol where they perform their function of host evasion. The most studied secretion systems of *M. tuberculosis* are the ESX system, Tat system, and Sec system.

The ESX system, so called because of a 6 kDa early secretory antigenic target (ESAT6) protein, is also known as type VII secretion systems. It is used by the *M. tuberculosis* to export secretory proteins that helps the pathogen to evade the host immunity. It has myriad functions to play in the survival of the pathogen inside macrophages. Some of these functions are in membrane disruption, in escape of the bacteria from the phagosomes, and in unabated growth of bacteria in the granuloma [59–61]. Therefore, it involves various signaling pathways to perform its function that may be targeted for the elimination of bacteria and hindering its survival in the host.

The Sec system is common to all bacteria. It is involved in the secretion of unfolded proteins, which have an N-terminal signal sequence. *M. tuberculosis* expresses a form of SecA protein called SecA2 which is required for virulence as well as host evasion by modulation of phagolysosomal fusion [62]. This makes the components of this system as potential drug targets to avoid pathogen survival in the macrophages and to ensure their early clearance.

The Tat pathway (twin-arginine translocase) is involved in the export of proteins with N-terminal signal peptides. The Tat pathway is found in many Gram-positive and Gram-negative bacteria and, along with being involved in virulence, is essential for mycobacterial survival in the host [63]. As mammals do not possess this system, targeting Tat pathway will provide no off-target effects in the host.

Proteasomes are cellular machinery responsible for degradation of cellular proteins that have been marked for degradation by addition of ubiquitin polypeptide. Ubiquitin-like proteosomal activity has been found in *M. tuberculosis* through various structural and biochemical studies [64]. These proteasomes target ubiquitinylated peptides and proteins via programmed targeting by the proteasomes. Recent studies have identified inhibitors of mycobacterial proteosomal machinery, which is not effective for host proteasomes [65, 66]. Two molecules of oxathiazol-2-one class have been reported to effectively target the bacterial proteosomal machinery without targeting the host [67]. These selective proteosomal inhibitors are very promising to be developed into drugs of future generation.

16.10 Adenylate Cyclase/cAMP Pathway

Phagolysosomal fusion is regulated by the modulation of ATP and cyclic AMP (adenosine monophosphate). Initial experiments using latex bead phagosomes show that phagosomes have an intrinsic adenylate cyclase activity which converted ATP into cAMP, which prevents lysosomal fusion and thus helps in the long-term survival of the pathogen in the host. Inhibiting this adenylate cyclase activity induces phagolysosomal fusion which was halted. A knockout of adenylate cyclase gene lost its ability to grow inside the host cells [68, 69], confirming the essential role this pathway plays in mycobacterial proliferation. Thus, this pathway can be targeted by inhibitors of adenylate cyclase or downstream proteins for potential anti-TB drug.

16.11 Elimination of *M. tuberculosis* by Targeting of Host Signaling Pathways

The successful clearance of bacteria by macrophages requires maturation of the phagosome leading to formation of phagolysosomes and clearance of the invading pathogen. However, *M. tuberculosis* has evolved several strategies for its long-lasting survival in the host. Although macrophages are very effective in phagocytosing and clearing most of the bacteria, *M. tuberculosis* has developed numerous schemes to evade the host immune system and for its effective survival in the host cells. These strategies include inhibition of phagolysosomal fusion; prevention of phagosome acidification; recruitment and maintenance of tryptophan-aspartate-containing coat protein (TACO) on phagosomes which inhibits the phagosome maturation; and expression of members of the host-induced repetitive glycine-rich protein (PE-PGRS) family of proteins. These pathways or strategies used by the pathogen can be targeted for development of novel drugs or therapies wherein evasion of the host immune system would be reversed which would lead to the successful clearance of the bacteria.

The key pathways which can be targeted are:

16.11.1 Blocking Survival Within Macrophages by Targeting the Host Factors Which Prevent Phagolysosomal Fusion

16.11.1.1 Coronin 1/Calcineurin Pathway

Pathogenic mycobacteria survive within macrophages by inhibiting lysosomal killing and residing in the phagosomes along with different other host evasion strategies [70–72]. For this, the host protein coronin 1 (TACO, tryptophan-aspartate-containing coat protein or P57) is recruited and retained toward the cytosolic side of the phagosome. Coronin 1 belongs to the highly conserved family of proteins, which is present in eukaryotes in all hematopoietic cells and is characterized by multiple tryptophan-aspartate repeat domains [73, 74]. Coronin 1-mediated inhibition of lysosomal trafficking has been found to be exclusive for phagosomes infected with mycobacteria [75]. Coronin 1 protein belongs to a large family of proteins, which have multiple tryptophan-aspartate repeat sequences [76]. During the maturation of lysosomes, coronin 1 is released from the phagosome, which leads to the fusion of the phagosome with the lysosome and eventually degradation of the cargo. However, during *M. tuberculosis* infection, coronin 1 stays on the cytosolic surface of the phagocytic vesicle blocking its degradation. It has been reported that the presence of coronin 1 on the phagosome leads to the activation of downstream calcium/calcineurin pathway that is responsible for inhibition of phagolysosome fusion and thus elimination of the internalized mycobacteria [77]. Furthermore, in coronin 1-deficient mice-derived macrophages, it has been observed that the mycobacteria containing phagocytic vesicles fuse readily with the lysosomes and lead to the degradation of the bacilli [77]. These coronin 1-lacking macrophages were found unable to support intracellular survival of the mycobacteria but were fully functional in other roles [77]. Coronin 1 activates calcineurin signaling pathway. Calcineurin is a calcium-sensitive phosphatase [78] which upon inhibition by its blockers results in bacterial killing [75]. Therefore, both coronin 1 and calcineurin signaling pathways and their downstream substrates can be targeted for development of a novel drug.

16.11.1.2 Voltage-Gated Channels

Calcium from the extracellular environment enters into the cells by means of the cell membrane calcium channels, such as voltage-gated calcium channels (VGCCs), which are selective calcium channels, or nonselective calcium channels, such as purine receptor (P2X7), cyclic nucleotide-gated (CNG) ion channels, and canonical transient receptor potential channels (TRPCs). These channels are stimulated by membrane potentials and ligands. VGCC is activated by changes in membrane potential, while the nonselective channels such as P2X7, CNG, and TRPC are stimulated by adenosine triphosphate (ATP), cAMP, and PLC or DAG, respectively [79, 80]. The role of these channels has gained importance in TB drug development recently. VGCCs are multimeric transmembrane channels that can affect the function of the immune system by activating the expression of pro-inflammatory genes and modulating the phagocytic capacity of the macrophages and DCs [81]. Use of VGCC blockers has resulted in the increase in calcium influx [81–83]. This increase in calcium influx results in enhanced secretion of pro-inflammatory cytokines. It has also been reported that inhibition of VGCCs in DCs leads to the activation of T lymphocytes and increases the ability of macrophages to perform their bactericidal functions [81, 84]. Mice experiments also prove that blocking the expression of VGCC with antibodies resulted in decreased *M. tuberculosis* number with a simultaneous increase in calcium level [81, 85, 86]. Under normal uninhibited state, there is a low influx of calcium and signaling in macrophages and DCs mainly due to high upregulation of VGCC. Moreover, peripheral blood mononuclear cells (PBMCs) isolated from TB-infected patients show higher VGCC expression level compared to healthy control [81], making VGCCs an important indicator of TB infection.

Another voltage-gated channel, which has been targeted for efficient control of TB, is the purine receptor (P2X7). P2X7 is a ligand-gated ion channel [87]. In macrophages and myeloid cells, P2X7 gets activated by ATP and leads to the downstream secretion of cytokines, which ultimately lead to interferon-gamma production [88, 89]. Other voltage-gated channels being studied as drug target for *M. tuberculosis* infection are NaV1.6 and KV1.3 [90, 91]. These channels or their downstream substrates can be explored for elimination of TB by using inhibitors against them.

Therefore, a combination of conventional anti-TB drug regimen along with voltage-gated channel blockers is a novel method of chemotherapy against TB and may help to shorten the length of treatment and prevent emergence of drug resistance.

The different possible calcium signaling pathways, which can be targeted for drug development, have been elucidated in Fig. 16.4.

16.11.1.3 Immunity-Related GTPase Family M Proteins

Immunity-related GTPase family M proteins, IRGM (also known as LRG47), are members of a family of IFN-gamma-regulated P47 GTPases [92]. These proteins have gained huge importance due to their role being recently highlighted in inducing autophagy upon infections by interacting with autophagy-linked proteins such as ATG5, ATG10, and LC3. These genes in mice are regulated by IFN-γ, as treatment

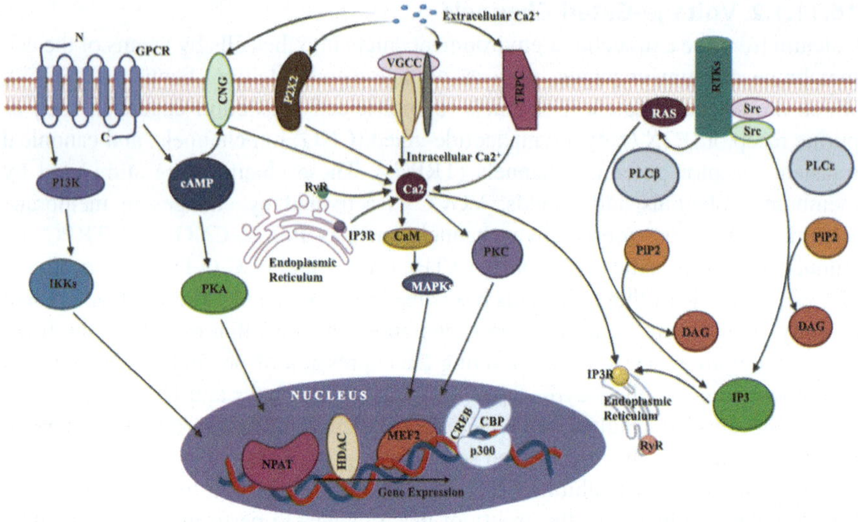

Fig. 16.4 All the possible calcium signaling pathways, which can be targeted in the immune cells for ensuring effective phagolysosomal fusion

of macrophages (derived from mice) with IFN-γ leads to killing of *M. tuberculosis* population, by induction of autophagy via IRGM1 (murine analogue of IRGM) [93, 94]. LRG47-mediated autophagy also protects against other infections such as *Escherichia coli* and *Salmonella typhimurium* [94–98]. However, IRGM-dependent autophagy does not function in the absence of infection, suggesting pathogen specificity of these proteins in the induction of autophagy [99, 100]. In the absence of IRGM protein, the phagosomes have low levels of V-ATPases, which help in the acidification and fusion of phagocytic vesicles [101]. Therefore, these GTPases along with other factors and genes regulating them can be targeted for effective anti-mycobacterial therapies.

16.11.1.4 Blocking the Persistence of the Bacteria Within the Host

After the treatment of infected patients with DOTS therapy, these patients start feeling better within 3–4 weeks of the treatment. These patients with initial recovery often discontinue the treatment, and this leads to the survival of few bacteria in the host, which hide in the host for a long time feeding themselves by taking nutrition from the host and shielding themselves from the host immune system. These surviving bacteria are known as persisters, and they become tolerant to front-line anti-TB drugs. This persister *M. tuberculosis* population may give rise to drug resistance and lead to the relapse of the disease in a more severe form. Total control of *M. tuberculosis* survival in the host requires effective targeting of these persisters along with the actively replicating bacteria. Persisters feed themselves by utilizing the lipid sources present within the host cells. The enzyme isocitrate lyase is responsible for using the lipids as nutrient sources. The expression level of this enzyme is elevated upon *M. tuberculosis* infection [102]. It is the key enzyme of the glyoxylate shunt pathway that dodges the carbon dioxide-producing steps of the Krebs pathway, allowing the bacteria to use lipids/fatty acids as substrates [103]. It has been studied that *M. tuberculosis* devoid of two of the genes coding for isocitrate lyase has difficulty in surviving and replicating in the host [104]. Therefore, these enzymes are attractive targets for anti-mycobacterial therapy.

Recently, it has been reported that persisters invade mesenchymal stem cells (MSCs) of the host and utilize the hypoxic environment and lipids of these cells for their survival. These MSC-residing bacteria have been shown to be eliminated effectively by a dual strategy of using lipid inhibitor and autophagy inducer along with conventional antibiotics to attain sterile clearance of the disease [3].

Other mechanisms involve increasing the IFN-γ immune response in the host for the effective clearance of the pathogen and thus increasing the host pro-inflammatory response by using immune modulators. Moreover, Toll-like receptors (TLRs) also help in the killing of the pathogen from the immune cells. TLR2, TLR4, and TLR9 signaling are mainly responsible for their function in the eradication of the *M. tuberculosis* [105, 106]. Therefore, it is quite reasonable to develop strategies involving the signaling pathways upstream or downstream of the TLR signaling pathways or IFN-γ signaling for the sterile clearance of persisters of *M. tuberculosis* population.

The possible mechanism of targeting host and bacterial factors for the survival of *M. tuberculosis* is shown schematically in Fig. 16.5.

16.12 Conclusions and Future Prospects

The past decade has been involved in extensive search for new drug targets for TB. This is evident from the discovery of bedaquiline and delamanid. Scientists all over the globe are working to understand both the bacterial and host factors which help in the growth and survival of *M. tuberculosis*. Many novel drug candidates have been identified which include kinases, phosphatases, and secreted proteins, which have been studied to block the lysosomal degradation of the bacteria. Using inhibitors against these promising drug candidates will help us decipher new pathways and molecules, which can be effective for drug development. However, it would be challenging to convert these studies into therapeutics for the treatment of patients owing to the side effects involved, which are still mostly unknown. We also need to study if we can target multiple kinases, which perform similar functions with a multiple kinase inhibitor. Therefore, the need is to study all the side effects and bactericidal potential of these drugs to convert them into drugs of the future.

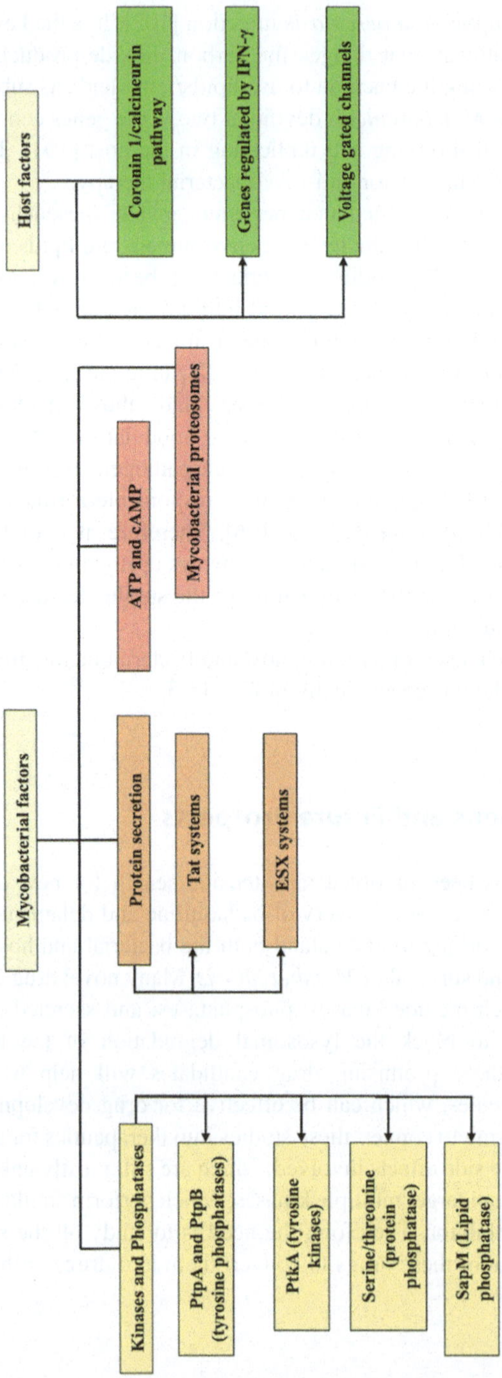

Fig. 16.5 A schematic diagram of host and bacterial factors which can be targeted for future drug therapy

References

1. WHO Report, 2019
2. Schlesinger LS (1993) Macrophage phagocytosis of virulent but not attenuated strains of Mycobacterium tuberculosis is mediated by mannose receptors in addition to complement receptors. J Immunol 150:2920–2930
3. Fatima S, Kamble SS, Dwivedi VP et al (2020) Mycobacterium tuberculosis programs mesenchymal stem cells to establish dormancy and persistence. J Clin Invest 130(2):655–661
4. Stutz MD, Clark MP, Doerflinger M, Pellegrini M (2018) Mycobacterium tuberculosis: rewiring host cell signaling to promote infection. J Leukoc Biol 103(2):259–268
5. D'Ambrosio L, Centis R, Tiberi S et al (2017) Delamanid and bedaquiline to treat multidrug-resistant and extensively drug-resistant tuberculosis in children: a systematic review. J Thorac Dis 9(7):2093–2101
6. Smith I (2003) Mycobacterium tuberculosis pathogenesis and molecular determinants of virulence. Clin Microbiol Rev 16(3):463–496
7. Bloom BR, Atun R, Cohen T et al (2017) Chapter 11: Tuberculosis. In: Holmes KK, Bertozzi S, Bloom BR et al (eds) Major infectious diseases, 3rd edn. The International Bank for Reconstruction and Development/The World Bank, Washington, DC
8. Guirado E, Schlesinger LS, Kaplan G (2013) Macrophages in tuberculosis: friend or foe. Semin Immunopathol 35(5):563–583
9. Ernst JD (1998) Macrophage receptors for Mycobacterium tuberculosis. Infect Immun 66 (4):1277–1281
10. Sia JK, Rengarajan J (2019) Immunology of *Mycobacterium tuberculosis* infections. Microbiol Spectr 7(4). https://doi.org/10.1128/microbiolspec.GPP3-0022-2018
11. Zhai W, Wu F, Zhang Y, Fu Y, Liu Z (2019) The immune escape mechanisms of *Mycobacterium Tuberculosis*. Int J Mol Sci 20(2):340
12. Ndlovu H, Marakalala MJ (2016) Granulomas and inflammation: host-directed therapies for tuberculosis. Front Immunol 7:434
13. Cooper AM (2009) Cell-mediated immune responses in tuberculosis. Annu Rev Immunol 27:393–422
14. Fattorini L, Piccaro G, Mustazzolu A, Giannoni F (2013) Targeting dormant bacilli to fight tuberculosis. Mediterr J Hematol Infect Dis 5(1):e2013072
15. Sotgiu G, Centis R, D'ambrosio L, Migliori GB (2015) Tuberculosis treatment and drug regimens. Cold Spring Harb Perspect Med 5(5):a017822
16. Meena LS (2015) An overview to understand the role of PE_PGRS family proteins in Mycobacterium tuberculosis H37 Rv and their potential as new drug targets. Biotechnol Appl Biochem 62(2):145–153
17. Bachhawat N, Singh B (2007) Mycobacterial PE_PGRS proteins contain calcium-binding motifs with parallel beta-roll folds. Geno Prot Bioinfo 5:236–241
18. Tian C, Jian-Ping X (2010) Roles of PE_PGRS family in Mycobacterium tuberculosis pathogenesis and novel measures against tuberculosis. Microb Pathog 49:311–314
19. Koul A, Choidas A, Treder M et al (2000) Cloning and characterization of secretory tyrosine phosphatases of Mycobacterium tuberculosis. J Bacteriol 182(19):5425–5432
20. Bach H, Papavinasasundaram KG, Wong D, Hmama Z, Av-Gay Y (2008) Mycobacterium tuberculosis virulence is mediated by PtpA dephosphorylation of human vacuolar protein sorting 33B. Cell Host Microbe 3(5):316–322
21. Beresford N, Patel S, Armstrong J, Szöor B, Fordham-Skelton AP, Tabernero L (2007) MptpB, a virulence factor from Mycobacterium tuberculosis, exhibits triple-specificity phosphatase activity. Biochem J 406(1):13–18
22. Koch MA, Waldmann H (2005) Protein structure similarity clustering and natural product structure as guiding principles in drug discovery. Drug Discov Today 10(7):471–483

23. Beresford NJ, Mulhearn D, Szczepankiewicz B et al (2009) Inhibition of MptpB phosphatase from Mycobacterium tuberculosis impairs mycobacterial survival in macrophages. J Antimicrob Chemother 63(5):928–936
24. Nören-Müller A, Reis-Corrêa I Jr, Prinz H et al (2006) Discovery of protein phosphatase inhibitor classes by biology-oriented synthesis. Proc Natl Acad Sci U S A 103 (28):10606–10611
25. Chow K, Ng D, Stokes R, Johnson P (1994) Protein tyrosine phosphorylation in Mycobacterium tuberculosis. FEMS Microbiol Lett 124:203–207
26. Walker JE, Saraste M, Runswick MJ, Gay NJ (1982) Distantly related sequences in the α- and β-subunits of ATP synthase, myosin, kinases and other ATP-requiring enzymes and a common nucleotide binding fold. EMBO J 1(8):945–951
27. Bach H, Wong D, Av-Gay Y (2009) Mycobacterium tuberculosis PtkA is a novel protein tyrosine kinase whose substrate is PtpA. Biochem J 20(2):155–160
28. Sajid A, Arora G, Gupta M, Upadhyay S, Nandicoori VK, Singh Y (2011) Phosphorylation of Mycobacterium tuberculosis Ser/Thr phosphatase by PknA and PknB. PLoS One 6(3):e17871
29. Boitel B, Ortiz-Lombardía M, Durán R et al (2003) PknB kinase activity is regulated by phosphorylation in two Thr residues and dephosphorylation by PstP, the cognate phospho-Ser/Thr phosphatase, in Mycobacterium tuberculosis. Mol Microbiol 49(6):1493–1508
30. Molle V, Brown AK, Besra GS, Cozzone AJ, Kremer L (2006) The condensing activities of the Mycobacterium tuberculosis type II fatty acid synthase are differentially regulated by phosphorylation. J Biol Chem 281(40):30094–30103
31. Wehenkel A, Bellinzoni M, Grana M, Duran R, Villarino A et al (2008) Mycobacterial Ser/Thr protein kinases and phosphatases: physiological roles and therapeutic potential. Biochim. Biophys Acta 1784:193–202
32. Fernandez P, Saint-Joanis B, Barilone N et al (2006) The Ser/Thr protein kinase PknB is essential for sustaining mycobacterial growth. J Bacteriol 188(22):7778–7784
33. Thakur M, Chaba R, Mondal AK, Chakraborti PK (2008) Interdomain interaction reconstitutes the functionality of PknA, a eukaryotic type Ser/Thr kinase from Mycobacterium tuberculosis. J Biol Chem 283(12):8023–8033
34. Thakur M, Chakraborti PK (2008) Ability of PknA, a mycobacterial eukaryotic-type serine/threonine kinase, to transphosphorylate MurD, a ligase involved in the process of peptidoglycan biosynthesis. Biochem J 415(1):27–33
35. Khan S, Nagarajan SN, Parikh A et al (2010) Phosphorylation of enoyl-acyl carrier protein reductase InhA impacts mycobacterial growth and survival. J Biol Chem 285 (48):37860–37871
36. Parikh A, Verma SK, Khan S, Prakash B, Nandicoori VK (2009) PknB-mediated phosphorylation of a novel substrate, N-acetylglucosamine-1-phosphate uridyltransferase, modulates its acetyltransferase activity. J Mol Biol 386(2):451–464
37. Veyron-Churlet R, Molle V, Taylor RC et al (2009) The Mycobacterium tuberculosis β-ketoacyl-acyl carrier protein synthase III activity is inhibited by phosphorylation on a single threonine residue. J Biol Chem 284(10):6414–6424
38. Veyron-Churlet R, Zanella-Cléon I, Cohen-Gonsaud M, Molle V, Kremer L (2010) Phosphorylation of the Mycobacterium tuberculosis β-ketoacyl-acyl carrier protein reductase MabA regulates mycolic acid biosynthesis. J Biol Chem 285(17):12714–12725
39. Sureka K, Hossain T, Mukherjee P et al (2010) Novel role of phosphorylation-dependent interaction between FtsZ and FipA in mycobacterial cell division. PLoS One 5(1):e8590
40. Wehenkel A, Fernandez P, Bellinzoni M et al (2006) The structure of PknB in complex with mitoxantrone, an ATP-competitive inhibitor, suggests a mode of protein kinase regulation in mycobacteria. FEBS Lett 580(13):3018–3022
41. Hatzios SK, Baer CE, Rustad TR, Siegrist MS, Pang JM et al (2013) Osmosensory signaling in Mycobacterium tuberculosis mediated by a eukaryotic-like Ser/Thr protein kinase. Proc Natl Acad Sci 110:E5069–E5077

42. Greenstein AE, Echols N, Lombana TN, King DS, Alber T (2007) Allosteric activation by dimerization of the PknD receptor Ser/Thr protein kinase from Mycobacterium tuberculosis. J Biol Chem 282(15):11427–11435
43. Vanzembergh F, Peirs P, Lefevre P et al (2010) Effect of PstS sub-units or PknD deficiency on the survival of Mycobacterium tuberculosis. Tuberculosis (Edinb) 90(6):338–345
44. Be NA, Bishai WR, Jain SK (2012) Role of Mycobacterium tuberculosis pknD in the pathogenesis of central nervous system tuberculosis. BMC Microbiol 12:7
45. Jayakumar D, Jacobs WR Jr, Narayanan S (2008) Protein kinase E of Mycobacterium tuberculosis has a role in the nitric oxide stress response and apoptosis in a human macrophage model of infection. Cell Microbiol 10(2):365–374
46. Walburger A, Koul A, Ferrari G et al (2004) Protein kinase G from pathogenic mycobacteria promotes survival within macrophages. Science 304(5678):1800–1804
47. Houben EN, Walburger A, Ferrari G et al (2009) Differential expression of a virulence factor in pathogenic and non-pathogenic mycobacteria. Mol Microbiol 72(1):41–52
48. Av-Gay Y, Everett M (2000) The eukaryotic-like Ser/Thr protein kinases of *Mycobacterium tuberculosis*. Trends Microbiol 8(5):238–244
49. Scherr N, Honnappa S, Kunz G et al (2007) Structural basis for the specific inhibition of protein kinase G, a virulence factor of Mycobacterium tuberculosis. Proc Natl Acad Sci 104 (29):12151–12156
50. Tiwari D, Singh RK, Goswami K, Verma SK, Prakash B, Nandicoori VK (2009) Key residues in Mycobacterium tuberculosis protein kinase G play a role in regulating kinase activity and survival in the host. J Biol Chem 284(40):27467–27479
51. Scherr N, Müller P, Perisa D, Combaluzier B, Jenö P, Pieters J (2009) Survival of pathogenic mycobacteria in macrophages is mediated through autophosphorylation of protein kinase G. J Bacteriol 191(14):4546–4554
52. Santhi N, Aishwarya S (2011) Insights from the molecular docking of withanolide derivatives to the target protein PknG from Mycobacterium tuberculosis. Bioinformation 7(1):1–4
53. Molle V, Reynolds RC, Alderwick LJ et al (2008) EmbR2, a structural homologue of EmbR, inhibits the Mycobacterium tuberculosis kinase/substrate pair PknH/EmbR. Biochem J 410 (2):309–317
54. Malhotra V, Arteaga-Cortés LT, Clay G, Clark-Curtiss JE (2010) Mycobacterium tuberculosis protein kinase K confers survival advantage during early infection in mice and regulates growth in culture and during persistent infection: implications for immune modulation. Microbiology (Reading, Engl) 156(9):2829–2841
55. Napier RJ, Rafi W, Cheruvu M et al (2011) Imatinib-sensitive tyrosine kinases regulate mycobacterial pathogenesis and represent therapeutic targets against tuberculosis. Cell Host Microbe 10(5):475–485
56. Saleh MT, Belisle JT (2000) Secretion of an acid phosphatase (SapM) by Mycobacterium tuberculosis that is similar to eukaryotic acid phosphatases. J Bacteriol 182(23):6850–6853
57. Vergne I, Chua J, Lee HH, Lucas M, Belisle J, Deretic V (2005) Mechanism of phagolysosome biogenesis block by viable Mycobacterium tuberculosis. Proc Natl Acad Sci U S A 102(11):4033–4038
58. Vergne I et al (2005) Mechanism of phagolysosome biogenesis block by viable Mycobacterium tuberculosis. Proc Natl Acad Sci U S A 102:4033–4038
59. Brodin P, Rosenkrands I, Andersen P, Cole ST, Brosch R (2004) ESAT-6 proteins: protective antigens and virulence factors? Trends Microbiol 12(11):500–508
60. van der Wel N, Hava D, Houben D et al (2007) M. tuberculosis and M. leprae translocate from the phagolysosome to the cytosol in myeloid cells. Cell 129(7):1287–1298
61. Volkman HE, Pozos TC, Zheng J, Davis JM, Rawls JF, Ramakrishnan L (2010) Tuberculous granuloma induction via interaction of a bacterial secreted protein with host epithelium. Science 327(5964):466–469

62. Sullivan JT, Young EF, McCann JR, Braunstein M (2012) The Mycobacterium tuberculosis SecA2 system subverts phagosome maturation to promote growth in macrophages. Infect Immun 80(3):996–1006
63. Dilks K, Rose RW, Hartmann E, Pohlschröder M (2003) Prokaryotic utilization of the twin-arginine translocation pathway: a genomic survey. J Bacteriol 185(4):1478–1483
64. Becker SH, Jastrab JB, Dhabaria A, Chaton CT, Rush JS, Korotkov KV, Ueberheide B, Darwin KH (2019) The Mycobacterium tuberculosis pup-proteasome system regulates nitrate metabolism through an essential protein quality control pathway. Proc Natl Acad Sci U S A 116(8):3202–3210
65. Cromm PM, Crews CM (2017) The proteasome in modern drug discovery: second life of a highly valuable drug target. ACS Cent Sci 3(8):830–838
66. Lin G, Li D, de Carvalho LP et al (2009) Inhibitors selective for mycobacterial versus human proteasomes. Nature 461(7264):621–626
67. Cheng Y, Pieters J (2010) Novel proteasome inhibitors as potential drugs to combat tuberculosis. J Mol Cell Biol 2(4):173–175
68. Shenoy AR, Visweswariah SS (2006) New messages from old messengers: cAMP and mycobacteria. Trends Microbiol 14(12):543–550
69. Agarwal N, Lamichhane G, Gupta R, Nolan S, Bishai WR (2009) Cyclic AMP intoxication of macrophages by a Mycobacterium tuberculosis adenylate cyclase. Nature 460(7251):98–102
70. Rohde K, Yates RM, Purdy GE, Russell DG (2007) Mycobacterium tuberculosis and the environment within the phagosome. Immunol Rev 219:37–54
71. Pieters J (2008) Mycobacterium tuberculosis and the macrophage: maintaining a balance. Cell Host Microbe 3:399–407
72. Warner DF, Mizrahi V (2007) The survival kit of Mycobacterium tuberculosis. Nat Med 13:282–284
73. Eckert C, Hammesfahr B, Kollmar M (2011) A holistic phylogeny of the coronin gene family reveals an ancient origin of the tandem-coronin, defines a new subfamily, and predicts protein function. BMC Evol Biol 11:268
74. Pieters J, Muller P, Jayachandran R (2013) On guard: coronin proteins in innate and adaptive immunity. Nat Rev Immunol 13:510–518
75. Jayachandran R, Sundaramurthy V, Combaluzier B, Mueller P, Korf H, Huygen K et al (2007) Survival of mycobacteria in macrophages is mediated by coronin 1-dependent activation of calcineurin. Cell 130:37–50
76. Neer EJ, Schmidt CJ, Nambudripad R, Smith TF (1994) The ancient regulatory-protein family of WD-repeat proteins. Nature 371(6495):297–300
77. Ferrari G, Langen H, Naito M, Pieters J (1999) A coat protein on phagosomes involved in the intracellular survival of mycobacteria. Cell 97:435–447
78. Winslow MM, Neilson JR, Crabtree GR (2003) Calcium signalling in lymphocytes. Curr Opin Immunol 15(3):299–307
79. Wiley JS, Sluyter R, Gu BJ, Stokes L, Fuller SJ (2011) The human P2X7 receptor and its role in innate immunity. Tissue Antigens 78:321e32
80. Song L, Cui R, Yang Y, Wu X (2015) Role of calcium channels in cellular antituberculosis effects: potential of voltage-gated calcium-channel blockers in tuberculosis therapy. J Microbiol Immunol Infect 48(5):471–476
81. Gupta S, Salam N, Srivastava V, Singla R, Behera D, Khayyam KU et al (2009) Voltage gated calcium channels negatively regulate protective immunity to Mycobacterium tuberculosis. PLoS One 4:e5305
82. Lammas DA, Stober C, Harvey CJ, Kendrick N, Panchalingam S, Kumararatne DS (1997) ATP-induced killing of mycobacteria by human macrophages is mediated by purinergic P2Z (P2X7) receptors. Immunity 7:433e44

83. Adams KN, Szumowski JD, Ramakrishnan L (2014) Verapamil, and its metabolite norverapamil, inhibit macrophage-induced, bacterial efflux pump-mediated tolerance to multiple antitubercular drugs. J Infect Dis 210:456e66
84. Sharma D, Tiwari BK, Mehto S et al (2016) Suppression of protective responses upon activation of L-type voltage gated calcium channel in macrophages during Mycobacterium bovis BCG infection. PLoS One 11(10):e0163845
85. Gupta S, Tyagi S, Almeida DV, Maiga MC, Ammerman NC, Bishai WR (2013) Acceleration of tuberculosis treatment by adjunctive therapy with verapamil as an efflux inhibitor. Am J Respir Crit Care Med 188:600e7
86. Chen C, Gardete S, Jansen RS et al (2018) Verapamil Targets Membrane Energetics in Mycobacterium tuberculosis. Antimicrob Agents Chemother 62(5):e02107–e02117
87. Caseley EA, Muench SP, Roger S, Mao HJ, Baldwin SA, Jiang LH (2014) Non-synonymous single nucleotide polymorphisms in the P2X receptor genes: association with diseases, impact on receptor functions and potential use as diagnosis biomarkers. Int J Mol Sci 15 (8):13344–13371
88. Tominaga K, Yoshimoto T, Torigoe K, Kurimoto M, Matsui K, Hada T et al (2000) IL-12 synergizes with IL-18 or IL-1beta for IFN-gamma production from human T cells. Int Immunol 12:151–160
89. Savio LEB, de Andrade Mello P, da Silva CG, Coutinho-Silva R (2018) The P2X7 receptor in inflammatory diseases: angel or demon? Front Pharmacol 9:52
90. Carrithers LM, Hulseberg P, Sandor M, Carrithers MD (2011) The human macrophage sodium channel NaV1.5 regulates mycobacteria processing through organelle polarization and localized calcium oscillations. FEMS Immunol Med Microbiol 63(3):319–327
91. Wulff H, Castle NA, Pardo LA (2009) Voltage-gated potassium channels as therapeutic targets. Nat Rev Drug Discov 8(12):982–1001
92. Gilly M, Wall R (1992) The IRG-47 gene is IFN-γ induced in B cells and encodes a protein with GTP-binding motifs. J Immunol 148(10):3275–3281
93. Gutierrez MG, Master SS, Singh SB, Taylor GA, Colombo MI, Deretic V (2004) Autophagy is a defense mechanism inhibiting BCG and Mycobacterium tuberculosis survival in infected macrophages. Cell 119(6):753–766
94. Singh SB, Ornatowski W, Vergne I et al (2010) Human IRGM regulates autophagy and cell-autonomous immunity functions through mitochondria. Nat Cell Biol 12(12):1154–1165
95. Petkova DS, Viret C, Faure M (2013) IRGM in autophagy and viral infections. Front Immunol 3:426
96. McCarroll SA, Huett A, Kuballa P, Chilewski SD, Landry A, Goyette P et al (2008) Deletion polymorphism upstream of IRGM associated with altered IRGM expression and Crohn's disease. Nat Genet 40:1107–1112
97. Lapaquette P, Glasser AL, Huett A, Xavier RJ, Darfeuille-Michaud A (2010) Crohn's disease-associated adherent-invasive E. coli are selectively favoured by impaired autophagy to replicate intracellularly. Cell Microbiol 12(1):99–113
98. Brest P, Lapaquette P, Souidi M, Lebrigand K, Cesaro A, Vouret-Craviari V et al (2011) A synonymous variant in IRGM alters a binding site for miR-196 and causes deregulation of IRGM-dependent xenophagy in Crohn's disease. Nat Genet 43:242–245
99. Gregoire IP, Rabourdin-Combe C, Faure M (2012) Autophagy and RNA virus interactomes reveal IRGM as a common target. Autophagy 8:1136–1137
100. Matsuzawa T, Kim BH, Shenoy AR, Kamitani S, Miyake M, Macmicking JD (2012) IFN-gamma elicits macrophage autophagy via the p38 MAPK signaling pathway. J Immunol 189:813–818
101. MacMicking JD, Taylor GA, McKinney JD (2003) Immune control of tuberculosis by IFN-γ-inducible LRG-47. Science 302(5645):654–659
102. Lee YV, Wahab HA, Choong YS (2015) Potential inhibitors for isocitrate lyase of Mycobacterium tuberculosis and non-M. tuberculosis: a summary. Biomed Res Int 2015:895453

103. Bishai W (2000) Lipid lunch for persistent pathogen. Nature 406(6797):683–685
104. Muñoz-Elías EJ, McKinney JD (2005) Mycobacterium tuberculosis isocitrate lyases 1 and 2 are jointly required for in vivo growth and virulence. Nat Med 11(6):638–644
105. Stamm CE, Collins AC, Shiloh MU (2015) Sensing of Mycobacterium tuberculosis and consequences to both host and bacillus. Immunol Rev 264(1):204–219
106. Basu J, Shin DM, Jo EK (2012) Mycobacterial signaling through toll-like receptors. Front Cell Infect Microbiol 2:145

Targeting Molecular and Cellular Mechanisms in Pulmonary Hypertension

17

Stephen A. Adefegha, Olorunfemi R. Molehin, and Aderonke E. Fakayode

Abstract

Pulmonary hypertension (PH) is described as an elevated mean pulmonary artery pressure (mPAP) of 25 mmHg or above, measured at rest by right-heart catheterization. The precise worldwide occurrence of PH is not easily measured, majorly as a result of its complex etiology, and its progression is likely to be underestimated. Extensive reports on the pathophysiology of PH and incidence etiology at the cellular and molecular level have been well documented. In addition, basic clinical research studies have shown promising potentials of popular and widely known cardiovascular biomarkers, but with limited clinical significance in the management and diagnosis of PH as a result of decreased specificity as well as several other cardiovascular complications of patients with PH. Conversely, a large panel of experimental research studies reveal novel cellular and molecular mechanisms, drug targets, and biomarkers following the principle of the evidence-based medicine. Unfortunately, the basic extrapolation of these finding results to clinical practice is not straightforward because of large complex nature of the pathophysiology of PH. Hence, there is the need for more translational medicine research to fully comprehend the pathophysiological

S. A. Adefegha
Functional Foods and Nutraceuticals Unit, Department of Biochemistry, Federal University of Technology, Akure, Nigeria
e-mail: saadefegha@futa.edu.ng

O. R. Molehin (✉)
Department of Biochemistry, Faculty of Science, Ekiti State University, Ado-Ekiti, Nigeria
e-mail: olorunfemi.molehin@eksu.edu.ng

A. E. Fakayode
Department of Biochemistry and Molecular Biology, Faculty of Science, Obafemi Awolowo University, Ile-Ife, Nigeria
e-mail: afakayode@pg-student.oauife.edu.ng

K. Dua et al. (eds.), *Targeting Cellular Signalling Pathways in Lung Diseases*,
https://doi.org/10.1007/978-981-33-6827-9_17

classification of PH and define accurately its biomarkers and therapeutic approach for PH treatment. We discuss in this chapter the types of PH, the novel therapeutic options available, and the molecular mechanism behind PH and its possible drug targets for treatments at the cellular and molecular level. Several factors like oxidative stress, cell signaling, inflammation, and mechanisms of immune systems are highlighted that may be responsible for the pathophysiology of PH and its complications. Each of these processes is discussed separately, for clarity purpose, but most importantly, crucial cross functions will occur among the pathways in PH patients.

Keywords

Pulmonary hypertension · Mean pulmonary artery pressure · Cardiovascular disease · Pathophysiology · Lung disease

17.1 Introduction

Hypertension (HP) is a global health burden because of its rising observed cases and its increased risk factor for cardiovascular diseases (CVDs). HP occurs when there is persistent elevation of artery's blood pressure, thus bringing about damages of organs and an increase in the world mortality rate [1]. When there is a rise in cardiac output and systemic vascular resistance, it results in blood pressure. Blood pressure has been found to elevate the risk of cardiovascular disease globally. Blood pressure is reported as two numbers which are obtained from the diastolic and systolic pressure [2]. The highest pressure in the arteries when the heart beats and fills the arteries is termed the systolic pressure, while the least pressure in the arteries when the heart rests between beats is termed the diastolic pressure.

During aging process, there is stiffness of blood vessels; this prevents the free movement of blood from the heart, thus bringing about an increase in systolic pressure as aging occurs. "In most cases, increased cardiac output occurs in younger age while in adults, systemic vascular resistance is observed; this may be a result of elevated α-adrenoreceptor stimulation; it could be that peptides like angiotensin are constantly released. Elevated increase of calcium in the cytosolic medium of vascular smooth muscle results in vasoconstriction. Other factors such as growth factors can also increase vascular smooth muscle, thus causing vasoconstriction, stiffening of the aorta and arteries which results in pulse pressure increase" [3]. It has been established that CVD is one of the major causes of global deaths; thus, the treatment and management of several CVDs have been put into consideration to reduce mortality rate worldwide.

Hypertension has been considered to be one of these CVDs and it occurs when there is a persistent elevation in the blood pressure of arteries. "Hypertension contributes majorly to increased risks for coronary heart disease, stroke, and several other heart-related diseases. It has been confirmed to be one of the major contributors to global deaths. In recent times, lots of studies and researches were carried out to combat the occurrence of hypertension worldwide. Hypertension was classified into primary hypertension which occurs as a result of genetic factors or nonspecific lifestyles such as excess salt in diet, excess body weight, physical inability, smoking, and secondary hypertension described as persistent diseased state such as chronic kidney disease. Treating of hypertension has been linked to a significant reduction in the occurrence of certain disease conditions, namely heart diseases, stroke, myocardial infarction, and other cardiovascular related diseases [2]. The autonomic nervous system is well documented to play a crucial function in the regulation of blood pressure as patients with hypertension have elevated peripheral sensitivity to certain hormones like norepinephrine; in addition, increased responsiveness to stressful stimuli is also observed in hypertensive patients" [2, 3].

17.2 Consequences and Complications of Hypertension

There is no gauge for complications to arise, as an increase in blood pressure is linked with increased morbidity across the whole blood pressure ranges of measurement. "Persistent increase in blood pressure brings about increase in muscle mass, thickening of the artery wall resulting in left ventricular hypertrophy (a condition that impairs diastolic function and slows ventricular relaxation) which is a separate risk factor for CVDs. Arterial hypertension is one of the major factors that increase the occurrence of coronary artery disease. It has been established that poorly controlled or untreated hypertensive patients are prone to having myocardial ischemia and myocardial infarction as a result of pressure-related increased demand for oxygen or a coronary oxygen supply depletion. Other complications of hypertension are heart failure, stroke, etc. which could be from intracranial hemorrhage or thrombosis" [2].

17.2.1 Treatment of Hypertension

Hypertension has greatly increased the world's mortality rate; therefore, there is a quick response to the search for its treatment and management. The first line of treatment and management of HP is modification in the way of living, which includes restriction in sodium intake, weight reduction, decreased alcohol intake, and frequent exercises. "The use of drug therapy has also been introduced in the treatment and management of hypertension. The mechanism of action of most hypertensive drugs differs from each other but every antihypertensive drug acts majorly by reducing the cardiac output, resistance of vascular tissues, or both. Some of the most common class of antihypertensive drugs include β-blockers, inhibitors of angiotensin-converting enzyme (ACE), thiazide diuretics, antagonists of the angiotensin II receptors, blockers (calcium channel, α-adrenoceptor, combined α- and β-blockers), etc." [3, 4].

17.3 Pulmonary Hypertension (PH)

PH occurs when the resting mean pulmonary artery pressure is 25 mmHg or above at the resting stage [1]. "It could occur due to pulmonary vascular disease, left-heart disease, lung disease or hypoxia, chronic thromboembolic disease, and a variety of disorders, including hemolytic anemias and sarcoidosis. Globally, it is estimated that over 100 million people have been affected. Currently, there exists no curative treatment for PH despite advancement in the study of the disease pathology but there are improved diagnostic markers and therapeutic approaches for patients. Patients with PH tend to show the following pathogenic features: prolonged pulmonary vasoconstriction, remodeled small pulmonary artery vasculature, in situ thrombosis, and elevated stiffness of the vascular wall, leading to increased pulmonary arterial pressure (Fig. 17.1) [1, 2]. One major feature of PH is an increased resistance of the vascular walls resulting in progressive elevations in pulmonary artery pressure [1]. Clinically, these pulmonary vascular changes come with some symptoms, such as syncope, chest pain, unexplained dyspnea on exertion, and fatigue.

17.3.1 Pulmonary Arterial Hypertension (PAH)

Pulmonary arterial hypertension (PAH) is a primary subdivision of PH which has a prominent characteristic of a gradual increase in resistance from the pulmonary vascular tissues due to unregulated remodeling of pulmonary vasculature, prolonged vasoconstriction, and in situ thrombosis (Fig. 17.2) [3]. The advancement in extensive clinical classification of PH, diagnostic indices, and novel therapeutic approach produced improved survival rate in PH in the past years [4].

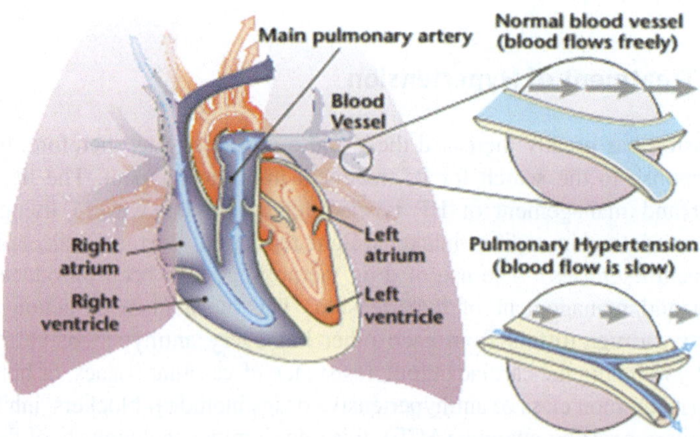

Fig. 17.1 The flow of blood in pulmonary hypertension [2]

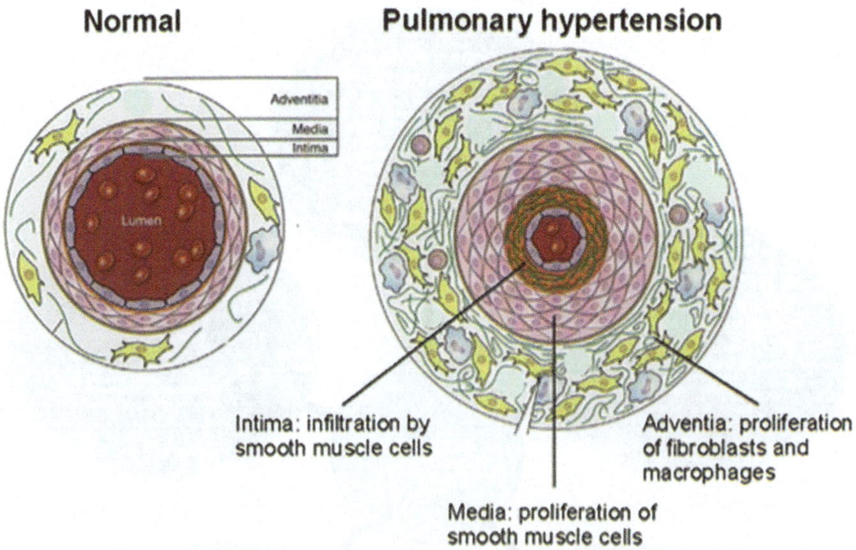

Fig. 17.2 Vascular framework in pulmonary hypertension. Cross-sectional diagram of a normal pulmonary arteriole and a pulmonary arteriole in pulmonary hypertension

17.3.2 Clinical Classification of Pulmonary Hypertension

PH was subdivided into groups based on the report of World Health Organization (WHO) that anorexigen is capable of inducing PAH. PH was then classified into five groups (Fig. 17.3).

17.3.2.1 Group 1: Pulmonary Arterial Hypertension (PAH)

This occurs when arteries in the lungs become narrowed, thickened off, or stiff. The right side of the heart must work harder to push blood through these narrowed arteries [4]. The major feature of PH in this category is a persistent rise in pulmonary vascular resistance (PVR) and mean pulmonary artery pressure (mPAP) due to an obstruction within the pulmonary vasculature [5]. The identified causes of PAH include idiopathic, heritable such as human immunodeficiency virus (HIV) infection, bone morphogenetic protein receptor 2 (BMPR2), toxin- and drug-induced and associated disorders, portal hypertension, and congenital heart diseases. Many mutations in the gene like BMPR2 predispose people to the incidence of PAH [3].

17.3.2.2 Group 2: PH Due to Left-Heart Disease (PH-LHD)

The PH in this category is a result of left-heart diseases. It is defined as mPAP $\geq$25 mmHg and pulmonary artery wedge pressure (PAWP) >15 mmHg. The prominent characteristics of this class of PH are a rise in mPAP, declining pulmonary vascular remodeling, failure of the right ventricle, and death [6]."

Fig. 17.3 Classification of pulmonary hypertension

17.3.2.3 Group 3: PH Due to Lung Diseases and/or Hypoxia

Group 3 class of PH is linked with hidden conditions like developmental lung abnormalities, interstitial lung diseases (ILD), chronic obstructive pulmonary disease (COPD), obstructive sleep apnea (OSA), and other pulmonary diseases [7]. Hypoxic vasoconstriction and extermination of the pulmonary vascular bed are the two main pathophysiologic bottom-line features of PH ascribed with hypoxia and COPD [6]. It has been documented that hypoxia is a major player in the incidence of endothelial cell damage, molecule release like endothelin giving birth to spasm reactions, and proliferation in the neighboring smooth muscle cells [3, 7]. The major therapeutic strategy to the management of group 3 PH is addressing the hidden disease process.

17.3.2.4 Group 4: Chronic Thromboembolic PH (CTEPH)

PH occurs as a result of chronic thromboembolic disease, giving birth to protracted occlusion of the pulmonary vasculature, thus resulting in abnormal mechanisms of fibrinolysis and autoimmune disorders. This condition greatly contributes to poor resolution of thrombi [5].

17.3.2.5 Group 5: PH with Unclear Multifactorial Mechanisms

PH with unclear many-sided mechanisms is classified as group 5 PH. It is sometimes called the orphan disease. Group 5 PH is a significant diversified group of diseases that comprises PH secondary to multifaceted mechanisms. In this diseased state, the precise incidence, etiology, and therapy remain unclear.

17.4 Pathophysiology of Pulmonary Hypertension

The classification of PH into groups has enhanced the gradual understanding of its mechanism of action; most especially, the pathophysiological mechanism of group 1 PAH has been greatly understood leading to diverse discoveries of potential drug targets; however, groups 2, 3, 4, and 5 have not been fully understood. However, there is a dearth of information on the mechanism behind the other groups of PH [2]. Moreover, they share some common denominators in their mechanism of action among all groups of PH [8]. The foundational mechanisms of increases in PVR in PAH include prolonged vasoconstriction, uncontrolled pulmonary vascular remodeling, and in situ thrombosis. The incidence and progression of PAH involve multiple factors and a plethora of several cell types inside the pulmonary artery vessel wall like primary pulmonary artery smooth muscle cells (PASMC), pulmonary artery endothelial cells (PAECs), fibroblasts, inflammatory cells, and platelets that are contributory factors in the disease process (Fig. 17.4).

17.4.1 Genetics and Genomics of PAH

17.4.1.1 Transcript Mapping and Positional Cloning of the Gene Underlying PAH

Genes with similar biological properties with PAH are characterized by direct Sanger sequencing; this leads to the recognition of the gene bone morphogenetic receptor type II (BMPR2) [10]. "BMPR2 is an approximately large gene, comprising 13 coding exons displaying over 190 kb of genomic DNA. At the transcriptional level, its start site is at base pair position 1148 very close to the initiation codon adenine. It has a remarkably long 3'UTR of an estimated size of 11 kb [9]. BMPR2 encodes the transmembrane bone morphogenetic receptor type II of the TGF-β superfamily of signaling molecules [10]. TGF-β molecules perform important functions in cellular activities such as migration, differentiation, proliferation, and apoptosis." (Fig. 17.5).

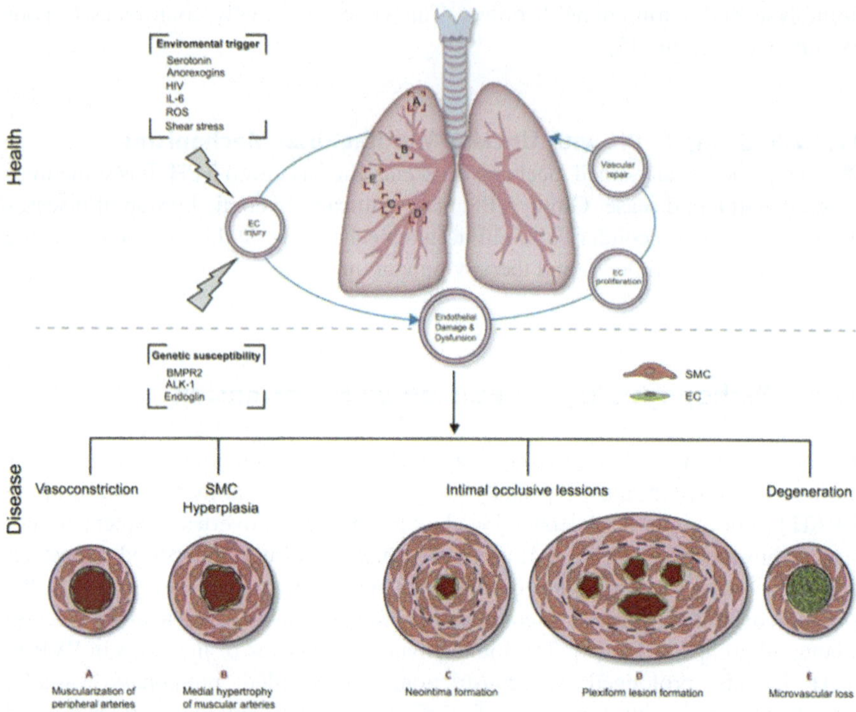

Fig. 17.4 Pathophysiological mechanisms of PAH

17.4.2 Role of Inflammation in PH

Inflammatory mechanisms appear to perform a crucial function in some types of PH including cases of rats induced with monocrotaline and PAH of several sources in humans. This inflammation has also been implicated in connective tissue diseases and human immunodeficiency virus infection. "Remarkably, some patients with serious PAH linked with systemic lupus erythematosus got better with immunosuppressive therapy, highlighting the importance of inflammation in this subset of patients [11]. Some immunological disturbances have been observed with idiopathic PAH patients which further supports the assertion that inflammation is implicated in the incidence and development of this disease. Undoubtedly, a cross section of PAH patients have circling autoantibodies which include antinuclear antibodies, together with increased rotating levels of proinflammatory cytokines IL-1 and IL-6. Histopathology of the lungs also shows inflammatory infiltrates (macrophages and lymphocytes) in the spectrum of plexiform lesions in severe PAH together with an elevated expression of chemokines RANTES and fractalkine" [12].

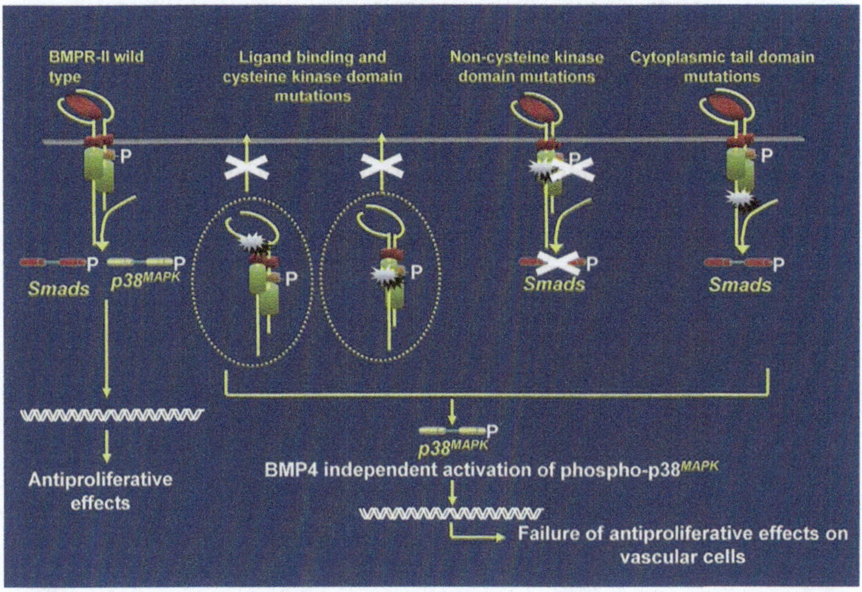

Fig. 17.5 Consequences of bone morphogenetic protein type II (BMPR2) receptor mutations on signaling [10]

17.4.3 Caveolin-1 Mutation in PH

Caveolin-1 is usually the point where many signaling cascades begin due to their interaction with G proteins, TGF-β-receptor 1, endothelial nitric oxide synthase (eNOS), and nitric oxide synthase 2A [13]. Caveolin-1 is a scaffolding protein and a major component of caveolae that connects with many signaling molecules including the ones implicated in PH and it is involved in their modulation [14]. The interruption and ongoing loss of endothelial caveolin-1 with reciprocal activation of proliferative pathways occur prior to the inception of PH, and the recapture of caveolin-1 impedes proliferative pathways and mitigates PH [15]. A comprehensive loss to the endothelial cells happens during the development of PH with resultant enhancement of caveolin-1 expression in the smooth muscle cells. In the smooth muscle cells, caveolin-1 moves from having an antiproliferative role to a proproliferative one and engages in cell proliferation and cell movement, probably going towards an irreversible process of PH [13, 15]. However, the interruption of endothelial caveolin-1 is not noticed in the reversible form of PH known as hypoxia. In contrast, proliferative pathways are activated in this model, suggesting a dysfunction in caveolin-1 function. Hence a dysfunction or disruption of endothelial caveolin-1 promotes PH, and the status of caveolin-1 may determine whether PH is reversible or irreversible [15, 16].

17.5 Metabolism Alterations in Pulmonary Hypertension

17.5.1 Energetic Metabolism in PH

Warburg effect, a phenomenon involving the substitution of energy acquisition in the Krebs cycle with glycolysis, is a characteristic of cancer cells and cells in PH patients (Fig. 17.6). Pyruvate, a key product of glycolysis in aerobic conditions, is metabolized into acetyl-CoA by the enzyme pyruvate dehydrogenase (PDH) which then enters into the tricarboxylic acid cycle (TCA cycle). The reduction in PDH activity as well as the conversion of pyruvate to lactate by the enzyme lactate dehydrogenase A (LDHA) in PAH PAECs has been proved and reported [17]. Increased fatty acid metabolism is the second metabolic deviation observed in PAH cells [17]. The third metabolic change is the glutaminolysis [18, 19]. The

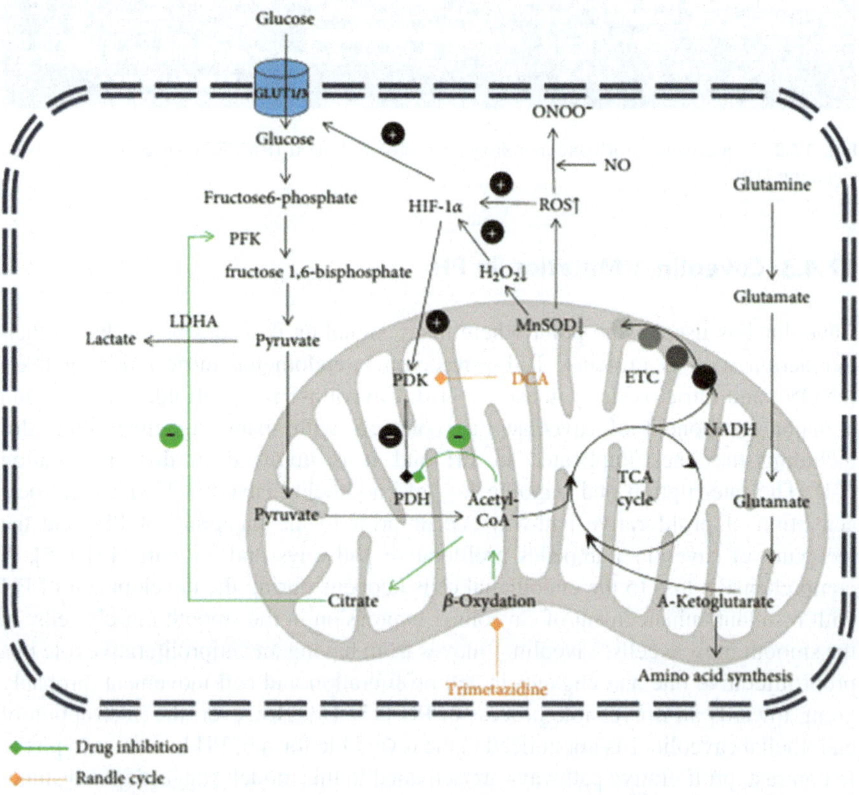

Fig. 17.6 Metabolism alteration in pulmonary hypertension [20]. *DCA* dichloroacetate, *ETC* electron transport chain, *HIF-1α* hypoxia inducible factor, *LDHA* lactate dehydrogenase A, *MnSOD* mitochondrial superoxide dismutase, *PDH* pyruvate dehydrogenase, *PFK-6* 6-phosphofructo-1-kinase, *ROS* reactive oxygen species, *TCA cycle* tricarboxylic acid cycle. Orange arrow: drug inhibition; green arrow: Randle cycle

additional significant changes relate to mitochondrial dysfunction. The function of mitochondria includes generation of ATP; they also act as oxygen sensors [20, 21].

17.6 Conclusion

To date, basic clinical research studies have shown promising potentials of popular and widely known cardiovascular biomarkers, but with limited clinical significance in the management and diagnosis of PH resulting from their decreased precision as well as several other cardiovascular complications of patients with PH [22, 23]. Conversely, a large panel of experimental research studies reveal novel cellular and molecular mechanisms, drug targets, and biomarkers following the principle of the proof-based medicine. Regrettably, the plain extrapolation of these findings to clinical application is not straightforward because of large complex nature of the pathophysiology of PH. Hence, there is the need for more translational medicine research to better understand the pathophysiological classification of PH and define accurately its biomarkers and therapeutic strategy for PH treatment.

References

1. Simonneau G, Gatzoulis MA, Adatia I (2013) Updated clinical classification of pulmonary hypertension. J Am Coll Cardiol 62:D34–D41. https://doi.org/10.1016/j.jacc.2013.10.029
2. Tuder RM, Archer SL, Dorfmüller P (2013) Relevant issues in the pathology and pathobiology of pulmonary hypertension. J Am Coll Cardiol 62:D4–D12. https://doi.org/10.1016/j.jacc.2013.10.025
3. Hoeper MM, Bogaard HJ, Condliffe R (2013) Definitions and diagnosis of pulmonary hypertension. J Am Coll Cardiol 62(Suppl):D42–D50. https://doi.org/10.1016/j.jacc.2013
4. Savale L, Chaumais MC, Cottin V (2012) Pulmonary hypertension associated with benfluorex exposure. Eur Respir J 40:1164–1172
5. Kim NH, Delcroix M, Jenkins DP (2013) Chronic thromboembolic pulmonary hypertension. J Am Coll Cardiol 62(25 Suppl):D92–D99
6. Thenappan T, Ryan JJ, Archer SL (2012) Evolving epidemiology of pulmonary arterial hypertension. Am J Respir Crit Care Med 186:707–709
7. Kosanovic D, Herrera EA, Sydykov A, Orfanos SE, El Agha E (2017) Pulmonary hypertension due to lung diseases and/or hypoxia: what do we actually know? Can Respir J 2017:9598089. https://doi.org/10.1155/2017/9598089
8. Hassoun PM, Mouthon L, Barbera JA (2009) Inflammation, growth factors, and pulmonary vascular remodeling. J Am Coll Cardiol 54:S10–S19
9. Aldred MA, Machado RD, James V, Morrell NW, Trembath RC (2007) Characterization of the BMPR2 5`-untranslated region and a novel mutation in pulmonary hypertension. Am J Respir Crit Care Med 176(8):819–824
10. Davies RJ, Holmes AM, Deighton J, Long L, Yang X, Barker L, Walker C, Budd DC, Upton PD, Morrell NW (2012) BMP type II receptor deficiency confers resistance to growth inhibition by TGF-b in pulmonary artery smooth muscle cells: role of proinflammatory cytokines. Am J Physiol Lung Cell Mol Physiol 302:L604–L615
11. Dorfmuller P, Perros F, Balabanian K, Humbert M (2003) Inflammation in pulmonary arterial hypertension. Eur Respir J 22:358–363

12. Balabanian K, Foussat A, Dorfmuller P (2002) CX (3) C chemokine fractalkine in pulmonary arterial hypertension. Am J Respir Crit Care Med 165:1419–1425
13. Mathew R (2014) Pathogenesis of pulmonary hypertension: a case for caveolin-1 and cell membrane integrity. Am J Physiol Heart Circ Physiol 306:H15–H25
14. Zhao YY, Zhao YD, Mirza MK et al (2009) Persistent eNOS activation secondary to caveolin-1 deficiency induces pulmonary hypertension in mice and humans through PKG nitration. J Clin Investig 119(7):2009–2018
15. Achcar RO, Demura Y, Rai PR, Taraseviciene-Stewart L, Kasper M, Voelkel NF, Cool CD (2006) Loss of caveolin and heme oxygenase expression in severe pulmonary hypertension. Chest 129:696–705
16. Zhao YY, Liu Y, Stan R (2002) Defects in caveolin-1 cause dilated cardiomyopathy and pulmonary hypertension in knockout mice. Proc Natl Acad Sci 99(17):11375–11380
17. Chen C, Luo F, Wu P (2020) Metabolomics reveals metabolite changes of patients with pulmonary arterial hypertension in China. J Cell Mol Med 24(4):2484–2496
18. Shahbazi R, Baradaran B, Khordadmehr M (2020) Targeting ROCK signaling in health, malignant and nonmalignant diseases. Immunol Lett 219:15–26
19. Simonneau G, Rubin LJ, Galiè N (2008) Addition of sildenafil to long-term intravenous epoprostenol therapy in patients with pulmonary arterial hypertension: a randomized trial. Ann Intern Med 149(8):521–530
20. Ryan TJ, Roy DS, Pignatelli M, Arons A, Tonegawa S (2015). Engram Cells Retain Memory Under Retrograde Amnesia. Sci 348(6238):1007–1013
21. Gajecki D, Gawrys J, Szahidewicz-Krupska E, Doroszko A (2020) Novel molecular mechanisms of pulmonary hypertension: a search for biomarkers and novel drug targets-from bench to bed site. Oxidative Med Cell Longev 2020:7265487. https://doi.org/10.1155/2020/7265487
22. Ryan JJ, Archer SL (2015) Emerging concepts in the molecular basis of pulmonary arterial hypertension. Part I: metabolic plasticity and mitochondrial dynamics in the pulmonary circulation and right ventricle in pulmonary arterial hypertension. Circulation 131(19):1691–1702
23. Stacher E, Graham BB, Hunt JM (2012) Modern age pathology of pulmonary arterial hypertension. Am J Respir Crit Care Med 186:261–272

Targeting Molecular and Cellular Mechanisms of Pulmonary Arterial Hypertension

18

Md Khadem Ali, Jay C. Horvat, and Edda F. Spiekerkoetter

Abstract

Pulmonary arterial hypertension (PAH) is a devastating disease of the pulmonary circulation, characterized by pulmonary vascular remodeling leading to elevated pulmonary arterial pressure, increased pulmonary vascular resistance, and right heart failure. Unfortunately, up until now, no definite cure exists for this disease. Currently available drugs focus on pulmonary vasodilation, anti-proliferation, and augmentation of endothelial function by targeting nitric oxide, endothelin, voltage-gated calcium channels, and prostacyclin signaling pathways. However, these drugs only partially improve survival and quality of life as they do not address the underlying pulmonary vascular remodeling. Over the past few years, attempts have been made to identify effective therapies that target different, anti-remodeling mechanisms and signaling pathways. Targets for these therapies include genetic and epigenetic modifications, growth factors and proliferation, inflammation and immunomodulation, endothelial-mesenchymal transition, and metabolic abnormalities. In this chapter, we outline and discuss promising novel therapeutic approaches that target diverse molecular and cellular signaling mechanisms involved in PAH.

M. K. Ali (✉) · E. F. Spiekerkoetter
Division of Pulmonary, Allergy and Critical Care Medicine, Stanford Medical School, Stanford, CA, USA

Vera Moulton Wall Center for Pulmonary Vascular Disease, Stanford University, Stanford, USA

J. C. Horvat
Priority Research Centre for Healthy Lungs, Hunter Medical Research Institute and School of Biomedical Sciences and Pharmacy, University of Newcastle, Newcastle, NSW, Australia

K. Dua et al. (eds.), *Targeting Cellular Signalling Pathways in Lung Diseases*,
https://doi.org/10.1007/978-981-33-6827-9_18

Keywords

Pulmonary arterial hypertension · Pulmonary vascular remodeling · Right ventricle remodeling · Inflammation

18.1 Introduction

Pulmonary hypertension is a life-threatening cardiopulmonary condition of abnormally elevated pressure in the pulmonary arteries. Pulmonary arterial hypertension (PAH) is defined by the World Health Organization as group I pulmonary hypertension, characterized by an increased mean pulmonary artery pressure >20 mmHg, a pulmonary artery capillary wedge pressure of <15 mmHg, and an increased pulmonary vascular resistance of 3 Wood units. A key pathological feature of PAH is pulmonary vascular remodeling, which leads to elevated pulmonary arterial pressure, pulmonary vascular resistance, and eventually right ventricular hypertrophy (RVH) and, if untreated, RV failure. Histopathological analysis suggests that dysfunction of key cellular components of the pulmonary vasculature, namely, endothelial and smooth muscle cells, pericytes, inflammatory cells, and adventitial fibroblasts, induces pulmonary vascular remodeling (e.g., medial hypertrophy, neointima formation, plexiform lesion) (Fig. 18.1). This results in narrowing of the vessel lumen and triggers complex vessel lesions, raising pulmonary vascular resistance, contributing to increased pulmonary arterial pressure to place a hemodynamic load on the right ventricle. Although PAH is a rare disease affecting only about one to two of every one million individuals annually, the mortality and morbidity rate is high, and, if untreated, PAH quickly leads to right ventricle failure and death after 2–3 years [1, 2]. PAH may be heritable (with a family history of PAH), idiopathic (without family history, unknown cause), or associated (linked to interstitial lung disease, congenital heart disease, autoimmune disease, etc.). While the exact cause of PAH is not known, genetic defects (mutations or epigenetic changes), environmental factors (e.g., hypoxia, viral infections, anorectic agents, etc.), and immune or inflammatory triggers may contribute to the cause or progression of the disease [3]. Importantly, there is no cure for PAH. Existing drugs target pulmonary vasodilation, proliferation, and endothelial function by increasing nitric oxide (NO), by inhibiting endothelin and voltage-gated calcium channels, and by augmenting prostacyclin signaling pathways. However, these drugs only partially increase survival and improve quality of life, while the majority of patients ultimately become resistant to medication and succumb to the disease. Therefore, development of new effective therapies is urgently needed. To discover novel therapies for PAH, it is important to understand the molecular mechanisms underlying the pathogenesis of the disease. Over the past two decades, many cellular and molecular mechanisms have been described as playing key roles in the pathogenesis of disease in preclinical and clinical settings. In this chapter, we summarize the most

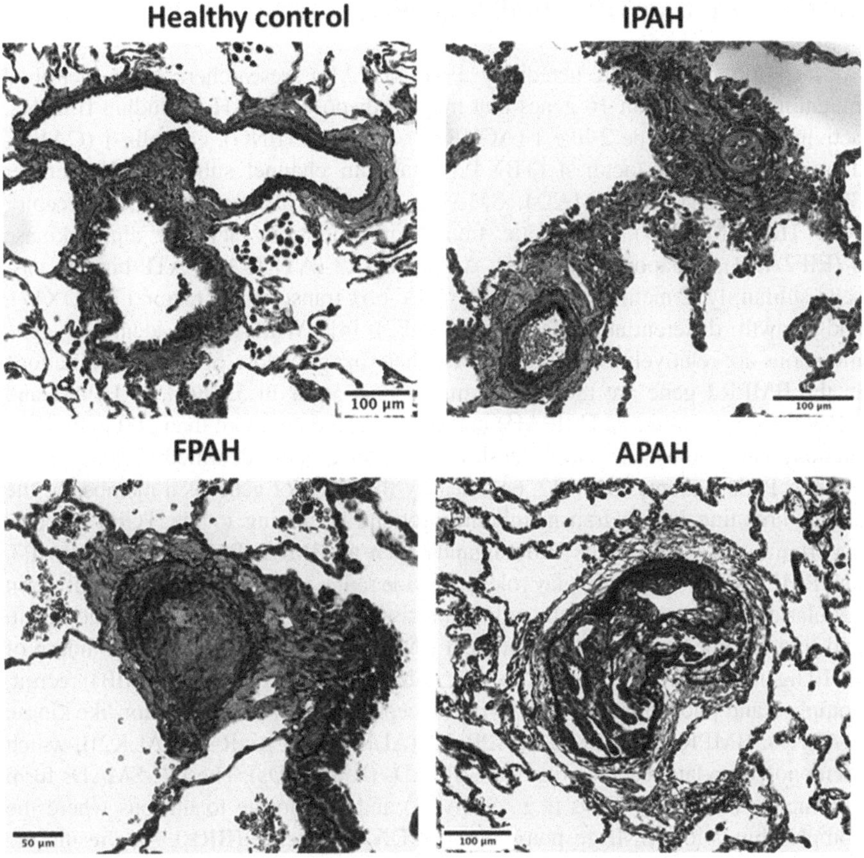

Fig. 18.1 Representative histological manifestations typically detected in lungs of patients with pulmonary arterial hypertension. Healthy control image shows a normal pulmonary artery (top) adjacent to an airway. Idiopathic pulmonary arterial hypertension (IPAH); the left lower vessel shows medial hypertrophy and some neointimal proliferation. The right upper lesion is a vessel totally occluded by neointima. Familial pulmonary arterial hypertension (FPAH) with a BMPR2 mutation image shows medial proliferation and total occlusion of the vessel with neointima. Associated pulmonary arterial hypertension (APAH) patient with congenital heart disease image shows a plexiform lesion

promising novel therapies that target key pathological mechanistic pathways involved in PAH. In particular, we focus on modulation of bone morphogenic protein receptor 2 (BMPR2) signaling/genetically determined targets, epigenetic modification, DNA damage and DNA damage response, inflammation and immunomodulation, growth factors and proliferation, metabolism, and endothelial-to-mesenchymal transition (EndMT).

18.2 Targeting TGFβ/BMP Signaling Activity

So far, in patients with hereditary PAH (HPAH), researchers have identified mutations in more than 16 genes that may predispose to PAH, including BMPR2, activin A receptor type 2-like 1 (ACVRL1), endoglin (ENG), caveolin-1 (CAV1), T-box transcription factor 4 (TBX4), potassium channel subfamily K member 3 (KCNK3), SMAD1, SMAD4, SMAD9, bone morphogenetic protein receptor type 1B (BMPR1B), eukaryotic translation initiation factor 2 alpha kinase 4 (EIF2AK4), aquaporin 1 (AQP1), ATPase 13A3 (ATP13A3), ATP-binding cassette subfamily C member 8 (ABCC8), SRY-box transcription factor 17 (SOX17), and growth differentiation factor 2 (GDF2) [4]. While most identified gene mutations are relatively rare (1–3% cases), heterozygous loss-of-function mutations in the BMPR2 gene are the most common and occur in 53–86% of HPAH and 14–35% of idiopathic PAH (IPAH) patients [5]. To date, more than 300 mutations, predominantly nonsense and frameshift types, have been identified in the BMPR2 gene in PAH patients. BMPR2, encoded by the *BMPR2* gene, is a member of the serine/threonine kinase transmembrane proteins belonging to the TGFβ receptor superfamily. BMPR2 binds BMP ligands such as BMP2, BMP4, BMP6, BMP7, and BMP9. BMPs typically play roles in a wide range of signal pathways involved in cellular differentiation, growth, and apoptosis and in embryogenesis, development, and tissue homeostasis. In the canonical BMP signaling pathway, upon binding of BMP ligands, BMP type 2 receptors (e.g., BMPR2 (ActRIIA) and ActRIIB)) recruit, complex, and phosphorylate BMP type 1 receptors (e.g., activin receptor-like kinase 1(ALK1), BMPR-1A (ALK3), BMPR-1B (ALK6), and ActR-1A (ALK2)), which then phosphorylate receptor-regulated SMADs (R-SMADs). These R-SMADs form a complex with co-SMADs (e.g., SMAD4) and translocate to nucleus where the complex binds to a BMP response element DNA sequence (BRE). As a result, the complex acts as a transcriptional regulator of target gene expression including inhibitor of DNA binding 1, 2, and 3 (ID1, ID2, ID3), which play a critical role in cell proliferation, apoptosis, and migration. In addition to the canonical Smad-mediated signaling pathway, several non-canonical BMP signaling pathways are also activated by BMPR2, including p38 mitogen-activated protein kinase (MAPK), extracellular signal-regulated kinase (ERK), wingless (Wnt), and NOTCH signaling [6].

Despite the high frequency of BMPR2 mutations in PAH patients, the disease penetrance rate is ~20% of the mutation carriers, suggesting that in addition to the BMPR2 mutations, other unidentified genetic, epigenetic, or environmental factors are involved in the development of the disease, potentially by decreasing BMPR2 expression and signaling activity below a specific threshold required to cause disease.

In PAH patients with and without BMPR2 mutations, BMPR2 expression and signaling activity is impaired in the pulmonary vasculature [7, 8], suggesting that dysfunction of BMPR2 signaling is a key common feature in PAH patients.

Pulmonary endothelial-specific deletion of BMPR2 in mice recapitulates human PAH features [9]. PAH manifestations are also observed in mice expressing a dominant-negative BMPR2 gene in pulmonary smooth muscle cells [10, 11]. Impaired BMPR2 signaling is associated with aberrant vascular cell phenotypes, including pulmonary arterial endothelial cell (PAEC) apoptosis, hyperproliferation and apoptosis resistance of pulmonary arterial smooth muscle cells (PASMCs), and inflammation. These findings suggest that targeting BMPR2 expression and signaling could be an effective therapeutic approach for treating PAH.

Over the past few years, several pharmaceutical drugs/ligands and approaches have been suggested to improve the BMPR2 expression and signaling in PAH. These include gene therapy (targeted delivery of adenoviral vectors containing BMPR2 gene to pulmonary endothelium) [12–14]; BMPR2 protein trafficking by chemical chaperones such as sodium 4-phenylbutyrate and probenecid [15]; read-through of premature STOP codons with ataluren [16]; inhibition of lysosomal degradation by chloroquine, hydroxychloroquine, and elafin [17]; activation of BMPR2 expression (e.g., paclitaxel, mercaptopurine) [18–20]; activation of BMP signaling (enzastaurin, FK506 (tacrolimus)) [21, 22]; TGFβ ligand trap (TGFBRII-Fc, ActRIIa-Fc) [23]; and exogenous BMP ligand delivery, e.g., BMP9 [24]. Of these therapies, FK506 and ActRIIa-Fc have progressed to clinical trials. FK506, an immunosuppressant targeting calcineurin function, is commonly used to prevent the rejection of organ transplants. Using a high-throughput screening of FDA-approved drugs, Spiekerkoetter and colleagues identified FK506 as a strong activator of the BMPR2 signaling pathway. Low-dose FK506 rescued endothelial dysfunction in PAECs from patients with IPAH and reversed monocrotaline (MCT)-and Sugen/hypoxia-induced experimental PH. Mechanistically, FK506 increased BMPR2 signaling twofold: by removing the TGFβ pathway inhibitor FKBP12 from the type 1 receptors ALK1, ALK2, and ALK3 and by inhibiting the phosphatase calcineurin, which both facilitated phosphorylation of the type 1 receptors and subsequent downstream SMAD signaling [22]. These findings led to a 16-week double-blind, placebo-controlled single-center phase IIa randomized clinical trial (RCT) (NCT01647945) to assess the safety and tolerability of FK506 in stable patients with PAH. Three severe end-stage patients that did not qualify for the trial because of severity of illness were treated with compassionate use of low-dose FK506 and continued receiving stable doses of PAH medication and diuretics. All three patients stabilized after 12 months of low-dose FK506 treatment in terms of symptoms, 6-minute walk distance (6MWD), N-terminal (NT)-pro hormone BNP (NT-proBNP), and RV function [25]. The RCT of 23 stable PAH patients revealed that low-dose FK506 is generally well tolerated. Although the focus of the trial was not to assess efficacy, the drug improved BMPR2 expression in peripheral blood mononuclear cell (PBMC) of some patients, supporting the further study of FK506 in PAH for efficacy and long-term safety in a phase IIb trial [26].

While PAH-associated BMPR2 mutations lead to an impairment of BMP signaling, TGFβ signaling activity has been shown to increase in the lung of patients with PAH and rodent models of PAH [27–30]. Imbalances of TGFβ/BMP signaling have been shown to contribute to pulmonary vascular remodeling, endothelial dysfunction, inflammation, and impaired angiogenesis in PAH [31]. Using a TGFβ ligand trap is an approach that inhibits TGFβ activity and rebalances BMPR2 signaling in PAH. A selective TGFβ1/3 ligand trap, TGFBRII-Fc (immunoglobulin Fc fusion protein of TGFβ), has been shown to improve PAH and vascular remodeling in preclinical PH models (Sugen/hypoxia and monocrotaline) [23]. Recently, Acceleron has announced ActRIIa-Fc (sotatercept) (another selective TGFβ ligand trap) as a breakthrough therapy for PAH based on a phase II PULSAR clinical trial results [32]. Acceleron showed that sotatercept reduces PVR (primary endpoint) and improves 6MWD, NT-proBNP, and WHO functional class (secondary endpoints) in a phase II, double-blind, placebo-controlled RCT in 106 adult patients with PAH, a majority of whom were on a stable combination background therapy with PAH-specific therapies (NCT03496207). In general, sotatercept was well tolerated, with very few adverse events in the trial, similar to those seen in other sotatercept trial for different disease. These exciting clinical findings provide the opportunity to deliver significant benefits over and beyond the treatments currently available. In addition, a phase II SPECTRA trial for evaluating the use of sotatercept in just NYHA FC III is actively recruiting PAH patients (SPECTRA; NCT03738150).

18.3 Targeting DNA Methylation and Histone Posttranslational Modifications

18.3.1 Targeting DNA Methylation

DNA methylation is a process of covalent addition of a methyl group to the cytosine-phosphate-guanine (CpG)-rich DNA sequences in the genome via the enzyme DNA methyltransferases (DNMTs). DNA methylation is involved in repressing and silencing of gene expression and in regulating gene expression and various cellular processes in development and human diseases. In PAH, DNA methylation was found to regulate expression of superoxide dismutase 2 (SOD2), which is a key mitochondrial antioxidant enzyme. Expression of SOD2 was reduced in plexiform lesions of PAH patients and in PASMC of fawn-hooded rats (FHR), a spontaneous and heritable model of experimental PH [33]. As sequencing revealed no mutations in the SOD2 gene, the authors examined possible epigenetic modification by bisulfite sequencing and identified hypermethylation located in an enhancer region within intron 2 and the promoter of the SOD2 gene in PASMC isolated from FHR [33]. In addition, there was an increase in expression of DNMT1 and DNMT3B in the lung and PASMC of the FHR [33]. Mechanistically, downregulation of SOD2 impaired H_2O_2-mediated redox signaling, activated HIF-1α, downregulated Kv1.5 channels, and produced a proliferative, apoptosis-resistant state [33]. In contrast, augmentation of SOD2 inhibited HIF-1α and restored Kv1.5 expression in FHR PASMCs. Importantly, treatment with 5-aza-2′-deoxycytidine, a DNMT inhibitor, reversed methylation, restored expression of SOD2, reduced proliferation, and enhanced apoptosis in FHR-derived PASMCs [33], suggesting that targeting DNA methylation to restore SOD2 could be an effective therapeutic option against PAH.

Another study revealed that the hypermethylation of BMPR2 promoter leads to decreased BMPR2 expression in HPAH patients. The reduced levels of BMPR2 expression promoted PAH development. Additionally, the results suggested that specific allelic methylation of the WT strand in HPAH patients may contribute to the penetrance of HPAH and may trigger the early onset of disease. Drugs that target the methylation of the BMPR2 promoter therefore might be a promising novel therapeutic approach for treating HPAH patients [34, 35].

Furthermore, a hypermethylation of the ATP-binding cassette subfamily A member 1 (ABCA1) gene was identified in PAH, which might be relevant to PAH development given the importance of a downregulation of ABCA1 gene in lipid metabolism. Furthermore, it was suggested that the methylation status of ABCA1 could be a useful biomarker for separating patients at high risk of PAH [34].

18.3.2 Targeting Histone Modifications

Emerging evidence suggests that elevated levels of histone deacetylase (HDAC) expression and activity are present in the lungs of PAH patients and animal models [36]. Several HDAC inhibitors have shown promise for suppressing the development of disease by inducing anti-proliferative and anti-inflammatory effects in rodent models of PAH. Preclinical studies have shown that pan-HDAC inhibitors reduce fibrosis, inflammation, and restenosis [36]. Furthermore, HDAC inhibitors have been tested in animal models of left ventricular (LV) dysfunction and found to be effective, which suggested that HDAC inhibitors might have the potential for treating RV failure in PAH [37]. Hypoxia-induced PAH in rats has been shown to be attenuated by the HDAC inhibitors, valproic acid (VPA, Class I HDAC inhibitor) and suberoylanilide hydroxamic acid (SAHA, Class I, II, and IV HDAC inhibitor) [38]. In line with this, VPA has been shown to prevent, and partially reverse, MCT- and chronic hypoxia-induced PAH in rats [39]. The HDAC inhibitor sodium valproate has been shown to attenuate pulmonary artery banding and MCT-induced RV hypertrophy [40]. MGCD0103, a Class I HDAC inhibitor, has also been shown to reduce pulmonary arterial wall thickening and to maintain RV function in experimental PH in rats [37].

Kim et al. demonstrated that increased nuclear accumulation of HDAC4 and HDAC5 reduced the activity of the transcription factor myocyte enhancer factor 2 (MEF2), leading to decreased levels of miR-424 and miR-503, which promotes endothelial cell proliferation. Importantly, a selective Class IIa HDAC inhibitor, MC1568, rescued MCT- and Sugen/hypoxia-induced impairment of MEF2 and PH and improved right ventricular systolic pressure (RVSP) [41]. In the MCT and Sugen/hypoxia-induced PH models, pharmacological inhibition of HDAC6 with tubastatin A (a selective HDAC6 inhibitor) was shown to improve PH and reduce pulmonary vascular remodeling [42]. Inactivation of SIRT1 occurs in PASMCs in PAH patients, disrupting the normal acetylation/deacetylation balance. It promotes proliferation of the PASMC, resulting in vascular remodeling, and increases vascular resistance in the pulmonary vasculature. Zurlo et al. demonstrated that sirtuin

1 (SIRT1)-inducible knockout mice exacerbated chronic hypoxia-induced vascular remodeling [43]. The SIRT1 activator resveratrol demonstrated a beneficial effect, in part by SIRT1 activation, which improved RVSP and decreased RVH. By inducing mitochondrial permeability transition dysfunction, SIRT1 activation also increased PASMC apoptosis [44], a strategy that could form the basis for a new future treatment for PAH.

Increasing numbers of studies are showing that bromodomain-containing protein 4 (BRD4) plays an important role in driving PAH pathogenesis. Meloche et al. showed that expression of BRD4 is enhanced in lungs, distal PAs, and PASMCs of patients with PAH. Mechanistically, BRD4 expression in PAH is miR-204-dependent, and overexpression of BRD4 promotes proliferation of PASMC and resistance to apoptosis [45]. In vitro studies suggested that the BRD4 antagonist, RVX208, normalized the hyperproliferative, inflammatory, apoptosis-resistant phenotype of smooth muscle cells and microvascular endothelial cells collected from patients with PAH [46]. Orally administered RVX208 has been shown to improve pulmonary hemodynamics and reverse pulmonary vascular remodeling in two separate models of monocrotaline + shunt and Sugen + hypoxia-induced PH [46]. Furthermore, RVX208 treatment has also been reported to support the pressure-loaded RV in PAB rats [46]. The findings suggested that inhibition of BRD4 appeared to be another promising drug target to treat PAH.

18.4 Targeting DNA Damage and DNA Repair Responses

Elevated DNA damage and impaired DNA repair responses are linked with disease progression of PAH. Veliparib (ABT-888), an inhibitor of poly-ADP-ribose polymerase-1 (PARP1, a crucial protein of DNA damage/repair system), has been shown to reverse PH in both Sugen/hypoxia- and monocrotaline-induced experimental PH in rats [47]. Another PARP1 inhibitor, olaparib, that has already been approved for the treatment of BRCA (breast cancer gene)-related breast cancer), is currently being tested in patients with PAH in a phase IB clinical trial (NCT03782818). A selective BET inhibitor, apabetalone (RVX208), has been reported to reverse PAH phenotypes in isolated PAH vascular cells and in two PH rat models [46]. The same study shows that RVX208 supported the RV in the PAB rat model [46]. These findings have recently led to the initiation of a pilot clinical trial of the use of RVX208 in ten PAH patients as a basis for a phase II clinical trial assessing RVX208 (NCT03655704) for the treatment of PAH.

18.5 Targeting Noncoding RNAs

Noncoding RNAs (ncRNAs) are RNA molecules that are not translated into proteins but play a role in a broad range of biological functions, including the regulation of gene expression at transcriptional and posttranscriptional levels as well as epigenetic modifications. The most studied ncRNAs in PAH are miRNAs. Altered levels of microRNAs (miRs) in the blood, lung tissues, and in particular PASMC, PAEC, fibroblasts, and cultured vascular cells have been shown to be associated with aberrant gene expression and signaling, metabolism, DNA damage, and vasoconstriction in PAH [48]. Also, miRs have been shown to regulate proliferation, migration, and apoptosis of PAEC and PASMC [49]. These findings suggested that dysregulated miR levels may play a role in the pathogenesis of PAH. Furthermore, targeting imbalanced microRNA levels, to restore their normal expression in the body, could be an effective treatment for PAH. A list of some key miRs involved in PAH pathogenesis as well as their associated molecular mechanisms is summarized in Table 18.1. Low plasma levels of miR-150 are correlated with an increased mortality in PAH [111], suggesting that expression of miRs could be beneficial as biomarker for patient survival. Targeting miRs with either miR-mimics or antagomirs has been shown to prevent or reverse PAH in mice and rat models of PH [61–63]. For example, Brock et al. showed that antagomirs targeted against miR-20a restored functional BMPR2 signaling and prevented vascular remodeling in hypoxia-induced PH models [96]. Antisense oligonucleotides against miR-145 treatment improved RVSP and reduced vascular remodeling [84]. Treatment with miR-140-5p mimics decreased pulmonary vascular remodeling and RVSP via targeting SMURF1 (SMAD-specific E3 ubiquitin protein ligase 1) [118]. miR-27b regulated PPARγ-dependent Hsp90-eNOS and NO signaling; and inhibition of miR-27b ameliorated MCT-induced endothelial dysfunction and PH [78]. Restoration of miR-223 expression in the lung of MCT-treated rats reversed established PH and improved vascular remodeling, right ventricle hypertrophy, pulmonary resistance, and survival [61]. However, as of now, it is too premature to determine whether these miR-based therapies could be used as treatment in PAH patients. Since miR-based therapies regulate a range of genes, which could be beneficial as multiple pathogenetic pathways could be targeted at once, this strategy is also potentially worrisome as several adverse off-target effects could be observed, limiting their translation into humans. Furthermore, of additional importance is the fact that many miR expressions show a high cell-type specificity; thus delivery of these miRs targeting mimics or antagomirs to the specific vascular cells would be important in order to avoid off-target effects on other cells. Thus, the development of vascular cell-specific delivery methods as well as techniques that provide deliberate miR release would be required before those strategies could be successfully translated into the clinic.

Recently, several long noncoding RNAs (lncRNAs) have also been implicated in pulmonary vascular remodeling and PAH pathogenesis. Using RNAseq data analyses of PASMCs and lung pericytes from IPAH patients and hypoxia-exposed PASMC and pericytes, Zehendner et al. have shown that tyrosine kinase receptor-inducing lncRNA (TYKRIL) is significantly increased in all hyperproliferative conditions [125]. Under these hyperproliferative conditions, TYKRIL has been shown to promote proliferation and inhibit apoptosis in PASMCs and pericytes by p53/PDGF signaling axis [125]. Since TYKRIL has poor conservation in animals, the authors have performed studies in ex vivo precision-cut lung slices (PCLS) collected from lungs of IPAH patients and demonstrated that GapmeR-mediated

Table 18.1 List of key miRs involved in PAH

miRNA	Expression levels	Phenotypes/function	Mechanism(s)/targets	Ref.
miR-1	↑ in plasma, ↓ in PASMC of patients	Induces endothelial dysfunction	Targets Kv1.5, TGFβR1, SphK1	[50–53]
miR-143/145	↑ in patients	Promotes PASMC proliferation	ABCA1	[54]
miR-124	↓ in PAEC	Reduces PAEC glycolysis and proliferation, inhibits PASMC proliferation	PTPB1, PKM1/PKM2, NFATc1	[55–58]
miR-29b	↓ in PA of hypoxia mice exposed to hypoxia and in hypoxia-treated PASMC	Inhibits proliferation and promotes the apoptosis of PASMCs	Mcl-1, CCND2	[59]
miR-204	↓ in rat PA intima and HPAECs by hypoxia	Improves hypoxia-induced PAH	Autophagy	[60]
miR-223	↓ in lung and PASMC of PAH patients	Prevents and reverses experimental PAH	ITGB3, RhoB/MLC2	[61–63]
miR-181b, miR-181a/b-5p	↓ in PAH	Inhibits EndMT, improves inflammatory response	Endocan, TGFBR1	[64, 65]
miR-195-5p	Secreted by anti-apoptosis EC	Promotes proliferation and migration of PASMC	HIF-1a/miR-195-5p/Smad7	[66]
miR-27a	↑ in the lung of hypoxia-exposed rats	miR-27a induces EndMT	BMP signaling	[67]
miR-15a-5p	↑ in the lung of MCT-induced PAH	Induces PASMC apoptosis in a PAH model	VEGF/p38/MMP-2 signaling pathway	[68]
miR-135a	↑ in the lung of MCT-, OVA-, and PM-induced PAH	miR-135a inhibition improves experimental PAH	β-Catenin/GSK-3β signaling pathway	[69–71]
miR-125a	↓ in the PA and PASMC of MCT rats	Improves MCT-induced PAH	TGFβ1, IL-6/STAT3, HK-II, mitofusin 1	[72–74]
miR-371b-5p	↓ in MCT PAH rats	Inhibits EC apoptosis in MCT-induced PAH	PTEN/PI3K/Akt signaling pathways	[75]
miR-455-3p-1	↓ in PAH patients	Represses FGF7 expression to inhibit PAH	RAS/ERK signaling	[76]
miR-19a	↑ in hypoxia-treated PASMC	Regulates hypoxia-mediated proliferation and migration	PTEN	[77]
miR-27b	↑ in the lung of MCT-induced PAH rats	miR-27b inhibition improves MCT-induced PAH	PPARγ-dependent Hsp90-eNOS and NO	[78]

miR-214	↑ in PASMC of PAH patients	Regulates PAH SMC phenotypes	MEF2C-MYOCD-leiomodin 1 (LMOD1)	[79]
miR-23a	↑ in serum of IPAH patients and PASMC exposed to hypoxia	Regulates PAH PASMC phenotypes	BMPR2/Smad1 signaling, HIF-1	[80–82]
miR-34a	↓ in the lung and PA of hypoxia-treated rats and PASMC	Regulates PASMC proliferation miR-34a	PDGFRA	[83]
miR-145	↑ in the lung of PAH patients and BMPR2-deficient mice	miR-145 inhibition protects and reverses PAH in rodent models of PAH	–	[84–86]
miR-100	↓ in the lung of hypoxia-treated rats	Regulates PASMC proliferation in hypoxia-treated rats	mTOR signaling	[87]
miR-190	↑ in pulmonary arteries of hypoxia-treated rats	Promotes hypoxic pulmonary vasoconstriction	Voltage-gated K^+ channel	[88]
miR-21	↑ in PA of hypoxia-treated mice	Regulates PASMC proliferation	–	[89]
miR-483	↓ in serum and PAEC of IPAH patients	Overexpression of miR-483 improves PAH in rats		[90]
miR-322	↑ in plasma of PH/PAH patients	Promotes PASMC proliferation and migration under hypoxia	BMP signaling	[91–93]
miR-103/107	↓ in the lung of hypoxia-exposed rats	Regulates hypoxia-induced PASMC proliferation	HIF-1β	[94]
miR-18a-5p	↑ in PAH patients	Increases proliferation and migration of PASMCs	Notch2	[95]
miR-20a	–	miR-20a inhibition prevents hypoxia-induced PAH	BMPR signaling	[96]
miR-30a-5p	↑ in plasma of patients	Promotes proliferation and inhibits apoptosis of PAEC under hypoxia	YKL-40	[97]
miR-138	↑ in hypoxic PASMC	miR-138 inhibition reduces MCT-induced PAH and negatively regulates PASMC apoptosis	KCNK3, SLC45A3, Mst1, TASK-1	[98–100]
miR-17/20	–	Regulate PASMC proliferation	PHD2/HIF-1	[101]
miR-17 ~ 92	↓ in PASMC from PAH patients	Regulates PASMC phenotypes, attenuates hypoxia-induced PAH	PDZ, LIM domain 5	[102] [103]
miR-328	↓ in PA	Regulates hypoxic PAH	IGF1R, L-type calcium channel α1C	[104]

(continued)

Table 18.1 (continued)

miRNA	Expression levels	Phenotypes/function	Mechanism(s)/targets	Ref.
miR-629	↑ in hypoxia-treated PASMC	Regulates hypoxic pulmonary vascular remodeling	PERP, FOXO3	[105]
miR-593-5p	↓ in the lung of hypoxia-treated rats	Promotes hypoxia-induced PAH development	PLK1	[106]
miR-361-5p	↑ in hypoxia-treated PASMC	miR-361-5p inhibition suppresses PASMC migration and survival	ABCA1, JAK2/STAT3 pathway	[107]
miR-132	↑ in MCT-induced PAH rats	Facilitates proliferation and migration of PASMC	PTEN	[108]
miR-205-5p	↓ in hypoxia-induced PAH in mice	Suppresses PASMC proliferation	Erk1/2 signaling	[109]
miR-30c	↓ in hypoxia-treated PASMC	Promotes hypoxia-induced PAH development	PDGFβ	[110]
miR-150	↓ in plasma of patients	Inhibits PASMC proliferation	HIF-1α	[111–113]
miR-17	↑ in blood of patients	Regulates hypoxia-induced PASMC proliferation	p21, MFN2, PTEN	[114–117]
miR-140	↓ in blood of patients	miR-140 mimic prevents and reverses PAH in rats	TNF-α, DNMT1	[118–120]
miR-221	↑ in the lung and PASMCs of patients	Inhibition attenuates Sugen/hypoxia-induced PAH	TIPP3 and AXIN2	[121, 122]
miR-193	↓ in the lung and serum of PAH patients	miR-193 mimic treatment ameliorates MCT- and hypoxia-induced PAH	IGF1R, lipoxygenases, PPAR-RXRα	[123, 124]

knockdown of TYKRIL in PCLS reversed pulmonary vascular remodeling [125], suggesting that TYKRIL has therapeutic potential for PAH treatment. Furthermore, PAXIP1 antisense RNA 1 (PAXIP1-AS1) expression has been demonstrated to be upregulated in small pulmonary arteries, adventitial fibroblasts, and PASMCs of IPAH patients [126]. Mechanistically, PAXIP1-AS1 has been shown to affect focal adhesions by regulating their downstream target paxillin [126]. However, from this study, it is not clear whether PAXIP1-AS1 plays a casual role in disease development or whether it is just a consequence of the remodeling processes. Lei et al. demonstrated that expression of smooth muscle-induced long noncoding RNA (SMILR) was significantly increased in PAH patients, in the MCT-induced PH model, as well as in hypoxia-exposed PASMC [127]. Furthermore, silencing of SMILR inhibited hypoxia-induced proliferation and migration via targeting miR-141 in PASMCs. Importantly, SMILR shRNA delivery in the MCT-induced PH rat model ameliorated PH and pulmonary vascular remodeling by targeting the Rho/ROCK/miR-141 axis [127], indicating that knocking down SMILR has great potential for the treatment of PAH. Sun and colleagues showed that expression of MEG3 is decreased in the lung and PAs of PAH patients and silencing of MEG3 expression triggered PASMC proliferation and migration by p53 signaling pathway in vitro [128]. In addition to these lncRNAs, dysregulation of maternally expressed 3 (MEG3), cancer susceptibility 2 (CASC2), taurine upregulated 1 (TUG1), H19, antisense noncoding RNA in the INK4 locus (ANRIL), HOXA cluster antisense RNA 3 (HOXAAS3), metastasis-associated lung adenocarcinoma transcript 1 (MALAT1), urothelial cancer-associated 1 (UCA1), and ribosomal protein S4-like (RPS4L) and lncRNA regulated by PDGF and TGFβ (LnRPT) and TCONS_00034812 has been shown to be associated with PAH pathogenesis. Their expression in PAH samples, function in pulmonary vascular cells, and mechanisms are summarized in Table 18.2. All these studies show that targeting lncRNAs could be a promising avenue to treat PAH. However, several limitations need to be overcome to consider these lncRNAs as potential therapeutics in PAH. Further studies are needed to explore the secondary structure of lncRNAs, their delivery into the human body, the speed of onset and duration of their action, as well as the prevention of off-target effects.

18.6 Targeting Metabolic Regulation

Metabolic abnormalities are associated with abnormal signaling pathways involved in the pulmonary vasculature and RV remodeling in PAH pathogenesis. Over the past few years, several metabolic and signaling pathways have been recognized as promising therapeutic targets for PAH [3]. These include the inhibition of pyruvate dehydrogenase kinase (PDK), an inhibitor of the mitochondrial enzyme pyruvate dehydrogenase (PDH, the gatekeeping enzyme of glucose oxidation), and regulation of p53, HIF-1, mTOR, and HIPPO signaling pathways and mitochondrial phosphatase [3].

Table 18.2 List of long noncoding RNAs involved in PAH

lncRNA	Expression in PAH	Key function in vascular cells	Mechanisms/targets	Ref.
TYKRIL	↑ in PASMC, pericytes	Promotes proliferation, inhibits apoptosis of PASMC	p53/PDGFRβ axis	[125]
PAXIP1-AS1	↑ in small PA, PASMC	Promotes proliferation and migration, inhibits apoptosis of PASMC	Paxillin	[126]
MEG3	↓ in the lung, PA, PASMC	Inhibits proliferation and migration of PASMC	miR-21/PTEN; p53 pathway	[128, 129]
MEG3	↑ in PASMC	Hyperactivates cell cycle progression, enhances proliferation of PASMC	miR-328-3p/IGF1R	[130]
CASC2	↓ in PA, PASMC	Inhibits cell proliferation, migration, and phenotypic switch of PASMC	α-SMA expression	[131]
TUG1	↑ in PA	Promotes proliferation and migration of HPASMC	miR-374c/Foxc1, notch signaling	[132, 133]
H19	↑ in serum, lung	Promotes proliferation of PASMC	miRNA let-7b/AT1R	[134]
MANTIS	↓ in the lung	Facilitates endothelial angiogenic function, promotes apoptosis	BRG1	[135]
ANRIL	↓ in PASMC	Promotes proliferation and migration of PASMC	Unknown	[136]
Hoxaas3	↑ in the lung, PASMC	Promotes proliferation and regulates the cell cycle in PASMC	HOXA3	[137]
MALAT1	↑ in PA, PASMC	Promote pulmonary vascular remodeling and cell cycle progression	hsa-miR-124-3p.1/KLF5	[138]
UCA1	↑ in PASMC	Promotes proliferation and inhibits apoptosis in PASMC	hnRNP I	[139]
SMILR	↑ in serum	Regulates vascular remodeling and PAH	RhoA/ROCK/miR-141 signaling	[127]
TCONS_00034812	↓ in PA, PASMC	Promotes proliferation and inhibits apoptosis of PASMC	Stox1/MAPK signaling	[140]
LincRNA-Cox2	↑ in blood, PASMCs	Promotes PASMC proliferation	LincRNA-COX2/miR-let-7a/STAT3 axis	[141]
Rps4l	↓ in PASMC	Modulates proliferation, migration, and cell cycle progression of PASMC	ILF3/HIF-1α	[142]
LnRPT	↓ in PASMC	Promotes PASMC proliferation	Notch signaling pathway	[143]
Lnc-Ang362	↑ in the lung	Promotes proliferation and migration of PASMCs	Lnc-Ang362/miR-221/miR-222	[144]

Dichloroacetate (DCA), an inhibitor of PDK, prevented and reversed MCT-induced PH by inducing apoptosis and restoring PDH activity [145]. DCA also prevented and reversed chronic hypoxia-induced PH by restoration of expression and function of Kv channels [146]. These results led to an open-label phase I clinical trial that revealed that chronic oral administration of DCA improved mean pulmonary arterial pressure (mPAP), pulmonary vascular resistance (PVR), and 6MWD in a subset of patients with IPAH [147]. Interestingly, this study showed that patients with inactivating mutations in two mitochondrial proteins SIRT3 and UCP2 were less responsive to DCA [147], demonstrating the importance of precision medicine.

Targeting fatty acid oxidation to increase the Randle cycle may be another possible approach to regulate the metabolism and improve the RV function in chronic PAH. Ranolazine pharmacologically targets this process by inhibiting fatty acid oxidation and enhancing glucose oxidation. Several clinical trials have already been completed studying the effect of ranolazine in PH/PAH (NCT01757808, NCT01174173, NCT02829034, NCT01839110, NCT02133352, NCT01917136). Ranolazine was found to be safe and well tolerated, associated with an improvement of functional class, a reduction in RV size, an improvement in RV function, and a trend toward improved exercise capacity, but did not improve hemodynamics in 11 patients with PAH in a phase III clinical trial (NCT01174173) [148].

Metformin, an antidiabetic drug, has also been reported to reverse PH in experimental rodent models via inhibition of the aromatase and estrogen synthesis potentially via adenosine monophosphate-activated protein kinase (AMPK) [149]. One study by Liu et al. reported that metformin prevented the exacerbation of experimental PH via activating AMPK and inhibiting autophagy [150]. Metformin has also been shown to reduce myocardial lipid levels and improve RV function in mice models of heritable PH [151]. Currently, two clinical trials are ongoing to assess the effects of metformin on insulin resistance, oxidant stress, RV lipid accumulation, exercise capacity, and WHO functional class in PAH (NCT03617458, NCT01884051).

Collectively, these findings underscore how metabolic modulation could be a promising therapeutic strategy in PAH. Future studies exploring the interaction between chronic hypoxia, inflammation, hormones, high shear stress, and metabolism may lead to the development of new effective metabolic-based interventions in PAH.

18.7 Targeting Growth Factors and Proliferation

Dysregulation of different growth factors (e.g., platelet-derived growth factor (PDGF), epidermal growth factor (EGF), fibroblast growth factor 2 (FGF2), vascular endothelial growth factor, nerve growth factor) and/or their respective receptor tyrosine kinases together with inflammatory mediators (e.g., cytokines, chemokines, immune complexes, and circulating autoantibodies) plays an important role in pulmonary vascular remodeling in PAH. Therapeutic targeting of these factors has shown promising results in PAH in both preclinical and clinical settings.

Activation and aberrant expression of PDGF and its receptors have been shown in the remodeled small PAs of experimental and clinical PAH. Inhibition of PDGF receptor by tyrosine kinase inhibitors such as imatinib, nilotinib, and dasatinib has been shown to improve PAH and pulmonary vascular remodeling in rodent models [152]. A phase III clinical trial of imatinib mesylate in patients with advanced PAH revealed that imatinib improved exercise capacity and hemodynamics, yet unexpected serious adverse events discouraged further clinical use of this kinase for PAH [153]. Dasatinib has been reported to cause PAH in humans [154, 155], and nilotinib has been associated with both, cause and treatment for PAH [156]. These findings impede the development of these kinases for PAH therapy.

Activation of EGF receptor (EGFR) signaling has been associated with PAH pathogenesis. Several EGFR tyrosine kinase inhibitors such as erlotinib, gefitinib, and lapatinib have been shown to ameliorate MCT-induced PH in rats [157].

FGF2 is believed to be important for the development of PAH by promoting PASMC proliferation [158]. Izikki et al. demonstrated that overproduction of FGF2 by endothelial cells contributed to PASMC hyperplasia [158]. Moreover, pharmacological inhibition of FGF receptor 1 with SU5402 or siRNA-mediated silencing of FGF2 reversed MCT-induced PH, suggesting FGF2 to be a promising target for new treatments.

Activation of the Rho-kinase (RhoA/ROCK1) pathway mediates PASMC proliferation and is involved in vasoconstriction and vascular remodeling in PAH. Fasudil, a Rho-kinase inhibitor, has been shown to improve PH manifestations in several preclinical rodent PH models [159–162] and in patients with PAH [163–165], suggesting that Rho-kinase could be a potential therapeutic target in PAH.

Activation of mammalian target of rapamycin (mTOR) is involved in PASMC proliferation and pulmonary vessel remodeling and RV dysfunction in PAH [166]. Everolimus, an inhibitor of mTOR, was shown to improve PVR and 6MWD in eight of the ten patients with PAH or chronic thromboembolic pulmonary hypertension in a prospective open-label pilot study [167]. Another mTOR inhibitor, nanoparticle albumin-bound rapamycin (ABI-009), is currently evaluated in a phase I clinical safety trial of 25 severe PAH patients (NCT02587325). Intravenous administration of rapamycin-loaded nanoparticles prevented development of MCT-induced experimental PH as evidenced by attenuated pulmonary arteriole hypertrophy, a decrease in RVSP and RV remodeling, and activation of downstream targets of the mTOR pathway [168]. Intraperitoneal injection of imatinib completely reversed established PH in Rictor KO mice, whereas the combination therapy with rapamycin and a lower dose of imatinib dramatically reverses the established severe PAH in rats (Sugen/hypoxia-induced PH); rapamycin alone only caused partial inhibition [169]. Rapamycin partially reversed the protein expression patterns of EndMT, improved experimental PH, and decreased the migration of human PAECs, providing the proof of concept that EndMT is druggable [170]. A recent study by Tsutsumi et al. has demonstrated that nintedanib (a tyrosine kinase inhibitor) ameliorated experimental PAH through inhibition of endothelial-mesenchymal transition and smooth muscle cell proliferation [171].

18.8 Targeting Inflammation and Immunomodulation

Chronic inflammation and altered immune responses are crucial contributors to the pathogenesis of PAH and pulmonary vascular remodeling [172]. Rituximab (anti-CD20 monoclonal antibody) (NCT01086540) and tocilizumab (interleukin-6 receptor antagonist) (NCT02676947) have recently completed clinical trials in patients with PAH (results not published yet). A recent completed single-group, open-label phase IB/II pilot study data suggested that IL-1 blockade with anakinra is feasible and safe in patients with PAH with RV failure [173]. However, a larger placebo-controlled RCT with the longer treatment period is required in order to further confirm these outcomes. A hybrid compound of bardoxolone methyl and the NO donor isosorbide 5-mononitrate has been reported to attenuate vascular remodeling (pulmonary artery medial thickness, vascular muscularization) and reduce PH (mean pulmonary arterial pressure, RVSP, RVH, and fibrosis) in MCT-treated rats [174], suggesting that this hybrid compound might be an effective treatment in PH. Bardoxolone methyl (an antioxidant inflammation modulator that activates NRF2 signaling pathway) improved 6MWD (primary endpoint) in two clinical trials in PAH (NCT02036970 (LARIAT) and NCT02657356 (CATALYST)). An expanded clinical trial for assessing long-term safety and tolerability of bardoxolone methyl is underway in eligible PAH patients who previously engaged in controlled bardoxolone methyl study (NCT03068130). CXA-10, an oral nitrated fatty acid affecting fibrotic and inflammatory signaling, is being tested to evaluate long-term safety, efficacy, and pharmacokinetics in PAH patients with a focus on RVEF and hemodynamics as endpoints on stable background therapy in a phase II multicenter double-blind, placebo-controlled RCT (NCT03449524). A pilot study assessing effectiveness of spironolactone (a drug with anti-inflammatory, immunomodulatory, and antioxidant properties and the potential to improve blood vessel function and endothelial dysfunction) in treating PAH patients is ongoing (NCT01712620). Two other clinical trials of spironolactone have been completed (NCT02253394, NCT01468571). Rabinovitch and colleagues have demonstrated that elafin, a naturally occurring elastase inhibitor, reversed obliterative changes in pulmonary arteries by elastase inhibition and caveolin-1-mediated amplification of BMPR2 signaling [175]. The same study also showed that elafin is pro-apoptotic and decreased neointimal lesions in PAH lung explant culture [175]. Currently, a clinical phase I trial is in progress to assess safety of elafin in healthy subjects (NCT03522935).

18.9 Targeting EndMT

EndMT, a process of phenotypic conversion of endothelial cells to mesenchymal cells, has been shown to play a critical role in experimental and clinical PAH [170, 176, 177]. Quo et al. showed co-expression of endothelial marker CD31 or von Willebrand antigen with smooth muscle α-actin in the neointimal lesions of PAH patients [178]. Ranchoux et al. demonstrated evidence of EndMT by showing presence of luminal endothelial cells exhibiting a mixed EC/mesenchymal phenotypes, loss of EC cell-cell junction, acquisition of SMA fibers, vimentin phosphorylation, and a marked invagination of transitional ECs into the subendothelial space in the lung of PAH patients as well as Sugen/hypoxia- and MCT-induced experimental PH models [170]. Another study reported by Hopper et al. showed that BMPR2 deficiency leads to EndMT in cultured PAECs by showing spindle SM-like morphology, decreased expression of CD31, and increased expression of αSMA, SM22α, calponin, and phospho-vimentin and Slug expression. Similar gene and protein expression changes were also observed in mice with loss of pulmonary EC-specific BMPR2 [179], suggesting that EndMT could be a cause rather than a consequence of PAH associated with alterations of BMPR2 axis.

A recent study suggested that mesenchymal stem cell (MSC) therapy ameliorated RV systolic pressure, pulmonary vascular muscularization, fibrosis, and inflammation, associated with the reversal of EndMT markers (FN, VE-Cad, CD31, Snail, vimentin (VIM), Twist1, co-expression of vWF and α-SMA in cells within the pulmonary vascular wall) in chronic hypoxia- and Sugen-hypoxia-induced PH models in rats [180]. In favor of these in vivo findings, the authors also showed that MSC-conditioned cell culture media ameliorated abnormal cellular morphology and migratory phenotypes and EndMT markers in cultured human primary microvascular endothelial cells exposed to hypoxia [180]. These findings suggest that MSC transplantation could be effective in targeting PH-associated EndMT. However, further studies are needed to explore more causative link supporting MSC therapy on targeting EndMT in PAH.

Another possible approach for targeting EndMT in PAH is to inhibit dipeptidyl peptidase-4 (DPP-4, CD26), a serine protease which regulates the activity of secreted polypeptides (e.g., cytokines, chemokines, and vasoactive peptides). Sitagliptin (SG), a typical inhibitor of DPP-4, has been shown to suppress EndMT in MCT-induced PH models, as evidenced by enhanced expression of vWF and VE-cadherin; attenuated expression of α-SMA, VIM, and FN; and reduced abundance of CD31 and α-SMA double-stained cells [181].

In addition, several other PAH-associated EndMT targeting therapeutics and approaches have been suggested in preclinical PAH models, which include hydrogen sulfide [182], nintedanib [171], paeoniflorin [183], CD44 [184], high-mobility group AT-Hook 1 (HMGA1) [179], microRNAs [60, 64, 67, 185, 186], restoration of BMPR2 signaling [12, 187], and rapamycin [170] (Table 18.3). Collectively, these approaches have exciting potential as new therapies targeting EndMT in PH/PAH.

Table 18.3 List of emerging therapies targeting EndMT in PAH

Approach/drug	PAH models	EndMT markers	Therapeutic effects	Ref.
H_2S (NaHS)	MCT rats	↑ expression of VE-cadherin; ↓ expression of α-SMA and Snail; morphological features of EndMT by EM	↓ RVSP, mPAP, RV/LV + S, TAPSE, medial wall thickness	[182]
Nintedanib	SuHx rats	↓ Twist1 expression	↓ RVSP, RV/LV + S, medial wall thickness, intimal occlusive lesions	[171]
Paeoniflorin	SuHx rats	↑ expression of VE-cadherin, ↓ expression of FN and Vim	↓ RVSP, RV/LV + S, medial wall thickness, CSA, pulmonary vascular adventitial fibrosis, RV fibrosis	[183]
Sulfasalazine	SuHx rats	↓ abundance of α-SMA and vWF co-expressing cells, ↓ expression of TNF-α, IL-1b, IL-6	↓ RVSP, RV/LV + S, % of vessels stained with CD44v + cells, vascular muscularization, CD44v mRNA	[184]
Rapamycin	MCT rats	↑ p-120-catenin, ↓ Twist1	Partially improve PAH	[170]
MSC therapy	Chronic hypoxia, SuHx rats	↑ expression of VE-Cad and CD31, ↓ expression of FN, Snail, Vim, and Twist1, ↓ abundance of vWF and α-SMA co-expressed cells	↓ RVSP, pulmonary vascular muscularization, fibrosis, and inflammation	[180]
Sitagliptin	MCT rats	↑ expression of vWF and VE-cadherin; ↓ expression of α-SMA, VIM, and FN; ↓ abundance of CD31 and α-SMA double-stained cells	↓ RVSP, hypertrophy of PA medial layer, RV remodeling	[181]

18.10 Conclusions

Over the past few years, extensive efforts have been made to develop and test novel effective strategies that target pathogenic pathways in PAH, including BMPR2/TGFβ signaling, epigenetic modification, DNA damage, immunity, inflammation, metabolism, endothelial-mesenchymal transition, and RV function and remodeling. So far, several potentially exciting targets have been identified in both preclinical and clinical studies. These targets either are currently being examined in preclinical studies or have been proposed or progressed to clinical trials with the intention of increasing the availability of effective therapies for patients with PAH that address the underlying pathology of vascular remodeling in addition to current therapies targeting pulmonary vasodilation.

Acknowledgments Authors are thankful to Dr. Adam Andruska, Division of Pulmonary, Allergy, and Critical Care Medicine, Stanford University, USA, for providing guidance, for editing the chapter, and also for providing the lung section images of PAH patients and healthy controls.

References

1. Benza RL et al (2012) An evaluation of long-term survival from time of diagnosis in pulmonary arterial hypertension from the REVEAL Registry. Chest 142(2):448–456
2. Peacock AJ et al (2007) An epidemiological study of pulmonary arterial hypertension. Eur Respir J 30(1):104–109
3. Humbert M et al (2019) Pathology and pathobiology of pulmonary hypertension: state of the art and research perspectives. Eur Respir J 53(1)
4. Southgate L et al (2020) Molecular genetic framework underlying pulmonary arterial hypertension. Nat Rev Cardiol 17(2):85–95
5. Machado RD et al (2015) Pulmonary arterial hypertension: a current perspective on established and emerging molecular genetic defects. Hum Mutat 36(12):1113–1127
6. Andruska A, Spiekerkoetter E (2018) Consequences of BMPR2 deficiency in the pulmonary vasculature and beyond: contributions to pulmonary arterial hypertension. Int J Mol Sci 19(9)
7. Atkinson C et al (2002) Primary pulmonary hypertension is associated with reduced pulmonary vascular expression of type II bone morphogenetic protein receptor. Circulation 105 (14):1672–1678
8. Chen NY et al (2016) Macrophage bone morphogenic protein receptor 2 depletion in idiopathic pulmonary fibrosis and Group III pulmonary hypertension. Am J Physiol Lung Cell Mol Physiol 311(2):L238–L254
9. Hong KH et al (2008) Genetic ablation of the BMPR2 gene in pulmonary endothelium is sufficient to predispose to pulmonary arterial hypertension. Circulation 118(7):722–730
10. West J et al (2004) Pulmonary hypertension in transgenic mice expressing a dominant-negative BMPRII gene in smooth muscle. Circ Res 94(8):1109–1114
11. West J et al (2008) Mice expressing BMPR2R899X transgene in smooth muscle develop pulmonary vascular lesions. Am J Physiol Lung Cell Mol Physiol 295(5):L744–L755
12. Reynolds AM et al (2012) Targeted gene delivery of BMPR2 attenuates pulmonary hypertension. Eur Respir J 39(2):329–343
13. Reynolds AM et al (2007) Bone morphogenetic protein type 2 receptor gene therapy attenuates hypoxic pulmonary hypertension. Am J Physiol Lung Cell Mol Physiol 292(5):L1182–L1192
14. Harper RL et al (2016) BMPR2 gene therapy for PAH acts via Smad and non-Smad signalling. Respirology 21(4):727–733
15. Dunmore BJ et al (2020) 4PBA restores signalling of a cysteine-substituted mutant BMPR2 receptor found in patients with PAH. Am J Respir Cell Mol Biol 63(2):160–171
16. Drake KM et al (2013) Correction of nonsense BMPR2 and SMAD9 mutations by ataluren in pulmonary arterial hypertension. Am J Respir Cell Mol Biol 49(3):403–409
17. Klionsky DJ et al (2016) Guidelines for the use and interpretation of assays for monitoring autophagy (3rd edition). Autophagy 12(1):1–222
18. Savai R et al (2014) Pro-proliferative and inflammatory signaling converge on FoxO1 transcription factor in pulmonary hypertension. Nat Med 20(11):1289–1300
19. Botros L et al (2020) The effects of mercaptopurine on pulmonary vascular resistance and BMPR2 expression in pulmonary arterial hypertension. Am J Respir Crit Care Med 202 (2):296–299
20. Kurakula K et al (2019) Prevention of progression of pulmonary hypertension by the Nur77 agonist 6-mercaptopurine: role of BMP signalling. Eur Respir J 54(3)
21. Dannewitz Prosseda S et al (2019) FHIT, a novel modifier gene in pulmonary arterial hypertension. Am J Respir Crit Care Med 199(1):83–98

22. Spiekerkoetter E et al (2013) FK506 activates BMPR2, rescues endothelial dysfunction, and reverses pulmonary hypertension. J Clin Invest 123(8):3600–3613
23. Yung LM et al (2016) A selective transforming growth factor-beta ligand trap attenuates pulmonary hypertension. Am J Respir Crit Care Med 194(9):1140–1151
24. Long L et al (2015) Selective enhancement of endothelial BMPR-II with BMP9 reverses pulmonary arterial hypertension. Nat Med 21(7):777–785
25. Spiekerkoetter E et al (2015) Low-dose FK506 (tacrolimus) in end-stage pulmonary arterial hypertension. Am J Respir Crit Care Med 192(2):254–257
26. Spiekerkoetter E et al (2017) Randomised placebo-controlled safety and tolerability trial of FK506 (tacrolimus) for pulmonary arterial hypertension. Eur Respir J 50(3)
27. Botney MD, Bahadori L, Gold LI (1994) Vascular remodeling in primary pulmonary hypertension. Potential role for transforming growth factor-beta. Am J Pathol 144(2):286–295
28. Zaiman AL et al (2008) Role of the TGF-beta/Alk5 signaling pathway in monocrotaline-induced pulmonary hypertension. Am J Respir Crit Care Med 177(8):896–905
29. Chen YF et al (2006) Dominant negative mutation of the TGF-beta receptor blocks hypoxia-induced pulmonary vascular remodeling. J Appl Physiol (1985) 100(2):564–571
30. Sheares KK et al (2004) Differential effects of TGF-beta1 and BMP-4 on the hypoxic induction of cyclooxygenase-2 in human pulmonary artery smooth muscle cells. Am J Physiol Lung Cell Mol Physiol 287(5):L919–L927
31. Tielemans B et al (2019) TGFbeta and BMPRII signalling pathways in the pathogenesis of pulmonary arterial hypertension. Drug Discov Today 24(3):703–716
32. ATS 2020 virtual: PULSAR phase 2 trial results of sotatercept in PAH. ATS 2020 Virtual, June 24, 2020
33. Archer SL et al (2010) Epigenetic attenuation of mitochondrial superoxide dismutase 2 in pulmonary arterial hypertension: a basis for excessive cell proliferation and a new therapeutic target. Circulation 121(24):2661–2671
34. Napoli C, Benincasa G, Loscalzo J (2019) Epigenetic inheritance underlying pulmonary arterial hypertension. Arterioscler Thromb Vasc Biol 39(4):653–664
35. Liu D et al (2017) Hypermethylation of BMPR2 promoter occurs in patients with heritable pulmonary arterial hypertension and inhibits BMPR2 expression. Am J Respir Crit Care Med 196(7):925–928
36. Chelladurai P, Seeger W, Pullamsetti SS (2016) Epigenetic mechanisms in pulmonary arterial hypertension: the need for global perspectives. Eur Respir Rev 25(140):135–140
37. Cavasin MA et al (2012) Selective class I histone deacetylase inhibition suppresses hypoxia-induced cardiopulmonary remodeling through an antiproliferative mechanism. Circ Res 110 (5):739–748
38. Zhao L et al (2012) Histone deacetylation inhibition in pulmonary hypertension: therapeutic potential of valproic acid and suberoylanilide hydroxamic acid. Circulation 126(4):455–467
39. Lan B et al (2015) Therapeutic efficacy of valproic acid in a combined monocrotaline and chronic hypoxia rat model of severe pulmonary hypertension. PLoS One 10(1):e0117211
40. Cho YK et al (2010) Sodium valproate, a histone deacetylase inhibitor, but not captopril, prevents right ventricular hypertrophy in rats. Circ J 74(4):760–770
41. Kim J et al (2015) Restoration of impaired endothelial myocyte enhancer factor 2 function rescues pulmonary arterial hypertension. Circulation 131(2):190–199
42. Boucherat O et al (2017) HDAC6: a novel histone deacetylase implicated in pulmonary arterial hypertension. Sci Rep 7(1):4546
43. Zurlo G et al (2018) Sirtuin 1 regulates pulmonary artery smooth muscle cell proliferation: role in pulmonary arterial hypertension. J Hypertens 36(5):1164–1177
44. Yu L et al (2017) Resveratrol protects against pulmonary arterial hypertension in rats via activation of silent information regulator 1. Cell Physiol Biochem 42(1):55–67
45. Meloche J et al (2015) Bromodomain-containing protein 4: the epigenetic origin of pulmonary arterial hypertension. Circ Res 117(6):525–535
46. Van der Feen DE et al (2019) Multicenter preclinical validation of BET inhibition for the treatment of pulmonary arterial hypertension. Am J Respir Crit Care Med 200(7):910–920

47. Meloche J et al (2014) Role for DNA damage signaling in pulmonary arterial hypertension. Circulation 129(7):786–797
48. Chun HJ, Bonnet S, Chan SY (2017) Translational advances in the field of pulmonary hypertension. translating microRNA biology in pulmonary hypertension. It will take more than "miR" words. Am J Respir Crit Care Med 195(2):167–178
49. Zhou G, Chen T, Raj JU (2015) MicroRNAs in pulmonary arterial hypertension. Am J Respir Cell Mol Biol 52(2):139–151
50. Mondejar-Parreno G et al (2019) miR-1 is increased in pulmonary hypertension and downregulates Kv1.5 channels in rat pulmonary arteries. J Physiol 597(4):1185–1197
51. Mondejar-Parreno G et al (2019) miR-1 induces endothelial dysfunction in rat pulmonary arteries. J Physiol Biochem 75(4):519–529
52. Sysol JR et al (2018) Micro-RNA-1 is decreased by hypoxia and contributes to the development of pulmonary vascular remodeling via regulation of sphingosine kinase 1. Am J Physiol Lung Cell Mol Physiol 314(3):L461–L472
53. Connolly M et al (2020) miR-1-5p targets TGF-betaR1 and is suppressed in the hypertrophying hearts of rats with pulmonary arterial hypertension. PLoS One 15(2): e0229409
54. Yue Y et al (2018) miR-143 and miR-145 promote hypoxia-induced proliferation and migration of pulmonary arterial smooth muscle cells through regulating ABCA1 expression. Cardiovasc Pathol 37:15–25
55. Caruso P et al (2017) Identification of microRNA-124 as a major regulator of enhanced endothelial cell glycolysis in pulmonary arterial hypertension via PTBP1 (polypyrimidine tract binding protein) and pyruvate kinase M2. Circulation 136(25):2451–2467
56. Zhang H et al (2017) Metabolic and proliferative state of vascular adventitial fibroblasts in pulmonary hypertension is regulated through a microRNA-124/PTBP1 (polypyrimidine tract binding protein 1)/pyruvate kinase muscle axis. Circulation 136(25):2468–2485
57. Wang D et al (2014) MicroRNA-124 controls the proliferative, migratory, and inflammatory phenotype of pulmonary vascular fibroblasts. Circ Res 114(1):67–78
58. Kang K et al (2013) MicroRNA-124 suppresses the transactivation of nuclear factor of activated T cells by targeting multiple genes and inhibits the proliferation of pulmonary artery smooth muscle cells. J Biol Chem 288(35):25414–25427
59. Chen J et al (2018) Effect of miR-29b on the proliferation and apoptosis of pulmonary artery smooth muscle cells by targeting Mcl-1 and CCND2. Biomed Res Int 2018:6051407
60. Liu T et al (2019) Down-regulation of miR-204 attenuates endothelial-mesenchymal transition by enhancing autophagy in hypoxia-induced pulmonary hypertension. Eur J Pharmacol 863:172673
61. Meloche J et al (2015) miR-223 reverses experimental pulmonary arterial hypertension. Am J Physiol Cell Physiol 309(6):C363–C372
62. Liu A et al (2019) Role of miR-223-3p in pulmonary arterial hypertension via targeting ITGB3 in the ECM pathway. Cell Prolif 52(2):e12550
63. Zeng Y et al (2016) MicroRNA-223 attenuates hypoxia-induced vascular remodeling by targeting RhoB/MLC2 in pulmonary arterial smooth muscle cells. Sci Rep 6:24900
64. Zhao H et al (2020) miR-181b-5p inhibits endothelial-mesenchymal transition in monocrotaline-induced pulmonary arterial hypertension by targeting endocan and TGFBR1. Toxicol Appl Pharmacol 386:114827
65. Zhao H et al (2020) miR-181a/b-5p ameliorates inflammatory response in monocrotaline-induced pulmonary arterial hypertension by targeting endocan. J Cell Physiol 235 (5):4422–4433
66. Zeng Z et al (2018) Anti-apoptosis endothelial cell-secreted microRNA-195-5p promotes pulmonary arterial smooth muscle cell proliferation and migration in pulmonary arterial hypertension. J Cell Biochem 119(2):2144–2155
67. Liu T et al (2019) miR-27a promotes endothelial-mesenchymal transition in hypoxia-induced pulmonary arterial hypertension by suppressing BMP signaling. Life Sci 227:64–73

68. Zhang W et al (2020) MicroRNA15a5p induces pulmonary artery smooth muscle cell apoptosis in a pulmonary arterial hypertension model via the VEGF/p38/MMP2 signaling pathway. Int J Mol Med 45(2):461–474
69. Yu RH, Wang LM, Hu XH (2019) MiR-135a inhibitor alleviates pulmonary arterial hypertension through beta-Catenin/GSK-3beta signaling pathway. Eur Rev Med Pharmacol Sci 23 (21):9574–9581
70. Lee HW, Park SH (2017) Elevated microRNA-135a is associated with pulmonary arterial hypertension in experimental mouse model. Oncotarget 8(22):35609–35618
71. Lee HW, Park SH (2017) Correction: elevated microRNA-135a is associated with pulmonary arterial hypertension in experimental mouse model. Oncotarget 8(35):59999
72. Cai Z et al (2018) MiR-125a-5p ameliorates monocrotaline-induced pulmonary arterial hypertension by targeting the TGF-beta1 and IL-6/STAT3 signaling pathways. Exp Mol Med 50 (4):45
73. Luo L et al (2020) miR-125a-5p inhibits glycolysis by targeting hexokinase-II to improve pulmonary arterial hypertension. Aging (Albany NY) 12(10):9014–9030
74. Ma C et al (2017) MiR-125a regulates mitochondrial homeostasis through targeting mitofusin 1 to control hypoxic pulmonary vascular remodeling. J Mol Med (Berl) 95(9):977–993
75. Zhu G et al (2018) miR371b5p inhibits endothelial cell apoptosis in monocrotaline-induced pulmonary arterial hypertension via PTEN/PI3K/Akt signaling pathways. Mol Med Rep 18 (6):5489–5501
76. Zhou C et al (2019) Mir-455-3p-1 represses FGF7 expression to inhibit pulmonary arterial hypertension through inhibiting the RAS/ERK signaling pathway. J Mol Cell Cardiol 130:23–35
77. Zhao M et al (2019) MiR-19a modulates hypoxia-mediated cell proliferation and migration via repressing PTEN in human pulmonary arterial smooth muscle. Life Sci 239:116928
78. Bi R et al (2015) MicroRNA-27b plays a role in pulmonary arterial hypertension by modulating peroxisome proliferator-activated receptor gamma dependent Hsp90-eNOS signaling and nitric oxide production. Biochem Biophys Res Commun 460(2):469–475
79. Sahoo S et al (2016) MEF2C-MYOCD and leiomodin1 suppression by miRNA-214 promotes smooth muscle cell phenotype switching in pulmonary arterial hypertension. PLoS One 11(5): e0153780
80. Sarrion I et al (2015) Role of circulating miRNAs as biomarkers in idiopathic pulmonary arterial hypertension: possible relevance of miR-23a. Oxidative Med Cell Longev 2015:792846
81. Yan L et al (2017) Effect of miR-23a on anoxia-induced phenotypic transformation of smooth muscle cells of rat pulmonary arteries and regulatory mechanism. Oncol Lett 13(1):89–98
82. Zhang Y, Peng B, Han Y (2018) MiR-23a regulates the proliferation and migration of human pulmonary artery smooth muscle cells (HPASMCs) through targeting BMPR2/Smad1 signaling. Biomed Pharmacother 103:1279–1286
83. Wang P et al (2016) miRNA-34a promotes proliferation of human pulmonary artery smooth muscle cells by targeting PDGFRA. Cell Prolif 49(4):484–493
84. Ogorodnikova N, Arenz C (2015) MicroRNA-145-targeted drug and its preventive effect on pulmonary arterial hypertension (patent WO2012153135 A1). Expert Opin Ther Pat 25 (6):723–727
85. Caruso P et al (2012) A role for miR-145 in pulmonary arterial hypertension: evidence from mouse models and patient samples. Circ Res 111(3):290–300
86. McLendon JM et al (2015) Lipid nanoparticle delivery of a microRNA-145 inhibitor improves experimental pulmonary hypertension. J Control Release 210:67–75
87. Wang AP et al (2015) miR-100 suppresses mTOR signaling in hypoxia-induced pulmonary hypertension in rats. Eur J Pharmacol 765:565–573
88. Li SS et al (2014) MicroRNA-190 regulates hypoxic pulmonary vasoconstriction by targeting a voltage-gated K(+) channel in arterial smooth muscle cells. J Cell Biochem 115 (6):1196–1205
89. Yang S et al (2012) miR-21 regulates chronic hypoxia-induced pulmonary vascular remodeling. Am J Physiol Lung Cell Mol Physiol 302(6):L521–L529

90. Zhang J et al (2020) MicroRNA-483 amelioration of experimental pulmonary hypertension. EMBO Mol Med 12(5):e11303
91. Connolly M et al (2018) miR-322-5p targets IGF-1 and is suppressed in the heart of rats with pulmonary hypertension. FEBS Open Bio 8(3):339–348
92. Zeng Y et al (2015) Hypoxia inducible factor-1 mediates expression of miR-322: potential role in proliferation and migration of pulmonary arterial smooth muscle cells. Sci Rep 5:12098
93. Baptista R et al (2018) MicroRNA-424(322) as a new marker of disease progression in pulmonary arterial hypertension and its role in right ventricular hypertrophy by targeting SMURF1. Cardiovasc Res 114(1):53–64
94. Deng B et al (2016) MicroRNA-103/107 is involved in hypoxia-induced proliferation of pulmonary arterial smooth muscle cells by targeting HIF-1beta. Life Sci 147:117–124
95. Miao R et al (2020) MiR-18a-5p contributes to enhanced proliferation and migration of PASMCs via targeting Notch2 in pulmonary arterial hypertension. Life Sci 257:117919
96. Brock M et al (2014) AntagomiR directed against miR-20a restores functional BMPR2 signalling and prevents vascular remodelling in hypoxia-induced pulmonary hypertension. Eur Heart J 35(45):3203–3211
97. Tan H et al (2019) MicroRNA30a5p promotes proliferation and inhibits apoptosis of human pulmonary artery endothelial cells under hypoxia by targeting YKL40. Mol Med Rep 20 (1):236–244
98. Le Ribeuz H et al (2020) In vivo miR-138-5p inhibition alleviates monocrotaline-induced pulmonary hypertension and normalizes pulmonary KCNK3 and SLC45A3 expression. Respir Res 21(1):186
99. Li S et al (2013) MicroRNA-138 plays a role in hypoxic pulmonary vascular remodelling by targeting Mst1. Biochem J 452(2):281–291
100. Liu JJ et al (2018) MicroRNA138 promotes proliferation and suppresses mitochondrial depolarization in human pulmonary artery smooth muscle cells through targeting TASK1. Mol Med Rep 17(2):3021–3027
101. Chen T et al (2016) miR-17/20 controls prolyl hydroxylase 2 (PHD2)/hypoxia-inducible factor 1 (HIF1) to regulate pulmonary artery smooth muscle cell proliferation. J Am Heart Assoc 5 (12)
102. Chen T et al (2018) PAI-1 is a novel component of the miR-17~92 signaling that regulates pulmonary artery smooth muscle cell phenotypes. Am J Physiol Lung Cell Mol Physiol 315 (2):L149–L161
103. Chen T et al (2015) Loss of microRNA-17 approximately 92 in smooth muscle cells attenuates experimental pulmonary hypertension via induction of PDZ and LIM domain 5. Am J Respir Crit Care Med 191(6):678–692
104. Guo L et al (2012) The microRNA-328 regulates hypoxic pulmonary hypertension by targeting at insulin growth factor 1 receptor and L-type calcium channel-alpha1C. Hypertension 59(5):1006–1013
105. Zhao M et al (2019) MiR-629 regulates hypoxic pulmonary vascular remodelling by targeting FOXO3 and PERP. J Cell Mol Med 23(8):5165–5175
106. Zhao TF et al (2019) MiR-593-5p promotes the development of hypoxic-induced pulmonary hypertension via targeting PLK1. Eur Rev Med Pharmacol Sci 23(8):3495–3502
107. Zhang X et al (2018) Inhibition of miR-361-5p suppressed pulmonary artery smooth muscle cell survival and migration by targeting ABCA1 and inhibiting the JAK2/STAT3 pathway. Exp Cell Res 363(2):255–261
108. Zeng ZH et al (2019) MicroRNA132 mediates proliferation and migration of pulmonary smooth muscle cells via targeting PTEN. Mol Med Rep 19(5):3823–3830
109. Tao W et al (2019) miR-205-5p suppresses pulmonary vascular smooth muscle cell proliferation by targeting MICAL2-mediated Erk1/2 signaling. Microvasc Res 124:43–50
110. Xing Y et al (2015) MicroRNA-30c contributes to the development of hypoxia pulmonary hypertension by inhibiting platelet-derived growth factor receptor beta expression. Int J Biochem Cell Biol 64:155–166

111. Rhodes CJ et al (2013) Reduced microRNA-150 is associated with poor survival in pulmonary arterial hypertension. Am J Respir Crit Care Med 187(3):294–302
112. Li Y et al (2019) MicroRNA-150 relieves vascular remodeling and fibrosis in hypoxia-induced pulmonary hypertension. Biomed Pharmacother 109:1740–1749
113. Chen M et al (2017) MicroRNA-150 attenuates hypoxia-induced excessive proliferation and migration of pulmonary arterial smooth muscle cells through reducing HIF-1alpha expression. Biomed Pharmacother 93:861–868
114. Lu Z et al (2016) Upregulated miR-17 regulates hypoxia-mediated human pulmonary artery smooth muscle cell proliferation and apoptosis by targeting mitofusin 2. Med Sci Monit 22:3301–3308
115. Liu G et al (2018) Upregulation of microRNA-17-5p contributes to hypoxia-induced proliferation in human pulmonary artery smooth muscle cells through modulation of p21 and PTEN. Respir Res 19(1):200
116. Pullamsetti SS et al (2012) Inhibition of microRNA-17 improves lung and heart function in experimental pulmonary hypertension. Am J Respir Crit Care Med 185(4):409–419
117. Li H et al (2020) MicroRNA-17 as a potential diagnostic biomarker in pulmonary arterial hypertension. J Int Med Res 48(6):300060520920430
118. Rothman AM et al (2016) MicroRNA-140-5p and SMURF1 regulate pulmonary arterial hypertension. J Clin Invest 126(7):2495–2508
119. Zhang Y, Xu J (2016) MiR-140-5p regulates hypoxia-mediated human pulmonary artery smooth muscle cell proliferation, apoptosis and differentiation by targeting Dnmt1 and promoting SOD2 expression. Biochem Biophys Res Commun 473(1):342–348
120. Zhu TT et al (2019) MicroRNA-140-5p targeting tumor necrosis factor-alpha prevents pulmonary arterial hypertension. J Cell Physiol 234(6):9535–9550
121. Yan Y et al (2020) MicroRNA-221 promotes proliferation and migration of pulmonary arterial smooth muscle cells (PASMCs) by targeting tissue inhibitor of metalloproteinases-3 (TIMP3). Cardiovasc Diagn Ther 10(4):646–657
122. Nie X et al (2019) MicroRNA-221-3p promotes pulmonary artery smooth muscle cells proliferation by targeting AXIN2 during pulmonary arterial hypertension. Vasc Pharmacol 116:24–35
123. Sharma S et al (2014) Apolipoprotein A-I mimetic peptide 4F rescues pulmonary hypertension by inducing microRNA-193-3p. Circulation 130(9):776–785
124. Xiao T et al (2017) Differential expression of microRNA in the lungs of rats with pulmonary arterial hypertension. Mol Med Rep 15(2):591–596
125. Zehendner CM et al (2020) Long noncoding RNA TYKRIL plays a role in pulmonary hypertension via the p53-mediated regulation of PDGFRbeta. Am J Respir Crit Care Med 202(10):1445–1457
126. Jandl K et al (2019) Long non-coding RNAs influence the transcriptome in pulmonary arterial hypertension: the role of PAXIP1-AS1. J Pathol 247(3):357–370
127. Lei S et al (2020) LncRNA-SMILR modulates RhoA/ROCK signaling by targeting miR-141 to regulate vascular remodeling in pulmonary arterial hypertension. Am J Physiol Heart Circ Physiol 319(2):H377–H391
128. Sun Z et al (2017) Long non-coding RNA MEG3 downregulation triggers human pulmonary artery smooth muscle cell proliferation and migration via the p53 signaling pathway. Cell Physiol Biochem 42(6):2569–2581
129. Zhu B et al (2018) Down-regulation of lncRNA MEG3 promotes hypoxia-induced human pulmonary artery smooth muscle cell proliferation and migration via repressing PTEN by sponging miR-21. Biochem Biophys Res Commun 495(3):2125–2132
130. Xing Y et al (2019) Long noncoding RNA-maternally expressed gene 3 contributes to hypoxic pulmonary hypertension. Mol Ther 27(12):2166–2181
131. Gong J et al (2019) Long non-coding RNA CASC2 suppresses pulmonary artery smooth muscle cell proliferation and phenotypic switch in hypoxia-induced pulmonary hypertension. Respir Res 20(1):53

132. Yang L et al (2019) LncRNA Tug1 involves in the pulmonary vascular remodeling in mice with hypoxic pulmonary hypertension via the microRNA-374c-mediated Foxc1. Life Sci 237:116769
133. Wang S et al (2019) TUG1 regulates pulmonary arterial smooth muscle cell proliferation in pulmonary arterial hypertension. Can J Cardiol 35(11):1534–1545
134. Su H et al (2018) LncRNA H19 promotes the proliferation of pulmonary artery smooth muscle cells through AT1R via sponging let-7b in monocrotaline-induced pulmonary arterial hypertension. Respir Res 19(1):254
135. Leisegang MS et al (2017) Long noncoding RNA MANTIS facilitates endothelial angiogenic function. Circulation 136(1):65–79
136. Wang S, Zhang C, Zhang X (2020) Downregulation of long noncoding RNA ANRIL promotes proliferation and migration in hypoxic human pulmonary artery smooth muscle cells. Mol Med Rep 21(2):589–596
137. Zhang H et al (2019) Long noncoding RNA Hoxaas3 contributes to hypoxia-induced pulmonary artery smooth muscle cell proliferation. Cardiovasc Res 115(3):647–657
138. Wang D et al (2019) Long noncoding RNA MALAT1 sponges miR1243p.1/KLF5 to promote pulmonary vascular remodeling and cell cycle progression of pulmonary artery hypertension. Int J Mol Med 44(3):871–884
139. Zhu TT et al (2019) Long noncoding RNA UCA1 promotes the proliferation of hypoxic human pulmonary artery smooth muscle cells. Pflugers Arch 471(2):347–355
140. Liu Y et al (2018) LncRNA-TCONS_00034812 in cell proliferation and apoptosis of pulmonary artery smooth muscle cells and its mechanism. J Cell Physiol 233(6):4801–4814
141. Cheng G, He L, Zhang Y (2020) LincRNA-Cox2 promotes pulmonary arterial hypertension by regulating the let-7a-mediated STAT3 signaling pathway. Mol Cell Biochem
142. Liu Y et al (2020) Long noncoding RNA Rps4l mediates the proliferation of hypoxic pulmonary artery smooth muscle cells. Hypertension 76(4):1124–1133
143. Chen J et al (2018) The long noncoding RNA LnRPT is regulated by PDGF-BB and modulates the proliferation of pulmonary artery smooth muscle cells. Am J Respir Cell Mol Biol 58(2):181–193
144. Wang H, Qin R, Cheng Y (2020) LncRNA-Ang362 promotes pulmonary arterial hypertension by regulating miR-221 and miR-222. Shock 53(6):723–729
145. McMurtry MS et al (2004) Dichloroacetate prevents and reverses pulmonary hypertension by inducing pulmonary artery smooth muscle cell apoptosis. Circ Res 95(8):830–840
146. Michelakis ED et al (2002) Dichloroacetate, a metabolic modulator, prevents and reverses chronic hypoxic pulmonary hypertension in rats: role of increased expression and activity of voltage-gated potassium channels. Circulation 105(2):244–250
147. Michelakis ED et al (2017) Inhibition of pyruvate dehydrogenase kinase improves pulmonary arterial hypertension in genetically susceptible patients. Sci Transl Med 9(413)
148. Khan SS et al (2015) Effects of ranolazine on exercise capacity, right ventricular indices, and hemodynamic characteristics in pulmonary arterial hypertension: a pilot study. Pulm Circ 5 (3):547–556
149. Dean A et al (2016) Metformin reverses development of pulmonary hypertension via aromatase inhibition. Hypertension 68(2):446–454
150. Liu Y et al (2019) Metformin prevents progression of experimental pulmonary hypertension via inhibition of autophagy and activation of adenosine monophosphate-activated protein kinase. J Vasc Res 56(3):117–128
151. Hemnes AR et al (2014) Evidence for right ventricular lipotoxicity in heritable pulmonary arterial hypertension. Am J Respir Crit Care Med 189(3):325–334
152. Pullamsetti SS et al (2014) Novel and emerging therapies for pulmonary hypertension. Am J Respir Crit Care Med 189(4):394–400
153. Hoeper MM et al (2013) Imatinib mesylate as add-on therapy for pulmonary arterial hypertension: results of the randomized IMPRES study. Circulation 127(10):1128–1138

154. Montani D et al (2012) Pulmonary arterial hypertension in patients treated by dasatinib. Circulation 125(17):2128–2137
155. Orlikow E, Weatherald J, Hirani N (2019) Dasatinib-induced pulmonary arterial hypertension. Can J Cardiol 35(11):1604.e1–1604.e3
156. McGee M et al (2018) Drug-associated pulmonary arterial hypertension. Clin Toxicol (Phila) 56(9):801–809
157. Dahal BK et al (2010) Role of epidermal growth factor inhibition in experimental pulmonary hypertension. Am J Respir Crit Care Med 181(2):158–167
158. Izikki M et al (2009) Endothelial-derived FGF2 contributes to the progression of pulmonary hypertension in humans and rodents. J Clin Invest 119(3):512–523
159. Abe K et al (2004) Long-term treatment with a Rho-kinase inhibitor improves monocrotaline-induced fatal pulmonary hypertension in rats. Circ Res 94(3):385–393
160. Abe K et al (2006) Long-term inhibition of Rho-kinase ameliorates hypoxia-induced pulmonary hypertension in mice. J Cardiovasc Pharmacol 48(6):280–285
161. Mouchaers KT et al (2010) Fasudil reduces monocrotaline-induced pulmonary arterial hypertension: comparison with bosentan and sildenafil. Eur Respir J 36(4):800–807
162. Qi L et al (2019) Fasudil dichloroacetate (FDCA), an orally available agent with potent therapeutic efficiency on monocrotaline-induced pulmonary arterial hypertension rats. Bioorg Med Chem Lett 29(14):1812–1818
163. Ruan H et al (2019) The acute effects of 30 mg vs 60 mg of intravenous Fasudil on patients with congenital heart defects and severe pulmonary arterial hypertension. Congenit Heart Dis 14(4):645–650
164. Ruan HY, Zhang YG, Liu R (2018) [Acute effects of intravenous fasudil with different dosage on patients with congenital heart defects and severe pulmonary arterial hypertension]. Zhonghua Yi Xue Za Zhi 98(9):678–681
165. Xiao JW et al (2015) Acute effects of Rho-kinase inhibitor fasudil on pulmonary arterial hypertension in patients with congenital heart defects. Circ J 79(6):1342–1348
166. Houssaini A, Adnot S (2017) mTOR: a key to both pulmonary vessel remodeling and right ventricular function in pulmonary arterial hypertension? Am J Respir Cell Mol Biol 57 (5):509–511
167. Seyfarth HJ et al (2013) Everolimus in patients with severe pulmonary hypertension: a safety and efficacy pilot trial. Pulm Circ 3(3):632–638
168. Segura-Ibarra V et al (2017) Rapamycin nanoparticles localize in diseased lung vasculature and prevent pulmonary arterial hypertension. Int J Pharm 524(1–2):257–267
169. Tang H et al (2018) Pathogenic role of mTORC1 and mTORC2 in pulmonary hypertension. JACC Basic Transl Sci 3(6):744–762
170. Ranchoux B et al (2015) Endothelial-to-mesenchymal transition in pulmonary hypertension. Circulation 131(11):1006–1018
171. Tsutsumi T et al (2019) Nintedanib ameliorates experimental pulmonary arterial hypertension via inhibition of endothelial mesenchymal transition and smooth muscle cell proliferation. PLoS One 14(7):e0214697
172. Rabinovitch M et al (2014) Inflammation and immunity in the pathogenesis of pulmonary arterial hypertension. Circ Res 115(1):165–175
173. Trankle CR et al (2019) IL-1 blockade reduces inflammation in pulmonary arterial hypertension and right ventricular failure: a single-arm, open-label, phase IB/II pilot study. Am J Respir Crit Care Med 199(3):381–384
174. Cheng Y et al (2018) Identification of a novel hybridization from isosorbide 5-mononitrate and bardoxolone methyl with dual activities of pulmonary vasodilation and vascular remodeling inhibition on pulmonary arterial hypertension rats. J Med Chem 61(4):1474–1482
175. Nickel NP et al (2015) Elafin reverses pulmonary hypertension via caveolin-1-dependent bone morphogenetic protein signaling. Am J Respir Crit Care Med 191(11):1273–1286
176. Good RB et al (2015) Endothelial to mesenchymal transition contributes to endothelial dysfunction in pulmonary arterial hypertension. Am J Pathol 185(7):1850–1858

177. Ranchoux B et al (2018) Endothelial dysfunction in pulmonary arterial hypertension: an evolving landscape (2017 Grover Conference Series). Pulm Circ 8(1):2045893217752912
178. Qiao L et al (2014) Endothelial fate mapping in mice with pulmonary hypertension. Circulation 129(6):692–703
179. Hopper RK et al (2016) In pulmonary arterial hypertension, reduced BMPR2 promotes endothelial-to-mesenchymal transition via HMGA1 and its target slug. Circulation 133 (18):1783–1794
180. Huang J et al (2020) Transplantation of mesenchymal stem cells attenuates pulmonary hypertension by normalizing the endothelial-to-mesenchymal transition. Am J Respir Cell Mol Biol 62(1):49–60
181. Xu J et al (2018) Dipeptidyl peptidase IV (DPP-4) inhibition alleviates pulmonary arterial remodeling in experimental pulmonary hypertension. Lab Investig 98(10):1333–1346
182. Zhang H et al (2019) Protective effect of hydrogen sulfide on monocrotaline-induced pulmonary arterial hypertension via inhibition of the endothelial mesenchymal transition. Int J Mol Med 44(6):2091–2102
183. Yu M et al (2020) Paeoniflorin ameliorates chronic hypoxia/SU5416-induced pulmonary arterial hypertension by inhibiting endothelial-to-mesenchymal transition. Drug Des Devel Ther 14:1191–1202
184. Isobe S et al (2019) Endothelial-mesenchymal transition drives expression of CD44 variant and xCT in pulmonary hypertension. Am J Respir Cell Mol Biol 61(3):367–379
185. Li L et al (2017) NF-kappaB mediated miR-130a modulation in lung microvascular cell remodeling: implication in pulmonary hypertension. Exp Cell Res 359(1):235–242
186. Xu YP et al (2017) MiR-126a-5p is involved in the hypoxia-induced endothelial-to-mesenchymal transition of neonatal pulmonary hypertension. Hypertens Res 40(6):552–561
187. Zhang H et al (2018) Bone morphogenetic protein-7 inhibits endothelial-mesenchymal transition in pulmonary artery endothelial cell under hypoxia. J Cell Physiol 233(5):4077–4090

Potential Cellular Targets Associated with the Signaling of the Pulmonary Hypertension

19

Meemansha Sharma, Thakur Uttam Singh,
Madhu Cholanhalli Lingaraju, and Subhashree Parida

Abstract

The mean value of normal pulmonary arterial pressure in human beings is 12–16 mmHg. Pulmonary arterial pressure beyond 25 mmHg is associated with the condition of pulmonary hypertension, and is related to right heart failure. It has been noticed that the pathophysiology of pulmonary hypertension is a multifactor process that involves both structural and functional changes in the pulmonary vasculature and is responsible for the increase in the pulmonary vascular resistance. There are several factors which are associated with the alterations in pulmonary pressure and release from the vascular endothelium such as nitric oxide, prostacyclin, and endothelium-derived hyperpolarizing factors. Disturbances in these factors lead to pulmonary hypertension. There are several important potential cellular targets which are associated with the pulmonary hypertension signaling such as TGF-β, BMR2, Rho, ROCK, CypA, Bsg, and AMPK. This chapter review all these potential cellular targets.

Keywords

Pulmonary · Hypertension · Vascular · Therapeutics · Lung

19.1 Introduction

Pulmonary arterial hypertension (PAH) is a life-threatening condition in which the mean arterial pressure becomes equal or above of 25 mmHg with normal pulmonary capillary wedge pressure equal or less than 15 mmHg [1]. PAH is a chronic disease

M. Sharma · T. U. Singh (✉) · M. C. Lingaraju · S. Parida
Division of Pharmacology and Toxicology, ICAR-Indian Veterinary Research Institute, Izatnagar, Uttar Pradesh, India

K. Dua et al. (eds.), *Targeting Cellular Signalling Pathways in Lung Diseases*,
https://doi.org/10.1007/978-981-33-6827-9_19

in which small pulmonary arteries get obstructed due to endothelial dysfunction and vascular remodeling, ultimately leading to right ventricular failure and death [2]. The estimated prevalence of pulmonary hypertension ranges from 11 to 26 cases per million in adults, and further, it is noted that the incidence of disease is more in women compared to men, which is 2–7.6 cases per million [2]. Abnormal proliferation of endothelial cells or smooth muscles, oxidative stress, and inflammation are the major factors which are involved in the pathogenesis of PAH [3]. There is an imbalance between vasoactive mediators of pulmonary arteries which leads to dysregulation of vascular tone and further contributes to pulmonary vascular remodeling. This vascular remodeling plays an important role in the development of PAH and also involves angiogenesis and vasoconstriction due to thrombosis and hypoxia [4]. Right ventricular hypertrophy and pressure overload are common pathological findings reported in PAH due to obstruction of blood vessels and increased pulmonary resistance. PAH has been divided into five categories of disorders [5]. Type 1 includes pulmonary hypertension which could be idiopathic or heritable and drug- and toxin-induced and also associated with some diseases such as human immunodeficiency virus (HIV) and schistosomiasis infections. Pulmonary veno-occlusive disease and pulmonary capillary hemangiomatosis are also included in the separate categories of type 1 pulmonary hypertension. Type 2 includes pulmonary hypertension which is caused by left heart disease. Alveolar hypoxia or respiratory disorder, associated pulmonary hypertension comes under type 3 category. Type 4 includes pulmonary hypertension that occurs due to chronic obstruction of pulmonary arteries by thrombi known as chronic thromboembolic pulmonary hypertension (CTEPH). Type 5 includes pulmonary hypertension that is caused by rare diseases with unknown mechanism (Table 19.1). Over the past decades, significant advances have been made in the treatment of pulmonary hypertension including endothelin receptor antagonists, prostaglandin analogues, and PDE-5 inhibitors [6]. Owing to the high morbidity and mortality associated with PAH, it is necessary to elucidate the novel therapeutic intervention which could be directly targeted toward the molecular mechanisms that are involved in pulmonary artery remodeling. Therefore, here we have summarized the pathophysiology, genetic phenomenon, and cellular and molecular mechanisms of pulmonary hypertension along with its novel therapeutic targets.

19.2 Pathophysiological Phenomenon Associated with PAH

The complex pathogenesis of PAH involves the imbalance between vasoconstrictors and vasodilatory mediators in pulmonary arteries [3]. Reduced concentration of vasodilatory mediators such as prostaglandin I_2 [7], nitric oxide, and cyclic guanosine monophosphate (cGMP) [8] and increased level of vasoconstrictors like thromboxane, endothelin, and serotonin are observed in pulmonary hypertensive patients [9]. These imbalances cause vasoconstriction in pulmonary arteries and contribute to vascular dysfunction in PAH [3]. In the development of pulmonary hypertension, the attention is being focused on inflammation in pulmonary vasculature. Mononuclear

Table 19.1 Classification of pulmonary hypertension

Group 1	(a) Idiopathic or heritable PAH (b) Drug- and toxin-induced (c) PAH associated with identified diseases (HIV and schistosomiasis)
Group 1'	Pulmonary veno-occlusive diseases (a) Pulmonary capillary hemangiomatosis
Group 2	Pulmonary hypertension resulting from left heart disease (a) Systolic dysfunction (b) Diastolic dysfunction (c) Valvular dysfunction
Group 3	Respiratory disorders associated with PAH (a) Chronic obstructive pulmonary disease (b) Alveolar hypoventricular disorders (c) Exposure to high-altitude disorder (d) Sleep-disordered breathing
Group 4	Chronic thromboembolic pulmonary hypertension (CTEPH)
Group 5	PH caused by rare diseases of unknown mechanism (a) Hematological disorders (b) Myeloproliferative disorders (c) Lymphangioleiomyomatosis (d) Vasculitis (e) Chronic renal failure (f) Tumor obstruction

This classification was adopted from Simonneau et al. [5]

fibrocytes and leukocytes are leaked from the blood vessel walls and contribute in remodeling of pulmonary vasculature [10]. The high level of circulating cytokines and their expressed receptors have been reported in idiopathic pulmonary hypertensive patients [11]. Among various cytokines, overexpression of IL-6 gene may be responsible for the pulmonary vascular disease in rodents [12], and resistance against hypoxia-induced pulmonary hypertension has been found in IL-6 knockout mice [13].

Dysfunctions are reported in potassium (K^+) and calcium (Ca^{+2}) channels of vascular smooth muscle cells (VSMC) in PAH [4]. Decreased expression of voltage-gated potassium (K^+) channels [14] and increased expression of transient receptor potential (TRP) ion channels [15] have been reported previously in idiopathic PAH. Further, these changes in ion channels produce proliferation in VSMC and dysregulate cellular homeostasis. In addition, extracellular matrix remodeling may also contribute in the pathophysiology of PAH [2]. Some studies have shown that the disruption of internal elastic lamina of pulmonary arteries is due to increased activity of serine elastases [16] and matrix metalloproteinases [17] in pulmonary hypertension. CTEPH is a different form of pulmonary hypertension. Pathogenesis of CTEPH is similar to idiopathic PAH which includes obstruction of central *pulmonary arteries* by organized thrombus and severe pulmonary *vascular remodeling* [18]. Reduced production of angiogenic vascular endothelial growth factors and decreased leukocyte recruitment are likely to delay the resolution of organized thrombus [19] (Fig. 19.1).

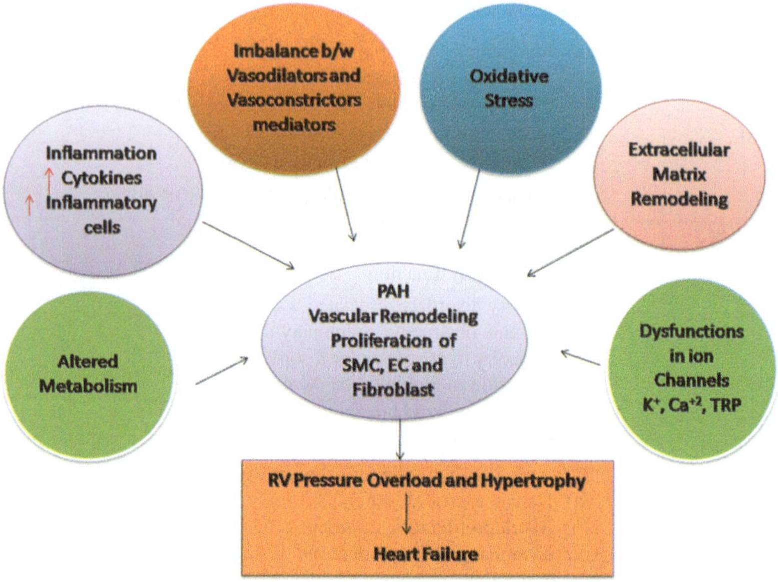

Fig. 19.1 Overview of pathophysiological mechanisms that are discussed in the text and involved in progression of pulmonary hypertension. These factors are contributing in remodeling of pulmonary vasculature and lumen occlusion which ultimately lead to right ventricular hypertrophy and heart failure. *SMC* smooth muscle cell, *EC* endothelial cell, K^+ potassium, Ca^{+2} calcium channels, *TRP* transient receptor potential ion channels

19.3 Potential Cellular Targets Associated with the Signaling of the Pulmonary Hypertension

19.3.1 TGF-β Signaling in Pulmonary Hypertension

TGF-β receptor family and their associated proteins are involved in the regulation of cell physiological activity such as proliferation, differentiation, and inflammation through signal transduction in pulmonary vascular endothelium [20]. Disruption of this signaling pathway is associated with vascular remodeling, atherosclerosis, and myocardial fibrosis [21]. Altered activity of activin-like kinase 1 and 5 (ALK1 and ALK5) and bone morphogenetic protein receptor-2 (BMPR2) due to mutation may be implicated in the hereditary and other form of PAH [4]. It has been observed that 75% of cases of hereditary PAH have been reported due to mutation which is shown in BMPR2, and this mutation is also responsible for idiopathic PAH which accounts for 25% of this phenomenon [22]. The loss of gene BMPR2 and its mutation has enhanced apoptosis and dysfunction in endothelial cells and also causes pulmonary vascular remodeling which are the contributing factors of pulmonary hypertension [23]. The previous report revealed that targeted delivery of wild-type BMPR2 gene into the rat lung endothelium has enhanced BMPR2 expression and prevents

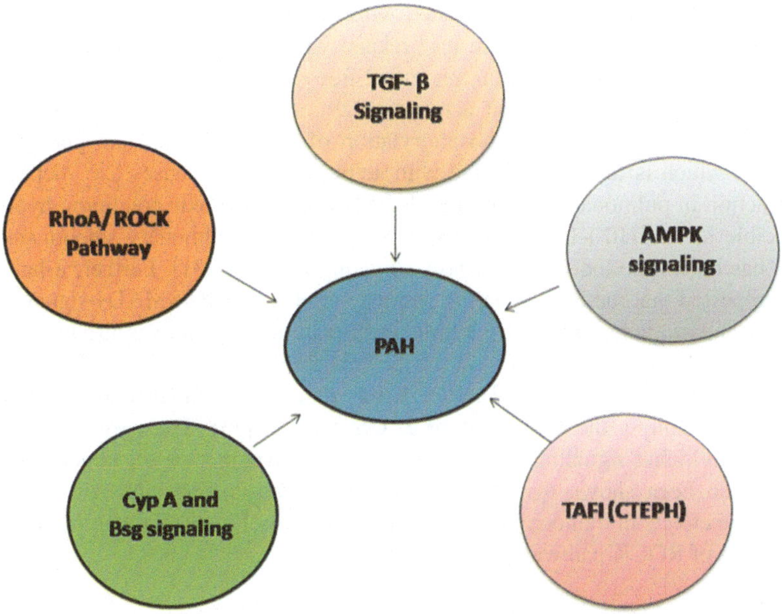

Fig. 19.2 Overview of potential cellular targets associated with the signaling of the pulmonary hypertension. *PAH* pulmonary arterial hypertension, *TGF-β* transforming growth factor beta, *AMPK* AMP-activated protein kinase, *ROCK* Rho-associated protein kinase, *CypA* cyclophilin A, *Bsg* basigin, *TAFI* thrombin-activatable fibrinolytic inhibitor, *CTEPH* chronic thromboembolic pulmonary hypertension

pulmonary hypertension [23]. Another report found that FK506 (tacrolimus) which is an activator of BMPR2 signaling has prevented the development of PAH in mice delivered by osmotic pump via continuous subcutaneous infusion [24]. 1–3% of PAH cases have been observed due to mutations in other gene members of transforming growth factor superfamily such as activin receptor-like kinase type 1 (ACVRL1 or ALK1) or endoglin genes [25] (Fig. 19.2).

The other genetic variations responsible for pulmonary hypertension have occurred in autosomal recessive manner in some genes such as caveolin-1 (CAV1), which regulates SMAD 2/3 phosphorylation, and potassium channel subfamily K member 3 (KCNK3), which is related to the proliferation of pulmonary artery smooth muscle cells. Eukaryotic translation initiation factor 2 alpha kinase 4 (EIF2AK4) is also associated with pulmonary vaso-occlusive disease [26, 27].

19.3.2 Involvement of RhoA/ROCK Pathway in Pulmonary Hypertension Signaling

Rho-associated protein kinase (ROCK) works as the effector for G protein Ras homolog family member A (RhoA) and contains serine-threonine kinase activity

[28]. Rho-kinase pathway is associated with the development and progression of pulmonary hypertension, because it is related with hypoxic exposure, endothelial dysfunction, VSMC proliferation, increased free radical production, chemotaxis of inflammatory cells, and platelet activation [29]. Activation of ROCK protein by RhoA is responsible for the phosphorylation of myosin light chain phosphatase (MLCP) which is further interacted with actin protein and causes Ca^{2+}-dependent contraction in pulmonary smooth muscle cells [30]. In hypoxic condition, hypoxia-inducible factor (HIF)-1α-activated vasoconstriction in pulmonary smooth muscle cells has been reported through RhoA signaling pathway [31]. Further, it has also been reported that the Rho signaling pathway inhibits BMP2/SMAD1 signal transduction and promotes pulmonary VSMC proliferation [32]. RhoA is also involved in endothelial dysfunction which is one of the contributing factors for the development of pulmonary hypertension [33]. ROCK has reduced the bioavailability of nitric oxide by inhibiting the expression of eNOS enzyme [33]. Thus, we can say that RhoA-Rho-kinase signaling pathway are potential novel therapeutic target for PAH. It has also been shown that the combination of fasudil and sildenafil produced protective effects in monocrotaline-induced pulmonary hypertension in rats through inhibition of ROCK pathway [34] (Fig. 19.2).

19.3.3 CypA (Cyclophilin A) and Bsg Signaling in Pulmonary Hypertension

Intracellular CypA protein related to immunophilin family acts as a receptor for cyclosporine drug which is used for immunosuppression [35]. Activated Rho kinase induces secretion of CypA from VSMC [36]. CypA binds to its Bsg receptor present in circulating cells; thereby, contributes in the progression of hypoxia-induced pulmonary hypertension. Further, this interaction causes vascular smooth muscle proliferation, disrupts nitric oxide metabolism, and enhances expression of inflammatory cytokines [37–39]. Amelioration of hypoxia-induced pulmonary hypertension is reported in $Bsg^{+/-}$ mice in comparison with $Bsg^{+/+}$ mice [38]. The higher expression of Bsg protein in pulmonary arteries as well as in inflammatory cells of animal models [40] and the high concentration of CypA protein in serum are reported from pulmonary arterial hypertensive patients [41]. Hence, inhibition of CypA secretion could be a new therapeutic target to prevent PAH. In addition, it is also found that statin drugs and Rho-kinase inhibitors could ameliorate PAH and prevent the secretion of CypA from VSMC [41] (Fig. 19.2).

19.3.4 AMPK Signaling in Pulmonary Hypertension

AMP-activated protein kinase acts as an energy sensor and has serine-threonine kinase activity [42]. Inhibition of Rho kinase leads to activation of AMP-activated protein kinase pathway [43]. Anti-apoptotic effects in endothelial cells [44] and pro-apoptotic effects in VSMC are produced by AMPK pathway [45], which are the crucial factors for remodeling in pulmonary hypertension. Apart from nitric oxide production in endothelial cells, AMPK signaling promotes proliferation of smooth muscle cells and vascular remodeling and reduces intracellular signaling pathways and secretion of growth factors [40]. The downregulation of AMPK in endothelial cells showed protection against pulmonary hypertensive patients and also hypoxia-induced pulmonary hypertension in mice [46]. Indeed, activation of AMPK via various drugs such as statins and metformin produces curative effects in PAH [46, 47]. Thus, these drugs could be novel therapeutic targets for amelioration of pulmonary hypertension. In addition, cytokines, chemokines, and various growth factors are also involved in regulation of endothelial dysfunction and participate in progression of pulmonary hypertension [40]. Thus, some effective therapeutic interventions are required which could prevent the generation of cytokines or inflammation in pulmonary hypertension (Fig. 19.2).

19.3.5 Alteration in Metabolic Pathways in PAH

Normally, pyruvate is formed in the cell by glycolysis, and further utilized in TCA cycle after formation of acetyl coenzyme A through pyruvate dehydrogenase enzyme (PDH) [21]. However, decreased activity of pyruvate dehydrogenase enzyme has been reported in PAH, and enhanced activities of lactate dehydrogenase have been reported in lungs and hearts of monocrotaline-induced rats [48]. In pulmonary hypertension, glycolytic end product pyruvate converts into lactate by the enzyme lactate dehydrogenase and ultimately induces acidosis; as a result, right ventricular function is impaired [49]. Some recent studies illustrated mitochondrial dynamics, in which mitochondrial structures were fragmented inside the pulmonary arterial smooth muscle cells and were involved in pathogenesis of pulmonary hypertension [50]. Mitochondrial dynamics is able to activate HIF-1α in hypoxic condition which further promotes glycolytic shift and depletion of energy in the right ventricle via defective ATP synthesis [50]. The high level of HIF-1α protein expression has been found in idiopathic pulmonary hypertension which is responsible for proliferative vasculopathy [51]. In addition, reactive oxygen species and free radicals can also induce the production of HIF-1α proteins, because the superoxide dismutase enzyme level is reduced due to mitochondrial dysfunction which scavenges free radicals and generates hypoxic conditions [22].

19.3.6 Thrombin-Activatable Fibrinolytic Inhibitor (TAFI) in CTEPH

The injury in lung vasculature activates several factors such as pro-coagulant and anticoagulant proteins, inflammatory cytokines, and various growth factors which are involved in the development of CTEPH [52]. Moreover, coagulation system becomes hyperactive as a result of lung injury which leads to fibrin deposition [52]. TAFI is a glycoprotein synthesized from the liver and activated by thrombin, plasmin, and activated platelets [53, 54]. TAFI prevents fibrinolysis by removing

C-terminal lysine amino acid from fibrin, thereby inhibiting fibrin binding to plasminogen activators and conversion of plasminogen to plasmin [55]. Thus, we could say that TAFI may be involved in pathogenesis of CTEPH and delayed fibrinolysis of thrombi. The high level of TAFI in plasma has been found in patients with CTEPH which directly indicates the correlation with impaired fibrinolysis [29]. Similarly, ex vivo plasma clot lysis assay has been performed in plasma from CTEPH patients which has shown resistance to fibrinolysis [29]. Inhibition of activated TAFI by using CPI-2KR (an inhibitor of activated TAFI) compound was found effective against resistant fibrinolysis and ameliorated CTEPH condition in patients [29]. Prostaglandin E1 has improved fibrinolysis impairment by inhibiting platelet activation and further secretion of TAFI [29]. Thus, inhibitor of TAFI could be a novel therapeutic target in CTEPH patients.

19.4 Conclusion

Pulmonary hypertension is considered as a complex disease with fatal outcomes. Till date, at molecular level, various studies have been conducted to identify new molecular mechanisms and pathophysiology involved in pulmonary hypertension, but detailed mechanisms still remain to be elucidated. Some important signaling identified are TGF-β, Rho/ROCK, CypA, Bsg, and AMPK. It is necessary to prioritize and understand these novel molecular pathways in detail and link to personalized therapeutic interventions for the treatment of the disease.

References

1. Hoeper MM, Ghofrani HA, Grünig E, Klose H, Olschewski H, Rosenkranz S (2017) Pulmonary hypertension. Dtsch Arztebl Int 114(5):73
2. Thenappan T, Chan SY, Weir EK (2018) Role of extracellular matrix in the pathogenesis of pulmonary arterial hypertension. Am J Phys Heart Circ Phys 315(5):H1322–H1331
3. Prins KW, Thenappan T (2016) WHO Group I pulmonary hypertension: epidemiology and pathophysiology. Cardiol Clin 34(3):363
4. Schermuly RT, Ghofrani HA, Wilkins MR, Grimminger F (2011) Mechanisms of disease: pulmonary arterial hypertension. Nat Rev Cardiol 8(8):443
5. Simonneau G, Montani D, Celermajer DS, Denton CP, Gatzoulis MA, Krowka M, Williams PG, Souza R (2019) Haemodynamic definitions and updated clinical classification of pulmonary hypertension. Eur Respir J 53:1
6. Buckley MS, Staib RL, Wicks LM (2013) Combination therapy in the management of pulmonary arterial hypertension. Int J Clin Pract 67:13–23
7. Ichida F, Uese KI, Hamamichi Y, Hashimoto I, Tsubata SI, Fukahara K, Murakami A, Miyawaki T (1998) Chronic effects of oral prostacyclin analogue on thromboxane a and prostacyclin metabolites in pulmonary hypertension. Pediatr Int 40(1):14–19
8. Ghofrani HA, Osterloh IH, Grimminger F (2006) Sildenafil: from angina to erectile dysfunction to pulmonary hypertension and beyond. Nat Rev Drug Discov 5(8):689–702
9. Farber HW, Loscalzo J (2004) Pulmonary arterial hypertension. N Engl J Med 351 (16):1655–1665

10. Rabinovitch M, Guignabert C, Humbert M, Nicolls MR (2014) Inflammation and immunity in the pathogenesis of pulmonary arterial hypertension. Circ Res 115(1):165–175
11. Soon EH (2010) Elevated levels of inflammatory cytokines predict survival in idiopathic and familial pulmonary arterial hypertension. Circulation 122(9):920r927
12. Chaouat A, Savale L, Chouaid C, Tu L, Sztrymf B, Canuet M, Maitre B, Housset B, Brandt C, Le Corvoisier P, Weitzenblum E (2009) Role for interleukin-6 in COPD-related pulmonary hypertension. Chest 136(3):678–687
13. Savale L, Tu L, Rideau D, Izziki M, Maitre B, Adnot S, Eddahibi S (2009) Impact of interleukin-6 on hypoxia-induced pulmonary hypertension and lung inflammation in mice. Respir Res 10(1):6
14. Moudgil R, Michelakis ED, Archer SL (2006) The role of K+ channels in determining pulmonary vascular tone, oxygen sensing, cell proliferation, and apoptosis: implications in hypoxic pulmonary vasoconstriction and pulmonary arterial hypertension. Microcirculation 13 (8):615–632
15. Yu Y, Fantozzi I, Remillard CV, Landsberg JW, Kunichika N, Platoshyn O, Tigno DD, Thistlethwaite PA, Rubin LJ, Yuan JXJ (2004) Enhanced expression of transient receptor potential channels in idiopathic pulmonary arterial hypertension. Proc Natl Acad Sci 101 (38):13861–13866
16. Cowan KN, Heilbut A, Humpl T, Lam C, Ito S, Rabinovitch M (2000) Complete reversal of fatal pulmonary hypertension in rats by a serine elastase inhibitor. Nat Med 6(6):698–702
17. Chelladurai P, Seeger W, Pullamsetti SS (2012) Matrix metalloproteinases and their inhibitors in pulmonary hypertension. Eur Respir J 40(3):766–782
18. Sacks RS, Remillard CV, Agange N, Auger WR, Thistlethwaite PA, Yuan JXJ (2006) Molecular biology of chronic thromboembolic pulmonary hypertension. Semin Thorac Cardiovasc Surg 18(3):265–276
19. Alias S, Redwan B, Panzenböck A, Winter MP, Schubert U, Voswinckel R, Frey MK, Jakowitsch J, Alimohammadi A, Hobohm L, Mangold A (2014) Defective angiogenesis delays thrombus resolution: a potential pathogenetic mechanism underlying chronic thromboembolic pulmonary hypertension. Arterioscler Thromb Vasc Biol 34(4):810–819
20. Jeffery TK, Morrell NW (2002) Molecular and cellular basis of pulmonary vascular remodeling in pulmonary hypertension. Prog Cardiovasc Dis 45(3):173–202
21. Gajecki D, Gawrys J, Szahidewicz-Krupska E and Doroszko A (2020) Novel Molecular Mechanisms of Pulmonary Hypertension: A Search for Biomarkers and Novel Drug Targets—From Bench to Bed Site. *Oxidative Medicine and Cellular Longevity, 2020*
22. Maron BA, Leopold JA (2015) Emerging concepts in the molecular basis of pulmonary arterial hypertension: part II: neurohormonal signaling contributes to the pulmonary vascular and right ventricular pathophenotype of pulmonary arterial hypertension. Circulation 131 (23):2079–2091
23. Lai YC, Potoka KC, Champion HC, Mora AL, Gladwin MT (2014) Pulmonary arterial hypertension: the clinical syndrome. Circ Res 115(1):115–130
24. Thompson AR, Lawrie A (2017) Targeting vascular remodeling to treat pulmonary arterial hypertension. Trends Mol Med 23(1):31–45
25. Montani D, Günther S, Dorfmüller P, Perros F, Girerd B, Garcia G, Jaïs X, Savale L, Artaud-Macari E, Price LC, Humbert M (2013) Pulmonary arterial hypertension. Orphanet J Rare Dis 8 (1):97
26. Austin ED, Ma L, LeDuc C, Berman Rosenzweig E, Borczuk A, Phillips JA III, Palomero T, Sumazin P, Kim HR, Talati MH, West J (2012) Whole exome sequencing to identify a novel gene (caveolin-1) associated with human pulmonary arterial hypertension. Circ Cardiovasc Genet 5(3):336–343
27. Eyries M, Montani D, Girerd B, Perret C, Leroy A, Lonjou C, Chelghoum N, Coulet F, Bonnet D, Dorfmüller P, Fadel E (2014) EIF2AK4 mutations cause pulmonary veno-occlusive disease, a recessive form of pulmonary hypertension. Nat Genet 46(1):65–69

28. Julian L, Olson MF (2014) Rho-associated coiled-coil containing kinases (ROCK) structure, regulation, and functions. Small GTPases 5(2):e29846
29. Yaoita N, Satoh K, Satoh T, Sugimura K, Tatebe S, Yamamoto S, Aoki T, Miura M, Miyata S, Kawamura T, Horiuchi H (2016) Thrombin-activatable fibrinolysis inhibitor in chronic thromboembolic pulmonary hypertension. Arterioscler Thromb Vasc Biol 36(6):1293–1301
30. Amano M, Nakayama M, Kaibuchi K (2010) Rho-kinase/ROCK: a key regulator of the cytoskeleton and cell polarity. Cytoskeleton 67(9):545–554
31. Wang XY, Mo D, Tian W, Liu XX, Zhou YG, Sun Y, Feng YD, Xiao X, Hao XW, Zhang HN, Li C (2019) Inhibition of RhoA/ROCK signaling pathway ameliorates hypoxic pulmonary hypertension via HIF-1α-dependent functional TRPC channels. Toxicol Appl Pharmacol 369:60–72
32. Wei H, Zhang D, Liu L, Xia W, Li F (2019) Rho signaling pathway enhances proliferation of PASMCs by suppressing nuclear translocation of Smad1 in PAH. Exp Ther Med 17(1):71–78
33. Yang G, Caldwell RB, Yao L, Romero MJ, Toque HA, Caldwell RW (2010) The role of RhoA/rho kinase pathway in endothelial dysfunction. J Cardiovasc Dis Res 1(4):165–170
34. Antoniu SA (2012) Targeting RhoA/ROCK pathway in pulmonary arterial hypertension. Expert Opin Ther Targets 16(4):355–363
35. Xue C, Sowden M, Berk BC (2017) Extracellular cyclophilin a, especially acetylated, causes pulmonary hypertension by stimulating endothelial apoptosis, redox stress, and inflammation. Arterioscler Thromb Vasc Biol 37(6):1138–1146
36. Satoh K (2015) Cyclophilin a in cardiovascular homeostasis and diseases. Tohoku J Exp Med 235(1):1–15
37. Miller LH, Ackerman HC, Su XZ, Wellems TE (2013) Malaria biology and disease pathogenesis: insights for new treatments. Nat Med 19(2):156–167
38. Satoh K, Satoh T, Kikuchi N, Omura J, Kurosawa R, Suzuki K, Sugimura K, Aoki T, Nochioka K, Tatebe S, Miyamichi-Yamamoto S (2014) Basigin mediates pulmonary hypertension by promoting inflammation and vascular smooth muscle cell proliferation. Circ Res 115 (8):738–750
39. Sunamura S, Satoh K, Kurosawa R, Ohtsuki T, Kikuchi N, Elias-Al-Mamun M, Shimizu T, Ikeda S, Suzuki K, Satoh T, Omura J (2018) Different roles of myocardial ROCK1 and ROCK2 in cardiac dysfunction and postcapillary pulmonary hypertension in mice. Proc Natl Acad Sci 115(30):E7129–E7138
40. Satoh K, Kikuchi N, Satoh T, Kurosawa R, Sunamura S, Siddique MAH, Omura J, Yaoita N, Shimokawa H (2018) Identification of novel therapeutic targets for pulmonary arterial hypertension. Int J Mol Sci 19(12):4081
41. Satoh K, Nigro P, Berk BC (2010) Oxidative stress and vascular smooth muscle cell growth: a mechanistic linkage by cyclophilin a. Antioxid Redox Signal 12(5):675–682
42. Fisslthaler B, Fleming I (2009) Activation and signaling by the AMP-activated protein kinase in endothelial cells. Circ Res 105(2):114–127
43. Noda K, Nakajima S, Godo S, Saito H, Ikeda S, Shimizu T, Enkhjargal B, Fukumoto Y, Tsukita S, Yamada T, Katagiri H (2014) Rho-kinase inhibition ameliorates metabolic disorders through activation of AMPK pathway in mice. PLoS One 9(11):e110446
44. Nagata D, Kiyosue A, Takahashi M, Satonaka H, Tanaka K, Sata M, Nagano T, Nagai R, Hirata Y (2009) A new constitutively active mutant of AMP-activated protein kinase inhibits anoxia-induced apoptosis of vascular endothelial cell. Hypertens Res 32(2):133–139
45. Igata M, Motoshima H, Tsuruzoe K, Kojima K, Matsumura T, Kondo T, Taguchi T, Nakamaru K, Yano M, Kukidome D, Matsumoto K (2005) Adenosine monophosphate-activated protein kinase suppresses vascular smooth muscle cell proliferation through the inhibition of cell cycle progression. Circ Res 97(8):837–844
46. Omura J, Satoh K, Kikuchi N, Satoh T, Kurosawa R, Nogi M, Otsuki T, Kozu K, Numano K, Suzuki K, Sunamura S (2016) Protective roles of endothelial AMP-activated protein kinase against hypoxia-induced pulmonary hypertension in mice. Circ Res 119(2):197–209

47. Satoh K, Fukumoto Y, Nakano M, Sugimura K, Nawata J, Demachi J, Karibe A, Kagaya Y, Ishii N, Sugamura K, Shimokawa H (2009) Statin ameliorates hypoxia-induced pulmonary hypertension associated with down-regulated stromal cell-derived factor-1. Cardiovasc Res 81 (1):226–234
48. Rafikova O, Meadows ML, Kinchen JM, Mohney RP, Maltepe E, Desai AA, Yuan JXJ, Garcia JG, Fineman JR, Rafikov R, Black SM (2016) Metabolic changes precede the development of pulmonary hypertension in the monocrotaline exposed rat lung. PLoS One 11(3):e0150480
49. Chen C, Luo F, Wu P, Huang Y, Das A, Chen S, Chen J, Hu X, Li F, Fang Z, Zhou S (2020) Metabolomics reveals metabolite changes of patients with pulmonary arterial hypertension in China. J Cell Mol Med 24(4):2484–2496
50. Ryan JJ, Archer SL (2015) Emerging concepts in the molecular basis of pulmonary arterial hypertension: part I: metabolic plasticity and mitochondrial dynamics in the pulmonary circulation and right ventricle in pulmonary arterial hypertension. Circulation 131(19):1691–1702
51. Fijalkowska I, Xu W, Comhair SA, Janocha AJ, Mavrakis LA, Krishnamachary B, Zhen L, Mao T, Richter A, Erzurum SC, Tuder RM (2010) Hypoxia inducible-factor1α regulates the metabolic shift of pulmonary hypertensive endothelial cells. Am J Pathol 176(3):1130–1138
52. Qin L, D'Alessandro-Gabazza CN, Aoki S, Gil-Bernabe P, Yano Y, Takagi T, Boveda-Ruiz D, Ramirez Marmol AY, San Martin Montenegro VT, Toda M, Miyake Y (2010) Pulmonary hypertension is ameliorated in mice deficient in thrombin-activatable fibrinolysis inhibitor. J Thromb Haemost 8(4):808–816
53. Antovic JP, Blombäck M (2002) Thrombin-activatable fibrinolysis inhibitor antigen and TAFI activity in patients with APC resistance caused by factor V Leiden mutation. Thromb Res 106 (1):59–62
54. Miljić P, Heylen E, Willemse J, Đorđević V, Radojković D, Čolović M, Elezović I, Hendriks D (2010) Thrombin activatable fibrinolysis inhibitor (TAFI): a molecular link between coagulation and fibrinolysis. Srp Arh Celok Lek 138(suppl. 1):74–78
55. Bouma BN, Meijers JCM (2003) Thrombin-activatable fibrinolysis inhibitor (TAFI, plasma procarboxypeptidase B, procarboxypeptidase R, procarboxypeptidase U). J Thromb Haemost 1 (7):1566–1574

Targeting Molecular and Cellular Mechanism of Influenza A Virus

20

Gayathri Gopal, Shibi Muralidar, and Senthil Visaga Ambi

Abstract

The respiratory epithelium is the erudite barrier that serves as the interface between the environmental factors and the host immune system. It maintains homeostasis and shields the lungs against foreign antigens and pathogens. Influenza is one of the highly contagious viral infections of the host respiratory tract mucosa caused by influenza A virus (IAV). IAV is considered a major life-threatening human pathogen that potentially causes epidemics and pandemics with high morbidity and mortality. IAV is capable of infecting a broad range of birds and mammals including humans. Due to this broad range of infectivity on both avian and mammalian organisms, zoonotic spillovers from any of these infected organisms can potentially end up in a pandemic with numerous detrimental consequences for the world population. Further, the emergence of H1N1, H5N1, and H7N9 (avian-origin) influenza viruses causing notable lethal cases has demonstrated the limitations in the existing drugs and strategies against IAV. Thus, there is an imperative need for novel influenza therapeutics and a new mechanism of action to combat the tenacious threat of IAV on the world population. In this review, we focus on the defense machinery of the respiratory epithelium and its response to IAV. We also attempted to enlighten the reader's knowledge on the underlying molecular and cellular mechanisms of IAV in lung

Gayathri Gopal and Shibi Muralidar contributed equally with all other contributors.

G. Gopal · S. Muralidar · S. V. Ambi (✉)
Biopharmaceutical Research Lab, Anusandhan Kendra-1, SASTRA Deemed-to-be-University, Thanjavur, India

Department of Bioengineering, School of Chemical and Biotechnology, SASTRA Deemed-to-be-University, Thanjavur, India
e-mail: senthilvisagaambi@scbt.sastra.edu

K. Dua et al. (eds.), *Targeting Cellular Signalling Pathways in Lung Diseases*,
https://doi.org/10.1007/978-981-33-6827-9_20

injury highlighting the potential drug candidates with their mechanism of action toward the end of this chapter.

Keywords

IAV life cycle · Lung injury · Alveolar epithelial cells · Cytokines · Apoptosis · Pharmacological treatment

20.1 Introduction

Influenza is one of the highly contagious viral infections caused by influenza A virus (IAV) through infecting the host respiratory tract mucosa. Being the major cause of seasonal or pandemic flu, IAV has imposed an enormous burden on both the economy and public health throughout the world. IAV is considered a major life-threatening human pathogen that potentially causes epidemics and pandemics with high morbidity and mortality [1–4]. With wild waterfowl as its primary reservoir, IAV is capable of infecting a broad range of birds and mammals including humans. Due to this broad range of infectivity on both avian and mammalian organisms, zoonotic spillovers from any of these infected organisms can potentially end up in a pandemic with numerous detrimental consequences for the world population. Moreover, the continuous evolvement of IAV due to its adaptation, antigenic mutation, and reassortment leads to the unexpected emergence of highly pathogenic virulent strains ultimately resulting in local epidemics or global pandemics. Some examples of such outbreaks include H1N1 (Spanish flu, 1918), H2N2 (Asian flu, 1957), H3N2 (Hong Kong flu, 1968), H5N1 (bird flu, 2005), H1N1 (swine flu, 2009), and H7N9 (bird flu, 2013) [3, 5].

Seasonal epidemics caused by the different strains of IAV infect 3–5 million people and kill about 250,000–500,000 people annually. In case of pandemic caused by the same IAV, the number of deaths can surge up to millions, especially during the pandemic years, 1957–1958 (H2N2, Asian flu) and 1918–1919 (H1N1, Spanish flu), where the deaths ranged from 1 million to 50 million, respectively [2, 6, 7]. Further, the emergence of H1N1, H5N1, and H7N9 (avian-origin) influenza viruses causing notable lethal cases has demonstrated the limitations in the existing drugs and strategies against IAV. Currently, the two main strategies available for controlling IAV include vaccination and small molecule anti-influenza drugs. However, selecting an apposite viral strain for the production of annual trivalent or quadrivalent vaccines is considered to be a formidable task due to rapid antigenic shift and drift in IAV [2, 3]. Thus, there is an imperative need for novel influenza therapeutics and a new mechanism of action to combat the tenacious threat of IAV on the world population. This chapter will enlighten the reader's knowledge on the underlying molecular and cellular mechanisms of IAV in lung injury.

20.2 A Glance on IAV and Its Replication

20.2.1 Biology of Influenza A Virus

Influenza viruses are enveloped RNA viruses that fall under the family *Orthomyxoviridae* and have a single-stranded, negative-sense, segmented RNA genome. They are classified into three types, namely, influenza A virus (IAV), influenza B virus (IBV), and influenza C virus (ICV), of which IAV was found to be highly pathogenic compared to the other two types [6, 8–10]. Classification of these types (IAV, IBV, and ICV) is due to their immunologically diverse matrix protein and nucleoprotein antigens. The annual influenza epidemics and pandemics are mainly contributed by IAV and IBV through antigenic variation triggered by random mutation (antigenic drift) and reassortment (antigenic shift), whereas ICV are only responsible for very mild infections. Further, the host range of ICV is restricted to humans and pigs, and in the case of IBV, it is restricted to humans and seals. However, in the case of IAV, it infects a variety of animals apart from humans which include horses, seals, pigs, dogs, ducks, chickens, and whales [8, 11–14].

Both IAV and IBV are pleomorphic with a diameter of 80–120 nm and contain eight negative-sense RNA segments. Apart from the influenza virus classification, IAV is also classified into subtypes based on its genetic and antigenic dissimilarities in their two membrane glycoproteins, namely, hemagglutinin (HA) and neuraminidase (NA). The 8 single-stranded negative-sense RNA segments encode a total of 12 proteins which include 8 structural proteins, namely, HA, NA, M1 (matrix protein), M2 (ion channel protein), 3 polymerases (PB1, PB2, PA), and nucleoprotein (NP), and 4 nonstructural proteins (NS), namely, NS1, NS2, PB1-F2, and N40 [6, 9, 15, 16].

In IAVs, so far 9 NA (N1–9) and 16 HA (H1–16) subtypes have been isolated from waterfowls. The sequence identity between the H1–16 and N1–9 subtypes was found to lie between 30% and 60%. Subtypes H17 and H18 of HA and N10 and N11 of NA are two additional subtypes that are recently identified in bats. Apart from the numerous possible combinations in the available subtypes, only three subtype combinations (H1N1, H2N2, and H3N2) have been constantly infecting the human population [8, 17–20]. All eight gene segments of IAV are encapsidated by their nucleoprotein, and the polymerase complexes containing three polymerase proteins PA, PB1, and PB2 are placed at the nucleocapsid ends. Further, the helical capsids are surrounded by M1 proteins which are then encircled by a host-derived lipid bilayer where the membrane glycoproteins (HA and NA) and M2 proteins are embedded. Despite the virion shape, HA was found to be the most abundant membrane glycoprotein situated in the viral envelope which is followed by NA and M2. The typical structure of IAV is illustrated in Fig. 20.1 [21, 22].

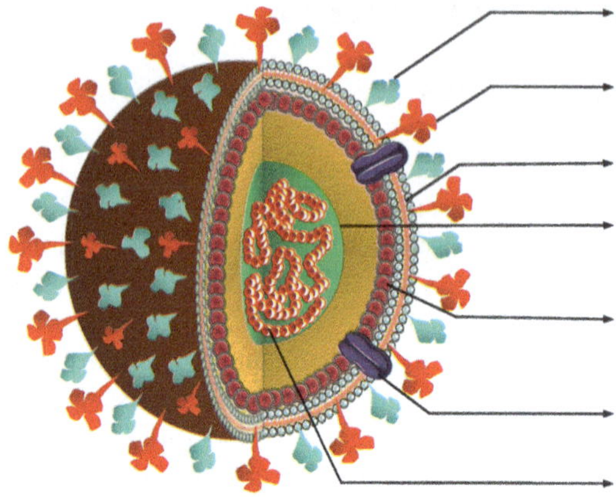

Fig. 20.1 Structure of IAV

20.2.2 Life Cycle of Influenza A Virus

Infection in the lower respiratory tract of humans can result in the deluging of alveolar compartments leading to the development of ARDS (acute respiratory distress syndrome) which may ultimately result in death by respiratory failure. IAV primarily targets and infects the airway and alveolar epithelial cells that are located on the surface of the respiratory tract. The viruses target alveolar epithelial cells due to the presence of sialic acid residues (SA residues) which is known to be the functional receptor of the virus. Infection in these alveolar epithelial cells can lead to the induction of necrotizing bronchitis and bronchiolitis along with interstitial pneumonia leading to epithelial damage [23, 24]. Infection of the influenza virus in the pulmonary mucosa can play a major influencing role on the airway functioning, and further, the replication of influenza virus in the airway epithelium can possibly disrupt the normal cellular morphology and functions. Also, the exclusive viral entry and release through the apical surface of airway epithelial cells and the great number of progeny virions that are released from the apical membrane can easily alter the optimal state and composition of the apical membrane [25–27].

20.2.2.1 Viral Entry

IAV's first step in its life cycle is the initiation of infection by binding to the potential host cells. HA protein present in the viral envelope is a homotrimer which is responsible for the formation of spike in the viral lipid membrane. HA0 (a precursor of HA) contains two subunits, namely, HA1 and HA2. HA1 subunit contains a receptor-binding domain, whereas the HA2 subunit comprises the fusion peptide. Both the subunits HA1 and HA2 are linked by disulfide bonds. Upon reaching the host cell, the spikes of HA use its receptor-binding site to get attached to the surface glycoconjugates which contain terminal SA residues [15, 28, 29]. After this attachment, IAV scans the cell surface using the sialidase function of NA to find suitable sialylated receptor, removes local SAs, and also unshackles the nonproductive associations of HA. Generally, HAs from human IAVs recognize α-2,6-linked SAs (predominantly found on epithelial cells (EC) of human) that results in a "bent" presentation, whereas HAs from avian IAVs recognize α-2,3-linked SAs (predominantly found on EC of duck intestine) resulting in a "linear" presentation. Although the binding preference of HA receptor to SA linkages is not crucial for infection, it is the most indispensable factor for IAV's host range and transmission [15, 28, 30–33].

20.2.2.2 Endocytosis and Fusion

Following the binding of HA to the host cell's SA residues, IAVs enter the host cell through the process of receptor-mediated endocytosis in which the virus gets into the cell in an endosome. The primary internalization pathway for IAV is through clathrin-mediated endocytosis which involves dynamin and epsin-1 (adaptor protein); however, clathrin-independent endocytosis and macropinocytosis are also found to be a possible mechanism for IAV internalization [15, 28, 29]. Once inside the host cell, the low pH (around 5 to 6) in the late endosome initiates the fusion of viral and endosomal membranes. This induction of large irreversible conformational change in HA due to the low pH dismisses the N-terminus of HA2 subunit (fusion peptide) and makes the HA2 fusion peptide exposed [34–38]. Further, the exposed HA2 fusion peptide gets itself inserted into the endosomal membrane which in turn leads to the fusion of viral and endosomal membranes by bringing both into contact with each other. The acidic environment in the late endosome is not only crucial to induce conformational changes and fusion of viral-endosomal membranes but also activates the M2 ion channel (a type III transmembrane protein). Opening of M2 ion channel leads to the acidification of the viral core leading to the disassociation of M1 protein from viral ribonucleoprotein complexes (vRNPs), and this acidic environment releases the packaged vRNPs to enter freely into the host cell's cytoplasm [15, 28, 36, 39–41]. Further, this disassociation of vRNPs from M1 protein subsequently results in the nuclear import of vRNPs facilitated by the cellular nuclear import factors like importin-α and importin-β [28].

20.2.2.3 Viral Replication, Transcription, and Translation

Unlike other negative-sense RNA viruses, the replication and transcription of IAVs take place in the host cell nucleus. This process is facilitated by the trimeric viral polymerase complex which comprises PB1, PB2, and PA subunits. The viral RNA replication initiates with the synthesis of (+) ssRNA (positive-sense copy of vRNA) which is also known as complementary RNA (cRNA). Further, the synthesized cRNA is then copied in order to produce a huge amount of vRNA. The synthesis of genomic (−) ssRNA takes place by using (+) ssRNA as a template strand. This synthesized (−) ssRNA is then packed in new virions or transported to the host cytosol where they can act as viral mRNA. The viral RNA transcription begins with the binding of PB2 protein to the 5′-cap of host mRNAs which is followed by the endonucleolytic cleavage activity of PA. Further, this endonuclease activity of PA leads to the production of cellular capped primer for viral mRNA synthesis [28, 42–

45]. Synthesis of viral mRNAs is mediated by the polymerase activity of PB1 protein, and then cooperation between the crucial proteins such as PB1, PB2, NP, M1, and PA leads to the formation of new vRNPs. These vRNPs are further transported from the host nucleus to the cytoplasm by the viral NS2/NEP protein and cellular CRM1 proteins for the purpose of nuclear export and incorporation of vRNPs into the virions [46, 47].

Influenza viruses use the host cell translation machinery to translate their viral mRNAs, and as a result, multiple interactions of viral mRNAs with the cellular transcriptional factors like eukaryotic initiation factor-4A (eIF4A), eIF4E, and eIF4G take place. Upon infection, the host cell protein synthesis will be suppressed, and IAV's viral mRNAs will be preferentially translated. Influenza virus depends on the host cell's splicing machinery to express its proteins, and at the same time, it precludes the host cell from using its splicing machinery and thereby blocks the procession of host cell mRNA. The interaction of NS1 with cleavage and polyadenylation specificity factor (CPSF) and poly (A)-binding protein II (PABII), and the interaction of viral polymerase complexes to C-terminal domain of cellular DNA-dependent RNA polymerase II (Pol II) are known to be the possible reason for the aforementioned inhibition of host cell mRNA synthesis. Subsequently, the NP and viral polymerase subunit proteins are imported into the nucleus through the nuclear localization signals. Further, M1, NS1, and NEP/NS2 are also imported into the nucleus to implement their respective roles in nuclear export of vRNPs [28, 29].

20.2.2.4 Export of vRNPs, Assembly, and Budding

Generally, negative-sense vRNPs are exported from nucleus to cytoplasm. This export of vRNPs is mediated by the CRM1-dependent pathway where the vRNPs are exported through the nuclear pores. Not only cellular CRM1 protein but IAV also depends on viral NS2/NEP and M1 proteins to export the newly synthesized vRNP complexes. M1 protein was found to directly interact with the vRNPs through the C-terminal protein end. Also, the M1 protein's N-terminal portion can bind to NEP, and NEP can potentially bind to CRM1 which is accompanied by GTP hydrolysis (normally happens in the CRM1-dependent export pathway). Binding of M1 protein to the negative-sense vRNPs and NEP, is followed by the binding of NEP to CRM1, and this double binding event ultimately leads to a "daisy-chain" complex that results in the vRNPs' export out of the nucleus. Further, the YB-1 (Y-box binding protein 1) is also found to be associated with vRNPs in the nucleus which is likely to be exported from the nucleus along with vRNPs. This YB-1 protein facilitates the association of vRNPs with microtubules for the further transportation of vRNPs to the plasma membrane [48–52]. On the other hand, M2, NA, and HA proteins are transported via the Golgi apparatus to the surface of the host cell in which the preparation of viral envelope takes place. The cleavage of HA proteins into HA1 and HA2 subunit mediated by the cellular furin-like proteases in the trans-Golgi network plays a crucial role in determining the virulence of influenza viruses [42, 53–55].

Once the vRNPs are out of the nucleus, the final process in the life cycle is to form viral particles and exit the cell. IAV being an enveloped virus depends on the host cell plasma membrane for the formation of viral particles that exit the cell to infect neighboring cells. Not only the vRNPs but also the viral proteins like HA, NA, and M2 that are present within the viral lipid bilayer are necessary to form a viral particle. The M1 protein present underneath the lipid bilayer plays a significant role in the assembly process due to its notable interactions with lipid membranes, NEP/NS2, and vRNPs. Further, the M2 protein also plays a crucial role in mediating the membrane scission and particle release. At last, the enzymatic activity of NA proteins removes the sialic acids from host cells, thereby allowing the virion release from the plasma membrane [28, 29, 42, 56]. A detailed pictorial representation of the aforementioned life cycle of IAV is given in Fig. 20.2.

20.3 Molecular Pathogenesis of Influenza Virus Infection

The interaction of viral proteins with the host immune cells is responsible for the pathogenesis of influenza virus infection. Both the viral factors and the host immune system indicate the molecular pathogenesis [57]. In recent years, it has been stated that host microflora also plays a significant role in activating the pulmonary immune cells to induce virus-specific immune responses. Hence, the immune-rich environment which surrounds the host and virus is the primary factor that contributes to viral pathogenesis. Influenza A virus infection in the respiratory epithelial cells destroys the host cell pre-mRNAs and further inhibits the translation of mRNAs leading to the death of the host cells either by cytolytic or apoptotic mechanisms [58]. Virus-infected cells adapt different immune response mechanisms to limit the speed of viral replication. As a response to the immune mechanism triggered by the incoming virus, various transcription factors are activated which is accompanied by the production of chemokines and pro-inflammatory and anti-inflammatory cytokines. The recruitment of other immune cells by cytokines, to the site of infection, induces inflammatory responses which on later stages initiate anti-inflammatory effects against the virus [59].

When IAV infects the epithelial cells of the respiratory tract or the alveolar macrophages, the Toll-like receptors (TLR) and retinoic acid-inducible gene I (RIG-I) of the host cell recognize the single-stranded RNA of the influenza virus. The signals produced by TLR and RIG-I induce the production of interferon, cytokines, and transcription factors which in turn activate the antiviral immune responses [60]. Conceptually, during the course of IAV infection, there are three stages of events, with respect to the responses that occur simultaneously throughout infection period. The first is the spread of the virus in the alveolar epithelium; the second is the immediate response of the host immune cells through innate and adaptive immune mechanisms to eliminate or neutralize the virus which can prevent the severity of the infection and help viral clearance. On the other hand, the immune responses to the virus entry can also induce significant damage to the infected respiratory epithelium and endothelium. The third is the long-term immunity developed against the viral strain which is also accompanied by the regeneration of damaged lung tissues [61].

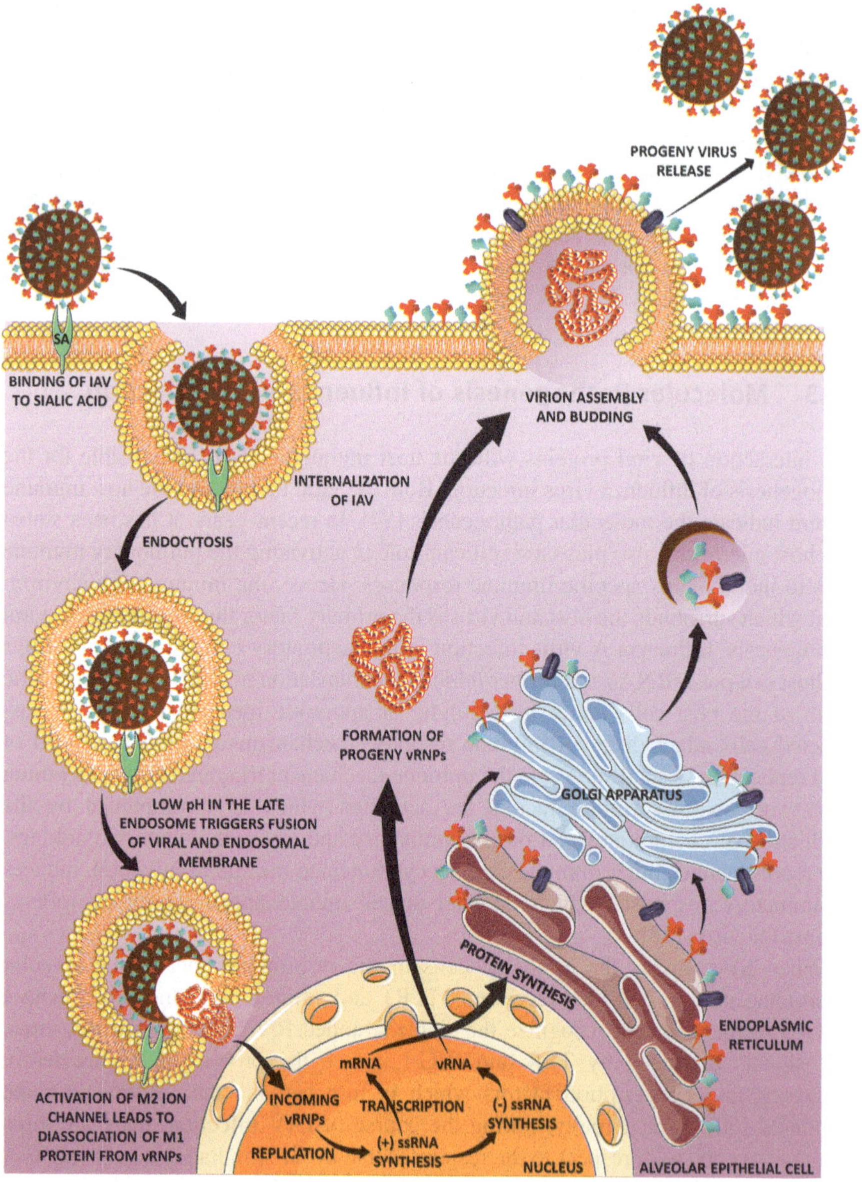

Fig. 20.2 Life cycle of IAV

20.3.1 Host Cell Immune Response Upon IAV Infection

Host immune cells start functioning after the detection of the virus. The first line of defense mechanism against the invading virus is provided by the host mucosal immune system which can prevent the adhesion of the virus to the susceptible immune cells. Host innate immunity includes phagocytic cells, interferons (IFNs), and pro-inflammatory cytokines and is capable of multiple mechanisms in fighting the invading virus [62]. Adaptive immunity in the host system is mediated by B lymphocytes and T lymphocytes, in the motive of neutralizing or eliminating the virus particles. In contrast, IAVs use a plethora of strategies to demonstrate a successful infection in the host cell and to escape the detection and clearance by immune cells. Upon detection of IAV infection, natural killer (NK) cells, neutrophils, dendritic cells, and macrophages infiltrate to the site of infection [57].

NK cells interact with the viral protein carrying dendritic cells and macrophages to secrete cytokines and chemokine and restrict the infection of IAV by lysing the infecting cells. Neutrophils are important innate immune cells that are hired by the pro-inflammatory cytokines to the site of infection via vascular endothelium. They are involved in the viral clearance through phagocytosis and degranulation. Further, they are helpful in adaptive immunity to regulate influenza-specific $CD8^+$ cells to the site of infection. The role of dendritic cells (DCs) is to recognize the virus. During IAV infection, DCs transport from the lungs to the lymph nodes to present the antigens to the T cells. T cells and B cells are key elements in elucidating immune response against IAV infection. Cytotoxic T lymphocytes (CTLs) defend IAV infection through activated cytokines and effectors. CTLs are also involved in class I MHC-mediated lysis of virus-infected cells. B cells are activated when $CD4^+$ cells interact with IAV-infected epithelial cells via class II MHC binding and thus promote antibody production [63].

When IAV attacks the host cell, the innate immune response provides a barrier to the incoming viral protein. It is the first line of defense mechanism which triggers the pro-inflammatory response. On the other hand, adaptive immunity has a crucial role in eliminating the pathogen during the later stages of infection. In addition to the two typical immune mechanisms acquired for the viral infection, respiratory mucosal immunity is triggered in the infected mucosal tissues. Despite several immune mechanisms adapted to restrict viral replication in the host cell, IAVs have developed various strategies to escape host immune responses and to exhibit successful infection [64].

20.3.1.1 Innate Immune Response Against IAV Infection

The innate immune response is the primary rapid response. During IAV infection, viral conserved components called pathogen-associated molecular patterns (PAMPs) are recognized by host pathogen recognition receptors (PRRs), such as RIG-I and TLR. This leads to the innate immune signaling resulting in the production of various cytokines and activation of antiviral molecules. The invading viral protein is recognized by the host receptor and discriminates them from the self-molecules. Among various PRRs, RIG-I is one of the primary receptors that recognizes the intracellular ssRNA and transcriptional intermediates of IAVs in the infected host cells. This in turn triggers the activation of caspase, leading to the induction of transcription factors including IRFs and NF-κB [57, 63–65].

In addition to the above transcription factors, TLRs expressed on the endosomes and lysosomes also recognize nucleic acid derived from IAVs. Further, it was demonstrated that TLR3 may be responsible for recognizing unidentified RNA structures present in the phagocytized cells during IAV infection [66]. IAV infection induces the activation of lung epithelial-associated immune cells and related soluble factors. More of ciliated cells and less of secretory cells and basal cells make up the human respiratory epithelial layer. Secretory cells, especially goblet cells, secrete mucus components and antiviral factors to evade primary barrier toward the viral protein. Similarly, ciliated cells aid in the removal of viral proteins and virus-engulfed debris via the mucociliary elevator to expel them from the respiratory tract. As a response to viral infection, NOD-like receptor family, pyrin domain-containing 3 (NLRP3) and NLR apoptosis inhibitory protein 5 were activated [17, 58, 65].

After the identification of viral components by the immune cells, the transcription factors including NF-κB and IRFs are activated, and this readily triggers the transcription of IFNs and other pro-inflammatory cytokines to eliminate the viral antigens. In response to the stimulation of NLRP3 inflammasome by IAV-associated M2 ion channel and PB1-F2, IL-1 secretion is stimulated. IFNs bind to receptors, resulting in upregulation of multiple interferon-stimulated genes (ISGs). These ISGs can modify the different steps of IAV life cycle, for instance, viral entry into host cells can be restricted via the involvement of Mx family, interferon-induced transmembrane (IFITM) protein family, cholesterol 25-hydroxylase (CH25H), and TRIM proteins [66].

20.3.1.2 Adaptive Immune Response Against IAV Infection

On detecting the influenza infection, DCs induce the IFNs and cytokines for their maturation into APCs (antigen-presenting cells), to initiate the T cell-mediated immune responses. The viral protein binds to the DCs, and the protein-bearing DCs get activated to recruit naive CD4+ T cells which get differentiated into Th1, Th2, Th7, regulatory T cells (Treg cells), follicular helper T cells, and killer cells. The signals produced by DCs present the antigen to the MHC class II molecules and induce the differentiation of Th2 cells that is regulated by CD4+ T cell. The mature Th2 cells trigger B cell activation and thereby maintain immunological memory [67]. During the early stages of infection, T cells expand independent of the TCR mechanism to efficiently eliminate the IAV-infected airway epithelial cells. CD8+ T cells are key components for virus clearance in adaptive immunity. When the DCs present the virus particles to the MHC class I molecule, CD8+ T cells get activated to undergo rapid expansion, differentiation, and migration to the infected sites. Generally, CD8+ T cells play a significant role in establishing adaptive immune responses through cytotoxic effects on the virus particles by cytotoxic T lymphocytes (CTLs) [68]. CTLs are responsible for the destruction of virus into cytotoxic granules containing perforin and granzymes (GrA and GrB) through apoptosis thereby inhibiting further IAV replication in the host cell. This process is regulated by the production of cytokines, such as TNF, FASL, and TRAIL. The recruitment of death receptors by cytokines will help enhance the cytotoxicity of IAV. Furthermore, CD8 + T cells are capable of restoring the IAV-specific memory CTLs against specific epitopes of IAV. In addition to the cytotoxic effects, CTLs also prevent the spread of

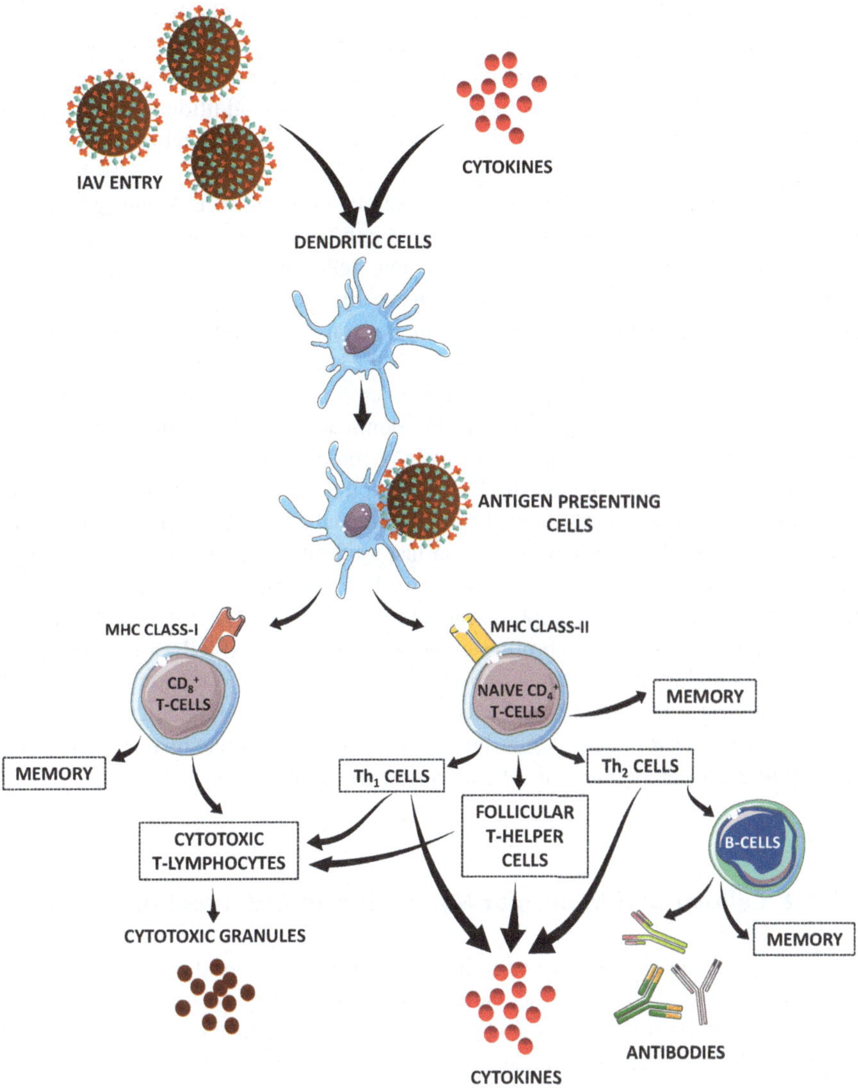

Fig. 20.3 Immune response against IAV infection

the virus from URT to the lungs. It is also evident that the process of autophagy has an important role in establishing memory CD8+ T cells [69].

CD4+ cells induce the Th2 cells to activate B cells. B cells generate specific antibodies against the viral proteins to eliminate them from the host cells and restore the memory for secondary infection. The overall mechanism of adaptive immune response against IAV infection is illustrated in Fig. 20.3. The function of antibodies could differ from each other: IgG is responsible for inhibition of IAV pathogenesis, while IgA blocks IAV transmission. Further, IAV-specific antibody-dependent cell-mediated cytotoxicity (CDCC) finds importance in providing cross-protection against IAV infection. However, mutation of HA and/or NA through antigenic shift and drift can help the virus to escape the host. Also, additional glycosylation on H5 HA is responsible for the viral inhibition of neutralization by specific antibodies [57, 63, 68].

With mucosal tissues being the primary site of virus entry, IgA and IgM in the mucosal tissue are the predominant antibodies exhibited to neutralize the pathogens and to prevent the viral entry and replication. IgA antibodies which are specific against HA and NA proteins of influenza virus are the primary neutralizing antibodies [70]. Generally, primary immune response for viral infections occurs in the lymphoid tissues, whereas the periphery is the site of secondary responses. After the viral entry, IgM is produced dominantly to exhibit primary responses against the invading viral proteins followed by the dominance of IgG during secondary responses. Production of higher levels of IgM could help in the rapid clearance of virus. Initial systematic response by IgM against the virus is usually required for the subsequent IgG antibody response. Generally the half-life of antibodies against any pathogen is short, but antibody titers can last a lifetime with the help of antibody-secreting cells (ASCs) which are long-lived. Furthermore, calcium modulator, transmembrane activator, cyclophilin ligand interactor (TACI) cytokines, B lymphocyte stimulator (BLyS), and a proliferation-inducing ligand (APRIL) are also responsible in producing the required immune responses and protection and against secondary viral infection. These factors can be targeted to maintain the antibody titer by increased survival of ASCs during the secondary immune responses, thereby leading to enhanced counterpart of subsequent infections [8, 65, 70].

20.3.2 Cellular and Molecular Mechanism of IAV Infection

When the virus infects the host cell, the goal is to replicate itself and attack a more number of cells. A viral enzyme, called polymerase, serves as the key to produce more copies in the host system [71]. It copies the genetic material of the virus and navigates the host cell mechanism toward the synthesis of viral proteins. This is made possible by utilizing the "cap" from the host cell wherein the viral polymerase binds to the mRNA of the host cell. The cap is cut off from the host cell mRNA and is added to the beginning of the viral mRNA by a process called "cap snatching." But the mechanism behind the enzyme subunits performing this action remains unclear [9]. The major mystery behind the influenza virus is whether the viral specificity toward the host cell receptor can generate a prevalent strain of virus [62]. The viral HA of the glycoprotein binds to the SA residues of the host cells. Also, specific amino acid changes in HA have been found remarkable in the specificity of the sialic acid receptor and the pathogenicity. The HA of the IAV virus typically binds to the α2,6-linked sialic acids found in the upper respiratory tract of humans which is the primary site of infection [61].

20.3.2.1 Host Cell Protein Synthesis

IAV infection in host cells is efficient enough to suppress the gene expression of the host cell. The viral polymerase protein complex binds to the 5′ ends of the host mRNAs and cleaves it. Now, the 5′ cap of the host mRNAs is used as a primer for viral mRNA synthesis [71]. IAV-encoded NS1 protein inhibits the splicing of cellular pre-mRNAs and blocks the nuclear export of host mRNAs. IAV-specific mRNAs are translated efficiently to the cytoplasm and ensure viral protein synthesis. A crucial element in the translation of IAV mRNAs is the sequences of 5′ untranslated end of viral mRNAs. Another important mechanism for the maintenance of viral mRNA translation is the downregulation of protein kinase receptor (PKR) activity. PKR phosphorylates eukaryotic initiation factor-2α to reduce the translation of viral mRNAs to the cytoplasm. IAV demonstrates two different ways to escape the inhibitory effects by PKR activation. First is the interaction of NS1 protein with PKR that disturbs the activation of PKR. Second is the activation of chaperone-associated protein (p58IPK) upon IAV infection which can interfere with the dimerization and activation of PKR. Controlling the viral protein synthesis and arrest of host protein synthesis leads to cell death within 20–40 h of infection [9].

20.3.2.2 Production of Cytokines and Activation of Transcription Factors in IAV Infection

As a response to the IAV infection, lung epithelial cells and leukocytes infection produce chemokines and cytokines. The respiratory epithelial cells and the macrophages are the targets of IAV infection. During the course of infection, a limited number of cytokines are produced by infected epithelial cells [63]. The cytokines include IFN-α/β, IL-1β, IL-6, and TNF-α. IAV-infected macrophages produce MIP-1α, MIP-1β, RANTES, MCP-1, MCP-3, and IP-10 and limited IL-8. Epithelial cells secrete IL-8, RANTES, and MCP as a response to IAV infection. The chemokines help in the recruitment of neutrophils to the site of infection. Dendritic cells have been found to secrete large quantities of IL-12 in response to viral infection. IAV infection also induces IL-15 production by peripheral blood mononuclear cells (PBMCs) [7]. It is elucidated that antigen-presenting cells can produce high levels of antiviral and immunostimulatory cytokines during the virus entry. Viral infection also initiates the activation of different transcription factors which induces the expression of cytokine and chemokine. NF-κB, AP-1, IRFs, STATs, and IL-6 are predominantly activated during the infection. Activation of NF-κB is detected within 1 hour of infection and activates the replication of the virus and viral protein synthesis [72]. The production of IAV-induced cytokines and chemokines is given in Fig. 20.4.

NF-κB is found to be activated due to ER stress caused by the expression of a single IAV gene. MAP kinases are important regulators of the gene expression of cytokines, and the Map kinase superfamily ERK and c-Jun-NH2 terminal kinase are activated during the IAV infection [9]. AP-1 activation requires viral replication as the expression of a single IAV gene HA or NP is not sufficient to induce the activation of AP-1, but NF-κB is readily activated. Interferon regulatory factor (IRF) activation is important for virus-induced response to the host cells. IRF is

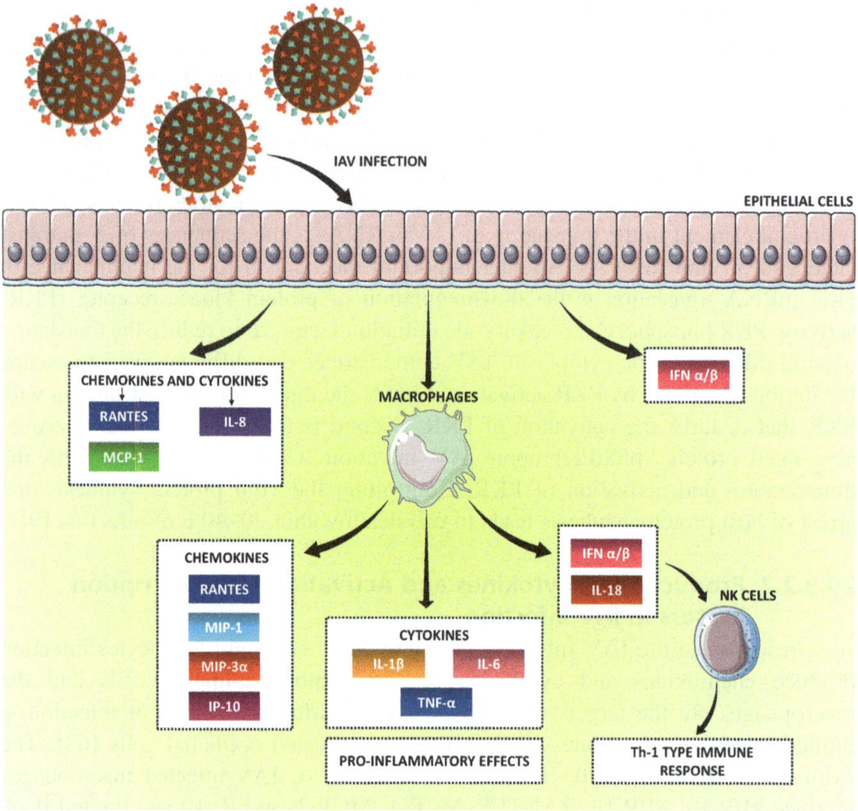

Fig. 20.4 Activation of cytokines and chemokines upon IAV infection

indirectly activated by the viral infection via the expression of IFN-α/β genes and regulates the activation of many other interferons. STATs are also activated by IAV indirectly by the upregulation of IFN-α/β. In addition to the upregulation of certain chemokine and cytokine genes, IAV infection can also regulate the post-translational mechanism involved in the production of cytokines [58]. IAV-infected macrophages produce IL-1β and IL-18 after 9–12 days of infection. Expression of proIL-18 by macrophages and osteoclasts occurs several hours after the accumulation of viral mRNAs and proteins. Caspase cascade activation regulates apoptosis during IAV infection. Multiple apoptosis signals are linked with the activation of caspase-9 and caspase-8 wherein IAV-infected macrophages activate caspase-1 and caspase-3. Activated caspase-1 helps in the cleavage of proIL-18 to mature into IL-18 and IL-1β [59].

20.3.2.3 Apoptosis in IAV Infection

Infection with influenza A virus (IAV) leads to significant cell death in the URT and LRT including lung parenchyma. High levels of cell death are observed in case of severe infections which can aggravate inflammation leading to respiratory failure. Several studies have been attempted to demonstrate the aid of IAV induces cell death that could either restore the lung homeostasis or progression of lung pathologies. There are two primary pathways involved in the activation of apoptosis: intrinsic and extrinsic pathways.

Intrinsic Pathway

The intrinsic pathway is triggered by intracellular stress and is mitochondria-associated. Host factors such as nitric oxide, cytochrome-C, and mitochondria-mediated activation caspases are responsible for the activation of this pathway [73]. The Bcl-2 protein family is significant in activating and mediating the intrinsic pathway. The intracellular mechanism of apoptosis is triggered by the catalytic activation of caspases. Initiator caspases get activated upon IAV infection and undergo homophilic dimerization resulting in autocatalytic cleavage to induce effector caspases. The activated caspases trigger the stimulation of caspase-activated DNase (CAD) for fragmentation of the genomic DNA of the host cells [73].

Effector caspase-8 cleaves pro-apoptotic B cell lymphoma 2 (Bcl-2) family proteins including BH3-interacting domain (Bid). The subsets of the Bcl-2 family include proteins such as Bid, Bad, Bik, NOXA, and PUMA. Some of these proteins act as cell death agonists, while some of them act as an antagonist to cell death. The intracellular signals by effector caspase-9 activate the Bax and Bak proteins that help in the pore formation in the outer membrane of mitochondria and disrupt the mitochondrial membrane potential. The activated Bid is responsible for mitochondrial damage and membrane potential loss which stimulates the release of cytochrome-C to the cytosol from mitochondria. The cytochrome-C binds to the apoptotic protease-activating factor 1 (Apaf1) to form a protein complex called the "apoptosome." The cascade of events proceeded by effector caspases amplify the apoptotic signal [74].

Extrinsic Pathway

The activation of death receptors on the cell surface which are derived from the TNFR superfamily regulates the extrinsic pathway of apoptosis. A few well-known death receptors that can induce apoptosis include TNFR1, TNFRSF6, DR3 (TNFRSF25), DR4 (TNFRSF25), and DR6 (TNFRSF21). When the viral protein binds to these death receptors, the protein-receptor complex ligates and activates apoptotic signals [67]. In the early infection stage, these receptors trigger the apoptosis via formation of a death-inducing signaling complex (DISC) which often deals with caspase-8. Stimulation of caspase-8 can activate the effector caspases which are caspase-3 and caspase-7. This cascade of events by caspase activation commits the cells to undergo apoptosis. During lethal infection of IAV, expression of FasL is abundant in the lung epithelial cells, macrophages, dendritic cells, and lymphocytes. It is also demonstrated that the time course of gene expression of FasL during virus infection is directly proportional to the body weight loss in mice models. Functional deficiency in the production of type I IFN during IAV infection correlates with the FasL expression on the cell surface. FasL-mediated apoptosis is highly associated with the severity of illness including lung inflammation and body weight loss in the host [70].

Activation of TRAIL is also associated with influenza pathogenicity. Expression of TRAIL gene during severe influenza infection induces apoptosis in airway epithelial cells and alveoli. Functional drawbacks in the expression of TRAIL genes result in increased multiplication of the virus. Hence, TRAIL is crucial for the control of IAV infection and propagation of the virus. The expression of all the TRAIL gene during IAV infection, on host survival, depends on the severity of the disease after infection [59, 70].

20.3.2.4 Autophagy

Autophagy is an important innate immune mechanism in which dysfunctional cellular components are removed or degraded to maintain the cellular metabolic function. The mTOR pathway is one of the highly conserved autophagic pathways required for the proper cellular homeostasis. Autophagy is significant in the replication of many viruses including IAV. But the autophagy mechanisms remain unclear [75]. It has demonstrated that nuclear protein- and M2 ion channel-mediated autophagy is responsible for promoting the replication of the virus in the host system. This is achieved by regulating the AKT-mTOR signaling pathway. The signals from mTOR complex 1 (MTORC1) negatively associates with the ULK1 kinase activity, thus inducing autophagy [66]. The interaction between the viral proteins (polymerase basic protein 2) and the host autophagy receptors (microtubule-associated protein light chain 3 (LC3)) promotes the production of viral particles and accelerates the production of viral progeny. Autophagy is also induced by the phosphorylation of Bcl-2/Beclin-1 complex through JNK1 [76]. Autophagy pathways for cellular metabolism and homeostasis have an impact for many viruses; however, the role of autophagy mechanism is known less in IAV infection. To limit infection of IAVs, TRIM23 is essential to mediate autophagy via its RING prostaglandin E3 ligase and ADP-ribosylation factor (ARF) GTPase activity [67]. In contrast to other host factors, Beclin-1- and TUFM-regulated autophagy also inhibited IAV replication. It has been demonstrated that in HeLa cells and A549 cells, activation of JNK1 after IAV infection induces autophagosome formation, while TGF-activated kinase 1 contributes to the process. Furthermore, the memory B cells after IAV infection are maintained by the process of autophagy to counteract the infection. To escape autophagy during infection, IAV utilizes autophagy to complete its life cycle. The viral protein NS1 can suppress JNK1-mediated autophagy induction, and the viral M2 proton ion channel could also block the maturation of autophagosome and mediates the LC3-bound membrane redistribution, thereby allowing the filamentous budding of IAV in the host cell [65, 68].

20.4 Pharmacological Management of IAV Infection

Anti-influenza drugs are the key players in the treatment and management of IAV-infected patients especially at the early pandemic period in which an effective vaccine is absent. Despite broad protective vaccines for flu, anti-influenza drugs play an indispensable role for the treatment of patients who show poor response toward vaccination. The anti-influenza drugs that are currently in use directly target the virus in different stages of viral life cycle. As of now, only two classes of anti-influenza drugs are available for the treatment and management of influenza infection which include M2 ion channel blockers and NA inhibitors. The first class, i.e., M2 ion channel inhibitors, comprises adamantane derivatives like rimantadine and amantadine. These drugs target the M2 ion channel protein and inhibit the proton conductivity of the same which in turn disrupts the viral-endosomal membrane fusion. Disruption of this process prevents the uncoating and subsequent release of vRNPs into the host cell cytoplasm in the viral life cycle. However, the emerging development of high resistance of influenza viruses toward the M2 ion channel inhibitors (rimantadine and amantadine) results in their large discontinuation which is then replaced by the NA inhibitors [77, 78].

NA inhibitors, being the competitive analogs of SA residues, interfere the NA protein's sialidase enzymatic activity toward sialic acid that is present in the host cell's surface. By disrupting this crucial process of IAV, the progeny viruses released from the infected cell get inhibited and thereby prevent the further spread of infection to other healthy cells. Oseltamivir, zanamivir, and peramivir are the three licensed NA inhibitors that are currently in use for the management of influenza infection. Baloxavir marboxil acts as a selective inhibitor of cap-dependent endonuclease activity of PA polymerase which is responsible for the cap-snatching activity of influenza viruses. Apart from these drugs, favipiravir which targets the viral RNA-dependent RNA polymerase is currently under clinical evaluation for its usage and effectiveness in management of influenza infections [77, 78]. Table 20.1 summarizes the available drugs, their target, and mechanism of action for the management and treatment of IAV infection.

20.5 Conclusion

Infection in the lower respiratory tract of humans can result in the deluging of alveolar compartments leading to the development of ARDS which may ultimately result in death by respiratory failure. IAV primarily targets and infects the airway and alveolar epithelial cells that are located on the surface of the respiratory tract. The respiratory epithelium elicits the innate and adaptive immune response against IAV infection. Viral infection also initiates the activation of different transcription factors which induce the expression of cytokines and chemokines. The continuous evolvement of IAV due to its adaptation, antigenic mutation, and reassortment leads to the unexpected emergence of highly pathogenic virulent strains ultimately resulting in local epidemics or global pandemics. A detailed understanding of the

Table 20.1 Potential drug candidates and their mechanism of action against IAV infection

Antiviral drug	Drug target	Mode of action	References
Amantadine	M2 proton ion channel of the virus	Blocks the IAV M2 ion channel and thereby disrupts the viral-endosomal membrane fusion which is much needed for the uncoating and vRNP release	Medication and Manufacturer [79]
Rimantadine	M2 proton ion channel of the virus	Blocks the IAV M2 ion channel and thereby disrupts the viral-endosomal membrane fusion which is much needed for the uncoating and vRNP release	Medication and Manufacturer [79]
Oseltamivir	Neuraminidase (NA)	Inhibits the binding of NA to the sialic acid residues present in the host lung epithelial cell surface	Amarelle et al. [80]
Zanamivir			
Peramivir			
Laninamivir			
Favipiravir	Viral protease	Selective inhibition of viral RNA-dependent RNA polymerase	Denney and Ho [66]
Umifenovir (Arbidol)	Viral components	Prevents the contact between the virus and the host cell and thereby inhibits the membrane fusion of the virus	Wang et al. [81]
Ribavirin	RNA transcription	Selective inhibition of viral RNA-dependent RNA polymerase	Mc Mahon and Martin-Loeches [82]
Glycyrrhizin	Pro-inflammatory gene expression	Inhibits the prostaglandin E2 production by activated macrophages; suppresses the formation of superoxide in macrophages	Wolkerstorfer et al. [83]
Ouabain	Protein translation	Inhibits the vRNP replication in alveolar epithelial cells by decreasing intracellular potassium levels which in turn inhibits protein translation	Amarelle et al. [84]
Baloxavir marboxil	Viral PA polymerase	Inhibits the cap-dependent endonuclease and interferes with the replication of viral mRNA	Shie and Fang [78]

pathobiology of IAV infection may be beneficial for suggesting novel therapies that could intervene the host signaling pathways crucial for viral entry and replication, reduce the inflammatory response, and decrease lung injury and enhance tissue regeneration. Further research and new therapeutic strategies will pave way to target the IAV infection and develop biomarkers that can identify patients with severe disease for effective treatment.

Acknowledgment The authors express their gratitude to SASTRA Deemed to be University, Tamil Nadu, India, for infrastructure and financial support. Authors also extend their appreciation for the contribution of Biopharmaceutical research lab members, SASTRA Deemed to be University.

Author Contribution **Gayathri Gopal:** Writing—Original Draft, Collection of data; **Shibi Muralidar:** Writing—Original Draft, Collection of data; **Senthil Visaga Ambi:** Conceptualization, Visualization, Supervision, and Writing—Review and Editing.

Declarations of Interest None

References

1. Brydon EWA, Morris SJ, Sweet C (2005) Role of apoptosis and cytokines in influenza virus morbidity. FEMS Microbiol Rev 29:837–850. https://doi.org/10.1016/j.femsre.2004.12.003
2. Kadam RU, Wilson IA (2017) Structural basis of influenza virus fusion inhibition by the antiviral drug Arbidol. Proc Natl Acad Sci U S A 114:206–214. https://doi.org/10.1073/pnas.1617020114
3. Shen Z, Lou K, Wang W (2015) New small-molecule drug design strategies for fighting resistant influenza A. Acta Pharm Sin B 5:419–430. https://doi.org/10.1016/j.apsb.2015.07.006
4. Thomas PG, Keating R, Hulse-Post DJ, Doherty PC (2006) Cell-mediated protection in influenza infection. Emerg Infect Dis 12:48–54. https://doi.org/10.3201/eid1201.051237
5. Edinger MO, Pohl MO, Stertz S (2014) Entry of influenza A virus: host factors and antiviral targets. J Gen Virol 95:263–277. https://doi.org/10.1099/vir.0.059477-0
6. Garman EF (2015) Antiviral adhesion molecular mechanisms for influenza: W. G. Laver's lifetime obsession. Philos Trans R Soc B Biol Sci 370. https://doi.org/10.1098/rstb.2014.0034
7. Keshavarz M, Solaymani-Mohammadi F, Namdari H et al (2020) Metabolic host response and therapeutic approaches to influenza infection. Cell Mol Biol Lett 25. https://doi.org/10.1186/s11658-020-00211-2
8. Böhm R, Haselhorst T, Itzstein M Von (2014) Influenza virus , overview: structures, infection mechanisms and antivirals viral pathogenesis of influenza A virus. 1–18. https://doi.org/10.1007/978-4-431-54836-2
9. Julkunen I, Sareneva T, Pirhonen J et al (2001) Molecular pathogenesis of influenza A virus infection and virus-induced regulation of cytokine. Gene Expr 12:171–180
10. Szewczyk B, Bieńkowska-szewczyk K, Król E (2014) Introduction to molecular biology of influenza A. Viruses 61:397–401
11. Parrish CR, Murcia PR, Holmes EC (2015) Influenza virus reservoirs and intermediate hosts: dogs, horses, and new possibilities for influenza virus exposure of humans. J Virol 89:2990–2994. https://doi.org/10.1128/jvi.03146-14
12. Simonsen L, Spreeuwenberg P, Lustig R et al (2013) Global mortality estimates for the 2009 influenza pandemic from the GLaMOR project: a Modeling study. PLoS Med 10. https://doi.org/10.1371/journal.pmed.1001558
13. Taubenberger JK, Kash JC (2010) Influenza virus evolution, host adaptation, and pandemic formation. Cell Host Microbe 7:440–451. https://doi.org/10.1016/j.chom.2010.05.009
14. Webby RJ, Webster RG, Richt JA (2007) Influenza viruses in animal wildlife populations. Curr Top Microbiol Immunol 315:67–83. https://doi.org/10.1007/978-3-540-70962-6_4
15. Dou D, Revol R, Östbye H et al (2018) Influenza A virus cell entry, replication, virion assembly and movement. Front Immunol 9:1–17. https://doi.org/10.3389/fimmu.2018.01581
16. Roberts PC, Lamb RA, Compans RW (1998) The M1 and M2 proteins of influenza A virus are important determinants in filamentous particle formation. Virology 240:127–137. https://doi.org/10.1006/viro.1997.8916
17. Yoon S-W, Webby RJ, Webster RG. Evolution and ecology of influenza a viruses. Curr Top Microbiol Immunolo. https://doi.org/10.1007/82_2014_396
18. Morens DM, Taubenberger JK, Anthony SF (2009) The Persistent Legacy of the 1918 Influenza Virus. N Engl J Med 361:225–229. https://doi.org/10.1056/NEJMp0904819

19. Tong S, Li Y, Rivailler P et al (2012) A distinct lineage of influenza A virus from bats. Proc Natl Acad Sci U S A 109:4269–4274. https://doi.org/10.1073/pnas.1116200109
20. Tong S, Zhu X, Li Y et al (2013) New World Bats Harbor diverse influenza a viruses. PLoS Pathog 9. https://doi.org/10.1371/journal.ppat.1003657
21. Blümel J, Burger R, Drosten C et al (2009) Influenza virus. Transfus Med Hemotherapy 36:32–39. https://doi.org/10.1159/000197314
22. Hutchinson EC, Charles PD, Hester SS et al (2014) Conserved and host-specific features of influenza virion architecture. Nat Commun:5. https://doi.org/10.1038/ncomms5816
23. Herold S, Becker C, Ridge KM, Budinger GRS (2015) Influenza virus-induced lung injury: pathogenesis and implications for treatment. Eur Respir J 45:1463–1478. https://doi.org/10.1183/09031936.00186214
24. Janke BH (2014) Influenza A virus infections in swine: pathogenesis and diagnosis. Vet Pathol 51:410–426. https://doi.org/10.1177/0300985813513043
25. Londino JD, Lazrak A, Collawn JF et al (2017) Influenza virus infection alters ion channel function of airway and alveolar cells: mechanisms and physiological sequelae. Am J Physiol Lung Cell Mol Physiol 313:L845–L858. https://doi.org/10.1152/ajplung.00244.2017
26. Tucker SP, Compans RW (1993) Virus infection of polarized epithelial cells. Adv Virus Res 42:187–247. https://doi.org/10.1016/S0065-3527(08)60086-X
27. Tucker SP, Thornton CL, Wimmer E, Compans RW (1993) Vectorial release of poliovirus from polarized human intestinal epithelial cells. J Virol 67:4274–4282. https://doi.org/10.1128/jvi.67.7.4274-4282.1993
28. Matsuoka Y, Matsumae H, Katoh M. et al (2013) A comprehensive map of the influenza A virus replication cycle. BMC Syst Biol 7:97. https://doi.org/10.1186/1752-0509-7-97
29. Samji T (2009) Influenza A: understanding the viral life. Cycle 82:153–159
30. Gambaryan AS, Tuzikov AB, Piskarev VE et al (1997) Specification of receptor-binding phenotypes of influenza virus isolates from different hosts using synthetic sialylglycopolymers: non-egg-adapted human H1 and H3 influenza A and influenza B viruses share a common high binding affinity for 6′-sialyl (N-ace). Virology 232:345–350. https://doi.org/10.1006/viro.1997.8572
31. Matrosovich M, Tuzikov A, Bovin N et al (2000) Early alterations of the receptor-binding properties of H1, H2, and H3 avian influenza virus hemagglutinins after their introduction into mammals. J Virol 74:8502–8512. https://doi.org/10.1128/jvi.74.18.8502-8512.2000
32. Nobusawa E, Aoyama T, Kato H et al (1991) Comparison of complete amino acid sequences and receptor-binding properties among 13 serotypes of hemagglutinins of influenza A viruses. Virology 182:475–485. https://doi.org/10.1016/0042-6822(91)90588-3
33. Sakai T, Nishimura SI, Naito T, Saito M (2017) Influenza A virus hemagglutinin and neuraminidase act as novel motile machinery. Sci Rep 7:1–11. https://doi.org/10.1038/srep45043
34. Bullough PA, Hughson FM, Skehel JJ, Wiley DC (1994) Structure of influenza haemagglutinin at the pH of membrane fusion. Nature 371:37–43
35. Huang Q, Sivaramakrishna RP, Ludwig K et al (2003) Early steps of the conformational change of influenza virus hemagglutinin to a fusion active state: stability and energetics of the hemagglutinin. Biochim Biophys Acta Biomembr 1614:3–13. https://doi.org/10.1016/S0005-2736(03)00158-5
36. Skehel JJ, Wiley DC (1999) Receptor binding and membrane fusion in virus entry: the influenza hemagglutinin. Annu Rev Biochem 1999:777–810
37. White J, Helenius A, Gething MJ (1982) Haemagglutinin of influenza virus expressed from a cloned gene promotes membrane fusion. Nature 300:658–659
38. Yoshimura A, Ohnishi S (1984) Uncoating of influenza virus in endosomes. J Virol 51:497–504. https://doi.org/10.1128/jvi.51.2.497-504.1984
39. Lakadamyali M, Rust MJ, Babcock HP, Zhuang X (2003) Visualizing infection of individual influenza viruses. Proc Natl Acad Sci U S A 100:9280–9285. https://doi.org/10.1073/pnas.0832269100

40. Pinto LH, Lamb RA (2006) The M2 proton channels of influenza A and B viruses. J Biol Chem 281:8997–9000. https://doi.org/10.1074/jbc.R500020200
41. Rust MJ, Lakadamyali M, Zhang F, Zhuang X (2004) Assembly of endocytic machinery around individual influenza viruses during viral entry. Nat Struct Mol Biol 11:567–573. https://doi.org/10.1038/nsmb769
42. Bavagnoli L, Maga G (2013) Identification of host cell factors involved in influenza A virus infection. 8:195–208
43. Braam J, Ulmanen I, Krug RM (1983) Molecular model of a eucaryotic transcription complex: functions and movements of influenza P proteins during capped RNA-primed transcription. Cell 34:609–618. https://doi.org/10.1016/0092-8674(83)90393-8
44. Dias A, Bouvier D, Crépin T et al (2009) The cap-snatching endonuclease of influenza virus polymerase resides in the PA subunit. Nature 458:914–918. https://doi.org/10.1038/nature07745
45. Resa-Infante P, Jorba N, Zamarreño N et al (2008) The host-dependent interaction of α-importins with influenza PB2 polymerase subunit is required for virus RNA replication. PLoS One 3. https://doi.org/10.1371/journal.pone.0003904
46. Digard P, Elton D, Simpson-Holley M, Medcalf E (2001) Interaction of the influenza virus nucleoprotein with F-actin. Int Congr Ser 1219:503–512. https://doi.org/10.1016/S0531-5131(01)00628-8
47. O'Neill RE, Talon J, Palese P (1998) The influenza virus NEP (NS2 protein) mediates the nuclear export of viral ribonucleoproteins. EMBO J 17:288–296. https://doi.org/10.1093/emboj/17.1.288
48. Akarsu H, Burmeister WP, Petosa C et al (2003) Crystal structure of the M1 protein-binding domain of the influenza A virus nuclear export protein (NEP/NS2). EMBO J 22:4646–4655. https://doi.org/10.1093/emboj/cdg449
49. Baudin F, Petit I, Weissenhorn W, Ruigrok RWH (2001) In vitro dissection of the membrane and RNP binding activities of influenza virus M1 protein. Virology 281:102–108. https://doi.org/10.1006/viro.2000.0804
50. Boulo S, Akarsu H, Ruigrok RWH, Baudin F (2007) Nuclear traffic of influenza virus proteins and ribonucleoprotein complexes. Virus Res 124:12–21. https://doi.org/10.1016/j.virusres.2006.09.013
51. Kawaguchi A, Matsumoto K, Nagata K (2012) YB-1 functions as a porter to Lead influenza virus ribonucleoprotein complexes to microtubules. J Virol 86:11086–11095. https://doi.org/10.1128/jvi.00453-12
52. Shapiro GI, Gurney T, Krug RM (1987) Influenza virus gene expression: control mechanisms at early and late times of infection and nuclear-cytoplasmic transport of virus-specific RNAs. J Virol 61:764–773. https://doi.org/10.1128/jvi.61.3.764-773.1987
53. Horimoto T, Kawaoka Y (1994) Reverse genetics provides direct evidence for a correlation of hemagglutinin cleavability and virulence of an avian influenza A virus. J Virol 68:3120–3128. https://doi.org/10.1128/jvi.68.5.3120-3128.1994
54. Stieneke-Grober A, Vey M, Angliker H et al (1992) Influenza virus hemagglutinin with multibasic cleavage site is activated by furin, a subtilisin-like endoprotease. EMBO J 11:2407–2414. https://doi.org/10.1002/j.1460-2075.1992.tb05305.x
55. Webster RG, Rottt R (1967) Influenza virus a pathogenicity: the pivotal role of hemagglutinin minireview. Cell 50:665–666
56. Nayak DP, Balogun RA, Yamada H et al (2009) Influenza virus morphogenesis and budding. Virus Res 143:147–161. https://doi.org/10.1016/j.virusres.2009.05.010
57. Gounder AP, Boon ACM (2019) Influenza pathogenesis: the effect of host factors on severity of disease. J Immunol 202:341–350. https://doi.org/10.4049/jimmunol.1801010
58. El-Sayed I, Bassiouny K, Nokaly A et al (2016) Influenza A virus and influenza b virus can induce apoptosis via intrinsic or extrinsic pathways and also via NF-κB in a time and dose dependent manner. Biochem Res Int 2016. https://doi.org/10.1155/2016/1738237

59. Fukuyama S, Kawaoka Y (2011) The pathogenesis of influenza virus infections: the contributions of virus and host factors. Curr Opin Immunol 23:481–486. https://doi.org/10.1016/j.coi.2011.07.016
60. Troy NM, Bosco A (2016) Respiratory viral infections and host responses; insights from genomics. Respir Res 17:1–12. https://doi.org/10.1186/s12931-016-0474-9
61. Zhang Y, Xu Z, Cao Y (2020) Host – virus interaction: how host cells defend. Viruses 12:1–23
62. Ramos I, Smith G, Ruf-Zamojski F et al (2019) Innate immune response to influenza virus at single-cell resolution in human epithelial cells revealed paracrine induction of interferon lambda 1. J Virol 93:1–22. https://doi.org/10.1128/jvi.00559-19
63. Chen X, Liu S, Goraya MU et al (2018) Host immune response to influenza A virus infection. Front Immunol 9:1–13. https://doi.org/10.3389/fimmu.2018.00320
64. Salomon R, Webster RG (2009) The influenza virus enigma. Cell 136:402–410. https://doi.org/10.1016/j.cell.2009.01.029
65. Allen EK, Thomas PG (2015) Immunity to influenza. Preventing infection and regulating disease. Am J Respir Crit Care Med 191:248–251. https://doi.org/10.1164/rccm.201412-2245ED
66. Denney L, Ho LP (2018) The role of respiratory epithelium in host defence against influenza virus infection. Biom J 41:218–233. https://doi.org/10.1016/j.bj.2018.08.004
67. Herold S, Ludwig S, Pleschka S, Wolff T (2012) Apoptosis signaling in influenza virus propagation, innate host defense, and lung injury. J Leukoc Biol 92:75–82. https://doi.org/10.1189/jlb.1011530
68. Lee HY, Topham DJ, Park SY et al (2009) Simulation and prediction of the adaptive immune response to influenza A virus infection. J Virol 83:7151–7165. https://doi.org/10.1128/jvi.00098-09
69. Kim TS, Sun J, Braciale TJ (2011) T cell responses during influenza infection: getting and keeping control. Trends Immunol 32:225–231. https://doi.org/10.1016/j.it.2011.02.006
70. Bahadoran A, Lee SH, Wang SM et al (2016) Immune responses to influenza virus and its correlation to age and inherited factors. Front Microbiol 7:1–11. https://doi.org/10.3389/fmicb.2016.01841
71. Atkin-Smith GK, Duan M, Chen W, Poon IKH (2018) The induction and consequences of Influenza A virus-induced cell death. Cell Death Dis:9. https://doi.org/10.1038/s41419-018-1035-6
72. Meseko CA, Oluwayelu DO (2019) Avian influenza, transboundary animal diseases in sahelian africa and connected regions, 345–374. https://doi.org/10.1007/978-3-030-25385-1_17
73. Fujikura D, Miyazaki T (2018) Programmed cell death in the pathogenesis of influenza. Int J Mol Sci 19:1–14. https://doi.org/10.3390/ijms19072065
74. Mehrbod P, Ande SR, Alizadeh J et al (2019) The roles of apoptosis, autophagy and unfolded protein response in arbovirus, influenza virus, and HIV infections. Virulence 10:376–413. https://doi.org/10.1080/21505594.2019.1605803
75. Zhou Z, Jiang X, Liu D et al (2009) Autophagy is involved in influenza A virus replication. Autophagy 5:321–328. https://doi.org/10.4161/auto.5.3.7406
76. Wang R, Zhu Y, Zhao J et al (2018) Autophagy promotes replication of influenza A virus in vitro. J Virol 93:1–17. https://doi.org/10.1128/jvi.01984-18
77. Pizzorno A, Padey B, Terrier O, Rosa-Calatrava M (2019) Drug repurposing approaches for the treatment of influenza viral infection: reviving old drugs to fight against a long-lived enemy. Front Immunol:10. https://doi.org/10.3389/fimmu.2019.00531
78. Shie JJ, Fang JM (2019) Development of effective anti-influenza drugs: congeners and conjugates–a review. J Biomed Sci 26:1–20. https://doi.org/10.1186/s12929-019-0567-0
79. Medication P, Manufacturer D (2020) Anti-influenza a drugs mechanism of action: indications: major drug interactions: notes: anti-influenza A & B drugs. Inhibitors of Viral 2020:1–17
80. Amarelle L, Lecuona E, Sznajder JI (2017) Anti-influenza treatment: drugs currently used and under development. Arch Bronconeumol 53:19–26. https://doi.org/10.1016/j.arbr.2016.11.020

81. Wang X, Cao R, Zhang H et al (2020) The anti-influenza virus drug, arbidol is an efficient inhibitor of SARS-CoV-2 in vitro. Cell Discov 6:4–8. https://doi.org/10.1038/s41421-020-0169-8
82. McMahon A, Martin-Loeches I (2017) The pharmacological management of severe influenza infection–'existing and emerging therapies'. Expert Rev Clin Pharmacol 10:81–95. https://doi.org/10.1080/17512433.2017.1255550
83. Wolkerstorfer A, Kurz H, Bachhofner N, Szolar OHJ (2009) Glycyrrhizin inhibits influenza A virus uptake into the cell. Antivir Res 83:171–178. https://doi.org/10.1016/j.antiviral.2009.04.012
84. Amarelle L, Katzen J, Shigemura M et al (2019) Cardiac glycosides decrease influenza virus replication by inhibiting cell protein translational machinery. Am J Physiol Lung Cell Mol Physiol 316:L1094–L1106. https://doi.org/10.1152/ajplung.00173.2018

Understanding the Biology of Non-typeable *Haemophilus influenzae* in Chronic Obstructive Pulmonary Disease Through the Lens of Genomics

21

Rajendra KC and Ronan F. O'Toole

Abstract

The genus *Haemophilus* contains a number of species of medical importance. In particular, *Haemophilus influenzae* causes a range of invasive diseases including meningitis, septicaemia, epiglottis, cellulitis and arthritis, as well as non-invasive infections that can present as sinusitis, otitis media, chronic bronchitis and pneumonia. Conventional laboratory biotyping and genotyping lack the resolution required to distinguish between isolates of across these different clinical phenotypes. However, advancements in whole-genome sequencing have made it possible to detect subtle differences in the bacterium's genome content. In this chapter, we explore the potential that genomics and bioinformatics offer to the management of *H. influenzae* infection. In particular, we discuss specific examples of their application in the study of COPD-related infections caused by non-typeable strains of *H. influenzae*.

Keywords

Non-typeable *Haemophilus influenzae* · Chronic obstructive pulmonary disease · Whole-genome sequencing

R. KC
Menzies Institute for Medical Research, University of Tasmania, Hobart, TAS, Australia

R. F. O'Toole (✉)
Department of Pharmacy and Biomedical Sciences, School of Molecular Sciences, College of Science, Health and Engineering, La Trobe University, Albury-Wodonga, VIC, Australia
e-mail: r.otoole@latrobe.edu.au

K. Dua et al. (eds.), *Targeting Cellular Signalling Pathways in Lung Diseases*,
https://doi.org/10.1007/978-981-33-6827-9_21

Abbreviations

ChoP	Phosphorylcholine
COPD	Chronic obstructive pulmonary disease
DNA	Deoxyribonucleic acid
HGT	Horizontal gene transfer
LOS	Lipooligosaccharide
MLEE	Multilocus enzyme electrophoresis
MLST	Multilocus sequence typing
NTHi	Non-typeable *Haemophilus influenzae*
PAFR	Platelet-activating factor receptor
pan-GWAS	Pan-genome-wide association studies
REA	Restriction endonuclease analysis
SNPs	Single-nucleotide polymorphisms
WGS	Whole-genome sequencing

21.1 Introduction to *Haemophilus* and Human Colonisation

Haemophilus was first described by Richard Pfeiffer as 'influenza bacillus' during the influenza pandemic of 1889–1892, when it was misidentified as a causative agent of human influenza [1]. The term *Haemophilus* is derived from the Greek words *haima* and *philus* for blood and lover, respectively. This acknowledges the fact that *Haemophilus* requires the inclusion of blood in in vitro culture medium, in particular, haem (factor X) and NAD (factor V). Species of the genus *Haemophilus* are nonmotile, aerobic or facultatively anaerobic, Gram-negative, non-spore-forming coccobacilli measuring less than 1 μm in diameter [2]. *Haemophilus* species when grown on nutrient-rich media generally produce non-pigmented or slightly yellowish, flat and convex colonies. They are nitrate reducers and are capable of fermenting carbohydrates [2].

In 1921, the genus *Haemophilus* was incorporated into the family *Pasteurellaceae*, which includes commensals and opportunistic pathogens of both humans and animals. *Haemophilus* species colonise mucosal surfaces of their specific hosts. Most *Haemophilus* species exhibit host specificity for humans with the exception of four animal-adapted species, i.e. *H. felis*, *H. haemoglobinophilus*, *H. paracuniculus* and *H. parasuis* [3]. Currently, there are nine human-specific species of the genus *Haemophilus* that are broadly classified into three groups based on phenotypic traits: the *H. influenzae* group which contains three X factor- and V factor-dependent species, i.e. *H. influenzae*, *H. aegyptius* and *H. haemolyticus*; the *H. parainfluenzae* group which consists of five X factor-independent and V factor-dependent species, i.e. *H. parainfluenzae*, *H. parahaemolyticus*, *H. paraphrohaemolyticus*, *H. pittmaniae* and *H. sputorum*; and, lastly, *H. ducreyi*, which requires X factor but not V factor [3].

Among the human-colonising *Haemophilus* species, *H. influenzae* is of highest clinical importance. It is a common inhabitant of the healthy nasopharynx and is generally a harmless commensal. However, when it penetrates into nasopharyngeal mucosa and reaches the bloodstream, or spreads to other regions of the respiratory tract, it can cause significant invasive or non-invasive infections [4]. Invasive infections due to *H. influenzae* include meningitis, septicaemia, epiglottis, cellulitis and arthritis [5], while *H. influenzae* non-invasive infections can be manifested as sinusitis, otitis media, chronic bronchitis and pneumonia. Generally, encapsulated *H. influenzae* is considered more virulent than their non-capsulated (non-typeable) counterparts. Among capsulated strains, *H. influenzae* serotype b (Hib) strains are the most pathogenic and were commonly associated with invasive disease prior to the introduction of Hib vaccine programmes from the 1980s onwards [6]. However, since the introduction of Hib vaccines, non-typeable *H. influenzae* (NTHi) has emerged as the leading cause of both invasive and non-invasive *Haemophilus* diseases [7].

H. aegyptius is associated with acute conjunctivitis (pink eye) as well as an invasive disease known as Brazilian purpuric fever [8, 9]. *H. haemolyticus*, which was once considered non-pathogenic, has occasionally been reported to cause invasive diseases that include bacteraemia, peritonitis and arthritis [10, 11]. *H. parainfluenzae* is present in the normal microbiota of the mouth and pharynx, and is a rare cause of meningitis, dental abscess, liver abscess, septic arthritis and endocarditis [12–15].

H. parahaemolyticus has been reported from patients with pharyngitis, and also from subacute endocarditis, brain abscess and septic shock [16–18]. The other three species of the *H. parainfluenzae* group, i.e. *H. paraphrohaemolyticus*, *H. pittmaniae* and *H. sputorum*, occur in the mouth and oropharynx of healthy individuals and are less commonly associated with lower respiratory tract infections and invasive disease such as liver abscess [19–21]. Finally, *H. ducreyi* is a causative agent of chancroid which is a major sexually transmitted infection in developing countries of Africa and Southeast Asia that manifests as genital ulcerative disease [22]. In addition, *H. ducreyi* has been implicated in chronic skin ulcers in children [23].

21.2 Non-typeable *Haemophilus influenzae* and Human Disease

NTHi is a major bacterial pathogen associated with mucosal infections, including otitis media, sinusitis, conjunctivitis and exacerbations of chronic obstructive pulmonary disease (COPD) [24–27]. In addition, there have been increasing reports of invasive disease caused by NTHi isolates [28]. Langereis and de Jonge have put forward four possible explanations for the emergence of NTHi [28]. Firstly, introduction of the Hib vaccine is considered to have caused a major reduction in nasopharyngeal colonisation by Hib leaving NTHi to become more dominant. Secondly, improved diagnostic techniques have likely resulted in an increase in the number of previously undetectable NTHi isolates now being identified in clinical specimens. Thirdly, NTHi is believed to have acquired through horizontal gene transfer additional virulence and survival factors that have augmented its pathogenicity and immune evasion properties. Finally, the epidemiology of *H. influenzae* infection, which was once primarily associated with paediatric disease, appears to have shifted more towards the elderly, a population group which is growing faster than other age groups globally [28].

Unlike encapsulated *H. influenzae*, which possesses a carbohydrate capsule as a major virulence factor, for NTHi there does not appear to be a single feature that is characteristic of all disease-associated strains [29]. Moreover, the extent to which disease isolates of NTHi are phenotypically or genotypically similar has not been studied extensively. NTHi possesses an array of adhesins and host immune evasion molecules that help them colonise the respiratory epithelium and establish infection therein [30, 31]. Fimbriae encoded by the *hif* locus [32], high molecular weight adhesins (HMW1 and HMW2) encoded by the *hmw1* and *hmw2* loci [33], Hia (for *H. influenzae* adherence protein) and Hap (for *Haemophilus* adhesion and penetration protein) encoded by the *hia* and *hap* loci, respectively [34], are utilised by NTHi to bind to human airway epithelial cell receptors or extracellular matrix proteins to initialise colonisation. Other surface adhesins include outer membrane proteins 2 (P2) and P5 [35], outer membrane lipoprotein (PCP) [36], protein E [37] and a phosphorylcholine (ChoP) moiety on lipooligosaccharide (LOS) [38]. It is conceivable that NTHi strains which have similar repertoires of virulence factors could be associated with very different clinical phenotypes [39]; however, this question requires high-resolution analysis of the genomes of a wide range of NTHi strains.

Several environmental factors and disease conditions can predispose individuals to higher susceptibility of respiratory infection [40–42]. Exposure to tobacco and biomass smoke induces lung inflammation and impairs the innate immune response by altering alveolar phagocytosis, resulting in increased susceptibility to respiratory infection [43–45]. In addition, exposure of airway epithelial cells to tobacco and biomass smoke upregulates the expression of host cell surface receptors, including platelet-activating factor receptor (PAFR), which is subsequently utilised by respiratory bacteria such as NTHi for adherence and invasion [30, 46, 47]. Upregulated PAFR is believed to be bound by an NTHi adhesin, a ChoP moiety which is presented on cell wall LOS [38, 48–50]. This may provide an opportunity for NTHi to establish infection and cause exacerbations in COPD patients.

21.3 Current Techniques for the Genotyping of NTHi

NTHi strains are highly diverse, and various typing methods have been proposed for epidemiological purposes [51–53]. NTHi are naturally transformable, and comparative genomic studies based on whole-genome sequencing have demonstrated the existence of substantial genomic plasticity among NTHi strains [53, 54]. This results mainly from horizontal gene transfer (HGT) processes, which cause nonuniform genetic exchanges among bacteria [55]. The high rate of HGT in NTHi has been demonstrated by the presence of multiple unique strains co-localised within the biofilms and by the simultaneous existence of polyclonal NTHi populations during chronic infections, such as COPD and cystic fibrosis [56, 57]. This genetic diversity is believed to be important for ensuring survival of the population. Environmental challenge may selectively enrich for resilient strains and against susceptible strains while maintaining survival of the population [58]. Thus, evolutionary pressures select beneficial mechanisms that generate diversity across the NTHi population.

Biotyping is one of the traditional methods of determining phenotypic diversity of NTHi isolates [59]. Classically, *H. influenzae* strains are phenotypically grouped into eight (I–VIII) biotypes based primarily on three biochemical reactions: ornithine decarboxylation, indole production and urea hydrolysis [60]. Unlike capsulated strains, NTHi isolates are found to be more diverse in biotype [60]. NTHi strains inhabiting a specific niche, such as the middle ear, sinuses, blood or the respiratory tract, or those associated with a certain clinical source, are indistinguishable from others based on their biotypes. For example, strains belonging to all five biotypes I, II, III, IV and V have been isolated from the middle ear of patients with chronic otitis media [61]. Similarly, biotypes II, III and V are found to be associated with both invasive strains and commensals of the upper respiratory tract [62, 63].

Some studies have reported a correlation between virulence-related phenotypes of NTHi and their clinical sources. For instance, Chin et al. reported a higher ability of NTHi strains isolated from COPD patients with exacerbations to induce a inflammatory response and to adhere to primary human tracheobronchial epithelial cells than strains isolated from patients with stable conditions [64]. Similarly, Bresser et al. demonstrated differences between persisting and non-persisting NTHi strains with regard to their ability to induce secretion of pro-inflammatory cytokines [65]. In contrast to the former study, persisting strains triggered lower levels of pro-inflammatory cytokines IL-6 and IL-8 than non-persisting strains, and both types of strains exhibited similar levels of adherence to lung cells [65]. Later, Bresser et al. showed that NTHi strains from the same clinical source exhibit diverse phenotypes, for example, strains from chronic bronchitis exhibited significantly different levels of adherence to extracellular matrix proteins [66]. The weak correlation of virulence-related phenotype with clinical source has been attributed to reversible phase variation in the expression of virulence-related genes in NTHi [67].

In terms of genetic diversity, for many years this has been characterised using the techniques ribotyping, restriction endonuclease analysis (REA), multilocus enzyme electrophoresis (MLEE) and multilocus sequence typing (MLST) [68–70]. REA involves digestion of bacterial genome with a restriction endonuclease followed by visualisation of the fragments, whereas in ribotyping only fragments that hybridise specifically with a ribosomal RNA are probed [71]. REA and ribotyping provide sufficient discriminatory power to be useful for taxonomic studies, however they are not very suitable for routine diagnostic and epidemiological studies as the procedures are time-consuming and labour-intensive due to the need for Southern blotting and probing [68].

MLEE typing is based on the comparison of the electrophoretic mobility of housekeeping enzymes [72]. The resolution of MLEE is low as it considers only limited variations within a small number of enzyme loci. The resolution can be increased by including more loci in the analysis; however, this adds technical complexity and is more time consuming.

MLST is an improved version of MLEE, where the nucleotide sequence of an internal fragment from each of seven housekeeping genes is analysed [73]. The major advantages of MLST are high-resolution discriminatory power and the production of data that can be compared between different laboratories [74]. In addition, the method has high throughput capability and is scalable for population-based studies [75]. However, MLST typing does not provide a sufficient level of resolution to discriminate between the NTHi isolates collected from different clinical sources [76–78].

More recently, whole-genome sequencing (WGS) allows the examination of the entire genome of an NTHi isolate and hence provides the opportunity for a higher level of discrimination between strains, including those from different disease phenotypes [79]. With the recent advancement in sequencing technologies, high-throughput sequencing of entire bacterial genomes with short turn-around times is becoming more affordable [80]. The sequencing can be coupled with appropriate bioinformatics workflows for robust handling and analysis of big sequence data. This facilitates the detection of small genetic differences and the distinction of closely related isolates.

21.4 Differential Genomic Content of NTHi Strains Associated with COPD

COPD is emerging as the third leading cause of mortality worldwide, claiming approximately 3.2 million lives in 2017 [81]. Bacterial infection is a frequent feature in the lower airways of COPD patients with NTHi being a predominant bacterium in that clinical setting [26, 82, 83]. NTHi colonises damaged airways in COPD patients and contributes to further inflammation and airway injury, particularly during acute exacerbations [84, 85]. It is therefore important to understand the underlying mechanisms that make NTHi proficient in establishing an infection of the COPD airways. Furthermore, elucidation of differences between NTHi strains associated with COPD versus other clinical phenotypes can aid the development of targeted therapy against NTHi-induced exacerbations of COPD.

There are a number of important studies that investigated genomic changes that occur in NTHi during persistence in COPD airways [78, 86]. Pettigrew and colleagues assessed 101 longitudinal pairs of NTHi that were collected over a period of 15 years. They reported polymorphisms in certain cell surface proteins (antigens) as well as changes in simple sequence repeats of genes that regulate vital virulence functions, such as adherence, nutrient uptake and modification of LOS [78]. Moleres and colleagues examined 92 NTHi isolates from 13 COPD patients collected over 1–9 years and found genetic changes in phase-variable genes as reported by Pettigrew and colleagues. In addition, they identified loss-of-function mutations in the *ompP1* (*fadL*) gene which encodes a bifunctional membrane protein OmpP1 that enables NTHi to adhere to airway epithelia as well as transport long-chain fatty acids into the cell [86].

In our recent work, we performed an in-depth comparative analysis on a set of 568 NTHi genomes that were isolated from different clinical phenotypes, including otitis media, conjunctivitis, meningitis and COPD [77]. We classified the collection, based on the MLST profile, into 174 unique sequence types (STs). Thirty-four of the STs contained both COPD and non-COPD strains of NTHi, indicating a limited ability of MLST genotyping to differentiate COPD strains of NTHi from other clinical phenotypes. We then expanded the analysis further by considering all of the polymorphic sites from 853 core genes. This led to the identification of eight (I–VIII) distinct subpopulations of NTHi. However, the discriminatory power of phylogeny based on core genome SNPs was still not sufficient to separate the NTHi isolates according to their clinical source. This result was consistent with previous findings by De Chiara et al. and Pettigrew et al. [76, 78]. On the other hand, when comparative gene content analysis was performed based on pan-genome accessory genes of the NTHi collection using discriminant analysis, a clear separation of NTHi isolates from COPD compared to other disease phenotypes emerged. Furthermore, using a pan-genome-wide association studies (pan-GWAS) approach, a set of genes were found to be significantly associated with COPD strains, and these included genes involved in adherence, immune evasion and nutrient uptake. In addition, variant forms of genes that play important roles in transformation competence and recombination, such as *tfoX*, *pilA* and *recJ* encoding TfoX (regulates transformation *via* competence-regulon genes), PilA (regulates DNA uptake *via* type IV pili) and RecJ (DNA repair and recombination), were found specifically in COPD isolates of NTHi. These COPD-associated accessory genes may provide a competitive advantage to NTHi in the colonisation of COPD airways.

21.5 Future Applications of Genomics in NTHi Diagnostics and Treatment

WGS, relative to older genotyping methods, is unique in offering the highest resolution to differentiate bacterial species [87]. Application of WGS in clinical microbiology has transformed the field of diagnostic and public health microbiology through rapid, culture-independent and accurate characterisation of pathogens and for the surveillance of infectious diseases, respectively [88, 89].

WGS has been useful in the detection and characterisation of NTHi from a wide spectrum of diseases, including otitis media, conjunctivitis, meningitis and COPD exacerbations. In addition, it has been successfully applied to elucidate the phylogeny of NTHi isolates and to study the evolution of NTHi genome during persistence in the human airways [76–78]. Furthermore, it can be employed to detect accessory genome content, and hence strains, with a predicted higher propensity to be associated with acute exacerbations of COPD, and other clinical disease phenotypes [77].

Recurrent infection with NTHi is a common clinical problem in children with otitis media and in patients with COPD [90, 91]. Recurrent infections are caused either through infection with new exogenous strains of NTHi or by failure of antibiotic therapy to effectively clear an existing infection. Precise differentiation between new and persisting isolates of NTHi can be achieved with the high resolution of WGS. The ability to distinguish between new and relapse NTHi infections in COPD would assist in guiding effective antibiotic selection in the treatment of patients.

NTHi is constantly evolving and modifying its genome to adapt to dynamic changes in the human host during chronic and recurrent infections [92]. The emergence of hypervirulent and antibiotic-resistant strains of bacteria, including NTHi, and their global dissemination is a potential threat to public health [93]. WGS provides the ability to inform public health management of NTHi by mapping the origin and transmission network of hypervirulent and antibiotic resistant strains.

21.6 Conclusions

The *Haemophilus* genus contains both commensal and pathogenic species of bacteria for which humans are their primary host. While disease is mostly associated with a small number of taxa including *H. influenzae*, *H. aegyptius* and *H. ducreyi*, it is apparent that a diversity of clinical phenotypes can be caused by the same species. This is particularly evident with respect to infection with *H. influenzae* which can manifest as sinusitis, otitis media, chronic bronchitis and pneumonia or cause invasive disease including cellulitis, epiglottis, septicaemia, meningitis and arthritis.

This raises the question as to whether certain biotypes or genotypes within the *H. influenzae* classification are more specialised than others for specific disease sequelae. It is apparent that classical biochemical and genetic techniques for differentiating isolates of *H. influenzae* are limited in their ability to answer this question and a higher resolution method is required. The emergence of whole-genome sequencing provides a platform for detecting subtle inherited and acquired differences between individual isolates of *H. influenzae* and correlating these with clinical disease phenotypes. Its application has led to the discovery that NTHi strains, which are associated with acute exacerbations of COPD, can now be differentiated genetically from non-COPD strains. A set of genes in the accessory genome of COPD isolates of NTHi encode adherence and cellular metabolism functions which are believed to enable NTHi to adapt to the microenvironment of COPD airways. In addition, NTHi strains harbour competence- and recombination-related genes which may facilitate incorporation of further horizontally acquired survival and virulence genes into their genomes. Knowledge of the distinct gene content in COPD isolates of NTHi has the potential to open up new avenues for the development of targeted therapeutics in the management of NTHi infections and related exacerbations in COPD.

Competing Interests The authors declare that they have no financial or other conflicts of interest.

Ethical Approval Not required.

Funding Not applicable.

References

1. Pfeiffer R (1893) Die aetiologie der influenza. Z Hyg Infekt 13:357–386
2. Kilian M (2015) *Haemophilus*. Bergey's manual of systematics of archaea and bacteria. 1–47
3. NØrskov-Lauritsen N (2014) Classification, identification, and clinical significance of *Haemophilus* and *Aggregatibacter* species with host specificity for humans. Clin Microbiol Rev 27:214–240
4. High NJ (2002) *Haemophilus influenzae*. Molecular medical microbiology. Elsevier, Amsterdam
5. McConnell A, Tan B, Scheifele D, Halperin S, Vaudry W, Law B, Embree J (2007) Invasive infections caused by *Haemophilus influenzae* serotypes in twelve Canadian IMPACT centers, 1996–2001. Pediatr Infect Dis J 26:1025–1031
6. Robbins JB, Schneerson R, Anderson P, Smith DH (1996) The 1996 Albert Lasker Medical Research Awards. Prevention of systemic infections, especially meningitis, caused by *Haemophilus influenzae* type b. Impact on public health and implications for other polysaccharide-based vaccines. JAMA 276:1181–1185
7. Campos J, Hernando M, Román F, PÉrez-VÁzquez M, Aracil B, Oteo J, LÁzaro E, De Abajo F (2004) Analysis of invasive *Haemophilus influenzae* infections after extensive vaccination against *H. influenzae* type b. J Clin Microbiol 42:524–529
8. Barbosa SF, Hoshino-Shimizu S, Alkmin MDGA, Goto H (2003) Implications of *Haemophilus influenzae* biogroup aegyptius hemagglutinins in the pathogenesis of Brazilian purpuric fever. J Infect Dis 188:74–80
9. Bengtson IA (1933) Seasonal acute conjunctivitis occurring in the southern states. Public Health Reports 1896–1970:917–926
10. Anderson R, Wang X, Briere EC, Katz LS, Cohn AC, Clark TA, Messonnier NE, Mayer LW (2012) *Haemophilus haemolyticus* isolates causing clinical disease. J Clin Microbiol 50:2462–2465
11. Morton DJ, Hempel RJ, Whitby PW, Seale TW, Stull TL (2012) An invasive *Haemophilus haemolyticus* isolate. J Clin Microbiol 50:1502–1503
12. Cardines R, Giufrè M, Ciofi Degli Atti M, Accogli M, Mastrantonio P, Cerquetti M (2009) *Haemophilus parainfluenzae* meningitis in an adult associated with acute otitis media. New Microbiol 32:213
13. Chattopadhyay B, Silverstone P, Winwood R (1983) Liver abscess caused by *Haemophilus parainfluenzae*. Postgrad Med J 59:788–789
14. Chunn CJ, Jones SR, Mccutchan JA, Young EJ, Gilbert DN (1977) *Haemophilus parainfluenzae* infective endocarditis. Medicine 56:99–113
15. Khor B-S, Liaw S-J, Liao H-C, Yang C-C, Lee M-H (2007) Bone and joint infections caused by*Haemophilus parainfluenzae*: report of two cases and literature review. Infect Dis Clin Pract 15:206–208
16. Chen CC, Wang SJ, Fuh JL (2001) Isolation of *Haemophilus parahaemolyticus* in a patient with cryptogenic brain abscess. Scand J Infect Dis 33:385–386
17. Le Floch AS, Cassir N, Hraiech S, Guervilly C, Papazian L, Rolain JM (2013) *Haemophilus parahaemolyticus* septic shock after aspiration pneumonia, France. Emerg Infect Dis 19:1694–1695

18. Pittman M (1953) A classification of the hemolytic bacteria of the genus haemophilus: haemophilus haemolyticus Bergey et al. and *Haemophilus parahaemolyticus* nov spec. J Bacteriol 65:750
19. Boucher MB, Bedotto M, Couderc C, Gomez C, Reynaud-Gaubert M, Drancourt M (2012) *Haemophilus pittmaniae* respiratory infection in a patient with siderosis: a case report. J Med Case Rep 6:120
20. Douglas GW, Buck LL, Rosen C (1979) Liver abscess caused by *Haemophilus paraphrohaemolyticus*. J Clin Microbiol 9:299–300
21. Nørskov-Lauritsen N, Bruun B, Andersen C, Kilian M (2012) Identification of haemolytic *Haemophilus* species isolated from human clinical specimens and description of *Haemophilus sputorum* sp. nov. Int J Med Microbiol 302:78–83
22. Trees DL, Morse SA (1995) Chancroid and *Haemophilus ducreyi*: an update. Clin Microbiol Rev 8:357–375
23. Mitjà O, Lukehart SA, Pokowas G, Moses P, Kapa A, Godornes C, Robson J, Cherian S, Houinei W, Kazadi W (2014) *Haemophilus ducreyi* as a cause of skin ulcers in children from a yaws-endemic area of Papua New Guinea: a prospective cohort study. Lancet Glob Health 2: e235–e241
24. Faden H, Bernstein J, Brodsky L, Stanievich J, Krystofik D, Shuff C, Hong JJ, Ogra PL (1989) Otitis media in children. I. The systemic immune response to nontypable *Haemophilus influenzae*. J Infect Dis 160:999–1004
25. Groenewegen KH, Wouters EF (2003) Bacterial infections in patients requiring admission for an acute exacerbation of COPD; a 1-year prospective study. Respir Med 97:770–777
26. Murphy T, Brauer AL, Schiffmacher AT, Sethi S (2004) Persistent colonization by *Haemophilus influenzae* in chronic obstructive pulmonary disease. Am J Respir Crit Care Med 170:266–272
27. Sunakawa K, Takeuchi Y, Iwata S (2011) Nontypeable *Haemophilus influenzae* (NTHi) epidemiology. Kansenshogaku Zasshi 85:227–237
28. Langereis JD, De Jonge IM (2015) Invasive disease caused by nontypeable *Haemophilus influenzae*. Emerg Infect Dis 21:1711
29. Moxon ER, Vaughn KA (1981) The type b capsular polysaccharide as a virulence determinant of *Haemophilus influenzae*: studies using clinical isolates and laboratory transformants. J Infect Dis 143:517–524
30. Kc R, Shukla SD, Walters EH, O'Toole RF (2017) Temporal upregulation of host surface receptors provides a window of opportunity for bacterial adhesion and disease. Microbiology 163:421–430
31. Rodriguez CA, Avadhanula V, Buscher A, Smith AL, Geme JW ST 3rd, Adderson EE (2003) Prevalence and distribution of adhesins in invasive non-type b encapsulated *Haemophilus influenzae*. Infect Immun 71:1635–1642
32. Krasan G, Cutter D, Block SL, Geme JWS (1999) Adhesin expression in matched nasopharyngeal and middle ear isolates of nontypeable *Haemophilus influenzae* from children with acute otitis media. Infect Immun 67:449–454
33. Vuong J, Wang X, Theodore JM, Whitmon J, De Leon PG, Mayer LW, Carlone GM, Romero-Steiner S (2013) Absence of high molecular weight proteins 1 and/or 2 is associated with decreased adherence among non-typeable *Haemophilus influenzae* clinical isolates. J Med Microbiol 62:1649–1656
34. St Geme JW 3rd (2002) Molecular and cellular determinants of non-typeable *Haemophilus influenzae* adherence and invasion. Cell Microbiol 4:191–200
35. Reddy MS, Bernstein JM, Murphy TF, Faden HS (1996) Binding between outer membrane proteins of nontypeable *Haemophilus influenzae* and human nasopharyngeal mucin. Infect Immun 64:1477–1479
36. Punturieri A, Copper P, Polak T, Christensen PJ, Curtis JL (2006) Conserved nontypeable *Haemophilus influenzae*-derived TLR2-binding lipopeptides synergize with IFN-beta to

increase cytokine production by resident murine and human alveolar macrophages. J Immunol 177:673–680
37. Ikeda M, Enomoto N, Hashimoto D, Fujisawa T, Inui N, Nakamura Y, Suda T, Nagata T (2015) Nontypeable *Haemophilus influenzae* exploits the interaction between protein-E and vitronectin for the adherence and invasion to bronchial epithelial cells. BMC Microbiol 15:1–11
38. Swords WE, Buscher BA, Ver Steeg II K, Preston A, Nichols WA, Weiser JN, Gibson BW, Apicella MA (2000) Non-typeable *Haemophilus influenzae* adhere to and invade human bronchial epithelial cells via an interaction of lipooligosaccharide with the PAF receptor. Mol Microbiol 37:13–27
39. Erwin AL, Smith AL (2007) Nontypeable *Haemophilus influenzae*: understanding virulence and commensal behavior. Trends Microbiol 15:355–362
40. Sonego M, Pellegrin MC, Becker G, Lazzerini M (2015) Risk factors for mortality from acute lower respiratory infections (ALRI) in children under five years of age in low and middle-income countries: a systematic review and meta-analysis of observational studies. PLoS One 10:e0116380
41. Trosini-Desert V, Germaud P, Dautzenberg B (2004) Tobacco smoke and risk of bacterial infection. Rev Mal Respir 21:539–547
42. Zalacain R, Sobradillo V, Amilibia J, Barrón J, Achótegui V, Pijoan JI, Llorente JL (1999) Predisposing factors to bacterial colonization in chronic obstructive pulmonary disease. Eur Respir J 13:343–348
43. Phipps JC, Aronoff D, Curtis JL, Goel D, O'brien E, Mancuso P (2010) Cigarette smoke exposure impairs pulmonary bacterial clearance and alveolar macrophage complement-mediated phagocytosis of *Streptococcus pneumoniae*. Infect Immun 78:1214–1220
44. Rylance J, Fullerton DG, Scriven J, Aljurayyan A, Mzinza D, Barrett S, Wright AKA, Wootton DG, Glennie SJ, Baple K, Knott A, Mortimer K, Russell DG, Heyderman RS, Gordon SB (2015) Household air pollution causes dose-dependent inflammation and altered phagocytosis in human macrophages. Am J Respir Cell Mol Biol 52:584–593
45. Van Der Vaart H, Postma DS, Timens W, Ten Hacken NHT (2004) Acute effects of cigarette smoke on inflammation and oxidative stress: a review. Thorax 59:713–721
46. Kc R, Hyland IK, Smith JA, Shukla SD, Hansbro PM, Zosky GR, Karupiah G, O'toole RF (2020) Cow dung biomass smoke exposure increases adherence of respiratory pathogen nontypeable *Haemophilus influenzae* to human bronchial epithelial cells. Exposure Health 2020:1–13
47. Kc R, Zosky GR, Shukla SD, O'toole RF (2018) A cost-effective technique for generating preservable biomass smoke extract and measuring its effect on cell receptor expression in human bronchial epithelial cells. Biol Methods Protocols 3:bpy010
48. O'Toole RF, Shukla SD, Walters EH (2016) Does upregulated host cell receptor expression provide a link between bacterial adhesion and chronic respiratory disease? J Transl Med 14:304
49. Shukla SD, Fairbairn RL, Gell DA, Latham RD, Sohal SS, Walters EH, O'Toole RF (2016) An antagonist of the platelet-activating factor receptor inhibits adherence of both nontypeable *Haemophilus influenzae* and *Streptococcus pneumoniae* to cultured human bronchial epithelial cells exposed to cigarette smoke. Int J Chron Obstruct Pulmon Dis 11:1647–1655
50. Shukla SD, Muller HK, Latham R, Sohal SS, Walters EH (2016) Platelet-activating factor receptor (PAFr) is upregulated in small airways and alveoli of smokers and COPD patients. Respirology 21:504–510
51. Erwin AL, Nelson KL, Mhlanga-Mutangadura T, Bonthuis PJ, Geelhood JL, Morlin G, Unrath WC, Campos J, Crook DW, Farley MM, Henderson FW, Jacobs RF, Muhlemann K, Satola SW, Van Alphen L, Golomb M, Smith AL (2005) Characterization of genetic and phenotypic diversity of invasive nontypeable *Haemophilus influenzae*. Infect Immun 73:5853–5863
52. Farjo RS, Foxman B, Patel MJ, Zhang L, Pettigrew MM, McCoy SI, Marrs CF, Gilsdorf JR (2004) Diversity and sharing of *Haemophilus influenzae* strains colonizing healthy children attending day-care centers. Pediatr Infect Dis J 23:41–46

53. Shen K, Antalis P, Gladitz J, Sayeed S, Ahmed A, Yu S, Hayes J, Johnson S, Dice B, Dopico R, Keefe R, Janto B, Chong W, Goodwin J, Wadowsky RM, Erdos G, Post JC, Ehrlich GD, Hu FZ (2005) Identification, distribution, and expression of novel genes in 10 clinical isolates of nontypeable *Haemophilus influenzae*. Infect Immun 73:3479–3491
54. Hogg JS, Hu FZ, Janto B, Boissy R, Hayes J, Keefe R, Post JC, Ehrlich GD (2007) Characterization and modeling of the *Haemophilus influenzae* core and supragenomes based on the complete genomic sequences of Rd and 12 clinical nontypeable strains. Genome Biol 8:R103
55. Dutta C, Pan A (2002) Horizontal gene transfer and bacterial diversity. J Biosci 27:27–33
56. Molin S, Tolker-Nielsen T (2003) Gene transfer occurs with enhanced efficiency in biofilms and induces enhanced stabilisation of the biofilm structure. Curr Opin Biotechnol 14:255–261
57. Murphy TF, Sethi S, Klingman KL, Brueggemann AB, Doern G (1999) Simultaneous respiratory tract colonization by multiple strains of nontypeable *Haemophilus influenzae* in chronic obstructive pulmonary disease: implications for antibiotic therapy. J Infect Dis 180:404–409
58. Lozupone C, Stombaugh JI, Gordon JI, Jansson JK, Knight R (2012) Diversity, stability and resilience of the human gut microbiota. Nature 489:220–230
59. Barenkamp SJ, Munson RS, Granoff D (1982) Outer membrane protein and biotype analysis of pathogenic nontypable *Haemophilus influenzae*. Infect Immun 36:535–540
60. Kilian M (1976) A taxonomic study of the genus *Haemophilus*, with the proposal of a new species. J Gen Microbiol 93:9–62
61. Demaria TF, Lim DJ, Barnishan J, Ayers LW, Birck HG (1984) Biotypes of serologically nontypable *Haemophilus influenzae* isolated from the middle ears and nasopharynges of patients with otitis media with effusion. J Clin Microbiol 20:1102–1104
62. Long SS, Teter MJ, Gilligan PH (1983) Biotype of *Haemophilus influenzae*: correlation with virulence and ampicillin resistance. J Infect Dis 147:800–806
63. Wallace Jr R, Musher D, Septimus E, McGowan Jr J, Quinones F, Wiss K, Vance P, Trier P (1981) *Haemophilus influenzae* infections in adults: characterization of strains by serotypes, biotypes, and β-lactamase production. J Infect Dis 144:101–106
64. Chin CL, Manzel L, Lehman EE, Humlicek A, Shi L, Starner TD, Denning GM, Murphy TF, Sethi S, Look DC (2005) *Haemophilus influenzae* from patients with chronic obstructive pulmonary disease exacerbation induce more inflammation than colonizers. Am J Respir Crit Care Med 172:85–91
65. Bresser P, Van Alphen L, Habets FJ, Hart AA, Dankert J, Jansen HM, Lutter R (1997) Persisting *Haemophilus influenzae* strains induce lower levels of interleukin-6 and interleukin-8 in H292 lung epithelial cells than nonpersisting strains. Eur Respir J 10:2319–2326
66. Bresser P, Out TA, Van Alphen L, Jansen HM, Lutter R (2000) Airway inflammation in nonobstructive and obstructive chronic bronchitis with chronic haemophilus influenzae airway infection. Comparison with noninfected patients with chronic obstructive pulmonary disease. Am J Respir Crit Care Med 162:947–952
67. Power PM, Sweetman W, Gallacher N, Woodhall M, Kumar G, Moxon E, Hood D (2009) Simple sequence repeats in *Haemophilus influenzae*. Infect Genet Evol 9:216–228
68. Bruce KD, Jordens J (1991) Characterization of noncapsulate *Haemophilus influenzae* by whole-cell polypeptide profiles, restriction endonuclease analysis, and rRNA gene restriction patterns. J Clin Microbiol 29:291–296
69. Meats E, Feil EJ, Stringer S, Cody AJ, Goldstein R, Kroll JS, Popovic T, Spratt BG (2003) Characterization of encapsulated and noncapsulated *Haemophilus influenzae* and determination of phylogenetic relationships by multilocus sequence typing. J Clin Microbiol 41:1623–1636
70. Van Alphen L, Caugant DA, Duim B, O'Rourke M, Bowler LD (1997) Differences in genetic diversity of nonencapsulated *Haemophilus influenzae* from various diseases. Microbiology 143:1423–1431
71. Maslow JN, Mulligan ME, Arbeit RD (1993) Molecular epidemiology: application of contemporary techniques to the typing of microorganisms. Clin Infect Dis 17:153–162. quiz 163-4

72. Selander RK, Caugant DA, Ochman H, Musser JM, Gilmour MN, Whittam TS (1986) Methods of multilocus enzyme electrophoresis for bacterial population genetics and systematics. Appl Environ Microbiol 51:873–884
73. Ibarz Pavón BA, Maiden MCJ (2009) Multilocus sequence typing. Methods Mol Biol (Clifton, NJ) 551:129–140
74. Maiden MCJ, Bygraves JA, Feil E, Morelli G, Russell JE, Urwin R, Zhang Q, Zhou J, Zurth K, Caugant DA, Feavers IM, Achtman M, Spratt BG (1998) Multilocus sequence typing: a portable approach to the identification of clones within populations of pathogenic microorganisms. Proc Natl Acad Sci 95:3140–3145
75. Boers SA, Van Der Reijden WA, Jansen R (2012) High-throughput multilocus sequence typing: bringing molecular typing to the next level. PLoS One 7:e39630–e39630
76. De Chiara M, Hood D, Muzzi A, Pickard DJ, Perkins T, Pizza M, Dougan G, Rappuoli R, Moxon E, Soriani M, Donati C (2014) Genome sequencing of disease and carriage isolates of nontypeable *Haemophilus influenzae* identifies discrete population structure. Proc Natl Acad Sci U S A 111:5439–5444
77. Kc R, Leong KWC, Harkness NM, Lachowicz J, Gautam SS, Cooley LA, McEwan B, Petrovski S, Karupiah G, O'toole RF (2020) Whole-genome analyses reveal gene content differences between nontypeable *Haemophilus influenzae* isolates from chronic obstructive pulmonary disease compared to other clinical phenotypes. Microb Genom 6. https://doi.org/10.1099/mgen.0.000405
78. Pettigrew MM, Ahearn CP, Gent JF, Kong Y, Gallo MC, Munro JB, D'Mello A, Sethi S, Tettelin H, Murphy TF (2018) *Haemophilus influenzae* genome evolution during persistence in the human airways in chronic obstructive pulmonary disease. Proc Natl Acad Sci U S A 115: E3256–e3265
79. Salipante SJ, Sengupta DJ, Cummings LA, Land TA, Hoogestraat DR, Cookson BT (2015) Application of whole-genome sequencing for bacterial strain typing in molecular epidemiology. J Clin Microbiol 53:1072–1079
80. Goodwin S, McPherson JD, McCombie WR (2016) Coming of age: ten years of next-generation sequencing technologies. Nat Rev Genet 17:333–351
81. Roth GA, Abate D, Abate KH, Abay SM, Abbafati C, Abbasi N, Abbastabar H, Abd-Allah F, Abdela J, Abdelalim A (2018) Global, regional, and national age-sex-specific mortality for 282 causes of death in 195 countries and territories, 1980–2017: a systematic analysis for the Global Burden of Disease Study 2017. The Lancet 392:1736–1788
82. Moghaddam SJ, Ochoa CE, Sethi S, Dickey BF (2011) Nontypeable *Haemophilus influenzae* in chronic obstructive pulmonary disease and lung cancer. Int J Chron Obstruct Pulmon Dis 6:113
83. Sriram KB, Cox AJ, Clancy RL, Slack MPE, Cripps AW (2018) Nontypeable *Haemophilus influenzae* and chronic obstructive pulmonary disease: a review for clinicians. Crit Rev Microbiol 44:125–142
84. Clancy RL, Dunkley M (2011) Acute exacerbations in COPD and their control with oral immunization with non-typeable *Haemophilus influenzae*. Front Immunol 2:7–7
85. Sethi S, Murphy TF (2001) Bacterial infection in chronic obstructive pulmonary disease in 2000: a state-of-the-art review. Clin Microbiol Rev 14:336–363
86. Moleres J, Fernandez-Calvet A, Ehrlich RL, Marti S, Perez-Regidor L, Euba B, Rodriguez-Arce I, Balashov S, Cuevas E, Linares J, Ardanuy C, Martin-Santamaria S, Ehrlich GD, Mell JC, Garmendia J (2018) Antagonistic pleiotropy in the bifunctional surface protein FadL (OmpP1) during adaptation of *Haemophilus influenzae* to chronic lung infection associated with chronic obstructive pulmonary disease. MBio 9
87. Didelot X, Bowden R, Wilson DJ, Peto TEA, Crook DW (2012) Transforming clinical microbiology with bacterial genome sequencing. Nat Rev Genet 13:601–612
88. Kao R, Haydon D, Lycett SJ, Murcia PR (2014) Supersize me: how whole-genome sequencing and big data are transforming epidemiology. Trends Microbiol 22:282–291
89. Kwong J, McCallum N, Sintchenko V, Howden BP (2015) Whole genome sequencing in clinical and public health microbiology. Pathology 47:199–210

90. Hurst JR, Vestbo J, Anzueto A, Locantore N, Müllerova H, Tal-Singer R, Miller B, Lomas DA, Agusti A, Macnee W, Calverley P, Rennard S, Wouters EF, Wedzicha JA (2010) Susceptibility to exacerbation in chronic obstructive pulmonary disease. N Engl J Med 363:1128–1138
91. Kaur R, Casey JR, Pichichero ME (2013) Relationship with original pathogen in recurrence of acute otitis media after completion of amoxicillin/clavulanate: bacterial relapse or new pathogen. Pediatr Infect Dis J 32:1159–1162
92. Harrison A, Hardison RL, Fullen AR, Wallace RM, Gordon DM, White P, Jennings RN, Justice SS, Mason KM (2020) Continuous microevolution accelerates disease progression during sequential episodes of infection. Cell Rep 30(e3):2978–2988
93. Beceiro A, Tomás M, Bou G (2013) Antimicrobial resistance and virulence: a successful or deleterious association in the bacterial world? Clin Microbiol Rev 26:185–230

Targeting Molecular and Cellular Mechanism in Rhinovirus Infection

22

Manju and Pranav Kumar Prabhakar

Abstract

More than 50% of the respiratory system infection is caused by rhinovirus, one of the most important members of *Picornaviridae* family. Since the discovery of rhinovirus by Dr. Price in 1956, it has been found to be the commonest cause of common cold. It is ubiquitous in prevalence, even though we have very limited knowledge and understanding of its mechanism of pathogenicity. The rhinovirus infection ultimately results in many different types of complications and abnormalities such as sinusitis, asthma and otitis media and other types of pulmonary diseases. An RT-PCR-based methodology has provided a way to identify the rhinovirus and its role in upper and lower respiratory systems. Here in this chapter we are going to discuss about the rhinovirus and immunological responses against rhinovirus. Finally, we are also going to explain various possible therapeutic approaches for the management of rhinovirus.

Keywords

Asthma · Rhinovirus · Pathogenesis · Viral load · Exacerbation

22.1 Introduction

Rhinoviruses which are a genus of Picornavirus family are the main cause of many human diseases. Some of the most common human diseases are common cough and cold (also known as virus-induced respiratory infections), viral meningitides, encephalitis and sepsis-like syndrome in neonates [1, 2]. More than 50% of the

Manju · P. K. Prabhakar (✉)
Department of Medical Laboratory Sciences, Lovely Professional University, Phagwara, Punjab, India

K. Dua et al. (eds.), *Targeting Cellular Signalling Pathways in Lung Diseases*,
https://doi.org/10.1007/978-981-33-6827-9_22

respiratory tract infection in humans are caused by the rhinoviruses. Rhinovirus, a member of *Picornaviridae* family, was isolated by and purified by Dr. Watson Price and his group at the Johns Hopkins University and also confirmed as the main cause of common cold in humans [3].

22.2 Structure of Rhinovirus (Picornavirus)

More than 100 serotypes of rhinovirus are included in the *Picornaviridae* family. The structures of these viruses are small (around 24–30 nm in diameter), are nonenveloped and contain a small nuclear capsid and a positive-sense single-stranded RNA of 7200–8500 nucleotides [4, 5]. The 5′ end of ssRNA of rhinovirus genome is protected with a protein of viral origin, and at 3′ end, it contains a poly (A) tail like the eukaryotic mRNA. The 5′ side of genome is responsible for the expression of the structural proteins, and non-structural proteins are encoded from the 3′ end of genome. After the entry of virus into the host cell, the viral genome is expressed; whole viral proteins are expressed as a single long stretch of polypeptide which is later cleaved into structural and non-structural proteins of virus [4, 6]. There are 4 different types of capsid proteins, namely, VP1, VP2, VP3 and VP4, arranged in 60 repeating protomeric icosahedral units (Fig. 22.1). Three out of four capsid proteins, VP1, VP2 and VP3, are present on the surface of capsid unit and possess antigenicity which is responsible for the host-pathogen immunological responses. The antigenic diversity among the rhinovirus serotypes is due to the variation in the capsid proteins VP1, VP2 and VP3. On the basis of the genetic homology, these serotypes can be divided into three categories of HRV-A, HRV-B and HRV-C. The

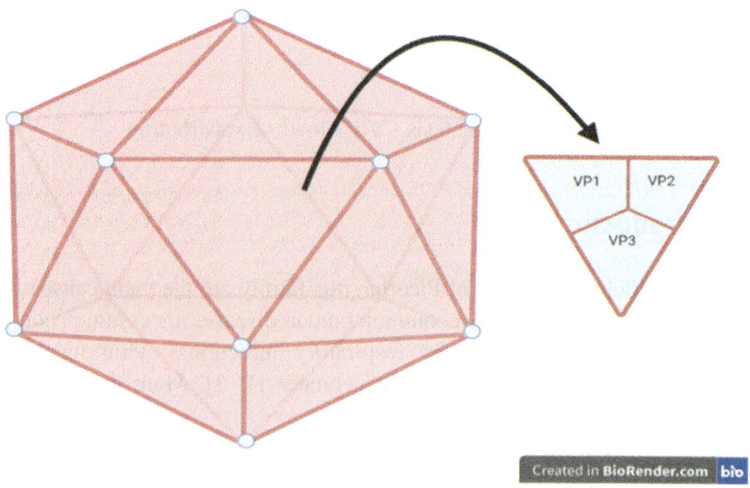

Fig. 22.1 Nonenveloped rhinovirus is a spherical icosahedral virus. Three types of viral capsid protein (VP1, VP2 and VP3) cover the viral single-stranded positive-sense RNA. The fourth capsid protein connects capsid with the genetic material

fourth capsid protein VP4 is present inside the virus and helps in anchoring ssRNA with the viral capsid. Overall, these capsid proteins protect the viral genome [3–5].

On the basis of the host cellular receptor, human rhinoviruses are classified into two categories. More than 90% of the HRV-A and HRV-B rhinovirus families enter into host cell through the attachment with the intercellular adhesion molecule-1 (ICAM-1), and the remaining rhinoviruses enter via the low-density lipoprotein receptor [4, 5]. On the capsid protein VP1, a hydrophobic pocket (known as canyon) is present which is the point where ICAM-1 at which virus gets connected with the host cells [7]. The fourth viral capsid protein VP4 is located on the internal portion and is required during the assembly of new viral particles at the time of replication and infection to the fresh healthy host cells [8]. In recent times, HRV-C has gained more interest due to rhinovirus-associated exacerbation of asthma [9]. Even the genomic sequencing of HRV-C has not able to elucidate the exact mechanism of infection and the receptor involved in the pathogenesis especially in the case of epithelial cells. The structural modelling analysis reveals that HRV-C neither using ICAM-1 nor LDL receptor. HRV-C has not been studied extensively, and hence very less information is available regarding the pathogenesis process.

22.3 Pathogenesis of Rhinovirus

The main route for the entry of rhinovirus is upper respiratory tract organs like the mouth and nasal epithelium. Viral load can be detected in the nasal secretion after 24 h of infection, and the viral load reaches to its peak on second or third day of infection. After the third day, the declination in the viral titre can be seen, and mostly on the fifth day, it is undetectable. Symptom for the common cold starts on the second day of infection and reaches to its peak on third or fourth day. The exact mechanism of rhinitis is not very clear that it is due to the cytocidal properties of virus or due to the release of chemical messengers. Histamine has not shown any role in the case of the occurrence of rhinitis, whereas kinins have been seen in elevated amount.

22.3.1 Pathogenesis of Rhinovirus in the Respiratory Tract (Upper and Lower)

In non-asthmatic people, manifestations of rhinovirus contamination are commonly restricted to the upper part of the respiratory tract. Rhinorrhoea and the congestion of nasal chamber due to mucus, the foremost noticeable indications, are related with a neutrophilic incendiary reaction that's related with expanded vascular permeability and induction of bodily fluid hypersecretion. Cough is an uncommon however irksome sign of rhinovirus URI. The pathogenesis of cough may include aggravation from back pharyngeal seepage or direct disease of the large airway's routes. Gwaltney et al. [10] showed sinus inclusion in numerous people with ordinary regular cold manifestations. The sinus infection settled without intercession proposing that these upper respiratory abnormalities ought to be more precisely described as a viral rhinosinusitis. Nonetheless, the irritation related with hindrance of sinus openings and optional Eustachian tube brokenness can incline to intense bacterial sinusitis and otitis media, individually.

Conversely, lower respiratory side effects related with rhinovirus contamination are generally conspicuous in patients who have basic asthma or other ongoing lung infection. These manifestations incorporate hack, windedness, chest snugness and wheezing [11–14]. The reason for these lower respiratory side effects has been a wellspring of contention concerning components of rhinovirus pathogenesis. In particular, the basic discussion focuses on the degree to which rhinovirus can contaminate cells of the lower respiratory tract and, thusly, regardless of whether bronchial disease shapes the reason for respiratory manifestations, rather than reflecting backhanded impacts identified with the safe reaction to the upper aviation route contamination. There is an assortment of likely hindrances to disease of lungs by rhinovirus including the temperature affectability of replication of the infection. Rhinovirus imitates ideally at 33 °C, a temperature essentially lower than that of bronchial aviation route epithelium [15]. It is imperative that rhinovirus has been correspondingly disconnected with bacterial microorganisms in 24–54% of youngsters and 10–18% of grown-ups with pneumonia [16–18]. In this manner, more imperative to respiratory tract irresistible ailment pathogenesis during rhinovirus contamination, might be the limit of rhinovirus to incline to accompanying or resulting disease with other respiratory microorganisms. For example, human tracheal epithelial cells at the same time tainted with RV14 and *Strept. pneumoniae* show expanded adherence of the *Strept.* [19]. So also, macrophages presented to rhinovirus demonstrated debilitated responsiveness of Pattern Recognition Receptors (PRRs) following presentation to bacterial cost like receptors (TLR) agonists for example lipopolysaccharide and lipoteichoic corrosive [20]. Moreover, considers embroiling rhinovirus contribution in lower respiratory tract disease dependent on the identification of rhinovirus antigens (or genomes) in lower respiratory tract routes are jumbled by the powerlessness to bar upper respiratory tract inferred rhinovirus tainting of bronchial samples. This is unquestionably risky with sputum investigation, yet even bronchoscopically acquired examples can be tainted during the bronchoscope's section through the upper aviation route. Notwithstanding, a few investigations uphold the presence of rhinovirus in the lower aviation route [21, 22] including work demonstrating rhinovirus by in situ hybridisation after exploratory RV16 contamination [23]. Work by this group likewise shows that while rhinovirus serotypes reproduce ideally at 33°, i.e. the temperature of the upper respiratory lot, the higher temperature of the lower aviation routes isn't a flat-out boundary to rhinovirus replication [24]. The prevalence of current feeling subsequently underpins the idea that rhinovirus probably can gainfully contaminate cells of the lower aviation routes.

22.3.2 Pathogenic Influences of Rhinovirus on the Epithelium

Rhinovirus can be transmitted from a carrier through direct contact or fomite contact or through aerosol. The virus enters into a healthy person through intranasal or conjunctival inoculation. Within 24 h of conjunctival inoculation, virus is transported into the nasopharynx via the lacrimal duct [25]. The nasal mucosa is the main site for rhinovirus infection. Based upon functional, ultrastructural and biochemical criteria, nasal epithelial cells are classified as ciliated, secretory and basal cells [26]. Other cells include dendritic cells, macrophages, mast cells and lymphocytes. The terminal processes of cholinergic and sensory nerves are also observed [27]. The mucous secreted by goblet cells and submucosal glands traps the inhaled foreign particles such as fungal spores, bacteria and viruses [28]. Mucous contains ions, water, immunoglobulins (such as IgA) and glycoproteins [29, 30]. Cilia transport the entrapped infectious particle along with the mucous to the oral cavity where they are swallowed and digested, resulting in the destruction of the foreign particle. However, if the balance between the composition and volume of mucus, ciliary beat, and periciliary fluid is disturbed, it may lead to the increased susceptibility to pathogens [31].

Majority of known RV serotypes, i.e. HRV-A and HRV-B, enter the airway pathway by attachment with the bound ICAM-1 (intercellular adhesion molecule-1) present on the surface of airway epithelial cell, while other minor serotypes enter through LDL (low-density lipoproteins) [32–34]; recently identified CDHR3 might serve as receptor for HRV-C [35]. The RV infection and allergic reactions upregulate the expression of membrane-bound ICAM-1 by NF-κβ-dependent mechanism. Respiratory viruses including respiratory synovial virus or influenza virus are associated with respiratory epithelial cell destruction. However, the epithelial cell structure and composition remain intact during the rhinovirus infection and are seldom associated with cytopathology effects [25, 36]. However, it causes dissociation of zona occludens 1 that leads to the disruption of epithelial cell barrier function; henceforth transmigration of pathogens and their soluble products and exposure of basolateral epithelial cell receptors such as TLR take place [7]. The RV infections are mostly associated with the upper respiratory tract; however, some asthmatic individuals may develop lower respiratory tract infection as well.

Virus uptake takes place through clathrin-independent or clathrin-dependent endocytosis or through micropinocytosis [8]. In case of a minor group like HRV-A2, after binding with LDLR present at the surface of ciliated epithelial cell, virus is internalised into early endosome. The mildly acidic environment present in this compartment triggers the virus-receptor dissociation. The virus-free LDL receptor through perinuclear recycling endosomes returned to the plasma membrane, and virus enters late endosomes. Under the low pH (≤ 5.6), viruses are converted into subviral A particles that contain viral genome but are devoid of VP4 (innermost protein coat). The viral RNA genome is released into the cytoplasm, and degradation of subviral B particles (empty capsid) takes place in lysosomes (Papadopoulos et al., 2000). In case of a major group like HRV-A89, both virus and ICAM-1 are sorted into recycling pathway. The acidic condition in perinuclear recycling endosome

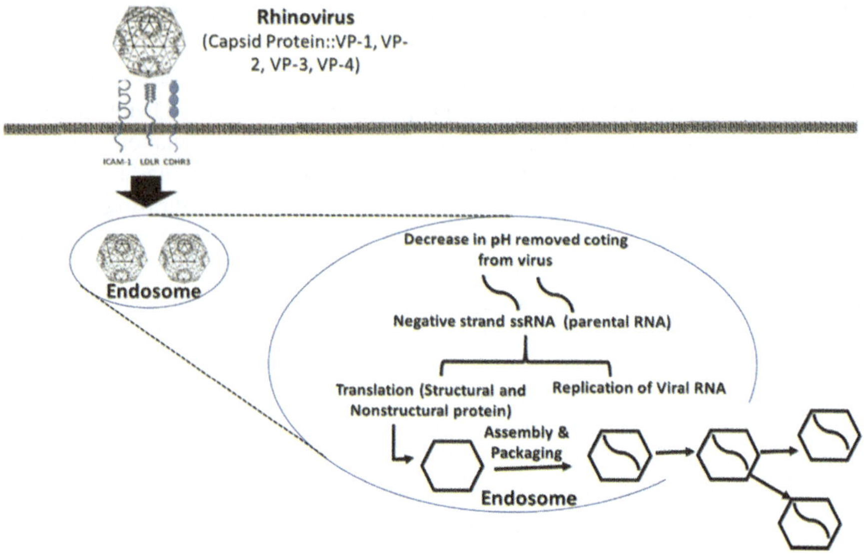

Fig. 22.2 Replication of rhinovirus in the host epithelial cells. Virus is taken inside the cells via different modes like clathrin-coated vesicle or clathrin-independent (low-density lipoprotein receptor (LDLR) or intercellular adhesion molecule-1 (ICAM-1). After endocytosis, there is a decrease on endocyte's pH which facilitates uncoating of viral particles. Single-stranded negative strand of viral RNA is translated into structural and non-structural viral protein which later spliced and replicated. The virion particles are assembled into viral particles, and packed viral particles are exported once cell lysed

leads to the conversion of virus into subviral A and B particles. Once viral RNA genome enters the cytosol, the empty capsid returns to apical mucous layer [37]. The release of RNA genome into the cytosol takes place either by rupturing the endosomal membrane or by forming pores in the membrane. As (+) ssRNA genome enters the cytosol, translation of viral genome into a polyprotein begins with the help of host cell ribosomes. After the autocatalytic cleavage of polyprotein into structural (VP1, VP2, VP3 and VP4 capsid proteins) and non-structural proteins takes place, viral polymerase initiates RNA replication. Biosynthesis is followed by the assembly of virus and then release of viral progeny into the nasal cavity (Fig. 22.2). In contrast to the cell lysis for the release of viral progeny, cell-to-cell spread of rhinovirus takes place via virus-carrying macrovesicles [38].

There are a number of viral pathogens such as influenza virus and respiratory syncytial virus which damage and destroy the nasal epithelial barrier, but the studies conducted on rhinovirus have confirmed that it doesn't damage epithelial cells and not cause cytopathology. This was studied by using monolayer adenoid tissue, and infected with rhinovirus and at the point of peak secreted viral titres, no damages to the adenoid tissue monolayer have been observed, and there was no cytopathic effect as well [25]. This is steady with the inability to watch cytopathology in rhinovirus-tainted nasal or bronchial biopsy tissue. Contamination does, notwithstanding, upset epithelial hindrance work. The impacts of rhinovirus to increment vascular spillage and bodily fluid discharge reflect partially this capacity of the rhinovirus to upset the epithelial obstruction, explicitly the interruption of tight intersections. Studies using refined human nasal epithelial cells demonstrated diminished zona occludens 1, claudin-1 and E-cadherin mRNA and protein levels after contamination with rhinovirus [39]. This is reliable with perceptions in regard to the interruption of aviation route epithelial apical intersections by poly(dI:dC) [40]. Notwithstanding expanding porousness, this disturbance of the epithelial obstruction will encourage movement of microorganisms (counting non-rhinovirus microbes) and their solvent items, and uncover basolateral epithelial receptors, where TLR and different PRRs are conspicuously found.

22.4 Immune Response to Rhinovirus

Without a capacity to credit the presence and degree of side effects to either virus titre or cytopathology, we recommend that it is the qualities of the host immune response against rhinovirus that are the essential determinant of indications. The host reaction to the infection incorporates those interceded by the innate immunity, humoral immunity and cell-mediated immunity. Somewhat these particular reactions speak to a continuum with the reformist advancement of more serious (and more indicative) reactions, in spite of the fact that the particular succession of this continuum may differ from patient to tolerant.

(a) *Innate immunity against rhinovirus infection*: When the host lacks pre-sensitised humoral immune component or some other mucosal epithelial surface-associated components, rhinovirus is able to infect the host epithelium and initiates the innate immune responses. The induction and initiation of innate immune response is very fast as it was evident in many studies. The studies revealed that the rhinovirus infection induces type-I interferon production and reduction in the nasal epithelial pH in less than 24 h after the induced infection [41]. The identification or detection of rhinovirus by innate immune components solely depends on the recognition of rhinovirus-associated PRR such as TLR, retinoic acid-inducible gene-1 (RIG-1) and melanoma differentiation-associated gene-5 (MDA-5). Host TLR-2, present on the cell surface of epithelial cells, is able to recognise rhinovirus capsid protein, and once rhinovirus has been internalised and rhinovirus-induced translation started, the single-stranded RNA and double-stranded DNA of rhinovirus are identified by the endosomal THR-3, TLR-7 and TLR-8. MDA-5 and RIG-1 also able to detect the double-stranded DNA (Fig. 22.3) [42]. This recognition of rhinovirus PAMPs with the host PRR induces the expression of a number of cytokines and releases cytokines such as type-I interferon (interferon alpha and beta), type-III interferons (interleukins 28-A, 28-B and 29) and some other cytokines like IL-6, IL-12 and IL-15. The antiviral mechanism of action of interferons is mediated by the inhibition of viral replication, whereas the mode of action of

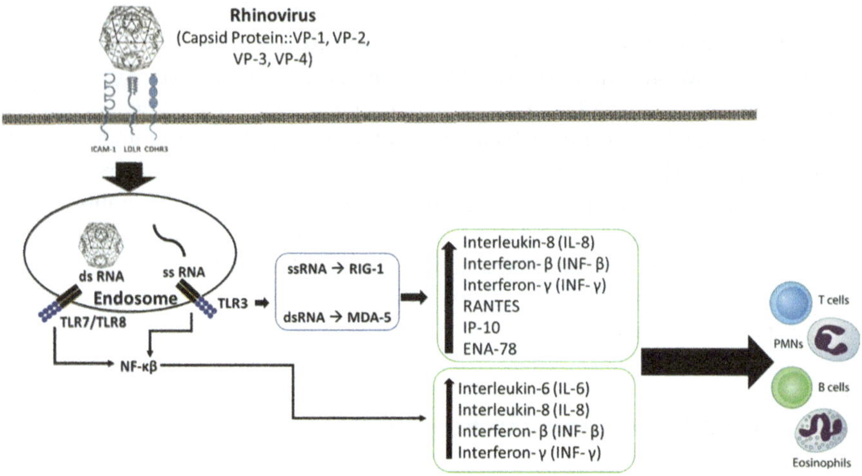

Fig. 22.3 Induction of innate immune system and activation of various cellular components of innate immunity. Viral particles enter into host cells through the binding to the host cell receptor. Once viral particle is inside the endosome inside host cell, decrease in pH uncoated viral particles. Viral dsRNA and ssRNA genetic material and newly produced genetic material are detected by the toll-like receptor-3 (TLR-3) and toll-like receptor-7/8 (TLR-7/8). The recognition of ssRNA and dsRNA by TLR-3 induces the expression of retinoic acid-inducible gene-1 (RIG-1) and melanoma differentiation-associated protein-5 (MDA-5), respectively. These RIG-1 and MDA-5 induce the expression and production of various cytokines such as RANTES (regulated on activation, normal T-cell expressed and secreted), IP-10 (IFN-γ-induced protein 10) and ENA-78 (epithelial-derived neutrophil-activating peptide 78). Other than these, the interaction of TLR-7 and TLR-8 induces the expression of IFN-β- and IFN-γ mediated through NF-κβ

other cytokines like IL-12 and IL-15 is through the induction of proliferation and differentiation of cytotoxic T-cell and natural killer cells, their survival and recruitment at the site of action [43]. The increased natural killer cells are the primary and early source of interferon-gamma secretion. Interleukin-6 also plays an important role in the induction of different aspects of innate immune components which is responsible for the removal of rhinovirus [44], and a single-nucleotide polymorphism in the interleukin-6 gene leads to serious illness [45]. Rhinovirus-infected epithelial cells also release some cytokines such as interleukin-1ß and interleukin-11, which also modulated immune system for the fight against rhinovirus. Apparently the most significant determinants of the clinical result of rhinovirus disease involve a number of growth factors such as G-CSF and GM-CSF, and various chemokines such as interleukin-8 (IL-8 or chemokine C-X-C motif ligand 8, CXCL8), chemokine (C-X-C motif) ligand 5 (CXCL5) or epithelial-derived neutrophil-activating peptide 78 (ENA-78), C-X-C motif chemokine ligand 10 (CXCL10) or interferon-gamma-induced protein 10 (IP-10) and chemokine (C-C motif) ligand 5 (CCL5) or regulated on activation, normal T-cell expressed and secreted (RANTES) that together drive granulocyte enlistment, survival and enactment. These granulocytes are mostly neutrophils, reflecting particularly the activities of CXCL8 and CXCL10. These mediators rapidly increase in the nasal secretion and serum of rhinovirus-infected person, and its concentration parallelly increases in the peripheral blood neutrophils. The exact role of polymorphonuclear leukocytes in the case immunity against rhinovirus is not very clear; however the resulting neutrophil-loaded nasal exudate is one of the more trademark highlights of 'colds', and the early articulation of CXCL8 and CXCL10 connects to the presence of indicative RV diseases [44]. A neutrophilic secretion has also been related with increments in kallikrein, which drives the creation of kallidin and bradykinin [45]. These kinins are raised in the nasal washes of the patients with rhinovirus infection and have sufficient symptoms, especially in those with sensitivities and asthma [46, 47]. Eosinophils can likewise be powerfully communicated [48]. The acceptance of eosinophilia may impact the capacity of RV to create nasal side effects by upgrading observer unfavourably susceptible responses (talked about underneath). Interestingly, eosinophils, to a limited extent through their capacity to discharge various strong RNAses, seem to advance infection annihilation [49].

(b) *Humoral immunity against rhinovirus*: The role of humoral immune system against the rhinovirus in the rhinovirus-induced patients is increasingly understood. In an experimental study with the rhinovirus infection, B-cell responses have been detected in the form of mucosal rhinovirus-specific immunoglobulin on third day of infection and immunoglobulin G on seventh to eighth days of infection [50]. A part for this humoral reaction is proposed by perceptions that the presence of serotype-explicit killing IgG antibodies blocks resulting challenge disease following exploratory vaccination with a RV of that serotype [51]. It ought to be underlined that given the requirement for killing neutraliser to be available at the nasal mucosa limit, almost certainly, secretory IgA would be the genuine determinant of defensive humoral insusceptibility. Antibodies could add to viral freedom by going about as killing antibodies, e.g. stopping cellular-associated ligands, opsonising the infection for introduction to phagocytic cells or starting NK cell-interceded counteracting agent subordinate cell cytotoxicity. Notwithstanding immediate infection balance, prior antibodies may likewise serve to intercede neutraliser-encouraged antigen uptake and promote quicker and more active cell-based immunological responses [52]. The idea that humoral immunological reactions have a prime role in the forestalling and destroying contamination is additionally derived from perceptions with respect to the expanded recurrence and seriousness of disease in patients with humoral safe disappointment (e.g. normal variable insusceptible inadequacy). In these conditions, RV was the most widely recognised infection delivering respiratory contaminations [21]. This was not rectified with substitution immunoglobulin, further involving the requirement for serotype-explicit antibodies, which could be deficient in some random business immunoglobulin arrangement.

(c) *T-cell-based immunity against rhinovirus*: When there is an unavailability of active innate immunity and neutralising immunoglobulin, the significance of rhinovirus-associated T-cell-mediated damage of the rhinovirus becomes very important. The viral titres start declining which gives an indication that there might be de novo activation of rhinovirus-specific T-cell just after 72 h of infection. This observed watched time period is just reliable with enactment of prior effector T-cell or memory T-cells, which should therefore react to shared epitope(s) shown by the infective rhinovirus. Due to rhinovirus infection as a consequence, both CD-4- and CD-8-specific T-cell responses get induced. CD4 cells are mainly Th-1 type of T-cells, and the interferon gamma released by them is responsible for the antiviral immunological responses, but these CD-4-specific T-cells also induce the development of humoral immune system components [21]. CD-8-specific T-cells are one of the centre points in the adaptive immune system-associated removal of rhinovirus from the host body; however, the availability and their significance have not been studied in detail. These T-cells can be detected after the rhinovirus infection, and the main characteristic feature is the production of interferon gamma. An extra molecular mechanism which may produce the induction of some rhinovirus-associated symptoms—like different respiratory infection focusing on resistant reactions—the inclination of these cells to associatively communicate IL-10 and subsequently moderate observer safe intervened harm [53, 54].

22.5 Therapeutic Approaches against Rhinovirus

Even after the tremendous scientific efforts by scientist and researcher, there is neither approved medicine nor any vaccine available for the management of rhinovirus infection. However, the clinical trials for the vaccines have been started in the 1960s. Scientific community has achieved some mixed success in these strategies. Various strategies have been used by the scientific communities to target rhinovirus.

22.5.1 Strategies to Targeting Virus

Capsid Binding Agents Rhinovirus is a nonenveloped virus which has ssRNA as the genetic material. There are four viral capsid proteins (VP1–VP4) which form an icosahedral structure and seven non-structural proteins. Some of these viral proteins and enzymes have been used to target rhinovirus such as capsid proteins, protease and RNA-dependent RNA polymerase ([55], [56], [57]). Among all these proteins, viral capsid proteins were the first viral components used as a therapeutic target for the inhibition of rhinovirus replication. These categories of compounds have affinity for the hydrophobic binding pocket present on the capsid of rhinovirus and result in a modulation in the conformation of capsid and increase their stability which ultimately reduced the binding capacity of viral capsid with host target protein [58]. Inhibition of these viral components shows an advantage to reducing the pathogenicity, but the disadvantage is that it leads to mutation-induced resistance.

The viral pathogenesis required the binding of viral capsid proteins with the host cell receptors such as ICAM-1 (by HRV-A HRV-B), LDL receptor (by HRV-A) and cadherin-related family member 3 (by HRV-C). This host receptor binds on the specific site on the viral capsid called canyon, and the canyon is the site for drug molecules which block virus entry into host cell through binding with the capsid. The first-generation drugs of capsid binder are WIN compounds, a drug with narrow strain specificity. One of the derivatives of these compounds, Picovir or pleconaril, reached clinical trial, but later the FDA has rejected due to some serious safety concern. The bioavailability of this pleconaril is almost 70% and has a large half-life.

Only one capsid binding drug which results in positive clinical data was Vapendavir molecule [59]. The Vapendavir is an experimental medicine administered orally, and the binding site for this is present on the VP1 viral capsid protein. This binding prevents the release of viral ssRNA genetic material into the host target cell. Vapendavir has shown some good result against HRV-A and HRV-B serotypes, whereas against HRV-C the result is not very conclusive.

Protease Inhibitors Another important target to block the rhinovirus pathogenesis is through blocking two proteases expressed by viral genome, namely, $2A^{pro}$ and $3C^{pro}$. These proteases are structurally similar to trypsin but differ in the active site nucleophile where serine is replaced with cysteine. The $2A^{pro}$ cleaves the polyprotein and separates structural and non-structural proteins. This has very limited activities for viral protein, but it works on host transcription factor (eIF4G) and blocks the host protein synthesis but let viral protein synthesis continue. Most of the drugs which target viral protease target $3C^{pro}$. One of the $3C^{pro}$ protease inhibitor is Rupintrivir, manufactured by Agouron and Pfizer in the form of nasal spray [60].

22.5.2 Strategies to Targeting Host Protein

There are some host proteins which can also be targeted for the management of rhinovirus infection. As during infection, these viruses hijack most of the cellular activities of cells by taking control over the cellular components. Virus uses host cellular component for the replication and viral protein synthesis. Inhibition of these host protein activities may also inhibit the replication and viral protein synthesis. One of the important and considerable drawbacks associated with inhibiting and blocking host protein is the toxicity it causes. As one of the large groups of human rhinovirus uses ICAM as the viral receptor for the entry into host cell, Boehringer Ingelheim designed soluble ICAM with the name of Tremacamra. Experimental data shows the reduction in the common cold symptom, but it has been stopped and discontinued [61].

Not only rhinovirus but all Picornavirus family uses cytosolic side of host's Golgi body and endoplasmic reticulum as a surface for their genetic material replication, and the structure and chemical composition of this membrane are modulated by the virus. According to studies, the non-structural protein of picornaviruses, 2-B, 2-C and 3-A, played an important role in the modulation of composition and morphology of ER membrane after associating with them. A detailed study of the interaction of these viral proteins and host proteins revealed a novel therapeutic target for the management of the designing of antiviral drugs. Enviroxime and other similar type of compounds were the first antiviral drug designed for the influencing of membrane remodelling feature [62]. Enviroxime has been failed due to some of the physico-chemical properties of drug such as weak exposure, its toxic effect and lacking efficacy when used through both oral and nasal routes [63].

One of the important enzyme Phosphatidylinositol-4-kinase IIIβ (PI4KIIIβ) play very significant role in the membrane trafficking in the Golgi body rhinovirus interact with PI4KIIIβ through its non-structural 3-A protein. The detailed mechanism of this interaction is not clear [64]. Viral RNS-dependent RNS polymerase (3-D^{pol}) hooks up on the phosphatidylinositol-4-phosphate (PtdIns4P)-rich membrane, and the PI4KIIIβ blockers like PIK93 successfully stop viral replication [65]. And hence a number of pharmaceutical industries targeted this protein complex; however there was no very much success.

It has been reported that another rhinovirus protein 3-A interacts with the host protein Arf guanosine triphosphate exchange factor GBF1 [66, 67]. The guanosine triphosphate binding protein ARF-1 plays a significant role in the endoplasmic reticulum and Golgi body-mediated membrane trafficking and activates PI4KIIIβ. This also affects the replicative process of rhinovirus through the inhibition of DNA replication by GBF1-suppressive protein Brefeldin A [68]. Oxysterol binding protein (OBP) joins PtdIns4P and Arf-1 and works as regulator for cholesterol homoeostasis. Some of the drugs against picornavirus and rhinovirus target this oxysterol binding protein like enviroxime compounds [69].

22.6 Summary

Still it is a part of discussion that the symptoms coned during the rhinovirus infection are due to the pathogenicity of it or it is due to the microenvironment in which the virus replicates and proliferates. We have seen that the many features of our immune system are responsible in regulating and restricting rhinovirus infections but at the same time reducing symptoms as well, as the virus itself is not cytopathic in nature. There are many immunological components which play their role in the rhinovirus-induced immunity which are innate immune system, adaptive immune system and cell-based immunological responses. As we will explore the molecular mechanism much more, it will open many future directions for the management and treatment strategies for rhinovirus, especially in those asthmatic patients where asthma goes severe exacerbation due to rhinovirus infections.

References

1. Kennedy JL, Pham S, Borish L (2019) Rhinovirus and asthma exacerbations. Immunol Allergy Clin N Am 39:335–344

2. Rotbart HA, Hayden FG (2000) Picornavirus infections: a primer for the practitioner. Arch Fam Med 9:913
3. Peltola V, Waris M, Österback R, Susi P, Hyypiä T, Ruuskanen O (2008) Clinical effects of rhinovirus infections. J Clin Virol 43:411–414
4. Pitkäranta A, Hayden FG (1998) Trends in clinical practice rhinoviruses: important respiratory pathogens. Ann Med 30:529–537
5. San-Juan-Vergara H, Peeples ME (2019) Importance of virus characteristics in respiratory syncytial virus-induced disease. Immunol Allergy Clin N Am 39:321–334
6. Han M, Rajput C, Hershenson MB (2019) Rhinovirus attributes that contribute to asthma development. Immunol Allergy Clin N Am 39:345–359
7. Kennedy JL, Turner RB, Braciale T, Heymann PW, Borish L (2012) Pathogenesis of rhinovirus infection. Curr Opin Virol 2:287–293
8. Fuchs R, Blaas D (2010) Uncoating of human rhinoviruses. Rev Med Virol 20:281–297
9. Warner SM, Wiehler S, Michi AN, Proud D (2019) Rhinovirus replication and innate immunity in highly differentiated human airway epithelial cells. Respir Res 20:150
10. Gwaltney JM, Phillips CD, Miller RD, Riker DK (1994) Computed tomographic study of the common cold. N Engl J Med 330:25–30
11. Collinson J, Nicholson KG, Cancio E, Ashman J, Ireland DC, Hammersley V, Kent J, O'callaghan C (1996) Effects of upper respiratory tract infections in patients with cystic fibrosis. Thorax 51:1115–1122
12. Folkerts G, Busse WW, Nukamp FP, Sorkness R, Gern JE (1998) Virus-induced airway hyperresponsiveness and asthma. Am J Respir Crit Care Med 157:1708–1720
13. Heras JL, Swanson VL (1983) Sudden death of an infant with rhinovirus infection complicating bronchial asthma: case report. Pediatr Pathol 1:319–323
14. Nicholson KG, Kent J, Hammersley V, Cancio E (1996) Risk factors for lower respiratory complications of rhinovirus infections in elderly people living in the community: prospective cohort study. BMJ 313:1119–1123
15. Killington RA, Stott EJ, Lee D (1977) The effect of temperature on the synthesis of rhinovirus type 2 RNA. J Gen Virol 36:403–411
16. Jennings LC, Anderson TP, Beynon KA, Chua A, Laing RTR, Werno AM, Young SA, Chambers ST, Murdoch DR (2007) Incidence and characteristics of viral community-acquired pneumonia in adults. Thorax 61:42–48
17. Juvén T, Mertsola J, Waris M, Leinonen M, Ruuskanen O (2004) Clinical response to antibiotic therapy for community-acquired pneumonia. Eur J Pediatr 163:140–144
18. Templeton KE, Scheltinga SA, Van Den Eeden WCJFM, Graffelman WA, Van Den Broek PJ, Claas ECJ (2005) Improved diagnosis of the Etiology of community-acquired pneumonia with real-time polymerase chain reaction. Clin Infect Dis 41:345–351
19. Ishizuka S, Yamaya M, Suzuki T, Takahashi H, Ida S, Sasaki T, Inoue D, Sekizawa K, Nishimura H, Sasaki H (2003) Effects of rhinovirus infection on the adherence of *Streptococcus pneumoniae* to cultured human airway epithelial cells. J Infect Dis 188:1928–1939
20. Oliver BGG, Lim S, Wark P, Laza-Stanca V, King N, Black JL, Burgess JK, Roth M, Johnston SL (2008) Rhinovirus exposure impairs immune responses to bacterial products in human alveolar macrophages. Thorax 63:519–525
21. Gern JE, Galagan DM, Jarjour NN, Dick EC, Busse WW (1997) Detection of rhinovirus RNA in lower airway cells during experimentally induced infection. Am J Respir Crit Care Med 155:1159–1161
22. Halperin SA, Eggleston PA, Hendley JO, Suratt PM, Gröschel DHM, Jack M, Gwaltney J (1983) Pathogenesis of lower respiratory tract symptoms in experimental rhinovirus infection. Am Rev Respir Dis 128:806–810
23. Papadopoulos NG, Bates PJ, Bardin PG, Papi A, Leir SH, Fraenkel DJ, Meyer J, Lackie PM, Sanderson G, Holgate ST, Johnston SL (2000) Rhinoviruses infect the lower airways. J Infect Dis 181:1875–1884

24. Papadopoulos NG, Sanderson G, Hunter J, Johnston SL (1999) Rhinoviruses replicate effectively at lower airway temperatures. J Med Virol 58:100–104
25. Winther B, Gwaltney JM Jr, Mygind N, Turner RB, Hendley JO (1986) Sites of rhinovirus recovery after point inoculation of the upper airway. JAMA 256:1763–1767
26. Tam A, Wadsworth S, Dorscheid D, Man SFP, Sin DD (2011) The airway epithelium: more than just a structural barrier. Ther Adv Respir Dis 5:255–273
27. Knight DA, Holgate ST (2003) The airway epithelium: structural and functional properties in health and disease. Respirology 8:432–446
28. Jeffery PK (1983) Morphologic features of airway surface epithelial cells and glands. Am Rev Respir Dis 128:S14–S20
29. Jeffery PK (1991) Morphology of the airway wall in asthma and in chronic obstructive pulmonary disease. Am Rev Respir Dis 143:1152–1158
30. Yoshida M, Kobayashi K, Kuo TT, Bry L, Glickman JN, Claypool SM, Kaser A, Nagaishi T, Higgins DE, Mizoguchi E, Wakatsuki Y, Roopenian DC, Mizoguchi A, Lencer WI, Blumberg RS (2006) Neonatal Fc receptor for IgG regulates mucosal immune responses to luminal bacteria. J Clin Invest 116:2142–2151
31. Song Y, Salinas D, Nielson DW, Verkman AS (2006) Hyperacidity of secreted fluid from submucosal glands in early cystic fibrosis. Am J Phys Cell Phys 290:C741–C749
32. Bella J, Rossmann MG (2000) ICAM-1 receptors and cold viruses. Pharm Acta Helv 74:291–297
33. Jacobs SE, Lamson DM, Soave R, Guzman BH, Shore TB, Ritchie EK, Zappetti D, Satlin MJ, Leonard JP, Van Besien K, Schuetz AN, Jenkins SG, George KS, Walsh TJ (2015) Clinical and molecular epidemiology of human rhinovirus infections in patients with hematologic malignancy. J Clin Virol 71:51–58
34. Vlasak M, Roivainen M, Reithmayer M, Goesler I, Laine P, Snyers L, Hovi T, Blaas D (2005) The minor receptor group of human rhinovirus (HRV) includes HRV23 and HRV25, but the presence of a lysine in the VP1 HI loop is not sufficient for receptor binding. J Virol 79:7389–7395
35. Bochkov YA, Watters K, Ashraf S, Griggs TF, Devries MK, Jackson DJ, Palmenberg AC, Gern JE (2015) Cadherin-related family member 3, a childhood asthma susceptibility gene product, mediates rhinovirus C binding and replication. Proc Natl Acad Sci 112:5485–5490
36. Turner RB, Hendley JO, Gwaltney JM Jr (1982) Shedding of infected ciliated epithelial cells in rhinovirus colds. J Infect Dis 145:849–853
37. Jurgeit A, Moese S, Roulin P, Dorsch A, Lötzerich M, Lee WM, Greber UF (2010) An RNA replication-center assay for high content image-based quantifications of human rhinovirus and coxsackievirus infections. Virol J 7:264
38. Inal JM, Jorfi S (2013) Coxsackievirus B transmission and possible new roles for extracellular vesicles. Biochem Soc Trans 41:299–302
39. Yeo NK, Jang YJ (2010) Rhinovirus infection-induced alteration of tight junction and adherens junction components in human nasal epithelial cells. Laryngoscope 120:346–352
40. Rezaee F, Meednu N, Emo JA, Saatian B, Chapman TJ, Naydenov NG, De Benedetto A, Beck LA, Ivanov AI, Georas SN (2011) Polyinosinic: polycytidylic acid induces protein kinase D–dependent disassembly of apical junctions and barrier dysfunction in airway epithelial cells. J Allergy Clin Immunol 128(e11):1216–1224
41. Triantafilou K, Vakakis E, Richer EA, Evans GL, Villiers JP, Triantafilou M (2011) Human rhinovirus recognition in non-immune cells is mediated by toll-like receptors and MDA-5, which trigger a synergetic pro-inflammatory immune response. Virulence 2:22–29
42. Slater L, Bartlett NW, Haas JJ, Zhu J, Walton RP, Sykes A, Dahdaleh S, Clarke DL, Belvisi MG, Kon OM (2010) Co-ordinated role of TLR3, RIG-I and MDA5 in the innate response to rhinovirus in bronchial epithelium. PLoS Pathog 6:e1001178
43. Shimasaki N, Jain A, Campana D (2020) NK cells for cancer immunotherapy. Nat Rev Drug Discov 2020:1–19

44. Gern JE, Vrtis R, Grindle KA, Swenson C, Busse WW (2000) Relationship of upper and lower airway cytokines to outcome of experimental rhinovirus infection. Am J Respir Crit Care Med 162:2226–2231
45. Lauredo IT, Forteza RM, Botvinnikova Y, Abraham WM (2004) Leukocytic cell sources of airway tissue kallikrein. Am J Phys Lung Cell Mol Phys 286:L734–L740
46. Igarashi Y, Skoner DP, Doyle WJ, White MV, Fireman P, Kaliner MA (1993) Analysis of nasal secretions during experimental rhinovirus upper respiratory infections. J Allergy Clin Immunol 92:722–731
47. Naclerio RM, Proud D, Lichtenstein LM, Kagey-Sobotka A, Hendley JO, Sorrentino J, Gwaltney JM (1988) Kinins are generated during experimental rhinovirus colds. J Infect Dis 157:133–142
48. Proud D, Leigh R (2011) Epithelial cells and airway diseases. Immunol Rev 242:186–204
49. Avila PC, Abisheganaden JA, Wong H, Liu J, Yagi S, Schnurr D, Kishiyama JL, Boushey HA (2000) Effects of allergic inflammation of the nasal mucosa on the severity of rhinovirus 16 cold. J Allergy Clin Immunol 105:923–932
50. Johnston S (2001) The immunology of virus infection in asthma. Eur Respir J 18:1013–1025
51. Alper CM, Doyle WJ, Skoner DP, Buchman CA, Cohen S, Gwaltney JM (1998) Prechallenge antibodies moderate disease expression in adults experimentally exposed to rhinovirus strain hanks. Clin Infect Dis 27:119–128
52. Yazdani R, Habibi S, Sharifi L, Azizi G, Abolhassani H, Olbrich P, Aghamohammadi A (2020) Common variable immunodeficiency: epidemiology, pathogenesis, clinical manifestations, diagnosis, classification, and management. J Investig Allergol Clin Immunol 30:14–34
53. Hwang HS, Chang M, Kim YA (2020) Influenza–host interplay and strategies for universal vaccine development. Vaccine 8:548
54. Sun J, Cardani A, Sharma AK, Laubach VE, Jack RS, Müller W, Braciale TJ (2011) Autocrine regulation of pulmonary inflammation by effector T-cell derived IL-10 during infection with respiratory syncytial virus. PLoS Pathog 7:e1002173
55. Barnard DL (2006) Current status of anti-picornavirus therapies. Curr Pharm Des 12:1379–1390
56. Bauer L, Lyoo H, Van Der Schaar HM, Strating JR, Van Kuppeveld FJ (2017) Direct-acting antivirals and host-targeting strategies to combat enterovirus infections. Curr Opin Virol 24:1–8
57. Morley A, Tomkinson N, Cook A, Macdonald C, Weaver R, King S, Jenkinson L, Unitt J, McCrae C, Phillips T (2011) Effect of lipophilicity modulation on inhibition of human rhinovirus capsid binders. Bioorg Med Chem Lett 21:6031–6035
58. Thibaut HJ, De Palma AM, Neyts J (2012) Combating enterovirus replication: state-of-the-art on antiviral research. Biochem Pharmacol 83:185–192
59. Smith TJ, Kremer MJ, Luo M, Vriend G, Arnold E, Kamer G, Rossmann MG, Mckinlay MA, Diana GD, Otto MJ (1986) The site of attachment in human rhinovirus 14 for antiviral agents that inhibit uncoating. Science 233:1286–1293
60. Hayden FG, Turner RB, Gwaltney JM, Chi-Burris K, Gersten M, Hsyu P, Patick AK, Smith GJ, Zalman LS (2003) Phase II, randomized, double-blind, placebo-controlled studies of ruprintrivir nasal spray 2-percent suspension for prevention and treatment of experimentally induced rhinovirus colds in healthy volunteers. Antimicrob Agents Chemother 47:3907–3916
61. Turner RB, Wecker MT, Pohl G, Witek TJ, McNally E, George RS, Winther B, Hayden FG (1999) Efficacy of tremacamra, a soluble intercellular adhesion molecule 1, for experimental rhinovirus infection: a randomized clinical trial. JAMA 281:1797–1804
62. Wikel J, Paget C, Delong D, Nelson J, Wu C, Paschal J, Dinner A, Templeton R, Chaney M (1980) Synthesis of syn and anti isomers of 6-[[(hydroxyimino) phenyl] methyl]-1-[(1-methylethyl) sulfonyl]-1H-benzimidazol-2-amine. Inhibitors of rhinovirus multiplication. J Med Chem 23:368–372
63. Hayden FG, Gwaltney J (1982) Prophylactic activity of intranasal enviroxime against experimentally induced rhinovirus type 39 infection. Antimicrob Agents Chemother 21:892–897

64. Dorobantu CM, Ford-Siltz LA, Sittig SP, Lanke KH, Belov GA, Van Kuppeveld FJ, Van Der Schaar HM (2015) GBF1-and ACBD3-independent recruitment of PI4KIIIβ to replication sites by rhinovirus 3A proteins. J Virol 89:1913–1918
65. Hsu N-Y, Ilnytska O, Belov G, Santiana M, Chen Y-H, Takvorian PM, Pau C, Van Der Schaar H, Kaushik-Basu N, Balla T (2010) Viral reorganization of the secretory pathway generates distinct organelles for RNA replication. Cell 141:799–811
66. Belov GA, Altan-Bonnet N, Kovtunovych G, Jackson CL, Lippincott-Schwartz J, Ehrenfeld E (2007) Hijacking components of the cellular secretory pathway for replication of poliovirus RNA. J Virol 81:558–567
67. Wessels E, Duijsings D, Lanke KH, Van Dooren SH, Jackson CL, Melchers WJ, Van Kuppeveld FJ (2006) Effects of picornavirus 3A proteins on protein transport and GBF1-dependent COP-I recruitment. J Virol 80:11852–11860
68. Arita M, Philipov S, Galabov AS (2015) Phosphatidylinositol 4-kinase III beta is the target of oxoglaucine and pachypodol (Ro 09-0179) for their anti-poliovirus activities, and is located at upstream of the target step of brefeldin A. Microbiol Immunol 59:338–347
69. Arita M, Kojima H, Nagano T, Okabe T, Wakita T, Shimizu H (2013) Oxysterol-binding protein family I is the target of minor enviroxime-like compounds. J Virol 87:4252–4260

Targeting Molecular and Cellular Mechanisms in Respiratory Syncytial Virus (RSV) Infection

23

Pooja Mittal, Ramit Kapoor, Anjali Saharan, and Rupesh K. Gautam

Abstract

RSV is a negatively stranded RNA virus from the paramyxovirus family. It sprouts from the apical surface of epithelial cells and creates a filamentous structure which contains both viral proteins and genomic RNA. This is the leading cause of bronchitis, asthma, and pneumonia which can lead to ventilation and respiratory failure in infants. RSV encodes for only 11 proteins, but the protein ratios are capable of causing virus replication. In the virion capsid, the most prominent proteins are F and G proteins which are present on the surface of cells from which the virus starts budding. The F protein is responsible for the fusion of virion to the cells and the modification or deletion of which is responsible for the drastic change in virus infection, while the modification/deletion of G protein has the modest effect. The F protein is conserved in all the strains of RSV. RSV F is necessary for the viral entry into the cells and facilitates the pH-independent fusion of virus to the host cell plasma membrane which leads to infection of the host. Following the viral fusion, viral genome- and vRNA-dependent RNA polymerase gets released into the cytoplasm which will lead to transcription of each viral protein and result in translation of the same. Targeting F protein or the related polymerases can lead to prevention of RSV infections. Here, we will discuss the recent review, therapies, and targeting mechanisms to prevent the RSV infection.

P. Mittal
School of Pharmaceutical Sciences, RIMT University, Mandi Gobindgarh, Punjab, India

R. Kapoor
Amity Institute of Pharmacy, Amity University, Noida, India

A. Saharan · R. K. Gautam (✉)
MM School of Pharmacy, Mahrishi Markandeshwar University, Sadopur, Ambala, Haryana, India

K. Dua et al. (eds.), *Targeting Cellular Signalling Pathways in Lung Diseases*,
https://doi.org/10.1007/978-981-33-6827-9_23

Keywords

Fusion protein · Respiratory syncytial virus · Paramyxovirus · Viral filaments

23.1 Introduction

Respiratory syncytial virus is a common respiratory problem in which the lungs are on high risk of infections and leads to breathing problem in the upper respiratory tracts. The interference in the lungs and breathing problems leads to a rise in mild cold and sneezing. This virus propagates in infants mainly, resulting in bronchitis which is the leading cause of narrowing of air pathway and causes obstruction in breathing [1, 2]. This can be severely increased which leads to pneumonia, respiratory failure, and apnea and results in death. This can be also propagated in elderly patients with severe lung damage. The ease of communication of this virus is well evident that it spreads from person to person. The virus enters into the body via different pathways such as the nose, eyes, and mouth from the sneezing of the infected person or to get in touch with the infected person having RSV infection. This RSV virion can also survive on the hard surfaces for some short span of time. The main streamway for the treatment of RSV infections is passive although with immunization. The main focusing targets of this virus were the infants and adults having a previous history of cardiac, pulmonary, and neuromuscular diseases. In the current era, a single line of treatment is available which is also not much effective in the treatment of RSV. From the geographical data results, it was observed that in 2015, 33.1 million episodes of acute RSV infection were out of which, 3.2 million patients were hospitalized and the mortality rate was about 59,600 which was mainly observed in the children under the age of 5 years [2, 3].

23.1.1 Etiology

From the research studies, it was concluded that RSV is a negatively charged single-stranded virus which belongs to the virion family such as *Paramyxoviridae* in the genus *Pneumovirus*. It was observed that in the era of 1955, it was first discovered from chimpanzees from where it gets transmitted to human pathogen.

RSV is composed of a bilipid layer envelope surrounded by a ribonucleic protein molecule along with several functional membrane proteins which help in the attachment of moiety to host cells and fusion of the same.

The RSV was also found to configure into two stereotype strains such as A and B. The human RSV is among one of the stereotypes. The two stereotypes can be differentiated on the basis of their several membrane proteins where they are attached or on the basis of attachment of membrane proteins [4].

23.1.2 Symptoms of RSV Infection

RSV was found to be characterized by runny nose and fever that start to begin in 3–5 days after the initiation of infection. It was observed that in the young children, initiation of cough and sneezing involves the lower respiratory tract only. In infants the first symptom was depicted by apnea that is not a period of not breathing. Due to this majority of the infants got developed respiratory distress, loss of appetite, fever, and death. In adults, the symptoms were mild and may manifest in the form of common cold only [3–5].

23.1.3 Diagnosis

Infections caused by RSV pathogen were determined using various virology tests such as viral culture in which a sample was collected from the infected person and was kept in various cell lines for which the virus was going to be tested. If the cell shows cytopathic changes, then the culture is said to be positive. The other rapid antigen detection test for the detection includes enzyme immunoassay which founds the antibodies with respect to a partiuclar disease. This test was used to determine the antibodies related to a particular infectious disease [6–10]. This method involves the amplification of target DNA by using polymerase chain reaction (PCR). It is used to measure the amount of a specific RNA which was monitored by amplification reaction under fluorescence. This phenomenon was known as RT-PCR or qPCR. These methods are used for the analysis of gene expressions and quantification of viral ribonucleic acid in research and clinical determination [11].

23.1.4 Structure of RSV

Human respiratory syncytial virus (RSV) is a single-stranded virion belonging to a *Paramyxoviridae* family. The genome of this virus contains 10 genes which encodes 11 proteins. Two surface proteins are present on its superficial surface entitled as F protein, and the G attachment which is a glycol protein helps and plays a major function as an antigen of virus and also plays a critical role for the virulence of the virus. Attachment to the host cell is carried out by the G protein. To permit the entry of the viral particle inside the host cell, F protein plays a role by enabling the fusion of the viral particle to the cell membrane of the host cell. Clumping of multinucleated cells is mediated by F protein after the fusion of the plasma membrane which is utilized to form the syncytia which helps in transmission of viral particles from one cell to another [1, 5, 11]. RSV is also composed of two different antigenic subtypes like A and B. The etiology of infection is not defined which is caused by which subtype of infection [9, 10].

23.2 Pathogenesis of RSV Infection

The spread of RSV infections happens by the nasopharyngeal or conjunctival mucosa after their contact with the infected person. It can cause pneumonia and bleeding of the bronchiolitis due to which alveoli get contacted and caused problem in breathing. The virus survives for 6 h on surfaces, on about 90 min on gloves and about 20 min on skin. That is why it is strictly advised to keep washing your hands frequently to avoid the contamination and contact precautions [12–14].

23.2.1 Incubation Period

The incubation of RSV can last between 2 and 8 days after coming in contact with the infected person. But symptoms may remain up to 3 weeks. It occurs very often during winters and spring. It arises in all infants, but the level of severity was more in the children with lung disease and in premature birth. RSV inoculation takes place in the upper respiratory tract followed by degradation of epithelial cells. It envelop the negatively stranded ribonucleic acid molecule which gets RNA and convert that into positive stranded of RNA with the help of RNAPCR reaction, then it further undergoes translation and forms a virus moiety which further forms a viral factory on replication [2, 7, 11–15]. The new virus invades the neighboring cell causing multinucleated syncytia and destroying other cells. The syncytia attract the new immune cells and destroy them. In response to this reaction, various chemokines gets released which further releases more epithelial cells which will further releases more mucous. As more number of immune cells gets released, they will leads to the release of more mucous and also leads to the damage of more blood vessels which cause inflammation and swelling. The extra fluid in the walls makes them thicker and causes narrowing of airway passage. In addition dead cells in mucous slide in the alveoli and cause over there a mucus plug formation which traps air. Over the time the trapped air slowly diffuses into the bloodstream and causes the airways to collapse called atelectasis. On the other hand, some tiny mucous plugs are formed which will act as a one way valve as they allow the air to enter the lungs but not leaving the same outside. It further leads to overinflatation of lungs which is also known as air trapping. As a result, both air trapping and atelectasis reduce the ability to carry oxygen and remove carbon dioxide from the lungs which will cause hypoxemia in the long run. [2, 7, 11, 15].

23.3 Immune Response to RSV

The information regarding the interaction of virus with the immune class is depicted well versed in the animal models and in vitro studies. It was evident from the studies that with the onset of infection in the body, the host cell interacts with the immune system and leads to the production of virus-neutralizing antibodies and T-cell-specific immunity products. On interaction with the immune cells, the host produced mainly two types of action coined as systemic and pulmonary response to RSV and cellular [10, 16–18].

23.3.1 Systemic Response to RSV

23.3.1.1 Neutrophils

The predominant systemic effect was initiated by the neutrophils. Neutrophils are the preexisting class type of bronchoalveolar lavage (BAL) fluid which exist in the inflated lungs of infants who are suffering from RES. During the early phase of pathogenesis, the neutrophils get activated if RSV is in the lower respiratory tract where they produced neutrophil elastase and activation markers. As a result, the virus directly interact with the neutrophilic cells and RSV protein which creates the turbulence between the intracellular virions resulting in phagocytosis [6, 9, 19].

23.3.1.2 Natural Killer (NK) Cells

Prevalence of RSV infection leads to a lesser count number of natural killer cells. These killer cells have the strong impact of the inhibition on the leukocyte immunoglobulin-like receptor which are responsible for the inflammation caused during the infection. As the number of NK cells decreases, the chances of infection rises up which leads to fatality. The rise in fatality is due to the gathering of granzyme B expressing NK cells in the respiratory tracts of infants who were already suffering from bronchitis [19, 20].

23.3.1.3 Dendritic Cells

Dendritic cells such as conventional dendritic cells and plasmacytoid dendrimeric cells are produced and mobilized in the nasal mucosa during the onset of infection. In this the RSV fusion protein which was present in the mucosa helps in the migration of dendrimers rather than monocytes. The movement of these dendrimers plays their role as primary antigen-presenting cells during the infection [10, 16, 17].

23.3.1.4 Monocytes and Macrophages

The mucous of the alveoli which is obtained from the bronchoalveolar lavage fluid of the RSV-infected adults and infants was found to contain surface glycoproteins of RSV such as HLA-DR molecules, IL-1β, and cytoplasmic TNF-α, which depicts a local immunoregulation and antigen-presenting role [7, 21].

23.3.1.5 T Lymphocytes

T lymphocytes developed from the stem cells of the bone marrow and form the interference potential immune evasion strategy in which the pathogenic microbes and tumors start invading the host cells by maximizing the rate of transmittance of disease from the fresh host to another one with constant growth. RSV infection leads to the rise of initial T systemic lymphopenia due to the increase in level of disease. RSV is also involved in the apoptosis caused by T cells which induces the expressions of series cell death [5, 8, 9, 11, 15, 19, 20, 22, 23].

23.3.2 Cellular Response in Term and Preterm Infants

The cellular response in terms of cellular level was analyzed using the neutrophil counts; lymphocytic counts in bronchoalveolar lavage (BAL) fluid were increased in infants ventilated on RSV bronchiolitis; all are higher in preterm infants which leads to maturation of immune system [5–7, 14, 21, 24].

23.4 Current Therapies for RSV Infection

In terms of treatment, a few drugs and treatment options are available for RSV. The target genes and proteins are widely used for the RSV infections for prevention and cure. The prophylactic and curative mode of the drugs determines their mechanism of action. On the basis of the proposed cycle of mechanism, only a few practical options are used for the cure of RSV. Various processes such as DNA replication, synthesis of mRNA, and protein synthesis can be targeted for the development of drugs. According to the FDA till date, ribavirin is the only approved inhaled therapy for the treatment of RSV. A guanosine analogue ribavirin possesses a broad antiviral activity and was approved in 1986 by the FDA for the treatment of bronchitis in infants and children [1, 25, 26]. The various treatment options for the RSV are mentioned in Table 23.1.

23.4.1 Treatments in Clinical Trials

Two major antibodies such as RI-001 and ALX-0171 are currently in clinical trial phases I and II, respectively. Out of which, RI-001 is a polyclonal antibody for the neutralization of RSV and is derived from the plasma component of immunocompromised adults and children. However, it was not found to be much suitable for the treatment of RSV has never been observed; it is hoped that RI-001 cts in immunocompromised population [1–3, 9, 10, 17, 25–31]. The Various treatments of RSV in clinical trials are mentioned in Table 23.1 and management strategies of RSV are mentioned in Table 23.2.

23.5 Molecular and Cellular Targets to Prevent RSV Infection

RSV belongs to family *Paramyxoviridae*, subfamily *Pneumovirinae*, and order *Mononegavirales*. It is composed of 15kb negatively charged single-stranded RNA which contains ten genes. These genes are responsible for the encoding of 11 proteins. The whole of the RNA genome is packed with nucleoprotein (N) which provides the helical symmetry. This helical symmetry helps in the attachment of polymerase (L) protein, phosphoprotein (P), and M2-1 protein. All these proteins are necessary for the transcription and translation processes.

Table 23.1 Management of RSV infection

Therapy	Components	Explanation
Management of symptoms	Bronchodilators	Should be used under careful monitoring and should not be used routinely
	Corticosteroids	Should not be used routinely
	Antibiotics	Use for children only who have confirmed with infection
	Oxygen supply	Use only if the saturation concentration of oxygen in blood is below 90%
Vaccination	Palivizumab	This is a monoclonal antibody with human origin. It targets the specific protein of RSV. Can be administrated i.m. of by i.v. route
	Motavizumab	20-fold more potent than palivizumab. Derived from palivizumab
	Medi-534	This is a live-attenuated virus vaccine and is effective against RSV
Therapeutics	Ribavirin	Only FDA-approved treatment for RSV. This is the antiviral drug which inhibits both DNA and RNA viruses. This inhibition is based on polymerase inhibition. It can be administered i.v., orally and by aerosols also
	RSV-IVIG	Adequately provide passive immunity and decrease the RSV-related hospitalizations to 41–65%
Alternative therapies	Vitamin D	It reduces acute respiratory symptoms
	Pre-probiotics	The role of gut microflora in prevention of RSV disease was evaluated some years ago
Phytotherapy	Curcumin	It has pleotropic effects which include anti-inflammatory, anticancer, antibacterial, antiviral, autophagy, etc. some studies suggested that it works excellently for the prevention of disease severity
	Modified Dingchuan decoction	This is a mixture of Chinese herbs *Salviae miltiorrhizae radix, Scutellariae radix, Farfarae flos, and Ephedrae* herbs. This is useful in the treatment of asthma, cough, and lung-related problems

As RSV contains an envelope, so to progress the infection, the attachment of the envelope to the cell membrane is a prerequisite to transfer its genome to the cytoplasm of the host cell. RSV contains three transmembrane glycoproteins, out of which two, i.e., G and F proteins, are responsible for the fusion process. G protein is the important one which is responsible for the viral attachment, while F protein is responsible for the fusion of the virus to the cell membrane. [2, 3, 32–35]. The structure of RSV is shown in Fig. 23.1.

To enter the target cell, the virus made a contact with the host cell through G protein which will activate the F protein through its engagement with the secondary protein. Extensive research was carried out to study the mechanism of the viral fusion to target cell. It was concluded by some researchers that heparin sulfate is needed for the fusion of the viral cell, while others have concluded that CX3CR 1 is sufficient to carry out this process. Studies by using human airway epithelial (HAE)

Table 23.2 Treatments of RSV in clinical trials

Therapy	Mechanism of action	Example of candidates	Conclusion
Antiviral drugs	Replication inhibition	RSV 604	
	RSV codon mutation	Ribavirin, viramidine, merimepodib	RSV treatment but can affect the host adversely
	Inhibitors of inosine monophosphate dehydrogenase enzyme	Ribavirin, mycophenolate, mizoribine	Promising anti-RSV drug
	Immunostimulation	Ribavirin	Studies are under trials
Fusion inhibitors	Inhibition of attachment to the cell membrane by F protein	Chemical, BMS-433771, RFI-641; Peptide, HR121, HR212, RhoA	Chemical fusion inhibitors comprise some serious side effects, while peptide fusion inhibitors are the promising one
Nanoparticles	Inhibits cell attachment	Silver and gold nanoparticles	Studies are under trials
Antisense therapy	Interference with RNA	siRNA-ALN-RSV01Phosphorodiamidate morpholino oligomers	Effective and found to be safe to the host
Ethnobotanicals	Can be the inhibitors of fusion	Plant extracts—*Cinnamomum cassia*, *Cimicifuga foetida*, Sheng-Ma-Ge-Gen-Tang, Ginger, etc. water extract of Licorice	They seem to be promising but still studies are under trials

cell lines have shown that heparin sulfate is found in very low quantity on the apical surface of the cells which suggests the clinically less relevance of heparin sulfate. A few information is available for the role of SH protein on the viral structure; however, it was found to be responsible for the virulence and possess no role in the fusion process [4, 7, 11, 36–38].

23.5.1 G Protein

A huge and important fusion protein on the cell surface of RSV which is G is responsible for the viral entry into the target cell and is known to bind heparin sulfate present on the cells while it binds to CX3CR 1 on ciliated cells. Indeed, heparin sulfate and CX3CR 1 are the potential targets for the prevention of viral fusion and thereby the RSV infection. It was also shown that the antibodies which prevent the interaction of G protein with CX3CR 1 reduce RSV infection in mice. In infants, a

Fig. 23.1 Structure of respiratory syncytial virus (RSV) showing all its proteins

few amount of CX3CR was found to be present inside the respiratory tract, and RSV is known to interact preferably with the CX3CR 1-expressing cells. Further, glycosylation strategy is mainly used for the shielding of antigenic protein from recognition via antibody. When the G protein is heavily glycosylated, it will change the glycan structure and will alter the macrostructure of G protein which will mask the protein backbone from the antibodies and will limit their affinity. Also, the antigenic variation is also generated across RSV strains on a huge variation in the oligosaccharide arrangement. This antigenic rearrangement will weaken the immune response as the antibodies generated previously have poor attraction with the new variants with G protein. Additionally, frameshift mutations, premature stops, and point mutations can be seen in the new variants. In this way, they get protected from the immune system. These strategies of immune protection reduce the possibility of neutralizing antibodies to protect the cells and effectively render the immunological memory to act as surplus against RSV infection. In the ectoderm layer, the G protein further contains two hypervariable regions which make them suitable for mutations and prevent the recognition by normal antibodies generated during the previous infection. RSV variants can be divided into two more categories: RSV A and RSV B based on these variations [8, 19, 28, 39–42]. These variations in strains lead to enhancement of viral pathogenicity along with the suppression of humoral immunity response. Antigen presentation to the T cell can be prevented on heavy glycosylation of the G protein which also suppresses the antibody response. The soluble form of RSV G (sG) protein will be produced which will behave as an immune "decoy." Generally, during the initial stages of viral life cycle, the sG protein production is high, which flooded the nearby areas and reduces the effectiveness of G protein-specific antibodies. The creation of sG protein can be achieved by initiation of RNA synthesis at the second AUG codon which will remove the 65 amino acids from N terminal which is responsible for the fixation of G protein in the viral envelope structure. Also, the release of G-specific antibodies can be prevented by the combination of variable glycosylation, and sG release can be prevented by sG. sG also uses multiple mechanisms to protect against antibodies. The modulatory effects of sG protein can also change the cytokine responses which will change the Th1 response. Also, the RSV deficient of sG protein will enhance the pro-inflammatory modulators like cytokines and IL-8. This concludes the role of sG in blocking the conscription of the immune cells to the site of infection. It also has an impact on the complement system which is involved in the clearance of the virus particles at the site of infection. The G protein also contains a CX3C motif which is also responsible for the modulation of immune cell response by limiting the immune cell recruitment. Through this motif, RSV cell acts as a mimic to the CX3CL1 cytokine and allows the attachment of the viral protein through this which will help in increasing the RSV infectivity. It can be concluded from the above discussion that RSV G protein is involved in immunomodulation by various means that can allow the virus to prevent the role of G-specific antibodies and can impact the normal functioning of the immune system against RSV. So, G protein of the RSV can said to provide the favorable and major target for the prevention of infection [9, 16, 22, 43–45].

23.5.1.1 Targeting of G Protein with Monoclonal Antibodies (mAb)

In RSV structure, the G protein is the major protein which can impact the viral replication, and targeting of G protein can abolish the RSV infection to a greater extent. G protein targeting has attracted an increased attention to provide the RSV prophylaxis in infants and can provide the posttreatment also [28, 42].

23.5.2 F Protein

The F protein is mandatory for the fusion process which renders the fusion of the virus to the host cells and is important for viral multiplication. Its way of positioning the RSV envelope and the way of sequencing make the F protein a primary target for the treatment of RSV infection. The vaccine development targeting F proteins are under trials. One mAb palivizumab is designed to target F protein, thus neutralizing the replication of the virus and can provide treatment toward RSV to the high-risk patients. However, it can provide treatment against the majority of RSV strains; one mutated strain which is carrying N276S mutation is resistant to palivizumab. The mechanism of cell attachment by F protein includes a conformational shift in the F protein structure when the virion came closer to the target cell, and as a result to this, it will anchor its N terminal hydrophobic head into the target cell membrane, and further it will fold back so that the virus as well as target cell can get attached to each other. So all the forms of F proteins can be a good target for RSV infection as mentioned in Fig. 23.2. Initially, the work on targeting of F protein in RSV infection was hindered by the numerous confirmations of F protein, while later on, it was found that the targeting of postfusion confirmation may have a modest therapeutic

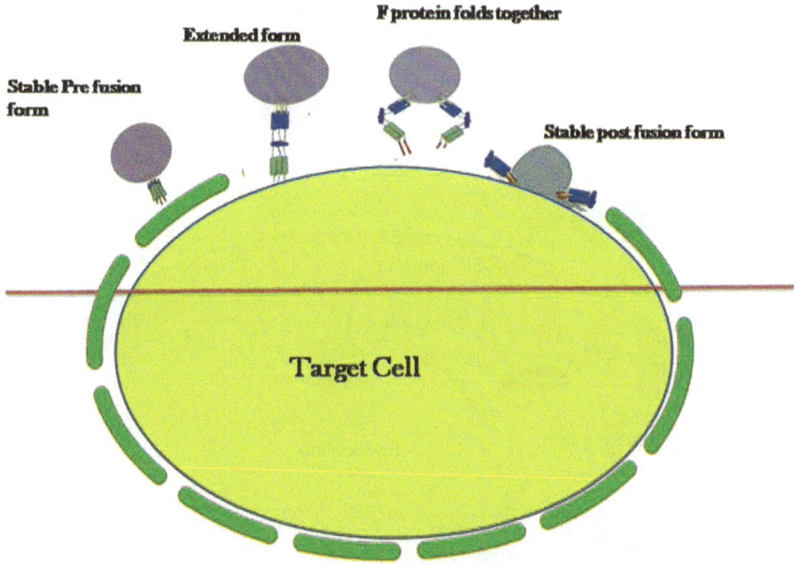

Fig. 23.2 The image showing the working of RSV F protein which helps in fusion of RSV to the target cell

benefit. Now, the crystal structures of prefusion F protein and postfusion F protein have been solved and will lead to the development of treatment therapy for the RSV infection. Small peptides were also developed previously against RSV, but the need of frequent injections and high cost of the therapy had ruled out their use [7, 19, 27].

Even though the G protein acts as a main attachment protein, F protein is also responsible for the cell attachment; it generates the functions which will defend the loss-of-function mutations in the G protein which is less stable than the F protein.

F protein was found to act on neonatal B regulatory cells (nBreg). It attaches to nBreg through B cell receptor (BCR) which will cause the upregulation of CX3CR1 which is further bonded to G protein through CX3C domain. Moreover, the RSV infection of nBreg will increase the production of IL-10 (anti-inflammatory) which will suppress the Th1 activity. This nBreg-specific activity is observed in the cord as well as in infant blood just after birth, but it decreases along with age [11, 32, 40, 44]. Thus, infants are more prone to RSV infection. Targeting of F protein can thus provide a better target for the treatment of RSV infection. Various targets for the prevention of RSV infection are shown in Fig. 23.3.

23.5.3 SH Protein

The small hydrophobic (SH) is the type II transmembrane capsid protein present on the surface of RSV envelope. Structural studies have shown that it is viroporin which is responsible for the formation of pores inside the cell membranes which alters the

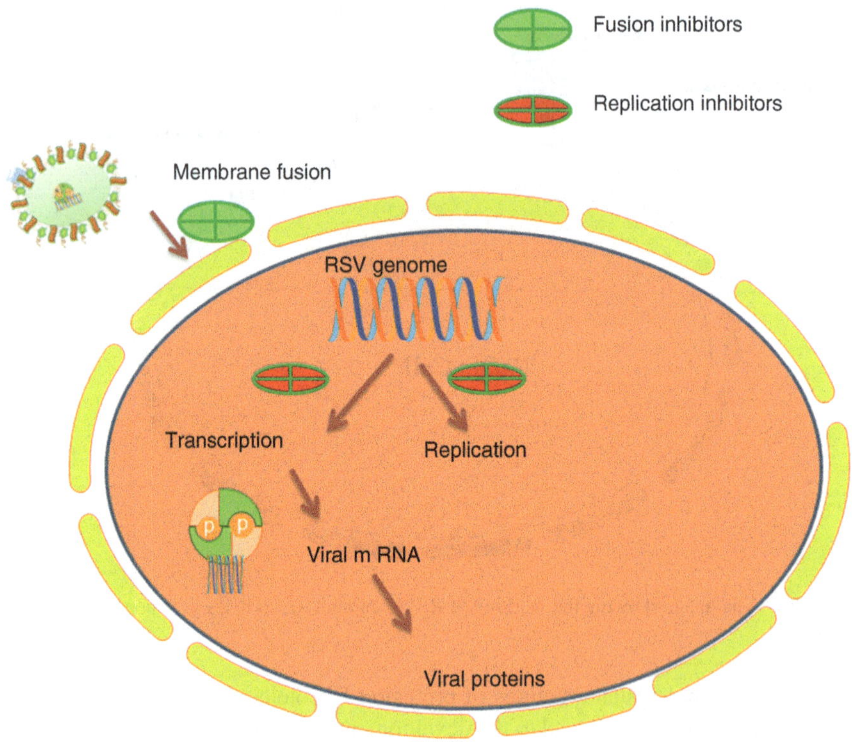

Fig. 23.3 The structure of RSV showing various targets for the treatment of infection

membrane permeability. SH deletion mutants still have the ability to enter the cells and can replicate. However, the deletion strains are less virulent than the wild strains of RSV SH. The role of SH is not fully understood yet; however, the research suggested that SH helps in prolongation of the life of the infected cells and makes them insensitive to apoptosis which will further help in viral replication. The presence of SH influences the cytokine production and increased production of IL-1B, and TNF expression was observed with deletion strains. That's why the deletion strains of SH were explored as a live-attenuated vaccine though, till date, nothing is present inside the market [5, 15, 28, 46].

23.5.4 N Protein

The nucleoprotein (N) along with the phosphoprotein (P) coats the RNA genome of RSV to protect the same from degradation. Structural studies had evaluated that it forms a left-handed helix around the viral genome. Along with its key roles, the N protein was also found to have their role in immune invasion functions. Nucleoprotein was also found to be present on the surface of infected epithelial cells along with their presence inside the nucleus. The presence of the N protein in a bilipid membrane leads to the reduction of T cell activation. This ability of RSV to modulate the T cell activation affects the antibody production so it can also be a good target for the treatment of RSV infection. Scientist had evaluated that after 6 hr of RSV infection, the N protein was colocalized with RIG-1 and MIDA5 and later on during infection, the viral inclusion bodies were observed. The interaction of N protein with RIG pathway components restricts the subsequent antiviral IFN response and can enable a more replication of RSV. Hence, it can act as an important target for the prevention of infection [7, 17, 21, 31, 36, 41, 47].

We have studied the various pathways by which G, F, SH, and N proteins act in the virus particle and how they are responsible for the modulation of the viral replication process. By causing the blockade of the cellular pathways, they can be the potential targets for the treatment of RSV infection.

23.6 Future Prospective

With 3.2 million RSV-related hospitalizations, each year, the development of vaccines is the key point which can prevent its occurrence as well as commencement. The complex nature of RSV capsule as well as its immune invasion strategies had hindered the development of vaccines. Further the immune profiles of the group of people who are at higher risk (old, neonatal, children, pregnant women) have added more complexity in the vaccine development stages. Till date, the prophylaxis treatment for children is available in the hospitals by oxygen, nebulization, and fluids to mitigate the bronchitis-like symptoms of RSV infection. Ribavirin is also offered in some countries as a treatment though it possesses certain side effects like high-cost treatment, low effectiveness, etc. The vaccine development for the protection against RSV can be useful as the primary purpose of vaccine is to develop the long-lasting antibodies against RSV and stimulation of B cell and T cell response. The antibodies generated during the first RSV infection are active for a shorter period of time, and consecutively two infections are needed to generate the long-lasting antibodies which can also decline rapidly. The antibodies gained through the mother's placenta also start declining at the age of 2–3 months, and the peak of RSV infection is at the age of 2.5 months. However, one study had also concluded that high tidal volume of RSV antibodies in mother's placenta close to birth does not reduce the chances of newborn RSV infection. Moreover, males are at higher risk of developing RSV infection as compared to females as the antibody transfer rate is lower in males as compared to females. Thus, a promising treatment for the prevention of RSV infection is needed certainly, and this target can be achieved with the help of target-specific vaccine development. An ideal curing treatment will be the one which provides the action for a longer period of time along with the direct action on the target. It should be able to conserve the immune evasion mechanism of RSV and thus enable the natural immunity of the body to fight against RSV. To develop a better design of the therapeutics, research workers must work together toward defining the role of all RSV proteins and define all the process that suppress the internal immunity and should also monitor the virus over the period of time for the genetic variations. Armed with this knowledge, we will be able to develop the better vaccine therapeutics which will protect our populations from this deadly viral disease.

References

1. Crowe JE, Murphy BR, Chanock RM, Williamson RA, Barbas CF, Burton DR (1994) Recombinant human respiratory syncytial virus (RSV) monoclonal antibody Fab is effective therapeutically when introduced directly into the lungs of RSV-infected mice. Proc Natl Acad Sci 91 (4):1386–1390
2. Crowe JE Jr, Bui PT, London WT et al (1994) Satisfactorily attenuated and protective mutants derived from a partially attenuated cold-passaged respiratory syncytial virus mutant by introduction of additional attenuating mutations during chemical mutagenesis. Vaccine 12 (8):691–699
3. Eisenhut M (2006) Extrapulmonary manifestations of severe respiratory syncytial virus infection–a systematic review. Crit Care 10(4):R107
4. Hervás D, Reina J, Yañez A, Del Valle J, Figuerola J, Hervás J (2012) Epidemiology of hospitalization for acute bronchiolitis in children: differences between RSV and non-RSV bronchiolitis. Eur J Clin Microbiol Infect Dis 31(8):1975–1981
5. Jensen TO, Stelzer-Braid S, Willenborg C et al (2016) Outbreak of respiratory syncytial virus (RSV) infection in immunocompromised adults on a hematology ward. J Med Virol 88 (10):1827–1831
6. Kusuda S, Takahashi N, Saitoh T et al (2011) Survey of pediatric ward hospitalization due to respiratory syncytial virus infection after the introduction of palivizumab to high-risk infants in Japan. Pediatr Int 53(3):368–373
7. Madge P, Paton J, McColl J, Mackie P (1992) Prospective controlled study of four infection-control procedures to prevent nosocomial infection with respiratory syncytial virus. Lancet 340 (8827):1079–1083
8. Meissner HC (2003) Selected populations at increased risk from respiratory syncytial virus infection. Pediatr Infect Dis J 22(2):S40–S45
9. Thomas M, Bedford-Russell A, Sharland M (2000) Hospitalisation for RSV infection in ex-preterm infants—implications for use of RSV immune globulin. Arch Dis Child 83 (2):122–127
10. Weber MW, Weber MW, Mulholland EK, Mulholland EK, Greenwood BM (1998) Respiratory syncytial virus infection in tropical and developing countries. Tropical Med Int Health 3 (4):268–280
11. Kim L, Rha B, Abramson JS et al (2017) Identifying gaps in respiratory syncytial virus disease epidemiology in the United States prior to the introduction of vaccines. Clin Infect Dis 65 (6):1020–1025
12. Barney SG (1995) Pathogenesis of respiratory syncytial virus vaccine-augmented pathology. Am J Respir Crit Care Med 152:563–S66
13. Brandenburg A, Neijens H, Osterhaus A (2001) Pathogenesis of RSV lower respiratory tract infection: implications for vaccine development. Vaccine 19(20-22):2769–2782
14. Breary S, Smyth RL (2007) Pathogenesis of RSV in children. Respiratory Syncytial Virus 14:141–162
15. Van Schaik S, Welliver R, Kimpen J (2000) Novel pathways in the pathogenesis of respiratory syncytial virus disease. Pediatr Pulmonol 30(2):131–138
16. Welliver RC (1988) Detection, pathogenesis, and therapy of respiratory syncytial virus infections. Clin Microbiol Rev 1(1):27–39

17. Whitehead SS, Juhasz K, Firestone C-Y, Collins PL, Murphy BR (1998) Recombinant respiratory syncytial virus (RSV) bearing a set of mutations from cold-passaged RSV is attenuated in chimpanzees. J Virol 72(5):4467–4471
18. Wu H, Pfarr D, Losonsky G, Kiener P (2008) Immunoprophylaxis of RSV infection: advancing from RSV-IGIV to palivizumab and motavizumab. Human antibody therapeutics for viral disease. Springer, Cham, pp 103–123
19. Oshansky CM, Zhang W, Moore E, Tripp RA (2009) The host response and molecular pathogenesis associated with respiratory syncytial virus infection. Future Microbiol 4 (3):279–297
20. Peebles RS Jr, Graham BS (2005) Pathogenesis of respiratory syncytial virus infection in the murine model. Proc Am Thorac Soc 2(2):110–115
21. Mangodt TC, Van Herck MA, Nullens S et al (2015) The role of Th17 and Treg responses in the pathogenesis of RSV infection. Pediatr Res 78(5):483–491
22. Tripp RA (2004) Pathogenesis of respiratory syncytial virus infection. Viral Immunol 17 (2):165–181
23. Watkiss ER (2012) Pathogenesis of respiratory syncytial virus. Curr Opin Virol 2(3):300–305
24. Bont L (2009) Current concepts of the pathogenesis of RSV bronchiolitis. Hot topics in infection and immunity in children V. Springer, Cham, pp 31–40
25. Bresee JS, Fischer M, Dowell SF et al (1996) Vitamin A therapy for children with respiratory syncytial virus infection: a multicenter trial in the United States. Pediatr Infect Dis J 15 (9):777–782
26. Chiba Y, Higashidate Y, Suga K, Honjo K, Tsutsumi H, Ogra PL (1989) Development of cell-mediated cytotoxic immunity to respiratory syncytial virus in human infants following naturally acquired infection. J Med Virol 28(3):133–139
27. Mejias A, Garcia-Maurino C, Rodriguez-Fernandez R, Peeples ME, Ramilo O (2017) Development and clinical applications of novel antibodies for prevention and treatment of respiratory syncytial virus infection. Vaccine 35(3):496–502
28. Rezaee F, Linfield DT, Harford TJ, Piedimonte G (2017) Ongoing developments in RSV prophylaxis: a clinician's analysis. Curr Opin Virol 24:70–78
29. Richardson JY, Ottolini MG, Pletneva L et al (2005) Respiratory syncytial virus (RSV) infection induces cyclooxygenase 2: a potential target for RSV therapy. J Immunol 174 (7):4356–4364
30. Sáez-Llorens X, Moreno MT, Ramilo O et al (2004) Safety and pharmacokinetics of palivizumab therapy in children hospitalized with respiratory syncytial virus infection. Pediatr Infect Dis J 23(8):707–712
31. Weinberger M (2004) Respiratory infections and asthma: current treatment strategies. Drug Discov Today 9(19):831–837
32. De Clercq E (2001) Molecular targets for antiviral agents. J Pharmacol Exp Ther 297(1):1–10
33. Geerdink RJ, Pillay J, Meyaard L, Bont L (2015) Neutrophils in respiratory syncytial virus infection: a target for asthma prevention. J Allergy Clin Immunol 136(4):838–847
34. Hallak LK, Spillmann D, Collins PL, Peeples ME (2000) Glycosaminoglycan sulfation requirements for respiratory syncytial virus infection. J Virol 74(22):10508–10513
35. Hashimoto K, Ishibashi K, Ishioka K et al (2009) RSV replication is attenuated by counteracting expression of the suppressor of cytokine signaling (SOCS) molecules. Virology 391 (2):162–170
36. Haynes LM, Caidi H, Radu GU et al (2009) Therapeutic monoclonal antibody treatment targeting respiratory syncytial virus (RSV) G protein mediates viral clearance and reduces the pathogenesis of RSV infection in BALB/c mice. J Infect Dis 200(3):439–447
37. Le Calvez H, Yu M, Fang F (2004) Biochemical prevention and treatment of viral infections–a new paradigm in medicine for infectious diseases. Virol J 1(1):12
38. Costello HM, Ray WC, Chaiwatpongsakorn S, Peeples ME (2012) Targeting RSV with vaccines and small molecule drugs. Infect Disorders-Drug Targets 12(2):110–128

39. Malley R, DeVincenzo J, Ramilo O et al (1998) Reduction of respiratory syncytial virus (RSV) in tracheal aspirates in intubated infants by use of humanized monoclonal antibody to RSV F protein. J Infect Dis 178(6):1555–1561
40. Mazur NI, Martinón-Torres F, Baraldi E et al (2015) Lower respiratory tract infection caused by respiratory syncytial virus: current management and new therapeutics. Lancet Respir Med 3 (11):888–900
41. McLellan JS, Chen M, Leung S et al (2013) Structure of RSV fusion glycoprotein trimer bound to a prefusion-specific neutralizing antibody. Science 340(6136):1113–1117
42. Pemberton R, Cannon M, Openshaw P, Ball L, Wertz G, Askonas B (1987) Cytotoxic T cell specificity for respiratory syncytial virus proteins: fusion protein is an important target antigen. J Gen Virol 68(8):2177–2182
43. Russell CD, Unger SA, Walton M, Schwarze J (2017) The human immune response to respiratory syncytial virus infection. Clin Microbiol Rev 30(2):481–502
44. Tauro S, Su Y-C, Thomas S et al (2008) Molecular and cellular mechanisms in the viral exacerbation of asthma. Microbes Infect 10(9):1014–1023
45. Wang SZ, Forsyth KD (2000) The interaction of neutrophils with respiratory epithelial cells in viral infection. Respirology 5(1):1–9
46. Atto B, Eapen MS, Sharma P et al (2019) New therapeutic targets for the prevention of infectious acute exacerbations of COPD: role of epithelial adhesion molecules and inflammatory pathways. Clin Sci 133(14):1663–1703
47. Griffiths C, Drews SJ, Marchant DJ (2017) Respiratory syncytial virus: infection, detection, and new options for prevention and treatment. Clin Microbiol Rev 30(1):277–319

Targeting Molecular and Cellular Mechanisms in SARS-CoV-2 Novel Coronavirus Disease 2019 (COVID-19)

24

Sagarika Majhi, Kamal Dua, Parteek Prasher,
Harlokesh Narayan Yadav, and Lubhan Singh

Abstract

Coronavirus disease 2019 (COVID-19) represents a significant threat to global well-being and safety. Controlling it and reducing the rate of death as soon as possible is a pressing job for the healthcare sector. The possible therapeutic agents used in COVID-19 are from the past encounters in combating SARS, MERS, and other viruses. Assuming that broad-spectrum antiviral drugs have long been accepted on the market to treat various infectious diseases, their metabolic features, dosage, possible efficacy, and side effects are yet to be studied for the current scenario. Repurposing of clinically approved drugs may be a salient short-term approach for the treatment of novel coronavirus. But the downside is that these therapies are too "broad spectrum" to target COVID-19 directly. Therefore,

S. Majhi
I. T. S College of Pharmacy, Muradnagar, Ghaziabad, India

K. Dua
Discipline of Pharmacy, Graduate School of Health, University of Technology Sydney, Sydney, NSW, Australia

Priority Research Centre for Healthy Lungs, University of Newcastle & Hunter Medical Research Institute, New Lambton Heights, Newcastle, NSW, Australia

P. Prasher
Department of Chemistry, University of Petroleum & Energy Studies, Dehradun, India

Department of Chemistry, UGC Sponsored Centre for Advanced Studies, Guru Nanak Dev University, Amritsar, India

H. Narayan Yadav
All India Institute of Medical Sciences (AIIMS)-New Delhi, New Delhi, India

L. Singh (✉)
Kharvel Subharti College of Pharmacy, Swami Vivekanand Subharti University, Meerut, India

K. Dua et al. (eds.), *Targeting Cellular Signalling Pathways in Lung Diseases*,
https://doi.org/10.1007/978-981-33-6827-9_24

its side effects cannot be underrated. A number of clinical trials are underway to evaluate the effectiveness of other treatment options.

Active symptomatic support is still the key to treatment. Although stem cells, monoclonal antibodies, polypeptides, interferon, and plasma from recovered patients have been shown to be effective in treating COVID-19 patients, their safeties are still being evaluated, and the efficacy remains to be further confirmed. This chapter illuminates the structural proteins of SARS-CoV-2 and helps to investigate certain proteins and gain new insights into the drugs used for COVID-19. The features of future therapy must be multicomponent, multitarget, and multi-pathway for disease, care and it has tremendous potential for COVID-19 therapy. Because the pandemic is rapidly emerging, such observations will lead to better understanding for making effective therapeutics to curb COVID-19.

Keywords

Coronavirus disease 2019 (COVID-19) · SARS-CoV-2 · Virus · Repurposing

24.1 Introduction

A recurring trend mostly in case of infectious disease has emerged in the environment. Many pathogens since their initial exposure were found to reappear, mostly in highly virulent types. Viral illnesses remain a major threat to the health and living style of public in society. In the past years, the world has encountered numerous highly infectious viral outbreaks, including severe acute respiratory coronavirus syndrome (MERS-CoV, SARS-CoV-1), H1N1 influenza, etc. The World Health Organization (WHO) has recently recognized a pandemic of respiratory illness of uncertain pathogenesis occurred on December 31, 2019. Due to the striking similarities of its indications to those caused by SARS, the International Association on Virus Taxonomy named it SARS-CoV-2. The WHO later announced that the illness characterized by the new virus is "COVID-19" which is the abbreviation for "coronavirus disease 2019" [1].

A single-stranded RNA-containing coronavirus (CoV) is covered with an envelope which contains a spike-like glycoprotein, which provides it a crown-like impression when viewed on a microscope. They exhibit various clinical incidences of respiratory and extrapulmonary system with associated increase in mortality [2, 3] (Fig. 24.1).

The RNA virus, usually in a circular or elliptic form, has a diameter of 60–140 nm. Its gene codes 29,891 nucleotides, encrypting 9860 amino acids and sharing a sequence identity of 99.9%, proposing a very notable host transition into human beings. Like most other CoV, it is susceptible to the high temperatures and ultraviolet radiations. In addition, lipid solvents such as chloroform, ether (75%), ethanol, peroxyacetic acid, and chlorine-containing disinfectant may efficiently inhibit the activity of such viruses [4, 5].

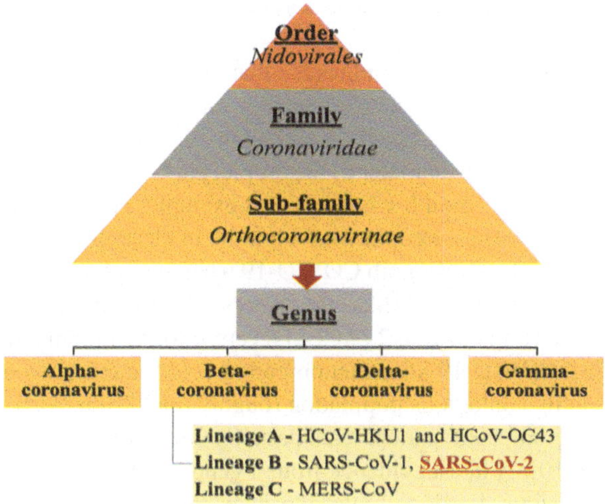

Fig. 24.1 Classification of SARS-CoV-2

SARS-CoV-2 is highly infectious but is usually mild and self-limiting. The symptomatic and asymptomatic carriers both can spread infection. According to current evidence, COVID-19 virus is highly contagious and gets communicated between persons through respiratory secretions and contact routes. Hence, washing hands frequently and keeping a distance of at least 1 meter are considered the main measures to prevent infection [6]. Personal protective equipment (PPE) kits may protect individuals from infection, especially in immune-compromised people. The WHO's approved PPE comprises of surgical masks, boots, gowns, or face masks, respirators, and aprons [7]. Newer possibilities of feco-oral transmission are also mentioned in some recent speculations. Rapid diagnostic tests will be helpful for screening and diagnosing COVID-19 patients.

Trials are currently underway on new antiviral drugs and vaccines to combat the pandemic, although isolation and containment is just a way to keep COVID-19 from spreading. The exact mechanism of the virus is still unknown so far, and there has been no effective drug produced for the virus also. Currently, it is necessary to monitor the source of infection, cut off the dissemination route, and make use of existing drugs and effectively manage the disease progression. New medications should also be developed, along with vaccine research and development, due to which disease morbidity and mortality will be reduced and people's lives would be better secured. Despite globalization, viral diseases are frequently increasing and can trigger a global outbreak. Biosafety and health security awareness will allow for the prevention and control of emerging infectious diseases.

24.2 Transmission Cycle

Current research indicates that COVID-19 passes by direct, by indirect (by contaminated items or object surfaces), or by mouth and nose secretions near infected persons. Such secretions involve blood, respiratory droplets, or droplets of secretion (the diameter of the droplet particles is >5–10 μm). They are expelled from the nose or mouth whenever, for example, an infected individual coughs, sneezes, talks, or sings. People who might be in close direct contact with an infected person (within 1 m) may grab COVID-19 when the contagious droplets get into their mouth, nose, or eyes. Unlike the others, SARS-CoV-2 also utilizes the angiotensin-converting enzyme 2 (ACE2) receptor for its binding and entry into the host system [8]. COVID-19's incubation period could fluctuate from 3 days to 14 days, and the highly contagious population roughly doubles in every 7 days on an average. Also, each infected patient can pass on the infection onto an additional 2.2 people [9, 10]. A latest evidence from a genome sequencing of SARS-CoV-2 showed that this virus developed into two main forms (assigned as L and S) with two distinct SNP. Although the L type is more widespread (~70%), is aggressive, and spread faster, as observed in the early stages of the outbreak, the S type (~30%) is an older and less violent form. Such ultrarapid advances in viral outbreaks clearly indicate an immediate need for awareness of these viral mechanisms in order to survive with this COVID-19 as a public health emergency.

Clinical Features: An individual with COVID-19 might show a range of symptoms. This may be categorized as asymptomatic or symptomatic, carrier or infectious. On the basis of the patient's immunity status, the disease can range from mild to profusely symptomatic. However, the rate of fatality ranged from 2.3% to 5% with an average of 3%. Feeble epidemiological risk factors for early diagnosis include older age, male gender, smokers, and related comorbidities comprising overweight, hypertension, diabetes, chronic lung problems, heart disease, and renal disease [11]. The clinical features of COVID-19 are depicted in Fig. 24.2.

Some diagnostic laboratory findings of COVID-19 along with its characteristic features are enlisted in Table 24.1.

24.3 Structure of SARS-CoV-2

The crown-like microstructure of SARS-CoV-2 is generated by transmembrane spike glycoproteins (S proteins). This further forms homotrimers on the virion surface [20]. An exopeptidase, namely, ACE2 (angiotensin-converting enzyme 2), is present on respiratory tract epithelial cells. It encompasses a therapeutic target for restricting SARS-CoV-2 entry into the cell. Viral S protein and host ACE2 facilitates endocytosis of pathogens that bind to the human host. S protein is sheared by host serine proteases, such as transmembrane serine protease 2 (TMPRSS2) which leads to the release of single-stranded RNA (ssRNA) (+) into the intracellular area. Host machinery converts the RNA to generate the replicase and the structural proteins (Fig. 24.3).

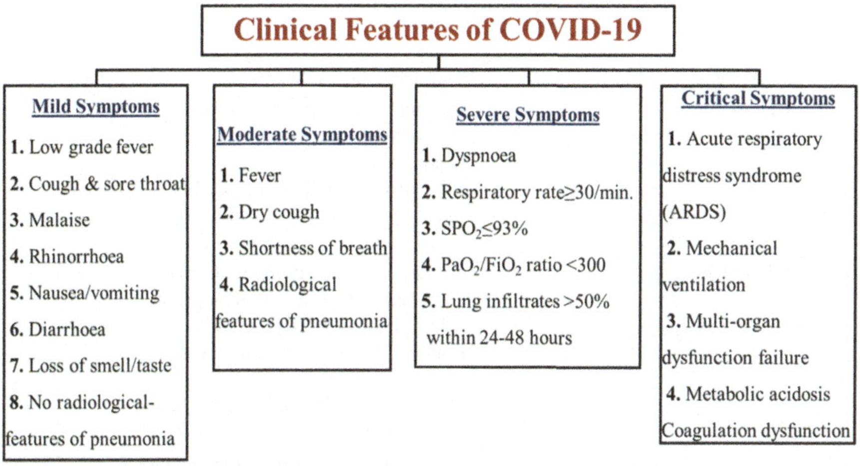

Fig. 24.2 Clinical symptoms of COVID-19 according to severity

Host and SARS-CoV-2 proteases split the replicase into nonstructural proteins including the RNA-dependent RNA polymerase (RdRp) protease. RdRp eases the replication and amplification of SARS-CoV-2 RNA. Two other proteases, 3CLpro and PLpro, are essential for replication and packaging of new virions. They process the translation of genomic RNA into structural and nonstructural protein. Both 3CLpro and PLpro are extremely important for replication and host cell regulation, and are also viable targets for antiviral agents (Fig. 24.4).

SARS-CoV-2 transmembrane proteins are transferred to the viral capsids formed with the help of endoplasmic reticulum and Golgi apparatus. It consists of spike [S], envelope [E], and membrane [M]. Viral assembly occurs when the viral RNA and nucleocapsid (N) protein are incorporated via viral transmembrane proteins. The S, N, M, and E form vital proteins which play a crucial role in the virus life cycle. The S protein promotes the binding of receptors and the fusion of membranes. The N protein strengthens the entry of pathogens and conducts cellular post-fusion processes in the host that are essential for viral survival. The E protein continues to promote virion formation and viral pathogenicity, while ribonucleoproteins are formed by the M protein which mediates inflammatory responses in hosts. The resulting exocytosis occurs with the release of newly synthesized virions.

24.4 Pharmacological Treatment

Numerous drugs have demonstrated benefit in COVID-19 by using chloroquine and hydroxychloroquine in patients with less serious illnesses. Many drugs currently being tested include lopinavir/ritonavir, remdesivir, favipiravir, oseltamivir, ribavirin, beta-interferon, tocilizumab, and Arbidol. Natural resources with antiviral properties promise candidates for selection as lead targets for drug design and

Table 24.1 Laboratory findings of COVID-19

Sr. no.	Test/marker	Characteristic features	References
1.	Hemogram	• Lymphopenia (in 80% of cases), mild thrombocytopenia, and leukocytosis • Neutrophil-to-lymphocyte ratio (NLR) $\geq$3.13 shows progression to severe illness	[8, 12]
2.	Inflammatory markers	• ↑ serum procalcitonin with severity • ↑C-reactive protein (CRP), lactate dehydrogenase (LDH), SGOT, troponin, D-dimer, ferritin, creatine kinase, and ESR • ↑interleukin (IL)-6, IL-4, IL-10, and tumor necrosis factor (TNF)-α	[13, 14]
3.	Serology (blood sampling is much easier than swab sampling from the oropharynx or nasopharynx)	**a. Enzyme-linked immune-sorbent assay (ELISA):** Sensitivity of 87.3% and specificity of 100%, based on Rp3 nucleoprotein to detect IgM and IgG against SARS-CoV-2	[13, 15]
		b. Immunochromatography (card test): Sensitivity of 82.4% and specificity of 100%, it is convenient, is cheaper, and offers a rapid turnover	
4.	Reverse-transcriptase polymerase chain reaction (RT-PCR)	• Sensitivity is 70% • Specimen collection from the upper/lower respiratory tract or sputum or bronchoalveolar lavage • Samples to be taken as early as symptom onset, to obtain high virus concentrations	[16, 17]
5.	Bronchoscopy	• Bronchoalveolar lavage (BAL) may be done when sputum sample cannot be obtained • Strict precautions are to be taken to avoid aerosol infections • The sensitivity for BAL 93%	[18]
6.	Radiology [chest X-ray, computed tomography, & lung sonography]	• They are nonspecific, normal in initial phases to patchy unilateral or bilateral involvement to lobar/multi-lobar/bilateral consolidation	[19]

production for COVID-19 therapy. During the present scenario of an international health emergency and shortage of care modalities, these available natural resources can save time as well as be cost-effective.

In the near future, the production of a biosensing material capable of detecting SARS-CoV-2 or a biomarker linked with COVID-19, wearable technology capable of detecting clinical symptoms related to the advent of SARS-CoV-2 (e.g., fever,

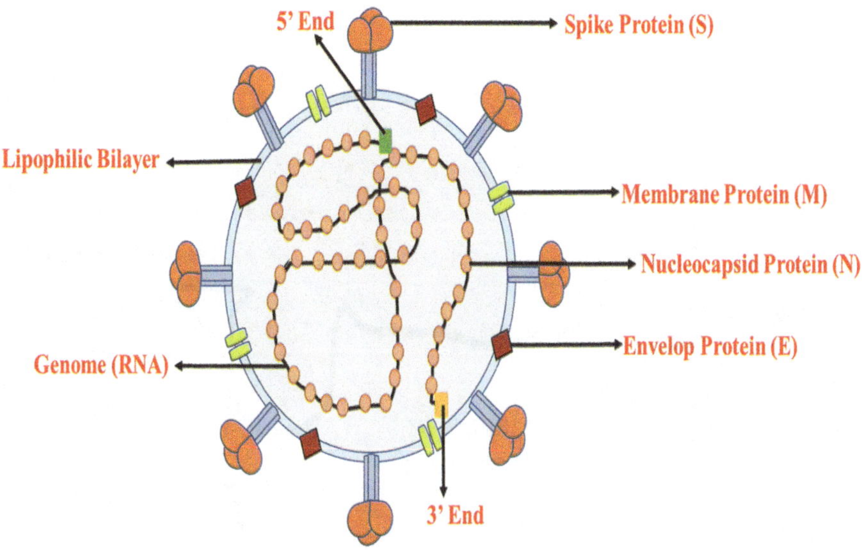

Fig. 24.3 Microstructure of SARS-CoV-2

cough, and fatigue), and vaccine production could help achieve early diagnosis and prevention of SARS-CoV-2.

Pharmaceutical industry faces several administrative impediments such as increased clinical trial design stringency and increased product licensing protection criteria. Given the many challenges in developing new antiviral substances, there has been an increasing initiative in several parts of the world to produce antivirals from natural products in both preclinical and clinical settings. Various synthetic drugs along with their molecular targets and natural bioactive compounds against coronavirus disease have been identified for their effective use and described in Tables 24.2 and 24.3, respectively.

24.5 Conclusion

In conclusion, the pervasiveness of SARS-CoV-2 and its progression depend on the relationship between the virus and the immune system of the person. Viral parameters include virus type, its tendency to mutate, viral load, and virus survival. The variables of immune system of the person include genes (such as HLA genes), age, sex, dietary habits, neuroendocrine-immune regulation, and physical status. All the above variables contribute to whether a person is infected with the virus, the infection period & severity, along with the re-infection propensity of COVID-19. Precise diagnosis in the preliminary stages of the outbreak helps monitor the spread of infectious virus. Developing modern, secure, reliable, fast, and easy SARS-CoV-2 detection technologies is imperative in the future. Physicians must actively

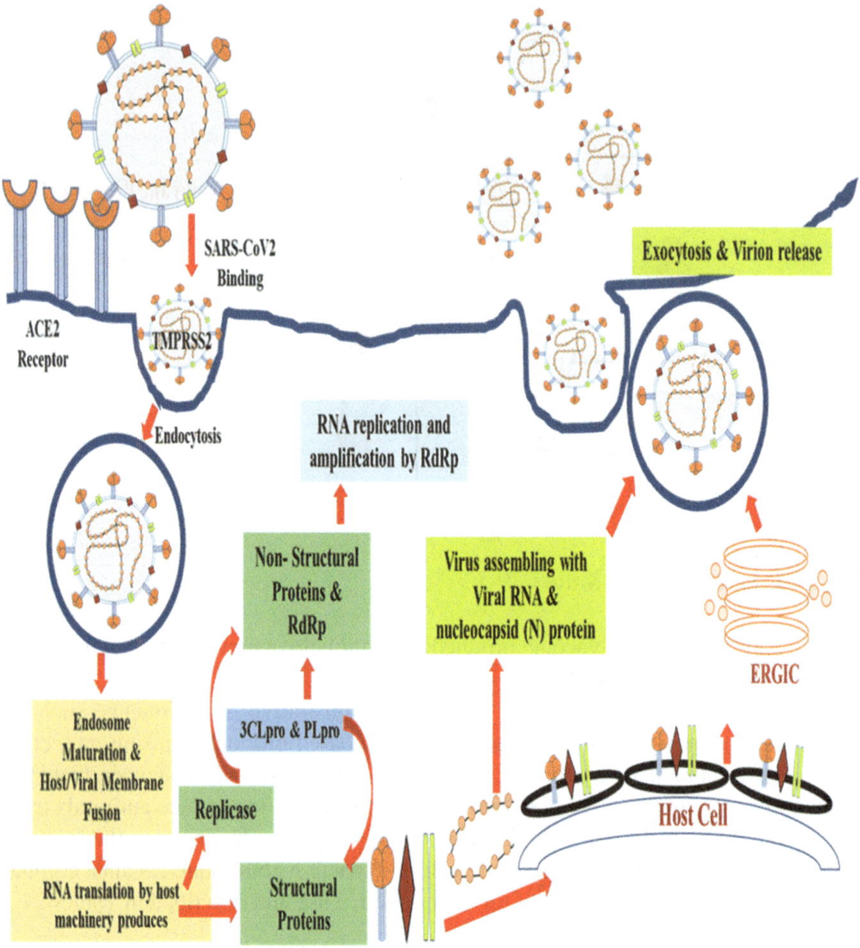

Fig. 24.4 Life cycle of SARS-CoV-2 in host cell

interfere in the two variables to make them grow into a direction that improves public well-being, and will help patients recover as soon as possible. Nevertheless, it must not be assumed that a 100% curative effect can be obtained by medical action.

Recent analysis offers a significant report on a range of natural products which have shown efficacy as anti-CoV drugs and potential for treatment. Fortunately, many researches on the antiviral impacts of natural molecules in this area are now only preliminary, and in-depth in vivo studies on suitable experimental animals are required to explore the cellular and molecular mechanisms. There are a few promising natural compounds on which studies should be performed to obtain pharmacokinetic profile comprising of absorption, distribution, metabolism, and excretion. In addition, clinical trials (phases I–III) are needed to test therapeutic

Table 24.2 Drugs/molecular targets against coronavirus disease

Sr. no.	Category	Drug/molecule	Molecular targets	Mechanism	Dosage	References
1.	Serine protease inhibitors	Camostat mesylate	TMPRSS2	SARS-CoV spike protein (S) cleavage and activation is required for membrane fusion and host cell entry. This is mediated by transmembrane serine protease 2 (TMPRSS2), an airway and alveolar cell serine protease. This partly blocks SARS-CoV and HCoV-NL63 infection in HeLa cell that expresses ACE2 and TMPRSS2; it has been shown to inhibit TMPRSS2 in human lung Calu-3 cells by camostat mesylate significantly which reduced infection with SARS-CoV-2	Three times daily dose of 100–300 mg	[21, 22, 23]
2.	Serine protease inhibitors	Nafamostat mesylate	TMPRSS2	Inhibit MERS-CoV S protein-mediated viral membrane fusion with TMPRSS2-host cells expressing the lung Calu-3 by inhibiting the function of TMPRSS2 protease. Since MERS-CoV and SARS-CoV-2 S proteins	Daily dose of 240 mg for 5 days	[24, 25]

(continued)

Table 24.2 (continued)

Sr. no.	Category	Drug/molecule	Molecular targets	Mechanism	Dosage	References
				share significant homology for the amino acid sequence. This can also impede SARS-CoV-2 cell entry at EC50 of 22.50 μM in the simian Vero E6 cells		
3.	Antimalarial drug	Chloroquine phosphate	ACE2	Chloroquine phosphate prevents terminal phosphorylation of ACE2, and hydroxychloroquine elevates the pH in endosomes involved in the entrance of virus cells, all of which represent important antiviral pathways for chloroquine and hydroxychloroquine. It has already been demonstrated that chloroquine phosphate prevents SARS-CoV infection and its spread via in vitro mechanism	250 mg daily until clinical convalescence, peroral	[26, 27]
4.	Antimalarial drug	Hydroxychloroquine	Endosome, pH elevation	In vivo, hydroxychloroquine is metabolized into chloroquine (mechanism as above)	400 mg loading dose twice daily at day 1, 200 mg twice daily for 4 days, or 600 mg for 6 days, or 400 mg for 5 days, peroral	[28]

5.	Anti-inflammatory alkaloid from *Stephania cepharantha* + anti-helminthic + antimalarial drug	Cepharanthine/selamectin/mefloquine hydrochloride	–	This prevents infection of simian Vero E6 cells with pangolin coronavirus GX P2V/2017/Guangxi (GX P2V), the S protein of which shares 92.2% of the amino acid identity with SARS-CoV-2. However, it was shown that GX P2V also uses ACE2 as the viral cell entry receptor	Repurposing clinical study	[29]
6.	Angiotensin-converting enzyme 2 inhibitors	• DX600, a metallocarboxypeptidase • TNF-α-converting enzyme (TACE) small-molecule inhibitor . TAPI-2 • Nicotianamine, a metal chelator	ACE2	ACE2 inhibition and thus blockade of SARS-CoV-2 cell entry	In vitro & in vivo preclinical studies	[30, 31, 32]
7.	Antiviral drug	Remdesivir	RNA-dependent RNA polymerase (RdRp)	Encourages proof-reading avoidance by viral exo-ribonuclease which leads to viral RNA synthesis inhibition. It works early in infection, and reduces dose-dependent levels of viral RNA that resembles in vitro viral load impairment	200 mg loading dose at day 1, 100 mg for 9–13 days, peroral or intravenous	[33, 34]

(continued)

Table 24.2 (continued)

Sr. no.	Category	Drug/molecule	Molecular targets	Mechanism	Dosage	References
8.	Antiviral drug	Favipiravir	RNA-dependent RNA polymerase (RdRp)	Selectively and potently inhibits the RdRp of RNA viruses and induces lethal RNA transversion mutations, thereby producing a nonviable virus phenotype	6000 mg loading dose at day 1, 2, 400 mg for days 2–10, peroral	[35, 36]
9.	Antiviral drug	Lopinavir/ritonavir	Viral proteases	Inhibits viral protease essential for intracellular assembly	400 mg lopinavir & 100 mg ritonavir twice daily, for 14 days, peroral	[37, 38]
10.	Antiviral drug	Arbidol/umifenovir	Membrane fusion, clathrin-mediated endocytosis	Umifenovir prevents viral host cell entry by inhibition of membrane fusion of viral envelope and host cell cytoplasmic membrane via inhibition of clathrin-mediated endocytosis, thereby preventing virus infection	200–400 mg three times daily, for 9 days, peroral	[39, 40]
11.	3Clpro protease inhibitors	–	3Clpro (also termed M^{pro}) protease	N3, a Michael acceptor inhibitor that can inhibit the 3Clpros of SARS-CoV and MERS-CoV, was shown to form a covalent bond and an irreversible inhibitor of SARS-CoV-2 3Clpro	Computer-aided drug design	[41]

12.	Anti-helminthic activity	Ivermectin (macrocyclic lactones)	Importin (IMP)-α/β heterodimer	Normally the IMP-α/β1 heterodimer binds to CoV's cargo protein in cytosol, and crosses the membrane via nuclear pore. As it reaches the nucleus, the complex dissociates and the cargo infects and declines host cell's antiviral response. This also prevents the cargo protein from crossing the membrane barrier to enter the nucleus, thereby preventing infection and improving the antiviral response	In vitro antiviral activity	[42]
13.	AP2-associated protein kinase-1 inhibitor	Baricitinib	BenevolentAI	Baricitinib is a high-affinity AP2-associated protein kinase-1 (AAK1)-inhibiting drug with inhibition Janus kinase 1/2 (an endocytosis regulator)	2 mg or 4 mg single dose	[43]
14.	Broad-spectrum antiviral agent	Interferon-α	Immune system	Through turning on inactive components and aligning them with the SARS-CoV-2 response mechanism, IFNs will strengthen the immune system	5 million U twice per day, vapor inhalation for <10 days	[44]

(continued)

Table 24.2 (continued)

Sr. no.	Category	Drug/molecule	Molecular targets	Mechanism	Dosage	References
15.	Passive immunization	Convalescent plasma therapy (CPT)	Neutralizing antibodies	The therapy theory is rather basic, and based on the assumption that a patient's plasma retrieved from COVID-19 produces highly specific antibodies that are capable of combating SARS-CoV-2. If antibodies acquired from recovering patients are administered to a patient receiving therapy, they may continue battle the infection more efficiently	Preventive measure with 200 mL CPT can improve condition within 3 days of treatment	[45]

Table 24.3 Natural compounds and their targets against coronavirus disease

Sr. no.	Natural/isolated compounds	Biological source	Target virus/mechanism of action	IC50 value	Reference
1.	Cepharanthine	*Stephania japonica*	SARS-CoV-2; ACE inhibitor	0.98 μmol/L	[46]
2.	Theaflavin	*Camellia sinensis*	SARS-CoV-2; inhibits RdRp activity	–	[47]
3.	Aescin	*Aesculus hippocastanum*	SARS-CoV	3.4 μmol/L	[48, 49]
4.	Reserpine	*Rauwolfia* species	SARS-CoV	6.0 μmol/L	[48, 49]
5.	Ginsenoside-Rb1	*Panax ginseng*	SARS-CoV; inhibits glycoprotein activity by disrupting the envelope	100 μmol/L	[49, 50]
6.	Leptodactylone	*Boenninghausenia sessilicarpa*	SARS-CoV	–	[51]
7.	Lycorine	*Lycoris radiata*	SARS-CoV	15.7 ± 1.2 nmol/L	[52]
8.	Cepharanthine	*Stephania japonica* (Qianjinteng)	2019-nCoV-related pangolin coronavirus GX_P2V infection	0.98 μmol/L	[46]
9.	Glycyrrhizin	*Glycyrrhiza glabra*	SARS-CoV; upregulates nitrous oxide synthase and nitrous oxide production	300 mg/L	[53, 54]
10.	Emodin	*Rheum palmatum*	SARS-CoV; blocks the binding of S protein to ACE2 in a dose-dependent manner	200 μmol/L	[55]
11.	Celastrol Pristimerin Tingenone Iguesterin	*Tripterygium regelii*	SARS-CoV; inhibits SARS-CoV-3CLpro	10.3 μmol/L 5.5 μmol/L 9.9 μmol/L 2.6 μmol/L	[56]
12.	Quercetin-3-b-galactoside	*Ginkgo biloba*	SARS-CoV; competitively inhibits SARS-CoV-3CLpro	42.79 ± 4.97 μmol/L	[57]
13.	Chalcones I–IX	*Angelica keiskei*	SARS-CoV; competitively inhibits SARS-CoV-3CLpro	11.4–129.8 μmol/L	[58]
14.	Tanshinones I–VII	*Salvia miltiorrhiza*	SARS-CoV; inhibits PLpro activity	0.7–30 μmol/L	[59]
15.	Hirsutenone	*Alnus japonica*	SARS-CoV; inhibits PLpro activity	4.1 μmol/L	[59, 60]
16.	Myricetin	*Myrica rubra*	SARS-CoV; inhibits ATPase activity	2.71 ± 0.19 μmol/L	[61]
17.	Scutellarein	*Scutellaria baicalensis*	SARS-CoV; inhibits ATPase activity	0.86 ± 0.48 μmol/L	[61]

(continued)

Table 24.3 (continued)

Sr. no.	Natural/isolated compounds	Biological source	Target virus/mechanism of action	IC50 value	Reference
18.	Saikosaponin B2	*Bupleurum chinense*	HCoV-229E; interferes with early viral entry and blocks viral penetration	1.7 ± 0.1 μmol/L	[52, 62]
19.	Tetrandrine	*Stephania tetrandra*	HCoV-OC43; targeting S protein, inhibits p38 MAPK pathway, suppress replication	0.33 ± 0.03 μmol/L	[63]
20.	Dihydrotanshinone	*Salviae Miltiorrhizae radix*	MERS-CoV	1 μg/mL	[64]
21.	Resveratrol	*Polygonum cuspidatum*	MERS-CoV; targeting N protein and prolonging cellular survival after viral infection	–	[65]
22.	Andrographolide	*Andrographis paniculata*	SARS-CoV-2; inhibitor of 3CLpro through in silico studies; also targets spike protein, ACE2, RdRp, & PLpro	–	[66, 67]
23.	Baicalin	*Scutellaria baicalensis Georgi*	SARS-CoV; inhibit ACE in vitro, strong binding effect with ACE2 and 3CLpro	12.5 μg/mL	[68]
24.	Patchouli alcohol (PA)	Traditional Chinese medicine *Patchouli*	SARS-CoV-2; binding with RdRp was satisfactory, modulates the levels of inflammatory cytokines (docking studies)	–	[69, 70]
25.	Luteolin	Chinese herbal medicine *Torreya nucifera*	SARS-CoV-2; binds with ACE2 and 3CLpro inhibitor, antiviral effect (docking studies)	–	[71]
26.	Curcumin	*Curcuma longa*	SARS-CoV; inhibits virus replication and 3CLpro	40 μmol/L	[72]
27.	Lignan (hinokinin, savinin)	*Chamaecyparis obtusa* var. *formosana*	SARS-CoV; inhibits virus replication and 3CLpro	25–100 μmol/L	[72]

efficacy and protection against CoV on human subjects. Most specifically, research will be conducted to investigate the possible associations of anti-CoV activity between natural products and usable antivirals. Optimizing any of the above lead compounds with established or new mode of action may contribute to the eventual production of new COVID-19 therapeutic agents.

References

1. WHO. https://www.who.int/emergencies/diseases/novel-coronavirus-2019 (Last accessed 08 September 2020)
2. Chan JF, Wang KK, Tse H, Jin DY, Yuen KY (2013) Interspecies transmission and emergence of novel viruses: lessons from bats and birds. Trends Microbiol 21(10):544–555
3. Chen Y, Liu Q, Guo D (2020) Emerging coronaviruses: genome structure, replication, and pathogenesis. J Med Virol 92(4):418–423
4. Lu R, Zhao X, Li J, Niu P, Yang B, Wu H et al (2020) Genomic characterisation and epidemiology of 2019 novel coronavirus: implications for virus origins and receptor binding. Lancet 395:565–574
5. Zhou P, Yang XL, Wang XG, Hu B, Zhang L, Zhang W et al (2020) A pneumonia outbreak associated with a new coronavirus of probable bat origin. Nature 579(7798):270–273
6. World Health Organization, Coronavirus disease (COVID-19) advice for the public. https://www.who.int/emergencies/diseases/novel-coronavirus-2019/advice-for-public (Last accessed 08 September 2020)
7. World Health Organization, Rational use of personal protective equipment for Coronavirus disease 2019 (COVID-19). https://apps.who.int/iris/bitstream/handle/10665/331215/WHO-2019-nCov-IPCPPE_use-2020.1-eng.pdf (Last accessed 08 September 2020)
8. Zhou P, Yang X-L, Wang X-G, Hu B, Zhang L, Zhang W et al (2020) Discovery of a novel coronavirus associated with the recent pneumonia outbreak in humans and its potential bat origin. bioRxiv. https://doi.org/10.1101/2020.01.22.914952
9. Li Q, Guan X, Wu P, Wang X, Zhou L, Tong Y et al (2020) Early transmission dynamics in Wuhan, China, of novel coronavirus-infected pneumonia. N Engl J Med 382:1199–1207
10. Bauch CT, Smith JOL, Coffee MP, Galvani AP (2005) Dynamically modelling SARS and other newly emerging respiratory illnesses: past, present, and future. Epidemiology 16(6):791–801
11. Cheng SC, Chang YC, Chiang YLF, Chien YC, Cheng M, Yang CH et al (2020) First case of coronavirus disease 2019 (COVID-19) pneumonia in Taiwan. J Formos Med Assoc 119 (3):747–751
12. Lai CC, Liu YH, Wang CY, Wang YH, Hsueh SC, Yen MY et al (2020) Asymptomatic carrier state, acute respiratory disease, and pneumonia due to severe acute respiratory syndrome coronavirus 2 (SARSCoV-2): facts and myths. J Microbiol Immunol Infect 53(3):404–412
13. Wang Yixuan WY, Chen Y, Qin Q (2020) Unique epidemiological and clinical features of the emerging 2019 novel coronavirus pneumonia (COVID-19) implicate special control measures. 92(6):568–576
14. Tanindi A, Cemri M (2011) Troponin elevation in conditions other than acute coronary syndromes. Vasc Health Risk Manag 7:597–603
15. Xiang J, Yan M, Li H, Liu T, Lin C, Huan S et al. (2020) Evaluation of enzyme-linked immunoassay and colloidal gold- immune-chromatographic assay kit for detection of novel coronavirus (SARS-Cov-2) causing an outbreak of pneumonia (COVID-19). doi: https://doi.org/10.1101/2020.02.27.20028787
16. Guo L, Ren L, Yang S, Xiao M, Chang D, Yang F et al. (2020) Profiling early humoral response to diagnose novel coronavirus disease (COVID-19), Clin Infect Dis; ciaa310. https://doi.org/10.1093/cid/ciaa310

17. Gao X, Zhou H, Wu C, Xiao Y, Ren L, Paranhos-Baccala G et al (2015) Antibody against nucleocapsid protein predicts susceptibility to human coronavirus infection. J Infect 71 (5):599–602
18. Wang W, XuY GR, Lu R, Han K, Wu G et al (2020) Detection of SARS-CoV-2 in different types of clinical specimens. J Am Med Assoc 323(18):1843–1844
19. Pan FY, Tianhe SP, Gui S, Bo L, Li L, Zheng D et al (2020) Time course of lung changes on chest CT during recovery from 2019 novel coronavirus (COVID-19) pneumonia. Radiology 77 (8):1–15
20. Tortorici MA, Veesler D (2019) Structural insights into coronavirus entry. Adv Virus Res 105:93–116
21. Ramsey ML, Nuttall J, Hart PA (2019) A phase 1/2 trial to evaluate the pharmacokinetics, safety, and efficacy of NI-03 in patients with chronic pancreatitis: study protocol for a randomized controlled trial on the assessment of camostat treatment in chronic pancreatitis (TACTIC). Trials 20(1):501
22. NCT02693093, ClinicalTrials.gov, (2016), Feb 26
23. Matsuyama S, Nao N, Shirato K, Kawase M, Saito S, Takayama I et al (2020) Enhanced isolation of SARS-CoV-2 by TMPRSS2-expressing cells. Proc Natl Acad Sci U S A 117 (13):7001–7003
24. Yamamoto M, Matsuyama S, Li X, Takeda M, Kawaguchi Y, Inoue JI et al (2016) Identification of nafamostat as a potent inhibitor of middle east respiratory syndrome coronavirus S protein-mediated membrane fusion using the split-protein based cell-cell fusion assay. Antimicrob Agents Chemother 60(11):6532–6539
25. Wang M, Cao R, Zhang L, Yang X, Liu J, Xu M et al (2020) Remdesivir and chloroquine effectively inhibit the recently emerged novel coronavirus (2019-nCoV) in-vitro. Cell Res 30 (3):269–271
26. Abdul M, Al-Bari A (2017) Targeting endosomal acidification by chloroquine analogs as a promising strategy for the treatment of emerging viral diseases. Pharmacol Res Perspect 5(1): e00293
27. Keyaerts E, Vijgen L, Maes P, Neyts J, Ranst MV (2004) In vitro inhibition of severe acute respiratory syndrome coronavirus by chloroquine. Biochem Biophys Res Commun 323 (1):264–268
28. Chen Z, Hu J, Zhang Z, Jiang S, Han S, Yan D et al., (2020) Efficacy of hydroxychloroquine in patients with COVID-19: results of a randomized clinical trial, medRXiv. https://doi.org/10.1101/2020.03.22.20040758
29. Fan HH, Wang LQ, Liu WL, An XP, Liu ZD, He XQ et al (2020) Repurposing of clinically approved drugs for treatment of coronavirus disease 2019 in a 2019-novel coronavirus (2019-nCoV) related coronavirus model. Chin Med J 133(9):1051–1056
30. Haga S, Nagata N, Okamura T, Yamamoto N, Sata T, Yamamoto N et al (2010) TACE antagonists blocking ACE2 shedding caused by the spike protein of SARS-CoV are candidate antiviral compounds. Antivir Res 85(3):551–555
31. Mohler KM, Sleath PR, Fitzner JN, Ceretti DP, Anderson M, Kerwar SS et al (1994) Protection against a lethal dose of endotoxin by an inhibitor of tumour necrosis factor processing. Nature 370(6486):218–220
32. Takahashi M, Terada Y, Nakai I, Nakanishi H, Yoshimura E, Mori S et al (2003) Role of nicotinamide in the intracellular delivery of metals and plant reproductive development. Plant Cell 15(6):1263–1280
33. Agostini ML, Andres EL, Sims AC, Graham RL, Sheahan TP, Lu X et al (2018) Coronavirus susceptibility to the antiviral remdesivir (GS-5734) is mediated by the viral polymerase and the proofreading exoribonuclease. MBio 9(2):e00221–e00218
34. Mulangu S, Dodd LE, Davey RT, Mbaya OT, Proschan M, Mukadi D et al (2019) A randomized, controlled trial of Ebola virus disease therapeutics. N Engl J Med 381 (24):2293–2303

35. Jin Z, Smith LK, Rajwanshi VK, Kim B, Deval J (2014) The ambiguous base-pairing and high substrate efficiency of T-705 (favipiravir) ribofuranosyl 5′-triphosphate towards influenza A virus polymerase. PLoS One 8(7):e68347
36. Sissoko D, Laouenan C, Folkesson E, M'Lebing AB, Beavogui AH, Baize S et al (2016) Experimental treatment with favipiravir for Ebola virus disease (the JIKI Trial): a historically controlled, single-arm proof-of-concept trial in Guinea. PLoS Med 13(3):e1001967
37. Chu CM, Cheng VCC, Hung IFN, Wong MML, Chan KH, Chan KS et al (2004) Role of lopinavir/ritonavir in the treatment of SARS: initial virological and clinical findings. Thorax 59 (3):252–256
38. Chan JF, Yao Y, Yeung ML, Deng W, Bao L, Jia L et al (2015) Treatment with lopinavir/ritonavir or interferon-β1b improves outcome of MERS-CoV infection in a nonhuman primate model of common marmoset. J Infect Dis 212(12):1904–1913
39. Leneva IA, Russell RJ, Boriskin YS, Hay AJ (2009) Characteristics of arbidol-resistant mutants of influenza virus: implications for the mechanism of anti-influenza action of arbidol. Antivir Res 81(2):132–140
40. Deng L, Li C, Zeng Q, Liu X, Li X, Zhang H et al (2020) Arbidol combined with LPV/r versus LPV/r alone against corona virus disease 2019: a retrospective cohort study. J Infect 81(1):e1–e5
41. Jin Z, Du X, Xu Y, Deng Y, Liu M, Zhao Y et al (2020) Structure of Mpro from COVID-19 virus and discovery of its inhibitors. Nature 582:289–293
42. Wagstaff KM, Rawlinson SM, Hearps AC, Jans DA (2011) An AlphaScreen®-based as- say for high-throughput screening for specific inhibitors of nuclear import. J Biomol Screen 16:192–200
43. Sorrell FJ, Szklarz M, Azeez KRA, Elkins JM, Knapp S (2016) Family-wide structural analysis of human numb-associated protein kinases. Structure 24(3):401–411
44. Jianping J. Randomized, open, blank control study on the efficacy and safety of recombinant human interferon α1 β; in the treatment of patients with new type of coronavirus infection in Wuhan. Source- US National Library of Medicine https://clinicaltrials.gov/ct2/show/NCT04293887 (Last accessed on 08 September 2020)
45. Duan K, Liu B, Li C, Zhang H, Yu T, Qu J et al (2020) Effectiveness of convalescent plasma therapy in severe COVID-19 patients. Proc Natl Acad Sci 117(17):9490–9496
46. Fan HH, Wang LQ, Liu WL, An XP, Liu ZD, He XQ et al (2020) Repurposing of clinically approved drugs for treatment of coronavirus disease 2019 in a 2019-novel coronavirus-related coronavirus model. Chin Med J 133:1051–1056
47. Lung J, Lin YS, Yang YH, Chou YL, Shu LH, Cheng YC et al (2020) The potential chemical structure of anti-SARS-CoV-2 RNA-dependent RNA polymerase. J Med Virol 92:693–697
48. Chen FE, Huang J (2005) Reserpine: a challenge for total synthesis of natural products. Chem Rev 105(12):4671–4706
49. Wu CY, Jan JT, Ma SH, Kuo CJ, Juan HF, Cheng YS et al (2004) Small molecules targeting severe acute respiratory syndrome human coronavirus. Proc Natl Acad Sci U S A 101:10012–10017
50. Jeong CS, Hyun JE, Kim YS (2003) Ginsenoside Rb1: the anti-ulcer constituent from the head of *Panax ginseng*. Arch Pharm Res 26(11):906–911
51. Yang QY, Tian XY, Fang WS (2007) Bioactive coumarins from *Boenninghausenia sessilicarpa*. J Asian Nat Prod Res 9(1):59–65
52. Li SY, Chen C, Zhang HQ, Guo HY, Wang H, Wang L et al (2005) Identification of natural compounds with antiviral activities against SARS-associated coronavirus. Antivir Res 67 (1):18–23
53. Hoever G, Baltina L, Michaelis M, Kondratenko R, Baltina L, Tolstikov GA et al (2005) Antiviral activity of glycyrrhizic acid derivatives against SARS-coronavirus. J Med Chem 48 (4):1256–1259

54. Cinatl J, Morgenstern B, Bauer G, Chandra P, Rabenau H, Doerr HW (2003) Glycyrrhizin, an active component of liquorice roots, and replication of SARS-associated coronavirus. Lancet 361(9374):2045–2046
55. Ho TY, Wu SL, Chen JC, Li CC, Hsiang (2007) CY Emodin blocks the SARS coronavirus spike protein and angiotensin-converting enzyme 2 interaction. Antivir Res 74(2):92–101
56. Ryu YB, Park SJ, Kim YM, Lee JY, Seo WD, Chang JS et al (2010) SARSCoV 3CLpro inhibitory effects of quinone-methide triterpenes from *Tripterygium regelii*. Bioorg Med Chem Lett 20(6):1873–1876
57. Chen L, Li J, Luo C, Liu H, XuW CG et al (2006) Binding interaction of quercetin-3-beta-galactoside and its synthetic derivatives with SARSCoV 3CL (pro): structure-activity relationship studies reveal salient pharmacophore features. Bioorg Med Chem 14(24):8295–8306
58. Park JY, Ko JA, Kim DW, Kim YM, Kwon HJ, Jeong HJ et al (2016) Chalcones isolated from *Angelica keiskei* inhibit cysteine proteases of SARS-CoV. J Enzyme Inhib Med Chem 31 (1):23–30
59. Park JY, Kim JH, Kim YM, Jeong HJ, Kim DW, Park KH et al (2012) Tanshinones as selective and slow-binding inhibitors for SARS-CoV cysteine proteases. Bioorg Med Chem 20 (19):5928–5935
60. Park JY, Jeong HJ, Kim JH, Kim YM, Park SJ, Kim D et al (2012) Diarylheptanoids from *Alnus japonica* inhibit papain-like protease of severe acute respiratory syndrome coronavirus. Biol Pharm Bull 35:2036–2042
61. Yu MS, Lee J, Lee JM, Kim Y, Chin YW, Jee JG et al (2012) Identification of myricetin and scutellarein as novel chemical inhibitors of the SARS coronavirus helicase, nsP13. Bioorg Med Chem Lett 22(12):4049–4054
62. Cheng PW, Ng LT, Chiang LC, Lin CC (2006) Antiviral effects of saikosaponins on human coronavirus 229E in vitro. Clin Exp Pharmacol Physiol 33(7):612–616
63. Kim DE, Min JS, Jang MS, Lee JY, Shin YS, Song JH et al (2019) Natural bis-benzylisoquinoline alkaloids-tetrandrine, fangchinoline, and cepharanthine, inhibit human coronavirus OC43 infection of MRC-5 human lung cells. Biomol Ther 9(11):696
64. Kim JY, Kim YI, Park SJ, Kim IK, Choi YK, Kim SH (2018) Safe, high throughput screening of natural compounds of MERS-CoV entry inhibitors using a pseudovirus expressing MERS-CoV spike protein. Int J Antimicrob Agents 52(5):730–732
65. Lin SC, Ho CT, Chuo WH, Li S, Wang TT, Lin CC (2017) Effective inhibition of MERS-CoV infection by resveratrol. BMC Infect Dis 17:144
66. Yu B, Dai CQ, Jiang ZY, Li EQ, Chen C, Wu XL et al (2014) Andrographolide as an anti-H1N1 drug and the mechanism related to retinoic acid-inducible gene-I-like receptors signaling pathway. Chin J Integr Med 20(7):540–545
67. Enmozhi SK, Raja K, Sebastine I, Joseph J (2020) Andrographolide as a potential inhibitor of SARS-CoV-2 main protease: an *in-silico* approach. J Biomol Struct Dyn:1–10
68. Deng YF, Aluko RE, Jin Q, Zhang Y, Yuan LJ (2012) Inhibitory activities of baicalin against renin and angiotensin-converting enzyme. Pharm Biol 50(4):401–406
69. Yu Y, Zhang Y, Wang S, Liu W, Hao C, Wang W (2019) Inhibition effects of patchouli alcohol against influenza a virus through targeting cellular PI3K/Akt and ERK/MAPK signaling pathways. Virol J:16–163
70. Li YC, Peng SZ, Chen HM, Zhang FX, Xu PP, Xie JH et al (2012) Oral administration of patchouli alcohol isolated from Pogostemonis Herba augments protection against influenza viral infection in mice. Int Immunopharmacol 12(1):294–301
71. Ryu YB, Jeong HJ, Kim JH, Kim YM, Park JY, Kim D et al (2010) Biflavonoids from Torreya nucifera displaying SARS-CoV 3CL(pro) inhibition. Bioorg Med Chem 18(22):7940–7947
72. Wen CC, Kuo YH, Jan JT, Liang PH, Wang SY, Liu HG et al (2007) Specific plant terpenoids and lignoids possess potent antiviral activities against severe acute respiratory syndrome coronavirus. J Med Chem 50(17):4087–4095

25 Underpinning the Rudimentary/Underlying Mechanisms Involved in the Pathogenesis of SARS-CoV-2 (COVID-19) in Human Lung Cells

Arnab Ghosh, Chandrika Bhattacharyya, Nidhan K. Biswas, and Amlan Das

Abstract

The COVID-19 pandemic is presently the major threat to human society and health due to its high infectivity and mortality rates. To date, this pandemic has resulted in more than 1.5 million deaths globally, affecting more than 200 countries. Phylogenetic analysis of the SARS-CoV-2 genome revealed its striking homology with the bat-derived coronavirus strains, thus confirming the zoonotic origin of the virus. SARS-CoV-2 binds to the angiotensin-converting enzyme 2 (ACE2) receptors expressed on the surface of the host cells, leading to endocytosis of the receptor, followed by the replication of the viral RNA, packaging, assembly, and release of the progeny viruses. This leads to the systemic infection in the host body and the shredding of the virus, causing its transmission to a new host. The extent of infection in the host cells depends on the expression of ACE2 expression and hyperactivation of the immune system to generate a cocktail of inflammatory cytokines, also referred to as the cytokine storm. This inflammatory response can cause severe damage to the lung tissues.

Keywords

Coronavirus disease-2019 (COVID-19) · Pandemic · Severe acute respiratory syndrome coronavirus 2 (SARS-CoV-2) · Receptor-binding domain (RBD) · Angiotensin-converting enzyme 2 (ACE2) · Cytokine storm

Arnab Ghosh, Chandrika Bhattacharyya and Amlan Das contributed equally to this work.

A. Ghosh · C. Bhattacharyya · N. K. Biswas · A. Das (✉)
National Institute of Biomedical Genomics, Kalyani, India
e-mail: ad4@nibmg.ac.in

K. Dua et al. (eds.), *Targeting Cellular Signalling Pathways in Lung Diseases*,
https://doi.org/10.1007/978-981-33-6827-9_25

25.1 Background

The coronavirus disease-2019 (COVID-19) pandemic has imposed a severe threat to humanity, affecting more than 200 countries globally (www.worldometers.info). COVID-19 is a highly infectious disease caused by SARS-CoV-2 (severe acute respiratory syndrome coronavirus 2), a novel coronavirus strain, also referred to as 2019 novel coronavirus (2019-nCoV) or human coronavirus 2019 (hCoV-19) [1]. SARS-CoV-2 is a positive-sense single-stranded RNA virus responsible for the highly contagious severe acute respiratory syndrome (SARS) in humans [2]. In December 2019, the initial cases of SARS-CoV-2 infection were reported in Wuhan City of Hubei Province, China [3]. Several cases were reported since 8 December 2019, revealing that many individuals employed or lived nearby the local Huanan Seafood Market were suffering from unidentified pneumonia [3]. With the increasing severity of the infection, the China Health Authority alerted the World Health Organization (WHO) about this situation on 31 December 2019. By the end of January 2020, there were 7736 confirmed and 12,167 suspected cases in China, and a further 82 confirmed cases were detected in 18 other countries [4]. Due to the increasing reports of COVID-19 infection and morbidity in China and other countries, the WHO declared this disease as a pandemic in March 2020 [5]. By the end of October 2020, the number of COVID-19 confirmed cases had surpassed 36 million, including 1 million 56 thousand deaths in more than 200 countries across the globe [**Weekly update of COVID-19 on 9 October 2020**].

25.2 Classification and Origin of SARS-CoV-2

SARS-CoV-2 belongs to the family *Coronaviridae*, which can be classified into two subfamilies, namely, *Coronavirinae* and *Torovirinae*. The members of the *Coronavirinae* subfamily are further subdivided into four genera such as (a) *Alphacoronavirus*, e.g., human coronavirus 229E (HCoV-229E) and HCoV-NL63; (b) *Betacoronavirus*, e.g., HCoV-OC43, severe acute respiratory syndrome coronavirus like SARS-CoV, HKU1, and Middle Eastern respiratory syndrome coronavirus (MERS-CoV); (c) *Gammacoronavirus* also known as avian coronavirus, responsible for the infection in birds; and (d) *Deltacoronavirus* including the viruses infecting pigs and birds [6]. The alpha and beta genera were derived from the bat gene pool, whereas the gamma and delta genera were derived from pig and avian sources [7]. Members of alphacoronavirus and betacoronavirus possess the pathogenic property, which may be mild or detrimental to humans. Four of these viruses, such as NL63, 229E, OC43, and HKU1, were known to cause mild respiratory problems in humans. In contrast, the other three strains, including MERS-CoV, SARS-CoV, and the newly emerged SARS-CoV-2, can cause severe respiratory syndromes [1]. SARS-CoV-2 belongs to the genera of *Betacoronavirus* and displays a crown-like morphology on the surface when observed under an electron microscope. The presence of this "crown-like structure" had influenced the nomenclature of this family of viruses as the coronaviruses (corona is the Latin term for crown).

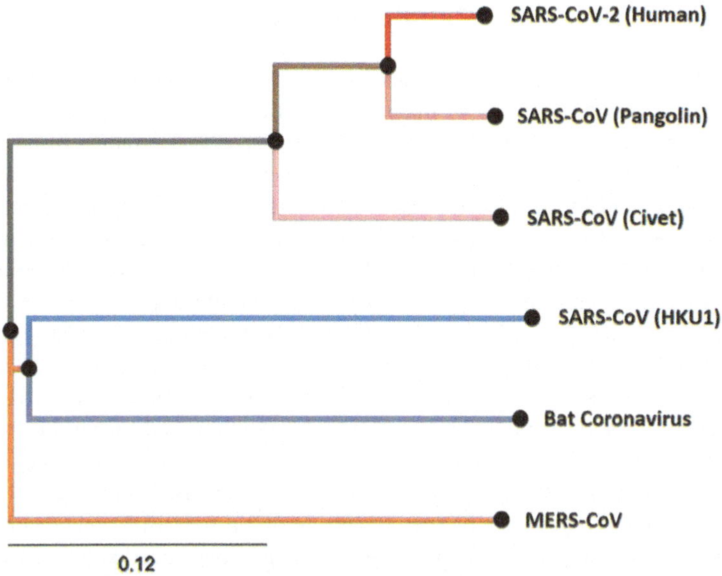

Fig. 25.1 Multiple sequence alignment of whole-genome sequences and phylogenetic tree of SARS-CoV-2 with human SARS-CoV, coronavirus from other species, and MERS-CoV (Clustal Omega). Novel coronavirus, i.e., SARS-CoV-2 showed more sequence similarity with coronaviruses from pangolin and civet. On the other hand, human SARS-CoV was placed more closer to coronavirus from bat

These "crown-like structures" represent the glycoprotein moieties present on the viral envelope, which facilitate the entry of the virus into the host cells.

The full-length genome sequences of SARS-CoV-2 obtained from early infected individuals related to the Wuhan meat market revealed that the genomic length of this novel coronavirus ranges between 29,891 and 29,903 nucleotides (nt). Phylogenetic analysis of the SARS-CoV-2 genome revealed the striking similarity of the virus with two bat-derived coronaviruses isolated in 2018 in eastern China, such as bat-SL-CoVZC45 and bat-SL-CoVZXC21 (>88% similarity) [8]. Furthermore, it was found to be genetically distinct from SARS-CoV (with about 79% similarity) and MERS-CoV. A further study revealed that SARS-CoV-2 was more related to the bat coronavirus RaTG13, which was previously detected in *Rhinolophus affinis*, an intermediate horseshoe bat (>92% sequence identity) from Yunnan Province, China. These findings suggest that SARS-CoV might have originated from bats and crossed the interspecies barrier to infect humans via zoonotic transmission [9] (Fig. 25.1).

25.3 Structural Organization of the Virus

The SARS-CoV-2 genome comprises of 30-kb-long single-stranded RNA that consists of a variable number of open reading frames (ORFs) between 6 and 11. The viral mRNA codes for 29 structural and nonstructural proteins (nsps), which include ORF1a/b polyprotein, spike (S) glycoprotein, envelope (E), membrane (M), and the nucleocapsid (N) protein. About two-thirds of the viral genome comprises overlapping replicase genes ORF1a and ORF1ab, which are directly translated into the polyprotein pp1a (nsp1–11) or pp1ab (nsp1–16), respectively, depending on a ribosomal (−1) frameshift [10]. The remaining one-third portion codes for the structural proteins, including S, E, M, and N proteins [11]. The translated polyproteins may undergo further proteolytic cleavage to generate 11 or 16 individual nsps. The viral nsps constitute the replication/transcription complex (RTC), a membrane-anchored, dynamic protein-RNA complex required for viral replication. The processing of viral polyproteins is mediated by two proteases, namely, the papain-like protease (PLpro or nsp3) encoded between nsp1 and 4 and the main protease or chymotrypsin-like protease (Mpro or 3CLpro or nsp5) encoded between nsp4 and 11/16. The papain-like proteinase is the first nsp encoded by ORF1ab and is an essential component of the RTC. The PLpro cleaves nsps 1–3 and blocks host innate immune response [12]. The viral Mpro facilitates the processing of the nsp7–10 region and liberates 4 small proteins nsp7, nsp8, nsp9, and nsp10, respectively [12]. SARS-CoV-2 also contains six accessory proteins, encoded by ORF3a, ORF6, ORF7a, ORF7b, and ORF8 genes. The structural proteins such as S protein, E protein, M protein, and N protein constitute the outer surface of SARS-CoV-2. The M and E proteins play crucial roles in the morphogenesis and assembly of viral particles, while the spike protein facilitates virus entry within the host cells [13]. The ~600 kDa S glycoprotein exists in a trimeric form, and each monomer in this complex is composed of two heterodimeric subunits, S1 and S2. The S1 subunit consists of the receptor-binding domain or RBD, which interacts with the peptidase domain (PD) of angiotensin-converting enzyme 2 (ACE2) receptor, expressed in the membrane of lung pneumocytes and other cell types [14]. RBDs of SARS-CoV and SARS-CoV-2 have 72% similarity in the amino acid sequence with highly similar tertiary structures. S protein needs to be cleaved at S1/S2 to get activated and further interaction of S1/ACE2, resulting in the fusion of S2 subunit with the host cell membrane [15]. This process is mediated by TMPRSS2 and furin-like proteases following a conformational change of S protein from a pre-fusion to post-fusion state. The replication of SARS-COV-2 in the host cells is mediated by a multi-subunit replication/transcription complex made of nsps such as nsp12, nsp7, and nsp8 [16]. The nsp12 acts as the catalytic subunit of the RNA-dependent RNA polymerase (RdRp), while nsp7 and nsp8 increase the RNA-binding ability and processivity of the enzyme [16]. Another nonstructural protein, nsp13, codes for the enzyme helicase, which catalyzes the unwinding of double-stranded RNA formed during replication and allows the next round of viral replication [11]. Thus, the coordinated functions of the structural and nonstructural proteins of SARS-CoV-2 facilitate the infection and replication of the virus in the host cells.

25.4 Human Transmission of SARS-CoV-2

In December 2019, many individuals were admitted to the local hospitals of Wuhan, the capital city of Hubei Province of China, with severe pneumonia of unknown cause. Many of the infected persons shared common exposure to the Hunan wholesale meat/seafood market. On 7 January 2020, the virus was identified as a coronavirus strain, which showed significant homology with the bat coronavirus. Furthermore, the environmental samples from the Hunan market also confirmed the origin of the virus from that place [17]. The number of cases then started to increase exponentially, and individuals who did not have exposure to the Hunan market were also found to be infected, thus revealing the contagious nature of SARS-CoV-2 infection in humans. The massive migration of Chinese people in and out of China during the Chinese New Year acted as a catalyst for the global transmission of the disease. Even the countries outside China, those with no history of travel to China, experienced a steady increase in the COVID-19-positive cases. These incidents confirmed that local human-to-human transmission was also occurring in those countries [18]. Close contacts between the individuals and exposure to large droplets generated during coughing and sneezing were identified as the main modes of transmission, and the disease was also found to be airborne since these droplets can spread 1–2 m in the air. Moreover, it was observed that a subset of COVID-19 patients may be asymptomatic but still can spread the virus via close contact with healthy individuals or transmission of droplets [19]. Approximately 40–45% of SARS-CoV-2 infections are accountable to asymptomatic persons. These individuals help in the transmission of the virus to other individuals for an extended period of time, which is typically longer than the usual 14 days. Computer tomography images showed that asymptomatic infection is associated with subclinical lung abnormalities [20]. Studies have also shown that viral loads are higher in the nasal cavity compared to the throat and there is no significant difference in viral burden between asymptomatic and symptomatic patients. For viral infections, there is a certain period of time, after which the symptoms become prominent in the hosts, which is known as the incubation time or incubation period. The mean incubation period for SARS-CoV-2 infection was found to vary from 2 to 14 days [median 5 days] [21]. The estimated necessary reproduction number (R0) and serial interval were estimated to be 2.2 and 5–6 days, respectively, with a doubling time of the number of infected individuals every 3 days [22].

25.5 The Temporal Spread of SARS-CoV-2 Subtypes

The availability of SARS-CoV-2 genome sequence data from almost every region of the world made the real-time tracking of its evolution possible. The SARS-CoV-2 virus infects humans with various degrees of severity; some regions of the world experienced much higher severity and mortality than the rest [23]. Like all other RNA viruses, with time, it accumulated mutations and evolved into multiple subtypes that spread nonuniformly among geographic locations. The first reported

SARS-CoV-2 sequences from Wuhan, China, in December 2019 are formally known as the "O" subtype and considered to be the ancestral one. Within 1 month, 9 subtypes were born, including the ancestral subtype with 2 of the O and B consisting of 86% of all viral sequences, and the rest were B2 (4.55%), A1a (2.69%), A3 (2.07%), A2 (1.66%), B1 (1.44%), B4 (1.66%), and A2a (0.41%) spread across Asia (13 countries), Europe (6 countries), North America (2 countries), and Oceania (1 country). From the subsequent month, rapid changes in frequencies of viral subtypes took place with the inclusion of another subtype A6 (consisting of 0.26% of viral sequences); the frequency of A2a subtype with Spike D614G and RdRp P323L (also known as ORF1b P314L) mutation gained 11.15% from less than 1% in the previous month. Also frequency of A1a gained from 2.69% to 11.41%. By the end of February, SARS-CoV-2 reaches greater portions of Asia (18 countries), Europe (17 countries), North America (4 countries), and Oceania (2 countries) followed by Africa (3 countries) and South America (1 country). From March, the A2a subtype started outcompeting all other subtypes gaining 63.58% frequency worldwide, which further rose to 76.60% till the end of May. The selective advantage of A2a subtype virus was predicted to be due to (1) an introduction of the additional cleavage site in S1/S2 junction of spike protein [24, 25] and (2) increased stability of spike protein by D614G mutation [26]. Further, A2a has evolved into three subtypes, namely, 20A, 20B, and 20C, among which 20B that harbors two consecutive mutations—G28881A and G28882A—attained the highest frequency. After May, the sequence data submission across geographical regions became parsed and nonuniform making it difficult to track real-time viral evolution. The current mutation of SARS-CoV-2 was estimated to be around 26.64 substitutions per year for each transmission line [www.nextstrain.org]. Thus, the temporal spreading of the virus is greatly influenced by the mutational potential of SARS-CoV-2 (Fig. 25.2).

25.6 Phases of Viral Life Cycle

The SARS-CoV-2 infection is initiated after its entry in the nasopharynx by getting attached to ACE2 receptor-enriched epithelial cells of the nasal and oral mucosa. This results in a reduced sense of smell and taste, and with time the virus is transmitted to the lungs, which then becomes the primary site of infection. After attachment, a chain of events occurs sequentially, including endocytosis, viral genome replication, transcription, assembly, and release of the progeny virus to the neighboring cells. As a result, the infection gradually gets transmitted to different organs throughout the human body.

Classically, the viral life cycle can be divided into five elements: attachment, entry via membrane fusion, replication and processing of the viral genome, viral assembly, and release. The cellular and molecular mechanisms of viral entry, replication, assembly, and release of the progeny virus particles have been summarized as follows.

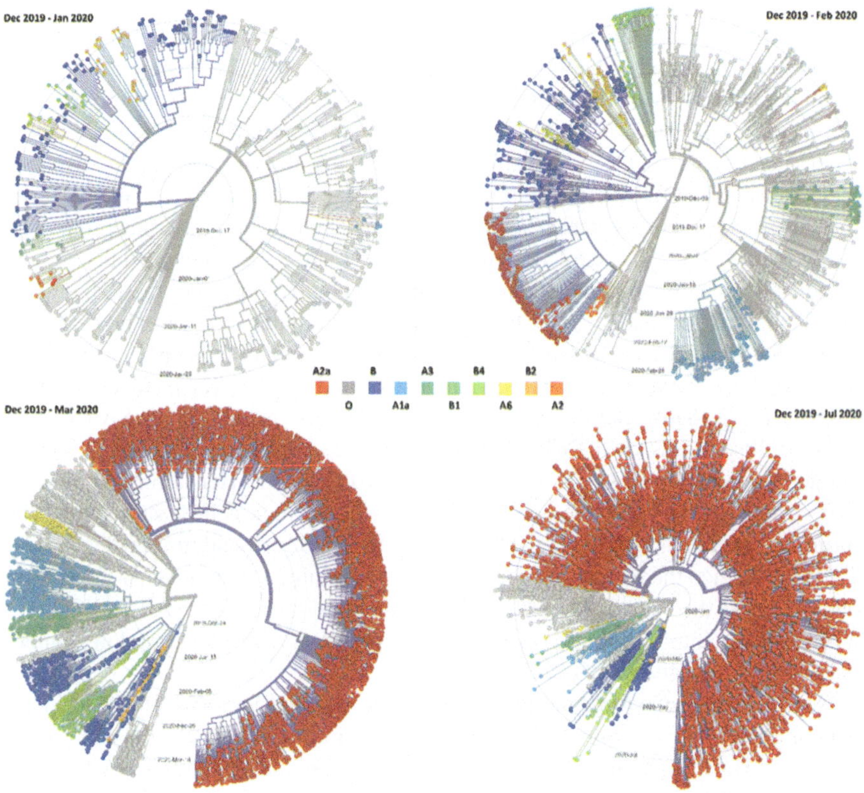

Fig. 25.2 Radial phylogenetic time tree based on more than 80,000 SARS-CoV-2 RNA sequences worldwide, showing the trajectory of the evolution of SARS-CoV-2 subtypes in different time points. Each concentric circle shows the date of sample collection; earlier dates are closer to the center. The A2a subtype (red), after originating in January 2020, rapidly outcompeted other subtypes within a few months and became the dominant subtype

25.6.1 Cleavage of S Protein

For cellular attachment, the S1 subunit of the viral spike protein binds angiotensin-converting enzyme 2 (ACE2) receptor, expressed in the membrane of lung pneumocytes. The S1 subunit consists of the receptor-binding domain or RBD, which interacts with the peptidase domain (PD) of the ACE2 protein [14]. An atypical interaction exists between SARS-CoV-2 spike protein RBD with hACE2 binding. Though SARS-CoV-2 RBD has a higher ACE2 binding affinity than SARS-CoV RBD, the SARS-CoV-2 spike has a lower ACE2 binding affinity than SARS-CoV spike protein. The cryo-electron microscopy of the SARS-CoV-2 spike protein revealed that its RBD is mostly in the lying-down state, a state associated with ineffective receptor binding. Therefore, compared to SARS-CoV, although SARS-CoV-2 RBD has a higher ACE2 binding affinity, it is less accessible, resulting in comparable or lower hACE2 binding affinity for SARS-CoV-2 spike. To maintain its high infectivity while keeping its RBD less accessible, SARS-CoV-2 relies on a second strategy—host protease activation. The fusion of the viral membrane with that of the host cell depends on the cleavage of S protein by host cell proteases at the S1/S2 and the S2′ site, which results in S protein activation. After attachment of the S protein to the host membrane, the second event is membrane fusion. This process is facilitated by the host proteases, namely, furin, transmembrane serine protease 2 (TMPRSS2), and cathepsin L. After the proteolytic cleavage of S at S1/S2 and S2′, the two subunits S1 and S2 are generated that remain associated via non-covalent interactions. The S1 subunit consists of the RBD, while the S2 subunit gets anchored to the membrane and initiates the fusion process.

The S1/S2 site in SARS-CoV-2 consists of an exposed loop that harbors a stretch of a polybasic sequence of 6–7 arginine (R) and lysine (K) residues such as RKKRKRYG. This multibasic amino acid sequence is unique to SARS-CoV-2 and is absent in SARSr-CoV (SARS-CoV-related coronaviruses), but present in human coronaviruses like OC43, HKU1, and MERS-CoV. Hence, in both SARS-CoV-2 and MERS-CoV, the entry of the virus in the lung cell phase required the furin-mediated pre-cleavage of the S protein at the S1/S2 site and further cleavage by TMPRSS2 since the expression of cathepsin L is minimal in lung cells. It has been observed that the furin preactivation step is required for the entry of SARS-CoV-2 pseudovirus into different types of ACE2-expressing cells, including lung epithelial and lung fibroblast cells [27].

Accumulating experimental evidence further revealed that TMPRSS2 independent of furin could also initiate the S protein activation and facilitate viral entry in the host cells. Hoffman et al. had shown that TMPRSS2 could also cleave SARS-CoV-2 S protein at S1/S2 and S′ sites and mediate the fusion of the free fusion peptide with the host cell membrane [28]. TMPRSS2 has been reported to cleave at single arginine or lysine residues (R/K↓) at S-cleavage sites. The arginine residues such as R682, R683, and R685 located in the S1/S2 cleavage sites of SARS-CoV-2 act as the recognition sequence for TMPRSS2-mediated cleavage. Moreover administration of camostat mesylate, a serine protease inhibitor, was found to block the SARS-CoV-2 infection in lung cells [28], proving the importance of TMPRSS2 in SARS-CoV-2 entry and transmission into the host cells. A cathepsin B/L-mediated auxiliary activation of SARS-CoV-2 S protein has been found to be operative in TMPRSS2-negative cells. However, the question of whether these cellular proteases act synergistically was yet to be solved until the report published by Shang et al. [27]. With the help of inhibitor study, they established the role of host cell proteases in SARS-CoV-2 infection. They observed that, in the presence of camostat, a TMPRSS2 inhibitor, entry of SARS-CoV-2 in three cell lines such as ACE2-transfected HeLa, Calu-3, and MRC-5 was significantly reduced. Moreover, with administration of E64d, a lysosomal cathepsin inhibitor, similar results were obtained. Additionally, pretreatment with PPCi, a furin inhibitor, enhanced the efficacy of camostat or E64d. These results clearly proved that a cumulative association of TMPSS-2 and cathepsin B/L with furin is required for the priming of the viral S protein before its entry into the host cells.

25.6.2 Binding of the Activated S Protein with ACE2 Receptor

The binding of the active SARS-CoV-2 S protein and ACE2 receptor occurs after the host proteases prime the virus to initiate the infection process. It has been observed that SARS-CoV-2 S protein binds to ACE2 with higher affinity (about 20 times higher) than the SARS-CoV counterpart. This increased specificity and affinity of SARS-CoV-2 S protein toward ACE2 may contribute to the higher rates of transmission and mortality of this virus in humans than its ancestral strain SARS-CoV or other coronaviruses. ACE2, on the other hand, is widely distributed throughout the body due to its high expression in the nasal epithelium, oral mucosa, lungs, heart, brain, vascular endothelium, small intestine, colon, and kidney, which facilitates the transmission of the virus into different organs.

25.6.3 Fusion of the Virus and Host Cell Membranes

The S2 subunit of SARS-CoV-2 mediates the fusion of viral and host cell membranes. The S2 subunit comprises five structural domains, namely, FP, HR1, HR2, TM, and CT domains, respectively, which are responsible for viral fusion and entry. FP domain acts as the fusion peptide, consisting of 15–20 conserved hydrophobic amino acids, such as glycine (G) or alanine (A). Due to its hydrophobicity, FP interacts with the lipid bilayer of the host cell membrane and anchors to the target membrane. HR1 and HR2 domains are composed of a repetitive heptapeptide sequence: HPPHCPC, where H is a hydrophobic residue, P is a hydrophilic residue, and C is any charged residue. HR1 and HR2 domains associate to form a six-helical bundle (6-HB), which is vital for the fusion of the S2 subunit. HR1 domain is located at the C-terminal end of FP, while HR2 is located at the N-terminal end of the TM domain. The TM domain anchors the S2 subunit protein to the viral membrane.

The viral fusion proteins may be classified into three distinct classes (class I, class II, and class III) based on their structure and mode of action. The coronavirus S protein belongs to the class I family due to the structural features of its fusion domain, the requirement of the proteolytic cleavage for activation, and the presence of heptad repeats that can fold into a 6-HB conformation. The fusion of class I fusion peptides with the host cell membrane involves the following sequential steps: (a) formation of a pre-fusion native state, (b) formation of a pre-fusion metastable state, (c) formation of a pre-hairpin intermediate state, and (d) formation of a post-fusion stable state. When the S protein is synthesized, it exists in the pre-fusion or native state. Proteolytic cleavage at the S1/S2 site results in the change of conformation of the S protein to a pre-fusion metastable state to generate separate S1 and S2 subunits that remain non-covalently associated. The transition of FP from the metastable state to the next pre-hairpin intermediate state required overcoming an energy barrier. The energy barrier may be overcome by FP only through the acquisition of a conformational change triggered by the neighboring environmental factors, such as pH, hydrophobicity, polarity, etc. This conformational change of FP then directs the three HR1 regions to assemble into a coiled-coil trimer. Immediately, the three HR2 regions then bind to the hydrophobic buckets of the HR1 trimer in an antiparallel orientation. This complex is known as the fusion core or 6-HB. When the FP attains this conformation, it comes in the close proximity of the host cell membrane so that both the viral and host cell membranes can fuse. The FP now exists in a stable post-fusion state.

The fusion of SARS-CoV-2 with the host cell membrane occurs by the abovementioned mechanism [29].

25.6.4 Replication of the Viral Genome

Like the other RNA viruses, immediately after the entry, the SARS-CoV-2 (+) ss-mRNA undergoes translation by using the host cell machinery to form important viral proteins that participate in viral replication, assembly, and release of progeny virus particles. The complete replication process of the viral genome includes RNA synthesis, proofreading of the template, and capping of the new RNA. Several SARS-CoV-2 nsps mediate this whole process by forming the replication/transcription complex (RTC), a membrane-anchored, dynamic protein-RNA complex required for viral replication. The nsp12 acts as the catalytic core of the RNA-dependent RNA polymerase (RdRp), while nsp7 and nsp8 enhance the RNA-binding ability and processivity of the enzyme [16]. Another nonstructural protein nsp13 codes for the enzyme helicase, which causes the unwinding of double-stranded RNA formed during replication and allows the next round of viral replication. Another nsp complex that is crucial for viral RNA replication is the nsp14–16 complex, of which nsp14 helps in RNA proofreading while nsp15 imparts uridylate-specific endoribonuclease activity. The nsp14 contains two domains, the N-terminal domain having an exoribonuclease activity participates in proofreading, while the C-terminal domain possesses the (N7 guanine)-methyl transferase activity and in capping the 5′end of the viral m RNA. The cap consists of an N7-methylated guanosine (GTP) molecule linked with the first transcribed nucleotide by 5′–5′ bond transcribed. The capping machinery apart from Nsp14 consists of Nsp13, 16, and cofactor Nsp10. Finally, nsp16 in complex with nsp10 mediates the termination of mRNA capping. A recent experimental study by Banerjee et al. showed that SARS-CoV-2 proteins interact with human RNAs. After SARS-CoV-2 infection, the viral NSP16 protein binds to the mRNA recognition domains of the splicing RNAs (U1 and U2) and downregulates global mRNA splicing. The other proteins, such as nsp1, bind to 18S ribosomal RNA and promote global inhibition of mRNA translation, while nsp8 and nsp9 regulate 7SL RNA and prevent protein trafficking to the cell membrane. Overall these SARS-CoV-2 proteins were also shown to downregulate host interferon response after viral infection so that progeny virus particles can evade the host immune responses [30].

25.6.5 Assembly and Release of the Progeny Virus Particles

The endoplasmic reticulum and Golgi apparatus of the host cells facilitate the assembly of the viral RNAs and associated proteins into the progeny virus particles. The M and E proteins interact together and also with N and S proteins to assemble new virus particles in the endoplasmic reticulum and Golgi apparatus.

After assembling, the release of the progeny virus particles may occur by budding, exocytosis, or host cell death. It has been observed that the incomplete virus particles generally are released through budding. In contrast, the mature or complete progeny viruses opted for either exocytosis or lysis of the host cells. These newly released viruses may either infect healthy neighboring cells or may be shredded into the environment via droplet transmission.

All these phases constitute the life cycle of SARS-CoV-2 within the host cells, including the lung pneumocytes (Fig. 25.3).

25.7 Pathogenicity of SARS-CoV-2 in Host

SARS-CoV-2 infection results in systemic and respiratory disorders in patients, including fever, ARDS, respiratory failure, multiple organ dysfunction, septic shock, coagulation dysfunction, and metabolic acidosis. In addition, gastrointestinal symptoms such as nausea, vomiting, abdominal pain, and diarrhea and other cardiovascular symptoms such as chest pain, heart palpitations, myocardial injury, acute coronary syndrome, arrhythmias, acute coronary syndrome, etc. are observed in COVID-19 patients. Moreover, reduced peripheral blood leukocyte and lymphocyte counts, abnormalities in liver enzymes, C-reactive protein, myocardial enzymes, and increased D-dimer and inflammatory factors are also reported in COVID-19 patients. The pathogenicity caused by SARS-CoV-2 in COVID-19 patients may be summarized as follows.

25.7.1 Downregulation of ACE2

ACE2 is an integral membrane protein with enzymatic activity that plays a key role in the regulation of the renin-angiotensin system (RAS). RAS-signaling pathway acts as a homeostatic regulator of vascular function. The major biochemical activity of ACE2 is the degradation of angiotensin II (Ang II). This crucial vasoactive peptide angiotensin II acts as a potent vasoconstrictor by binding to its receptor angiotensin II type-I receptor (AT1R). Binding of Ang II to AT1R results in angiotensin 1–7, which opposes the action of Ang II. It has been observed that the pathophysiological actions of Ang II are mediated via AT1 and AT2 receptors. In contrast, the activity of Ang (1–7) is mediated through the Mas oncogene and Mas-related G-protein-coupled D receptors. Thus, the RAS system comprises of two axes: (1) the ACE-Ang II-AT1 receptor axis that promotes vasoconstriction, fibrosis, proliferation, thrombosis, oxidative stress, and inflammation and (2) the

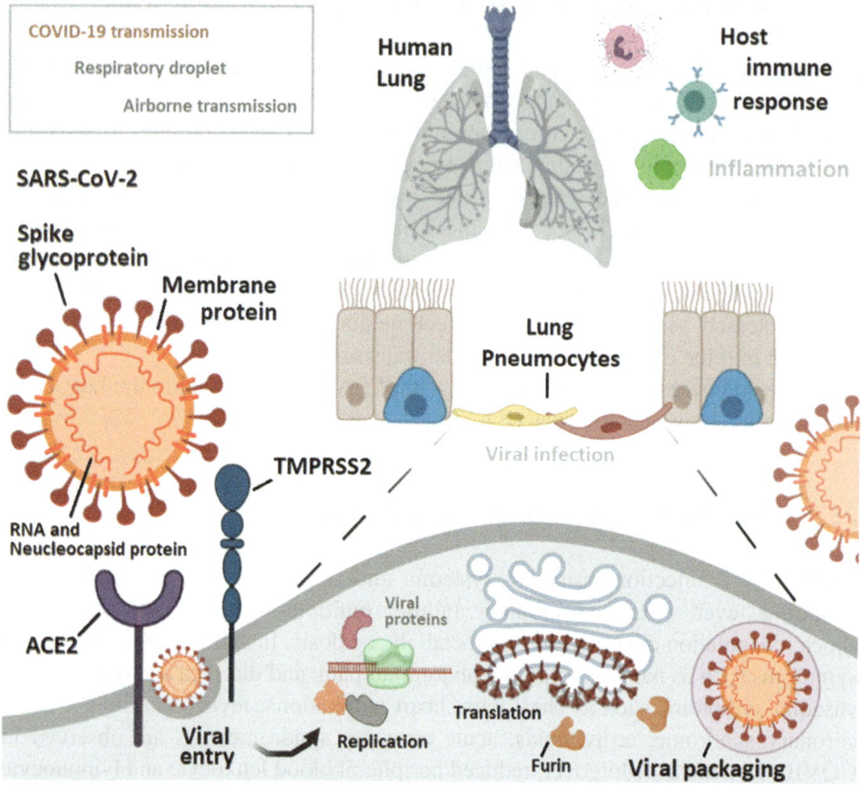

Fig. 25.3 Human-to-human SARS-CoV-2 transmission takes place through respiratory droplets. Lung pneumocytes are most susceptible for SARS-CoV-2 infection. The viral spike protein, which consists of S1 and S2 subunits, binds to the host transmembrane ACE2 receptor. Activation of SARS-CoV-2 spike protein through cleavage at S1/S2 by host protease TMPRSS2 has been shown to be indispensable. Upon cleavage, at S1/S2 site, the spike protein gets activated and attaches to the host ACE2 receptor, followed by fusion of viral membrane with the host cell membrane and release of viral RNA into the cytoplasm. Viral proteins (nsps, RdRp, etc.), along with host proteins, guide the replication followed by the translation of viral RNA. Upon translation, the viral proteins go through maturation and folding in the endoplasmic reticulum. Finally, viral packaging takes place within Golgi vesicle, followed by release from the host cell. SARS-CoV-2 acquires an additional furin cleavage site in spike protein over SARS-CoV, which results in half-cleaved spike protein at the time of the release of the viral particle that provides the virus with tissue tropism

ACE2-Ang (1–7)-Mas receptor axis that counterbalances the effects of Ang II which imparts anti-inflammatory, antithrombotic, cardioprotective, vasodilatory, and antiproliferative actions [31]. During the fusion of the viral membrane with the host cell membrane, the ACE2 receptor gets internalized, and this results in the downregulation of ACE2 on the host cell surface. Downregulation of ACEs leads to the dysregulation of both ACE2-Ang II-AT1 and the ACE2-Ang 1–7-Mas axes. Hence, the protective function of Ang II-controlled downstream signaling cascades is disrupted, and increased respiratory damage, cell damage, inflammation, thrombosis, etc., occur in the COVID-19 patients.

25.7.2 Impairment of the Immune System

A severe impairment of the immune system is generally observed in COVID-19 patients, associated with high levels of inflammatory cytokines, referred to as the "cytokine storm." The major immune dysfunctions caused by SARS-CoV-2 include the following events.

25.7.2.1 Cytokine Storm

Medical reports of COVID-19 patients have revealed the severity of this disease is associated with the massive production of inflammatory cytokines, also known as the cytokine storm. This uncontrolled inflammatory response in the lungs can lead to severe pneumonia in the patients and even death. The COVID-19 patients who have been hospitalized or kept under ICU had shown higher levels of interleukin (IL) such as IL-1β, IL-1Rα, IL-7, IL-8, IL-9, IL-10, G-CSF, GM-CSF, IFN-γ, CCL2, CCL3, CCL4, CXCL10, PDGF, VEGF, and TNFα, in the plasma compared to the healthy individuals [32]. This cytokine storm may be caused by Nlrp3γ inflammasome, a powerful pro-inflammatory regulatory system, found activated in COVID-19 patients. The activation of the Nlrp3γ inflammasome is known to trigger an immune response by activating the intracellular caspase-1. Activated caspase-1 further mediates the release of pro-inflammatory cytokines, such as IL-1β and IL-18, and the formation of gasdermin D (GSDMD) pore channels in plasma membranes. These events lead to the release of several biologically active damage-associated molecular patterns or DAMPs. These DAMPs result in the lysis and death of the lung cells by a process known as pyroptosis.

25.7.2.2 Activation of the Complement System

DAMP-mediated killing of SARS-CoV-2-infected cells leads to the activation of the innate immune response in the patients, thus activating the complement cascade (ComC). In addition to DAMPs, the ComC also can be directly activated by mannan-binding lectin (MBL), which can bind to SARS-CoV-2 proteins. MBL is generally present in serum that can interact with MBL-associated serine protease 2 (MASP-2). This MBL-MASP-2 complex can initiate the complement-activated lectin pathway via binding to sugar residues on the virus membrane. The N protein of SARS-CoV-2 was found to interact with MASP-2, activating the complement protein C4. Activation of the complement cascade further leads to the damage and lysis of alveolar lung cells.

25.7.2.3 Lymphocyte Dysfunction

Massive lymphocyte dysfunction, also known as lymphopenia, is commonly observed in COVID-19 patients and is responsible for the severity of the disease. This medical condition is featured by decreased CD4+ and CD8+ T cells in the peripheral blood and increased double-positive HLA-DR and CD38 populations. The later event indicates the activation of T cells. Also, a high population of pro-inflammatory Th17 and cytotoxic CD8+ T cells were observed in the COVID-19 patients suffering from severe immune dysfunction. Th17 cells act as the central executioner of the immune dysfunction since it produces a plethora of inflammatory interleukins, cytokines, and growth factors, which can cause severe damage to the immune system. Th17 cells can produce IL-17, GM-CSF (mainly associated with TH1 cells), IL-21, and IL-22. It can also regulate the pro-inflammatory cytokines such as G-CSF, IL-1β, IL-6, and TNFα; chemokines such as KC, MIP2A, MIP3A, IL-8, and IP10; and matrix metalloproteinases (MMPs). G-CSF is required for granulopoiesis, and recruitment of neutrophils, IL-1β, IL-6, and TNFα can lead to systemic inflammatory symptoms, including fever, the chemokines can recruit the immune infiltrates, and the MMPs can induce tissue damage. Thus systemic immune dysregulation can be induced by the Th17 cells via activation of the cytokine storm [33]. All these immune disorders lead to severe respiratory disorders in COVID-19 patients, including death.

25.8 Role of Host Genetics in Viral Pathogenesis

SARS-CoV-2 infection showed enormous inter-individual variability in clinical readouts that includes silent infection to severe disease. To understand the host genomic-guided underpinnings of viral pathogenesis, disease susceptibility, and varied clinical outcomes, multiple genome-scale studies were performed across different populations of the world. Host genetic variations in key members of viral entry genes (ACE2, TMPRSS2, and FURIN) can alter the expression and functionality of these proteins, which results in altered susceptibility to the SARS-CoV2 infection. Studies showed that ACE2 occurs at non-polymorphic frequencies in most human populations [34–36]. There are a few variants in the gene (rs73635825 and rs143936283) that showed different intermolecular interactions with the SARS-CoV-2 spike protein [37]. TMPRSS2 is another key gene that modulates cellular entry of SARS-CoV-2 by priming the spike protein. Our study and one recent study have identified an intergenic single nucleotide polymorphism, rs35074065 (deletion of C allele), to be associated with increased expression of both TMPRSS2 and the interferon-inducible gene MX1 in lung tissue [24, 38]. Individuals carrying this SNP-delC genotype have higher TMPRSS2 expression, which helps in increased viral entry. The frequency of this SNP-delC is extremely low in East Asian populations and significantly higher in European and North-American populations. Our study has further identified a neutrophil elastase (NE) cut-site on the 614G position on the SARS-CoV-2 A2a subtype. The A2a subtype swept through the world and outcompeted other subtypes after it emerged in late January. The increased expression of MX1 due to delC SNP leads to interferon-mediated neutrophil trafficking in lung cells. We speculated that the availability of NE either due to neutrophil infiltration or due to other mechanisms (such as availability of NE inhibitor protein—alpha 1 antitrypsin) has favored enhanced host cell entry of the

SARS-CoV-2614G subtype that has a specific NE cut-site in the 614G position [24] of the spike protein. SARS-CoV-2-pseudotyped lentiviral particles with this variant 614G were shown to infect multiple human cell types more efficiently (~3.5-fold), compared to the 614D variant [39].

Another study by Asselta et al. identified an exonic variant (p.Val160Met) and two distinct haplotypes in TMPRSS2, associated with increased gene expression [40]. Notably, one of these variants overlaps with a known androgen-responsive enhancer for TMPRSS2, indicating a plausible mechanism behind the sex bias in COVID-19 severity. Recently, a study attempted to explore the variations in the furin gene with the susceptibility of COVID-19 [41]. However, there was no association found between furin gene variations with SARS-CoV-2 infection.

Further, multiple studies explored the disease severity in the context of COVID infection. A genetic susceptibility locus in 3p21.31 (SLC6A20, LZTFL1, CCR9, FYCO1, CXCR6, and XCR1 gene clusters) was found to be associated with COVID-19 patients with respiratory failure and further confirmed a potential involvement of the ABO blood group system [42]. Further studies pointed out that the risk genomic segment of around 50 kb in size was inherited from Neanderthals and is carried by ~50% of people in south Asia and ~16% of people in Europe [43]. Rare loss of function variants in 13 human loci known to involve in TLR3- and IRF7-dependent type-I interferon (IFN) response has been shown to be associated with life-threatening COVID-19 disease [44]. These loci were prior found to be involved with influenza virus infection. Further studies have shown that about 10% of patients with severe COVID-19 lung disease have neutralizing autoantibodies against type-I IFNs [44].

25.9 Therapeutics Developments for Management of COVID-19

Since the emergence of the COVID-19 pandemic, quite a few therapeutic regimens have been explored based on the clinical data obtained from the epidemics caused by the ancestral strains of SARS-CoV-2, such as SARS-CoV and MERS-CoV. Undoubtedly, the vaccines will serve as the ultimate therapeutic strategy against this virus, but the timeline for their production is a matter of concern to combat the present pandemic situation. Alternatively, therapeutics such as small molecule inhibitors and neutralizing antibodies may be used to prevent SARS-CoV-2 infection. Many of these agents have already got approval for clinical trials, and a few have shown promising results. Among the plethora of therapeutic agents presently under clinical trials in different countries, remdesivir (Gilead Sciences), earlier repurposed for Ebola; hydroxychloroquine, the antimalarial agent; ritonavir and lopinavir, two anti-HIV drugs; ivermectin, an FDA-approved anti-parasite and anti-HIV drug, etc.; and the neutralizing antibodies such as an anti-SARS-CoV-2 antibody, anti-C5a antibody, etc., deserve special mention [45].

The ongoing therapies can be broadly divided into two categories depending on their mode of action. One group of drugs can affect the coronavirus directly by targeting the critical viral proteins. These drugs may either inhibit the enzymes responsible for the replication of the viral genome or inhibit the viral entry into the host cells. The second group includes those who can modulate the host immune system by boosting the innate immune responses. Some of the novel or repurposed therapeutic agents which are presently under clinical trial against SARS-CoV-2 are briefly summarized as follows.

25.9.1 Agents Targeting Viral Replication

25.9.1.1 Remdesivir (GS-5734)

Remdesivir, a structural analogue of adenosine, is so far the most promising drug that exhibits significant antiviral activities against RNA viruses, including SARS-CoV-2. Due to its structural resemblance with adenosine, it can readily incorporate into the nascent viral RNA, and further inhibits the polymerase activity of the viral RNA-dependent RNA polymerase (RdRp) [46]. Inhibition of viral RdRp results in the premature termination of the RNA chain, thereby inhibiting the replication of the SARS-CoV-2 genome. Remdesivir, originally developed by Gilead Sciences (USA), was first tested against the Ebola virus in the Democratic Republic of the Congo and also known to be effective against different coronaviruses, including SARS-CoV and MERS-CoV. In a very recent case study, it was reported that remdesivir treatment resulted in the reduction of the recovery time in patients hospitalized with COVID-19 by inhibiting the lower respiratory tract infection [47].

25.9.1.2 Favipiravir

Favipiravir, developed by Toyama Chemical (Fujifilm, Japan), is a structural analogue of guanine and effectively inhibits the activity of RdRp. It is an approved medication for influenza with minimal side effects. Early clinical data with favipiravir administered to COVID-19 patients showed promising outcomes. An open-label nonrandomized trial involving 80 patients in China reported a significant reduction in the viral clearance time in patients treated with favipiravir [48].

25.9.1.3 Ivermectin

The anti-parasite drug ivermectin is an FDA-approved agent that was also reported to be effective against both dengue virus and human immunodeficiency virus (HIV). A recent study has shown that ivermectin can reduce SARS-CoV-2 RNA up to 5000-fold in mammalian cells [49]. Ivermectin, in combination with the antibiotic doxycycline, has been approved for clinical trials against COVID-19 in several countries [clinicaltrials.gov].

25.9.1.4 Ribavirin

This drug is a guanosine analogue that is known to inhibit RNA-dependent RNA polymerase activity of SARS-CoV-2. Based on the clinical data of this drug against HCV, ribavirin was approved for clinical trials in COVID-19 patients alone or in combination. An earlier case report revealed that ribavirin in combination with lopinavir-ritonavir and interferon-β-1b improved the medical conditions of patients with mild COVID-19 symptoms [50].

25.9.2 Agents Blocking the Virus-Host Cell Membrane Fusion

25.9.2.1 Recombinant Human Angiotensin-Converting Enzyme 2 (APN01)

As described earlier, the ACE2 receptor plays an important role in the entry of SARS-CoV-2 in the lung cells, which highly express this receptor. Also, ACE2, being a regulator of the renin-angiotensin system, is known to protect multiple tissues from injury. The human recombinant ACE2 or hrsACE2 (APN01; APEIRON Biologics, Vienna, Austria) is a soluble protein that can be implicated in anti-COVID therapy due to its ability to bind to the viral spike protein. APN01 may have two modes of action that will be beneficiary to COVID-19 patients. Firstly, it can neutralize SARS-CoV-2 by binding to S protein, and secondly, it can minimize the injury to multiple organs, such as lungs, kidneys, and heart, to name a few. The in vitro studies revealed that hrsACE2 could reduce SARS-CoV-2 load in cell lines by a factor of 1000–5000, and a recently published case study revealed the efficacy of APN01 in COVID-19 patients [51].

25.9.2.2 Hydroxychloroquine

Hydroxychloroquine (HQ) is an FDA-approved antimalarial agent that can also inhibit viral infection by raising the endosomal pH required for membrane fusion between the virus and the host cell. Moreover, HQ was shown to block the replication of SARS-CoV by interfering with the glycosylation of ACE2 receptor, thereby reducing the viral copy number. Based on this evidence, the FDA issued an emergency. Although not such promising results were reported from the clinical studies, more randomized controlled trials are recommended for further evaluation.

25.9.2.3 Arbidol Hydrochloride (Umifenovir)

Arbidol is a well-known drug against influenza that targets the viral membrane-expressed glycoprotein hemagglutinin (HA) and blocks the fusion of the viral membrane with the host endosome after endocytosis. Presently, it is undergoing trials as a single agent (NCT04260594, NCT04255017). In vitro studies have revealed that Arbidol effectively inhibited the entry of SARS-CoV-2 in mammalian cells [52]. Also, in a recently completed clinical trial, Arbidol in combination with HQ had shown promising outcomes in patients hospitalized with COVID-19 symptoms [53].

25.9.3 Immunomodulatory Medications

25.9.3.1 Tocilizumab

It is a humanized monoclonal antibody targeting the interleukin-6 receptor (IL-6R). Tocilizumab has also been approved for the treatment of the cytokine storm associated with cancer immunotherapy and other lung disorders. Although it does not possess any direct antiviral effects, it significantly counters the massive cytokine storm syndrome, a severe symptom of COVID-19, by inhibiting the binding of IL-6 to its receptor. After its efficacy was tested in vivo systems, the FDA has approved for a phase III randomized, double-blind, placebo-controlled trial for this antibody-based drug in hospitalized COVID-19 patients.

25.9.3.2 Sarilumab and Emapalumab

Sarilumab is an anti-IL-6Rα monoclonal antibody antagonizing IL-6 in rheumatoid arthritis, while emapalumab targets interferon (IFN)-γ, and is used in the treatment of hemophagocytic lymphohistiocytosis. Like tocilizumab, both sarilumab and emapalumab can effectively block the massive cytokine release due to SARS-CoV-2 infection. The Italian media agency, Agenzia Italiana del Farmaco (AIFA), had approved the use of either sarilumab or emapalumab in phase II/III clinical trials involving hospitalized COVID-19 patients with pulmonary complications.

25.9.3.3 Bevacizumab

Bevacizumab is a recombinant humanized monoclonal antibody that can specifically bind to VEGF and is widely used in the treatment of different types of cancers. Presently, a clinical trial is under process to evaluate the effectiveness of bevacizumab to treat SARS-CoV-2 infection (NCT04275414).

These sets of medications described here either can directly bind to the essential SARS-CoV-2 proteins or inhibit the viral replication or entry into the host cells or those that can reduce the cytokine storm-generated in virus-infected lung pneumocytes due to high inflammation. The increasing viral load, coupled with the cytokine storm, can do irreversible damage to the lung cells, which may lead to severe pneumonia in COVID-19 patients. Since vaccine development is not feasible soon, these medications are playing a crucial role in checking the infection and reducing the death rate.

25.10 Conclusion

The COVID-19 pandemic has become a severe threat to human health due to the contagious nature of the infection and high mortality rates. Till October 2020, it has resulted in more than 1.5 million deaths globally, affecting more than 200 countries, and the numbers are increasing day by day. Genetic analysis of SARS-CoV-2 revealed its striking similarity with the bat-derived strains CoVZC45, CoVZXC21, and RaTG13 and thus confirmed the zoonotic origin of the virus. Starting from the first reported case in December 2019 at Wuhan, China, the SARS-CoV-2 strain slowly mutates into newer subtypes that spread across the world. Over time, one of the subtypes, A2a, with two signature mutations, Spike D614G and RdRp P323L, outcompeted the other subtypes and became the major pathogenic strain across Asia, Europe, and North America. Our group first revealed that the mutant subtype (614G) outcompeted the pre-existing type (614D), significantly faster in Europe and North America than in East Asia. Our hypothesis was based on the phylodynamic analysis of over 70,000 SARS-CoV-2 coronavirus genome sequences, available worldwide, until July 2020. Additionally, we also identified a novel neutrophil elastase (ELANE) cleavage site introduced in the G mutant, near the S1/S2 junction of the spike protein, which was absent in wild-type strain [24]. We hypothesized that elevation of neutrophil elastase level at the site of infection would enhance the activation of spike protein, thus facilitating host cell entry for 614G, but not the 614D, subtype. This finding will definitely help to understand the mechanism of viral entry and pathogenesis in host cells.

References

1. Gorbalenya AE et al (2020) The species severe acute respiratory syndrome-related coronavirus: classifying 2019-nCoV and naming it SARS-CoV-2. Nat Microbiol 5(4):536–544. https://doi.org/10.1038/s41564-020-0695-z
2. Chan JF-W et al (2020) A familial cluster of pneumonia associated with the 2019 novel coronavirus indicating person-to-person transmission: a study of a family cluster. Lancet (London, England) 395(10223):514–523
3. Lu H, Stratton CW, Tang Y-W (2020) Outbreak of pneumonia of unknown etiology in Wuhan, China: the mystery and the miracle. J Med Virol 92(4):401–402
4. Burki TK (2020) Coronavirus in China. Lancet Respir Med 8(3):238
5. WHO (2020) WHO Director-General's opening remarks at the media briefing on COVID-19, 11 Mar 2020. Available from: https://www.who.int/dg/speeches/detail/who-director-general-s-opening-remarks-at-the-media-briefing-on-covid-19%2D%2D-11-march-2020
6. Burrell CJ, Howard CR, Murphy FA (2016) Fenner and White's medical virology, 5th edn. Academic Press, Amsterdam
7. Woo PCY et al (2012) Discovery of seven novel mammalian and avian coronaviruses in the genus deltacoronavirus supports bat coronaviruses as the gene source of alphacoronavirus and betacoronavirus and avian coronaviruses as the gene source of gammacoronavirus and deltacoronavirus. J Virol 86(7):3995–4008. Available from: http://europepmc.org/abstract/MED/22278237
8. Wu F et al (2020) A new coronavirus associated with human respiratory disease in China. Nature 579(7798):265–269
9. Zhou P et al (2020) A pneumonia outbreak associated with a new coronavirus of probable bat origin. Nature 579(7798):270–273. https://doi.org/10.1038/s41586-020-2012-7
10. Wu A et al (2020) Genome composition and divergence of the novel coronavirus (2019-nCoV) originating in China. Cell Host Microbe 27(3):325–328
11. Tanner JA et al (2003) The severe acute respiratory syndrome (SARS) coronavirus NTPase/helicase belongs to a distinct class of 5′ to 3′ viral helicases. J Biol Chem 278 (41):39578–39582. Available from: http://europepmc.org/abstract/MED/12917423
12. Krichel B et al (2020) Processing of the SARS-CoV pp1a/ab nsp7-10 region. Biochem J 477 (5):1009–1019
13. Van Boheemen S et al (2012) Genomic characterization of a newly discovered coronavirus associated with acute respiratory distress syndrome in humans. MBio 3(6):e00473-12

14. Wang Q et al (2020) Structural and functional basis of SARS-CoV-2 entry by using human ACE2. Cell 181(4):894–904.e9. Available from: http://www.sciencedirect.com/science/article/pii/S009286742030338X
15. Nieto-Torres JL et al (2011) Subcellular location and topology of severe acute respiratory syndrome coronavirus envelope protein. Virology 415(2):69–82
16. Kirchdoerfer RN, Ward AB (2019) Structure of the SARS-CoV nsp12 polymerase bound to nsp7 and nsp8 co-factors. Nat Commun 10(1):2342. https://doi.org/10.1038/s41467-019-10280-3
17. China Center for Disease Control and Prevention (2020) China CDC detects a large number of new coronaviruses at Huanan Seafood Market in Wuhan. Available from: http://www.chinacdc.cn/yw_9324/202001/t20200127_211469.html. Accessed 20 Feb 2020
18. Rothe C et al (2020) Transmission of 2019-nCoV infection from an asymptomatic contact in Germany. N Engl J Med 382:970–971
19. Bai Y et al (2020) Presumed asymptomatic carrier transmission of COVID-19. JAMA 323 (14):1406–1407
20. Oran DP, Topol EJ (2020) Prevalence of asymptomatic SARS-CoV-2 infection. Ann Internal Med 173(5):362–367. https://doi.org/10.7326/M20-3012
21. Li Q et al (2020) Early transmission dynamics in Wuhan, China, of novel coronavirus-infected pneumonia. N Engl J Med 382(13):1199–1207
22. Netea MG et al (2020) Trained immunity: a tool for reducing susceptibility to and the severity of SARS-CoV-2 infection. Cell 181(5):969–977. Available from: http://www.sciencedirect.com/science/article/pii/S0092867420305079
23. Biswas N, Majumder P (2020) Analysis of RNA sequences of 3636 SARS-CoV-2 collected from 55 countries reveals selective sweep of one virus type. Indian J Med Res 151(5):450–458. Available from: http://www.ijmr.org.in/preprintarticle.asp?id=284484;type=0
24. Bhattacharyya C et al (2020) Global spread of SARS-CoV-2 subtype with spike protein mutation D614G is shaped by human genomic variations that regulate expression of *TMPRSS2* and *MX1* genes. BioRxiv. Available from: http://biorxiv.org/content/early/2020/05/05/2020.05.04.075911.abstract
25. Hu J et al (2020) The D614G mutation of SARS-CoV-2 spike protein enhances viral infectivity and decreases neutralization sensitivity to individual convalescent sera. BioRxiv. Available from: https://www.biorxiv.org/content/early/2020/06/20/2020.06.20.161323
26. Korber B et al (2020) Tracking changes in SARS-CoV-2 spike: evidence that D614G increases infectivity of the COVID-19 virus. Cell 182(4):812–827.e19. Available from: https://pubmed.ncbi.nlm.nih.gov/32697968
27. Shang J et al (2020) Cell entry mechanisms of SARS-CoV-2. Proc Natl Acad Sci 117 (21):11727–11734. Available from: https://www.pnas.org/content/117/21/11727
28. Hoffmann M et al (2020) SARS-CoV-2 cell entry depends on ACE2 and TMPRSS2 and is blocked by a clinically proven protease inhibitor. Cell 181(2):271–280.e8. Available from: http://www.sciencedirect.com/science/article/pii/S0092867420302294
29. Tang T et al (2020) Coronavirus membrane fusion mechanism offers a potential target for antiviral development. Antivir Res 178:104792
30. Banerjee AK et al (2020) SARS-CoV-2 disrupts splicing, translation, and protein trafficking to suppress host defenses. Cell. Available from: http://www.sciencedirect.com/science/article/pii/S0092867420313106
31. Jia HP et al (2009) Ectodomain shedding of angiotensin converting enzyme 2 in human airway epithelia. Am J Physiol Lung Cell Mol Physiol 297(1):L84–L96
32. Yang Y et al (2020) SARS-CoV-2: characteristics and current advances in research. Virol J 17 (1):117. https://doi.org/10.1186/s12985-020-01369-z
33. Wu D, Yang XO (2020) TH17 responses in cytokine storm of COVID-19: an emerging target of JAK2 inhibitor Fedratinib. J Microbiol Immunol Infect (Wei Mian Yu Gan Ran Za Zhi) 53 (3):368–370

34. Cao Y et al (2020) Comparative genetic analysis of the novel coronavirus (2019-nCoV/SARS-CoV-2) receptor ACE2 in different populations. Cell Discov 6(1):11. https://doi.org/10.1038/s41421-020-0147-1
35. Renieri A et al (2020) ACE2 variants underlie interindividual variability and susceptibility to COVID-19 in Italian population. MedRxiv. Available from: https://www.medrxiv.org/content/early/2020/04/06/2020.04.03.20047977
36. Stawiski EW et al (2020) Human ACE2 receptor polymorphisms predict SARS-CoV-2 susceptibility. BioRxiv. Available from: https://www.biorxiv.org/content/early/2020/04/10/2020.04.07.024752
37. Hussain M et al (2020) Structural variations in human ACE2 may influence its binding with SARS-CoV-2 spike protein. J Med Virol 92(9):1580–1586. https://doi.org/10.1002/jmv.25832
38. Russo R et al (2020) Genetic analysis of the coronavirus SARS-CoV-2 host protease TMPRSS2 in different populations. Front Genet 11:872
39. Ozono S et al (2020) Naturally mutated spike proteins of SARS-CoV-2 variants show differential levels of cell entry. BioRxiv. Available from: https://www.biorxiv.org/content/early/2020/06/26/2020.06.15.151779
40. Asselta R et al (2020) ACE2 and TMPRSS2 variants and expression as candidates to sex and country differences in COVID-19 severity in Italy. Aging 12(11):10087–10098. https://doi.org/10.18632/aging.103415
41. Torre-Fuentes L et al (2020) ACE2, TMPRSS2, and Furin variants and SARS-CoV-2 infection in Madrid, Spain. J Med Virol. https://doi.org/10.1002/jmv.26319
42. The Severe Covid-19 GWAS Group (2020) Genomewide Association Study of severe Covid-19 with respiratory failure. N Engl J Med 383(16):1522–1534. https://doi.org/10.1056/NEJMoa2020283
43. Zeberg H, Pääbo S (2020) The major genetic risk factor for severe COVID-19 is inherited from Neanderthals. Nature. https://doi.org/10.1038/s41586-020-2818-3
44. Bastard P et al (2020) Auto-antibodies against type I IFNs in patients with life-threatening COVID-19. Science:eabd4585. Available from: http://science.sciencemag.org/content/early/2020/09/23/science.abd4585.abstract
45. Tu Y-F et al (2020) A review of SARS-CoV-2 and the ongoing clinical trials. Int J Mol Sci 21(7):2657
46. Yin W et al (2020) Structural basis for inhibition of the RNA-dependent RNA polymerase from SARS-CoV-2 by remdesivir. Science 368(6498):1499–1504. Available from: http://science.sciencemag.org/content/368/6498/1499.abstract
47. Beigel JH et al (2020) Remdesivir for the treatment of Covid-19—final report. N Engl J Med. https://doi.org/10.1056/NEJMoa2007764
48. Cai Q et al (2020) Experimental treatment with favipiravir for COVID-19: an open-label control study. Engineering (Beijing, China) 6:1192–1198
49. Caly L et al (2020) The FDA-approved drug ivermectin inhibits the replication of SARS-CoV-2 in vitro. Antivir Res 178:104787. Available from: http://www.sciencedirect.com/science/article/pii/S0166354220302011
50. Hung IF-N et al (2020) Triple combination of interferon beta-1b, lopinavir–ritonavir, and ribavirin in the treatment of patients admitted to hospital with COVID-19: an open-label, randomised, phase 2 trial. Lancet 395(10238):1695–1704. https://doi.org/10.1016/S0140-6736(20)31042-4
51. Zoufaly A et al (2020) Human recombinant soluble ACE2 in severe COVID-19. Lancet Respir Med. https://doi.org/10.1016/S2213-2600(20)30418-5
52. Wang X et al (2020) The anti-influenza virus drug, arbidol is an efficient inhibitor of SARS-CoV-2 in vitro. Cell Discov 6(1):28. https://doi.org/10.1038/s41421-020-0169-8
53. Nojomi M et al (2020) Effect of arbidol on COVID-19: a randomized controlled trial. Available from: http://europepmc.org/abstract/PPR/PPR217076

Targeting Molecular and Cellular Mechanisms in SARS-CoV-2 Novel Corona (COVID-19) Virus Infection

26

Nabanita Roy Chattopadhyay, Nilanjana Das, and Tathagata Choudhuri

Abstract

SARS-CoV-2, or novel coronavirus, is causing the fatal and contagious coronavirus disease-2019 (COVID-19) affecting thousands of people every single day. Researchers are continuously searching for any possible cure and/or vaccine, but no conclusive report is available till date. SARS-CoV-2 is a positive-sense single-stranded RNA virus (+ssRNA virus) which uses its RNA as its genetic material as well as mRNA for the viral protein production. The severity and mode of infection of SARS-CoV-2 are attributed to the presence of a cleavage site for furin endoproteases in the junction of the S1 and S2 domains of the spike glycoprotein. The S1 part attaches the virion to host ACE2 (angiotensin-converting enzyme 2), a cell-surface receptor found in various tissue types including the alveolar epithelial cells, and is internalized inside the endosome. This primes the S2 by cleaving the S1/S2 junction by the cellular serine protease, transmembrane protease serine 2 (TMPRSS2), which actually is a furin endoprotease, and unmasks it for the fusion of the viral and cellular endosome membranes. This fusion facilitates the viral genome to enter into the cytoplasm of the host cell and establish infection. Immediately after entering, the viral RNA uses the host cell's protein synthesis machinery for producing its own proteins. SARS-CoV-2 does not lyse the host cell; instead, they "bud" off from the cell and infect nearby cells in the same way. As most of the human epithelial cells express ACE2, particularly the alveolar cells and cells of the intestine, those infectious virions eventually spread throughout the body and may pose the host toward a critical condition. The complex signaling required for activating innate and adaptive immunity needs a lot of proteins to be expressed by the host. But in

N. Roy Chattopadhyay · N. Das · T. Choudhuri (✉)
Department of Biotechnology, Visva-Bharati, Santiniketan, India
e-mail: tathagata.choudhuri@visva-bharati.ac.in

K. Dua et al. (eds.), *Targeting Cellular Signalling Pathways in Lung Diseases*,
https://doi.org/10.1007/978-981-33-6827-9_26

COVID-19, the initial stages of pathogenesis may suppress the immune responses of the host by suppressing the protein production, as stated earlier.

Keywords

COVID-19 · SARS-CoV-2 · Pandemic · ACE2 · RdRp · Spike glycoprotein · Signaling · Pathways · Immunity

26.1 Introduction

Severe acute respiratory syndrome coronavirus 2 (SARS-CoV-2), or novel coronavirus, is the causative agent of the fatal and contagious coronavirus disease-2019 (COVID-19), affecting thousands of people in every single day posing a global threat. The first incidence disclosed worldwide in December 2019 has reported an unknown disease with fever and pneumonia-like symptoms eventually resulting in acute respiratory distress syndrome and death. Later it was confirmed that the disease has been caused by a coronavirus, related to bat coronaviruses, and it may have originated from the animal market of the capital of Hubei province in China, named Wuhan [1, 2]. Since then it has continued, and still continuing, to spread throughout the world irrespective of the socioeconomic status, ethno-genetic diversity, and topographical differences [3, 4]. The disease has been declared as a Public Health Emergency of International Concern (PHEIC) on 30 January 2020 and a pandemic on 11 March 2020 by the World Health Organization (WHO) [4]. Common symptoms include dry cough, fever, breathing trouble, and other symptoms of common flu. Sore throat, diarrhea, loss of smell and taste, and sputum secretion are generally reported in less people [5, 6]. While immune-competent healthy people show recovery from this disease, severe pneumonia and multi-organ failure are noted in some and became the cause of their death [3, 7]. The average rate of deaths per hundred confirmed infections is about 4.5% [8]; still the situation is frightening due to a few reasons. Firstly, the basic reproduction number (R_0) of the virus is 1.4–3.9 in different countries, i.e., one infected person can infect 1.4–3.9 new people who come in contact with the infected one [9]. Therefore, the possible number of infections and deaths is quite scaring as the huge population of human beings and the universal susceptibility toward this virus can pose a serious threat to mankind. Secondly, no cure is known till date. A range of antivirals and antiparasitic drugs are being tested clinically and are being used in patients along with antibiotics like azithromycin; but no conclusive data is still available to declare that as a cure for COVID-19. Thirdly, no vaccine is also known. Again, a number of candidate vaccines are being tested clinically, and some show promising results, but it must be noted that approval of any vaccine needs at least 12–15 months to pass through all the stages of clinical trials. SARS-CoV-2 is a newly found virus, and so is the resultant disease. Therefore, determination of the molecular and biochemical characteristics of the virus and targeting one or more of those are time-taking, eventually delaying the process of vaccine generation. And finally, the situation is frightening and probably alarming due to the information gathered in the last few months. The mechanism of infection of this virus indicates that it might be present in other mammals also; indeed, it is confirmed that COVID-19 is a potent zoonotic disease [4, 9–14].

The virus usually infects people during close contact with other infected persons or objects, and through respiratory droplets during coughing and sneezing by these persons [4, 15, 16]. Diagnosis can be done by chest CT scan, antigen test, or antibody test in the suspected person, but the presence of the virus is mainly confirmed by reverse transcription polymerase chain reaction (RT-PCR) from a nasopharyngeal swab [5, 17]. No drug and vaccine are available for the disease till date, and the incubation period of the virus inside the infected person may vary from a few days to as long as 2 weeks or even more; therefore, social distancing, use of masks, frequent washing of hands with soap, maintaining personal hygiene and healthy lifestyle, and avoidance of touching the mouth and nose are the only methods suggested to prevent the infection [4, 5, 18].

26.2 Phylogeny and Molecular Mechanism of Infectivity of SARS-CoV-2

SARS-CoV-2 belongs to the realm *Riboviria*, order *Nidovirales*, family *Coronaviridae*, genus *Betacoronavirus*, subgenus *Sarbecovirus*, and species severe acute respiratory syndrome-related coronavirus [19, 20]. It is a positive-sense single-stranded RNA virus (+ssRNA virus); the RNA genome is surrounded by an envelope embedded with spike glycoprotein responsible for the pathogenicity. Though this virus shows significant homology with the bat coronaviruses and its homologue found in pangolins, the severity or mode of infection of SARS-CoV-2 is attributed to the presence of a cleavage site for furin endoproteases in the junction of the S1 and S2 domains of the spike glycoprotein (nucleotide number 21563.0.25384 of the reference sequence NC_045512). Western blot analysis of 293 T cells expressing S protein of SARS-CoV-2 with C terminal HA tag showed two bands. One band corresponded to the unprocessed S protein (S0) and another for the S2 subunit of the S protein, indicating efficient proteolytic processing of SARS-CoV-2S protein [20–25]. The S1 part attaches the virion to host ACE2 (angiotensin-converting enzyme 2), a cell-surface receptor found in various tissue types including the alveolar epithelial cells and intestinal cells. The internalization of the virus via endosome primes the S2 domain of spike glycoprotein by cleaving the S1/S2 junction with the help of the cellular serine protease, transmembrane protease serine 2 (TMPRSS2). This enzyme actually is a furin endoprotease, and unmasks S2 for the fusion of the viral envelope and cellular endosome membrane. This fusion facilitates the viral genome to enter into the cytoplasm of the host cell and start initiating the establishment of infection [20–29]. SARS-COV, the close relative of SARS-CoV-2, utilizes endosomal cathepsin B and L (Cat B/L) and the serine protease TMPRSS2 [26] for S protein priming. Ammonium chloride (Cat B/L activity blocker) treatment strongly inhibited the entry of SARS-CoV-2 in the TMPRSS2-cell, whereas it was less effective in TMPRSS2+ cell. The result was clinically proven where serine protease

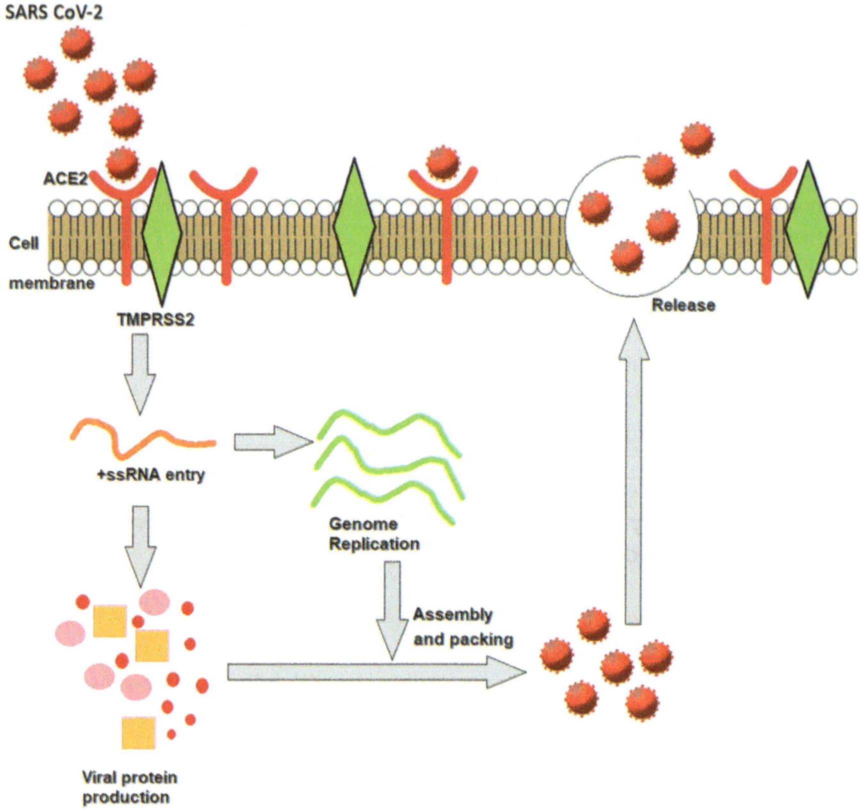

Fig. 26.1 SARS-CoV-2 and its mode of infection. The schematic representation of the steps of establishment of infection inside a host cell. Abbreviations: *+ssRNA* positive-sense single-stranded RNA, *ACE2* angiotensin-converting enzyme 2, *TMPRSS2* transmembrane protease serine 2

inhibitor camostat mesylate and E-64d, an inhibitor of Cat B/L, was used. The SARS-CoV-2S can use both Cat B/L and TMPRSS2 for the priming process [25–27]. Moreover, ACE2 expression correlates along with the expression of other human CoV receptors [alanyl aminopeptidase (ANPEP) and dipeptidyl peptidase-4 (DPP4)]. Thus, this finding indicates that the stated furin endoprotease may act as a co-receptor for S protein of SARS-CoV-2 [28, 29].

As the genetic material carries positive sense, the virus uses its RNA as its genetic material as well as mRNA for the viral protein production (Fig. 26.1) [11–13, 21–24]. The viral RNA enters the host cytoplasm and immediately starts using the host cell's protein synthesis machinery for the production of its own proteins needed for genome replication. The details of the nucleotide sequences, their respective peptide sequences, structures, and annotated functions are available in public databases [5, 21–23]. On a particular note for the most important mature proteins for infectivity, a few are mentioned hereafter. Two-thirds of the genome, called the ORF1ab, is responsible for producing the ORF1ab replicase polyprotein; this actually produces a set of 16 mature viral proteins needed for different aspects of establishment of infection [4, 5, 21–26]. The remaining parts encode the spike glycoprotein (S), small envelope protein (E), matrix protein (M), nucleocapsid protein (NC), and some other accessory proteins to evade the host immune system. The first 265 nucleotides comprise of the 5′UTR and then the sequence for ORF1ab starts. From nucleotide number 266–805 (nt 266–805), a protein called nsp1 ("nsp" stands for nonstructural protein) is synthesized. This protein binds the host 40S ribosomal subunit, and this binding pushes the host mRNAs toward degradation by cleaving them near their 5′UTRs. Thus, the productions of the host proteins are usually stopped and can make the host immune-compromised. Viral mRNAs are protected from such cleavage as they bear a 5′-leader sequence; and therefore, the productions of viral proteins are facilitated. Viral proteins nsp7 and nsp8 (nt11843.0.12091 and 12092.0.12685, respectively) interact with each other to facilitate viral replication, probably by acting as a primase complex. One of the most important mature viral proteins is nsp10 (nt13025.0.13441), which stimulates two other mature viral proteins—nsp14 (nt18040.0.19620) and nsp16 (nt20659.0.21552). Another viral protein called nsp15 can cleave the long viral sense RNA in some mature fragments, each with a 5′cap-like structure made of a 2′–3′-cyclic phosphate (Fig. 26.2). This structure might initially protect the viral RNAs from host-mediated destruction and marks them for further processing. After activation by nsp10, the N7-guanine methyltransferase activity of nsp14 adds the N7-methylguanosine cap to the viral mRNAs and recruits nsp16 to those. This protein then adds a methyl group to the 2′-*O*-ribose of that N7-methylguanosine. These methylations form a compact cap at the mature viral mRNAs and prepare those for protein production using the eukaryotic host-cell machinery as well as escape from host immune clearance. All these, and some other proteins, help the RNA-dependent RNA polymerase (RdRp), or nsp12 (nt13,442–16,236), to efficiently replicate and transcribe the viral genome to establish the infection securely and to generate new infective virions [21–23].

SARS-CoV-2 continues to replicate its genome required for the synthesis of new virions; alongside, it also synthesizes a nested set of subgenomic mRNAs for the production of viral proteins needed for packing and release of those virions. They do not lyse the host cell after the packing of even a large number of virions; rather, they "bud" off from the cell and infect nearby cells in the same way. As most of the human epithelial cells express ACE2, particularly the alveolar cells and those in the intestine, those infectious virions eventually spread throughout the body and push the host toward a critical condition if not treated timely [27]. But unfortunately, no specific and confirmed drug is available till date to treat SARS-CoV-2 properly [4–6]. According to current information, a few antiviral and antimalarial drugs are being used to manage, at least minimally, COVID-19; however, researchers are testing various other drugs and drug candidates virtually and in clinical trials also. Any affective vaccination is also not available till date [4–6, 30–32]. Therefore, immune-compromised people, for example, persons having other diseases, may sometimes face severe complications due to COVID-19 and may even die due to ADRD and severe pneumonia [4, 13, 17, 33, 34]. The immune response of the host plays the vital role for clearing any infection. But in COVID-19, the initial stages of

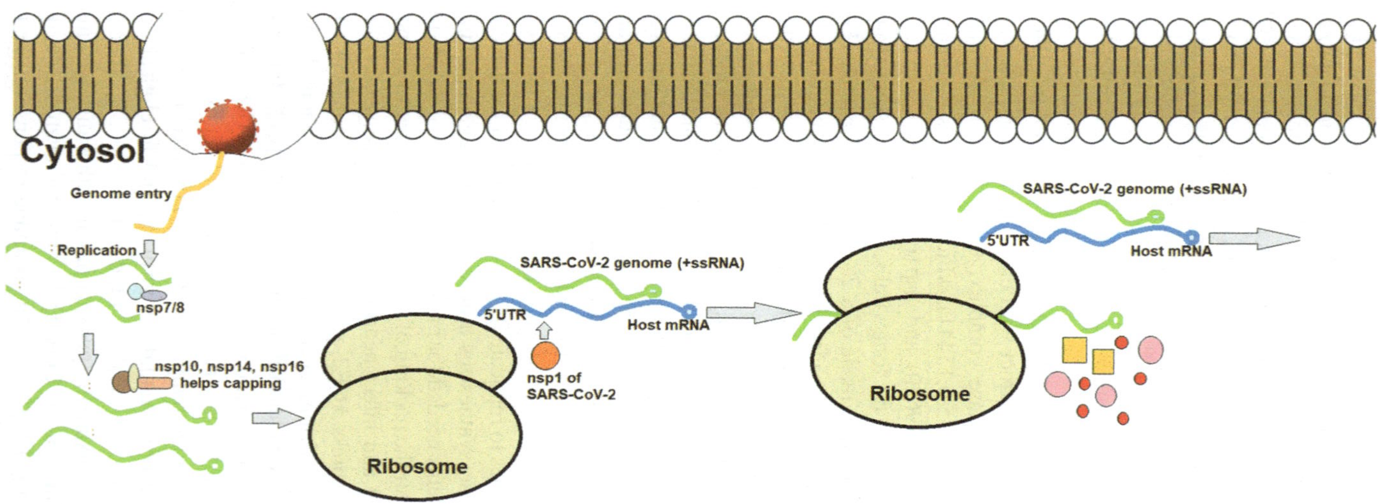

Fig. 26.2 Some viral proteins facilitate survival inside the host cell. A viral protein called nsp1 binds the host 40S ribosomal subunit and helps degrade host mRNAs. Viral proteins nsp7 and nsp8 interact with each other to facilitate viral replication. nsp10 stimulates nsp14 and nsp16 to form a protective cap on viral mRNA and protects it from degradation inside the host cytoplasm

pathogenesis try to suppress the immune responses of the infected person. The complex signaling required for activating innate as well as adaptive immunity needs a lot of proteins to be expressed by the host. But SARS-CoV-2 tries to suppress the production of host proteins at a very initial stage of its infection, as stated earlier [4, 18, 25, 27, 30, 33]. Though a few cytokines and chemokines are reported in COVID-19 patients, which molecules might trigger a series of immune responses and cytokine storm, it should be noted that such induction of immune responses is not universal. In addition, cytokine storm may lead to severe respiratory distress which is widely reported to be associated with COVID-19.

26.3 The Signaling Pathways and Key Molecules Tangled with SARS-CoV-2 Infections in the Lung

Early in this pandemic, little information about infection mechanism of SARS-CoV-2 was available in published reports. It took a long time to know about the similarities and differences of SARS-CoV-2 with other beta coronaviruses. The novelty of the mechanism of infectivity of this virus, as discussed earlier, makes it more perilous and hard to treat. SARS-CoV-2 enters into the human body through the nasal-paranasal and oral route. The virus binds to the epithelial cells of these cavities, which act as the portals for initial infection and transmission. The viral infection goes downward through the respiratory tract to conducting airways and spread into the lung alveoli. In severely affected individuals, ARDS and pneumonia are seen with the major symptom of low ratio of partial pressure arterial oxygen (PaO_2) against the fraction of inspired oxygen (FiO_2), i.e., low PaO_2/FiO_2 compared to normal individuals. In 18.9% of the cases, patients suffering from ARDS show PaO_2/FiO_2 ratio less than <100 mmHg. The increased vascular permeability, plasma leakage, and severe hypoxemia are the main characteristics of ARDS. Infiltration of active immune cells, mainly neutrophils and mononuclear cells, therefore develops hyperinflammatory state and hypercoagulation. Consequently, it develops disseminated microvascular coagulation and eventually becomes fatal [25, 33–36].

The surface epithelium of the alveoli, or the pneumocyte layer, is composed of two types of cells. Type I cells are complex branched cells containing cytoplasmic plates, whereas type II is the progenitor cell type for both type I and type II. The type II cells help in the renewal of the damaged area of the lung. They synthesize and discharge pulmonary surfactants and thus create a receptive condition for gas exchange. Also, the lungs contain functionally distinct types of other stem cells like basal cells, club cells, and bronchoalveolar stem cells, each present in specific anatomical locations. Type II pneumocytes expressing ACE2 are the prime target of SARS-CoV-2 infection of the human lung as detailed earlier. The Oct4+ pulmonary stem cells, large and small bronchial epithelial cells, goblet/club cells, and pulmonary endothelial cells also can express ACE2. Thus, the lung is the primary target for this fatal virus. Tissue-resident immune cells within the lung and type II pneumocytes are the first candidates to encounter SARS-CoV-2. As the host becomes immune-compromised by the viral interaction with protein production, these cells become the primary sites for viral proliferation and host inflammatory responses [25, 33–36].

26.4 Role of Angiotensin-Converting Enzyme 2 (ACE2) and Proteases Toward SARS-CoV-2 Infection

In previous sections, it has been discussed that ACE2 protein is the functional receptor for the spike glycoprotein of the SARS-CoV-2, and initially, the entry of SARS-CoV-2 needs the binding of the surface unit S1 of the spike protein to the ACE2 receptor. The renin-angiotensin system (RAS) has an important role in circulatory homeostasis as well as in injury/repair response. Stimulation of a local RAS within the pulmonary system (circulation and lung parenchyma) could promote the pathogenesis of lung injury or infection by increasing vascular permeability, vascular tone, and fibroblast activity, and by depleting alveolar epithelial cell survival. In a close relationship with ACE2, the RAS appears to have a critical role in SARS-CoV-2 infection [37]. Generally, after the CoV binds to its receptor, the proteases like trypsin and TMPRSS2 activate the early fusion pathway; if not, the virus becomes endocytosed and eventually destroyed. When furin cleaves the S protein in proper time of infection pathway, the exogenous and membrane-bound proteases track the early entry. Otherwise, the virus may again be endocytosed after the cleavage at the S1/S2 site. The low pH within the endosome activates cathepsin L; thus, it cleaves the S2 site and triggers the viral fusion facilitating the infection pathway. Likewise, SARS-CoV-2S protein also possesses a unique furin cleavage site at the S1/S2 region, and SARS-CoV-2 can utilize membrane-bound TMPRSS2 or endosomal cathepsin L for its entry [26, 38, 39].

In the ACE-Ang II-AT1R (ACE-angiotensin II-angiotensin II type 1 receptor) pathway, the classical RAS increases sympathetic nervous system tension. This results in vasoconstriction and increases blood pressure; thereafter it boosts inflammation, fibrosis, and myocardial hypertrophy (Fig. 26.3). On the other hand, ACE2-Ang 1–7-MasR-based pathway antagonizes these effects [40]. SARS-CoV-2 induces the downregulation of ACE2 in various tissues [41–43]. Thus, the ACE-Ang II-AT1R pathway takes over the command and increases vascular permeability through the JAK/STAT pathway [41]. Some cytokines like IL-6, IL-1β, tumor TNF-α, and IFN-γ have been testified to be commonly raised in COVID-19. Among them, IL-6, IL-1β, and IFN-γ were reported to hinder ACE2 expression, which ultimately alters the balance of the renin-angiotensin system (RAS) [44–47]. Since SARS-CoV-2 enters cells mainly via an ACE2, targeting this protein will be a beneficial therapeutic strategy in controlling the COVID-19 outbreak. The proposed and repurposed ACE2 inhibitors are listed in Table 26.1.

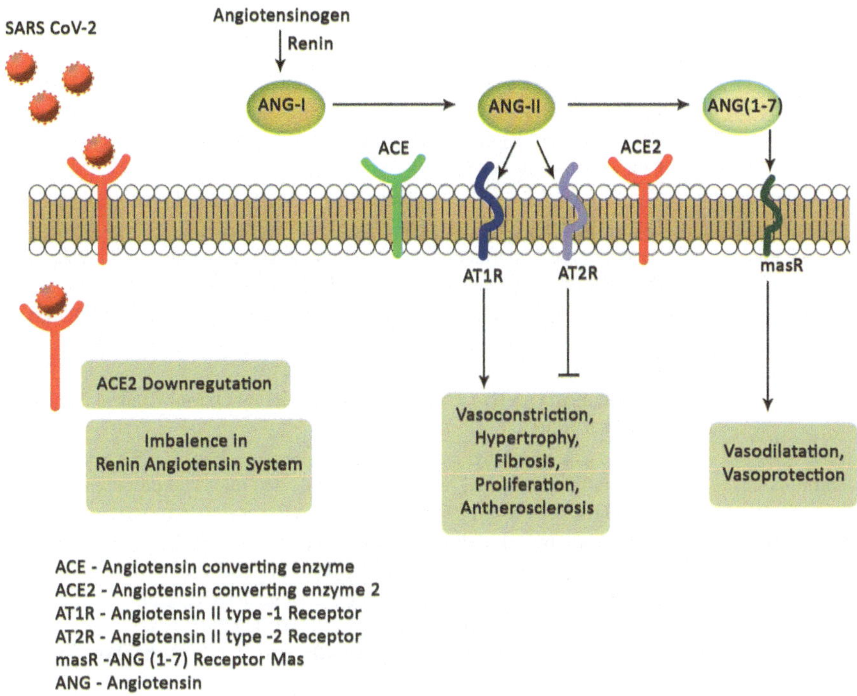

Fig. 26.3 Modulation of angiotensin II and its signaling pathway by SARS-CoV-2 complicates infection. As ACE2 is the main receptor for SARS-CoV-2 entry, targeting the ACE2-AngII-AT1R pathway has a great potential for COVID-19 therapeutic plans

26.5 Toll-Like Receptors (TLRs) and Activation of NF-κB and Interferon Regulatory Factor in the Lung

TLRs play an important role in the innate immune system by recognizing pathogen-associated molecular patterns (PAMPs). Upon recognition of PAMPs, TLRs signal by recruiting specific adaptor molecules which in turn activate transcription factors. The immune-resident and epithelial cells of the lung express TLRs such as TLR3 and TLR7/8, which can recognize pathogen-associated molecular patterns and viral nucleic acids and can induce downstream signal [57, 58]. TLRs activate mitogen-activated protein (MAP) kinases and a number of transcription factors like NF-κB, interferon regulatory factor 3 (IRF3), and IRF7; thereby initiating consequential activation of many pro-inflammatory cytokines and type I (α and β) and type III (λ) IFNs [59]. It was found in common respiratory viruses that TLR3 activates IRF3 via TIRF (TIR-domain-containing adapter-inducing interferon-β) and IRF7, and also activates NF-κbeta through MyD88. Like TLR3, other endosomal TLRs like TLR7/8 and TLR9 can detect viral nucleic acids resulting in activation of IRF7 and NF-κbeta via MyD88. Concerning the lung, SARS-CoV infection triggers an

Table 26.1 The proposed and repurposed drugs, targeting specific signaling molecules in COVID-19 treatment

Targets	Drug trials
Human ACE2 inhibitor	Lividomycin, burixafor, quisinostat, fluprofylline, pemetrexed, spirofylline, edotecarin, and diniprofylline [48] COL-3 and CGP-60474 [49]
IL-6 inhibitors	Tocilizumab [clinical trial in Italy, NCT04317092, and different countries in Europe and the USA, NCT04320615] [50]
IL-1 inhibitors	Anakinra [clinical trial in Italy (NCT04324021) and another trial in Italy and the USA, by Novartis (NCT04362813)] [50]
Application of IFNs	Interferon Beta-1A and Interferon Beta-1B [clinical trial NCT04343768] [51]
TNFα inhibitors	Infliximab [clinical trial—NCT04425538] [52] Adalimumab (AVID-CC) [under clinical trial by Oxford Clinical Trials Research Unit]
NF-k beta inhibitors	Caffeic acid phenethyl ester (CAPE), resveratrol, Bay11-7082, and parthenolide [validated in SARS-CoV-infected mice and proposed for COVID-19 treatment] [53, 54]
p38MAPK inhibitors	Silymarin [clinical trial—NCT04394208] [55] Losmapimod [clinical trial—NCT04511819] [56]
Janus kinase inhibitors	Baricitinib [13 different clinical studies (ClinicalTrials.gov)] Ruxolitinib [clinical trial by Novartis NCT04337359] Tofacitinib [clinical trial in Italy (NCT04332042, NCT04390061)] [50]

inflammatory response by augmenting the production of IL-6, IL-8, IFN-γ, inducible protein 10 (IP-10), and NF-κB in the differentiated Calu-3 cell line grown at the air-liquid interface (ALI). Besides these, two helicases called RIG-1 and MDA5 can recognize viral replication products from cytosol and activate IRF3 via IPS-1/MAVS/VISA/Cardif [57, 60].

26.6 Association of Interferons in COVID-19

Interferons (IFNs) are cytokines formed by cells during virus infections. They initiate antiviral signaling activity by binding to specific receptors (IFN alpha/beta) on the cell surface and consequently can activate the JAK/STAT signaling cascade (Fig. 26.4). Thus, the establishment of an antiviral state is achieved by the expression of hundreds of interferon-stimulated genes. COVID-19 is characterized by elevated serum cytokines (particularly IL-1β, IL-6, and TNF-α), weakened interferon responses, and peripheral lymphopenia [61]. COVID-19 is associated with a reduced IFN-I and IFN-III host response [62]. The ORF6 and ORF8 proteins and nucleocapsid of SARS-CoV-2 can hinder IFN-I signaling in vitro [63]. However, these pro-inflammatory molecules may restrict SARS-CoV-2 replication in vitro. Clinical trials regarding the administration of exogenous IFN-I and IFN-III against COVID-19 (NCT04343768 and NCT04354259, respectively) are still going on [64, 65] (Table 26.1).

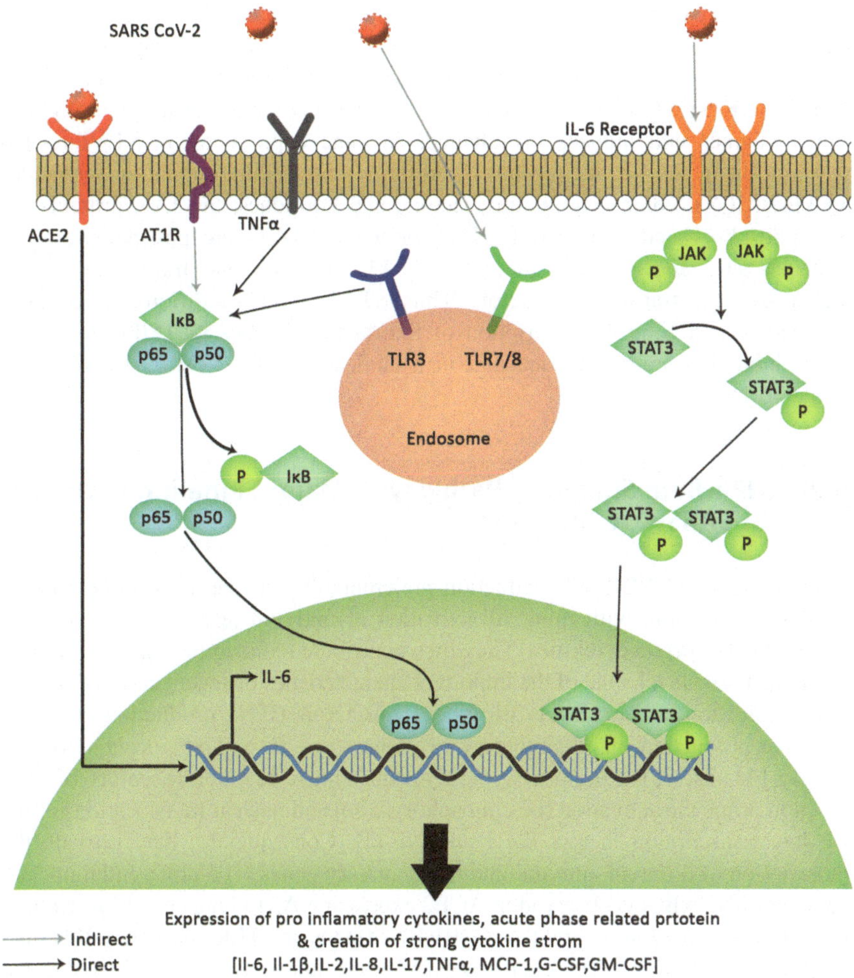

Fig. 26.4 SARS-CoV-2 infection affects a number of host signaling pathways. A number of important signaling pathways in the host are affected by SARS-CoV-2 infection, and thus each of the molecules of these pathways may be targeted by candidate drug molecules

26.7 Involvement of Il-6 and JAK/STAT Signaling During CoV Infection

Lung epithelial cells and macrophages respond to SARS-CoV-2 infections, activating NF-κB-and its targets like IL-6, IL-8, and TNF [66–68]. Furthermore, IL-6 shows pleiotropic effects concerning different lung-related diseases. It was observed that in murine models of acute lung injury, IL-6 initiates an inflammatory response through the STAT3 pathway within the lungs [69, 70]. The IL-6−/− mice showed a decreased inflammatory influx in the bronchoalveolar lavage [71]. The importance of IL-6 was also proved in COPD and asthma [72]. Especially IL-6-induced expression of Ang II creates a feedback loop by promoting the expression of IL-6 via JAK/STAT. JAK/STAT pathway involves the rapid transmission of extra-cellular signals and induces the changes in gene expression via STAT-related transcription factors. Upon receptor binding of extracellular signaling molecules cytokines, IFNs, colony-stimulating factors, and hormones, the JAK proteins are cross-phosphorylated and recruit STAT proteins. STATs are phosphorylated at particular sites and activate. Activated STATs dimerize and translocate to the nucleus to act as transcription factors. Thus, STATs bind at a different position in the genome and control transcription of thousands of genes [73]. Il-6 and some kinase inhibitors are repurposed and in a clinical trial for COVID-19 treatment (Table 26.1).

26.8 NF-κBeta Signaling Pathway: A Major Signaling Pathway in SARS-CoV Infection

Severe cases of SARS-CoV-2 infection preferentially activate a pro-inflammatory cytokine production, with minimal activation of antiviral response. The infected epithelial cells secrete cytokines. They then contribute to the initiation of inflammation at the tissue level. One of the important characteristics of severe cases of SARS-CoV-2 infection is low levels of type I interferon (IFN) production, whereas excessive production of inflammatory cytokines (i.e., IL-6, TNF) occurs simultaneously [44, 74, 75]. Commonly, the cytosolic innate immune receptors of viral infection prime the activation (i.e., phosphorylation and nuclear translocation) of the two key transcription factors IRF3 and NF-κB. Consequently, they activate the transcription of different antiviral and inflammatory genes. The preferential nuclear transport of NF-κB was observed in ACE2 expressing A549 lung cells [76]. Knockout and overexpression of viral RNAs (RIG-I, MDA5, and TLR3 as well as IFN) and receptors were done in ACE2 expressing A549 cells infected with SARS-CoV-2. No significant changes were observed in IFIT1 mRNA (IRF3 target) and TNF mRNA (NF-κB target) levels in those cells. The results indicate that RIG-I, MDA5, TLR3, and IFN receptors (cellular RNA sensors) are not involved in NF-κB activation. Commonly, cGAS and STING work as intracellular sensors for cytosolic DNA.

Upon DNA binding with the aid of ATP and GTP, the cGAS catalyzes the production of cGAMP. This cGAMP then acts as a second messenger, which in turn binds and triggers STING, thus leading to type I IFN production [77, 78]. Though cGAS mainly acts as a cytosolic DNA sensor, the induction of the cGAS-STING signaling axis and activation of downstream NF-κB and IRF3 have been shown for numerous RNA virus infections [79]. The cGAS knockout mice are more vulnerable to the infection of a positive-sense single-stranded RNA virus, West Nile virus (WNV) [80]. Likewise, STING-deficient mice had greater susceptibility to RNA vesicular virus stomatitis (VSV) infection, and STING-deficient cells could not produce an aggravated innate immune response against

RNA viruses such as VSV and SeV (Sendai virus) [81]. Upon infection of SARS-CoV-2, both cGAS and STING are re-localized to perinuclear clusters of infected cells. Therefore, the condition indicates the activation of cGAS and STING. The pharmacological inhibitor of STING H-151 significantly decreases in the levels of TNF mRNA in infected cells, both in ACE2 expressing A549 and Calu-3 lung cells. Generally, STING activation connects with NF-κB and IRF3 activation. But several reports showed that interference in ER to Golgi translocation of STING preferably stimulates the NF-κB pathway [82, 83]. The researcher observed that there is no significant colocalization of STING and Golgi markers in SARS-CoV-2-infected cells, indicating impaired STING translocation and activation of a specific NF-κB inflammatory response. Hence, SARS-CoV-2 infection activates the cGAS-STING pathway, which in turn induces pro-inflammatory cytokines through NF-κB, and the STING inhibitors can control this specific response efficiently. However, pharmacological STING inhibitors could not block TNF upregulation completely. It can be speculated that multiple pathways may be responsible for NF-κB pathway activation [76]. Thus, combinatorial therapeutic approach can be used to target multiple NF-κB activation pathways in SARS-CoV-2-infected cells (Table 26.1).

26.9 NLPR3 Inflammasomes

NLRP3 (NOD-, LRR-, and pyrin domain-containing protein 3) acts as an intracellular sensor that perceives a broad range signal along with PAMPs from viruses, leading to the formation of the inflammasome. Formation of NLRP3 inflammasome primes caspase 1-dependent release of IL-1β and IL-18. Among them, IL-1β can promote the production of IFNs and antiviral factors through IRF1/STAT1 signaling, helping virus elimination [84]. In the case of SARS-CoV infection, ORF8b and ORF3a are capable of triggering NLPR3 inflammasomes [85, 86]. The largest nonstructural protein of SASR-CoV-2 (nsp3) is composed of several domains. Among them, the SARS-CoV unique domain (SUD) (Nsp3c) is also found in MERS-CoV and SARS-CoV-2 having 15% and 75% identity with SARS-CoV, respectively. SUD can induce NLRP3 production and promote inflammation by inducing CXCL10 in the lung [87]. Therefore, a precise and regulated approach to modulate NLPR3 activity and its levels could be helpful for ARDS treatment in COVID-19 patients.

26.10 p38-MAP Kinase Pathway

Ang II shows its effects by activating the p38 MAPK pathway, whereas ACE2 converts Ang II to Ang 1–7, which in turn binds to the Mas receptor and consequently decreases p38 MAPK activation [88]. In SARS-CoV-2, upon viral entry, loss of ACE2 activity is very prominent. Therefore, it leads to the downregulation of Ang 1–7 production, accelerate Ang II-mediated activation of p38, and uncontrolled inflammation. Additionally, p38 activation produces a positive feedback loop with

ADAM17, which cleaves the ACE2 ectodomain, and further declines the ACE2 protective activity [89]. In COVID-19 patients, Ang II levels were positively related to the degree of lung injury and viral load, indicating RAS imbalance [90]. SARS-CoV can directly upregulate p38 activity with the help of its protein and promote its replication. It can be predicted that by possessing a high homology with SARS-CoV, the new virus SARS-CoV-2 may follow a similar mechanism [56]. Besides these, ERK, JNK, and p38 involve numerous aspects of COPD development [91]. Stress, LPS, and different inflammatory factors can activate the p38MAPK pathway, which produces inflammatory cytokines [92]. Among its four isoforms, p38α MAPK associates with the production of IL-8 and IL-6 in response to IL-1 and TNF-α, respectively [93, 94]. Additionally, p38MAPK can posttranscriptionally regulate TNF-α and COX-2 gene via AREs (AU-rich element), but the exact mechanisms of p38MAPK-mediated posttranscriptional regulations are still unclear [95].

The expression of intercellular cell adhesion molecule-1 (ICAM-1) increases in the early phase of acute lung injury (ALI), thus leading to the aggravation of disease [96]. p38MAPK can regulate the expression of ICAM-1 by human antigen R (HuR) or p53. Being a posttranscriptional regulatory factor, HuR affects the half-life and/or translation of different mRNAs, like ICAM-1, TNF-α, COX-2, and TLR4 through binding to AREs [97–99]. Therefore, pharmacological targeting of the MAPK pathway in ARDS will be a wise approach (Table 26.1).

26.11 Targeting Potential Molecules for Treatment of COVID-19

Designing a successful and universal treatment regimen for COVID-19 patients becomes critical due to these above-discussed mechanisms of infection of the virus and the respective host responses. Various signaling pathways and their functional proteins are used by the virus for successful establishment of infection. Potentially each of these molecules can be targeted for designing a therapeutic plan for COVID-19 patients. As mentioned earlier, a lot of these targets are indeed being tested in different laboratories throughout the world. A set of combinatorial approaches are also being investigated by some. Ideally, it is needed to attack the infection from, at least, three sides: a drug to kill or stop the virus; a chemical or biochemical blocker for any step of the establishment of the infection; and, last but not the least, a modulator for one or more important signaling molecules. An elaborate list of the drugs against signaling molecules has been presented in Table 26.1. Table 26.2 lists the widely used antiviral and antiparasitic drugs. Oseltamivir, peramivir, zanamivir, ganciclovir, acyclovir, and methylprednisolone are already reported to be almost ineffective against SARS-CoV-2. Some other antiviral drugs like umifenovir, galidesivir, ribavirin, triazavirin, ritonavir, nitazoxanide, favipiravir, lopinavir, ritonavir, nafamostat, darunavir, and remdesivir, and an antimalarial drug chloroquine, have been used separately or in combination to treat COVID-19 patients. Recently, a new line of attack is proposed depending upon the mechanism of entry of the virus. A blocker for ACE2 or TMPRSS2 is being tried to prevent the viral entry into the host cell. Interestingly, the WHO has announced a

Table 26.2 Selected antiviral and antiparasitic drugs for final screening and their mode of actions

Drug	Mode of action [5, 100–106]
Favipiravir	Selectively inhibits RNA polymerase and prevents replication of the viral genome
Remdesivir	Nucleoside analog that is expected to inhibit the action of RNA polymerase by incorporating those into RNA during replication and/or transcription
Ribavirin	Ribavirin triphosphate (RTP) is the predominant metabolite which directly inhibits viral mRNA polymerase by binding to the nucleotide binding site of the enzyme
Galidesivir	Binds to viral RNA polymerase at the binding site of natural nucleotides; thereby leads to structural change in the viral enzyme and disruption of the viral RNA polymerase activity resulting in premature termination of the elongating RNA strand
Umifenovir	Interacts at the plasma membrane to stabilize it and to prevent viral entry
Pirodavir	Binds and stabilizes the viral capsid
Chloroquine	Increases endosomal and lysosomal pH; thus it might prevent the release of viral genome, which may inhibit viral RdRp
Proguanil	Specifically inhibits parasitic dihydrofolate reductase
Quinine	Might act similarly like chloroquine
Artemether	Possibly creates oxidative and metabolic stress and accumulates intracellular calcium
Artesunate	Increases reactive oxygen species (ROS) and decreases glutathione on parasite
Nafamostat, camostat mesylate	Inhibits TMPRSS2
Losartan, olmesartan	Inhibits AT1R
Benazepril, captopril, enalapril, fosinopril, lisinopril, moexipril, perindopril, quinapril, ramipril, trandolapril	Inhibits ACE and helps increase ACE2

combinatorial approach using remdesivir, chloroquine, a blocker as said above, and an immune booster as said above [8, 31, 102]. A chloroquine derivative, hydroxychloroquine, is being widely used as a promising treatment strategy for COVID-19 and is also suggested by the WHO [103]. It is an effective antimalarial drug which increases lysosomal pH and reduces inflammatory responses by a complex immune-modulating process. But it should be noted that hydroxychloroquine can reduce both innate and adaptive immune responses significantly by suppressing Toll-like receptor signaling (TLR) pathways. In COVID-19, lack of strong and effective immunity plays one of the major roles for poor prognoses in patients [4, 7, 9, 18]. Therefore, the use of chloroquine/hydroxychloroquine molecule in COVID-19 patients can pose a serious threat on their cure, and survival also. This indeed is being reported very recently, and makes scientists bound to exclude this molecule from treatment proposals.

26.12 Conclusion

Novel coronavirus disease-2019 or COVID-19 has made people look back into many things. Each and every day, scientists as well as general public are reporting striking incidents and novel scientific findings. Probable zoonotic transfer of SARS-CoV-2, its scaring abilities for adaptation and infection, destruction of various species and climate change, effects of social distancing in public psychology, and even survival of man and economy are being studied by all and are really important indeed. But still it cannot be denied that the search for possible treatment of COVID-19 patients has outweighed all as the first and foremost task at present is to make people survive through this hell. Published reports are showing individual results for some known and/or unknown candidates for the treatments. Some are reporting about any antiviral; some are trying to establish the efficacy of antiparasitic drugs like the hydroxyl derivative of chloroquine and hydroxychloroquine; some are trying inhibitors to block host-cell entry by SARS-CoV-2; and even others are trying to help immunity of affected people by testing interferons. But as far people have learned about the characteristics of the viral genome and its mode of infection, it can be inferred that no one drug available at present would serve the purpose of curing a significant number of patients. Rather, a combination of molecules, each for targeting a specific step of establishment of infection, would be helpful as a full-proof treatment plan; and this has also been proposed by the WHO recently. Though drug interactions are checked thoroughly in clinical trials, it should be noted that the dose and dosage of any candidate molecule should be finalized very carefully by efficient and experienced healthcare professionals to prevent any unwanted adverse result(s) as very less is known about this fatal virus till now (Fig. 26.5). Like other pandemics which passed in earlier centuries, we hope that proper and widespread management of COVID-19 will also become possible and help survival of mankind.

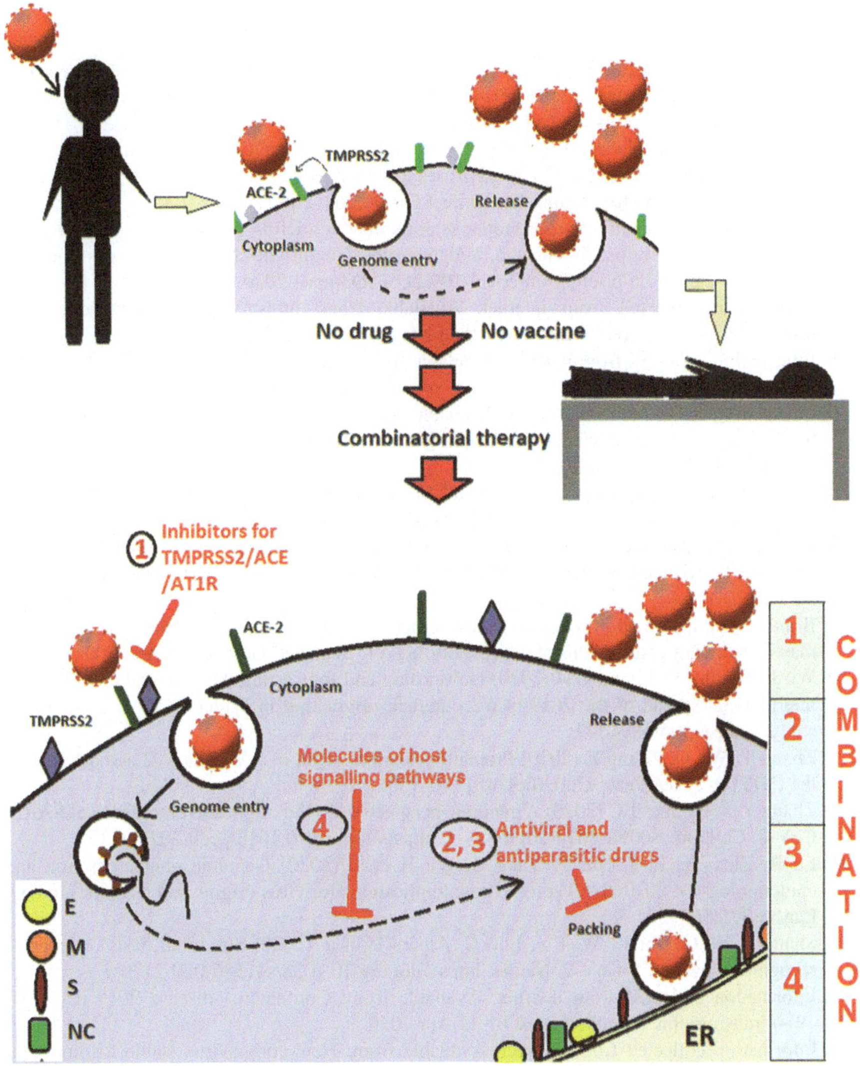

Fig. 26.5 Proposed treatment plans for fighting COVID-19. A set of possible combinatorial treatment plans including an antiviral, an antiparasitic drug, an inhibitor for preventing host-cell entry, and some molecular targets in the host signaling pathways might improve management of COVID-19. Adapted and modified from Chattopadhyay NR, Chatterjee K, Banerjee A, Choudhuri T. Combinatorial therapeutic trial plans for COVID-19 treatment armed up with antiviral, antiparasitic, cell-entry inhibitor, and immune-boosters. VirusDisease. (2020). 31 (4):479–489

References

1. Chen L, Liu W, Zhang Q, Xu K, Ye G, Wu W et al (2020) RNA based mNGS approach identifies a novel human coronavirus from two individual pneumonia cases in 2019 Wuhan outbreak. Emerg Microbes Infect 9:313–319
2. Hui DS, Azhar EI, Madani TA, Ntoumi F, Kock R, Dar O et al (2020) The continuing 2019-nCoV epidemic threat of novel coronaviruses to global health—the latest 2019 novel coronavirus outbreak in Wuhan, China. Int J Infect Dis 91:264–266
3. WHO Director-General's opening remarks at the media briefing on COVID-19. 3 Apr 2020. Available from: https://www.who.int/dg/speeches/detail/who-director-general-s-opening-remarks-at-the-media-briefing-on-covid-19%2D%2D3-april-2020
4. Information collected from internet. Available from: https://www.who.int/emergencies/diseases/novel-coronavirus-2019. Searched latest by 1 Apr 2020
5. Information collected from internet. Available from: www.cdc.gov. Searched latest by 1 Apr 2020
6. Information collected from internet. Available from: Coronavirus COVID-19 global cases by the Center for Systems Science and Engineering (CSSE) at Johns Hopkins University (JHU). Searched latest by 12 Apr 2020
7. Information collected from internet. Available from: Q&A on coronaviruses. World Health Organization (WHO). Searched latest by 27 Mar 2020
8. Information collected from internet. Available from: Naming the coronavirus disease (COVID-19) and the virus that causes it. World Health Organization. Searched latest by 30 Mar 2020
9. Branswell H. Information collected from internet. Available from: Limited data on coronavirus may be skewing assumptions about severity. STAT. Searched latest by 30 Mar 2020
10. Wu JT, Leung K, Leung GM (2020) Nowcasting and forecasting the potential domestic and international spread of the 2019-nCoV outbreak originating in Wuhan, China: a modelling study. Lancet 395:689–697
11. Zhang T, Wu Q, Zhang Z (2020) Probable pangolin origin of SARS-CoV-2 associated with the COVID-19 outbreak. Curr Biol 30:1346–1351
12. Zhang Y-Z, Holmes EC (2020) A genomic perspective on the origin and emergence of SARS-CoV-2. Cell. pii: S0092-8674(20)30328-7. https://doi.org/10.1016/j.cell.2020.03.035
13. Lu R, Zhao X, Li J, Niu P, Yang B, Wu H et al (2020) Genomic characterisation and epidemiology of 2019 novel coronavirus: implications for virus origins and receptor binding. Lancet 395:565–574
14. Shang J, Ye G, Shi K, Wan Y, Luo C, Aihara H et al (2020) Structural basis of receptor recognition by SARS-CoV-2. Nature. https://doi.org/10.1038/s41586-020-2179-y
15. Information collected from internet. Available from: Coronavirus disease 2019 (COVID-19)—transmission. Searched latest by 12 Apr 2020
16. Information collected from internet. Available from: New coronavirus stable for hours on surfaces. National Institutes of Health. Searched latest by 10 Apr 2020
17. Jin YH, Cai L, Cheng ZS, Cheng H, Deng T, Fan YP et al (2020) A rapid advice guideline for the diagnosis and treatment of 2019 novel coronavirus (2019-nCoV) infected pneumonia (standard version). Mil Med Res 7:4–26
18. Velavan TP, Meyer CG (2020) The COVID-19 epidemic. Tropical Med Int Health 25:278–280
19. Gorbalenya AE, Baker SC, Baric RS, de Groot RJ, Drosten C, Gulyaeva AA et al (2020) The species Severe acute respiratory syndrome-related coronavirus: classifying 2019-nCoV and naming it SARS-CoV-2. Nat Microbiol 5:536–544
20. Cui J, Li F, Shi ZL (2019) Origin and evolution of pathogenic coronaviruses. Nat Rev Microbiol 17:181–192
21. Information collected from internet. Available from: https://www.gisaid.org/. Searched latest by 4 Apr 2020

22. Information collected from internet. Available from: https://www.ncbi.nlm.nih.gov/nuccore/NC_045512. Searched latest by 1 Apr 2020
23. Information collected from internet. Available from: https://swissmodel.expasy.org/repository/species/2697049. Searched latest by 12 Apr 2020
24. Information collected from internet. Available from: https://www.niaid.nih.gov/news-events/novel-coronavirus-sarscov2-images. Searched latest by 4 Apr 2020
25. Hoffmann M, Kleine-Weber H, Schroeder S, Krüger N, Herrler T, Erichsen S et al (2020) SARS-CoV-2 cell entry depends on ACE2 and TMPRSS2 and is blocked by a clinically proven protease inhibitor. Cell. pii: S0092-8674(20)30229-4. https://doi.org/10.1016/j.cell.2020.02.052
26. Coutard B, Valle C, de Lamballerie X, Canard B, Seidah NG, Decroly E (2020) The spike glycoprotein of the new coronavirus 2019-nCoV contains a furin-like cleavage site absent in CoV of the same clade. Antivir Res 176:104742. https://doi.org/10.1016/j.antiviral.2020.104742
27. Glowacka I, Bertram S, Müller MA, Allen P, Soilleux E, Pfefferle S et al (2011) Evidence that TMPRSS2 activates the severe acute respiratory syndrome coronavirus spike protein for membrane fusion and reduces viral control by the humoral immune response. J Virol 85:4122–4134
28. Raj VS, Mou H, Smits SL, Dekkers DHW, Müller MA, Dijkman R et al (2013) Dipeptidyl peptidase 4 is a functional receptor for the emerging human coronavirus-EMC. Nature 495:251–254
29. Peck KM, Scobey T, Swanstrom J, Jensen KL, Burch CL, Baric RS et al (2017) Permissivity of dipeptidyl peptidase 4 orthologs to middle east respiratory syndrome coronavirus is governed by glycosylation and other complex determinants. J Virol. 91:e00534-17
30. Li G, De Clercq E (2020) Therapeutic options for the 2019 novel coronavirus (2019-nCoV). Nat Rev Drug Discov 19:149–150
31. Guo YR, Cao QD, Hong ZS, Tan YY, Chen SD, Jin HJ et al (2020) The origin, transmission and clinical therapies on coronavirus disease 2019 (COVID-19) outbreak—an update on the status. Mil Med Res. https://doi.org/10.1186/s40779-020-00240-0
32. Wu C, Liu Y, Yang Y, Zhang P, Zhong W, Wang Y et al (2020) Analysis of therapeutic targets for SARS-CoV-2 and discovery of potential drugs by computational methods. Acta Pharm Sin B. https://doi.org/10.1016/j.apsb.2020.02.008
33. Lin L, Lu L, Cao W, Li T (2020) Hypothesis for potential pathogenesis of SARS-CoV-2 infection-a review of immune changes in patients with viral pneumonia. Emerg Microbes Infect 9:727–732
34. Wang D, Hu B, Hu C, Zhu F, Liu X, Zhang J et al (2020) Clinical characteristics of 138 hospitalized patients with 2019 novel coronavirus-infected pneumonia in Wuhan, China. JAMA. https://doi.org/10.1001/jama.2020.1585
35. Li M-Y, Li L, Zhang Y, Wang X-S (2020) Expression of the SARS-CoV-2 cell receptor gene ACE2 in a wide variety of human tissues. Infect Dis Poverty 9:45
36. Devaux CA, Rolain J-M, Raoult D (2020) ACE2 receptor polymorphism: susceptibility to SARS-CoV-2, hypertension, multi-organ failure, and COVID-19 disease outcome. J Microbiol Immunol Infect. Available from: http://www.sciencedirect.com/science/article/pii/S1684118220301092
37. Morrell NW, Stenmark KR (2013) The renin-angiotensin system in pulmonary hypertension. Am J Respir Crit Care Med 187:1138–1139
38. Walls AC, Park Y-J, Tortorici MA, Wall A, McGuire AT, Veesler D (2020) Structure, function, and antigenicity of the SARS-CoV-2 spike glycoprotein. Cell. pii: S0092-8674(20)30262-2
39. Jaimes JA, André NM, Chappie JS, Millet JK, Whittaker GR (2020) Phylogenetic analysis and structural modeling of SARS-CoV-2 spike protein reveals an evolutionary distinct and proteolytically sensitive activation loop. J Mol Biol 432:3309–3325

40. Cheng H, Wang Y, Wang G-Q (2020) Organ-protective effect of angiotensin-converting enzyme 2 and its effect on the prognosis of COVID-19. J Med Virol 92:726–730
41. Marrero MB, Schieffer B, Paxton WG, Heerdt L, Berk BC, Delafontaine P et al (1995) Direct stimulation of Jak/STAT pathway by the angiotensin II AT1 receptor. Nature 375:247–250
42. Gheblawi M, Wang K, Viveiros A, Nguyen Q, Zhong J-C, Turner AJ et al (2020) Angiotensin-converting enzyme 2: SARS-CoV-2 receptor and regulator of the renin-angiotensin system. Circ Res 126:1456–1474
43. Perrotta F, Matera MG, Cazzola M, Bianco A (2020) Severe respiratory SARS-CoV2 infection: does ACE2 receptor matter? Respir Med 168:105996
44. Huang C, Wang Y, Li X, Ren L, Zhao J, Hu Y et al (2020) Clinical features of patients infected with 2019 novel coronavirus in Wuhan, China. Lancet 395:497–506
45. de Lang A, Osterhaus ADME, Haagmans BL (2006) Interferon-γ and interleukin-4 downregulate expression of the SARS coronavirus receptor ACE2 in Vero E6 cells. Virology 353:474–481
46. McGonagle D, Sharif K, O'Regan A, Bridgewood C (2020) The role of cytokines including interleukin-6 in COVID-19 induced pneumonia and macrophage activation syndrome-like disease. Autoimmun Rev 19:102537
47. Kuba K, Imai Y, Rao S, Gao H, Guo F, Guan B et al (2005) A crucial role of angiotensin converting enzyme 2 (ACE2) in SARS coronavirus-induced lung injury. Nat Med 11:875–879
48. Teral K, Baddal B, Gülcan HO (2020) Prioritizing potential ACE2 inhibitors in the COVID-19 pandemic: insights from a molecular mechanics-assisted structure-based virtual screening experiment. J Mol Graph Model 100:107697–107697
49. He B, Garmire L (2020) Prediction of repurposed drugs for treating lung injury in COVID-19 [version 2; peer review: 2 approved]. F1000Research 9. Available from: http://openr.es/mbm
50. Colafrancesco S, Scrivo R, Barbati C, Conti F, Priori R (2020) Targeting the immune system for pulmonary inflammation and cardiovascular complications in COVID-19 patients. Front Immunol 11:1439
51. Zeng Y-M, Xu X-L, He X-Q, Tang S-Q, Li Y, Huang Y-Q et al (2020) Comparative effectiveness and safety of ribavirin plus interferon-alpha, lopinavir/ritonavir plus interferon-alpha, and ribavirin plus lopinavir/ritonavir plus interferon-alpha in patients with mild to moderate novel coronavirus disease 2019: study protocol. Chin Med J (Engl) 133:1132–1134
52. Stallmach A, Kortgen A, Gonnert F, Coldewey SM, Reuken P, Bauer M (2020) Infliximab against severe COVID-19-induced cytokine storm syndrome with organ failure—a cautionary case series. Crit Care 24:444
53. DeDiego ML, Nieto-Torres JL, Regla-Nava JA, Jimenez-Guardeño JM, Fernandez-Delgado R, Fett C et al (2014) Inhibition of NF-κB-mediated inflammation in severe acute respiratory syndrome coronavirus-infected mice increases survival. J Virol 88:913
54. Catanzaro M, Fagiani F, Racchi M, Corsini E, Govoni S, Lanni C (2020) Immune response in COVID-19: addressing a pharmacological challenge by targeting pathways triggered by SARS-CoV-2. Signal Transduct Target Ther 5:84–84
55. Liu C-H, Jassey A, Hsu H-Y, Lin L-T (2019) Antiviral activities of silymarin and derivatives. Molecules 24:1552
56. Grimes JM, Grimes KV (2020) p38 MAPK inhibition: a promising therapeutic approach for COVID-19. J Mol Cell Cardiol 144:63–65
57. Vareille M, Kieninger E, Edwards MR, Regamey N (2011) The airway epithelium: soldier in the fight against respiratory viruses. Clin Microbiol Rev 24:210
58. Sabroe I, Parker L, Dower S, Whyte M (2008) The role of TLR activation in inflammation. J Pathol 214:126–135
59. Iwamura C, Nakayama T (2008) Toll-like receptors in the respiratory system: their roles in inflammation. Curr Allergy Asthma Rep 8:7
60. Sallenave J-M, Guillot L (2020) Innate immune signaling and proteolytic pathways in the resolution or exacerbation of SARS-CoV-2 in Covid-19: key therapeutic targets? Front Immunol 11:1229–1229

61. Chen G, Wu D, Guo W, Cao Y, Huang D, Wang H et al (2020) Clinical and immunological features of severe and moderate coronavirus disease 2019. J Clin Invest 130:2620–2629
62. Blanco-Melo D, Nilsson-Payant BE, Liu W-C, Uhl S, Hoagland D, Møller R et al (2020) Imbalanced host response to SARS-CoV-2 drives development of COVID-19. Cell 181:1036–1045.e9
63. Li J-Y, Liao C-H, Wang Q, Tan Y-J, Luo R, Qiu Y et al (2020) The ORF6, ORF8 and nucleocapsid proteins of SARS-CoV-2 inhibit type I interferon signaling pathway. Virus Res 286:198074
64. Felgenhauer U, Schoen A, Gad HH, Hartmann R, Schaubmar AR, Failing K et al (2020) Inhibition of SARS-CoV-2 by type I and type III interferons. J Biol Chem. Available from: http://www.jbc.org/content/early/2020/06/25/jbc.AC120.013788.abstract
65. Lee JS, Shin E-C (2020) The type I interferon response in COVID-19: implications for treatment. Nat Rev Immunol 20:585–586
66. Zhang X, Wu K, Wang D, Yue X, Song D, Zhu Y et al (2007) Nucleocapsid protein of SARS-CoV activates interleukin-6 expression through cellular transcription factor NF-κB. Virology 365:324–335
67. Wang W, Ye L, Ye L, Li B, Gao B, Zeng Y et al (2007) Up-regulation of IL-6 and TNF-α induced by SARS-coronavirus spike protein in murine macrophages via NF-κB pathway. Virus Res 128:1–8
68. Chang Y-J, Liu CY-Y, Chiang B-L, Chao Y-C, Chen C-C (2004) Induction of IL-8 release in lung cells via activator protein-1 by recombinant baculovirus displaying severe acute respiratory syndrome-coronavirus spike proteins: identification of two functional regions. J Immunol 173:7602
69. Pedroza M, Schneider DJ, Karmouty-Quintana H, Coote J, Shaw S, Corrigan R et al (2011) Interleukin-6 contributes to inflammation and remodeling in a model of adenosine mediated lung injury. PLoS One 6:e22667
70. Zhao J, Yu H, Liu Y, Gibson SA, Yan Z, Xu X et al (2016) Protective effect of suppressing STAT3 activity in LPS-induced acute lung injury. Am J Phys Lung Cell Mol Phys 311:L868–L880
71. Goldman JL, Sammani S, Kempf C, Saadat L, Letsiou E, Wang T et al (2014) Pleiotropic effects of interleukin-6 in a "two-hit" murine model of acute respiratory distress syndrome. Pulm Circ 4:280–288
72. Rincon M, Irvin CG (2012) Role of IL-6 in asthma and other inflammatory pulmonary diseases. Int J Biol Sci 8:1281–1290
73. O'Shea JJ, Schwartz DM, Villarino AV, Gadina M, McInnes IB, Laurence A (2015) The JAK-STAT pathway: impact on human disease and therapeutic intervention. Annu Rev Med 66:311–328
74. Wen W, Su W, Tang H, Le W, Zhang X, Zheng Y et al (2020) Immune cell profiling of COVID-19 patients in the recovery stage by single-cell sequencing. Cell Discov 6:31
75. Chua RL, Lukassen S, Trump S, Hennig BP, Wendisch D, Pott F et al (2020) COVID-19 severity correlates with airway epithelium–immune cell interactions identified by single-cell analysis. Nat Biotechnol 38:970–979
76. Neufeldt CJ, Cerikan B, Cortese M, Frankish J, Lee J-Y, Plociennikowska A et al (2020) SARS-CoV-2 infection induces a pro-inflammatory cytokine response through cGAS-STING and NF-κB. bioRxiv. https://doi.org/10.1101/2020.07.21.212639
77. Gao D, Wu J, Wu Y-T, Du F, Aroh C, Yan N et al (2013) Cyclic GMP-AMP synthase is an innate immune sensor of HIV and other retroviruses. Science 341:903
78. Sun L, Wu J, Du F, Chen X, Chen ZJ (2013) Cyclic GMP-AMP synthase is a cytosolic DNA sensor that activates the type I interferon pathway. Science 339:786
79. Ma Z, Damania B (2016) The cGAS-STING defense pathway and its counteraction by viruses. Cell Host Microbe 19:150–158

80. Schoggins JW, MacDuff DA, Imanaka N, Gainey MD, Shrestha B, Eitson JL et al (2014) Pan-viral specificity of IFN-induced genes reveals new roles for cGAS in innate immunity. Nature 505:691–695
81. Ishikawa H, Ma Z, Barber GN (2009) STING regulates intracellular DNA-mediated, type I interferon-dependent innate immunity. Nature 461:788–792
82. Stempel M, Chan B, Juranić Lisnić V, Krmpotić A, Hartung J, Paludan SR et al (2019) The herpesviral antagonist m152 reveals differential activation of STING-dependent IRF and NF-κB signaling and STING's dual role during MCMV infection. EMBO J 38:e100983
83. de Oliveira Mann CC, Orzalli MH, King DS, Kagan JC, Lee ASY, Kranzusch PJ (2019) Modular architecture of the STING C-terminal tail allows interferon and NF-κB signaling adaptation. Cell Rep 27:1165–1175.e5
84. Orzalli M, Smith A, Jurado K, Iwasaki A, Garlick J, Kagan J (2018) An antiviral branch of the IL-1 signaling pathway restricts immune-evasive virus replication. Mol Cell 71:825–840.e6
85. Shi C-S, Nabar NR, Huang N-N, Kehrl JH (2019) SARS-coronavirus open reading frame-8b triggers intracellular stress pathways and activates NLRP3 inflammasomes. Cell Death Discov 5:101
86. Siu K-L, Yuen K-S, Castano-Rodriguez C, Ye Z-W, Yeung M-L, Fung S-Y et al (2019) Severe acute respiratory syndrome coronavirus ORF3a protein activates the NLRP3 inflammasome by promoting TRAF3-dependent ubiquitination of ASC. FASEB J 33:8865–8877
87. Chang Y-S, Ko B-H, Ju J-C, Chang H-H, Huang S-H, Wen C-L (2020) SARS unique domain (SUD) of severe acute respiratory syndrome coronavirus induces NLRP3 inflammasome-dependent CXCL10-mediated pulmonary inflammation. Int J Mol Sci 21:3179
88. Park J-K, Fischer R, Dechend R, Shagdarsuren E, Gapeljuk A, Wellner M et al (2007) p38 Mitogen-activated protein kinase inhibition ameliorates angiotensin II–induced target organ damage. Hypertension 49:481–489
89. Scott AJ, O'Dea KP, O'Callaghan D, Williams L, Dokpesi JO, Tatton L et al (2011) Reactive oxygen species and p38 mitogen-activated protein kinase mediate tumor necrosis factor α-converting enzyme (TACE/ADAM-17) activation in primary human monocytes. J Biol Chem 286:35466–35476
90. Liu Y, Yang Y, Zhang C, Huang F, Wang F, Yuan J et al (2020) Clinical and biochemical indexes from 2019-nCoV infected patients linked to viral loads and lung injury. Sci China Life Sci 63:364–374
91. Renda T, Baraldo S, Pelaia G, Bazzan E, Turato G, Papi A et al (2008) Increased activation of p38 MAPK in COPD. Eur Respir J 31:62
92. Cuenda A, Rousseau S (2007) p38 MAP-kinases pathway regulation, function and role in human diseases. Biochim Biophys Acta Mol Cell Res 1773:1358–1375
93. Shapiro L, Dinarello CA (1995) Osmotic regulation of cytokine synthesis in vitro. Proc Natl Acad Sci U S A 92:12230–12234
94. Beyaert R, Cuenda A, Vanden Berghe W, Plaisance S, Lee JC, Haegeman G et al (1996) The p38/RK mitogen-activated protein kinase pathway regulates interleukin-6 synthesis response to tumor necrosis factor. EMBO J 15:1914–1923
95. Neininger A, Kontoyiannis D, Kotlyarov A, Winzen R, Eckert R, Volk H-D et al (2002) MK2 targets AU-rich elements and regulates biosynthesis of tumor necrosis factor and interleukin-6 independently at different post-transcriptional levels. J Biol Chem 277:3065–3068
96. Villar J, Muros M, Cabrera-Benítez NE, Valladares F, López-Hernández M, Flores C et al (2014) Soluble platelet-endothelial cell adhesion molecule-1, a biomarker of ventilator-induced lung injury. Crit Care 18:R41
97. Zucal C, D'Agostino V, Loffredo R, Mantelli B, Thongon N, Lal P et al (2015) Targeting the multifaceted HuR protein, benefits and caveats. Curr Drug Targets 16:499–515
98. Pullmann R, Rabb H (2014) HuR and other turnover- and translation-regulatory RNA-binding proteins: implications for the kidney. Am J Physiol Renal Physiol 306:F569–F576

99. Feng Y, Fang Z, Liu B, Zheng X (2019) p38MAPK plays a pivotal role in the development of acute respiratory distress syndrome. Clinics (Sao Paulo) 74:e509–e509
100. Drugbank White Paper (2020) COVID-19: finding the right fit identifying potential treatments using a data-driven approach. Information (pdf) collected from internet. Available from: https://blog.drugbankplus.com/data-driven-approaches-to-identify-potential-covid-19-therapies/?li=DB-banner/Covid-2/1&_ga=2.131289871.1097267322.1587024179-257168694.1586191970
101. Information collected from internet. Available from: https://www.ebi.ac.uk/chembl/. Searched latest by 12 Apr
102. Wang M, Cao R, Zhang L, Yang X, Liu J, Xu M et al (2020) Remdesivir and chloroquine effectively inhibit the recently emerged novel coronavirus (2019-nCoV) in vitro. Cell Res 30:269–271
103. Cortegiani A, Ingoglia G, Ippolito M, Giarratano A, Einav S (2020) A systematic review on the efficacy and safety of chloroquine for the treatment of COVID-19. J Crit Care. pii: S0883-9441(20)30390-7. https://doi.org/10.1016/j.jcrc.2020.03.005
104. Ablasser A, Goldeck M, Cavlar T, Deimling T, Witte G, Röhl I et al (2013) cGAS produces a 2′-5′-linked cyclic dinucleotide second messenger that activates STING. Nature 498:380–384
105. Lim J, Jeon S, Shin HY, Kim MJ, Seong YM, Lee WJ et al (2020) Case of the index patient who caused tertiary transmission of COVID-19 infection in Korea: the application of lopinavir/ritonavir for the treatment of COVID-19 infected pneumonia monitored by quantitative RT-PCR. J Korean Med Sci. https://doi.org/10.3346/jkms.2020.35.e79
106. Holshue ML, DeBolt C, Lindquist S, Lofy KH, Wiesman J, Bruce H et al (2020) First case of 2019 novel coronavirus in the United States. N Engl J Med 382:929–936

Special Features of Human Lung ACE2 Sensitivity to SARS-CoV-2 Spike Glycoprotein

27

Dipannita Santra, Amrita Banerjee, and Smarajit Maiti

Abstract

The present review explains some special features of lung tissues in relation to angiotensin (Ang) metabolism and function. The enzymes ACE and ACE2 remain in a balanced state to maintain the normal physiological vaso-status. In some pathological conditions, this balance is hampered, and the immediate effects are pressure deregulations and hypertension. A number of cardiac and renal anomalies are associated with hypertensive disorder. Adrenal functions and corticosteroids are also associated with physiological functions of angiotensin and peripheral salt regulation linking hyper- or hypotension. ACE2 is the most efficient target of SARS-CoV-2, the cause of the present pandemic that claimed millions of life globally. The death from this virus is mostly caused by higher rate of viral propagation with extreme fidelity and Ang-linked multi-organ failure. The special property of lung surface ACE2 and its efficacy in viral propagations have been briefly reviewed in this chapter with some therapeutic strategies related to Ang regulations.

Keywords

SARS CoV-2 · Angiotensin · ACE2 · Comorbid · Hypertensive vascular disorder

D. Santra · A. Banerjee
Cell and Molecular Therapeutics Laboratory, Department of Biochemistry and Biotechnology, Oriental Institute of Science and Technology, Midnapore, India

S. Maiti (✉)
Cell and Molecular Therapeutics Laboratory, Department of Biochemistry and Biotechnology, Oriental Institute of Science and Technology, Midnapore, India

Epidemiology and Human Health Division, Agricure Biotech Research Society, Midnapore, India

K. Dua et al. (eds.), *Targeting Cellular Signalling Pathways in Lung Diseases*,
https://doi.org/10.1007/978-981-33-6827-9_27

27.1 Introduction

In the last week of December 2019, a few cases of pneumonia were reported in Wuhan of China. After a careful analysis, experts of the Centers for Disease Control and Prevention, China declared that the symptoms were caused by novel coronavirus [1]. The World Health Organization (WHO) mentioned the disease as "COVID-19." The International Committee on Taxonomy of Viruses named the virus "severe acute respiratory syndrome coronavirus 2" (SARS-CoV-2). There are 49,664,999 confirmed cases of novel coronavirus disease 2019 (COVID-19) globally, with 1,248,759 deaths and reserved. The total fatality rate is 0.025%. Major thrust was delivered on generating research intelligence to guide evidence-based actions to control the virus as a worldwide response to manage the pandemic condition. Its genetic similarities with the SARS virus (outbreak 2003) and dissimilarities with its virulence and pathogenicity make it unique. Its clinical characteristics were very similar to those of viral pneumonia. The coronaviruses are single-stranded RNA viruses with a diameter of 80–120 nm. It belongs to the *Coronaviridae* family and the *Orthocoronavirinae* subfamily, which is divided into four genera: *Alphacoronavirus*, *Betacoronavirus*, *Gammacoronavirus*, and *Deltacoronavirus*. SARS-CoV, MERS-CoV, and SARS-CoV-2 are all betacoronaviruses, a genus that includes many viruses that infect humans, bats, and other domesticated and wild animals [2]. Bats are considered to be the natural hosts of SARS-CoV-2. Pangolins and snakes are known to be intermediate hosts. No definite intermediate host of SARS-CoV-2 has been suggested till date. A study from Wuhan Institute of Virology showed 96.2% similarity in the gene sequence alignment between SARS-CoV-2 and bat coronavirus [3]. This may suggest that bats are the potential source of SARS-CoV-2. Using macrogenomic sequencing, molecular biological detection, and electron microscopic analysis, Xu et al. [4] showed 99% similarity between SARS-CoV-2 isolated from pangolins and SARS-CoV-2 from infected humans. No studies to date have fully elucidated the potential natural and intermediate host of SARS-CoV-2. Coronaviruses exploit their homotrimeric spike glycoprotein (comprising an S1 subunit and S2 subunit in each monomer) on the envelope to bind to their cellular receptors, ACE2 present in lung cell surface. Such a binding triggers a cascade of events that results in the fusion between cell and viral membranes resulting in entry to lung cell. Previous cryo-electron microscopy studies of the SARS-CoV 2003 spike protein and its interaction with the cell receptor ACE2 have shown that receptor binding induces the dissociation of the S1 with ACE2. This event induces the S2 to transfer from a pre-fusion to a more-stable post-fusion state. This is essential for final membrane fusion [5] upon engagement of ACE2 by a receptor-binding domain (RBD) in S1 [6]. Efficient rearrangements occur that cause S1 shedding and cleavage of S2 by host proteases, and exposure of a fusion peptide adjacent to the S2′ proteolysis site activates the fusion [7]. It has been reported that angiotensin-converting enzyme 2 (ACE2) plays a crucial role in the entry of virus into the cell to cause the final infection and proliferation of the virus [8]. ACE2 protects the lung and heart from acute respiratory distress syndrome (ARDS) and acute myocarditis. Angiotensin II has several adverse effects in the lung and tissues.

SARS-CoV-2 is highly contagious and is able to infect humans not only through the mucous membranes of the nose and mouth and eyes [9]. These results explain the faster transmission capability of SARS-CoV-2 in humans compared with SARS-CoV. In this regard, soluble ACE2 could also be a possible potential candidate for the treatment of COVID-19.

27.2 Angiotensin and Its Physiological Functions

Angiotensin is a protein hormone that causes vasoconstriction and blood vessels to make it narrower. It is the part of the renin-angiotensin system, which helps to maintain blood pressure and fluid balance in the body. Angiotensin I is produced by the action of renin (released from the kidney juxtaglomerular cells) on a protein called angiotensinogen (an α-2-globulin), which is formed by the liver. Angiotensin I (Ang I) is further cleaved to angiotensin II (Ang II) by angiotensin-converting enzyme (ACE). ACE is an integral component of the renin-angiotensin system (RAS). It converts angiotensin I to angiotensin II (ATII). The renin-angiotensin system (RAS) is considered to be the major regulator of blood pressure and electrolyte balance. Several new members of the RAS have been identified so far. These new members include (pro)renin receptor [10], angiotensin (1–12) [11], ACE2 [12], angiotensin (1–7) [13], and the Mas receptor [14]. All the major components of the RAS including renin, angiotensinogen, ACE, ACE2, AT1R, AT2R, and the Mas receptor are expressed in the lung tissue [15]. After angiotensin I is converted to angiotensin II, it has effects on the kidney, adrenal cortex, arterioles, and brain tissues by binding to angiotensin II type I (AT1) and angiotensin II type II (AT2) receptors. AT1 receptor, a G-protein-coupled receptor, exerts its function in the adrenal gland, renal glomeruli and proximal tubules, vascular and cardiac muscles, and brain tissues. The AT2 receptor is coupled with various phosphatases and mediates protein dephosphorylation. These receptors are present in several tissues including the adrenal gland, heart, and brain. AT1 contributes to such pathological conditions by the stimulation of cellular growth via protein phosphorylation, which activates DNA transcription of several cytokines and growth mediators. AT2 functions to promote vascular remodeling and proliferation, mainly via the angiotensin II type I (AT1) receptor. This result increased capillary growth, vascular permeability, and oxidative stress responses. AT1 receptor blockers (ARBs) are non-peptide antagonists for the clinical use purposes in the treatment of hypertension [16]. Two chemical compounds ditrifluoroacetate and trifluoroacetate salt show the pharmacology and functions of AT2 receptor [17, 18]. The ACE cleaves the C-terminal dipeptide end from angiotensin I to generate a potent vasopressor, angiotensin II [19]. It inactivates the vasodilator bradykinin by the eventual removal of two C-terminal dipeptides [20]. The level of Ang II in the renal proximal tubules and interstitial fluid is approximately 10^3 times than the concentration in plasma. In the plasma, angiotensin II sustains with a half-life of 1–2 min. During this period peptidases degrade it into angiotensins III and IV. Angiotensin III stimulates an aldosterone effect of angiotensin II. Angiotensin IV has further decreased the systemic effect. The ultimate effect of angiotensin II is to increase blood pressure, body water, and sodium content.

Angiotensin deficiency can prevent the regulation of blood volume and pressure. High angiotensin levels can also cause cardiac anomalies and heart failure. Angiotensin blockers can help in these situations by blocking the receptor sites.

27.3 Role of Angiotensin-Metabolizing Proteins

The conversion of angiotensin I to angiotensin II is catalyzed by an enzyme called angiotensin-converting enzyme (ACE). ACE is located primarily in the vascular endothelium of the lungs and kidneys in human bodies (Fig. 27.1). Angiotensin-converting enzyme (ACE) is a zinc metallopeptidase [19]. ACE occurs in two isoforms in humans, encoded by a single gene located on chromosome 17 at q23. Both of the ACE isoforms hydrolyze the circulating peptides and catalyze the hydrolysis of substance P, Ang 1–9, *N*-acetyl-seryl-aspartyl-lysyl-proline (Ac-SDKP), cholecystokinin, and other proteins in addition to Ang I. It also affects the vasodilator peptides bradykinin and kallidin [21]. Its further action is also associated with sperm membrane and infertility. ACE has also been implicated during a range of physiological processes unrelated to blood pressure regulation,

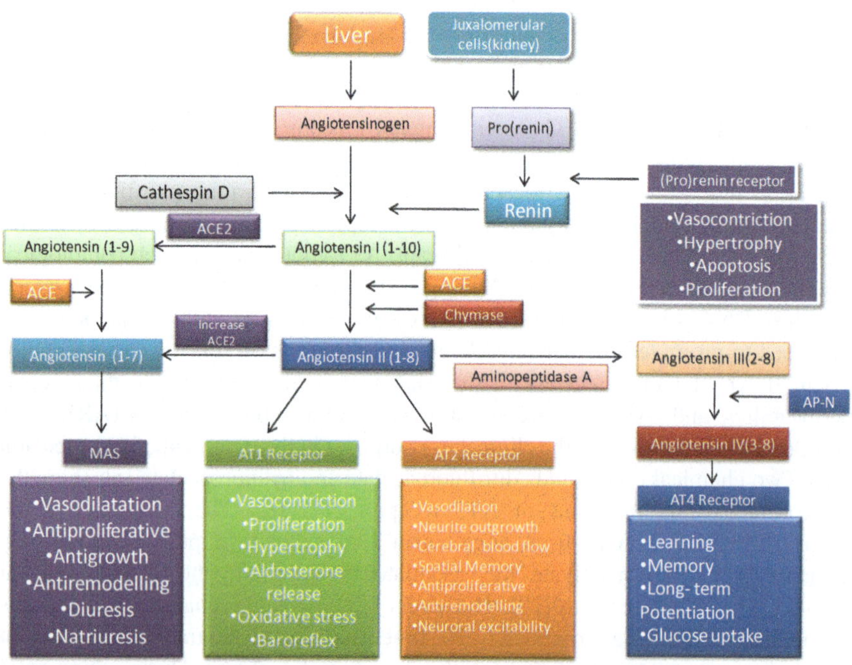

Fig. 27.1 Role of different angiotensin receptors in physiological and pathological functions is shown

like hematopoiesis, immunity, reproduction, and neuropeptide regulation (Fig. 27.1). Angiotensin-converting enzyme inhibitors (ACE inhibitors) are a therapeutic materials that slow down the activity of the enzyme ACE and eventually decrease the assembly of angiotensin II. Consequently, blood vessels enlarge or dilate, and blood pressure is significantly reduced. The medicines act on the renin-angiotensin-aldosterone system, but the ACE inhibitors inhibit the formation of angiotensin II and consequently the downstream effects via AT1 receptor (vasoconstriction, cell growth, sodium and water retention, and sympathetic activation). In contrast, ARBs were screened and designed to replace angiotensin II from the AT1 receptor. ARBs augment vasodilation and natriuresis (Fig. 27.1). However, this ARB-associated autocrine cascade with bradykinin, nitric oxide, and vasoactive prostaglandins is less important than that of ACE inhibitors [22]. Therefore, the potent effect of ACE inhibitors (CEI) on humans and on hypertensive animals is a significant decrease of Ang II formation and an interference with kinin metabolism [23, 24]. ACE inhibitors have been linked with the IgA nephropathy and particularly on its sustained modification. IgA nephropathy is related to the accumulation of extracellular matrix (ECM) in the kidney.

Tipnis et al. [25] cloned a human metalloproteinase for the first time [12]. On the other hand, angiotensin-converting enzyme 2 (ACE2) is a carboxymonopeptidase with a preference for hydrolysis between proline and carboxy-terminal hydrophobic residues in cardiovascular, neuronal, and reproductive organs [12].

27.4 Functions of ACE2 and Its Role in Lung Functions

Both ACE and ACE2 are endothelium-bound carboxypeptidases and work in coordination. The ACE2 protein is highly expressed in the heart and kidney. In the kidney, it is expressed in renal tubular in epithelium, vascular smooth muscle cells of the arteries, and the glomeruli [26]. ACE2 is a type I transmembrane zinc metallocarboxypeptidase with a variable homology to ACE which has been discussed earlier. The importance of the RAS in lung diseases has recently been focused in regard to the identification of ACE2 role in severe acute respiratory syndrome (SARS) coronavirus receptor. There is abundant expression of RAS components within the lung, including ACE and ACE2. In the recent pandemic situation by SARS-CoV-2, it has again highlighted that activation of the intrapulmonary RAS could influence the pathogenesis of lung injury causing the highest rate of morbidity [15]. The systemic ANG II is mainly generated in the lung [27]. Reports suggest that in bleomycin-induced pulmonary fibrosis in rats or mice, ACE inhibitors or AT1 receptor blockers can attenuate epithelial apoptosis, interstitial fibrosis, and collagen deposition [28]. This secreted protein catalyzes the cleavage of angiotensin I into angiotensin 1–9, and angiotensin II into the vasodilator angiotensin 1–7 (Fig. 27.1). The tissue-specific expression of this gene suggests its regulatory role in cardiovascular and renal function. In addition, this protein is a functional receptor for the spike glycoprotein of the earlier emerged human coronavirus HCoV-NL63, SARS-CoV and recently occurring human severe acute respiratory syndrome coronaviruses SARS-CoV-2 (COVID-19 virus). ACE2 functions with preference for hydrolysis between proline and a hydrophobic or basic C-terminal amino acid residue [29].

The primary function of ACE2 is maintaining the balance with angiotensin-converting enzyme (ACE) function. ACE2 successively cleaves the carboxyl-terminal amino acid phenylalanine from angiotensin II (Asp-Arg-Val-Tyr-Ile-His-Pro-Phe) and hydrolyzes it into the vasodilator angiotensin (1–7) (H-Asp-Arg-Val-Tyr-Ile-His-Pro-OH) [30]. ACE2 also cleaves other peptides including [des-Arg9]-bradykinin, apelin, neurotensin, dynorphin A, and ghrelin [30]. Report suggests that ACE2 regulates the membrane trafficking of the neutral amino acid transporter SLC6A19 and has been implicated in Hartnup's disease [31]. The importance of ACE2 in pulmonary physiology was appreciated when this enzyme was identified as a functional receptor for the SARS coronavirus (SARS-CoV) occurring in December, the causative agent of severe acquired respiratory syndrome (SARS) [32]. In vivo experimental studies revealed that lung ACE2 expression is significantly decreased upon SARS-CoV infection, which eventually results in respiratory failure and death [33]. It is suggested that ACE2 expression is reduced in established atherosclerotic plaques [34] and in proatherosclerotic states, such as diabetes [35].

Decreased expression of ACE2 is linked with lung fibrosis in both human and experimental animals [36]. Moreover, lung overexpression of ACE2 in lungs attenuates monocrotaline-induced pulmonary hypertension [37] and bleomycin-induced pulmonary fibrosis [38]. Again, ACE2 protects murine lungs from acute injury [39, 40]. On the contrary, mutant ACE2 mice exhibit enhanced vascular permeability with depressed lung function [41]. Administration of recombinant ACE2 has been shown to prevent the development of lung failure in ACE2 knockout mice [41]. An altered rennin-angiotensin system (RAS) has been implicated as a causative factor in the pathogenesis of pulmonary arterial hypertension (PAH). Activation of the classic ACE-AngII-AT1R axis of the RAS regulates the role of the vasoactive peptide angiotensin II (Ang II), and its receptor AT1R. That adversely affects pulmonary hemodynamics to cause PAH [15]. Besides this, modulation of matrix metalloproteinases (MMP-2 and MMP-9) by ACE2 may also be responsible for impaired tissue remodeling [42]. Furthermore, ACE2 overexpression has been shown to inhibit hypoxia-induced collagen production by cardiac fibroblasts. This suggests that the anti-fibrotic action of ACE2 has a strong physiological implication [43]. In acute respiratory distress syndrome (ARDS), ACE, Ang II, and AT1 receptor function as lung injury-promoting factors, while ACE2 protects the lung from injury as a negative regulator of the RAS [40]. The pathogenesis of pulmonary fibrosis includes endothelial and epithelial cell injury, invasion of inflammatory cells, and production of chemical mediators leading to the cellular proliferation and activation of fibroblasts [44]. So, manipulation of ACE2 could protect against pulmonary fibrosis as a therapeutic target.

27.5 ACE2 Polymorphism and Its Physiological Implications

The Study on Leeds Family by Rice et al. [45] is one of the first studies to explore the link of ACE2 polymorphisms and inheritance of hypertension. The levels of active circulating ACE2 have also been correlated with hypertensive symptoms. Studies have examined the connection between ACE2 genetic variants and disease severity; the risk of calculation developing hypertension in several ethnic populations is accountable. Higher levels of ACE2 expression are suggested to cause severity of coronavirus infection COVID-19 symptoms. A polymorphism of ACE2 gene was earlier documented in the Chinese population with three ACE2 variants (rs4240157, rs4646155, and rs4830542) associated with HT [46]. In a Nicotine Dependence in Teens Canadian cohort, rs2074192, rs233575, and rs2158083 mutations were significantly associated with clinical variations of blood pressure and hypertension [47]. In India, the study of 246 HT patients and 274 normotensive people indicated an association of HT with ACE2 rs21068809 mutation [48].

The renin-angiotensin-aldosterone system (RAAS) pathway can also be regulated by a polymorphism in ACE. In African-American population, hypertension with an ACE polymorphism was reported [49]. SARS-CoV and SARS-CoV-2 spike proteins interact very efficiently with human angiotensin-converting enzyme 2 (ACE2) as their receptor, whereas MERS-CoV spike protein attaches with dipeptidyl peptidase-4 (DPP4) [50]. COVID-19 severity depends on the host transmembrane serine protease-2 (TMPRSS2) for SARS-CoV-2 spike (S) protein priming [51]. During SARS-CoV-2 infection, not only the ACE2 is facilitated by the invasion of SARS virus for rapid replication, but also ACE2 is depleted from the cell membrane. Therefore, the damaging effects of Ang II are enhanced, resulting in the severe damage of lung tissues [52]. ACE2 promotes the entry of CoV into host cells in two different ways. The first one involves ACE2 receptor-mediated clathrin-dependent endocytosis. The RBD of the virus is recognized by the extracellular PD (peptidase domain) of ACE2 mainly through polar residues. When CoV is attached to ACE2, the ACE2 extracellular domain controlling the catalytic effect is cleaved off by specific proteases. As a result, the transmembrane domain is internalized. The other is the assistance of clathrin [53]. The viral particles and host cells fuse. The intracellular structure of ACE2 promotes viral transport from the cell membrane to the cytoplasm. In vitro studies have shown that ADAM17 inhibitors can attenuate virus replication [54].

The second way involves ACE2 receptor-mediated transmembrane serine protease-2 (TMPRSS2)-dependent membrane fusion. When the SARS-S protein binds to ACE2, processing by TMPRSS2 is suggested to permit fusion at the cell surface. Prevalent polymorphisms in *TMPRSS2,* including p.Val160Met (rs12329760), support potential explanations for differential genetic susceptibility to COVID-19. The viral receptor-binding domain (RBD) located in S1 has been narrowed at the lower portion to amino acid residues 318–510 [55]. Similar to SARS-CoV, RBD of SARS-CoV-2 spike protein comprises two subdomains: core and extended loop. Cocrystal structures of SARS-CoV and SARS-CoV-2 spike proteins complexed with ACE2 have shown that the extended loop of RBD directly interacts with loops flanked by α2- and α3-helices. Nevertheless a β-hairpin loop between β3- and β4-sheets of ACE2 is also involved in the process [56]. The strong binding mechanism between spike and the ACE2 has been explored. At the N-terminus of α1, Gln^{498}, Thr^{500}, and Asn^{501} of the RBD form a network of H bonds with Tyr^{41}, Gln^{42}, Lys^{353}, and Arg^{357} from ACE2. At the C-terminus of α1, Gln^{474} of the RBD is H-bonded to Gln^{24} of ACE2. The points presented so far propose that the link, involving ACE2, between the preexistent condition of hypertension and an elevated risk of SARS-CoV-2 infection (and the resultant mortality) is mostly due to the regulation of ACE2 expression. Almost all of these variants of ACE2 are diversely occurring (minor allele frequency R 5%) in one or the other population. Notwithstanding, it raises the possibility of genetic regulation of ACE2 expression being highly widespread.

27.6 Evolutionary Basis of ACE2 and SARS Spike Glycoprotein Binding

Evolution is a dynamic and continuous interplay where pathogens and hosts are interacted. SARS-CoV-2 is the seventh coronavirus known to infect humans. The other six coronaviruses are HCoV-229E, HCoV-NL63, HCoV-OC43, HCoV-HKU1, SARS-CoV, and Middle East respiratory syndrome coronavirus (MERS-CoV) [57]. This is expected that all these have some evolutionary relationship. HCoV-229E and HCoV-NL63 belong to the alphacoronavirus. And the others including SARS-CoV-2 belong to betacoronavirus. SARS-CoV and MERS-CoV are considered highly pathogenic and are known to be transmitted from bats to humans via some intermediate host [58]. Phylogenetic analysis determines the evolutionary relationship and host selection between spike glycoproteins within the human-specific betacoronavirus. To better understand the host selection of betacoronaviruses, the connection of spike glycoprotein between SARS-CoV-2 and other closely related betacoronaviruses has been analyzed (Fig. 27.2). The result showed that bat SARS-like CoV RaTG13 has 96.2% overall genome sequence identity [5], and those are joint neighbor of SARS-CoV-2. The phylogenetic analysis may help the understanding of the highly virulent nature of SARS-CoV-2.

27.7 Clinical Feature of ACE2 and Spike Binding

Current clinical reports and epidemiological data indicate that SARS-CoV-2 is associated with significant morbidity of cardiovascular diseases and complications (Fig. 27.2). These include hypertension (HTN), myocarditis, acute myocardial infarction, and increased heart and kidney failure [59]. In mouse models of acute respiratory distress syndrome, ACE2 knockout mice displayed more severe symptoms, while overexpression of ACE2 had some protective effects [60]. This suggests the ACE2 may have some adaptive role in COVID-19 pathogenesis (Fig. 27.2). In SARS-CoV infection of mice, both viral replication and viral spike

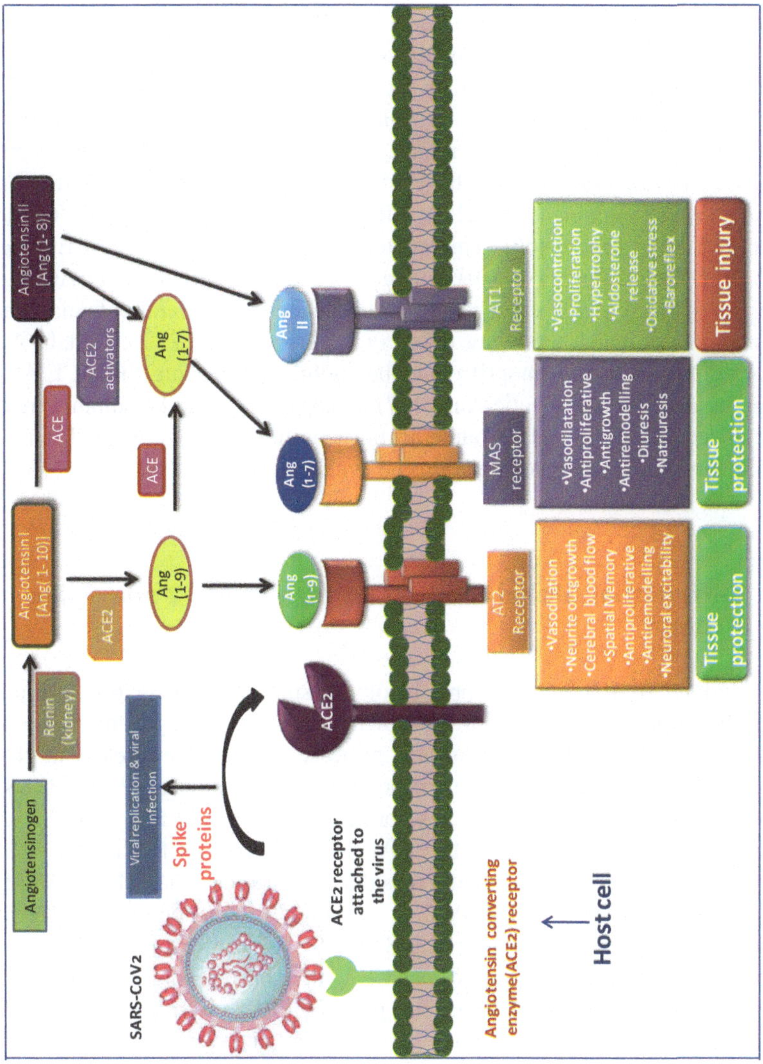

Fig. 27.2 SARS-CoV-2-mediated angiotensin deregulations and abnormal physiological functions

protein separately have been demonstrated to selectively reduce ACE2 but not ACE expression [33]. Nevertheless, SARS-CoV also induces rapid downregulation of ACE2 from the cell surface [61]. It promotes the release of catalytically active ACE2 ectodomains [62]. These results indicate the importance of physiological balance between ACE/ACE2 and Ang II/Ang (1–7), which is likely to be disrupted by SARS-CoV viral infection. In mouse models, recombinant ACE2 administration was shown to inhibit myocardial remodeling and terminate Ang II-induced cardiac hypertrophy [63]. Moreover, renal oxidative stress, inflammation, and fibrosis have also been terminated [64]. Exogenous recombinant ACE2 attenuates acute lung failure in ACE2 knockout also as in wild-type mice. This suggests the cross-interactivity and similarity between mouse and human proteins. Moreover, recombinant ACE2 protein was shown to have a therapeutic potential for the SARS-CoV infection [33]. Recombinant ACE2 is safe in healthy human subjects [65] and patients with lung disease [66]. This strategy has been evaluated in a European phase 2 clinical trials for COVID-19 pathogenesis. Peptide derivatives of ACE2 are also being exploited as cell entry inhibitors [67]. There are several potential therapeutic approaches which are being tested or developed (Fig. 27.2). Certain ACE inhibitors drugs such as lisinopril could be useful in balancing the ACE/ACE2 function. In addition, therapeutic Ang (1–7) heptapeptide could be administered to activate its receptor MAS and to counteract the activities of Ang II. Notwithstanding drugs blocking Ang II receptors can also be evaluated (Fig. 27.2).

27.8 Differential Global Binding Pattern of ACE2 and Spike Protein

Up on binding to ACE2, the S protein of the spike of SARS-CoV-2 requires cleavage by proteases to fuse with the cell membrane and trigger the process to enter the host cell [68]. A number of host proteases are capable of cleaving the S protein of SARS-CoV-2, including transmembrane serine protease-2 (TMPRSS2), cathepsins, and furin [69]. Among these proteases, TMPRSS2 is crucial for viral entry and spread in the host infected by SARS-CoV-2 [69]. ACE2 and TMPRSS2 together confer specificity of host cell types that the virus can enter. TMPRSS2 has been docked against the SARSCoV-2 S protein [70]. An approved TMPRSS2 protease inhibitor is capable of blocking SARS-CoV-2 cell entry [68]. Moreover, variations in TMPRSS2, as opposed to ACE2, have been predicted as associated with disease severity in COVID-19 patients [69]. TMPRSS2 is frequently mutated in prostate cancer, which may contribute to the higher rate of COVID-19 mortality observed in age-matched men compared to women [71]. Both factors likely contribute to the efficiency of virus transmission, making COVID-19 more contagious than infections by SARS-CoV.

27.9 Therapeutic Strategies Against SARS-CoV-2 Via ACE2

Presently, no specific antiviral agent is available against SARS-CoV-2 infection [72]. For studying SARS-CoV-2 pathogenicity, SARS-CoV animal models can be used because SARS-CoV-2 is 80% similar to SARS-CoV. Repurposing of previous drugs was the initial strategy to treat COVID-19. The newest guideline published by the National Health Commission (NHC) recommends IFN-α, lopinavir/ritonavir, ribavirin, chloroquine phosphate, and Arbidol as antiviral therapy [73]. In addition, the heptad repeat 1 (HR1) and heptad repeat 2 (HR2) on SARS-CoV-2 involved viral and cell membrane fusion [74]. Xia et al. reported that HR2-derived peptides (HR2P) and EK1 (a modified OC43-HR2P peptide) exhibited effective fusion-inhibitory activity toward SARS-CoV-2. AP-2-associated protein kinase 1 (AAK1) is a host kinase that regulate a clathrin-mediated endocytosis [75]. This can be therapeutically targeted. A group of approved drugs targeting AAK1 were searched out based on artificial intelligence (AI) technology [76]. Among them, the Janus kinase inhibitor baricitinib, an AAK1-binding drug, was supposed as a initially severed as suitable drug for COVID-19 [76]. Arbidol was shown to inhibit virus entry/fusion of viral membranes with cellular membranes [77]. Chloroquine, a traditional antimalarial drug, was shown to be effective against SARS-CoV-2 infection in vitro [78]. Recently, it is shown that HCQ treatment is significantly applied in COVID-19 patients, and its effect is reinforced by azithromycin [79]. A few combination of medications, all currently included with HCQ are being used in COVID-19 cases [80]. But this drug has severe range of toxicity and is used in systemic lupus erythematosus (SLE) and rheumatoid arthritis (RA) treatment. A lot of antiviral agents have been developed against viral proteases, polymerases, MTases, and entry proteins. A number of clinical trials of antiviral drugs are currently in the testing stage. These drugs are remdesivir (NCT04252664, NCT04257656), favipiravir (ChiCTR2000029600, ChiCTR2000029544), ASC09 (ChiCTR2000029603), lopinavir/ritonavir (ChiCTR2000029387, ChiCTR2000029468, ChiCTR2000029539), and Arbidol (ChiCTR2000029621). Remdesivir is a monophosphoramidate prodrug of an adenosine analog, and it is an inhibitor of viral RNA polymerase. Combined treatment of remdesivir with chloroquine has also been proposed to inhibit SARS-CoV-2. Research in the last several years has been conducted with tea (*Camellia sinensis*) flavonoids for their different therapeutic and disease protective roles. Tea flavonoids have been confirmed as strong antitoxicant, antioxidant, and anti-inflammatory agents. An antitumerigenic role of catechin derivatives especially epigallocatechin gallate (EGCG) has been shown decisively by several laboratories [81–83]. Tea flavonoids have been confirmed as strong antiinfective, antioxidant and anti-inflammatory agents [84]. Some of the traditional molecules as phytochemical sources have been tested and found that nigellidine from *Nigella sativa* L., black cuminseed, can be a great importance in COVID-19 treatment. Importantly SARS-CoV-2 infection ACE2 mediated and impairment with aldosterone system can be repaired by NS. Vasorelaxant and antihypertensive function of NS helps in the modulation of renin-angiotensin system (RAS) or the diuretic activity [85]. Other than nigellidine there are several compounds like thymoquinone groups of drugs and α-hederin, and these compounds are suggested to have important therapeutic activities against different types of pathogenic infections.

27.10 Conclusion

SARS-CoV-2 is highly ubiquitous and more infectious than previously experienced coronaviruses. The case fatality rate varies on geographical locations and ethnicity. The high affinity of SARS-CoV-2 spike (S) protein to lung ACE2 receptor contributes to the higher infectivity of virus and greater fidelity in its propagation. Vaso-dilation and constriction and peripheral vascular status is strictly related to the health of several organs like the heart, kidney, and brain. The physiological vaso-state is completely devastated during SARS-CoV-2 pathogenesis especially in aged individuals and again with comorbid condition. Diabetes, hypertensive disorder, and lung and kidney disorder are the fatal situation in COVID-19 pathogenesis. Asymptomatic contact is very often overlooked, but this causes super-spreading of the disease. Symptomatic and asymptomatic variants are possibly the outcome of interindividual variability in their metabolic and genetic profile. At present there are no effective treatments such as antiviral or passive immunization schemes to control this pandemic. Development of a safe and effective vaccine will take time. Recent studies have revealed some of the potential therapeutic strategies such as ACE2/TMPRSS2 receptor inhibition, immunotherapy, vaccination, application of antiviral drugs/peptides and plasma therapy that inhibit the inflammatory responses and interaction between ACE2 and SARS-CoV-2. Recently, teicoplanin, an antibiotic used to treat *Staphylococcus* infection, has been found to inhibit cathepsin-L in several coronaviruses, including SARS-CoV-2 [86]. It is shown in molecular modeling that nigellidine (from *Nigella sativa*, black cumin) can bind in the active sites of several important proteins of SARS-CoV-2 [87]. Our earlier study also proposed that *epigallocatechin gallate* (EGCG) and *theaflavin gallate* (TDG) from tea are the potent binder to the CoV-2 spike protein [88]. This has been demonstrated in our previous findings on some suitable epitope screening from SARS-CoV-2 spike [89]. Possible suitable epitope as screened in the current study may be helpful in this global pandemic situation.

Conflict of Interests None.

References

1. Huang C, Wang Y, Li X, Ren L, Zhao J, Hu Y (2020) Clinical features of patients infected with 2019 novel coronavirus in Wuhan, China. Lancet 395:497–506
2. International Committee on Taxonomy of Viruses (ICTV) (2018) Virus taxonomy: 2018b release
3. Zhou P, Yang XL, Wang XG, Hu B, Zhang L, Zhang W (2020) A pneumonia outbreak associated with a new coronavirus of probable bat origin. Nature 579(7798):270–273

4. Xu X, Chen P, Wang J, Feng J, Zhou H, Li X (2020) Evolution of the novel coronavirus from the ongoing Wuhan outbreak and modeling of its spike protein for risk of human transmission. Sci China Life Sci 63(3):457–460
5. Gui M et al (2017) Cryo-electron microscopy structures of the SARS-CoV spike glycoprotein reveal a prerequisite conformational state for receptor binding. Cell Res 27:119–129
6. Wong SK, Li W, Moore MJ, Choe H, Farzan M (2004) A 193-amino acid fragment of the SARS coronavirus S protein efficiently binds angiotensin-converting enzyme 2. J Biol Chem 279:3197–3201
7. Madu IG, Roth SL, Belouzard S, Whittaker GR (2009) Characterization of a highly conserved domain within the severe acute respiratory syndrome coronavirus spike protein S2 domain with characteristics of a viral fusion peptide. J Virol 83:7411–7421
8. Lu R, Zhao X, Li J, Niu P, Yang B, Wu H et al (2020) Genomic characterisation and epidemiology of 2019 novel coronavirus: implications for virus origins and receptor binding. Lancet 395:565–574. https://doi.org/10.1016/S0140-6736(20)30251-8
9. Lu CW, Liu XF, Jia ZF (2020) 2019-nCoV transmission through the ocular surface must not be ignored. Lancet 395(10224):e39. https://doi.org/10.1016/S0140-6736(20)30313-5
10. Nguyen G, Delarue F, Burckle C, Bouzhir L, Giller T, Sraer JD (2002) Pivotal role of the renin/prorenin receptor in angiotensin II production and cellular responses to renin. J Clin Invest 109:1417–1427
11. Nagata S, Kato J, Sasaki K, Minamino N, Eto T, Kitamura K (2006) Isolation and identification of proangiotensin-12, a possible component of the renin–angiotensin system. Biochem Biophys Res Commun 350:1026–1031
12. Donoghue M, Hsieh F, Baronas E, Godbout K, Gosselin M, Stagliano N, Donovan M, Woolf B, Robison K, Jeyaseelan R (2000) A novel angiotensin-converting enzyme-related carboxypeptidase (ACE2) converts angiotensin I to angiotensin 1–9. Circ Res 87:E1–E9
13. Schiavone MT, Santos RA, Brosnihan KB, Khosla MC, Ferrario CM (1988) Release of vasopressin from the rat hypothalamo-neurohypophysial system by angiotensin-(1–7) heptapeptide. Proc Natl Acad Sci U S A 85:4095–4098
14. Santos RA, Simoes e Silva AC, Maric C, Silva DM, Machado RP, de Buhr I, Heringer-Walther S, Pinheiro SV, Lopes MT, Bader M (2003) Angiotensin-(1–7) is an endogenous ligand for the G protein-coupled receptor Mas. Proc Natl Acad Sci U S A 100:8258–8263
15. Marshall RP (2003) The pulmonary renin–angiotensin system. Curr Pharm Des 9:715–722
16. Karnik SS, Unal H, Kemp JR, Tirupula KC, Eguchi S et al (2015) International Union of Basic and Clinical Pharmacology. XCIX. Angiotensin receptors: interpreters of pathophysiological angiotensinergic stimuli. Pharmacol Rev 67:754–819
17. Chiu AT, Herblin WF, McCall DE, Ardecky RJ, Carini DJ et al (1989) Identification of angiotensin II receptor subtypes. Biochem Biophys Res Commun 165:196–203
18. Chang RS, Lotti VJ (1990) Two distinct angiotensin II receptor binding sites in rat adrenal revealed by new selective nonpeptide ligands. Mol Pharmacol 37:347–351
19. Skeggs LT Jr, Khan JR, Shumway NP (1956) The preparation and function of the hypertension converting enzyme. J Exp Med 103(3):295–299
20. Yang HY, Erdos EG, Levin Y (1970) A dipeptidyl carboxy peptidase that converts angiotensin I and inactivates bradykinin. Biochem Biophys Acta 214(2):374–376
21. Fuchs S, Frenzel K, Xiao HD, Adams JW, Zhao H, Keshelava G, Teng L, Bernstein KE (2004) Newly recognized physiologic and pathophysiologic actions of the angiotensin-converting enzyme. Curr Hypertens Rep 6:124–128
22. Messerli FH, Weber MA, Brunner HR (1996) Angiotensin II receptor inhibition: a new therapeutic principle. Arch Intern Med 156:1957–1965
23. Erdös EG (1979) Inhibitors of kininases. Fed Proc 38:2774–2777
24. Krieger EM, Salgado HC, Assan CJ et al (1971) Potential screening test for detection of overactivity of renin-angiotensin. Lancet 2:269–275
25. Tipnis SR, Hooper NM, Hyde R, Karran E, Christie G, Turner AJ (2000) A human homolog of angiotensin-converting enzyme. Cloning and functional expression as a captopril-insensitive

carboxypeptidase. J Biol Chem 275(43):33238–33243. https://doi.org/10.1074/jbc.M002615200.
26. Crackower MA, Sarao R, Oudit GY et al (2002) Angiotensin-converting enzyme 2 is an essential regulator of heart function. Nature 417:822–828
27. Skeggs LT, Dorer FE, Levine M, Lentz KE, Kahn JR (1980) The biochemistry of the renin–angiotensin system. Adv Exp Med Biol 130:1–27
28. Ward WF, Sharplin J, Franko AJ, Hinz JM (1989) Radiation-induced pulmonary endothelial dysfunction and hydroxyproline accumulation in four strains of mice. Radiat Res 120 (1):113–120
29. Vickers C, Hales P, Kaushik V et al (2002) Hydrolysis of biological peptides by human angiotensin-converting enzyme-related carboxypeptidase. J Biol Chem 277:14838–14843
30. Turner AJ (2015) Chapter 25: ACE2 cell biology, regulation, and physiological functions. In: Unger T, Ulrike M, Steckelings UM, dos Santos RA (eds) The protective arm of the renin angiotensin system (RAS): functional aspects and therapeutic implications. Academic Press, London, pp 185–189
31. Kuba K, Imai Y, Penninger JM (2013) Multiple functions of angiotensin-converting enzyme 2 and its relevance in cardiovascular diseases. Circ J 77(2):301–308. https://doi.org/10.1253/circj.cj-12-1544
32. Li W, Moore MJ, Vasilieva N, Sui J, Wong SK, Berne MA et al (2003) Angiotensin-converting enzyme 2 is a functional receptor for the SARS coronavirus. Nature 426:450–454
33. Kuba K, Imai Y, Rao S, Gao H, Guo F, Guan B et al (2005) A crucial role of angiotensin converting enzyme 2 (ACE2) in SARS coronavirus-induced lung injury. Nat Med 11:875–879
34. Sluimer JC, Gasc JM, Hamming I et al (2008) Angiotensin-converting enzyme 2 (ACE2) expression and activity in human carotid atherosclerotic lesions. J Pathol 215(3):273–279
35. Tikellis C, Johnston CI, Forbes JM et al (2003) Characterization of renal angiotensin—converting enzyme 2 in diabetic nephropathy. Hypertension 41(3):392–397
36. Li X, Molina-Molina M, Abdul-Hafez A, Uhal V, Xaubet A, Uhal BD (2008) Angiotensin converting enzyme-2 is protective but downregulated in human and experimental lung fibrosis. Am J Physiol Lung Cell Mol Physiol 295:L178–L185
37. Yamazato Y, Ferreira AJ, Hong KH, Sriramula S, Francis J, Yamazato M, Yuan L, Bradford CN, Shenoy V, Oh SP (2009) Prevention of pulmonary hypertension by angiotensin-converting enzyme 2 gene transfer. Hypertension 54:365–371
38. Shenoy V, Ferreira AJ, Qi Y, Fraga-Silva RA, Diez-Freire C, Dooies A, Jun JY, Sriramula S, Mariappan N, Pourang D (2010) The angiotensin-converting enzyme 2/angiogenesis-(1–7)/Mas axis confers cardiopulmonary protection against lung fibrosis and pulmonary hypertension. Am J Respir Crit Care Med 182:1065–1072
39. Imai Y, Kuba K, Penninger JM (2008) The discovery of angiotensin-converting enzyme 2 and its role in acute lung injury in mice. Exp Physiol 93:543–548
40. Imai Y, Kuba K, Penninger JM (2007) Angiotensin-converting enzyme 2 in acute respiratory distress syndrome. Cell Mol Life Sci 64:2006–2012
41. Imai Y, Kuba K, Rao S, Huan Y, Guo F, Guan B, Yang P, Sarao R, Wada T, Leong-Poi H (2005) Angiotensin-converting enzyme 2 protects from severe acute lung failure. Nature 436:112–116
42. Kassiri Z, Zhong J, Guo D, Basu R, Wang X, Liu PP, Scholey JW, Penninger JM, Oudit GY (2009) Loss of angiotensin-converting enzyme 2 accelerates maladaptive left ventricular remodeling in response to myocardial infarction. Circ Heart Fail 2:446–455
43. Grobe JL, Der Sarkissian S, Stewart JM, Meszaros JG, Raizada MK, Katovich MJ (2007) ACE2 overexpression inhibits hypoxia-induced collagen production by cardiac fibroblasts. Clin Sci (Lond) 113:357–364
44. Marshall RP, McAnulty RJ, Laurent GJ (1997) The pathogenesis of pulmonary fibrosis: is there a fibrosis gene? Int J Biochem Cell Biol 29:107–120

45. Rice GI, Jones AL, Grant PJ, Carter AM, Turner AJ, Hooper NM (2006) Circulating activities of angiotensin-converting enzyme, its homolog, angiotensin-converting enzyme 2, and neprilysin in a family study. Hypertension 48:914–920
46. Yi L, Gu YH, Wang XL, An LZ, Xie XD, Shao W et al (2006) Association of ACE, ACE2 and UTS2 polymorphisms with essential hypertension in Han and Dongxiang populations from North-Western China. J Int Med Res 34:272–283
47. Malard L, Kakinami L, O'Loughlin J, Roy-Gagnon MH, Labbe A, Pilote L et al (2013) The association between the angiotensin-converting enzyme-2 gene and blood pressure in a cohort study of adolescents. BMC Med Genet 14:117
48. Patnaik M, Pati P, Swain SN, Mohaprata MK, Dwibedi B, Kar SK et al (2014) Association of angiotensin-converting enzyme and angiotensin-converting enzyme-2 gene polymorphisms with essential hypertension in the population of Odisha, India. Ann Hum Biol 41:143–150
49. Duru K, Farrow S, Wang JM, Lockette W, Kurtz T (1994) Frequency of a deletion polymorphism in the gene for angiotensin con-verting enzyme is increased in African-Americans with hypertension. Am J Hypertens 7(8):759–762
50. Li F (2015) Receptor recognition mechanisms of coronaviruses: a decade of structural studies. J Virol 89(4):1954–1964
51. Hoffmann M, Kleine-Weber H, Schroeder S, Kruger N, Herrler T, Erichsen S, Schiergens TS, Herrler G, Wu NH, Nitsche A et al (2020) SARS-CoV-2 cell entry depends on ACE2 and TMPRSS2 and is blocked by a clinically proven protease inhibitor. Cell 181(2):271–280
52. Wong MKS (2016) Angiotensin converting enzymes, subchapter 29D. In: Takei Y, Ando H, Tsutsui K (eds) Handbook of hormones. Comparative endocrinology for basic and clinical research. Elsevier, Amsterdam, pp 263–265. e29D-1–e29D-4.
53. Wang H, Yang P, Liu K, Guo F, Zhang Y, Zhang G, Jiang C (2008) SARS coronavirus entry into host cells through a novel clathrin- and caveolae-independent endocytic pathway. Cell Res 18(2):290–301. https://doi.org/10.1038/cr.2008.15
54. Haga S, Nagata N, Okamura T, Yamamoto N, Sata T, Yamamoto N, Sasazuki T, Ishizaka Y (2010) TACE antagonists blocking ACE2 shedding caused by the spike protein of SARS-CoV are candidate antiviral compounds. Antivir Res 85(3):551–555. https://doi.org/10.1016/j.antiviral.2009.12.001
55. Babcock GJ, Esshaki DJ, Thomas WD Jr, Ambrosino DM (2004) Amino acids 270 to 510 of the severe acute respiratory syndrome coronavirus spike protein are required for interaction with receptor. J Virol 78:4552–4560
56. Li F, Li W, Farzan M, Harrison SC (2005) Structure of SARS coronavirus spike receptor-binding domain complexed with receptor. Science 309(5742):1864–1868
57. Tang Q, Song Y, Shi M, Cheng Y, Zhang W, Xia XQ (2015) Inferring the hosts of coronavirus using dual statistical models based on nucleotide composition. Sci Rep 5:17155
58. Drosten C, Kellam P, Memish ZA (2014) Evidence for camel-to-human transmission of MERS coronavirus. N Engl J Med 371(14):1359–1360
59. Fried JA, Ramasubbu K, Bhatt R, Topkara VK, Clerkin KJ, Horn E, Rabbani LR, Brodie D, Jain SS, Kirtane AJ, Masoumi A, Takeda K, Kumaraiah D, Burkhoff D, Leon M, Schwartz A, Uriel N, Sayer G (2020) The variety of cardiovascular presentations of COVID-19. Circulation 141:1930–1936. https://doi.org/10.1161/circulationaha.120.047164
60. Imai Y, Kuba K, Rao S, Huan Y, Guo F, Guan B, Yang P, Sarao R, Wada T, Leong-Poi H, Crackower MA, Fukamizu A, Hui CC, Hein L, Uhlig S, Slutsky AS, Jiang C, Penninger JM (2005) Angiotensin-converting enzyme 2 protects from severe acute lung failure. Nature 436 (7047):112–116
61. Glowacka I, Bertram S, Herzog P, Pfefferle S, Steffen I, Muench MO, Simmons G, Hofmann H, Kuri T, Weber F, Eichler J, Drosten C, Pöhlmann S (2010) Differential downregulation of ACE2 by the spike proteins of severe acute respiratory syndrome coronavirus and human coronavirus NL63. J Virol 84:1198–1205
62. Haga S, Yamamoto N, Nakai-Murakami C, Osawa Y, Tokunaga K, Sata T, Yamamoto N, Sasazuki T, Ishizaka Y (2008) Modulation of TNF-alpha-converting enzyme by the spike

protein of SARSCoV and ACE2 induces TNF-alpha production and facilitates viral entry. Proc Natl Acad Sci U S A 105:7809–7814
63. Huentelman MJ et al (2005) Protection from angiotensin II-induced cardiac hypertrophy and fibrosis by systemic lentiviral delivery of ACE2 in rats. Exp Physiol 90:783–790
64. Zhong J et al (2011) Prevention of angiotensin II-mediated renal oxidative stress, inflammation, and fibrosis by angiotensin-converting enzyme 2. Hypertension 57:314–322
65. Haschke M, Schuster M, Poglitsch M, Loibner H, Salzberg M, Bruggisser M, Penninger J, Krähenbühl S (2013) Pharmacokinetics and pharmacodynamics of recombinant human angiotensin-converting enzyme 2 in healthy human subjects. Clin Pharmacokinet 52:783–792. https://doi.org/10.1007/s40262-013-0072-7pmid:23681967
66. Khan A, Benthin C, Zeno B, Albertson TE, Boyd J, Christie JD, Hall R, Poirier G, Ronco JJ, Tidswell M, Hardes K, Powley WM, Wright TJ, Siederer SK, Fairman DA, Lipson DA, Bayliffe AI, Lazaar AL (2017) A pilot clinical trial of recombinant human angiotensin-converting enzyme 2 in acute respiratory distress syndrome. Crit Care 21:234. https://doi.org/10.1186/s13054-017-1823-xpmid:28877748
67. Zhang G et al (2020) Investigation of ACE2 N-terminal fragments binding to SARS-CoV-2 spike RBD. bioRxiv. [Preprint]. 17 Jun 2020. https://doi.org/10.1101/2020.03.19.999318
68. Hoffmann M, Kleine-Weber H, Schroeder S, Kruger N, Herrler T, Erichsen S, Schiergens TS, Herrler G, Wu NH, Nitsche A, Muller MA, Drosten C, Pohlmann S (2020) SARS-CoV-2 cell entry depends on ACE2 and TMPRSS2 and is blocked by a clinically proven protease inhibitor. Cell 181:271–280
69. Lukassen S, Lorenz Chua R, Trefzer T, Kahn NC, Schneider MA, Muley T, Winter H, Meister M, Veith C, Boots AW, Hennig BP, Kreuter M, Conrad C, Eils R (2020) SARS-CoV-2 receptor ACE2 and TMPRSS2 are primarily expressed in bronchial transient secretory cells. EMBO J 39:e105114
70. Asselta R, Paraboschi EM, Mantovani A, Duga S (2020) ACE2 and TMPRSS2 variants and expression as candidates to sex and country differences in COVID-19 severity in Italy. medRxiv. https://doi.org/10.1101/2020.03.30.20047878
71. Stopsack KH, Mucci LA, Antonarakis ES, Nelson PS, Kantoff PW (2020) TMPRSS2 and COVID-19: serendipity or opportunity for intervention? Cancer Discov. https://doi.org/10.1158/2159-8290.CD-20-0451
72. de Wit E, van Doremalen N, Falzarano D et al (2016) SARS and MERS: recent insights into emerging coronaviruses. Nat Rev Microbiol 14(8):523–534
73. National Health Commission of the People's Republic of China. Notice on printing and distributing the diagnosis and treatment plan of pneumonia with new coronavirus infection (trial version 7). http://www.nhc.gov.cn/yzygj/s7653p/202001/f492c9153ea9437bb587ce2ffcbee1fa.shtml. Accessed 4 Mar 2020
74. Xia S, Zhu Y, Liu M, Lan Q, Xu W, Wu Y, Ying T, Liu S, Shi Z, Jiang S, Lu L (2020) Fusion mechanism of 2019-nCoV and fusion inhibitors targeting HR1 domain in spike protein. Cell Mol Immunol. https://doi.org/10.1038/s41423-020-0374-2
75. Neveu G, Ziv-Av A, Barouch-Bentov R, Berkerman E, Mulholland J, Einav S (2015) AP-2-associated protein kinase 1 and cyclin G-associated kinase regulate hepatitis C virus entry and are potential drug targets. J Virol 89:4387–4404
76. Richardson P, Griffin I, Tucker C, Smith D, Oechsle O, Phelan A, Stebbing J (2020) Baricitinib as potential treatment for 2019-nCoV acute respiratory disease. Lancet 395:e30–e31
77. Blaising J, Polyak SJ, Pecheur EI (2014) Arbidol as a broad-spectrum antiviral: an update. Antivir Res 107:84–94
78. Wang M, Cao R, Zhang L, Yang X, Liu J, Xu M, Shi Z, Hu Z, Zhong W, Xiao G (2020) Remdesivir and chloroquine effectively inhibit the recently emerged novel coronavirus (2019-nCoV) in vitro. Cell Res 30:269–271
79. Gautret P, Lagier J-C, Parola P, Hoang VT, Meddeb L, Mailhe M et al (2020) Hydroxychloroquine and azithromycin as a treatment of COVID-19: results of an open-label

non-randomized clinical trial. Int J Antimicrob Agents 56(1):105949. https://doi.org/10.1016/j.ijantimicag.2020.105

80. Wang M, Cao R, Zhang L, Yang X, Liu J et al (2020) Remdesivir and chloroquine effectively inhibit the recently emerged novel coronavirus (2019-nCoV) in vitro. Cell Res 30(3):269–271. https://doi.org/10.1038/s41422-020-0282-0
81. Acharyya N, Ali S-S, Deb B, Chattopadhyay S, Maiti S (2015) Green tea (*Camellia sinensis*) alleviates arsenic-induced damages to DNA and intestinal tissues in rat and in situ intestinal loop by reinforcing antioxidant system. Environ Toxicol 30(9):1033–1044. https://doi.org/10.1002/tox.21977
82. Acharyya N, Chattopadhyay S, Maiti S (2014) Chemoprevention against arsenic-induced mutagenic DNA breakage and apoptotic liver damage in rat via antioxidant and SOD1 upregulation by green tea (*Camellia sinensis*) which recovers broken DNA resulted from arsenic-H2O2 related in vitro oxidant stress. J Environ Sci Health C Environ Carcinog Ecotoxicol Rev 32(4):338–361. https://doi.org/10.1080/10590501.2014.96706
83. Maiti S, Acharyya N, Ghosh T-K, Ali S-S, Manna E, Nazmeen A, Sinha N-K (2017) Green tea (*Camellia sinensis*) protects against arsenic neurotoxicity via antioxidative mechanism and activation of superoxide dismutase activity. Cent Nerv Syst Agents Med Chem 17 (3):187–195. https://doi.org/10.2174/1871524917666170201
84. Li J, Song D, Wang S, Dai Y, Zhou J, Gu J (2020) Antiviral effect of epigallocatechin gallate via impairing porcine circovirus type 2 attachment to host cell receptor. Viruses 12(2):176. https://doi.org/10.3390/v12020176
85. Ajebli M, Eddouks M (2019) Phytotherapy of hypertension: an updated overview. Endocr Metab Immune Disord Drug Targets. [published online ahead of print, 2019 Dec 26]. https://doi.org/10.2174/1871530320666191227104648
86. Baron SA, Devaux C, Colson P, Raoult D, Rolain JM (2020) Teicoplanin: an alternative drug for the treatment of COVID-19? Int J Antimicrob Agents 55:105944. https://doi.org/10.1016/j.ijantimicag.2020.105944
87. Banerjee A, Nazmeen A, Das S, Kanwar M, Maiti S (2020) Active-site molecular docking of Nigellidine with nucleocapsid-NSP2-MPro of COVID-19 and to human IL1R-IL6R and strong antioxidant role of Nigella-sativa in experimental rats. J Drug Target. https://doi.org/10.1080/1061186X.2020.1817040
88. Banerjee A, Santra D, Maiti S Energetics based epitope screening in SARS CoV-2 (COVID 19) spike glycoprotein by immuno-informatic analysis aiming to a suitable vaccine development. bioRxiv. https://doi.org/10.1101/2020.04.02.021725
89. Maiti S, Banerjee A (2020) Epigallocatechin-gallate and theaflavin-gallate interaction in SARS CoV-2 spike-protein central-channel with reference to the hydroxychloroquine interaction: bioinformatics and molecular docking study. Drug Dev Res. [Preprints]. https://doi.org/10.20944/preprints202004.0247.v1

Implications of Phosphoinositide 3-Kinase (PI3K) Signalling in Cellular and Molecular Mechanisms of Respiratory Diseases

28

Biswarup Basu, Sandip Ghosh, Souvik Das, and Amlan Das

Abstract

Phosphoinositide 3-kinases (PI3Ks) are the central modulators of different cellular signalling pathways and play an important role in cell survival, proliferation, growth, metabolism, cell polarity, and activation of various immune cells, to mention a few. PI3Ks are classified into three main categories based on their structure and substrate specificity. In mammals and higher eukaryotes, there exist four isoforms of class I PI3K (PI3Kα, β, γ, δ), three isoforms of class II PI3K (PI3KC2α, C2β, C2γ), and a single class III PI3K. These isoforms have overlapping functions, but the stimulation for specific receptors required for their activation is also different. PI3Ks regulate a plethora of signalling cascades that are involved in pathway members which are involved in various physiological processes such as ROS generation, mast cell activation, neutrophil migration, etc. These events eventually lead to inflammation in various respiratory diseases like chronic obstructive pulmonary disease, asthma, emphysema, cystic fibrosis, etc. PI3K signalling has also been reported to be involved in SARS-CoV-2-mediated pneumonia in the lungs. Nowadays, inhibition of various members of PI3K pathway is the leading approach for the treatment of respiratory diseases.

Biswarup Basu, Sandip Ghosh and Souvik Das contributed equally to this work.

B. Basu · S. Ghosh · S. Das
Department of Neuroendocrinology and Experimental Hematology, Chittaranjan National Cancer Institute, Kolkata, West Bengal, India

A. Das (✉)
National Institute of Biomedical Genomics, Kalyani, Nadia, West Bengal, India
e-mail: ad4@nibmg.ac.in

K. Dua et al. (eds.), *Targeting Cellular Signalling Pathways in Lung Diseases*,
https://doi.org/10.1007/978-981-33-6827-9_28

Keywords

PI3K · Akt · mTOR · Lung disorders · Asthma · Inflammation

28.1 Introduction

Impairment of the normal lung function due to genetic mutations, infection, inflammation, lifestyle, and exposures to external agents such as cigarette smoke or pollutants can lead to severe respiratory diseases. Chronic inflammation is found to be associated with several respiratory diseases such as chronic obstructive pulmonary disease (COPD), asthma, cystic fibrosis, and idiopathic pulmonary fibrosis, to name a few [1]. The limited success of glucocorticoid-based treatment in these diseases, coupled with the lack of potent anti-inflammatory medications, had initiated the search for new therapeutic targets, including a plethora of protein or lipid kinases which may trigger the inflammatory responses leading to lung disorders [2]. The phosphoinositide 3-kinase (PI3K) has been identified as the central player of several respiratory disorders by its ability to regulate inflammatory responses in the lung via different downstream signalling proteins [3]. Several studies have shown that selective targeting of PI3K signalling may be a potential therapeutic strategy for controlling respiratory disorders [1]. PI3Ks belong to a family of intracellular signalling kinases that participate in orchestrating a plethora of vital cellular functions such as growth and differentiation, proliferation, cell cycle progression, migration, adhesion, and metabolism, to mention a few [4].

28.2 Historical Overview of PI3 Kinase

In the mid-1980s, a tyrosine-phosphorylated 85 kD protein, associated with unknown lipid kinase activity, was first reported in platelet-derived growth factor (PDGF)-stimulated cells or oncogenically transformed cells (such as the Tyr kinase SRC or polyomavirus middle T antigen-transformed cells). Researchers also identified a phosphorylated lipid in GPCR-stimulated neutrophils, which was found to be PI-3,4,5-P3 (phosphatidylinositol (3,4,5)-trisphosphate (PIP_3)) [5]. The same lipid PIP_3 was also found in neutrophils upon treatment with tyrosine kinase agonists. Phosphorylated forms of the phosphatidyl glyceride class of lipid, phosphatidylinositol (PI), are called phosphoinositides. The phosphoinositides are known to play crucial roles in diverse signalling pathways via phosphorylation by lipid kinases. The inositol ring of PI can be phosphorylated on the three, four, and five hydroxyl (–OH) groups to form seven different types of phosphoinositide species by a variety of kinases. Whitman et al. demonstrated that the enzymatic activity associated with oncoproteins (polyoma middle T antigen) could phosphorylate the 3′-OH group in the inositol ring of PI to form phosphatidylinositol-3-phosphate (PI-3-P). A year later, Auger et al. revealed that in PDGF-stimulated smooth muscle cells, this enzyme could produce phosphatidylinositol-3,4-bisphosphate (PI-3,4-P2) and phosphatidylinositol 3,4,5-trisphosphate (PI-3,4,5-P3) [6]. These discoveries led to the finding that the bioactive product of phosphoinositide 3-kinase (PI3K) is important for cellular responses to growth factors and also for malignant transformation. Further, Lewis Cantley, an American cell biologist, and his group made the surprising discovery that this oncoprotein-associated kinase is actually a "*PI-3-kinase*" which phosphorylates the 3-OH group of the inositol ring of phosphoinositides.

28.3 Classification and Overview of PI3K Signalling Cascades

28.3.1 Classification

PI3Ks exist as heterodimers composed of a regulatory subunit and a catalytic subunit and participate in the activation of receptor-mediated signalling cascades via phosphorylation of phosphoinositol class of lipids. The phosphorylated lipids, in turn, act as the secondary messengers and regulate a wide array of downstream signalling pathways [4]. On the basis of the structural organization, substrate-binding specificity, and downstream signalling ability, the PI3K family of kinases may be classified into three subtypes [classes I, II, and III]. The class I PI3K may be further subgrouped into class IA and IB based on their associated catalytic subunits and regulatory subunits. Class IA PI3K is activated by membrane-expressed receptor tyrosine kinases (RTKs) such as growth factors, cytokines, and insulin, while class IB PI3K is activated by G-protein-coupled receptors (GPCRs).

The class II PI3Ks consist of three isoforms (PI3KC2α, PI3KC2β, and PI3KC2γ), which are widely expressed in vertebrates. This family of PI3K can phosphorylate phosphoinositol and phosphoinositol 4-monophosphate in vitro to form PI-3-P and PI-3,4-P2, respectively. PI3KC2α and PI3KC2β isoforms consist of an N-terminal clathrin-binding (CB) region and are known to regulate clathrin-mediated endocytosis. Class II PI3Ks are known to be activated by various activators and growth factors, including chemokines like CC chemokine ligand 2 (CCL2), lysophosphatidic acid (LPA), cytokines such as leptin and tumor necrosis factor a (TNFa), insulin, and the growth factors such as epidermal growth factor (EGF), platelet-derived growth factor (PDGF), etc. All three types of PI3KC2 subclass possess a unique C-terminal region that contains a C2 domain and a PX domain. The PX domain helps in the binding of PI3KC2 with phosphatidylinositol 4,5-bisphosphate [PI-4,5-P_2].

The class III PI3K, also known as vacuolar protein sorting 34 (Vps34), is conserved in all eukaryotes. Vps34 binds to an adaptor protein Vps15 that regulates the intracellular membrane localization and activity of Vps34. Vps15 also recruits essential signalling proteins, such as Rab5 GTPase, to coordinate with Vps34 at endosomes. Class III PI3Ks are also known to regulate Toll-like receptor (TLR) signalling. Detailed nomenclature, classification, composition, activation, etc. of PI3Ks are documented in Table 28.1.

28.3.2 Downstream Signal Transduction Pathways of PI3 Kinase

Three classes of PI3K have different substrate specificities and downstream signalling events given in detail below. The signalling cascades of different classes of PI3Ks have been schematically represented in Fig. 28.1.

28.3.2.1 Class I PI3K

Upon binding to extracellular signals such as growth factors to the cell surface receptors such as RTKs, RAS, or G-protein-coupled receptor (GPCR), activation of the class I PI3K pathway is initiated. PI3K class IA (p110α, p110β, and p110δ) is activated via RTK activation, while the catalytic p110γ subunit of class IB PI3K gets activated by Gβγ subunit of GPCR and RAB5. In resting condition, the P110 catalytic subunit of PI3K IA exists as a heterodimer with p85 regulatory subunit and remains inactive by the inhibitory action exerted by the p85 subunit [3, 5, 8].

Three different pathways of activation of PI3K can be found through RTK activation. The first way of activation involves the direct binding of p85 to the activated RTK intercellular phospho-YXXM domain (Y, tyrosine; X, any amino acid; M, methionine) through the SH2 domain. p110 is now free from inhibition and converts its substrate phosphatidylinositol 4,5-bisphosphate (PI-4,5-P_2; also known as PIP_2) to phosphatidylinositol 3,4,5-trisphosphate (PI-3,4,5-P_3; also known as PIP_3). Another PI3K activation pathway depends on an adaptor protein growth factor receptor-bound protein 2 (GRB2), which binds to RTK YXN motif. Scaffolding protein Grb2-associated binder (GAB) then binds GRB2. Finally, GAB interacts with p85. The third way of PI3K activation is the interaction of GRB2 with SOS. SOS in turn activates RAS. RAS further activates p110 independently of p85 [7, 8].

As mentioned earlier, the activated PI3K converts PI-4,5-P_2 to PI-3,4,5-P3 for triggering the downstream signalling. Signalling proteins possessing the pleckstrin homology (PH) domains can bind to PI-3,4,5-P3 and get accumulated at sites of PI3K. Important signalling mediators having the PH domain include serine-threonine kinases AKT or protein kinase B (PKB) and phosphoinositide-dependent kinase-1 (PDK1). Association with the membrane-bound PI-3,4,5-P3 recruits these proteins into proximity and results in the phosphorylation of Akt by PDK1. PDK1 phosphorylates AKT at T308 and results in partial activation, while the mammalian target of rapamycin complex 2 (mTORc2) then further phosphorylates Akt at S473, resulting in its complete activation [8]. Activated AKT can phosphorylate up to 100 substrates, thereby modulating a variety of cellular functions, including cellular growth and differentiation, death, and apoptosis, migration, epithelial-to-mesenchymal transition, and glucose homeostasis to name a few. Hence, the PI3K/AKT axis is known as the canonical PI3K signalling.

Interestingly phosphorylation of most of the downstream target protein targets by Akt resulted in an inhibitory effect. For example, phosphorylation of the

Table 28.1 Different classes of PI3Ks

Types of PI3K	Subtypes	Catalytic subunit	Genes	Regulatory subunit/ adapter protein	Genes	Activated by	Product	Function	Expression	References
Class I PI3K	Class IA	p110α, p110β and p110δ	*PIK3CA*, *PIK3CB*, *PIK3CD* genes, respectively.	p85α, p85β, p55γ, p55α, p50α	PIK3R1, PIK3R2, PIK3R3	RTK, cytokine receptor, TLR, TCR, BCR, small GTPases (like Ras), GPCR, Rab5	Phosphorylate the phosphatidylinositol 4,5-bisphosphate to produce phosphatidylinositol 3,4,5-trisphosphate	Angiogenesis, metabolism, growth, proliferation, survival, protein synthesis, transcription, and apoptosis	Ubiquitous for p110α, p110β; leukocyte for p110δ	[7–10]
	Class IB	p110γ	*PIK3CG*	p101 or p84/p87	PIK3R5, PIK3R6	GPCR			Leukocyte, endothelium and heart	
Class II PI3K		PI3K-C2α	*PIK3C2A*	Clathrin/no regulator		CCL2, TNFα, LPA, insulin, EFG, PDGF, SCF	Phosphatidylinositol 4,5-bisphosphate and phosphatidylinositol 3-phosphate	PI3K-C2α specific GLUT 4 translocation to PLM, clathrin-mediated endocytosis; angiogenesis, vesicular trafficking, mitosis	Ubiquitous	
		PI3KC2β	*PIK3C2B*					Cell migration, HL 60 differentiation, stabilization of nucleotide-free RAS, mTOR signalling repression	Ubiquitous	

(continued)

Table 28.1 (continued)

Types of PI3K	Subtypes	Catalytic subunit	Genes	Regulatory subunit/ adapter protein	Genes	Activated by	Product	Function	Expression	References
		PI3KC2γ	*PIK3C2G*					Akt2 activation, glycogen storage in the liver.	Liver	
Class III PI3K		Vps34		P150 (or VSP 15)			Formation of phosphatidylinositol 3-phosphate from phosphatidylinositol	TLR signalling, generation of PtdIns3P (major protein of vesicular trafficking), autophagy, mTOR signalling regulation by crosstalking with PI3K I, autophagy		

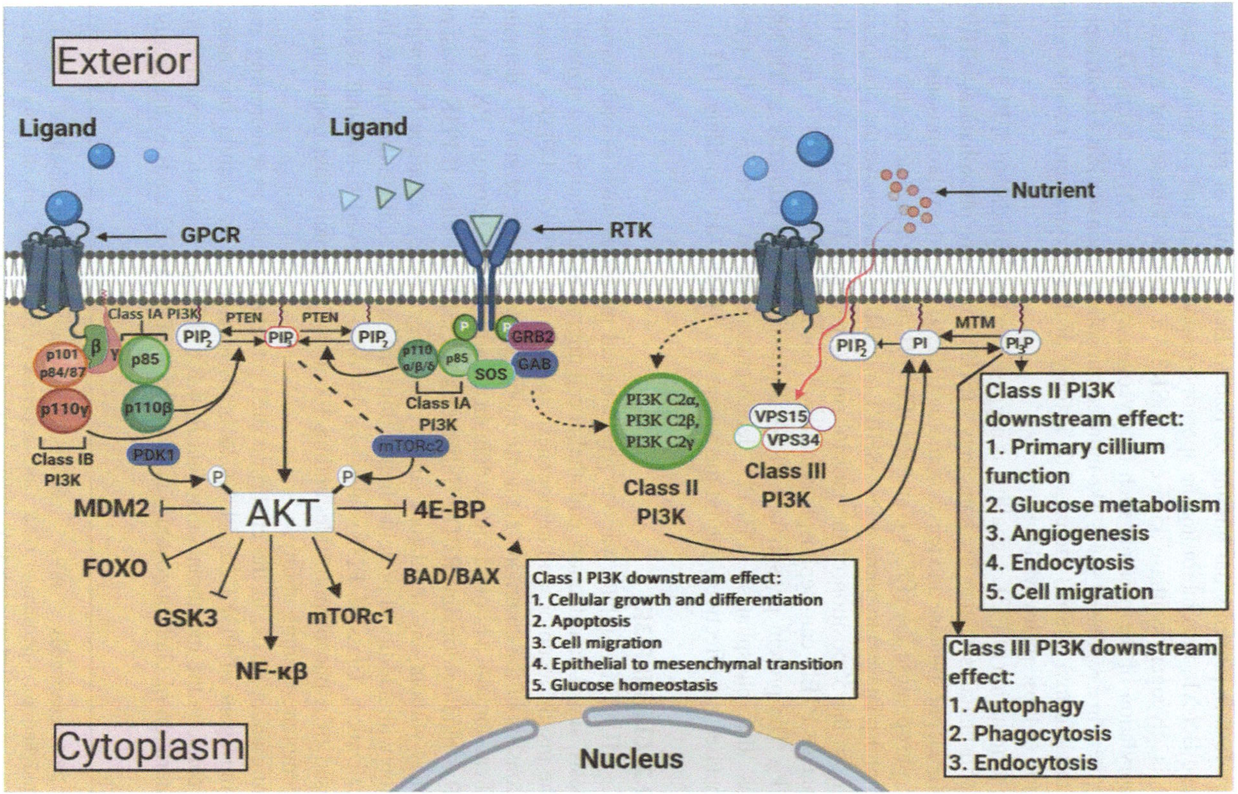

Fig. 28.1 Scheme of PI3K cellular signalling cascades and type-specific downstream cellular effects. *PI3K* phosphoinositide 3-kinase, *GPCR* G-protein-coupled receptor, *RTK* receptor tyrosine kinase, *PTEN* phosphatase and tensin homolog, *PIs* phosphoinositides, *PIP* PI 4-phosphate, PIP_2 PI 4,5-bisphosphate,

Forkhead-related transcription factor 1 (FKHR-L1) by Akt results inhibits the transcription of genes, which are stimulated by FKHR-L1. Similarly, phosphorylation of pro-apoptotic protein Bad by Akt creates a binding site for 14-3-3. This complex further prevents the binding of Bad to Bcl-2 family of proteins such as Bcl-2 and Bcl-XL, thus inhibiting the apoptotic response. Another prominent target of Akt is glycogen synthase kinase 3 (GSK3), a protein kinase that mediates phosphorylation of the proteins such as glycogen synthase, c-Myc, cyclin D, etc. to keep them in a repressed state. Akt-induced phosphorylation of GSK3 (both alpha and beta isoforms) inhibits the catalytic activity of this enzyme and results in the activation of those pathways [4, 9].

The termination of PI3K signalling is triggered by the degradation of PI-3,4,5-P3 by the phosphatases such as Src homology 2 (SH2)-containing phosphatases (SHIP1 and SHIP2), phosphatase and tensin homolog (PTEN) deleted on chromosome TEN, protein phosphatase 2A (PP2A), and PHLPP (PH domain and leucine-rich repeat protein phosphatases) to name a few. SHIP1 and SHIP2 can remove the phosphate group from the 5′-position of the inositol ring to generate PI-3,4-P2, whereas the phosphatase PTEN can dephosphorylate at 3′-position of PI-3,4,5-P3 to produce PI-4,5-P_2. The protein phosphatase PP2A dephosphorylates primarily at Thr308, whereas the phosphatase PHLPP dephosphorylates at Ser473. All these phosphatases can effectively inhibit the activation of PI3K/Akt signalling [11].

28.3.2.2 Class II PI3K

The class II PI3K regulates various physiological processes such as endocytosis, exocytosis, signal transduction, endothelial cell migration, activation and signalling from the insulin receptor, and angiogenesis to name a few. Like all PI3Ks, class II PI3K consists of the common "signature motif" containing the central C2 domain, one helical domain, a bilobed kinase domain, an additional C2 domain, PX domain, and a Ras-binding domain (RBD). Unlike class I PI3Ks, the class II PI3Ks exist as monomers without any regulatory subunit. But they possess a unique N-terminal extension with additional protein-binding domains, which facilitate their intracellular localization. Class II PI3Ks also consist of a conserved C-terminal region composed of a phosphoinositide-binding PX and C2 domains that contribute to their membrane-binding ability.

During endocytosis, in the clathrin-coated pit, clathrin acts as a recruiting and activating agent of PI3KC2α. PI3KC2α binds to clathrin by clathrin box motif, which converts the membrane-abundant PI-4-P to PI-3,4-P2 which is required for the maturation of clathrin-coated pits. Formation of PI-3,4-P2 leads to the recruit-

Fig. 28.1 (continued) PIP_3 phosphatidylinositol (3,4,5)-trisphosphate, *PI-3-P* phosphatidylinositol-3-phosphate, *GAB* Grb2-associated binder, *GRB2* growth factor receptor-bound protein 2, *MTM* myotubularin, *VPS* vacuolar protein sorting, *PDK1* phosphoinositide-dependent kinase-1, *MDM2* mouse double minute 2 homolog, *FOXO* Forkhead box protein O, *GSK* glycogen synthase kinase, *NF-kB* nuclear factor kappa-light-chain-enhancer of activated B cells, *mTOR* mammalian target of rapamycin, *4E-BP* 4E-binding protein

ment of the curvature inducing PX–BAR domain protein sorting nexin 9 (SNX9), which facilitates the internalization of the clathrin-coated pits via membrane invagination. This leads to the dynamin-mediated membrane scission and vesicle release, thus completing the endocytosis. PI-3,4-P2 is then rapidly converted to PI-3-P by type I inositol-3,4-bisphosphate 4-phosphatase (INPP4), resulting in the transition of clathrin-coated pits into internalized early endosomes. The PI-3,4-P2 generated by PI3KC2α activates the enzyme Rab1 at the base of the primary cilium-involved in sensing extracellular stimuli and transducing intracellular signalling.

Formation of PI-3,4-P2 by PI3KC2α at the membrane and intracellular vesicles triggers various intracellular signal transduction cascades. For example, the catalytic activity of PI3KC2α is related to increased PI-3,4-P2 levels, which facilitate the internalization of TGF-β receptor into SARA (Smad anchor for receptor activation) containing early endosomes. The endocytosis of the transforming growth factor beta (TGF-β) receptor leads to Smad2/3 phosphorylation, which in turn activates the angiogenic signalling pathways. Also, clustering of PI3KC2α takes place around the insulin-bound receptor leading to the local production of PI-3,4-P2, and subsequent activation of Akt1, which recruits the glucose transporter type 4 (GLUT4) to the plasma membrane and glucose uptake, is initiated. PI3KC2α is also responsible for the SNAP25-dependent fusion of exocytic vesicles with the membrane, regulating the exocytosis of both neurotransmitters and insulin granules in pancreatic B cells. Thus, PI3KC2α plays an important role in endocytosis, exocytosis, and intracellular trafficking.

PI3KC2β was initially identified as an activator of Ca^{2+} flux while later also found to be the downstream modulator of growth factors, such as EGF and stem cell factor (SCF). PI3KC2β is known to play important roles in IgE-dependent allergy, cell migration, and metabolic regulation. PI3KC2β was also found to be localized on lysosomes and late endosomes under nutrient deprivation or stress conditions. PI3KC2β is recruited in lysosomal compartments by PI-3,4-P2, where it forms a complex with mTORC1 and acts as the repressor. PI3KC2β is known to promote the gating of the Ca^{2+}-activated K^+ channel. It plays a vital role in IgE-mediated Ca^{2+} entry and subsequent degranulation of the mast cells.

PI3KC2γ is the least characterized among all three class II PI3Ks but known to play a crucial role in vesicular trafficking and glucose homeostasis by sustaining the Akt signalling. After binding to an agonist, the activated insulin receptor gets internalized into Rab5-positive endosomes, from which it may be either recycled to the membrane or degraded in lysosomes. In the early endosome, the endosomal signalling protein anaplastic large cell lymphoma (APPL1) facilitates Akt recruitment and activation. But the rapid degradation of PI3KC2β-produced PI-3,4,5-P3 into PI-3-P induces the release of APPL1 from endosomes, and the downstream signalling cascade is terminated. Interestingly, upon insulin stimulation, PI3KC2γ gets recruited by Rab5-GTP in early endosomes and promotes P-3,4-P2 accumulation. As a result, the release of APPL1 is inhibited, leading to long-term Akt activation. PI3KC2γ-mediated signalling has significant implications in several respiratory and inflammatory disorders [12].

28.3.2.3 Class III PI3K

As a lipid kinase, Vps34 generates PtdIns3P, and the generation of PtdIns3P is an important early event of autophagy. PtdIns3P pool is finely regulated inside the cell by phosphatases called myotubularin (MTM), which dephosphorylates PI-3-P to PIs [13].

PI3K class III can form two different complexes containing the catalytic subunit Vps34, protein kinase Vps15, and Beclin1 (also known as Atg6) as a common core part. Complex I contains Atg14/ATG14L along with core complex, and directs the complex I to phagophore initiation sites. The localization of this complex I to ER is important for the formation of autophagosome, which is controlled by the protein kinase ULK1. ULK1 phosphorylates several sites of beclin1 protein, hence activating the complex I promoting autophagy. When UVRAG (UV radiation resistance-associated gene) associates with the common core complex, it is known as complex II, directing endosome and autophagosome maturation, autophagosome lysosome fusion, and endocytic trafficking [14].

Vps34 also participates in endocytosis when forming a complex with Vps15, UVRAG, and Beclin1. Vps34 enters the early endosomes through the interaction between Vps15 and Rab5. In the early endosome membrane, Vps34 generates PI-3-P leading to the recruitment of effector proteins such as EEA1, Hrs (also known as HGS), rabankyrin-5, and rabenosyn-5. These effector proteins can regulate the docking and fusion of cargo-containing vesicles from the plasma membrane and also participate in sorting of the enclosed cargo, which can be recycled back to the membrane or delivered to the lysosome for further degradation [15].

During stress and glucose starvation, protein kinase MAPKAP2 (mitogen-activated protein kinase-activated protein kinase 2) and AMPK phosphorylates different sites of beclin1 to induce autophagy. DAPK (death-associated protein kinase) also promotes autophagy through beclin1 phosphorylation. Phosphorylation of specific tyrosine residues in beclin1 by epidermal growth factor receptor (EGFR) upon binding to growth factors, e.g., EGF (epidermal growth factor), and phosphorylation of several serine residues of ATG14L by mTORC1 lead to suppression of autophagy [14].

28.4 Molecular Mechanism of Different Respiratory Diseases Involving PI3K Pathway

The complex interplay between external factors like allergens, environmental pollution, etc. and internal factors like oxidative stress, hypoxia, autophagy, and immune regulation contributes to the development of pulmonary diseases where PI3K signalling has a pivotal role. The initiation of autophagy involves the activity of the PI3K-III/Beclin1 complex, whereas stabilization of HIF-1α under hypoxia is dependent on the PI3K/Akt pathway. ROS can activate autophagy by inhibiting the PI3K/Akt/mTOR pathway or by activating AMPK to inhibit the mTOR signalling pathway. Inactivation of PTEN can also increase mTORC1 activity via the PI3K/Akt pathway, thus suppressing autophagy [16]. Dysregulation of one or more of these pathways may result in different lung pathophysiologies as discussed below (Fig. 28.2).

28.4.1 Asthma

Asthma is one of the important chronic, heterogeneous diseases of the lung characterized mainly by denudation of the epithelium due to mucosal inflammation and epithelial shedding, increased submucosal mucus gland area in large airways, wheezing, coughing, and shortness of breath. Different types of lung cells, including epithelial cells and smooth muscle cells, communicate against inhaled stimuli (allergens, viruses, air pollutants, and proteases), which activate pattern recognition receptors (PRRs) expressed at airway epithelial cells. Increased collagens 1, 3, and 5, fibronectin, tenascin, versican, collagen I, and hyaluronan and decreased elastin are the critical attributes of a different layer of airway muscle. An increase in smooth muscle mass, angiogenesis and vascular remodelling, and thickening of the airway wall are the other significant symptoms that occur in individuals who have asthma [17]. Immediate hypersensitivity response, genetic and environmental factors, immunological factors, behavioral factors such as active or passive smoking, age, and sex also have a close influence on the severity of asthma. There may be a close association of PI3Kδ and PRR (pattern recognition receptor) signalling pathway. Activation of PI3Kδ results in the increase of oxidative stress by mtROS (mitochondrial reactive oxygen species) in the airway epithelium. PI3Kδ activation in the airway epithelium can regulate NLRP3 (nucleotide-binding domain and leucine-rich repeat-containing protein 3) inflammasome and ER (endoplasmic reticulum) stress. Activation of PI3Kδ in airway epithelium possibly helps in the release of cytokine (TSLP) and interleukins (IL-25, IL-33) from airway epithelium that acts on dendritic cell migration, mast cell, and ILC2 (type 2 innate lymphoid) cells to recruit both innate and adaptive hematopoietic cells and initiate the release of T helper cell 2 (Th2) cytokines (IL-4, IL-5, and IL-13). PI3Ks are activated in naïve $CD4^+$ T cell upon contact with antigen-presenting cells like dendritic cell (DC) [18]. PI3Kδ has an additional role in the activation of Th2 lymphocytes by C–C chemokine ligand (CCL)17 and CCL22 secreted from DCs [74]. Cytokines secreted from DCs attract Th2 cells via C–C chemokine receptor (CCR) type 4. IL-4 stimulates B cells that secrete IgE into circulation. IgE binds with FcεR1 surface receptors on mast cells. Binding of allergens to IgE helps in the secretion of histamine and prostaglandin D2 (PGD_2) that evokes bronchoconstriction. Mast cells also attract Th2 cells and eosinophils through the release of PGD_2 through CRTh2 (chemotactic receptors of Th2 cells). IL-5 and IL-13 production from both Th2 lymphocyte and ILC2 is necessary for the induction of type 2 eosinophilic inflammation that is crucial for asthma pathogenesis and relatively resistant to corticosteroids. IL-13 is an inducer of mucus hypersecretion [19]. The binding of IL-13 to the smooth muscle cell surface receptor results in the production of PIP_3 by PI3K. PDK1 activation by PIP_3 phosphorylates Akt that helps in RhoA-dependent smooth muscle cell contraction. Mitogens like epidermal growth factor (EGF), insulin-like growth factors (IGFs),

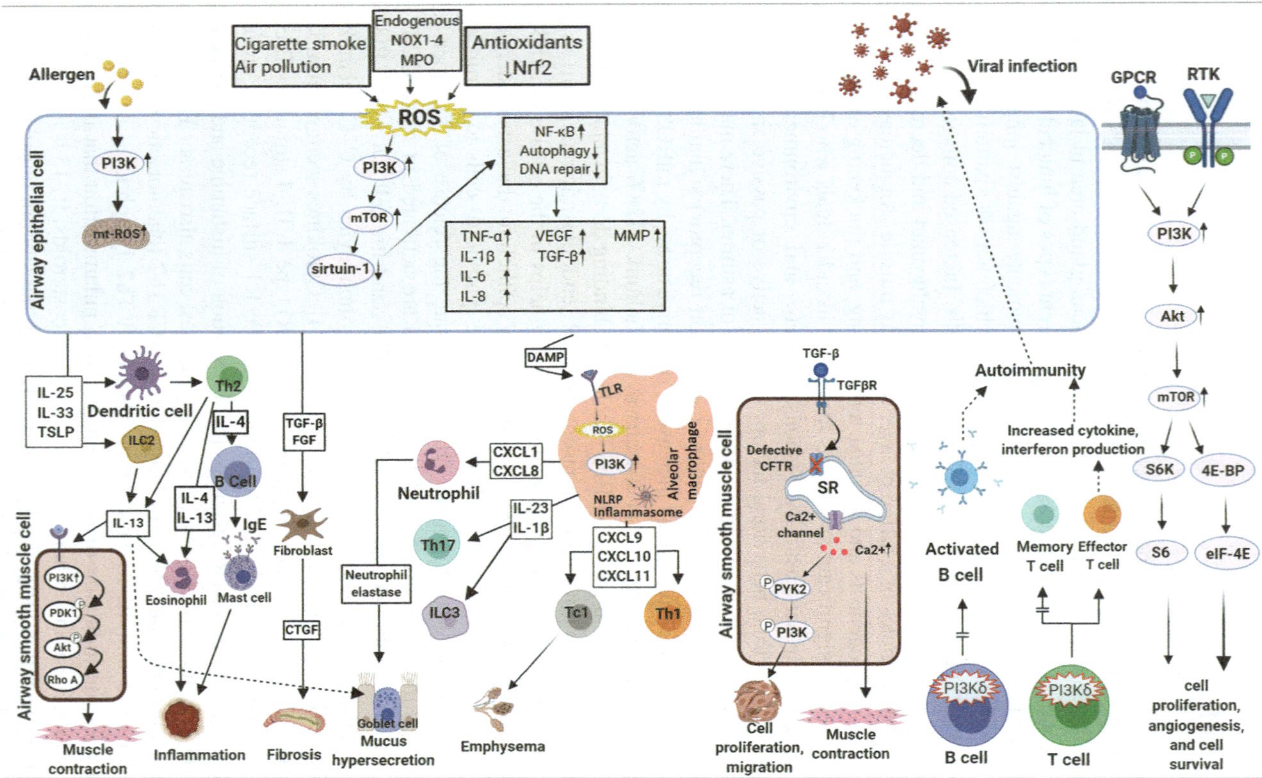

Fig. 28.2 Molecular and cellular mechanism(s) of respiratory disorders involving PI3K signalling. Different external and internal factors trigger PI3K signalling-mediated downstream cellular responses like secretion of cytokines and growth factors that get involved in the complex interplay of immune cells and pulmonary cell-mediated airway smooth muscle contraction, inflammation, fibrosis, mucus hypersecretion, emphysema, autoimmunity, and finally progression to several respiratory disease phenotypes

platelet-derived growth factor (PDGF), and fibroblast growth factor (FGF)-2 acts via tyrosine kinase receptors (TKR) and G-protein-coupled receptors (GPCRs) present on airway smooth muscle cell membrane. p21ras activates through these receptors and interacts with the downstream PI3K pathway. cAMP response element-binding (CREB) protein transcription factor phosphorylation occurs through PDK1, Akt, and S6 kinase beta-1 (S6K1, also known as $p70^{s6k}$), and activation of nuclear factor kappa-light-chain-enhancer of activated B cells (NF-κB) occurs through Ras-related C3 botulinum toxin substrate 1 (Rac1). These activated transcription factors help in cell proliferation by cyclin D_1 gene expression [20].

Neutrophilic inflammation such as Th1 (type 1) and Th17 (type 17) is critically involved in the development of non-type 2 asthma. In severe asthmatic patients, increased Th17 cells may regulate neutrophilic inflammation by CXCL8, released from airway epithelial cells. Increased release of IL-9 (Th9) from a population of $CD4^+$ cells occurs in asthma that plays a role in airways to maintain mast cells [19].

Epigenetic regulation is also observed in PI3K signalling and asthma. A decrease in the level of histone deacetylase 2 (HDAC2) in airway epithelium by activation of PI3K leads to an increase in the pro-inflammatory cytokines [18]. The epigenetic modulators such as microRNAs (miRNAs) also play a pivotal role in asthma. The miRNAs contribute to the phenotypes such as airway hypersensitivity, Th2-related inflammation, airway recruitment of eosinophils, and mucus secretion, associated with asthma. Some miRNAs have a contribution to asthma by regulating the PI3K pathway, including miR-21, miR-133a, miRNA-27b-3p, miR-19, miR-200a, miR-223, miR-221, MiR-138, and miR-10a (Table 28.2).

28.4.2 Chronic Obstructive Pulmonary Disease (COPD)

Chronic obstructive pulmonary disease, commonly referred to as COPD, is a progressive lung disease with most common features like emphysema and chronic bronchitis. Although smoking remains the main causative agent for COPD, many environmental factors like air pollution, fumes, and dust can result in developing COPD in the case of nonsmokers. In both, massive inflammation takes place with the generation of ROS [28]. Pesticide use and autoimmunity also have roles in COPD progression. The inflammatory response can be triggered by particulate matter (PM), nitrogen dioxide (NO_2), and ozone (O_3) exposure and increased IL-8 concentration. Lung injury and destruction of the parenchyma are caused by IL-8-stimulated neutrophil elastase that attacks lung elastin and alveolar wall. AAT (alpha-1 antitrypsin) inhibits neutrophil elastase, and AAT deficiency causes the development of COPD in a small number of individuals [29].

PI3K pathway regulates autophagy that leads to apoptosis of the alveolar epithelial cells and causes COPD. PI3K signalling regulates different other signalling pathways, e.g., p38MAPK, Erk, etc. and downstream responses (e.g., NF-κB-dependent transcription) within the immune cells. Nuclear erythroid 2-related factor-2 (Nrf2), a transcription factor regulating most antioxidants, is not properly activated in the lung due to high oxidative stress [30]. In the case of COPD, oxidative

Table 28.2 miRNA regulation in PI3K pathway dysregulation

miRNAs regulated through the PI3K pathway	Lung diseases	References
Asthma		
miR-21	Increase airway hyperresponsiveness, proliferation and migration of ASM cells. Play important role in airway remodelling.	[21]
miR-133a	Upregulation of miR-133a expression inhibits airway remodelling by inhibition of PI3K/AKT/mTOR pathway	[22]
miRNA-27b-3p	Regulate SYK gene expression in the pathogenesis of pediatric asthma	[23]
miR-19a	Downregulate TGF-β receptor 2 and increase severe asthma proliferation	[21]
miR-200a	Upregulation of miR-200a downregulates PI3K expression and suppress airway smooth muscle cell proliferation and airway remodelling in asthmatic mice	[24]
miR-223	Downregulate eosinophil progenitor cell proliferation	[25]
miR-221	Regulate airway smooth muscle cell proliferation and release of IL-6 in severe asthmatic patients	[21]
miR-138	Suppress PDK1 expression by targeting the 3′-UTR of the gene and inhibit airway smooth muscle cell proliferation	[21]
miR-10a	Regulate mitogen-induced ASM proliferation targeting PI3K pathway by downregulating PIK3CA expression	[21]
COPD		
miRNA-101-3p.1	Activates EGFR/PI3K/AKT signaling pathway and facilitates COPD progression	[26]
miR-34a	Overexpression can induce PI3Kα expression and reduce SIRT1/-6 expression resulting oxidative stress and COPD	[27]

stress may be increased by the reduction in the transcription factor Nrf2, activation of NADPH oxidases (NOX), myeloperoxidase (MPO), and reduced superoxide dismutase (SOD) [31]. PI3K activation by ROS also activates NF-κB that amplifies inflammatory responses. Sirtuin-1 (SIRT-1) deacetylates target proteins like Forkhead box protein O1 (FOXO1), NF-κB, and matrix metalloprotease 9 (MMP-9). ROS activates PI3K, followed by the activation of Akt and mTOR, which then inhibits SIRT-1. Reduced expression of SIRT-1 leads to cellular senescence in COPD due to acetylation of the target proteins. Accelerated lung aging and inflammation are caused by the acquisition of senescence-associated secretory phenotype (SASP) which is characterized by the upregulation of NF-κB, downregulation of autophagy, and decreased DNA repair process. p38 mitogen-activated protein kinase and Janus-activated kinase are activated by SASP, and this results in the release of inflammatory cytokines (IL-1β, IL-6, and TNF-α), growth factors like vascular endothelial growth factor (VEGF), TGF-β1, chemokines, and MMPs, downstream from NF-κB activation [32]. Accelerated aging of the lung in COPD includes senescent epithelial cells, endothelial cells, and fibroblast release inflammatory proteins, including TNF-α, IL-β, IL-6, CCL2, CXCL1, CXCL8, TGF-β, and MMP-9. ROS-mediated activation of PI3K activates NLR family pyrin domain-containing 3 (NLRP3) inflammasome by the induction of NF-κB signalling activating caspase-1-dependent pathway and interleukins like IL-1α, IL-1β, IL-33, and IL-18. Chemokines released from activated macrophage, including CXCL8, CXCL1, and leukotriene B4 (LTB4), attract neutrophils by the CXCR2 cell surface receptor. Neutrophil elastase, cathepsin G, and proteinase-3 stimulate submucosal glands and goblet cells that result in mucus hypersecretion. CXCL9, CXCL10, and CXCL11 interact via CXCR3 chemokine receptor of CD8+ cytotoxic T (Tc1) cells and CD4+ Th1 cells. The release of proteases (e.g. MMP-9) from macrophage causes elastin degradation. CD8+ cytotoxic T (Tc1) cells release perforin, granzyme B, and TNF-α that induce apoptosis of alveolar cells. These could result in alveolar destruction or emphysema. MMP-2, MMP-9, MMP-12, and cathepsins K, L, and S are the elastolytic enzymes released from alveolar macrophages that causes elastolysis in COPD patients. Epithelial cells and macrophages also release TGF-β, which stimulates fibroblast proliferation and the release of connective tissue growth factor (CTGF), which results in fibrosis around the small airways, along with neutrophilic inflammation [19]. Additionally, miRNA-101-3p.1 and miR-34a are reported to help activate the PI3K pathway and facilitate COPD progression (Table 28.2).

28.4.3 Cystic Fibrosis (CF)

Cystic fibrosis (CF) is caused by CFTR (cystic fibrosis transmembrane conductance regulator) gene mutation that leads to defective CFTR protein. ΔF508 CFTR (deletion of phenylalanine 508 of the cystic fibrosis transmembrane conductance regulator) is the most common form of the impaired CFTR protein. A thin layer of mucus and liquid over the airway surface helps to move bacteria toward the pharynx by ciliary beating, which gets compromised in cystic fibrosis leading to a chronic inflammatory response. CF causes chronic endobronchial infection, malnutrition, and decreased lung function. CF patients may develop pancreatic insufficiency, CF-related diabetes, and liver disease [33].

Different signalling pathways are involved in cystic fibrosis (CF) lung pathogenesis, such as hypoxia-inducible factor (HIF)-1α, NF-κB, PI3K, AKT, proline-rich tyrosine kinase (PYK)-2, mTOR, and mitogen-activated protein kinase (MAPK). PI3K signalling is responsible for elevated TGF-β-mediated airway remodelling in CF patients. The loss of CFTR can increase intracellular Ca^{2+} from the sarcoplasmic reticulum by TGF-β. TGF-β interacts with the sarcoplasmic reticulum via TGF-βR1 or TGF-βR2, which leads to the autophosphorylation of proline-rich tyrosine kinase 2 (PYK2). Activated PYK2 then further activates PI3K that stimulates cell proliferation and migration. Increased Ca^{2+} level can induce cell contraction independently or in conjugation with PYK2. The upstream and downstream targets of Ca^{2+}, PYK2, and PI3K play important roles in modulating cell contraction, proliferation, and migration in normal and CFTR-deficient airway smooth muscle cells [34].

The PI3K/Akt/mTOR pathway plays a vital role in the regulation of ΔF508 CFTR stability. Misfolding, ER accumulation, and an inability of trafficking to the plasma membrane are the main characteristics of ΔF508 CFTR. CFTR, those are unable to degrade by proteasome, can be stocked in the cytoplasm in the form of aggresome. Upregulation in mTORC1/2 signalling in ΔF508 CFBE cell shows a defective autophagy pathway and reduced aggresome clearance. Inhibitors of PI3K/AKT/mTOR pathway decrease in mTORC1/2 activation and increase in autophagic activity [35].

28.4.4 Activated PI3K Delta Syndrome (APDS)

APDS is denoted as a primary immunodeficiency disease caused by the gain-of-function mutations in PI3K signalling genes where the development and response of lymphocytes get altered. APDS patients having T cells in the circulation that are terminally differentiated, low IL-2 secreting, senescent, shortened telomeres, and poor proliferative capacity, whereas transitional B cells are mainly found in circulation in APDS patients. Patients with APDS also are susceptible to pathogenic virus and bacterial infections [36].

There are two different types of APDS. APDS I is due to a single amino acid substitution in the gene PIK3CD encoding for p110δ subunit marked by the hyperactivated PI3Kδ signalling pathway where it converts more PIP_2 to PIP_3. To date, there are ten such mutations reported, of which E1021K is most common. APDS II is caused by splice site mutation leading to the exclusion of exon 11 in the gene PIK3RI (encoding for p85α, p55α, p50α). This mutation results in the inhibition of the interaction between the regulatory part p85 and its catalytic counterpart [37]. Activation of PI3K in B lymphocytes results in phosphorylation of AKT, which, in turn, phosphorylates mTOR, FOXO, and BACH2 (BTB Domain And CNC Homolog 2) [38]. Unphosphorylated FOXO acts as a transcription factor when found inside the nucleus. FOXO upregulates several genes like IL-17 receptor alpha (Il-7rα), L-selectin (CD62L), recombination-activating gene 1 (RAG1) and RAG2, IKAROS, activation-induced cytidine deaminase (AID), and early B cell factor 1 (EBF1) to name a few. APDS is the hyperactivated situation of PI3K signalling, leading to increased phosphorylated FOXO, rendering its localization outside of the nucleus by interaction with 14-3-3 protein [39]. This results in impaired B cell population generation due to dysregulation of the abovementioned conditions. A fine regulation of B cell lymphoma 6 (BCL6) protein and Blimp-1 protein is required for plasma B cell generation and subsequent B cell formation. This regulation gets altered in hyperactivated PI3K signalling like APDS [40].

In T lymphocyte, PI3Kδ/AKT pathway is activated by signals from the T cell receptor, inducible T cell costimulator (ICOS), and IL2 receptor. Active unphosphorylated FOXO1 is required for naïve T cell survival by regulating the expression of IL-7 receptor-α (IL-7Rα), and homing toward secondary lymphoid organs. When FOXO is inactivated by the activated PI3K/AKT via phosphorylation, the process is altered. Inactivation of FOXO results in the downregulation of IL-7Rα and CD62L expression, which directs the T cells to exit the lymph nodes and circulate through the vascular systems and organs as found in APDS patients [36]. As a result, more Th1 cells are produced instead of Treg cells. Since Th1 cells produce a high level of IFNγ, autoimmunity occurs in APDS patients. Another important target of FOXO is C–C chemokine receptor 7 (CCR7), required for Treg cell homing. Activated PI3K/AKT leads to reduced Treg function through the downregulation of CCR7 [39]. Inactive FOXO in case of APDS causes more effector T cell pool than memory T cell. This change in T cell population is also influenced by phosphorylated EZH2, which promotes effector T cell formation. Unphosphorylated EZH2 is responsible for memory T cell formation [38]. These are the leading cause of the increased level of the effector T cell population in APDS patients [39]. PI3K downstream protein mTOR remains in a hyperactivated state in APDS patients. Activated mTOR convert naïve T cell to effector T cell by increasing the cellular metabolism. In a hyperactivated condition, T cell drives toward a senescence state rather than a quiescent state upon antigen exposure. This hyperproliferation of T cell may cause T cell senescence in APDS without a known mechanism [36].

28.4.5 Idiopathic Pulmonary Fibrosis (IPF)

Idiopathic pulmonary fibrosis (IPF) is a chronic, progressive, fatal interstitial lung disease characterized by fibroblast/myofibroblast accumulation in the alveolar wall by circulating fibrocytes, gradually replacing normal lung tissue by fibrotic scarring and honeycombing [1]. Enormous matrix deposition to alveoli leads to distortion of the alveolar architecture and consequent loss of respiratory function. Alterations in lung functions like reduced lung compliance (pulmonary surfactant alterations and lung extracellular matrix compliance reduction) and lung volumes, reduced diffusing capacity, increased dead space ventilation, chronic arterial hypoxemia, and altered airway physiology such as increased cough reflex and increased airway volume as well as pulmonary hemodynamics are characteristics of IPF.

Immune cells, like monocyte-derived alveolar macrophages, are crucial for the development of lung fibrosis. IPF is positively correlated with age, male sex, as well as cigarette smoking [41].

It has been observed that there are no significant differences between IPF and control tissues in p110α, p110β, and p110δ expression, whereas p110γ appeared to be overexpressed nearly 1.5-fold in IPF lung tissue. The fibrocyte influx in lung alveoli is influenced by stromal cell-derived factor-1 (SDF-1) signalling working through CXCR4, which is the predominant chemokine receptor on fibrocytes. CXCR4 expression is upregulated by growth factors and hypoxia (HIF1a), which is prevented by PI3K inhibition with pharmacological inhibitor LY2949002

[42]. Akt inhibitors can partly reverse the EMT process responsible for fibrosis, indicating the PI3K/Akt pathway contributes to fibrosis and EMT progression [43]. Besides, fibrocyte infiltration can be abolished by inhibition of mTOR into the lung, reducing collagen deposition and CXCR4 expression from fibrocyte in vivo [42].

28.4.6 Lung Cancer

As described earlier, PI3K signalling plays a vital role in cell proliferation, survival, and growth. Aberration in this pathway component, including amplification and mutations, results in tumor cell proliferation, apoptosis inhibition, metastasis, angiogenesis, and chemotherapy resistance. Genetic alteration of PI3K pathway components leads to increased PI3K signalling and causes lung cancer. Mutational activation of p110α subunit of PIK3CA gene by mutant Ras oncogene results in overexpression of Akt in non-small cell lung cancer cells (NSCLC). Akt activation in NSCLC tumors relates to poor prognosis and resistance to chemoradiotherapy. mTOR (mammalian target of rapamycin), the downstream regulator protein of the PI3K pathway, is reported to be mutated in lung adenocarcinomas and reported to be important for tumor progression metastasis along with KRAS mutation. These events are also responsible for the downstream activation of S6 and eIF-4E needed for cell proliferation, angiogenesis, and cell survival [44].

28.5 Role of PI3K Inhibitors in the Prevention of Respiratory Diseases

Many PI3K inhibitors are identified and synthesized for respiratory diseases, mostly targeting cancers of different organs, including the lung. Among them, few are pan-PI3K inhibitors (Wortmannin, LY294002), and few target different classes of PI3Ks (IC87114, YM-024, TGX-221) (Table 28.3). Many PI3K inhibitors are reported to be in the clinical trial for the treatment of respiratory diseases. Several inhibitors are in a clinical trial also used in combination with other drugs. The same drug can also be used in clinical trials for different respiratory diseases (Table 28.4).

28.6 Implications of PI3Kinase Signalling in COVID Infection

In recent times COVID-19 (coronavirus disease-2019) pandemic caused by the coronavirus SARS-CoV-2 (severe acute respiratory syndrome coronavirus 2) had unleashed a severe assault on the human health and society with 0.6 million deaths globally, affecting more than 200 countries [WHO Situation Report-185]. One of the primary reasons resulting in high mortality in the advanced-stage COVID-19 patients may be due to the unregulated inflammation in the lungs, which can further lead to acute respiratory distress syndrome (ARDS). As a central regulator, PI3K,

Table 28.3 PI3K inhibitors in lung diseases

Inhibitors	Role in lung disease	Reference
Wortmannin	Pan-PI3K inhibitor	[45]
LY294002	Pan-PI3K inhibitor inhibits OVA induced activation of Akt	[45]
IC87114	PI3Kδ specific inhibitor for p110δ subunit reduces CD4+ T cells and allergic responses of mast cells.	[45]
YM-024	Class I PI3K inhibitor for p110α, inhibit Akt and α-SMA expression	[46]
TGX-221	Class I PI3K inhibitor for p110β, inhibit cell proliferation.	[46]
AS-252424	Class IB p110γ inhibitor	[46]
Theophylline	PI3Kδ inhibitor, restore corticosteroid sensitivity in COPD	[47]
MK-2206	Increase CFTR stability and expression by restoring autophagy	[35]

along with its downstream effectors such as Akt or mTOR, controls multiple cellular processes that may be advantageous to the viral growth [48, 49]. It has been observed that several viruses such as RSV, HBV, HPV, and HCV have been reported to activate the PI3K/Akt pathway to reduce apoptosis and hence to facilitate prolonged cell survival required for the viral replication. Also, some viruses such as adenovirus, influenza A, and EBV can enhance the remodelling of the cellular cytoskeleton via activation of the PI3K/Akt pathway. As mentioned earlier, PI3K/Akt/mTOR pathway is also involved in the regulation of the immune responses of the host cells such as maturation, and trafficking of leukocytes, regulation of Toll-like receptor (TLR)-mediated signalling, and also the regulation of a wide array of transcription factors such as AP-1, NF-kB, CREB, GSK-3β, etc. and PI3K/Akt signalling may be represented as a two-faced player in host-virus infection. In a very recent publication, Klann et al. reported that multiple pathways activated by growth factor receptor signalling cascades, including the RAF/MEK/ERK MAPK signalling pathway and also PI3K/Akt/mTOR, are required for the replication of SARS-CoV-2 in mammalian cells [50]. They observed that the administration of PI3K inhibitor pictilisib and dual PI3K/mTOR inhibitor omipalisib inhibited the replication of the SARS-CoV-2 genome and production of the viral progeny in the host cells. Although the host PI3 kinase/Akt pathway acts as a beneficiary pathway to the viral replication and growth, its role in SARS-CoV-2 replication in lung pneumocytes has not been demonstrated yet. Nevertheless, an adverse effect of PI3K/Akt suppression in the patient lung cells cannot be ruled out. As mentioned earlier, pharmacological inhibition of the PI3K/Akt pathway by agents like triciribine, pictilisib, and MK2206, alone or in combination, can provide potential medication options for the COVID-19 patients with respiratory complications like ARDS [51], but extensive clinical trials are required to establish the importance of PI3K signalling as a potential drug target against COVID-19.

Table 28.4 PI3K pathway inhibitors in clinical trial for lung diseases (compiled from www.clinicaltrial.gov)

PI3K pathway inhibitors	Clinical trial identifier	Trial phase	Combination with other drug	Condition
BKM120	NCT01723800	Phase 1 completed	Carboplatin and pemetrexed	Stage IV non-small cell lung cancer
BKM120	NCT02194049	Phase 1 completed	Cisplatin/etoposide	Advanced solid tumors or small cell lung cancer
GDC-0941	NCT00974584	Phase 1 completed	Paclitaxel, carboplatin, bevacizumab, pemetrexed, and cisplatin	Advanced non-small cell lung cancer
Gedatolisib (PF-05212384)	NCT03065062	Phase 1	CDK4/6 inhibitor Palbociclib (PD-0332991)	Squamous cell lung cancer
IPI-145	NCT01653756	Phase 2	Alone	Asthma
Defactinib hydrochloride and vismodegib	NCT02465060	Phase 2	Alone	Lung carcinoma
PF-05212384	NCT02920450	Phase 2	Paclitaxel and carboplatin	Non-small cell lung cancer
Aspirin	NCT03532698	–	Osimertinib	Non-small cell lung cancer stage IIIB and IV
RAD001	NCT00124280	Phase 1	Alone	Non-small-cell lung carcinoma
Idelalisib	NCT03257722	Phase 2	Pembrolizumab	Non-small-cell lung carcinoma
SAR245409	NCT01390818	Phase 1	SAR245409	Non-small-cell lung carcinoma
Ipatasertib	NCT04467801	Phase 2	Docetaxel	NSCLC stage IV and stage IIIB
Serabelisib	NCT04073680	Phase 1b/2	Canagliflozin	Lung cancer
BGB-10188	NCT04282018	Phase 1/2	Zanubrutinib and tislelizumab	Non-small cell lung cancer
CHF6523	NCT04032535	Phase 1	Alone	COPD
IPI-145	NCT01653756	Phase 2a	Alone	Asthma
GSK2269557	NCT02522299	Phase 2 completed	Alone	COPD
RV1729	NCT02140346	Phase 1 completed	Alone	COPD
RV1729	NCT01813084	Phase 1 completed	Alone	Asthma
RV1729	NCT02140320	Phase 1 completed	Alone	Asthma
Rapamycin	NCT03383380	Phase 2	Alone	APDS

28.7 Conclusion

As discussed, PI3K signalling has an important role in the normal function of the human lung, and deregulation at any level may result in serious pulmonary malfunction, progressive degeneration, as well as mild to severe lung complications, including lung injuries. These features became important in the case of the COVID-19 pandemic, where the severity of the disease can be attributed to preexisting lung injuries as comorbidity features, although a clear link is yet to be established. The suitability of selective PI3K inhibitors, as with many potential therapeutic targets, depends on a balance of expression, function, disease relevance, toxicity profile, and tolerance within the proposed disease. Bottlenecks like a selective PI3K inhibitor are exemplified by the lack of progression from in vivo models and early-stage human trials to late-stage development. Present therapies are inadequate resulting in considerable unmet medical needs such as in severe asthma, COPD, CF, and IPF. Selective inhibition of PI3K inhibitors at controlling lung disease is promising, but potential side effects of selective inhibition will be an essential therapeutic determinant in terms of cost-benefit ratio. For conditions such as CF and IPF, which are generally progressive and where effective pharmacological therapy is almost nonavailable, side effect profiles may be clinically more acceptable compared to the administration of inhibitors at mild asthma. A short-term application may be suitable for those where toxicity profiles and tolerance are poor, such as during exacerbations of severe asthma and COPD. In the case of viral infection and lung injury, the immune system becomes unstable, and PI3K signalling may become an unlikely target as in the case of the recent COVID-19 pandemic or other lung disorders. It will take further development in preclinical and clinical trials for new selective therapeutic modality targeting PI3K signalling to be used in the near future.

References

1. Marwick J, Chung KF, Adcock I (2010) Phosphatidylinositol 3-kinase isoforms as targets in respiratory disease. Ther Adv Respir Dis 4(1):19–34
2. Lugogo N, Kraft M (2006) Epidemiology of asthma. Clin Chest Med 27(1):1–15
3. Medina-Tato D, Ward S, Watson M (2007) Phosphoinositide 3-kinase signalling in lung disease: leucocytes and beyond. Immunology 121(4):448–461
4. Cantley L (2002) The phosphoinositide 3-kinase pathway. Science 296(5573):1655–1657
5. Vanhaesebroeck B, Stephens L, Hawkins P (2012) PI3K signalling: the path to discovery and understanding. Nat Rev Mol Cell Biol 13(3):195–203
6. Vanhaesebroeck B, Vogt P, Rommel C (2010) PI3K: from the bench to the clinic and back. Curr Top Microbiol Immunol 347:1–19
7. Martini M, De Santis M, Braccini L, Gulluni F, Hirsch E (2014) PI3K/AKT signalling pathway and cancer: an updated review. Ann Med 46(6):372–383
8. Bilanges B, Posor Y, Vanhaesebroeck B (2019) PI3K isoforms in cell signalling and vesicle trafficking. Nat Rev Mol Cell Biol 20(9):515–534
9. Carracedo A, Pandolfi P (2008) The PTEN–PI3K pathway: of feedbacks and cross-talks. Oncogene 27(41):5527–5541
10. Castellano E, Downward J (2011) RAS interaction with PI3K: more than just another effector pathway. Genes Cancer 2(3):261–274

11. Fruman D, Chiu H, Hopkins B, Bagrodia S, Cantley L, Abraham R (2017) The PI3K pathway in human disease. Cell 170(4):605–635
12. Gulluni F, De Santis M, Margaria J, Martini M, Hirsch E (2019) Class II PI3K functions in cell biology and disease. Trends Cell Biol 29(4):339–359
13. Mruk D, Cheng C (2010) The myotubularin family of lipid phosphatases in disease and in spermatogenesis. Biochem J 433(2):253–262
14. Hurley J, Young L (2017) Mechanisms of autophagy initiation. Annu Rev Biochem 86 (1):225–244
15. Jaber N, Mohd-Naim N, Wang Z, DeLeon J, Kim S, Zhong H, Sheshadri N, Dou Z, Edinger A, Du G, Braga V, Zong W (2016) Vps34 regulates Rab7 and late endocytic trafficking through recruitment of the GTPase-activating protein Armus. J Cell Sci 129(23):4424–4435
16. Ornatowski W, Lu Q, Yegambaram M, Garcia A, Zemskov E, Maltepe E, Fineman J, Wang T, Black S (2020) Complex interplay between autophagy and oxidative stress in the development of pulmonary disease. Redox Biol 36:101679
17. Lemmon M, Schlessinger J (2010) Cell signalling by receptor tyrosine kinases. Cell 141 (7):1117–1134
18. Jeong J, Kim J, Kim S, Lee Y (2019) Defining bronchial asthma with phosphoinositide 3-kinase delta activation: towards endotype-driven management. Int J Mol Sci 20(14):3525
19. Barnes P (2017) Cellular and molecular mechanisms of asthma and COPD. Clin Sci 131 (13):1541–1558
20. Pelaia G, Renda T, Gallelli L, Vatrella A, Busceti M, Agati S, Caputi M, Cazzola M, Maselli R, Marsico S (2008) Molecular mechanisms underlying airway smooth muscle contraction and proliferation: implications for asthma. Respir Med 102(8):1173–1181
21. Szymczak I, Wieczfinska J, Pawliczak R (2016) Molecular background of miRNA role in asthma and COPD: an updated insight. Biomed Res Int 2016:1–10
22. Shao Y, Chong L, Lin P, Li H, Zhu L, Wu Q, Li C (2018) MicroRNA-133a alleviates airway remodeling in asthma through PI3K/AKT/mTOR signalling pathway by targeting IGF1R. J Cell Physiol 234(4):4068–4080
23. Dong X, Zhong N, Fang Y, Cai Q, Lu M, Lu Q (2018) MicroRNA 27b-3p modulates SYK in pediatric asthma induced by dust mites. Front Pediatr 6:301
24. Liu Y, Miao Y, Gao X, Wang Y, Wang H, Zheng Y, Zhao Z (2018) MicroRNA-200a affects the proliferation of airway smooth muscle cells and airway remodeling by targeting FOXC1 via the PI3K/AKT signalling pathway in ovalbumin-induced asthmatic mice. Cell Physiol Biochem 50 (6):2365–2389
25. Heffler E, Allegra A, Pioggia G, Picardi G, Musolino C, Gangemi S (2017) MicroRNA profiling in asthma: potential biomarkers and therapeutic targets. Am J Respir Cell Mol Biol 57(6):642–650
26. Chen S, Zhang Z, Chen L, Zhang J (2019) miRNA 101 3p.1 as an independent diagnostic biomarker aggravates chronic obstructive pulmonary disease via activation of the EGFR/PI3K/ AKT signalling pathway. Mol Med Rep 20:4293–4302
27. Baker J, Vuppusetty C, Colley T, Papaioannou A, Fenwick P, Donnelly L, Ito K, Barnes P (2016) Oxidative stress dependent microRNA-34a activation via PI3Kα reduces the expression of sirtuin-1 and sirtuin-6 in epithelial cells. Sci Rep 6(1):35871
28. Forbes L, Kapetanakis V, Rudnicka A, Cook D, Bush T, Stedman J, Whincup P, Strachan D, Anderson H (2009) Chronic exposure to outdoor air pollution and lung function in adults. Thorax 64(8):657–663
29. Wells A, Woods A, Hilleman D, Malesker M (2019) Alpha-1 antitrypsin replacement in patients with COPD. P T 44(7):412–415
30. Malhotra D, Thimmulappa R, Navas-Acien A, Sandford A, Elliott M, Singh A, Chen L, Zhuang X, Hogg J, Pare P, Tuder R, Biswal S (2008) Decline in NRF2-regulated antioxidants in chronic obstructive pulmonary disease lungs due to loss of its positive regulator, DJ-1. Am J Respir Crit Care Med 178(6):592–604

31. Barnes P (2014) Cellular and molecular mechanisms of chronic obstructive pulmonary disease. Clin Chest Med 35(1):71–86
32. Hikichi M, Mizumura K, Maruoka S, Gon Y (2019) Pathogenesis of chronic obstructive pulmonary disease (COPD) induced by cigarette smoke. J Thorac Dis 11(S17):S2129–S2140
33. Wilschanski M, Rivlin J, Cohen S, Augarten A, Blau H, Aviram M, Bentur L, Springer C, Vila Y, Branski D, Kerem B, Kerem E (1999) Clinical and genetic risk factors for cystic fibrosis-related liver disease. Pediatrics 103(1):52–57
34. Natarajan V (2020) Is PI3K a villain in cystic fibrosis? Am J Respir Cell Mol Biol 62 (5):552–553
35. Reilly R, Mroz M, Dempsey E, Wynne K, Keely S, McKone E, Hiebel C, Behl C, Coppinger J (2017) Targeting the PI3K/Akt/mTOR signalling pathway in cystic fibrosis. Sci Rep 7(1):7642
36. Lucas C, Chandra A, Nejentsev S, Condliffe A, Okkenhaug K (2016) PI3Kδ and primary immunodeficiencies. Nat Rev Immunol 16(11):702–714
37. Michalovich D, Nejentsev S (2018) Activated PI3 kinase delta syndrome: from genetics to therapy. Front Immunol 9:369
38. Preite S, Gomez-Rodriguez J, Cannons J, Schwartzberg P (2019) T and B-cell signalling in activated PI3K delta syndrome: from immunodeficiency to autoimmunity. Immunol Rev 291 (1):154–173
39. Cabrera-Ortega A, Feinberg D, Liang Y, Rossa C, Graves D (2017) The role of Forkhead box 1 (FOXO1) in the immune system: dendritic cells, T cells, B cells, and hematopoietic stem cells. Crit Rev Immunol 37(1):1–13
40. Crotty S, Johnston R, Schoenberger S (2010) Effectors and memories: Bcl-6 and Blimp-1 in T and B lymphocyte differentiation. Nat Immunol 11(2):114–120
41. Lederer D, Martinez F (2018) Idiopathic pulmonary fibrosis. N Engl J Med 378(19):1811–1823
42. Mehrad B, Burdick M, Strieter R (2009) Fibrocyte CXCR4 regulation as a therapeutic target in pulmonary fibrosis. Int J Biochem Cell Biol 41(8-9):1708–1718
43. Zhang X, Xing R, Chen L, Liu C, Miao Z (2016) PI3K/Akt signalling is involved in the pathogenesis of bleomycin-induced pulmonary fibrosis via regulation of epithelial-mesenchymal transition. Mol Med Rep 14(6):5699–5706
44. Sarris E, Saif M, Syrigos K (2012) The biological role of PI3K pathway in lung cancer. Pharmaceuticals 5(11):1236–1264
45. Ito K, Caramori G, Adcock I (2006) Therapeutic potential of phosphatidylinositol 3-kinase inhibitors in inflammatory respiratory disease. J Pharmacol Exp Ther 321(1):1–8
46. Conte E, Fruciano M, Fagone E, Gili E, Caraci F, Iemmolo M, Crimi N, Vancheri C (2011) Inhibition of PI3K prevents the proliferation and differentiation of human lung fibroblasts into myofibroblasts: the role of class I P110 isoforms. PLoS One 6(10):e24663
47. To Y, Ito K, Kizawa Y, Failla M, Ito M, Kusama T, Elliott W, Hogg J, Adcock I, Barnes P (2010) Targeting phosphoinositide-3-kinase-δ with theophylline reverses corticosteroid insensitivity in chronic obstructive pulmonary disease. Am J Respir Crit Care Med 182(7):897–904
48. Dunn E, Connor J (2012) HijAkt: the PI3K/Akt pathway in virus replication and pathogenesis. Prog Mol Biol Transl Sci 106:223–250
49. Ji W, Liu H (2008) PI3K-Akt signalling and viral infection. Recent Pat Biotechnol 2 (3):218–226
50. Klann K, Bojkova D, Tascher G, Ciesek S, Münch C, Cinatl J (2020) Growth factor receptor signalling inhibition prevents SARS-CoV-2 replication. Mol Cell 80(1):164–174.e4
51. Somanath P (2020) Is targeting Akt a viable option to treat advanced-stage COVID-19 patients? Am J Phys Lung Cell Mol Phys 319(1):L45–L47

29 The Role of the Cholinergic System in Lung Diseases

Vedran Milankovic and Tamara Lazarevic-Pasti

Abstract

Acetylcholinesterase is a crucial player in the catalytic hydrolysis of cholinergic neurotransmitters. Recent studies have indicated that acetylcholinesterase is potentially a marker and a regulator of apoptosis, as well as a tumor suppressor with great potential. It was shown that the gene for acetylcholinesterase expression is usually decreased in tumor tissues, regardless of the cellular origin of an enzyme. Acetylcholinesterase is also a newly established regulator in cell proliferation and cell death. Hence, for lung cancer diagnosis and prognosis, acetylcholinesterase could be used as a potential marker. Moreover, the neurotransmitter acetylcholine is an autocrine growth factor for human lung cancer. The released acetylcholine binds back to nicotinic and muscarinic receptors on lung cancer cells. That action accelerates the proliferation, migration, and invasion of cancer cells. The cholinergic system is also involved in asthma and chronic obstructive pulmonary disease. An improved understanding of this complicated regulation will yield new insights into apoptosis biology and pathways that might be strategic targets for designing different lung disease therapeutics.

Keywords

Acetylcholinesterase · Acetylcholine · Cancer · Asthma · COPD

V. Milankovic
Faculty of Physical Chemistry, University of Belgrade, Belgrade, Serbia

T. Lazarevic-Pasti (✉)
Department of Physical Chemistry, VINČA Institute of Nuclear Sciences – National Institute of the Republic of Serbia, University of Belgrade, Belgrade, Serbia; https://www.researchgate.net/profile/Tamara_Lazarevic-Pasti

K. Dua et al. (eds.), *Targeting Cellular Signalling Pathways in Lung Diseases*,
https://doi.org/10.1007/978-981-33-6827-9_29

29.1 Introduction

The cholinergic system is a neurotransmitter system dominantly involved in learning and memory processes [1]. Besides its main component, the neurotransmitter acetylcholine (ACh) (Fig. 29.1), the cholinergic system includes enzymes acetylcholinesterase (AChE), butyrylcholinesterase (BuChE), and choline acetyltransferase (ChAT), as well as the muscarinic and nicotinic acetylcholine receptors (mAChR and nAChR) and vesicular acetylcholine transporter (VAChT). All components of the system are present in the central nervous system (CNS). Nevertheless, some components are present in other parts of the human body, implicating the cholinergic system's less-known alternative roles.

Numerous studies showed the presence of ACh, AChE, or AChRs in immune [2], skin, skeletal muscle, cardiovascular, digestive, urinary, reproductive, and respiratory system cells [3]. It seems that almost all types of cells express acetylcholine receptors. Still, the cell should be able to synthesize and release ACh to be considered a part of the cholinergic system. It was shown that ACh could be released by and act on non-neuronal cells [3]. In these cells, the cholinergic system exerts its non-neuronal roles. In the last few decades, extensive research revealed the involvement and disfunction of cholinergic components in many diseases related to different organ systems. Still, the exact role of all non-neuronal cholinergic functions remained unknown.

Nowadays, the neuronal cholinergic system's modulation as a treatment for different diseases is often a mainstream choice. AChE inhibitors are well-known therapy for various neurological disorders, especially Alzheimer's disease [4–8]. Lately, some established therapies are targeting the non-neuronal cholinergic system [3]. As already mentioned, AChE inhibitors are also used in some cancer treatments [9]. Still, the expression and the role of the non-neuronal cholinergic system are not fully understood. The modulation of the non-neuronal cholinergic system has many potential applications in different medicinal treatments, once we elucidate the exact mechanism of its actions.

This chapter will discuss the components of the cholinergic system, their neuronal and non-neuronal roles, and diseases caused due to disruption in their functioning. Special attention will be given to the pulmonary cholinergic system, as well as its

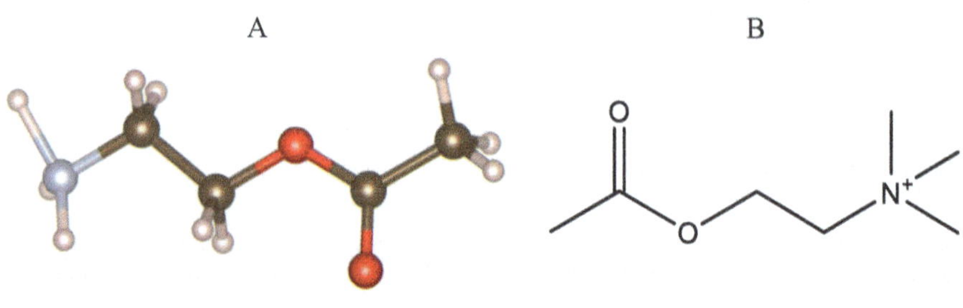

Fig. 29.1 (**a**) 3D structure of acetylcholine. Carbon atoms, brown; oxygen atoms, red; nitrogen atoms, blue; hydrogen atoms, white. (**b**) 2D structure of acetylcholine

role in lung cancer development, asthma, and chronic obstructive pulmonary disease.

29.2 Components of the Cholinergic System

Components of the cholinergic system are neurotransmitter acetylcholine, cholinergic enzymes (acetylcholinesterase, butyrylcholinesterase, and choline acetyltransferase), acetylcholine receptors (muscarinic and nicotinic), and vesicular acetylcholine transporters.

29.2.1 Acetylcholine

Acetylcholine (Fig. 29.1) is synthesized from choline and acetyl-coenzyme A (acetyl-CoA), supplied by mitochondria. The enzyme choline acetyltransferase catalyzes this process in the cytoplasm. After synthesis, ACh is transported by vesicular ACh transporters into synaptic vesicles of cholinergic neurons, localized throughout the peripheral and central nervous systems. The rate-limiting steps in ACh synthesis are the availability of choline and acetyl-CoA. Cholinergic neurons innervate almost all brain structures and regions. Cholinergic neurons are either a part of local circuits (interneurons) or neuron projections that spread throughout various parts of the brain and wire them up [10, 11]. Considering the abundance of cholinergic neurons across the brain, it is not surprising that the cholinergic system has an important place in the regulation of various physiological processes. Those processes are attention, learning, memory, stress response, wakefulness, sleep, emotional responses, and behavior [12, 13] and overall have a prominent role in developing many CNS diseases. Following synthesis, ACh undergoes exocytosis and release into the synaptic cleft, binding to two different types of receptors.

When classically viewed, acetylcholine is a neurotransmitter regulating cognitive and behavioral functions in the brain—autonomous ganglionic and parasympathetic postganglionic transmission. Within the respiratory tract, acetylcholine is in charge of mucus secretion from airway submucosal glands and a bronchomotor tone [14]. It is the dominant parasympathetic neurotransmitter in the lungs. There are some additional acetylcholine roles in the respiratory system, such as inflammation and remodeling in inflammatory lung diseases [15–17]. It was also proven that acetylcholine is synthesized in the airway epithelium cells and inflammatory cells [18–20], two types of cells that express both nicotinic and muscarinic receptors for acetylcholine. These receptors later could influence the inflammatory response [15, 19]. The presented facts led to the change in the traditional view of acetylcholine's physiological and pathophysiological role. It allowed us to open new possibilities for therapeutic targeting of the cholinergic system in the lungs [21].

29.2.2 Enzymes of the Cholinergic System

There are three enzymes included in the cholinergic system: acetylcholinesterase, butyrylcholinesterase, and choline acetyltransferase.

Cholinesterases are the enzymes from the serine hydrolase family. This class of enzymes has serine residue in their active site responsible for the hydrolysis of their substrate [22]. By the term cholinesterase, there are two types of enzymes known so far: acetylcholinesterases and butyrylcholinesterases [22]. Both of them can hydrolyze acetylcholine and some other choline esters but with different relative abilities. Besides cleaving acetylcholine, cholinesterase may also be involved in peptide hydrolysis [23] or may show aryl acylamidase activity [9, 24].

ACh present at the synaptic cleft is rapidly inactivated and hydrolyzed to choline and acetate by the enzyme acetylcholinesterase [25], an enzyme with the highest catalysis rate known in biology [26]. AChE has been secreted in cell bodies of cholinergic neurons and spread across the neuron by axoplasmic transport into the synaptic cleft attached to a plasma membrane [12, 27]. AChE's primary functional target is to quickly breakdown the ACh to choline and acetate in the synapse and neuromuscular junction, which ends neural response [13, 28].

Besides this well-known function of AChE, some novel non-cholinergic roles of this enzyme are proven. These include the involvement of AChE in cell adhesion, differentiation, and proliferation [29]. Some studies have recently indicated a potential role of AChE as a marker and regulator of apoptosis [29]. Also, there is evidence that AChE is a very promising tumor suppressor [29]. Having in mind the very diverse roles of AChE in the human organism, it is expected that there is more than one isoform present. The human organism expresses three different alternative spliced isoforms of AChE [30]. The synaptic isoform of AChE (S) is tetrameric and expressed in the brain and muscle tissue [29]. This form contains a hydrophobic subunit, which allows it to be anchored to the membrane [29]. The erythrocytic isoform of AChE (E) is dimeric and contains a hydrophobic region [29]. The read-through isoform of AChE (R) is monomeric and transforms into a hydrophilic monomer at the cholinergic synapses to respond to stress [31–33]. The first known function of AChE, termination of acetylcholine-mediated nerve impulses, is the most studied one.

Recently, there has been a revolution in AChE research, and additional roles of AChE became the subject of intensive investigation. It was motivated by the fact that some of the long-suspected nonclassical functions of this enzyme are confirmed [33]. The complexity of AChE gene regulation is emphasized and thoroughly studied in the twenty-first century. Despite that, our understanding of the additional roles of AChE is still incomplete. It is why the usage of AChE activity modulators is a topic that is very carefully considered. In terms of neurodegenerative diseases, such as Alzheimer's (AD) and Parkinson's (PD), the situation is clear—AChE needs to be inhibited to improve the patient's condition. On the contrary, the role of AChE in different cancer conditions is not so coherent, which is understandable, having in mind different etiologies of various cancers [9, 34, 35].

The first structure of AChE was determined in the enzyme isolated from electric ray *Torpedo californica* [36]. This result allowed the visualization of a binding pocket for acetylcholine at an atomic resolution. Surprisingly, it was shown that the active site of AChE is not on the surface of the protein [36]. It is located at the bottom of a 20 Å deep gorge lined with numerous aromatic residues. AChE isolated from *T. californica* is structurally similar to AChE in vertebrate nerves and muscles [37]. The 3D structure of recombinant human AChE revealed in 2010 confirmed this [38].

First kinetic studies showed that the active site (catalytic anionic site, CAS) of AChE contains two subsites named "esteratic" and "anionic" subsite [39]. While the "esteratic" site is accountable for the enzyme's catalytic activity, the "anionic" site is known to be a choline-binding pocket. Eventually, it was shown that serine and histidine residues located in the "esteratic" active site of the enzyme are crucial for the activity of AChE [40, 41]. "Anionic" subsite handles the interaction with a positively charged quaternary group of acetylcholine. Also, the "anionic" subsite is the binding site for some quaternary ligands that act as the inhibitors of AChE [42, 43]. Besides, some quaternary oximes, which act as reactivators of AChE after inhibition by organophosphates, bind to the "anionic" site. Besides two subsites of the AChE catalytic center, there is the "peripheral" anionic site (PAS) [44]. It is different from the choline-binding pocket of the active site. This site is involved in substrate inhibition of AChE. However, at low substrate concentrations, binding to the "peripheral" anionic site could accelerate the acylation step in the catalytic pathway [45]. AChE is the target molecule for various drugs, chemical weapons, pesticides, and snake venoms. Knowing the 3D structure of AChE is essential for an understanding of its catalytic activity and molecular basis of its interaction with muscarinic and nicotinic ACh receptors and rational drug design [46].

Besides the primary catalytic function of AChE in acetylcholine hydrolysis, many non-cholinergic functions of this enzyme have been shown. These include AChE involvement in cell growth, stem cell differentiation [47–49], cell recognition, cell signaling, synaptogenesis, activation of dopamine neurons [33], tumorigenesis, amyloid fiber assembly [33, 50, 51], regulation of neuritic growth, neuronal network formation, cell adhesion [52], hematopoiesis, and thrombopoiesis [33].

There is increasing evidence of apoptosis involvement in many diseases, such as cancer, diabetes, or Alzheimer's disease, which indicate the role of AChE in all these conditions [53–55]. The acetylcholine level is also critical for the successful control of inflammation and immune response in peripheral tissues [56]. An increase in acetylcholine levels above a certain threshold can suppress the production of pro-inflammatory cytokines [57]. Since AChE is responsible for acetylcholine level regulation, its role in modulating inflammation is evident [56, 57]. Furthermore, the process of inflammation is linked to various conditions [56], including cancer, so this is one more clue suggesting that AChE is involved in these conditions. In the last several years, a novel signaling system differing from the traditional cholinergic transmission system was proposed. The system shares the same components as the traditional ones, the alpha-7 nicotinic acetylcholine receptor (α7-nAChR) and AChE, and could be operational throughout the brain and body [58]. Namely,

AChE is found throughout the brain and body where its typical substrate ACh is relatively scarce, and no neuronal transmission occurs. In specific, AChE has been located in several non-neuronal cell types, including placenta, glia, endothelial, epithelial, cancer, and immune cells [58]. Moreover, non-hydrolytic actions of AChE have been commonly associated with developmental processes, stress responses, and regulation of apoptotic cell death [58]. Hence, a new "para-cholinergic" system, where the AChE peptide would serve as the signaling molecule, is characterized by (1) more generalized mechanism involving the body and brain; (2) dependence on and limitations to the availability of choline and the signaling via an allosteric site on the α7 nAChR linked to an intracellular cascade mediated by calcium; (3) involvement of the endocrine and immune system, besides nervous one; (4) slow action (seconds to minutes vs. milliseconds as in the cholinergic system); (5) long-term effects (neurite outgrowth); (6) inclusion of glia [59], folliculostellate endocrine cells [60], chromaffin cells [61], and cancer cells [62] in the system, besides only neurons. Nowadays, we are experiencing an increased number of evidence that AChE is linked with many diseases in somewhat different ways. Not all of its roles are clear, but it is evident that they are very complex. Hence, it is essential to study further this enzyme and its functions and the moderators of its activity to understand them better [9].

In vertebrates, besides the AChE, another cholinesterase efficiently hydrolyzes the ACh and is called butyrylcholinesterase [63]. Although these two enzymes have the same function, they use different substrates, tissue distribution, and sensitivity to inhibitors [47, 64]. Butyrylcholinesterase is well recognized for its role in metabolizing bioactive esters in food and medications. It is best known for inactivating succinylcholine, a muscle relaxant used to facilitate abdominal surgery but endangers patients with "atypical BChE" that hydrolyzes it poorly. Although some BChE genetic variants are associated with an elevated risk of cardiac death [65, 66] or early-onset Alzheimer's disease [67], this enzyme was not known to play a direct role in mammalian physiology. But in 2004, De Vriese reported that BChE is capable of hydrolyzing the acylated peptide known as "ghrelin," which stimulates hunger and food-seeking [68]. This deacylation reaction cleaves the octanoyl group essential for ghrelin activity at the growth hormone secretagogue receptor, "GHSR1a," which drives growth hormone release from the pituitary gland [69, 70]. Ghrelin affects a wide array of physiological functions and pathological states, including insulin release [71], adiposity [72], energy homeostasis [73], and the development and progression of several types of cancers [74]. Besides, ghrelin is also involved in psychosocial states, memory, and learning [75]. BChE looks to be a key regulator of ghrelin in the peripheral circulation with widespread impacts [76].

The gene that codes BuChE displays polymorphisms, resulting in several enzymes with different activity levels, including enzymatically silent variants. Moreover, BuChE exists in several molecular forms, including monomers and oligomers, consisting of identical catalytic subunits [77].

AChE is consistently associated with both cholinergic and cholinoceptive neurons. At the same time, BuChE immunoreactivity has been detected in all brain regions using an enzyme-linked immunosorbent assay [78]. AChE is expressed at exceptionally high levels in the hippocampus formation [79, 80] and the motor, premotor, and neocortical areas of the human cerebral cortex [81]. BuChE is also expressed in the hippocampus and temporal neocortex, still at lower levels than AChE [79]. Mesulam et al. [79] reported that hippocampal and neocortical AChE is localized in the axons and pyramidal neurons, while BuChE is associated with glial cells. However, Racké and Matthiesen [16] reported that, in hippocampal formation, AChE is present in both neurons and neuropil. At the same time, BuChE is only detected in neurons and suggested that these enzymes colocalize. In the amygdala, the number of BuChE-positive neurons is reported to exceed the number of AChE-positive neurons, with BuChE residing predominantly in the neurons and their dendritic processes and AChE residing in the neuropil [82]. The distinct distribution of AChE and BuChE within the brain suggests that these enzymes may both play important biological roles [83].

Evidence for the role of BuChE in cholinergic signaling in humans comes from a study that demonstrated that BuChE could hydrolyze acetylthiocholine in human brain tissue treated with the AChE inhibitor BW-284C51 [79]. Together, these studies in AChE-knockout mice and human brain tissue have shown that BuChE can hydrolyze ACh and can compensate for AChE when levels are depleted.

In addition to its role in the hydrolysis of ACh, BuChE is also known to have nonenzymatic functions. It has been suggested that AChE may accelerate amyloid deposition in Alzheimer's brain [51]. BuChE can associate with amyloid-β (Aβ) protein and may delay the onset and rate of neurotoxic Aβ fibril formation in vitro [84]. AChE and BuChE may also be involved in inflammatory pathways [83, 85].

Choline acetyltransferase is the enzyme in charge of the biosynthesis of acetylcholine. ChAT is presently the most specific indicator for monitoring the functional state of cholinergic neurons in CNS and peripheral nervous systems. Choline acetyltransferase is a globular protein composed of a single strand. ChAT is synthesized in the perikaryon of cholinergic neurons. From the synthesis position, it is later moved most likely by both slow and rapid axoplasmic flows to the terminals of nerve cells. At the terminals of cholinergic nerve cells, ChAT exists in either soluble or non-ionically membrane-bound form. Three species of mRNA, R, N, and M types of mRNA, are transcribed from various promoter regions in mice, rats, and humans. They are also produced by different splicing processes. In rodents, all transcripts encode the same protein. In humans, only M-type mRNA can generate both large and small forms of ChAT protein, while R- and N-type mRNAs generate only the small form of protein, which is the same as the rodent type of ChAT protein. There is a unique property related to the genomic structure of ChAT in comparison to other enzymes responsible for neurotransmitters synthesis. The ChAT gene has the first intron with the open reading frame encoding vesicular acetylcholine transporter, protein accountable for the ACh transport from the cytoplasm into the synaptic vesicles. It means that the expressions of choline acetyltransferase and vesicular acetylcholine transporter are coordinately regulated in cholinergic cells, probably by multiple regulatory elements. In the central nervous system, studies showed the exact localization of cholinergic neurons. They could be found in the basal nucleus of Meynert, the nucleus of the diagonal band of Broca, the medial septal nucleus, the caudate nucleus, the nucleus accumbens, the pedunculopontine and laterodorsal tegmental nucleus, the putamen, the medial habenular and parabigeminal nucleus, cranial nerve nuclei, and the anterior horn of the spinal cord. Cholinergic neurons project fibers to many areas in the CNS and the other areas of the human body, constructing a complicated cholinergic network with many vital roles in various physiological properties. On the other hand, this complex system's disturbances lead to many diseases, varying from neurodegenerative to different syndromes with completely non-related etiologies. The abnormalities of choline acetyltransferase in the brain are involved in schizophrenia and sudden infant death syndrome [86]. The abnormalities of choline acetyltransferase in the other parts of the human organism are still to be connected to the respective disorders.

29.2.3 ACh Receptors

The cholinergic system is major neurotransmitters systems involved in learning and memory [1].

The ACh receptor is an essential protein in the cell membrane, interacting with ACh and thus allowing it to act as a neurotransmitter. The cholinergic receptors are divided into two groups based on their exogenous agonists: muscarinic and nicotinic ACh receptors. In different regions of the brain, the expression of AChRs is diverse. Studies showed that nicotinic receptors could be found throughout the CNS, including the neuromuscular junction and autonomic ganglia. Muscarinic receptors are expressed in the brain both at pre- and postsynaptic nerve terminals, as well as parasympathetic effector organs [87].

29.2.3.1 The Muscarinic Acetylcholine Receptors

The muscarinic acetylcholine receptors are members of the family of seven transmembrane receptors coupled to G-proteins (GPCRs). They regulate different physiological processes [88]. The muscarinic acetylcholine receptor is consisted of a single polypeptide, forming seven transmembrane domains to build a central pore. Inside this pore is a binding site for acetylcholine. When ACh is bounded to the central pore, the signaling cascade via G-proteins is activated [89]. The prominent role of muscarinic acetylcholine receptors is exerted in CNS and includes the regulation of sensory, motor, and autonomic processes [90]. The involvement of this group of receptors in learning and memory is well established [91].

The muscarinic group of acetylcholine receptors is divided into five types, labeled as M1, M2, M3, M4, and M5, and encoded by five different genes [92]. The structure of different types of muscarinic receptors is similar apart from the third intracellular loop. Their signaling attributes are different, so they are further divided into two subgroups based on that [93–95]. Subgroups determine the specific coupling preferences of these receptors [96]. M1, M3, and M5 are coupled to Gαq/11 protein, which causes the activation of phospholipase C. Stimulation of these receptors as a consequence has the process of phosphorylation, allowing the regulation of different proteins and their functions by that process. On the other hand, M2 and M4 are coupled to Gi/o, inhibiting adenylate cyclase [97]. The stimulation of these receptors, as a result, has a reduced cAMP level in the cytosol [98]. The intracellular signaling due to the muscarinic stimulation includes the activation of protein kinases, phospholipases A2 (releasing arachidonic acid) and D (releasing choline), as well as the regulation of potassium and calcium channels [99]. In the human body, muscarinic acetylcholine receptors are distributed centrally and peripherally, and they are involved in both cognitive processes such as memory [100–102] and motor function [1, 103].

29.2.3.2 The Nicotinic Acetylcholine Receptors

The nicotinic acetylcholine receptors are ligand-gated ion channels. These receptors are formed as the congregation of five subunits gathered around one central pore. They can be found in homomeric or heteromeric conformation [104]. The common subunits found in these molecules are α2–α9 and β2–β4 [105]. The neural subunits eligible to form heteromeric receptor with αβ subunit combinations are α2–α6 and β2–β4. Functional homomeric receptors could be formed from subunits α7–α9 [104]. The β subunits alone are unable to form a functional receptor. At the same time, α2–α6 subunits alone can only make receptors with a weak response to ligand. Based on that, only the combination of α and β receptors can form a fully functional receptor and provide an adequate physiological role [106]. It was shown that α subunits are responsible for agonist recognition and contain a binding site. In contrast, β subunits can increase affinity towards agonists and stabilize the whole receptor in that manner [106]. Among various combinations of subunits' stoichiometries, the most common in CNS are $(\alpha 7)_5$, $(\alpha 4)_2(\beta 2)_3$, and $(\alpha 4)_3(\beta 2)_2$ nAChR types [107]. Each subunit consists of two hydrophilic extracellular segments, N- and C-terminals of the protein, four transmembrane domains labeled as M1–M4, and the intracellular loop between domains M3 and M4 [11]. It was shown that phosphorylation sites could be found within that loop [11] and that M2 forms the central pore [104]. For the central pore to be open, two molecules of ligand must be bind to the receptor. With that action, the permeation of cations Ca^{2+}, Na^{+}, and K^{+} is enabled [11].

In the mammalian brain, the α4β2- and α7-type receptors are most abundant [105]. In rodents, the α4β2 receptor is the most represented combination in all layers of the cerebral cortex and hippocampus [106], while the α7 receptor is mostly expressed in basal forebrain neurons [108, 109].

The function of the nicotinic acetylcholine receptors is regulated by the binding of ligands, nicotine, or ACh. ACh binds to the receptor's extracellular N-terminal domain at the boundary between α- and non-α-subunits [110]. The binding causes an influx of different cations, especially Ca^{2+} ions, inside the cell [111]. This nAChR mediated Ca^{2+} entry causes a marked increase in intracellular Ca^{2+} concentration, which is adequate to initiate Ca^{2+} sensitive processes [1, 11].

29.2.4 Vesicular Acetylcholine Transporter

The vesicular acetylcholine transporter mediates the storage of acetylcholine by synaptic vesicles. However, the release of acetylcholine independent from VAChT is found to be important during development. The vesicular acetylcholine transporter exchanges cytoplasmic acetylcholine for two vesicular protons [112, 113]. A decrease in VAChT expression has functional consequences for mammals. The study with mice with a 70% reduction in the expression levels of VAChT showed that mice are myasthenic and have cognitive deficits [114]. Hence, vesicular transport activity is rate-limiting for neurotransmission "in vivo" [114, 115]. Exocytosis of synaptic vesicle contents is the principal mechanism for the secretion of neurotransmitters and its regulation [116]. Still, there are some alternative mechanisms of neurotransmitters' secretion [117–119]. Like those in nerve terminals, ACh release in quants has also been detected in myocytes and fibroblasts, cells with no VAChT expressed [120, 121]. The result indicates that the release of acetylcholine during development is not dependent on VAChT. The question is whether the release is nonvesicular or vesicular storage can occur without VAChT [122]. Besides, decreased VAChT expression leads to decreased endogenous cholinergic signaling, resulting in pronounced allergic airway inflammation [123].

29.3 Pulmonary Cholinergic System

Besides the lung's primary function in gas exchange, there are some additional roles: lungs are a barrier and crucial defense when it comes to environmental contaminants and pathogens. These essential physiological functions are regulated by acetylcholine produced and released by neuronal and non-neuronal sources and their action on muscarinic receptors. Acetylcholine action provokes airway smooth muscle to contract and control the tone and regulates the openness of the conducting airways. Also, acetylcholine is responsible for smooth muscle relaxation and vasodilation in blood vessels. Acetylcholine was shown to regulate mucus secretion and clearance at mucosal glands and epithelial cells. Last but not least, acetylcholine has a role in inflammation modulation [124].

The main sources of acetylcholine in the lung are parasympathetic nerves. They synthesize and release acetylcholine in this part of the human body. They innervate all conducting airways and pulmonary blood vessels [125–127]. Due to the stimulation, parasympathetic nerves release acetylcholine, which acts on muscarinic receptors on airway smooth muscle and causes bronchoconstriction.

Like the other parts of the human body, acetylcholine synthesis and release are not restricted to cholinergic neurons in the lung. Both epithelial cells [128, 129] and endothelial cells [127, 130] in the pulmonary system can synthesize and release acetylcholine. Those cells express choline acetyltransferase, the enzyme in charge of the synthesis of acetylcholine; hemicholinium-3 sensitive choline transporters that transport choline into cells [131]; and vesicular acetylcholine transporters, responsible for packing acetylcholine into vesicles. Also, organic cation transporters OCT1 and OCT2 are expressed in the luminal membranes of ciliated epithelial cells. OCT1 and OCT2 are polyspecific organic cation transporters that can transport acetylcholine. They were shown to directly release acetylcholine into the lumen of the airway [132, 133]. In pulmonary arteries, acetylcholine induces vasodilation throughout its action at muscarinic receptors [134, 135].

The sources of acetylcholine are the endothelial cells. The expression of ChAT in these cells is mosaic. It suggests that acetylcholine production is related to local mechanical forces' differences due to blood flow [124, 130].

The studies of muscarinic receptor distribution in lung tissues by autoradiographic labeling showed that muscarinic receptor density is highest in parasympathetic ganglia, mucous glands, smooth muscle, and nerve fibers [136, 137]. Subtype-specific distribution and function were also determined using pharmacological analysis, in situ hybridization, RT-PCR, and knockout mice [124]. Muscarinic acetylcholine receptors are present on both preganglionic and postganglionic parasympathetic nerves in the lungs. They are densest at the ganglia [136, 137]. Receptors expressed on parasympathetic nerves control acetylcholine release by postganglionic nerves at target tissues due to the modulation of neurotransmission between the pre- and postganglionic nerves. Muscarinic receptors control neurotransmission across this synapse in the lungs [138]. Preganglionic neurons in the lungs express inhibitory M2 receptors at the synapse. Inhibitory M2 receptors on postganglionic parasympathetic nerves are very important for proper lung functioning. M2 receptors are activated by acetylcholine, and that activity inhibits additional acetylcholine release. In that way, mucus secretion and bronchoconstriction in a healthy organism are controlled [139, 140]. Receptors from the M1 group are present in postganglionic nerves. The importance of these receptors in terms of synaptic neurotransmission modulation varies in different species. In allergic humans, the role of M1 receptors in facilitating parasympathetic neurotransmission and bronchoconstriction decrease was proven.

On the other hand, the role of M1 receptors in healthy humans is still not clear [141]. Bronchoconstriction in the lungs caused by acetylcholine is enabled via smooth muscle contraction [142]. Studies from the 1990s proved M2 and M3 muscarinic receptors' presence on airway smooth muscle [143–146]. From a physiological point of view, the role of M3 receptors is key in smooth muscle contraction. M2 receptors are more represented than M3 receptors, but their airway smooth muscle contraction role is indirect. They can inhibit induced smooth muscle relaxation on the airway [143]. Neuronal acetylcholine does not relax precontracted human pulmonary arteries, but exogenous acetylcholine will if the endothelium is intact [134].

Acetylcholine acts on muscarinic receptors on endothelial cells and causes nitric oxide production. Due to nitric oxide production, the smooth muscle relaxes [134, 135, 147]. M3 receptors are important for vasodilatation and blood pressure decrease in vivo [148]. It was shown that active muscarinic receptors in epithelial cells in the lungs increase intracellular calcium [149] and ciliary beat frequency [149–151]. That increases the transport of mucus and particulates out of the lung. M3 receptors are dominant and only required for the full increase in ciliary beat frequency and particle transport speed [149], while M1 and M2 receptors can indirectly contribute to the overall increase of ciliary transport speed [150]. Also, in the last few decades, the role of M1 and M3 receptors in immune response in the lungs has been demonstrated in alveolar macrophages, mast cells, and airway epithelial cells. The investigation of the role of inhibitory M1 receptors in human mast cells showed that acetylcholine inhibits evoked histamine release [152].

On the other hand, M3 receptors exert their action in alveolar macrophages, causing acetylcholine to induces the release of leukotriene B_4, affecting inflammation in airway epithelial cells [153, 154]. Acetylcholine is also able to contribute to airway remodeling. By its action on muscarinic receptors, fibroblasts, and smooth muscle, cell proliferation is increased [155, 156]. Human lung fibroblasts express M1, M2, M3, and M4 muscarinic receptors, but M2 is dominant [157, 158].

Besides its well-known primary role as a neurotransmitter, acetylcholine can be released by and act on non-neuronal cells. In non-neuronal cells, the system in charge of synthesis, transport, reception, and acetylcholine degradation is labeled as the non-neuronal cholinergic system (NNCS). It was shown many times that the NNCS is dysregulated in various diseases. Those alterations influence the pathology of the mentioned diseases. Still, we don't know much about the expression and function of the NNCS in many organ systems. It would be extremely useful to dig deep in that area and establish new promising tools for a series of disorders connected with pulmonary and other NNCS.

While ACh was considered only as a part of CNS in humans until the 1990s, in the view of phylogenesis, the non-neuronal cholinergic system already existed before the nervous system was developed in non-neuronal cells like bacteria, algae, and protozoa [159]. After the rediscovery of the NNCS in the 1990s, the research was focused on the distribution, the functions, the molecular components, and its involvement in pathological conditions. Although almost all cell types express cholinergic receptors, that alone is not enough to label them as part of NNCS. They should also synthesize and release acetylcholine to be called the cells of NNCS, and usually, they contain the same machinery for the complement of that task as neurons. As already mentioned, acetylcholine is synthesized by choline acetyltransferase in most of the neurons and non-neuronal cells [159, 160]. Today, it is known that acetylcholine could also be a product of the enzyme carnitine acetyltransferase (CarAT) in some non-neuronal cells, such as skeletal muscle cells and the urothelium [161, 162]. Again, the rate-limiting step in this synthesis is the reuptake of choline, as the essential nutrient. In neurons and some other cells, the high-affinity choline transporter-1 (CHT1) is in charge of that [163–166]. In other cases, choline transporter-like proteins (CTL1-5) [167, 168] or organic cation transporters (OCTs) are responsible for choline reuptake. Also previously mentioned, acetylcholine is stored and released by the vesicular acetylcholine transporter in neurons [169] and some specific non-neuronal cells [170]. Some cells do not express vesicular acetylcholine transporter, and they usually do not store acetylcholine, but they release it directly via OCTs [133, 171]. The studies acknowledged the presence of mediatophore [171], a protein of 220 kDa consisting of 15-kDa proteolipid subunits of the vacuolar H^+-ATPase, responsible for acetylcholine exocytosis in some cells, enabling extracellular acetylcholine to exerts its effect on a variety of nicotinic and muscarinic receptors.

The respiratory system includes the nose, nasopharynx, trachea, and lung. The mentioned organs are required for the transport and exchange of the respiratory gases oxygen and CO_2. The respiratory epithelium lines up the surface in the lumen of the airways. The respiratory epithelium consists of at least 12 cell types. Besides epithelial cells, airway pathophysiology involves other types of the cell, such as fibroblasts and inflammatory cells. Besides, autonomic nerve fibers take place close to the airways. It is hard to make a difference between the effects caused by the neuronal or the non-neuronal cholinergic system. It was previously discussed that neuronal acetylcholine triggers mucus secretion and bronchoconstriction via muscarinic receptors [172]. Nicotinic acetylcholine receptors were also found in the airways, precisely on fibroblasts, immune cells, and the respiratory system [3, 173].

Nicotinic acetylcholine receptor labeled as α7nAChR is responsible for regulating lung inflammation and cytokine release in acute models of inflammation related to lung pathophysiology. Both neuronal and non-neuronal acetylcholine can induce anti-inflammatory effects via α7nAChR in different models of inflammation [174–176]. According to one model, stimulation of α7nAChR provoke the expression of the SOCS3 protein. This action further leads to the downregulation of the JAK-2/STAT-3 pathway [177]. The consequence of this event is a decrease in pro-inflammatory cytokine production, especially TNF-α [123, 178, 179].

29.4 Cholinergic System and Lung Cancer

The cholinergic system is directly involved in apoptosis. Apoptosis represents a homeostatic mechanism that involves genetically determined elimination of cells throughout development, differentiation, and aging. Dysregulation of apoptosis necessarily leads to pathology, mainly malignancies.

Apoptosis occurs via two distinct, precisely controlled biochemical pathways: the extrinsic or death receptor-mediated pathway and the intrinsic or mitochondria-mediated pathway [180]. The extrinsic pathway involves ligation of cell surface death receptors by their respective ligands initiating the caspase-8 activation, which triggers apoptosis. In contrast, the intrinsic pathway, which is largely controlled by the Bcl-2 family of pro- and anti-apoptotic proteins, starts inside the cell as a response to death signals caused by different types of cellular stress with additional activation of caspase-9. Both pathways initiate caspase cascade, which leads to the activation of executioner caspases (caspase-3) and thus triggers the irreversible apoptotic program [180]. Although there is a wide range of physiological and pathological stimuli that may trigger apoptosis, not all cells will necessarily die; abnormalities in cell death regulation pathway can cause different types of abnormalities and disorders, primarily cancer [9].

Studies in vitro and in vivo have shown that besides its pivotal role in neurotransmission, acetylcholinesterase is involved in regulating cell proliferation, differentiation, hematopoiesis, and apoptosis [64]. Moreover, irregular expression of AChE has been found in several types of tumors, suggesting the involvement of AChE in the tumorigenesis [30, 181]. As mentioned, AChE is encoded by a single gene, but there are three isoforms of this enzyme, which differ in their carboxy-terminal sequences; synaptic AChE (AChE-S), erythrocytic AChE (AChE-E), and read-through AChE (AChE-R) [182]. All AChE isoforms selectively contribute to the processes involved in promoting or attenuating cell death [183]. However, since AChE-S has an important role in forming an apoptosome and its expression may be elicited in different cell types during cell death processes, that isoform has been considered a potential marker regulator of apoptosis [33].

In contrast to AChE-S that promotes cell death, the AChE-R has the opposite effect and positively regulates cell proliferation [183]. In general, the AChE displays a pro-apoptotic function. The AChE is upregulated in response to different apoptotic stimuli, thus promoting apoptosis. In contrast, knockdown of its expression either by antisense RNA (asRNA), small interfering ribonucleic acids (siRNAs), or heterozygous deletion of the AChE gene attenuates apoptosis [184]. These data indicate the possibility that the pro-apoptotic function of AChE may play a role in tumor suppression.

At the early stage of apoptosis, AChE occurs in the cytoplasm and then translocates to the nucleus, while at late stages, AChE is only present in the fragmented nuclei [64]. These findings indicate that AChE participates in modulating nuclear components, causing the chromatin condensation and fragmentation, which confirms its function in apoptosis. According to Zhang et al., the AChE is expressed during apoptosis in several cell lines. They have noticed that in some cells, such as PC12 cells that normally express a low level of AChE or in cells that normally do not express AChE (human lung fibroblasts (HLF) cells and rat kidney (NRK) cells), an increase of AChE levels can be observed during apoptosis [64]. Although the inhibition of AChE expression by siRNAs, for instance, may rescue cells from apoptosis, it is believed that overexpression of AChE itself cannot initiate the apoptotic process [185]. In other words, for cell death initiation, the translocation of AChE triggered by internal/external stimuli is required. Recent studies indicate an important role of the Ran-binding protein microtubule-organizing center (RanBPM) in translocation processes of AChE during apoptosis. The RanBPM protein makes a platform for the interaction of different signaling proteins such as cell surface receptors, nuclear receptors, nuclear transcription factors, and tyrosine kinases [186].

Additionally, RanBPM was considered a pro-apoptotic protein due to its ability to activate caspase-3 and its modulating effects on the function and stability of many proteins that regulate cell death [187]. Moreover, it has been found that the RanBPM binds AChE-S during translocation to the nucleus [186]. Although the translocation mechanism remains to be clarified, the fact that RanBPM is an AChE interacting protein and is considered a crucial regulator of cell proliferation confirms the importance of this protein in the cell death process and tumorigenesis.

Some recent studies have demonstrated a novel role of AChE-S as a DNase. Considering that AChE-S is a bifunctional enzyme with acetylcholine hydrolysis and DNA cleavage domains, Du et al. assumed that translocation could be a crucial step that figured out the function's change of AChE-S as a cholinesterase to a function as a DNase function [188]. In this study, AChE-S was found to hold intrinsic endonuclease activity between amino acids 32–138 and promote apoptosis only after translocation into the nucleus. These findings implicate the pro-apoptotic functions of AChE independent of its cholinesterase function [188]. Overall, these results represent a very important platform for further research on the role and mechanisms by which AChE participates in cell death.

In recent years, it has been demonstrated that AChE participates in the formation of apoptosome by influencing the interaction between apoptotic protease activating factor-1 (Apaf-1) and cytochrome-c, which consequently activate procaspase-9. Generally, mitochondrial apoptotic signaling leads to the release of cytochrome-c that binds to apoptotic protease-activating factor 1 (Apaf-1) and procaspase-9, making an apoptosome formation. The apoptosome activates caspase-9 and proceeds down the same common pathway as the extrinsic pathway, which leads to the activation of executioner caspases (caspase-3) and thus triggers apoptosis. Park et al. tried to elucidate the molecular role of AChE in apoptosome formation [189]. They have shown that AChE plays a pivotal role in the oligomerization of Apaf-1 and that the interaction between AChE and caveolin-1 and, subsequently, cytochrome-c is indispensable for the interaction between cytochrome-c and Apaf-1 and apoptosome formation [189]. Interestingly, the silencing of the AChE gene by small interfering ribonucleic acids (siRNAs) abolishes the expression of AChE and disrupts the interaction of Apaf-1 and cytochrome-c, subsequently blocking the activation of caspase-9 and preventing apoptosis [190].

The pathology's development and progression depend on a rate of disturbance in the balance between cell proliferation, survival, and cell death (apoptosis). As mentioned above, the expression and activity of AChE vary between different types of tumors. Moreover, a significant difference in AChE expression levels between tumor tissue and normal tissue was confirmed. In general, the alteration of AChE expression levels depends on the cell type, which confirms the complex but important role of AChE in apoptosis. Although the involvement of AChE in tumorigenesis remains unclear, current findings indicate that AChE is indeed involved in this process [191].

In recent years, it has been demonstrated that the phosphatidylinositol 3-kinase (PI3K)/Akt signaling pathway regulates numerous cellular functions such as proliferation, growth, survival, and tumor progression in response to different cellular stimuli or toxic injuries. To preserve cell viability, activated PI3K activates the Akt pathway via its downstream kinase cascade, maintaining the phosphorylation status of Akt proteins [192]. Although this pathway plays a key role in various physiological processes, it also represents one of the main pathways constitutively activated in several human tumors. The inhibition of Akt signaling is known to induce apoptosis. These findings indicate that AChE is capable of reducing cell differentiation [193] and inhibiting signal transduction via the PI3K/Akt pathway [194]. Notably, recent studies indicate an important role of a novel form of AChE proteins in apoptosis [195]. The 55-kDa AChE protein originated from the cleavage of 68 kDa AChE, expressed after the activation of caspases, is present during apoptosis progression.

On the contrary, the 55-kDa AChE protein's decreased expression might contribute to angiogenesis and be related to tumorigenesis [196]. According to the study of Xie et al., considering the relationship between the activation of the PI3K/Akt pathway and the formation of the 55 kD AChE protein, the activation of the Akt cascade suppresses the formation of the 55 kDa AChE protein. It is observed that PI3K promotes this inhibitory effect via the activation of the endogenous Akt pathway [180, 196]. Tumor suppressor, P53, is a regulating apoptotic protein with a key role in cell apoptosis and tumorigenesis [180, 197]. Ye et al. have demonstrated that AChE and p53 are associated with the cell apoptotic pathway suggesting that AChE could be a downstream component of the p53 capable of inducing apoptosis [9, 53].

As mentioned, the lung tissue expresses the cholinergic system, including nicotinic acetylcholine receptors, which is included in many physiologic and pathologic processes [198].

Lung cancer is the most frequently diagnosed cancer and one of the leading causes of death worldwide. It was established that lung cancer is mostly caused by smoking [199]. Other factors contributing to lung cancer development are various occupational and environmental contaminants and numerous unknown factors [199]. Lung cancers are called small-cell carcinomas (SCLC) and non-small-cell carcinomas (NSCLC). The division is formed due to the morphological differences in tissues and clinical outcomes [200]. SCLC cover about a quarter of all lung cancer cases. It is closely related to smoking and metastasizes very quickly [201–203]. NSCLC are classified as large-cell carcinomas, squamous cell carcinomas, and adenocarcinomas. Regardless of the great advance in diagnostic test development, chemotherapeutic treatments, and surgical procedures, the survival rate for patients suffering from lung cancer remains low because of a lack of specific biomarkers for early screening [204].

The investigation showed that normal human lung fibroblasts do not express AChE. On the other hand, it was proven that the expression of AChE is upregulated during apoptosis [64, 185]. The AChE could be considered as a marker of early cancer differentiation. Different studies showed that the AChE gene could be either deleted, mutated, or amplified in various cancers [181, 205]. There are many examples of decreased AChE activity in cancer patients, but the opposite was found [30, 181, 206, 207]. Hence, it is necessary to further thoroughly examine the role of AChE in various types and stages of cancer development.

It was shown that AChE activity was significantly lower in different subtypes of lung cancer patients compared to control [208, 209]. As a consequence of decreased expression and activity of AChE, the acetylcholine level in lung cancer tissue is elevated [209]. It was shown that acetylcholine promotes lung cancer cell growth [210]. Due to the unusually high level of acetylcholine, cell proliferation is enhanced, proving the involvement of AChE in the regulation of lung cancer cell growth. The exact role of AChE in this process is still not fully understood. Recent studies showed that AChE gene expression is always decreased in lung cancer tissues. Due to that, AChE is recognized as a regulator in cell proliferation and death, advocating the use of AChE as a potential marker for lung cancer diagnosis and prognosis. In the next decades, AChE enhancers could be applied as drugs, alone or in combination with conventional cytostatics, and be the future of lung cancer treatment [29].

It was proven that α7nAChR is responsible for the proliferative, pro-angiogenic, and pro-metastatic effects of nicotine in lung cancer [211]. Nicotine exerts its biological effects through nAChRs in human lung cancer cells [212], and cancer development is initiated due to the activation of the α7 subunit of nAChRs [213–216]. α7nAChR is expressed in squamous cell lung cancer cells, lung adenocarcinoma, and SCLC [217–221].

29.5 The Cholinergic System in Asthma and Chronic Obstructive Pulmonary Disease

Asthma is a chronic inflammatory disorder of the airways involving many different cell types and cellular elements. Rather than being a single disease, it consists of related, overlapping syndromes. Asthma includes three domains of symptoms: (1) variable airway obstruction, (2) airway hyperresponsiveness (or bronchial hyperreactivity), and (3) airway inflammation, no one of which is essential for diagnosis [222]. Airway hyperresponsiveness associated with asthma leads to recurrent wheezing episodes, shortness of breath, chest tightness, and coughing [223]. The etiology of asthma remains undefined despite advances in understanding this disease's pathogenesis, genetics, and clinical characteristics [224]. Causes of asthma could be both genetic and environmental, each contributing to about 50% of the risk of the disease [225]. Environmental factors that cause asthma induce airway inflammation with eosinophils (more common) or neutrophils and airway hyperresponsiveness. Chronic exposure to environmental allergens (such as dust) initiates immunoglobulin E (IgE)—a mediated reaction that is probably the cause of the lower airway inflammation [222]. The genetic bases of asthma are undeniable, but not yet well explained. There is evidence that the development of asthma could be genetically predisposed. The researcher identified a number of chromosomal regions associated with asthma susceptibility. Some of them are associated with the assembly of IgE antibodies and the mediators of inflammation production. Still, more investigation is necessary to identify the exact genes implicated in asthma development. Besides, the gene-environment interactions that may lead to the disease's expression should also be taken into account [226]. Variability in the clinical phenotype caused by the multiple molecular mechanisms underlying the complex pathological processes involved in disease development and progression makes it hard to explain polymorphisms, which are believed that cause the disease [227]. Like other atopic conditions, asthma is related to T helper cell type-2 (Th2) immune response.

Triggers for asthma development may be allergic stimuli, such as dust, dander, mold, and pollen, and nonallergic stimuli, including viral infections, cold air, exercise, and tobacco smoke. All those stimuli cause events that, as a result, have chronic airway inflammation. Th2 cells in airways release specific cytokines—interleukins like IL-4, IL-5, IL-9, and IL-13. Mentioned interleukins lead to eosinophilic inflammation and immunoglobulin E production. Increased IgE levels, as a result, have the release of histamine and cysteinyl leukotrienes, inflammatory mediators that cause bronchospasm, edema, and increased mucous secretion. All these manifestations of inflammation promotion lead to the development of characteristic asthma symptoms [228]. The inflammatory response is further propagated, leading to the late-phase asthmatic response characterized by progressive airway inflammation and bronchial hyperreactivity [228]. Ultimately, the airway remodeling caused by recurrent asthma episodes, as a result, has further lung function decline and airway obstruction [226, 229].

Chronic obstructive pulmonary disease (COPD) is generally defined as slowly progressive airflow obstruction, which is only partially reversible [230]. COPD implies a set of conditions that develop progressively due to a number of different disease processes, most commonly in patients with chronic bronchitis and emphysema [231]. Clinical symptoms usual for chronic bronchitis are excessive cough and sputum production. On the other hand, emphysema is related to chronic dyspnea, and it is a consequence of distended air spaces and destroyed lung tissue [232]. Many COPD patients have both of these conditions. The main cause of COPD is long-term exposure to lung irritants that damage the lungs and the airways. Cigarette smoking is the usual cause of COPD. Up to 75% of people who have COPD smoke or used to smoke. Other irritants include secondhand smoke, air pollution, chemical fumes, and dust from the environment and workplaces [231]. The cause of COPD can also be genetic. The genetic condition called alpha-1 antitrypsin deficiency is rare but should be considered if a person under the age of 40 develops COPD. A low level of alpha-1 antitrypsin can lead to lung damage and COPD development. Suppose the patient diagnosed with this condition smokes or is exposed to smoke. In that case, COPD can worsen quickly [232]. Cigarette smoking or exposure to harmful agents in most cases is leading to the inflammation of lung tissue, bronchial airway disease, and the destruction of parenchyma. The destruction of alveoli is a characteristic of emphysema. As a result of the lungs' inability to drain, the air is trapped inside, which has dyspnea as a consequence. In the late stages of COPD, the diaphragm is flattened, the rib cage is enlarged, and hypoxemia and pulmonary hypertension develop. The overall result of pathological processes underlying COPD has increased resistance to airflow and decreased expiratory flow rate. At that stage, the removal of inflammatory stimulus does not reduce the inflammation [232].

The cholinergic system plays a crucial role in asthma and COPD. ACh, a neurotransmitter present in the nerve cells, is synthesized from choline and acetyl—CoA, a thioester used in metabolic reactions and an acceptor and a donor of acetyl groups catalyzed by ChAT, an enzyme found only in cholinergic cells, as already mentioned. ACh is released as a result of the brain signal to the lung in response to the lung's reaction to tobacco smoke or environmental pollutants. ACh has many roles in the lungs. It contracts smooth muscle to control the tone and regulate the patency of conducting airways. It causes vasodilatation and smooth muscle relaxation in blood vessels. It regulates mucus secretion and mucus clearance mucosal glands and epithelial cells, and it modulates inflammation. When produced and released, ACh acts through muscarinic receptors to regulate the lung's primary role – gas exchange. It also helps regulate the lung's other physiological functions, as the lung serves as the barrier against pathogens and environmental contaminants [124]. Airway neurons and non-neuronal cells, such as airway epithelial cells and inflammatory cells, release ACh, which binds to airway muscarinic receptors and triggers smooth muscle contraction and mucus secretion [233]. Preganglionic fibers release ACh at the level of peribronchial ganglia, where postganglionic fibers are generated, which leads to the release of ACh in the bronchial wall. Airway smooth muscle and mucous bronchial glands in the medium-to-large airway are stimulated by this, which results in bronchoconstriction and mucus secretion [233].

The stimulation of nerve endings by tobacco smoke, cold air, dust, or other asthma triggers can instigate acetylcholine release [234]. Abnormal sensitivity of cholinergic receptors often causes narrowing of the airways, resulting in contracting airway smooth muscles when they should not.

Two major mechanisms that lead to bronchoconstriction and mucus secretion in asthma and COPD are:

1. Increased expression and enhanced function of signaling molecules essential for muscarinic receptor-mediated airway smooth muscle contraction.
2. Exaggerated release of neuronal ACh due to neuronal mechanisms associated with inflammation [235].

29.5.1 Treatment of Asthma and COPD with Anticholinergics

Anticholinergics are compounds that are blocking acetylcholine actions. They are used in the treatment of chronic obstructive pulmonary disease and asthma for a long time. The timeline of using anticholinergics in the treatment of asthma and COPD is given in Fig. 29.2 [236].

Anticholinergics' origins are from botanical preparations of the deadly nightshade family (Solanaceae), which was used for centuries in many cultures worldwide. *Datura* is a genus of the Solanaceae family. Present worldwide is the inhalation of the aerosol that these plant roots, stems, and seeds release. At the same time, burning was a treatment for obstructive airway disease for centuries because the antimuscarinic compound, atropin (Fig. 29.3), was released this way. Other potent alkaloids that are released are scopolamine and hyoscyamine, and they lead to intoxication, hallucinations, and poisoning [237, 238].

In the 1800s, British military officers from India introduced *Datura stramonium* to Western medicine. Its use was controversial, but it soon became popular. It was

Fig. 29.2 The timeline of using anticholinergics in the treatment of asthma and COPD

Fig. 29.3 Structure of atropin

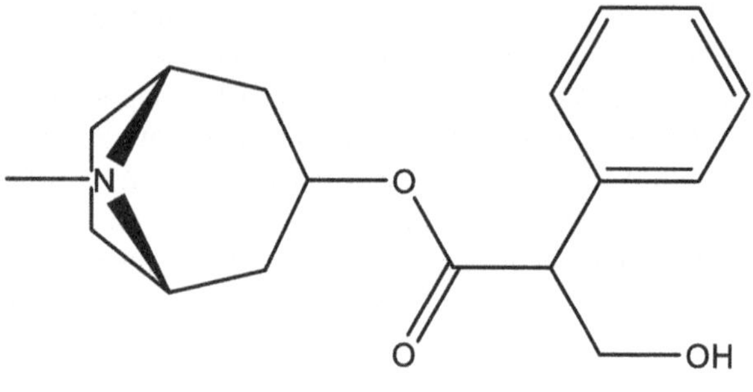

taken in the form of cigarettes or pipe tobacco. Atropin-based agents were in use for almost a century. As atropine is well-absorbed into the systematic circulation and penetrates the blood-brain barrier, many side effects took place during its use. In the 1900s, adrenaline replaced these agents as the first-line treatment for respiratory disorders [239].

As the need to replace β-agonist agents, anticholinergics are back in use in the 1970s. The development of less toxic alternatives of atropin gave them a wider clinical use [240].

The quaternization of tertiary nitrogen of the tropine moiety of atropine and the scopine moiety of scopolamine is accomplished that systemic adsorption of these molecules is prevented and their potent anticholinergic activity is retained. This way, molecules with highly charged quaternary ammonium salts that are poorly adsorbed across membranes and had low oral and systemic bioavailability and low blood-brain barrier penetration are still highly potent antagonists, and muscarinic receptors are synthesized. These compounds have a relatively short duration of action, because of the short residence time in the lungs or at muscarinic receptors, because of which they have to be taken multiple times per day [241].

Anticholinergic therapy is anti-bronchoconstrictor and not bronchodilator therapy as β-agonists are, so the time needed for anticholinergics to start acting is assumed to be longer [241].

Depending on the time of action, anticholinergics are divided into short-acting muscarinic anticholinergics (SAMA) and long-acting muscarinic anticholinergics (LAMA) [223].

29.5.2 Short-Acting Muscarinic Anticholinergics (SAMA)

SAMAs are nonselective anticholinergic agents that block both prejunctional M2 and postjunctional M3 receptors. SAMAs include atropine, ipratropium bromide, and oxitropium bromide. Atropine is rarely used nowadays because it is easily absorbed across the oral and respiratory mucosa, which leads to many side effects. Unlike atropine, ipratropium bromide and oxitropium bromide have low lipid solubility. They do not penetrate the blood-brain barrier, which makes them effective quaternary SAMAs that can be used to treat respiratory diseases [236].

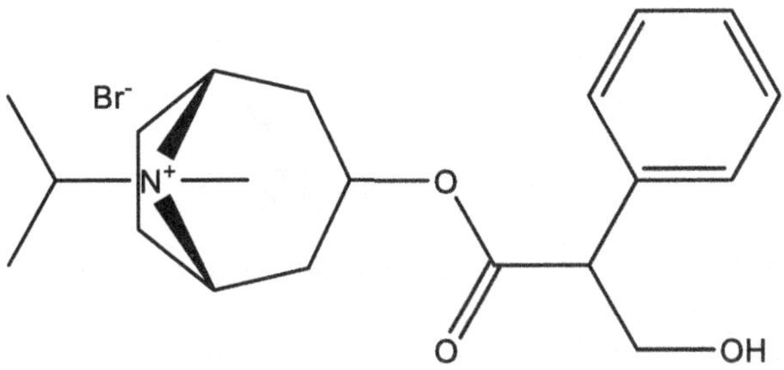

Fig. 29.4 Structure of ipratropium bromide

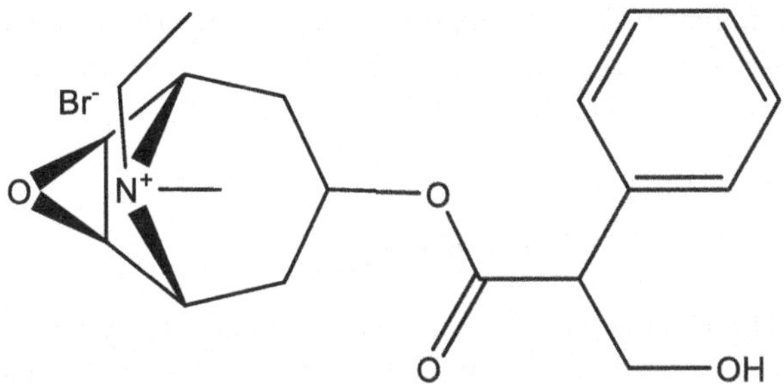

Fig. 29.5 Structure of oxitropium bromide

Ipratropium bromide (Fig. 29.4) starts acting within 15–30 min, but the bronchodilator effect peak may take up to 90 min after inhalation. Ipratropium bromide is useful for patients that cannot tolerate β-agonists because the side effects are very low. The time of action of inhaled ipratropium bromide is 4–6 h, so it is supposed to be taken 4–6 times a day, which is inconvenient [242].

Oxitropium bromide (Fig. 29.5) is, unlike ipratropium bromide, based on a scopolamine molecule instead of atropine. It shows a bit longer duration than ipratropium bromide, which is 5–8 h, and its peak bronchodilation is 60–90 min after inhalation [236].

29.5.3 Long-Acting Muscarinic Anticholinergics

LAMAs take more time to start acting than ipratropium bromide, but last longer, 12–24 h. The most used LAMA is tiotropium bromide.

Tiotropium bromide (Fig. 29.6) also contains quaternary ammonium like ipratropium. Tiotropium bromide binds to all muscarinic receptors, but its most interesting property is its significantly greater duration of binding to M1 and M3 receptors than M2 receptors, which makes this drug selective for M1 and M3 receptors. Binding to muscarinic M2 receptors occurs early after administration, which increases the release of ACh, but as the M3 receptor is also blocked, bronchoconstriction does not occur. Neuronal release of ACh returns to the baseline within 2 h, when M2 receptors are no longer blocked. As the muscarinic M3 receptor's function only begins to return after 7 h, the M2 receptors' blockade is not significant. Bronchodilation in humans reaches a peak in 3–4 h after the

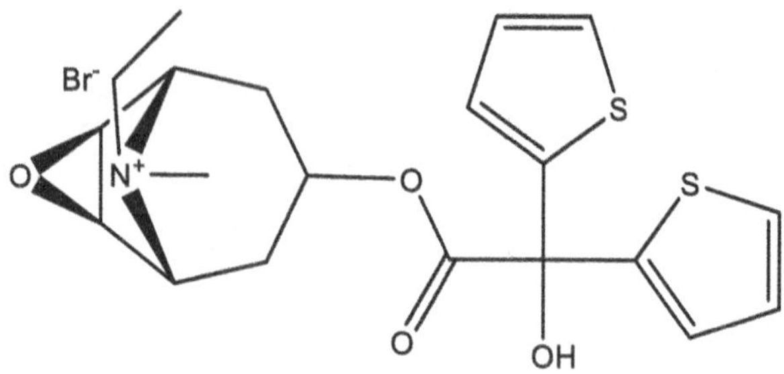

Fig. 29.6 Structure of tiotropium bromide

administration of tiotropium bromide, which is slow and makes this drug inappropriate for rescue medication. The duration of its action is 1–2 days, making it a great once-daily bronchodilator [223, 243].

29.6 Future Perspective

The modulation of the neuronal cholinergic system as a treatment for different diseases is one of today's mainstream choices. On the other hand, targeting the non-neuronal cholinergic system as a therapeutic approach in disease control is still not common. Expression and the non-neuronal cholinergic system's role are still not fully understood, but there are some AChE inhibitors in use for the treatment of lung cancer. The mechanism underlining airway hyperresponsiveness in asthma and chronic obstructive pulmonary disease is still indefinite, but the malfunctioning of the pulmonary cholinergic system is thought to be a contributing mechanism. Concerning that, anticholinergic therapy is used widely and with a high degree of success in treating this condition. This field is highly promising for novel therapy design and should be further investigated. Crucial is to expand our knowledge regarding the exact role of every component of the non-neuronal cholinergic system in the lungs. With that knowledge gathered, we could establish novel approaches in lung disease treatment.

References

1. Ahmed T, Zahid S, Mahboob A, Farhat SM (2017) Cholinergic system and post-translational modifications: an insight on the role in Alzheimer's disease. Curr Neuropharmacol 15 (4):480–494. https://doi.org/10.2174/1570159X14666160325121145
2. Fujii T, Mashimo M, Moriwaki Y, Misawa H, Ono S, Horiguchi K, Kawashima K (2017) Expression and function of the cholinergic system in immune cells. Front Immunol 8(1085). https://doi.org/10.3389/fimmu.2017.01085
3. Beckmann J, Lips KS (2013) The non-neuronal cholinergic system in health and disease. Pharmacology 92(5–6):286–302. https://doi.org/10.1159/000355835
4. Bondžić AM, Lazarević-Pašti TD, Leskovac AR, Petrović SŽ, Čolović MB, Parac-Vogt TN, Janjić GV (2020) A new acetylcholinesterase allosteric site responsible for binding voluminous negatively charged molecules—the role in the mechanism of AChE inhibition. Eur J Pharm Sci 151:105376. https://doi.org/10.1016/j.ejps.2020.105376

5. Lazarević-Pašti TD, Bondžić AM, Pašti IA, Vasić VM (2012) Indirect electrochemical oxidation of organophosphorous pesticides for efficient detection via acetylcholinesterase test. Pestic Biochem Physiol 104(3):236–242. https://doi.org/10.1016/j.pestbp.2012.09.004
6. Lazarević-Pašti T, Nastasijević B, Vasić V (2011) Oxidation of chlorpyrifos, azinphos-methyl and phorate by myeloperoxidase. Pestic Biochem Physiol 101(3):220–226. https://doi.org/10.1016/j.pestbp.2011.09.009
7. Lazarević-Pašti T, Čolović M, Savić J, Momić T, Vasić V (2011) Oxidation of diazinon and malathion by myeloperoxidase. Pestic Biochem Physiol 100(2):140–144. https://doi.org/10.1016/j.pestbp.2011.03.001
8. Lazarević Pašti T, Momić T, Onjia A, Vujisić L, Vasić V (2010) Myeloperoxidase-mediated oxidation of organophosphorus pesticides as a pre-step in their determination by AChE based bioanalytical methods. Microchim Acta 170(3):289–297. https://doi.org/10.1007/s00604-010-0324-2
9. Lazarevic-Pasti T, Leskovac A, Momic T, Petrovic S, Vasic V (2017) Modulators of acetylcholinesterase activity: from Alzheimer's disease to anti-cancer drugs. Curr Med Chem 24(30):3283–3309. https://doi.org/10.2174/0929867324666170705123509
10. Armstrong DM, Saper CB, Levey AI, Wainer BH, Terry RD (1983) Distribution of cholinergic neurons in rat brain: demonstrated by the immunocytochemical localization of choline acetyltransferase. J Comp Neurol 216(1):53–68. https://doi.org/10.1002/cne.902160106
11. Lucas-Meunier E, Fossier P, Baux G, Amar M (2003) Cholinergic modulation of the cortical neuronal network. Pflugers Arch 446(1):17–29. https://doi.org/10.1007/s00424-002-0999-2
12. Ferreira-Vieira TH, Guimaraes IM, Silva FR, Ribeiro FM (2016) Alzheimer's disease: targeting the cholinergic system. Curr Neuropharmacol 14(1):101–115. https://doi.org/10.2174/1570159x13666150716165726
13. Singh M, Kaur M, Kukreja H, Chugh R, Silakari O, Singh D (2013) Acetylcholinesterase inhibitors as Alzheimer therapy: from nerve toxins to neuroprotection. Eur J Med Chem 70:165–188. https://doi.org/10.1016/j.ejmech.2013.09.050
14. Belmonte KE (2005) Cholinergic pathways in the lungs and anticholinergic therapy for chronic obstructive pulmonary disease. Proc Am Thorac Soc 2(4):297–304. https://doi.org/10.1513/pats.200504-043SR
15. Gosens R, Zaagsma J, Meurs H, Halayko AJ (2006) Muscarinic receptor signaling in the pathophysiology of asthma and COPD. Respir Res 7(1):73. https://doi.org/10.1186/1465-9921-7-73
16. Racké K, Matthiesen S (2004) The airway cholinergic system: physiology and pharmacology. Pulm Pharmacol Ther 17(4):181–198. https://doi.org/10.1016/j.pupt.2004.03.001
17. Racké K, Juergens UR, Matthiesen S (2006) Control by cholinergic mechanisms. Eur J Pharmacol 533(1):57–68. https://doi.org/10.1016/j.ejphar.2005.12.050
18. Proskocil BJ, Sekhon HS, Jia Y, Savchenko V, Blakely RD, Lindstrom J, Spindel ER (2004) Acetylcholine is an autocrine or paracrine hormone synthesized and secreted by airway bronchial epithelial cells. Endocrinology 145(5):2498–2506. https://doi.org/10.1210/en.2003-1728
19. Wessler I, Kirkpatrick CJ (2008) Acetylcholine beyond neurons: the non-neuronal cholinergic system in humans. Br J Pharmacol 154(8):1558–1571. https://doi.org/10.1038/bjp.2008.185
20. Wessler IK, Kirkpatrick CJ (2001) The non-neuronal cholinergic system: an emerging drug target in the airways. Pulm Pharmacol Ther 14(6):423–434. https://doi.org/10.1006/pupt.2001.0313
21. Kolahian S, Gosens R (2012) Cholinergic regulation of airway inflammation and remodelling. J Allergy 2012:681258. https://doi.org/10.1155/2012/681258
22. Colović MB, Krstić DZ, Lazarević-Pašti TD, Bondžić AM, Vasić VM (2013) Acetylcholinesterase inhibitors: pharmacology and toxicology. Curr Neuropharmacol 11(3):315–335. https://doi.org/10.2174/1570159x11311030006

23. Small DH, Ismael Z, Chubb IW (1986) Acetylcholinesterase hydrolyses chromogranin a to yield low molecular weight peptides. Neuroscience 19(1):289–295. https://doi.org/10.1016/0306-4522(86)90022-9
24. Jaganathan L, Boopathy R (2000) A direct method to visualise the aryl acylamidase activity on cholinesterases in polyacrylamide gels. BMC Biochem 1:3. https://doi.org/10.1186/1471-2091-1-3
25. Augustinsson KB, Nachmansohn D (1949) Distinction between acetylcholine-esterase and other choline ester-splitting enzymes. Science (New York, NY) 110(2847):98–99. https://doi.org/10.1126/science.110.2847.98
26. Potter PE, Hadjiconstantinou M, Meek JL, Neff NH (1984) Measurement of acetylcholine turnover rate in brain: an adjunct to a simple HPLC method for choline and acetylcholine. J Neurochem 43(1):288–290. https://doi.org/10.1111/j.1471-4159.1984.tb06712.x
27. McMahan UJ, Sanes JR, Marshall LM (1978) Cholinesterase is associated with the basal lamina at the neuromuscular junction. Nature 271(5641):172–174. https://doi.org/10.1038/271172a0
28. Johnson G, Moore SW (2006) The peripheral anionic site of acetylcholinesterase: structure, functions and potential role in rational drug design. Curr Pharm Des 12(2):217–225. https://doi.org/10.2174/138161206775193127
29. Xi HJ, Wu RP, Liu JJ, Zhang LJ, Li ZS (2015) Role of acetylcholinesterase in lung cancer. Thoracic Cancer 6(4):390–398. https://doi.org/10.1111/1759-7714.12249
30. Ruiz-Espejo F, Cabezas-Herrera J, Illana J, Campoy FJ, Muñoz-Delgado E, Vidal CJ (2003) Breast cancer metastasis alters acetylcholinesterase activity and the composition of enzyme forms in axillary lymph nodes. Breast Cancer Res Treat 80(1):105–114. https://doi.org/10.1023/a:1024461108704
31. Massoulié J, Bon S, Perrier N, Falasca C (2005) The C-terminal peptides of acetylcholinesterase: cellular trafficking, oligomerization and functional anchoring. Chem Biol Interact 157–158:3–14. https://doi.org/10.1016/j.cbi.2005.10.002
32. Montenegro MF, Ruiz-Espejo F, Campoy FJ, Muñoz-Delgado E, de la Cadena MP, Rodríguez-Berrocal FJ, Vidal CJ (2006) Cholinesterases are down-expressed in human colorectal carcinoma. Cell Mol Life Sci 63(18):2175–2182. https://doi.org/10.1007/s00018-006-6231-3
33. Soreq H, Seidman S (2001) Acetylcholinesterase—new roles for an old actor. Nat Rev Neurosci 2(4):294–302. https://doi.org/10.1038/35067589
34. Deng R, Li W, Guan Z, Zhou JM, Wang Y, Mei YP, Li MT, Feng GK, Huang W, Liu ZC, Han Y, Zeng YX, Zhu XF (2006) Acetylcholinesterase expression mediated by c-Jun-NH2-terminal kinase pathway during anticancer drug-induced apoptosis. Oncogene 25 (53):7070–7077. https://doi.org/10.1038/sj.onc.1209686
35. Hyatt JL, Tsurkan L, Morton CL, Yoon KJP, Harel M, Brumshtein B, Silman I, Sussman JL, Wadkins RM, Potter PM (2005) Inhibition of acetylcholinesterase by the anticancer prodrug CPT-11. Chem Biol Interact 157–158:247–252. https://doi.org/10.1016/j.cbi.2005.10.033
36. Sussman JL, Harel M, Frolow F, Oefner C, Goldman A, Toker L, Silman I (1991) Atomic structure of acetylcholinesterase from Torpedo californica: a prototypic acetylcholine-binding protein. Science (New York, NY) 253(5022):872–879. https://doi.org/10.1126/science.1678899
37. Bon S, Vigny M, Massoulié J (1979) Asymmetric and globular forms of acetylcholinesterase in mammals and birds. Proc Natl Acad Sci U S A 76(6):2546–2550. https://doi.org/10.1073/pnas.76.6.2546
38. Dvir H, Silman I, Harel M, Rosenberry TL, Sussman JL (2010) Acetylcholinesterase: from 3D structure to function. Chem Biol Interact 187(1–3):10–22. https://doi.org/10.1016/j.cbi.2010.01.042
39. Nachmansohn D, Wilson IB (1951) The enzymic hydrolysis and synthesis of acetylcholine. Adv Enzymol Relat Subj Biochem 12:259–339. https://doi.org/10.1002/9780470122570.ch5

40. MacPhee-Quigley K, Taylor P, Taylor S (1985) Primary structures of the catalytic subunits from two molecular forms of acetylcholinesterase. A comparison of NH2-terminal and active center sequences. J Biol Chem 260(22):12185–12189
41. Wilson IB, Bergmann F (1950) Acetylcholinesterase. VIII Dissociation constants of the active groups. J Biol Chem 186(2):683–692
42. Mooser G, Sigman DS (1974) Ligand binding properties of acetylcholinesterase determined with fluorescent probes. Biochemistry 13(11):2299–2307. https://doi.org/10.1021/bi00708a010
43. Wilson IB, Quan C (1958) Acetylcholinesterase studies on molecular complementariness. Arch Biochem Biophys 73(1):131–143. https://doi.org/10.1016/0003-9861(58)90248-0
44. Taylor P, Lappi S (1975) Interaction of fluorescence probes with acetylcholinesterase. The site and specificity of propidium binding. Biochemistry 14(9):1989–1997. https://doi.org/10.1021/bi00680a029
45. Johnson JL, Cusack B, Davies MP, Fauq A, Rosenberry TL (2003) Unmasking tandem site interaction in human acetylcholinesterase. Substrate activation with a cationic acetanilide substrate. Biochemistry 42(18):5438–5452. https://doi.org/10.1021/bi027065u
46. Dougherty DA, Stauffer DA (1990) Acetylcholine binding by a synthetic receptor: implications for biological recognition. Science (New York, NY) 250(4987):1558–1560. https://doi.org/10.1126/science.2274786
47. Imamura O, Arai M, Dateki M, Ogata T, Uchida R, Tomoda H, Takishima K (2015) Nicotinic acetylcholine receptors mediate donepezil-induced oligodendrocyte differentiation. J Neurochem 135(6):1086–1098. https://doi.org/10.1111/jnc.13294
48. Sogorb MA, Fuster E, Del Río E, Estévez J, Vilanova E (2016) Effects of mipafox, paraoxon, chlorpyrifos and its metabolite chlorpyrifos-oxon on the expression of biomarker genes of differentiation in D3 mouse embryonic stem cells. Chem Biol Interact 259(Pt B):368–373. https://doi.org/10.1016/j.cbi.2016.04.017
49. Sperling LE, Steinert G, Boutter J, Landgraf D, Hescheler J, Pollet D, Layer PG (2008) Characterisation of cholinesterase expression during murine embryonic stem cell differentiation. Chem Biol Interact 175(1–3):156–160. https://doi.org/10.1016/j.cbi.2008.05.034
50. Alvarez A, Opazo C, Alarcón R, Garrido J, Inestrosa NC (1997) Acetylcholinesterase promotes the aggregation of amyloid-beta-peptide fragments by forming a complex with the growing fibrils. J Mol Biol 272(3):348–361. https://doi.org/10.1006/jmbi.1997.1245
51. Inestrosa NC, Alvarez A, Pérez CA, Moreno RD, Vicente M, Linker C, Casanueva OI, Soto C, Garrido J (1996) Acetylcholinesterase accelerates assembly of amyloid-beta-peptides into Alzheimer's fibrils: possible role of the peripheral site of the enzyme. Neuron 16 (4):881–891. https://doi.org/10.1016/s0896-6273(00)80108-7
52. Paraoanu LE, Layer PG (2008) Acetylcholinesterase in cell adhesion, neurite growth and network formation. FEBS J 275(4):618–624. https://doi.org/10.1111/j.1742-4658.2007.06237.x
53. Ye X, Zhang C, Chen Y, Zhou T (2015) Upregulation of acetylcholinesterase mediated by p53 contributes to cisplatin-induced apoptosis in human breast cancer cell. J Cancer 6(1):48–53. https://doi.org/10.7150/jca.10521
54. Zhang B, Yang L, Yu L, Lin B, Hou Y, Wu J, Huang Q, Han Y, Guo L, Ouyang Q, Zhang B, Lu L, Zhang X (2012) Acetylcholinesterase is associated with apoptosis in β cells and contributes to insulin-dependent diabetes mellitus pathogenesis. Acta Biochim Biophys Sin Shanghai 44(3):207–216. https://doi.org/10.1093/abbs/gmr121
55. Toiber D, Berson A, Greenberg D, Melamed-Book N, Diamant S, Soreq H (2008) N-acetylcholinesterase-induced apoptosis in Alzheimer's disease. PLoS One 3(9):e3108. https://doi.org/10.1371/journal.pone.0003108
56. Hernández AF, Menéndez P (2016) Linking pesticide exposure with pediatric leukemia: potential underlying mechanisms. Int J Mol Sci 17(4):461–461. https://doi.org/10.3390/ijms17040461

57. Gilboa-Geffen A, Hartmann G, Soreq H (2012) Stressing hematopoiesis and immunity: an acetylcholinesterase window into nervous and immune system interactions. Front Mol Neurosci 5:30–30. https://doi.org/10.3389/fnmol.2012.00030
58. Garcia-Ratés S, Greenfield S (2017) Cancer and neurodegeneration: two sides, same coin? Oncotarget 8(14):22307–22308. https://doi.org/10.18632/oncotarget.16190
59. Bond CE, Zimmermann M, Greenfield SA (2009) Upregulation of alpha7 nicotinic receptors by acetylcholinesterase c-terminal peptides. PLoS One 4(3):e4846–e4846. https://doi.org/10.1371/journal.pone.0004846
60. Fauquier T, Guérineau NC, McKinney RA, Bauer K, Mollard P (2001) Folliculostellate cell network: a route for long-distance communication in the anterior pituitary. Proc Natl Acad Sci U S A 98(15):8891–8896. https://doi.org/10.1073/pnas.151339598
61. Appleyard ME, Smith AD (1989) Secretion of acetylcholinesterase and butyrylcholinesterase from the guinea-pig isolated ileum. Br J Pharmacol 97(2):490–498. https://doi.org/10.1111/j.1476-5381.1989.tb11977.x
62. Onganer PU, Djamgoz MBA, Whyte K, Greenfield SA (2006) An acetylcholinesterase-derived peptide inhibits endocytic membrane activity in a human metastatic breast cancer cell line. Biochim Biophys Acta 1760(3):415–420. https://doi.org/10.1016/j.bbagen.2005.12.016
63. Falugi C, Castellani P, Raffanti S, Borsi L, Zardi L (1986) Acetylcholinesterase in normal and malignant human cells. Basic Appl Histochem 30(4):433–446
64. Zhang XJ, Yang L, Zhao Q, Caen JP, He HY, Jin QH, Guo LH, Alemany M, Zhang LY, Shi YF (2002) Induction of acetylcholinesterase expression during apoptosis in various cell types. Cell Death Differ 9(8):790–800. https://doi.org/10.1038/sj.cdd.4401034
65. Goliasch G, Haschemi A, Marculescu R, Endler G, Maurer G, Wagner O, Huber K, Mannhalter C, Niessner A (2012) Butyrylcholinesterase activity predicts long-term survival in patients with coronary artery disease. Clin Chem 58(6):1055–1058. https://doi.org/10.1373/clinchem.2011.175984
66. Arbel Y, Shenhar-Tsarfaty S, Waiskopf N, Finkelstein A, Halkin A, Revivo M, Berliner S, Herz I, Shapira I, Keren G, Soreq H, Banai S (2014) Decline in serum cholinesterase activities predicts 2-year major adverse cardiac events. Mol Med 20(1):38–45. https://doi.org/10.2119/molmed.2013.00139
67. Duysen EG, Li B, Lockridge O (2009) The butyrylcholinesterase knockout mouse a research tool in the study of drug sensitivity, bio-distribution, obesity and Alzheimer's disease. Expert Opin Drug Metab Toxicol 5(5):523–528. https://doi.org/10.1517/17425250902915555
68. De Vriese C, Gregoire F, Lema-Kisoka R, Waelbroeck M, Robberecht P, Delporte C (2004) Ghrelin degradation by serum and tissue homogenates: identification of the cleavage sites. Endocrinology 145(11):4997–5005. https://doi.org/10.1210/en.2004-0569
69. Kojima M, Hosoda H, Date Y, Nakazato M, Matsuo H, Kangawa K (1999) Ghrelin is a growth-hormone-releasing acylated peptide from stomach. Nature 402(6762):656–660. https://doi.org/10.1038/45230
70. Peino R, Baldelli R, Rodriguez-Garcia J, Rodriguez-Segade S, Kojima M, Kangawa K, Arvat E, Ghigo E, Dieguez C, Casanueva F (2000) Ghrelin-induced growth hormone secretion in humans. Eur J Endocrinol 143(6):R11
71. Tong J, Prigeon RL, Davis HW, Bidlingmaier M, Kahn SE, Cummings DE, Tschöp MH, Alessio D (2010) Ghrelin suppresses glucose-stimulated insulin secretion and deteriorates glucose tolerance in healthy humans. Diabetes 59(9):2145. https://doi.org/10.2337/db10-0504
72. Tschöp M, Smiley DL, Heiman ML (2000) Ghrelin induces adiposity in rodents. Nature 407 (6806):908–913. https://doi.org/10.1038/35038090
73. Wynne K, Stanley S, McGowan B, Bloom S (2005) Appetite control. J Endocrinol 184 (2):291. https://doi.org/10.1677/joe.1.05866
74. Chopin L, Walpole C, Seim I, Cunningham P, Murray R, Whiteside E, Josh P, Herington A (2011) Ghrelin and cancer. Mol Cell Endocrinol 340(1):65–69. https://doi.org/10.1016/j.mce.2011.04.013

75. Gahete MD, Córdoba-Chacón J, Kineman RD, Luque RM, Castaño JP (2011) Role of ghrelin system in neuroprotection and cognitive functions: implications in Alzheimer's disease. Peptides 32(11):2225–2228. https://doi.org/10.1016/j.peptides.2011.09.019
76. Brimijoin S, Chen VP, Pang Y-P, Geng L, Gao Y (2016) Physiological roles for butyrylcholinesterase: a BChE-ghrelin axis. Chem Biol Interact 259(Pt B):271–275. https://doi.org/10.1016/j.cbi.2016.02.013
77. Darvesh S, Hopkins DA, Geula C (2003) Neurobiology of butyrylcholinesterase. Nat Rev Neurosci 4(2):131–138. https://doi.org/10.1038/nrn1035
78. Brimijoin S, Hammond P (1988) Butyrylcholinesterase in human brain and acetylcholinesterase in human plasma: trace enzymes measured by two-site immunoassay. J Neurochem 51 (4):1227–1231. https://doi.org/10.1111/j.1471-4159.1988.tb03091.x
79. Mesulam M, Guillozet A, Shaw P, Quinn B (2002) Widely spread butyrylcholinesterase can hydrolyze acetylcholine in the normal and Alzheimer brain. Neurobiol Dis 9(1):88–93. https://doi.org/10.1006/nbdi.2001.0462
80. Green RC, Mesulam MM (1988) Acetylcholinesterase fiber staining in the human hippocampus and parahippocampal gyrus. J Comp Neurol 273(4):488–499. https://doi.org/10.1002/cne.902730405
81. Mesulam MM, Geula C (1991) Acetylcholinesterase-rich neurons of the human cerebral cortex: cytoarchitectonic and ontogenetic patterns of distribution. J Comp Neurol 306 (2):193–220. https://doi.org/10.1002/cne.903060202
82. Darvesh S, Grantham DL, Hopkins DA (1998) Distribution of butyrylcholinesterase in the human amygdala and hippocampal formation. J Comp Neurol 393(3):374–390
83. Nordberg A, Ballard C, Bullock R, Darreh-Shori T, Somogyi M (2013) A review of butyrylcholinesterase as a therapeutic target in the treatment of Alzheimer's disease. Prim Care Companion CNS Disord 15(2):PCC.12r01412. https://doi.org/10.4088/PCC.12r01412
84. Diamant S, Podoly E, Friedler A, Ligumsky H, Livnah O, Soreq H (2006) Butyrylcholinesterase attenuates amyloid fibril formation in vitro. Proc Natl Acad Sci U S A 103(23):8628–8633. https://doi.org/10.1073/pnas.0602922103
85. Das UN (2007) Acetylcholinesterase and butyrylcholinesterase as possible markers of low-grade systemic inflammation. Med Sci Monit 13(12):Ra214–Ra221
86. Oda Y (1999) Choline acetyltransferase: the structure, distribution and pathologic changes in the central nervous system. Pathol Int 49(11):921–937. https://doi.org/10.1046/j.1440-1827.1999.00977.x
87. Kato G, Agid Y (1979) [Acetylcholine receptors (author's transl)]. La Nouvelle presse medicale 8(29):2407–2411
88. Lefkowitz RJ (2007) Seven transmembrane receptors: a brief personal retrospective. Biochim Biophys Acta 1768(4):748–755. https://doi.org/10.1016/j.bbamem.2006.11.001
89. Hulme EC, Curtis CA, Wheatley M, Aitken A, Harris AC (1989) Localization and structure of the muscarinic receptor ligand binding site. Trends Pharmacol Sci Suppl:22–25
90. Koshimizu H, Leiter LM, Miyakawa T (2012) M4 muscarinic receptor knockout mice display abnormal social behavior and decreased prepulse inhibition. Mol Brain 5(1):10. https://doi.org/10.1186/1756-6606-5-10
91. Coyle JT, Price DL, DeLong MR (1983) Alzheimer's disease: a disorder of cortical cholinergic innervation. Science (New York, NY) 219(4589):1184–1190. https://doi.org/10.1126/science.6338589
92. Bonner TI, Buckley NJ, Young AC, Brann MR (1987) Identification of a family of muscarinic acetylcholine receptor genes. Science (New York, NY) 237(4814):527–532. https://doi.org/10.1126/science.3037705
93. Hulme EC, Birdsall NJ, Buckley NJ (1990) Muscarinic receptor subtypes. Annu Rev Pharmacol Toxicol 30:633–673. https://doi.org/10.1146/annurev.pa.30.040190.003221
94. Wess J (1996) Molecular biology of muscarinic acetylcholine receptors. Crit Rev Neurobiol 10 (1):69–99. https://doi.org/10.1615/critrevneurobiol.v10.i1.40

95. Wess J, Liu J, Blin N, Yun J, Lerche C, Kostenis E (1997) Structural basis of receptor/G protein coupling selectivity studied with muscarinic receptors as model systems. Life Sci 60 (13–14):1007–1014. https://doi.org/10.1016/s0024-3205(97)00041-6
96. Wess J (1993) Molecular basis of muscarinic acetylcholine receptor function. Trends Pharmacol Sci 14(8):308–313. https://doi.org/10.1016/0165-6147(93)90049-p
97. Matsui M, Yamada S, Oki T, Manabe T, Taketo MM, Ehlert FJ (2004) Functional analysis of muscarinic acetylcholine receptors using knockout mice. Life Sci 75(25):2971–2981. https://doi.org/10.1016/j.lfs.2004.05.034
98. Caulfield MP, Birdsall NJ (1998) International Union of Pharmacology. XVII Classification of muscarinic acetylcholine receptors. Pharmacol Rev 50(2):279–290
99. Lanzafame AA, Christopoulos A, Mitchelson F (2003) Cellular signaling mechanisms for muscarinic acetylcholine receptors. Receptors Channels 9(4):241–260
100. Bartus RT, Dean RL 3rd, Beer B, Lippa AS (1982) The cholinergic hypothesis of geriatric memory dysfunction. Science (New York, NY) 217(4558):408–414. https://doi.org/10.1126/science.7046051
101. Bartus RT, Johnson HR (1976) Short-term memory in the rhesus monkey: disruption from the anti-cholinergic scopolamine. Pharmacol Biochem Behav 5(1):39–46. https://doi.org/10.1016/0091-3057(76)90286-0
102. Everitt BJ, Robbins TW (1997) Central cholinergic systems and cognition. Annu Rev Psychol 48:649–684. https://doi.org/10.1146/annurev.psych.48.1.649
103. Howe AR, Surmeier DJ (1995) Muscarinic receptors modulate N-, P-, and L-type Ca2+ currents in rat striatal neurons through parallel pathways. J Neurosci 15(1 Pt 1):458–469. https://doi.org/10.1523/jneurosci.15-01-00458.1995
104. Dani JA, Bertrand D (2007) Nicotinic acetylcholine receptors and nicotinic cholinergic mechanisms of the central nervous system. Annu Rev Pharmacol Toxicol 47:699–729. https://doi.org/10.1146/annurev.pharmtox.47.120505.105214
105. Dani JA (2001) Overview of nicotinic receptors and their roles in the central nervous system. Biol Psychiatry 49(3):166–174. https://doi.org/10.1016/s0006-3223(00)01011-8
106. Tohgi H, Utsugisawa K, Yoshimura M, Nagane Y, Mihara M (1998) Age-related changes in nicotinic acetylcholine receptor subunits alpha4 and beta2 messenger RNA expression in postmortem human frontal cortex and hippocampus. Neurosci Lett 245(3):139–142. https://doi.org/10.1016/s0304-3940(98)00205-5
107. Pohanka M (2012) Alpha7 nicotinic acetylcholine receptor is a target in pharmacology and toxicology. Int J Mol Sci 13(2):2219–2238. https://doi.org/10.3390/ijms13022219
108. Hernandez CM, Dineley KT (2012) α7 nicotinic acetylcholine receptors in Alzheimer's disease: neuroprotective, neurotrophic or both? Curr Drug Targets 13(5):613–622. https://doi.org/10.2174/138945012800398973
109. Wang HY, Lee DH, D'Andrea MR, Peterson PA, Shank RP, Reitz AB (2000) beta-Amyloid (1-42) binds to alpha7 nicotinic acetylcholine receptor with high affinity. Implications for Alzheimer's disease pathology. J Biol Chem 275(8):5626–5632. https://doi.org/10.1074/jbc.275.8.5626
110. Resende RR, Adhikari A (2009) Cholinergic receptor pathways involved in apoptosis, cell proliferation and neuronal differentiation. Cell Commun Signal 7:20. https://doi.org/10.1186/1478-811x-7-20
111. Itier V, Bertrand D (2001) Neuronal nicotinic receptors: from protein structure to function. FEBS Lett 504(3):118–125. https://doi.org/10.1016/s0014-5793(01)02702-8
112. Nguyen ML, Cox GD, Parsons SM (1998) Kinetic parameters for the vesicular acetylcholine transporter: two protons are exchanged for one acetylcholine. Biochemistry 37 (38):13400–13410. https://doi.org/10.1021/bi9802263
113. Parsons SM (2000) Transport mechanisms in acetylcholine and monoamine storage. FASEB J 14(15):2423–2434. https://doi.org/10.1096/fj.00-0203rev
114. Prado VF, Martins-Silva C, de Castro BM, Lima RF, Barros DM, Amaral E, Ramsey AJ, Sotnikova TD, Ramirez MR, Kim HG, Rossato JI, Koenen J, Quan H, Cota VR, Moraes MF,

Gomez MV, Guatimosim C, Wetsel WC, Kushmerick C, Pereira GS, Gainetdinov RR, Izquierdo I, Caron MG, Prado MA (2006) Mice deficient for the vesicular acetylcholine transporter are myasthenic and have deficits in object and social recognition. Neuron 51 (5):601–612. https://doi.org/10.1016/j.neuron.2006.08.005

115. Edwards RH (2007) The neurotransmitter cycle and quantal size. Neuron 55(6):835–858. https://doi.org/10.1016/j.neuron.2007.09.001
116. Sudhof TC (2004) The synaptic vesicle cycle. Annu Rev Neurosci 27:509–547. https://doi.org/10.1146/annurev.neuro.26.041002.131412
117. Falk-Vairant J, Corrèges P, Eder-Colli L, Salem N, Roulet E, Bloc A, Meunier F, Lesbats B, Loctin F, Synguelakis M, Israel M, Dunant Y (1996) Quantal acetylcholine release induced by mediatophore transfection. Proc Natl Acad Sci U S A 93(11):5203–5207. https://doi.org/10.1073/pnas.93.11.5203
118. Sun YA, Poo MM (1985) Non-quantal release of acetylcholine at a developing neuromuscular synapse in culture. J Neurosci 5(3):634–642. https://doi.org/10.1523/jneurosci.05-03-00634.1985
119. Vyskocil F, Illés P (1977) Non-quantal release of transmitter at mouse neuromuscular junction and its dependence on the activity of Na+-K+ ATP-ase. Pflugers Arch 370(3):295–297. https://doi.org/10.1007/bf00585542
120. Dan Y, Poo MM (1992) Quantal transmitter secretion from myocytes loaded with acetylcholine. Nature 359(6397):733–736. https://doi.org/10.1038/359733a0
121. Girod R, Popov S, Alder J, Zheng JQ, Lohof A, Poo MM (1995) Spontaneous quantal transmitter secretion from myocytes and fibroblasts: comparison with neuronal secretion. J Neurosci 15(4):2826–2838. https://doi.org/10.1523/jneurosci.15-04-02826.1995
122. de Castro BM, De Jaeger X, Martins-Silva C, Lima RDF, Amaral E, Menezes C, Lima P, Neves CML, Pires RG, Gould TW, Welch I, Kushmerick C, Guatimosim C, Izquierdo I, Cammarota M, Rylett RJ, Gomez MV, Caron MG, Oppenheim RW, Prado MAM, Prado VF (2009) The vesicular acetylcholine transporter is required for neuromuscular development and function. Mol Cell Biol 29(19):5238–5250. https://doi.org/10.1128/MCB.00245-09
123. Pinheiro NM, Miranda CJCP, Santana FR, Bittencourt-Mernak M, Arantes-Costa FM, Olivo C, Perini A, Festa S, Caperuto LC, Tibério IFLC, Prado MAM, Martins MA, Prado VF, Prado CM (2020) Effects of VAChT reduction and α7nAChR stimulation by PNU-282987 in lung inflammation in a model of chronic allergic airway inflammation. Eur J Pharmacol 882:173239. https://doi.org/10.1016/j.ejphar.2020.173239
124. Buels KS, Fryer AD (2012) Muscarinic receptor antagonists: effects on pulmonary function. Handb Exp Pharmacol 208:317–341. https://doi.org/10.1007/978-3-642-23274-9_14
125. Canning BJ, Fischer A (1997) Localization of cholinergic nerves in lower airways of guinea pigs using antisera to choline acetyltransferase. Am J Phys 272(4 Pt 1):L731–L738. https://doi.org/10.1152/ajplung.1997.272.4.L731
126. Cavallotti C, D'Andrea V, Cavallotti C, Cameroni M (2005) Distribution of acetylcholinesterase and cholineacetyl-transferase activities in the human pulmonary vessels of younger and older adults. Geriatr Gerontol Int 5(4):286–292. https://doi.org/10.1111/j.1447-0594.2005.00308.x
127. Haberberger R, Schemann M, Sann H, Kummer W (1997) Innervation pattern of guinea pig pulmonary vasculature depends on vascular diameter. J Appl Physiol (Bethesda, MD: 1985) 82(2):426–434. https://doi.org/10.1152/jappl.1997.82.2.426
128. Proskocil BJ, Sekhon HS, Jia Y, Savchenko V, Blakely RD, Lindstrom J, Spindel ER (2004) Acetylcholine is an autocrine or paracrine hormone synthesized and secreted by airway bronchial epithelial cells. Endocrinology 145(5):2498–2506. https://doi.org/10.1210/en.2003-1728
129. Reinheimer T, Münch M, Bittinger F, Racké K, Kirkpatrick CJ, Wessler I (1998) Glucocorticoids mediate reduction of epithelial acetylcholine content in the airways of rats and humans. Eur J Pharmacol 349(2-3):277–284. https://doi.org/10.1016/s0014-2999(98)00185-x

130. Haberberger RV, Bodenbenner M, Kummer W (2000) Expression of the cholinergic gene locus in pulmonary arterial endothelial cells. Histochem Cell Biol 113(5):379–387. https://doi.org/10.1007/s004180000153
131. Ferguson SM, Savchenko V, Apparsundaram S, Zwick M, Wright J, Heilman CJ, Yi H, Levey AI, Blakely RD (2003) Vesicular localization and activity-dependent trafficking of presynaptic choline transporters. J Neurosci 23(30):9697–9709. https://doi.org/10.1523/jneurosci.23-30-09697.2003
132. Kummer W, Wiegand S, Akinci S, Wessler I, Schinkel AH, Wess J, Koepsell H, Haberberger RV, Lips KS (2006) Role of acetylcholine and polyspecific cation transporters in serotonin-induced bronchoconstriction in the mouse. Respir Res 7(1):65. https://doi.org/10.1186/1465-9921-7-65
133. Lips KS, Volk C, Schmitt BM, Pfeil U, Arndt P, Miska D, Ermert L, Kummer W, Koepsell H (2005) Polyspecific cation transporters mediate luminal release of acetylcholine from bronchial epithelium. Am J Respir Cell Mol Biol 33(1):79–88. https://doi.org/10.1165/rcmb.2004-0363OC
134. Greenberg B, Rhoden K, Barnes PJ (1987) Endothelium-dependent relaxation of human pulmonary arteries. Am J Phys 252(2 Pt 2):H434–H438. https://doi.org/10.1152/ajpheart.1987.252.2.H434
135. McMahon TJ, Kadowitz PJ (1992) Methylene blue inhibits neurogenic cholinergic vasodilator responses in the pulmonary vascular bed of the cat. Am J Phys 263(5 Pt 1):L575–L584. https://doi.org/10.1152/ajplung.1992.263.5.L575
136. van Koppen CJ, Blankesteijn WM, Klaassen AB, Rodrigues de Miranda JF, Beld AJ, van Ginneken CA (1988) Autoradiographic visualization of muscarinic receptors in human bronchi. J Pharmacol Exp Ther 244(2):760–764
137. van Koppen CJ, Blankesteijn WM, Klaassen AB, Rodrigues de Miranda JF, Beld AJ, van Ginneken CA (1987) Autoradiographic visualization of muscarinic receptors in pulmonary nerves and ganglia. Neurosci Lett 83(3):237–240. https://doi.org/10.1016/0304-3940(87)90092-9
138. Myers AC, Undem BJ (1996) Muscarinic receptor regulation of synaptic transmission in airway parasympathetic ganglia. Am J Phys 270(4 Pt 1):L630–L636. https://doi.org/10.1152/ajplung.1996.270.4.L630
139. Ayala LE, Ahmed T (1989) Is there loss of protective muscarinic receptor mechanism in asthma? Chest 96(6):1285–1291. https://doi.org/10.1378/chest.96.6.1285
140. Minette PA, Lammers JW, Dixon CM, McCusker MT, Barnes PJ (1989) A muscarinic agonist inhibits reflex bronchoconstriction in normal but not in asthmatic subjects. J Appl Physiol (Bethesda, MD: 1985) 67(6):2461–2465. https://doi.org/10.1152/jappl.1989.67.6.2461
141. Fujimura M, Kamio Y, Matsuda T (1992) Effect of a M1-selective muscarinic receptor antagonist (pirenzepine) on basal bronchomotor tone in young women. Respiration 59(2):102–106. https://doi.org/10.1159/000196036
142. Struckmann N, Schwering S, Wiegand S, Gschnell A, Yamada M, Kummer W, Wess J, Haberberger RV (2003) Role of muscarinic receptor subtypes in the constriction of peripheral airways: studies on receptor-deficient mice. Mol Pharmacol 64(6):1444–1451. https://doi.org/10.1124/mol.64.6.1444
143. Fernandes LB, Fryer AD, Hirshman CA (1992) M2 muscarinic receptors inhibit isoproterenol-induced relaxation of canine airway smooth muscle. J Pharmacol Exp Ther 262(1):119–126
144. Haddad EB, Landry Y, Gies JP (1991) Muscarinic receptor subtypes in guinea pig airways. Am J Phys 261(4 Pt 1):L327–L333. https://doi.org/10.1152/ajplung.1991.261.4.L327
145. Mak JC, Barnes PJ (1990) Autoradiographic visualization of muscarinic receptor subtypes in human and guinea pig lung. Am Rev Respir Dis 141(6):1559–1568. https://doi.org/10.1164/ajrccm/141.6.1559
146. Mak JC, Baraniuk JN, Barnes PJ (1992) Localization of muscarinic receptor subtype mRNAs in human lung. Am J Respir Cell Mol Biol 7(3):344–348. https://doi.org/10.1165/ajrcmb/7.3.344

147. Furchgott RF, Zawadzki JV (1980) The obligatory role of endothelial cells in the relaxation of arterial smooth muscle by acetylcholine. Nature 288(5789):373–376. https://doi.org/10.1038/288373a0
148. Fisher JT, Vincent SG, Gomeza J, Yamada M, Wess J (2004) Loss of vagally mediated bradycardia and bronchoconstriction in mice lacking M2 or M3 muscarinic acetylcholine receptors. FASEB J 18(6):711–713. https://doi.org/10.1096/fj.03-0648fje
149. Salathe M, Lipson EJ, Ivonnet PI, Bookman RJ (1997) Muscarinic signaling in ciliated tracheal epithelial cells: dual effects on Ca2+ and ciliary beating. Am J Phys 272(2 Pt 1): L301–L310. https://doi.org/10.1152/ajplung.1997.272.2.L301
150. Klein MK, Haberberger RV, Hartmann P, Faulhammer P, Lips KS, Krain B, Wess J, Kummer W, König P (2009) Muscarinic receptor subtypes in cilia-driven transport and airway epithelial development. Eur Respir J 33(5):1113–1121. https://doi.org/10.1183/09031936.00015108
151. Seybold ZV, Mariassy AT, Stroh D, Kim CS, Gazeroglu H, Wanner A (1990) Mucociliary interaction in vitro: effects of physiological and inflammatory stimuli. J Appl Physiol (Bethesda, MD: 1985) 68(4):1421–1426. https://doi.org/10.1152/jappl.1990.68.4.1421
152. Reinheimer T, Möhlig T, Zimmermann S, Höhle KD, Wessler I (2000) Muscarinic control of histamine release from airways. Inhibitory M1-receptors in human bronchi but absence in rat trachea. Am J Respir Crit Care Med 162(2 Pt 1):534–538. https://doi.org/10.1164/ajrccm.162.2.9911094
153. Koyama S, Rennard SI, Robbins RA (1992) Acetylcholine stimulates bronchial epithelial cells to release neutrophil and monocyte chemotactic activity. Am J Phys 262(4 Pt 1):L466–L471. https://doi.org/10.1152/ajplung.1992.262.4.L466
154. Koyama S, Sato E, Nomura H, Kubo K, Nagai S, Izumi T (1998) Acetylcholine and substance P stimulate bronchial epithelial cells to release eosinophil chemotactic activity. J Appl Physiol (Bethesda, MD: 1985) 84(5):1528–1534. https://doi.org/10.1152/jappl.1998.84.5.1528
155. Gosens R, Nelemans SA, Grootte Bromhaar MM, McKay S, Zaagsma J, Meurs H (2003) Muscarinic M3-receptors mediate cholinergic synergism of mitogenesis in airway smooth muscle. Am J Respir Cell Mol Biol 28(2):257–262. https://doi.org/10.1165/rcmb.2002-0128OC
156. Krymskaya VP, Orsini MJ, Eszterhas AJ, Brodbeck KC, Benovic JL, Panettieri RA Jr, Penn RB (2000) Mechanisms of proliferation synergy by receptor tyrosine kinase and G protein-coupled receptor activation in human airway smooth muscle. Am J Respir Cell Mol Biol 23 (4):546–554. https://doi.org/10.1165/ajrcmb.23.4.4115
157. Haag S, Matthiesen S, Juergens UR, Racké K (2008) Muscarinic receptors mediate stimulation of collagen synthesis in human lung fibroblasts. Eur Respir J 32(3):555–562. https://doi.org/10.1183/09031936.00129307
158. Matthiesen S, Bahulayan A, Kempkens S, Haag S, Fuhrmann M, Stichnote C, Juergens UR, Racké K (2006) Muscarinic receptors mediate stimulation of human lung fibroblast proliferation. Am J Respir Cell Mol Biol 35(6):621–627. https://doi.org/10.1165/rcmb.2005-0343RC
159. Wessler I, Kirkpatrick CJ, Racké K (1999) The cholinergic 'pitfall': acetylcholine, a universal cell molecule in biological systems, including humans. Clin Exp Pharmacol Physiol 26 (3):198–205. https://doi.org/10.1046/j.1440-1681.1999.03016.x
160. Wessler I, Kilbinger H, Bittinger F, Unger R, Kirkpatrick CJ (2003) The non-neuronal cholinergic system in humans: expression, function and pathophysiology. Life Sci 72 (18–19):2055–2061. https://doi.org/10.1016/s0024-3205(03)00083-3
161. Tucek S (1982) The synthesis of acetylcholine in skeletal muscles of the rat. J Physiol 322:53–69. https://doi.org/10.1113/jphysiol.1982.sp014022
162. Lips KS, Wunsch J, Zarghooni S, Bschleipfer T, Schukowski K, Weidner W, Wessler I, Schwantes U, Koepsell H, Kummer W (2007) Acetylcholine and molecular components of its synthesis and release machinery in the urothelium. Eur Urol 51(4):1042–1053. https://doi.org/10.1016/j.eururo.2006.10.028

163. Okuda T, Haga T (2003) High-affinity choline transporter. Neurochem Res 28(3–4):483–488. https://doi.org/10.1023/a:1022809003997
164. Haberberger RV, Pfeil U, Lips KS, Kummer W (2002) Expression of the high-affinity choline transporter, CHT1, in the neuronal and non-neuronal cholinergic system of human and rat skin. J Invest Dermatol 119(4):943–948. https://doi.org/10.1046/j.1523-1747.2002.00182.x
165. Pfeil U, Haberberger RV, Lips KS, Eberling L, Grau V, Kummer W (2003) Expression of the high-affinity choline transporter CHT1 in epithelia. Life Sci 72(18–19):2087–2090. https://doi.org/10.1016/s0024-3205(03)00088-2
166. Lips KS, Pfeil U, Reiners K, Rimasch C, Kuchelmeister K, Braun-Dullaeus RC, Haberberger RV, Schmidt R, Kummer W (2003) Expression of the high-affinity choline transporter CHT1 in rat and human arteries. J Histochem Cytochem 51(12):1645–1654. https://doi.org/10.1177/002215540305101208
167. Yamada T, Inazu M, Tajima H, Matsumiya T (2011) Functional expression of choline transporter-like protein 1 (CTL1) in human neuroblastoma cells and its link to acetylcholine synthesis. Neurochem Int 58(3):354–365. https://doi.org/10.1016/j.neuint.2010.12.011
168. Traiffort E, Ruat M, O'Regan S, Meunier FM (2005) Molecular characterization of the family of choline transporter-like proteins and their splice variants. J Neurochem 92(5):1116–1125. https://doi.org/10.1111/j.1471-4159.2004.02962.x
169. Erickson JD, Varoqui H, Schäfer MK, Modi W, Diebler MF, Weihe E, Rand J, Eiden LE, Bonner TI, Usdin TB (1994) Functional identification of a vesicular acetylcholine transporter and its expression from a "cholinergic" gene locus. J Biol Chem 269(35):21929–21932
170. Kummer W, Lips KS, Pfeil U (2008) The epithelial cholinergic system of the airways. Histochem Cell Biol 130(2):219–234. https://doi.org/10.1007/s00418-008-0455-2
171. Israël M, Dunant Y (1998) Acetylcholine release. Reconstitution of the elementary quantal mechanism. J Physiol Paris 92(2):123–128. https://doi.org/10.1016/s0928-4257(98)80149-6
172. Buels KS, Fryer AD (2012) Muscarinic receptor antagonists: effects on pulmonary function. Handb Exp Pharmacol 208:317–341. https://doi.org/10.1007/978-3-642-23274-9_14
173. Arias HR, Richards VE, Ng D, Ghafoori ME, Le V, Mousa SA (2009) Role of non-neuronal nicotinic acetylcholine receptors in angiogenesis. Int J Biochem Cell Biol 41(7):1441–1451. https://doi.org/10.1016/j.biocel.2009.01.013
174. Pinheiro NM, Santana FPR, Almeida RR, Guerreiro M, Martins MA, Caperuto LC, Câmara NOS, Wensing LA, Prado VF, Tibério IFLC, Antônio M, Prado M, Prado CM (2017) Acute lung injury is reduced by the α7nAChR agonist PNU-282987 through changes in the macrophage profile. FASEB J 31(1):320–332. https://doi.org/10.1096/fj.201600431r
175. Su X, Lee JW, Matthay ZA, Mednick G, Uchida T, Fang X, Gupta N, Matthay MA (2007) Activation of the alpha7 nAChR reduces acid-induced acute lung injury in mice and rats. Am J Respir Cell Mol Biol 37(2):186–192. https://doi.org/10.1165/rcmb.2006-0240OC
176. Su X, Matthay MA, Malik AB (2010) Requisite role of the cholinergic alpha7 nicotinic acetylcholine receptor pathway in suppressing Gram-negative sepsis-induced acute lung inflammatory injury. J Immunol 184(1):401–410. https://doi.org/10.4049/jimmunol.0901808
177. de Jonge WJ, van der Zanden EP, The FO, Bijlsma MF, van Westerloo DJ, Bennink RJ, Berthoud HR, Uematsu S, Akira S, van den Wijngaard RM, Boeckxstaens GE (2005) Stimulation of the vagus nerve attenuates macrophage activation by activating the Jak2-STAT3 signaling pathway. Nat Immunol 6(8):844–851. https://doi.org/10.1038/ni1229
178. Chatterjee PK, Al-Abed Y, Sherry B, Metz CN (2009) Cholinergic agonists regulate JAK2/STAT3 signaling to suppress endothelial cell activation. Am J Physiol Cell Physiol 297(5):C1294–C1306. https://doi.org/10.1152/ajpcell.00160.2009
179. Peña G, Cai B, Deitch EA, Ulloa L (2010) JAK2 inhibition prevents innate immune responses and rescues animals from sepsis. J Mol Med (Berl) 88(8):851–859. https://doi.org/10.1007/s00109-010-0628-z
180. Elmore S (2007) Apoptosis: a review of programmed cell death. Toxicol Pathol 35(4):495–516. https://doi.org/10.1080/01926230701320337

181. Ruiz-Espejo F, Cabezas-Herrera J, Illana J, Campoy FJ, Vidal CJ (2002) Cholinesterase activity and acetylcholinesterase glycosylation are altered in human breast cancer. Breast Cancer Res Treat 72(1):11–22. https://doi.org/10.1023/a:1014904701723
182. Grisaru D, Sternfeld M, Eldor A, Glick D, Soreq H (1999) Structural roles of acetylcholinesterase variants in biology and pathology. Eur J Biochem 264(3):672–686. https://doi.org/10.1046/j.1432-1327.1999.00693.x
183. Jiang H, Zhang XJ (2008) Acetylcholinesterase and apoptosis. A novel perspective for an old enzyme. FEBS J 275(4):612–617. https://doi.org/10.1111/j.1742-4658.2007.06236.x
184. Ye W, Gong X, Xie J, Wu J, Zhang X, Ouyang Q, Zhao X, Shi Y, Zhang X (2010) AChE deficiency or inhibition decreases apoptosis and p53 expression and protects renal function after ischemia/reperfusion. Apoptosis 15(4):474–487. https://doi.org/10.1007/s10495-009-0438-3
185. Jin QH, He HY, Shi YF, Lu H, Zhang XJ (2004) Overexpression of acetylcholinesterase inhibited cell proliferation and promoted apoptosis in NRK cells. Acta Pharmacol Sin 25 (8):1013–1021
186. Gong X, Ye W, Zhou H, Ren X, Li Z, Zhou W, Wu J, Gong Y, Ouyang Q, Zhao X, Zhang X (2009) RanBPM is an acetylcholinesterase-interacting protein that translocates into the nucleus during apoptosis. Acta Biochim Biophys Sin Shanghai 41(11):883–891. https://doi.org/10.1093/abbs/gmp082
187. Suresh B, Ramakrishna S, Baek KH (2012) Diverse roles of the scaffolding protein RanBPM. Drug Discov Today 17(7–8):379–387. https://doi.org/10.1016/j.drudis.2011.10.030
188. Du A, Xie J, Guo K, Yang L, Wan Y, OuYang Q, Zhang X, Niu X, Lu L, Wu J, Zhang X (2015) A novel role for synaptic acetylcholinesterase as an apoptotic deoxyribonuclease. Cell Discov 1:15002. https://doi.org/10.1038/celldisc.2015.2
189. Park SE, Jeong SH, Yee S-B, Kim TH, Soung YH, Ha NC, Kim ND, Park J-Y, Bae HR, Park BS, Lee HJ, Yoo YH (2008) Interactions of acetylcholinesterase with caveolin-1 and subsequently with cytochrome c are required for apoptosome formation. Carcinogenesis 29 (4):729–737. https://doi.org/10.1093/carcin/bgn036
190. Pick M, Perry C, Lapidot T, Guimaraes-Sternberg C, Naparstek E, Deutsch V, Soreq H (2006) Stress-induced cholinergic signaling promotes inflammation-associated thrombopoiesis. Blood 107(8):3397–3406. https://doi.org/10.1182/blood-2005-08-3240
191. Halliday AC, Greenfield SA (2012) From protein to peptides: a spectrum of non-hydrolytic functions of acetylcholinesterase. Protein Pept Lett 19(2):165–172. https://doi.org/10.2174/092986612799080149
192. Zhou H, Li XM, Meinkoth J, Pittman RN (2000) Akt regulates cell survival and apoptosis at a postmitochondrial level. J Cell Biol 151(3):483–494. https://doi.org/10.1083/jcb.151.3.483
193. Xiang AC, Xie J, Zhang XJ (2008) Acetylcholinesterase in intestinal cell differentiation involves G2/M cell cycle arrest. Cell Mol Life Sci 65(11):1768–1779. https://doi.org/10.1007/s00018-008-8016-3
194. Bleau AM, Hambardzumyan D, Ozawa T, Fomchenko EI, Huse JT, Brennan CW, Holland EC (2009) PTEN/PI3K/Akt pathway regulates the side population phenotype and ABCG2 activity in glioma tumor stem-like cells. Cell Stem Cell 4(3):226–235. https://doi.org/10.1016/j.stem.2009.01.007
195. Day T, Greenfield SA (2003) A peptide derived from acetylcholinesterase induces neuronal cell death: characterisation of possible mechanisms. Exp Brain Res 153(3):334–342. https://doi.org/10.1007/s00221-003-1567-5
196. Zhang X-J, Greenberg DS (2012) Acetylcholinesterase involvement in apoptosis. Front Mol Neurosci 5:40–40. https://doi.org/10.3389/fnmol.2012.00040
197. Levine AJ, Hu W, Feng Z (2006) The P53 pathway: what questions remain to be explored? Cell Death Differ 13(6):1027–1036. https://doi.org/10.1038/sj.cdd.4401910
198. Hajiasgharzadeh K, Sadigh-Eteghad S, Mansoori B, Mokhtarzadeh A, Shanehbandi D, Doustvandi MA, Asadzadeh Z, Baradaran B (2019) Alpha7 nicotinic acetylcholine receptors

in lung inflammation and carcinogenesis: friends or foes? J Cell Physiol 234(9):14666–14679. https://doi.org/10.1002/jcp.28220
199. Field RW, Withers BL (2012) Occupational and environmental causes of lung cancer. Clin Chest Med 33(4):681–703. https://doi.org/10.1016/j.ccm.2012.07.001
200. Brambilla E, Travis WD, Colby TV, Corrin B, Shimosato Y (2001) The new World Health Organization classification of lung tumours. Eur Respir J 18(6):1059–1068. https://doi.org/10.1183/09031936.01.00275301
201. Khuder SA (2001) Effect of cigarette smoking on major histological types of lung cancer: a meta-analysis. Lung Cancer (Amsterdam, Netherlands) 31(2–3):139–148. https://doi.org/10.1016/s0169-5002(00)00181-1
202. Ezzati M, Lopez AD (2003) Estimates of global mortality attributable to smoking in 2000. Lancet (London, England) 362(9387):847–852. https://doi.org/10.1016/s0140-6736(03)14338-3
203. Ezzati M, Henley SJ, Lopez AD, Thun MJ (2005) Role of smoking in global and regional cancer epidemiology: current patterns and data needs. Int J Cancer 116(6):963–971. https://doi.org/10.1002/ijc.21100
204. Molina JR, Yang P, Cassivi SD, Schild SE, Adjei AA (2008) Non-small cell lung cancer: epidemiology, risk factors, treatment, and survivorship. Mayo Clin Proc 83(5):584–594. https://doi.org/10.4065/83.5.584
205. Bartha E, Szelényi JG, Hollán SR (1982) Acetylcholinesterase (AchE) activity of lymphocytes in chronic lymphoid leukemia (CLL). Leuk Res 6(6):861–864. https://doi.org/10.1016/0145-2126(82)90071-6
206. Motamed-Khorasani A, Jurisica I, Letarte M, Shaw PA, Parkes RK, Zhang X, Evangelou A, Rosen B, Murphy KJ, Brown TJ (2007) Differentially androgen-modulated genes in ovarian epithelial cells from BRCA mutation carriers and control patients predict ovarian cancer survival and disease progression. Oncogene 26(2):198–214. https://doi.org/10.1038/sj.onc.1209773
207. Nieto-Cerón S, Vargas-López H, Pérez-Albacete M, Tovar-Zapata I, Martínez-Hernández P, Rodríguez-López JN, Cabezas-Herrera J (2010) Analysis of cholinesterases in human prostate and sperm: implications in cancer and fertility. Chem Biol Interact 187(1–3):432–435. https://doi.org/10.1016/j.cbi.2010.03.038
208. Martínez-López de Castro A, Nieto-Cerón S, Aurelio PC, Galbis-Martínez L, Latour-Pérez J, Torres-Lanzas J, Tovar-Zapata I, Martínez-Hernández P, Rodríguez-López JN, Cabezas-Herrera J (2008) Cancer-associated differences in acetylcholinesterase activity in bronchial aspirates from patients with lung cancer. Clin Sci (Lond) 115(8):245–253. https://doi.org/10.1042/cs20070393
209. Martínez-Moreno P, Nieto-Cerón S, Torres-Lanzas J, Ruiz-Espejo F, Tovar-Zapata I, Martínez-Hernández P, Rodríguez-López JN, Vidal CJ, Cabezas-Herrera J (2006) Cholinesterase activity of human lung tumours varies according to their histological classification. Carcinogenesis 27(3):429–436. https://doi.org/10.1093/carcin/bgi250
210. Song P, Sekhon HS, Fu XW, Maier M, Jia Y, Duan J, Proskosil BJ, Gravett C, Lindstrom J, Mark GP, Saha S, Spindel ER (2008) Activated cholinergic signaling provides a target in squamous cell lung carcinoma. Cancer Res 68(12):4693–4700. https://doi.org/10.1158/0008-5472.can-08-0183
211. Wang S, Hu Y (2018) α7 nicotinic acetylcholine receptors in lung cancer. Oncol Lett 16(2):1375–1382. https://doi.org/10.3892/ol.2018.8841
212. Improgo MR, Scofield MD, Tapper AR, Gardner PD (2010) The nicotinic acetylcholine receptor CHRNA5/A3/B4 gene cluster: dual role in nicotine addiction and lung cancer. Prog Neurobiol 92(2):212–226. https://doi.org/10.1016/j.pneurobio.2010.05.003
213. Wessler I, Kirkpatrick CJ (2008) Acetylcholine beyond neurons: the non-neuronal cholinergic system in humans. Br J Pharmacol 154(8):1558–1571. https://doi.org/10.1038/bjp.2008.185

214. Cattaneo MG, D'Atri F, Vicentini LM (1997) Mechanisms of mitogen-activated protein kinase activation by nicotine in small-cell lung carcinoma cells. Biochem J 328(Pt 2):499–503. https://doi.org/10.1042/bj3280499
215. Song P, Sekhon HS, Proskocil B, Blusztajn JK, Mark GP, Spindel ER (2003) Synthesis of acetylcholine by lung cancer. Life Sci 72(18–19):2159–2168. https://doi.org/10.1016/s0024-3205(03)00078-x
216. Song P, Sekhon HS, Lu A, Arredondo J, Sauer D, Gravett C, Mark GP, Grando SA, Spindel ER (2007) M3 muscarinic receptor antagonists inhibit small cell lung carcinoma growth and mitogen-activated protein kinase phosphorylation induced by acetylcholine secretion. Cancer Res 67(8):3936–3944. https://doi.org/10.1158/0008-5472.can-06-2484
217. Schaal C, Chellappan S (2016) Nicotine-mediated regulation of nicotinic acetylcholine receptors in non-small cell lung adenocarcinoma by E2F1 and STAT1 transcription factors. PLoS One 11(5):e0156451. https://doi.org/10.1371/journal.pone.0156451
218. Dasgupta P, Kinkade R, Joshi B, Decook C, Haura E, Chellappan S (2006) Nicotine inhibits apoptosis induced by chemotherapeutic drugs by up-regulating XIAP and survivin. Proc Natl Acad Sci U S A 103(16):6332–6337. https://doi.org/10.1073/pnas.0509313103
219. Dasgupta P, Rastogi S, Pillai S, Ordonez-Ercan D, Morris M, Haura E, Chellappan S (2006) Nicotine induces cell proliferation by beta-arrestin-mediated activation of Src and Rb-Raf-1 pathways. J Clin Invest 116(8):2208–2217. https://doi.org/10.1172/jci28164
220. Lam DC, Girard L, Ramirez R, Chau WS, Suen WS, Sheridan S, Tin VP, Chung LP, Wong MP, Shay JW, Gazdar AF, Lam WK, Minna JD (2007) Expression of nicotinic acetylcholine receptor subunit genes in non-small-cell lung cancer reveals differences between smokers and nonsmokers. Cancer Res 67(10):4638–4647. https://doi.org/10.1158/0008-5472.can-06-4628
221. Brown KC, Perry HE, Lau JK, Jones DV, Pulliam JF, Thornhill BA, Crabtree CM, Luo H, Chen YC, Dasgupta P (2013) Nicotine induces the up-regulation of the α7-nicotinic receptor (α7-nAChR) in human squamous cell lung cancer cells via the Sp1/GATA protein pathway. J Biol Chem 288(46):33049–33059. https://doi.org/10.1074/jbc.M113.501601
222. Barnes KC (2011) Genetic studies of the etiology of asthma. Proc Am Thorac Soc 8 (2):143–148. https://doi.org/10.1513/pats.201103-030MS
223. Quirce S, Domínguez-Ortega J, Barranco P (2015) Anticholinergics for treatment of asthma. J Investig Allergol Clin Immunol 25(2):84–93; quiz 94–95
224. Sears MR (1997) Descriptive epidemiology of asthma. Lancet 350:S1–S4. https://doi.org/10.1016/S0140-6736(97)90028-3
225. Cookson W, Moffatt M (2004) Making sense of asthma genes. N Engl J Med 351 (17):1794–1796. https://doi.org/10.1056/NEJMe048232
226. Quirt J, Hildebrand KJ, Mazza J, Noya F, Kim H (2018) Asthma. Allergy Asthma Clin Immunol 14(Suppl 2):50–50. https://doi.org/10.1186/s13223-018-0279-0
227. Cockcroft DW (2018) Environmental causes of asthma. Semin Respir Crit Care Med 39 (1):12–18. https://doi.org/10.1055/s-0037-1606219
228. Lemanske RF Jr, Busse WW (2010) Asthma: clinical expression and molecular mechanisms. J Allergy Clin Immunol 125(2 Suppl 2):S95–S102. https://doi.org/10.1016/j.jaci.2009.10.047
229. Bai TR, Vonk JM, Postma DS, Boezen HM (2007) Severe exacerbations predict excess lung function decline in asthma. Eur Respir J 30(3):452–456. https://doi.org/10.1183/09031936.00165106
230. Standards for the diagnosis and care of patients with chronic obstructive pulmonary disease (COPD) and asthma. This official statement of the American Thoracic Society was adopted by the ATS Board of Directors, November 1986, 1987. Am Rev Respir Dis 136(1):225–244. https://doi.org/10.1164/ajrccm/136.1.225
231. Barnes PJ (2000) Chronic obstructive pulmonary disease. N Engl J Med 343(4):269–280. https://doi.org/10.1056/nejm200007273430407
232. Devine JF (2008) Chronic obstructive pulmonary disease: an overview. Am Health Drug Benef 1(7):34–42

233. Gosens R, Gross N (2018) The mode of action of anticholinergics in asthma. Eur Respir J 52 (4):1701247. https://doi.org/10.1183/13993003.01247-2017
234. Evans CM, Kim K, Tuvim MJ, Dickey BF (2009) Mucus hypersecretion in asthma: causes and effects. Curr Opin Pulm Med 15(1):4–11. https://doi.org/10.1097/MCP.0b013e32831da8d3
235. Gosens R, Zaagsma J, Meurs H, Halayko AJ (2006) Muscarinic receptor signaling in the pathophysiology of asthma and COPD. Respir Res 7(1):73. https://doi.org/10.1186/1465-9921-7-73
236. Rodrigo GJ (2018) Anticholinergics for asthma: a long history. Curr Opin Allergy Clin Immunol 18(1):38–43. https://doi.org/10.1097/aci.0000000000000408
237. Gandevia B (1975) Historical review of the use of parasympatholytic agents in the treatment of respiratory disorders. Postgrad Med J 51(7 Suppl):13–20
238. Miraldi E, Masti A, Ferri S, Barni Comparini I (2001) Distribution of hyoscyamine and scopolamine in Datura stramonium. Fitoterapia 72(6):644–648. https://doi.org/10.1016/s0367-326x(01)00291-x
239. Anderson PJ (2005) History of aerosol therapy: liquid nebulization to MDIs to DPIs. Respir Care 50(9):1139–1150
240. Dessanges JF (2001) A history of nebulization. J Aerosol Med 14(1):65–71. https://doi.org/10.1089/08942680152007918
241. Belmonte KE (2005) Cholinergic pathways in the lungs and anticholinergic therapy for chronic obstructive pulmonary disease. Proc Am Thorac Soc 2(4):297–304; discussion 311–312. https://doi.org/10.1513/pats.200504-043SR
242. Soler X, Ramsdell J (2014) Anticholinergics/antimuscarinic drugs in asthma. Curr Allergy Asthma Rep 14(12):484. https://doi.org/10.1007/s11882-014-0484-y
243. Alagha K, Palot A, Sofalvi T, Pahus L, Gouitaa M, Tummino C, Martinez S, Charpin D, Bourdin A, Chanez P (2014) Long-acting muscarinic receptor antagonists for the treatment of chronic airway diseases. Ther Adv Chronic Dis 5(2):85–98. https://doi.org/10.1177/2040622313518227

The Keap1-Nrf2 Signaling Pathway in Lung Cancer

30

Azhwar Raghunath, Raju Nagarajan, and Kiruthika Sundarraj

Abstract

The survival and proliferation of lung cancers widely depend on various signaling pathways; Keap1-Nrf2-ARE is one of the primary pathways in lung cancers. The principal function of Nrf2 is to offer protection to cells from stress induced by xenobiotics through the activation of cytoprotective transcriptional programs. Nrf2 is a Janus-faced transcription factor that can act both as a preneoplastic or an antineoplastic in cancers. In lung cancers, persistent activation of Nrf2 offers resistance to chemotherapeutic drugs and radiotherapy thereby facilitates proliferation. The sustainable Nrf2 actuation impacts the different hallmarks of cancers such as proliferation, metastasis, angiogenesis, and apoptosis. In aberrantly activated Nrf2 lung cancers, inhibitors of Nrf2 can act as therapeutic agents. In this chapter, we highlight the relationship between lung cancers and Keap1-Nrf2-ARE signaling with an emphasis on the dysregulation of Keap1-Nrf2 signaling, Keap1 and Nrf2 mutations, microRNAs in Keap1-Nrf2 signaling, smoking and Nrf2 activation in lung cancer, and Keap1-Nrf2 pathway as a therapeutic target.

Keywords

Keap1 · Lung cancer · MicroRNA · Mutation · Nrf2 · Oxidative stress

Azhwar Raghunath and Raju Nagarajan contributed equally to this work.

A. Raghunath (✉)
Department of Pediatrics, University of Michigan, Ann Arbor, MI, USA
e-mail: razh@umich.edu

R. Nagarajan
Department of Biotechnology, Indian Institute of Technology Madras, Chennai, Tamil Nadu, India

K. Sundarraj
Department of Biotechnology, Kongu Arts and Science College, Erode, Tamil Nadu, India

K. Dua et al. (eds.), *Targeting Cellular Signalling Pathways in Lung Diseases*,
https://doi.org/10.1007/978-981-33-6827-9_30

30.1 Introduction

Lung cancer contributes to one-fourth of all cancer death and is the second most common cancer in both sexes. Mostly older people over 65 years of age and a small group of individuals less than 45 years are diagnosed with lung cancer. Among the 2.1 million estimated lung cancer cases, 1.8 million deaths are reported worldwide [1]. Globally, the primary cause of lung cancer is tobacco use; air pollution, occupational agents, radon, and secondhand smoke are the other risk factors. Based on the pathological features, the two main types of lung cancer are non-small-cell lung cancer (NSCLC) and small-cell lung cancer (SCLC), which constitutes 80% and 20% of lung cancers, respectively. Efficacious therapies are available for a subset of lung cancer patients. The Cancer Genome Atlas (TCGA) revealed that the modulation of the Keap1-Nrf2 pathway occurs in one-third of squamous cell lung cancer [2]. Nrf2 signal initiates and progresses the lung cancer. Cancer cells experience severe insult from both exogenous and endogenous agents. This dynamic defense signaling pathway safeguards the cancer cells from these agents. The master transcription factor called nuclear factor erythroid 2-related factor 2 (Nrf2) transcribed from the gene *nuclear factor, erythroid 2 like 2* (*NFE2L2*) offers cytoprotection against the oxidative and xenobiotic stresses [3]. A repertoire of genes involved in the defense system is under the control of Nrf2. The modulation of genes governed by the Keap1-Nrf2 pathway promotes the progression of lung tumors.

30.2 Keap1-Nrf2 Signaling Pathway

Under normal physiological conditions, Kelch-like erythroid cell-derived protein with CNC homology (ECH)-associated protein 1 (Keap1) accommodates Cul3-containing E3 ubiquitin ligase complex, thereby acting as a substrate adaptor protein [4]. This complex retains the master regulator Nrf2 within the cytoplasm and marks for ubiquitination and degradation [5]. By these processes, Nrf2 is maintained at an optimal level through sequestration by Keap1. Under electrophilic or stressed conditions, Keap1 undergoes modifications and is no longer having the potential to sequester newly synthesized Nrf2 within the cytoplasm. Eventually, the phosphorylation of Nrf2 translocates it into the nucleus and partners with a diverse array of proteins and binds the antioxidant response element (ARE) of target genes to activate their transcription [3, 6]. The protein products of these myriad genes defend the cell from electrophilic and oxidative insults. The target antioxidant proteins and detoxification enzymes include glutamate-cysteine ligase, glutathione reductase, heme oxygenase, NADPH quinone oxidoreductase 1, peroxidase, thioredoxin, etc. [7]. Besides the above canonical activation by Keap1, Nrf2 can also be activated through Keap1-independent mechanisms (Fig. 30.1). The major players involved in this noncanonical Keap1-independent activation of Nrf2 are protein kinase C (PKC), p62, p53-induced p21, glycogen synthase kinase-3β (GSK3β), mitogen-activated protein kinase (MAPK), extracellular regulated protein kinase (ERK), and pAMPK

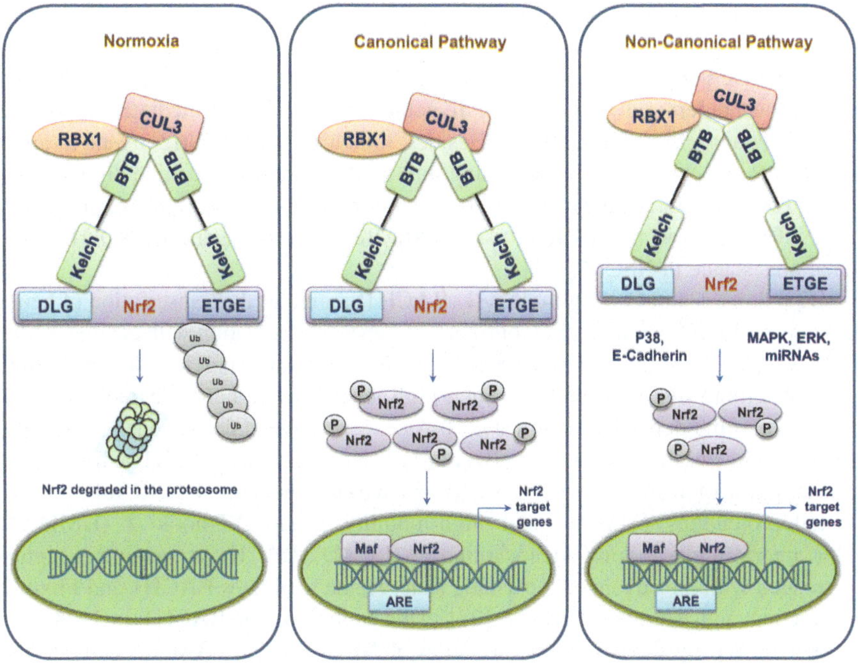

Fig. 30.1 Keap1-Nrf2 signaling pathway. Under normal physiological conditions, Keap1 retains Nrf2 in the cytoplasm and directs it to proteolysis in the proteasome. Nrf2 gets activated through canonical and noncanonical modes. When activated, Nrf2 heterodimerizes with Maf, binds AREs, and activates the transcription of the Nrf2 target genes

[8–12]. Genetically engineered mouse models (GEMM) uncovered that Nrf2 stabilization, accumulation, and activation are not a lung cancer driver [13]. But Keap1 inactivation along with Nrf2 activation and oncogenic mutations transform lung epithelium to malignancy. Though the primary role of the Keap1-Nrf2 signaling cascade is to protect cells from redox imbalance, it is also found to modulate in lung cancer.

30.3 Regulation and Dysregulation of Nrf-Keap1 Signaling Pathway in Lung Cancer

The defective Nrf2 pathway increases the susceptibility to lung tumorigenesis. The activation of Nrf2 offers protection to cells from carcinogens. In contrast, persistent Nrf2 activity promotes the progression phase of lung cancer. NSCLCs generate reactive oxygen species as a result of enhanced metabolic output, to maintain redox homeostasis Nrf2 activates the transcription of antioxidant genes. The stress experienced by the lung cancer cells is overcome by the activation of Nrf2. Alterations such as mutations and DNA methylation activate Nrf2, which regulates NSCLC development [14–16]. These changes suggest that Nrf2 activation is one of the key mechanisms in NSCLC [17, 18]. Modifications in the Keap1-Nrf2 axis deregulate Nrf2 ultimately triggering tumorigenesis leading to lung adenocarcinomas. In human lung mucoepidermoid cancer cells, hyperactivation of Nrf2-mediated oxidative stress response is observed [19]. Besides tolerating stress, Nrf2 overactivation diminishes the chemosensitivity in lung tumors. Nicotinamide adenine dinucleotide (NADPH) Q oxidoreductase 1 (NQO1), a target gene of Nrf2, is highly expressed in the NSCLC than in normal lung tissue [20]. Besides tumorigenesis, NQO1 offers chemoresistance in NSCLC [21].

BTB domain and CNC homology 1 (BACH1) regulates the transcription of heme oxygenase 1 (HO1) and other cytoprotective genes in association with Nrf2 and Maf [22]. Oxidative insult releases heme from hemoproteins and free heme in turn generates more free radicals [23]. Such high levels of free heme degrade BACH1 in a proteasome-dependent manner [24]. As excess heme is deleterious to the cells, HO1 catabolizes free heme [25]. Besides HO1 expression, BACH1 activates the transcription of metastatic genes—CXC-chemokine receptor 4 (CXCR4), matrix metalloproteinases (MMPs), and high-mobility group AT-hook 2 (HMGA2) [26]. Interestingly, in lung adenocarcinoma (LUAD), BACH1 expression is highly correlated with the advanced metastatic state. Heme facilitates BACH1 and Fbxo22 interaction and degrades BACH1 [27]. This high level of BACH1 in LUAD is overshadowed by the sustained activation of Nrf2. Nrf2 and BACH1 activate the transcription of antioxidant and pro-metastatic genes, respectively. Nrf2 stabilized the BACH1 through the inhibition of Fbxo22-dependent BACH1 degradation and expression of HO1, which curtails the free heme. In LUAD, elevated expression of Nrf2, BACH1, and HO1 was observed. The pharmacological inhibition of Nrf2, BACH1, and HO1 could impair lung cancer cell proliferation and invasion.

Anomalous conditions of the Keap1-Nrf2 signaling pathway persist in lung cancer. Such disruption occurs due to mutations in Keap1 and Nrf2, DNA methylations in the promoter of Keap1 and Nrf2, and loss of heterozygosity [28]. Lung cancers exhibited alterations in Keap1 leading to the deviant Keap1 which no longer holds Nrf2 for ubiquitination and degradation, which eventually accumulates Nrf2 and activates Nrf2 target genes [17]. A few of the Keap1 mutations augmented the interaction and ubiquitination between Keap1 and Nrf2 but diminished the Nrf2 degradation [29]. Several of these somatic mutations reside within the conserved regions of the Kelch and IVR domain of the Keap1, which disrupted the Keap1 function. Hypermethylation in the Keap1 promoter was observed in lung cancers, which resulted in the feeble Keap1 expression leading to Nrf2 activation [16]. Besides Keap1 mutations, somatic mutations within Nrf2 also contribute to the dysregulation of the Keap1-Nrf2 signaling pathway in lung cancer [30]. Mutations in the DLG and ETGE motifs of Nrf2 are unresponsive to Keap1-mediated Nrf2 regulation, which led to the abnormal accumulation of Nrf2 and persistent activation of Nrf2 target genes.

30.4 Mutations in Keap1 and Nrf2 in Lung Cancer

Among all the cancer types, the greatest enrichment in the mutations of Nrf2 and Keap1 is observed in lung cancer which directly influences the increased Nrf2 activity [31]. Generally, the mutations in Keap1 lead to the loss of its ability to regulate Nrf2 termed as the loss-of-function (LOF) mutations, and the mutations (mostly missense mutations) in Nrf2 help to retain its function, termed as the gain-of-function (GOF) mutations. Several studies investigated and reported that the occurrence of mutations in Keap1 and Nrf2 is mutually exclusive, particularly in NSCLC, and the distribution of mutation frequency is unequal and varies in different histological lung cancer types [2, 17, 32–34]. Significant enrichment of GOF mutations in Nrf2 (59.2%) and LOF mutations in Keap1 (72.2%) was observed in squamous cell carcinoma (SqCC) and adenocarcinoma (AC), respectively [2, 18, 30–32, 35–38]. TCGA studies performed in the genomic characterization of 178 lung squamous cell carcinomas reported that the Keap1/Nrf2 mutations were observed in one-third (34%) of the patients. [2]. In a large scale clinical study of 2455 patients, alterations including mutations observed in Keap1 and Nrf2 in 15% and 2.9% of patients, respectively [32, 39–42]. Very recently, 3.5% and 11.3% of patients identified with Nrf2 and Keap1 mutations among the 1391 NSCLC patients. Interestingly, many studies reported the higher mutation rate of Keap1 and Nrf2 in smokers than nonsmokers [19, 43, 44]

In Nrf2, almost all the observed mutations in AC and SqCC lung tumors were located in or around the coding regions of Keap1-interacting regions such as DLG (43%) and ETGE (57%) motifs, which makes Nrf2 insensitive to Keap1-mediated regulation and thus leads to the Nrf2 accumulation [2, 15, 17, 28, 45]. Genomic profiling analysis reported the two hot spots of Nrf2 mutation in ~10% of SqCC patients which enable the escape from the Keap1-mediated suppression [46, 47]. As Keap1 is the prime regulator of Nrf2, the mutations in Keap1 which were located across all the domains directly lifts the repression of Nrf2 and increases its abundance and nuclear translocation, which activates Nrf2 target genes to facilitate the tumor growth in lung cancer. Keap1 mutations were first reported in lung cancer, and the mutations were observed in 19% of patients in all lung cancer types. As the third most commonly mutated gene in ADC, the mutations were observed in 26% of ADC patients [17, 32, 35]. In NSCLC patients, somatic mutations in Keap1 increased Nrf2 abundance and conferred the worse progress-free survival [14, 48]. Glycine-to-cysteine gene mutations (G430C and G364C, in Kelch-repeat domain) observed in Keap1 activates Nrf2 due to the reduced binding affinity with Nrf2 in human lung adenocarcinoma cell lines [5, 49]. Among the 18 characterized mutations in SqCC, 5 mutations which include N469fs, G333C, R554Q, P318fs, and W544C were promoted in Nrf2 activation, 9 involved in the hypomorphic Nrf2 suppression, and the remaining 4 mutations did not affect the Nrf2 suppression [2, 29]. Several other somatic mutations were observed in the highly conserved Kelch binding or intervening linker (IVR) domains that include NL20, A549, H460, H1435, H23, H358, H1993, H1395, H838, H1299, and H292 mutations in NSCLC patients [15, 17].

30.5 MicroRNAs Associated with the Keap1-Nrf2 Signaling Pathway in Lung Cancer

MicroRNAs (miRNA) are a type of noncoding RNA molecules that regulate the protein synthesis and posttranscriptional expression of targeted genes by the complete or partial binding with the 3′ untranslated region (3′-UTR) of the mRNA targets [50–52]. Many investigations were performed to understand the dual role of miRNA, tumor suppressor as well as oncogenic, in the regulation of various cancer types particularly in lung cancer [53]. Among the many miRNAs involved in the regulation of lung cancer, some miRNAs disrupt the Nrf2-keap1 signaling pathway by regulating the expression level of Nrf2 and keap1 by disrupting the stability and degradation of Nrf2 and Keap1 mRNA [54, 55]. Overexpression of miR-155 facilitates the malignant transformation of lung cells and increased the resistance of arsenic trioxide by upregulating the Nrf2 level in arsenic-transformed bronchial epithelial cells in adenocarcinoma [56–58]. In contrast, miR-34a directly inhibits the expression of Nrf2 in NSCLC cells that leads to the decreased synthesis of antioxidants such as glutathione [59]. MiR-141 facilitates the degradation of Keap1 mRNA by directly binding to its 3′-UTR and promotes the proliferation of cancer cells in NSCLC [60, 61]. Recent studies showed that miR-421 is associated with the poor prognosis in NSCLC patients that promotes cancer progression by targeting the 3′ UTR of keap1 and downregulating its expression, which reduces the drug resistance level [62, 63]. In NSCLC cells, activation of Nrf2 inhibits the expression of miR-1 and miR-206 and promotes the pentose phosphate pathway, which leads to a reduced survival rate [64].

30.6 Smoking and Nrf2 Activation in Lung Cancer

Lung cancer is one of the three major cigarette smoking (CS)-related diseases with the highest rates of mortality and morbidity [65]. CS is responsible for the direct exposure of more than 5000 different chemicals which includes reactive oxygen/nitrogen species and >60 International Agency of Cancer Research (IARC)-listed carcinogens to the tissues in the respiratory tract [66]. The formation of many additional reactive toxicants, with the interaction of host-derived molecules, together, causes major tissue damage in the lungs that leads to chronic inflammation and disease development [67–69] particularly in non-small-cell lung cancer [70] and squamous cell lung cancer [71]. Also, CS contains numerous components from different classes that are accountable for the thiol oxidation of host protein and non-protein molecules which induce oxidative stress and act as Nrf2 activators [66, 72, 73]. CS-induced Nrf2 activation is majorly carried out by the thiol modification of the cysteine residues present in the Keap1 protein [66, 72–74]. Although all the cysteine residues may act as redox sensors, residues in the position 151, 273, and 288 are the major targets for this covalent modification [75–79]. These covalent modifications in keap1 perturb the physical interaction between keap1 and Nrf2 that lifts the cytoplasmic repression of Nrf2 resulting in the release of Nrf2 from the Keap1-Nrf2 complex [74]. Alternative to the thiol modification in keap1, CS-induced Nrf2 activation also carried out by the complementary mechanism involving the phosphorylation of the residues at serine-40 and threonine-80 located in the Neh2 domain of the Nrf2, which directly affects the binding of Nrf2 to Keap1 [80–84]. Many investigations carried out in different experimental models concluded that the various chemical components present in CS act as Nrf2 activators, either thiol modification in Keap1 or Nrf2 phosphorylation, particularly nitric oxide [85–87], CS condensate [88], heavy metals [89, 90], α,β-unsaturated aldehydes [91, 92], diphenols [93, 94], peroxides [95], polyenes, and phase I inducers [96]. In both Nrf2-activation mechanisms, the CS components responsible for the direct activation of Nrf2 are termed as monofunctional, whereas CS contains many bifunctional components that indirectly activate Nrf2 [97, 98]. One of the bifunctional components, xenobiotics, present in the CS activates another transcription factor, the aryl hydrocarbon (Ah) receptor. Ah receptor activates the production of Nrf2 by binding to the xenobiotic response element (XRE) present in the promoter region of Nrf2 [97, 99, 100]. Yet other investigations sequenced Nrf2 sequences from lung cancer patients and showed the Nrf2 mutation is common in smokers and the mutation rate is higher in smokers than nonsmokers [44, 101].

30.7 Nrf2-Keap1 Signaling Pathway as a Therapeutic Target in Lung Cancer

Hyperactivation of Nrf2 augments chemoresistance and radioresistance in lung cancer [102]. Multidrug resistance-associated protein 1 (MRP1) is an efflux transporter that expels chemotherapeutic agents from cancer cells [103]. In lung cancer, Nrf2 upregulated MRP1 thereby mediates chemoresistance [104]. Nrf2 silencing inhibited colony formation in NSCLC, and Nrf2 depletion in NSCLC repressed the formation of tumor in athymic mice [15, 105]. The interaction of cyclin-dependent kinase 20 with Keap1 conferred chemo- and radioresistance in lung cancer cells through Nrf2 activation [106]. Nrf2 inhibitors inhibited the tumor activity in the Keap1-lacking NSCLC [107]. Hence, Nrf2 mitigation can be a therapeutic target in lung cancer. Evidence supports that Nrf2 inhibition offers merit in the treatment of lung cancer patients. Though considerable evidence displayed the efficacy of Nrf2 inhibitors augmenting the chemotherapeutic agents against lung cancer, there is an immense search for an improved Nrf2 inhibitor.

30.8 Inhibitors of Nrf2 in Lung Cancer

In many cancers, including lung cancer, the elevated level of Nrf2 facilitates cancer progression and increased resistance to cancer therapy. Given the oncogenic role of Nrf2 overexpression, the strategy of inhibiting Nrf2 is becoming a promising approach in lung cancer therapy. Several natural chemical components derived from small molecules and repurposed drugs have been studied for the inhibitory properties of Nrf2 for lung cancer therapy (Table 30.1). Cinnamomi cortex extract (CCE) procyanidin inhibited the Nrf2 expression and reduced Nrf2 mRNA levels that suppress cancer cell proliferation in A549 NSCLC cells [128–130]. Quinoid brusatol enhanced the Nrf2 degradation via increased ubiquitination, suppressed the colony formation, reduced tumor growth, and improved survival in murine A549 xenograft models [131–133]. In both in vitro and in vivo experiments using murine A549 xenograft models, luteolin flavonoid suppressed the growth of A549 cells via inhibiting Nrf2 and improved antitumor efficacy by sensitizing A549 cells for several chemotherapeutics such as bleomycin, doxorubicin, cisplatin, and oxaliplatin [120, 134]. Other Nrf2 inhibitors such as convallatoxin, quinazoline alkaloid halofuginone, camptothecin, and 4-methoxy-chalcone (4-MC) sensitized A549 cells and increased the efficacy of drugs like cisplatin [135–138]. High-throughput screening of small-molecule libraries by several groups revealed the potent, high A549 NSCLC-specific Nrf2 inhibitors such as ARE expression modulator 1 (AEM1) [139] and ML385 [107]. With the list of novel Nrf2 inhibitors, several repurposed drugs including auranofin [140], glucocorticoid clobetasol propionate [141], metformin [142], and camptothecin [143] exhibited promising NRF2 inhibiting ability by reducing Nrf2 nuclear accumulation and enhanced its degradation. Even though the promising Nrf2 inhibiting ability of the abovementioned molecules was tested in lung cancer cell lines, no clinical study results were available except for the recent clinical trial in which the combination of luteolin and ascorbic acid potentially modulated Nrf2 dysregulation in a patient with Nrf2 GOF mutations [144].

30.9 Conclusion

Research in the last decade revealed the Keap1-Nrf2 pathway as an oncogenic signaling pathway in lung cancer. Nrf2 inhibitors and agonists tested in preclinical studies displayed both tumor suppression and initiation activities. In NSCLC patients, Nrf2 inhibitors offer promise as a therapeutic agent. A comprehensive understanding of this dynamic complexity of the Keap1-Nrf2-ARE signaling pathway in lung cancer is warranted as Nrf2 cross-talks with other signaling networks. Such networks will control unique metabolic reprogramming in lung cancers and identification of Nrf2-regulated genes and offer diagnostic and prognostic markers in lung cancers.

Table 30.1 Compounds/phytochemicals that modulate the Keap1-Nrf2 signaling pathway in lung cancer

Compounds	Chemosensitization/ radiosensitization	Cell lines	Dose and duration	Molecular targets	Molecular mechanism	Role of the compound	References
Acetaminophen	Chemosensitizes erastin	A549 and H1299 cells	Combination of 20 μM/mL erastin and 160 μm/mL acetaminophen for 24 h	↓Nrf2, ↓HO-1	Acetaminophen sensitized erastin-induced ferroptosis by regulating the Nrf2/HO-1 signaling pathway	Combination treatment of erastin and acetaminophen increased lipid peroxidation, loss of mitochondrial membrane potential leads to the inhibition of Nrf2/HO-1 redox regulation	[108]
Coroglaucigenin	Enhanced radiosensitivity	A549, NCI-H460, and NCI-H446 cells	1 μM for 24 h	↓Nrf2, ↓NQO-1, ↓HO-1	Combination treatment of coroglaucigenin and X-ray irradiation reduced the expression of Nrf2 and NQO-1	Coroglaucigenin treatment increases the cellular ROS and eventually causes DNA damage involving in an Nrf2-dependent pathway	[109]
Cryptotanshinone	Chemosensitizes cisplatin	A549 and A549/DDP cells	10 μM for 24 h	↓Nrf2, ↓p-JNK, ↓p-ERK, ↓p-p38, ↓p-Akt, ↓p-STAT3	Cryptotanshinone treatment induced the sensitivity of A549/DDP cell to cisplatin and blocked Nrf2 signaling pathway	Cryptotanshinone treatment down-regulated the protein levels of p-JNK, p-ERK, p-P38, p-Akt, and p-STAT3 and contributed to	[110]

(continued)

Table 30.1 (continued)

Compounds	Chemosensitization/ radiosensitization	Cell lines	Dose and duration	Molecular targets	Molecular mechanism	Role of the compound	References
						enhanced expression of Keap1 downregulating Nrf2	
Diosmetin	Chemosensitizes paclitaxel	A549 and H1299 cells	20 μM for 12 h	↓Nrf2, ↓p-Akt, ↓p-GSK-3β	Diosmetin treatment increased ROS accumulation and apoptosis by interfering with Nrf2 antioxidant defense via the disruption of the PI3K/Akt/GSK-3β pathway	Combination treatment of diosmetin and paclitaxel reduced the phosphorylation of Akt and GSK-3β expression and efficiently suppressed Nrf2 pathway	[111, 112]
Epigallocatechin gallate	Chemosensitizes etoposide	A549 and NCIH23 cells	0.5 μM Egigallacatechin gallate and 1 μM etoposide for 48 h	↓Nrf2, ↑Keap1, ↑p53, ↓p21, ↑RXR, ↑RAR	EGCG treatment sensitized A549 cells towards etoposide chemotherapy noncanonical regulation of Nrf2	Combination treatment disrupted redox homeostasis, enhanced RXR and RAR, increased p53 and decreased p21, and inhibited constitutive Nrf2	[113]

Erastin and Sorafenib	Chemosensitizes with a small amount of cisplatin (CDDP)	N5CP cells	20 μM sorafenib and 10 μM erastin for 12 h and 10 μM CDDP for 12 h	↓Nrf2, ↓xCT	Cultured cells were treated with CDDP and erastin or sorafenib increased ROS accumulation and decreased cell survival rate by modulating the Nrf2/xCT pathway	Erastin or sorafenib combination with a small dose of CDDP effectively induces ferroptosis in CDDP-resistant NSCLC cells through increased ROS accumulation and the inhibition of the Nrf2 and its downstream target gene *xCT*	[114]
Flumethasone	Chemosensitizes cisplatin, doxorubicin, and 5-FU	A549 and H460 cells	100 nM for 24 h	↓Nrf2	Flumethasone inhibited Nrf2 signaling in A549 and H460 cells by promoting Nrf2 degradation	Flumethasone enhanced cisplatin, doxorubicin, and 5-FU through the inhibition of Nrf2 signaling	[115]
Ganoderic acid	–	H460 cells	5, 10, 20, 50, 80 μM for 24 and 48 h	↓Nrf2	Ganoderic acid treatment inhibits cancer cell proliferation, viability, and ROS and effectively inhibited the Nrf2 expression	Ganoderic acid targets Nrf2 in H460 cells	[116]

(continued)

Table 30.1 (continued)

Compounds	Chemosensitization/radiosensitization	Cell lines	Dose and duration	Molecular targets	Molecular mechanism	Role of the compound	References
Genistein	Radiosensitization	A549 cells	10 μM for 48 h	↓Nrf2, ↓NQO1, ↓HO-1, ↑cleaved caspase3, ↑cytochrome-c	Genistein treatment on A549 cells inhibited the methylation in the Keap1 promoter region leading to the transcription of Keap1 inhibiting Nrf2	Genistein treatment induced ROS production and mitochondrial cytochrome c-mediated caspase-3 activation and selectively inhibited the methylation of Keap1 gene promoter region and Nrf2 expression	[117]
Ginsenoside Rd	Reverses cisplatin resistance	A549 and A549/DDP cells	80 μmol/L Ginsenoside Rd. for 48 h	↓PCNA, ↓Nrf2, ↓HO1, ↓NQO1, ↓GCLC, ↓MDR1, ↓MRP1	Ginsenoside Rd. treatment on A549 and A549/DDP cells blocks activation of the Nrf2 signaling pathway and restores its sensitivity to chemotherapy drugs	Ginsenoside Rd. suppressed Nrf2 activation, reversed chemoresistance, and induced G0/G1 phase arrest	[118]
2-Undecanone	–	BEAS-2B cells	25, 50, 100 μM 2-undecanone for 24 h	↑Nrf2, ↑NQO1, ↑HO-1	2-Undecanone treatment suppressed benzo	2-Undecanone activated the Nrf2-HO-1/NQO-	[119]

					(a)pyrene-induced ROS, DNA damage, and inflammation by the activation of the Nrf2 pathway to induce HO-1, NAD(P)H, NQO-1	1 signaling pathway and prevented benzo (a)pyrene-induced lung tumorigenesis	
Luteolin	Chemosensitizes oxaliplatin, bleomycin, and doxorubicin	A549 cells	1, 10, 20 μM for 24 and 48 h	↓Nrf2, ↓NQO1, ↓AKR1C1, ↓HO-1, ↓AKR1C	Luteolin inhibited Nrf2 activity resulting in the downregulation of ARE-gene batteries	Luteolin enhanced a dramatic reduction in Nrf2 leading to decreased Nrf2 binding to AREs and enhance the responsiveness of cancer cells to chemotherapeutic drugs	[120]
Metformin	Reverses cisplatin resistance	A549 and A549/DDP cells	1, 5, 10 mM for 24 and 48 h	↓Nrf2, ↓PI3K, ↓pAkt, ↓pERK1/2, ↑p38MAPK, ↑pJNK, ↓GSTA1, ↓ABCC1	Metformin inhibited the Nrf2, GSTA1, and ABCC1 to reduce proliferation and induced the apoptosis	Metformin inhibited the expression of Nrf2 by inhibiting PI3K/Akt and ERK1/2 signaling	[121]

(continued)

Table 30.1 (continued)

Compounds	Chemosensitization/ radiosensitization	Cell lines	Dose and duration	Molecular targets	Molecular mechanism	Role of the compound	References
Metformin	Sensitizes epigallocatechin-3-gallate (EGCG)	A549, H1299, and H460 cells	0.4 mM metformin for 48 h and 80 and 100 μM EGCG for 24 h	↓Nrf2, ↓HO-1, ↑SIRT1	Metformin increased intracellular ROS and upregulated SIRT1 expression through the NF-KB pathway	Metformin treatment-induced SIRT1 expression by NFkB pathway and activated deacetylation of the Nrf2 and suppressed the EGCG-activated Nrf2 signaling pathway	[122]
Nobiletin	Potentiates paclitaxel	A549/T xenograft model	15 mg/kg paclitaxel and 12.5, 25, and 50 mg/kg nobiletin	↓Nrf2, ↓p-Akt, ↓p-ERK	Nobiletin treatment increase the therapeutic effects of paclitaxel and inhibit the MDR tumor sizes in A549/T xenograft nude mice model and modulating Nrf2/Akt/ERK pathways	Combination treatment of nobiletin and paclitaxel inhibits the phosphorylation of Akt, and ERK pathways	[123]
Indolyl-chalcone derivates (compound 3d)	–	A549 cells	0.1, 1, 2.5, 5, 20 μM for 24 and 48 h	↑Nrf2, ↑HO-1,	Compound 3d treatment elevated Nrf2 activity and induced apoptosis	Compound 3d increased induced the level of ROS which leads to the activation and nuclear transportation of Nrf2, HO-1 expression	[124]

Triptolide	Sensitizes cisplatin, etoposide, and epirubicin	A549, 3LL cells, and C57BL/6 mice	0.1, 0.5 μM for 24 h in A549 cells, 0.05, 0.1 μM for 24 h in 3LL cells and 0,25 and 1 mg/kg in C57BL/6 mice	↓Nrf2, ↓GCLC, ↓GCLM, ↓AKR1C1	Triptolide efficiently suppressed the Nrf2-ARE pathway and increases the sensitivity of cancer cells	Triptolide treatment sensitized chemotherapeutic drugs-induced cytotoxicity by suppressing the Nrf2-ARE pathway	[125]
(+)-Usnic acid	Enhances paclitaxel efficacy	H520 and Calu-1 cells	0, 10, 20, 40 μM for 8/24 h	↓Nrf2, ↓HO1, ↓Nqo1, ↓p-Akt	(+)-Usnic acid induces ROS accumulation and disrupts the mitochondrial respiratory chain and the PI3K/Akt/Nrf2 pathway	(+)-Usnic acid treatment effectively induces ROS dependent apoptosis by inhibition of Nrf2 expression and PI3K/Akt pathway	[126]
Wentilactone	–	NCI-H446, NCI-HI688, and LTEPsm cells	48 h IC50 value of Wentilactone was 3.44 0.38 μmol/L for NCI-H446, 0.41 0.18 μmol/L for NCI-HI688 and 0.57 0.10 μmol/L for LTEPSM cell	↑c-Caspase-3, ↓c-FLIP, ↓AKR1C1	Wentilactone treatment inhibited the expression of AKR1C1 and attenuated cancer cell proliferation via the apoptosis mechanism	Wentilactone promoted apoptosis via IGF-1R/IRS1/PI3K/Akt/Nrf2/FLIP/Caspase-3 signaling pathways	[127]

References

1. Bray F, Ferlay J, Soerjomataram I et al (2018) Global cancer statistics 2018: GLOBOCAN estimates of incidence and mortality worldwide for 36 cancers in 185 countries. CA Cancer J Clin 68:394–424
2. Cancer Genome Atlas Research Network (2012) Comprehensive genomic characterization of squamous cell lung cancers. Nature 489:519–525
3. Li W, Yu S, Liu T et al (2008) Heterodimerization with small Maf proteins enhances nuclear retention of Nrf2 via masking the NESzip motif. Biochim Biophys Acta 1783:1847–1856
4. Zhang DD, Lo SC, Cross JV et al (2004) Keap1 is a redox-regulated substrate adaptor protein for a Cul3-dependent ubiquitin ligase complex. Mol Cell Biol 24:10941–10953
5. Kobayashi A, Kang MI, Okawa H (2004) Oxidative stress sensor Keap1 functions as an adaptor for Cul3-based E3 ligase to regulate proteasomal degradation of Nrf2. Mol Cell Biol 24:7130–7139
6. Raghunath A, Sundarraj K, Nagarajan R et al (2018) Antioxidant response elements: discovery, classes, regulation and potential applications. Redox Biol 17:297–314
7. Baird L, Dinkova-Kostova AT (2011) The cytoprotective role of the Keap1-Nrf2 pathway. Arch Toxicol 85:241–272
8. Buelna-Chontal M, Guevara-Chávez JG, Silva-Palacios A et al (2014) Nrf2-regulated antioxidant response is activated by protein kinase C in postconditioned rat hearts. Free Radic Biol Med 74:145–156
9. Chen W, Sun Z, Wang XJ et al (2009) Direct interaction between Nrf2 and p21(Cip1/WAF1) upregulates the Nrf2-mediated antioxidant response. Mol Cell 34:663–763
10. Joo MS, Kim WD, Lee KY et al (2016) AMPK facilitates nuclear accumulation of Nrf2 by phosphorylating at serine 550. Mol Cell Biol 36:1931–1942
11. Komatsu M, Kurokawa H, Waguri S et al (2010) The selective autophagy substrate p62 activates the stress responsive transcription factor Nrf2 through inactivation of Keap1. Nat Cell Biol 12:213–223
12. Salazar M, Rojo AI, Velasco D et al (2006) Glycogen synthase kinase-3beta inhibits the xenobiotic and antioxidant cell response by direct phosphorylation and nuclear exclusion of the transcription factor Nrf2. J Biol Chem 281:14841–14851
13. Best SA, De Souza DP, Kersbergen A et al (2018) Synergy between the Keap1/Nrf2 and PI3K pathways drives non-small-cell lung cancer with an altered immune microenvironment. Cell Metab 27:935–943
14. Muscarella LA, Parrella P, D'Alessandro V et al (2011) Frequent epigenetics inactivation of KEAP1 gene in non-small cell lung cancer. Epigenetics 6:710–719
15. Ohta T, Iijima K, Miyamoto M et al (2008) Loss of Keap1 function activates Nrf2 and provides advantages for lung cancer cell growth. Cancer Res 68:1303–1309
16. Wang R, An J, Ji F et al (2008) Hypermethylation of the Keap1 gene in human lung cancer cell lines and lung cancer tissues. Biochem Biophys Res Commun 373:151–154
17. Singh A, Misra V, Thimmulappa RK et al (2006) Dysfunctional KEAP1-NRF2 interaction in non-small-cell lung cancer. PLoS Med 3:e420
18. Solis LM, Behrens C, Dong W et al (2010) Nrf2 and Keap1 abnormalities in non-small cell lung carcinoma and association with clinicopathologic features. Clin Cancer Res 16:3743–3753
19. Sekine T, Sakaguchi C, Fukano Y (2015) Investigation by microarray analysis of effects of cigarette design characteristics on gene expression in human lung mucoepidermoid cancer cells NCI-H292 exposed to cigarette smoke. Exp Toxicol Pathol 67:143–151
20. Saigusa D, Motoike IN, Saito S et al (2020) Impacts of NRF2 activation in non-small-cell lung cancer cell lines on extracellular metabolites. Cancer Sci 111:667–678
21. Kolesar JM, Pritchard SC, Kerr KM et al (2002) Evaluation of NQO1 gene expression and variant allele in human NSCLC tumors and matched normal lung tissue. Int J Oncol 21:1119–1124

22. Sun J, Brand M, Zenke Y et al (2004) Heme regulates the dynamic exchange of Bach1 and NF-E2-related factors in the Maf transcription factor network. Proc Natl Acad Sci U S A 101:1461–1466
23. Gozzelino R, Jeney V, Soares MP (2010) Mechanisms of cell protection by heme oxygenase-1. Annu Rev Pharmacol Toxicol 50:323–534
24. Zenke-Kawasaki Y, Dohi Y, Katoh Y et al (2007) Heme induces ubiquitination and degradation of the transcription factor Bach1. Mol Cell Biol 27:6962–6971
25. Li C, Stocker R (2009) Heme oxygenase and iron: from bacteria to humans. Redox Rep 14:95–101
26. Zhang X, Guo J, Wei X et al (2018) Bach1: function, regulation, and involvement in disease. Oxid Med Cell Longev 2018:1347969
27. Lignitto L, LeBoeuf SE, Homer H et al (2019) Nrf2 activation promotes lung cancer metastasis by inhibiting the degradation of Bach1. Cell 178:316–329
28. Shibata T, Ohta T, Tong KI et al (2008) Cancer related mutations in NRF2 impair its recognition by Keap1-Cul3 E3 ligase and promote malignancy. Proc Natl Acad Sci U S A 105:13568–13573
29. Hast BE, Cloer EW, Goldfarb D et al (2014) Cancer-derived mutations in KEAP1 impair NRF2 degradation but not ubiquitination. Cancer Res 74:808–817
30. Kim YR, Oh JE, Kim MS et al (2010) Oncogenic NRF2 mutations in squamous cell carcinomas of oesophagus and skin. J Pathol 220:446–451
31. Kerins MJ, Ooi A (2018) A catalogue of somatic NRF2 gain-of-function mutations in cancer. Sci Rep 8:12846
32. Cancer Genome Atlas Research Network (2014) Comprehensive molecular profiling of lung adenocarcinoma. Nature 511:543–550
33. Cancer Genome Atlas Research Network (2016) Distinct patterns of somatic genome alterations in lung adenocarcinomas and squamous cell carcinomas. Nat Genet 48:607–616
34. Li QK, Singh A, Biswal S et al (2011) KEAP1 gene mutations and NRF2 activation are common in pulmonary papillary adenocarcinoma. J Hum Genet 56:230–234
35. Campbell JD, Alexandrov A, Kim J et al (2016) Distinct patterns of somatic genome alterations in lung adenocarcinomas and squamous cell carcinomas. Nat Genet 48:607–616
36. Kim Y, Hammerman PS, Kim J et al (2014) Integrative and comparative genomic analysis of lung squamous cell carcinomas in east Asian patients. J Clin Oncol 32:121–128
37. Sanchez-Vega F, Mina M, Armenia J et al (2018) Oncogenic signaling pathways in the cancer genome atlas. Cell 173:e10
38. Sasaki H, Shitara M, Yokota K et al (2012) Increased NRF2 gene (NFE2L2) copy number correlates with mutations in lung squamous cell carcinomas. Mol Med Rep 6:391–394
39. Ding L, Getz G, Wheeler DA et al (2008) Somatic mutations affect key pathways in lung adenocarcinoma. Nature 455:1069–1075
40. Imielinski M, Berger AH, Hammerman PS et al (2012) Mapping the hallmarks of lung adenocarcinoma with massively parallel sequencing. Cell 150:1107–1120
41. Jordan EJ, Kim HR, Arcila ME et al (2017) Prospective comprehensive molecular characterization of lung adenocarcinomas for efficient patient matching to approved and emerging therapies. Cancer Discov 7:596–609
42. Rizvi NA, Hellmann MD, Snyder A et al (2015) Cancer immunology. Mutational landscape determines sensitivity to PD-1 blockade in non-small cell lung cancer. Science 348:124–128
43. Frank R, Scheffler M, Merkelbach-Bruse S et al (2018) Clinical and pathological characteristics of KEAP1- and NFE2L2-mutated non-small cell lung carcinoma (NSCLC). Clin Cancer Res 24:3087–3096
44. Hu Y, Ju Y, Lin D et al (2012) Mutation of the Nrf2 gene in non-small cell lung cancer. Mol Biol Rep 39:4743–4747
45. Hayes JD, McMahon M (2009) NRF2 and KEAP1 mutations: permanent activation of an adaptive response in cancer. Trends Biochem Sci 34:176–188

46. Goldstein LD, Lee J, Gnad F et al (2016) Recurrent loss of NFE2L2 exon 2 is a mechanism for Nrf2 pathway activation in human cancers. Cell Rep 16:2605–2617
47. Hayes DN, Kim J, Cho S et al (2014) Integrative and comparative genomic analysis of lung squamous cell carcinomas in east Asian patients. J Clin Oncol 32:121–128
48. Takahashi T, Sonobe M, Menju T et al (2010) Mutations in Keap1 are a potential prognostic factor in resected non-small cell lung cancer. J Surg Oncol 101:500–506
49. Padmanabhan B, Tong KI, Ohta T et al (2006) Structural basis for defects of Keap1 activity provoked by its point mutations in lung cancer. Mol Cell 21:689–700
50. Bartel DP (2004) MicroRNAs: genomics, biogenesis, mechanism, and function. Cell 116:281–297
51. Garzon R, Marcucci G, Croce CM (2010) Targeting microRNAs in cancer: rationale, strategies and challenges. Nat Rev Drug Discov 9:775–789
52. Guarnieri DJ, Dileone RJ (2008) MicroRNAs: a new class of gene regulators. Ann Med 40:197–208
53. Bartel DP (2009) MicroRNA target recognition and regulatory functions. Cell 136:215–233
54. Cortez MA, Valdecanas D, Zhang X et al (2014) Therapeutic delivery of miR-200c enhances radiosensitivity in lung cancer. Mol Ther 22:1494–1503
55. Zhang B, Ma Z, Tan B et al (2019) Targeting the cell signaling pathway Keap1-Nrf2 as a therapeutic strategy for adenocarcinomas of the lung. Expert Opin Ther Targets 23:41–250
56. Chen C, Jiang X, Gu S et al (2017) MicroRNA-155 regulates arsenite-induced malignant transformation by targeting Nrf2-mediated oxidative damage in human bronchial epithelial cells. Toxicol Lett 278:38–47
57. Eren E, Tufekci KU, Isci KB et al (2018) Sulforaphane inhibits lipopolysaccharide-induced inflammation, cytotoxicity, oxidative stress, and miR-155 expression and switches to mox phenotype through activating extracellular signal-regulated kinase 1/2-nuclear factor erythroid 2-related factor 2/antioxidant response element pathway in murine microglial cells. Front Immunol 9:36
58. Gu S, Lai Y, Chen H et al (2017) miR-155 mediates arsenic trioxide resistance by activating Nrf2 and suppressing apoptosis in lung cancer cells. Sci Rep 7:12155
59. Trivedi M, Singh A, Talekar M et al (2017) MicroRNA-34a encapsulated in hyaluronic acid nanoparticles induces epigenetic changes with altered mitochondrial bioenergetics and apoptosis in non-small-cell lung cancer cells. Sci Rep 7:3636
60. Fu WF, Chen WB, Dai L et al (2016) Inhibition of miR-141 reverses cisplatin resistance in non-small cell lung cancer cells via upregulation of programmed cell death protein 4. Eur Rev Med Pharmacol 20:2565–2572
61. Mei Z, He Y, Feng J et al (2014) MicroRNA-141 promotes the proliferation of non-small cell lung cancer cells by regulating expression of PHLPP1 and PHLPP2. FEBS Lett 588:3055–3061
62. Duan F, Wang M, Cao Y et al (2019) MicroRNA-421 confers paclitaxel resistance by binding to the KEAP1 3′UTR and predicts poor survival in non-small cell lung cancer. Cell Death Dis 10:821
63. Li YX, Cui XM, Li YD et al (2018) Upregulated expression of miR-421 is associated with poor prognosis in non-small-cell lung cancer. Cancer Manag Res 10:2627–2633
64. Singh A, Happel C, Manna SK et al (2013) Transcription factor NRF2 regulates miR-1 and miR-206 to drive tumorigenesis. J Clin Invest 123:2921–2934
65. Müller T, Hengstermann A (2012) Nrf2: friend and foe in preventing cigarette smoking-dependent lung disease. Chem Res Toxicol 25:1805–1824
66. Rodgman A, Perfetti TA (2009) The chemical compounds of tobacco and tobacco smoke. CRC Press, Boca Raton
67. Müller T, Haussmann HJ, Schepers G (1997) Evidence for peroxynitrite as an oxidative stress-inducing compound of aqueous cigarette smoke fractions. Carcinogenesis 18:295–301
68. Nakayama T, Kaneko M, Kodama M et al (1985) Cigarette smoke induces DNA single-strand breaks in human cells. Nature 314:462–464

69. Squadrito GL, Pryor WA (1998) Oxidative chemistry of nitric oxide: the roles of superoxide, peroxynitrite, and carbon dioxide. Free Radic Biol Med 25:392–403
70. Bryant A, Cerfolio RJ (2007) Differences in epidemiology, histology, and survival between cigarette smokers and never-smokers who develop non-small-cell lung cancer. Chest 132:185–192
71. Landi MT, Dracheva T, Rotunno M et al (2008) Gene expression signature of cigarette smoking and its role in lung adenocarcinoma development and survival. PLoS One 3:e1651
72. MacNee W (2010) Oxidative stress and chronic obstructive pulmonary disease. Eur Resp Monogr 46:100–129
73. Wooten JB, Chouchane S, McGrath TE (2006) Tobacco smoke constituents affecting oxidative stress. In: Halliwell B, Poulsen H (eds) Cigarette smoke and oxidative stress. Springer, Heidelberg, pp 5–46
74. Itoh K, Wakabayashi N, Katoh Y et al (1999) Keap1 represses nuclear activation of antioxidant responsive elements by Nrf2 through binding to the amino-terminal Neh2 domain. Genes Dev 13:76–86
75. Eggler AL, Liu G, Pezzuto JM et al (2005) Modifying specific cysteines of the electrophile-sensing human Keap1 protein is insufficient to disrupt binding to the Nrf2 domain Neh2. Proc Natl Acad Sci 102:10070–10075
76. Holland R, Fishbein JC (2010) Chemistry of the cysteine sensors in kelch-like ech-associated protein 1. Antioxid Redox Signal 13:1749–1761
77. Levonen AL, Landar A, Ramachandran A et al (2004) Cellular mechanisms of redox cell signalling: role of cysteine modification in controlling antioxidant defences in response to electrophilic lipid oxidation products. Biochem J 378:373–382
78. Wakabayashi N, Dinkova-Kostova AT, Holtzclaw WD et al (2004) Protection against electrophile and oxidant stress by induction of the phase 2 response: fate of cysteines of the Keap1 sensor modified by inducers. Proc Natl Acad Sci 101:2040–2045
79. Zhang DD, Hannink M (2003) Distinct cysteine residues in Keap1 are required for Keap1-dependent ubiquitination of Nrf2 and for stabilization of Nrf2 by chemopreventive agents and oxidative stress. Mol Cell Biol 23:8137–8151
80. Bloom DA, Jaiswal AK (2003) Phosphorylation of Nrf2 at Ser40 by protein kinase C in response to antioxidants leads to the release of Nrf2 from INrf2, but is not required for Nrf2 stabilization/accumulation in the nucleus and transcriptional activation of antioxidant response element-mediated NAD(P)H:quinone oxidoreductase-1 gene expression. J Biol Chem 278:44675–44682
81. Bobrovnikova-Marjon E, Grigoriadou C, Pytel D et al (2010) PERK promotes cancer cell proliferation and tumor growth by limiting oxidative DNA damage. Oncogene 29:3881–3895
82. Cullinan SB, Zhang D, Hannink M et al (2003) Nrf2 is a direct PERK substrate and effector of PERK-dependent cell survival. Mol Cell Biol 23:7198–7209
83. Huang HC, Nguyen T, Pickett CB (2002) Phosphorylation of Nrf2 at Ser-40 by protein kinase C regulates antioxidant response element-mediated transcription. J Biol Chem 277:42769–42774
84. Jaiswal AK (2004) Nrf2 signaling in coordinated activation of antioxidant gene expression. Free Radic Biol Med 36:1199–1207
85. Cheng SE, Lee IT, Lin CC et al (2010) Cigarette smoke particle-phase extract induces HO-1 expression in human tracheal smooth muscle cells: role of the c-Src/NADPH oxidase/MAPK/Nrf2 signaling pathway. Free Radic Biol Med 48:1410–1422
86. McMahon M, Lamont DJ, Beattie KA et al (2010) Keap1 perceives stress via three sensors for the endogenous signaling molecules nitric oxide, zinc, and alkenals. Proc Natl Acad Sci 107:18838–18843
87. Um HC, Jang JH, Kim DH et al (2011) Nitric oxide activates Nrf2 through S-nitrosylation of Keap1 in PC12 cells. Nitric Oxide 25:161–168

88. Goven D, Boutten A, Leçon-Malas V et al (2009) Prolonged cigarette smoke exposure decreases heme oxygenase-1 and alters Nrf2 and Bach1 expression in human macrophages: roles of the MAP kinases ERK1/2 and JNK. FEBS Lett 583:3508–3518
89. Alam J, Wicks C, Stewart D (2000) Mechanism of heme oxygenase-1 gene activation by cadmium MCF-7 mammary epithelial cells. Role of p38 kinase and Nrf2 transcription factor. J Biol Chem 275:27694–27702
90. He X, Ma Q (2009) NRF2 cysteine residues are critical for oxidant/electrophile-sensing, kelch-like ech-associated protein-1-dependent ubiquitination-proteasomal degradation, and transcription activation. Mol Pharmacol 76:1265–1278
91. Lee SE, Jeong SI, Kim GD et al (2011) Upregulation of heme oxygenase-1 as an adaptive mechanism for protection against crotonaldehyde in human umbilical vein endothelial cells. Toxicol Lett 201:240–248
92. Zhang H, Forman HJ (2007) Acrolein induces heme oxygenase-1 through PKC-δ and PI3K in human bronchial epithelial cells. Am J Respir Cell Mol Biol 38:483–490
93. Dinkova-Kostova AT, Wang XJ (2011) Induction of the Keap1/Nrf2/ARE pathway by oxidizable diphenols. Chem Biol Interact 192:101–106
94. Wang XJ, Hayes JD, Higgins LG et al (2011) Activation of the NRF2 signaling pathway by copper-mediated redox cycling of para- and orthohydroquinones. Chem Biol 17:75–85
95. Fourquet S, Guerois R, Biard D et al (2010) Activation of NRF2 by nitrosative agents and H_2O_2 involves KEAP1 disulfide formation. J Biol Chem 285:8463–8471
96. Cederbaum AI (2009) Nrf2 and antioxidant defense against CYP2E1 toxicity. Expert Opin Drug Metab Toxicol 5:1223–1244
97. Nguyen T, Sherratt PJ, Pickett CB (2003) Regulatory mechanisms controlling gene expression mediated by the antioxidant response element. Annu Rev Pharmacol Toxicol 43:233–260
98. Prochaska HJ, Talalay P (1988) Regulatory mechanisms of monofunctional and bifunctional anticarcinogenic enzyme inducers in murine liver. Cancer Res 48:4776–4782
99. Miao W, Hu L, Scrivens PJ et al (2005) Transcriptional regulation of NF-E2 p45-related factor (NRF2) expression by the aryl hydrocarbon receptor-xenobiotic response element signaling pathway: direct cross-talk between phase I and II drug-metabolizing enzymes. J Biol Chem 280:20340–20348
100. Wakabayashi N, Slocum SL, Skoko JJ et al (2010) When Nrf2 talks, who's listening? Antioxid Redox Signal 13:1649–1663
101. Sasaki H, Suzuki A, Shitara M et al (2013) Genotype analysis of the NRF2 gene mutation in lung cancer. Int J Mol Med 31:1135–1138
102. Bai X, Chen Y, Hou X et al (2016) Emerging role of NRF2 in chemoresistance by regulating drug-metabolizing enzymes and efflux transporters. Drug Metab Rev 48:541–567
103. Munoz M, Henderson M, Haber M et al (2007) Role of the MRP1/ABCC1 multidrug transporter protein in cancer. IUBMB Life 59:752–757
104. Ji L, Li H, Gao P et al (2013) Nrf2 pathway regulates multidrug-resistance-associated protein 1 in small cell lung cancer. PLoS One 8:e63404
105. Singh A, Boldin-Adamsky S, Thimmulappa RK et al (2008) RNAi-mediated silencing of nuclear factor erythroid-2-related factor 2 gene expression in non-small cell lung cancer inhibits tumor growth and increases efficacy of chemotherapy. Cancer Res 68:7975–7984
106. Wang Q, Ma J, Lu Y et al (2017) CDK20 interacts with KEAP1 to activate NRF2 and promotes radiochemoresistance in lung cancer cells. Oncogene 36:5321–5330
107. Singh A, Venkannagari S, Oh KH et al (2016) Small molecule inhibitor of NRF2 selectively intervenes therapeutic resistance in KEAP1-deficient NSCLC tumors. ACS Chem Biol 11:3214–3225
108. Gai C, Yu M, Li Z et al (2020) Acetaminophen sensitizing erastin-induced ferroptosis via modulation of Nrf2/heme oxygenase-1 signaling pathway in non-small-cell lung cancer. J Cell Physiol 235:3329–3339
109. Sun M, Pan D, Chen Y et al (2017) Coroglaucigenin enhances the radiosensitivity of human lung cancer cells through Nrf2/ROS pathway. Oncotarget 8:32807–32820

110. Xia C, Bai X, Hou X et al (2015) Cryptotanshinone reverses cisplatin resistance of human lung carcinoma A549 cells through down-regulating Nrf2 pathway. Cell Physiol Biochem 37:816–824
111. Chen LH, Liao CY, Lai LC et al (2019) Semaphorin 6A attenuates the migration capability of lung cancer cells via the NRF2/HMOX1 axis. Sci Rep 9:13302
112. Chen X, Wu Q, Chen Y et al (2019) Diosmetin induces apoptosis and enhances the chemotherapeutic efficacy of paclitaxel in non-small cell lung cancer cells via Nrf2 inhibition. Br J Pharmacol 176:2079–2094
113. Datta S, Sinha D (2019) EGCG maintained Nrf2-mediated redox homeostasis and minimized etoposide resistance in lung cancer cells. J Funct Foods 62:103553
114. Li Y, Yan H, Xu X et al (2020) Erastin/sorafenib induces cisplatin-resistant non-small cell lung cancer cell ferroptosis through inhibition of the Nrf2/xCT pathway. Oncol Lett 19:323–333
115. Zhou Y, Zhou Y, Wang K et al (2020) Flumethasone enhances the efficacy of chemotherapeutic drugs in lung cancer by inhibiting Nrf2 signaling pathway. Cancer Lett 474:94–105
116. Gill BS, Kumar S, Navgeet (2017) Ganoderic acid targeting nuclear factor erythroid 2-related factor 2 in lung cancer. Tumour Biol 39:1010428317695530
117. Liu X, Sun C, Liu B (2016) Genistein mediates the selective radiosensitizing effect in NSCLC A549 cells via inhibiting methylation of the keap1 gene promoter region. Oncotarget 7:27267–27279
118. Chian S, Zhao Y, Xu M et al (2019) Ginsenoside Rd reverses cisplatin resistance in non-small-cell lung cancer A549 cells by downregulating the nuclear factor erythroid 2-related factor 2 pathway. Anti-Cancer Drugs 30:838–845
119. Lou Y, Guo Z, Zhu Y et al (2019) Houttuynia cordata Thunb. and its bioactive compound 2-undecanone significantly suppress benzo(a)pyrene-induced lung tumorigenesis by activating the Nrf2-HO-1/NQO-1 signaling pathway. J Exp Clin Res 38:242
120. Tang X, Wang H, Fan L et al (2011) Luteolin inhibits Nrf2 leading to negative regulation of the Nrf2/ARE pathway and sensitization of human lung carcinoma A549 cells to therapeutic drugs. Free Radic Biol Med 50:1599–1609
121. Zhang J, Jiao K, Liu J et al (2018) Metformin reverses the resistance mechanism of lung adenocarcinoma cells that knocks down the Nrf2 gene. Oncol Lett 16:6071–6080
122. Yu C, Jiao Y, Xue J et al (2017) Metformin sensitizes non-small cell lung cancer cells to an epigallocatechin-3-gallate (EGCG) treatment by suppressing the Nrf2/HO-1 signaling pathway. Int J Biol Sci 13:1560–1569
123. Feng SL, Tian Y, Huo S et al (2020) Nobiletin potentiates paclitaxel anticancer efficacy in A549/T xenograft model: pharmacokinetic and pharmacological study. Phytomedicine 67:153141
124. Zhao X, Dong WL, Gao YD et al (2017) Novel indolyl-chalcone derivatives inhibit A549 lung cancer cell growth through activating Nrf-2/HO-1 and inducing apoptosis in vitro and in vivo. Sci Rep 7:3919
125. Zhu J, Wang H, Chen F et al (2018) Triptolide enhances chemotherapeutic efficacy of antitumor drugs in non-small-cell lung cancer cells by inhibiting Nrf2-ARE activity. Toxicol Appl Pharmacol 358:1–9
126. Qi W, Lu C, Huang H et al (2020) (+)-Usnic acid induces ROS-dependent apoptosis via inhibition of mitochondria respiratory chain complexes and Nrf2 expression in lung squamous cell carcinoma. Int J Mol Sci 21:876
127. Jiang W, Meng L, Xu G et al (2018) Wentilactone A induces cell apoptosis by targeting AKR1C1 gene via the IGF-1R/IRS1/PI3K/AKT/Nrf2/FLIP/Caspase-3 signaling pathway in small cell lung cancer. Oncol Lett 16:6445–6457
128. Ohnuma T, Anzai E, Suzuki Y et al (2015) Selective antagonization of activated Nrf2 and inhibition of cancer cell proliferation by procyanidins from Cinnamomi cortex extract. Arch Biochem Biophys 585:17–24

129. Ohnuma T, Matsumoto T, Itoi A et al (2011) Enhanced sensitivity of A549 cells to the cytotoxic action of anticancer drugs via suppression of Nrf2 by procyanidins from Cinnamomi cortex extract. Biochem Biophys Res Commun 413:623–629
130. Ohnuma T, Sakamoto K, Shinoda A et al (2017) Procyanidins from Cinnamomi cortex promote proteasome-independent degradation of nuclear Nrf2 through phosphorylation of insulin-like growth factor-1 receptor in A549 cells. Arch Biochem Biophys 635:66–73
131. Olayanju A, Copple IM, Bryan HK et al (2015) Brusatol provokes a rapid and transient inhibition of Nrf2 signaling and sensitizes mammalian cells to chemical toxicity implications for therapeutic targeting of Nrf2. Free Radic Biol Med 78:202–212
132. Ren D, Villeneuve NF, Jiang T et al (2011) Brusatol enhances the efficacy of chemotherapy by inhibiting the Nrf2-mediated defense mechanism. Proc Natl Acad Sci U S A 108:1433–1438
133. Tao S, Wang S, Moghaddam SJ et al (2014) Oncogenic KRAS confers chemoresistance by upregulating NRF2. Cancer Res 74:7430–7441
134. Chian S, Thapa R, Chi Z et al (2014) Luteolin inhibits the Nrf2 signaling pathway and tumor growth in vivo. Biochem Biophys Res 447:602–608
135. Ahmad F, Dixit D, Sharma V et al (2016) Nrf2-driven TERT regulates pentose phosphate pathway in glioblastoma. Cell Death Dis 7:e2213
136. Lee J, Kang JS, Nam LB et al (2018) Suppression of NRF2/ARE by Convallatoxin sensitises A549 cells to 5-FU-mediated apoptosis. Free Radic Res 52:1416–1423
137. Tsuchida K, Tsujita T, Hayashi M et al (2017) Halofuginone enhances the chemo-sensitivity of cancer cells by suppressing NRF2 accumulation. Free Radic Biol Med 103:236–247
138. Tsujita T, Peirce V, Baird L et al (2014) Transcription factor Nrf1 negatively regulates the cystine/glutamate transporter and lipid-metabolizing enzymes. Mol Cell Biol 34:3800–3816
139. Bollong MJ, Yun H, Sherwood L et al (2015) A small molecule inhibits deregulated NRF2 transcriptional activity in cancer. ACS Chem Biol 10:2193–2198
140. Xiaobo C, Majidi M, Feng M et al (2016) TUSC2(FUS1)-erlotinib induced vulnerabilities in epidermal growth factor receptor(EGFR) wildtype non-small cell lung cancer(NSCLC) targeted by the repurposed drug Auranofin. Sci Rep 6:35741
141. Choi EJ, Jung BJ, Lee SH et al (2017) A clinical drug library screen identifies clobetasol propionate as an NRF2 inhibitor with potential therapeutic efficacy in KEAP1 mutant lung cancer. Oncogene 37:5285–5295
142. Do MT, Kim HG, Khanal T et al (2013) Metformin inhibits heme oxygenase-1 expression in cancer cells through inactivation of Raf-ERK-Nrf2 signaling and AMPK-independent pathways. Toxicol Appl Pharmacol 271:229–238
143. Chen F, Wang H, Zhu J et al (2017) Camptothecin suppresses NRF2-ARE activity and sensitises hepatocellular carcinoma cells to anticancer drugs. Br J Cancer 117:1495–1506
144. Huppke P, Weissbach S, Church JA et al (2017) Activating de novo mutations in NFE2L2 encoding NRF2 cause a multisystem disorder. Nat Commun 8:818

Role of Toll-Like Receptors in Molecular and Cellular Mechanisms of Respiratory Diseases

31

Indu Lata Kanwar, Tanweer Haider, Vikas Pandey, Prem N. Gupta, and Vandana Soni

Abstract

The family of toll-like receptors (TLRs) is receiving considerable attention as potential regulators and controllers of the immune response through their ability to recognize and defend against invading pathogens. TLRs are key components of the innate immune system. Inappropriate or unregulated activation of TLR signaling can lead to chronic inflammation and autoimmune disorders. TLRs have been implicated in a number of lung-associated immune responses and pathogenesis of some respiratory diseases including asthma, chronic obstructive pulmonary disease, lung cancer, and infections. This chapter details the different TLRs, signaling transduction of TLR, defines their possible role in the pathogenesis of the main respiratory diseases, and finally, speculates over the therapeutic possibilities by targeting TLR signaling.

Keywords

Toll like receptors · Asthma · Lung cancer · COPD · TLR signalling

I. L. Kanwar · T. Haider · V. Pandey · V. Soni (✉)
Department of Pharmaceutical Sciences, Dr. Harisingh Gour University, Sagar, Madhya Pradesh, India
e-mail: Vsoni@dhsgsu.edu.in

P. N. Gupta
Formulation and Drug Delivery Division, CSIR-Indian Institute of Integrative Medicine, Jammu, Jammu and Kashmir, India
e-mail: pngupta@iiim.res.in

K. Dua et al. (eds.), *Targeting Cellular Signalling Pathways in Lung Diseases*,
https://doi.org/10.1007/978-981-33-6827-9_31

31.1 Introduction

The immune system comprises a complex system of cells and proteins, which defend the body from infections. Principally, immunity can be of two types: innate (natural) and acquired (adaptive) immunity. Innate immunity is the vanguard of defense and is characterized by unspecificity and a lack of memory. In contrast, adaptive immunity is characterized by specificity, adaptability to the antigens, and memory to previous infections [1]. Activated innate immunity is an essential factor that is required in developing specific acquired immunity against antigens. With innate immunity, the primary response to a pathogen mediated by pattern recognition receptors (PRRs), which recognize the pathogen-associated molecular patterns (PAMP), present in a wide range of microbes [1, 2]. Some examples of PRRs are RNA helicase retinoic acid-inducible gene I (RIG-I)-like receptors, C-type lectin receptors, DNA sensors, nucleotide-binding oligomerization domain (NOD)-like receptors (NLRs), and Toll-like receptors (TLRs). TLRs are associated with the immune systems microbial recognition as an innate and adaptive immune response in higher vertebrates. The TLR protein is type-I transmembrane receptors with -COOH terminal intracellular tails containing a conserved region (Toll/interleukin (IL)-1) and -NH2 terminal with leucine-rich extracellular repeat domains. The extracellular domain with many leucine-rich extracellular repeat domains is involved in ligands binding and necessary for the TLR dimerization. The intracellular domain (82 amino acids region) is highly variable and involved in the protein-protein interaction between TLR proteins and downstream signal transduction components [3–5]. TLRs recognize a large number of varied and complex PAMP, which represent conserved molecular features of many microbial classes such as Gram-negative lipopolysaccharides (LPS) (TLR4 ligands), dsRNA (TLR3 ligands), viral ssRNA (TLR7 and TLR8 ligands), bacterial flagellin (TLR5 ligands), and cytosine-phosphate-guanosine (CpG) DNA (TLR9 ligands) [3, 6, 7]. When these microbes rupture the physical barriers like skin, they are then recognized by the TLR, which activates the immune cell response. The TLR protein engagement leads to the inflammatory reaction caused by the activation of transcription factor nuclear factor kappa-light-chain-enhancer of activated B (NF-κB) regulating the expression of cytokines (ILs, interferons (INFs), tumor necrosis factor (TNF)-α), and chemokines along with reactive oxygen and nitrogen species [3, 8]. TLRs are involved in the activation of the immune response against the pathogens as well as in tumor progression [9, 10], hematopoietic malignancies [11], myocardial inflammation [12], the pathogenesis of the autoimmune disease, rheumatoid arthritis, myocarditis [13], atherosclerosis [14], noninfectious pulmonary diseases [15], etc.

The TLR is also involved in several lung-associated pathologies and immune responses in lung injury and chronic inflammatory lung diseases, e.g., cystic fibrosis, chronic obstructive pulmonary disease (COPD), airway hyperreactivity, and asthma [3, 16]. TLR stimulation in human lung macrophages provides first-line defense against inhaled pathogens. Human lung macrophages induce an intense production of cytokines and chemokines that are characteristic of the pro-inflammatory M1 macrophage phenotype [17]. TLR2 and TLR4 play a critical role in COPD and contribute to the development of T helper-17 immune response and finally help in COPD [18]. TLR also modulates the functions of nonimmune cells like endothelial cells and pulmonary hypertension. Pulmonary hypertension is recognized as abnormal vasoconstriction and remodeling of the pulmonary arteries followed by right heart failure and then death [19]. The involvement of the different TLRs in various other pulmonary diseases and their molecular and cellular aspect is essential. The activation and signaling of TLRs and their contribution to lung diseases may better understand the treatment aspects. Targeting at the specific site, i.e., at the molecular level and/or cellular level, may help develop therapeutics in treating different TLR associated lung diseases. In this chapter, we will discuss the types of TLR, function, the activation of the signaling of TLR, the role of TLR in pulmonary diseases, and targeting of TLR signaling for the therapeutic of lung diseases.

31.2 Types and Functions of TLR

There are some well-known TLRs present in humans and animals. TLRs are extensively articulated in both resident cells and lymphoid and myeloid cells of lung tissue. Some examples of TLRs such as TLR1 to TLR10 present in humans, whereas TLR1 to TLR9, TLR11, TLR12, and TLR13 present in mice and TLR21 in apes [6, 20, 21]. Some of the human TLRs such as TLR1, TLR2, TLR4 to TLR6, and TLR10 are present on the cell surface, whereas TLR3 and TLR7 to TLR9 are expressed on the intracellular vesicles mainly in the endosome and on cell sub organelle, i.e., endoplasmic reticulum (ER) (Table 31.1) [1, 20]. Each TLR has a specific recognition ability for specific molecule groups. For example, TLR2 is associated with TLR1/TLR6 for Gram-positive bacteria; TLR3 for dsRNA; TLR4 for LPS of Gram-negative bacteria; TLR5 for bacterial flagella; TLR7, TLR8, and TLR9 for viral ssRNA, dsRNA, and nucleic acids of viruses, etc. [1, 3].

TLRs are commonly available in both resident cells and lymphoid and myeloid cells of lung tissues [23] where the expression of TLR1, TLR2, TLR4, TLR7, and TLR8 at high and TLR3, TLR5, and TLR9 at low level was observed on the surface of human alveolar macrophages [24]. The expression of TLR3 has been found in lung tissue and endothelial cells of pulmonary arterial hypertension and also in human lung epithelial during the respiratory syncytial virus infection [25, 26]. Increased expression of the surface protein TLR4 has been observed in LPS lung injury and acute lung injury [27, 28]. TLR4 is also expressed on the lung's endothelial cells and is necessary for the occlusion of capillaries and the accumulation of neutrophils after systemic administration of LPS [29]. These TLRs are also overexpressed in different cancers and played a role in cancer prognosis. Lung cancers have significant expression of TLRs. The expression of TLR2, TLR3, TLR7, and TLR9 has been found in non-small-cell lung cancer (NSCLC) patients, whereas TLR3 and TLR9 significantly elevated in idiopathic pulmonary fibrosis [30]. Numerous studies suggested that TLR expression was observed in pro-tumor activities as well as in antitumor activities. The high expression of TLR5 has been significantly associated with the better prognosis of NSCLC [31].

Table 31.1 TLRs, their adaptors, and ligands [1, 20, 22]

TLRs	Pathogen detected	Adapter molecules	Pathways	PAMPS	Regulation
Cell surface-localized TLRs					
TLR1	Bacteria	MyD88, TIRAP	MyD88	LPS, porin, modulin, peptidoglycan, lipoteichoic acid, lipopeptides, zymosan, glycophosphatidylinositols	IL-6, and by elevated concentrations of IFN-αβ, IL-10, and TNF-α
TLR2	Bacteria, protozoa, fungi	MyD88, TIRAP	MyD88	Lipopeptides, porin, peptidoglycan zymosan, GIPL, β-Glycan, lipoproteins	IL-6 and TNF-α, IL-1β and IL-10
TLR4	Bacteria and virus	MyD88, TIRAF, TRIF, TRAM	MyD88, TRIF	LPS, RSV, F protein, VSV, Env-prot, MMTV	IFN-δ and IL-1β
TLR5	Bacteria	MyD88	MyD88	Flagellin	IL-6, IL-10, TNF-α and IFN-δβ
TLR6	Bacteria	MyD88, TIRAP	MyD88	Macrophage-activating lipopeptide-2, LPS, porin, modulin, peptidoglycan, lipoteichoic acid, lipopeptides, zymosan, glycophosphatidylinositols	IFN-δ and IL-1β
TLR10	Bacteria	MyD88, TIRAP	MyD88	Ac3LP, lipopeptides, porin, peptidoglycan zymosan, GIPL, β-glycan, lipoproteins	IFN-δ, IL-1β, IL-6, IL-10, and TNF-α
Intracellular localized TLRs					
TLR3	Virus	TRIF	TRIF	Analog of dsRNA	IFN-δ, IL-1β, IL-6, IL-10, and TNF-α
TLR7	Virus	MyD88	MyD88	ssRNA,	TLR7: IL-6, IFN-δ and IL-1β
TLR8	Virus	MyD88	MyD88	ssRNA	TLR8: IL-1β, IL-6, IL-10, and TNF-α Expression enhanced by exposure to IFN-δ
TLR9	Bacteria, virus	MyD88	MyD88	Bacterial DNA and viral RNA. Cytosine guanine dinucleotide-DNA	IFN-δ, IL-1β, IL-6, IL-10 and TNF-α

The expression of TLR1, TLR2, TLR4, TLR5, and TLR9 also are found on the recruited neutrophils [20, 32]. The significant elevation of TLR7 and TLR9 has expressed on plasmacytoid dendritic cells in the lungs and has been involved in suppression in allergic responses by the T-cell regulation [33]. The recognition of TLR ligands present in dendritic cells and macrophages leads to the rapid production of cytokines and chemokines that indicate the pathogen's presence. The production of cytokines and chemokines initiates the rapid recruitment of cells because of the immunity system to the site of infection. It activates them by immediately initiating antigen-presenting cells, as they induce expression of CD40, CD80, and CD86 as costimulatory molecules for the activation of T lymphocytes [34]. The TLRs induce the differentiation of T helper lymphocytes (CD4 +) or cytotoxic lymphocytes (CD8 +) from the T cells. T helper-1 stimulated by IL-12 produces IL-2, IFN-γ, and TNF, whereas T helper-2 lymphocytes stimulated by IL-12 produce IL-4 to IL-6 [1, 35]. Other lymphocytes, i.e., T-regulator and T helper-17, are essential for developing and regulating the immune response [18]. Thus, different types of TLRs are available on surfaces of lung diseases, macrophages, lung tissue, endothelial cells, etc. and activate the immune system.

31.3 Activation and Signaling of TLR

Activation of TLRs causes acute inflammation and regulates the adapted immunity at various levels. The activation-specific transcription factors, for the immunity and inflammatory responses, are activated by the partially overlapping of the intracellular pathways of every TLR. The TLR signaling cascades initiate from the toll-interleukin-1 receptor (TIR) domain. The TIR is a cytoplasmic domain of TLR signaling, which includes the four types of adaptor molecules such as TIR-domain-containing adapter-inducing interferon-β (TRIF), TRIF-related adaptor molecule (TRAM), myeloid differentiation primary response protein (MyD88), and MyD88 adaptor-like (Mal) or TIR domain-containing adaptor protein (TIRAP). After the ligand binding of cytoplasmic TIR domains of TLR, the dimerization of TIR adaptor molecules such as MyD88, TIRAP, TRIF, and TRAM takes place. The dimerization of adaptor molecules is recruited by TIR, which initiates the series of events and finally activates and translocates the transcriptor factors. The TLR signaling (Fig. 31.1) is categorized into two categories, such as MyD88 (critical adaptor molecules in single transduction)-dependent or MyD88-independent [16]. MyD88-dependent pathways occur in TLR 2, TLR5, TLR7, TLR8, TLR9, and TLR 11 signaling [16, 22]. After the pathogenesis, PAMPs bind the TLRs, which lead to the changes of cytoplasmic domains, which later recruits MyD88 to TLRs by the TIR domain interactions. After recruiting serine/threonine IL-1 receptor-associated kinase-4 (IRAK-4) to MyD88 downstream signal transduction and after binding with MyD88, IRAK-4 recruits the IRAK-1 and phosphorylases the IRAK-1. After the phosphorylation of IRAK-1, it autophosphorylates and generates the new docking sites. These new docking sites enable the tumor necrosis factor receptor-associated factor 6 (TRAF 6) to bind to the MyD88/IRAK-4/IRAK-1 complex. The

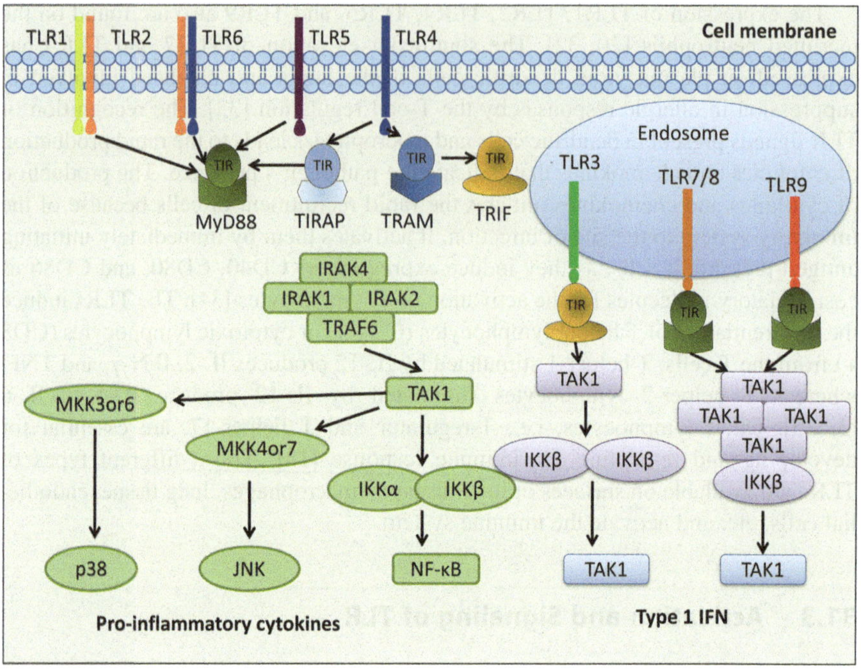

Fig. 31.1 TLR signaling pathways [38]

dissociation of TRAF 6 and IRAK-1 from the complex leads to activate the c-Jun N terminal kinase (JNK) and inhibitor of NF-κB kinase (INK) by the interaction with complex transforming growth factor (TGF)-β-activated kinase 1/TGF-β-activated kinase 1/2/3. These active JNK and INK lead to the activation of the activator protein 1 (AP1) and NF-κB. NF-κB moves to the nucleus and acts as a transcription factor for gene-coded pro-inflammatory chemokines and cytokines such as TNF-α, IL-6, IL-8, and IL-1β [22, 36].

Another TLR signaling pathway is the MyD88-independent or TRIF pathway. There are several reports that suggested that the TLR3 and TLR4 are stimulated by the MyD88-independent signal transduction or TRIF pathway. The TIRF pathways activate by the pathogens, resulting in interferon regulatory factor 3 (IRF3) activation, an important transcription factor needed to transcription antiviral genes like IFN-β [20, 37]. In the TRIF pathways, TRIF binds with the TIR domain via TRAM (a bridge domain molecule). TRIF binds with the TRAF 3 and TRAF 6. The MyD88 pathways activated by the binding of TRIF to TRAF 6. Nevertheless, if TRIF binds to TRAF 3, it activates the TGF-β-activated kinase 1 and then phosphorylates and activates the IRF3 and IRF7. Their homodimers enter into the nucleus and attach to their binding segment on DNA and transcribe IFN-α and IFN-β [20].

31.4 Implications of TLR in Respiratory Diseases

31.4.1 TLR in Asthma

Asthma is a fatal disease that can cause chronic airway inflammation, which manifests as bronchoconstriction, goblet cell hyperplasia, excessive mucus secretion, and tissue remodeling that occurs in childhood. Potential immune response to asthma targeting environmental antigens, including pollen or dust particles, constitutes the habitat of antigen-specific Th2 cells that secrete antigen-specific IgE in the lungs [39]. Innate immunity plays a vital role in disease pathogenesis because infections, either viral or bacterial, are related to induction or resisting asthma. Various human, animal, and epidemiological studies have reported that the time and extent of exposure to LPS and probably TLR4 activation seem to determine either a defensive Th1 response or tolerance Th2 response develops in the lungs [40]. In research, it was reported that intranasal administration of low-dose LPS could cause a Th2 reaction in the lung, while LPS in other parts of the body can produce a robust Th1 response [41]. However, experimental treatment of mice with TLR agonists or microbes represses airway hyperresponsiveness, allergic sensitization, and eosinophilic inflammation. Recent research also shows that *Acinetobacter lwoffii* F78 can prevent ovalbumin-induced asthma of the offspring when administered intranasally to pregnant mice. The protective effect depends on the expression pattern of maternal TLRs, and the recognition of microbes by maternal TLRs during pregnancy triggers the fetal lung environment, which produces a subsequent Th1 response to these antigens. Lung resident cells expressing TLR4 recognize house dust mite (a ubiquitous indoor allergen) and induce Th2 cell response to which produce IL-25, IL-33, granulocyte-macrophage colony-stimulating factor, and stromal lymphopoietin thymus [42]. These cytokines cause the activation and polarization of innate lymphocytes. Besides, eosinophil-derived neurotoxins stimulate the activation of TLR2, which ultimately leads to the large secretion of IL-6, IL-10, and Th2 polarization. Similarly, basophils also cause Th2 cell activation. As many genetic association studies have shown, TLR is also associated with increased allergic asthma [43]. For example, TLR7 and TLR8 are associated with asthma progression, and their ligands can protect airway remodeling in experimentally induced asthma. TLR10 single-nucleotide polymorphism is also associated with asthma in two separate samples, although the ligands for TLR10 are not yet defined. TLR4 and TLR9 are connected with breathlessness, and TLR4 is associated with allergen-specific IgE secretion. Based on this observation, TLR9 ligand is currently being used in clinical trials to treat or prevent asthma [44].

Besides, bacterial infections such as *Mycoplasma pneumoniae*, *Chlamydia pneumoniae*, and *Streptococcal pneumoniae* may worsen the disease condition. It was shown that MyD88-deficient mice infected with *C. pneumoniae* showed reduced expression of cytokines and chemokines, delayed the recruitment of CD4 + and CD8+ cells, and failed to clear lung bacteria, resulting in severe chronic inflammation and increase mortality. However, the secretion of IL-1β, IFN-γ, and other inflammatory mediators may be enhanced through MyD88-independent pathways later, but the mortality associated with *C. pneumoniae* cannot be ruled out. In such infections, the TLR2 and TLR4 work synergistically, or in some cases, other PRRs also involved [16].

Excessive secretion of airway mucin and upregulation of TLR2 resulted from the *M. pneumoniae* infection, whereas during allergic inflammation caused by *M. pneumoniae* infection, the TLR2 is inhibited due to the production of IL-6 and other mediators. As a result, bacterial clearance is reduced. Antibiotic therapy is highly effective in asthma patients with *M. pneumoniae* infection because they improve lung function and suggest the role of bacteria and host immune system interaction in asthma exacerbation and death. One study showed that other PRRs also play a part in the infection of *S. pneumoniae*, in which high and low infectious doses of *S. pneumoniae* are cleared, and the production of inflammatory mediators is moderately reduced in TLR2-deficient mice [45]. With the lack of MyD88, the severity of response to *S. pneumoniae* is significantly increased, instead of deleting TLR2 and TLR4 individually or in combination. Lower respiratory tract viral infections may also lead to the development and worsening of asthma. Respiratory syncytial virus (RSV) is an incredibly vital cause of acute bronchiolitis and wheezing in children and may lead to the subsequent development of asthma. After getting severe RSV in early life, wheezing is related to elevated blood lipids Th2 response, eosinophilia, and IL-10 production [46]. During RSV infection, the virus attaches to lung epithelial cells through the G protein, and then, the virus envelope fuses with the plasma membrane of the host cell-mediated by F protein. TLR4 recognizes the viral F protein and stimulates NK cells. Therefore, TLR4-deficient mice have higher viral loads and defective NK cells. According to a study, insufficient TLR4 signal is also associated with the worsening of preterm infants. The dsRNA formed during RSV replication stimulates TLR3-mediated signal transduction in fibroblasts and the human lung epithelial cells, resulting in excessive secretion of RANTES and IP-10 chemokines. The lack of TLR3 signaling leads to increased mucus secretion, increased airway eosinophilia, and also increased levels of IL-5 and IL-13. It is observed that the retinoic acid-inducible gene I (RIG-I) induces the expression of IFN-β during RSV, which triggers TLR3 activation, and that TLR3 mediates the secondary immune signaling pathway [26]. In contrast, RSV virus clearance is completely mediated by the TLR2/TLR6 heterodimer. Therefore, allergic asthma can be mediated by infection, and genetic susceptibility can make it worse. These infections can cause protective inflammatory reactions in some cases, acute allergic reactions in other cases, or can allow long-term Th2 reactions.

31.4.2 TLR in Lung Cancer

Lung cancer is widespread cancer, accounting for more than a quarter of cancer-related deaths. Its development and progress are related to the immune system's inability to establish an effective antitumor immune response. As previously reported, TLR also drives innate immunity. Therefore, the exogenous modulation of immunity through TLRs may be a strategy to fight cancer, but the underlying chronic inflammation will aggravate the disease and further promote lung cancer [20].

Airway epithelial cells express elevated TLR1, TLR2, TLR3, TLR4, TLR5, TLR6, and TLR9 in lung carcinoma compared with the normal lung. Similarly, TLR is expressed by infiltrating myeloid and lymphoid origin's immune cells, and its activation can determine adaptive immune polarization. Activation of TLRs in stromal cells and hematopoietic cell types can promote the immune stimulation or tolerance mechanism of cells to the lung, where they can participate in carcinogenic processes. The activation of TLRs in epithelial cells stimulates the production of chemokines, such as CXCL-8 (IL-8), a potent neutrophil chemoattractant, and the release of growth factors, such as vascular endothelial-derived growth factor (VEGF), which recruits blood vessels to the inflamed and damaged airway epithelium [30]. Similarly, activating TLR on antigen-presenting cells (innate immune cells) activates adaptive immunity through various mechanisms, including antigen processing and presentation, upregulation of costimulatory molecules, and modulation of T regulatory cells. The synthesis/release of pro-inflammatory cytokines induced by TLR can determine the adaptive immune polarization. For example, TLR binding can lead to the synthesis of IL-2 and IL-12p35 or IL-12p19, which promote Th1 or Th17 immunity, respectively. Th1-type bias may affect immune surveillance for lung cancer with effective antitumor immune responses [47].

TLR is essential to the humoral immune system carefully arranged by B cells. B cells undergo maturation, proliferation, and immunoglobulin conversion after TLR activation. Under physiological conditions, the number of B cells in the lung is minimal (0.1%), but lung cancer, especially when there are no tolerogenic cells, increases three times. B cells refer to both as antitumor and tumor cells in controversial research. B cells express high levels of TLR7 and TLR9. Some authors believe that B cells can resist T cell-mediated immunity, thereby enhancing tumor cell growth. However, the regulation of B cell-derived immunity by other TLRs in lung cancer needs advanced research [48]. Similarly, compared with TLR7 and TLR9, lung macrophages express high TLR2, TLR3, TLR4, and TLR6. The M2 phenotype is characterized by the production of large amounts of IL-10 and TGF-β, immunosuppressive cytokines, anti-inflammatory activity, and tissue repair function. Cancer tissue is rich in M2, which can promote angiogenesis and tissue remodeling. MyD88 deficiency reduces tumor growth due to less macrophage incursion into the tumor microenvironment. MyD88-mediated signaling activates the M2 phenotype, while TLR3 activation strongly induces TRIF-dependent signaling to stimulate macrophages M1 phenotype [49]. TLR activation on macrophages and myeloid-derived suppressor cells (MDSC) inhibits immune surveillance of tumor cells. A recent study confirmed that the activation of TLR2 on MDSC promotes tumor growth in lung adenocarcinoma, lymphoma, and colon cancer via immune evasion.

NK cells are also involved as primary effector cells for antitumor, and the MyD88-dependent pathway hardly activates them, but the TRIF-dependent pathway attracts their cytotoxicity. It has been found that after the administration of poly I:C (ligand for TLR3), the absence of TRIF mediates tumor progression. However, in prostate cancer, poly I:C administration can degenerate tumor growth because the activation of TLR3 may also lead to the apoptotic cascade. Signal transduction of TLR7 and TLR9 is reduced after a viral infection, which reduces the innate immune response during tumorigenesis. Activation of TLR2-MyD88 in MDSC also leads to the activation of NK cells. However, due to NK cell activation after the administration of TLR3, TLR7/8 ligand, a decrease in lung metastasis has been observed [47]. Also, mast cells are found in several tumor types. The activation of mast cell receptors and TLRs can stimulate various growth-stimulating mediators and pro-angiogenic factors, promoting tumor spread. In contrast, in the mouse model, TLR2 activation on mast cells reduced lung cancer growth after LLC1 cells were implanted subcutaneously [50]. Oldford et al. have shown that TLR2 activation on mast cells promotes the release of IL-6 and chemokine ligand 1 (CCL1), which leads to increased antitumor activity and recruitment of NK and CD8+ cytotoxic T cells [51]. Although lung cancer has Th2-based pathological features, mast cells' role in tumor progression is still elusive.

31.4.3 TLR in COPD

TLR in COPD can be associated with abnormal inflammation and limited expiratory airflow that cannot be completely reversed. The prevalence of the disease is related to smoking and age. Although COPD's pathogenesis is unclear, all aspects of innate lung immunity, including mucociliary clearance, alveolar macrophage function, and airway antimicrobial polypeptide expression, are weakened. That is why these microbial pathogens settle in the lower region of the respiratory tract and cause severe inflammation and infection, leading to a gradual loss of lung function [20].

There is increasing evidence that impaired innate immunity may be the cause of COPD. Recently, the vital role of TLR was described in maintaining lung structural homeostasis under environmental conditions. Studies in the rat model indicated that the overexpression of TLR4 in the lung's structural cells is necessary to maintain its normal structure and prevent oxidative stress. The absence of TLR4 leads to high concentrations of Nox3 (an intracellular oxidant derived from the reduced nicotinamide adenine dinucleotide phosphate oxidase system) endothelial cells, which can produce inflammation during/in COPD. Furthermore, TLR4 is believed to act as an inhibitor of the endogenous activity of Nox3 in the lungs, and its presence allows the protection of lung integrity by regulating the oxidative system [52]. Based on this finding, it has been hypothesized that tobacco smoke's free radicals and oxidative properties can destroy innate immunity; thus, cell necrosis and tissue damage of the lungs happened. Besides, acute exposure to cigarette smoke (two cigarettes, twice a day for 3 days) induces acute lung inflammation in rats, dependent on the signal of TLR4/MyD88 and IL-1R1/MyD88 [53].

It has been observed that C3H/HeJ mice with naturally impaired TLR4 signaling had less chronic inflammation after being exposed to cigarette smoke for 5 weeks. Long-term exposure to cigarette smoke can cause strain-dependent emphysema in mice, and the link with TLR has not been established. Several studies have discussed the relationship between the function and expression of TLR in smokers, nonsmokers, and patients with COPD. A study showed that after ex vivo ligand stimulation, the expression of TLR2 decreased in alveolar macrophages in smokers and COPD patients. Other studies have reported that the expression levels of TLR2, TLR4, CD14, or MD2 are comparable, but stimulation of alveolar macrophages with ligands of TLR2 or TLR4 can reduce the mRNA and levels of inflammatory cytokines (such as IL-6, TNF-α, (IL) protein expression-1β) and chemokines (such as RANTES and IL-8) can exhibit defective NF-κB activation, IRAK-1, and p38 phosphorylation in smokers [20]. Therefore, the authors suggest that the inflammatory response mediated by TLR2 and TLR4 depends on the exposure time of LPS that selectively reprograms alveolar macrophages through smoking. Additionally, cigarette smoke and particle phases contain 4500 or more components, including harmful particles, reactive chemicals, free radicals, and overly researched risk factors for COPD. The components of cigarette smoke stimulate and strengthen immune cells, such as macrophages. Lung tissue damage is caused by excessive mucus secretion, mucociliary insufficiency, and inflammation. The cigarette irritation response accumulates neutrophils, monocytes, dendritic cells, pro-inflammatory mediators, ROS, and factors that attract T lymphocytes and proteolytic enzymes [16]. These response functions play vital roles in COPD.

Toll-like receptors associated with cigarette smoke-induced inflammation are TLR2, TLR4, and TLR9. Karimi et al. conducted extensive studies on the function of TLR4 and found that TLR4 is involved in the production of cytokines induced by cigarette smoke [54]. Sarir et al. revealed that brief exposure to cigarette smoke reduces TLR4 surface performance [49]. Pace et al. used epithelial cells to confirm that TLR4 is involved in inducing cigarette smoke-induced CXCL8 lung inflammation (Pace, Ferraro et al. 2008). Maes et al. studied that in subacute cigarette exposure neutrophils, compared with the wild-type mice in the control group, the level of lymphocytes in the BALF of TLR4 knockout mice was reduced [55]. To find out the relationship between TLR4 and COPD, a study was conducted on 240 heavy smokers. It was observed that the development of COPD is related to the TLR4 polymorphisms (TLRD299G, TLR2R753Q, TLR4T3991). TLR4T3991 polymorphism worsening contributes to the development of COPD in smokers, which is exposure to increased susceptibility to infection, and the D299G TLR4 polymorphism has been attributed to the lower severity of COPD due to low reactivity to LPS. To fully clarify the role of TLR polymorphism in COPD, more research is needed. TLR4 deficiency in mice causes pulmonary emphysema [20]. It indicates that TLR4 is involved in healthy tissue homeostasis. In summary, the above knowledge summarizes the inflammation mediated by TLR4 caused by environmental components and the uncontrolled TLR4 in diseases.

31.4.4 TLR in Pulmonary Infection

TLR is composed of a group of proteins associated with recognizing and triggering the various pathogens and particular responses against these pathogenic attacks. The myeloid and stromal cells in the lung express TLRs, which recognize the PAMP and damage-associated molecular patterns (DAMPs) and are associated with signaling in host defense against the infectious agents [20]. On getting repeated exposure to various infectious agents in the alveolus and respiratory tract, the role of TLR becomes important, and other receptors expressed on lung cells like PRRs, NLRs, and RIG-I also play a key role in managing and protecting against pathogens [15]. The majority of TLR are present both intracellularly within endosomes (TLR3, TLR7/8, TLR9) and at the cell surface (TLR1-6, 10). TLRs have a unique role in inflammatory lung diseases, ranging from acute respiratory distress syndrome (ARDS) to asthma and COPD. Thus, the molecules are under trials, which can target these receptors [56].

TLR signaling is the critical factor playing an important role in host defense against acute infections caused by bacteria, viruses, and fungi. Humans experiencing genetic mutations in TLR2 and TLR4 are probably more susceptible to these infections. Nevertheless, an increase in mortality in pneumonia models has not been observed on the deletion of an individual TLR, as recorded in a study about the TLR2 role in pneumococcal pneumonia [57]. Both TLR2 and TLR4 are involved in immune response in tuberculosis, while human polymorphism in TLR2 is associated with enhanced susceptibility to tuberculosis and leprosy. TLR2- and TLR4-knockout mice probably lead and result in enhanced mortality in tuberculosis, whereas the deficiency in TLR and IL-1R signaling adapter (MyD88) could show defective innate immune response against tuberculosis [58].

The maximum threat is observed in lower respiratory tract infections, predominantly in pneumonia and pneumonia-associated complications, and is a significant threat and global burden to human health, particularly in children. The bacterial infection is a common cause of severe pneumonia, which could be superimposed in inflammatory conditions like COPD or influenza. Bacteria often violate the host defense mechanism even in the presence of the pulmonary immune system and its sophisticated mechanisms, helping in the development of innate and acquired immune responses against various infections. It causes rapid bacterial duplication in the lungs and spreading to remote organs. Consequently, a significant lung's immune response is significantly required to eliminate bacteria and other infectious agents successfully [59].

31.5 Targeting of TLR Signaling

The role of adaptive immunity is a subject of important consideration as dominated research in lung diseases is producing a new area of research, thus focused on developing novel therapeutics approaches for the treatment of inflammatory diseases. The explosion of a family of proteins called TLRs in the research is being processed for its treatment. These TLRs have vital roles in adaptive and innate immunity. Thus targeting TLRs signaling offers significant new opportunities to interfere in lung diseases [58]. TLRs are expressed on nearly all innate immune cells (circulating leukocytes like neutrophils and monocytes, macrophages, dendritic cells, NK cells), adaptive immune cells (B and T lymphocytes), and nonimmune cells like endothelial and epithelial cells and fibroblasts [60].

The ligands of TLR have been classified as exogenous and endogenous ligands of TLRs. The exogenous ligands include PAMPs, like flagellin protein from bacterial flagella, LPS from Gram-negative bacteria, peptidoglycan (PGN) and lipoteichoic acid (LTA) from Gram-positive bacteria, lipopeptides, lipoarabinomannan (LAM), lipomannans and lipoglycans from mycobacteria, zymosan from yeast, dsRNA of viruses, and DNA from viruses and bacteria [38, 61]. The endogenous ligands or host-derived DAMPs are generally formed on account of injury and nonphysiological cell death. They include fibrinogen and hyaluronan as an extracellular matrix component, constituents of the plasma membrane, nuclear and cytosolic proteins such as high-mobility group box protein 1 (HMGB1) and heat shock proteins (HSPs), and damaged and/or fragmented organelle elements like mitochondrial DNA (mtDNA) [38, 62].

Apart from these ligands, there are five types of adaptor proteins for the TIR domain in mammals, which include [63] TIRAP or Mal, TRIF, MyD88, sterile α- and armadillo-motif-containing protein (SARM), TRAM [63].

The precise intracellular downstream signaling cascade is initiated by binding various ligands to TLR, which stimulates and initiates host defense reactions; for example, PAMP-PRR interactions cause the production of pro-inflammatory cytokines and type-1 INF initiating the immune responses against particular microbes. The nature of the stimulus controls the TLR signaling that is of following two distinct pathways (Fig. 31.1). One is the MyD88-dependent pathway, which leads to inflammatory cytokine production by using TLRs, except TLR3. Second is the TRIF-dependent pathway associated with the stimulation of INF type-1 and utilized by TLR3 and 4 [64, 65].

TLRs are indispensable fundamentals of the innate immune system and play a significant function in the host-defensive mechanism against microbes. However, TLR overactivation disrupts the immune system's homeostasis and leads to higher production of pro-inflammatory cytokines involved in the pathogenesis of various inflammatory and autoimmune diseases. Thus, the inhibition of signaling pathways of TLR could be forecasted as an efficient therapeutic approach to suppress unwanted surplus disease-associated inflammatory reactions [66].

The inhibition of TLRs can be achieved in two major ways; firstly, by blocking the binding of TLR ligands to the receptors and secondly by preventing the signal transduction through the interference of intracellular signaling pathways. It can be achieved by using various therapeutic agents that inhibit the TLR signaling from controlling excessive inflammation. These agents can be grouped as small-molecule inhibitors, oligonucleotides, antibodies, microRNAs, lipid-A analogs, and other new emerging nano-inhibitors [38, 66] (Table 31.2).

TLRs help in developing the self-defense ability through the identification of endogenous molecules and invading pathogens in damaged tissues. However, too much activation of TLR may disturb the immune homeostasis by producing pro-inflammatory cytokines and chemokines. These contribute to the development

Table 31.2 Various TLR inhibitors

TLR inhibitors	Mechanism	Examples	Targeted TLR	References
Small-molecule inhibitors (SMIs)	• Suppress autoantigen presentation by getting accumulate in the acidic intracellular space endosomes and lysosomes • Blockade of endosomal TLR7, 8, and 9 signaling • Decrease in cytokine production	CpG-52364, SM934, hydroxychloroquine sulfate, chloroquine	TLR7, TLR8, TLR9	[38, 67]
Antibodies	Block the binding of ligands to the specific TLRs	OPN-305 T2.5 NI-0101 1A6	TLR7, TLR8, TLR9	[38, 68]
Oligonucleotides	Interfere the binding of ligands to endosomal TLRs and hence block the TLR signal transduction	IRS954 INH-ODN-24888	TLR7 TLR9	[38, 69]
Lipid A analogs	Prevents LPS to TLR4 binding Reduce activation of LPS-induced NF-κB and pro-inflammatory cytokine production	Eritoran	TLR4	[38, 70]
MicroRNA inhibitors (miRNAs)	Reduction in the production of IL-6 Protein translation repression or mRNA degradation	miR-146a, miR-155, and miR-21	TLR2, TLR4	[38, 71]
Nano-inhibitors	Suppress LPS-induced MyD88-dependent NF-κB activation Inhibit subsequent cytokines production in mouse macrophages	High-density lipoprotein (HDL)-like nanoparticle	TLR4, TLR2, TLR3, TLR5	[38, 72]

of many inflammatory diseases in the lung, like asthma. Thus, inhibitors/antagonists that target the TLR signals could be useful for managing inflammatory diseases.

For the treatment of COPD, a combination of inhaled corticosteroids and bronchodilators has been used. These corticosteroids enhance β-adrenergic responses and repress inflammatory responses in the airway. But in severe COPD and asthma cases, corticosteroids do not produce desired effects, which could be attributed to devastating oxidative stress and subsequent DNA damage. Thus, the antioxidant use may decrease DAMP production by scavenging ROS and inhibiting the activation of TLRs.

Drugs that target TLRs have been classified as agonists or antagonists. TLR agonists, as the name suggests, augment receptors' and antagonists' response usually lessens the responses through attenuating inflammation. Chronic inflammation in allergic asthma is caused due to allergen exposure. Antagonists in asthma treatment target the muscarinic receptors to promote bronchodilation through the airway's smooth muscle relaxation. The Protollin (which contains both TLR2 and TLR4 ligands) is useful in lowering allergic responses to intranasal administration [73]. Similarly, the Resiquimod (R-848) is a TLR7 agonist and was advantageous in suppressing allergic airway diseases [74]. TLR7 possesses anti-inflammatory properties. The activation of TLR7 controls the inflammation by preventing the entry of leukocytes into the airways. R-848 reduces leukocyte recruitment and IL-5 and IL-13 production [75]. The mechanisms behind TLRs' protective action are thought to be the reorientation of the immune system reducing the Th2 function, including Th2 cytokine production, eosinophils, and bronchial hypersensitivity, which all leading to reduced airway inflammation [76].

In asthma, COPD, and inflammatory airway diseases, TLR signaling pathways have been closely associated with the pathophysiology of the disease. In allergic asthma, the house dust mites are one of the main risk factors that trigger the airway monocytes and neutrophils through a TLR4-dependent mechanism. Besides, LPS (TLR4 agonist) induces Th1 or Th2 immune response in a dose-dependent manner in asthmatic conditions. In COPD, TLR2, TLR4, and TLR9 participate in inflammatory responses induced by cigarette smoking. The excessive lung tissue demolition and inflammatory responses during the airway diseases' exacerbation are caused through the overactive TLR signaling [66].

31.6 Conclusion

Innate immunity is the primary mechanism for maintaining the dynamic balance of lung tissue despite continuous exposure to pathogenic microorganisms and environmental irritants. TLRs are mainly involved in the activation of the innate immune system, and many studies have been conducted in past years to find out the role of TLRs in host defense and tissue homeostasis. The insights generated by this work allow people to come up with some general principles about lung innate immunity. These principles relate to the development of acute lung diseases, often leading to chronic inflammatory diseases such as fibroproliferative ARDS or severe RSV infection that precedes the development of asthma. Secondly, innate immune deficiency leads to the development of the chronic obstructive pulmonary disease. These diseases may be due to the host's direct or indirect tendency to infection, for example, in chronic *S. pneumoniae* infection, leading to mild aggravation of COPD. Similarly, TLR is also involved in the occurrence and development of lung cancer. Contemporary research has proved that TLR signal transduction can promote and inhibit tumors. On the new side, these receptors participate in the immune system activation to produce an antitumor effect. TLRs also help in tumor treatment by activating tumor microenvironment immune cells, but more research is needed to elucidate the complete role of TLRs in tumorigenicity. Although TLR activation plays an essential role in developing lung in pulmonary infection, many questions need to be answered. Also, the use of TLR as a therapeutic target involves the use of TLR agonists or antagonists. Few TLR antagonists and agonists are used as treatments for airway diseases, but they have not been successful due to various reasons involving dose toxicity and inappropriate administration time and route. Therefore, more experimental research is required to explore its potential as a treatment strategy. Understanding TLR biology will provide more opportunities to clarify the link between innate immunity and acute and chronic lung infections and diseases. This knowledge will help us identify TLRs as new therapeutic targets that can reduce the burden of lung diseases.

References

1. Crespo-Lessmann A, Juárez-Rubio C, Plaza-Moral V (2010) Role of toll-like receptors in respiratory diseases. Arch Bronconeumol 46(3):135–142
2. Duez C, Gosset P, Tonnel A-B (2006) Dendritic cells and toll-like receptors in allergy and asthma. Eur J Dermatol 16(1):12–16
3. Basu S, Fenton MJ (2004) Toll-like receptors: function and roles in lung disease. Am J Phys Lung Cell Mol Phys 286(5):L887–L892
4. O'Neill LA, Dinarello CA (2000) The IL-1 receptor/toll-like receptor superfamily: crucial receptors for inflammation and host defense. Immunol Today 21(5):206–209
5. Smirnova I, Poltorak A, Chan EK et al (2000) Phylogenetic variation and polymorphism at the toll-like receptor 4 locus (TLR4). Genome Biol 1(1):research002. 1
6. Lai C-Y, Yu G-Y, Luo Y et al (2019) Immunostimulatory activities of CpG-oligodeoxynucleotides in teleosts: toll-like receptors 9 and 21. Front Immunol 10:179
7. Yang J, Yan H (2017) TLR5: beyond the recognition of flagellin. Cell Mol Immunol 14 (12):1017–1019
8. Blaser H, Dostert C, Mak TW et al (2016) TNF and ROS crosstalk in inflammation. Trends Cell Biol 26(4):249–261
9. Haider T, Tiwari R, Vyas SP et al (2019) Molecular determinants as therapeutic targets in cancer chemotherapy: an update. Pharmacol Ther 200:85–109
10. Shchebliakov D, Logunov DY, Tukhvatulin A et al (2010) Toll-like receptors (TLRs): the role in tumor progression. Acta Nat 2(3):6
11. Monlish DA, Bhatt ST, Schuettpelz LG (2016) The role of toll-like receptors in hematopoietic malignancies. Front Immunol 7:390
12. Yang Y, Lv J, Jiang S et al (2016) The emerging role of toll-like receptor 4 in myocardial inflammation. Cell Death Dis 7(5):e2234–e2234
13. Li M, Zhou Y, Feng G et al (2009) The critical role of toll-like receptor signaling pathways in the induction and progression of autoimmune diseases. Curr Mol Med 9(3):365–374
14. Curtiss LK, Tobias PS (2009) Emerging role of toll-like receptors in atherosclerosis. J Lipid Res 50(Supplement):S340–S345
15. Kovach MA, Standiford TJ (2011) Toll like receptors in diseases of the lung. Int Immunopharmacol 11(10):1399–1406
16. Lafferty EI, Qureshi ST, Schnare M (2010) The role of toll-like receptors in acute and chronic lung inflammation. J Inflamm 7(1):57

17. Grassin-Delyle S, Abrial C, Salvator H et al (2020) The role of toll-like receptors in the production of cytokines by human lung macrophages. J Innate Immun 12(1):63–73
18. Sidletskaya K, Vitkina T, Denisenko Y (2020) The role of toll-like receptors 2 and 4 in the pathogenesis of chronic obstructive pulmonary disease. Int J Chron Obstruct Pulmon Dis 15:1481
19. Bhagwani A, Thompson AR, Farkas L (2020) When innate immunity meets angiogenesis—the role of toll-like receptors in endothelial cells and pulmonary hypertension. Front Med 7
20. Arora S, Ahmad S, Irshad R et al (2019) TLRs in pulmonary diseases. Life Sci 233:116671
21. Christmas P (2010) Toll-like receptors: sensors that detect infection. Nat Educ 3(9):85
22. Sharma S, Garg I, Ashraf MZ (2016) TLR signalling and association of TLR polymorphism with cardiovascular diseases. Vasc Pharmacol 87:30–37
23. Sha Q, Truong-Tran AQ, Plitt JR et al (2004) Activation of airway epithelial cells by toll-like receptor agonists. Am J Respir Cell Mol Biol 31(3):358–364
24. Hoppstädter J, Diesel B, Zarbock R et al (2010) Differential cell reaction upon toll-like receptor 4 and 9 activation in human alveolar and lung interstitial macrophages. Respir Res 11(1):124
25. Farkas D, Thompson AR, Bhagwani AR et al (2019) Toll-like receptor 3 is a therapeutic target for pulmonary hypertension. Am J Respir Crit Care Med 199(2):199–210
26. Wang MM, Lu M, Zhang CL et al (2018) Oxidative stress modulates the expression of toll-like receptor 3 during respiratory syncytial virus infection in human lung epithelial A549 cells. Mol Med Rep 18(2):1867–1877
27. Esawy MM, Shabana MA, Baioumy SA et al (2019) Diagnostic and prognostic roles of peripheral blood toll-like receptor-4 and stanniocalcin-1 genes expression in acute lung injury. Immunobiology 224(6):734–738
28. Tang S-E, Wu S-Y, Chu S-J et al (2019) Pre-treatment with ten-minute carbon dioxide inhalation prevents lipopolysaccharide-induced lung injury in mice via down-regulation of toll-like receptor 4 expression. Int J Mol Sci 20(24):6293
29. Andonegui G, Bonder CS, Green F et al (2003) Endothelium-derived toll-like receptor-4 is the key molecule in LPS-induced neutrophil sequestration into lungs. J Clin Invest 111 (7):1011–1020
30. Samara KD, Antoniou KM, Karagiannis K et al (2012) Expression profiles of toll-like receptors in non-small cell lung cancer and idiopathic pulmonary fibrosis. Int J Oncol 40(5):1397–1404
31. Zhou H, Chen JH, Hu J et al (2014) High expression of toll-like receptor 5 correlates with better prognosis in non-small-cell lung cancer: an antitumor effect of TLR5 signaling in non-small cell lung cancer. J Cancer Res Clin Oncol 140(4):633–643
32. Koller B, Kappler M, Latzin P et al (2008) TLR expression on neutrophils at the pulmonary site of infection: TLR1/TLR2-mediated up-regulation of TLR5 expression in cystic fibrosis lung disease. J Immunol 181(4):2753–2763
33. Plantinga M, Hammad H, Lambrecht BN (2010) Origin and functional specializations of DC subsets in the lung. Eur J Immunol 40(8):2112–2118
34. Lee K, Kim SH, Yoon HJ et al (2011) Bacillus-derived poly-gamma-glutamic acid attenuates allergic airway inflammation through a toll-like receptor-4-dependent pathway in a murine model of asthma. Clin Exp Allergy 41(8):1143–1156
35. Arancibia SA, Beltrán CJ, Aguirre IM et al (2007) Toll-like receptors are key participants in innate immune responses. Biol Res 40(2):97–112
36. Chen ZJ (2005) Ubiquitin signalling in the NF-κB pathway. Nat Cell Biol 7(8):758–765
37. Collins SE, Noyce RS, Mossman KL (2004) Innate cellular response to virus particle entry requires IRF3 but not virus replication. J Virol 78(4):1706–1717
38. El-Zayat SR, Sibaii H, Mannaa FA (2019) Toll-like receptors activation, signaling, and targeting: an overview. Bull Natl Res Cent 43(1):187
39. Banno A, Reddy AT, Lakshmi SP et al (2020) Bidirectional interaction of airway epithelial remodeling and inflammation in asthma. Clin Sci 134(9):1063–1079
40. Zakeri A, Yazdi FG (2017) Toll-like receptor-mediated involvement of innate immune cells in asthma disease. Biochim Biophys Acta 1861(1):3270–3277

41. Herrick CA, Bottomly K (2003) To respond or not to respond: T cells in allergic asthma. Nat Rev Immunol 3(5):405–412
42. Conrad ML, Ferstl R, Teich R et al (2009) Maternal TLR signaling is required for prenatal asthma protection by the nonpathogenic microbe Acinetobacter lwoffii F78. J Exp Med 206 (13):2869–2877
43. Møller-Larsen S, Nyegaard M, Haagerup A et al (2008) Association analysis identifies TLR7 and TLR8 as novel risk genes in asthma and related disorders. Thorax 63(12):1064–1069
44. Genuneit J, Cantelmo J, Weinmayr G et al (2009) A multi-centre study of candidate genes for wheeze and allergy: the International Study of Asthma and Allergies in Childhood Phase 2. Clin Exp Allergy 39(12):1875–1888
45. Brooks LR, Mias GI (2018) Streptococcus pneumoniae's virulence and host immunity: aging, diagnostics, and prevention. Front Immunol 9:1366
46. Acosta PL, Caballero MT, Polack FP (2016) Brief history and characterization of enhanced respiratory syncytial virus disease. Clin Vaccine Immunol 23(3):189–195
47. Pinto A, Morello S, Sorrentino R (2011) Lung cancer and toll-like receptors. Cancer Immunol Immunother 60(9):1211
48. Ren T, Xu L, Jiao S et al (2009) TLR9 signaling promotes tumor progression of human lung cancer cell in vivo. Pathol Oncol Res 15(4):623–630
49. Sarir H, Mortaz E, Karimi K et al (2009) Cigarette smoke regulates the expression of TLR4 and IL-8 production by human macrophages. J Inflamm 6(1):12
50. Welsh TJ, Green RH, Richardson D et al (2005) Macrophage and mast-cell invasion of tumor cell islets confers a marked survival advantage in non–small-cell lung cancer. J Clin Oncol 23 (35):8959–8967
51. Oldford SA, Haidl ID, Howatt MA et al (2010) A critical role for mast cells and mast cell-derived IL-6 in TLR2-mediated Inhibition of tumor growth. J Immunol 185(11):7067–7076
52. Zhang X, Shan P, Jiang G et al (2006) Toll-like receptor 4 deficiency causes pulmonary emphysema. J Clin Invest 116(11):3050–3059
53. Doz E, Noulin N, Boichot E et al (2008) Cigarette smoke-induced pulmonary inflammation is TLR4/MyD88 and IL-1R1/MyD88 signaling dependent. J Immunol 180(2):1169–1178
54. Karimi K, Sarir H, Mortaz E et al (2006) Toll-like receptor-4 mediates cigarette smoke-induced cytokine production by human macrophages. Respir Res 7(1):66
55. Maes T, Bracke KR, Vermaelen KY et al (2006) Murine TLR4 is implicated in cigarette smoke-induced pulmonary inflammation. Int Arch Allergy Immunol 141(4):354–368
56. Chaudhuri N, Dower SK, Whyte MK et al (2005) Toll-like receptors and chronic lung disease. Clin Sci (Lond) 109(2):125–133
57. Knapp S, Wieland CW, van't Veer C et al (2004) Toll-like receptor 2 plays a role in the early inflammatory response to murine pneumococcal pneumonia but does not contribute to antibacterial defense. J Immunol 172(5):3132–3138
58. Chaudhuri N, Dower SK, Whyte MK et al (2005) Toll-like receptors and chronic lung disease. Clin Sci 109(2):125–133
59. Baral P, Batra S, Zemans RL et al (2014) Divergent functions of toll-like receptors during bacterial lung infections. Am J Respir Crit Care Med 190(7):722–732
60. Delneste Y, Beauvillain C, Jeannin P (2007) Immunité naturelle-Structure et fonction des toll-like receptors. Médecine/Sciences 23(1):67–74
61. Ayres JS, Schneider DS (2012) Tolerance of infections. Annu Rev Immunol 30:271–294
62. Murad S (2014) Toll-like receptor 4 in inflammation and angiogenesis: a double-edged sword. Front Immunol 5:313
63. Akira S, Takeda K (2004) Toll-like receptor signalling. Nat Rev Immunol 4(7):499–511
64. Yamamoto M, Takeda K (2010) Current views of toll-like receptor signaling pathways. Gastroenterol Res Pract 2010
65. Kawai T, Akira S (2011) Toll-like receptors and their crosstalk with other innate receptors in infection and immunity. Immunity 34(5):637–650

66. Gao W, Xiong Y, Li Q et al (2017) Inhibition of toll-like receptor signaling as a promising therapy for inflammatory diseases: a journey from molecular to nano therapeutics. Front Physiol 8:508
67. Kužnik A, Benčina M, Švajger U et al (2011) Mechanism of endosomal TLR inhibition by antimalarial drugs and imidazoquinolines. J Immunol 186(8):4794–4804
68. Daubeuf B, Mathison J, Spiller S et al (2007) TLR4/MD-2 monoclonal antibody therapy affords protection in experimental models of septic shock. J Immunol 179(9):6107–6114
69. Zhu F-G, Jiang W, Dong Y et al (2012) IMO-8400, a novel TLR7, TLR8 and TLR9 antagonist, inhibits disease development in lupus-prone NZBW/F1 mice (119.12). Am Assoc Immnol
70. Opal SM, Laterre P-F, Francois B et al (2013) Effect of eritoran, an antagonist of MD2-TLR4, on mortality in patients with severe sepsis: the ACCESS randomized trial. JAMA 309 (11):1154–1162
71. Shen N, Liang D, Tang Y et al (2012) MicroRNAs—novel regulators of systemic lupus erythematosus pathogenesis. Nat Rev Rheumatol 8(12):701–709
72. Babazada H, Yamashita F, Yanamoto S et al (2014) Self-assembling lipid modified glycol-split heparin nanoparticles suppress lipopolysaccharide-induced inflammation through TLR4–NF-κ B signaling. J Control Release 194:332–340
73. Shalaby KH, Jo T, Nakada E et al (2012) ICOS-expressing CD4 T cells induced via TLR4 in the nasal mucosa are capable of inhibiting experimental allergic asthma. J Immunol 189 (6):2793–2804
74. Xirakia C, Koltsida O, Stavropoulos A et al (2010) Toll-like receptor 7–triggered immune response in the lung mediates acute and long-lasting suppression of experimental asthma. Am J Respir Crit Care Med 181(11):1207–1216
75. G Drake M, H Kaufman E, D Fryer A et al (2012) The therapeutic potential of toll-like receptor 7 stimulation in asthma. Inflamm Allergy-Drug Targets 11(6):484–491
76. Zuo L, Lucas K, Fortuna CA et al (2015) Molecular regulation of toll-like receptors in asthma and COPD. Front Physiol 6:312

Biomarkers of Oxidative Stress

32

Jia-fu Feng

Abstract

Free radicals (FRs) and/or reactive oxygen species (ROS) are bioactive substances generated inevitably during the metabolic process of organisms. To combat excessive free radical and/or reactive oxygen production, living organisms have evolved many sophisticated peroxide-antioxidant defense systems. These systems are located in a dynamic equilibrium state under normal physiological conditions, while the body antioxidant system could be unbalanced and lead to oxidative stress in pathological states. Oxidative stress is closely related to the occurrence and development of various diseases, including cancer. Therefore, FRs and/or ROS involved in pathological reactions can be used as markers of oxidative stress. Although most oxidation-antioxidant markers are not difficult to be measured by modern medical detection technology separately, the detection of each oxidation-antioxidant substance is not only time- and energy-consuming, but also inaccurate. One of the reasons for inaccuracy is the incomplete understanding and detection of oxidation-antioxidant substances in the organism. The other is the superposition effect produced by various oxidation-antioxidant substances which have a synergistic effect in the same system. In view of this, only combined total oxidant status (TOS) with total antioxidant status (TAS) and oxidant stress index (OSI) can accurately assess the oxidant stress status of subjects. In this chapter, the species of single oxidation or antioxidant, TOS, TAs and OSI, and the determination method of end products of lipid hydroxides (malondialdehyde and 4-hydroxynonene) are introduced.

J.-f. Feng (✉)
Mianyang Central Hospital, School of Medicine, University of Electronic Science and Technology of China, Mianyang, Sichuan, China

K. Dua et al. (eds.), *Targeting Cellular Signalling Pathways in Lung Diseases*,
https://doi.org/10.1007/978-981-33-6827-9_32

Keywords

Reactive oxygen species · Free radical · Oxidative Stress; Biomarker; Tumour; Cancer · Biomarker · Tumour · Cancer

Free radicals (FR) and reactive oxygen species (ROS) are naturally produced in all aerobic organisms [1]. On the one hand, FR and ROS perform many normal physiological functions in the body. On the other hand, excessive generation of FR and ROS may attack biological macromolecules, resulting in oxidative damage to the body. It has been confirmed that intracellular oxidative damage is mainly caused by FR and ROS [2]. FR can inhibit the function of many components in normal cells, react with unsaturated bonds in membrane lipids, denature proteins, and damage nucleic acids, etc. The metabolic form of ROS and the scavenging rate of the body's antioxidants constitute the peroxidation-antioxidant system of the body. Under physiological conditions, the system is in a state of dynamic equilibrium. But in the case of diseases (such as inflammation, trauma, infection, and cancer), the former's generation is faster than the latter's clearance, disordering the dynamic balance [3]. The subsequent body response leads to oxidative stress (OxS) state, resulting in oxidative damage of biological macromolecules such as proteins, lipids, and nucleic acids, interfering with normal life activities. Therefore, OxS refers to when the body encounters various harmful stimuli, highly active molecules in the body, such as ROS and reactive nitrogen species (RNS) excessively produced, and at the same time, the degree of oxidation exceeds the removal of oxides, causing an imbalance of oxidation-antioxidant system, which leads to oxidative damage of biological macromolecules such as proteins, lipids, and/or nucleic acids, and interferes with normal life activities, resulting in a serious state of antioxidant stress.

32.1 Single Oxidation or Antioxidant

Aerobic organisms have evolved an antioxidant defense system that removes FR and ROS. The system can be divided into three levels [4]: ① primary antioxidants, such as superoxide dismutase (SOD), glutathione peroxidase (GSH-Px), catalase (CAT), glutathione S-transferase (GSH), paraoxonase 1 (PON1) enzyme, whose function is to prevent the production of new FR and/or ROS; ② secondary antioxidants, such as vitamins (Vit) A, Vit C, Vit E, uric acid, glutathione, α-lipoic acid, carotene, trace elements copper, zinc, manganese, and selenium, whose function is to clear FR and/or ROS before FR and/or ROS trigger lipid peroxidation chain reaction; ③ tertiary antioxidants, such as DNA repair enzymes and methionine oxysulfide reductase, whose function is to repair nucleic acid chains damaged by FR and/or ROS oxidation and maintain the normal physiological function of cells. However, in some pathological conditions, the body cannot defense against the increase of oxidations or decrease of antioxidants, and the balance between oxidation and antioxidation is transformed to the oxidative state, which will inevitably lead to OxS reaction.

All biochemical antioxidants involved in the body's antioxidant defense system are biomarkers of OxS. In order to evaluate the antioxidant status in vivo, it is necessary to detect the antioxidants in the body. These antioxidants such as SOD, GSH-Px, CAT, GSH, PON1, Vit A, Vit C, Vit E, and uric acid in the oxidation defense system all can be detected separately. In the past, many researchers have often used these indicators to reflect the body's antioxidant status. However, due to the tremendous number of different antioxidants in plasma, serum, urine, or other biological samples, it is difficult to implement a single determination of various antioxidants.

In the body's oxidation-antioxidation system, the opposite of antioxidants is oxidations, which are mainly FR and oxidants. The FR that can be generated in organisms by enzymatic and/or nonenzymatic reactions include ① ROS, such as $\cdot O_2$-, $\cdot OH_2$; ② RNS, such as ·NO, $\cdot NO_2$, and ·ONOO-. Commonly mentioned oxidants are $^1O_2$, H_2O_2, etc. The FR and oxidants often mentioned in the study are shown in Table 32.1. Strictly speaking, $^1O_2$ and H_2O_2 are not oxygen free radicals, but active oxygen.

The following oxidizing substances are all OxS biomarkers (Table 32.2), but oxidizing substances are not easier to detect than anti-oxidizing substances, and are more difficult to detect individually.

Tables 32.3 and 32.4 list the biomarkers of oxidative stress commonly used in clinical or scientific research or reported in the literature (except for the nitro-oxidative stress class). Among them, many markers have been commonly recognized as OxS biomarkers.

32.2 Total Oxidant Status and Total Antioxidant Status

There are two types of antioxidant systems in the human body. One is the enzyme antioxidant system, including superoxide dismutase (SOD), catalase (CAT), and glutathione peroxidase (GSH-Px), and the other is the nonenzyme antioxidant

Table 32.1 Oxyradical abbreviation and terminology

Abbreviation	Terminology
$O_2^{\cdot -}$	Superoxide anion radical
$HO_2^{\cdot}$	Hydrogen peroxide radical
H_2O_2	Hydrogen peroxide
HO·	Hydroxyl radical
RO·	Oxygen organic free radical
ROO·	Organic peroxide radical
$^1O_2$	Singlet oxygen
ROS	Reactive oxygen species
L·	Lipid free radical
LOO·	Lipid peroxide radical

Table 32.2 Reactive oxygen species with the significance of oxidative stress

Species	Terminology	Characteristics
$O_2^{\cdot-}$	Superoxide anion	The single electron reduction state; formed by many oxygen reactions (such as flavin protein, redox cycle)
$HO_2^{\cdot}$	Hydrogen peroxy	Formed by the protonation of $O_2^{\cdot-}$; enhanced fat solubility
H_2O_2	Hydrogen peroxide	Two-electron reduced state; formed by disproportionation of $O_2^{\cdot-}(HO_2^{\cdot})$, or directly formed by O_2
$HO\cdot$	Hydroxyl radical	Three-electron reduction state; formed by Fenton reaction and metal-catalyzed Haber–Weiss reaction; is highly active
$RO\cdot$	R-oxygen radical, alkoxy radical	Oxygen organic free radicals (such as lipids)
$ROO\cdot$	R-Peroxy radical, alkyl peroxy radical	Formed from organic hydroperoxide (ROOH), such as lipid, by hydrogen extraction (or homolysis)
ROOH	R-hydroperoxide	Organic hydroperoxides (such as fatty acids and thymine hydroperoxides)
O_2^* or $^1O_2$	Singlet oxygen	First excitation; higher than ground state oxygen (O); red (bimolecular) or infrared (monomolecular) light emission
^{3}R'R"CO (R'R"CO*)	Triplet carbonyl	Excited carbonyl compounds, blue-green light emission (i.e. via dioxane intermediates)

system, including Vit C, Vit E, glutathione, melatonin, alpha-lipoic acid, carotenoids, trace elements copper, zinc, selenium, etc. With existing biochemical and/or molecular biology technologies, most of the antioxidants in the antioxidant system can be independently detected. However, detection of each antioxidant separately is time-consuming, labor-intensive, expensive, complicated, and inaccurate. The reason for this inaccuracy is that the antioxidants have a synergistic effect in the same system and will produce a superimposed effect. Moreover, both the oxidizing substance and the anti-oxidizing substance have what we know and we have not yet known. Therefore, the determination of one and/or several oxidative or antioxidant substances or their metabolites alone cannot correctly evaluate the oxidative or antioxidant status of the body. In addition, oxidative/antioxidant substances can be divided into known and unknown. Using existing inspection methods, the known oxidation/antioxidants can be detected, but it is time-consuming and laborious, and the unknown ones still cannot be detected. Otherwise, the effects of different oxidation/antioxidants can be superimposed, and the detection of only a few oxidative/anti-oxidant substances does not represent a change in overall levels, because the changes of other oxidative/anti-oxidant substances are not clear.

In view of this, the concepts of total antioxidant status (TAS) and total oxidant status (TOS) came into being. TAS represents the overall level of enzymes and nonenzyme antioxidants in the organism. It is also called total antioxidant capacity (TAC), total antioxidant activity (TAA), total antioxidant power (TAOP), total antioxidant response (TAR), total reactive antioxidant potential (TRAP), etc. TAS

Table 32.3 OxS biomarkers commonly used in clinical practice or research (antioxidants)

Abbreviation	Full name	Abbreviation	Full name
AOPP	Advanced oxidation protein products	Mel	Melatonin
ALA	α-Lipoic acid	Myase	Myeloperoxidase
apoA-I	Apolipoprotein A-I	OMP	Oxidatively modified protein
ADMA	Asymmetric dimethyl-L-arginine	DHN	1,4-dihydroxynonene
CbP	Carbonylproteine	Anti-oxLDL	Ox-LDL antibody
ACR	Carotene	PHPA	Para-hydroxyphenylacetic acid
CAT	Catalase	PON1	Paraoxonase 1
CoQ10	Coenzyme Q10	PMN-Elae	Polymorphonuclear leukocyte elastase
Cu	Copper	GSH	Reduced glutathione
CRP	C-reactive protein	Se	Selenium
COX	Cyclooxygenase	SOD	Superoxide dismutase
8-OHdG	8-hydroxy-20-deoxyguanosine	SDMA	Symmetrisches Dimethylarginin
F_2-IsoP	F_2-isoprostane	Try	Tryptophan
FB	Free biotin	DNP	2,4-Dinitrophenylhydrazine
GSSG	Glutathione disulfide	Ubi	Ubiquinone
GSHPx	Glutathione peroxidase	UA	Uric acid
GSH	Glutathione S-transferase	VitA	Vitamin A
Hp	Haptoglobin	VitB	Vitamin B6
IDO	Indolamin-2,3-Dioxygenase	VitB12	Vitamin B12
Kyn	Kynurenin	VitAC	Vitamin C
Lyso	Lysozym	Vit/E	Vitamin E
Mn	Manganese	Zn	Zinc

is synonymous with the body's total antioxidant level. It not only represents the sum of enzymes and nonenzyme antioxidants in the body but also reflects the relationship of mutual connection and synergism between the antioxidants. There is a close relationship between the strength of the body's antioxidant defense system and its health and disease status. When it decreases, it will inevitably cause inflammation, tumors, and immune system diseases. Therefore, the TAS level reflects the comprehensive information of the body's antioxidant capacity in different states.

Compared to TAS, total oxidant status (TOS) represents the overall level of all oxidants in the oxidation-antioxidant system that maintains the body's antioxidant defense capabilities. It is also named total peroxide (TP), serum oxidation activity (SOA), reactive oxygen metabolites (ROM), oxygen radical absorbance capacity (ORAC), or some other synonyms. TOS is synonymous with the total oxidation level of the body. Like the TAS, it not only represents the sum of oxidations in the body but also reflects the relationship of mutual connection and synergism between oxidations. Oxidation is an essential component of the human body's antioxidant

Table 32.4 OxS biomarkers commonly used in clinical practice or research (oxidants)

Classification	Abbreviation	Full name
Oxidant	ALE	Advanced lipoxidation end product
	8-iso-$PGF_2\alpha$	8-iso-prostaglandin $F_2\alpha$
	Fe^{2+}	Ferrous ion
	4-HNE	4-hydroxy-2-nonenal
	HNA	4-hydroxynonenoic acid
	HydrP	Hydroperoxide
	DDG	7,8-dihydro-8-oxo-20-Deoxyguanosine
	LO	Lipid alkoxyl radical
	LOOH	Lipid hydroperoxide
	LOO	Lipid peroxyl radical
	MDA	Malondialdehyd
	Nrf2	Nuclear factor-like 2
	OMP	Oxidatively modified proteins
	oxLDL	Oxidierte LDL
	di-Tyr	O,o'-dityrosine
	ProC	Protein carbonyl
	SNST	S-nitrosothiols
	NHPA	3-nitro-4-hydroxyphenylacetic acid
	Cl-Tyr	3-chlorotyrosine
Integral	OSI	oxidant stress index
	TAS	Total antioxidant status
	TOS	Total oxidant status

defense system. There is a close relationship between the strength of the body's antioxidant defense system and health or disease states. When oxidation is elevated, it will inevitably cause inflammation, tumor, and immune system diseases. Therefore, the TOS level reflects comprehensive information about oxidizing ability in different states.

It can be seen that TAS and TOS are necessary detection indicators to fully reflect the body's antioxidant effect. At present, both TAS and TOS can realize automatic detection, with high precision and good reproducibility, and can be used for the detection of any biological sample, which is easy to popularize.

Taking Hitachi automatic biochemical analyzer as an example, the instrument setting parameters for automatic detection of TAS (Table 32.3) and TOS (Table 32.4) are given here. Other brands and models of automatic biochemical analyzers can refer to this parameter and instrument performance, which can be easily modified settings.

32.2.1 Measurement of Total Antioxidant Status

The following is a brief introduction of the TAS full-automatic detection method.

Table 32.5 TAS test parameter setting of Hitachi LAbOSPECT 008AS automatic biochemical analyzer

Test parameters	Instrument setting
Method	2-Point End
Sample volume	4 μL
R1	160 μL
R2	16 μL
Reaction time	10 min
Temperature	37 °C
Read points	21, 38
Primary wavelength	405 nm
Secondary wavelength	546 nm
Calibration type	Linear
Unit	mmol/L

32.2.1.1 Detection Principle

The TAS assay relies on the ability of antioxidants in the sample to inhibit the oxidation of the peroxidase methemoglobin from ABTS (2,2'-azino-di-3-ethylbenzthiazoline sulfonate) to $ABTS^+$. The amount of $ABTS^+$ produced can be monitored at 600 nm using an automatic biochemical analyzer. Under the reaction conditions used, the antioxidants in the sample suppress the absorbance at 600 nm to an extent proportional to their concentration.

32.2.1.2 Reagent Composition

Reagent 1 (R1): 0.1 mol/L citric acid/trisodium citrate dihydrate buffer (pH = 5.8).

Reagent 2 (R2): 0.1 mol/L citric acid/trisodium citrate dihydrate buffer (pH = 3.6), 8.0 mmol/L ABTS [2, 2'-azino-bis(3-ethylbenzothiazoline-6-sulfonic acid)] and 2.0 mmol/L H_2O_2. After the reagent is prepared, it will be kept away from light and at room temperature for at least 1 h, and it can be used only after ABTS is fully oxidized.

pH Correction: Use 0.5 mol/L citric acid or 0.1 mol/L NaOH to accurately adjust the pH = 3.6 of R2.

The reagent can be stored for 2 weeks at room temperature (20–25 °C) and at least 6 months at 4 °C.

Standard solution: According to different biological samples, a kind of water-soluble vitamin E analogue 6-Hydroxy-2,5,7,8-tetramethylchromane-2-carboxylic acid (Trolox) (AR) was used to prepare different concentrations as standard.

32.2.1.3 Instrument Setting

Take Hitachi LAbOSPECT 008AS automatic biochemical analyzer as an example, the instrument parameters of TAS automatic detection are listed in Table 32.5. It is easy to adjust the test parameters according to the principle of the experiment, reagent composition, and instrument performance. The overall antioxidant levels in the samples are calculated using a known concentration of antioxidant. TAS values are expressed as mmol Trolox equivalent/L (mmol Trolox equiv./L).

32.2.2 Measurement of Total Oxidant Status

The following is a brief introduction of TOS full-automatic detection method.

32.2.2.1 Detection Principle

The TAS assay relies on the oxidation of ferrous ions to ferric ions in an acidic medium in the presence of various oxidative species. Ferric ion concentration was measured by xylenol orange. The assay was calibrated with a hydrogen peroxide standard (unit: μmol/L). Results were expressed in μmol H_2O_2 equivalent/L (μmol H_2O_2 equiv./L).

32.2.2.2 Reagent Composition

Reagent 1 (R1): 150.0 μmol/L xylenol orange, 140.0 mmol/L NaCl, 1.35 mol/L glycerol, and 16.0 mmol/L HCl.

Reagent 2 (R2): 5.0 mmol/L ferrous ammonium sulfate, 10.0 mmol/L o-dianisidine dihydrochloride, and 17.8 mmol/L HCl.

pH Correction: Use 0.1 mol/L HCl or 0.1 mol/L NaOH to accurately adjust the R1 and R2 reagents to pH = 1.75.

The reagent can be stored for 2 weeks at room temperature (20–25 °C) and at least 6 months at 4 °C.

Standard solution: 30% hydrogen peroxide (GR), according to the different biological samples, different standard concentrations will be prepared.

32.2.2.3 Instrument Setting

Take Hitachi LAbOSPECT 008AS automatic biochemical analyzer as an example; the instrument parameters of TAS automatic detection are listed in Table 32.6. It is easy to adjust the test parameters according to the principle of the experiment, reagent composition, and instrument performance. Likewise, the overall oxidant levels in the samples are calculated using a known concentration

Table 32.6 TOS test parameter setting of Hitachi LAbOSPECT 008AS automatic biochemical analyzer

Test parameters	Instrument setting
Method	2-Point end
Sample volume	21 μL
R1	162 μL
R2	12 μL
Reaction time	10 min
Temperature	37 °C
Read points	19, 38
Primary wavelength	570 nm
Secondary wavelength	800 nm
Calibration type	Linear
Unit	μmol/L

of hydrogen peroxide. TOS values are expressed as μmol H_2O_2 equivalent/L (μmol H_2O_2 equiv./L).

32.3 Calculation of Oxidant Stress Index

The oxidant stress index (OSI) is an index that reflects the state of redox balance in the human body. It can be calculated by the following formula:

$$\mathrm{OSI} = \mathrm{TOS}/\mathrm{TAS}.$$

When the TOS unit is μmoL H_2O_2equiv/L and the TAS unit is μmoL Troloxequiv/L, the above formula can be converted into:

$$\mathrm{OSI}(\text{arbitrary unit}) = [(\mathrm{TOS}, \mu\mathrm{moL}\,\mathrm{H_2O_2equiv/L})/(\mathrm{TAS}, \mu\mathrm{moL}\,\mathrm{Troloxequiv/L}] \times 100$$

Whether in a physiological state or pathological conditions, the results of the individual detection of one or several oxidized and/or antioxidant substances alone may rise or decrease compared to the levels of healthy individuals. The most puzzling thing is that TAS or TOS may also change (rise or fall) to varying degrees. However, when TAS and TOS simultaneously increase or decrease proportionally, the body will not produce OxS at this time if the ratio OSI does not change significantly (that is, there is a small fluctuation within the allowable range). At this moment, if only one or several oxidants and/or antioxidants, even TAS or TOS levels, changed significantly compared with those in healthy individuals, the observer is likely to judge that OxS occurred. However, OxS does not occur-if OSI remains relatively stable.

It can be seen that OSI is the key indicator for judging whether the oxidation-antioxidation balance of the body is disordered, which leads to the occurrence of OxS.

32.4 End Products of Lipid Hydroperoxide

The main target of reactive oxygen species is polyunsaturated fatty acids on the cell membrane, which triggers lipid peroxidation and leads to cell structure and function damage. In addition, the decomposition of lipid hydroperoxide produces many end products, such as compounds containing aldehyde groups (malondialdehyde, MDA), keto groups and hydroxyl (4-hydroxynonene, 4-HNE), organic hydrocarbons alkane, alkene and new oxygen free radicals, etc., which can accelerate biological oxidation in cells.

Lipid peroxidation is a free radical chain reaction. The formation of lipid peroxides has the following two types:

1. Enzymatic reactions: Some lipoxygenases can promote the reaction of oxygen with polyunsaturated fatty acids to form lipid peroxides. For example, 5-lipoxygenase and 12-lipoxygenase can promote the carbon atoms in the fifth and twelfth sites of arachidonic acid to be oxygenated to form 5-hydrogen peroxy-arachidonic acid and 12-hydrogen peroxy-arachidonic acid.
2. Nonenzymatic reaction: Polyunsaturated fatty acids have multiple double bonds, and more active hydrogen atoms are located on the methylene group between the two double bonds. For example, the C-H bond located in the methylene but not affected by the double bond has a dissociation energy of 393.56 kJ/mol, while the C-H bond located in the methylene and affected by two double bonds has a dissociation energy of only 355.85 kJ/mol. Therefore, when subjected to the action of light, radiation, and free radicals, etc., polyunsaturated fatty acids can easily remove hydrogen atoms from the methylene group located between two double bonds to form lipid radicals, and then double bonds and unpaired electron sites transfer to form relatively stable conjugated double bonds and then react with oxygen to form products such as lipid peroxy radicals and lipid peroxides.

Under the action of light, radiation, or free radicals, lipid molecules (LH) remove 1 hydrogen atom to form lipid free radicals (L·). Lipid free radicals react with oxygen to form lipid peroxyl radicals (LOO·); then LOO· radicals attack other lipid molecules and seize their hydrogen atoms to generate lipid radicals (L·) and lipid hydrogen peroxide (LOOH). This reaction is repeated in this manner, resulting in continuous consumption of lipids and large production of lipid peroxides.

RO, RO_2·, and ROOH are lipid peroxidation products; however, the content of these lipid peroxidation products in the human body is extremely low in normal physiological conditions, and their products will be converted to harmless substances, even if there is a chance of lipid peroxidation. Lipid peroxide can be decomposed into aldehydes, ketones, alcohols, ethers, carboxylic acids, and alkanes, among which malondialdehyde is the most representative lipid peroxidation product. Consequently, many researchers have tested malondialdehyde to determine whether a lipid peroxidation reaction has occurred in a system. But in terms of the human body, it is extremely one-sided to determine whether OxS occurs in the human body with this method [5].

The products of lipid peroxidation are commonly used as biomarkers of OxS or oxidative stress/damage. Lipid peroxidation generates a variety of relatively stable end products for decomposition, which can then be measured as an indirect biomarker of OxS in biological samples.

It is possible to estimate antioxidant activity in vivo by detecting the changes of lipid, protein, and/or DNA oxidative damage markers in biological samples. However, most of these markers are nonspecific, and their detection may also be interfered with by compounds of non-peroxidative origin [6]. There are many methods available for detecting oxidative damage to human lipid, protein, and DNA. A series of peroxidative products involved in the methods have been applied, including thiobarbital acid reactive substances (TBARS), converged dienes, hydrocarbons, lipid peroxides, F_2-isostandards, protein carbonyls, 8-hydrodeoxyguanosine, etc.

Among them, the method of detecting MDA based on the TBARS reaction principle has been widely used with its simpleness in technology. However, it is interfered with by compounds of non-peroxidative origin in human biological samples. It is also affected by Fe content in buffers and reagents. There are significant differences in the values of healthy subjects between different laboratories. High-performance liquid chromatography has improved specificity, but it is not easy to be popularized because of the limitation of instrument price and technical difficulty. In view of these factors, only the spectrophotometry of MDA is introduced here.

32.5 Measurement of Malondialdehyde

In this book, a modified spectrophotometric method of the thiobarbituric acid (TBA) test is introduced for the determination of MDA.

32.5.1 Detection Principle

MDA in LPO degradation products can combine with thiobarbituric acid to form a red complex TBARS, which has a maximum peak at 532 nm. The concentration of MDA in the sample can be calculated by comparing it with the standard of equivalent tests.

32.5.2 Reagent Composition

Reagent 1 (R1): 10% trichloroacetic acid.
Reagent 2 (R2): 0.67% thiobarbituric acid.

Standard solution: 6.0 μmol/L 1,1,3,3-tetramethoxypropane. It can be prepared into a standard storage concentrated solution of 1.0 mmol/L and can be stored in a refrigerator at 4 °C for at least 3 months.

32.5.3 Manipulation Steps

Step 1: Deproteination.
Take a clean 5 mL test tube, first add 200 μL of the sample, then add 400 μL of reagent R1, and shake vigorously or use a micro shaker to make it thoroughly mixed. Then centrifuge at 5000 rpm for 10 min.

Step 2: Color reaction.

Take 300 μL of supernatant and add 300 μL of reagent R2. Then boil the sample in an open water bath for 10 min, remove, and cool to room temperature.

Step 3: Colorimetric determination.

Read the absorbance using a spectrophotometer at the wavelength of 532 nm. The concentration of MDA in the sample can be obtained by calculating the standard tube operated at the same time.

Although the method is simple and easy to be popularized. it cannot be automatically detected because of the need for centrifugation and boiling, which can be seen from the **Manipulation** steps.

32.6 Conclusions

Oxidative stress refers to the imbalance between the oxidation and antioxidant system in the body, which causes a pathological process of oxidative damage of cells and/or tissues. When OxS occurs, the oxidation-antioxidant system tends to be unbalanced in the direction of oxidation, resulting in inflammatory infiltration of neutrophils, increased secretion of proteases, and the production of a large amount of oxidation intermediates [7]. Therefore, OxS is a negative effect from the free radicals in the body. In physiological conditions, it can not only promote the aging of the body but also promote the occurrence and development of diseases in pathological conditions. More than 95% of the free radicals in the body are oxygen free radicals, with the characteristics as follows: (1) the human body can not only produce free radicals but also scavenge them to keep the dynamic balance, so that the body can protect cells, tissues, and organs from oxidative damage; (2) oxygen free radicals can not only cause damage to the body but also promote certain physiological functions of the body; (3) the production and removal of oxygen free radicals are in a dynamic balance. If this dynamic balance is broken, it will cause damages to cells, tissues, and/or organs leading to the occurrence and development of disease.

Antioxidants are the substances that the body fights against oxygen free radicals (oxidants). At present, there are many kinds of related biomarkers (i.e., OxS biomarkers) used to reflect the oxidation/oxidation status of the body. But so far, no matter what kind of disease, there is still no widely accepted and highly specific biomarkers of OxS as an evaluation indicator of diagnosis, risk prediction, and prognosis for clinical application. Many oxidative damages may be a cascade reaction, which not only has complicated disease course but also involves a special tissue structure. Thus, the use of a single biomarker of oxidative stress is very limited, because it can only reflect a certain stage or aspect of damage to cells or tissues. TAS, TOS, and OSI could reflect the state of OxS in a system (cell, tissue, organ, or whole body), and the combined detection of them is the best choice to evaluate the OxS system. However, these three indexes have no tissue specificity, and they can only reflect the overall level of the body. With the continuous development of science and technology, a more understanding of proteomics, metabonomics, and bioinformatics will promote the development of OxS biomarkers with tissue organ specificity, high accuracy, and sensitivity to provide reliable clinical evidence for disease prevention and treatment.

References

1. Chakraborti S, Parinandi NL, Ghosh R et al (2019) Oxidative stress in lung diseases [M]. Springer, Singapore
2. Sies H (2019) Oxidative stress: eustress and distress[M]. Elsevier Science
3. Maurya PK, Chandra P (2017) Oxidative stress: diagnostic methods and applications in medical science[M]. Springer, Singapore
4. Armstrong D, Stratton RD (2016) Oxidative stress and antioxidant protection: the science of free radical biology and disease[M]. Wiley
5. Gelpi RJ, Boveris A, Poderoso JJ (2016) Biochemistry of oxidative stress: physiopathology and clinical aspects[M]. Springer International Publishing
6. Gowder SJT (2015) Basic principles and clinical significance of oxidative stress[M]. IntechOpen
7. Gilmour N (2015) Oxidative stress: biological aspects[M]. Callisto Reference

Targeting Chronic Lung Diseases Using Advanced Drug Delivery Systems

33

Himanshu Gandhi, Shikha Mahant, Amit Bansal, and Deepak N. Kapoor

Abstract

Targeting chronic lung diseases using advanced drug delivery systems explores the development of novel therapeutics and diagnostics to improve pulmonary disease management. Chronic lung diseases comprise a variety of serious diseases, including asthma, chronic obstructive pulmonary disease (COPD), idiopathic pulmonary fibrosis (IPF), pulmonary cystic fibrosis, and lung cancers. Here, the chapter will discuss advanced drug delivery systems developed for the management of various lung diseases.

Keywords

Chronic lung diseases · Targeted drug delivery · Pulmonary delivery · Advanced drug delivery · Treatment of pulmonary diseases

33.1 Introduction

Pulmonary diseases are among the leading cause of death worldwide. According to the WHO, millions of people are affected by pulmonary diseases and the population continues to increase [1]. These range from cough, common cold, asthma, and bronchiolitis to more severe conditions such as chronic obstructive pulmonary disease (COPD), lung cancer, idiopathic pulmonary fibrosis (IPF), and pulmonary hypertension [2]. Some of these diseases are irreversible and often fatal, and no

H. Gandhi · S. Mahant · D. N. Kapoor (✉)
School of Pharmaceutical Sciences, Shoolini University of Biotechnology and Management Sciences, Solan, Himachal Pradesh, India

A. Bansal
Formulation Research & Development, Perrigo Company PLC, Allegan, MI, USA

K. Dua et al. (eds.), *Targeting Cellular Signalling Pathways in Lung Diseases*,
https://doi.org/10.1007/978-981-33-6827-9_33

effective treatment has been reported to restore full lung function. Currently, there are more than 300 million people worldwide living with asthma, of which approximately 30 million are in China [3]. The prevalence of asthma and COPD is still increasing, unlike some other major diseases [4]. COPD is already the third leading cause of death from chronic disease worldwide, a major threat to public health [5]. Pulmonary tuberculosis spreads from one person to another through air. WHO estimates that 10.0 million (range: 9.0–11.1) people were diagnosed with tuberculosis and it caused 1.5 million (range: 1.4–1.6) deaths in 2018 [6]. IPF is one of the most common fatal interstitial lung diseases and is reported to affect more than 2.1 million patients [7]. This disease produces a strong inflammatory component of the lung, as well as the production of inflammatory cytokines, cellular infiltration, and airway hyperreactivity [8]. Traditional pharmacotherapy for chronic lung disease can be divided into several categories, depending on the type of therapeutic agent. Various chemical drugs, peptides, antibodies, and genetic molecules (eg, siRNA, shRNA, and miRNA) have been used to treat chronic lung disease [7]. Unfortunately, most chronic lung diseases cannot be completely cured by pharmacotherapy alone. In asthma, symptom control is the only option currently available. Likewise, steroids, bronchodilators, pirfenidone, and nintedanib are currently used to treat COPD or IPF, but no effective treatment is available to cure this disease completely. However, pharmacotherapy remains important for chronic lung disease, and patients should receive appropriate treatment throughout their life or until lung transplantation [9].

In recent decades, advances in drug delivery systems have made it possible to develop new pharmaceutical systems such as liposomes, niosomes, microemulsions, lipid micelles, solid lipid nanoparticles, polymer micelles, dendrimers, polymeric nanoparticles, nanogels, and nano-capsules. The goal of these advanced drug delivery systems is to modify pharmacokinetics, pharmacodynamics, nonspecific toxicity, immunogenicity, and biologic system recognition to achieve increased efficacy [10]. These systems are used for therapeutic purposes to transport the drug from the application site to the therapeutic target site in a controlled manner [11].

Traditional pharmacotherapy often provides several limitations, for example, poor pharmacokinetic profile and low drug diffusion, which results in poor response to the treatment [7]. For gene therapy, an effective vector is a prerequisite for successful gene delivery into the cell because genetic molecules are not easily delivered into cells without carriers and frequently degrade in the biological fluids [10]. This chapter discusses the treatment of various chronic lung diseases using advanced drug delivery systems.

33.2 Targeting of Various Chronic Lung Diseases Using Advanced Drug Delivery Systems

Chronic lung diseases comprise a variety of diseases including asthma, chronic obstructive pulmonary disease (COPD), idiopathic pulmonary fibrosis (IPF), pulmonary cystic fibrosis, and lung cancers [2]. The drug delivery system is designed to

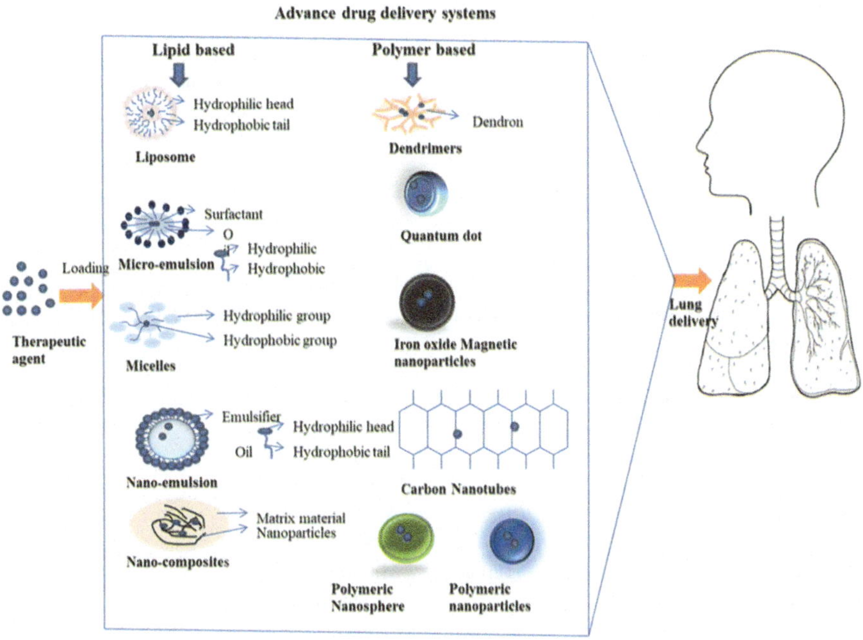

Fig. 33.1 Advanced drug delivery systems used for lung targeting

keep the drug in a pharmacologically active state until it is effectively delivered to the lungs where it is needed and shows the desired effect. Retention of the drug in the lungs for the desired time and avoidance of its clearance from the site of action is an important property desired by the advanced pulmonary drug delivery system [9]. Various advanced drug delivery systems used for drug targeting in the lungs are shown in Fig. 33.1.

33.2.1 Asthma

Asthma is a chronic lung disease that causes the airways to become inflamed and narrowed. The main features of asthma are bronchial hyperreactivity, reversible airflow obstruction, and tissue remodeling [12]. Asthmatic patients often experience symptoms such as shortness of breath, chest tightness, coughing, and periods of wheezing. Asthma affects a person of any age but is most frequent in childhood [13]. Various advanced drug delivery systems used for the treatment of asthma are listed in Table 33.1.

Ahmad et al. developed chitosan-coated nanoparticles (NPs) loaded with budesonide (BUD-NPs) to improve bioavailability, lung deposition, and pharmacokinetic profile of the BUD-NPs [14]. The prepared BUD-NPs increase bioavailability and in vivo deposition in the lungs of animal models in comparison to i.v. or oral route. Liquid crystal nanoparticles loaded with Celastrol were formulated by Chan

Table 33.1 Advanced drug delivery systems for asthma

Delivery systems	Therapeutic drugs	Particle size	Animal model	Cell lines	References
Liquid crystalline nanoparticles	Quercetin	268.7 nm	–	BCi NS1.1	[19]
Nanoparticles	Andrographolide	205 nm	Murine asthma mice model	–	[18]
Nanoparticles	Atropine	88.30 nm	Rat model	–	[16]
Liquid crystalline nanoparticles	Celastrol	194.1 nm	–	BCi-NS1.1	[15]
Nanoparticles	Budesonide	196.4 nm	Rat model	–	[14]
Nanoparticles	Betamethasone phosphate	116 nm	Murine mice asthma model	–	[24]
Nanoparticles	Bilirubin	100 nm	Asthma mice model	–	[22]
PLGA nanoparticles	Caryota mitis profilin	180 nm	Murine asthma mice model	–	[25]
Liposomes	*CpG* oligodeoxy nucleotide	–	Murine model of asthma using mice	–	[17]
PLGA-PEG nanoparticles	Bavachinin	196 nm	Mice model	Caco-2 cells, HeLa cells and NIH-3 T3	[21]
Chitosan–hyaluronic acid nanoparticles	Heparin	152 nm	–	–	[26]
Microspheres	–	–	Rat asthma model	–	[20]
Solid lipid nanoparticles	Curcumin	190 nm	Rat astma model	–	[23]
Microparticle	Salbutamol	8.24 μm	Rabbit model	A549 cells	[27]

et al. using the ultrasonication method [15]. These nanoparticles reduced the production of IL-1b (pro-inflammatory marker) in immortalized human bronchial epithelial cell lines (BCi-NS1.1), which improves asthma symptoms and causes no toxicity. Chattopadhyay et al. synthesized atropine-loaded nanoparticles (ANP) coated with polymers poly (lactic-co-glycolic acid) and polyethylene glycol [16]. ANP shows an excellent pharmacokinetic profile compared to dry powder. Further, 18 days of ANP treatment reduced collagen deposition and increased progressive airway obstruction. Thus, ANP nanoparticles delivered to the lungs increased the surfaces of pulmonary airways and increased the lung hyperresponsiveness, obstruction, and inflammation.

The major disadvantages of allergen-specific immunotherapy are the inability to modulate immune responses, long-term treatment, nonadherence, and the ability to develop life-threatening anaphylaxis. Custodio et al. investigated the allergen-specific immunotherapy with a liposomal formulation containing a low dose of allergens (OVA) combine with CpG-ODN (synthetic TLR9 agonist) in a murine model of asthma that depends on MyD88 signaling in dendritic cells [17]. The liposome composition containing the co-encapsulated allergen plus CpGODN was found to show short-term treatment, but not allergen and CpG-ODN alone. This prepared liposomal formulation was also effective and safe against allergens derived from *Blomia tropicalis* mite extract. In 2019, Chakraborty et al. prepared andrographolide (AG)-loaded nanoparticles to evaluate the anti-asthmatic efficacy on murine asthma model by oral/pulprolonged in vitro release and improved AG bioavailability upon oral/pulmonary administration [18]. The number of cells, levels of IL-4, IL-5, and IL-13 in bronchoalveolar lavage fluid, and serum IgE levels were significantly reduced after administration of AG nanoparticles compared to treatment with free AG. It was observed that the pulmonary route was more efficient than the oral administration. Cherkyong et al. formulated quercetin-loaded liquid crystalline nanoparticles (LCN) and surface-modified liquid crystalline nanoparticles (sm-LCN) to study their anti-inflammatory activity using lipolysis-induced primary human bronchial epithelial cell lines (BCi-NS1.1) [19]. Quercetin containing LCN and sm-LCN significantly ($p < 0.05$) reduced the production of pro-inflammatory markers namely IL-1β, IL-6, and IL-8 compared to the LPS-only group. In another study, She et al. investigated the effects of nebulized inhalation of anti-NGF microspheres (NANM) on ovalbumin (OVA)-induced airway remodeling [20]. It was observed that NANM significantly enhanced OVA-induced remodeling through regulation of the TGF-b1/Smad3 pathway.

Feng and colleagues prepared PEG5000-PLGA nanoparticles containing bavachinin for the treatment of asthma in a murine model by selectively inhibiting Th2 cytokine production by oral administration [21]. Through oral administration, these prepared nanoparticles exhibited excellent anti-asthma therapeutic effects in a murine allergic asthma model as assessed by histological section analysis, local and systemic cytokine expression, and T cell differentiation. Bilirubin-based nanoparticles (BRNPs) were prepared by Kim et al. for allergic lung inflammatory diseases [22]. The antihistaminic effect of prepared nanoparticles was assessed in a mouse model. Compared with unconjugated bilirubin (UCB), BRNP treatment suppressed the symptoms of experimental allergic asthma and Th2-related allergic lung inflammation. In another study, Wang et al. formulated curcumin-loaded solid lipid nanoparticles (curcumin-SLNs) to increase their therapeutic effectiveness in a model of ovalbumin-induced allergic asthma (OVA) in mice [23]. The prepared formulation showed entrapment efficiency of up to 75%. The curcumin plasma concentration was significantly higher than those obtained with curcumin alone. In the animal study, curcumin-loaded SLNs effectively suppressed airway hyperresponsiveness and inflammatory cell infiltration and as well as inhibit Th2 cytokine expression.

33.2.2 Chronic Obstructive Pulmonary Disease

Chronic obstructive pulmonary disease (COPD) is the most common poorly reversible obstructive disease of lungs with a high global incidence, which is characterized by decreased lung function. It is associated with chronic inflammation, mucus hypersecretion, oxidative stress, airway obstruction, cellular senescence, and increased autophagy. Cigarette smoke is one of the most important risk factors for the development of the pathogenesis of COPD. Other factors are chronic exposure to ambient air pollutants and commercial gases. Exposure to biomass smoking in many developing countries also contributes to the development and progression of COPD [28]. Several advanced delivery systems used in targeted therapeutic deliveries for COPD are shown in Table 33.2.

Ainali and co-workers fabricated fluticasone propionate (FLU) and salmeterol xinafoate (SX) dual drug-loaded microparticles by using the ionotropic gelation technique [29]. In vitro release studies revealed that FLU and SX showed a great amelioration in drug dissolution profiles from all modified chitosan microparticles compared to pure drugs. Xu et al. investigated the mechanism of lentivirus-mediated overexpression of transmembrane regulators of cystic fibrosis conduction under oxidative stress and inflammatory reactions in lung tissue of a COPD-induced mouse model [30]. Overexpression of CFTR can inhibit apoptosis of pulmonary endothelial cells, which inhibits reactive oxygen species (ROS), low glutathione (GSH), and malondialdehyde (MDA) levels and increases superoxide dismutase (SOD) as well as the total antioxidant capacity (T-AOC), and GSH peroxidase (GSH-Px Excess CFTR) can protect lung tissue from injury due to oxidative stress and the inflammatory reactions in mouse models of COPD. The mechanism involves

Table 33.2 Advanced drug delivery systems for chronic obstructive pulmonary disease (COPD)

Dosage form	Therapeutic drugs	Particle size	Animal model	Cell line	References
Solid lipid microparticles	Salmeterol xinafoate	3.5–4 μm	–	16HBE cell line	[33]
Magnetic nanoparticles	–	–	Mice model	–	[36]
Lipid-polymer nanoparticles	–	124 nm	–	A549 cells	[34]
Solid lipid microparticles	Fluticasone propionate	4.0 μm	–	16-HBE, cell	[35]
Chitosan microparticles	fluticasone propionate and Salmeterol Xinafoate	2.2 μm	–	–	[29]
Nanosphere	Amikacin	200 nm	Mice model	A549, L929, BEAS-2B	[31]
Nanoparticle	–	244 nm	–	A549 cells	[32]

in this can be attributed to the repression of MAPK/NF-κB p65 signaling pathway. Li et al. prepared amikacin-loaded black quantum dots (BPQDs) conjugated with PEGylated chitosan nanospheres (PEG@CS/BPQDs-AM NPs) for the efficient treatment of COPD [31]. Rapid release of therapeutic amikacin after BPQD erosion and more than 70% release of amikacin within 7 days increased the therapeutic effect of amikacin. In animal models, the airflow barrier was significantly reduced after injection of PEG @ CS/BPQDs-AM NPs, and a marked antibacterial effect against *Pseudomonas aeruginosa* was observed. Therapy based on RNA interference (RNAi) is seen as an endogenous mechanism for gene suppression to modulate gene expression, which includes the degradation of the mRNA sequence mediated by certain RNAs. In addition, microRNAs (miRNAs) are diagnostic and therapeutic targets for treatment of COPD. Mohammed et al. prepared nanoparticles by incorporating cationic lipid dioleoyltrimethylammoniumpropane (DOTAP) and surface modified with the adsorption of miR-146a for delivery to lung epithelial cells [32]. The results revealed that miR146a, supplied as miR-146a nanoparticles, had a dose-dependent effect on the highest nanoparticle concentrations (0.321 and 0.625 mg/mL) and reduced expression of the IRAK1 target gene by up to 40%. The MiR-146a nanoparticles decreased the production of the IL-8 promoter-reporter (GFP). Mucoadhesive solid lipid nanoparticles (SLM) encapsulated with salmeterol xinafoate have been developed by Amore et al. for therapy of COPD [33]. The technique has been shown to allow the preparation of microparticles loaded with SX with suitable dimensions (3.5–4 μm) to release the drug into the secondary bronchi, avoid systemic side effects, and increase the retention time and therapeutic effect of SX in the pulmonary epithelium. Reactive oxygen species (ROS) and epigenetic abnormalities are strongly associated with the pathology of COPD, and excess production of ROS leads to glucocorticoid resistance and deficiency of histone deacetylase 2 (HDAC2) for the reduction of ROS level and glucocorticoid resistance. Chikuma et al. developed core-shell-type lipid-polymer nanoparticles (LPNs), in which potent antioxidant Mn-porphyrin dimer (MnPD) and HDAC2-encoding plasmid DNA (pHDAC2) were successfully encapsulated [34]. The MnPD antioxidant activity and transfection of HDAC2 expression levels work synergistically to reduced the excess production of IL-8, a chemokine related to COPD, and to improve mitochondrial dysfunction. In another study, Amore et al. prepared solid lipid nanoparticles loaded with drug fluticasone propionate (FP) for COPD treatment [35]. The result demonstrated that the FP-loaded solid lipid nanoparticles were more effective in controlling lung inflammation than FP alone.

33.2.3 Lung Cancer

Lung cancer is considered to be the most chronic form of cancer worldwide. Conventional treatments for lung cancer include a variety of medical interventions such as surgical removal, chemotherapy, and radiation therapy. However, this type of approach lacks specificity and also destroys neighboring healthy cells [37, 38]. Recently, the targeted drug delivery approach received more attention to addressing this problem [39]. Drug carriers can transport drugs to the lungs, prolong their time duration, adjust their therapeutic doses, and reduce the complications and risk of toxic drug side effects in patients. This targeted approach transports the therapeutic agent to the site of action and reduces their proliferation to adjacent nontarget or healthy tissues and organs [39, 40]. Advanced drug delivery systems for the treatment of lung cancer are listed in Table 33.3.

Almuqbil et al. developed a doxorubicin-loaded nanocarrier-based drug delivery system (nanoDDS) to increase the penetration of the therapeutic drug doxorubicin in lung tumors and to compare these results with its efficacy [41]. This work demonstrated the conjugation of DOX to G4SA, which promotes DOX penetration into the extracellular matrix and produces a co-culture spheroids lung cancer model. The G4SA-GFLG-DOX penetration into the spheroid core was 3.1-fold higher than that of free DOX, which was associated with an increased efficiency measured by the caspase 3/7 test. Growth inhibition studies exhibited that the released DOX maintains its activity and reduced tumors to similar levels of free DOX. In another study, Bai et al. prepared docetaxel (DTX) loaded with pH-sensitized nanoparticles made up of alendronate (ALN) and polyamidoamine (PAMAM) for the treatment of bone metastasis in lung cancer [42]. The results revealed that resulting nanoparticles considerably enhanced the anticancer activity of DTX and inhibited osteoclast formation.

Chen and co-workers synthesized nanoparticles containing dual drugs, cisplatin and epigallocatechin gallate (EGCG), to chemically eliminate cisplatin and destroy cancerous cells of lungs [43]. It was observed that low concentration of prepared nanoparticles (EGCG 5 μg/mL: cisplatin 2 μg/mL) exhibited significant cytotoxicity as compared to cisplatin alone. The prepared cisplatin nanoparticles were freely absorbed by the cells through endocytosis and increase the intracellular cisplatin concentration to therapeutic levels. Curcumin-loaded poly (ester-thioether) microspheres co-loaded with α-tocopherol succinate and erlotinib have been described by Cheng et al. for non-small-cell lung cancer combination therapy [44]. The poly (ester-thioether) microspheres extensively increased the bioavailability of both erlotinib and α-tocopheryl succinate, in contrast to the free drug combination and obtained synergistic effect on inhibition of A549 cell proliferation in both in vitro and in vivo studies. The porous microspheres exhibit faster erosion and drug release compared to nonporous analogues and thus show increased effectiveness against cancer. Coban et al. synthesized liposomes coated with bovine serum albumin containing erlotinib or Zn (II) phthalocyanine-containing ferrocene groups (Pc-Zn) to treat lung cancer by using photodynamic therapy [45]. Liposomes loaded with Pc-Zn had a significant cytotoxic effect under light irradiation compared to liposomes loaded with erlotinib. Liposomes coated with BSA did not show any significant effect on cytotoxicity. Guo et al. reported the preparation of liposome formulation with GIE11 for the targeted delivery of doxorubicin [46]. The results exhibited a significant reduction in tumor size with prepared liposome formulation (312 mm^3) as compared to the free drug (540 mm^3).

Table 33.3 Advanced drug delivery systems for lung cancer

Therapeutic drugs	Dosage form	Particle size	Animal model	IC_{50}	Cell line	References
Phthalocyanine	Liposomes	187.2 nm	–	–	A549 cells	[45]
Cisplatin or epigallocatechin gallate	Nanoparticles	75 nm	–	–	A549 cells	[43]
Erlotinib α-tocophery and poly (ester-thioether)	Microsphers	13.6 μm	Mice model	17.4 μg/mL	NIH/3 T3 and A549 cells	[44]
Docetaxel	Nanoparticles	100 nm	Mice model	–	A549 and MDA-MB-231 cells	[42]
Irinotecan	Liposomes	127 nm	Mice model	–	A549 cells	[47]
Doxorubicin	Liposomes	485 nm	–	–	A549, Calu-3, 16HBE14o cells	[57]
Doxorubicin	Dendrimers	10 nm	–	0.58 μM	A549 and 3 T3 cells	[41]
Tel and docetaxel	Liposomes	151 nm	Mice model	–	H460 WT and CD133+ H460 stem cells	[58]
Doxorubicin	Liposomes	120 nm	Mouse model	–	H 1299 cells	[46]
Euphorbia fischeriana root	Nanoparticles	40 nm	–	14.5 μg/mL	A549 cells	[59]
Cisplatin	Nanoparticles	245 nm	Mice model	–	AT1 cells	[48]
Doxorubicin and celecoxib	Nanoparticles	152 nm	–	0.396 μM	A549-Luc, NCI-H1650-Luc, and PC-9 cells	[60]
Docetaxel	Tocopheryl polyethylene glycol 1000 succinate-coated liposomes	140 nm	Mice model	24.54ug/mL	A549 and A549/DDP	[61]
Gemcitabine	Nanoparticles	302 nm	Mice model	–	LL/2 and BEAS-2B cells	[55]

(continued)

Table 33.3 (continued)

Therapeutic drugs	Dosage form	Particle size	Animal model	IC_{50}	Cell line	References
Pirfenidone	Liposomes	214 nm	–	0.2 mg/mL	A549, H157, H460, and H4006 cells	[62]
Paclitaxel	Liposomes	254 nm	–	0.085 μM	NCI-H460 cells	[54]
Paclitaxel, vinorelbine	Liposomes	150 nm	–	–	A549, H1299, and NSCLC cells	[50]
Betulinic acid, parthenolide, honokiol, and ginsenoside	Liposomes	100 nm	Mice model	–	A549 cells	[49]
Quinacrine	Nanoparticles	216 nm	–	1.3 μM	A549 NSCLC cells	[53]
Doxorubicin	Dendrimers	–	Mice model	109.90 μM	B16-F10 cells	[63]
Erlotinib	Nanoparticles	112.5 nm	Mice model	–	A549 cells	[64]

Another study reported the development of irinotecan loaded liposome formulation, surface modified with NF-κB inhibitor (CB5005) for non-small-cell lung cancer treatment [47]. The prepared system showed a significant anti-tumor effect in mice carrying the A549 xenografts model. Iyer et al. prepared cisplatin-loaded glutathione (GSH)-coated polyurethane nanoparticles (GPUs) for treatment of lung cancer [48]. The resulting nanoparticle formulation showed significant tumor reduction in the in vitro survival fraction (A549 lung cancer cells) as compared to free cisplatin. It was observed that the cisplatin-loaded nanoparticles demonstrated inhibition of tumor growth in mouse xenograft A549 lung tumor compared to the free cisplatin. Jin et al. designed a system for the treatment of lung cancer by combining four natural products, i.e., betulinic acid, parthenolide, honokiol, and ginsenoside Rh2 in the liposomal formulation [49]. The findings of in vitro and in vivo studies revealed that combination of four natural products exhibited a synergistic effect and more efficient and safer treatment for lung cancer as compared to cisplatin alone. The folate-targeted, co-drug paclitaxel- and vinorelbine-loaded radiolabel theranostic liposomal formulations were prepared by Karpuz et al. for the diagnosis and therapy of lung cancer [50]. It was observed that in vitro anticancer activity of liposome formulations containing the drug is significantly higher than the activity of single-drug liposomal formulations. Superparamagnetic iron oxide nanoparticles (SPIONs) are found to be a promising drug delivery system and hyperthermia agent in both cancer therapy and diagnosis due to their magnetic nature. SPIONs coated with silica layers of different types, namely, mesoporous (@mSiO_2), nonporous (@SiO_2), or with a combination of d mesoporous layers and nonporous (@mSiO_2@SiO_2), were prepared by Recznka et al. using the sol-gel method [51]. The outcomes of the study revealed that all types of silica coating significantly decreased iron release (tenfolds) as compared to uncoated SPIONs. In vitro findings of the study showed that a delay in proliferation was observed in BEAS-2B cells compared to AAS49 cells. Tie et al. prepared a folate-modified liposome complex for targeting folate receptor β positive tumor-associated macrophages in lung cancer treatment [52]. It was observed that prepared liposomal formulation (F-PLP/pBIM) significantly reduced lung cancer growth. Inhalable quinacrine (QA)-loaded bovine serum albumin (BSA)-modified cationic nanoparticles were prepared by Vaidya et al. for lung cancer treatment [53]. The resulting nanoparticles showed increased therapeutic potential for NSCLC cells, cell cycle arrest in the G2/M phase, and increased apoptosis. The resulting particles also showed increased therapeutic efficacy in cell culture 3D models. In 2019, Zhang et al. designed paclitaxel (PTX) and siRNA-loaded liposomes for lung cancer treatment [54]. The results of in vitro studies on NCI-H460 lung cancer cells showed the highest cellular uptake, lowest cell viability, and strong apoptosis A system for targeted drug delivery of silk fibroin nanoparticle (SFNP)-loaded with gemcitabine (Gem) was developed by Motaghitlab et al. for the treatment of lung cancer [55]. The targeted Gem-loaded SFNPs showed higher cellular uptake, cytotoxicity, and accumulation in the targeted lung tissue as compared to nontargeted SFNPs. The in vivo study found lower mortality, higher survival rate, and no sign of metastasis. Li et al. investigated the synergistic effect of cold atmospheric plasma (CAP) and iron oxide-based magnetic nanoparticles (IOMNPs) on epidermal growth factor receptor (EGFR), cellular bioactivity, and the downstream EGFR signaling pathway [56]. The results revealed that the synergistic effect of CAP and IOMNPs significantly inhibited cell proliferation and decrease EGFR regulation, whereas CAP suppressed lung cancer cells by suppressing pAKT and pERK.

33.2.4 Idiopathic Pulmonary Fibrosis

Idiopathic pulmonary fibrosis (IPF) is a chronic non-neoplastic lung disease characterized by the damage of alveolar epithilium cells and scarring without any known stimulation. Shortness of breath due to exercise and chronic dry cough are major symptoms [65, 66]. Various advanced drug delivery systems for the treatment of pulmonary fibrosis are shown in Table 33.4. In 2020, Ghodake et al. fabricated a cefoperazone sodium-loaded liposome formulation to alleviate *P. aeruginosa* biofilm in cystic fibrosis infection [67]. There was 17% reduction in *P. aeruginosa* biofilm after treatment with 0.42 μg/mL of the prepared formulation. Amikacin-loaded liposomal suspension for inhalation was prepared by Bilton et al. for *P. aeruginosa* infection in cystic fibrosis [68]. The results showed that the cyclic dosing of once-daily amikacin liposomal inhaled suspension was non-inferior to the twice-daily doses of tobramycin-loaded liposomal inhaled suspension to reduce infection in cystic fibrosis and improve lung functions.

Table 33.4 Advanced drug delivery systems for pulmonary fibrosis

Therapeutic drugs	Dosage form	Particle size	Animal model	Cell line	References
Amikacin	Liposomes	–	–	–	[68]
Cefoperazone sodium	Liposomes	362 nm	–	–	[67]
Pirfenidone	Nanoparticles	80 nm	–	HDF cells	[71]
–	Nanoparticles	59.3 nm	Mouse model	THP-1(ATCC) cells	[75]
Ciprofloxacin	Nanoparticles	190.4 nm	–	Calu-3 and CFBE41o cells	[70]
Gadolinium	Nanoparticles	–	Mice model	–	[76]
Tacrolimus	Nanoparticles	182 nm	Mouse model	–	[72]
Pirfenidone	Nanoparticles	278.5-241 nm	Mice model	C57Bl/6 cells	[77]
Fluorofenidone	Nanoparticles	172.5 nm	Rat model	A549 cells	[69]
Bleomycin	Chitosan-based nanoparticles	189 nm	Rat model	(MRC-5) cells	[78]
Thymoquinone	PLGA-PVA Nanoparticles	20 nm	Rat model	–	[73]

In another study, fluorofenidone-loaded poly(lactic-*co*-glycolic acid) (PLGA) nanoparticles modified with spermidine (Spd) were fabricated by Tang et al. for the treatment of idiopathic pulmonary fibrosis [69]. It was observed that the A549 cellular uptake of prepared nanoparticles was reduced significantly with an increase in spermidine concentration. Biodistribution of prepared formulation increased 3.62–4.66 times compared to plain fluorofenidone-loaded nanoparticles. The application of nanoparticles with a high antibiotic content can easily penetrate the biofilm and become an adventure in overcoming treatment obstacles. Tureli et al. developed ciprofloxacin-loaded PLGA nanoparticles for the treatment of cystic fibrosis [70]. The findings of the study demonstrated enhanced antibacterial activity with the prepared nanoparticles and decreased mucus turbidity when incubated with nanoparticles.

Abnoos and co-workers prepared pirfenidone-loaded chitosan-coated sodium alginate nanoparticles by using the pre-gelation method [71]. The findings of the in vitro drug release study represented sustained release at 24 h with 50 and 94% of drug loading efficiency. The penetration of the prepared nanoparticles through the skin was significantly increased in comparison to the pirfenidone solution. Seo et al. prepared inhaled tacrolimus-bound albumin-coated nanoparticle formulation for the treatment of bleomycin-induced cystic fibrosis [72]. The authors found that the prepared inhaled nanoparticle demonstrated significant anti-fibrotic efficacy in a bleomycin-induced cystic fibrosis mice model as compared to intraperitoneal administration of tacrolimus (60 μg/mouse). The benefits of thymoquinone (TQ) are still partially problematic due to poor water solubility. Therefore, TQ-loaded PLGA-PVA nanoparticles were fabricated by Saghir et al. for evaluating the potential effect on bleomycin-induced pulmonary fibrosis in the albino rat model [73]. The designed nanoparticles showed an encapsulation efficiency of about 80%. The findings of the study have concluded that the formulated nanoparticles significantly reduce bleomycin-induced pulmonary fibrosis through the regulation of GF-β1 and IL10 and downregulation of iNOS regulation in lung tissue. The inhaled tacrolimus (TAC)-loaded chitosan-coated PLGA nanoparticles were fabricated by Lee et al. by using the W/O emulsion diffusion method for the treatment of bleomycin-induced pulmonary fibrosis [74]. The nanoparticles exhibited a significantly high TAC entrapment efficiency of 37.7%. It was observed that the direct inhalation of prepared nanoparticles (TAC 180 mg/mouse) twice weekly in mice leads to remarkable anti-fibrotic efficacy as compared to daily oral administration (TAC 300 mg/ mouse). Imaging of the lung deposits showed that the chitosan-coated TAC-loaded PLGA nanoparticles were well localized in the lungs and gradually faded within 96 h.

33.2.5 Pulmonary Arterial Hypertension

Pulmonary arterial hypertension (PAH) is a chronic and progressive disease with increased pulmonary arterial blood pressure on average from 25 mm Hg due to an increase in pulmonary vascular resistance. The main consequence of the PAH is, if untreated, it may lead to right ventricular failure, fluid retention that causes peripheral edema, and eventually death [79]. There is no known treatment yet available for the PAH, but recent studies on molecular mechanisms responsible for disease development and progression resulted in new therapeutic options improving quality of life and chances of extended survival [80]. Several advanced drug delivery systems for pulmonary arterial hypertension are shown in Table 33.5.

In 2020, Zerong and co-workers formulated resveratrol-loaded dipalmitoylphosphatidylcholine-coated lipid nanoparticles (DPPC-LNs) by using the film hydration-ultrasonic dispersion technique for the treatment of PAH [81]. The prepared nanoparticle formulation showed high drug encapsulation efficiency of 94.40% and 80% of cumulative drug release over 48 h. Results suggested that the prepared formulation could enhance time-dependent cellular uptake by pulmonary arterial smooth muscle cells (PASMCs). Analysis of plasma and lung tissue data showed greater accumulation of nanoparticles after intratracheal administration compared with intravenous administration of resveratrol solution. Nafee et al. prepared sildenafil-loaded nanostructured lipid carriers (NLCs) and solid lipid nanoparticles (SLNs) containing precirol and PVA for the treatment of pulmonary hypertension via nebulization [82]. The prepared NLCs demonstrated a high drug release profile and cell viability of A549 cells (increases threefold IC50 over SLNs). In vivo administration of a nano-encapsulated drug, formulation confirmed the normal lung tissue with minimal evidence of bleeding.

Metal-organic framework (MOF) nano-medicine containing sildenafil was designed by Nura et al. for the future treatment of PAH [83]. Outcomes of the study revealed that the prepared nano-system was nontoxic when incubated with a human endothelial cell for 24 h. The nano-formulation of sildenafil retained its crystallinity, delayed sildenafil release, and induced vasodilation activity after a lag phase of more than 4 h. Gupta et al. investigated the effectiveness of fasudil-loaded CARSKNKDC (CAR) peptide-DSPE-PEG micelle hybrid nanoparticles for the targeted delivery to the pulmonary artery to treat pulmonary arterial hypertension [84]. The encapsulation efficiency of CAR micelles containing fasudil was found to be ~58%. The CAR peptide exhibited a ~1.7-fold higher cellular uptake in pulmonary artery smooth muscle (PASM) cells in comparison to plain micelles. The formulation was more likely to accumulate in the pulmonary vessel than in peripheral blood, which is evident by the ratio of decreased pulmonary and systemic arterial pressures.

Cerivastatin containing nano-liposomes was prepared by Young et al. to explore potential treatment for pulmonary arterial hypertension [85]. The developed nanoliposomal formulation demonstrated significantly lower cellular cytotoxicity than free cerivastatin by inhibiting the proliferation of PASM cells. The formulation showed a significant decrease in pulmonary artery pressure (55.13 ± 9.82 to 35.56 ± 6.59 mm Hg) in monocrotalin-induced PAH in a rat model. Another study reported the development of spray-dried hydrogel microparticles containing sildenafil citrate for the treatment of PAH [86]. The results demonstrated the highest entrapment efficiency of more than 80%. The in vitro release profile of the drug sildenafil citrate showed a sustained release of over 24 h. In vivo pharmacokinetic

Table 33.5 Advanced drug delivery system for pulmonary arterial hypertension

Therapeutic drugs	Dosage form	Particle size	Animal model	Cell line	References
Sildenafil	Nanomedicine	50 nm	Mice model	Human blood outgrowth endothelial cells (BOECs)	[83]
Fasudil	Peptide-micelle hybrid nanoparticles	14.47 nm	Rat model	PASM cells	[84]
Tadalafil	Nanocomposites as a dry powder	3.2 μm	Rat model	–	[87]
Sildenafil citrate	Microparticles	2–5 μm	Rat model	RAW 264.7 cells	[86]
Cerivastatin	Nanoliposomes	98 nm	Rat model	HPASMC cells	[85]
Sildenafil citrate	Nanostructured lipid carriers	760 nm	Rat model	A549 cells	[82]
Resveratrol	Dipalmitoylphosphatidylcholine-coated lipid nanoparticles	123.7 nm	Rat model	PASMC cells	[81]

studies have shown that hydrogel microparticles increase drug deposition in the lungs and have a longer elimination half-life than commercial sildenafil citrate tablets. Teymourirad et al. prepared an inhalable dry powder nanocomposite formulation containing tadalafil using spray drying techniques for increased treatment efficacy of PAH [87]. The prepared nanocomposites demonstrated a 13.7-fold increased solubility compared to the pure drug tadalafil. In addition, the in vivo evaluation showed that intratracheal administration of prepared tadalafil nanocomposites achieves 2.3 and 3.7 times the highest mean residence time, respectively, compared to conventional oral administration.

33.2.6 Pulmonary Tuberculosis

Tuberculosis (TB) is an ancient multisystemic infectious disease commonly caused by *Mycobacterium tuberculosis* which causes significant health challenges in modern days. One-third of the worldwide human population is infected with the deadly bacteria *M. tuberculosis* and more than nine million new cases of TB are reported annually [88, 89]. Early diagnosis and prompt initiation of delayed diagnosis may increase disease severity, increase the risk of death, and increase community transmission of tuberculosis [90]. Advanced drug delivery systems for the treatment of pulmonary tuberculosis are shown in Table 33.6. Rifabutin-loaded nanostructured lipid carriers (NLCs) were synthesized by Pinheiro et al. by using high-shear

Table 33.6 Advanced drug delivery system for pulmonary tuberculosis

Therapeutic drugs	Dosage form	Particle size	Animal model	Cell line	References
Rifabutin	Nanostructured lipid carrier	175–213 nm	–	Calu-3, A549, and RAW 264.7 cells,	[91]
Ethambutol	Solid lipid nanoparticles	56.79 nm	–	A549 cells	[94]
Curcumin	Nanoparticles	400 nm	–	RAW 264.7 cells	[96]
Moxifloxacin	Moxifloxacin PLGA nanoparticles	430–920 nm	–	H37Ra cells	[97]
Isoniazid	Boron nitride nanoclusters, nanoparticles, and nanotubes	–	–	–	[98]
Isoniazid	Magnetic nanoparticles	12.93 nm	–	–	[95]
Rifampicin	Solid lipid nanoparticles	315 nm	Mice model	–	[93]
Rifampicin	Solid lipid nanoparticles	520 nm	Mice model	–	[92]

homogenization and ultrasonic technique for the treatment of TB [91]. The drug-loaded NLCs exhibited significant cytotoxicity toward Raw 264.7, A549, and Calu-3, cells with an IC_{50} of 238.9, 185.7, and 108.7 μg/mL, respectively. The prepared NLCs exhibited the highest entrapment efficiency of 80%. In addition, rifabutin-loaded NLCs follow a pH-sensitive drug release profile, whereas rapid drug release was at acidic pH than neutral pH.

The active targeting of alveolar macrophages (AM) in the lungs is a promising approach to improve the therapeutic efficiency of "old" drugs currently available in clinical practice for the treatment of TB. Truzzi et al. developed a rifampicin (RIF)-loaded solid lipid nanoparticles (SLNs) conjugated with mannose-based surfactant for the treatment of TB [92]. The results revealed that the prepared drug-loaded nanocarrier system effectively targets the AM through the mannose-receptor mediated pathway with a poor systemic biodistribution. Vieira et al. synthesized a rifampicin-loaded mannosylated nanostructured lipid carrier (NLC) for the active treatment of tuberculosis [93]. The developed nanoparticles exhibited higher efficiency of 90% and higher cellular uptake by bone marrow-derived macrophages.

Another study reports the development of solid lipid nanoparticles (SLNs) loaded with ethambutol by spray drying method as a dry powder inhaler for tuberculosis therapy [94]. The entrapment efficiency was found to be more than 98%. MTT assay confirms the biocompatibility and lack of toxicity of drug-loaded SLNs. Zargarnezhad et al. developed isoniazid conjugated magnetic nanoparticles coated with lipoamino acid, which evaluate the antimicrobial activity against *M.* tuberculosis and Gram-positive and Gram-negative non-mycobacterial strain [95]. The outcomes of the study revealed that 44.8% and 16.7% of isoniazid amount decreased against *M. tuberculosis* when conjugated to naked and surface-modified magnetic nanoparticles, respectively.

Jahangirdhar and the group developed rifampicin- and curcumin-loaded polymeric nanoparticles for active intra-macrophage delivery and clearance of *M. tuberculosis* [96]. The prepared nanoparticles exhibited 1.5 folds higher drug internalization than free drugs and were nontoxic to RAW 264.7 macrophages. The formulation shows high efficacy against *M. tuberculosis*-infected macrophages at 25 MIC (98.03) 2.5%) with complete clearance over 50 MIC. Abdelghany et al. synthesized dual drug (amikacin and moxifloxacin)-loaded PLGA nanoparticles modified with alginate for treatment of tuberculosis [97]. The in vitro anti-mycobacterial activity of the prepared dual drug-loaded nanoparticles was evaluated by using macrophages infected with *M. tuberculosis*, which showed significant inhibition in viable bacterial count compared to single drug-loaded formulations. The bacterial viability of 0.6% was confirmed by dually encapsulated drug formulation relative to 6.49% for only amikacin and 3.27% for only moxifloxacin nanoparticles.

33.3 Future Perspectives

Pharmaceutical technology has been playing an essential role in medicine as it allows us to study various materials and their combinations to develop drug carriers, support them with superior properties, achieve desired goals, and end with therapeutic success. In lung drug delivery, the primary goal is to keep the drug in the lungs as is the case with local delivery approaches, thereby reducing systemic absorption. In other cases, a systemic effect is the desired result, and the carrier is designed to avoid retention. Most of the current research in advanced drug delivery systems is focused on developing new materials and manufacturing polymer responsive systems with their macro- and microstructure and chemical profile. Special attention is drawn to the design of the hydrophilic/hydrophobic copolymer segment and the production of star-shaped polymer and dendrimers as nanocarriers for bioactive compounds. Not only should future researches focus on discovering new potential targets for lung recovery and regeneration, but also more research should be needed for better-understood disease models, which could further accelerate research to develop new therapeutic approaches for chronic lung diseases. In the near future, the emergence of new, effective, and specially developed ADDS can be traced back to interdisciplinary research knowledge leading to personalized medical health solutions.

33.4 Conclusion

It is becoming increasingly difficult to develop effective controlled release delivery systems for the diagnosis and treatment of chronic lung disease due to the complex physiological and pathological mechanisms. Extensive development in the area of drug delivery systems has helped to overcome the problems of traditional medical interventions such as low therapeutic efficacy, drug resistance, nonspecific targeting, and unexpected side effects. These advanced drug delivery systems have become ideal carriers for targeted therapies for various lethal diseases. Various advanced drug delivery systems like polymeric nanoparticles, liposomes, polymeric micelles, carbon nanotubes, dendrimers, quantum dots, nanostructured lipid carriers, solid lipid nanoparticles, and magnetic nanoparticles have been developed and used with a variety of medical intervention to track and target the therapeutic agents at the infected area of the body.

References

1. Varkey B, Maier LA (2015) Chronic respiratory diseases: challenges in diagnosis and prevention. Curr Opin Pulm Med 21:111–113. https://doi.org/10.1097/MCP.0000000000000146
2. Thakur AK, Chellappan DK, Dua K et al (2020) Patented therapeutic drug delivery strategies for targeting pulmonary diseases. Expert Opin Ther Pat 30:375–387. https://doi.org/10.1080/13543776.2020.1741547

3. Li M, Jin R, Peng Y, Wang C, Ren W, Lv F, Gong S, Feng F et al (2020) Generation of antibodies against COVID-19 virus for development of diagnostic tools. medRxiv 11:39. https://doi.org/10.1101/2020.02.20.20025999
4. Cruz FF, Rocco PRM (2020) The potential of mesenchymal stem cell therapy for chronic lung disease. Expert Rev Respir Med 14:31–39. https://doi.org/10.1080/17476348.2020.1679628
5. Singh D, Agusti A, Anzueto A et al (2019) Global strategy for the diagnosis, management, and prevention of chronic obstructive lung disease: the GOLD science committee report 2019. Eur Respir J 53:1900164. https://doi.org/10.1183/13993003.00164-2019
6. Ding C, Wang S, Shangguan Y et al (2020) Epidemic trends of tuberculosis in China from 1990 to 2017: evidence from the global burden of disease study. Infect Drug Resist 13:1663–1672. https://doi.org/10.2147/IDR.S249698
7. Yhee JY, Im J, Nho RS (2016) Clinical medicine advanced therapeutic strategies for chronic lung disease using nanoparticle-based drug delivery. J Clin Med 5(9):82. https://doi.org/10.3390/jcm5090082
8. Chiang PC, Alsup JW, Lai Y et al (2009) Evaluation of aerosol delivery of nanosuspension for pre-clinical pulmonary drug delivery. Nanoscale Res Lett 4:254–261. https://doi.org/10.1007/s11671-008-9234-1
9. Sharma D, Goyal K (2018) Recent approaches for novel treatment for pulmonary diseases. Int J Pul Res Sci 2(4):1–12. https://doi.org/10.19080/IJOPRS.2018.02.555593
10. Smola M, Vandamme T, Sokolowski A (2008) Nanocarriers as pulmonary drug delivery systems to treat and to diagnose respiratory and non respiratory diseases. Int J Nanomedicine 3(1):1–19. https://www.ncbi.nlm.nih.gov/pmc/articles/PMC2526354
11. Courrier HM, Butz N, Vandamme F (2002) Pulmonary drug delivery systems: recent developments and prospects. Crit Rev Ther Drug Carrier Syst 19(4–5):425–498. https://doi.org/10.1615/CritRevTherDrugCarrierSyst.v19.i45.40
12. Zuo L, Pannell BK, Liu Z (2016) Characterization and redox mechanism of asthma in the elderly. Oncotarget 7:25010–25021. https://doi.org/10.18632/oncotarget.7075
13. Urso DL (2009) Asthma in the elderly. Curr Gerontol Geriatr Res 2009:858415. https://doi.org/10.1155/2009/858415
14. Ahmad N, Ahmad R, Almakhamel Z et al (2020) A comparative pulmonary pharmacokinetic study of budesonide using polymeric nanoparticles targeted to the lungs in treatment of asthma. Artif Cells Nanomed Biotechnol 48:749–762. https://doi.org/10.1080/21691401.2020.1748640
15. Chan Y, Ng SW, Chellappan DK et al (2020) Celastrol-loaded liquid crystalline nanoparticles as an anti-inflammatory intervention for the treatment of asthma. Int J Polym Mater Polym Biomater 3:1–10. https://doi.org/10.1080/00914037.2020.1765350
16. Chattopadhyay P, Pathak MP, Patowary P et al (2020) Synthesized atropine nanoparticles ameliorate airway hyperreactivity and remodeling in a murine model of chronic asthma. J Drug Deliv Sci Technol 56:101507. https://doi.org/10.1016/j.jddst.2020.101507
17. Alberca-Custodio RW, Faustino LD, Gomes E et al (2020) Allergen-specific immunotherapy with liposome containing CpG-ODN in murine model of asthma relies on MyD88 signaling in dendritic cells. Front Immunol 11:692. https://doi.org/10.3389/fimmu.2020.00692
18. Chakraborty S, Ehsan I, Mukherjee B et al (2019) Therapeutic potential of andrographolide-loaded nanoparticles on a murine asthma model. Nanomed Nanotechnol Biol Med 20:102006. https://doi.org/10.1016/j.nano.2019.04.009
19. Cherk Yong DO, Saker SR, Wadhwa R et al (2019) Preparation, characterization and in-vitro efficacy of quercetin loaded liquid crystalline nanoparticles for the treatment of asthma. J Drug Deliv Sci Technol 54:101297. https://doi.org/10.1016/j.jddst.2019.101297
20. She W, Mei Z, Zhao H et al (2019) Nebulized inhalation of anti-nerve growth factor microspheres inhibits airway remodeling in an ovalbumin-induced rat asthma model. J Aerosol Med Pulm Drug Deliv 32:70–77. https://doi.org/10.1089/jamp.2018.1453
21. Feng Y, Hong Z, Wang K et al (2018) Oral delivery of bavachinin-loaded PEG-PLGA nanoparticles for asthma treatment in a murine model. Artic J Biomed Nanotechnol 14:1806–1815. https://doi.org/10.1166/jbn.2018.2618

22. Kim D-E, Medicine S, Lee Y et al (2017) Bilirubin nanoparticles ameliorate allergic lung inflammation in a mouse model of asthma. Biomaterials 140:37–44. https://doi.org/10.1016/j.biomaterials.2017.06.014
23. Wang W, Zhu R, Xie Q et al (2012) Enhanced bioavailability and efficiency of curcumin for the treatment of asthma by its formulation in solid lipid nanoparticles. Int J Nanomedicine 7:3667–3677. https://doi.org/10.2147/IJN.S30428
24. Matsuo Y, Ishihara T, Ishizaki J et al (2009) Effect of betamethasone phosphate loaded polymeric nanoparticles on a murine asthma model. Cell Immunol 260:33–38. https://doi.org/10.1016/j.cellimm.2009.07.004
25. Xiao X, Zeng X, Zhang X et al (2013) Effects of Caryota mitis profilin-loaded PLGA nanoparticles in a murine model of allergic asthma. Int J Nanomedicine 8:4553–4562. https://doi.org/10.2147/IJN.S51633
26. Oyarzun-Ampuero FA, Brea J, Loza MI et al (2009) Chitosan-hyaluronic acid nanoparticles loaded with heparin for the treatment of asthma. Int J Pharm 381:122–129. https://doi.org/10.1016/j.ijpharm.2009.04.009
27. Sreeharsha N, Venugopala KN, Nair AB et al (2019) An efficient, lung-targeted, drug-delivery system to treat asthma via microparticles. Drug Des Devel Ther 13:4389–4403. https://doi.org/10.2147/DDDT.S216660
28. Passi M, Shahid S, Chockalingam S et al (2020) Conventional and nanotechnology based approaches to combat chronic obstructive pulmonary disease: implications for chronic airway diseases. Int J Nanomedicine 15:3803–3826. https://doi.org/10.2147/IJN.S242516
29. Ainali NM, Xanthopoulou E, Michailidou G et al (2020) Microencapsulation of fluticasone propionate and salmeterol xinafoate in modified chitosan microparticles for release optimization. Molecules 25:3888. https://doi.org/10.3390/molecules25173888
30. Xu X, Huang H, Yin X et al (2020) Effect of lentivirus-mediated CFTR overexpression on oxidative stress injury and inflammatory response in the lung tissue of COPD mouse model. Biosci Rep 40(1):1–12. https://doi.org/10.1042/BSR20193667
31. Li Z, Luo G, Hu W-P et al (2020) Mediated drug release from nano-vehicles by black phosphorus quantum dots for efficient therapy of chronic obstructive pulmonary disease. Angew Chemie Int Ed 59(46):20568–20576. https://doi.org/10.1002/anie.202008379
32. Mohamed A, Kunda NK, Ross K et al (2019) Polymeric nanoparticles for the delivery of miRNA to treat Chronic Obstructive Pulmonary Disease (COPD). Eur J Pharm Biopharm 136:1–8. https://doi.org/10.1016/j.ejpb.2019.01.002
33. Amore E, Manca ML, Ferraro M et al (2019) Salmeterol Xinafoate (SX) loaded into mucoadhesive solid lipid microparticles for COPD treatment. Int J Pharm 562:351–358. https://doi.org/10.1016/j.ijpharm.2019.03.059
34. Chikuma K, Arima K, Asaba Y et al (2019) The potential of lipid-polymer nanoparticles as epigenetic and ROS control approaches for COPD. Free Radic Res 1:1–12. https://doi.org/10.1080/10715762.2019.1696965
35. Amore E, Manca ML, Gjomarkaj M (2017) Mucoadhesive solid lipid microparticles for controlled release of a corticosteroid in the chronic obstructive pulmonary disease treatment. Nanomedicine 12:2287–2302. https://doi.org/10.2217/nnm-2017-0072
36. Al FA, Shaik AS, Afzal S et al (2014) MR imaging and targeting of a specific alveolar macrophage subpopulation in LPS-induced COPD animal model using antibody-conjugated magnetic nanoparticles. Int J Nanomedicine 9:1491–1503. https://doi.org/10.2147/IJN.S59394
37. Awasthi R, Rathbone MJ, Hansbro PM et al (2018) Therapeutic prospects of microRNAs in cancer treatment through nanotechnology. Drug Deliv Transl Res 8:97–110
38. Huang CY, Ju DT, Chang CF et al (2017) A review on the effects of current chemotherapy drugs and natural agents in treating non-small cell lung cancer. Biomed 7:12–23. https://doi.org/10.1051/bmdcn/2017070423
39. Sharma P, Mehta M, Dhanjal DS et al (2019) Emerging trends in the novel drug delivery approaches for the treatment of lung cancer. Chem Biol Interact 309:108720. https://doi.org/10.1016/j.cbi.2019.06.033

40. Badrzadeh F, Rahmati-Yamchi M, Badrzadeh K et al (2016) Drug delivery and nanodetection in lung cancer. Artif Cells Nanomed Biotechnol 44:618–634. https://doi.org/10.3109/21691401.2014.975237
41. Almuqbil RM, Heyder RS, Bielski ER et al (2020) Dendrimer conjugation enhances tumor penetration and efficacy of doxorubicin in extracellular matrix-expressing 3D lung cancer models. Mol Pharm 17:1648–1662. https://doi.org/10.1021/acs.molpharmaceut.0c00083
42. Bai SB, Cheng Y, Liu DZ et al (2020) Bone-targeted PAMAM nanoparticle to treat bone metastases of lung cancer. Nanomedicine 15:833–849. https://doi.org/10.2217/nnm-2020-0024
43. Chen Y-J, Wang Z-W, Lu T-L et al (2020) The synergistic anticancer effect of dual drug-(cisplatin/epigallocatechin gallate) loaded gelatin nanoparticles for lung cancer treatment. Hindawi J Nanomater 2020:15. https://doi.org/10.1155/2020/9181549
44. Cheng F, Peng X, Meng G et al (2020) Poly(ester-thioether) microspheres co-loaded with erlotinib and α-tocopheryl succinate for combinational therapy of non-small cell lung cancer. J Mater Chem B 8:1728–1738. https://doi.org/10.1039/c9tb02840d
45. Çoban Ö, Barut B, Yalçın CÖ et al (2020) Development and in vitro evaluation of BSA-coated liposomes containing Zn (II) phthalocyanine-containing ferrocene groups for photodynamic therapy of lung cancer. J Organomet Chem 925:121469. https://doi.org/10.1016/j.jorganchem.2020.121469
46. Guo Y, Zhu Q, Shi J et al (2020) Doxorubicin loaded liposomes for lung tumor targeting: in vivo cell line study in human lung cancer. JBUON 25:1504–1511
47. Hu Y, Zhang Y, Wang X et al (2020) Treatment of lung cancer by peptide-modified liposomal irinotecan endowed with tumor penetration and NF-κB inhibitory activities. Mol Pharm 17 (10):3685–3695. https://doi.org/10.1021/acs.molpharmaceut.0c00052
48. Iyer R, Nguyen T, Padanilam D et al (2020) Glutathione-responsive biodegradable polyurethane nanoparticles for lung cancer treatment. J Control Release 321:363–371. https://doi.org/10.1016/j.jconrel.2020.02.021
49. Jin X, Yang Q, Cai N, Zhang Z (2020) A cocktail of betulinic acid, parthenolide, honokiol and ginsenoside Rh2 in liposome systems for lung cancer treatment. Nanomedicine 1:41–54. https://doi.org/10.2217/nnm-2018-0479
50. Karpuz M, Silindir-Gunay M, Kursunel MA et al (2020) Design and in vitro evaluation of folate-targeted, co-drug encapsulated theranostic liposomes for non-small cell lung cancer. J Drug Deliv Sci Technol 57:101707. https://doi.org/10.1016/j.jddst.2020.101707
51. Reczyńska K, Marszałek M, Zarzycki A et al (2020) Superparamagnetic iron oxide nanoparticles modified with silica layers as potential agents for lung cancer treatment. Nano 10:1076. https://doi.org/10.3390/nano10061076
52. Tie Y, Zheng H, He Z et al (2020) Targeting folate receptor β positive tumor-associated macrophages in lung cancer with a folate-modified liposomal complex. Signal Transduct Target Ther 5(6). https://doi.org/10.1038/s41392-020-0115-0
53. Vaidya B, Kulkarni NS, Shukla SK et al (2020) Development of inhalable quinacrine loaded bovine serum albumin modified cationic nanoparticles: repurposing quinacrine for lung cancer therapeutics. Int J Pharm 577:118995. https://doi.org/10.1016/j.ijpharm.2019.118995
54. Zhang C, Zhang S, Zhi D et al (2020) Co-delivery of paclitaxel and survivin siRNA with cationic liposome for lung cancer therapy. Colloids Surf A Physicochem Eng Asp 585:124054. https://doi.org/10.1016/j.colsurfa.2019.124054
55. Mottaghitalab F, Kiani M, Farokhi M et al (2017) Targeted Delivery System Based on Gemcitabine-Loaded Silk Fibroin Nanoparticles for Lung Cancer Therapy. ACS Appl Mater Interfaces 9:31600–31611. https://doi.org/10.1021/acsami.7b10408
56. Li W, Ding D, Chen Z et al (2018) Cold atmospheric plasma and iron oxide-based magnetic nanoparticles for synergetic lung cancer therapy. Free Redical Biol Med 130:71–81. https://doi.org/10.1016/j.freeradbiomed.2018.10.429
57. Anabousi S, Bakowsky U, Schneider M et al (2006) In vitro assessment of transferrin-conjugated liposomes as drug delivery systems for inhalation therapy of lung cancer. Eur J Pharm Sci 29:367–374. https://doi.org/10.1016/j.ejps.2006.07.004

58. Arthur P, Patel N, Surapaneni SK et al (2020) Targeting lung cancer stem cells using combination of Tel and Docetaxel liposomes in 3D cultures and tumor xenografts. Toxicol Appl Pharmacol 401:115112. https://doi.org/10.1016/j.taap.2020.115112
59. Zhang H, Liang Z, Zhang J et al (2020) Zinc oxide nanoparticle synthesized from Euphorbia fischeriana root inhibits the cancer cell growth through modulation of apoptotic signaling pathways in lung cancer cells. Arab J Chem 13:6174–6183. https://doi.org/10.1016/j.arabjc.2020.05.020
60. Lee R, Choi YJ, Jeong MS et al (2020) Hyaluronic acid-decorated glycol chitosan nanoparticles for pH-sensitive controlled release of doxorubicin and celecoxib in nonsmall cell lung cancer. Bioconjug Chem 31:923–932. https://doi.org/10.1021/acs.bioconjchem.0c00048
61. Li N, Mai Y, Liu Q et al (2020) Docetaxel-loaded D-α-tocopheryl polyethylene glycol-1000 succinate liposomes improve lung cancer chemotherapy and reverse multidrug resistance. Drug Deliv Transl Res 12:1–1. https://doi.org/10.1007/s13346-020-00720-9
62. Parvathaneni V, Kulkarni NS, Shukla SK et al (2020) Systematic development and optimization of inhalable pirfenidone liposomes for non-small cell lung cancer treatment. Pharmaceutics 12:206. https://doi.org/10.3390/pharmaceutics12030206
63. Zhong Q, Bielski ER, Rodrigues LS et al (2016) Conjugation to poly(amidoamine) dendrimers and pulmonary delivery reduce cardiac accumulation and enhance antitumor activity of doxorubicin in lung metastasis. Mol Pharm 13:2363–2375. https://doi.org/10.1021/acs.molpharmaceut.6b00126
64. Shen Y, Li W (2018) HA/HSA co-modified erlotinib–albumin nanoparticles for lung cancer treatment. Drug Des Devel Ther 12:2285–2292. https://doi.org/10.2147/DDDT.S169734
65. Thannickal VJ, Toews GB, White ES et al (2004) Mechanisms of pulmonary fibrosis. Annu Rev Med 55:395–417. https://doi.org/10.1146/annurev.med.55.091902.103810
66. Meltzer EB, Noble PW (2008) Idiopathic pulmonary fibrosis. Orphanet J Rare Diseases 3:8. https://doi.org/10.1186/1750-1172-3-8
67. Ghodake V, Vishwakarma J, Vavilala SL, Patravale V (2020) Cefoperazone sodium liposomal formulation to mitigate P. aeruginosa biofilm in Cystic fibrosis infection: a QbD approach. Int J Pharm 587:119696. https://doi.org/10.1016/j.ijpharm.2020.119696
68. Bilton D, Pressler T, Fajac I et al (2020) Amikacin liposome inhalation suspension for chronic Pseudomonas aeruginosa infection in cystic fibrosis. J Cyst Fibros 19:284–291. https://doi.org/10.1016/j.jcf.2019.08.001
69. Tang J, Li JM, Li G et al (2017) Spermidine-mediated poly(Lactic-co-glycolic acid) nanoparticles containing fluorofenidone for the treatment of idiopathic pulmonary fibrosis. Int J Nanomedicine 12:6687–6704. https://doi.org/10.2147/IJN.S140569
70. Günday Türeli N, Torge A, Juntke J et al (2017) Ciprofloxacin-loaded PLGA nanoparticles against cystic fibrosis P. aeruginosa lung infections. Eur J Pharm Biopharm 117:363–371. https://doi.org/10.1016/j.ejpb.2017.04.032
71. Abnoos M, Mohseni M, Mousavi SAJ et al (2018) Chitosan-alginate nano-carrier for transdermal delivery of pirfenidone in idiopathic pulmonary fibrosis. Int J Biol Macromol 118:1319–1325. https://doi.org/10.1016/j.ijbiomac.2018.04.147
72. Seo J, Lee C, Hwang HS et al (2016) Therapeutic advantage of inhaled tacrolimus-bound albumin nanoparticles in a bleomycin-induced pulmonary fibrosis mouse model. Pulm Pharmacol Ther 36:53–61. https://doi.org/10.1016/j.pupt.2016.01.001
73. Saghir SAM, Al-Gabri NA, Khafaga AF et al (2019) Thymoquinone-PLGA-PVA nanoparticles ameliorate bleomycin-induced pulmonary fibrosis in rats via regulation of inflammatory cytokines and iNOS signaling. Animals 9:951. https://doi.org/10.3390/ani9110951
74. Lee C, Seo J, Hwang HS et al (2016) Treatment of bleomycin-induced pulmonary fibrosis by inhaled tacrolimus-loaded chitosan-coated poly(lactic-co-glycolic acid) nanoparticles. Biomed Pharmacother 78:226–233. https://doi.org/10.1016/j.biopha.2016.01.027
75. Huang J, Huang J, Ning X et al (2020) CT/NIRF dual-modal imaging tracking and therapeutic efficacy of transplanted mesenchymal stem cells labeled with Au nanoparticles in silica-induced pulmonary fibrosis. J Mater Chem B 8:1713–1727. https://doi.org/10.1039/c9tb02652e

76. Tassali N, Bianchi A, Lux F et al (2016) MR imaging, targeting and characterization of pulmonary fibrosis using intra-tracheal administration of gadolinium-based nanoparticles. Contrast Media Mol Imaging 11:396–404. https://doi.org/10.1002/cmmi.1703
77. Trivedi R, Redente EF, Thakur A et al (2012) Local delivery of biodegradable pirfenidone nanoparticles ameliorates bleomycin-induced pulmonary fibrosis in mice. Nanotechnology 23. https://doi.org/10.1088/0957-4484/23/50/505101
78. Zhang G, Mo S, Fang B et al (2019) Pulmonary delivery of therapeutic proteins based on zwitterionic chitosan-based nanocarriers for treatment on bleomycin-induced pulmonary fibrosis. Int J Biol Macromol 133:58–66. https://doi.org/10.1016/j.ijbiomac.2019.04.066
79. Farber HW, Loscalzo J (2004) Mechanisms of disease: pulmonary arterial hypertension. N Engl J Med 351:1655–1665. https://doi.org/10.1056/NEJMra035488
80. Chan SY, Loscalzo J (2009) Pathogenic mechanisms of pulmonary arterial hypertension. J Mol Cell Cardiol 44(1):14–30. https://doi.org/10.1016/j.yjmcc.2007.09.006
81. Li Z, Qiao W, Wang C et al (2020) DPPC-coated lipid nanoparticles as an inhalable carrier for accumulation of resveratrol in the pulmonary vasculature, a new strategy for pulmonary arterial hypertension treatment. Drug Deliv 27:736–744. https://doi.org/10.1080/10717544.2020.1760962
82. Nafee N, Makled S, Boraie N (2018) Nanostructured lipid carriers versus solid lipid nanoparticles for the potential treatment of pulmonary hypertension via nebulization. Eur J Pharm Sci 125:151–162. https://doi.org/10.1016/j.ejps.2018.10.003
83. Mohamed NA, Saleh HA, Kameno Y et al (2019) Metal-organic framework (MOF) nanomedicine preparations of sildenafil designed for the future treatment of pulmonary arterial hypertension. BioRxiv 1:718478. https://doi.org/10.1101/718478
84. Gupta N, Ibrahim HM, Ahsan F (2014) Peptide-micelle hybrids containing fasudil for targeted delivery to the pulmonary arteries and arterioles to treat PAH. J Pharm Sci 103:3743–3753. https://doi.org/10.1002/jps.24193
85. Lee Y, Pai SB, Bellamkonda RV et al (2018) Cerivastatin nanoliposome as a potential disease modifying approach for the treatment of pulmonary arterial hypertension. J Pharmacol Exp Ther 366:66–74. https://doi.org/10.1124/jpet.118.247643
86. Shahin HI, Vinjamuri BP, Mahmoud AA et al (2019) Design and evaluation of novel inhalable sildenafil citrate spray-dried microparticles for pulmonary arterial hypertension. J Control Release 302:126–139. https://doi.org/10.1016/j.jconrel.2019.03.029
87. Teymouri Rad R, Dadashzadeh S, Vatanara A et al (2019) Tadalafil nanocomposites as a dry powder formulation for inhalation, a new strategy for pulmonary arterial hypertension treatment. Eur J Pharm Sci 133:275–286. https://doi.org/10.1016/j.ejps.2019.04.001
88. Ravimohan S, Kornfeld H, Weissman D, Bisson GP (2018) Tuberculosis and lung damage: from epidemiology to pathophysiology. Eur Respir Rev 27:147. https://doi.org/10.1183/16000617.0077
89. Banerjee S, Roy S, Bhaumik KN, Pillai J (2020) Mechanisms of the effectiveness of lipid nanoparticle formulations loaded with anti-tubercular drugs combinations toward overcoming drug bioavailability in tuberculosis. J Drug Target 28:55–69. https://doi.org/10.1080/1061186X.2019.1613409
90. Sreeramareddy CT, Panduru KV, Menten J, Van Den Ende J (2009) Time delays in diagnosis of pulmonary tuberculosis: a systematic review of literature. BMC Infect Dis 9:91. https://doi.org/10.1186/1471-2334-9-91
91. Pinheiro M, Ribeiro R, Vieira A et al (2016) Design of a nanostructured lipid carrier intended to improve the treatment of tuberculosis. Drug Des Devel Ther 10:2467–2475. https://doi.org/10.2147/DDDT.S104395
92. Truzzi E, Nascimento TL, Iannuccelli V et al (2020) In vivo biodistribution of respirable solid lipid nanoparticles surface-decorated with a mannose-based surfactant: a promising tool for pulmonary tuberculosis treatment? Nano 10:568. https://doi.org/10.3390/nano10030568

93. Vieira ACC, Magalhães J, Rocha S et al (2017) Targeted macrophages delivery of rifampicin-loaded lipid nanoparticles to improve tuberculosis treatment. Nanomedicine 12:2721–2736. https://doi.org/10.2217/nnm-2017-0248
94. Mokhtarzadeh A, Mohammadi A, Hamishehkar H et al (2019) Ethambutol-loaded solid lipid nanoparticles as dry powder inhalable formulation for tuberculosis therapy. AAPS PharmSciTech 20:120. https://doi.org/10.1208/s12249-019-1334-y
95. Zargarnezhad S, Gholami A, Khoshneviszadeh M et al (2020) Antimicrobial activity of isoniazid in conjugation with surface-modified magnetic nanoparticles against mycobacterium tuberculosis and nonmycobacterial microorganisms. J Nanomater 2020:7372531. https://doi.org/10.1155/2020/7372531
96. Jahagirdar PS, Gupta PK, Kulkarni SP, Devarajan PV (2020) Intramacrophage delivery of dual drug loaded nanoparticles for effective clearance of mycobacterium tuberculosis. J Pharm Sci 109:2262–2270. https://doi.org/10.1016/j.xphs.2020.03.018
97. Abdelghany S, Parumasivam T, Pang A et al (2019) Alginate modified-PLGA nanoparticles entrapping amikacin and moxifloxacin as a novel host-directed therapy for multidrug-resistant tuberculosis. J Drug Deliv Sci Technol 52:642–651. https://doi.org/10.1016/j.jddst.2019.05.025
98. El-Mageed HRA, Mustafa FM, Abdel-Latif MK (2020) Boron nitride nanoclusters, nanoparticles and nanotubes as a drug carrier for isoniazid anti-tuberculosis drug, computational chemistry approaches. J Biomol Struct Dyn 28:1–10. https://doi.org/10.1080/07391102.2020.1814871

Plant-Based Chemical Moieties for Targeting Chronic Respiratory Diseases 34

Yinghan Chan, Sin Wi Ng, Kamal Dua, and Dinesh Kumar Chellappan

Abstract

The World Health Organization (WHO) has reported that chronic respiratory diseases such as asthma, chronic obstructive pulmonary disease, interstitial lung disease, and lung cancer are among the major chronic human diseases that posed a huge challenge to public health and socioeconomic growth. Pharmacotherapy is crucial in the management of these diseases; however, the utilization of conventional treatments is found to be futile, as most patients remained poorly controlled with low quality of life. This has prompted the discovery and development of novel therapeutic agents to improve treatment outcomes. Over the years, researchers have studied a vast range of natural products for their potential in managing chronic respiratory diseases. It has been demonstrated that chemical moieties obtained from plant sources improved pharmacokinetic and toxicological profiles, with a robust multi-prolonged action. Hence, they are held in high regard as possible replacements to address the limitations faced by current therapies. In this chapter, such a phytochemical approach with respect to their

Y. Chan (✉)
School of Pharmacy, International Medical University, Kuala Lumpur, Malaysia

Department of Pharmacology, Faculty of Medicine, University of Malaya, Kuala Lumpur, Malaysia

S. W. Ng
School of Pharmacy, International Medical University, Kuala Lumpur, Malaysia

Cancer Research Malaysia, Subang Jaya, Selangor, Malaysia

K. Dua
Discipline of Pharmacy, Graduate School of Health, University of Technology Sydney, Sydney, NSW, Australia

D. K. Chellappan
Department of Life Sciences, School of Pharmacy, International Medical University, Kuala Lumpur, Malaysia

K. Dua et al. (eds.), *Targeting Cellular Signalling Pathways in Lung Diseases*,
https://doi.org/10.1007/978-981-33-6827-9_34

molecular mechanisms targeting signalling pathways involved in various chronic respiratory diseases will be discussed. We have also summarized some of the experimental evidence that supports the use of plant-based chemical moieties in chronic respiratory diseases.

Keywords

Plants · Herbs · Natural products · Molecular mechanisms · Chronic respiratory disease

34.1 Introduction

Chronic respiratory diseases (CRDs) such as asthma, chronic obstructive pulmonary disease (COPD), cystic fibrosis, and lung cancer are the most common inflammatory diseases and notable causes of death worldwide, posing significant and enormous economic and social burden in terms of the cost incurred with treatment, hospitalization, as well as reduced productivity. The burden induced by such diseases has also severely affected the livelihood of suffering individuals [1, 2]. According to a report by the World Health Organization (WHO), COPD alone has exhibited three million annual mortalities, whereas approximately 334 million people, which includes 14% children, are suffering from asthma. Besides, pulmonary tuberculosis has also claimed the lives of 1.4 million people annually. In addition, lung cancer too represents a fatal neoplasm that claimed more than 2 million lives each year [1, 3, 4]. To make things worse, the prevalence of these CRDs is still growing rapidly, and it is predicted that a huge number of global populations will be shredded in the near future due to CRDs. Inflammatory CRDs are further exacerbated by several prime mediators, such as worsening air pollution, increased exposure to occupational allergens, and a growing population of smokers [5–7].

Typically, conventional management strategies rely on the use of therapeutic drugs, whereby research and medical advancements have led to the development of a wide range of chemical compounds and biomolecules for treating CRDs. Nevertheless, these efforts have been futile as most of the therapeutic drugs are unable to fully cure CRDs [8–10]. For instance, controlling symptoms is the only management option in asthma. In certain cases where patients are not adequately controlled, they may require therapeutic intervention using more than one type of drug, thereby exposing them to a greater risk of adverse reactions. These include the undesirable development of dyspepsia, tremors, dizziness, and sour throat as observed in the prolonged use of corticosteroids and bronchodilators [11, 12]. Likewise, in COPD, although current treatment options such as leukotriene antagonists and phosphodiesterase 4 (PDE4) inhibitors demonstrated a certain extent of efficacies, they are unable to reverse the pathophysiological process of the disease and have no significant positive effects on mortality rate and disease progression [13, 14]. Thus, there is a pressing need for the development of novel safe and effective therapeutics that could be utilized as alternative treatment options to combat CRDs, as well as to complement existing conventional therapeutics for enhancing treatment outcomes in patients with CRDs.

Plants and herbal-based remedies have been widely employed as therapeutic tools for a variety of diseases due to their extended spectrum of pharmacological effects and improved toxicological profiles [15, 16]. As reported by the WHO, 80% of the world population utilizes phyto-therapeutic agents for their primary health-related needs, whereas 11% of marketed drugs are formulated using natural plant-based moieties [17]. In terms of CRDs, several plant extracts such as *Tripterygium wilfordii*, *Echinacea purpurea*, and *Curcurma longa* contain active constituents, namely, flavonoids, iridoids, and triterpenoids, whereby several studies have proven that these active constituents exert complex mechanisms of action that targets the cellular pathways of CRDs [1]. In this chapter, we provide an insight into the potential and challenges of various plant-based chemical moieties in the management of CRDs, with particular focus on asthma, COPD, and lung cancer.

34.2 Plant-Based Chemical Moieties in the Management of Chronic Respiratory Diseases: From Bench to Bedside

The prevalence of CRDs has been rapidly increasing and has affected more than millions of people around the globe. Such a growing threat of CRDs to public health has urged medical researchers to design and develop efficient and safe novel therapeutic agents with regard to the present scenario. Even though current conventional agents demonstrated a certain extent of potent efficacy, the therapeutic outcomes remained unsatisfactory, as most of them are linked with certain clinical consequences, such as high treatment costs, limited effectiveness, as well as side effects [3, 8, 18–20]. Therefore, researchers have shifted their attention to the usage of natural sources for the development of novel therapeutics in the management of CRDs.

Bioactive compounds derived from plants and herbal origins have been an integral part of culture and history, whereby plant products have been used as the backbone of traditional healing systems since the inception of human beings [21]. Although many plants were documented in the early years of having remarkable medicinal properties and were used for treating various pathological conditions, the use of traditional medicine has been overshadowed by modern therapeutics as the means for managing human diseases [21, 22]. Nonetheless, over recent years, there has been a substantial increase in the use of medicinal plants for the promotion of health and management of diseases in many countries. In contrary to the traditional use of plant-based therapies which utilizes unmodified whole plant for preserving the original composition as well as the integrity of the source plant, modern therapeutics are focused on identifying biologically active compounds from plant extracts and extracting them for specific therapeutic applications [23, 24]. For this instance, numerous studies have proven that plant-based chemical moieties are effective in targeting various cell signalling pathways for combating human diseases. As they are naturally occurring, they also exhibited lower systemic toxicities as compared to chemically synthesized compounds. In addition, the cost of conventional drugs is usually more expensive in comparison to plant-based products. Therefore, the use of conventional therapies for patients residing in rural areas may be unaffordable [25, 26]. Subsequently, a paradigm shift from the use of chemical drugs towards the application of phytochemicals in the management of CRDs was observed.

Throughout these years, a diversity of natural and herbal products has been discovered and explored for their potential in managing CRDs (Table 34.1). They have attracted considerable interest as novel therapeutic agents in CRDs due to the structural diversity of compounds present in plants, as well as their ability to target multiple biological pathways that modulate the complex pathogenesis of these diseases. Typically, the pharmacological activities of these plants are attributed to the presence of bioactive constituents from various classes, such as alkaloids, flavonoids, iridoids, and triterpenoids [27, 28]. The following section offers a brief overview of the pathogenesis of various CRDs and how plant-based chemical moieties can help manage CRDs by targeting the underlying molecular mechanisms of these diseases, evidenced by some of the latest studies conducted in this field of research. The findings are compiled in Table 34.1, along with several other examples of plant-based chemical moieties that presented potential anti-CRD effects but are not discussed in the following text.

34.2.1 Asthma

34.2.1.1 Overview of Asthma

Asthma is a disorder of the airways characterized by chronic airway inflammation, tissue remodeling, as well as declining airway function. It is one of the most common CRDs and affects more than 339 million people worldwide, yet its pathophysiology remained poorly understood. Indoor pollutants and outdoor allergens, tobacco smoke, and air pollution are among the major risk factors for asthma [29, 30]. Asthma can manifest at any age, and the common symptoms are coughing, wheezing, and dyspnoea. It is a complex disease that could manifest as 'episodic', referring to the periods where symptoms appear and resolve upon therapy, or as 'persistent', which can be identified by the presence of characteristic asthma clinical symptoms [31].

An in-depth understanding of asthma pathogenesis is essential for the effective management of the disease. Typically, most cases of asthma are linked with the sensitization of airways to allergens. Upon exposure to these allergens, it produces a cascade of events which eventually lead to chronic airway inflammation. T helper cell (Th) type-2 (Th2) immune responses induce the release of pro-inflammatory cytokines, such as interleukin (IL)-4, IL-5, IL-9, as well as IL-13 [32, 33]. Among these, IL-4 is responsible for the hyperproduction of immunoglobulin E (IgE) via IgE switch in B cells. IgE-mediated activation of airway mast cells then leads to rapid airway narrowing, followed by the release of rapidly acting granule-associated preformed mediators tryptase and histamine [19, 34]. On the other hand, IL-5 is involved in the recruitment of eosinophils that results in the development of allergic rhinitis in the upper respiratory airways. Along with prostaglandin D_2, cysteinyl

Table 34.1 Summary of the therapeutic effects and mechanisms of action of various plant-based chemical moieties against chronic respiratory diseases

Plant	Active constituent (s)/extract(s)	Chronic respiratory disease(s)	Model(s)	Mechanisms of action/therapeutic effects	Reference
Allium sativum	Diallyl disulfide	COPD	In vivo emphysema model	– Inhibits the activity of NF-κB and activates the NRF2 pathway – Downregulates the expressions of MMP-9 and TIMP-1	[109]
Alstonia scholaris	Total alkaloids	Asthma, COPD	In vivo LPS-induced airway inflammation model	– Decreases the percentage of neutrophils – Inhibits the production of TNF-α and IL-8.	[110]
Andrographis paniculata	Andrographolide	COPD	In vivo model of inflammation	– Reduces lung cellular infiltration – Decreases lung levels of TNF-α, IL-1β, MMP-8, and MMP-9 – Increases activation of NRF2 and decreases the function of Keap1 – Enhances gene expression of antioxidant enzymes such as HO-1	[111]
Artemisia annua L.	Artemisinin	Lung cancer	In vitro NSCLC cell lines and in vivo NSCLC model	– Suppresses cell viability and relative tumour growth – Activates caspases-3, -8, and -9, stimulating apoptotic pathways – Phosphorylates histone H2AX, inducing DNA damage	[112]
Astragalus membranaceus	Astralagus extract	Asthma	In vivo asthmatic model	– Attenuates lung inflammation, goblet cell hyperplasia, and airway hyperresponsiveness – Decreases total eosinophils and lymphocytes – Reduces expressions of Th2 cytokines	[113]

(continued)

Table 34.1 (continued)

Plant	Active constituent (s)/extract(s)	Chronic respiratory disease(s)	Model(s)	Mechanisms of action/therapeutic effects	Reference
				– Inhibits NF-κB and its translocation	
	Astragaloside IV	Asthma	In vivo asthmatic model	– Relieves airway hyperresponsiveness and reduces inflammatory cells – Decreases IL-4 and IL-5 and increases IFN-γ – Inhibits the synthesis of GATA-3 encoding mRNA – Increases the synthesis of T-bet encoding mRNA	[114]
Azadirachta indica	Leaves extract	Lung cancer	In vivo mice model	– Normalizes the balance between carcinogen activation and detoxification.	[115]
Berberis vulgaris	Berberine	COPD	In vivo model of airway inflammation	– Inhibits ERK and p38 signalling proteins – Decreases levels of TNF-α, IL-1β, MCP-1, and MUC5AC – Attenuates goblet cell hyperplasia, airway inflammation, and mucus hypersecretion	[116]
Brucea javanica	Bruceine D	Lung cancer	In vitro NSCLC cells	– Suppresses proliferation of wild-type NSCLC cells in a dose- and time-dependent manner – Decreases colony-forming ability and migration of tumour cells – Effectively induces apoptosis that is associated with G0-G1 cell	[102]

				cycle arrest, accumulation of intracellular ROS, and disruption of mitochondrial membrane potential – Suppresses the expressions of anti-apoptotic proteins Bcl-xL and Bcl-2 – Enhances the expressions of pro-apoptotic proteins Bax and Bak – Inhibits the expressions of pro-caspase-3 and pro-caspase-8	
Callicarpa japonica	Methanol extract	COPD	In vitro model of human pulmonary mucoepidermoid cell line and *in vivo* model of inflammation	– Effectively suppresses the infiltration of neutrophils and mucus hypersecretion – Decreases the production of ROS – Significantly inhibits the production of IL-6 and TNF-α – Decreases the expression of MUC5AC – Attenuates the phosphorylation of ERK	[117]
Camellia sinensis	L-Theanine	Asthma	In vivo asthmatic model	– Inhibits ovalbumin-induced mucus hyperproduction and inflammatory cell infiltration – Decreases the production of IgE, MCP-1, IL-4, IL-5, IL-13, TNF-α, and IFN-γ – Reduces the generation of ROS – Reduces the activation of NF-κB and MMP-9	[118]
	Epigallocatechin-3-gallate	COPD	In vitro model of human ECV304 endothelial cells	– Regulates the expression of MMP-9 via suppression of ROS, NF-κB, and AP-1	[119]

(continued)

Table 34.1 (continued)

Plant	Active constituent (s)/extract(s)	Chronic respiratory disease(s)	Model(s)	Mechanisms of action/therapeutic effects	Reference
Centella asiatica	Asiatic acid	COPD	In vivo lung inflammation model	– Attenuates recruitment of inflammatory cells and mucus overproduction – Decreases the activation of MAPKs and NF-κB – Reduces the expression of MCP-1	[120]
Chilli pepper	Capsaicin	Lung cancer	In vitro NSCLC cells	– Induces p53-mediated degradation of HIF-1α as well as SMAR1-induced downregulation of COX-2 – Restrains HIF-1α nuclear localization – Downregulates VEGF expression	[98]
Cinnamomum chartophyllum	Ethanol extract	COPD	In vitro model of human bronchial epithelial cells	– Activates Nrf2 downstream genes NQO-1 and γ-GCS which enhances nuclear translocation and stabilization of Nrf2	[68]
Citrus fruits	Narigenin	Asthma, COPD	Unspecified	– Inhibits the production of TNF-α, IL-8, and MMP9 – Inhibits the NF-κB pathway by reduction of NF-κB and IκB phosphorylation	[121]

Citrus tachibana	Leaves ethanol extract	Asthma	In vivo asthmatic model	– Reduces airway inflammation, and ovalbumin specific IgE and IgG1 levels – Restores Th1/Th2 homeostasis via increase of TNF-α, IL-4, and IL-6 and decrease of IFN-γ and IL-12 – Inhibits phosphorylation of NF-κB and IκB-α	[122]
Cnidii monnieri	Osthol	Asthma	In vitro model of human bronchial epithelial cell line	– Suppresses the expression of eotaxin in a dose-dependent manner – Suppresses IL-4-induced STAT6 in a dose-dependent manner	[123]
Crataegus pinnatifida	Ursolic acid	COPD	In vivo emphysema model	– Alleviates emphysema-related lung pathology and improves cigarette smoke-induced oxidative stress – Downregulates expression of PERK pathway proteins – Induces the expression of Bcl-2 and inhibits expression of Bax – Upregulates expression of the NRF2 pathway	[76]
Curcuma longa	Curcumin	Lung cancer	In vitro A549 and H1299 lung cancer cell lines	– Decreases viability of human lung cancer cell lines – Suppresses colony formation capacities of tumour cells – Increases cell apoptosis and the number of autophagosomes – Reduces phosphorylation of	[124]

(continued)

Table 34.1 (continued)

Plant	Active constituent (s)/extract(s)	Chronic respiratory disease(s)	Model(s)	Mechanisms of action/therapeutic effects	Reference
				mTOR, ribosomal protein S6, AKT, and phosphoinositide 3-kinase	
		Asthma	In vivo asthmatic model	– Reduces the phosphorylation of JNK, ERK, and p38 via the MAPK pathway modulation – Inhibits the expression of NF-κB	[44]
		COPD	In vivo model of airway inflammation and emphysema	– Upregulates PGC-1α/SIRT3 signalling pathway – Attenuates skeletal muscle mitochondrial impairment – Decreases oxidative stress and inflammation	[125]
Ebenus boissieri Barbey	Hydroalcoholic extract	Lung cancer	In vitro A549 human lung cancer cell line	– Increases activity in all caspases, inducing apoptosis. – Remarkable induction of caspase-3. – Increases release of TNF-α and IFN-γ.	[126]
Echinacea purpurea	*Echinacea* complex	Asthma	In vivo model of allergic asthma	– Significantly reduces the concentration of Th2 cytokines. – Exerts bronchodilatory effect but the exact mechanism unclear.	[127]
Embelia ribes	Embelin	Lung cancer	In vitro A549 lung cancer cell line	– Enhances the levels of phosphorylated p38 and JNK. – Induces the activation of caspase-3. – Induces apoptotic effects.	[128]

Eucalyptus globulus	Eucalyptol	COPD	In vivo lung inflammation model	– Attenuates the activation of p65 subunit of NF-κB. – Reduces total leucocyte and macrophages. – Attenuates cigarette smoke-induced lung histopathological alterations.	[129]
Euterpe oleracea	Açai stone extract	COPD	In vivo lung inflammation model	– Significantly reduces alveolar macrophage, neutrophil numbers, myeloperoxidase, superoxide dismutase, catalase, glutathione peroxidase activity, and nitrite.	[130]
Forsythia suspensa	Forsythiaside	COPD	In vivo lung inflammation model	– Attenuates inflammatory cells infiltration and suppresses the production of TNF-α, IL-1β, and IL-6. – Inhibits NF-κB signalling pathway and upregulates Nrf2 expression.	[69]
Gardenia jasminoides	Genipin	Lung cancer	In vitro human NSCLC cells	– Decreases cell viability through apoptosis. – Induces mitochondrial execution pathway via the activation of caspase-9 and caspase-3, release of cytochrome c, and increase in Bax/Bcl-2 ratio. – Increases phosphorylation of p38 MAPK.	[131]
Ginkgo biloba	Leaves extract	Asthma, COPD	In vivo lung inflammation model	– Inhibits protein leakage and neutrophil infiltration. – Suppresses myeloperoxidase	[132]

(continued)

Table 34.1 (continued)

Plant	Active constituent (s)/extract(s)	Chronic respiratory disease(s)	Model(s)	Mechanisms of action/therapeutic effects	Reference
				activity, lipid peroxidation, and MMP-9 activity. – Inhibits the phosphorylation of NF-κB and prevents the degradation of NF-κB inhibitor, IκB.	
Glossogyne tenuifolia	Glossogin	Lung cancer	In vitro A549 human lung cancer cell line	– Releases cytochrome c and activates caspase-9 and caspase-3. – Decreases protein levels of Bcl-2 and Bcl-xL. – Increases protein expression of Bad.	[133]
Hevea brasiliensis	γ-tocotrienol	COPD	In vivo lung inflammation model	– Reduces neutrophil count and levels of cytokines, chemokines, pro-inflammatory gene expressions, and oxidative damage biomarkers. – Inhibits nuclear translocation of STAT3 and NF-κB. – Upregulates NRF2 in the lungs.	[134]
Houttuynia cordata	Sodium new houttuyfonate	Lung cancer	In vitro NSCLC cells	– Suppresses the expression of Linc00668, inhibiting the migration and invasion of NSCLC cells. – Acts as ceRNA by sponging miR-147a that regulates Slug mRNA levels.	[135]
Hydrastis canadensis	Berberine	Lung cancer	In vitro NSCLC cell lines	– Inhibits AP-2α, AP-2β, and hTERT. – Downregulates expressions of	[136]

				HIF-1α and VEGF. – Inhibits phosphorylation of Erk and Akt. – Triggers the release of cytochrome-c. – Promotes caspase cleavage and PARP, affecting the expression of Bax and Bcl-2.	
Kanahia laniflora	Methanolic extract	Lung cancer	In vitro A549 human NSCLC cells	– Inhibits proliferation of tumour cells. – Increases the percentage of apoptotic cells.	[137]
Ligusticum chuanxiong	Ligustrazine	Asthma	In vivo asthmatic model	– Reduces influx of neutrophils and eosinophils via the regulation of Tregs/Th17 cells. – Regulates the homeostasis between FOXP3 and RORγT transcription factors. – Regulates the expressions of T-bet and GATA-3	[50]
Magnolia fargesii	Lignans	COPD	In vitro model of airway epithelial cells and in vivo model of inflammation	– Remarkably inhibits both Erk and Akt phosphorylation levels, along with a dose-dependent suppression of EGFR. – Attenuates the infiltration of neutrophils and macrophages. – Suppresses the secretion of TNF-α and IL-6.	[82]
Magnolia officinalis	Magnolol and polyphenol mixture	Lung cancer	Lung cancer A549 xenograft model	– Significantly suppresses class I HDACs, leading to cell apoptosis via cell cycle arrest.	[106]

(continued)

Table 34.1 (continued)

Plant	Active constituent (s)/extract(s)	Chronic respiratory disease(s)	Model(s)	Mechanisms of action/therapeutic effects	Reference
				– Activates pro-apoptotic factors such as DR5, Bax, and cleaved caspase 3. – Reduces tumour growth.	
Moringa oleifera	β-sitosterol	Asthma	In vivo asthmatic model	– Lowers the level of eosinophils and neutrophils. – Suppresses TNF-α, IL-4, and IL-5. – Protects airway against ovalbumin-induced lung tissue histopathological changes.	[138]
Panax ginseng	Ginsan	Asthma	In vivo asthmatic model	– Reduces airway hyperresponsiveness, remodelling, and eosinophilia. – Decreases IL-5 level. – Upregulates mRNA expressions of COX-1, COX-2, and PGE2.	[139]
Pistacia weinmannifolia	Root extract	Asthma	In vitro and in vivo asthmatic models	– Downregulates the MAPK and NF-κB signalling pathways. – Reduces the levels of eosinophils and Th2 cytokines. – Inhibits the influx of inflammatory cells and mucus hypersecretion.	[45]
Platycodon grandiflorum	Platycodin D	COPD	In vivo lung inflammation model	– Attenuates cigarette smoke-induced lung pathological changes. – Reduces inflammatory cell	[140]

				infiltration and the production of TNF-α and IL-1β. – Activates NRF2 signalling pathway and inhibits the activation of NF-κB.	
Polygonum japonicum	Resveratrol	COPD	In vitro model of human bronchial epithelial cells	– Blocks the PI3K/Akt signalling pathway. – Decreases the expressions of MMP-2 and MMP-9.	[141]
Pourthiaea villosa	Baicalin	Asthma	In vitro and in vivo inflammatory models	– Inhibits the activities of PDE4A and PDE4B. – Suppresses the expression of TNF-α. – Reduces the infiltration of inflammatory cells.	[142]
Pseudolysimachion rotundum	Piscroside C	COPD	In vitro model of human bronchial epithelial cells	– Inhibits the TNF-α/NF-κB pathway by suppressing the activity of PKCδ.	[143]
	3-Methoxy-catalposide	Asthma, COPD	In vitro LPS-induced inflammatory model	– Inhibits the expression of COX-2 and iNOS. – Suppresses the release of NO and PGE2. – Reduces the levels of pro-inflammatory genes, such as IL-1β, IL-6, and TNF-α. – Inhibits the activation of Erk1/2, JNK, and p38 MAPKs. – Inhibits nuclear translocation of AP-1 and NF-κB.	[144]
Psoralea corylifolia L.	Psoralen	Asthma	In vivo asthmatic model	– Significantly suppresses the infiltration of inflammatory cells.	[40]

(continued)

Table 34.1 (continued)

Plant	Active constituent (s)/extract(s)	Chronic respiratory disease(s)	Model(s)	Mechanisms of action/therapeutic effects	Reference
				– Decreases mucus secretion. – Inhibits GATA-3 protein expression, thereby reducing Th2 cells response.	
	Bavachinin	Asthma	In vivo asthmatic model	– Inhibits the production of Th2 cytokines, including IL-4, IL-5, and IL-13. – Selectively influences the level of GATA-3 via modulation of GATA-3 mRNA.	[145]
Rabdosia rubescens	Oridonin	Lung cancer	In vitro lung cancer cell lines	– Increases the level of Bax and decreases the level of Bcl-2. – Inhibits the proliferation in a dose-dependent manner.	[146]
Radix et Rhizoma Leonticsi	Taspine	Lung cancer	In vitro A549 cell line and in vivo lung cancer model	– Inhibits tumour cell proliferation and tumour growth. – Inhibits the secretion of VEGF. – Inhibits tube formation and tissue vascularization.	[147]
Rhodiola rosea	Salidroside	COPD	In vivo lung inflammation model	– Inhibits the generation of pro-inflammatory cytokines including TNF-α, IL-1β, and IL-6. – Inhibits the protein levels of the MAPK/NF-κB pathway.	[148]
Rhus chinensis Mill.	Galla Chinensis extract	Acute lung injury, COPD	In vivo model of inflammation	– Reduces the production of MCP-1 in a dose-dependent manner. – Significantly inhibits	[149]

				inflammatory cells infiltration into the lung. – Downregulates TNF-α, IL-6, and MCP-1 mRNA expression in lung tissue.	
Salvia miltiorrhiza	Tanshinone IIA	COPD	In vivo model	– Attenuates decline in lung function, airspace enlargement, mucus hyperproduction, bronchial collagen deposition, as well as cigarette smoke- and LPS-induced inflammatory and oxidative stress responses. – Reduces secretion of IL-6 and IL-8. – Inhibits the activation of Erk1/2 and NF-κB.	[150]
	Salvianolic acid B	COPD	In vivo lung inflammation model	– Inhibits cigarette smoke-induced lung pathological changes. – Inhibits the infiltration of inflammatory cells, as well as the production of TNF-α, IL-1β, IL-6, and MCP-1. – Upregulates the expression of NRF2 and suppresses NF-κB.	[151]
Sanguinaria canadensis	Sanguinarine	Lung cancer	In vitro human A549 cancer cell line	– Suppresses tube formation and migration of tumour cells. – Decreases the secretion of VEGF and inhibits VEGF-mediated activation of Akt and p38.	[97]
Saururus chinensis		Asthma	In vivo asthmatic model	– Significantly lowers the production of IL-4, IL-5, IL-13,	[152]

(continued)

Table 34.1 (continued)

Plant	Active constituent (s)/extract(s)	Chronic respiratory disease(s)	Model(s)	Mechanisms of action/therapeutic effects	Reference
	meso-Dihydroguaiaretic acid			eotaxin, MCP-1, TNF-α, and IgE. – Inhibits ovalbumin-induced infiltration of inflammatory cells and mucus hyperproduction. – Attenuates NF-κB, Erk1/2, and p38 MAPK.	
Schisandra chinensis	Lignans	Cough hypersensitivity syndrome	In vivo model exposed to cigarette smoke	– Attenuates the infiltration of pulmonary neutrophils and total inflammatory cells. – Attenuates TNF-α and IL-8. – Inhibits the expressions of TRPV1 and TRPA1.	[153]
Scutellaria baicalensis	Oroxylin A	Asthma	In vivo asthmatic model	– Attenuates airway hyperresponsiveness and reduces the number of inflammatory cells. – Inhibits the expressions of IL-4, IL-5, IL-13, and IgE. – Inhibits the activation of NF-κB signalling pathway.	[154]
	Skullcapflavone II	Asthma	In vivo asthmatic model	– Reduces airway hyperresponsiveness, airway eosinophilia, and production of Th2 cytokines. – Increases the level of TGF-β1. – Suppresses subepithelial collagen deposition and goblet cell hyperplasia.	[55]

				– Elevates Smad7 expression but suppresses pSmad2/3 expression.	
Scutellariae radix	Wogonin	Asthma	In vivo asthmatic model	– Inhibits allergen-induced eosinophilic inflammation in lung tissues. – Downregulates IL-4 induced STAT6 translocation and activation, thereby suppressing Th2-mediated inflammation.	[41]
Silybum marianum	Silymarin	Lung cancer	In vitro NSCLC cell line	– Suppresses cell migration in a concentration-dependent manner. – Inhibits HDAC activity and decreases levels of class I HDACs. – siRNA knockdown of ZEB1, downregulating the expression of HDACs in NSCLC cells,	[108]
		COPD	In vivo model of airway inflammation	– Attenuates cigarette smoke-induced thickening of airway epithelium and inflammatory cell infiltration. – Decreases total cells, macrophages, and neutrophils. – Decreases the secretion of TNF-α, IL-1β, and IL-8. – Attenuates phosphorylation of ERK and p38 MAPKs.	[155]
	Silibinin	Lung cancer	In vivo model of lung adenocarcinoma	– Decreases both tumour number and tumour size. – Decreases tumour expression of TNF-α and IL-13. – Decreases HIF-1α expression	[156]

(continued)

Table 34.1 (continued)

Plant	Active constituent (s)/extract(s)	Chronic respiratory disease(s)	Model(s)	Mechanisms of action/therapeutic effects	Reference
				and nuclear localization. – Decreases phosphorylation of p65 NF-κB and STAT3 in tumour cells. – Upregulates angiogenic inhibitors Ang-2 and Tie-2.	
Sophora flavescens	Matrine	Asthma	In vivo asthmatic model	– Significantly reduces airway hyperresponsiveness. – Suppresses goblet cell hyperplasia and the infiltration of eosinophils and decreases inflammatory response. – Reduces the level of Th2 cytokines. – Suppresses the production of IgE, pro-inflammatory cytokines, and eotaxins. – Suppresses the expression of ICAM-1.	[157]
Stemona tuberosa	Aqueous extract	Asthma, COPD	In vivo lung inflammation model	– Decreases total macrophages, neutrophils, and lymphocytes. – Reduces the levels of TNF-α and IL-6. – Ameliorates bronchiole epithelial hyperplasia.	[158]
Taraxacum officinale	Taraxasterol	Asthma	In vivo asthmatic model	– Significantly decreases total inflammatory cell count.	[159]

				– Reduces the production of Th2 cytokines, including IL-4, IL-5, and IL-13. – Suppresses the infiltration of inflammatory cells into lung tissues.	
Trichosanthes kirilowii Maxim	Methylene chloride fraction	Lung cancer	Xenograft NSCLC tumour model	– Suppresses proliferation, migration, invasion, tumorigenicity, and tumour growth. – Induces G2-M arrest, necrosis, and apoptosis.	[160]
Tripterygium wilfordii	Celastrol	Asthma	In vitro model of human bronchial epithelial cells	– Suppresses the expression of IL-1β, suggesting that celastrol possesses an anti-inflammatory potential for attenuation of asthma.	[18]
Unspecified	Rutin	Asthma	In vivo *asthmatic model*	– Inhibits cellular infiltration in the airways and suppresses Th2 and Th17 cytokines. – Inhibits airway hyperresponsiveness. – Downregulates NF-κB and iNOS inflammatory pathways. – Inhibits the expression of MMP-9.	[161]
Unspecified	Quercetin	Lung cancer	In vitro HCC827 NSCLC cell line and in vivo NSCLC model	– Inhibits proliferation and anchorage-independent growth. – Suppresses the expression of Src, inhibiting Fn14/NF-κB signalling pathway. – Inhibits growth of solid tumours.	[162]
Unspecified		COPD	In vivo emphysema model	– Reduces the expressions and activities of MMP-9 and MMP-12	[79]

(continued)

Table 34.1 (continued)

Plant	Active constituent (s)/extract(s)	Chronic respiratory disease(s)	Model(s)	Mechanisms of action/therapeutic effects	Reference
				by increasing the expression of SIRT1. – Alleviates lung inflammation and goblet cell metaplasia and inhibits pro-inflammatory cytokines mRNA expressions.	
Viburnum grandiflorum	Plant extract	Lung cancer	In vitro NSCLC cells	– Significantly enhances rate of apoptosis. – Cleaves pro-caspase-8, pro-caspase-9, and pro-caspase-3. – Diminishes protein expression of MCL-1.	[101]
Vitex rotundifolia	Casticin	Asthma, COPD	In vitro model of human pulmonary epithelial cells	– Reduces the levels of TNF-α, IL-6, IL-8, and PGE2. – Suppresses the phosphorylation of Akt, PI3K, and MAPK. – Blocks the translocation of p65, which is the protein subunit of NF-κB.	[83]
Zingiber officinale	Ethanol and aqueous extracts	Asthma	In vivo asthmatic model	– Significantly reduces goblet cell hyperplasia and infiltration of inflammatory cells in airways. – Significantly reduces eosinophils and neutrophils count. – Inhibits Th2-mediated immune response and decreases mRNA expressions of IL-4 and IL-5. – Suppresses serum IgE level.	[94]

leukotrienes, adenosine, and other inflammatory mediators, they act synergistically to promote airway smooth muscle contraction and vascular leakage, resulting in mucus hypersecretion, bronchospasm, and elevated immune cells influx, which are the characteristic symptoms of asthma. Studies have shown that the airway epithelium has a crucial role in modulating Th2 responses via the production of master regulators, such as thymic stromal lymphopoietin, IL-33, and IL-25 [3, 19, 34, 35]. Furthermore, these cytokines and mediators that are produced during the early phase of an immune response can further propagate, leading to the late-phase asthmatic response that is characterized by progressive airway inflammation and hypersensitivity of the bronchus. As a result, patients may experience frequent asthma exacerbations that greatly damages lung function due to severe obstruction of airways [11].

Despite inhaled corticosteroids and long-acting β_2-adrenoceptor agonists being the mainstay of asthma therapeutics, these therapies failed to prevent severe exacerbations or to prevent the progression of the disease. This is because the response to these therapeutics is likely to vary by the characteristics of patients, namely, adherence and techniques of using medications [36, 37]. Besides, most of the drugs currently used for asthma management are also linked with unwanted adverse events. For example, prolonged steroid treatment may lead to systemic side effects, including candidiasis, dysphonia, osteoporosis, and adrenal suppression [8, 35]. In addition, with the increasing severity and chronicity of the disease, fibrosis of the airway wall leads to permanent narrowing of the airway, thereby reducing the effectiveness of bronchodilators. As a result, the aim to achieve good asthma control has become increasingly important with the prominent increase in the global prevalence of the disease, particularly in those with difficult-to-treat asthma where the disease cannot be controlled even with a high-dose combination of multiple therapeutic drugs [37].

34.2.1.2 Evidence of Plant-Based Chemical Moieties in the Management of Asthma

Th1/Th2 imbalance is widely known to induce the development of allergic asthma. Interferon (IFN)-γ, IL-12, and tumour necrosis factor (TNF)-β are secreted by Th1 cells, whereas IL-4, IL-5, and IL-13 are mostly secreted by Th2 cells. Therefore, bioactive compounds that can modulate the expression of T-bet and GATA-3 transcription factors may be beneficial in the management of asthma as these transcription factors influence the balance of Th1/Th2 cells [10, 38]. T-bet and GATA-3 can also be modulated by IFN-γ, IL-4, and IL-12 via the signal transducers and activators of transcription (STAT) pathways, namely, STAT1, STAT6, and STAT4, respectively [38, 39]. Jin et al. in a study evaluated the effects of psoralen, a major constituent found in the dried and ripe fruit of *Psoralea corylifolia* L., on Th2 responses in an asthmatic in vivo rat model. Results demonstrated significant suppression of inflammatory cell infiltration as well as decreased mucus secretion in lung tissues. This is attributed to the inhibition of GATA-3 protein expression that reduced Th2 response, resulting in alleviation of airway hyperresponsiveness and airway inflammation in asthma [40]. In another study, wogonin obtained from

Scutellariae radix was investigated for its anti-inflammatory effects in an asthmatic mouse model. The researchers reported that wogonin remarkably inhibited allergen-induced eosinophilic inflammation in lung tissues. It was also found that such an effect is attributed to downregulation of IL-4-induced STAT6 translocation and activation, thereby suppressing Th2-mediated inflammation [41].

Mitogen-activated protein kinases (MAPK) comprising of extracellular signal-regulated kinase (ERK) 1/2, p38, and c-Jun-N-terminal kinase (JNK) 1/2 are critical modulating factors for the activation of inflammatory cells in asthma pathogenesis. Besides, nuclear factor kappa-B (NF-κB) is another transcription factor involved in the upregulation of multiple pro-inflammatory genes, whereby increased activation of NF-κB can be observed upon allergen challenge and can be found in the airway epithelium cells of asthmatic patients [42, 43]. Therefore, plant-based chemical moieties that are capable of regulating both MAPK and NF-κB signalling pathways can be exploited for their potential in treating asthma. Chauhan et al. evaluated the therapeutic potential of curcumin, a polyphenol found in *Curcuma longa*, on ovalbumin-induced asthmatic mice models. Significant reduction of JNK, ERK, and p38 phosphorylation was documented, along with inhibited NF-κB expression in lung tissues, suggesting that curcumin can protect against asthma via MAPK/NF-κB signalling pathways [44]. Similarly, Lee et al. also reported downregulation of MAPK and NF-κB signalling pathways by *Pistacia weinmannifolia* root extract. In vivo model demonstrated reduction in the levels of eosinophils and Th2 cytokines, which eventually led to the inhibition of inflammatory cells influx and mucus hypersecretion, indicating that it can possibly be used as an adjuvant to existing therapies for the effective management of asthma [45].

T-regulatory cell (Tregs) and Th17 cell targeting is another strategy that can be exploited in the management of asthma. Driven by forkhead box P3 (FOXP3) transcription factor, Tregs are responsible for regulating the homeostasis of pulmonary immunity through the suppression of excessive immune responses deleterious to the host or by establishing immune tolerance to non-harmful antigens [46–48]. On the other hand, RORγT is the transcription factor for Th17 and it plays a crucial role in the activation of IL-17 production in Th17 cells. In asthmatic patients, IL-17 affects airway smooth muscle via the upregulation of airway hyperresponsiveness triggered by allergens [10, 49]. Thus, plants presenting FOXP3 and RORγT targeted activity may be further investigated for their potential in alleviating asthma. One study was conducted by Ji et al. to evaluate the feasibility of ligustrazine in asthma treatment, which is a bioactive compound extracted from the Chinese herb *Ligusticum chuanxiong*. In the mouse asthmatic model, it was found that ligustrazine reduced influx of neutrophils and eosinophils via the regulation of Tregs/Th17, by achieving a balance between FOXP3 and RORγT transcription factors. In addition, ligustrazine also regulates the expressions of T-bet and GATA-3 in asthma. Synergistically, ligustrazine can potentially be developed as a novel therapeutic agent for asthma [50].

Several other disease-targeting mechanisms by plant-based chemical moieties were also reported by researchers. Choi et al. reported that petatewalide B, a compound isolated from the leaves of *Petasites japonicus*, inhibited antigen-induced degranulation of β-hexosaminidase in mast cells. Strong inhibitions of eosinophils, macrophages, and lymphocytes accumulation were also observed in an ovalbumin-induced asthmatic model [51]. As mast cell degranulation leads to the release of mediators that attract leukocytes to inflammatory sites which further amplifies inflammatory response, inhibition of mast cell degranulation may be a helpful strategy in the management of asthma [52]. As contraction of the airway is the distinctive feature of asthma, strategies that help to relax airway smooth muscle, such as anticholinergic, antihistamine, β2-adrenoceptors stimulation, and blockade of Ca^{2+} signalling may also be beneficial to treat asthma. For this instance, Mokhtari-Zaer et al. in a study demonstrated the relaxant effect of *Crocus sativus* L. on smooth muscles via the activation of β2-adrenoceptors, as well as the suppression of histamine H1, muscarinic receptors, and calcium channels. This suggests that the plant may help in the alleviation of asthma [10, 53]. Platelet-derived growth factor (PDGF) and transforming growth factor (TGF)-β are involved in asthma airway remodelling, whereby TGF-β1/Smad is the major signalling pathway that induces airway remodelling. Xu et al. has proven the effectiveness of *Astragali radix* in improving the symptoms of allergic airway remodelling in a chronic asthmatic mouse model. It was found that eosinophilic airway inflammation was greatly suppressed, attributed to the inhibition of Th2 cytokines and TGF-β1 [10, 54]. Similarly, another study by Jang et al. demonstrated that skullcapflavone II derived from *Scutellaria baicalensis* reduced several primary pathophysiological features of asthma such as Th2 cytokines production, airway eosinophilic inflammation and airway hyperresponsiveness by modulating the TGF-β1/Smad signalling pathway. Subepithelial collagen deposition and goblet cell hyperplasia were remarkably inhibited, accompanied by decreased TGF-β1 levels, upregulated Smad7 expression, and downregulated pSmad2/3 expressions [55].

34.2.2 Chronic Obstructive Pulmonary Disease

34.2.2.1 Overview of Chronic Obstructive Pulmonary Disease

COPD is a common type of CRD which is characterized by functional and structural alterations resulted from long-term inhalation of harmful particles. Continuous exposure to infectious agents, pollutants, as well as cigarette smoke are the major risk factors of COPD. As it is currently the fourth leading cause of death worldwide with more than 250 million people affected by this disease, exacerbations of COPD pose a substantial burden on global healthcare systems [19, 56, 57]. The clinical presentations of COPD include chronic bronchitis and emphysema. In chronic bronchitis, the bronchial tubes are inflamed, narrow, and cilia-free, thereby leading to mucus hypersecretion in the lungs. On the other hand, in emphysema, small sacs which are present in the alveoli wall become larger, thereby decreasing its oxygen absorption capacity and damaging the alveoli. These manifestations ultimately deem COPD an incurable progressive disease [58]. Patients with COPD usually exhibit symptoms such as breathlessness, tightness of the chest, frequent coughing, as well as wheezing. All these symptoms are such that the disease can progress for years without the patient noticing them [19, 58, 59].

A rise in the number of macrophages, neutrophils, and CD8+ lymphocytes in alveolar and peripheral spaces are among the major characteristics of chronic inflammation in COPD. Inflammation in COPD involves components from both the adaptive and innate immune systems, which majorly affects the bronchoalveolar walls of both small and large airways [19, 60]. Alveolar macrophages release matrix metallopeptidase (MMP)-9 which results in inflammation and associated pathology of the disease. They also generate and induce the local burden of reactive oxygen species (ROS) in the lungs [61]. Airway fibrosis that is presented in COPD is mainly associated with the elevated expression of transforming growth factor (TGF)-β via the release of connective tissue growth factor (CTGF), resulting in the deposition of collagen [62]. Progressive lung damage is linked with the release of leukotriene B4, IL-8, interferon (IFN)-γ, MMP-2, MMP-9, MMP-12, tumour necrosis factor (TNF)-α, and TGF-β. NF-κB inflammatory pathway can be stimulated by ROS and reactive nitrogen species, which are induced by stimulated macrophages and neutrophils from cigarette smoke or other noxious gases [19, 62]. Oxidative stress that is present in COPD leads to increased production of hydrogen peroxide and carbon monoxide in the airways [62]. Nevertheless, these responses of the immune system were found to differ across different subtypes of COPD. Generally, immune responses in chronic bronchitis are associated with secreted serine proteases and elevated ROS levels from inhaled smoke and activated macrophages, whereas emphysema is directly caused by cigarette smoke [19].

Along with smoking cessation, current therapies are focused on the improvement of symptoms and prevention of exacerbations; however, disease-modifying treatment is yet to be available. Complete recovery is currently only possible via transplantation of the lung, which is not a practical approach in most patients. Existing drugs such as anticholinergics, phosphodiesterase inhibitors, long-acting β2 agonists, and corticosteroids are also linked with severe adverse reactions and cause susceptibility to other diseases [9, 59]. Thus, there is a compelling demand for the development of safe, novel, and potent alternatives for managing COPD.

34.2.2.2 Evidence of Plant-Based Chemical Moieties in the Management of Chronic Obstructive Pulmonary Disease

It has been widely established that cigarette smoke is the primary cause for the development of COPD, whereby it is the primary source of ROS in the lungs of COPD patients. The presence of ROS in the lungs further leads to the development of oxidative stress and inflammatory responses which contributes to the progression of COPD [63, 64]. This is due to the upregulation of NF-κB and p38 MAPK pathways by ROS, which results in the activation of pro-inflammatory cytokines in COPD pathogenesis. Besides, TGF-β is also induced by ROS which causes fibrosis in the lung [65, 66]. Thus, modulation of oxidative stress via boosting endogenous levels of antioxidants is a promising strategy to treat and manage COPD. As oxidative stress can be regulated by the NRF2 transcription factor, the NRF2 pathway can be targeted to mitigate inflammation and antioxidant defence mechanism in the lungs [64, 67]. Zhou et al. investigated the antioxidative effect of *Cinnamomum chartophyllum* extract on human bronchial epithelial cells. It was documented that Nrf2 downstream genes NQO-1 and γ-GCS were activated, leading to significant enhancement of nuclear translocation and stabilization of Nrf2. No observed adverse events were observed at Nrf2-inducing doses. These findings suggest that *C. chartophyllum* can potentially be utilized in the management of COPD attributed to its ability to prevent oxidative insults in human lungs [68]. Likewise, Cheng et al. in a study demonstrated the anti-inflammatory and antioxidant effects of forsythiaside isolated from *Forsythia suspensa* in a cigarette smoke-induced COPD mice model. The study showed that forsythiaside attenuated inflammatory cells infiltration and suppressed the production of TNF-α, IL-1β, and IL-6. Further investigation demonstrated that such effects are associated with the inhibition of the NF-κB signalling pathway and upregulation of Nrf2 expression [69].

In the pathogenesis of COPD, oxidative stress in the lung may further cause oxidative damage to various lung macromolecules. One such example is lung proteins, whereby oxidatively damaged proteins are generally non-functional and may be cytotoxic. These damaged proteins are accumulated in the endoplasmic reticulum (ER), resulting in a condition known as ER stress. Unfolded protein response (UPR) is a compensatory cellular response that refers to the transcriptional, translational, and post-translational cascades that facilitate the reversal of ER stress via reducing the flow of new proteins into the ER, thereby enhancing the capacity of ER for folding and processing proteins. This eventually degrades misfolded proteins in the ER. However, when the reversal of ER stress fails, UPR induces programmed cell death. Thus, UPR signalling pathways are the determinants of whether cells undergo apoptosis or successfully restore proteostasis [70–73]. PRK-like ER kinase (PERK), inositol requiring protein (IRE)-1, and activated transcription factor (ATF)-6 are the three signalling pathways that form the UPR and are involved in COPD-associated emphysema [74]. Specifically, the upregulation of the PERK pathway reduces protein synthesis by phosphorylating eukaryotic translation initiator factor (eIF)-2α and eventual translation of ATF4. Prolonged expression of ATF4 upregulates pro-apoptotic proteins, such as Bax, and downregulates anti-apoptotic proteins, such as Bcl-2 proteins [75]. Lin et al. explored the effects and mechanisms of ursolic acid, a pentacyclic triterpenoid commonly found in *Crataegus pinnatifida* and other edible plants, on a cigarette smoke-induced rat emphysema model. It was reported that ursolic acid alleviated emphysema-related pathology and improved cigarette smoke-induced oxidative stress in rat lungs. The researchers determined that ursolic acid downregulated the expression of PERK pathway proteins, along with the induced expression of Bcl-2 and inhibited the expression of Bax, which led to reduced apoptotic cells in rat lungs. Besides, protein expression of the NRF2 pathway was upregulated, thus reducing oxidative stress [76].

Considerable evidence has also been presented that matrix metalloproteinases (MMP) can have modulatory effects in COPD-induced emphysema, due to their involvement in the proteolytic attacks on the alveolar wall matrix. In addition, MMPs can modify the secretion of fibrogenic growth factors that leads to various lesions associated with COPD [77]. Sirtuin 1 (SIRT1) is a type 3 histone deacetylase

(HDAC) that regulates various cellular responses such as ageing and apoptosis. The loss of SIRT1 has been reported in COPD, and it is correlated with the acetylation of tissue inhibitor of metalloproteinase (TIMP)-1 lysine, leading to subsequent degradation of the anti-MMP protease and increased levels of MMP-9 in lung tissue [64, 78]. Ganesan et al. demonstrated the potential of quercetin, a plant flavonoid, in preventing the progression of COPD-induced emphysema in an in vivo model. It was shown that quercetin treatment reduced the expressions and activities of MMP-9 and MMP-12 by increasing the expression of SIRT1. This resulted in the alleviation of lung inflammation, goblet cell metaplasia, and inhibition of pro-inflammatory cytokines mRNA expressions. The results correspond to another finding where co-treatment with sirtinol, a SIRT1 inhibitor, blocked the effects of quercetin in LPS-exposed mice [79].

Another prominent feature of COPD is neutrophilic inflammation. Excessive neutrophilic inflammation will lead to the production of ROS and the release of serine proteases, MMPs, and myeloperoxidase, thereby causing collateral damage as neutrophils infiltrate into lung tissues. Neutrophil influx can be inhibited by targeting chemokine receptors CXCR1, CXCR2, as well as PI3Kγ, whereas neutrophil function can be modified by using selective PI3Kδ inhibitors [64, 80]. Furthermore, the PI3K signalling pathway has been associated with corticosteroid insensitivity in COPD patients, mediated by reduced HDAC2 expression. Total PI3K activity can also be determined by the expression of its downstream target protein, Akt [64, 81]. Lee et al. isolated lignans from the flower buds of *Magnolia fargesii* and elucidated their mechanisms of action underlying their COPD-protective effects. The researchers highlighted that lignans remarkably inhibited both Erk and Akt phosphorylation levels, along with a dose-dependent suppression of EGFR. Infiltration of neutrophils and macrophages was attenuated, and the secretion of TNF-α and IL-6 was suppressed both in vivo and in vitro [82]. Similarly, a study by Liou et al. presented promising anti-inflammatory effects of casticin, a compound isolated from *Vitex rotundifolia*. Results showed that casticin reduced the levels of TNF-α, IL-6, IL-8 and PGE2. Furthermore, casticin suppressed the phosphorylation of Akt, PI3K, and MAPK. The translocation of p65, the protein subunit of NF-κB, was also attenuated. These findings proved that casticin suppressed inflammation in pulmonary epithelial cells via inhibitory effects on PI3K/Akt, NF-κB, and MAPK signalling pathways [83].

34.2.3 Lung Cancer

34.2.3.1 Overview of Lung Cancer

Lung cancer has been the leading cause of cancer-related mortalities worldwide, posing extensive healthcare and economic burden in many countries. This is most likely due to the disease being undetected until there has been substantial progression of the disease, resulting in remarkable reduction in the quality of life of patients [84]. Various risk factors for lung cancer have been identified, including active cigarette smoking and exposure to second-hand smoke, prolonged exposure to pollutants and radiation, as well as occupational exposure to agents including asbestos, nickel, and chromium [85]. Smoking cessation is widely regarded as effective prevention of lung cancer as the tobacco-induced susceptibility of the disease is said to be influenced by competitive gene-enzyme interactions at the procarcinogen level and the resulting extent of DNA damage [19].

Lung cancer can generally be classified into two different subtypes, namely, non-small-cell lung cancer (NSCLC) and small-cell lung cancer (SCLC), whereby approximately 85–90% of lung cancers are determined to be NSCLC and are mostly diagnosed only at the advanced stage of metastasis. At this stage, the curative effects of surgical intervention are reduced drastically. Along with the emergence of drug resistance, these lead to therapeutic failure in NSCLC and eventual tumour recurrence and disease progression [86, 87]. NSCLC can be further categorized into adenocarcinoma, large-cell carcinoma, and squamous cell carcinoma [88]. The development of malignant cancer cells is associated with genetic alterations in DNA and mutations that affect the regulation of normal cells. For example, epidermal growth factor receptor (EGFR), B-Raf proto-oncogene (BRAF), echinoderm microtubule-associated protein-like (EMAPL)-4 anaplastic lymphoma kinase (ALK) fusion oncogene, c-ROS oncogene 1 (ROS1) fusions, and p53 gene mutations are some of the driver mutation genes that are found to be associated with the development of NSCLC [89, 90]. CD44 gene mutations have also been observed in both NSCLC and SCLC [91–93]. Besides, programmed death ligand 1 (PD-L1) is found overexpressed in NSCLC patients without driver mutations [93].

Typically, treatment of cancer aims to eradicate cancerous cells without any harmful effects on normal cells. Traditional management strategies of lung cancer include surgery, radiotherapy, and chemotherapy, which can be used either individually or in combination with each other. Although chemotherapy is mostly utilized in lung cancer, there are several issues associated with its use, such as severe adverse reactions and limited efficacy due to multidrug resistance [84, 88]. As poor prognosis remains a challenge for the effective management of lung cancer, further research and development of novel therapies are urgently warranted.

34.2.3.2 Evidence of Plant-Based Chemical Moieties in the Management of Lung Cancer

Angiogenic pathways are frequently regarded as the primary target in the treatment of NSCLC as tumour cells presented the development and generation of new blood vessels in the desmoplastic stroma for its growth and progression [19, 94]. Therefore, targeting pro-angiogenic factors is a pivotal strategy to inhibit angiogenic required for tumour growth. Vascular endothelial growth factor (VEGF) is the major factor that initiates angiogenesis, whereby the VEGF family of growth factors exert their effects via vascular endothelial growth factor receptor (VEGFR) interaction [95, 96]. When this signalling pathway is upregulated, it results in the proliferation of endothelial cells, degradation of the extracellular matrix, and migration of endothelial cells, followed by the formation of new blood vessels [95]. A study conducted by Xu et al. evaluated the therapeutic potential of sanguinarine, an alkaloid isolated from *Sanguinaria canadensis*, in lung cancer. It was observed that sanguinarine remarkably suppressed tube formation of human microvascular endothelial cells (HMVEC), as well as the migration of human A549 lung cancer cells. The findings were attributed to decreased secretion of VEGF in both HMVECs and human A549 lung cancer cells in a dose-dependent manner. Besides, VEGF-mediated Akt and p38 activation was also inhibited. These results suggest that sanguinarine may be beneficial in lung cancer treatment due to its antiangiogenic and anti-invasive properties [97]. In another study, Chakraborty et al. studied the anticancer effects of capsaicin, an active constituent found in chilli pepper, on NSCLC in a hypoxic environment. It was reported that capsaicin-induced p53-mediated degradation of HIF-1α as well as SMAR1-induced downregulation of COX-2 restrained HIF-1α nuclear localization, which eventually led to the downregulation of VEGF expression. The findings indicate that capsaicin can achieve efficient NSCLC therapy via its antiangiogenic effect in tumour cells [98].

Typically, all types of cancer cells evade normal apoptotic pathways through the dysregulation of apoptotic factors [99]. Therefore, a great understanding of these apoptotic pathways may allow the development of novel strategies that enhance lung cancer cell death. Caspases are ubiquitously expressed cysteine proteases that play a crucial role in the regulation of apoptosis, whereby aspartic residue cleavage of caspases and removal of their N-terminal inhibitory domain induced by death-inducing stimuli usually result in the demolition phase of apoptosis [100]. Apoptosis involves two major pathways, namely, the extrinsic death receptor pathway and the intrinsic mitochondria pathway. The extrinsic pathway can be induced by the activation of several death receptors including Fas, TNFs, as well as TNF-related apoptosis-inducing ligand receptors (TRAILR). On the other hand, the intrinsic apoptotic pathway is tightly regulated by Bcl-2 family of proteins, which include pro-apoptotic members such as Bax, and anti-apoptotic members such as Bcl-2 and MCL-1 [99, 100]. For example, NSCLC is often associated with irregularities in Bcl-2 family proteins, whereas clinical drug resistance is usually associated with the amplification of anti-apoptotic Bcl-2 and MCL-1 proteins [99]. A recent study by Han et al. evaluated the anticancer effect of *Viburnum grandiflorum* extract (VGE) against lung cancer cells. It was found that the rate of apoptosis was significantly enhanced, most likely due to cleavage of pro-caspase-8, -9, and -3. Besides, the protein expression of MCL-1 was also diminished. These findings suggest that VGE inhibited cell viability of lung carcinoma via the caspase-dependent pathway of apoptosis [101]. Xie et al. also reported a similar pro-apoptotic effect in a study utilizing bruceine D, a component extracted from *Brucea javanica*, against NSCLC cells. The researchers found that bruceine D suppressed the proliferation of wild-type NSCLC cells in a dose- and time-dependent manner, thereby decreasing the colony-forming ability and migration of human lung cancer A549 cells. In addition, bruceine D also effectively induced apoptosis in A549 cells that is associated with G0-G1 cell cycle arrest, accumulation of intracellular ROS, as well as disruption of mitochondrial membrane potential. These correspond with the findings that bruceine D suppressed the expressions of anti-apoptotic proteins, namely, Bcl-xL and Bcl-2, and enhanced the expressions of pro-apoptotic proteins, namely Bax and Bak. The expressions of pro-caspase-3 and pro-caspase-8 were also inhibited. Hence, these results suggest that bruceine D may be effective in the treatment of NSCLC due to its role in modulating the signalling of ROS-mitochondrial-mediated cell death [102].

Another strategy in targeting lung cancer is via the inhibition of histone deacetylases (HDAC), as they are known to regulate the activities and expressions of various proteins involved in the initiation and progression of cancer [103]. Studies have shown that HDAC inhibitors impact cell proliferation, survival, and angiogenesis by modulating molecular chaperones and signal transduction proteins, as well as inhibiting hypoxia-inducible factors and VEGF [104, 105]. Furthermore, HDAC inhibitors enhance immune responses and upregulate major histocompatibility complex (MHC) class I and II protein, as well as other co-stimulatory molecules such as CD80 and CD86 [104]. Studies have also indicated that HDAC inhibitors facilitate tumour cell apoptosis via expression induction of surface TRAIL receptors 1 (DR4) and TRAIL receptors 2 (DR5), which eventually leads to the initiation of the caspase cascade. As such, HDAC inhibitors are potential anticancer agents due to their ability to induce autophagy and apoptosis in tumour cells via the activation of intrinsic mitochondrial pathways [106, 107]. Liu et al. in their study has investigated the anticancer activity of magnolol and polyphenol mixture obtained from *Magnolia officinalis* in lung cancer. Results demonstrated significant suppression of class I HDACs, leading to cell apoptosis via cell cycle arrest and activation of pro-apoptotic factors such as DR5, Bax, and cleaved caspase 3. Tumour growth reduction was also observed in the lung cancer A549 xenograft model. These indicate that the magnolol and polyphenol extract induced tumour cell apoptosis by activating DR5 through the inhibition of class I HDACs [106]. In another study, Singh et al. evaluated the anticancer effect of silymarin obtained from *Silybum marianum* in a metastatic human NSCLC cell line. They reported a concentration-dependent suppression of cell migration that is associated with inhibition of HDAC activity, along with decreased levels of class I HDACs. Such HDAC inhibitory effects were comparable to those of approved synthetic HDAC inhibitors including trichostatin A and sodium butyrate. siRNA knockdown of ZEB1 was also observed, leading to downregulated expression of HDACs in NSCLC cells. These findings suggest that silymarin is effective in inhibiting the migration of lung cancer cells and can be developed as a potential agent to prevent lung cancer metastasis [108].

34.3 Conclusion and Future Directions

Chronic respiratory diseases have brought enormous social, economic, and healthcare burden, and they are increasingly prevalent globally due to the lack of effective measures that can be utilized to combat these chronic respiratory diseases. Therefore, the use of plant-based chemical moieties may be a promising approach in the development of novel therapeutic agents for treating these diseases. Over the years, many multifaceted approaches that involve the combination of biological, chemical, and pharmacological aspects in discovering the potential benefits of plant-based chemical moieties are performed both in vivo and in vitro. It is undoubtedly that a diversity of plants has demonstrated remarkable potential in managing chronic respiratory diseases, attributed to their ability in targeting various signalling pathways underlying the pathogenesis of these diseases. Nonetheless, the findings obtained in those studies are still insufficient for an in-depth understanding of their therapeutic potential in a scientific discipline. Further studies are needed to elucidate their precise disease-targeting mechanisms at the molecular level, along with accurate identification of the phytochemical constituent(s) responsible for the documented pharmacological effects in chronic respiratory diseases. Ethnopharmacological studies are also crucial to establish the parameters for safety, toxicity, quality, and effectiveness of these plant-based chemical moieties. It is anticipated that the field of phytotherapeutics will be of particular interest in the future, as there is still a diversity of plants that are yet to be discovered and remain unstudied across the world. Hence, scientific research and interdisciplinary collaborations should be continued to explore further opportunities where plant-based products can be directly extrapolated on human and to establish sufficient evidence in which plant-based products can be used as effective therapeutic agents in chronic respiratory diseases.

References

1. Prasher P, Sharma M, Mehta M, Paudel KR, Satija S, Chellappan DK, Dureja H, Gupta G, Tambuwala MM, Negi P, Wich PR, Hansbro NG, Hansbro PM, Dua K (2020) Plants derived therapeutic strategies targeting chronic respiratory diseases: chemical and immunological perspective. Chem Biol Interact 325:109125
2. Ambrosino N, Bertella E (2018) Lifestyle interventions in prevention and comprehensive management of COPD. Breathe 14:186–194
3. Barnes PJ, Bonini S, Seeger W, Belvisi MG, Ward B, Holmes A (2015) Barriers to new drug development in respiratory disease. Eur Respir J 45:1197–1207
4. World Health Organization Chronic respiratory diseases. https://www.who.int/health-topics/chronic-respiratory-diseases#tab=tab_1. Accessed 15 Oct 2020
5. Duncan D (2016) Chronic obstructive pulmonary disease: an overview. Br J Nurs 25:360–366. https://doi.org/10.12968/bjon.2016.25.7.360
6. Burney P (2017) Chronic respiratory disease—the acceptable epidemic? Clin Med J R Coll Phys Lond 17:29–32
7. Hatipoglu U (2018) Chronic obstructive pulmonary disease: more than meets the eye. Ann Thorac Med 13:1–6
8. Lin BF, Chiang BL, Ma Y, Lin JY, Chen ML (2015) Traditional herbal medicine and allergic asthma. Evid Based Complement Altern Med 2015:510989
9. Santana FPR, Pinheiro NM, Mernak MIB, Righetti RF, Martins MA, Lago JHG, Lopes FDTQDS, Tibério IFLC, Prado CM (2016) Evidences of herbal medicine-derived natural products effects in inflammatory lung diseases. Mediat Inflamm 2016:2348968
10. Liu F, Xuan NX, Ying SM, Li W, Chen ZH, Shen HH (2016) Herbal medicines for asthmatic inflammation: from basic researches to clinical applications. Mediat Inflamm 2016:6943135
11. Yu J (2008) Inflammatory mechanisms in the lung. J Inflamm Res 2:1. https://doi.org/10.2147/jir.s4385
12. Durham AL, Caramori G, Chung KF, Adcock IM (2016) Targeted anti-inflammatory therapeutics in asthma and chronic obstructive lung disease. Transl Res 167:192–203
13. Gross NJ, Barnes PJ (2017) New therapies for asthma and chronic obstructive pulmonary disease. Am J Respir Crit Care Med 195:159–166

14. Kim HP, Lim H, Kwon YS (2017) Therapeutic potential of medicinal plants and their constituents on lung inflammatory disorders. Biomol Ther 25:91–104
15. Ng SW, Chan Y, Chellappan DK, Madheswaran T, Zeeshan F, Chan YL, Collet T, Gupta G, Oliver BG, Wark P, Hansbro N, Hsu A, Hansbro PM, Dua K, Panneerselvam J (2019) Molecular modulators of celastrol as the keystones for its diverse pharmacological activities. Biomed Pharmacother 109:1785–1792
16. Chan Y, Ng SW, Xin Tan JZ, Gupta G, Tambuwala MM, Bakshi HA, Dureja H, Dua K, Ishaq M, Caruso V, Chellappan DK (2020) Emerging therapeutic potential of the iridoid molecule, asperuloside: a snapshot of its underlying molecular mechanisms. Chem Biol Interact 315:108911
17. Alamgeer YW, Asif H, Sharif A, Riaz H, Bukhari IA, Assiri AM (2018) Traditional medicinal plants used for respiratory disorders in Pakistan: a review of the ethno-medicinal and pharmacological evidence Milen Georgiev, Ruibing Wang. Chin Med (United Kingdom) 13:1–29
18. Chan Y, Ng SW, Chellappan DK, Madheswaran T, Zeeshan F, Kumar P, Pillay V, Gupta G, Wadhwa R, Mehta M, Wark P, Hsu A, Hansbro NG, Hansbro PM, Dua K, Panneerselvam J (2020) Celastrol-loaded liquid crystalline nanoparticles as an anti-inflammatory intervention for the treatment of asthma. Int J Polym Mater Polym Biomater. https://doi.org/10.1080/00914037.2020.1765350
19. Shukla SD, Swaroop Vanka K, Chavelier A, Shastri MD, Tambuwala MM, Bakshi HA, Pabreja K, Mahmood MQ, O'Toole RF (2020) Chronic respiratory diseases: an introduction and need for novel drug delivery approaches. In: Targeting chronic inflammatory lung diseases using advanced drug delivery systems. Elsevier, San Diego, pp 1–31
20. Sadikot RT (2018) The potential role of nanomedicine in lung diseases. Med Res Arch 6:1–9. https://doi.org/10.18103/MRA.V6I5.1723
21. Veeresham C (2012) Natural products derived from plants as a source of drugs. J Adv Pharm Technol Res 3:200–201
22. Thomford NE, Senthebane DA, Rowe A, Munro D, Seele P, Maroyi A, Dzobo K (2018) Natural products for drug discovery in the 21st century: innovations for novel drug discovery. Int J Mol Sci 19:1578
23. Wright GD (2019) Unlocking the potential of natural products in drug discovery. Microb Biotechnol 12:55–57. https://doi.org/10.1111/1751-7915.13351
24. Falzon CC, Balabanova A (2017) Phytotherapy: an introduction to herbal medicine. Prim Care Clin Off Pract 44:217–227
25. Karunamoorthi K, Jegajeevanram K, Vijayalakshmi J, Mengistie E (2013) Traditional medicinal plants: a source of phytotherapeutic modality in resource-constrained health care settings. J Evid Based Complement Altern Med 18:67–74. https://doi.org/10.1177/2156587212460241
26. Atanasov AG, Waltenberger B, Pferschy-Wenzig EM, Linder T, Wawrosch C, Uhrin P, Temml V, Wang L, Schwaiger S, Heiss EH, Rollinger JM, Schuster D, Breuss JM, Bochkov V, Mihovilovic MD, Kopp B, Bauer R, Dirsch VM, Stuppner H (2015) Discovery and resupply of pharmacologically active plant-derived natural products: a review. Biotechnol Adv 33:1582–1614
27. Nderitu KW, Mwenda NS, Macharia NJ, Barasa SS, Ngugi MP (2017) Antiobesity activities of methanolic extracts of Amaranthus dubius, Cucurbita pepo, and Vigna unguiculata in progesterone-induced obese mice. Evid Based Complement Altern Med 2017:4317321. https://doi.org/10.1155/2017/4317321
28. Sun NN, Wu TY, Chau CF (2016) Natural dietary and herbal products in anti-obesity treatment. Molecules 21:1351
29. Horak F, Doberer D, Eber E, Horak E, Pohl W, Riedler J, Szépfalusi Z, Wantke F, Zacharasiewicz A, Studnicka M (2016) Diagnosis and management of asthma—statement on the 2015 GINA Guidelines. Wien Klin Wochenschr 128:541–554. https://doi.org/10.1007/s00508-016-1019-4

30. Castillo JR, Peters SP, Busse WW (2017) Asthma exacerbations: pathogenesis, prevention, and treatment. J Allergy Clin Immunol Pract 5:918–927. https://doi.org/10.1016/j.jaip.2017.05.001
31. Ray A, Oriss TB, Wenzel SE (2015) Emerging molecular phenotypes of asthma. Am J Physiol Lung Cell Mol Physiol 308:L130–L140
32. Pavord ID, Beasley R, Agusti A, Anderson GP, Bel E, Brusselle G, Cullinan P, Custovic A, Ducharme FM, Fahy JV, Frey U, Gibson P, Heaney LG, Holt PG, Humbert M, Lloyd CM, Marks G, Martinez FD, Sly PD, von Mutius E, Wenzel S, Zar HJ, Bush A (2018) After asthma: redefining airways diseases. Lancet 391:350–400
33. Broide DH (2008) Immunologic and inflammatory mechanisms that drive asthma progression to remodeling. J Allergy Clin Immunol 121:560–570
34. Holgate ST (2013) Mechanisms of asthma and implications for its prevention and treatment: a personal journey. Allergy Asthma Immunol Res 5:343–347
35. Fahy JV (2015) Type 2 inflammation in asthma-present in most, absent in many. Nat Rev Immunol 15:57–65
36. Zoratti EM, O'Connor GT (2020) New therapeutic strategies for asthma. JAMA J Am Med Assoc 323:517–518
37. Agbetile J, Green R (2011) New therapies and management strategies in the treatment of asthma: patient-focused developments. J. Asthma Allergy 4:1–12
38. Agrawal DK, Shao Z (2010) Pathogenesis of allergic airway inflammation. Curr Allergy Asthma Rep 10:39–48
39. van Rijt L, von Richthofen H, van Ree R (2016) Type 2 innate lymphoid cells: at the crossroads in allergic asthma. Semin Immunopathol 38:483–496
40. Jin H, Wang L, Xu C, Li B, Luo Q, Wu J, Lv Y, Wang G, Dong J (2014) Effects of Psoraleae fructus and its major component psoralen on Th2 response in allergic asthma. Am J Chin Med 42:665–678. https://doi.org/10.1142/S0192415X14500438
41. Ryu EK, Kim TH, Jang EJ, Choi YS, Kim ST, Hahm KB, Lee HJ (2015) Wogonin, a plant flavone from Scutellariae radix, attenuated ovalbumin-induced airway inflammation in mouse model of asthma via the suppression of IL-4/STAT6 signaling. J Clin Biochem Nutr 57:105–112. https://doi.org/10.3164/JCBN.15-45
42. Arthur JSC, Ley SC (2013) Mitogen-activated protein kinases in innate immunity. Nat Rev Immunol 13:679–692
43. Gras D, Chanez P, Vachier I, Petit A, Bourdin A (2013) Bronchial epithelium as a target for innovative treatments in asthma. Pharmacol Ther 140:290–305
44. Chauhan PS, Singh DK, Dash D, Singh R (2018) Intranasal curcumin regulates chronic asthma in mice by modulating NF-κB activation and MAPK signaling. Phytomedicine 51:29–38. https://doi.org/10.1016/j.phymed.2018.06.022
45. Lee JW, Min JH, Kim MG, Kim SM, Kwon OK, Oh TK, Lee JK, Kim TY, Lee SW, Choi S, Li WY, Ryu HW, Ahn KS, Oh SR (2019) Pistacia weinmannifolia root exerts a protective role in ovalbumin-induced lung inflammation in a mouse allergic asthma model. Int J Mol Med 44:2171–2180. https://doi.org/10.3892/ijmm.2019.4367
46. Ramsdell F, Ziegler SF (2014) FOXP3 and scurfy: how it all began. Nat Rev Immunol 14:343–349
47. Josefowicz SZ, Lu LF, Rudensky AY (2012) Regulatory T cells: mechanisms of differentiation and function. Annu Rev Immunol 30:531–564
48. Ohkura N, Kitagawa Y, Sakaguchi S (2013) Development and maintenance of regulatory T cells. Immunity 38:414–423
49. Chung KF (2015) Targeting the interleukin pathway in the treatment of asthma. Lancet 386:1086–1096
50. Ji NF, Xie YC, Zhang MS, Zhao X, Cheng H, Wang H, Yin KS, Huang M (2014) Ligustrazine corrects Th1/Th2 and Treg/Th17 imbalance in a mouse asthma model. Int Immunopharmacol 21:76–81. https://doi.org/10.1016/j.intimp.2014.04.015

51. Choi YW, Lee KP, Kim JM, Kang S, Park SJ, Lee JM, Moon HR, Jung JH, Lee YG, Im DS (2016) Petatewalide B, a novel compound from Petasites japonicus with anti-allergic activity. J Ethnopharmacol 178:17–24. https://doi.org/10.1016/j.jep.2015.12.010
52. Krystel-Whittemore M, Dileepan KN, Wood JG (2016) Mast cell: a multi-functional master cell. Front Immunol 6:620
53. Mokhtari-Zaer A, Khazdair MR, Boskabady MH (2015) Smooth muscle relaxant activity of Crocus sativus (saffron) and its constituents: possible mechanisms. Avicenna J Phytomed 5:365–375
54. Xu S, Tian BP, Zhang LH, Hua W, Xia LX, Chen ZH, Li W, Shen HH (2013) Prevention of allergic airway hyperresponsiveness and remodeling in mice by Astragaliradix Antiasthmatic decoction. BMC Complement Altern Med 13:369. https://doi.org/10.1186/1472-6882-13-369
55. Jang HY, Ahn KS, Park MJ, Kwon OK, Lee HK, Oh SR (2012) Skullcapflavone II inhibits ovalbumin-induced airway inflammation in a mouse model of asthma. Int Immunopharmacol 12:666–674. https://doi.org/10.1016/j.intimp.2012.01.010
56. Qureshi H, Sharafkhaneh A, Hanania NA (2014) Chronic obstructive pulmonary disease exacerbations: latest evidence and clinical implications. Ther Adv Chronic Dis 5:212–227
57. Vijayan VK (2013) Chronic obstructive pulmonary disease. Indian J Med Res 137:251–269
58. Zhou HX, Ou XM, Tang YJ, Wang L, Feng YL (2015) Advanced chronic obstructive pulmonary disease: innovative and integrated management approaches. Chin Med J 128:2952–2959
59. Ram A, Balachandar S, Vijayananth P, Singh VP (2011) Medicinal plants useful for treating chronic obstructive pulmonary disease (COPD): current status and future perspectives. Fitoterapia 82:141–151
60. Suissa S, Dell'Aniello S, Ernst P (2012) Long-term natural history of chronic obstructive pulmonary disease: severe exacerbations and mortality. Thorax 67:957–963. https://doi.org/10.1136/thoraxjnl-2011-201518
61. Boukhenouna S, Wilson MA, Bahmed K, Kosmider B (2018) Reactive oxygen species in chronic obstructive pulmonary disease. Oxidative Med Cell Longev 2018:5730395
62. Wadhwa R, Shukla SD, Chellappan DK, Gupta G, Collet T, Hansbro N, Oliver B, Williams K, Hansbro PM, Dua K, Maurya PK (2019) Phytotherapy in inflammatory lung diseases: an emerging therapeutic interventional approach. In: Phytochemistry: an in-silico and in-vitro update. Springer, Singapore, pp 331–347
63. Barnes PJ (2017) Cellular and molecular mechanisms of asthma and COPD. Clin Sci 131:1541–1558
64. Gonçalves PB, Romeiro NC (2019) Multi-target natural products as alternatives against oxidative stress in Chronic Obstructive Pulmonary Disease (COPD). Eur J Med Chem 163:911–931
65. Barnes PJ (2013) New anti-inflammatory targets for chronic obstructive pulmonary disease. Nat Rev Drug Discov 12:543–559
66. Barnes PJ (2015) Identifying molecular targets for new drug development for chronic obstructive pulmonary disease: what does the future hold? Semin Respir Crit Care Med 36:508–522. https://doi.org/10.1055/s-0035-1555611
67. Boutten A, Goven D, Artaud-Macari E, Boczkowski J, Bonay M (2011) NRF2 targeting: a promising therapeutic strategy in chronic obstructive pulmonary disease. Trends Mol Med 17:363–371
68. Zhou MX, Li GH, Sun B, Xu YW, Li AL, Li YR, Ren DM, Wang XN, Wen XS, Lou HX, Shen T (2018) Identification of novel Nrf2 activators from Cinnamomum chartophyllum H.W. Li and their potential application of preventing oxidative insults in human lung epithelial cells. Redox Biol 14:154–163. https://doi.org/10.1016/j.redox.2017.09.004
69. Cheng L, Li F, Ma R, Hu X (2015) Forsythiaside inhibits cigarette smoke-induced lung inflammation by activation of Nrf2 and inhibition of NF-κB. Int Immunopharmacol 28:494–499. https://doi.org/10.1016/j.intimp.2015.07.011

70. Kelsen SG (2016) The unfolded protein response in chronic obstructive pulmonary disease. In: Annals of the American Thoracic Society. American Thoracic Society, New York, pp S138–S145
71. Pathinayake PS, Hsu ACY, Waters DW, Hansbro PM, Wood LG, Wark PAB (2018) Understanding the unfolded protein response in the pathogenesis of asthma. Front Immunol 9:1
72. Hetz C, Chevet E, Oakes SA (2015) Proteostasis control by the unfolded protein response. Nat Cell Biol 17:829–838
73. Brewer JW (2014) Regulatory crosstalk within the mammalian unfolded protein response. Cell Mol Life Sci 71:1067–1079
74. Gan G, Hu R, Dai A, Tan S, Ouyang Q, Fu D, Jiang D (2011) The role of endoplasmic reticulum stress in emphysema results from cigarette smoke exposure. Cell Physiol Biochem 28:725–732. https://doi.org/10.1159/000335766
75. Chang J, Wang H, Wang X, Zhao Y, Zhao D, Wang C, Li Y, Yang Z, Lu S, Zeng Q, Zimmerman J, Shi Q, Wang Y, Yang Y (2015) Molecular mechanisms of polyphyllin I-induced apoptosis and reversal of the epithelial-mesenchymal transition in human osteosarcoma cells. J Ethnopharmacol 170:117–127. https://doi.org/10.1016/j.jep.2015.05.006
76. Lin L, Yin Y, Hou G, Han D, Kang J, Wang Q (2017) Ursolic acid attenuates cigarette smoke-induced emphysema in rats by regulating PERK and Nrf2 pathways. Pulm Pharmacol Ther 44:111–121. https://doi.org/10.1016/j.pupt.2017.03.014
77. Churg A, Zhou S, Wright JL (2012) Matrix metalloproteinases in COPD. Eur Respir J 39:197–209
78. Yanagisawa S, Papaioannou AI, Papaporfyriou A, Baker JR, Vuppusetty C, Loukides S, Barnes PJ, Ito K (2017) Decreased serum sirtuin-1 in COPD. Chest 152:343–352. https://doi.org/10.1016/j.chest.2017.05.004
79. Ganesan S, Faris AN, Comstock AT, Chattoraj SS, Chattoraj A, Burgess JR, Curtis JL, Martinez FJ, Zick S, Hershenson MB, Sajjan US (2010) Quercetin prevents progression of disease in elastase/LPS-exposed mice by negatively regulating MMP expression. Respir Res 11:131. https://doi.org/10.1186/1465-9921-11-131
80. Mårdh CK, Root J, Uddin M, Stenvall K, Malmgren A, Karabelas K, Thomas M (2017) Targets of neutrophil influx and weaponry: therapeutic opportunities for chronic obstructive airway disease. J Immunol Res 2017:5273201
81. Barnes PJ (2016) Kinases as novel therapeutic targets in asthma and chronic obstructive pulmonary disease. Pharmacol Rev 68:788–815. https://doi.org/10.1124/pr.116.012518
82. Lee SU, Ryu HW, Lee S, Shin IS, Choi JH, Lee JW, Lee J, Kim MO, Lee HJ, Ahn KS, Hong ST, Oh SR (2018) Lignans isolated from flower buds of Magnolia fargesii attenuate airway inflammation induced by cigarette smoke in vitro and in vivo. Front Pharmacol 9. https://doi.org/10.3389/fphar.2018.00970
83. Liou CJ, Huang WC (2017) Casticin inhibits interleukin-1β-induced ICAM-1 and MUC5AC expression by blocking NF-κB, PI3K-Akt, and MAPK signaling in human lung epithelial cells. Oncotarget 8:101175–101188. https://doi.org/10.18632/oncotarget.20933
84. De Sousa Monteiro L, Bastos KX, Barbosa-Filho JM, De Athayde-Filho PF, De Fátima Formiga Melo Diniz M, Sobral MV (2014) Medicinal plants and other living organisms with antitumor potential against lung cancer. Evid Based Complement Altern Med 2014:604152
85. Alberg AJ, Brock MV, Ford JG, Samet JM, Spivack SD (2013) Epidemiology of lung cancer: diagnosis and management of lung cancer, 3rd ed: American College of Chest Physicians evidence-based clinical practice guidelines. Chest 143. https://doi.org/10.1378/chest.12-2345
86. Matsuda T, Machii R (2015) Morphological distribution of lung cancer from cancer incidence in five continents vol. x. Jpn J Clin Oncol 45:404. https://doi.org/10.1093/jjco/hyv041
87. Iglesias VS, Giuranno L, Dubois LJ, Theys J, Vooijs M (2018) Drug resistance in non-small cell lung cancer: a potential for NOTCH targeting? Front Oncol 8:267
88. Sharma P, Mehta M, Dhanjal DS, Kaur S, Gupta G, Singh H, Thangavelu L, Rajeshkumar S, Tambuwala M, Bakshi HA, Chellappan DK, Dua K, Satija S (2019) Emerging trends in the

novel drug delivery approaches for the treatment of lung cancer. Chem Biol Interact 309:108720
89. Gariani J, Martin SP, Hachulla AL, Karenovics W, Adler D, Soccal PM, Becker CD, Montet X (2018) Noninvasive pulmonary nodule characterization using transcutaneous bioconductance: preliminary results of an observational study. Medicine (United States) 97:e11924. https://doi.org/10.1097/MD.0000000000011924
90. Herbst RS, Heymach JV, Lippman SM (2008) Lung cancer. N Engl J Med 359:1367–1380. https://doi.org/10.1056/NEJMra0802714
91. Mulvihill MS, Kratz JR, Pham P, Jablons DM, He B (2013) The role of stem cells in airway repair: implications for the origins of lung cancer. Chin J Cancer 32:71–74
92. Li G, Gao Y, Cui Y, Zhang T, Cui R, Jiang Y, Shi J (2016) Overexpression of CD44 is associated with the occurrence and migration of non-small cell lung cancer. Mol Med Rep 14:3159–3167. https://doi.org/10.3892/mmr.2016.5636
93. Overview of the initial treatment and prognosis of lung cancer—UpToDate
94. Khan AM, Shahzad M, Raza Asim MB, Imran M, Shabbir A (2015) Zingiber officinale ameliorates allergic asthma via suppression of Th2-mediated immune response. Pharm Biol 53:359–367. https://doi.org/10.3109/13880209.2014.920396
95. Alasvand M, Assadollahi V, Ambra R, Hedayati E, Kooti W, Peluso I (2019) Antiangiogenic effect of alkaloids. Oxidative Med Cell Longev 2019:9475908
96. Zhong W, Qin Y, Chen S, Sun T (2017) Antitumor effect of natural product molecules against lung cancer. In: A global scientific vision—prevention, diagnosis, and treatment of lung cancer. InTech, Rijeka
97. Xu J-Y, Meng Q-H, Chong Y, Jiao Y, Zhao L, Rosen EM, Fan S (2013) Sanguinarine is a novel VEGF inhibitor involved in the suppression of angiogenesis and cell migration. Mol Clin Oncol 1:331–336. https://doi.org/10.3892/mco.2012.41
98. Chakraborty S, Adhikary A, Mazumdar M, Mukherjee S, Bhattacharjee P, Guha D, Choudhuri T, Chattopadhyay S, Sa G, Sen A, Das T (2014) Capsaicin-induced activation of p53-SMAR1 auto-regulatory loop down-regulates VEGF in non-small cell lung cancer to restrain angiogenesis. PLoS One 9:e99743. https://doi.org/10.1371/journal.pone.0099743
99. Haura EB, Cress WD, Chellappan S, Zheng Z, Bepler G (2004) Antiapoptotic signaling pathways in non-small-cell lung cancer: biology and therapeutic strategies. Clin Lung Cancer 6:113–122. https://doi.org/10.3816/CLC.2004.n.025
100. Liu G, Pei F, Yang F, Li L, Amin AD, Liu S, Ross Buchan J, Cho WC (2017) Role of autophagy and apoptosis in non-small-cell lung cancer. Int J Mol Sci 18:367
101. Han B, Wu J, Huang L (2020) Induction of apoptosis in lung cancer cells by viburnum grandiflorum via mitochondrial pathway. Med Sci Monit 26:e920265-1. https://doi.org/10.12659/MSM.920265
102. Xie JH, Lai ZQ, Zheng XH, Xian YF, Li Q, Ip SP, Xie YL, Chen JN, Su ZR, Lin ZX, Yang XB (2019) Apoptosis induced by bruceine D in human non-small-cell lung cancer cells involves mitochondrial ROS-mediated death signaling. Int J Mol Med 44:2015–2026. https://doi.org/10.3892/ijmm.2019.4363
103. Capelletto E, Novello S (2012) Emerging new agents for the management of patients with non-small cell lung cancer. Drugs 72:37–52
104. Ansari J, Shackelford RE, El-Osta H (2016) Epigenetics in non-small cell lung cancer: from basics to therapeutics. Transl Lung Cancer Res 5:155–171. https://doi.org/10.21037/tlcr.2016.02.02
105. Mamdani H, Jalal SI (2020) Histone deacetylase inhibition in non-small cell lung cancer: hype or hope? Front Cell Dev Biol 8:1126. https://doi.org/10.3389/fcell.2020.582370
106. Liu Y, Tong Y, Yang X, Li F, Zheng L, Liu W, Wu J, Ou R, Zhang G, Hu M, Liu Z, Lu L (2016) Novel histone deacetylase inhibitors derived from Magnolia officinalis significantly enhance TRAIL-induced apoptosis in non-small cell lung cancer. Pharmacol Res 111:113–125. https://doi.org/10.1016/j.phrs.2016.05.028

107. Fulda S (2012) Histone deacetylase (HDAC) inhibitors and regulation of TRAIL-induced apoptosis. Exp Cell Res 318:1208–1212
108. Singh T, Prasad R, Katiyar SK (2016) Therapeutic intervention of silymarin on the migration of non-small cell lung cancer cells is associated with the axis of multiple molecular targets including class 1 HDACs, ZEB1 expression, and restoration of miR-203 and E-cadherin expression. Am J Cancer Res 6:1287–1301
109. Liu Y, Li A, Feng X, Sun X, Zhu X, Zhao Z (2018) Pharmacological investigation of the anti-inflammation and anti-oxidation activities of diallyl disulfide in a rat emphysema model induced by cigarette smoke extract. Nutrients 10:79. https://doi.org/10.3390/nu10010079
110. Zhao YL, Shang JH, Pu SB, Wang HS, Wang B, Liu L, Liu YP, Hong-Mei S, Luo XD (2016) Effect of total alkaloids from Alstonia scholaris on airway inflammation in rats. J Ethnopharmacol 178:258–265. https://doi.org/10.1016/j.jep.2015.12.022
111. Tan WSD, Peh HY, Liao W, Pang CH, Chan TK, Lau SH, Chow VT, Wong WSF (2016) Cigarette smoke-induced lung disease predisposes to more severe infection with nontypeable haemophilus influenzae: protective effects of andrographolide. J Nat Prod 79:1308–1315. https://doi.org/10.1021/acs.jnatprod.5b01006
112. Rassias DJ, Weathers PJ (2019) Dried leaf artemisia annua efficacy against non-small cell lung cancer. Phytomedicine 52:247–253. https://doi.org/10.1016/j.phymed.2018.09.167
113. Yang ZC, Qu ZH, Yi MJ, Wang C, Ran N, Xie N, Fu P, Feng XY, Lv ZD, Xu L (2013) Astragalus extract attenuates allergic airway inflammation and inhibits nuclear factor κb expression in asthmatic mice. Am J Med Sci 346:390–395. https://doi.org/10.1097/MAJ.0b013e3182753175
114. Qiu YY, Zhu JX, Bian T, Gao F, Qian XF, Du Q, Yuan MY, Sun H, Shi LZ, Yu MH (2014) Protective effects of astragaloside IV against ovalbumin-induced lung inflammation are regulated/mediated by T-bet/GATA-3. Pharmacology 94:51–59. https://doi.org/10.1159/000362843
115. Koul A, Kapoor N, Bharati S (2012) Histopathological, enzymatic, and molecular alterations induced by cigarette smoke inhalation in the pulmonary tissue of mice and its amelioration by aqueous azadirachta indica leaf extract. J Environ Pathol Toxicol Oncol 31:7–15. https://doi.org/10.1615/JEnvironPatholToxicolOncol.v31.i1.20
116. Xu D, Wan C, Wang T, Tian P, Li D, Wu Y, Fan S, Chen L, Shen Y, Wen F (2015) Berberine attenuates cigarette smoke-induced airway inflammation and mucus hypersecretion in mice. Int J Clin Exp Med 8:8641–8647
117. Lee JW, Shin NR, Park JW, Park SY, Kwon OK, Lee HS, Hee Kim J, Lee HJ, Lee J, Zhang ZY, Oh SR, Ahn KS (2015) Callicarpa japonica Thunb. attenuates cigarette smoke-induced neutrophil inflammation and mucus secretion. J Ethnopharmacol 175:1–8. https://doi.org/10.1016/j.jep.2015.08.056
118. Hwang YP, Jin SW, Choi JH, Choi CY, Kim HG, Kim SJ, Kim Y, Lee KJ, Chung YC, Jeong HG (2017) Inhibitory effects of L-theanine on airway inflammation in ovalbumin-induced allergic asthma. Food Chem Toxicol 99:162–169. https://doi.org/10.1016/j.fct.2016.11.032
119. Khoi PN, Park JS, Kim JH, Xia Y, Kim NH, Kim KK, Do JY (2013) (−)-Epigallocatechin-3-gallate blocks nicotine-induced matrix metalloproteinase-9 expression and invasiveness via suppression of NF-κB and AP-1 in endothelial cells. Int J Oncol 43:868–876. https://doi.org/10.3892/ijo.2013.2006
120. Lee JW, Park HA, Kwon OK, Jang YG, Kim JY, Choi BK, Lee HJHJ, Lee S, Paik JH, Oh SR, Ahn KS, Lee HJHJ (2016) Asiatic acid inhibits pulmonary inflammation induced by cigarette smoke. Int Immunopharmacol 39:208–217. https://doi.org/10.1016/j.intimp.2016.07.010
121. Chin LH, Hon CM, Chellappan DK, Chellian J, Madheswaran T, Zeeshan F, Awasthi R, Aljabali AA, Tambuwala MM, Dureja H, Negi P, Kapoor DN, Goyal R, Paudel KR, Satija S, Gupta G, Hsu A, Wark P, Mehta M, Wadhwa R, Hansbro PM, Dua K (2020) Molecular mechanisms of action of naringenin in chronic airway diseases. Eur J Pharmacol 879:173139
122. Bui TT, Piao CH, Kim SM, Song CH, Shin HS, Lee CH, Chai OH (2017) Citrus tachibana leaves ethanol extract alleviates airway inflammation by the modulation of Th1/Th2 imbalance

via inhibiting NF-κB signaling and histamine secretion in a mouse model of allergic asthma. J Med Food 20:676–684. https://doi.org/10.1089/jmf.2016.3853

123. Chiu PR, Te Lee W, Te Chu Y, Lee MS, Jong YJ, Hung CH (2008) Effect of the Chinese herb extract osthol on IL-4-induced eotaxin expression in BEAS-2B cells. Pediatr Neonatol 49:135–140. https://doi.org/10.1016/S1875-9572(08)60028-5
124. Wang A, Wang J, Zhang S, Zhang H, Xu Z, Li X (2017) Curcumin inhibits the development of non-small cell lung cancer by inhibiting autophagy and apoptosis. Exp Ther Med 14:5075–5080. https://doi.org/10.3892/etm.2017.5172
125. Zhang M, Tang J, Li Y, Xie Y, Shan H, Chen M, Zhang J, Yang X, Zhang Q, Yang X (2017) Curcumin attenuates skeletal muscle mitochondrial impairment in COPD rats: PGC-1α/SIRT3 pathway involved. Chem Biol Interact 277:168–175. https://doi.org/10.1016/j.cbi.2017.09.018
126. Aydemir E, Simsek E, Imir N, Göktürk R, Yesilada E, Fiskin K (2015) Cytotoxic and apoptotic effects of Ebenus boissieri Barbey on human lung cancer cell line A549. Pharmacogn Mag 11:37. https://doi.org/10.4103/0973-1296.157679
127. Šutovská M, Capek P, Kazimierová I, Pappová L, Jošková M, Matulová M, Fraňová S, Pawlaczyk I, Gancarz R (2015) Echinacea complex—chemical view and anti-asthmatic profile. J Ethnopharmacol 175:163–171. https://doi.org/10.1016/j.jep.2015.09.007
128. Avisetti DR, Babu KS, Kalivendi SV (2014) Activation of p38/JNK pathway is responsible for embelin induced apoptosis in lung cancer cells: transitional role of reactive oxygen species. PLoS One 9:e87050. https://doi.org/10.1371/journal.pone.0087050
129. Kennedy-Feitosa E, Okuro RT, Pinho Ribeiro V, Lanzetti M, Barroso MV, Zin WA, Porto LC, Brito-Gitirana L, Valenca SS (2016) Eucalyptol attenuates cigarette smoke-induced acute lung inflammation and oxidative stress in the mouse. Pulm Pharmacol Ther 41:11–18. https://doi.org/10.1016/j.pupt.2016.09.004
130. De Moura RS, Ferreira TS, Lopes AA, Pires KMP, Nesi RT, Resende AC, Souza PJC, Da Silva AJR, Borges RM, Porto LC, Valenca SS (2012) Effects of Euterpe oleracea Mart. (AÇAÍ) extract in acute lung inflammation induced by cigarette smoke in the mouse. Phytomedicine 19:262–269. https://doi.org/10.1016/j.phymed.2011.11.004
131. Yang X, Yao J, Luo Y, Han Y, Wang Z, Du L (2013) P38 MAP kinase mediates apoptosis after genipin treatment in non-small-cell lung cancer H1299 cells via a mitochondrial apoptotic cascade. J Pharmacol Sci 121:272–281. https://doi.org/10.1254/jphs.12234FP
132. Huang CH, Yang ML, Tsai CH, Li YC, Lin YJ, Kuan YH (2013) Ginkgo biloba leaves extract (EGb 761) attenuates lipopolysaccharide-induced acute lung injury via inhibition of oxidative stress and NF-κB-dependent matrix metalloproteinase-9 pathway. Phytomedicine 20:303–309. https://doi.org/10.1016/j.phymed.2012.11.004
133. Hsu HF, Houng JY, Kuo CF, Tsao N, Wu YC (2008) Glossogin, a novel phenylpropanoid from Glossogyne tenuifolia, induced apoptosis in A549 lung cancer cells. Food Chem Toxicol 46:3785–3791. https://doi.org/10.1016/j.fct.2008.09.068
134. Peh HY, Tan WSD, Chan TKK, Pow CW, Foster PS, Wong WSF (2017) Vitamin E isoform γ-tocotrienol protects against emphysema in cigarette smoke-induced COPD. Free Radic Biol Med 110:332–344. https://doi.org/10.1016/j.freeradbiomed.2017.06.023
135. Jiang R, Hu C, Li Q, Cheng Z, Gu L, Li H, Guo Y, Li Q, Lu Y, Li K, Chen M, Zhang X (2019) Sodium new houttuyfonate suppresses metastasis in NSCLC cells through the Linc00668/miR-147a/slug axis. J Exp Clin Cancer Res 38. https://doi.org/10.1186/s13046-019-1152-9
136. Fu L, Chen W, Guo W, Wang J, Tian Y, Shi D, Zhang X, Qiu H, Xiao X, Kang T, Huang W, Wang S, Deng W (2013) Berberine targets AP-2/hTERT, NF-κB/COX-2, HIF-1α/VEGF and cytochrome-c/caspase signaling to suppress human cancer cell growth. PLoS One 8:e69240. https://doi.org/10.1371/journal.pone.0069240
137. Alfaifi MY (2016) Kanahia Laniflora methanolic extract suppressed proliferation of human non-small cell lung cancer A549 cells. Asian Pac J Cancer Prev 17:4755. https://doi.org/10.22034/APJCP.2016.17.10.4755

138. Mahajan SG, Mehta AA (2011) Suppression of ovalbumin-induced Th2-driven airway inflammation by β-sitosterol in a guinea pig model of asthma. Eur J Pharmacol 650:458–464. https://doi.org/10.1016/j.ejphar.2010.09.075
139. Lim YJ, Na HS, Yun YS, Choi IS, Oh JS, Rhee JH, Cho BH, Lee HC (2009) Suppressive effects of ginsan on the development of allergic reaction in murine asthmatic model. Int Arch Allergy Immunol 150:32–42. https://doi.org/10.1159/000210378
140. Gao W, Guo Y, Yang H (2017) Platycodin D protects against cigarette smoke-induced lung inflammation in mice. Int Immunopharmacol 47:53–58. https://doi.org/10.1016/j.intimp.2017.03.009
141. Guan Z, Shen L, Liang H, Yu H, Hei B, Meng X, Yang L (2017) Resveratrol inhibits hypoxia-induced proliferation and migration of pulmonary artery vascular smooth muscle cells by inhibiting the phosphoinositide 3-kinase/protein kinase B signaling pathway. Mol Med Rep 16:1653–1660. https://doi.org/10.3892/mmr.2017.6814
142. Park K, Lee JS, Choi JS, Nam YJ, Han JH, Byun HD, Song MJ, Oh JS, Kim SG, Choi Y (2016) Identification and characterization of baicalin as a phosphodiesterase 4 inhibitor. Phytother Res 30:144–151. https://doi.org/10.1002/ptr.5515
143. Lee SU, Lee S, Ro H, Choi JH, Ryu HW, Kim MO, Yuk HJ, Lee J, Hong ST, Oh SR (2018) Piscroside C inhibits TNF-α/NF-κB pathway by the suppression of PKCδ activity for TNF-RSC formation in human airway epithelial cells. Phytomedicine 40:148–157. https://doi.org/10.1016/j.phymed.2018.01.012
144. Ryu HW, Lee SU, Lee S, Song HH, Son TH, Kim YU, Yuk HJ, Ro H, Lee CK, Hong ST, Oh SR (2017) 3-Methoxy-catalposide inhibits inflammatory effects in lipopolysaccharide-stimulated RAW264.7 macrophages. Cytokine 91:57–64. https://doi.org/10.1016/j.cyto.2016.12.006
145. Chen X, Wen T, Wei J, Wu Z, Wang P, Hong Z, Zhao L, Wang B, Flavell R, Gao S, Wang M, Yin Z (2013) Treatment of allergic inflammation and hyperresponsiveness by a simple compound, Bavachinin, isolated from Chinese herbs. Cell Mol Immunol 10:497–505. https://doi.org/10.1038/cmi.2013.27
146. Li S, Shi D, Zhang L, Yang F, Cheng G (2018) Oridonin enhances the radiosensitivity of lung cancer cells by upregulating bax and downregulating bcl-2. Exp Ther Med 16:4859–4864. https://doi.org/10.3892/etm.2018.6803
147. Lu W, Dai B, Ma W, Zhang Y (2012) A novel taspine analog, HMQ1611, inhibits growth of non-small cell lung cancer by inhibiting angiogenesis. Oncol Lett 4:1109–1113. https://doi.org/10.3892/ol.2012.855
148. Luo F, Liu J, Yan T, Miao M (2017) Salidroside alleviates cigarette smoke-induced COPD in mice. Biomed Pharmacother 86:155–161. https://doi.org/10.1016/j.biopha.2016.12.032
149. Lee H, Yu SR, Lim D, Lee H, Jin EY, Jang YP, Kim J (2015) Galla Chinensis attenuates cigarette smoke-associated lung injury by inhibiting recruitment of inflammatory cells into the lung. Basic Clin Pharmacol Toxicol 116:222–228. https://doi.org/10.1111/bcpt.12308
150. Li D, Wang J, Sun D, Gong X, Jiang H, Shu J, Wang Z, Long Z, Chen Y, Zhang Z, Yuan L, Guan R, Liang X, Li Z, Yao H, Zhong N, Lu W (2018) Tanshinone IIA sulfonate protects against cigarette smoke-induced COPD and down-regulation of CFTR in mice. Sci Rep 8:376. https://doi.org/10.1038/s41598-017-18745-5
151. Zhang DF, Zhang J, Li R (2015) Salvianolic acid B attenuates lung inflammation induced by cigarette smoke in mice. Eur J Pharmacol 761:174–179. https://doi.org/10.1016/j.ejphar.2015.05.003
152. Song JW, Seo CS, Cho ES, Kim TI, Won YS, Kwon HJ, Son JK, Son HY (2016) Meso-dihydroguaiaretic acid attenuates airway inflammation and mucus hypersecretion in an ovalbumin-induced murine model of asthma. Int Immunopharmacol 31:239–247. https://doi.org/10.1016/j.intimp.2015.12.033
153. Zhong S, Nie YC, Gan ZY, Liu XD, Fang ZF, Zhong BN, Tian J, Huang CQ, Lai KF, Zhong NS (2015) Effects of Schisandra chinensis extracts on cough and pulmonary inflammation in a

cough hypersensitivity guinea pig model induced by cigarette smoke exposure. J Ethnopharmacol 165:73–82. https://doi.org/10.1016/j.jep.2015.02.009
154. Zhou DG, Diao BZ, Zhou W, Feng JL (2016) Oroxylin A inhibits allergic airway inflammation in ovalbumin (OVA)-induced asthma murine model. Inflammation 39:867–872. https://doi.org/10.1007/s10753-016-0317-3
155. Li D, Xu D, Wang T, Shen Y, Guo S, Zhang X, Guo L, Li X, Liu L, Wen F (2015) Silymarin attenuates airway inflammation induced by cigarette smoke in mice. Inflammation 38:871–878. https://doi.org/10.1007/s10753-014-9996-9
156. Tyagi A, Singh RP, Ramasamy K, Raina K, Redente EF, Dwyer-Nield LD, Radcliffe RA, Malkinson AM, Agarwal R (2009) Growth inhibition and regression of lung tumors by silibinin: modulation of angiogenesis by macrophage-associated cytokines and nuclear factor-κB and signal transducers and activators of transcription 3. Cancer Prev Res 2:74–83. https://doi.org/10.1158/1940-6207.CAPR-08-0095
157. Huang WC, Chan CC, Wu SJ, Chen LC, Shen JJ, Kuo ML, Chen MC, Liou CJ (2014) Matrine attenuates allergic airway inflammation and eosinophil infiltration by suppressing eotaxin and Th2 cytokine production in asthmatic mice. J Ethnopharmacol 151:470–477. https://doi.org/10.1016/j.jep.2013.10.065
158. Lee H, Jung KH, Park S, Kil YS, Chung EY, Jang YP, Seo EK, Bae H (2014) Inhibitory effects of Stemona tuberosa on lung inflammation in a subacute cigarette smokeinduced mouse model. BMC Complemnt Altern Med 14. https://doi.org/10.1186/1472-6882-14-513
159. Zhang X, Liu J, Xiong H, Cheng Y, Cui C, Zhang X, Xu L (2013) Effects of taraxasterol on ovalbumin-induced allergic asthma in mice. J Ethnopharmacol 148:787–793. https://doi.org/10.1016/j.jep.2013.05.006
160. Ni L, Zhu X, Gong C, Luo Y, Wang L, Zhou W, Zhu S, Li Y (2015) Trichosanthes kirilowii fruits inhibit non-small cell lung cancer cell growth through mitotic cell-cycle arrest. Am J Chin Med 43:349–364. https://doi.org/10.1142/S0192415X15500238
161. Liu LL, Zhang Y, Zhang XF, Li FH (2018) Influence of rutin on the effects of neonatal cigarette smoke exposure-induced exacerbated MMP-9 expression, Th17 cytokines and NF-kB/iNOS-mediated inflammatory responses in asthmatic mice model. Korean J Physiol Pharmacol 22:481–491. https://doi.org/10.4196/kjpp.2018.22.5.481
162. Dong Y, Yang J, Yang L, Li P (2020) Quercetin inhibits the proliferation and metastasis of human non-small cell lung cancer cell line: the key role of Src-mediated fibroblast growth factor-inducible 14 (Fn14)/nuclear factor kappa B (NF-κB) pathway. Med Sci Monit 26: e920537-1. https://doi.org/10.12659/MSM.920537

Role of Phytoconstituents in Targeting Cytokines for Managing Pathophysiology of Lung Diseases

35

Anand Maurya, Anurag Kumar Singh, Harsh Yadav, Gaurav Mishra, Rajendra Awasthi, Sunil Kumar Mishra, and Sushil Kumar Chaudhary

Abstract

Normal lungs can be distressed by internal as well as external factors which can be reason for producing diseases. The lungs respond against these factors via a protective mechanism known as inflammation. Cytokines, small secreted proteins, play a crucial role to control these inflammations, but overproductions of cytokines create many respiratory diseases (responsible for high mortality rates). These types of respiratory diseases are treated with the help of anti-inflammatory agents to minimize the pulmonary inflammation. Medicinal plants and drugs derived from plants have numerous therapeutic effects when utilized by patients. Occurrence of several active phytoconstituents in plants is useful to treat many types of inflammatory diseases. The use of medicinal plants and

A. Maurya · G. Mishra
Department of Medicinal Chemistry, Faculty of Ayurveda, Institute of Medical Sciences, Banaras Hindu University, Varanasi, Uttar Pradesh, India

A. K. Singh
Centre of Experimental Medicine and Surgery, Institute of Medical Sciences, Banaras Hindu University, Varanasi, Uttar Pradesh, India

H. Yadav
Department of Pharmacy, Indira Gandhi National Tribal University, Amarkantak, Madhya Pradesh, India

R. Awasthi
Amity Institute of Pharmacy, Amity University Uttar Pradesh, Noida, Uttar Pradesh, India

S. K. Mishra
Department of Pharmaceutical Engineering and Technology, Indian Institute of Technology (Banaras Hindu University), Varanasi, Uttar Pradesh, India

S. K. Chaudhary (✉)
Faculty of Pharmacy, DIT University, Dehradun, Uttarakhand, India

K. Dua et al. (eds.), *Targeting Cellular Signalling Pathways in Lung Diseases*,
https://doi.org/10.1007/978-981-33-6827-9_35

phytoconstituents (alkaloids, flavonoids, terpenes, and others) as curative tool in pulmonary inflammation is increasing significantly. In the current book chapter, the role of medicinal plants and phytoconstituents for the managing pathophysiology of lung diseases is appraised.

Keywords

Phytoconstituents · Cytokines · Lung diseases · Interleukins · COPD

Abbreviations

CCL17	Thymus and activation-regulated chemokine (also known as TARC)
COPD	Chronic obstructive pulmonary disease
CTGF	Connective tissue growth factor
GM-CSF	Granulocyte-macrophage colony-stimulating factor
IL	Interleukin
MCP-1	Monocyte chemotactic protein (also known as CCL2)
PARC	Pulmonary and activation-regulated chemokine (also known as CCL18)
PDGF	Platelet-derived growth factor
RANTES	Regulated and normal T-cell expressed and secreted
SDF-1/CXCL12	Stromal cell-derived factor-1
TGF-β	Transforming growth factor beta
TNF-α	Tumor necrosis factor alpha
TSLP	Thymic stromal lymphopoietin
VEGF	Vascular endothelial growth factor

35.1 Introduction

Lung disease frequency has been expanding, and it is identified with reasons for death for quite a long time. Among the lung maladies, asthma, chronic obstructive pulmonary disease (COPD), and acute respiratory distress syndrome (ARDS) are the well-known and responsible for high mortality and morbidity rates [1–3]. Most of the lung diseases commonly consist of lung inflammation which may be chronic or acute [4]. Entry of foreign materials in the lung caused inflammation that is part of cellular response. These responses are significant for the living cell, but chronic inflammation might be injurious to the lung. Lung inflammation includes the actuation of various inflammatory cells, such as lymphocytes, eosinophils, macrophages, and neutrophils. These cells are a wellspring of various inflammatory agents, for example, tumor necrosis factor (TNF-α), histamine, interleukins (IL-1β, IL-4, IL-5, and IL-6), leukotrienes, prostaglandins, and nitric oxide. Secretion of inflammatory agents associated with the signals and symptoms detected in lung ailments, for example, lung function loss, obstruction and hyperresponsiveness in airway, edema in airway, hypersecretion of mucus, and lung restoration [3, 4]. Chronic diseases of the lungs like chronic obstructive pulmonary disease, asthma, pulmonary fibrosis, airflow limitation, chronic inflammation, loss of elasticity, emphysema, mucus hypersecretion, and bronchoconstriction have different etiologies but also share usual characters in that they often involve repeated cycles of injury to the respiratory epithelium (due to inhaled pollutants such as pathogens or tobacco smoke), shared with dysregulated epithelial repair pathways and chronic activation of inflammatory processes, collectively leading to incongruous production of airway mucus, increased myofibroblast differentiation, fibroblast activation, smooth muscle proliferation, etc., thereby subsequently resulting in airway remodeling and decline in lung function. While differing in etiology, COPD and asthma share clinical symptoms including coughing, wheezing, shortness of breath, and sputum production [5]. Asthma, COPD, acute lower respiratory tract infections, TB, and lung cancer are among the major five public causes of severe illness and death globally [6].

Chronic obstructive pulmonary disease (COPD) is a disease of the airways and other structures of the lung and causes a foremost health-care problem. Although COPD usually exhibits at an older age as part of multi-disease, there is growing indication that events early in life add to diminished lung function in adults, which proposes that risk factors other than those already known (gases from cigarette smoking and inhaled particles and biomass fuel) are essential in the disease's etiology [7]. COPD is not one single disease but an umbrella term used to describe chronic lung diseases that cause restrictions in lung airflow. The most common symptoms of COPD are breathlessness, or a "need for air," excessive sputum production, and a chronic cough. The WHO Global Alliance against Chronic Respiratory Diseases was established with the goal of reducing the load of chronic respiratory diseases, toward a world in which all individuals breathe freely, and focuses on the needs of people with chronic respiratory diseases in low-income and middle-income countries [8]. Worldwide, greater than 500 million people are travailing from different lung diseases [9]. About 333 million people are affected by asthma, the most common chronic disease of childhood affecting 14% of all children globally. Millions of people are killed by pneumonia annually which is a foremost cause of mortality among children below 5 years old. Each year over ten million people develop tuberculosis (TB), and 1.4 million die from it, making it the most common lethal infectious disease. Further each year lung cancer kills 1.6 million people and is one of the most deadly cancers [10]. Pulmonary inflammation is a trademark of many respiratory diseases such as asthma, chronic obstructive pulmonary disease (COPD), and acute respiratory syndrome distress (ARDS). Most of these diseases are cured with anti-inflammatory therapy in order to avoid or to reduce the pulmonary inflammation.

In the modern world, approximately 81% of the present worldwide population depends on the use of herbal medicines for their principal health care and other necessities which signifies the possibility of herbal medicines. Besides, herbal medicines have a specific range of therapeutic actions and are less toxic with rarer adverse effects. According to the World Health Organization (WHO), all over the globe, the use of herbal medicines surpasses conventional drugs by a factor two to three times [11]. Currently, herbal medicines have regained their status for treatment of lung diseases with their effectiveness and safety feature being strongly sustained by controlled clinical trials [12]. In recent years herbal medicine-derived natural products have been used in traditional medicine, and scientific studies to evaluate the value of these compounds have developed. Various constituents obtained from plant origin have biological effects in vitro and in vivo, comparatively flavonoids, alkaloids, and terpenoids. The respiratory tract is a vital site of immune regulation, required to allow protective immunity counter to pathogens, while avoiding irregular inflammatory responses to inhaled allergens and reducing tissue damage. Numerous cell types work in a group to control pulmonary immune responses and preserve tolerance in the respiratory tract, together with regulatory and effector T cells, airway and interstitial macrophages, dendritic cells, and the airway epithelium. The cytokines transforming growth factor β, interleukin (IL-) 10, IL-27, and IL-35 are main coordinators of immune regulation in tissues, for example, the lung. Numerous cell types are associated in the regulation of immune responses in the lung, including FoxP3+ and FoxP3- regulatory T-cell (Treg) subsets [13], resident airway macrophages (AMs) [14], interstitial macrophages (IMs) [15], dendritic cells (DCs) [16], and the alveolar epithelia and conducting airway [17], emphasizing the significance of cell-cell communication in directing pulmonary immunity. This kind of cellular interactions in the immune system is subject to signaling mediated by cytokines. Various natural constituents from different plants can target the cell-signaling pathway showing valuable activity in contrast to respiratory disease (Figs. 35.1 and 35.2). The natural products which serve as a storehouse of essential chemotherapeutics contain alkaloids, flavonoids, and terpenes that produce desired effects against chronic respiratory diseases. These also hasten the development of novel drug systems by providing appropriate pharmacophores for generating optimum effect against the target pathways connected with the manifestation of respiratory ailments. This book chapter presents a brief discussion on the effects of various phytoconstituents like alkaloids, flavones, and terpenes for stopping the conventional and emerging respiratory disorders and their role in chronic lung disease and cytokines. The pathophysiology of COPD reveals chronic inflammation in the lung parenchyma intervened by macrophages, neutrophils, and cytotoxic (CD8+) T lymphocytes [18]. Plant-derived products have been utilized traditionally in various periods of civilization from ancient time, and the consumption of medicinal plant-based products has been expanding significantly [19]. Notwithstanding, various plant-based moieties considerably affect cellularly, and proof of the beneficial effects of plant-based moiety in inflammatory lung diseases has been growing [19].

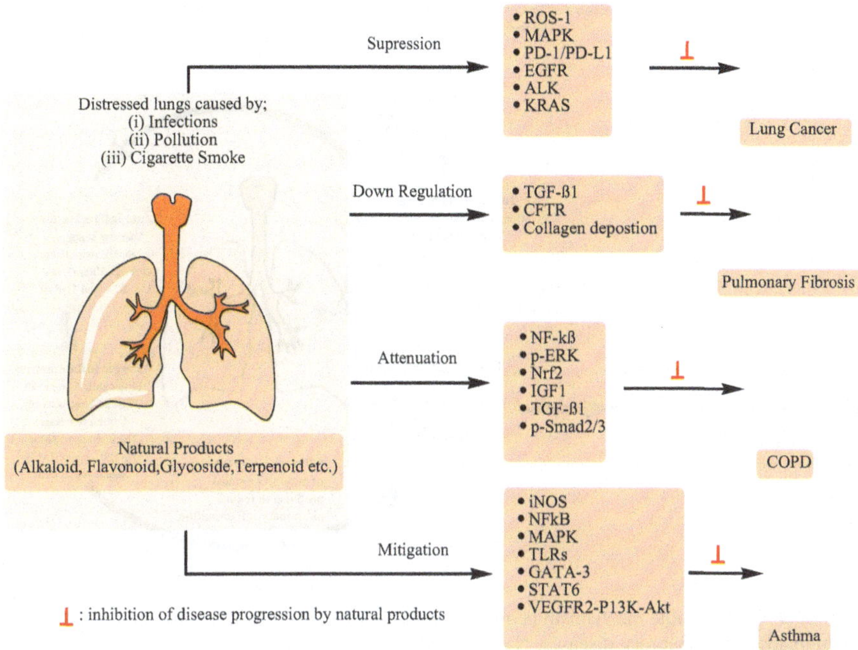

Fig. 35.1 Disease progression inhibition by targeting different cell-signaling pathways by natural products

35.2 Cytokines and Type of Cytokines

Cytokines are lesser, secreted monitoring proteins that act a significant role in immune habitat. Cytokines are involved in cell-cell communication and control several activities with cell existence, cell growth, and orientation of gene appearance. Elevated levels of cytokines are produced during an adaptive immune reaction by CD4+ "helper T cells" (TH) for different purposes. These helper T cells can develop TH1 cells creating high levels of interferon (IFN); TH2 cells producing high levels of interleukin (IL)-13, IL-5, and IL-4; or TH17 cells making high levels of IL-17 [20]. Cytokines are involved variously in COPD, asthma, and pulmonary fibrosis in different ways, so give significant targets for therapeutic action (Table 35.1).

35.2.1 Etiology and Therapeutics of COPD

The obsessive components influencing COPD are different and unpredictably connected. In the falling apart advancement of COPD, different incendiary middle components are delivered from epithelial cells and also invaded incendiary cells in

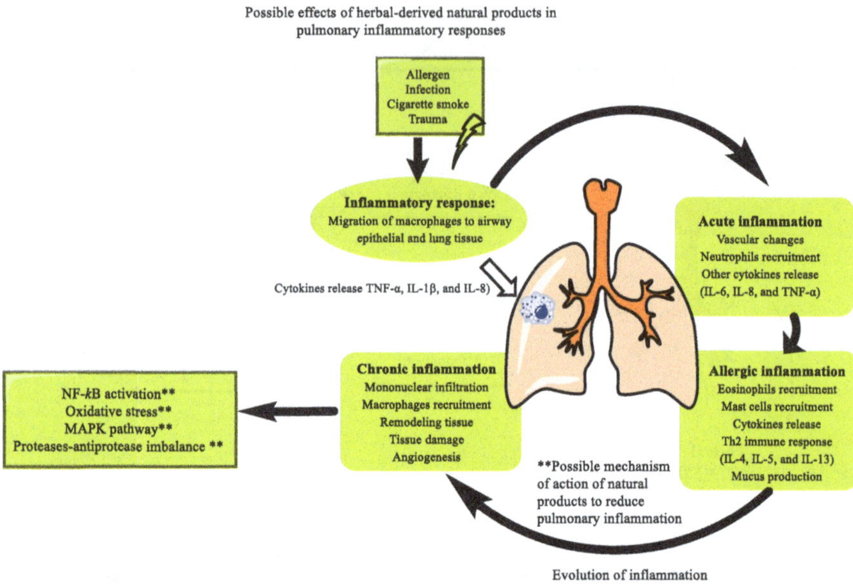

Fig. 35.2 The possible mechanism behind the effects of natural products is to reduce pulmonary inflammation in respiratory diseases. Natural products reduce the pulmonary inflammation by the inhibition of the transcription of NF-κB to the nucleus, hence stopping the progress of all the inflammatory processes triggered by an allergen, cigarette smoke, virus, or bacteria. Thus, these natural products can reduce inflammatory cytokine release and oxidative stress

Table 35.1 List of cytokines implied in COPD, pulmonary fibrosis, and asthma pathogeneses

COPD	Pulmonary fibrosis	Asthma
IL-6, IL-8, IL-32, IL-18, IL-17, TNF-α, TGF-β, IL-1, TSLP	IL-1, IL-17,IL-10, CTGF, IL-4, IL-13, MCP-1, GM-CSF CCL17, PARC/CCl18 SDF-1/CXCL12, oncostatin M PDGF	IL-17, IL-4, IL-5, IL-13, IL-25, IL-33, TSLP eotaxin, VEGF, IL-1, IL-8, IL-18, TNF-α, RANTES, GH-CSF

the lungs, comprising neutrophils, macrophages, and T lymphocytes. Tumor necrosis factor-α (TNF-α), interleukin-1 (IL-1) and IL-6 (proinflammatory cytokines), and chemokines containing IL-8 initiate and pull in the surrounding cells during obsessive measure. Transforming growth factor-β (TGF-β) has been accounted for to cause route fibrosis, prompting airway obliteration. Many hypotheses have been established for inhibition of cytokines or their receptors. The IL-1β and IL-18 (major component of the inflammasome) are considered as active targets toward different inflammasome constituents [21, 22]. Reactive oxygen species (ROS) are also involved in provoking COPD. High ratio of oxidants is obtained in tobacco-smoking person which facilitates a range of free radicals with ROS. Oxidative hassle with an additional production of ROS increases the inflammatory responses and progresses the pathological steps involved in COPD. However, many molecules attached like NADPH oxidase, nuclear erythroid-2-related factor 2 (Nrf2), myeloperoxidase, and superoxide dismutase can be viewed as targets for COPD remedy. Moreover the disproportion between proteases as well as anti-proteases leads to alveolar wall annihilation. Mainly, matrix metalloproteinase (MMP) and neutrophil elastase are involvedly controlled in COPD remedy. Various studies denote that the initiation and/or raised expression of matrix metalloproteinases such as MMP-2, MMP-9, and MMP-12 are closely linked to the development of COPD [23]. In other studies, sirtuins were indicated to be deeply attached in COPD. The behavior of sirtuin 1 expression is reduced in the lungs of COPD patients. The stimulation of sirtuin 1 and 6 has been shown to have defensive effects against COPD [24], and sirtuin activators maybe will be used for COPD treatment.

35.3 Effect of Various Phytoconstituents in Chronic Lung Disease and Cytokines

3.1 Alkaloid
3.2 Flavonoid
3.3 Terpenes
3.4 Miscellaneous.

35.3.1 Alkaloids

The alkaloids have played a historical role in the management of various lung diseases (e.g., burnt *Atropa belladonna* fumes) in different cultures. Some alkaloids which show bronchodilatory effects like l-vasicinone, vasicinolone, vasicine, deoxyvasicine, vasicinal, moringine, maiontone, and trigonelline reduced blocking in airways; piperine and dergamine act as a respiratory stimulant and several with a potential role in the effective management of numerous lung diseases [25]. For the treatment of respiratory disorder, some structure-based alkaloid medicinal compounds are classified as follows.

35.3.1.1 Iso-steroid Alkaloids

Liu et al., 2020, reported the beneficial potential of six iso-steroid alkaloids (1–6, Fig. 35.3) found from *Fritillaria cirrhosa* bulbus contrary to the CS-induced oxidative stress in RAW264.7 macrophages. Reportedly, the recognized six test alkaloids pointedly mitigated the production of reactive oxygen species (ROS), increased the level of antioxidant molecule glutathione (GSH), and stimulated the Nrf2-induced expression of HO-1 protein. Primarily, glucoside moiety at carbon-3 position in alkaloid **2** and the lack of β-OH at carbon-17 position in alkaloid **5** caused a higher GSH/GSSG ratio. Also, the occurrence of β-CH_3 substituent at carbon-20 position in alkaloid **3** favored its efficacy and acceptable cytotoxicity. The presence of −OH substitution at carbon-3 position in all the test alkaloids confirmed the importance for

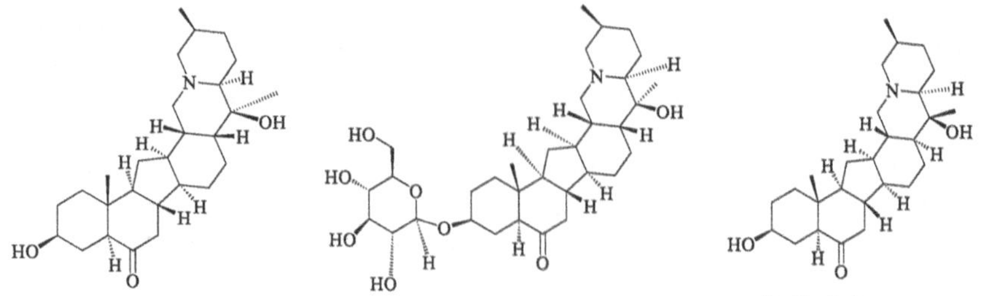

1. Imperialine 2. Imperialine-3-b-D-glucoside 3. Verticinone

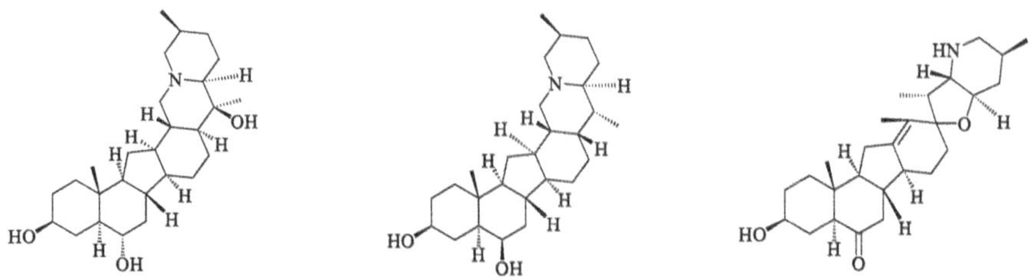

4. Verticine 5. Delavine 6. Peimisine

Fig. 35.3 Compilation of iso-steroid alkaloids

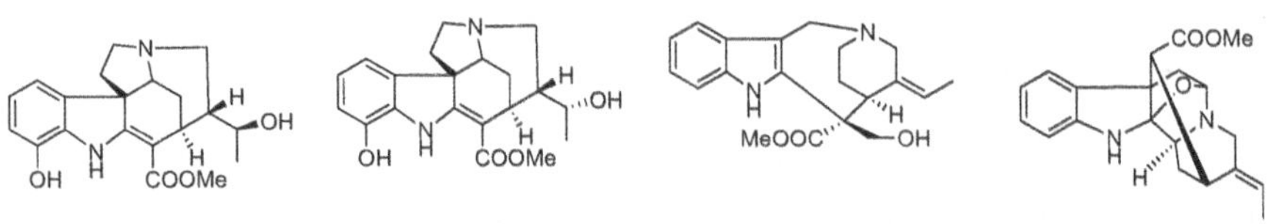

7. Scholaricine 8. 19-epischolaricine 9. Vallesamine 10. Picrinine

Fig. 35.4 List of indole and quinolone alkaloids

inducing HO-1 expression, which debilitated in the presence of C=O group at the same position [26].

35.3.1.2 Indole and Quinolone Alkaloids

Zhao et al., 2017, reported the pharmacokinetics and therapeutic effects of alkaloids (7–10, Fig. 35.4), found from *Alstonia scholaris*, on ovalbumin-induced airway allergic inflammatory model [27]. These compounds downregulated the stages of eosinophils and leukocytes, which was established by histopathological analysis of lungs. Significantly, the test alkaloids reduced the secretion of proinflammatory cytokine IL-4, a significant mediator of allergic responses, ultimately resulting in a significant reduction of pulmonary eosinophils and balancing the increase of immunoglobulin E (IgE) in serum. Besides, these alkaloids exhibited usefulness for the treatment of postinfectious symptoms in animal models by alleviating the levels of inflammatory cytokines followed by downregulation of the expression of IL-6 [28].

35.3.1.3 Quinazoline Alkaloids

Liu et al., 2015, isolated quinazoline from the aerial parts of *Peganum harmala* L alkaloids (11–13, Fig. 35.5) and confirmed their expectorant, antitussive, and bronchodilating properties in animal models. The alkaloids excellently depressed the symptoms accompanying capsaicin-induced acute pulmonary inflammation at doses 5, 15, and 45 mg/kg, associated with codeine phosphate administered at 30 mg/kg. Additionally, the bronchodilating test confirmed that the test alkaloids extensively prolonged the preconvulsive times in animal models, greater to the standard drug aminophylline, thus confirming the potency of these alkaloids for treating bronchial asthma [29].

35.3.1.4 Bisbenzylisoquinoline

Alkaloids isolated from *Cissampelos sympodialis* "warifteine" (14, Fig. 35.6) show distinguished efficacy in airway hyperreactivity in the animal model of asthma [30]. Oral pretreatment with the test alkaloid warifteine in animal models expressively reduced the allergen-induced airway hyperreactivity (AHR) to inhaled methacholine. Furthermore, it also reduced the IL-13 levels in bronchoalveolar lavage, which supports as the key regulator of AHR [31]. Kim et al., 2019, recognized natural bisbenzylisoquinoline **alkaloids 15, 16, and 17** (Fig. 35.5) from *Stephania tetrandra*, which inhibited human coronavirus OC43 infection of MRC-5 human lung cells at its early stages [32]. Coronaviruses infect the respiratory system, thus exhibiting severe conditions such as bronchiolitis and pneumonia [33]. The test alkaloids reportedly controlled the replication of human coronavirus OC43 and controlled the expression of viral protein and the virus-induced response of the host MRC-5 cells. Notably, alkaloid 15 (Fig. 35.5) activated the p38 mitogen-activated protein kinase (MAPK) pathway in the virus-infected cells, which ultimately improved their feasibility with the least signs of cytotoxicity. The MRC-5 cells exposed to the test alkaloids exhibited negligible expression of proinflammatory cytokines, which are then upregulated by the virus infection [34].

35.3.1.5 Benzophenanthridine

Kim et al., 2015, reported the alkaloid "chelidonine" isolated from *Chelidonium majus* decreased the IL-4- and eotaxin-2-mediated eosinophilic airway inflammation

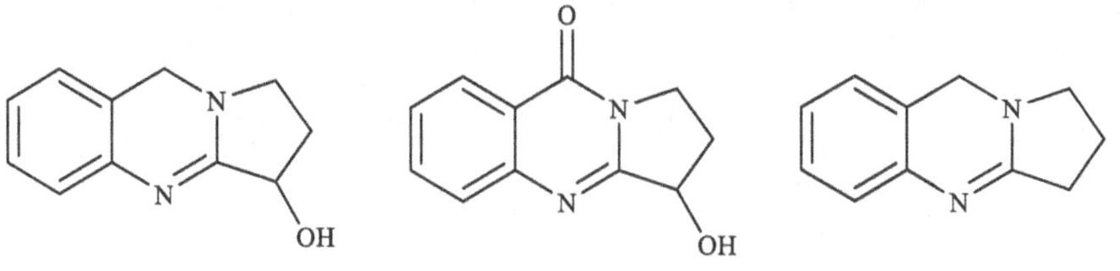

11. Vasicine 12. Vasicinone 13. Deoxyvasicine

Fig. 35.5 Structure quinazoline alkaloids

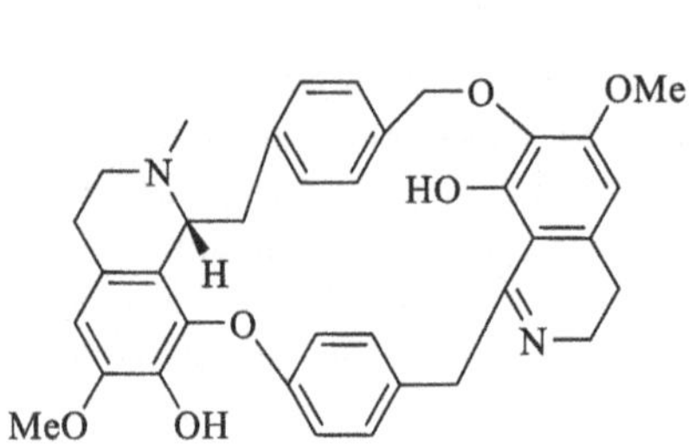

14. Warifteine

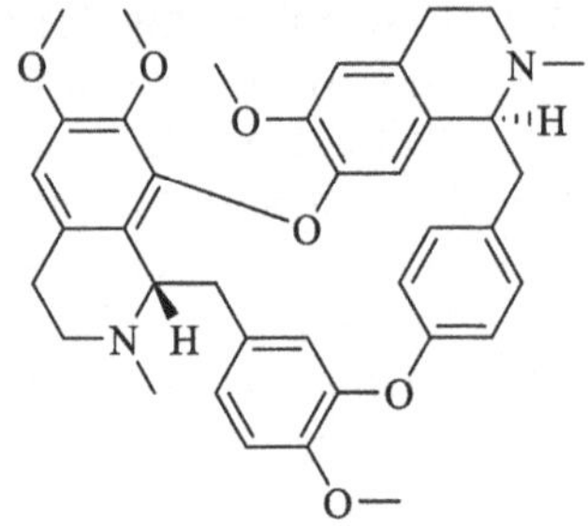

15. Tetrandrine

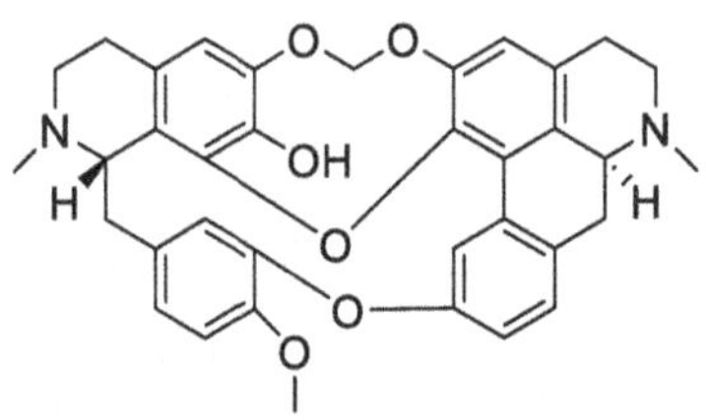

16. Fangchinoline

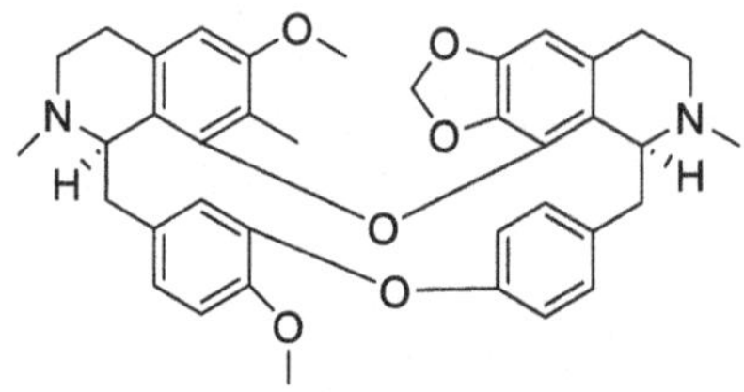

17. Cepharanthine

Fig. 35.6 Bisbenzylisoquinoline derivatives

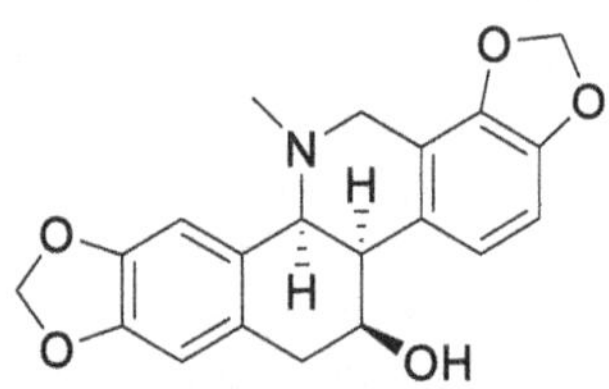

18-Chelidonine

Fig. 35.7 derivatives of benzophenanthridine

in asthmatic animal models (18 Fig. 35.7) [35]. The test alkaloid considerably suppressed the level of eosinophils in the airways, additionally downregulating eotaxin-2, interleukins, and cytokines in the bronchoalveolar lavage fluid (BALF). Markedly, chelidonine-treated animal models revealed a significant decrease of CD4 + and CD8+ T cells, Gr-1+/CD11b+, and 351 CD3/CCR3+ positive cells, which otherwise exaggerate inflammation process by secreting Th2 cytokines and degranulation of eosinophils [36].

35.3.1.6 Acridone Quinolone Alkaloids

Acridone quinolone alkaloids (19 and 20, Fig. 35.8) found from *Pericarpium Citri Reticulatae* reportedly show significant anti-asthmatic activity. The administration

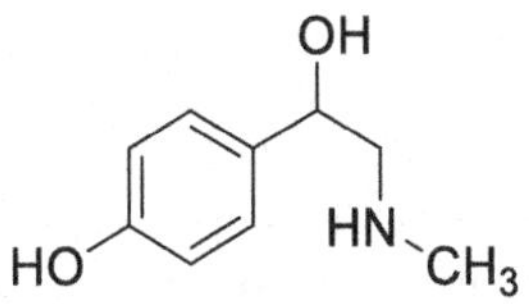

19. Synephrine

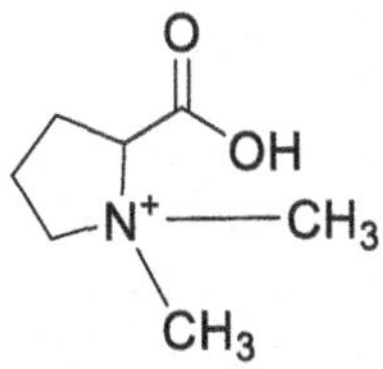

20. Stachydrine

Fig. 35.8 List of acridone quinolone alkaloids

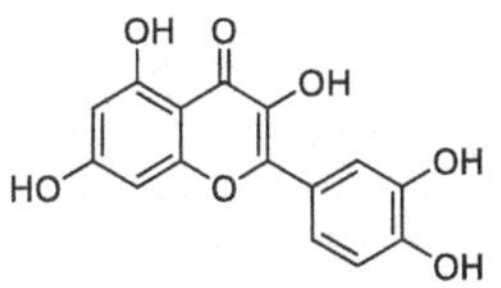

21. Quercetin

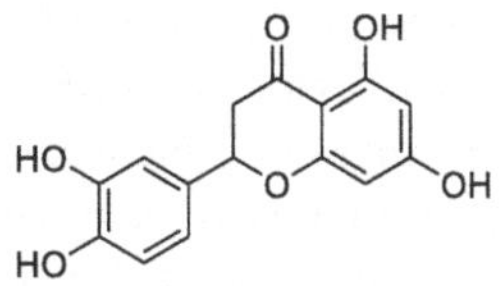

22. Eriodictyol

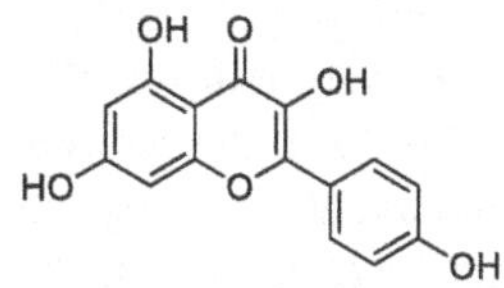

23. Kaempferol

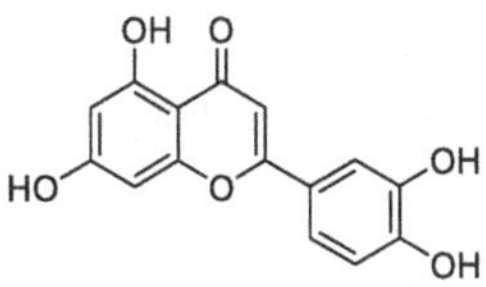

24. Luteolin

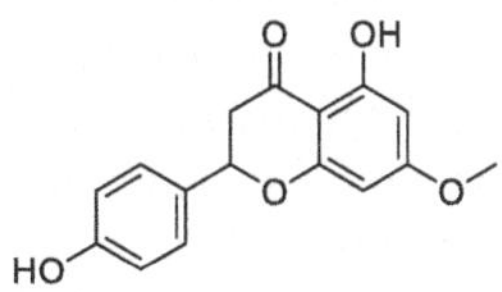

25. Sakuranetin

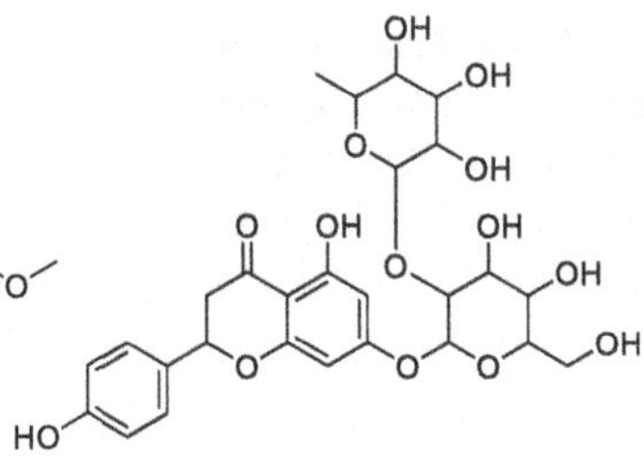

26. Naringin

Fig. 35.9 List of flavonoids

of acridone quinolone alkaloids in animal models with histamine-induced asthma downregulated eosinophil expression in bronchoalveolar lavage fluid and serum accompanied by prominent attenuation of IgE, IL-4, and IL-5. The alkaloid confirmed spasmolytic effects on acetylcholine chloride-induced contractions in the animal trachea [37].

35.3.2 Flavonoids

List is depicted in Fig. 35.9.

35.3.2.1 Quercetin

Quercetin is isolated from a variety of berry plants. It is polyphenolic flavonoid (3, 3′, 4′, 5, 7-pentahydroxyflvanone), brilliant citron yellow in color, isolated from fruits and leafy vegetables, and vital part of the individual diet (approximately daily intake; 50–800 mg/day). Despite attaining significance for its exceptional variety of health benefits, which ensures it is an essential flavonoid for the enhancement of novel and effective functional remedies and foods. Numerous investigational trials have indicated that quercetin exerts many therapeutic effects against a wide range of diseases [38]. A recent investigation suggested quercetin as having promising anti-inflammatory properties in the management of airway allergic inflammation [39]. Its prominent therapeutic potential is hindered by its high lipophilicity, low aqueous solubility (0.48 μg/mL), and low oral bioavailability (2%) [40, 41]. Quercetin effectively reduces different serum cytokines TNF-α, IL1-β, IL-6, and nitric oxide (NO), and the mechanism involves an increase in IL-10 secretion, an anti-inflammatory cytokine. It also reduced lung permeability, the number of macrophages and neutrophils, and the myeloperoxidase activity. Additionally quercetin is effective in reducing COX-2, iNOS expression, HMGB1, and p65-nuclear factor kappa B (NF-κB) [42].

35.3.2.2 Eriodictyol

Eriodictyol is a Chinese herb flavonoid isolated from *Dracocephalum rupestre*. Eriodictyol has long been used as an antioxidant and anti-inflammatory agent. Eriodictyol effectively improves lipopolysaccharide (LPS)-induced acute lung injury (ALI) in mice which is shown by stopping the expression of inflammatory cytokines in macrophages and regulating the transcription factor nuclear erythroid-2-related factor 2 (Nrf2) pathway [43].

35.3.2.3 Kaempferol

Kaempferol is a naturally occurring flavonoid [44]. Kaempferol is the main glycoside found in family Zingiberaceae *Kaempferia* rhizomes, present in several plants and plant-derived foods while evaluated in ALI models induced by LPS, appearing effective in decreasing pulmonary edema as well as the bleeding and width of the alveolar wall. Kaempferol also inhibited the inflammatory cells and complete protein in the BALF and cytokines, IL-1β and IL-6, and TNF-α. Despite the increase in superoxide dismutase action, the mechanism of action is controlled by signaling pathways MAPK and NF-κB [45].

35.3.2.4 Luteolin

Luteolin is a naturally occurring flavonoid biphenol (3′,4′,5,7-tetrahydroxyflavone) and is found in a variety of dietary fruits, spices, medicinal herbs, and vegetable sources, including celery, peppers, carrots, oregano, olive oil, thyme, rosemary, and peppermint; it possesses a range of pharmacological activities. Luteolin has properties characteristically to prompt accumulation of O_2; however, it reduces the H_2O_2 concentration in lung cancer cells. Many scientific pieces of evidence reported that luteolin has anti-inflammatory, antioxidant, neuroprotective, and anticancer effects. Luteolin has its anti-inflammatory activity by suppressing IL-17, IL-6, IL-1β, and TNF-α [46–48]. Many studies suggested that pretreatment with flavonoid luteolin reduced neutrophilic inflammation, pulmonary hemorrhage, and interstitial edema. The mechanism to regulate pulmonary inflammation was because of the reduction of cytokines such as KC, ICAM-1, and TNF-α content in the bronchoalveolar lavage fluid (BALF). Management of oxidative damage and lipid peroxidation was recognized by reduced activity of catalase and superoxide dismutase. The mechanism behind the effects of luteolin is the inhibition of NF-κB and the MAPK activity [49].

35.3.2.5 Sakuranetin

Sakuranetin (4′,5-dihydroxy-7-methoxyflavanone) a flavonoid is mainly isolated from the species *Baccharis retusa* (family: Asteraceae) [50]. Sakuranin is the chief glycoside isolated from the bark of *Prunus pseudo-cerasus* for the first time by Asahina et al. The reported asthma model suggested that sakuranetin flavonoid prompts a reduction of the Th2 cytokines such as IL-5, eotaxin, and RANTES in sensitized mice. Additionally, this flavonoid reduced the number of inflammatory cells in the lung, mainly IgE and eosinophils in the ovalbumin-sensitized animal model. Sakuranetin effects appear to be associated with the inhibition of NF-κB in the lung [51].

35.3.2.6 Naringin

Naringin is a constituent of the dried unripe or ripe fruit peel of the *Citrus grandis* "Tomentosa" (*Exocarpium Citri Grandis*); the BALF in a model of asthma using guinea pigs has confirmed its effects by airway hyperresponsiveness and reduced enhanced cough formation in the lungs and inhibited the increases in the leukocytes, IL-4, IL-5, and IL-13 [52]. It was confirmed that after naringin administration, inhibition of IL-4 and INF-δ levels, ovalbumin-induced increased airway resistance (Raw), and eosinophil infiltration occurs [53].

35.3.3 Terpenes

List is depicted in Fig. 35.10.

35.3.3.1 Limonene

Limonene is present naturally in bushes (e.g., citrus and other fruits), vegetables, spices, certain trees, and meats [54]. Limonene is a potential monoterpene in treating lung inflammation. Reportedly, *Dermatophagoides farinae* animal model treatment with limonene reduces the allergic airway inflammation, mainly by alleviating the reactive oxygen species. The limonene exposure significantly decreases the levels of TGF-β, interleukins-3/5, eotaxin, and MCP-1 in the bronchoalveolar lavages. Besides, the limonene administration revoked airway fibrosis, goblet cell metaplasia, and thickness of the lung's smooth muscles. These outcomes confirmed the candidature of limonene as a preventive agent in the asthmatic treatment [55].

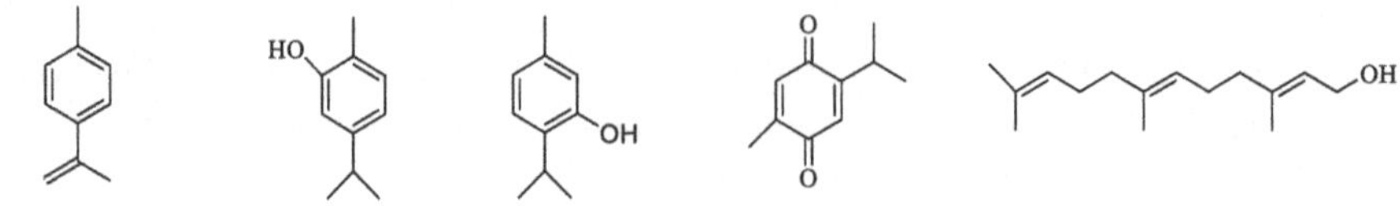

27. Limonene 28. Carvacrol 29. Thymol 29. Thymoquinone 30. Famesol

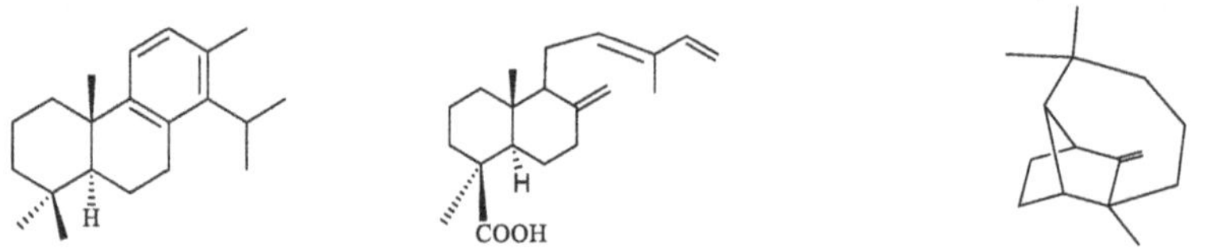

31. Totarol 32. trans-communic acid 33. Longifolene

Fig. 35.10 Derivatives of terpenes

35.3.3.2 Carvacrol

Carvacrol (5-isopropyl-2-methylphenol) is a monoterpenoid phenol, mainly isolated from some plant species like thyme, oregano, wild bergamot, as well as pepperwort [56, 57]. Reports suggested that bronchoalveolar lavage fluid of the sensitized animal models of carvacrol exerts a relaxant effect on the trachea's smooth muscles by decreasing the total number of WBC, monocyte, neutrophil, and eosinophil count in the human blood [58].

35.3.3.3 Thymol

Thymol (2-isopropyl-5-methylphenol, IPMP) is a naturally occurring phenol extracted from various plant sources, such as *Thymus*, *Satureja*, *Euphrasia rostkoviana*, *Lippia*, and *Coridothymus*. Additionally, it is also present in different amounts in the essential oil of some plant sources belonging to the genera *Thymus* (thyme), *Carum* (ajowan), and *Origanum* (*Origanum*) [59]. Thymol has reported several pharmacological activities and biological properties such as antioxidant, anti-infectious, antifungal, anesthetic, analgesic, antibacterial, and antiparasitic properties; it has potential as an immunomodulator and growth promoter. Thymol is the chief constituent of thyme oil and other natural plant sources with a low water solubility and phenolic structure [60]. Some thymols are terpene alcohols, terpenes, aldehydes, ketones, esters, ethers, and phenolic derivatives [61, 81]. The therapeutic effects in lung tissues of thymol in animal models with LPS-induced acute lung injury showed improved pathological changes. Reportedly, thymol administration caused the LPS-induced influx of the inflammatory cells, and metabolites, TNF-α, and interleukins decreased in bronchoalveolar lavage fluid [55].

In addition, thymol also controlled the LPS-mediated upsurge of MDO and MPS levels and significantly depressed the activity of SOD by this means, pausing the activation of NF-κB in the lungs. These localized enzymes mainly affect the adhesion and margination of neutrophils in the lung, thereby causing severe lung injury [62].

35.3.3.4 Thymoquinone

Thymoquinone (2-methyl-5-isopropyl-1,4-benzoquinone), a monoterpene molecule, is found in seeds of *Nigella sativa* L. and belongs to the Ranunculaceae family. The ameliorative protective effects of thymoquinone on the LPS-induced pulmonary vascular damage appear because of the presence of thymoquinone, which causes the downregulation of proinflammatory cytokines and interleukins. Furthermore, thymoquinone administration reduced the expression of NF-κB, TNF-α, and IL-1β in the respiratory airways produced due to the LPS sensitization [63]. Thymoquinone also guards against respiratory damage caused by cigarette smoke by alleviating the expression of proinflammatory leukotrienes, thromboxanes, prostaglandins, and prominently IL-1β, the prime biomarker cytokine in cigarette smoker's lung [64].

The protective effects of thymoquinone reportedly attenuated the expression of IL-4/5 and enhanced the expression of platelet endothelial cell adhesion molecule1 (CD31) and α-smooth muscle actin (α-SMA) in ovalbumin-sensitized asthma animal models. Moreover, thymoquinone for asthma amelioration deactivated VEGFR2-PI3K-Akt pathway and upregulated the expression of Slit glycoprotein-2 (Slit-2), which validates its anti-neoangiogenesis effect [65].

35.3.3.5 Farnesol

Farnesol (sesquiterpenols) contains 15 carbon atoms that exist extensively in fruits like peaches, vegetables like tomatoes and corn, herbs like lemongrass and chamomile, and in the volatile oils of ambrette seeds and citronella [66, 67]. It has been reported that farnesol is active against oxidative stress, inflammation, and lung injury brought by intratracheal instillation of cigarette smoke extract in rats [68]. Farnesol inhibits inflammation in the lungs and airways in asthmatic mice by restoring the discharge capability of peritoneal macrophages and decreased TNF-α/IL-10 cytokine secretion ratios, which indicates that farnesol might enhance general immunity [69]. The isoprenoid farnesol holds outstanding antinociceptive and chemopreventive strength and proves protection against chronic lung inflammation, oxidative stress, and lung injury produced by cigarette smoke. Farnesol showed protective action of the lung by dropping LDH levels and lowered activity of reduced glutathione reductase (GR), glutathione (GSH), glutathione peroxidase (GPx), and catalase enzymes. The lowered H_2O_2 content in lung cells further confirmed the lung cells' cytoprotective effects against cigarette smoke. Farnesol was found effective in the alleviation of benzopyrene-induced respiratory stress in animal models. In many animal models (lung tissue), farnesol was able to sustain optimal levels of phospholipids and transformed the catalytic activity of benzopyrene enzymes NADPH-cytochrome P450 reductase, glutathione S-transferase (GST), and microsomal epoxide hydrolase (mEH). All results proved farnesol's defensive action against benzopyrene-induced lung inflammation, edema, and epithelial damage [70].

35.3.4 Miscellaneous

List is depicted in Fig. 35.11.

35.3.4.1 Sulforaphane

Sulforaphane (SFN) is a natural dietary phytonutrient; isothiocyanate group of organosulfur compounds presents in many cruciferous family and is the maximum in broccoli sprouts, cabbage, broccoli, and kale. Numerous studies have shown that sulforaphane retains strong antioxidant properties. The inflammatory properties of particulate, such as smoke and tobacco, were blocked by SFN; airborne pollutants produce inflammatory properties which are controlled by SFN by upregulation of the phase II enzymes in the respiratory epithelial cells. Sulforaphane binds to the cysteine residues of NF-κB and thus prevents NF-κB activation; SFN is the effective inducer of PII enzymes and is supposed to perform through stimulation of the Nrf2 transcription factor and the antioxidant response element (ARE) [71]. Inflammatory effects of oxidative stress in respiratory airway passages in human subjects are reduced by SFN. Additionally, it provokes the symptoms of mucosal phase II enzymes in the upper airway passage [72].

35.3.4.2 Curcumin

Curcumin a polyphenol is an active secondary metabolite obtained from in *Curcuma longa*. Curcumin I plays an immunomodulatory role as it regulates the inflammatory expression of cytokines (IL-1, IL-6, TNF-α, TGF-β) in alveolar cells. In COPD conditions, it prevents the harmful cardiovascular events by decreasing the athero-sclerotic AT-LDL levels [73].

35.3.4.3 Caffeic Acid

Caffeic acid (CA) (group of hydroxycinnamic acid) [74, 75] is made up of two functional groups phenolic and acrylic and is present in the bark of *Eucalyptus globulus* [76, 77]. CA is used in various food supplements to improve exercise-related fatigue, increase athletic performance, encourage weight loss, and avoid cancer due to its antioxidant properties [78, 79]. Khayyal et al. 2003 reported CA considerable inhibition in the improvement of ventilatory functions in human asthma subjects by lessening in proinflammatory factors TNF-α, IL-6, IL-8, ICAM-1, prostaglandins E2 and F2α, leukotriene D4, and the upsurge in IL-10 [80].

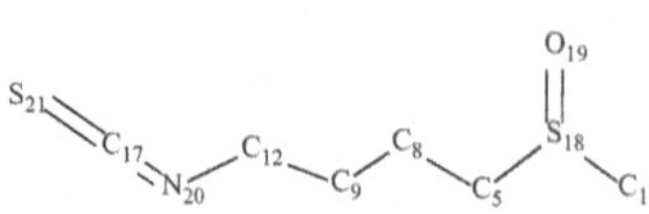

34. Sulforaphane

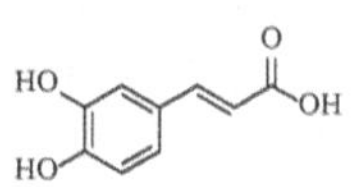

35. Caffeic acid

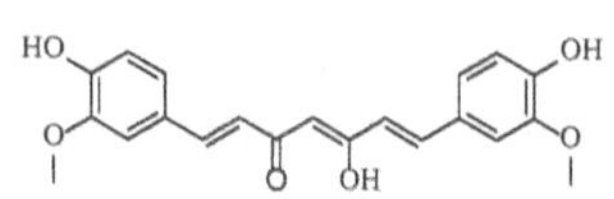

36. Curcumin

Fig. 35.11 Compilation of miscellaneous derivatives

35.4 Conclusion

Lung disease is a usual and serious reason of sickness and death worldwide. Medicinal plant-based drugs may be considered as important remedies for management of lung diseases. Various compounds present in plants including alkaloid (e.g., imperialine, vasicine, picrinine), flavonoids (e.g., quercetin, eriodictyol, kaempferol), terpenes (e.g., limonene, thymol, carvacrol), etc. show potent activity in lungs diseases by different pathways.

Conflict of Interest The authors confirm that this article has no conflict of interest.

References

1. Bateman ED, Hurd SS, Barnes PJ, Bousquet J, Drazen JM, FitzGerald M, Gibson P, Ohta K, O'byrne P, Pedersen SE, Pizzichini E (2008) Global strategy for asthma management and prevention: GINA executive summary. Eur Respir J 31(1):143–178
2. Ware LB, Matthay MA (2000) The acute respiratory distress syndrome. N Engl J Med 342 (18):1334–1349
3. Barnes PJ (2000) Chronic obstructive pulmonary disease. N Engl J Med 343(4):269–280
4. Murphy DM, O'Byrne PM (2010) Recent advances in the pathophysiology of asthma. Chest 137(6):1417–1426
5. Mehta P, Bothiraja C, Mahadik K, Kadam S, Pawar A (2018) Phytoconstituent based dry powder inhalers as biomedicine for the management of pulmonary diseases. Biomed Pharmacother 108:828–837
6. Grote L (2019) The global burden of sleep apnoea. Lancet Respir Med 7(8):645–647
7. Britto CJ, Brady V, Lee S, Cruz CS (2017) Respiratory viral infections in chronic lung diseases. Clin Chest Med 38(1):87–96
8. WHO. Global alliance against chronic respiratory disease-about GARD. https://www.who.int/gard/en/. Accessed 22 Oct 2020
9. Carkeek EC (2018) The burden of chronic respiratory disease in the Western Cape. Doctoral dissertation, University of Cape Town
10. Forum of International Respiratory Societies (2017) The global impact of respiratory disease, 2nd edn. European Respiratory Society, Sheffield
11. Mali RG, Dhake AS (2011) A review on herbal antiasthmatics. Orient Pharm Exp Med 11 (2):77–90
12. Mehta P, Shah R, Lohidasan S, Mahadik KR (2015) Pharmacokinetic profile of phytoconstituent (s) isolated from medicinal plants-a comprehensive review. J Tradit Complement Med 5(4):207–227
13. Lloyd CM, Hawrylowicz CM (2009) Regulatory T cells in asthma. Immunity 31(3):438–449
14. Hussell T, Bell TJ (2014) Alveolar macrophages: plasticity in a tissue-specific context. Nat Rev Immunol 14(2):81–93
15. Bedoret D, Wallemacq H, Marichal T, Desmet C, Calvo FQ, Henry E, Closset R, Dewals B, Thielen C, Gustin P, de Leval L (2009) Lung interstitial macrophages alter dendritic cell functions to prevent airway allergy in mice. J Clin Invest 119(12):3723–3738
16. Cook PC, MacDonald AS (2016) Dendritic cells in lung immunopathology. In Seminars in immunopathology 38: 4, 449–460. Springer Berlin Heidelberg
17. Whitsett JA, Alenghat T (2015) Respiratory epithelial cells orchestrate pulmonary innate immunity. Nat Immunol 16(1):27–35
18. Franconi F, Campesi I, Romani A (2020) Is extra virgin olive oil an ally for women's and men's cardiovascular health? Therapeutics, Cardiovascular

19. Slader CA, Reddel HK, Jenkins CR, Armour CL (2006) BOSNIC-ANTICEVICH SZ. Complementary and alternative medicine use in asthma: who is using what? Respirology 11(4):373–387
20. Zhu J, Paul WE (2008) CD4 T cells: fates, functions, and faults. Blood J Am Society Hematol 112(5):1557–1569
21. Rovina N, Dima E, Gerassimou C, Kollintza A, Gratziou C, Roussos C (2009) Interleukin-18 in induced sputum: association with lung function in chronic obstructive pulmonary disease. Respir Med 103(7):1056–1062
22. Zhang H (2011) Anti-IL-1β therapies. Recent Pat DNA Gene Seq 5(2):126–135
23. Churg A, Zhou S, Wright JL (2012) Matrix metalloproteinases in COPD. Eur Respir J 39 (1):197–209
24. Chun P (2015) Role of sirtuins in chronic obstructive pulmonary disease. Arch Pharm Res 38 (1):1–0
25. Khan H (2015) Alkaloids: potential therapeutic modality in the management of asthma. J Ayurvedic Herbal Med 1(3)
26. Liu S, Yang T, Ming TW, Gaun TK, Zhou T, Wang S, Ye B (2020) Isosteroid alkaloids from *Fritillaria cirrhosa* bulbus as inhibitors of cigarette smoke-induced oxidative stress. Fitoterapia 140:104434
27. Zhao YL, Cao J, Shang JH, Liu YP, Khan A, Wang HS, Qian Y, Liu L, Ye M, Luo XD (2017) Airways antiallergic effect and pharmacokinetics of alkaloids from *Alstonia scholaris*. Phytomedicine 27:63–72
28. Zhao YL, Yang ZF, Shang JH, Huang WY, Wang B, Wei X, Khan A, Yuan ZW, Liu YP, Wang YF, Wang XH (2018) Effects of indole alkaloids from leaf of *Alstonia scholaris* on post-infectious cough in mice. J Ethnopharmacol 218:69–75
29. Liu W, Wang Y, He DD, Li SP, Zhu YD, Jiang B, Cheng XM, Wang ZT, Wang CH (2015) Antitussive, expectorant, and bronchodilating effects of quinazoline alkaloids (±)-vasicine, deoxyvasicine, and (±)-vasicinone from aerial parts of *Peganum harmala* L. Phytomedicine 22 (12):1088–1095
30. Bezerra-Santos CR, Vieira-de-Abreu A, Vieira GC, Jaime Filho R, Barbosa-Filho JM, Pires AL, Martins MA, Souza HS, Bandeira-Melo C, Bozza PT, Piuvezam MR (2012) Effectiveness of *Cissampelos sympodialis* and its isolated alkaloid warifteine in airway hyperreactivity and lung remodeling in a mouse model of asthma. Int Immunopharmacol 13(2):148–155
31. Corren J (2013) Role of interleukin-13 in asthma. Curr Allergy Asthma Rep 13(5):415–420
32. Kim DE, Min JS, Jang MS, Lee JY, Shin YS, Park CM, Song JH, Kim HR, Kim S, Jin YH, Kwon S (2019) Natural bis-benzylisoquinoline alkaloids-tetrandrine, fangchinoline, and cepharanthine, inhibit human coronavirus OC43 infection of MRC-5 human lung cells. Biomol Ther 9(11):696
33. Huang C, Wang Y, Li X, Ren L, Zhao J, Hu Y, Zhang L, Fan G, Xu J, Gu X, Cheng Z (2020) Clinical features of patients infected with 2019 novel coronavirus in Wuhan, China. Lancet 395 (10223):497–506
34. Prasher P, Sharma M, Mehta M, Paudel KR, Satija S, Chellappan DK, Dureja H, Gupta G, Tambuwala MM, Negi P, Wich PR (2020) Plants derived therapeutic strategies targeting chronic respiratory diseases: chemical and immunological perspective. Chem Biol Interact 2020:109125
35. Kim SH, Hong JH, Lee YC (2015) Chelidonine, a principal isoquinoline alkaloid of Chelidonium majus, attenuates eosinophilic airway inflammation by suppressing IL-4 and eotaxin-2 expression in asthmatic mice. Pharmacol Rep 67(6):1168–1177
36. Lourenço O, Fonseca AM, Taborda-Barata L (2016) Human CD8+ T cells in asthma: possible pathways and roles for NK-like subtypes. Front Immunol 7:638
37. Fu M, Zou B, An K, Yu Y, Tang D, Wu J, Xu Y, Ti H (2019) Anti-asthmatic activity of alkaloid compounds from *Pericarpium Citri Reticulatae* (*Citrus reticulata* 'Chachi'). Food Funct 10 (2):903–911

38. De Oliveira MR, Nabavi SM, Braidy N, Setzer WN, Ahmed T, Nabavi SF (2016) Quercetin and the mitochondria: a mechanistic view. Biotechnol Adv 34(5):532–549
39. Scalia S, Haghi M, Losi V, Trotta V, Young PM, Traini D (2013) Quercetin solid lipid microparticles: a flavonoid for inhalation lung delivery. Eur J Pharm Sci 49(2):278–285
40. Hatahet T, Morille M, Hommoss A, Devoisselle JM, Müller RH, Bégu S (2016) Quercetin topical application, from conventional dosage forms to nanodosage forms. Eur J Pharm Biopharm 108:41–53
41. D'Andrea G (2015) Quercetin: a flavonol with multifaceted therapeutic applications? Fitoterapia 106:256–271
42. Wang L, Chen J, Wang B, Wu D, Li H, Lu H, Wu H, Chai Y (2014) Protective effect of quercetin on lipopolysaccharide-induced acute lung injury in mice by inhibiting inflammatory cell influx. Exp Biol Med 239(12):1653–1662
43. Zhu GF, Guo HJ, Huang Y, Wu CT, Zhang XF (2015) Eriodictyol, a plant flavonoid, attenuates LPS-induced acute lung injury through its antioxidative and anti-inflammatory activity. Exp Ther Med 10(6):2259–2266
44. Pan X, Liu X, Zhao H, Wu B, Liu G (2020) Antioxidant, anti-inflammatory and neuroprotective effect of kaempferol on rotenone-induced Parkinson's disease model of rats and SH-S5Y5 cells by preventing loss of tyrosine hydroxylase. J Funct Foods 74:104140
45. Chen X, Yang X, Liu T, Guan M, Feng X, Dong W, Chu X, Liu J, Tian X, Ci X, Li H (2012) Kaempferol regulates MAPKs and NF-κB signaling pathways to attenuate LPS-induced acute lung injury in mice. Int Immunopharmacol 14(2):209–216
46. Lamy S, Moldovan PL, Saad AB, Annabi B (2015) Biphasic effects of luteolin on interleukin-1β-induced cyclooxygenase-2 expression in glioblastoma cells. Biochim Biophys Acta Mol Cell Res 1853(1):126–135
47. Shi F, Zhou D, Ji Z, Xu Z, Yang H (2015) Anti-arthritic activity of luteolin in Freund's complete adjuvant-induced arthritis in rats by suppressing P2X4 pathway. Chem Biol Interact 226:82–87
48. Xu Y, Zhang J, Liu J, Li S, Li C, Wang W, Ma R, Liu Y (2015) Luteolin attenuate the D-galactose-induced renal damage by attenuation of oxidative stress and inflammation. Nat Prod Res 29(11):1078–1082
49. Kuo MY, Liao MF, Chen FL, Li YC, Yang ML, Lin RH, Kuan YH (2011) Luteolin attenuates the pulmonary inflammatory response involves abilities of antioxidation and inhibition of MAPK and NFκB pathways in mice with endotoxin-induced acute lung injury. Food Chem Toxicol 49(10):2660–2666
50. Stompor M (2020) A Review on Sources and Pharmacological Aspects of Sakuranetin. Nutrients 12(2):513
51. Toledo AC, Sakoda CP, Perini A, Pinheiro NM, Magalhães RM, Grecco S, Tibério IF, Câmara NO, Martins MA, Lago JH, Prado CM (2013) Flavanone treatment reverses airway inflammation and remodelling in an asthma murine model. Br J Pharmacol 168(7):1736–1749
52. Jiao HY, Su WW, Li PB, Liao Y, Zhou Q, Zhu N, He LL (2015) Therapeutic effects of naringin in a guinea pig model of ovalbumin-induced cough-variant asthma. Pulm Pharmacol Ther 33:59–65
53. Guihua X, Shuyin L, Jinliang G, Wang S (2016) Naringin protects ovalbumin-induced airway inflammation in a mouse model of asthma. Inflammation 39(2):891–899
54. Nikfar S, Behboudi AF (2014) Limonene. 78–82
55. Hirota R, Nakamura H, Bhatti SA, Ngatu NR, Muzembo BA, Dumavibhat N, Eitoku M, Sawamura M, Suganuma N (2012) Limonene inhalation reduces allergic airway inflammation in Dermatophagoides farinae-treated mice. Inhal Toxicol 24(6):373–381
56. Singh T, Chittenden C (2010) Efficacy of essential oil extracts in inhibiting mould growth on panel products. Build Environ 45(10):2336–2342
57. De Martino L, De Feo V, Formisano C, Mignola E, Senatore F (2009) Chemical composition and antimicrobial activity of the essential oils from three chemotypes of Origanum vulgare L. ssp. hirtum (Link) Ietswaart growing wild in Campania (Southern Italy). Molecules 14 (8):2735–2746

58. Alavinezhad A, Khazdair MR, Boskabady MH (2018) Possible therapeutic effect of carvacrol on asthmatic patients: a randomized, double blind, placebo-controlled, Phase II clinical trial. Phytother Res 32(1):151–159
59. Anand N, Sharma S (eds) (1997) Approaches to design and synthesis of antiparasitic drugs. Elsevier
60. Najafloo R, Behyari M, Imani R, Nour S (2020) A mini-review of thymol incorporated materials: applications in antibacterial wound dressing. J Drug Delivery Sci Technol 2020:101904
61. Hudaib M, Speroni E, Di Pietra AM, Cavrini V (2002) GC/MS evaluation of thyme (Thymus vulgaris L.) oil composition and variations during the vegetative cycle. J Pharm Biomed Anal 29(4):691–700
62. Wan L, Meng D, Wang H, Wan S, Jiang S, Huang S, Wei L, Yu P (2018) Preventive and therapeutic effects of thymol in a lipopolysaccharide-induced acute lung injury mice model. Inflammation 41(1):183–192
63. Al-Gabri NA, Qaid MM, El-Shaer NH, Ali MH, Abudabos AM (2019) Thymoquinone ameliorates pulmonary vascular damage induced by Escherichia coli–derived lipopolysaccharide via cytokine downregulation in rats. Environ Sci Pollut Res 26 (18):18465–18469
64. Yetkin NA, Büyükoğlan H, Sönmez MF, Tutar N, Gülmez I, Yilmaz I (2020) The protective effects of thymoquinone on lung damage caused by cigarette smoke. Biotech Histochem 95 (4):268–275
65. Su X, Ren Y, Yu N, Kong L, Kang J (2016) Thymoquinone inhibits inflammation, neoangiogenesis and vascular remodeling in asthma mice. Int Immunopharmacol 38:70–80
66. Duncan RE, Archer MC (2008) Farnesol decreases serum triglycerides in rats: identification of mechanisms including up-regulation of PPARα and down-regulation of fatty acid synthase in hepatocytes. Lipids 43(7):619–627
67. Khan R, Sultana S (2011) Farnesol attenuates 1, 2-dimethylhydrazine induced oxidative stress, inflammation and apoptotic responses in the colon of Wistar rats. Chem Biol Interact 192 (3):193–200
68. Qamar W, Sultana S (2008) Farnesol ameliorates massive inflammation, oxidative stress and lung injury induced by intratracheal instillation of cigarette smoke extract in rats: an initial step in lung chemoprevention. Chem Biol Interact 176(2–3):79–87
69. Ku CM, Lin JY (2015) Farnesol, a sesquiterpene alcohol in herbal plants, exerts anti-inflammatory and antiallergic effects on ovalbumin-sensitized and-challenged asthmatic mice. Evid Based Complement Alternat Med 2015:387357
70. Qamar W, Khan AQ, Khan R, Lateef A, Tahir M, Rehman MU, Ali F, Sultana S (2012) Benzo (a) pyrene-induced pulmonary inflammation, edema, surfactant dysfunction, and injuries in rats: alleviation by farnesol. Exp Lung Res 38(1):19–27
71. Riedl MA, Nel AE (2008) Importance of oxidative stress in the pathogenesis and treatment of asthma. Curr Opin Allergy Clin Immunol 8(1):49–56
72. Riedl MA, Saxon A, Diaz-Sanchez D (2009) Oral sulforaphane increases Phase II antioxidant enzymes in the human upper airway. Clin Immunol 130(3):244–251
73. Funamoto M, Sunagawa Y, Katanasaka Y, Miyazaki Y, Imaizumi A, Kakeya H, Yamakage H, Satoh-Asahara N, Komiyama M, Wada H, Hasegawa K (2016) Highly absorptive curcumin reduces serum atherosclerotic low-density lipoprotein levels in patients with mild COPD. Int J Chron Obstruct Pulmon Dis 11:2029
74. Zhang N, Chen Y, Zhao Y, Fan D, Li L, Yan B, Tao G, Zhao J, Zhang H, Wang M (2020) Caffeic acid assists microwave heating to inhibit the formation of mutagenic and carcinogenic PhIP. Food Chem 317:126447
75. Yamatani K, Kawatani R, Ajiro H (2020) Synthesis of glucosamine derivative with double caffeic acid moieties at N–and 6-O-positions for developments of natural based materials. J Mol Struct 1206:127689

76. Neiva DM, Araújo S, Gominho J, de Cássia Carneiro A, Pereira H (2018) Potential of Eucalyptus globulus industrial bark as a biorefinery feedstock: chemical and fuel characterization. Ind Crop Prod 123:262–270
77. Rodrigues VH, de Melo MM, Portugal I, Silva CM (2019) Simulation and techno-economic optimization of the supercritical CO2 extraction of *Eucalyptus globulus* bark at industrial scale. J Supercrit Fluids 145:169–180
78. Spagnol CM, Assis RP, Brunetti IL, Isaac VL, Salgado HR, Corrêa MA (2019) In vitro methods to determine the antioxidant activity of caffeic acid. Spectrochim Acta A Mol Biomol Spectrosc 219:358–366
79. Wang G, Lei Z, Zhong Q, Wu W, Zhang H, Min T, Wu H, Lai F (2017) Enrichment of caffeic acid in peanut sprouts and evaluation of its in vitro effectiveness against oxidative stress-induced erythrocyte hemolysis. Food Chem 217:332–341
80. Khayyal MT, El-Ghazaly MA, El-Khatib AS, Hatem AM, De Vries PJ, El-Shafei S, Khattab MM (2003) A clinical pharmacological study of the potential beneficial effects of a propolis food product as an adjuvant in asthmatic patients. Fundam Clin Pharmacol 17(1):93–102
81. Lee SJ, Umano K, Shibamoto T, Lee KG (2005) Identification of volatile components in basil (Ocimum basilicum L.) and thyme leaves (Thymus vulgaris L.) and their antioxidant properties. Food Chem 91(1):131–137

Targeting Cellular Signaling Pathways in Lung Cancer and Role of Phytochemicals as Novel Therapeutic Approach

36

Anchal Dubey, Abhishek Kumar, and Bechan Sharma

Abstract

In present times cancer is among one of the deadliest diseases of the world. Mutations in DNA lead to uncontrolled cell growth and cell proliferation resulting in cancer. Lung cancer not only contributes to an estimated 1.2 million deaths per year but is also known to have the lowest 5-year survival rate. Genomic mutations severely affect the signaling pathways which lead to the formation of tumors and progression of disease. A deep study of these pathways would help in the development of specific drugs which would help in targeting dysregulated pathways. However, current ongoing chemotherapy and other treatments are accompanied by a lot of side effects. A number of phytochemicals have antitumor potential and also are nontoxic toward other body cells. Utilization of these beneficial properties could play a role in providing a cost-effective and environment friendly way of cancer treatment. However, many previous supplementation trials have also shown inconsistent results. This chapter evaluates the limitation of current anticancer drugs available and the potential of phytochemicals as novel therapeutic against lung cancer.

Keywords

SCLC · NSCLC · Apoptosis · Cell proliferation · Genomic mutation · Phytochemical

A. Dubey · A. Kumar · B. Sharma (✉)
Department of Biochemistry, University of Allahabad, Prayagraj, Uttar Pradesh, India

K. Dua et al. (eds.), *Targeting Cellular Signalling Pathways in Lung Diseases*,
https://doi.org/10.1007/978-981-33-6827-9_36

36.1 Introduction

In present times cancer is among one of the deadliest diseases of the world. Mutations in DNA lead to uncontrolled cell growth and cell proliferation resulting in cancer. Lung cancer not only contributes to an estimated 1.2 million deaths per year but is also known to have the lowest 5-year survival rate [1]. Genomic mutations severely affect the signaling pathways which lead to formation of tumors and progression of disease. A deep study of these pathways would help in development of specific drugs which would help in targeting dysregulated pathways. However current ongoing chemotherapy and other treatments are accompanied by a lot of side effects. A number of phytochemicals have antitumor potential and also are nontoxic toward other body cells. Utilization of these beneficial properties could play a role in providing a cost-effective and environment friendly way of cancer treatment. However, many previous supplementation trials have also shown inconsistent results. This chapter will evaluate the limitation of current anticancer drugs available and the potential of phytochemicals as novel therapeutic against lung cancer.

36.2 Causes of Lung Cancer

Lung cancer is of two types, small cell lung cancer (SCLC) and non-small cell lung cancer (NSCLC). Among the two, NSCLC is more common and also develops slowly, while SCLC is rarer with frequent growth and is usually well developed with metastasis even at the time of diagnosis. Smoking is the most common cause of lung cancer responsible for about 85% of NSCLC and SCLC. More than 70 carcinogens are known to be present in tobacco smoke, most common being them nitrosamine ketones and polycyclic aromatic hydrocarbons. These carcinogens lead to genomic mutation by DNA adduct formation [2]. This formation occurs as a result of the metabolism of the carcinogens by cytochrome P450, CYP family enzymes, and glutathione-S-transferases (GSTs). The function of these enzymes is changed in most of the signaling pathways involved in lung cancer. Oncogenes lead these changes in the enzymes resulting in uncontrolled proliferation, malignant phenotypes, and evasion of apoptosis [3]. Persistent presence of these DNA adduct formations causes severe transversion mutation in p53 and Ras gene. In nonsmokers these mutations are more of transition type. Genes susceptible to lung cancer are generally those involved in metabolism of carcinogens and repair of DNA. However, less than 20% of smokers develop lung cancer indicating that susceptibility of these genes is determined by genetic changes. In the past incidences of increased lung cancer risk have also been associated with homozygous deletion of GSTM1 and number of polymorphisms that occurs at cytochrome P450 1A1 gene [4, 5].

Most of the molecular changes that occur during lung cancer are related with changes in oncogenes and tumor suppressor genes. Point mutations, loss or gain of complete chromosome, loss of heterozygosity, and changes in microsatellite DNA are some of the events which contributes to previously mentioned molecular changes [6, 7].

Among the SCLC, NSCLC, and other cases of tumors recognized by different clinical outcomes, molecular differences have been observed [8]. In case of SCLC, overexpression of tyrosine kinase receptors, multiple chromosome aberrations, and deletion of site-specific chromosome-containing tumor suppressor p53 gene have been observed. Amplification of 1p, 2p, and 3q and deletion of 18q have also been observed in SCLC cell line contributing to more aggressive phenotype of disease [9]. Loss of 3p allele is considered to be an early event in LC pathogenesis and is recorded in almost 90% of SCLC cases [10].

Although tobacco consumption is still considered the major lung cancer-related risk, family history has been known to contribute 2.5-fold increased risk [11] indicating a major susceptibility risk at 6q23–25 locus [12].

36.3 Signaling Pathways in Lung Cancer

Most of the genetic abnormalities of lung cancer can be linked to the alterations occurring in various signaling pathways. These alterations are now targeted for therapy. Oncogenes stimulate the pathways leading the cells toward a malignant phenotype. Oncogene addiction is a unique property exhibited by tumor cells in which tumor cells get addicted to the abnormal functions due to mutated oncogenic protein. Once the targeted drugs inhibit this function, cells die, providing a chance for better pharmacogenetic progress. Normal cells being resistant to targeted drugs are not harmed. This resistance is due to the lack of mutant proteins [13].

36.3.1 Epidermal Growth Factor Receptor Pathway

Receptor of tyrosine kinases (RTKs) forms a group of 58 surface growth factor receptor with ligand-mediated tyrosine kinase activity [14]. EGFR (ERBB1, HER1), ERBB2 (HER2, Neu), ERBB3 (HER3), and ERBB4 (HER4) constitute the family of four RTKs known as HER/ErbB family. TK domain of the four genes shows the greatest sequence homology. Different ligands cause activation of different family member. For EGFR, epidermal growth factor transforms growth factor-α and amphiregulin. Once the ligand binding is complete, formation of homodimer or heterodimer occurs resulting in activation of tyrosine kinases and receptor transphosphorylation. This causes activation of a number of other pathways including Ras, PI3K. Major cause of this activation is the formation of docking sites for various cytoplasmic signaling molecules. In NSCLC EGFR deregulation has been observed. Overexpression of EGFR protein was observed in NSCLC of squamous cells and ADC subtypes by Hirsch et al. [15]. Overexpression of EGFR is also related with adverse prognosis of the disease [16].

EGFR activation is often linked with cell proliferation and survival through various pathways including RAS/MAPK and PI3K/AKT pathway [17]. In approximately 10–30% of NSCLC cases, KRAS and EGFR mutations have been observed [18]. Both EGFR and KRAS mutations are associated with ADC histology but differ in their association with East Asian ethnicity, male-female sex, and smoking status. EGFR mutations are more frequently seen in East Asians, female sex, and nonsmokers, while that of KRAS is very rare in East Asians and targets smokers and male [19]. Critical regions of TK domain associated with downstream signaling are major target of mutations including insertion, deletion, and point mutations. However, deletion in exon 19 and point mutation L858R in exon 21 constitute 85% of mutation.

EGFR was among the first selected pathway for target therapy. Initially monoclonal antibody was used for blocking the ligand receptor interaction. Recently small molecule reversible TK inhibitors are used. A number of inhibitors have already been used in preclinical trials. Afatinib is one such inhibitor which blocks ErbB family irreversibly and has shown preclinical activity in NSCLC with EGFR mutation. In phase 2 trial, in patients of advanced lung carcinoma with EGFR mutations, afatinib showed antitumor activity [20]. In another phase 2 trial, dacomitinib (another irreversible EGFR inhibitor) was compared with erlotinib (reversible EGFR inhibitor) in advanced NSCLC patients. Dacomitinib shows improvement in progression-free survival with acceptable toxicity in comparison with erlotinib [21]. The illustrated mechanism of EGFR signaling pathway is explained in Fig. 36.1.

36.3.2 Estrogen Receptor Pathway

NSCLC has been recognized as an estrogen receptor-positive cancer; particularly estrogen receptor alpha (ERα) and estrogen receptor beta (ERβ) have been identified in NSCLC cell lines and tissues [22, 23]. Among the two receptors, estrogen receptor beta is more prominent in NSCLC with an overexpression of about 60–80% in lung cancer tissue. Its levels were also found to be upregulated in lung tumor when compared with normal lung tissues of same patient [24, 25]. ERβ expression has been also linked with prognosis in patients. Interestingly studies have shown that ERβ expression in men correlates to good prognosis while in women it indicates poor prognosis [26–28]. ERα expression is less frequent than ERβ and is mostly restricted to cases of lung tumor related with EGFR mutation [29]. In patients of advanced lung disease, expression of ERα is associated with good prognosis [30].

Recently signaling pathways are being targeted to develop a better understanding of underlying pathways and the role of estrogen in promoting lung carcinogenesis. Once estrogen receptors (ERs) are activated, it can lead to lung carcinogenesis through genomic or non-genomic pathway. In genomic pathway ER translocates to the nucleus leading to transcription of genes, and in non-genomic pathway, translocation of ER occurs to the cell membrane leading to activation of second messengers and ion channels. The role of estrogen in NSCLC was studied, and it was found that estrogen causes cell proliferation in NSCLC and also causes increased growth of tumor, while antiestrogen drugs are found to reduce cell proliferation and

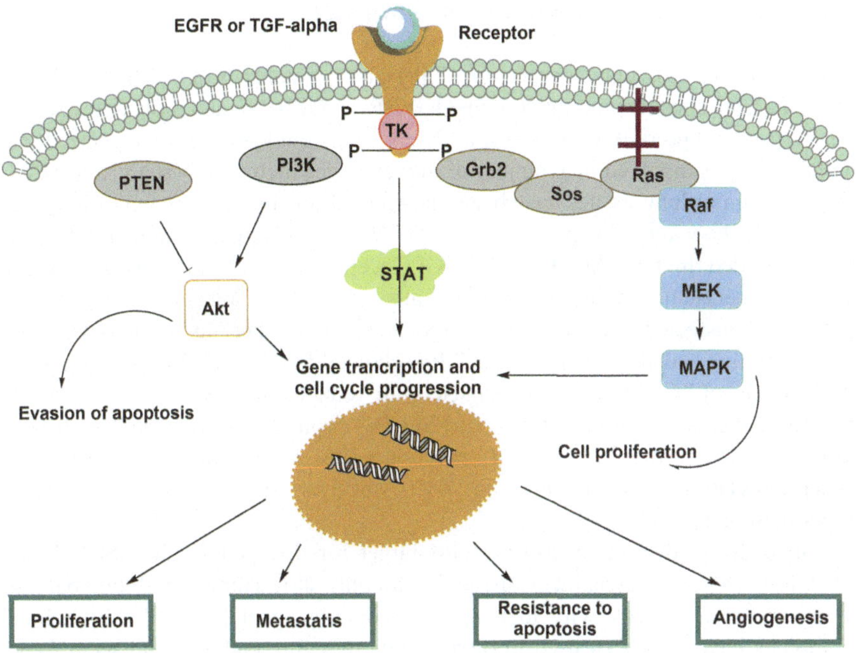

Fig. 36.1 A schematic representation of mechanism of EGFR signaling pathway in cancer progression

tumor size [31, 32]. Estrogen through non-genomic pathways causes activation of cAMP, MAPK, and AKT signaling pathways inducing cell proliferation. Estradiol also causes cell cycle progression and proliferation through genomic pathway promoting expression of c-myc, cyclin D, and Id proteins [33, 34].

Since estrogen plays a significant role in alteration of significant pathways, estrogen receptor inhibitors are used in a number of clinical trials in preventing malignancies. Fulvestrant is one such example which was used in NSCLC preclinical study. It is a selective estrogen receptor degrader causing the destabilization and degradation of estrogen receptors. In a mice model, the use of fulvestrant not only inhibited tumor growth by 32% but also ERβ expression [35].

Co-treatment of fulvestrant and vandetanib (which is a multi-target inhibitor of both EGFR and VEGFR) in NSCLC cell and xenograft model showed increase in apoptosis, inhibition of tumor growth, and decrease in cell growth. This indicates that fulvestrant enhances vandetanib effect by causing deactivation of EGFR pathway by blocking estrogen-driven activation [36].

36.3.3 Insulin-Like Growth Factor Pathway

The insulin-like growth factor receptor (IGF) pathway consists of two ligands IGF-1 and IGF-2, six specific binding proteins (IGFBP-1–IGFBP-6) and proteases, and two cell surface receptors IGF-1R and IGF-2R. Various evidences have suggested that IGF pathway is involved in a number of malignancies including NSCLC and SCLC [37, 38]. Tumor lung tissues have been known to show high expression of IGF, and tumor cells show autocrine production of IGF [39–41]. Overexpression of IGF-2 has been recorded in both SCLC and NSCLC cell lines [42]. In a mouse model, increased metastatic activity was reported after intrasplenic injection of lung cancer cell lines transfected with IGF-1R receptor [43]. In another study one of the four SCLC cell line and all four NSCLC cell line showed low IGFBP-3 expression level. High levels of IGFBP-3 are associated with reduced lung cancer risk [44]. In NSCLC patients, hypermethylation of IGFBP-3 promoter is associated with poor survival rate, with 38% 5-year survival rate in patients with IGFBP-3 hypermethylation as compared to 64% survival rate in patients without hypermethylation [45].

IGF pathway disruption also presents target for therapeutics. In NSCLC cell, A549 IGF-1R signaling pathway disruption not only shows tumor inhibition but also shows increase in sensitivity to apoptotic inducing agent [46, 47]. Monoclonal antibodies have also been used to disrupt IGF signaling pathway and seem to act by downregulation of IGF-1R [48]. In an in vitro study of 6 NSCLC cell line, the effect of anti-IGF-1R monoclonal antibody was studied in blockage of IGF-1R function. It was found that in one cell line it showed synergistic effect and increases the cytotoxic effect of radiation; in another cell line, an additive effect was observed; and in another cell line, subadditive effect was observed [49]. In an animal model, complete inhibition of tumor growth was recorded when anti-IGF-1R monoclonal antibody was combined with vinorelbine or an anti-epidermal growth factor receptor antibody [47].

36.3.4 Hedgehog Signaling Pathway

HH pathway during embryogenesis regulates morphogenesis of various organs, and in even adults, HH pathway regulates renewal of stem cells and maintenance of organ homeostasis [50, 51]. Sonic hedgehog (SHH), Indian hedgehog (IHH), and desert hedgehog (DHH) are three ligands recognized in canonical HH signaling pathway. Each HH ligand has their own specific spatial and temporal expression patterns and binds to a 12-pass spanning membrane receptor, PTCH, activating HH signaling. However, in the absence of ligands, PTCH does not activate HH signaling and continues to suppress activity of SMO which is a member of G-protein-coupled receptor family, a 7-pass membrane spanning protein [52]. HH signaling pathway is found to have critical roles during embryonic lung development as well as postnatal lung development [53].

In many types of cancer including lung, colon, and stomach, constitutive action of HH signaling has been observed and is supposed to promote cancer cell proliferation and metastasis [54]. In SCLC cases cells involved in HH pathway are not found to be mutated; however, in many SCLC cases, HH pathway was found to be activated [55]. In a xenograft SCLC model of nude mice, it was observed that in SHH-producing SCLC cells activation of HH pathway was observed and in surrounding non-SHH-producing cancer cells activation of HH was not observed indicating autocrine or juxtacrine activation of HH pathway in SCLC. Also, in a mouse model, HH pathway activation causes progression in SCLC; SMO suppression reduced SCLC initiation and progression [56]. In NSCLC too HH activation is found. In NSCLC cells epithelial to mesenchymal transition is associated with aggressive nature of cells [57]. In A549 cells with mesenchymal phenotype, upregulation of SHH ligand and GLI1 expression were observed as compared with A549 cells. In A549 mesenchymal cells, HH pathway is supposed to be activated by autocrine signaling. Also, suppression of HH pathway leads to suppressed cancer cell migration and metastasis induced by TGF-β signaling [58]. In 5 of 10 human NSCLC, reduced expression of HIP has been observed; HIP is a natural antagonist of HH signaling [59].

In an experiment conducted by Park et al. in mouse xenograft model, a combined treatment of etoposide and LDE225 (a SMO inhibitor) reduced tumor recurrence in SCLC. Even in TKI-resistant NSCLC cell line, LDE225 reduces tumor growth [56, 60].

36.3.5 Vesicular Endothelial Growth Factor Pathway (VEGF)

In the 1930s, an idea of tumor angiogenesis was proposed which explained the idea of tumor having its own blood supply [61]. VEGF stimulates tumor angiogenesis. VEGF constitutes family of growth factors which include VEGF-A, VEGF-B, VEGF-C, VEGF-D, VEGF-E, and placenta growth factor (PIG). All these growth factors promote cell proliferation and survival, cause inhibition of apoptosis, and regulate endothelial cell permeability [62]. In both SCLC and NSCLC cases, high levels of VEGF are related with poor prognosis [63, 64]. Different VEGF isoforms show different results in clinical trials. In a study involving NSCLC, VEGF-A_{189} levels correlate with intra-tumor mean vascular density, disease recurrence, and overall survival, while most abundant VEGF-A_{165} form did not show these clinical correlation [65]). A recombinant VEGF-neutralizing antibody bevacizumab which binds with all forms of VEGF-A isoform combinations has been tested along with chemotherapy in a number of clinical trials of NSCLC patients. In colorectal cancer use of bevacizumab along with chemotherapy resulted in both progression-free and overall survival improvement [66, 67]. In 2004, in a phase II clinical trial of patients having recurring NSCLC, the effect of bevacizumab was observed. It was found that patients who received bevacizumab along with chemotherapy showed better recovery rate than patients with only chemotherapy. While overall survival was also improved, some side effects were also observed. Most severe was that the bleeding frequency was increased [68].

36.4 Limitations of Ongoing Cancer Therapy

The most common methods for cancer therapy include surgical removal of tumor, chemotherapy, radiation therapy, immunotherapy, or combination of these treatments. However, all of these treatments include side effects like restricted metastasis, limited bioavailability, toxicity, nonspecific behavior, and fast clearance [69–71].

Chemotherapeutic agents include cytotoxic and cytostatic drugs involving topoisomerase inhibitors like irinotecan and doxorubicin, alkylating agents like cisplatin and cyclophosphamide, and microtubule-acting agent like vincristine, vinblastine, etc. Although these chemotherapeutic agents are highly successful in treating cancer, they have also been accompanied by severe side effects like neutropenia, sensory neuropathy and diarrhea (irinotecan), cardiotoxicity (doxorubicin), nephrotoxicity, cardiotoxicity, gastrointestinal toxicity, and pulmonary toxicity (cisplatin, carboplatin) [71, 72]. These drugs are also not cost-effective and non-ecofriendly.

The main target of anticancer drugs are abnormally dividing cancer cells; however, in the body, a number of highly dividing normal cells are also present like hair follicles, digestive tract cells, bone marrow cells, etc. These cells also get targeted by these drugs and leads to GIT inflammation, hair loss, immunosuppression, decrease in blood loss, and nervous disorders.

Sometimes after undergoing mutation, cancer cells start getting resistant to these drugs. Like in human breast cancer cell MCF7, docetaxel application showed overexpression of drug-resistant genes (ABCA 4 and ABCA 12). However, application of curcumin along with docetaxel caused downregulation of drug resistance [73]. This shows that monotherapy of cancer is not that successful. The use of phytochemicals or their derived analogues along with these therapies can not only be more effective but will also be less toxic as has been proven in extensive research findings [74].

36.5 Phytochemicals as Novel Therapeutic Approach

Since ancient times, plants have been used as medicines. It has been estimated that of all the plant species present, only 10% of them have been studied for medicinal purposes. In different parts of the plants like flower, leaf, embryo, fruits, seeds, etc., various phytochemicals and their derived analogues are present having a number of pharmacological functions. It has been found in various studies that phytochemicals have the property to inhibit the progression of cancer [73]. Plant products shows anticancer property by either inhibiting cancer cell-activating proteins, enzymes, and signaling pathways or by activating DNA repair mechanism or by inducing

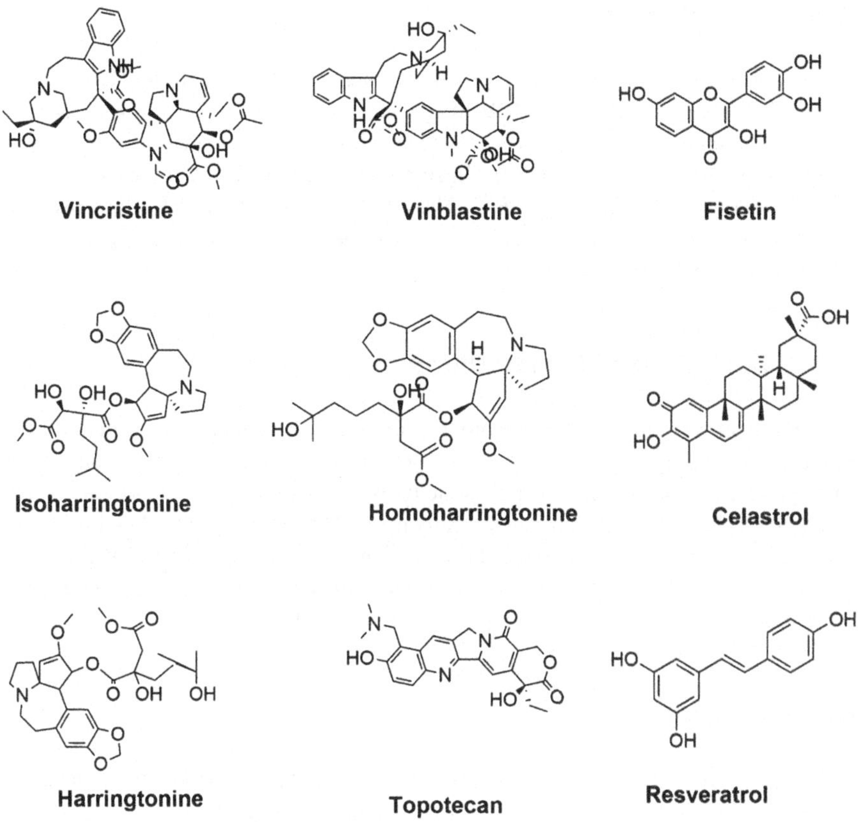

Fig. 36.2 Structure of some anticancer phytochemicals used in lung cancer treatment

antioxidant action [75, 76]. Some most potent phytochemicals and their derivative's structures are given in Fig. 36.2.

36.5.1 Vinca Alkaloids

Isolated from *Catharanthus roseus* of family Apocynaceae, vinca alkaloids are the versatile phytochemicals which are used in therapeutic treatment of various types of cancers including lung cancer. These phytochemicals (vincristine and vinblastine) arrest the cell cycle at metaphase by binding with specific sites named tubulin heterodimers and disrupt function of microtubules [77]. Vinorelbine, vindesine, and vinfosiltine are semisynthetic derivatives of vinca alkaloids and are being used currently alone or with some phytochemical combination against a large number of cancers [78].

36.5.2 Cephalotaxus

Cephalotaxus alkaloids have been known to possess antitumor activity. This phytochemical is also used against A549 lung cancer cell line by targeting molecular events involved in protein synthesis as well as inhibiting protein synthesis, however, not effecting protein elongation [79]. Harringtonine and isoharringtonine are isolated from *Cephalotaxus harringtonia.* The FDA has approved the use of homoharringtonine against chronic myelogenous leukemia in China, Japan, Pakistan, the USA, and Germany [80].

36.5.3 Derivatives of Camptothecin

Camptothecin was first isolated from *Camptotheca acuminata.* A water-soluble derivative of camptothecin is topotecan (TPT) and is known to have antitumor activity against lung cancer. TPT is actually topoisomerase 1 inhibitor which prevents DNA replication ultimately causing cell death [81]. The US Food and Drug Administration has approved the use of intravenous formulation of this drug for the treatment of SCLC patient at a recommended dose of 1.5 mg/m^2 for 5 continuous days with treatment repeated after 3 weeks [82].

36.5.4 Celastrol

Celastrol isolated from the bark of *Tripterygium wilfordii* is another anticancer compound. In 95-D lung cancer cells, celastrol inhibits heat-shock protein blocking its interaction with Cdc37 and inducing apoptosis via caspase-3 enzyme [83]. In another study, the effect of celastrol was observed in H1650 and H1975 two gefitinib-resistant NSCLC cell lines. It was found that celastrol exerted an apoptotic effect on cell lines in a time- and dose-dependent manner. Bax/Bcl 2 ratio was also increased. Caspases were also activated. The level of EGFR and AKT was downregulated; both are Hsp90 client proteins. These results proved the efficiency of celastrol in treating NSCLC cells by causing degradation of Hsp90 client protein and inducing caspase-dependent apoptotic pathway [84].

36.5.5 Resveratrol

Found in peanut, grapes, and mulberry, resveratrol is a naturally occurring polyphenol. It has shown anticancer property against various types of cancers including lung cancer. It mainly functions by upregulating P53- and Bcl-2-associated X proteins; downregulation of MMPs, NF-kB, AP-1, Bcl-2, cyclins, cyclin-dependent kinases, cytokines, and COX-2 proteins was also observed [85, 86]. The mechanism of action of resveratrol also includes suppression of VEGF protein by reduction of MAP kinase phosphorylation and inhibition of angiogenesis [87].

36.5.6 Fisetin

A bioactive flavone molecule found in various fruits and vegetables such as grapes, onion, apple, cucumber, etc. Fisetin inhibits both PI3K/Akt signaling pathway and is used in the treatment of lung cancer [88]. Fisetin inhibits more than one type of cancer. Fisetin induces apoptosis in both lung cancer and oral cancer. In lung cancer, apoptosis occurs by inhibition of MAPK signaling pathway and in oral cancer by inhibition of production of reactive oxygen species. In human renal carcinoma caki cells, p53-mediated upregulation of DR5 expression induces apoptosis [89].

36.6 Conclusion

Cancer being a global menace requires immediate attention for better therapy, as ongoing therapies are not only accompanied by severe effects but are also expensive and not very reliable. An approach toward greener therapy would be very beneficial. Phytochemicals would be the best alternative for this. Already a number of phytochemicals are in use for the same. More research is required for the full utilization of phytochemicals in revolutionizing cancer therapeutics. Not only as alternate, phytochemicals can also be used along with ongoing therapeutics as it can lead to a more effective treatment and risk of toxicity could also be reduced.

Acknowledgments AD and AK are grateful to the University Grants Commission-New Delhi for providing financial support in the form of research fellowships. The authors acknowledge DST-FIST, New Delhi, and UGC-SAP, New Delhi, for providing grants to generate research facilities at the Department of Biochemistry of the University of Allahabad, India.

References

1. Kasala ER, Bodduluru LN, Barua CC, Sriram CS, Gogoi R (2015) Benzo(a)pyrene induced lung cancer: role of dietary phytochemicals in chemoprevention. Pharmacol Rep 67 (5):996–1009
2. Hecht SS (1999) Tobacco smoke carcinogens and lung cancer. J Natl Cancer Inst 91 (14):1194–1210
3. Brambilla E, Gazdar A (2009) Pathogenesis of lung cancer signalling pathways: roadmap for therapies. Eur Respir J 33(6):1485–1497
4. Schwartz AG, Prysak GM, Bock CH, Cote ML (2007) The molecular epidemiology of lung cancer. Carcinogenesis 28(3):507–518
5. Paz-Elizur T, Sevilya Z, Leitner-Dagan Y, Elinger D, Roisman LC, Livneh Z (2008) DNA repair of oxidative DNA damage in human carcinogenesis: potential application for cancer risk assessment and prevention. Cancer Lett 266(1):60–72
6. Massion PP, Carbone DP (2003) The molecular basis of lung cancer: molecular abnormalities and therapeutic implications. Respir Res 4(1):12
7. Park IW, Wistuba II, Maitra A, Milchgrub S, Virmani AK, Minna JD, Gazdar AF (1999) Multiple clonal abnormalities in the bronchial epithelium of patients with lung cancer. J Natl Cancer Inst 91(21):1863–1868. https://doi.org/10.1093/jnci/91.21.1863

8. Zöchbauer-Müller S, Gazdar AF, Minna JD (2002) Molecular pathogenesis of lung cancer. Annu Rev Physiol 64:681–708
9. Sattler M, Salgia R (2003) Molecular and cellular biology of small cell lung cancer. Semin Oncol 30(1):57–71
10. Balsara BR, Testa JR (2002) Chromosomal imbalances in human lung cancer. Oncogene 21 (45):6877–6883
11. Amos CI, Xu W, Spitz MR (1999) Is there a genetic basis for lung cancer susceptibility? Recent Results Cancer Res 151:3–12
12. Bailey-Wilson JE, Amos CI, Pinney SM, Petersen GM, De Andrade M, Wiest JS, Fain P, Schwartz AG, You M, Franklin W, Klein C, Gazdar A, Rothschild H, Mandal D, Coons T, Slusser J, Lee J, Gaba C, Kupert E, Perez A, Zhou X, Zeng D, Liu Q, Zhang Q, Seminara D, Minna J, Anderson MW (2004) A major lung cancer susceptibility locus maps to chromosome 6q23-25. Am J Hum Genet 75(3):460–474. https://doi.org/10.1086/423857
13. Weinstein IB, Joe A (2008) Oncogene addiction. Cancer Res 68(9):3077–3080; discussion 3080
14. Robinson DR, Wu YM, Lin SF (2000) The protein tyrosine kinase family of the human genome. Oncogene 19(49):5548–5557
15. Hirsch FR, Varella-Garcia M, Bunn PA, Di Maria MV, Veve R, Bremnes RM, Barón AE, Zeng C, Franklin WA (2003) Epidermal growth factor receptor in non-small-cell lung carcinomas: correlation between gene copy number and protein expression and impact on prognosis. J Clin Oncol 21(20):3798–3807. https://doi.org/10.1200/JCO.2003.11.069
16. Nicholson RI, Gee JMW, Harper ME (2001) EGFR and cancer prognosis. Eur J Cancer 37 (Suppl 4):S9–S15. https://doi.org/10.1016/s0959-8049(01)00231-3
17. Sharma SV, Bell DW, Settleman J, Haber DA (2007) Epidermal growth factor receptor mutations in lung cancer. Nat Rev Cancer 7(3):169–181
18. Eberhard DA, Johnson BE, Amler LC, Goddard AD, Heldens SL, Herbst RS, Ince WL, Jänne PA, Januario T, Johnson DH, Klein P, Miller VA, Ostland MA, Ramies DA, Sebisanovic D, Stinson JA, Zhang YR, Seshagiri S, Hillan KJ (2005) Mutations in the epidermal growth factor receptor and in KRAS are predictive and prognostic indicators in patients with non-small-cell lung cancer treated with chemotherapy alone and in combination with erlotinib. J Clin Oncol 23 (25):5900–5909. https://doi.org/10.1200/JCO.2005.02.857
19. Gazdar AF, Shigematsu H, Herz J, Minna JD (2004) Mutations and addiction to EGFR: the Achilles "heal" of lung cancers? Trends Mol Med 10(10):481–486. https://doi.org/10.1016/j.molmed.2004.08.008
20. Yang JCH, Shih JY, Su WC, Hsia TC, Tsai CM, Ou SHI, Yu CJ, Chang GC, Ho CL, Sequist LV, Dudek AZ, Shahidi M, Cong XJ, Lorence RM, Yang PC, Miller VA (2012) Afatinib for patients with lung adenocarcinoma and epidermal growth factor receptor mutations (LUX-lung 2): a phase 2 trial. Lancet Oncol 13(5):539–548. https://doi.org/10.1016/S1470-2045(12)70086-4
21. Ramalingam SS, Blackhall F, Krzakowski M, Barrios CH, Park K, Bover I, Heo DS, Rosell R, Talbot DC, Frank R, Letrent SP, Ruiz-Garcia A, Taylor I, Liang JQ, Campbell AK, O'Connell J, Boyer M (2012) Randomized phase II study of dacomitinib (PF-00299804), an irreversible pan-human epidermal growth factor receptor inhibitor, versus erlotinib in patients with advanced non-small-cell lung cancer. J Clin Oncol 30(27):3337–3344. https://doi.org/10.1200/JCO.2011.40.9433
22. Niikawa H, Suzuki T, Miki Y, Suzuki S, Nagasaki S, Akahira J, Honma S, Evans DB, Hayashi SI, Kondo T, Sasano H (2008) Intratumoral estrogens and estrogen receptors in human non-small cell lung carcinoma. Clin Cancer Res 14(14):4417–4426. https://doi.org/10.1158/1078-0432.CCR-07-1950
23. Siegfried JM, Hershberger PA, Stabile LP (2009) Estrogen receptor signaling in lung cancer. Semin Oncol 36(6):524–531. https://doi.org/10.1053/j.seminoncol.2009.10.004
24. Chakraborty S, Ganti AK, Marr A, Batra SK (2010) Lung cancer in women: role of estrogens. Expert Rev Respir Med 4(4):509–518

25. Siegfried JM (2014) Smoking out reproductive hormone actions in lung cancer. Mol Cancer Res 12(1):24–31
26. Skjefstad K, Grindstad T, Khanehkenari MR, Richardsen E, Donnem T, Kilvaer T, Andersen S, Bremnes RM, Busund LT, Al-Saad S (2016) Prognostic relevance of estrogen receptor α, β and aromatase expression in non-small cell lung cancer. Steroids 113:5–13. https://doi.org/10.1016/j.steroids.2016.05.008
27. Baik CS, Eaton KD (2012) Estrogen signaling in lung cancer: an opportunity for novel therapy. Cancers (Basel) 4(4):969–988
28. Schwartz AG, Prysak GM, Murphy V, Lonardo F, Pass H, Schwartz J, Brooks S (2005) Nuclear estrogen receptor β in lung/cancer: expression and survival differences by sex. Clin Cancer Res 11(20):7280–7287. https://doi.org/10.1158/1078-0432.CCR-05-0498
29. Mazières J, Rouquette I, Lepage B, Milia J, Brouchet L, Guibert N, Beau-Faller M, Validire P, Hofman P, Fouret P (2013) Specificities of lung adenocarcinoma in women who have never smoked. J Thorac Oncol 8(7):923–929. https://doi.org/10.1097/JTO.0b013e3182904dfb
30. Li W, Tse LA, Wang F (2015) Prognostic value of estrogen receptors mRNA expression in non-small cell lung cancer: a systematic review and meta-analysis. Steroids 104:129–136. https://doi.org/10.1016/j.steroids.2015.09.005
31. Márquez-Garbán DC, Chen HW, Goodglick L, Fishbein MC, Pietras RJ (2009) Targeting aromatase and estrogen signaling in human non-small cell lung cancer. Ann N Y Acad Sci 1155:194–205
32. Stabile LP, Gaither Davis AL, Gubish CT, Hopkins TM, Luketich JD, Christie N, Finkelstein S, Siegfried JM (2002) Human non-small cell lung tumors and cells derived from normal lung express both estrogen receptor α and β and show biological responses to estrogen. Cancer Res 62(7):2141–2150
33. Siegfried JM, Stabile LP (2014) Estrogenic steroid hormones in lung cancer. Semin Oncol 41 (1):5–16. https://doi.org/10.1053/j.seminoncol.2013.12.009
34. Tang H, Liao Y, Zhang C, Chen G, Xu L, Liu Z, Fu S, Yu L, Zhou S (2014) Fulvestrant-mediated inhibition of estrogen receptor signaling slows lung cancer progression. Oncol Res 22 (1):13–20. https://doi.org/10.3727/096504014X14077751730315
35. Hershberger PA, Vasquez AC, Kanterewicz B, Land S, Siegfried JM, Nichols M (2005) Regulation of endogenous gene expression in human non-small cell lung cancer cells by estrogen receptor ligands. Cancer Res 65(4):1598–1605. https://doi.org/10.1158/0008-5472.CAN-04-2694
36. Siegfried JM, Gubish CT, Rothstein ME, Henry C, Stabile LP (2012) Combining the multitargeted tyrosine kinase inhibitor vandetanib with the antiestrogen fulvestrant enhances its antitumor effect in non-small cell lung cancer. J Thorac Oncol 7(3):485–495. https://doi.org/10.1097/JTO.0b013e31824177ea
37. Hofmann F, García-Echeverría C (2005) Blocking the insulin-like growth factor-I receptor as a strategy for targeting cancer. Drug Discov Today 10(15):1041–1047
38. Tao Y, Pinzi V, Bourhis J, Deutsch E (2007) Mechanisms of disease: signaling of the insulin-like growth factor 1 receptor pathway—therapeutic perspectives in cancer. Nat Clin Pract Oncol 4(10):591–602
39. Yu H, Spitz MR, Mistry J, Gu J, Hong WK, Wu X (1999) Plasma levels of insulin-like growth factor-I and lung cancer risk: a case-control analysis. J Natl Cancer Inst 91(2):151–156. https://doi.org/10.1093/jnci/91.2.151
40. Minuto F, Del Monte P, Barreca A, Alama A, Cariola G, Giordano G (1988) Evidence for autocrine mitogenic stimulation by somatomedin-C/insulin-like growth factor I on an established human lung cancer cell line. Cancer Res 48(13):3716–3719
41. Catrambone G, Del Monte P, Barreca A, Fortini P, Cariola G, Giordano G (1986) Evidence for an increased somatomedin-c/Insulin-like growth factor I content in primary human lung tumors. Cancer Res 46(2):985–988. https://doi.org/10.1016/s0169-5002(87)80440-3

42. Reeve JG, Morgan J, Schwander J, Bleehen NM (1993) Role for membrane and secreted insulin-like growth factor-binding protein-2 in the regulation of insulin-like growth factor action in lung tumors. Cancer Res 53:4680–4685. https://doi.org/10.1016/0169-5002(94)90316-6
43. Long L, Rubin R, Brodt P (1998) Enhanced invasion and liver colonization by lung carcinoma cells overexpressing the type 1 insulin-like growth factor receptor. Exp Cell Res 238 (1):116–121. https://doi.org/10.1006/excr.1997.3814
44. Renehan AG, Zwahlen M, Minder C, O'Dwyer ST, Shalet SM, Egger M (2004) Insulin-like growth factor (IGF)-I, IGF binding protein-3, and cancer risk: systematic review and meta-regression analysis. Lancet 363(9418):1346–1353. https://doi.org/10.1016/S0140-6736(04)16044-3
45. Chang YS, Gong K, Sun S, Liu D, El-Naggar AK, Khuri FR, Hong WK, Lee HY (2002) Clinical significance of insulin-like growth factor-binding protein-3 expression in stage I non-small cell lung cancer. Clin Cancer Res 8(12):3796–3802
46. Jiang Y, Rom WN, Yie TA, Chi CX, Tchou-Wong KM (1999) Induction of tumor suppression and glandular differentiation of A549 lung carcinoma cells by dominant-negative IGF-I receptor. Oncogene 18(44):6071–6077. https://doi.org/10.1038/sj.onc.1202984
47. Goetsch L, Gonzalez A, Leger O, Beck A, Pauwels PJ, Haeuw JF, Corvaia N (2005) A recombinant humanized anti-insulin-like growth factor-receptor type I antibody (h7C10) enhances the antitumor activity of vinorelbine and anti-epidermal growth factor receptor therapy against human cancer xenografts. Int J Cancer 113(2):316–328. https://doi.org/10.1002/ijc.20543
48. Cohen BD, Baker DA, Soderstrom C, Tkalcevic G, Rossi AM, Miller PE, Tengowski MW, Wang F, Gualberto A, Beebe JS, Moyer JD (2005) Combination therapy enhances the inhibition of tumor growth with the fully human anti-type 1 insulin-like growth factor receptor monoclonal antibody CP-751,871. Clin Cancer Res 11(5):2063–2073. https://doi.org/10.1158/1078-0432.CCR-04-1070
49. Cosaceanu D, Carapancea M, Castro J, Ekedahl J, Kanter L, Lewensohn R, Dricu A (2005) Modulation of response to radiation of human lung cancer cells following insulin-like growth factor 1 receptor inactivation. Cancer Lett 222(2):173–181. https://doi.org/10.1016/j.canlet.2004.10.002
50. Ingham PW, McMahon AP (2001) Hedgehog signaling in animal development: paradigms and principles. Genes Dev 15(23):3059–3087
51. Beachy PA, Karhadkar SS, Berman DM (2004) Tissue repair and stem cell renewal in carcinogenesis. Nature 432(7015):324–331
52. Rohatgi R, Milenkovic L, Scott MP (2007) Patched1 regulates hedgehog signaling at the primary cilium. Science 317(5836):372–376. https://doi.org/10.1126/science.1139740
53. Kugler MC, Joyner AL, Loomis CA, Munger JS (2015) Sonic hedgehog signaling in the lung: from development to disease. Am J Respir Cell Mol Biol 52(1):1–13. https://doi.org/10.1165/rcmb.2014-0132TR
54. Rubin LL, de Sauvage FJ (2006) Targeting the hedgehog pathway in cancer. Nat Rev Drug Discov 5(12):1026–1033
55. Voortman J, Lee JH, Killian JK, Suuriniemi M, Wang Y, Lucchi M, Smith WI, Meltzer P, Wang Y, Giaccone G (2010) Array comparative genomic hybridization-based characterization of genetic alterations in pulmonary neuroendocrine tumors. Proc Natl Acad Sci U S A 7 (29):13040–13045. https://doi.org/10.1073/pnas.1008132107
56. Park KS, Martelotto LG, Peifer M, Sos ML, Karnezis AN, Mahjoub MR, Bernard K, Conklin JF, Szczepny A, Yuan J, Guo R, Ospina B, Falzon J, Bennett S, Brown TJ, Markovic A, Devereux WL, Ocasio CA, Chen JK, Stearns T, Thomas RK, Dorsch M, Buonamici S, Watkins DN, Peacock CD, Sage J (2011) A crucial requirement for hedgehog signaling in small cell lung cancer. Nat Med 17(11):1504–1508. https://doi.org/10.1038/nm.2473
57. Thomson S, Buck E, Petti F, Griffin G, Brown E, Ramnarine N, Iwata KK, Gibson N, Haley JD (2005) Epithelial to mesenchymal transition is a determinant of sensitivity of non-small-cell

lung carcinoma cell lines and xenografts to epidermal growth factor receptor inhibition. Cancer Res 65(20):9455–9462. https://doi.org/10.1158/0008-5472.CAN-05-1058
58. Maitah MY, Ali S, Ahmad A, Gadgeel S, Sarkar FH (2011) Up-regulation of sonic hedgehog contributes to TGF-β1-induced epithelial to mesenchymal transition in NSCLC cells. PLoS One 6(1):e16068. https://doi.org/10.1371/journal.pone.0016068
59. Olsen CL, Hsu PP, Glienke J, Rubanyi GM, Brooks AR (2004) Hedgehog-interacting protein is highly expressed in endothelial cells but down-regulated during angiogenesis and in several human tumors. BMC Cancer 4:43. https://doi.org/10.1186/1471-2407-4-43
60. Della Corte CM, Bellevicine C, Vicidomini G, Vitagliano D, Malapelle U, Accardo M, Fabozzi A, Fiorelli A, Fasano M, Papaccio F, Martinelli E, Troiani T, Troncone G, Santini M, Bianco R, Ciardiello F, Morgillo F (2015) SMO gene amplification and activation of the hedgehog pathway as novel mechanisms of resistance to anti-epidermal growth factor receptor drugs in human lung cancer. Clin Cancer Res 21(20):4686–4697. https://doi.org/10.1158/1078-0432.CCR-14-3319
61. Ferrara N (2002) VEGF and the quest for tumour angiogenesis factors. Nat Rev Cancer 2 (10):795–803
62. Ferrara N, Gerber HP, LeCouter J (2003) The biology of VEGF and its receptors. Nat Med 9 (6):669–676
63. Stefanou D, Batistatou A, Arkoumani E, Ntzani E, Agnantis NJ (2004) Expression of vascular endothelial growth factor (VEGF) and association with microvessel density in small-cell and non-small-cell lung carcinomas. Histol Histopathol 19(1):37–42. https://doi.org/10.14670/HH-19.37
64. Herbst RS, Sandler AB (2004) Non-small cell lung cancer and antiangiogenic therapy: what can be expected of bevacizumab? Oncologist 9(Suppl 1):19–26. https://doi.org/10.1634/theoncologist.9-suppl_1-19
65. Yuan A, Yu CJ, Kuo SH, Chen WJ, Lin FY, Kwen-Tay L, Yang PC, Lee YC (2001) Vascular endothelial growth factor 189 mRNA isoform expression specifically correlates with tumor angiogenesis, patient survival, and postoperative relapse in non-small-cell lung cancer. J Clin Oncol 19(2):432–441. https://doi.org/10.1200/JCO.2001.19.2.432
66. Hurwitz H, Fehrenbacher L, Novotny W, Cartwright T, Hainsworth J, Heim W, Berlin J, Baron A, Griffing S, Holmgren E, Ferrara N, Fyfe G, Rogers B, Ross R, Kabbinavar F (2004) Bevacizumab plus irinotecan, fluorouracil, and leucovorin for metastatic colorectal cancer. N Engl J Med 350(23):2335–2342. https://doi.org/10.1056/NEJMoa032691
67. Giantonio BJ, Catalano PJ, Meropol NJ, O'Dwyer PJ, Mitchell EP, Alberts SR, Schwartz MA, Benson AB (2007) Bevacizumab in combination with oxaliplatin, fluorouracil, and leucovorin (FOLFOX4) for previously treated metastatic colorectal cancer: results from the eastern cooperative oncology group study E3200. J Clin Oncol 25(12):1539–1544. https://doi.org/10.1200/JCO.2006.09.6305
68. Johnson DH, Fehrenbacher L, Novotny WF, Herbst RS, Nemunaitis JJ, Jablons DM, Langer CJ, DeVore RF, Gaudreault J, Damico LA, Holmgren E, Kabbinavar F (2004) Randomized phase II trial comparing bevacizumab plus carboplatin and paclitaxel with carboplatin and paclitaxel alone in previously untreated locally advanced or metastatic non-small-cell lung cancer. J Clin Oncol 2(11):2184–2191. https://doi.org/10.1200/JCO.2004.11.022
69. Patra CR, Mukherjee S, Kotcherlakota R (2014) Biosynthesized silver nanoparticles: a step forward for cancer theranostics? Nanomedicine 9(10):1445–1448
70. Mukherjee S, Patra CR (2016) Therapeutic application of anti-angiogenic nanomaterials in cancers. Nanoscale 8(25):12444–12470
71. Weaver BA (2014) How Taxol/paclitaxel kills cancer cells. Mol Biol Cell 25(18):2677–2681
72. Caruso M, Colombo AL, Fedeli L, Pavesi A, Quaroni S, Saracchi M, Ventrella G (2000) Isolation of endophytic fungi and actinomycetes taxane producers. Ann Microbiol 50:3–13. https://doi.org/10.1023/A:1013374417961
73. Aung TN, Qu Z, Kortschak RD, Adelson DL (2017) Understanding the effectiveness of natural compound mixtures in cancer through their molecular mode of action. Int J Mol Sci 18(3):656

74. Singh S, Sharma B, Kanwar SS, Kumar A (2016) Lead phytochemicals for anticancer drug development. Front Plant Sci 7:1667
75. Thakore P, Mani RK, Kavitha DJS (2011) A brief review of plants having anticancer property. Ijprd 3(11):129–136
76. Tariq A, Sadia S, Pan K, Ullah I, Mussarat S, Sun F, Abiodun OO, Batbaatar A, Li Z, Song D, Xiong Q, Ullah R, Khan S, Basnet BB, Kumar B, Islam R, Adnan M (2017) A systematic review on ethnomedicines of anti-cancer plants. Phytother Res 31(2):202–264
77. Moudi M, Go R, Yien CYS, Nazre M (2013) Vinca alkaloids. Int J Prev Med 4(11):1231–1235
78. Almagro L, Fernández-Pérez F, Pedreño MA (2015) Indole alkaloids from Catharanthus roseus: bioproduction and their effect on human health. Molecules 20(2):2973–3000
79. Heiblig M, Sobh M, Nicolini FE (2014) Subcutaneous omacetaxine mepesuccinate in patients with chronic myeloid leukemia in tyrosine kinase inhibitor-resistant patients: review and perspectives. Leuk Res 38(10):1145–1153
80. Li FS, Weng JK (2017) Demystifying traditional herbal medicine with modern approaches. Nat Plants 3:17109
81. Kim SH, Kaplan JA, Sun Y, Shieh A, Sun HL, Croce CM, Grinstaff MW, Parquette JR (2015) The self-assembly of anticancer camptothecin-dipeptide nanotubes: a minimalistic and high drug loading approach to increased efficacy. Chem Eur J 21(1):101–105. https://doi.org/10.1002/chem.201404520
82. Vennepureddy A, Atallah J-P, Terjanian T (2015) Role of topotecan in non-small cell lung cancer: a review of literature. World J Oncol 6(5):429–436. https://doi.org/10.14740/wjon950e
83. Zhu H, Ding WJ, Wu R, Weng QJ, Lou JS, Jin RJ, Lu W, Yang B, He QJ (2010) Synergistic anti-cancer activity by the combination of TRAIL/APO-2L and celastrol. Cancer Investig 28 (1):23–32. https://doi.org/10.3109/07357900903095664
84. Fan XX, Li N, Wu JL, Zhou YL, He JX, Liu L, Leung ELH (2014) Celastrol induces apoptosis in gefitinib-resistant non-small cell lung cancer cells via caspases-dependent pathways and hsp90 client protein degradation. Molecules 19(3):3508–3522. https://doi.org/10.3390/molecules19033508
85. Patel KR, Brown VA, Jones DJL, Britton RG, Hemingway D, Miller AS, West KP, Booth TD, Perloff M, Crowell JA, Brenner DE, Steward WP, Gescher AJ, Brown K (2010) Clinical pharmacology of resveratrol and its metabolites in colorectal cancer patients. Cancer Res 70 (19):7392–7399. https://doi.org/10.1158/0008-5472.CAN-10-2027
86. Varoni EM, Lo Faro AF, Sharifi-Rad J, Iriti M (2016) Anticancer molecular mechanisms of resveratrol. Front Nutr 3:8
87. Brâkenhielm E, Cao R, Cao Y (2001) Suppression of angiogenesis, tumor growth, and wound healing by resveratrol, a natural compound in red wine and grapes. FASEB J 15 (10):1798–1800. https://doi.org/10.1096/fj.01-0028fje
88. Khan N, Afaq F, Khusro FH, Mustafa Adhami V, Suh Y, Mukhtar H (2012) Dual inhibition of phosphatidylinositol 3-kinase/Akt and mammalian target of rapamycin signaling in human nonsmall cell lung cancer cells by a dietary flavonoid fisetin. Int J Cancer 130(7):1695–1705. https://doi.org/10.1002/ijc.26178
89. Jin MK, Nam JO, Kwon TK (2017) Fisetin induces apoptosis through p53-mediated up-regulation of dr5 expression in human renal carcinoma caki cells. Molecules 22(8):1285. https://doi.org/10.3390/molecules22081285

Natural Compounds Targeting Major Signaling Pathways in Lung Cancer

37

Sunil Kumar, Abhishek Kumar Sharma, H. Lalhlenmawia, and Deepak Kumar

Abstract

Among all types of human cancer, lung cancer is one of the most common and leading causes of cancer mortality worldwide. In the past few years, the growing understanding of molecular and tumor biology has dramatically changed the paradigms of cancer treatment. This involves designing more efficient and less toxic treatments, such as immunotherapy, targeted therapy, and cancer vaccines, as well as improving decades-old therapies, viz., radiation therapy, chemotherapy, and surgery. Various signaling pathways have been identified to be responsible in the pathophysiology of lung cancer. Advancements in technology and cancer biology and different oncogenic targets for cell signaling pathways are now understood, and these targeted therapies are being considered and established. Specific therapy or molecular-based therapy for each type of lung cancer provides improved outcomes as genomic changes, origin, and growth patterns vary in each subtype of lung cancer. Several *natural compounds* extracted from plants have been investigated for effective *cancer treatment*. This chapter reviews the recent developments and current potential of natural compounds derived from plant sources to target various signaling pathways and treatment of lung cancer.

S. Kumar · D. Kumar (✉)
Department of Pharmaceutical Chemistry, School of Pharmaceutical Sciences, Shoolini University, Solan, Himachal Pradesh, India

A. K. Sharma
Department of Pharmaceutics, School of Pharmaceutical Sciences, Shoolini University, Solan, Himachal Pradesh, India

H. Lalhlenmawia
Department of Pharmacy, Regional Institute of Paramedical and Nursing Sciences, Aizawl, Mizoram, India

K. Dua et al. (eds.), *Targeting Cellular Signalling Pathways in Lung Diseases*, https://doi.org/10.1007/978-981-33-6827-9_37

Keywords

Targeted therapies · Signaling pathways · Chemotherapeutic agents · Natural compounds

37.1 Introduction

Lung cancer is a severe and potentially life-threatening ailment and is still the leading cause of death worldwide [1]. Non-small cell lung cancers (NSCLCs) account nearly 80–85% of lung cancers, and small cell lung cancers (SCLCs) are about 10–15%. NSCLCs are *subclassified* into large cell carcinoma (LCC), adenocarcinoma (AD), and squamous cell carcinoma (SCC) [2]. Globally, a very large number of new NSCLC cases are diagnosed every year, around 17.8% 5-year survival [3]. As a result, the NSCLC is regarded to be very harmful with both high incidence and a significantly *low survival rate*. NSCLC is commonly found in older people, and smoking persons are considered the most significant risk factor. Further risk issues include exposure to organic chemicals and environmental toxicants. SCLC is almost entirely *linked* to *cigarette smoking* and contributes for around 15% of all lung cancer cases [4]. SCLC patients have a poor prognosis, with a low 5-year survival rate of less than 5%, although the primary response to chemotherapy might be positive [5]. The detection of various molecular abnormalities has contributed to the progress of targeted therapies that use compounds to target particular genes and proteins linked with cancer growth and progression [6]. Despite the immense variety of genes involved in tumorigenesis, the p53 inactivation suppress tumor and is common in tumorigenesis [7]. ADs often include loss of function mutations and deletions in tumor suppressor genes like p53, STK11, SMARCA4, CDKN2A, RB1, NF1, and KEAP1 [8]. The mutation within the p53 gene is the most prevalent alteration in lung SCC, followed by PIK3CA amplification. In addition to the PIK3CA and P53 alterations, MLL-2, CDKN2A, *CDH8*, *PTEN*, *ADCY8*, *CALCR*, *PTPRT*, *FBXW 7*, *GRM8*, *RB1*, NFE2L2, and KEAP1 mutations are among the most frequent genetic changes in lung SCC [9]. The p53 inactivation and retinoblastoma (Rb) proteins have been revealed to account for up to 90% of SCLC cases, and inactivation of both proteins can be triggered by mutation. Numerous studies have reported the potential of natural compounds to provide anticancer effects in lung cancer with different underlying mechanisms. This chapter presents a discussion on the major signaling pathways in lung cancer and the potential of natural compounds derived from plant sources for the therapy of lung cancer.

37.2 Signaling Pathways in Lung Cancer

The key signaling pathways that could provide roadmaps for lung cancer therapy are discussed below.

37.2.1 Receptor Tyrosine Kinase (RTK) Pathway

Mutations influencing RTK signaling also contribute to malignant *transformation of cells*, which is seen in various malignancies [10]. Such mutations affect RTKs or their downstream pathway elements (MAPK and PI3K/AKT), leading to increase in cell growth, invasion, survival, and metastasis. For researchers and clinicians working in the field of cancer, targeting RTK signaling pathways remains a formidable challenge. In order to target the RTK, PI3K/AKT, and the MAPK pathways, numerous antibodies and small compound inhibitors are being clinically developed [11].

37.2.2 Epidermal Growth Factor Receptor (*EGFR*)

EGFR is an *RTK* that gets activated by binding to specific ligands such as epidermal growth factor [12]. When activation, EGFR tyrosine kinase triggers downstream pathways (MAPK and PI3K), leading to DNA synthesis and cell proliferation [13]. It has been reported to be involved in tumor development and cancer progression. EGFR overexpression accounts for up to 60% of NSCLC cases and is often linked with poor prognosis [14].

37.2.3 ALK Fusion Proteins

Oncogenic translocations including anaplastic lymphoma kinase (ALK), a tyrosine kinase, are often associated with the pathogenesis of various types of *cancers* [15]. Echinoderm microtubule-linked protein-like 4 (EML4)-ALK is implicated about 3–7% of NSCLC patients [16]. EML4-ALK activates a variety of intracellular downstream pathways, such as PI3K/Akt and JAK/STAT pathways, leading to reduced apoptosis and increased proliferation of transformed cells [17]. The NSCLCs have also been found associated with several additional oncogenic ALK fusion proteins including KIF5B-ALK [18] and TFG-ALK [19] that bear oncogenic functions.

37.3 RAS Pathway

Twenty to Forty percent of lung ADs are KRAS mutations [20]. It interacts with a PI3 kinase subunit (p110), and the disruption of this interaction induces regression of tumor growth and development, particularly in combination with MEK inhibition [21]. Preliminary investigations indicate that MEK inhibition might be an effective approach in the treatment of KRAS tumors, and several clinical trials are presently studying this strategy. Inhibition of Hsp90 is another approach to clinically target KRAS-mutated tumors [22]. *GTPase HRAS* belongs to the RAS family which activates the RAS-RAF-MEK-ERK pathway. HRAS mutations are observed very rarely in *lung cancers*. Processes like cell differentiation, cell proliferation, apoptosis, cell adhesion, cell migration, and actin cytoskeletal integrity are controlled by Ras-regulated signal pathways through the MAPK and PI3K pathways.

37.4 BRAF/MAPK Pathway

An oncogenic driver gene is BRAF mutation in NSCLC that activates the downstream effectors MEK and ERK, ultimately *promoting cell progression* and survival [23]. The rate of BRAF mutations in NSCLC is around 3.5–4%. V600 mutations account for ~50% of melanomas, and the rest of the cases harbor non-V600 mutations [24–26]. Both inhibitors have shown remarkable efficiency in progressive-stage NSCLC patients with BRAF V600E mutation [27].

37.5 PI3K Pathway

In several cancer types, including NSCLC, the phosphoinositide-3-kinase (PI3K) signaling pathway is activated and has a vital part in cell proliferation and survival. PI3K/AKT/mTOR pathway alterations can result in PI3K activation and malignant transformation [28]. PI3K pathway is often deregulated owing to genetic changes that affect one of its components, leading to the rise in PI3K signaling in lung cancer. PI3K activation mostly occurs as a result of mutations or amplification in RTK's, PI3K amplification, overexpression of downstream kinase AKT, inactivating mutations or deletions of PTEN, mutational initiation of PIK3CA gene encoding the p110a catalytic subunit, and activation by oncogenic RAS [29].

37.6 LKB1/AMPK Pathway

The LKB1 (also called STK11) is a tumor suppressor gene often inactivated and mutated in NSCLC. STK11 acts directly upon AMPK and regulates the action of mTORC by phosphorylating TSC1 in AMPK-governed pathways [30]. Loss of LKB1 activity can lead to abnormal *differentiation* and facilitates metastasis [31, 32]. AMPK has a crucial part in cell progress regulation, proliferation, and autophagy by regulating mTOR activity, which is frequently deregulated by cancerous cells. AMPK/mTOR targeting is a promising strategy for developing therapeutic agents against NSCLC [33].

37.7 TP53 Pathway

Mutations in tumor suppressor gene TP53 are the most common somatogenic variations in NSCLC. *TP53* has been reported to be mutated in around 50% of NSCLC and more than 70% of SCLC case [34]. Abnormalities in the TP53 tumor suppressor gene are among the most significant events in lung cancer and found an important role in lung epithelial cell oncogenesis [35]. TP53 mutations endow "gain-of-function" (GOF) activity, and this mutant TP53 protein can hasten tumor progression or enhance resistance to anticancer treatment. Most clinical trials indicate that NSCLC with TP53 abnormalities has a poor prognosis and could be comparatively more resistant to radiation and chemotherapy [36].

37.8 RB1 Pathway

It is a tumor suppressor that finds a significant part in regulating the cell cycle and is often inactivated in many cancer types. Tumor-suppressive action of RB1 is attributed to its efficiency to disrupt the G1/S transition by repressing the target genes of E2F which are involved in DNA synthesis and progression of the cell cycle [37]. RB1 mutations are rare in AD, common in SCC (7%), and prevalent for SCLC in more than 90% of tumors.

37.9 MYC Pathway

Myc is an oncogene known to be downregulated in several cancer types, especially lung cancer, where it contributes to tumorigenic progression and development. Elevated amounts of Myc have already been linked with drug resistance [38]. Oncogene alterations are also prevalent in SCLC, with approximately 20% of patients exhibiting amplification or transcriptional upregulation of a member of the Myc family, MYC (6%), MYCN (4%), or MYCL (9%) [39, 40]. MYC is elevated in approximately 33% of NSCLC patients and is a prognostic marker of early-stage tumors [41, 42].

37.10 Developmental Pathways

Several developmental pathways, like hedgehog, Notch, and Wnt, play a vital role in controlling the proliferation, cell death, motility, migration, and stemness of cancer cells. Notch is among the most important cell signaling pathways and regulates many genes including Hes1, cyclinD1, c-Myc, and Akt via association with Delta and/or Jagged/Serrate families. Notch demonstrates both cancer-promoting effects and suppressive functions in lung cancer. Overexpression of Wnt signaling is prevalent in many hematological malignancies and also in solid tumors. Experimental and clinical evidence suggests that activation of Wnt/β-catenin is essential for cancer development, *angiogenesis*, migration, and invasion. Wnt signaling activates multiple transduction cascades in a cell, such as Wnt/β-catenin-dependent pathways and β-catenin-independent pathways. Alterations and deregulation of the Wnt pathway can lead to different types of cancers.

Similar to Wnt and Notch pathways, hedgehog is a developmental pathway that can regulate tumorigenic and physiological aspects of postnatal developmental events such as proliferation, differentiation, cell death, motility, migration, and invasion. The hedgehog signaling pathway is linked to three ligands: sonic hedgehog (SHh), Indian hedgehog (IHh), and desert hedgehog (DHh), and additional

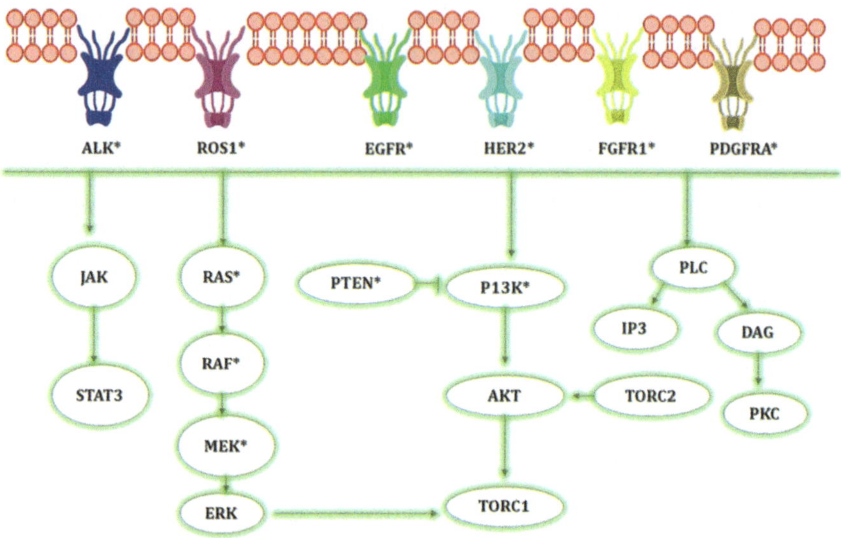

Fig. 37.1 Major drug targets for lung cancer

components of the pathway include two receptors (PTCH1 and PTCH2), three transcription factors (GLI1, GLI2, GLI3), and a smoothened receptor (SMO) that mediates signal transduction (Fig. 37.1) [43].

37.11 Natural Compounds as Anticancer Agent in Lung Cancer

Natural products derived from a variety of plant *sources have long been used as anticancer agents*. For several decades, orthodox Chinese medicines have been utilized in China for cancer therapy. In clinical practice, herbal medicines are widely used owing to their low cost, ease of availability, and very few complications or allergic events. As each expression of lung cancer varies in terms of genetic changes, cell origin, and growth patterns, the emphasis on new therapies or molecular-targeted therapy for each form of lung cancer promises improved outcomes. Numerous research studies have been conducted with a particular focus on possible natural products to determine the best chemotherapeutic agents that can reduce the risk of lung cancer. Alkaloids, terpenes, and flavonoids are natural botanical extracts that serve as potential chemotherapeutic agents and have beneficial effects against respiratory infections [44]. Some of the natural compounds that have shown remarkable potential for the treatment of lung cancer are discussed below.

37.11.1 Curcumin

Curcumin, the primary compound extracted from the turmeric rhizome (*Curcuma longa*) in the ginger family (Zingiberaceae), is a yellow polyphenol. Previous studies have shown that curcumin activity has been promising against different cancer types. Curcumin can suppress the signaling pathways of cancer, suppress metastases and angiogenesis, cause apoptosis, and induce tumor cell cancer sensitization [45]. Numerous research studies have shown that in a wide range of cancer therapy forms, such as lung cancer, curcumin can affect cancer stem cells (CSCs). By causing DNA damage or blocking DNA repair mechanisms, curcumin may reduce the ability of lung CSCs to self-regenerate [46]. The impact of curcumin on circulating cancer stem cells shown by a sphere formation assay has been documented, and important inhibitory effects of sphere formation have been observed. Curcumin has been identified as interfering with the signaling pathway of JAK2/STAT3 leading to a reduction in in vitro and in vivo viability of H460 lung cancer cells. Curcumin has also been shown to decrease the expression of CSC markers (Oct4, Nanog, ALDHA1, CD44, and CD133), suppress spread, and provoke apoptosis in H1299 and A549 cells via the downregulation of Wnt/β-catenin and sonic hedgehog pathway [47].

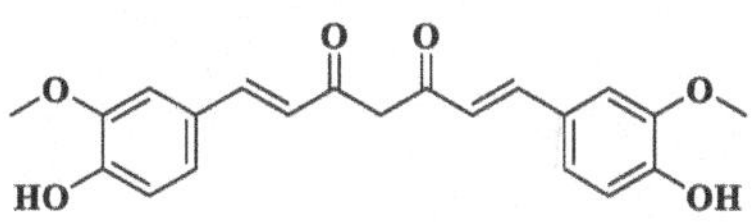

Curcumin

37.11.2 Wortmannin and Roscovitine

Roscovitine is a purine derivative which, through specific interaction at the ATP-binding site, suppresses the function of cyclin-dependent kinases (CDKs). Roscovitine is especially active toward Cdk5, Cdk1 (Cdc2), and Cdk2, which provoke cell cycle arrests of G1 and G2-M. Various studies have demonstrated the efficacy of roscovitine as an antitumor agent in various cancer cell lines. Likewise, in A549 cells, roscovitine induces dose-dependent apoptosis [48]. Similarly, wortmannin, a fungal metabolite, is a prominent antagonist specific to PI3K that attaches to the PI3K catalytic subunit p110 and irreversibly suppresses the enzyme that can chemosensitize various lung cancer cell lines (HeLa cells, HCT116, and A549). Wortmannin increases dose-dependent roscovitine-induced apoptosis associated with PKB/Act phosphorylate suppression in A549 cells [49].

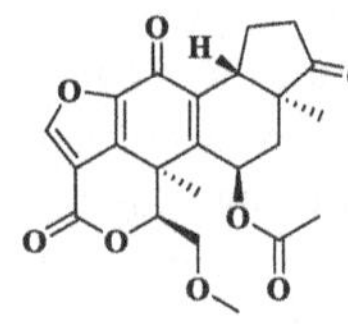

Wortmannin

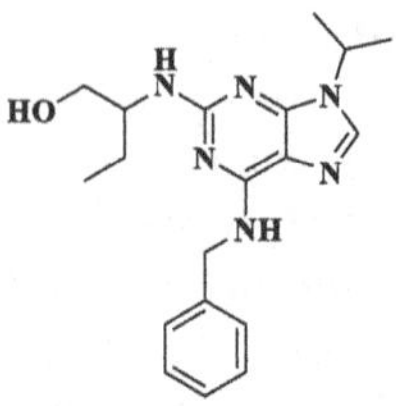

Roscovitine

37.11.3 Gigantol

Bhummaphan et al. demonstrated that gigantol, isolated from *Dendrobium draconis*, displayed CSC-inhibiting action on human lung carcinoma. The results of the study showed that the treatment of cancer cells with gigantol leads to decrease in lung CSC markers (ALDH1A1 and CD133) suggesting a decrease in cancer cell growth and reduced spheroid formation [50].

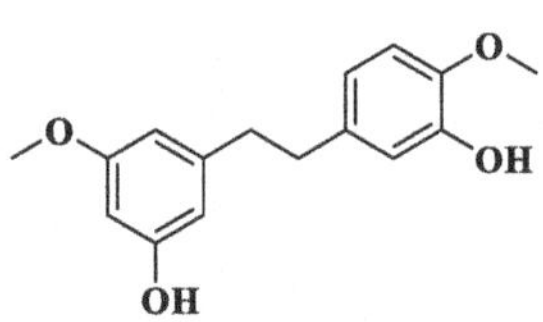

Gigantol

37.11.4 Chrysotoxine

Chrysotoxine, obtained from *Dendrobium pulchellum*, is documented to anchorage independently, sensitize anoikis, and suppress lung cancer cell metastases [51]. It has been reported that nontoxic doses ($\leq$20 μM) of chrysotoxine suppressed CSC-like phenotypes and reduced markers of ABCG2, CD133, ALDH1A1, and CD44 mediated by the Src-AKT-Sox2-dependent pathways [52].

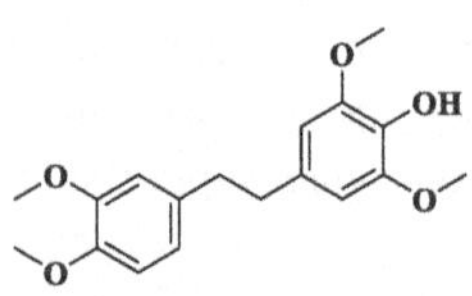

Chrysotoxine

37.11.5 Vanillin

Vanillin derived from *Vanilla planifolia* seed is commonly utilized as a flavor enhancer in cosmetics and food. Vanillin has been reported to suppress cell proliferation and promote apoptosis in several forms of cancer, such as lung carcinomas. Nontoxic doses (less than 100 μM) of vanillin can inhibit the formation of the spheroid and reduce ALDH1A1 and CD133 CSC markers and associated transcription factors, Nanog and Oct4, in H460 cells by reducing AKT and downstream CSC transcription factors [53].

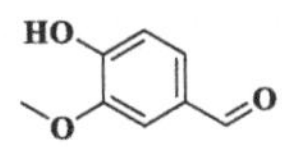

Vanillin

37.11.6 Silibinin

Silibinin is isolated from the seed of milk thistle (*Silybum marianum*) a polyphenolic flavonoid that can reduce different forms of cancer, particularly lung cancer. To evaluate inhibitory effect of silibinin on lung cancer stem cells, erlotinib cells (PC-9/Erl-R cells) have been reported in a routine culture medium containing a high dose of erlotinib (1 μM) by developing NSCLC PC9 cells expressing EGFR exon 19 deletion mutations. Flow cytometry and ALDEFLUOR reagent findings demonstrated that silibinin decreased CSC-like aldehyde dehydrogenase in erlotinib refractory cells and prevented dose-dependent lung cancer cell development [54].

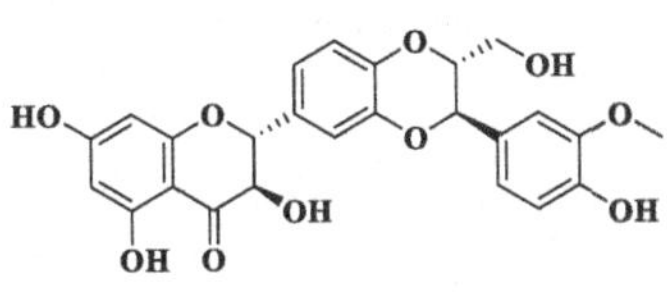

Silibinin

37.11.7 Parthenolide

Parthenolide, derived from *Tanacetum parthenium*, a sesquiterpene lactone has an anticancer effect on CSCs. Parthenolide was reported to selectively suppress CSCs in A549/shCDH1 cells in which shRNA was suppressed by CDH1/E-cadherin through the ER stress and apoptosis signaling pathway. Its basic function is to regulate the expression of stimulating transcription factor 4 (ATF4) and DNA damage-inducible transcript 3 (DDIT3), leading to the upregulation of the expression of polymerase-1 (PMAIP1) polymerase (ADP-ribose) [55].

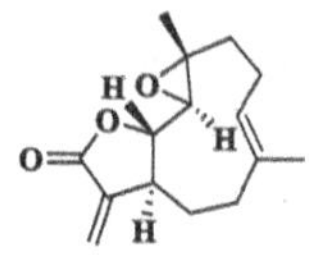

Parthenolide

37.11.8 Renieramycin M

Renieramycin M (RM) obtained from the *Xestospongia* sp. induces apoptosis in lung cancer cells. Its treatment at nontoxic concentrations substantially suppressed the colony and spheroids in H460 cells. In contrast, RM can also monitor CD133, ALDH1A1, and CD44 markers for CSC-enriched H460 cells [56].

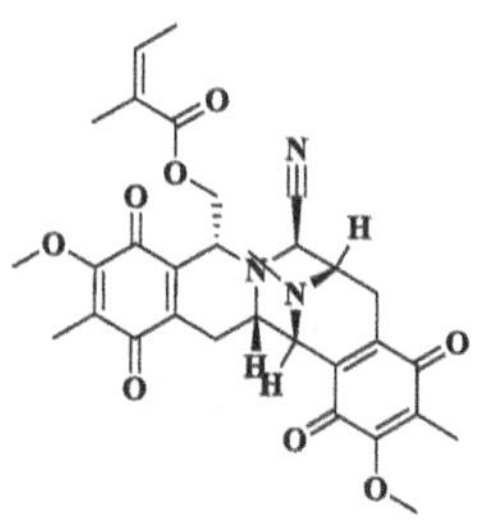

Renieramycin M

37.11.9 Salinomycin

It is derived from *Streptomyces albus* a polyether ionophoric antibiotic. Salinomycin demonstrated time-and dose-dependent antitumor activity assessed by sulforhodamine B and colony-forming assays in lung cancer cells (A549 and LNM35). Treatment with salinomycin has been shown to significantly inhibit the development of tumor spheroids and decrease the expression of OCT-4, NANOG, and SOX2 stem cells in ALDH A549 lung cell lines. Salinomycin nanoparticles and gefitinib nanoparticles have been developed using an emulsion/solvent evaporation method to kill both lung CSCs and lung cancer cells (A549 and A431). Moreover, the combination of salinomycin nanoparticles and gefitinib nanoparticles has a greater suppression effect on tumor growth than the combination of gefitinib and salinomycin or single salinomycin nanoparticles or gefitinib nanoparticles alone [57].

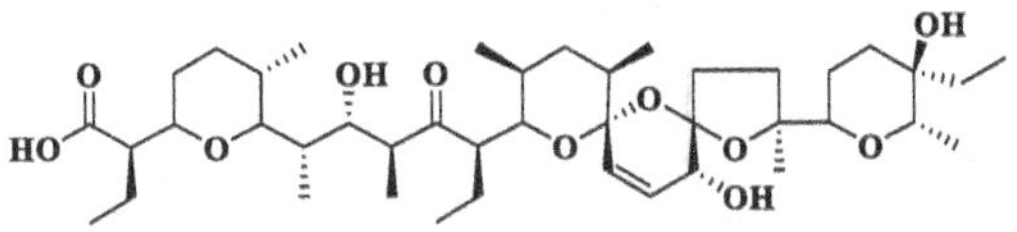

Salinomycin

37.11.10 Cordyceps militaris

Cordyceps militaris is a well-known ancient medicinal mushroom and a primary target for carcinoma therapy. The water extract from the *C. militaris* (WECM) promotes apoptosis of A549 cells via a signaling cascade of death-receptor-mediated extrinsic and mitochondrial-mediated intestinal caspase pathways. Apoptotic events due to WECM have also been identified to induce reduced telomerase activity by inhibition of hTERT transcriptional activity [58].

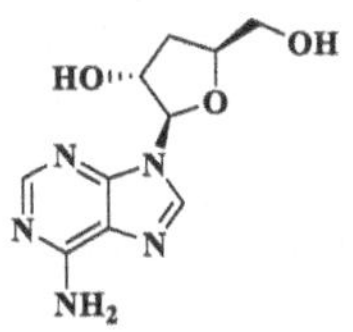

Cordyceps militaris

37.11.11 Resveratrol

It has been reported that resveratrol inhibits platelet aggregation and has antioxidant properties. Lung cancer has been shown to have inhibitory effects; it modifies a large number of genes and proteins and prevents the proliferation of A549 cells by inducing cell arrest, triggering apoptosis, and altering the TGF-β intracellular signaling pathway. It has already been shown in human A549 lung cancer cells as an antiproliferative agent and suppresses the Rb protein's phosphorylation and transcription factors like activator protein-1 and nuclear factor-kB (NF-kB) [59].

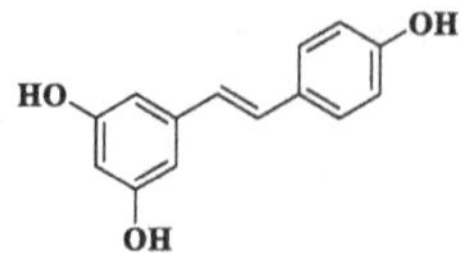

Resveratrol

37.11.12 Myricetin

It is a flavonoid compound widely present in wines, tea, berries, fruits, and medicinal plants and has antioxidant, antiproliferative, and anti-inflammatory properties. Earlier results have shown that myricetin has an antiproliferative impact on lung, esophageal, leukemia, and carcinoma cells of the prostate. Myricetin could serve as a strong antioxidant to scavenge or quench oxygen-free radicals and as an indirect antioxidant to shield cells from H_2O_2-induced cell damage by inducing antioxidant enzymes [60].

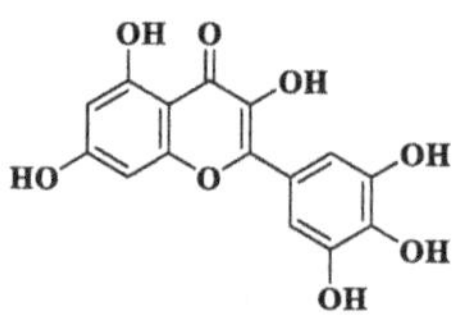

Myricetin

37.11.13 Berberine

Berberine is an isoquinoline alkaloid contained in medicinal plants that include *Hydrastis canadensis*, *Rhizoma coptidis*, and *Cortex phellodendri*. It has been described as having different pharmacological activities, viz., inhibition of DNA and protein synthesis, suppression of cell cycle progression, inhibition of tumor cell proliferation, and anticancer activity. Berberine has been reported to reduce motility and invasion of NSLC cells by improving the activation of c-Fos, c-Jun, and NF-£B, thus inhibiting uPA and MMP2 proteins [61].

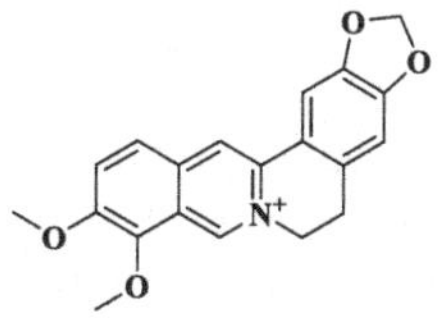

Berberine

37.11.14 Antroquinonol

Antroquinonol, a derivative of ubiquinone, is derived from mycelia. Antroquinonol is documented to reduce the viability of lung cancer and liver cancer cells by controlling AMP-activated protein kinase (AMPK) or phosphatidylinositol-3-kinase (PI3K)/rapamycin target mammalian (mTOR) pathways [62].

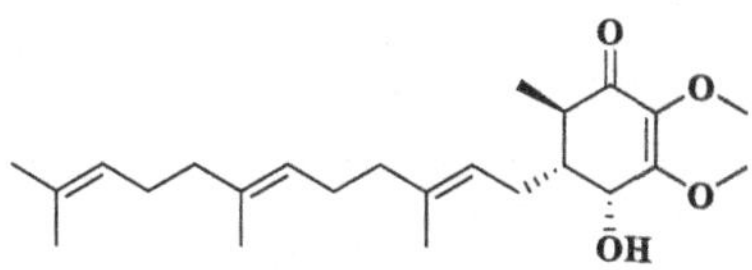

Antroquinonol

37.11.15 Honokiol

Honokiol is a naturally occurring biologically active agent derived from *Magnolia* spp. Compared to radiation alone, Lewis lung carcinoma cells (LL/2) treated with liposomal honokiol for 24 h demonstrated a higher radiation enhancement ratio (~twofold), showing increased immunity to radiation-induced cytotoxicity when co-treated with honokiol [63]. Combined treatment with radiation therapy resulted in a larger decrease in tumor volume compared to radiotherapy alone in the animal model (Lewis lung carcinoma-bearing C57BL/6 mice) [64]. In the lung cancer cell line xenograft lung cancer model, honokiol alone has been found to have antitumor effects, and the combined effect of honokiol and standard drug (cisplatin) decreased tumor volume (3.59-fold) as compared to standard drug alone [65].

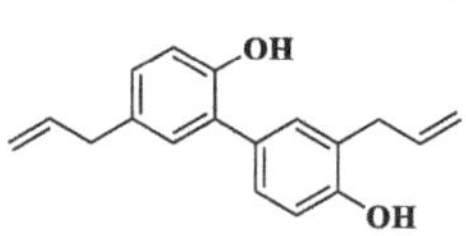

Honokiol

37.11.16 Emodin

Emodin, an important anticancer agent found in roots and rhizomes of various plant species, is used to treat tuberculosis, gallstones, hepatitis, arthritis, and osteomyelitis. MAPK was prevented by emodin's antiproliferative activity before emodin was identified, and a study showed that emodin can antagonize the signal. Both SK-MES-1 and A549 emodin-treated cells have been shown to have vacuolar cytoplasm degeneration [66]. There may be some other signaling pathways for emodin to prevent the spread of lung cancer cells. Deregulation is one of the molecular mechanisms for the proliferation of emodin suppression cells in human ERCC1 and Rad51 pulmonary carcinoma cells, and the ERK1/2 signaling pathway in human pulmonary carcinoma cells is inactivated. As a result, emodin can be used for the therapy of lung cancer as an alternative target [67].

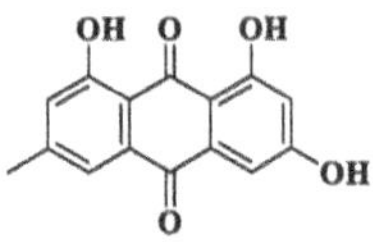

Emodin

37.11.17 Dioscin

Dioscin, a steroid of natural saponin, has an anticancer activity and is unclear in human lung cancer cells. Dioscin's inhibitory properties were tested, and the findings proved that dioscin inhibited human cancer cells A549, NCI-H446, and NCI-H460.0 from spreading. Single-cell gel electrophoresis and in situ terminal deoxynucleotidyl transferase nick-end mark assays have observed DNA disruption and cell apoptosis in dioscin-treated cells. By comparison, on the basis of the electron transmission microscope and the flow cytometry examination, dioscin has induced changes in the mitochondrial structure and disrupted the S-phase cell cycle. In fact, dioscin therapy has resulted in cytochrome c being moved from mitochondria to cytosol [68].

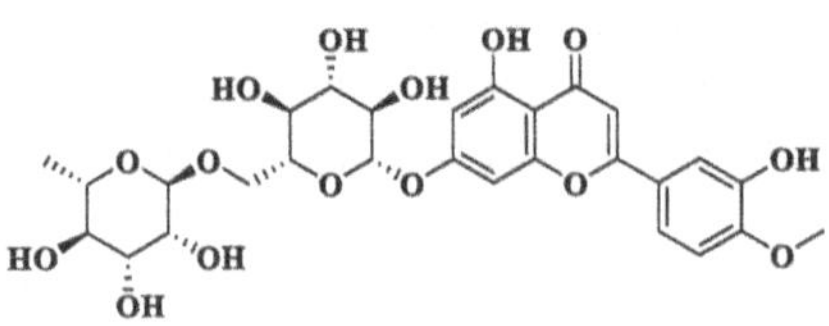

Diosein

37.11.18 Piperine

Piperine produces cytotoxic and apoptotic effects on human A549 lung cancer cells and the discovery of novel pathways. Piperine was shown to have the largest dose-dependent cytotoxic effect on A549 cells, while it did not have an effect on human lung fibroblasts in WI38. Cell DNA damage and cytotoxic activity may be due to this cell growth-inhibiting effect. In particular, in the G2/M process, piperine was able to induce cell cycle arrest and in the A549 cells to activate caspase-3 and caspase-9 cascades. In fact, piperine-mediated apoptosis may have been blocked by the broad caspase inhibitor z-VAD-fmk [69]. Compared to control piperine therapy lowered the expression of Bcl-2 protein and boosted the expression of Bax protein in A549 cells with increased p53 expression [70].

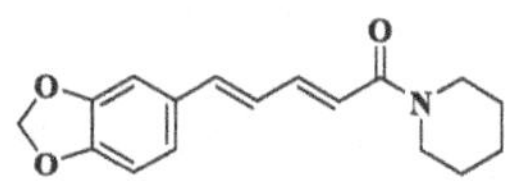

Piperine

37.11.19 Plumbagin

Anticancer activity has been demonstrated by plumbagin (PLB), but the mechanism is not clear. PLB has a potent proapoptotic and pro-autophagic effect on A549 and H23 cells; this study recorded. In the G2/M step, PLB arrests cells and raises the intracellular level of the reactive oxygen species across all cells [71]. As demonstrated by decreased phosphorylation of Akt and mTOR, dose-dependent PLB induces autophagy by suppression of the PI3 K/Akt/mTOR pathway. Autophagia suppression or implantation increases PLB-induced apoptosis. A cross talk between apoptosis caused by PLB and autophagia occurs. The results in NSCLC cells showed that PLB has organized pathways of both apoptosis and autophagy [72].

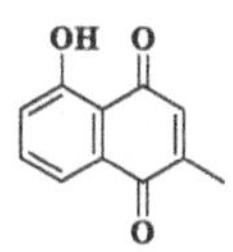

Plumbagin

37.11.20 Epigallocatechin-3-gallate (EGCG)

In various cancer types, the most prevalent polyphenol obtained from the green tea extract, epigallocatechin-3-gallate, was found possessing anticancer effects. In biology, the mechanics are not fully obvious. The impact of EGCG on migration, invasion, angiogenesis, and nicotine-induced epithelial mesenchymal transformation (EMT) was examined in A549 NSCLC cells, and molecular mechanisms were preliminarily examined [73]. The findings reported that various amounts of EGCG significantly suppressed nicotine-induced migration and invasion [74].

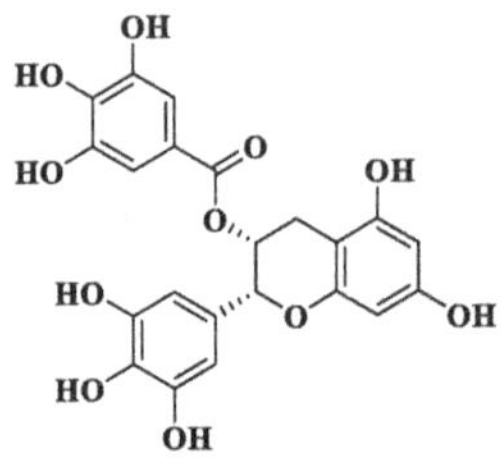

Epigallocatechin-3-gallate

37.11.21 Picropodophyllin

Picropodophyllin (PPP) acts as a IGF-1R antagonist in many tumors. However, the chemopreventive role of PPP has not been studied in lung tumorigenesis. In a mouse lung tumor model, the chemopreventive function of PPP has been studied. To cause lung tumors, benzo(a)pyrene was used, and PPP was given to female A/J mice through nasal inhalation. Tumor multiplicity and tumor load are measured for lung tumorigenesis. PPP reduced tumor multiplicity and tumor load substantially. The tumor multiplicity and load of 4 mg/mL of aerosolized PPP were lowered. Analysis of pharmacokinetics demonstrated strong bioavailability of lung and plasma PPP. This therapy improved staining of cleaved caspase-3 and decreased Ki-67 in lung tumors, indicating that PPP's lung tumor inhibitory effects were partly due to proliferation inhibition and apoptosis induction. PPP impaired cell proliferation in human lung cancer cell lines and also impaired phosphorylation of IGF-1R downstream markers, AKT and MAPK, eventually resulting in enhanced apoptosis. In lung cancer cell lines, PPP also decreased cell invasion [75].

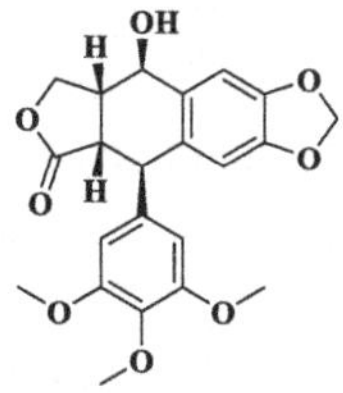

Picropodophyllin

37.11.22 Alantolactone

There are a number of pharmacological properties of alantolactone, a sesquiterpene lactone drug, including anti-inflammatory and anticancer effect. It blocked the cancer cell proliferation, apoptotic cells, and cell cycle distribution [76]. Apoptosis and induced cell cycle G1/G0 step arrest were caused by alantolactone. In addition, caspase-8, caspase-9, caspase-3, PARP, and Bax expressions were substantially upregulated, whereas expression of the antiapoptotic factor Bcl-2 was inhibited. In addition, alantolactone downregulated the expressions of cyclin-dependent kinase 4 (CDK4), CDK6, cyclin D3, and cyclin D1. Therefore, results have shown that alantolactone plays an antiproliferative function in the treatment of lung cancer cells and may be promising chemotherapy drugs for SK-MES-1 squamous lung cancer cells [77].

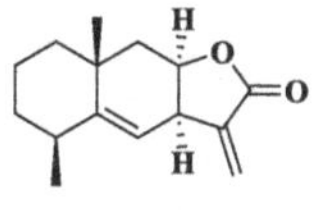

Alantolactone

37.11.23 Costunolide

Costunolide, a lactone of sesquiterpene, is considered to have anticancer effects. Costunolide's anticancer activity toward the NSCLC cell line of H1299 was investigated in this study. Costunolide suppression of viability of cell was identified by performing MTT assay. Furthermore, using annexin V/propidium iodide marking, the apoptotic rate was observed. To examine the antiproliferative effects of costunolide, forming a colony of cell-based assay was conducted. To evaluate the inhibitory action of the costunolide on migration and invasion, respectively, transwell and wound healing assays were conducted. In order to identify protein expression, Western blot analysis was undertaken and reverse transcription semi-quantitative PCR conducted to evaluate expression levels of mRNA. The findings reported costunolide, with IC_{50} value of 23.93 ± 1.67μM, inhibited the viability of H1299 cells and promoted cellular apoptosis. These results demonstrate the potential for the treatment of NSCLC with costunolide [78].

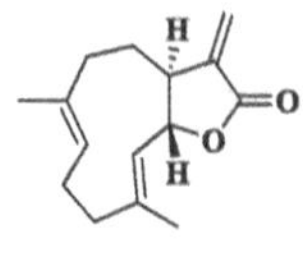

Custonolide

37.11.24 Catechins

The antagonist activity of catechins toward lung cancer has been reported through numerous experiments using chemically treated and transgenic mouse system. The introduction of EGCG or EGC through injection reduced lung tumorigenesis in mice and rats [79]. The formation of lung adenomas of adenocarcinomas in tumor-bearing A/J mice for 32 weeks was avoided by an oral route of 0.5% PPE or 0.044% caffeine in drinking water [80]. Study of the IHC found that PSA and caffeine therapy prevent enhanced apoptosis and cell proliferation and decreased c-Jun and phospho-ERK1/2 adenocarcinoma levels. In the lung tissues, neither agent has had a substantial effect on cell proliferation or apoptosis, indicating that the function is unique toward tumor cells [81]. Various studies reported the inhibitory activity of catechins toward lung tumorigenesis as well as the cancer prevention actions of caffeine. However, the mechanisms of action of these agents are yet to be further investigated [82].

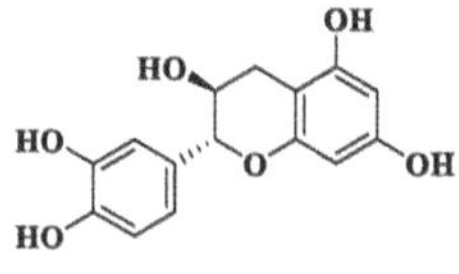

Catechins

37.11.25 Wentilactone A

The recently isolated marine fungus *Aspergillus dimorphicus* has developed the possible antitumor agent wentilactone. The deep-sea sediment was extracted from this fungus and characterized through a polyphasic method, incorporating extrolite, molecular, and phenotypic profiles. However, the production of wentilactone in static cultures reported with very low concentrations [83]. In attempt to increase the development of wentilactone, using the reaction surface technique, culture conditions were optimized. The experimental values were closely in line with the prediction model according to the optimum static fermentation circumstances. Wentilactone B and A yields were raised by nearly 13.4-fold to 11 and 6.5 mg/L, respectively. The fermentation scale-up for the manufacture of wentilactone further confirmed the result. In addition, specific small-molecule elicitors were identified to have the ability to promote the development of wentilactone [84].

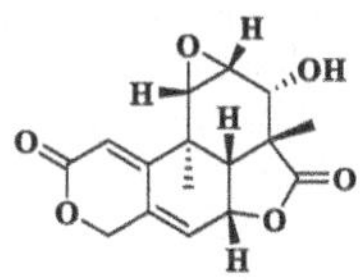

Wentilactone A

37.11.26 Evodiamine

Evodiamine is an antitumor alkaloid present in *E. rutaecarpa* and can turn into a therapeutic antitumor agent [88]. This study investigated the function of evodiamine in the proliferation of human A549 pulmonary cancer cells and the mechanism behind these effects [89]. The findings showed that the proliferation of A549 lung cancer cells with evodiamine was significantly blocked, apoptosis was induced and reactive oxygen species expressed, the cell cycle was halted, and evodiamines were due to its capacity to trigger oxidative cell destruction, induce apoptosis and cell cycle intervention, and regulate AKT-/NF- βB and SHH/GLI1 routes (Tables 37.1 and 37.2) [85].

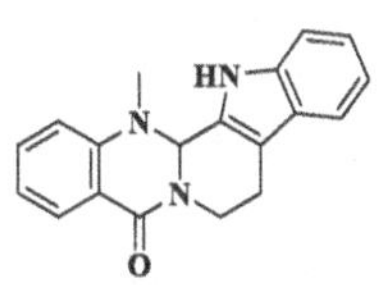

Evodiamine

37.12 Conclusion

The aim of this chapter was to summarize the present understanding and knowledge of molecular mechanisms in pancreatic cancer, especially in relation to their therapeutic targeting potential. Cancer of the lung is basically a set of multiple diseases, but in case of large cell carcinoma, different histologies do not generally suggest different molecular blueprints of tumors. The latter, dependent on the molecular

Table 37.1 Natural molecules targeting and their major signaling pathways in lung cancer

Major signaling pathways	Compounds	Target	Responses	References
Notch	Diallyl trisulfide Curcumin	Targets Notch-1 intracellular domain Downregulates transcription and translation Notch-1 and downstream genes Hes-1, Hey-1, and Hey-2 mRNA levels	Decreases the expression of Notch downstream genes Increases the tumor suppressor microRNA (miR143 and miR-145) expression and decreases tumor-promoting microRNA miR-21 Encourages apoptosis by elevating ROS	[43]
Wnt/β-catenin	Apigenin Resveratrol	β-catenin β-catenin; histone H2AX	Apoptosis of OS cells by decreasing mRNA and protein expression of β-catenin and c-Myc Histone H2AX phosphorylation causes telomere instability and DNA damage	[43]
Hedgehog	Cyclopamine	Binds to SMO	Inhibits the signal transduction to GLIS	[43]
PI3/AKT	Sulforaphane	ERK and AKT	Suppresses ERK and AKT phosphorylation, produces apoptosis through G2/M phase arrest	[62]

changes occurring in these tumors, has recently been identified to contribute to either the SCLC or AC community. Tumor molecular templates are emerging as conclusive indicators of the path of therapy to be selected, but not all driver variations in lung cancer have presumably been detected, and many of the driver mutations detected are actually not targetable. Lung cancer requires many oncogenic mechanisms, but most genetic mutations (with the omission ofTP53) are not strongly recurring within each histological form in most cases. Translocations and mutations and of receptor tyrosine kinases, for instance, do not contribute for the majority of adenocarcinoma cases. However, the advancement of selective receptor tyrosine kinase inhibition therapies has changed the diagnosis and management of adenocarcinoma, although comparable progress has not yet been made in exploring different oncogenic pathways for other forms of lung cancer. In the cancer of the lung, tumor inhibitors play a superficially universal function, but most of them remain indefinable therapeutic intervention targets. Today, in various clinical trials, many of the mechanisms implicated in the progression and development in the cancers of lung are currently targeted. In particular, immune therapies and immune checkpoint blockade have provided highly promising clinical trial results and will remain the

Table 37.2 Natural compound targeting lung cancer stem cells

Compound	Source	Proteins/signals	References
Curcumin	*Curcuma longa*	DNA damage, DNA repair, JAK1STAT3/ WnV -catenin, sonic hedgehog CD133, CD44, ALDHAI, Nanog, Oct4	[47]
Gigantol	*Dendrobium draconis*	AKT, CD133, ALDH1A1	[50]
Chrysotoxine	*Dendrobium pulchellum*	Src-AKT-Sox2, CD133, CD44, ABCG2, ALDH1A1	[51]
Vanillin	*Vanilla planifolia*	AKT-proteasomal degradation, CD133, ALDH1A1, Oct4, Nanog	[53]
Silibinin	*Silybum marianum*	ALDH activity	[54]
Parthenolide	*Tanacetum parthenium*	ER stress, apoptosis, ATF4, DDIT3, PMAIP	[55]
Renieramycin	*Xestospongia* sp.	CD133, CD44, ALDHIAI	[56]
Salinomycin	*Streptomyces albus*	OCT-4, Nanog, Sox2	[57]

focus of intensive clinical investigations. Investigational treatments and reasonable combinations of these are expected to lead to significant strides in improving outcomes for lung cancer patients. Treatment and diagnosis of lung cancer have now changed the development of targeted and immune therapies. Molecular examination has become a standard diagnostic technique in AC for KRAS, EML4-ALK, and EGFR. In terms of the advancement of targeted treatments, other types of lung cancer lag behind, but this is bound to change with several clinical trials underway in SCC and SCLC.

Acknowledgments Department of Pharmaceutical Chemistry, School of Pharmaceutical Sciences, Shoolini University of Biotechnology and Management Sciences, Post Box No. 9, Solan, Himachal Pradesh 173229, India.

Conflicts of Interest The authors declare no conflict of interest.

References

1. Bray F, Ferlay J, Soerjomataram I, Siegel RL, Torre LA, Jemal A (2018) Global cancer statistics 2018: GLOBOCAN estimates of incidence and mortality worldwide for 36 cancers in 185 countries. CA Cancer J Clin 68(6):394–424. https://doi.org/10.3322/caac.21492
2. Zheng MJ (2016) Classification and pathology of lung cancer. Surg Oncol Clin N Am 25 (3):447–468
3. Zappa C, Mousa SAJ (2016) Non-small cell lung cancer: current treatment and future advances. Transl Lung Cancer Res 5(3):288
4. Torre LA, Siegel RL, Jemal A (2016) Lung cancer statistics. In: Lung cancer and personalized medicine. Springer, pp 1–19

5. Stinchcombe TE, Gore EM (2010) Limited-stage small cell lung cancer: current chemoradiotherapy treatment paradigms. Oncologist 15(2):187
6. Kris MG, Johnson BE, Berry LD, Kwiatkowski DJ, Iafrate AJ, Wistuba II, Varella-Garcia M, Franklin WA, Aronson SL, Su P-F (2014) Using multiplexed assays of oncogenic drivers in lung cancers to select targeted drugs. JAMA 311(19):1998–2006
7. Rivlin N, Brosh R, Oren M, Rotter VJ (2011) Mutations in the p53 tumor suppressor gene: important milestones at the various steps of tumorigenesis. Genes Cancer 2(4):466–474
8. Imielinski M, Berger AH, Hammerman PS, Hernandez B, Pugh TJ, Hodis E, Cho J, Suh J, Capelletti M, Sivachenko AJ (2012) Mapping the hallmarks of lung adenocarcinoma with massively parallel sequencing. Cell 150(6):1107–1120
9. Choi M, Kadara H, Zhang J, Parra E, Rodriguez-Canales J, Gaffney S, Zhao Z, Behrens C, Fujimoto J, Chow CJ (2017) Mutation profiles in early-stage lung squamous cell carcinoma with clinical follow-up and correlation with markers of immune function. Ann Oncol 28 (1):83–89
10. Regad TJ (2015) Targeting RTK signaling pathways in cancer. Cancers (Basel) 7 (3):1758–1784
11. Gocek E, Moulas AN, Studzinski GP (2014) Non-receptor protein tyrosine kinases signaling pathways in normal and cancer cells. Crit Rev Clin Lab Sci 51(3):125–137
12. Wieduwilt M, Moasser MJC, Sciences ML (2008) The epidermal growth factor receptor family: biology driving targeted therapeutics. Cell Mol Life Sci 65(10):1566–1584
13. Montor WR, Salas AROSE, de Melo FHM (2018) Receptor tyrosine kinases and downstream pathways as druggable targets for cancer treatment: the current arsenal of inhibitors. Mol Cancer 17(1):55
14. Sharma SV, Bell DW, Settleman J, Haber DA (2007) Epidermal growth factor receptor mutations in lung cancer. Nat Rev Cancer 7(3):169–181
15. Ducray SP, Natarajan K, Garland GD, Turner SD, Egger G (2019) The transcriptional roles of ALK fusion proteins in tumorigenesis. Cancers (Basel) 11(8):1074
16. Koivunen JP, Mermel C, Zejnullahu K, Murphy C, Lifshits E, Holmes AJ, Choi HG, Kim J, Chiang D, Thomas R (2008) EML4-ALK fusion gene and efficacy of an ALK kinase inhibitor in lung cancer. Clin Cancer Res 14(13):4275–4283
17. Heuckmann JM, Balke-Want H, Malchers F, Peifer M, Sos ML, Koker M, Meder L, Lovly CM, Heukamp LC, Pao W (2012) Differential protein stability and ALK inhibitor sensitivity of EML4-ALK fusion variants. Clin Cancer Res 18(17):4682–4690
18. Takeuchi K, Choi YL, Togashi Y, Soda M, Hatano S, Inamura K, Takada S, Ueno T, Yamashita Y, Satoh Y (2009) KIF5B-ALK, a novel fusion oncokinase identified by an immunohistochemistry-based diagnostic system for ALK-positive lung cancer. Clin Cancer Res 15(9):3143–3149
19. Ducray SP, Natarajan K, Garland GD, Turner SD, Egger G (2019) The transcriptional roles of ALK fusion proteins in tumorigenesis. Cancers 11(8):1074
20. Ricciuti B, Leonardi GC, Metro G, Grignani F, Paglialunga L, Bellezza G, Baglivo S, Mencaroni C, Baldi A, Zicari D (2016) Targeting the KRAS variant for treatment of non-small cell lung cancer: potential therapeutic applications. Expert Rev Respir Med 10 (1):53–68
21. Murillo MM, Rana S, Spencer-Dene B, Nye E, Stamp G, Downward J (2018) Disruption of the interaction of RAS with PI 3-kinase induces regression of EGFR-mutant-driven lung cancer. Cell Rep 25(13):3545–3553. e3542
22. Chatterjee S, Bhattacharya S, Socinski MA, Burns TF (2016) HSP90 inhibitors in lung cancer: promise still unfulfilled. Clin Adv Hematol Oncol 14(5):346–356
23. Mu Y, Yang K, Hao X, Wang Y, Wang L, Liu Y, Lin L, Li J, Xing PJ (2020) Clinical characteristics and treatment outcomes of 65 patients with BRAF-mutated non-small cell lung cancer. Front Oncol 10:603

24. Marchetti A, Felicioni L, Malatesta S, Grazia Sciarrotta M, Guetti L, Chella A, Viola P, Pullara C, Mucilli F, Buttitta FJ (2011) Clinical features and outcome of patients with non-small-cell lung cancer harboring BRAF mutations. J Clin Oncol 29(26):3574–3579
25. Paik PK, Arcila ME, Fara M, Sima CS, Miller VA, Kris MG, Ladanyi M, Riely GJ (2011) Clinical characteristics of patients with lung adenocarcinomas harboring BRAF mutations. J Clin Oncol 29(15):2046
26. Luk PP, Yu B, Ng CC, Mercorella B, Selinger C, Lum T, Kao S, O'Toole SA, Cooper WA (2015) BRAF mutations in non-small cell lung cancer. Transl Lung Cancer Res 4(2):142
27. O'Leary CG, Andelkovic V, Ladwa R, Pavlakis N, Zhou C, Hirsch F, Richard D, O'Byrne KJ (2019) Targeting BRAF mutations in non-small cell lung cancer. Transl Lung Cancer Res 8 (6):1119
28. Tsay J-CJ, Wu BG, Badri MH, Clemente JC, Shen N, Meyn P, Li Y, Yie T-A, Lhakhang T, Olsen E (2018) Airway microbiota is associated with upregulation of the PI3K pathway in lung cancer. Am J Respir Crit Care Med 198(9):1188–1198
29. Gustafson AM, Soldi R, Anderlind C, Scholand MB, Qian J, Zhang X, Cooper K, Walker D, McWilliams A, Liu G (2010) Airway PI3K pathway activation is an early and reversible event in lung cancer development. Sci Transl Med. In: 2 (26):26ra25-26ra25
30. Alessi DR, Sakamoto K, Bayascas JR (2006) LKB1-dependent signaling pathways. Annu Rev Biochem 75:137–163
31. Facchinetti F, Bluthgen MV, Tergemina-Clain G, Faivre L, Pignon J-P, Planchard D, Remon J, Soria J-C, Lacroix L, Besse B (2017) LKB1/STK11 mutations in non-small cell lung cancer patients: Descriptive analysis and prognostic value. Lung Cancer 112:62–68
32. Chan KT, Asokan SB, King SJ, Bo T, Dubose ES, Liu W, Berginski ME, Simon JM, Davis IJ, Gomez SM (2014) LKB1 loss in melanoma disrupts directional migration toward extracellular matrix cues. J Cell Biol 207(2):299–315
33. Han D, Li S-J, Zhu Y-T, Liu L, Li M-X (2013) LKB1/AMPK/mTOR signaling pathway in non-small-cell lung cancer. Asian Pac J Cancer Prev 14(7):4033–4039
34. Toyooka S, Tsuda T, Gazdar AF (2003) The TP53 gene, tobacco exposure, and lung cancer. Hum Mutat 21(3):229–239
35. Mogi A, Kuwano H (2011) TP53 mutations in nonsmall cell lung cancer. J Biomed Biotechnol 2011:583929
36. Zhao J, Han Y, Li J, Chai R, Bai C (2019) Prognostic value of KRAS/TP53/PIK3CA in non-small cell lung cancer. Oncol Lett 17(3):3233–3240
37. Indovina P, Pentimalli F, Casini N, Vocca I, Giordano A (2015) RB1 dual role in proliferation and apoptosis: cell fate control and implications for cancer therapy. Oncotarget 6(20):17873
38. Massó-Vallés D, Beaulieu M-E, Soucek L (2020) MYC, MYCL, and MYCN as therapeutic targets in lung cancer. Expert Opin Ther Targets 24(2):101–114
39. George J, Lim JS, Jang SJ, Cun Y, Ozretić L, Kong G, Leenders F, Lu X, Fernández-Cuesta L, Bosco G (2015) Comprehensive genomic profiles of small cell lung cancer. Nature 524 (7563):47–53
40. Mollaoglu G, Guthrie MR, Böhm S, Brägelmann J, Can I, Ballieu PM, Marx A, George J, Heinen C, Chalishazar MD (2017) MYC drives progression of small cell lung cancer to a variant neuroendocrine subtype with vulnerability to aurora kinase inhibition. Cancer Cell 31 (2):270–285
41. Iwakawa R, Kohno T, Kato M, Shiraishi K, Tsuta K, Noguchi M, Ogawa S, Yokota J (2011) MYC amplification as a prognostic marker of early-stage lung adenocarcinoma identified by whole genome copy number analysis. Clin Cancer Res 17(6):1481–1489
42. Schaub FX, Dhankani V, Berger AC, Trivedi M, Richardson AB, Shaw R, Zhao W, Zhang X, Ventura A, Liu Y (2018) Pan-cancer alterations of the MYC oncogene and its proximal network across the cancer genome atlas. Cell Syst 6(3):282–300. e282
43. Giroux-Leprieur E, Costantini A, Ding VW, He B (2018) Hedgehog signaling in lung cancer: from oncogenesis to cancer treatment resistance. Int J Mol Sci 19(9):2835

44. Istudor V (2004) Pharmacologically active natural compounds for lung cancer. Altern Med Rev 9(4):402–419
45. Mehta HJ, Patel V, Sadikot RT (2014) Curcumin and lung cancer—a review. Target Oncol 9 (4):295–310
46. Chen H-W, Lee J-Y, Huang J-Y, Wang C-C, Chen W-J, Su S-F, Huang C-W, Ho C-C, Chen JJ, Tsai M-F (2008) Curcumin inhibits lung cancer cell invasion and metastasis through the tumor suppressor HLJ1. Cancer Res 68(18):7428–7438
47. Pillai GR, Srivastava AS, Hassanein TI, Chauhan DP, Carrier E (2004) Induction of apoptosis in human lung cancer cells by curcumin. Cancer Lett 208(2):163–170
48. Zhang F, Zhang T, Jiang T, Zhang R, Z-h T, Li C, Gu Z-P, Mei Q (2009) Wortmannin potentiates roscovitine-induced growth inhibition in human solid tumor cells by repressing PI3K/Akt pathway. Cancer Lett 286(2):232–239
49. Zhang F, Zhang T, Gu Z-P, Zhou Y-A, Han Y, Li X-F, Wang X-P, Cheng Q-S, Mei Q-B (2008) Enhancement of radiosensitivity by roscovitine pretreatment in human non-small cell lung cancer A549 cells. J Radiat Res 49(5):541–548
50. Charoenrungruang S, Chanvorachote P, Sritularak B, Pongrakhananon V (2014) Gigantol, a bibenzyl from Dendrobium draconis, inhibits the migratory behavior of non-small cell lung cancer cells. J Nat Prod 77(6):1359–1366
51. Maiuthed A, Chantarawong W, Chanvorachote P (2018) Lung cancer stem cells and cancer stem cell-targeting natural compounds. Anticancer Res 38(7):3797–3809
52. Chanvorachote P, Chamni S, Ninsontia C, Phiboonchaiyanan PP (2016) Potential anti-metastasis natural compounds for lung cancer. Anticancer Res 36(11):5707–5717
53. Lirdprapamongkol K, Kramb J-P, Suthiphongchai T, Surarit R, Srisomsap C, Dannhardt G, Svasti J (2009) Vanillin suppresses metastatic potential of human cancer cells through PI3K inhibition and decreases angiogenesis in vivo. J Agric Food Chem 57(8):3055–3063
54. Chu SC, Chiou HL, Chen PN, Yang SF, Hsieh Y (2004) Silibinin inhibits the invasion of human lung cancer cells via decreased productions of urokinase-plasminogen activator and matrix metalloproteinase-2. Mol Carcinog 40(3):143–149
55. Zhao X, Liu X, Su L (2014) Parthenolide induces apoptosis via TNFRSF10B and PMAIP1 pathways in human lung cancer cells. J Exp Clin Cancer Res 33(1):3. https://doi.org/10.1186/1756-9966-33-3
56. Halim H, Chunhacha P, Suwanborirux K, Chanvorachote P (2011) Anticancer and antimetastatic activities of Renieramycin M, a marine tetrahydroisoquinoline alkaloid, in human non-small cell lung cancer cells. Anticancer Res 31(1):193–201
57. Wang Y (2011) Effects of salinomycin on cancer stem cell in human lung adenocarcinoma A549 cells. Med Chem 7(2):106–111
58. Bizarro A, Ferreira IC, Soković M, Van Griensven LJ, Sousa D, Vasconcelos MH, Lima RT (2015) Cordyceps militaris (L.) link fruiting body reduces the growth of a non-small cell lung cancer cell line by increasing cellular levels of p53 and p21. Molecules 20(8):13927–13940
59. Wang H, Zhang H, Tang L, Chen H, Wu C, Zhao M, Yang Y, Chen X, Liu G (2013) Resveratrol inhibits TGF-β1-induced epithelial-to-mesenchymal transition and suppresses lung cancer invasion and metastasis. Toxicology 303:139–146
60. Shih Y-W, Wu P-F, Lee YC, Shi M-D, Chiang T-A (2009) Myricetin suppresses invasion and migration of human lung adenocarcinoma A549 cells: possible mediation by blocking the ERK signaling pathway. J Agric Food Chem 57(9):3490–3499. https://doi.org/10.1021/jf900124r
61. Peng P-L, Hsieh Y-S, Wang C-J, Hsu J-L, Chou F-P (2006) Inhibitory effect of berberine on the invasion of human lung cancer cells via decreased productions of urokinase-plasminogen activator and matrix metalloproteinase-2. Toxicol Appl Pharmacol 214(1):8–15. https://doi.org/10.1016/j.taap.2005.11.010
62. Ho C-L, Wang J-L, Lee C-C, Cheng H-Y, Wen W-C, Cheng HH-Y, Chen MC-M (2014) Antroquinonol blocks Ras and Rho signaling via the inhibition of protein isoprenyltransferase activity in cancer cells. Biomed Pharmacother 68(8):1007–1014. https://doi.org/10.1016/j.biopha.2014.09.008

63. Arora S, Singh S, Piazza GA, Contreras CM, Panyam J, Singh AP (2012) Honokiol: a novel natural agent for cancer prevention and therapy. Curr Mol Med 12(10):1244–1252. https://doi.org/10.2174/156652412803833508
64. Yang S-E, Hsieh M-T, Tsai T-H, Hsu S-L (2002) Down-modulation of Bcl-XL, release of cytochrome c and sequential activation of caspases during honokiol-induced apoptosis in human squamous lung cancer CH27 cells. Biochem Pharmacol 63(9):1641–1651. https://doi.org/10.1016/S0006-2952(02)00894-8
65. Singh T, Prasad R, Katiyar SK (2013) Inhibition of class I histone deacetylases in non-small cell lung cancer by honokiol leads to suppression of cancer cell growth and induction of cell death in vitro and in vivo. Epigenetics 8(1):54–65
66. Ko J-C, Su Y-J, Lin S-T, Jhan J-Y, Ciou S-C, Cheng C-M, Lin Y-W (2010) Suppression of ERCC1 and Rad51 expression through ERK1/2 inactivation is essential in emodin-mediated cytotoxicity in human non-small cell lung cancer cells. Biochem Pharmacol 79(4):655–664. https://doi.org/10.1016/j.bcp.2009.09.024
67. He L, Bi J-J, Guo Q, Yu Y, Ye X-F (2012) Effects of emodin extracted from Chinese herbs on proliferation of non-small cell lung cancer and underlying mechanisms. Asian Pac J Cancer Prev 13(4):1505–1510
68. Wei Y, Xu Y, Han X, Qi Y, Xu L, Xu Y, Yin L, Sun H, Liu K, Peng J (2013) Anti-cancer effects of dioscin on three kinds of human lung cancer cell lines through inducing DNA damage and activating mitochondrial signal pathway. Food Chem Toxicol 59:118–128. https://doi.org/10.1016/j.fct.2013.05.054
69. Selvendiran K, Singh JPV, Krishnan KB, Sakthisekaran D (2003) Cytoprotective effect of piperine against benzo[a]pyrene induced lung cancer with reference to lipid peroxidation and antioxidant system in Swiss albino mice. Fitoterapia 74(1):109–115. https://doi.org/10.1016/S0367-326X(02)00304-0
70. Lin Y, Xu J, Liao H, Li L, Pan L (2014) Piperine induces apoptosis of lung cancer A549 cells via p53-dependent mitochondrial signaling pathway. Tumour Biol 35(4):3305–3310
71. Li Y-C, He S-M, He Z-X, Li M, Yang Y, Pang J-X, Zhang X, Chow K, Zhou Q, Duan W, Zhou Z-W, Yang T, Huang G-H, Liu A, Qiu J-X, Liu J-P, Zhou S-F (2014) Plumbagin induces apoptotic and autophagic cell death through inhibition of the PI3K/Akt/mTOR pathway in human non-small cell lung cancer cells. Cancer Lett 344(2):239–259. https://doi.org/10.1016/j.canlet.2013.11.001
72. Shieh J-M, Chiang T-A, Chang W-T, Chao C-H, Lee Y-C, Huang G-Y, Shih Y-X, Shih Y-W (2010) Plumbagin inhibits TPA-induced MMP-2 and u-PA expressions by reducing binding activities of NF-κB and AP-1 via ERK signaling pathway in A549 human lung cancer cells. Mol Cell Biochem 335(1):181–193. https://doi.org/10.1007/s11010-009-0254-7
73. Relat J, Blancafort A, Oliveras G, Cufí S, Haro D, Marrero PF, Puig T (2012) Different fatty acid metabolism effects of (−)-Epigallocatechin-3-Gallate and C75 in Adenocarcinoma lung cancer. BMC Cancer 12(1):280. https://doi.org/10.1186/1471-2407-12-280
74. Sadava D, Whitlock E, Kane SE (2007) The green tea polyphenol, epigallocatechin-3-gallate inhibits telomerase and induces apoptosis in drug-resistant lung cancer cells. Biochem Biophys Res Commun 360(1):233–237. https://doi.org/10.1016/j.bbrc.2007.06.030
75. Zhang Q, Pan J, Lubet RA, Wang Y, You M (2015) Targeting the insulin-like growth factor-1 receptor by picropodophyllin for lung cancer chemoprevention. Mol Carcinog 54(S1):E129–E137
76. Maryam A, Mehmood T, Zhang H, Li Y, Khan M, Ma T (2017) Alantolactone induces apoptosis, promotes STAT3 glutathionylation and enhances chemosensitivity of A549 lung adenocarcinoma cells to doxorubicin via oxidative stress. Sci Rep 7(1):6242. https://doi.org/10.1038/s41598-017-06535-y
77. Zhao P, Pan Z, Luo Y, Zhang L, Li X, Zhang G, Zhang Y, Cui R, Sun M, Zhang X (2015) Alantolactone induces apoptosis and cell cycle arrest on lung squamous cancer SK-MES-1 cells. J Biochem Mol Toxicol 29(5):199–206

78. Wei M, Li J, Qiu J, Yan Y, Wang H, Wu Z, Liu Y, Shen X, Su C, Guo Q (2020) Costunolide induces apoptosis and inhibits migration and invasion in H1299 lung cancer cells. Oncol Rep 43 (6):1986–1994
79. Yang CS, Wang H (2016) Cancer preventive activities of tea catechins. Molecules 21(12):1679
80. Xu Y, Ho C-T, Amin SG, Han C, Chung F-L (1992) Inhibition of tobacco-specific nitrosamine-induced lung tumorigenesis in A/J mice by green tea and its major polyphenol as antioxidants. Cancer Res 52(14):3875–3879
81. Seeram NP, Zhang Y, Nair MG (2003) Inhibition of Proliferation of Human Cancer Cells and Cyclooxygenase Enzymes by Anthocyanidins and Catechins. Nutr Cancer 46(1):101–106. https://doi.org/10.1207/S15327914NC4601_13
82. Suganuma M, Saha A, Fujiki H (2011) New cancer treatment strategy using combination of green tea catechins and anticancer drugs. Cancer Sci 102(2):317–323. https://doi.org/10.1111/j.1349-7006.2010.01805.x
83. Jiang W, Meng L, Xu G, Lv C, Wang H, Tian H, Chen R, Jiao B, Wang B, Huang C (2018) Wentilactone A induces cell apoptosis by targeting AKR1C1 gene via the IGF-1R/IRS1/PI3K/AKT/Nrf2/FLIP/Caspase-3 signaling pathway in small cell lung cancer. Oncol Lett 16 (5):6445–6457. https://doi.org/10.3892/ol.2018.9486
84. Lv C, Hong Y, Miao L, Li C, Xu G, Wei S, Wang B, Huang C, Jiao B (2013) Wentilactone A as a novel potential antitumor agent induces apoptosis and G2/M arrest of human lung carcinoma cells, and is mediated by HRas-GTP accumulation to excessively activate the Ras/Raf/ERK/p53-p21 pathway. Cell Death Dis 4(12):e952–e952. https://doi.org/10.1038/cddis.2013.484
85. Hong J-Y, Park SH, Min H-Y, Park HJ, Lee SK (2014) Anti-proliferative effects of evodiamine in human lung cancer cells. J Cancer Prev 19(1):7–13. https://doi.org/10.15430/jcp.2014.19.1.7

Drug Delivery in Respiratory Diseases: Current Opportunities, Molecular and Cellular Mechanism, and Future Challenges

38

Debabrata Ghosh Dastidar, Dipanjan Ghosh, Swarnasree Ghosh, and Gopal Chakrabarti

Abstract

Respiratory diseases are a major cause of concern throughout the world. Lung cancer, chronic obstructive pulmonary disease (COPD), and asthma deserve special mention among them. According to the WHO, 235 million people suffer from asthma, and above three million people die annually from COPD, which is about 6% of all deaths across the globe. These diseases do not have a full-proof cure, and the treatment strategies are only symptomatic medications for temporary relief. Moreover, the respiratory system has a number of potential barriers to drug delivery. Secretion of mucus, ciliary action, and presence of macrophages are few among them. This chapter focusses on chronic diseases like COPD, tuberculosis, emphysema, asthma, and lung cancer and acute disease like pneumonia. Tuberculosis and lung cancer are the deadliest form of respiratory diseases that cause high annual death rates in the world. Pulmonary oedema, a commonly observed clinical manifestation, and cystic fibrosis, an autosomal recessive disorder, have also been discussed. The pathophysiology of diseases at the molecular level, currently available therapies, and the recent research on improvement of clinical efficiency have been elaborated. Different novel drug delivery systems

Dipanjan Ghosh and Swarnasree Ghosh contributed equally with all other contributors.

D. G. Dastidar
Department of Biotechnology, Dr. B.C. Guha Centre for Genetic Engineering and Biotechnology, University of Calcutta, Kolkata, WB, India

Guru Nanak Institute of Pharmaceutical Science & Technology, Kolkata, West Bengal, India
e-mail: debabrata.ghoshdastidar@gnipst.ac.in

D. Ghosh · S. Ghosh · G. Chakrabarti (✉)
Department of Biotechnology, Dr. B.C. Guha Centre for Genetic Engineering and Biotechnology, University of Calcutta, Kolkata, WB, India
e-mail: gcbcg@caluniv.ac.in

K. Dua et al. (eds.), *Targeting Cellular Signalling Pathways in Lung Diseases*,
https://doi.org/10.1007/978-981-33-6827-9_38

for pulmonary route and delivery device for pulmonary administration are discussed in detail. The mechanism of nanoparticle internalization by the epithelial cells and macrophages has been detailed together with the mechanism behind penetrating the mucus barrier. Finally, the future challenges in treating respiratory diseases are outlined.

Keywords

Respiratory disease · Pulmonary administration · Lung cancer · COPD · Tuberculosis · Porous microparticle · Inhalation · Dry powder inhaler · Nanoparticle

Abbreviations

ACP	Acyl carrier protein
AIDS	Acquired immunodeficiency syndrome
ALI	Acute lung injury
BCL2	B cell lymphoma 2
BMP	Bone morphogenetic protein
cAMP	Cyclic adenosine monophosphate
CCL	Chemokine (C-C motif) ligand
CCR	Chemokine receptor
CD	Cluster of differentiation
CF	Cystic fibrosis
CNS	Central nervous system
CXCL	Chemokine (C-X-C motif) ligand
GLUT1	Glucose transporter 1
GSH	Glutathione
hCAP-18	Human cationic antimicrobial protein- 18
HFC	Hydrofluorocarbon
HIV	Human immunodeficiency virus
ICAM- 1	Intercellular adhesion molecule 1
Ig	Immunoglobulin
IL	Interleukin
InhA	Inhibin α
MAPK	Mitogen-activated protein kinase
MMP	Matrix metalloproteinase
NAD+	Nicotinamide adenine dinucleotide (oxidized form)
NADH	Nicotinamide adenine dinucleotide (reduced form)
NF-κB	Nuclear factor kappa B
NOX2	NADPH oxidase 2
NSAIDS	Non-steroidal anti-inflammatory drugs
PavA	Pneumococcal adherence and virulence factor A
PD- L1	Programmed death-ligand 1

PDE- 4	Phosphodiesterase-4
PI3K	Phosphoinositide 3-kinase
PKA	Protein kinase A
PlGF	Placental growth factor
PsaA	Pneumococcal surface adhesin A
ROS	Reactive oxygen species
ROS1	c-ros oncogene 1
SNX9	Sorting nexin 9
TGF-β	Transforming growth factor- beta
TME	Tumour microenvironment
TNF-α	Tumour necrosis factor-α
USPtNs	Ultra-small platinum nanoparticles
WHO	World Health Organization

38.1 Introduction

38.1.1 Anatomy

The human respiratory system consists primarily of the nose, oropharynx, larynx, trachea, bronchi, bronchioles, and the lungs. Lungs are internal sac-shaped elastic respiratory organs lying ventral to the digestive tract. They are present in thoracic cavity beside the mediastinum and connected to the outside environment through the trachea. Trachea branches into two bronchi, one to each lung. Each bronchus branches into successively smaller bronchioles which supply air to the respiratory surfaces within the lung. The respiratory epithelium of the lungs is thin and well vascularized. It is divided into a number of small units to provide the vast surface area sufficient for exchange of gases between the lung and the blood. The lungs are further divided into five separate lobes, two on the left and three on the right by pleural membranes. These lobes are further divided into smaller pyramidal-shaped sections known as the bronchopulmonary segments that ultimately consist of over 300 million alveoli [1]. Alveoli being the structural and functional unit of the respiratory system are wrapped by a network of superfine capillaries which perform the exchange of oxygen into the arterial system for tissue perfusion in return of carbon dioxide [2]. This complex branching network looks like a tree, known as the tracheobronchial tree. The lymphoid tissue of the tracheobronchial system contains diffused, aggregated, and solitary lymphatic nodules. This is known as bronchus-associated lymphoid tissue (BALT), responsible for the hypersensitive reactions during respiration.It provides defence against microbes present in inhaled air [3]. Diaphragm is the primary respiratory muscle. The external intercostal muscles are inspiratory muscles used mainly during exercise and respiratory distress.

38.1.2 Functioning of the Lung in the Human Body

38.1.2.1 Respiratory Functions

Lungs are the primary organs for the process of respiration. The alveolar-capillary bed is responsible for the exchange of oxygen and carbon dioxide by diffusion. The conducting airway tract is responsible for maintaining the temperature, humidity, and sterility of inhaled air. It is known as air conditioning [4].

38.1.2.2 Non-respiratory Functions of the Lungs

The vascular endothelial cells of the lungs perform various metabolic activities. It converts angiotensin-I into angiotensin-II [5] and degrades vasoactive mediators like serotonin, bradykinin, and norepinephrine. Small cluster of neuroendocrine cells are present in bronchial mucosa. They are called *Kulchitsky cells*. They are responsible for the secretion of different important chemical mediators, including catecholamine, hormones (calcitonin, serotonin), and gastrin-releasing factors (bombesin) [6]. Pulmonary epithelium acts as the first line immunological barrier to the foreign materials present in inspired air. The bronchial surfaces are covered with a layer of mucus generated by the submucous glands and "goblet cells" that are present abundantly on the bronchial surface. Ciliated columnar cells are also present on the airway tract. The foreign material, including microorganisms that land on the bronchial surface, is entrapped into the mucus layer and transported towards the mouth due to the coordinated ciliary movement [7].

The lungs also play important roles both in innate and adaptive immune system. Alveolar macrophage, neutrophils, innate lymphoid cells (ILC) type I and II, and mucosal-associated invariant T cells (MAIT) take part in innate immune system. However, strict regulation of macrophage activation is required as airway cells are constantly exposed to airway endotoxins and their hypo-activity is essential for normal macrophage functioning. IL-10, TGF-β, CD200, and different surfactant proteins like SPA and SPD help in reducing those pro-inflammatory signalling [8]. The leukocytes together with the soluble proteins (IgG, complement factors, surfactants, and surfactant-associated proteins) present in alveolar fluids detect the microbes with pathogen-associated-molecular patterns (PAMPs) present on their cell surface. The airway aqueous fluid also contains lysozyme, lactoferrin, IgA, and antimicrobial peptides, e.g. defensins [9]. All these components help in encountering the pathogens. In pneumonia and sarcoidosis, antimicrobial peptides like cathelicidins LL-37 and hCAP-18 are expressed by the epithelial cells and neutrophils [10]. Alveolar fluid contains high proportions of lipopolysaccharide (LPS)-binding proteins (LBP) and soluble CD14 (sCD14). Alveolar macrophages recognize LPS with these LBP and sCD14 [11, 12]. TNF-α produced during bacterial and viral infection upregulates mucin gene expression [13]. This stimulates B lymphocytes to start producing antibodies. Antibodies have prime role in the process of opsonization and thus help in destroying the microbes by leukocytes present in tissue and lymph nodes.

38.2 Effect of Particle Size on Pulmonary Administration of Drug

Delivery of drugs locally into lungs has less chances of dose-dependent systemic toxicity. Sometimes the drugs undergoing extensive hepatic first-pass metabolism are intentionally administered through pulmonary route for direct systemic absorption [14]. Modern therapy of respiratory diseases involves different biomolecules like proteins (monoclonal antibodies) and peptides. These biopharmaceuticals are not suitable for oral delivery system as they are susceptible to degradation in the acidic environment of stomach and hepatic first-pass metabolism [15]. So, delivery through pulmonary route is a good alternative of choice. Moreover, in the treatment of disease like tuberculosis, drugs are to be specifically delivered to the alveolar macrophages [16]. The pulmonary system can be functionally divided into the conducting zone (consisting of trachea, bronchi, and bronchioles) and the respiratory zone (constituting the airways and alveoli). The total volume of alveolar septal tissue is 230 ± 38 cm^3. Five hundred million alveoli actively participate in the gaseous exchange process [17]. The adult human lungs have approximately 75–140 m^2surface area [18]. Macrophages (with cell diameter of 15–22 μm) uptake particles, preferentially of 1–3 μm in diameter [19]. Particles, smaller than 0.26 μm diameter, might escape phagocytosis by macrophages [20] and get in contact with the epithelial cells followed by clathrin/caveolae-mediated endocytosis [21].

The distribution of drug molecules after pulmonary administration primarily depends on the size of carrier particles (Fig. 38.1). The mechanisms behind distribution of particles are (i) inertial impaction, (ii) gravitational sedimentation, and (iii) Brownian diffusion. The larger particles having aerodynamic diameter > 5 μm pass through the oropharynx, trachea, and other large-sized airways [22]. The particles having aerodynamic diameter 1–5 μm are more prone to sedimentation in the smaller airways and the respiratory bronchioles. The small-sized nanoscaled drug carrier particles are mainly distributed by the Brownian motion of diffusion [23–25]. But, administered particles with sizes of <1 μm are exhaled out because of their low inertia [26] whereas mucociliary action is the mechanism behind clearance of larger particles. The rate of clearance depends on the particle diameter [27].

38.3 Barriers to Drug Delivery at the Respiratory Tract

The mucus, lining the inner surface of pulmonary tract, is the innate immunological barrier that contributes most to the drug delivery through pulmonary route [28]. The specialized goblet cells are responsible for the secretion of mucus consisting primarily of heavily glycosylated mucins and 95–99.5% water [29]. Due to viscoelasticity, mucus lining entraps the foreign pathogen and protects the epithelium from infection [30]. Due to impairment of mucociliary clearance and hypersecretion of mucus, the average thickness of airway mucus lining is found to be increased significantly in different respiratory diseases like asthma [31]. Drug-loaded carrier particles when delivered through pulmonary route are trapped and eliminated by this non-selective

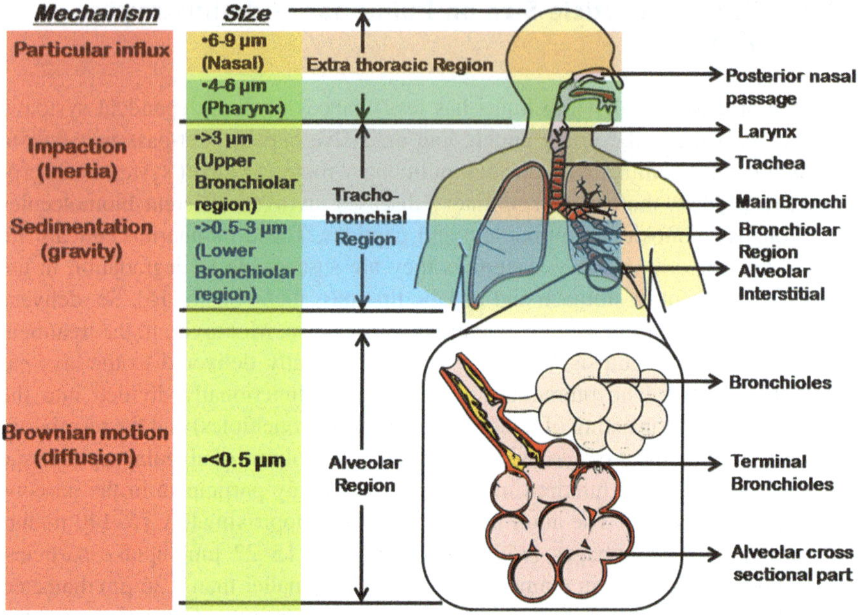

Fig. 38.1 Schematic representation of human respiratory system. Particles are distributed and deposited into the various regions of respiratory system based on their aerodynamic diameter. Various types of mechanisms are also involved for deposition depending on the size of particles

protective mucus layer. In general, particles larger than the mucus mesh spacing get trapped [32]. Thus, the entry of particles to the target site gets hindered. Presence of glycans makes mucin negatively charged. As a result, particles of cationic polymers, e.g. chitosan, polymethacrylate, and polyethylenimine, get captured in the mucus lining. Moreover, hydrophobic interactions are also responsible behind entrapment of nanoparticle by respiratory mucus layer [33].

A cluster of nanoparticles forming micron size particles are suitable for pulmonary administration. In the pulmonary fluid, the nanoparticles are reconstituted. The nanoparticles, if not stealth, are rapidly recognized and subsequently cleared by the alveolar macrophages [34]. This largely affects the half-life and therapeutic efficacy of the loaded drug [35].

38.4 Respiratory Barriers to Drug Administration

Polymeric nanoparticles can be administered as intravenous injection [36], aerosol formulation, and dry powder inhaler (DPI) through pulmonary route [37, 38]. Polymeric nanoparticles are usually hydrophilic in nature. The surface charge of the particle is a very important factor for lung cell targeting. Amine-modified polystyrene nanoparticles having neutral surface charge at physiological pH penetrate better than carboxylated polystyrene nanoparticles in the human sputum [39]. Similarly,

De Smedt et al. reported that positively and negatively charged particles are less mobile than PEGylated neutral particles in sputum [40]. Actually, positively charged nanoparticles get cross-linked with the mucin network and form the viscous mucous [41]. Negatively charged carboxylated particles create more hydrogen bonds due to chelation. Thus, viscosity of mucus is further increased. Electrically neutral PEGylated polymeric nanoparticles are hydrophilic in nature and able to mobilize the mucosal layer due to favourable hydrophilic–hydrophobic interactions with the mucin network [42, 43].

Foreign particles that get deposited in the central and intermediate airway zones undergo mucociliary clearance. Thus, the pulmonary absorption remains incomplete. The particles may be coated with mucoadhesive polymers to prolong the airway retention time resulting in enhanced pulmonary bioavailability [44]. Paradoxically Schneider et al. found that mucoadhesive particles are very much susceptible to entrapment in the mucus lining. As a result, they cannot reach the underlying airway epithelium [45]. Particles smaller than 300 nm diameter and without mucoadhesive coating were found to be uniformly distributed and retained for prolonged period of time. Nanoparticles that are smaller than mucus mesh spacing and have surface properties similar to viruses have good mucus-penetrating efficiency [46].

Alveolar macrophages are big barrier to the targeting of epithelial cells with nanoparticles. Macrophages can easily recognize and scavenge the foreign nanoparticles. PEGylation of nanoparticle surface makes them stealth to this scavenging process [47]. These "stealth" nanoparticles are reported to have considerably improved residence time and delayed clearance in vivo [48, 49].

38.5 Pathophysiology of Respiratory Diseases

Respiratory diseases in adults are a serious cause of concern. It is the primary cause of morbidity and mortality round the globe. Respiratory diseases can be classified into acute diseases (e.g. pneumonia and influenza), chronic diseases (e.g. chronic obstructive pulmonary disease and asthma), occupational lung diseases (e.g. byssinosis, asbestosis, and coal worker's pneumoconiosis), and other parenchymal lung diseases (hypersensitivity pneumonia). Lung cancer, tuberculosis, and AIDS-related pulmonary distress deserve special mention as these are some of the deadliest diseases responsible for annual high death rates globally [50]. Respiratory diseases can further be classified into obstructive and restrictive disease. Obstructive diseases are associated with impaired expiration leading to significant volume of air entrapment in the lungs after expiration. Thus, the value of functional residual capacity (FRC) increases. COPD is a typical example. Restrictive lung disease, on the other hand, exhibits restricted lung expansion causing decreased lung volumes. Idiopathic pulmonary fibrosis, pneumoconiosis, sarcoidosis, and lung parenchymal diseases are the typical examples [2]. Apart from these, some bacterial infections which occur in the upper respiratory tract include epiglottitis and laryngotracheitis caused likely by *Haemophilus influenzae* type b and bacterial pharyngitis caused by

Streptococcus pyogenes. Common cold is the most common viral infection mostly caused by rhinoviruses. Bronchitis, bronchiolitis, and pneumonia are manifestations of lower respiratory tract infection, caused by *H. influenzae*, respiratory syncytial virus, and *S. pneumoniae*, respectively [51].

38.5.1 Pneumonia

Pneumonia is associated with inflammation of the lung parenchymatous tissue. Lobar pneumonia is the alveolar infection of entire lobe of lung. Bronchopneumonia involves an alveolar process with a patchy distribution. Cough, fever, chest pain, tachycardia, and sputum production are the common symptoms of pneumonia. Depending on the age and the aetiological agents, some patients may suffer from headache, abdominal pain, nausea, vomiting, and diarrhoea [51]. Some common sources of pneumonia are bacterial pneumonia, aspiration pneumonia, and atypical pneumonia. *Streptococcus pneumoniae* is the major aetiologic agent that causes community-acquired acute bacterial pneumonia [50]. *Streptococcus pyogenes* causes haemorrhagic pneumonitis and empyema. Elderly patients over 50 years old and suffering from COPD or alcoholism are susceptible to *Haemophilus influenzae* and *Klebsiella pneumoniae* infection. *Mycobacterium tuberculosis* can also cause pneumonia. Patients with periodontal disease or decreased consciousness are susceptible to aspiration pneumonia caused by *Actinomyces, Bacteroides, Peptostreptococcus, Veillonella, Propionibacterium, Eubacterium*, and *Fusobacterium* species. Atypical pneumonias do not show typical bacterial lobar pneumonia symptoms. It can be caused by *Mycoplasma pneumoniae* in young people. Legionella including *L. pneumophila* causes atypical pneumonia. The mode of transmission is inhalation from tap water aerosols, respiratory devices, and air conditioners. *Chlamydia* causes pneumonia in bird handlers, especially in neonates and young infants with occupational pneumonitis. There are other viruses, *Actinomyces, Nocardia*, and fungus which can cause pneumonia or pneumonitis in various cohorts of healthy or immunocompromised patients [51].

38.5.1.1 Pathophysiology and Cellular Mechanism

S. pneumonia infects upper mucosal cells by means of binding of adhesion proteins like RrgA to TLR 2, lipoprotein PsaA to the E-cadherin receptor, and PavA protein to fibronectin and integrin. Following adhesion, the bacterial phosphorylcholine or choline-binding protein A (CbpA) binds either to the platelet-activating factor receptor (PAF-R) or to the polymeric immunoglobulin receptor (pIgR). This ultimately transports bacteria to the basement membrane of the host, helping in further invasion and disease progression [52].

Aqp5 and Mak6 genes play pivotal role in pathogenesis of pneumonia. In a recent study by Weiping Zhou et al. (2020) [53], differential expression of these genes were found to be significant in alveolar epithelial cells of mice infected with *Streptococcus pneumoniae*. microRNA-181a-5p, Stat3, and Sp3 were also identified as potential molecules that significantly regulate dysfunction modules in the development of

Table 38.1 Different genes responsible for pneumonia

Sl. no.	Expression	Candidate genes	Functions	References
1.	Differentially expressed	Aqp5, Mak6	Assist in the development of pneumonia	[53]
2.		microRNA-181a-5p, Stat3, and Sp3	Responsible for pathogenesis of pneumonia	[53]
3.	Downregulated	ND1, ND3, ND4L, and ND6	Responsible for immune response in pneumonia	[54, 57]
4.	Upregulated	MIR449A	Modulates inflammatory response	[54, 55]
5.		TAS2R43	Involved in immune response	[54, 56]

pneumonia. In another study [54], *ND1, ND3, ND4L, and ND6* genes that encode subunits of NADH dehydrogenase have been found to be downregulated in patients with severe pneumonia alone and severe pneumonia combined with COPD. Further, MIR449A genes, encoding miRNA-449aA, and TAS2R43 genes, encoding bitter taste 2 receptor member 43, were found to be upregulated [54]. miRNA-449a targets the NOTCH signalling pathway and alters the expression of the inflammatory marker YKL40 [55]. The genes of bitter taste receptors are involved in pneumonia pathogenesis by modulating inflammation and immunity. Amarogentin, an agonist for bitter taste receptor, causes IL-8 and MMP-1 expression in human mast cells and keratinocytes [56]. Various types of responsible genes are given in Table 38.1.

38.5.1.2 Treatment Strategies

Standard monotherapy/combinatorial treatment regimen with conventional antibiotics like amoxicillin, doxycycline, macrolides, clavulanate or cephalosporin, and fluoroquinolone [58] have become ineffective.Different adverse drug reactions, inefficient delivery of drugs into lungs, and development of multidrug resistance are the major reasons behind it. Moreover, development of gastrointestinal intolerance with macrolides are some pertinent drawbacks in pneumonia treatment [59]. Newer and novel treatment strategies of pneumonia are listed in Table 38.2.

38.5.1.3 Novel Drug Delivery Strategies

Pneumonia is an inflammatory disorder of respiratory system due to viral and bacterial infections. Zhang et al. developed stimuli-responsive nanoparticle to target inflammatory lung cells. The nanoparticle had a core-shell structure. The core containing acid-sensitive poly(β-amino esters) was covered with a layer of polyethylene glycol (PEG). TPCA-1, as model drug, was loaded within the core. The PEG molecules on the surface were biotinylated (conjugated with the biotin molecules) and coated with anti-ICAM-1 antibody for facilitating bioconjugation, lung targeting, and extended circulation [80].

N-Fumaroylated diketopiperazine (FDKP) is an FDA-approved excipient for pulmonary administration. Wang et al. developed azithromycin-loaded FDKP microparticle as dry powder for pulmonary administration (intratracheal

Table 38.2 Different strategies for the treatment of pneumonia

Sl. no.	Class	Drugs	General mechanism of action	References
1.	Beta-lactam + beta-lactamase inhibitors	Ceftolozane–tazobactam, Ceftazidime–avibactam, Meropenem–vaborbactam, Imipenem–relebactam, Aztreonam–avibactam	Increases efficiency of beta lactam antibiotics by blocking beta lactamase enzymes	[59–64]
2.	Cephalosporins (a type of beta-lactam antibiotics)	Cefiderocol (S-649266), Ceftobiprole	Inhibits the in bacterial cell wall synthesis	[59, 65]
3.	Macrolide	Solithromycin (fourth generation)	Binds to the 50S ribosomal subunit and inhibit protein synthesis	[59, 66, 67]
4.	Newer aminoglycoside	Plazomicin	Protein synthesis inhibitor	[59, 68]
5.	Currently developed fluoroquinolones	Levonadifloxacin (WCK 771 and WCK 2349)	Inhibits DNA topoisomerases involved in bacterial DNA replication	[59, 69, 70]
6.	Newer tetracycline	Eravacycline	Binds to the 30S ribosomal subunit and inhibit protein synthesis	[59, 71, 72]
7.	Novel pleuromutilin	Lefamulin	Binds to 50S ribosome and inhibits bacterial protein synthesis	[59, 73]
8.	Newer oxazolidinone	Tedizolid	Binds to the P site at the ribosomal 50S subunit and inhibit protein synthesis	[59, 74, 75]
9.	Novel lipoglycopeptide	Telavancin	Inhibit bacterial cell wall synthesis by inhibiting polymerization of N-acetyl muramic acid and N-acetyl glucosamine	[59, 76, 77]
10.	Newer outer membrane protein targeting antibiotics	Murepavadin	Hinders lipopolysaccharide (LPS) transport in gram-negative bacteria, causing cell death	[59, 78, 79]

insufflations) [81]. The aerosolization performance, moisture resistance, and exceptional deep lung deposition capability of this product were exceptionally good. As per preclinical study in mice model, this novel formulation led to higher local concentration and prolonged retention time of drug in comparison to intravenous injection and oral administration.

In another study, Hsu et al. developed nanovesicles (phosphatiosomes) for lung-targeted delivery of ciprofloxacin through intravenous (femoral vein) route in rat [82]. The soya-ethyl-morpholinium-ethosulphate (SME) intercalated on the surface of these PEGylated phosphatidylcholine (PC)-rich nanovesicles was responsible for lung targeting. It had superior anti-methicillin-resistant Staphylococcus aureus (MRSA) activity in comparison to plain phosphatiosomes and free ciprofloxacin. They evaluated the efficacy of this nanoformulation to eradicate extracellular and intracellular MRSA. These nanovesicles were readily taken up by alveolar macrophages, and the intracellular MRSA were killed. In a rat model of lung infection, the MRSA burden in the lungs was decreased by eightfold after administration of SME phosphatiosomes [82].

Cevher et al. developed doripenem-loaded chitosan microparticle using ionotropic gelation and spray-drying technique for the treatment of pneumonia caused by *Pseudomonas aeruginosa.* Spray-drying technique had production yield as high as ~98% with mean diameter of particles ranging from 3.8 to 6.9 μm. Particles of diameter 1–5 μm were suitable for controlled release of doripenem after pulmonary administration [83].

38.5.2 Tuberculosis

Mycobacterial infection is mainly responsible for tuberculosis in human. *Mycobacterium tuberculosis* is the most common. The types of tuberculosis and associated pathological manifestations are shown in Table 38.3. The persons infected with pulmonary or laryngeal TB generate infectious airborne droplet nuclei while coughing, sneezing, singing, or talking [84]. These 1–5 μm droplets containing the bacteria is the medium for transmission to a healthy person via inhalation. The

Table 38.3 Classification of tuberculosis

Sl. no.	Types of tuberculosis	Pathological manifestations
1.	Pulmonary TB	Affects the (alveolar) region
2.	Extrapulmonary TB	Found among non-HIV patients, particularly lymphatic tuberculosis; more common in women and young children [100]
3.	Pleural tuberculosis	Infects the pleural fluid, results from primary progression or reactivation of latent infection
4.	TB in the CNS	Causes meningitis resulting from infection and host immune response. Neurological symptoms include cranial nerve palsies and motor, sensory, and cerebellar defects and even seizures [101]
5.	Bone/joint tuberculosis	The thoracic section of the spine affected mostly
6.	Genitourinary tuberculosis	It is uncommon; the kidney and genitals are affected. In women, it may cause infertility [102]
7.	Disseminated tuberculosis	Involves many organs simultaneously and can occur due to primary progression of the disease or reactivation of suppressed infection

alveolar macrophages engulf the pathogen and trigger a series of events resulting in either the containment of infection or the manifestation of disease. *M. tuberculosis* replicates slowly and steadily within the alveolar macrophages and spreads to hilar lymph nodes via lymphatic system. Initially the infected macrophages produce interleukins 12 and 18 that stimulate T lymphocytes (predominantly CD4 − + T cells) for interferon γ (IFN-γ) secretion. This accelerates the phagocytosis of *M. tuberculosis* in the macrophage [85]. IFN-γ also stimulates the macrophage to release tumour necrosis factor α. It has important role in granuloma formation and controlling the extent of infection [86].

Helper T cells get matured into two functionally different populations TH1 and TH2. The TH1 response causes granuloma formation and protection.The TH2 response is responsible for tissue-necrotizing hypersensitivity and disease progression. In general, after two to eight weeks of infection the patient develops enough helper T cells for positive tuberculin skin test. Actually, formation of granulomas limits the further replication and spreading of the bacilli.Without further defect in cell-mediated immunity, the infection remains restricted to the granulomas with no active manifestation. But failure of host immunity to inhibit the replication of *M. tuberculosis* leads to progression of the infection most commonly to the parenchymatous tissue of the middle and lower lung, and in the hilar lymph nodes. HIV-affected immunocompromised individuals show increased chances of disease progression. Defects in the production of interferon γ or tumour necrosis factor α as well as in the interferon γ receptor and interleukin-12 receptor have also been reported for disease progression [84, 87].

Mycobacterial virulence factors such as components of the ESX-1 secretion system [88], protein tyrosine phosphatase PtpA [89], and SecA2 transport ATPase of the secretory system [90] interfere with lysosomes to kill *M. tuberculosis* phagocytosed by macrophages (Fig. 38.2).The intracellular pathogen then blocks NF-κB and MAPK signal transduction pathways, lowering the production of cytokines [91]. The bacteria can then burst out from phagosomes and damage mitochondrial DNA. This or the bacterial DNA itself can be detected by cytosolic DNA sensor proteins, which induces type 1 IFN expression, responsible for disease progression [92]. The bacterial cell wall components, via TLR-2 signalling, cause lipid bodies to accumulate in the macrophages. The pathogen either stays within these lipid bodies without further manifestation or hydrolyses these lipids by secreting a membrane-associated hydrolytic Msh1 protein to derive nutrients [93]. The bacterial virulence factor, early secreted antigenic target 6 (ESAT-6), increases the uptake of glucose. ESAT-6 translocates GLUT1 transporters to the cell surface and stimulates glycolytic enzymes to produce dihydroxyacetone phosphate (DHAP) and acetyl-CoA in excess, which are utilized as raw materials for the synthesis of triacylglycerol (TAG) [94]. *M. tuberculosis* induces miR-33 expression which also prevents autophagic degradation of lipid bodies [95]. Mycobacterial protein PE-PGRS47A obstructs autophagosomal breakdown of mycobacteria and subsequent bacterial antigenic peptide presentation by MHC class II molecules [96]. Other virulence factors also interfere with the antigen presentation process. *M. tuberculosis* is also reported to inhibit macrophage apoptosis [97], which would

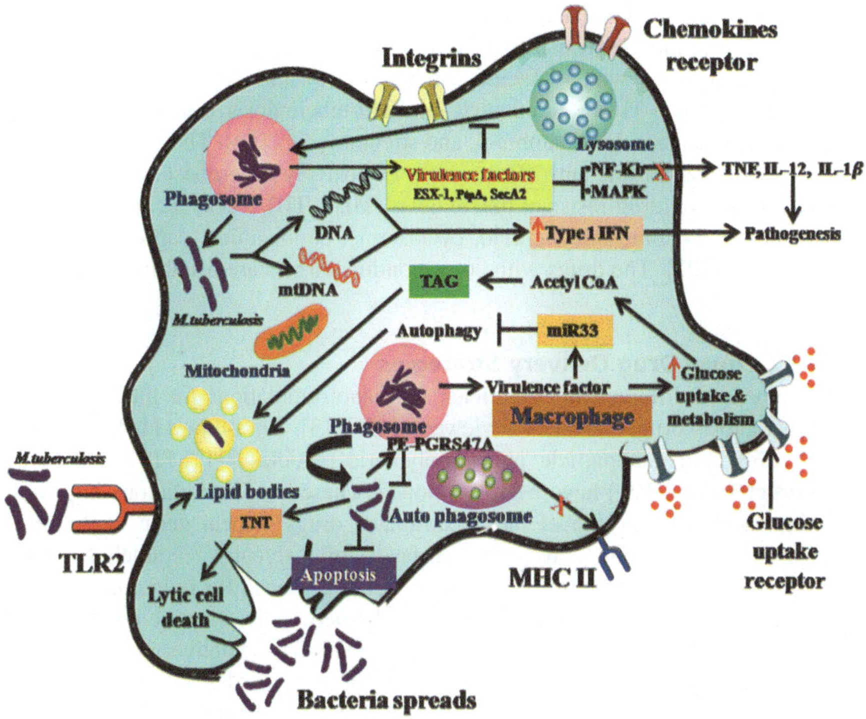

Fig. 38.2 Interplay between various cellular and molecular events in an *M. tuberculosis*-infected macrophage. Mycobacterial virulence factors ESX-1, PtpA, and SecA2 prevent lysosomal fusion with phagosomes containing the pathogen. They also block the NF-κB and MAPK signalling to lower protective cytokines. Bacteria, on bursting out from phagosome, damage mitochondria. Damaged mtDNA or bacterial DNA can upregulate type I IFN production, which helps in disease progression. In a TLR2- dependent manner, bacterial cell wall components cause lipid bodies to accumulate, which provides nutrients/shelter to the bacteria. Virulence factors also increase glucose uptake and metabolism, providing substrates like acetyl CoA for TAG synthesis. Expression of microRNA 33 has been found to be upregulated, which prevents autophagy of lipid bodies. Mycobacterial PE-PGRS47A protein prevents autophagosomal destruction of bacteria and its processing and presentation by class II MHC molecules. *M. tuberculosis* also prevents macrophage apoptosis. It secretes tuberculosis-necrotizing toxin (TNT), which helps in host cell lysis and bacterial spread (see text for details)

otherwise result in bacterial lysis and initiate adaptive immunity. The bacteria exit the host cell and spread to the neighbouring cells and tissues by inducing host cell lysis [98], probably enhanced by the secretion of tuberculosis-necrotizing toxin, which hydrolyses the coenzyme NAD+ and causes macrophage necrosis (TNT) [99].

38.5.2.1 Treatment Strategies

Treatment of TB with multiple antibiotics such as isoniazid, rifampicin, pyrazinamide, and ethambutol is required for a prolonged time period to attain bactericidal activity. However, development of antibiotic resistance is a concerning problem in multidrug-resistant tuberculosis (MDR-TB) infections. The World Health Organization (WHO) has devised a novel strategy called DOT (directly observed therapy), in which specific combination of anti-TB drugs is prescribed over a short course [103]. Some first-line anti-tuberculosis drugs are isoniazid, rifampicin, pyrazinamide, ethambutol, and streptomycin [104]. The next-generation drugs include second-line anti-TB drugs for extensively drug-resistant tuberculosis (XDR-TB) or multidrug-resistant tuberculosis (MDR-TB). These include amikacin, kanamycin, para-aminosalicylic acid, cycloserine, ethionamide, capreomycin, and ciprofloxacin [105]. The drugs with corresponding targets are listed in Table 38.4.

38.5.2.2 Novel Drug Delivery Strategies

Development of different microparticles and nanoformulations for the delivery of anti-tuberculosis drugs has been reviewed [112, 113]. Liposome [114], niosomes [115], solid lipid nanoparticle [116], polymeric particles [117–119], and metal nanoparticles [120, 121] have been developed. These nanoformulations improved the pharmacokinetic profile of the encapsulated drugs. Wheat germ agglutinin's receptors are present on the surface of alveolar epithelial cells. WSA-functionalized different nanoparticles were developed for pulmonary administration. Khullar et al. developed WGA-coated poly (lactide-co-glycolide) nanoparticles (PLG-NPs) as bioadhesive drug carriers for the delivery of rifampicin, isoniazid, and pyrazinamide [122]. These 350–400 nm particles, having encapsulation efficiency of 54–66%, get adhered to alveolar epithelial cell surface and maintained plasma concentration for prolonged period of time [122]. Khuller et al. tested inhalable multilamellar liposome-encapsulating rifampicin and isoniazid in guinea pigs [114]. The liposome was aerosolized with a nebulizer. 94% droplets had mass median aerodynamic diameter < 6um. It maintained the plasma level of drugs up to 48 h. The mean residence time (MRT) of intravenously administered rifampicin and isoniazid were ~2 h and ~2.5 h, respectively. Whereas the aerosolized liposome enhanced the MRT rifampicin to ~20 h and isoniazid ~18 h, respectively.

The frequency of administration of dose and amount of total dose administered are important regarding manifestation of dose-dependent side effects and patient compliance. In a preclinical study using *M. tuberculosis* H37Rv-infected mice, Khullar et al. compared the therapeutic efficacy of orally administered rifampicin, isoniazid, and pyrazinamide with that of solid lipid nanoparticle (SLN) containing equivalent doses of the same drugs [116]. Delivery of drugs as encapsulated in SLN enhanced the bioavailability of rifampicin, isoniazid, and pyrazinamide by ~11, ~30, and ~13 times. The MRT of rifampicin-, isoniazid-, and pyrazinamide-loaded SLN was ~92 h, ~ 98 h, and ~100 h, respectively, whereas free drugs had MRT values of ~6 h, ~5.5 h, and ~6.6 h. As a result, the therapeutic effect of 5 oral doses of drug-loaded SLNs was comparable to 46 daily doses of free drugs.

These three molecules rifampicin, isoniazid, and pyrazinamide were also formulated as niosomes with biocompatible surfactant tyloxapol [115]. These 150 nm niosomes had drug loading efficiency of more than 95% and sustained the release of drugs for prolonged period of time.

Table 38.4 Different drugs and their targets for tuberculosis

Generation	Sl. no.	Broad class	Drug	Cellular target	References
First generation (high rates of resistance)	1.	Isonicotinic acid	Isoniazid	Enoyl-ACP reductase, mycolic acid elongation	[104]
	2.	Rifamycin	Rifampicin	DNA-dependent RNA polymerase	[103]
	3.	Pyrazine	Pyrazinamide	Fatty acid biogenesis or membrane depolarization or ribosomal protein S1 (RpsA), protein translation and the ribosome-sparing process of trans-translation	[104]
	4.	Ethylenediamine	Ethambutol	Arabinan deposition in the cell wall	[103]
	5.	Aminoglycosides	Streptomycin	Inhibits protein synthesis by binding to the 16S rRNA of the 30S subunit of the bacterial ribosome	[104]
Second generation (less effective than the first-line drugs, may have toxic side effects or unavailability in many developing countries)	6.	Aminoglycosides	Amikacin	Misreading of mRNA and protein synthesis inhibition	[103]
			Kanamycin	Mistranslation and indirect inhibition of translocation during protein synthesis	[105]
	7.	Aminophenol	Para-Aminosalicylic acid	Inhibition of folic acid biosynthesis and iron uptake	[103]
	8.	D-alanine	Cycloserine	Inhibition of peptidoglycan synthesis by inhibiting the enzymes d-alanine racemase (AlrA) and d-alanine:d-alanine ligase (Ddl)	[105]
	9.	Thioamide	Ethionamide	ACP reductase InhA, mycolic acid synthesis	[103]

(continued)

Table 38.4 (continued)

Generation	Sl. no.	Broad class	Drug	Cellular target	References
	10.	Polypeptides	Capreomycin	Inhibits protein synthesis by changing ribosomal structures at the 16S rRNA	[105]
	11.	Fluoroquinolones	Ciprofloxacin	Trapping gyrase and topoisomerase IV on DNA-forming ternary complexes, blocking replication fork movement and transcription complexes	[103]
Latest drugs	12.	Nitroimidazoles	OPC-67683 (Delamanid)	Inhibition of methoxy- and keto-mycolic acid biosynthesis at a much lower dose	[103, 105, 106]
	13.	Quinolines	TMC-207 (Bedaquiline)/J compound	Inhibits ATP synthase	[103, 105, 107]
	14.	Antibiotics	Fluoroquinolones	Binds to DNA gyrase and topoisomerase IV and inhibits bacterial DNA synthesis	[69, 103, 105]
			Rifapentine	Inhibits mycobacterial DNA-dependent RNA polymerase	[103, 105]
	15.	Ethambutol derivatives	SQ- 109	Inhibits cell wall synthesis by targeting MmpL3, a transmembrane transporter for trehalose monomycolate; blocks ATP synthesis as an uncoupler	[103, 105, 108]
	16.	Oxazolidinones	Linezolid	Binds to the ribosomal 50S subunit, inhibiting protein synthesis	[103, 105, 109]
	17.		Sutezolid (PNU-100480)	Same mechanism as linezolid, with more efficacy	[103, 105]
	18.	Tetrabenzothiophenes	AX20017	Inhibits protein kinase G, enzyme responsible for blocking lysosomal degradation of mycobacteria in lysosomes	[103, 110]

	19.	Piperidine derivatives	BTZ043	Inhibits the enzyme decaprenylphosphoryl-β-D-ribose 2-epimerase	[103]
	20.	Pyrimidinediones	SQ641	Inhibits bacterial translocase I enzyme responsible for peptidoglycan biosynthesis	[103]
	21.	Spectinomycin	Spectinamide 1599	Selectively inhibits ribosomal activity and hence protein synthesis	[103, 111]

Hussein et al. developed magnetic nanoparticle-loaded chitosan nanoparticle as potential tool for both diagnosis and treatment of tuberculosis [117]. Streptomycin was loaded on the surface of the nanoparticle.

Recently, the therapeutic potential of antimicrobial nanoparticles like silver nanoparticle has nicely been reviewed by Mocan et al. [121]. Silver nanoparticles synthesized from different sources and capped with peptides or polymers like chitosan have good antimycobacterial effects. But macrophagic internalization of mycobacteria limits the effect of nanoparticle. Combination of AgNPs with classical anti-TB therapeutics has proven synergistic effect on antimycotic activity both extra- and intracellularly.

38.5.3 Emphysema

Emphysema is a disease condition, normally found in the age group of 45–60 years where the air space of adult lung is enlarged. Alveolar walls are ruptured; adjoining air spaces are associated to form the vesicles. The primary cause of emphysema is the chronic inhalation of obnoxious agents from cigarette smoke and polluted air. Mechanical forces like hyperinflation; triggering of apoptosis and cellular senescence; and failure in lung tissue repairment are responsible clinical manifestation for emphysema [123].

38.5.3.1 Cellular Mechanism

Genes regulating inflammatory pathways, proteolysis, and ROS generation are primarily involved in the pathogenesis of emphysema [124–126]. Proteolytic degradation of alveolar capillary cells is one of the major causes behind the destructive air-space enlargement. A balance between proteases and anti-proteases plays the pivotal role. Infiltration of neutrophils, macrophages, and lymphocytes in alveolar septae trigger immunological responses like inflammation. This inflammatory response causes release of proteases against lungs matrix proteins. The neutrophil elastase and MMP-12 released by activated macrophages enzymatically destroy the elastin substrate of the alveolar spaces. The α1-antitrypsin (AAT) is a neutrophil elastase inhibitor. Defect of gene regulating the expression of AAT is associated with emphysema [127]. Collagenases have been described for collagen breakdown in emphysema and emphysema models [128, 129]. Cathepsin S-dependent epithelial cell apoptosis is a crucial event in the pathogenesis of IFN-γ-induced alveolar remodelling and emphysema [130]. The fate of lung alveolar cells in emphysema is shown in Fig. 38.3.

Lung innate immunity is exposed to a plethora of exogenous microorganisms. Microbial LPS activate TLR4 and subsequent production of cytokines and ROS. In mice model of emphysema, overexpression of lung-specific TNF-α was observed. It was also associated with hypertension [131]. Moreover, IL-1β, IL-4, IL-13, and IL-γ, generally associated with chronic inflammation, were also overexpressed [123, 132]. Glutathione peroxidase-II is an isoform of glutathione peroxidase (GPX). Cigarette smoke induces glutathione peroxidase-II in lungs. This process is

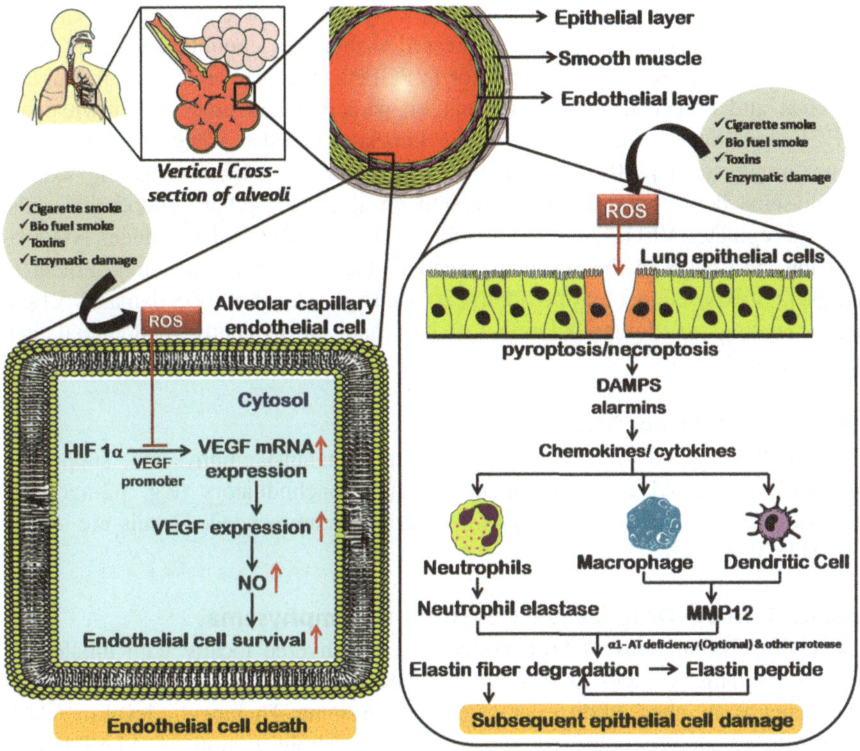

Fig. 38.3 The fate of lung alveolar cells in emphysema. Different factors like exposure to parasites, toxins, oxidants, particulate matters in cigarette/biofuel smoke, enzymatic damage induce the secretion of DAMPs by the epithelial cells undergoing necroptosis. A cascade of immunological reactions is triggered following recognition of DAMPs by innate immune cells. Resident/infiltrating neutrophils, macrophages, and dendritic cells release proteases like neutrophil elastase and MMP-12 that cause degradation of elastin fibres releasing elastin peptides. Such peptides are chemotactic in nature, recruiting further inflammatory cells and repeating the process. Other proteases are also known to degrade collagen and fibronectin (not shown here).Emphysema has been associated with familial α1AT deficiency (see text). Further inflammation eventually leads to epithelial cell death. In alveolar capillary endothelial cells, HIF-1α regulates VEGF gene expression.Activation of VEGRF induce NO production. NO helps in survival of endothelial cells. Higher level of ROS damage VEGF gene leading to downregulation of NO and subsequently endothelial cell death

regulated by the antioxidant transcription factor nuclear erythroid-related factor 2 (Nrf2) [133]. On knocking down Nrf2, mouse models have been reported to develop airspace enlargement and enhanced inflammatory responses. The alveolar septal epithelial and endothelial cells were killed by apoptosis [134]. Deficiency of TLR4 causes pulmonary emphysema as evidenced from TLR4-knockout mice (Tlr4−/− mice) [135]. TLR4 contributes to protection from ROS and has also been reported to help in helper T cell (TH2) activation [136]. Other knockout models of genes like VEGF conditional KO [137, 138] and surfactant-associated protein D

(SftpD) [123] have been reported to show symptoms of emphysema. Vascular endothelial growth factor (VEGF), expressed in abundant amounts in the lung tissue [138], is recognized by VEGF receptor (VEGFR). Alveolar septal endothelial cell survival and homeostasis is dependent on paracrine and autocrine downstream VEGF signalling [139]. Not only VEGF knockout but also VEGFR blockage can cause emphysema [140], indicating a crucial role for VEGF signalling axis in emphysema development and progression. In alveolar capillary endothelial cells, HIF-1a regulates VEGF gene expression. Activation of VEGRF induces NO production. NO helps in the survival of endothelial cells. The lung environment in smokers is exposed to a lot of oxidative stress. Higher level of ROS damages VEGF gene leading to downregulation of NO and subsequently endothelial cell death [141].

38.5.3.2 Treatment Strategies

Though there is no specific treatment strategy against emphysema, supportive medicines are used for symptomatic relief. Bronchodilators (e.g. β2-agonists, anticholinergics) and corticosteroids are mainly recommended. Details are shown in Table 38.5.

38.5.3.3 Novel Drug Delivery Systems for Emphysema

Generally, formulations of bronchodilators are delivered locally with inhalers or nebulizers. This provides local and immediate effect. Alpha-1 antitrypsin (A1AT) deficiency is an inherited disorder causing emphysema. Therapeutic level of A1AT in the lungs can be achieved with pulmonary administration of A1AT-loaded nanoparticle. Ghasemi et al. developed A1AT-loaded chitosan-genipin nanohydrogel [153]. The nanoformulation (particle size ~30–100 nm) had more than 80% entrapment efficiency. It provided sustained release of A1AT for 12 days in simulated lung fluid [153].

38.5.4 Pulmonary Oedema

Pulmonary oedema can be defined as the accumulation of excess fluid in the alveolar walls and alveolar spaces of the lungs. The basic forces that maintain fluid balance between the microcirculation and interstitium are known as "Starling Forces" [154]. An imbalance in "Starling Forces" causes cardiogenic (hydrostatic) pulmonary oedema (non-inflammatory type). The damage of the lymphatic drainage system or direct injury to lung parenchyma/vasculature increases the permeability. As a result, fluid may accumulate causing non-cardiogenic pulmonary oedema [154].

38.5.4.1 Pathophysiology

The pulmonary capillary pressure ranges from 6 to 13 mm Hg. The pressure is 10 mm Hg under normal conditions. Under different coronary artery disease conditions (myocardial infarction), congestive heart failure, and cardiomyopathy,

Table 38.5 Different drugs and their mechanism of actions against emphysema

Class	Drugs	Mechanism of action	Indications	References
Inhaled bronchodilators				
β2 agonists [mimic catecholamines and relax airway smooth muscles]	Albuterol	Binds to β2 adrenergic receptors > G-protein signalling > adenylyl cyclase increases intracellular [cAMP] > PKA activation > phosphorylation ofGq-coupled receptors > reduction of intracellular [Ca^{2+}] > inhibition of myosin light chain phosphorylation>prevention of airway smooth muscle contraction [142]	Short-acting β2 agonists (SABA), for intermittent dyspnoea	[143]
	Formoterol, salmeterol, indacaterol, olodaterol, vilanterol		Long-acting beta2 agonists (LABA), for increasing or chronic dyspnoea	[144, 145]
Anticholinergic medications (inhibit bronchoconstriction induced by acetylcholine neurotransmitter)	Ipratropium, oxitropium	Acetylcholine stimulates G protein-coupled M3 muscarinic receptors. Anticholinergics are competitive antagonists of acetylcholine; reduce/prevent the effects of cholinergic neurotransmission in the CNS and peripheral tissue [146]	Short-acting muscarinic antagonists (SAMA), for intermittent dyspnoea	[145]
	Tiotropium		Long-acting muscarinic antagonists (LAMA) for increasing or chronic dyspnoea	[147]
Inhaled corticosteroid (ICS)	Beclomethasone, budesonide and fluticasone	Potent glucocorticoid activity; reverses capillary permeability and lysosomal stability [148]	Reduces inflammation	[144, 145, 149]
Oral Phosphodiesterase-4 inhibitors	Roflumilast	PDE-4, found in lung cells, degrades intracellular cAMP. PDE-4 inhibitors stop this, increasing [cAMP], causing bronchial muscle relaxation. Also,	Reduces inflammation	[151]

(continued)

Table 38.5 (continued)

Class	Drugs	Mechanism of action	Indications	References
		they decrease pro-inflammatory mediators [150]		
Triple inhaled therapy (LABA+ LAMA+ ICS)	Budesonide (glucocorticoid) + glycopyrrolate (LAMA) + formoterol (LABA)	Long-acting beta2 agonists relax muscle tone, long-acting muscarinic antagonists inhibit bronchoconstriction, inhaled corticosteroids relax muscle and reduce inflammation	Combined efficiency of bronchodilator + corticosteroids	[145, 152]
Enzymatic therapy	Alpha1 antitrypsin	A1AT protein balances the effect of neutrophil elastase enzyme, released from leukocytes to fight infection. In its absence, neutrophil elastase destroys the alveoli, causing emphysema. Hence, A1AT is administered exogenously	For patients with familial alpha1 antitrypsin deficiency, responsible for emphysema pathogenesis	[127, 145]

the pressure is increased to cause pulmonary oedema [155, 156]. The extra-alveolar interstitial spaces have negative pressure to keep the alveoli dry. In patients with cardiovascular diseases, increased pressure in pulmonary capillary causes fluid accumulation in the interstitial spaces. Non-cardiogenic pulmonary oedema may also results from injury to the endothelial and epithelial layers of lungs [156].

38.5.4.2 Treatment Strategies

Therapies mainly make the best use of symptomatic drugs and medications which target underlying causes. They include diuretics [157], vasodilators [158, 159], calcium channel blockers [160, 161], inotropes [162], and morphine [163]. Details are shown in Table 38.6.

Table 38.6 Different categories of drugs and their mechanism of actions against pulmonary oedema

Sl. no.	Class	Drugs	Mechanism of action	References
1.	Diuretics	Furosemide	Improves dyspnoea by reducing pressure caused by fluid overload	[157, 164]
2.	Vasodilators	Nitroglycerin, nesiritide (administered intravenously), serelaxin (a recombinant human form of relaxin, inducing nitric oxide activation)	Causes dilatation of blood vessels, reducing pressure build-up	[158, 159]
3.	Calcium channel blockers	Clevidipine, nifedipine (prophylactically used to treat hypoxia-mediated pulmonary vasoconstriction in high-altitude pulmonary oedema, HAPE)	Help in lowering pressure	[160, 161]
4.	Inotropes	Dobutamine, dopamine, milrinone (administered intravenously, with vasodilatory activity)	Used during pulmonary congestion	[162]
5.	Analgesics	Morphine	Reduces systemic resistance of the blood vessels and anxiety; lowers dyspnoea and cause venous dilatation; used previously in acute pulmonary oedema, stopped because of adverse effects	[157, 163]

38.5.5 Lung Cancer

Lung cancer refers to tumours arising in the lung parenchymatous tissue or within the bronchi. It is the most commonly diagnosed cancer (12.4%) around the globe [165]. In the year 2018, as per the WHO, the death of 23.1% cancer patients was because of lung cancer [166].Tobacco smoking, specially passive smoking, remains predominantly the most common cause of lung cancer, with the risk being more acute in males who were former smokers [167]. The chance of mortality is higher in US population of males as surveyed by the American Cancer Society and the National Cancer Institute [168]. The risk is further increased with exposure to radiation therapy and carcinogens like asbestos, chromium, nickel, cadmium, chromium, silica, arsenic, and polycyclic aromatic hydrocarbons [165].

Approximately 10% of lung cancer patients suffer from small cell lung cancers (SCLCs). According to the WHO, SCLCs are of three cell subtypes: oat cell, intermediate cell, and combined cell (SCLC with components of non-small cell lung cancer, squamous cell carcinoma, or adenocarcinoma). Smoking is the most common cause of SCLC. The thyroid transcription factor-1, CD56, synaptophysin, and chromogranin are the popular markers of SCLC [165].

The non-small cell lung cancer (NSCLC) accounts for ~85% of lung cancers. It is traditionally divided into 3 major cell types: adenocarcinoma (~50%), squamous cell carcinoma (~35%), and large cell carcinoma (~15%) [169]. Adenocarcinoma is mainly associated with non-smokers, whereas squamous cell carcinoma is associated with smoking. Napsin A, Cytokeratin-7, and thyroid transcription factor-1 are the well-proven markers.

Moreover, NSCLC may include adeno squamous carcinoma, carcinoid tumours, and unconventional forms of non-small cell lung cancer. Giant cell carcinoma and sarcomatoid carcinoma are such unconventional form of NSCLC [165].

38.5.5.1 Pathogenesis

Environmental carcinogens cause abnormal growth of cells in the originating tissue of lung epithelium. It is known as dysplasia of lung. The carcinogens are mostly mutagens. On persistent exposure, they cause aberrant genetic mutations leading to defective protein synthesis. The cell-cycle homeostasis is lost resulting in carcinogenesis. The most common mutations are MYC, BCL2, and p53 for SCLC. Whereas mutations of EGFR, KRAS, ALK, ROS-1 [165, 170], BRAF [171], and p16 are responsible for NSCLC[165, 169, 172]. Squamous cell carcinoma involves mutations in components in PI3K pathway, fibroblast growth factor receptor, and discoidin domain receptor [169]. Cancer stem cell regulation is largely dependent on Hedgehog, Wnt, and Notch pathways and hence have crucial roles in lung cancer development [173].

Lungs do have microbiotica. In healthy person, they provide pulmonary immunity. But in person with respiratory disease, they interact with the pathogens and act as co-factor to help in progression of the disease [174]. Again, the gut microflora has important roles in the immune regulatory functions of human body. It modulates the treatment of different respiratory diseases including lung cancer. Recently Bingula et al. (2018) studied the role of the gut, lung, and upper airway microbiota in the treatment of NSCLC patients [175]. The findings suggest that the life style of patients (nutrition, profession, smoking, drinking, etc.) have direct effect on the microflora and subsequently on the progression of lung cancer [175].

38.5.5.2 Treatment Strategies

Treatment of NSCLC

The chemotherapeutic agents used in the treatment of lung cancer are summarized in Table 38.7. Surgery is the most sought-after strategy for treating stage 1 NSCLC. The patients of stage 2 NSCLC are treated with radiotherapy first. Thereafter, surgical removal of tumour is followed by adjuvant/neo-adjuvant chemotherapy, preferably with etoposide and cisplatin [165].

Stage 3 NSCLC tumours show a lot of heterogeneity with respect to tumour cell microenvironment and invasion. The usual procedure is surgery followed by adjuvant chemotherapy or chemoradiotherapy, depending on patient health status. In some cases with unrespectable tumours, surgery post-chemotherapy can be

Table 38.7 Summary of different chemotherapeutic agents used in the treatment of lung cancer

Class	Drugs	Mechanism of action	References
Podophyllotoxin derivative	Etoposide	Topoisomerase II inhibitor	[182]
Platinum compounds	Cisplatin	Forms DNA adducts, induces DNA damage and cell death	[183]
Immunotherapy	Bevacizumab	Vascular endothelial growth factor (VEGF) inhibitor	[176]
	Nivolumab	Anti-PD1 antibody	[165]
	Pembrolizumab (often used in combination with pemetrexed and carboplatin)	Anti-PD1 antibody	
	Atezolizumab	Anti- PDL1 antibody	
Tyrosine kinase inhibitors (TKIs)	First-generation EGFR-TKIs: Erlotinib, gefitinib	Inhibits EGFR (epidermal growth factor receptor) wild-type variant or activating mutations	[172, 177]
	Second generation: ErbB family blockers like afatinib, dacomitinib	Irreversibly blocks receptors of ErbB family, including EGFR	[178, 179]
	ALK (anaplastic lymphoma kinase) inhibitors like crizotinib, ceritinib, and alectinib	Targets and inhibits ALK gene rearrangements, responsible for NSCLC progression	[165]
	Crizotinib	Targets ROS1 translocation observed in lung adenocarcinoma	[180]

followed. The patients of stage 4 NSCLC have very less chance of survival. A combination of chemotherapeutic agents is generally preferred.

Bevacizumab is a well-known inhibitor of vascular endothelial growth factor (VEGF).This clinically approved anti-angiogenic agent has shown favourable results in patients of non-squamous NSCLC who do not have brain metastasis or haemoptysis [176]. The other emerging approaches for targeted therapies of NSCLCs focus on inhibiting the pathways of "driver mutations" that lead to advanced NSCLCs. The epidermal growth factor receptor (EGFR) is blocked with first-generation tyrosine kinase inhibitors (TKIs) like erlotinib and gefitinib [172, 177]. If the cancer is resistant to the first-generation TKIs, second-generation EGFR-TKIs like afatinib [178] and dacomitinib [179] are preferred. Crizotinib, ceritinib, and alectinib are used as inhibitors of ALK (anaplastic lymphoma kinase) [165]. Crizotinib has also been approved for another type of translocation observed in lung adenocarcinoma, ROS1 [180].

In malignant cells, programmed death receptor 1 (PD-1) is found important in downregulating T cells and promoting self-tolerance. PD-L1 and PD-L2 are the ligands that bind to PD-1 and inactivate T cells. Using antibodies against PD-1 or PD-L1 is a good strategy to block this pathway. Nivolumab (antibody against PD-1), pembrolizumab (antibody against PD-1, used often in combination with pemetrexed and carboplatin), and atezolizumab (antibody against PD-L1) are widely used for NSCLC treatment [165].

Treatment of Small Cell Lung Cancer (SCLC)

Treatment regimen includes surgery, chemotherapy, and radiation therapy [165]. Potential drugs under investigation include the mTOR inhibitors everolimus and temsirolimus [181].

38.5.5.3 Delivery Systems

Since anticancer drugs are cytotoxic, targeted drug delivery systems play a key role in anticancer therapy. The targeting strategies are of two types: passive targeting [184, 185] and active targeting [186, 187].

In the case of lung cancer, cancer cells are generally developed in the blood vessel-enriched pulmonary alveolar region. The epithelial tissues are directly connected with the blood vessels. Tumour vasculature is the key factor for passive targeting [188]. Tumour vasculature is largely different from normal blood vessel [189, 190]. Tumour tissue proliferates rapidly. As the size is increased, the demands of oxygen and other key nutrients are also increased. At a critical size of 1–2 mm^3, tumour cells start suffering from starvation of oxygen and nutrients because of lack of supplying blood vessels. Thus, the further growth of tumour is retarded [191]. Tumour vasculature is highly heterogeneous in distribution; microvessels are discontinuous and leaky. Depending on the location of tumour, the intra-cellular gap of endothelial cells ranges from 100 to 780 nm [192]. So, drug molecules like paclitaxel [193], gemcitabine [194], and cisplatin [195] can easily diffuse out and are not accumulated in the tumour microenvironment at the required concentration. This problem is solved when anticancer agents are delivered as encapsulated in nanoparticles [196, 197]. Solid tumours have poor lymphatic drainage system. Moreover, the interstitial fluid pressure is also high. These two factors slow down the clearance of drug encapsulated nanoparticles entered passively into the leaky tumour microenvironment. This mechanism of targeting a solid tumour was first explained by Matsumura and Maedain 1986 [198]. It is known as the enhanced permeation and retention (EPR) effect.

In active targeting, the surface of nanocarriers is decorated with ligand corresponding to receptors overexpressed on to the tumour cells. Thus, the drug is selectively accumulated in cancer cells, and the normal tissues are escaped from toxicity.

Different biocompatible and biodegradable polymers have been reported for the development of polymeric nanoparticle, micelle, and polymer-lipid hybrid nanoparticle to deliver anticancer drugs against lung cancer [199, 200]. Poly(D,L -lactic acid) (PLA), poly(D,L-lactic-co-glycolic acid) (PLGA), poly(ε-caprolactone) (PCL), poly-alkyl-cyanoacrylates, gelatin, albumin, chitosan, and their copolymers di-blocked or multi-blocked with poly(ethylene glycol) (PEG) are most popular. Different lipid-based nanoformulations like solid lipid nanoparticle liposomes, microemulsion, and niosomes have been well studied in lung cancer [201]. DOXIL™ (PEGylated liposomal formulation of doxorubicin) was the first FDA-approved clinically available nanoformulation for solid tumours [202, 203]. It is associated with passive targeting strategy only. Though many clinical trials had been conducted for the treatment of lung cancer patients with intravenously administered DOXIL in combination with other drugs like cisplatin, carboplatin, and cyclophosphamide, no promising results have been reported [204]. Abraxane®, the human serum albumin nanoparticle, carrying paclitaxel has been approved by FDA for first-line treatment of locally advanced or metastatic NSCLC. It is recommended in combination with carboplatin for patients who are not under curative surgery or radiation therapy [205].

Gelatin nanoparticles were grafted with biotinylated epithelial growth factor (EGF) molecules for targeted delivery of anticancer drugs to lung tumour. This enhanced the uptake of nanoparticles by A549 lung adenocarcinoma cells ex vivo and lung cancer cells in a mouse model after administration as aerosol [206]. Recently Dastidar et al. reported a simple strategy for targeting the lung cancer cells through oral route [201]. They developed paclitaxel-encapsulated core-shell nanoparticle of cetyl alcohol. These ~78 nm particles had ~98% drug encapsulation efficiency and oral bioavailability of ~95% in rats. In comparison to free paclitaxel, this nanoformulation was 6.6-folds more potent to A549 cells and 7.8-folds more potent to 10 nM paclitaxel-resistant A549 cells. This was due to the rapid and complete cellular uptake of nanoparticles by the A549 cells and paclitaxel-resistant A549 cells. Actually, cancer cells uptake more lipids as a fuel for their high metabolic activities and have higher activity of ADH and ALDH enzymes. ALDH is a marker protein of lung cancer stem cells. Since cetyl alcohol is metabolized by ADH and ALDH enzymes, cancer cells find it as good source of their metabolic needs and uptake rapidly from the surrounding medium. This was the simple strategy behind actively targeting the lung cancer cells by cetyl alcohol nanoparticles [201].

Pulmonary administration of anticancer drugs has also been investigated due to advantages like site-specific drug delivery, avoidance of first-pass metabolism, possibility of fewer side effects, and needle-free delivery devices leading to better patient compliance [207, 208]. Meenach et al. developed camptothecin-loaded porous microsphere of acetylated dextran (Ac-DEX) for the treatment of lung cancer [209]. The degradation kinetics of the polymeric matrix is a function of molecular weight of Ac-DEX. Hence, the rate of drug release can be controlled by using polymers of different molecular weight. Even duration of few hours to month can be achieved. Tafaghodi et al. developed paclitaxel encapsulated large porous microsphere of PLGA for the treatment of lung cancer [210]. In comparison to intravenous route, pulmonary administration of this microsphere was 11.9-fold more efficient in the delivery of drug to the target cells. Again, endotracheal administration in rats maintained effective plasma concentration of paclitaxel for fourfold longer period of time.

In a recent study, a group of scientists developed pH/redox dual stimuli-responsive clustered nanoparticle for targeted delivery of ultra-small platinum nanoparticles (USPtNs) and gemcitabine (GEM) simultaneous. It was tested in NSCLC [211]. Details of treatment strategy are shown in Fig. 38.4. The size of the nanocomposite particle was ~165 nm. The clustered nanoparticle is composed of GEM-grafted copolymers (PEG-b-P(LL-g-GEM)), pH-sensitive polypeptides (OAPI), and USPtNs. When present within cancer cells, this hybrid nano-system performs a number of programmed tasks to kill the cell. When encapsulated in lysosomal vesicle, USPtN is released due to acidic pH. When released in cytoplasm, GEM is released due to higher intracellular concentration of GSH in cancer cells.

38.5.6 Asthma

Asthma is defined as the chronic inflammation of the respiratory airways leading to bronchoconstriction and airway remodelling. Asthma is characterized by wheeze, shortness of breath, chest tightness, and cough. Influenced by genetic and environmental factors, asthma may have either early onset or late onset. Apart from allergic or atopic IgE-associated asthma, non-allergic asthma can also be induced by intolerance to NSAIDs, rhinosinusitis, and nasal polyps.

38.5.6.1 Pathogenesis

Schematic representation of allergen sensitization in pulmonary epithelium is shown in Fig. 38.5. Bacteria, virus, and pollutants invade the airway mucosal epithelium and get recognized due to the presence of pathogen-associated molecular patterns (PAMPs) on their surface. The downstream signalling causes the mucosal epithelial cells to secrete chemokines like CCL19, CCL20, and CCL27. They bind to chemokine receptors present on immature dendritic cells (DCs).The chemokines CCL19, CCL20, and CCL27 bind to the receptors CCR7, CCR6, and CCR10, respectively.

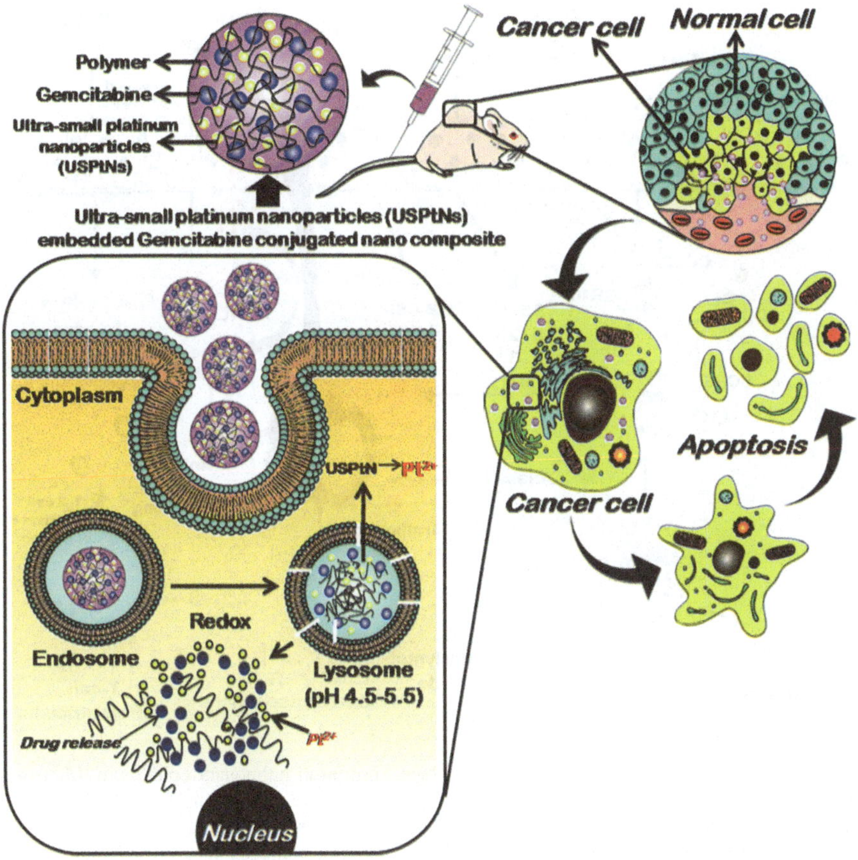

Fig. 38.4 Schematic representation of pH-responsive drug delivery system as cancer cell-targeting strategy. USPtNs and pH-responsive polymer-conjugated gemcitabine-combined hybrid nano-system complete multiple tasks inside cancer cells. This nanocomposite generates the cytotoxic Pt ions in response to lysosomal acidic environments into the cytoplasm of cancer cells due to increased reduced GSH and also subsequent release of gemcitabine in cytoplasmic environments. Abbreviations: *USPtNs* ultra small platinum nanoparticles, *GSH* glutathione

Thus, the immatured DCs matured into antigen-presenting myeloid-type DCs. They are called activated DCs. They detect and process the invading allergens into small peptides [212]. The processed peptides are then presented with the major histocompatibility complex (MHC) class II molecules to the T cell receptors (TCRs) of naive T cells, present in local lymph node [213]. Upon further interactions with co-stimulatory molecules, the T cells are differentiated into T helper (TH2) cells that promote IgE production. IgE are bound on mast cell surface. Upon further exposure to allergens, the mast cells, IgE, and eosinophils cause inflammatory responses. Mediators like histamine and bradykinin are released, resulting in constriction of the bronchial airways [214].

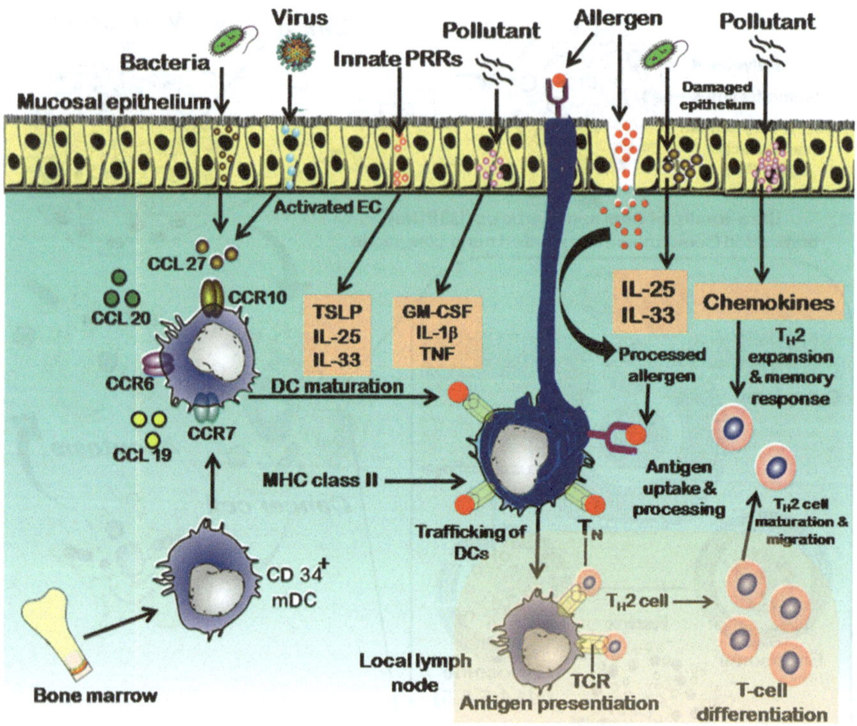

Fig. 38.5 Schematic representation of allergen sensitization in pulmonary epithelium. Microbes and pollutants initiate the activation of innate system. The airway epithelial cells (ECs) secrete chemokines and lead to trafficking of immature dendritic cells (DCs) to the mucosal epithelium. The DCs use their pattern recognition receptors (PRRs) to detect the pathogen/allergen and stimulate the cell signalling pathways leading to their maturation into competent antigen-presenting myeloid-type DCs. The pathogen/allergen-loaded DCs play as a leader for T cell differentiation. They migrate to the local lymph nodes and interact with naive T cells (TN) via the T cell receptor (TCR), Major Histocompatibility Complex (MHC) class II and co-stimulatory molecules. DCs activation and T-helper-2 (TH2) cell maturation and migration into the mucosa are influenced by additional epithelial-derived cytokines and chemokines, including IL-25, IL-33, CC-chemokines ligand 17 (CCL17), and CCL22. Abbreviations: *CCR* CC-chemokine receptor, *GM-CSF* granulocyte–macrophage colony-stimulating factor, *mDC* mucosal DC, *TNF* tumour necrosis facto, *TSLP* thymic stromal lymphopoietin

38.5.6.2 Treatment Strategies

38.5.6.3 Novel Drug Delivery Systems

Asthma is a chronic inflammatory disorder. It causes changes in the ultra-fine structure of airways and affects the remodelling process [218]. The common therapies have been summarized in Table 38.8. Generally, inhalation of steroids is prescribed. Though the damage is not completely reversed, systemic side effects are inevitable. Adrenocortical suppression, Cushing's syndrome, and osteoporosis are the most common among them. Only targeted delivery of drug can reduce the side

Table 38.8 Summary of drugs that are used in the treatment of asthma

Sl. no.	Class	Drugs	Mechanism of action	References
1.	Inhaled corticosteroids (ICSs)	Budesonide, fluticasone	Reduces inflammation	[215]
2.	Long-acting β2-adrenergic receptor agonists (LABAs)	Salmeterol, formoterol	Relaxes airway smooth muscles	[215]
3.	Long-acting muscarinic antagonists (LAMAs)	Tiotropium	Anticholinergic in nature (inhibit bronchoconstriction induced by acetyl-choline)	[216]
4.	Leukotriene receptor antagonists (LTRAs)	Montelukast, zafirlukast	Inhibit leukotriene- receptor activation and subsequent bronchoconstriction and chronic inflammation in chronic asthma	[215]
5.	Immunotherapy	Omalizumab	IgE-specific monoclonal antibody	[217]

effects. A recent study in athymic mice has shown that the systemic delivery of self-assembled dexamethasone nanoparticles (Dex-NP) targeted lung and was more effective than free dexamethasone (Dex) [219]. Intraperitoneal injection of ovalbumin on days 0 and 14 sensitizes the mice. After two weeks, daily 30 min exposure to ovalbumin aerosol at a frequency of three times per week induces asthma. This exposure to ovalbumin aerosol continues for the total duration of experiment. Dex-NP injection significantly reduced the count of total cells and eosinophils in the lung lavage. The level of the interleukin (IL)-4 and monocyte chemotactic protein-1 (MCP-1) was significantly reduced in comparison to untreated mice [219].

Theophylline, an anti-inflammatory drug effective in allergic asthma, was delivered intranasally, as complex with thiolated chitosan nanoparticles [220]. Since thiolated chitosan nanoparticles have enhanced mucoadhesiveness and epithelial cell permeability, this strategy significantly enhanced the anti-inflammatory effects of theophylline in ovalbumin-induced allergic mice model [220]. Thymulin is a serum thymus factor. It is a biologically inactive nonapeptide. It becomes activated upon coupling with zinc ions in vivo. It has an important role in modulation of intra- and extra-thymic T cell differentiation. Thus, it mediates anti-inflammatory and anti-fibrotic effects and is a good candidate drug for the treatment of allergic asthma. Morales et al. developed DNA nanoparticle of ~90 nm for the delivery of thymulin gene to lung cells. Single molecule of plasmid DNA was compacted with block copolymers of poly-L-lysine and polyethylene glycol linked by a cysteine residue (CK30PEG). The effect of intratracheal delivery (using a microsprayer) of this nanoformulation was studied in OVA-sensitized mice. The gene transfection was successful to inhibit the inflammatory and remodelling processes [221]. In the case of virus-induced chronic inflammatory lung diseases like asthma, COPD, or cystic fibrosis, stimulation of antiviral cytokine IFN-beta is a good strategy of choice. Polyinosinic-polycytidylic acid is well known for IFN-β induction. But it also causes the production of chemokine IL-8 leading to inflammation. So, a cytoplasmic delivery system is required that will efficiently stimulate the pathway for the production of IFN-β only. Dauletbayev et al. developed polyinosinic-polycytidylic acid encapsulated liposomal formulation that selectively stimulated the pathway for the upregulation of IFN-β in airway epithelial cells [222].

38.5.7 Cystic Fibrosis

Cystic fibrosis is an autosomal recessive lethal genetic disorder, caused by mutations in a gene on chromosome 7. The molecular mechanism of cystic fibrosis is shown in Fig. 38.6. The gene codes for cystic fibrosis transmembrane conductance regulator (CFTR) protein functioning as transmembrane cAMP-activated chloride channel [223]. It belongs to the ATP-binding cassette (ABC) superfamily [224]. Clinical manifestation takes place in a homozygous person. Different mutations in CFTR gene can lead to defective protein synthesis, misregulation of protein channel, defective chloride transport or enhanced degradation and subsequent unavailability of CFTR protein channel [225]. The most common mutation is Δ F508, causing abnormal CFTR protein folding and its early degradation [226]. In healthy individuals, Cl^- ions get actively transported via CFTR present in the apical membrane of airway epithelial cells, causing passive movement of Na^+ ions out of the cell. The increased electrolyte concentration on the luminal surface creates an osmotic force for water to be secreted via aquaporin channels [227]. This maintains the fluidity of the mucus required to clear the pathogen/foreign particles encountered through inhalation. However, in the case of defective CFTR protein, the Cl^- ion transport mechanism is impaired resulting in drying of mucus layer [228] and subsequent loss of innate defence mechanism. Apart from respiratory obstruction, the bowel, intestines, pancreas, and sweat glands, all of which are functionally dependent on chloride transport, get affected [227, 228].

38.5.7.1 Treatment Strategies

The choice of therapy depends on the type of mutation. It is summarized in Table 38.9.

38.5.7.2 Novel Drug Delivery Systems

Mucolytics [233] and few novel anti-inflammatory agents are good candidates for the treatment of cystic fibrosis (CF) [234]. N-Acetyl cysteine, Dornase alfa, gelsolin, and thymosin b4 therapy can be a good approach for CF [233]. Nanoparticles could be the delivery system of choice to overcome the barrier of thick mucus layer and bacterial biofilm [235]. The mucus of cystic fibrosis patients has very low water content than that of a healthy person. The healthy person mucus has a mesh size of 500 nm. In CF, there is formation of tough cross-linked network consisting of 70–80% mucin fibres, DNA (as a shell around the mucin core), actin, and other macromolecules. The mesh size is 100–300 nm [236, 237]. Depending on the disease stage of the patients, the viscoelasticity is significantly increased

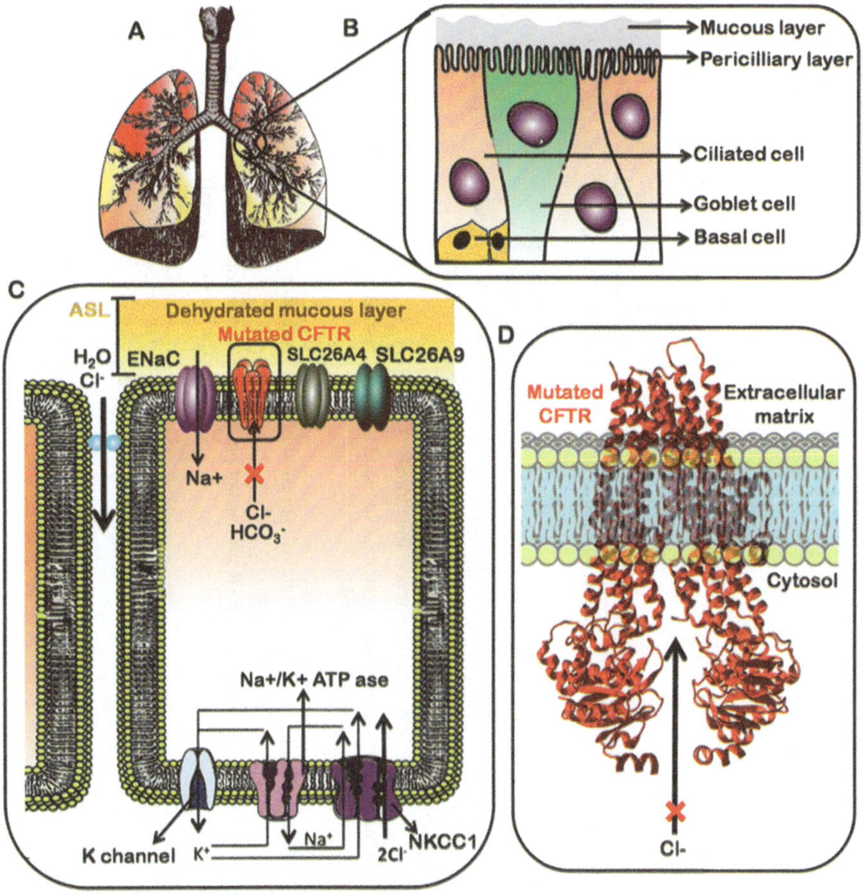

Fig. 38.6 Pathophysiology of cystic fibrosis. (**a**) Human Lung. (**b**) Detailed structure of conducting airways. Three types of cells are present—basal cells, ciliated cells, and goblet cells. The surface is covered with a mucus layer. (**c**) Molecular mechanism across basolateral membrane. The paracellular transport of Na^+ ions and water takes place following the electrochemical gradient. The intracellular movement of different ions is regulated by different transporter proteins like CFTR, NKCC, Na+/K+-ATPase, ENaC, and members of SLC26. NKCC1 is present on the basolateral membrane. It functions in co-ordination with Na+/K+-ATPase to accumulate Cl^- ions within the cells. Recycling of K^+ ion is associated with proper functioning of the Na+/K+-pump. SLC26A4 and SCL26A9 are anion exchangers. They are functionally important for mucus homeostasis. In CF, the mutated CFTR protein is not able to exchange of Cl^- ions. The movement of Na^+ ions stops. Secretion of water is reduced leading to dehydration of mucus. (**d**) Enlarged view of the mutated CFTR, taken from crystalline structure (PDB ID- 5UAK). Abbreviations: *ASL* airway surface liquid, *CFTR* cystic fibrosis transmembrane conductance regulator, *NKCC* Na-K-Cl cotransporter, *ENaC* epithelial sodium channel, *SLC 26* solute carrier family 26

[238]. This leads to the formation of strong hydrophobic and electrostatic barriers to the penetration of nanoparticles. There are two common strategies to overcome these barriers. The surface of nanoparticles is coated with a dense layer of low-molecular-

Table 38.9 Summary of treatment strategies for cystic fibrosis

Type of mutation	Molecular defect	Result	Class of drugs	Mechanism of action	Approved or in trial	References
I	Nonsense, frameshift, or splice-site mutation	Defective protein synthesis	Read-through agents	Override premature stop signals	SPX-101 (phase 2 completed), AZD5634 (phase 1 completed)	[229, 230]
II	Missense mutation	Defective post-translational processing and trafficking of the protein	Correctors and potentiators	Correctors repair defective protein folding and rescue trafficking, potentiators for mutant proteins expressed but non-functional	Orkambi, Symdeko	[225]
III	Missense mutation, channel gating	Decreased protein activity (non-functionality)	Potentiators	For mutant proteins expressed but non-functional	Ivacaftor	[231]
IV	Missense mutation	Faulty cl^{-} ion conductivity	Potentiators	For mutant proteins expressed but non-functional	Ivacaftor	[225]
V	Missense or frame-shift mutation	Reduced CFTR stability at plasma membrane with rapid turnover	Stabilizers	Stabilizes CFTR proteins at plasma membrane	Cavosonstat (N91115) (phase I completed)	[232]

weight polyethylene glycol (PEG). Since PEG is "muco-inert" polymer, both the electrostatic and hydrophobic interactions are eliminated. Mucolytic agent like N-acetylcysteine (NAC) is co-administered with the nanoformulation to reduce mucosal viscoelasticity [235, 239].

38.5.8 COPD (Chronic Obstructive Pulmonary Disease)

COPD is a chronic disorder, manifested by airway tract obstruction and parenchymatous tissue damage, caused by long-term inhalation of obnoxious chemicals/cigarette smoke. Other causes may be passive smoking, chronic exposure to hazardous environmental pollutants, or even familial alpha-1 antitrypsin deficiency (AATD). These induce downstream inflammatory reactions and result in lung damage and decreased lung elasticity. Symptoms include cough, dyspnoea, and sputum formation [240].

38.5.8.1 Pathophysiology

In patients with COPD, exposure to chronic lung irritants causes oxidative stress and disrupts equilibrium between proteases and anti-proteases. This causes enlargement of air sacs and loss of tissue elasticity due to elastin degradation (Fig. 38.7). Also, neutrophils and macrophages infiltrate at the site of irritation/damage. Persistent macrophage activation leads to the release of multiple inflammatory mediators causing an aggravated inflammatory response resulting in epithelial cell death [241]. Such events are more commonly observed in emphysema (discussed previously in Sect. 38.5.3). In patients with AATD, misfolded mutated protein accumulates in the liver, causing anti-protease insufficiency in the lung and subsequent protease-mediated cell damage. A decrease in the forced expiratory volume (FEV1) is observed along with excess inflation of the lungs. Gas exchange gets impaired in the course of disease progression. Incomplete exhalation and increase in physiologic dead space increase carbon dioxide (CO_2) levels. Lack of O_2 causes vasoconstriction, leading to pulmonary hypertension [240].

38.5.8.2 Treatment of Strategies

See Sect. 38.5.3.2 (Treatment strategy of emphysema).

38.5.8.3 Delivery Systems

Bronchodilators (β2-agonists and antimuscarinics) are the primary treatment strategy in patients with COPD. Dry powder inhalers, pressurized metered dose inhalers, and nebulizers are the most commonly used drug delivery devices to deliver the drugs locally in the respiratory tract for providing immediate relief. See Sect. 38.7 for details.

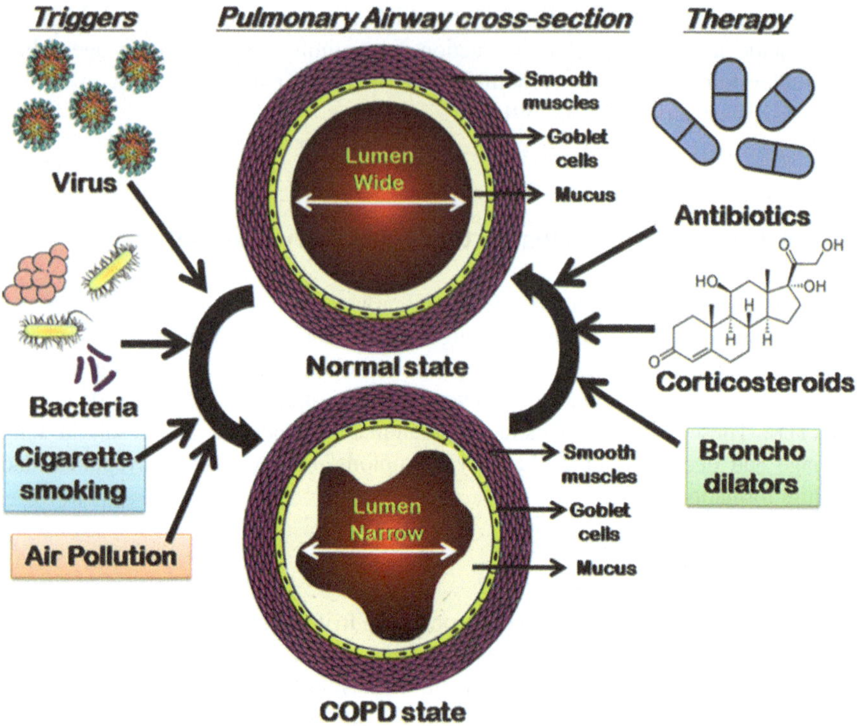

Fig. 38.7 Pathophysiology and treatment strategy of COPD. Invasion by pathogen, exposure to cigarette smoking, and different air pollutants cause induction of excessive mucous secretion by goblet cells. Thus, the airway is obstructed. So, antibiotic therapy, use of corticosteroids, and bronchodilators are the strategy of treatment

38.6 Mechanisms of Particle Internalization into Cell

Alveolar cells can be classified as macrophages and pneumocytes. Pneumocytes are again of two types—I and II. Type I cells are the larger flat epithelial lining cells involved in the exchange of oxygen and carbon dioxide in the alveoli. Type II cells are smaller in size but play big roles in defence mechanism of the alveoli. They synthesize and secret surfactants, keep the alveolar space relatively free from fluid, serve as progenitor cells, and boost the innate immune system. It is the macrophages and the type I epithelial cells that take part in the endocytosis process [242]. Different internalization mechanisms of nanoparticles are shown in Fig. 38.8 and discussed in detail below.

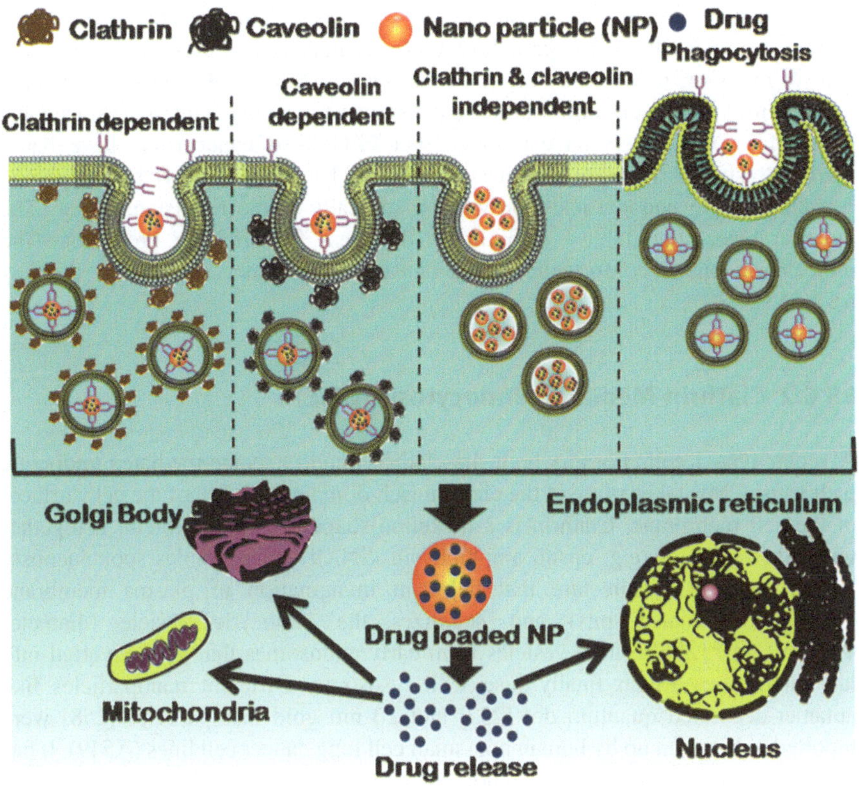

Fig. 38.8 Different mechanisms behind cellular uptake of nanoparticles

38.6.1 Phagocytosis

Phagocytosis is one of the major cellular internalization mechanisms. Different phagocytes, like macrophages, monocytes, neutrophils, and dendritic cells, are present in our body. They are responsible for engulfing foreign particles like bacteria and virus and stimulating the host's adaptive immune system. Fibroblasts and epithelial and endothelial cells are also phagocytic [243]. These phagocytic cells are the main barrier towards the delivery of nanoparticles, because they detect the nanoparticles as foreign and internalize them. The size, shape, and surface charge nanoparticles are the important factors that determine this phagocytosis process [244]. The very initial step of phagocytosis is known as opsonization [245, 246]. Different proteins like antibodies [247, 248], components of complement system [249], laminin, and fibronectin [250–252] coat the surface of the nanoparticles. This helps the phagocytic cells to detect and engulf the nanoparticles. Larger particles are generally internalized by this mechanism; particles of 460–2100 nm diameter have been found to be taken up by mouse peritoneal macrophages [253, 254]. Charge and hydrophobicity of nanoparticle's surface play important roles in opsonization and subsequent engulfment by phagocytic cells present in the reticuloendothelial system (RES). Hydrophobic particles are preferentially taken up by the phagocytic cells of the liver, followed by the spleen and lungs. Smaller nanoparticles (diameter < 35 nm) with hydrophilic surface are less prone to be taken up in the spleen and liver. Many researchers have reported that PEGylated nanocarriers have better barriers to surface adsorption of plasma proteins. The layer of polyethylene glycol, surrounding the nanoparticles, provides a sterically stable repulsive barrier. The polymer layer should have a minimum thickness for effective protection. The molecular weight, conformation, and density of polymer chain play the key role [255].

38.6.2 Clathrin-Mediated Endocytosis (CME)

Alveolar type I cells possess both the clathrin- and caveolae-mediated endocytic pathways. CME takes place at the clathrin-rich domain (0.5–2% of the cell surface) of the cell membrane. Clathrin is a triskelion-shaped scaffold protein. It together with other proteins (e.g. epsin, amphiphysin, SNX9) co-assembles spontaneously into a complex architecture that helps in invagination of plasma membrane (as clathrin coated pits) and stabilizes the endocytic vesicle (diameter 100–150 nm) [256]. These vesicles form early endosomes that get converted into late endosomes which finally fuse with lysosome. Different nanoparticles like aptamer-decorated quantum dot [257] and 20 nm gold nanoparticles [258] were reported to be taken up by human non-small cell lung cancer cell lines (A549). It has been shown that 50 nm siRNA nanoparticles follow passive diffusion to enter the alveolar cells and remain free in the cytoplasm. Whereas 100 nm nanoparticles enter primarily via clathrin- and caveolin-mediated endocytosis and are found in endosomes [242]. The same was observed for 50 nm and 100 nm amine-modified poly-styrene particle [242].

38.6.3 Caveolae-Dependent Endocytosis (CDE)

Caveolae are 50–80 nm flask-shaped membrane invaginations that are lined with a dimeric protein, caveolin-1 [256]. The other associated proteins are cavin, dynamin, vesicle-associated membrane protein (VAMP2), and synaptosome-associated protein (SNAP) [259]. Caveolae are involved in trans-endothelial albumin uptake. Again, in lung cancer cells caveolin-1 is overexpressed [259]. The commercially available albumin-bound form of paclitaxel (Abraxane®) is clinically used in the treatment of lung cancer. This nanoparticle is taken up by caveolae-mediated endocytosis [259]. This overexpression of caveolin-1 is the strategy behind targeted delivery of anticancer drug-loaded nanoparticles to lung cancer [260]. Sahay et al. developed doxorubicin-loaded pH-responsive polymeric micelles with cross-linked ionic cores of poly methacrylic acid and non-ionic shell of poly(ethylene oxide) [260]. This nanoparticle cannot enter into normal epithelial cells because of the presence of tight junctions. But, lung cancer cells uptake the nanoparticles by caveolae-mediated endocytosis. After internalization, drug-loaded micelles reach the lysosomes. The acidic environment causes the release of drug, linked to the polymer via pH-sensitive hydrazone bonds [260].

38.6.4 Clathrin/Caveolae-Independent Endocytosis

Non-small cell lung cancer (NSCLC), particularly lung adenocarcinomas, has overexpression of folate receptor-α (FRA) on their surface than that of normal tissue. Thus, conjugation of folic acid on the surface of anticancer drug-loaded nanoformulation is a good strategy to target the NSCLC [261]. The folic acid acts as ligands to the folate receptors on cancer cell surface. Due to this ligand-receptor interactions, the nanoparticles get internalized independent of the functioning of clathrin and caveolae [256]. Thus, the nanoparticles do bypass the trafficking into lysosomes. They are either retained in endocytic vacuoles or get released into cytoplasm [262].

38.7 Delivery Device

38.7.1 Dry Powder Inhaler (DPI)

Bell and colleagues are famous for their pioneer works on dry powder aerosols. They reported the first dry powder inhalation device in the year 1971 [263]. Nowadays, varieties of DPIs are available for clinical use. Some devices are breath actuated. They get activated and deliver consistent dose only when a threshold inspiratory flow rate is reached [264]. The more advanced delivery devices are electrically controlled. Battery is used as the source of power. They consist of an impellers and piezoelectric crystals. The impellers mix the content and the piezoelectric crystals vibrate to deliver the micronized dry solids [43]. As these devices are battery driven, the delivery of dose is reproducible and is independent of inspiratory flow rate [43]. Some devices are breath actuated. They get activated and deliver consistent dose only when a threshold inspiratory flow rate is reached [264]. These devices are very sophisticated and relatively expensive [18].

DPIs are suitable only for the administration of microparticles having proper size range. Therefore, the nanoparticles are formulated as nano-embedded microparticles (NEMs). These micron-sized particles get transported across the respiratory system. After dissolution of microparticle matrix in the respiratory fluid the nanoparticles are released and penetrate through the mucus barrier. Spray drying is the most common technique for converting nanoparticles suspension into stable, inhalable microparticles [265]. NEM powders should have good re-dispersibility. It can be assured with proper use of excipients and well-controlled manufacturing process [266]. The most efficient materials for the preparation of NEM are biocompatible and biodegradable polymers like PLGA (Polylactic-co-glycolic acid) [267, 268], chitosan [269], gelatine, and polyacrylate [270, 271].

38.7.2 Nebulizers

In nebulizers the solutions and suspensions are converted into droplets of appropriate size for pulmonary administration. In comparison to DPIs and pMDIs, nebulizers are suitable for aerosolization of lager doses of drugs [272]. Jet, vibrating, mesh, and ultrasonic are different types of nebulizers that are available in clinic [273]. Nebulizers are good alternative to inhalers. They are highly recommended for COPD and asthmatic patients who are unable to use inhalers [66].

But with nebulization technique, it is difficult to maintain aggregation free nanosuspension for prolonged period of time. Moreover, nebulizers have limited portability, and suffers from bacterial contamination. Therefore DPIs are preferred [274].

38.7.3 Soft Mist Inhaler (SMI)

"Soft mist" is inhalable aerosol cloud, generated from fine gets of drug solution. In general, a predetermined dose is pumped out through especially designed nozzle to generate the liquid jets converging at a pre-set angle. Respimat® is such a new-generation, propellant-free SMI [275, 276]. The aerosol cloud generated with MSI contains smaller particles than that of pMDIs and DPIs. Moreover, the particles have lower velocity. So, the cloud stays for prolonged period of time. This results in improved inhalation and higher deposition of drug molecules at the target site, leading to better therapeutic efficacy than pMDIs and DPIs [275].

38.7.4 Pressurized Metered Dose Inhalers (pMDI)

In pMDIs, propellants are used to deliver the content (solution or suspension), actively from a pressurized storage container. pMDI was introduced first by Riker Laboratories in the year 1956 [277]. Chlorofluorocarbon propellants were used that time. Nowadays, CFCs are banned and hydrofluoroalkanes (HFAs) like tetrafluoroethane (HFA-134a), heptafluoropropane (HFA-227) are used as environment friendly propellant [278].The formulation should be a homogenous mixture of API or API-loaded nanoparticles and the excipients. Generally, sorbitan trioleate or lecithin is used as surfactant to reduce particle aggregation [279]. It is advised to shake the inhaler before administration of dose [280]. Again, a good coordination between inspiration and actuation is necessary. The inspiration should be slow and steady followed by holding of breath for few seconds. Therefore, the elder patients find difficulties in using pMDI device [281].

Poor colloidal stability of microcrystals in hydrofluoroalkanes is a technical problem in using HFA propellants [282]. A new proprietary technology

Co-Suspension™ has been invented where spray-dried distearoyl-phosphatidylcholine is used as suspending agent [283]. In coming days microprocessor-based intelligent inhalers will be available in clinic [284].

38.7.5 Metered Dose Inhaler (MDI) with Spacer

This device delivers the required dose, precisely in the form of aerosolized droplets. It is associated with decreased deposition of API in the oropharyngeal area and subsequently efficient pulmonary administration. Spacers are tubes that act as a reservoir to hold the medication before spraying [39]. Thus, the velocity of aerosolized particles is reduced ensuring minimum fraction of dose to enter into gastrointestinal tract [41].

38.7.5.1 Breath-Actuated Metered Dose Inhaler (baMDI)

Patients having poor inhalation coordination with actuation of DPI or pMDI require breath-actuated devices [285]. In this case the coordination is automatic. The operation of breath-actuated device is based on spring operated optimized technology [286]. During inhalation, a forced is applied onto a valve causing the spring to operate. The formulation comes out in the form of aerosolized droplets of proper size distribution. Thus, this device has good patients compliance and preferred mostly by geriatric patients [287].

38.8 Future Challenges

Respiratory illnesses are the most critical health-related issues. There are a number of challenges to be addressed in clinical management of respiratory illness. The prime challenge is the targeted delivery of drugs in the case of lung cancer and microbial infections. Drug targeting is possible with properly designed nanoformulations, but only microparticles with proper density and particle size are suitable for pulmonary administration. Therefore, nanoparticle-incorporated microparticles (hybrid systems), especially porous microparticles, are deemed suitable. But these are difficult to synthesize and a lot of research scopes are there. Second challenge is stealing the nanoparticles/microparticles from dendritic cells in lungs. Though a few strategies are reported, more robust techniques are required. Another key challenge is to study the clearance of the matrix material from lungs. The presence of matrix material for prolonged period of time may be detrimental. Moreover, since the deposition of particles to a particular region of lung depends on a number of factors like particle size, surface properties, particle density, and disease condition of the patient, high stringency is required in controlling the quality of the product and development of robust delivery device. Another big challenge is the development of robust preclinical models for evaluating drugs administered through the pulmonary route, because it is very difficult to administer proper dose in animal via inhalation. Though device like microspray is available, more sophisticated delivery devices are required to assure accurate administration of dose. Some pulmonary diseases like cystic fibrosis are good candidate for gene therapy. So, development of efficient vector and device for targeted delivery of the vector is a major challenge. Though different types of delivery device for pulmonary administration are available in clinic, bioelectronic device and microprocessor-controlled devices are the future.

References

1. Brinkman JE, Sharma S (2020) Physiology, pulmonary. StatPearls Publishing, Treasure Island, FL
2. Haddad M, Sharma S (2019) Physiology, Lung. StatPearls Publishing, Treasure Island, FL
3. Marin ND, Dunlap MD, Kaushal D, Khader SA (2019) Friend or foe: the protective and pathological roles of inducible bronchus-associated lymphoid tissue in pulmonary diseases. J Immunol 202(9):2519
4. Brown K, Lynch DT (2018) Histology, lung. StatPearls Publishing, Treasure Island, FL
5. Fishman AP (1977) Nonrespiratory functions of the lungs. Chest 72(1):84–89
6. Jin L, Wang Z, Qi X (2018) Diffuse idiopathic pulmonary neuroendocrine cell hyperplasia: case series and a review of the literature. Medicine 97(52)
7. Effros RM (2006) Anatomy, development, and physiology of the lungs. GI Motility online.
8. Morgan DJ, Casulli J, Chew C, Connolly E, Lui S, Brand OJ, Rahman R, Jagger C, Hussell T (2018) Innate immune cell suppression and the link with secondary lung bacterial pneumonia. Front Immunol 9:2943
9. Lehrer RI Primate defensins. Nat Rev Microbiol 2:727–738. (1740–1526 (Print))
10. Bals R, Hiemstra PS (2004) Innate immunity in the lung: how epithelial cells fight against respiratory pathogens. Eur Respir J 23(2):327–333
11. Martin TR, Mathison JC, Tobias PS, Letúrcq DJ, Moriarty AM, Maunder RJ, Ulevitch RJ (1992) Lipopolysaccharide binding protein enhances the responsiveness of alveolar macrophages to bacterial lipopolysaccharide. Implications for cytokine production in normal and injured lungs. J Clin Invest 90:2209–2219. (0021-9738 (Print))
12. Martin TR, Rubenfeld GD, Ruzinski JT, Goodman RB, Steinberg KP, Leturcq DJ, Moriarty AM, Raghu G, Baughman RP, Hudson LD Relationship between soluble CD14, lipopolysaccharide binding protein, and the alveolar inflammatory response in patients with acute respiratory distress syndrome. Am J Respir Crit Care Med 155:937–944. (1073-449X (Print))
13. Galeas-Pena M, McLaughlin N, Pociask D (2019) The role of the innate immune system on pulmonary infections. Biol Chem 400:443–456
14. Patton JS, Fishburn CS, Weers JG (2004) The lungs as a portal of entry for systemic drug delivery. Proc Am Thorac Soc 1(4):338–344
15. Hamman JH, Enslin GM, Kotzé AF (2005) Oral delivery of peptide drugs: barriers and developments. BioDrugs 19(3):165–177
16. Chellat F, Merhi Y, Moreau A, Yahia L (2005) Therapeutic potential of nanoparticulate systems for macrophage targeting. Biomaterials 26(35):7260–7275
17. Stone KC, Mercer RR, Gehr P, Stockstill B, Crapo JD (1992) Allometric relationships of cell numbers and size in the mammalian lung. Am J Respir Cell Mol Biol 6(2):235–243
18. Groneberg DA, Witt C, Wagner U, Chung KF, Fischer A (2003) Fundamentals of pulmonary drug delivery. Respir Med 97(4):382–387
19. Chono S, Tanino T, Seki T, Morimoto K (2006) Influence of particle size on drug delivery to rat alveolar macrophages following pulmonary administration of ciprofloxacin incorporated into liposomes. J Drug Target 14(8):557–566

20. Lauweryns JM, Baert JH (1977) Alveolar clearance and the role of the pulmonary lymphatics. Am Rev Respir Dis 115(4):625–683
21. Rejman J, Oberle V, Zuhorn IS, Hoekstra D (2004) Size-dependent internalization of particles via the pathways of clathrin- and caveolae-mediated endocytosis. Biochem J 377 (Pt 1):159–169
22. Dhand C, Prabhakaran MP, Beuerman RW, Lakshminarayanan R, Dwivedi N, Ramakrishna S (2014) Role of size of drug delivery carriers for pulmonary and intravenous administration with emphasis on cancer therapeutics and lung-targeted drug delivery. RSC Adv 4 (62):32673–32689
23. Patton JS (1996) Mechanisms of macromolecule absorption by the lungs. Adv Drug Deliv Rev 19(1):3–36
24. Courrier HM, Butz N, Vandamme TF (2002) Pulmonary drug delivery systems: recent developments and prospects. Crit Rev Ther Drug Carrier Syst 19(4–5):425–498
25. Patton JS, Brain JD, Davies LA, Fiegel J, Gumbleton M, Kim KJ, Sakagami M, Vanbever R, Ehrhardt C (2010) The particle has landed--characterizing the fate of inhaled pharmaceuticals. J Aerosol Med Pulm Drug Deliv 23(Suppl 2):S71–S87
26. Heyder J, Rudolf G (1984) Mathematical models of particle deposition in the human respiratory tract. J Aerosol Sci 15(6):697–707
27. Stahlhofen W, Koebrich R, Rudolf G, Scheuch G (1990) Short-term and long-term clearance of particles from the upper human respiratory tract as function of particle size. J Aerosol Sci 21:S407–S410
28. Alberts BJA, Lewis J (2002) The airways and the gut. In: Molecular biology of the cell. Garland, New York
29. Fahy JV, Dickey BF (2010) Airway mucus function and dysfunction. N Engl J Med 363 (23):2233–2247
30. Naughton J, Duggan G, Bourke B, Clyne M (2014) Interaction of microbes with mucus and mucins: recent developments. Gut Microbes 5(1):48–52
31. Evans CM, Kim K, Tuvim MJ, Dickey BF (2009) Mucus hypersecretion in asthma: causes and effects. Curr Opin Pulm Med 15(1):4–11
32. Knowles MR, Boucher RC (2002) Mucus clearance as a primary innate defense mechanism for mammalian airways. J Clin Invest 109(5):571–577
33. Khanvilkar K, Donovan MD, Flanagan DR (2001) Drug transfer through mucus. Adv Drug Deliv Rev 48(2–3):173–193
34. Jain RK (2001) Delivery of molecular medicine to solid tumors: lessons from in vivo imaging of gene expression and function. J Control Rel 74(1–3):7–25
35. El-Sherbiny I, Elbaz N, Yacoub M (2015) Inhaled nano- and microparticles for drug delivery. Global Cardiol Sci Pract:2015
36. Lim YH, Tiemann KM, Hunstad DA, Elsabahy M, Wooley KL (2016) Polymeric nanoparticles in development for treatment of pulmonary infectious diseases. WIREs Nanomed Nanobiotechnol 8(6):842–871
37. Park CW, Li X, Vogt FG, Hayes D Jr, Zwischenberger JB, Park ES, Mansour HM (2013) Advanced spray-dried design, physicochemical characterization, and aerosol dispersion performance of vancomycin and clarithromycin multifunctional controlled release particles for targeted respiratory delivery as dry powder inhalation aerosols. Int J Pharm 455(1–2):374–392
38. Li X, Vogt FG, Hayes D Jr, Mansour HM (2014) Physicochemical characterization and aerosol dispersion performance of organic solution advanced spray-dried microparticulate/nanoparticulate antibiotic dry powders of tobramycin and azithromycin for pulmonary inhalation aerosol delivery. Eur J Pharm Sci 52:191–205
39. Dawson M, Wirtz D, Hanes J (2003) Enhanced viscoelasticity of human cystic fibrotic sputum correlates with increasing microheterogeneity in particle transport. J Biol Chem 278 (50):50393–50401
40. Forier K, Messiaen A-S, Raemdonck K, Deschout H, Rejman J, Baets FD, Nelis H, Smedt SCD, Demeester J, Coenye T, Braeckmans K (2013) Transport of nanoparticles in cystic

fibrosis sputum and bacterial biofilms by single-particle tracking microscopy. Nanomedicine 8 (6):935–949
41. Chen EYT, Wang Y-C, Chen C-S, Chin W-C (2010) Functionalized positive nanoparticles reduce mucin swelling and dispersion. PLoS One 5(11):e15434
42. Jokerst JV, Lobovkina T, Zare RN, Gambhir SS (2011) Nanoparticle PEGylation for imaging and therapy. Nanomedicine (Lond) 6(4):715–728
43. Boylan NJ, Kim AJ, Suk JS, Adstamongkonkul P, Simons BW, Lai SK, Cooper MJ, Hanes J (2012) Enhancement of airway gene transfer by DNA nanoparticles using a pH-responsive block copolymer of polyethylene glycol and poly-L-lysine. Biomaterials 33(7):2361–2371
44. Liu X-B, Ye J-X, Quan L-H, Liu C-Y, Deng X-L, Yang M, Liao Y-H (2008) Pulmonary delivery of scutellarin solution and mucoadhesive particles in rats. Eur J Pharm Biopharm 70 (3):845–852
45. Schneider CS, Xu Q, Boylan NJ, Chisholm J, Tang BC, Schuster BS, Henning A, Ensign LM, Lee E, Adstamongkonkul P, Simons BW, Wang SS, Gong X, Yu T, Boyle MP, Suk JS, Hanes J (2017) Nanoparticles that do not adhere to mucus provide uniform and long-lasting drug delivery to airways following inhalation. Sci Adv 3(4):e1601556
46. Lai SK, Wang YY, Hanes J (2009) Mucus-penetrating nanoparticles for drug and gene delivery to mucosal tissues. Adv Drug Deliv Rev 61(2):158–171
47. Li SD, Huang L (2009) Nanoparticles evading the reticuloendothelial system: role of the supported bilayer. Biochim Biophys Acta 1788(10):2259–2266
48. Yang M, Lai SK, Wang YY, Zhong W, Happe C, Zhang M, Fu J, Hanes J (2011) Biodegradable nanoparticles composed entirely of safe materials that rapidly penetrate human mucus. Angew Chem Int Ed Engl 50(11):2597–2600
49. Ensign LM, Tang BC, Wang YY, Tse TA, Hoen T, Cone R, Hanes J (2012) Mucus-penetrating nanoparticles for vaginal drug delivery protect against herpes simplex virus. Sci Transl Med 4(138):138ra179
50. Speizer FE, Horton S, Batt J, Slutsky AS (2006) Respiratory diseases of adults. Disease control priorities in developing countries. The International Bank for Reconstruction and Development/The World Bank, Washington, DC
51. Dasaraju PV, Liu C (1996) Infections of the respiratory system. In: Baron S (ed) Medical microbiology, 4th edn. University of Texas Medical Branch at Galveston
52. Novick S, Shagan M, Blau K, Lifshitz S, Givon-Lavi N, Grossman N, Bodner L, Dagan R, Mizrachi Nebenzahl Y (2017) Adhesion and invasion of Streptococcus pneumoniae to primary and secondary respiratory epithelial cells. Mol Med Rep 15(1):65–74
53. Weiping Zhou SS (2020) Exploring the molecular mechanisms of pneumonia induced by Streptococcus pneumoniae infected lung epithelial cells through comprehensive network analysis. Int J Clin Exp Med 13(5):3745–3753
54. Feng C, Huang H, Huang S, Zhai YZ, Dong J, Chen L, Huang Z, Zhou X, Li B, Wang LL, Chen W, Lv FQ, Li TS (2018) Identification of potential key genes associated with severe pneumonia using mRNA-seq. Exp Ther Med 16(2):758–766
55. Sarma NJ, Tiriveedhi V, Subramanian V, Shenoy S, Crippin JS, Chapman WC, Mohanakumar T (2012) Hepatitis C virus mediated changes in miRNA-449a modulates inflammatory biomarker YKL40 through components of the NOTCH signaling pathway. PLoS One 7(11): e50826
56. Wolfle U, Haarhaus B, Schempp CM (2015) Amarogentin displays immunomodulatory effects in human mast cells and keratinocytes. Mediat Inflamm 2015:630128
57. Miller JL, Velmurugan K, Cowan MJ, Briken V (2010) The type I NADH dehydrogenase of Mycobacterium tuberculosis counters phagosomal NOX2 activity to inhibit TNF-alpha-mediated host cell apoptosis. PLoS Pathog 6(4):e1000864
58. Metlay JP, Waterer GW, Long AC, Anzueto A, Brozek J, Crothers K, Cooley LA, Dean NC, Fine MJ, Flanders SA, Griffin MR, Metersky ML, Musher DM, Restrepo MI, Whitney CG (2019) Diagnosis and treatment of adults with community-acquired pneumonia. An official clinical practice guideline of the American Thoracic Society and Infectious Diseases Society of America. Am J Respir Crit Care Med 200(7):e45–e67

59. Nayar S, Hasan A, Waghray P, Ramananthan S, Ahdal J, Jain R (2019) Management of community-acquired bacterial pneumonia in adults: limitations of current antibiotics and future therapies. Lung India 36(6):525–533
60. Xiao AJ, Miller BW, Huntington JA, Nicolau DP (2016) Ceftolozane/tazobactam pharmacokinetic/pharmacodynamic-derived dose justification for phase 3 studies in patients with nosocomial pneumonia. J Clin Pharmacol 56(1):56–66
61. Shirley M (2018) Ceftazidime-avibactam: a review in the treatment of serious gram-negative bacterial infections. Drugs 78(6):675–692
62. Jorgensen SCJ, Rybak MJ (2018) Meropenem and vaborbactam: stepping up the battle against carbapenem-resistant enterobacteriaceae. Pharmacotherapy 38(4):444–461
63. Rizk ML, Rhee EG, Jumes PA, Gotfried MH, Zhao T, Mangin E, Bi S, Chavez-Eng CM, Zhang Z, Butterton JR (2018) Intrapulmonary pharmacokinetics of relebactam, a novel beta-lactamase inhibitor, dosed in combination with imipenem-cilastatin in healthy subjects. Antimicrob Agents Chemother 62(3)
64. Cies JJ, LaCoursiere RJ, Moore WS 2nd, Chopra A (2017) Therapeutic drug monitoring of prolonged infusion aztreonam for multi-drug resistant Pseudomonas aeruginosa: a case report. J Pediatr Pharmacol Ther 22(6):467–470
65. Kisgen J, Whitney D (2008) Ceftobiprole, a broad-spectrum cephalosporin with activity against methicillin-resistant Staphylococcus aureus (MRSA). P T 33(11):631–641
66. Donald BJ, Surani S, Deol HS, Mbadugha UJ, Udeani G (2017) Spotlight on solithromycin in the treatment of community-acquired bacterial pneumonia: design, development, and potential place in therapy. Drug Des Devel Ther 11:3559–3566
67. Schroeder MR, Stephens DS (2016) Macrolide resistance in Streptococcus pneumoniae. Front Cell Infect Microbiol 6:98–98
68. McKinnell JA, Connolly LE, Pushkin R, Jubb AM, O'Keeffe B, Serio AW, Smith A, Gall J, Riddle V, Krause KM (2017) Improved outcomes with plazomicin (PLZ) compared with colistin (CST) in patients with bloodstream infections (BSI) caused by carbapenem-resistant Enterobacteriaceae (CRE): results from the CARE study. In: Open forum infectious diseases, vol suppl_1. Oxford University Press, Cary, NC, pp S531–S531
69. Blondeau JM (2004) Fluoroquinolones: mechanism of action, classification, and development of resistance. Surv Ophthalmol 49(Suppl 2):S73–S78
70. Xue G, Crabb DM, Xiao L, Liu Y, Waites KB (2018) In vitro activities of the benzoquinolizine fluoroquinolone levonadifloxacin (WCK 771) and other antimicrobial agents against mycoplasmas and ureaplasmas in humans, including isolates with defined resistance mechanisms. Antimicrob Agents Chemother 62(11):e01348–e01318
71. Bassetti M, Vena A, Castaldo N, Righi E, Peghin M (2018) New antibiotics for ventilator-associated pneumonia. Curr Opin Infect Dis 31(2):177–186
72. Fabian N, Agata LS, Stefan A, Daniel S, Alexandra D, Daniel NW (2014) Tetracycline antibiotics and resistance mechanisms. Biol Chem 395(5):559–575
73. Veve MP, Wagner JL (2018) Lefamulin: review of a promising novel pleuromutilin antibiotic. Pharmacotherapy 38(9):935–946
74. Hall RG 2nd, Smith WJ, Putnam WC, Pass SE (2018) An evaluation of tedizolid for the treatment of MRSA infections. Expert Opin Pharmacother 19(13):1489–1494
75. Pandit N, Singla RK, Shrivastava B (2012) Current updates on oxazolidinone and its significance. Int J Med Chem 2012:159285
76. Bagheri H, Sohrabi B, Salarian S, Bagheri B (2018) Lipoglycopeptide antibiotics: reliable fighters against gram-positive pathogens. Arch Clin Infect Dis 13(6):e59526
77. Nnedu ON, Pankey GA (2015) Update on the emerging role of telavancin in hospital-acquired infections. Ther Clin Risk Manag 11:605–610
78. Martin-Loeches I, Dale GE, Torres A (2018) Murepavadin: a new antibiotic class in the pipeline. Expert Rev Anti-Infect Ther 16(4):259–268

79. Wach A, Dembowsky K, Dale GE (2018) Pharmacokinetics and safety of intravenous murepavadin infusion in healthy adult subjects administered single and multiple ascending doses. Antimicrob Agents Chemother 62(4):e02355–e02317
80. Zhang CY, Lin W, Gao J, Shi X, Davaritouchaee M, Nielsen AE, Mancini RJ, Wang Z (2019) pH-responsive nanoparticles targeted to lungs for improved therapy of acute lung inflammation/injury. ACS Appl Mater Interfaces 11(18):16380–16390
81. Wang Q, Mi G, Hickey D, Li Y, Tu J, Webster TJ, Shen Y (2018) Azithromycin-loaded respirable microparticles for targeted pulmonary delivery for the treatment of pneumonia. Biomaterials 160:107–123
82. Hsu C-Y, Sung CT, Aljuffali IA, Chen C-H, Hu K-Y, Fang J-Y (2018) Intravenous anti-MRSA phosphatiosomes mediate enhanced affinity to pulmonary surfactants for effective treatment of infectious pneumonia. Nanomedicine 14(2):215–225
83. Yildiz-Peköz A, Akbal O, Tekarslan SH, Sagirli AO, Mulazimoglu L, Morina D, Cevher E (2018) Preparation and characterization of doripenem-loaded microparticles for pulmonary delivery. J Aerosol Med Pulm Drug Deliv 31(6):347–357
84. Dye C, Floyd K (2006) Tuberculosis. Disease control priorities in developing countries. The International Bank for Reconstruction and Development/The World Bank, Washington, DC
85. Frieden TR, Sterling TR, Munsiff SS, Watt CJ, Dye C (2003) Tuberculosis. Lancet 362 (9387):887–899
86. Flynn JL, Goldstein MM, Chan J, Triebold KJ, Pfeffer K, Lowenstein CJ, Schreiber R, Mak TW, Bloom BR Tumor necrosis factor-alpha is required in the protective immune response against Mycobacterium tuberculosis in mice. Immunity 2:561–572. (1074–7613 (Print))
87. Altare F, Ensser A, Breiman A, Reichenbach J, Baghdadi JE, Fischer A, Emile JF, Gaillard JL, Meinl E, Casanova JL (2001) Interleukin-12 receptor beta1 deficiency in a patient with abdominal tuberculosis. J Infect Dis 184(2):231–236
88. MacGurn JA, Cox JS (2007) A genetic screen for Mycobacterium tuberculosis mutants defective for phagosome maturation arrest identifies components of the ESX-1 secretion system. Infect Immun 75(6):2668–2678
89. Wong D, Bach H, Sun J, Hmama Z, Av-Gay Y (2011) Mycobacterium tuberculosis protein tyrosine phosphatase (PtpA) excludes host vacuolar-H+-ATPase to inhibit phagosome acidification. Proc Natl Acad Sci U S A 108(48):19371–19376
90. Sullivan JT, Young EF, McCann JR, Braunstein M (2012) The Mycobacterium tuberculosis SecA2 system subverts phagosome maturation to promote growth in macrophages. Infect Immun 80(3):996–1006
91. Wang J, Li B-X, Ge P-P, Li J, Wang Q, Gao GF, Qiu X-B, Liu CH (2015) Mycobacterium tuberculosis suppresses innate immunity by coopting the host ubiquitin system. Nat Immunol 16(3):237–245
92. Berry MP, Graham CM, McNab FW, Xu Z, Bloch SA, Oni T, Wilkinson KA, Banchereau R, Skinner J, Wilkinson RJ, Quinn C, Blankenship D, Dhawan R, Cush JJ, Mejias A, Ramilo O, Kon OM, Pascual V, Banchereau J, Chaussabel D, O'Garra A (2010) An interferon-inducible neutrophil-driven blood transcriptional signature in human tuberculosis. Nature 466 (7309):973–977
93. Singh KH, Jha B, Dwivedy A, Choudhary E, Arpitha G, Ashraf A, Arora D, Agarwal N, Biswal BK (2017) Characterization of a secretory hydrolase from Mycobacterium tuberculosis sheds critical insight into host lipid utilization by M. tuberculosis. J Biol Chem 292 (27):11326–11335
94. Singh V, Kaur C, Chaudhary VK, Rao KV, Chatterjee S (2015) M. tuberculosis secretory protein ESAT-6 induces metabolic flux perturbations to drive foamy macrophage differentiation. Sci Rep 5:12906
95. Ouimet M, Koster S, Sakowski E, Ramkhelawon B, van Solingen C, Oldebeken S, Karunakaran D, Portal-Celhay C, Sheedy FJ, Ray TD, Cecchini K, Zamore PD, Rayner KJ, Marcel YL, Philips JA, Moore KJ (2016) Mycobacterium tuberculosis induces the miR-33 locus to reprogram autophagy and host lipid metabolism. Nat Immunol 17(6):677–686

96. Saini NK, Baena A, Ng TW, Venkataswamy MM, Kennedy SC, Kunnath-Velayudhan S, Carreno LJ, Xu J, Chan J, Larsen MH, Jacobs WR Jr, Porcelli SA (2016) Suppression of autophagy and antigen presentation by Mycobacterium tuberculosis PE_PGRS47. Nat Microbiol 1(9):16133
97. Keane J, Remold HG, Kornfeld H (2000) Virulent Mycobacterium tuberculosis strains evade apoptosis of infected alveolar macrophages. J Immunol 164(4):2016–2020
98. Mahamed D, Boulle M, Ganga Y, Mc Arthur C, Skroch S, Oom L, Catinas O, Pillay K, Naicker M, Rampersad S (2017) Intracellular growth of Mycobacterium tuberculosis after macrophage cell death leads to serial killing of host cells. elife 6:e22028
99. Sun J, Siroy A, Lokareddy RK, Speer A, Doornbos KS, Cingolani G, Niederweis M (2015) The tuberculosis necrotizing toxin kills macrophages by hydrolyzing NAD. Nat Struct Mol Biol 22(9):672–678
100. Stead WW (1970) Epidemiologic basis of tuberculosis eradication. Ann Intern Med 73 (1):135–136
101. Cairns H, Smith HV, Vollum RL (1950) Tuberculous meningitis. J Am Med Assoc 144 (2):92–96
102. Merchant S, Bharati A, Merchant N (2013) Tuberculosis of the genitourinary system-urinary tract tuberculosis: renal tuberculosis-part I. Indian J Radiol Imaging 23(1):46–63
103. Bansal R, Sharma D, Singh R (2018) Tuberculosis and its treatment: an overview. Mini Rev Med Chem 18(1):58–71
104. Dover LG, Coxon GD (2011) Current status and research strategies in tuberculosis drug development. J Med Chem 54(18):6157–6165
105. Kolyva AS, Karakousis PC (2012) Old and new TB drugs: mechanisms of action and resistance.
106. Matsumoto M, Hashizume H, Tomishige T, Kawasaki M, Tsubouchi H, Sasaki H, Shimokawa Y, Komatsu M (2006) OPC-67683, a nitro-dihydro-imidazooxazole derivative with promising action against tuberculosis in vitro and in mice. PLoS Med 3(11):e466
107. Palomino JC, Martin A (2013) TMC207 becomes bedaquiline, a new anti-TB drug. Future Microbiol 8(9):1071–1080
108. Igarashi M, Ishizaki Y, Takahashi Y (2017) New antituberculous drugs derived from natural products: current perspectives and issues in antituberculous drug development. J Antibiot (Tokyo). https://doi.org/10.1038/ja.2017.126
109. Sotgiu G, Centis R, D'Ambrosio L, Spanevello A, Migliori GB, International Group for the study of L (2013) Linezolid to treat extensively drug-resistant TB: retrospective data are confirmed by experimental evidence. Eur Respir J 42(1):288–290
110. Scherr N, Honnappa S, Kunz G, Mueller P, Jayachandran R, Winkler F, Pieters J, Steinmetz MO (2007) Structural basis for the specific inhibition of protein kinase G, a virulence factor of Mycobacterium tuberculosis. Proc Natl Acad Sci U S A 104(29):12151–12156
111. Lee RE, Hurdle JG, Liu J, Bruhn DF, Matt T, Scherman MS, Vaddady PK, Zheng Z, Qi J, Akbergenov R, Das S, Madhura DB, Rathi C, Trivedi A, Villellas C, Lee RB, Rakesh, Waidyarachchi SL, Sun D, McNeil MR, Ainsa JA, Boshoff HI, Gonzalez-Juarrero M, Meibohm B, Bottger EC, Lenaerts AJ (2014) Spectinamides: a new class of semisynthetic antituberculosis agents that overcome native drug efflux. Nat Med 20(2):152–158
112. Athar MT, Ullah Z (2017) Current approaches and future prospects of nanomedicine in tuberculosis therapy. Recent Pat Antiinfect Drug Discov 12(2):120–127
113. Nasiruddin M, Neyaz MK, Das S (2017) Nanotechnology-based approach in tuberculosis treatment. Tuberculosis Res Treat 2017:4920209
114. Pandey R, Sharma S, Khuller GK (2004) Nebulization of liposome encapsulated antitubercular drugs in guinea pigs. Int J Antimicrob Agents 24(1):93–94
115. Mehta SK, Jindal N (2013) Formulation of Tyloxapol niosomes for encapsulation, stabilization and dissolution of anti-tubercular drugs. Colloids Surf B Biointerfaces 101:434–441
116. Pandey R, Sharma S, Khuller GK (2005) Oral solid lipid nanoparticle-based antitubercular chemotherapy. Tuberculosis (Edinb) 85(5–6):415–420

117. El Zowalaty M, Hussein-Al-Ali S, Husseiny M, Geilich B, Webster T, Hussein M (2015) The ability of streptomycin-loaded chitosan-coated magnetic nanocomposites to possess antimicrobial and antituberculosis activities. Int J Nanomedicine 10:3269–3274
118. Asadi A (2014) Streptomycin-loaded PLGA-alginate nanoparticles: preparation, characterization, and assessment. Appl Nanosci 4(4):455–460
119. Ahmad Z, Khuller GK (2008) Alginate-based sustained release drug delivery systems for tuberculosis. Expert Opin Drug Deliv 5(12):1323–1334
120. Kreytsberg GN, Gracheva IE, Kibrik BS, Golikov IV (2011) Antituberculous effect of silver nanoparticles. J Phys Conf Ser 291:012030
121. Tăbăran AFMC, Mocan T, Tăbăran A, Mihaiu M, Iancu C, Mocan L (2020) Silver nanoparticles for the therapy of tuberculosis. Int J Nanomedicine 15:2231–2258
122. Sharma A, Sharma S, Khuller GK (2004) Lectin-functionalized poly (lactide-co-glycolide) nanoparticles as oral/aerosolized antitubercular drug carriers for treatment of tuberculosis. J Antimicrob Chemother 54(4):761–766
123. Taraseviciene-Stewart L, Voelkel NF (2008) Molecular pathogenesis of emphysema. J Clin Invest 118(2):394–402
124. Ferrarotti I, Zorzetto M, Beccaria M, Gilè LS, Porta R, Ambrosino N, Pignatti PF, Cerveri I, Pozzi E, Luisetti M (2003) Tumour necrosis factor family genes in a phenotype of COPD associated with emphysema. Eur Respir J 21(3):444–449
125. Gauldie J, Kolb M, Ask K, Martin G, Bonniaud P, Warburton D (2006) Smad3 signaling involved in pulmonary fibrosis and emphysema. Proc Am Thorac Soc 3(8):696–702
126. Ito I, Nagai S, Handa T, Muro S, Hirai T, Tsukino M, Mishima M (2005) Matrix metalloproteinase-9 promoter polymorphism associated with upper lung dominant emphysema. Am J Respir Crit Care Med 172(11):1378–1382
127. Eriksson S (1964) Pulmonary emphysema and alpha1-antitrypsin deficiency. Acta Med Scand 175:197–205
128. D'Armiento J, Dalal SS, Okada Y, Berg RA, Chada K (1992) Collagenase expression in the lungs of transgenic mice causes pulmonary emphysema. Cell 71(6):955–961
129. Wright JL, Churg A (1995) Smoke-induced emphysema in guinea pigs is associated with morphometric evidence of collagen breakdown and repair. Am J Phys Lung Cell Mol Phys 268(1):L17–L20
130. Zheng T, Kang MJ, Crothers K, Zhu Z, Liu W, Lee CG, Rabach LA, Chapman HA, Homer RJ, Aldous D, DeSanctis G, Underwood S, Graupe M, Flavell RA, Schmidt JA, Elias JA (2005) Role of cathepsin S-dependent epithelial cell apoptosis in IFN-γ-induced alveolar remodeling and pulmonary emphysema. J Immunol 175(3):2026
131. Fujita M, Shannon JM, Irvin CG, Fagan KA, Cool C, Augustin A, Mason RJ (2001) Overexpression of tumor necrosis factor-alpha produces an increase in lung volumes and pulmonary hypertension. Am J Physiol Lung Cell Mol Physiol 280(1):L39–L49
132. Ma B, Blackburn MR, Lee CG, Homer RJ, Liu W, Flavell RA, Boyden L, Lifton RP, Sun CX, Young HW, Elias JA (2006) Adenosine metabolism and murine strain-specific IL-4-induced inflammation, emphysema, and fibrosis. J Clin Invest 116(5):1274–1283
133. Singh A, Rangasamy T, Thimmulappa RK, Lee H, Osburn WO, Brigelius-Flohé R, Kensler TW, Yamamoto M, Biswal S (2006) Glutathione peroxidase 2, the major cigarette smoke-inducible isoform of GPX in lungs, is regulated by Nrf2. Am J Respir Cell Mol Biol 35 (6):639–650
134. Rangasamy T, Cho CY, Thimmulappa RK, Zhen L, Srisuma SS, Kensler TW, Yamamoto M, Petrache I, Tuder RM, Biswal S (2004) Genetic ablation of Nrf2 enhances susceptibility to cigarette smoke-induced emphysema in mice. J Clin Invest 114(9):1248–1259
135. Zhang X, Shan P, Jiang G, Cohn L, Lee PJ (2006) Toll-like receptor 4 deficiency causes pulmonary emphysema. J Clin Invest 116(11):3050–3059
136. Maes T, Bracke KR, Vermaelen KY, Demedts IK, Joos GF, Pauwels RA, Brusselle GG (2006) Murine TLR4 is implicated in cigarette smoke-induced pulmonary inflammation. Int Arch Allergy Immunol 141(4):354–368

137. Tang K, Rossiter HB, Wagner PD, Breen EC (2004) Lung-targeted VEGF inactivation leads to an emphysema phenotype in mice. J Appl Physiol (1985) 97(4):1559–1566. discussion 1549
138. Voelkel NF, Vandivier RW, Tuder RM (2006) Vascular endothelial growth factor in the lung. Am J Physiol Lung Cell Mol Physiol 290(2):L209–L221
139. Lee S, Chen TT, Barber CL, Jordan MC, Murdock J, Desai S, Ferrara N, Nagy A, Roos KP, Iruela-Arispe ML (2007) Autocrine VEGF signaling is required for vascular homeostasis. Cell 130:691–703. (0092-8674 (Print))
140. Kasahara Y, Tuder RM, Taraseviciene-Stewart L, Le Cras TD, Abman S, Hirth PK, Waltenberger J, Voelkel NF (2000) Inhibition of VEGF receptors causes lung cell apoptosis and emphysema. J Clin Invest 106(11):1311–1319
141. Ziel KA, Grishko V, Campbell CC, Breit JF, Wilson GL, Gillespie MN (2005) Oxidants in signal transduction: impact on DNA integrity and gene expression. FASEB J 19(3):387–394
142. Hsu E, Bajaj T (2020) Beta 2 agonists. StatPearls, Treasure Island, FL
143. Johnson DB, Merrell BJ, Bounds CG (2020) Albuterol. StatPearls, Treasure Island, FL
144. Calverley PMA, Anderson JA, Brook RD, Crim C, Gallot N, Kilbride S, Martinez FJ, Yates J, Newby DE, Vestbo J, Wise R, Celli BR, Investigators S (2018) Fluticasone furoate, vilanterol, and lung function decline in patients with moderate chronic obstructive pulmonary disease and heightened cardiovascular risk. Am J Respir Crit Care Med 197(1):47–55
145. Pahal P, Avula A, Sharma S (2019) Emphysema. StatPearls Publishing, Treasure Island, FL
146. Ghossein N, Kang M, Lakhkar AD (2020) Anticholinergic medications. StatPearls, Treasure Island, FL
147. Delgado BJ, Bajaj T (2019) Tiotropium. StatPearls Publishing, Treasure Island, FL
148. Liang TZ, Chao JH (2020) Inhaled corticosteroids. StatPearls, Treasure Island, FL
149. Wood-Baker RR, Gibson PG, Hannay M, Walters EH, Walters JA (2005) Systemic corticosteroids for acute exacerbations of chronic obstructive pulmonary disease. Cochrane Database Syst Rev 1:CD001288
150. Padda IS, Tripp J (2020) Phosphodiesterase inhibitors. StatPearls, Treasure Island, FL
151. Cilli A, Bal H, Gunen H (2019) Efficacy and safety profile of roflumilast in a real-world experience. J Thorac Dis 11(4):1100–1105
152. Rabe KF, Martinez FJ, Ferguson GT, Wang C, Singh D, Wedzicha JA, Trivedi R, St. Rose E, Ballal S, McLaren J, Darken P, Aurivillius M, Reisner C, Dorinsky P (2020) Triple inhaled therapy at two glucocorticoid doses in moderate-to-very-severe COPD. N Engl J Med 383 (1):35–48
153. Ghasemi A, Mohtashami M, Sheijani SS, Aliakbari K (2015) Chitosan-genipin nanohydrogel as a vehicle for sustained delivery of alpha-1 antitrypsin. Res Pharm Sci 10(6):523–534
154. Taylor A (1981) Capillary fluid filtration. Starling forces and lymph flow. Circ Res 49 (3):557–575
155. Iqbal MA, Gupta M Cardiogenic pulmonary edema. BTI - StatPearls, Treasure Island, FL
156. Murray JF (2011) Pulmonary edema: pathophysiology and diagnosis. Int J Tuberc Lung Dis 15(2):155–160. i
157. Purvey M, Allen G (2017) Managing acute pulmonary oedema. Aust Prescr 40(2):59–63
158. Metra M, Teerlink JR, Cotter G, Davison BA, Felker GM, Filippatos G, Greenberg BH, Pang PS, Ponikowski P, Voors AA, Adams KF, Anker SD, Arias-Mendoza A, Avendano P, Bacal F, Bohm M, Bortman G, Cleland JGF, Cohen-Solal A, Crespo-Leiro MG, Dorobantu M, Echeverria LE, Ferrari R, Goland S, Goncalvesova E, Goudev A, Kober L, Lema-Osores J, Levy PD, McDonald K, Manga P, Merkely B, Mueller C, Pieske B, Silva-Cardoso J, Spinar J, Squire I, Stepinska J, Van Mieghem W, von Lewinski D, Wikstrom G, Yilmaz MB, Hagner N, Holbro T, Hua TA, Sabarwal SV, Severin T, Szecsody P, Gimpelewicz C, Investigators R-A-C (2019) Effects of serelaxin in patients with acute heart failure. N Engl J Med 381(8):716–726
159. Publication Committee for the VI (2002) Intravenous nesiritide vs nitroglycerin for treatment of decompensated congestive heart failure: a randomized controlled trial. JAMA 287 (12):1531–1540

160. Bartsch P, Maggiorini M, Ritter M, Noti C, Vock P, Oelz O (1991) Prevention of high-altitude pulmonary edema by nifedipine. N Engl J Med 325(18):1284–1289
161. Peacock WF, Chandra A, Char D, Collins S, Der Sahakian G, Ding L, Dunbar L, Fermann G, Fonarow GC, Garrison N, Hu MY, Jourdain P, Laribi S, Levy P, Mockel M, Mueller C, Ray P, Singer A, Ventura H, Weiss M, Mebazaa A (2014) Clevidipine in acute heart failure: results of the a study of blood pressure control in acute heart failure-a pilot study (PRONTO). Am Heart J 167(4):529–536
162. Cuffe MS, Califf RM, Adams KF Jr, Benza R, Bourge R, Colucci WS, Massie BM, O'Connor CM, Pina I, Quigg R, Silver MA, Gheorghiade M (2002) Short-term intravenous milrinone for acute exacerbation of chronic heart failure: a randomized controlled trial. JAMA 287:1541–1547. (0098-7484 (Print))
163. Malek RSS (2020) Pulmonary edema. StatPearls Publishing, Treasure Island, FL
164. Chioncel O, Collins SP, Ambrosy AP, Gheorghiade M, Filippatos G (2015) Pulmonary oedema-therapeutic targets. Card Fail Rev 1(1):38–45
165. Siddiqui F, Siddiqui AH Cancer, lung. BTI – StatPearls, Treasure Island, FL
166. Wild CP, Stewart BW, Wild C (2014) World cancer report 2014. World Health Organization, Geneva, Switzerland
167. Burns DM (2000) Primary prevention, smoking, and smoking cessation: implications for future trends in lung cancer prevention. Cancer 89(11 Suppl):2506–2509
168. Miller KD, Siegel RL, Lin CC, Mariotto AB, Kramer JL, Rowland JH, Stein KD, Alteri R, Jemal A (2016) Cancer treatment and survivorship statistics, 2016. CA Cancer J Clin 66 (4):271–289
169. Cagle PT, Allen TC, Olsen RJ (2013) Lung cancer biomarkers: present status and future developments. Arch Pathol Lab Med 137(9):1191–1198
170. Janne PA, Meyerson M (2012) ROS1 rearrangements in lung cancer: a new genomic subset of lung adenocarcinoma. J Clin Oncol 30(8):878–879
171. Imielinski M, Cha S, Rejtar T, Richardson EA, Karger BL, Sgroi DC (2012) Integrated proteomic, transcriptomic, and biological network analysis of breast carcinoma reveals molecular features of tumorigenesis and clinical relapse. Mol Cell Proteomics 11(6):M111.014910
172. Lindeman NI, Cagle PT, Beasley MB, Chitale DA, Dacic S, Giaccone G, Jenkins RB, Kwiatkowski DJ, Saldivar JS, Squire J, Thunnissen E, Ladanyi M (2013) Molecular testing guideline for selection of lung cancer patients for EGFR and ALK tyrosine kinase inhibitors: guideline from the College of American Pathologists, International Association for the Study of Lung Cancer, and Association for Molecular Pathology. J Thorac Oncol 8(7):823–859
173. Gottschling S, Schnabel PA, Herth FJF, Herpel E (2012) Are we missing the target? – cancer stem cells and drug resistance in non-small cell lung cancer. Cancer Genomics Proteomics 9 (5):275–286
174. Chellappan DK, Sze Ning QL, Su Min SK, Bin SY, Chern PJ, Shi TP, Ee Mei SW, Yee TH, Qi OJ, Thangavelu L, Rajeshkumar S, Negi P, Chellian J, Wadhwa R, Gupta G, Collet T, Hansbro PM, Dua K (2019) Interactions between microbiome and lungs: paving new paths for microbiome based bio-engineered drug delivery systems in chronic respiratory diseases. Chem Biol Interact 310:108732
175. Bingula R, Filaire M, Radosevic-Robin N, Berthon JY, Bernalier-Donadille A, Vasson MP, Thivat E, Kwiatkowski F, Filaire E (2018) Characterisation of gut, lung, and upper airways microbiota in patients with non-small cell lung carcinoma: study protocol for case-control observational trial. Medicine (Baltimore) 97(50):e13676
176. Mulvenna P, Nankivell M, Barton R, Faivre-Finn C, Wilson P, McColl E, Moore B, Brisbane I, Ardron D, Holt T, Morgan S, Lee C, Waite K, Bayman N, Pugh C, Sydes B, Stephens R, Parmar MK, Langley RE Dexamethasone and supportive care with or without whole brain radiotherapy in treating patients with non-small cell lung cancer with brain metastases unsuitable for resection or stereotactic radiotherapy (QUARTZ): results from a phase 3, non-inferiority, randomised trial. Lancet 388:2004–2014. (1474-547X (Electronic))

177. Crequit P, Chaimani A, Yavchitz A, Attiche N, Cadranel J, Trinquart L, Ravaud P (2017) Comparative efficacy and safety of second-line treatments for advanced non-small cell lung cancer with wild-type or unknown status for epidermal growth factor receptor: a systematic review and network meta-analysis. BMC Med 15(1):193
178. Yang JC, Shih JY, Su WC, Hsia TC, Tsai CM, Ou SH, Yu CJ, Chang GC, Ho CL, Sequist LV, Dudek AZ, Shahidi M, Cong XJ, Lorence RM, Yang PC, Miller VA (2012) Afatinib for patients with lung adenocarcinoma and epidermal growth factor receptor mutations (LUX-lung 2): a phase 2 trial. Lancet Oncol 13(5):539–548
179. Ramalingam SS, Blackhall F, Krzakowski M, Barrios CH, Park K, Bover I, Seog Heo D, Rosell R, Talbot DC, Frank R, Letrent SP, Ruiz-Garcia A, Taylor I, Liang JQ, Campbell AK, O'Connell J, Boyer M (2012) Randomized phase II study of dacomitinib (PF-00299804), an irreversible pan-human epidermal growth factor receptor inhibitor, versus erlotinib in patients with advanced non-small-cell lung cancer. J Clin Oncol 30(27):3337–3344
180. Yasuda H, de Figueiredo-Pontes LL, Kobayashi S, Costa DB (2012) Preclinical rationale for use of the clinically available multitargeted tyrosine kinase inhibitor crizotinib in ROS1-translocated lung cancer. J Thorac Oncol 7(7):1086–1090
181. Schmid K, Bago-Horvath Z, Berger W, Haitel A, Cejka D, Werzowa J, Filipits M, Herberger B, Hayden H, Sieghart W (2010) Dual inhibition of EGFR and mTOR pathways in small cell lung cancer. Br J Cancer 103(5):622–628
182. Reyhanoglu G, Tadi P (2020) Etoposide. In: StatPearls. StatPearls Publishing, Treasure Island, FL
183. Makovec T (2019) Cisplatin and beyond: molecular mechanisms of action and drug resistance development in cancer chemotherapy. Radiol Oncol 53(2):148–158
184. Rosenblum D, Joshi N, Tao W, Karp JM, Peer D (2018) Progress and challenges towards targeted delivery of cancer therapeutics. Nat Commun 9(1):1410
185. Bazak R, Houri M, Achy SE, Hussein W, Refaat T (2014) Passive targeting of nanoparticles to cancer: a comprehensive review of the literature. Mol Clin Oncol 2(6):904–908
186. Yoo J, Park C, Yi G, Lee D, Koo H (2019) Active targeting strategies using biological ligands for nanoparticle drug delivery systems. Cancers (Basel) 11(5):640
187. Goel A, Baboota S, Sahni JK, Ali J (2013) Exploring targeted pulmonary delivery for treatment of lung cancer. Int J Pharm Investig 3(1):8–14
188. Chen F, Cai W (2014) Tumor vasculature targeting: a generally applicable approach for functionalized nanomaterials. Small 10(10):1887–1893
189. Nagy JA, Chang SH, Dvorak AM, Dvorak HF (2009) Why are tumour blood vessels abnormal and why is it important to know? Br J Cancer 100(6):865–869
190. Wehbe K, Pineau R, Eimer S, Vital A, Loiseau H, Déléris G (2010) Differentiation between normal and tumor vasculature of animal and human glioma by FTIR imaging. Analyst 135 (12):3052–3059
191. Ghosh Dastidar D, Ghosh D, Chakrabarti G (2020) Tumour vasculature targeted anti-cancer therapy. Vessel Plus 4:14
192. Deshpande PP, Biswas S, Torchilin VP (2013) Current trends in the use of liposomes for tumor targeting. Nanomedicine (Lond) 8(9):1509–1528
193. Weaver BA (2014) How Taxol/paclitaxel kills cancer cells. Mol Biol Cell 25(18):2677–2681
194. Chen Z, Zheng Y, Shi Y, Cui Z (2018) Overcoming tumor cell chemoresistance using nanoparticles: lysosomes are beneficial for (stearoyl) gemcitabine-incorporated solid lipid nanoparticles. Int J Nanomedicine 13:319–336
195. Duan X, He C, Kron SJ, Lin W (2016) Nanoparticle formulations of cisplatin for cancer therapy. Wiley Interdiscip Rev Nanomed Nanobiotechnol 8(5):776–791
196. Danhier F, Lecouturier N, Vroman B, Jérôme C, Marchand-Brynaert J, Feron O, Préat V (2009) Paclitaxel-loaded PEGylated PLGA-based nanoparticles: in vitro and in vivo evaluation. J Control Rel 133(1):11–17
197. Lu Z, Yeh TK, Tsai M, Au JL, Wientjes MG (2004) Paclitaxel-loaded gelatin nanoparticles for intravesical bladder cancer therapy. Clin Cancer Res 10(22):7677–7684

198. Matsumura Y, Maeda H (1986) A new concept for macromolecular therapeutics in cancer chemotherapy: mechanism of tumoritropic accumulation of proteins and the antitumor agent smancs. Cancer Res 46(12 Pt 1):6387–6392
199. Kim H, Griffith TS, Panyam J (2019) Poly(d,l-lactide-co-glycolide) nanoparticles as delivery platforms for TLR7/8 agonist-based cancer vaccine. J Pharmacol Exp Ther 370(3):715–724
200. Adhikari HS, Yadav PN (2018) Anticancer activity of chitosan, chitosan derivatives, and their mechanism of action. Int J Biomater 2018:2952085
201. Dastidar DG, Das A, Datta S, Ghosh S, Pal M, Thakur NS, Banerjee UC, Chakrabarti G (2019) Paclitaxel-encapsulated core–shell nanoparticle of cetyl alcohol for active targeted delivery through oral route. Nanomedicine (Lond) 14(16):2121–2150
202. Barenholz Y (2012) Doxil® — the first FDA-approved nano-drug: lessons learned. J Control Release 160(2):117–134
203. Patel J (1996) Liposomal doxorubicin: Doxil®. J Oncol Pharm Pract 2(4):201–210
204. In GK, Nieva JJTCR (2015) Emerging chemotherapy agents in lung cancer: nanoparticles therapeutics for non-small cell lung cancer. Transl Cancer Res 4(4):340–355
205. Hyeon T, Lee SS, Park J, Chung Y, Na HB (2001) Synthesis of highly crystalline and monodisperse maghemite nanocrystallites without a size-selection process. J Am Chem Soc 123(51):12798–12801
206. Pan C, Hu B, Li W, Sun Y, Ye H, Zeng X (2009) Novel and efficient method for immobilization and stabilization of β-d-galactosidase by covalent attachment onto magnetic Fe3O4–chitosan nanoparticles. J Mol Catal B Enzym 61(3–4):208–215
207. Carvalho TC, Carvalho SR, McConville JT (2011) Formulations for pulmonary administration of anticancer agents to treat lung malignancies. J Aerosol Med Pulm Drug Deliv 24(2):61–80
208. Madumani A, Kerry G, David WH, Vasantha HP (2019) Pulmonary nano-drug delivery systems for lung cancer: current knowledge and prospects. J Lung Health Dis 3(2):17
209. Meenach SA, Kim YJ, Kauffman KJ, Kanthamneni N, Bachelder EM, Ainslie KM (2012) Synthesis, optimization, and characterization of camptothecin-loaded acetalated dextran porous microparticles for pulmonary delivery. Mol Pharm 9(2):290–298
210. Alipour S, Montaseri H, Tafaghodi M (2015) Inhalable, large porous PLGA microparticles loaded with paclitaxel: preparation, in vitro and in vivo characterization. J Microencapsul 32 (7):661–668
211. Shi H, Xu M, Zhu J, Li Y, He Z, Zhang Y, Xu Q, Niu Y, Liu Y (2020) Programmed co-delivery of platinum nanodrugs and gemcitabine by a clustered nanocarrier for precision chemotherapy for NSCLC tumors. J Mater Chem B 8(2):332–342
212. Jacquet A (2011) Interactions of airway epithelium with protease allergens in the allergic response. Clin Exp Allergy 41(3):305–311
213. Holgate ST (2012) Innate and adaptive immune responses in asthma. Nat Med 18(5):673–683
214. Holgate ST, Wenzel S, Postma DS, Weiss ST, Renz H, Sly PD (2015) Asthma. Nat Rev Dis Primers 1(1):15025
215. Choi J, Azmat CE (2020) Leukotriene receptor antagonists. StatPearls, Treasure Island, FL
216. Puggioni F, Heffler EM, Canonica GW, Latorre M, Blasi F, Paggiaro P, Caminati M, Senna G (2019) Long-acting anti-muscarinic agents (LAMA) frequency of use and clinical features of patients with severe asthma in real-life setting: data from the severe asthma network in Italy (SANI) registry. Eur Respir J 54(suppl 63):PA4255
217. Godse K, Mehta A, Patil S, Gautam M, Nadkarni N (2015) Omalizumab-A review. Indian J Dermatol 60(4):381–384
218. Matsuo Y, Ishihara T, Ishizaki J, Miyamoto K, Higaki M, Yamashita N (2009) Effect of betamethasone phosphate loaded polymeric nanoparticles on a murine asthma model. Cell Immunol 260(1):33–38
219. Kenyon NJ, Bratt JM, Lee J, Luo J, Franzi LM, Zeki AA, Lam KS (2013) Self-assembling nanoparticles containing dexamethasone as a novel therapy in allergic airways inflammation. PLoS One 8(10):e77730–e77730

220. Lee D-W, Shirley SA, Lockey RF, Mohapatra SS (2006) Thiolated chitosan nanoparticles enhance anti-inflammatory effects of intranasally delivered theophylline. Respir Res 7(1):112
221. da Silva AL, Martini SV, Abreu SC, Samary Cdos S, Diaz BL, Fernezlian S, de Sá VK, Capelozzi VL, Boylan NJ, Goya RG, Suk JS, Rocco PR, Hanes J, Morales MM (2014) DNA nanoparticle-mediated thymulin gene therapy prevents airway remodeling in experimental allergic asthma. J Control Rel 180:125–133
222. Dauletbaev N, Cammisano M, Herscovitch K, Lands LC (2015) Stimulation of the RIG-I/MAVS pathway by polyinosinic: polycytidylic acid upregulates IFN-β in airway epithelial cells with minimal costimulation of IL-8. J Immunol 195(6):2829–2841
223. Riordan JR (2008) CFTR function and prospects for therapy. Annu Rev Biochem 77:701–726
224. Gadsby DC, Vergani P, Csanady L (2006) The ABC protein turned chloride channel whose failure causes cystic fibrosis. Nature 440(7083):477–483
225. Awatade NT, Wong SL, Hewson CK, Fawcett LK, Kicic A, Jaffe A, Waters SA (2018) Human primary epithelial cell models: promising tools in the era of cystic fibrosis personalized medicine. Front Pharmacol 9:1429
226. Lukacs GL, Verkman AS (2012) CFTR: folding, misfolding and correcting the DeltaF508 conformational defect. Trends Mol Med 18(2):81–91
227. Saint-Criq V, Gray MA (2017) Role of CFTR in epithelial physiology. Cell Mol Life Sci 74 (1):93–115
228. Yu E, Sharma S Cystic fibrosis. BTI - StatPearls, Treasure Island, FL
229. Couroux P, Farias P, Rizvi L, Griffin K, Hudson C, Crowder T, Tarran R, Tullis E (2019) First clinical trials of novel ENaC targeting therapy, SPX-101, in healthy volunteers and adults with cystic fibrosis. Pulm Pharmacol Ther 58:101819
230. Gardiner P, Malmgren A, Ersdal E, Goldwater R, Patel N (2017) ENaC inhibitor AZD5634 first in human trial reveals promising clinical profile for the treatment of cystic fibrosis. In: D94. Advances in cystic fibrosis and non-cystic fibrosis bronchiectasis. American Thoracic Society international conference abstracts. American Thoracic Society, New York. https://doi.org/10.1164/ajrccm-conference.2017.195.1_MeetingAbstracts.A730610.1164/ajrccm-conference.2017.195.1_MeetingAbstracts.A7306
231. Condren ME, Bradshaw MD (2013) Ivacaftor: a novel gene-based therapeutic approach for cystic fibrosis. J Pediatr Pharmacol Ther 18(1):8–13
232. Donaldson SH, Solomon GM, Zeitlin PL, Flume PA, Casey A, McCoy K, Zemanick ET, Mandagere A, Troha JM, Shoemaker SA, Chmiel JF, Taylor-Cousar JL (2017) Pharmacokinetics and safety of cavosonstat (N91115) in healthy and cystic fibrosis adults homozygous for F508DEL-CFTR. J Cyst Fibros 16(3):371–379
233. Henke MO, Ratjen F (2007) Mucolytics in cystic fibrosis. Paediatr Respir Rev 8(1):24–29
234. Mitri C, Xu Z, Bardin P, Corvol H, Touqui L, Tabary O (2020) Novel anti-inflammatory approaches for cystic fibrosis lung disease: identification of molecular targets and design of innovative therapies. Front Pharmacol 11:1096
235. Ong V, Mei V, Cao L, Lee K, Chung EJ (2019) Nanomedicine for cystic fibrosis. SLAS Technol 24(2):169–180
236. Kreda SM, Davis CW, Rose MC (2012) CFTR, mucins, and mucus obstruction in cystic fibrosis. Cold Spring Harb Perspect Med 2(9):a009589
237. Duncan GA, Jung J, Hanes J, Suk JS (2016) The mucus barrier to inhaled gene therapy. Mol Ther 24(12):2043–2053
238. Ibrahim BM, Park S, Han B, Yeo Y (2011) A strategy to deliver genes to cystic fibrosis lungs: a battle with environment. J Control Release 155(2):289–295
239. Savla R, Minko T (2013) Nanotechnology approaches for inhalation treatment of fibrosis. J Drug Target 21(10):914–925
240. Agarwal AK, Raja A, Brown BD (2020) Chronic obstructive pulmonary disease (COPD). StatPearls, Treasure Island, FL
241. Craig JM, Scott AL, Mitzner W (2017) Immune-mediated inflammation in the pathogenesis of emphysema: insights from mouse models. Cell Tissue Res 367(3):591–605

242. Thorley AJ, Ruenraroengsak P, Potter TE, Tetley TD (2014) Critical determinants of uptake and translocation of nanoparticles by the human pulmonary alveolar epithelium. ACS Nano 8 (11):11778–11789
243. Swanson JA (2008) Shaping cups into phagosomes and macropinosomes. Nat Rev Mol Cell Biol 9(8):639–649
244. Hillaireau H, Couvreur P (2009) Nanocarriers' entry into the cell: relevance to drug delivery. Cell Mol Life Sci 66(17):2873–2896
245. Rosales C, Uribe-Querol E (2017) Phagocytosis: a fundamental process in immunity. Biomed Res Int 2017:9042851–9042851
246. Hiemstra PS, Daha MR (1998) Opsonization. In: Delves PJ (ed) Encyclopedia of immunology, 2nd edn. Elsevier, Oxford, pp 1885–1888. https://doi.org/10.1006/rwei.1999.0475
247. Griffin FM (1977) Opsonization. In: Day NK, Good RA (eds) Biological amplification systems in immunology. Springer, Boston, MA, pp 85–113. https://doi.org/10.1007/978-1-4684-2544-4_5
248. Janeway CA Jr, Travers P, Walport M (2001) The distribution and functions of immunoglobulin isotypes. In: Immunobiology: the immune system in health and disease. Garland Science, New York
249. Janeway CA Jr, Travers P, Walport M (2001) The complement system and innate immunity. In: Immunobiology: the immune system in health and disease. Garland Science, New York
250. Vercellotti GM, McCarthy JB, Lindholm P, Peterson PK, Jacob HS, Furcht LT (1985) Extracellular matrix proteins (fibronectin, laminin, and type IV collagen) bind and aggregate bacteria. Am J Pathol 120(1):13–21
251. Proctor RA, Textor JA, Vann JM, Mosher DF (1985) Role of fibronectin in human monocyte and macrophage bactericidal activity. Infect Immun 47(3):629–637
252. Vaca DJ, Thibau A, Schütz M, Kraiczy P, Happonen L, Malmström J, Kempf VAJ (2020) Interaction with the host: the role of fibronectin and extracellular matrix proteins in the adhesion of gram-negative bacteria. Med Microbiol Immunol 209(3):277–299
253. Tabata Y, Ikada Y (1988) Effect of the size and surface charge of polymer microspheres on their phagocytosis by macrophage. Biomaterials 9(4):356–362
254. Liu Y, Ibricevic A, Cohen JA, Cohen JL, Gunsten SP, Fréchet JMJ, Walter MJ, Welch MJ, Brody SL (2009) Impact of hydrogel nanoparticle size and functionalization on in vivo behavior for lung imaging and therapeutics. Mol Pharm 6(6):1891–1902
255. Owens DE 3rd, Peppas NA (2006) Opsonization, biodistribution, and pharmacokinetics of polymeric nanoparticles. Int J Pharm 307(1):93–102
256. Behzadi S, Serpooshan V, Tao W, Hamaly MA, Alkawareek MY, Dreaden EC, Brown D, Alkilany AM, Farokhzad OC, Mahmoudi M (2017) Cellular uptake of nanoparticles: journey inside the cell. Chem Soc Rev 46(14):4218–4244
257. Engelberg S, Modrejewski J, Walter JG, Livney YD, Assaraf YG (2018) Cancer cell-selective, clathrin-mediated endocytosis of aptamer decorated nanoparticles. Oncotarget 9 (30):20993–21006
258. Ng CT, Tang FM, Li JJ, Ong C, Yung LL, Bay BH (2015) Clathrin-mediated endocytosis of gold nanoparticles in vitro. Anat Rec (Hoboken, NJ : 2007) 298(2):418–427
259. Chatterjee M, Ben-Josef E, Robb R, Vedaie M, Seum S, Thirumoorthy K, Palanichamy K, Harbrecht M, Chakravarti A, Williams TM (2017) Caveolae-mediated endocytosis is critical for albumin cellular uptake and response to albumin-bound chemotherapy. Cancer Res 77 (21):5925–5937
260. Sahay G, Kim JO, Kabanov AV, Bronich TK (2010) The exploitation of differential endocytic pathways in normal and tumor cells in the selective targeting of nanoparticulate chemotherapeutic agents. Biomaterials 31(5):923–933
261. Shi H, Guo J, Li C, Wang Z (2015) A current review of folate receptor alpha as a potential tumor target in non-small-cell lung cancer. Drug Des Devel Ther 9:4989–4996
262. Kelemen LE (2006) The role of folate receptor alpha in cancer development, progression and treatment: cause, consequence or innocent bystander? Int J Cancer 119(2):243–250

263. Bell JH, Hartley PS, Cox JS (1971) Dry powder aerosols. I A new powder inhalation device. J Pharm Sci 60(10):1559–1564
264. Corradi M, Chrystyn H, Cosio BG, Pirozynski M, Loukides S, Louis R, Spinola M, Usmani OS (2014) NEXThaler, an innovative dry powder inhaler delivering an extrafine fixed combination of beclometasone and formoterol to treat large and small airways in asthma. Expert Opin Drug Deliv 11(9):1497–1506
265. Sung JC, Pulliam BL, Edwards DA (2007) Nanoparticles for drug delivery to the lungs. Trends Biotechnol 25(12):563–570
266. Tomoda K, Ohkoshi T, Kawai Y, Nishiwaki M, Nakajima T, Makino K (2008) Preparation and properties of inhalable nanocomposite particles: effects of the temperature at a spray-dryer inlet upon the properties of particles. Colloids Surf B Biointerfaces 61(2):138–144
267. Lebhardt T, Roesler S, Uusitalo HP, Kissel T (2011) Surfactant-free redispersible nanoparticles in fast-dissolving composite microcarriers for dry-powder inhalation. Eur J Pharm Biopharm 78(1):90–96
268. Ungaro F, d'Angelo I, Coletta C, d'Emmanuele di Villa Bianca R, Sorrentino R, Perfetto B, Tufano MA, Miro A, La Rotonda MI, Quaglia F (2012) Dry powders based on PLGA nanoparticles for pulmonary delivery of antibiotics: modulation of encapsulation efficiency, release rate and lung deposition pattern by hydrophilic polymers. J Control Rel 157 (1):149–159
269. Grenha A, Seijo B, Remuñán-López C (2005) Microencapsulated chitosan nanoparticles for lung protein delivery. Eur J Pharm Sci 25(4–5):427–437
270. Chan HK (2006) Dry powder aerosol delivery systems: current and future research directions. J Aerosol Med 19(1):21–27
271. Hadinoto K, Phanapavudhikul P, Kewu Z, Tan RB (2007) Dry powder aerosol delivery of large hollow nanoparticulate aggregates as prospective carriers of nanoparticulate drugs: effects of phospholipids. Int J Pharm 333(1–2):187–198
272. Laube BL, Janssens HM, de Jongh FH, Devadason SG, Dhand R, Diot P, Everard ML, Horvath I, Navalesi P, Voshaar T, Chrystyn H (2011) What the pulmonary specialist should know about the new inhalation therapies. Eur Respir J 37(6):1308–1331
273. Hess DR (2000) Nebulizers: principles and performance. Respir Care 45(6):609–622
274. Newhouse MT, Hirst PH, Duddu SP, Walter YH, Tarara TE, Clark AR, Weers JG (2003) Inhalation of a dry powder tobramycin PulmoSphere formulation in healthy volunteers. Chest 124(1):360–366
275. Anderson P (2006) Use of Respimat Soft Mist inhaler in COPD patients. Int J Chron Obstruct Pulmon Dis 1:251–259
276. Laube BL, Dolovich MB (2014) 66—Aerosols and aerosol drug delivery systems. In: Adkinson NF, Bochner BS, Burks AW et al (eds) Middleton's allergy, 8th edn. Content Repository Only! London, pp 1066–1082. https://doi.org/10.1016/B978-0-323-08593-9.00067-X
277. Bonini M, Usmani OS (2015) The importance of inhaler devices in the treatment of COPD. COPD Res Pract 1(1):9
278. Hendeles L, Colice GL, Meyer RJ (2007) Withdrawal of albuterol inhalers containing chlorofluorocarbon propellants. N Engl J Med 356(13):1344–1351
279. Lavorini F, Fontana GA, Usmani OS (2014) New inhaler devices - the good, the bad and the ugly. Respiration 88(1):3–15
280. Byron PR (1997) Performance characteristics of pressurized metered dose inhalers in vitro. J Aerosol Med 10(s1):S-3–S-6
281. Yawn BP, Colice GL, Hodder R (2012) Practical aspects of inhaler use in the management of chronic obstructive pulmonary disease in the primary care setting. Int J Chron Obstruct Pulmon Dis 7:495–502
282. Vehring R, Lechuga-Ballesteros D, Joshi V, Noga B, Dwivedi SK (2012) Cosuspensions of microcrystals and engineered microparticles for uniform and efficient delivery of respiratory therapeutics from pressurized metered dose inhalers. Langmuir 28(42):15015–15023

283. Fabbri LM, Kerwin EM, Spangenthal S, Ferguson GT, Rodriguez-Roisin R, Pearle J, Sethi S, Orevillo C, Darken P, St Rose E, Fischer T, Golden M, Dwivedi S, Reisner C (2016) Dose-response to inhaled glycopyrrolate delivered with a novel Co-Suspension™ delivery technology metered dose inhaler (MDI) in patients with moderate-to-severe COPD. Respir Res 17 (1):109
284. Newman S (2014) Improving inhaler technique, adherence to therapy and the precision of dosing: major challenges for pulmonary drug delivery. Expert Opin Drug Deliv 11 (3):365–378
285. Newman SP, Weisz AW, Talaee N, Clarke SW (1991) Improvement of drug delivery with a breath actuated pressurised aerosol for patients with poor inhaler technique. Thorax 46 (10):712–716
286. (2005) Inhalation devices. CMAJ 173(6 Suppl):S39–S45
287. Chapman KR, Love L, Brubaker H (1993) A comparison of breath-actuated and conventional metered-dose inhaler inhalation techniques in elderly subjects. Chest 104(5):1332–1337

Future Prospects and Challenges in Targeting Cellular and Molecular Mechanisms in Respiratory Diseases

39

Nitin Verma, Komal Thapa, and Kamal Dua

Abstract

Respiratory disease is a global health issue causing increased death rate. Significant research has been done in understanding the fundamental mechanism of lung disease for managing patient's health with a range of respiratory diseases. Regardless of these researches, still there is inadequate disease prediction before the occurrence of symptoms, effective treatment, as well as effective drug target. This chapter emphasizes the prospect and challenges associated with genomics and molecular characterization of lung disease. It also highlights the mechanism of lung injury and repair and translational medicine for lung disease with cutting-edge technologies in the management of lung disease.

Keywords

Lung disease · Future prospects · Cellular mechanism · Molecular mechanism

N. Verma (✉) · K. Thapa
Chitkara University School of Pharmacy, Chitkara University, Himachal Pradesh, India
e-mail: nitin.verma@chitkarauniversity.edu.in

K. Dua
Discipline of Pharmacy, Graduate School of Health, University of Technology Sydney, Ultimo, NSW, Australia
e-mail: Kamal.Dua@uts.edu.au

K. Dua et al. (eds.), *Targeting Cellular Signalling Pathways in Lung Diseases*,
https://doi.org/10.1007/978-981-33-6827-9_39

39.1 Introduction

The world is facing immense health and economic burden due to respiratory diseases. 235 million people have been reported to be affected with asthma and more than 200 million people with chronic obstructive pulmonary disease (COPD), 65 million are affected with moderate-to-severe COPD [1], annually 8.7 million people develop tuberculosis (TB) [2], and more than one billion individuals are suffering from chronic respiratory conditions [3, 4]. Biofuel consumption releases toxic effects affecting at least two billion people. Every year 4 million people are dying from chronic respiratory disease [5]. Pneumonia is the foremost cause of death in young children [6]. Globally asthma has affected 14% [7]. Lung cancer is the most lethal cancer killing 1.4 million people each year [8]. The whole world is suffering from economic burden due to health cost for treating respiratory diseases. The annual cost for the treatment of asthma is estimated to be $18 billion [9]. This chapter is proposed to explore numerous challenges associated with molecular and genomic characterization of lung injury and repair. This chapter also emphasizes traditional research in lung medicine and the role of microbiome in lung disease along with cutting-edge technologies in the management of lung disease (Fig. 39.1).

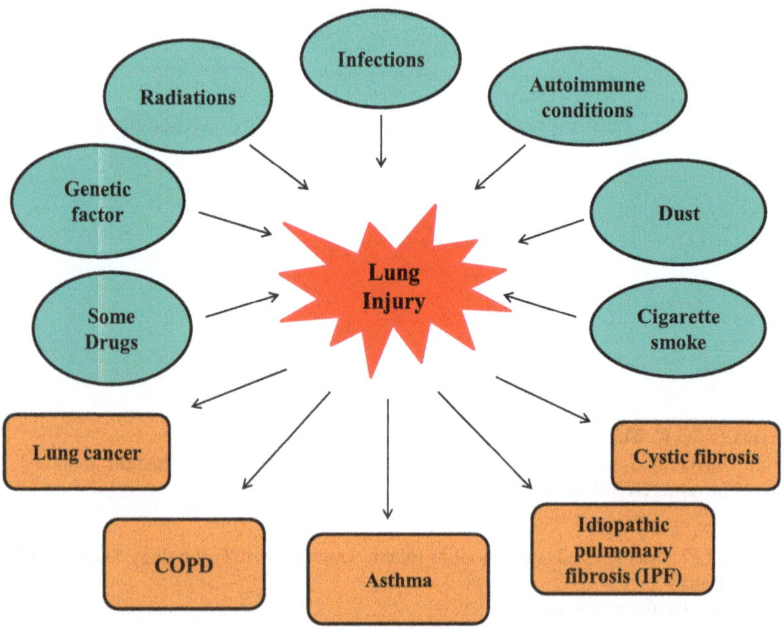

Fig. 39.1 Depiction of various factors causing lung injuries

39.2 Genomic and Molecular Characterization of Lung Disease

Genome-wide association studies (GWAS) have presented new approach into the molecular mechanisms of lung diseases and its function [10]. Eighty percent of families are found to have identified with familial pulmonary artery hypertension (FPAH) due to mutation of BMPR2 gene [11]. Alterations in transforming growth factor (TGF)-β have also been described to be associated with PAH [12]. Mice with β_6 gene have shown exaggerated inflammatory response in epithelial organ but after knockdown of β_6 provided protection in models against tissue fibrosis, bleomycin-induced pulmonary edema, or pulmonary fibrosis, suggesting its pathogenic role in acute lung injury [13]. Another gene is matrix metalloproteinase 12 (MMP-12) that showed inflammatory responses in mice induced with pulmonary emphysema via chronic exposure to tobacco smoke, whereas MMP-12 knockout mice offered anti-inflammatory effects with induced pulmonary emphysema [14, 15]. It has been reported that loss of α_1-antitrypsin (α_1-AT) gene in mice is lethal for the embryo during lung development [16]. TGF-β has also shown complex roles in COPD and has been used to discover important molecules and pathways in COPD [17]. Mice deficient in tyrosine kinase domain ($TK^{-/-}$) of the receptor (Mst1r) have shown an increased vulnerability to nickel (Ni)-induced acute lung injury (ALI) [18]. There are numerous transcription factors that may influence differentiation and formation of the respiratory epithelium, and these are GATA-6, TTF-1, Foxa1, NF-1, Foxa2, β-catenin, Foxj1, C/EBPα, Sox family members, p63, and others [19]. Studies related to gene deletion in mouse model demonstrate the significance of these transcription factors in lung function and homeostasis after birth. Many of these transcription factors regulate lung repair following either injury or unilateral pneumonectomy [20]. Together, these studies demonstrate that transcriptional programs play an important role in mediating epithelial cell differentiation and lung morphogenesis during injury and repair.

39.3 Mechanisms of Respiratory Diseases

39.3.1 Lung Injury Mechanism

39.3.1.1 Inflammatory Mechanism

Airway inflammation is mainly caused by exposure to pathogens such as bacteria, virus, pollutants, allergens, and irritants [21]. Toll-like receptors (TLRs) recognize alteration in the molecular patterns due to pathogen attack and activate inflammatory cells such as nuclear factor kappa-light-chain-enhancer of activated B cells (NF-κB) and generate growth factors such as pro-inflammatory cytokines and chemokines such as interleukin-8 (IL-8) and tumor necrosis factor alpha (TNF-α) [22, 23]. Various other proteins are associated with airway inflammation such as matrix metalloproteinase 9 (MMP-9), intercellular adhesion molecule-1 (ICAM-1), cyclooxygenase-2 (COX-2), cytosolic phospholipase A2 (cPLA2), and vascular cell adhesion molecule-1 (VCAM-1) [24]. A communication between neutrophils and epithelial cells influences signaling pathways to trigger immune cells for further inflammatory responses in the injured site [25–32]. Even though numerous studies have been conducted to understand inflammation mechanisms, still some parameters are not clear; therefore better understanding of the mechanism may help to develop new strategies to evaluate disease susceptibility and convincingly develop new approaches to prevent and treat chronic diseases.

39.3.1.2 The Role of Nitric Oxide in Lung Injury

The character of nitric oxide is associated with increased inflammatory and oxidative stress [33], as there are three distinct forms of NO• synthase (NOS-1, NOS-2 and NOS-3) which are responsible for generation of NO• and are present in almost every cell type, including epithelial cells of the airway, macrophages, neuronal cell, mast cells, neutrophils, and endothelial and smooth-muscle cells [34]. NOS isoforms NOS-1 and NOS-3 can be activated by any type of stress; NOS-2 is commonly activated by pro-inflammatory cytokines and bacterial products [35]. The increased activity of NOS-2 is mainly observed in inflammatory lung diseases such as acute respiratory distress syndrome (ARDS), bronchiectasis, and asthma that are described with increased NO• production almost certainly as host defense mechanism against viral and bacterial infections [35]. The disadvantage of excessive NO• production is harmful production of reactive nitrogen species (RNS), including peroxynitrite ($ONOO^-$) and nitrogen dioxide (NO_2 which is the prime etiology of inflammatory lung disease [36–38]. In several animal models of lung disease such as ischemia/ reperfusion, endotoxemia, and radiation, pharmacological NOS inhibitors have reduced oxidative injury [39–41]. However, in some cases, inhibition of NOS had found to worsen lung injury, revealing the protective and anti-inflammatory role of NO•. Therefore more researches and studies are needed to understand mechanistic aspects of NO• in respiratory diseases.

39.3.1.3 Apoptosis in Lung Injury

In vitro studies have revealed the role of macrophages and phagocytosed apoptotic PMNs in inflammatory response [42]. Also patients affected with acute respiratory disease (ARDS) showed increased number of PMN in BALF [43]. During epithelial injury of the lung's alveoli in ALI and ARDS pathway, Fas/FasL pathway plays an essential role in the apoptosis signaling system [44]. Increased level of BAL fluid in patients with ALI/ARDS is connected with increased mortality rates; decreased injury to lungs was observed in FasL-deficient mice [45, 46]. Further, caspase inhibition also inhibited PMN-induced acute lung injury in wild-type mice; thus this date indicates the role of Fas/FasL pathway in apoptosis for progression of ALI. Another factor for apoptosis is NF-κB that plays a critical role on ARDS as it increases the expression of many cytokines and can alter lung neutrophil apoptosis after endotoxemia [43, 47]. Also, G-CSF and its receptors were elevated in lung neutrophils after endotoxemia and may exacerbate the acute neutrophil-driven pulmonary inflammation independent of NF-κB [43, 47–49]. In LPS model of AKI, two tyrosine kinases, Src and Jak, have a dangerous role in the activation of multiple downstream signaling pathways such as cytokine signaling and

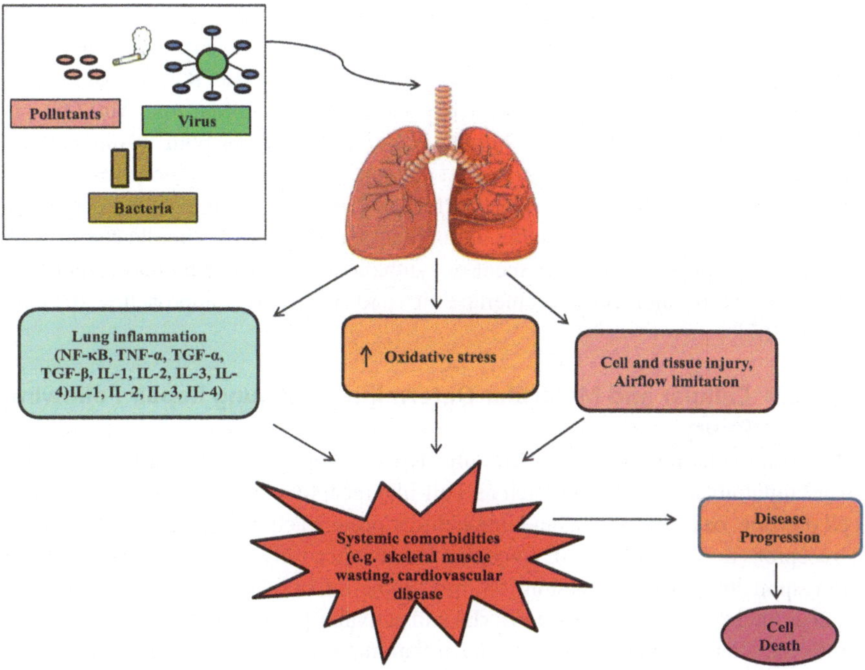

Fig. 39.2 Mechanism of lung inflammation in response to respiratory pathogens, pollutants, and cigarette smoke. Increased lung inflammation and oxidative stress cause wasting of skeletal muscle and worsening comorbid conditions such as cardiovascular disease and further progression leading to death

inflammatory responses [50–53]. In lung and serum animal model, inhibition of these kinases reduces production of various pro-inflammatory cytokines like TNF-a and IL-6 [54]. These tyrosine kinases activate downstream signaling effectors such as signal transducer and activator of transcription (STAT) factors that have similarly regulated gene expression and are responsible for inflammation and immune responses [55]. STAT3 has been recognized to play and essential role in increased expression of inflammatory chemokines, cytokines, and inflammatory mediator [56, 57]. In previous studies, Severgnini et al. evaluated that LPS stimulated STAT3 and Src and Jak activation produced reactive oxygen species (Fig. 39.2) [58, 59].

39.3.2 Lung Repair Mechanism

39.3.2.1 The Role of Antioxidant Enzymes

Antioxidant enzymes play a defensive role in reducing oxidative stress in lung disease [60, 61]. The endogenous levels of these antioxidant enzymes in the lung help to assess oxidant-induced cell damage [62]. In several stresses such as hyperoxia, mineral dusts, and paraquat, increased expression of superoxide dismutase (MnSOD) has been observed. The raised MnSOD expression serves as a biomarker of chronic lung inflammation in lung disease. Rats exposed with asbestos by inhalation showed increased mRNA levels of MnSOD in bronchioalveolar cells [63–65]. Numerous antioxidants exist both in intracellular and extracellular site of the lungs. Recent data proposed that glutathione redox cycle provided protection by detoxifying low levels of oxidants, with catalase. Development of vectors for genes encoding antioxidant enzymes can be implicated in gene therapy [66]. Moreover, development of synthetic ROS scavengers may target lung cells effectively and provide therapeutic and preventive approaches to lung disease [67].

39.3.2.2 Cellular and Molecular Determinants of Lung Repair Following Injury

Response of the lung to danger stimuli involves signal transduction that activates many inflammatory pathways such as Toll-like receptors (TLRs) that may stimulate cell damage, oxidative stress, and intracellular protein activation such as interleukin-1 receptor (IL-1R) and RAGE (receptor for advanced glycation end products) that are responsible for inflammation [68, 69]. Some inflammatory signaling molecules such as nuclear factor-kappa B [NF-κB] and activator protein [AP]-1 synthesize new molecules that ultimately mediate the inflammatory response and release a wide variety of mediators, such as eicosanoid, chemokines, and growth factors, into the extracellular space. Mechanism of tissue repair involves reabsorption of edema, reduced inflammation, and cell proliferation. Inflammatory response can be regulated by several immune mediators [70]. There are some cytokines that have anti-inflammatory effect; IL-10 has been studied, and when these pro-inflammatory pathways are downregulated due to stimuli, these anti-inflammatory mediators reduced the expression of other cytokine [71]. It has been documented that the level of some pro-survival signals such as granulocyte colony-stimulating factor (G-CSF) is reduced during apoptosis of inflammatory cells such as neutrophils [72]. Various growth factors can promote cell proliferation, and these involve epidermic [EGF] and hepatic growth factor [HGF]) and keratinocyte [KGF] that acts via tyrosine kinase receptors, promoting cell proliferation [73].

39.4 Challenges in Targeting Molecular Mechanisms in Respiratory Diseases

Nowadays lack of efficient drug is the most general reason of drug attrition as observed in preclinical animal models of respiratory diseases that poorly predict human state. In 2011 and 2012, 60% of drugs have failed in phase III and beyond because of lack of efficacy [74]. Studies relying on better understanding of the mechanism of disease will help to reduce failure of drugs by identifying its cause during development course and save future costs. Failure of drugs due to toxicity reasons can be improved by eliminating compound linked to toxicity. Better models of toxicity showed decreased risk of failure on toxicological grounds. For the prediction of drug's efficacy, better biomarkers are needed to develop drug into useful clinical treatment [75, 76]. Duration as well as cost of developing respiratory drugs is higher than for most therapeutic areas [77]. An interesting approach to speed up drug discovery is to evaluate existing drug against novel drugs [78, 79], for example, theophylline which is a bronchodilator for airway smooth muscles via inhibition of phosphodiesterase-3. But later it has been discovered that low dose of theophylline has shown to increase histone deacetylase-2 that got decreased by oxidative stress via inhibition of phosphoinositide-3-kinase-δ (PI3Kδ) [80, 81]. Theophylline at low dose in combination with low-dose oral corticosteroids is under clinical investigation for COPD treatment [82]. Another drug is macrolide which was developed as antibiotic, but it also possesses anti-inflammatory effects via suppression of the pro-inflammatory transcription factor nuclear factor-κB and improves responsiveness toward corticosteroid via PI3Kδ suppression [83, 84]. Numerous new drugs for respiratory disease have failed in clinical setting due to safety or efficacy issues but showed better results in preclinical animal models. The translation of potential drug applicants from animal models to humans has aroused question regarding its usefulness of in vivo studies, and there is a requirement for more predictive models depending on current expertise [85].

39.5 Translational Lung Medicine

Diverse range of lung diseases offers a great challenge in the field of respiratory medicine as the disease is more virulent and drug-resistant respiratory pathogens have been emerging such as multidrug-resistant *M. tb*, methicillin-resistant *Staphylococcus aureus*, carbapenem-resistant *Enterobacteriaceae*, and penicillin-resistant pneumococci [86]. The huge impact of SARS and H1N1 2009 influenza A globally has been highlighted, and now mankind remains vulnerable to ever-evolving microbial enemies [87]. Significant translational researches on lung diseases have been done. A study revealed that treatment with PI3K inhibitor in severe asthma mitigated insensitivity to glucocorticoid of peripheral blood mononuclear cells (PBMCs) [88]. Inhibition of JNK also reduced pulmonary fibrosis and lung remodeling in clinical setting [89]. The manifestations of CF (cystic fibrosis) in the lung involve the progressive development of bronchiectasis and obstructive lung disease. Studies of CF (cystic fibrosis) disease have been carried out using animal models or in vitro cell culture models of animal or human cells. CF is mainly caused due to alteration in the gene called cystic fibrosis transmembrane conductance regulator (CFTR), resulting in inflammation, unusual secretions, recurring infections. and untimely death. The US FDA approved use of Kalydeco (VX-809) for patients with CFTR mutations in May 2017. VX-809 restored lung function up to 25% in wild-type CFTR mice model and also improved pulmonary function in CFTR-mutated patients [90, 91]. Oxidative stress plays a major function in the development of various respiratory diseases such as acute respiratory distress syndrome (ARDS), chronic obstructive pulmonary disease (COPD), asthma, idiopathic pulmonary fibrosis (IPF), and lung cancer

[92]. Nrf2 is the master antioxidant transcription factor found abundantly in the lungs and is involved in the pathogenesis of various lung disorders, involving inflammation, apoptosis, oxidative stress, and carcinogenesis. Nrf2-deficient mice provide an efficient tool for investigating the role of Nrf2 in the disease model of oxidative pulmonary disease for better understanding of Nrf2 function in pulmonary diseases [92].

39.5.1 Lung Microbiome

The lung is regularly exposed to microbiota either by inhalation or subclinical microaspiration from birth [93]; however lung microbiota regulates homeostasis and immunity [94]. Hilty et al. reported that airways of asthmatic patients were enriched with *Proteobacteria* phylum [95]. COPD patient's sputum contains large amount of bacteria *Haemophilus*, *Pseudomonas*, and *Moraxella* that are associated with exacerbations [96]. In addition, the influence of viral exposure may trigger COPD exacerbations [97]. It has been reported in an experiment that patients were infected with rhinovirus that developed COPD clinical symptoms and the offender viruses were isolated from respiratory samples in 36–56% of patients [98–100]. COPD patients have been also reported with increased community of *Firmicutes* in severe disease that is responsible for further increase in the *Lactobacillus* genus [101, 102].

It has been demonstrated through animal models of respiratory syncytial virus (RSV) infection that antiviral response within the lung mucosa can be altered by the administration of *Lactobacillus rhamnosus* species former to infection [103]. Thus, alteration in microbiome may protect the lungs from respiratory viral infection. Studies have revealed that during asthma the airway microbiota composition gets altered when compared to controls, for example, the bronchial tree has more frequent *Proteobacteria* (particularly *Haemophilus*) compared to controls in asthma [95]. Predominance of *Proteobacteria* in severe asthma has been observed and is more diverse than non-asthmatic controls [104]. A connection between increased amount of *Proteobacteria* and bronchial hyperresponsiveness has been reported [105]. Microbiome alteration occurs in both severe and mild disease and is associated with traits of the disease. A study revealed that Th17-dependent neutrophil inflammation was mediated by resident microbiota in murine model of ovalbumin-induced asthma [106, 107]. Idiopathic pulmonary fibrosis (IPF) is a deadly disease of lung parenchyma of unknown cause, and studies have highlighted the pathogenetic role of bacterial as well as viral infection in IPF [108–113]. Alteration in microbiome due to increased number of *Streptococcus* and *Staphylococcus* bacteria may progress IPF [114, 115]. Further IPF patients had increased number of *Streptococcus*, *Neisseria*, *Veillonella*, and *Haemophilus* in BAL fluid as compared to controls [116–124]. Bronchiectasis and COPD are the manifestations of CF (cystic fibrosis) that are responsible for accelerated disease progression with significant mortality and morbidity [125]. Disease exacerbations were observed from increased number of pathogens that included *Staphylococcus aureus* and *Pseudomonas aeruginosa* in sputum of patients, and there was no alteration in bacterial density of sputum when antibiotics are administered [126–133]. The appearance of more new pathogens may be implicated in understanding the association of microbiome and lung disease. In CF, non-tuberculosis mycobacterium (NTM), particularly *abscessus* are linked with increased morbidity and mortality [133].

39.6 Genetic Consideration in Treatment of Lung Diseases

39.6.1 Gene Therapy for Acute and Acquired Lung Disease

39.6.1.1 Lung Injury

Bronchiolitis obliterans, a type of chronic graft rejection, is a serious problem in lung transplantation because of cell-mediated inflammation during the early stages of transplantation in severe acute lung injury [134]; therefore gene therapy for bronchiolitis obliterans may help to reduce inflammation in the very early stages [135]. DNA/liposome-based and adenovirus (Ad) vectors had been evaluated in rat lung transplantation models, with successful gene expression following transplantation [136, 137]. Numerous transgenes have been tested with purpose of decreasing acute rejection and decreasing inflammatory mediators such as interleukin (IL-10) [138], nitric oxide synthase [139], transforming growth factor beta-1 [140], Fas ligand [141], and CTLA4Ig [142]. Results obtained were encouraging, but still better efficiency in gene transfer is needed for therapeutic advantage in the grafted lung. Also feasibility of treatment with radioactive gene to avoid acute tissue damage and subsequent irradiation in the treatment of cancer has also been examined. Rodent lung was tested with transgene that encoded antioxidant manganese superoxide dismutase as delivered and carried by adenovirus and DNA/liposome [143]. Release of DNA/liposomes intratracheally into mouse lungs before irradiation leads to increased transient transgene expression in parenchymal and epithelial cell types [144].

Lung Cancer

Lung cancer is the result of mutations that lead to cell transformation and develop metastatic disease [145]. Gene therapy is considered when both chemotherapy and radiotherapy becomes resistant in malignant cells. One such approach is delivery of tumor suppressor genes such as p53 [146] and delivery of antisense such as K-ras oncogene or hammerhead ribozyme transgenes that can downregulate the expression of tumor cells [147, 148]. Another strategy is the RNA interference (RNAi) for downregulation of gene expression [149]. Introduction of p53 expressing-retroviral vectors in nine patients having non-small cell lung cancers showed degeneration in tumor in three patients and its stabilization in further three patients [150]. Similar effects were given by Ad vectors [151, 152]. New Ad vectors have the ability to replicate in p53-mutant cells that can be infused intravenously in patients with advanced lung metastasis; as a result intratumoral viral replication was observed in three out of four patients [153]. Monthly delivery of Ad with p53 expression showed better tolerance with only minor side effects [154]. Another vector is cationic polymer PEI that delivered p53 via aerosol in the lungs of rodent leading to increased transgene expression all through the lung causing noteworthy decrease in murine models with lung tumor [155–157]. Another approach is the inhibition of tumor angiogenesis using gene transfer [158]. Potent angiogenesis inhibitor endostatin expressed in recombinant Ad reduced the rate of breast growth and lung cancer and stopped development of lung micrometastases in animal models [159]. Another approach in the field of gene therapy is the prodrug or suicide gene therapy that can transmit genes and sensitize tumor cells toward other nontoxic drugs [160–162]. Delivery of DNA expressing herpes simplex virus thymidine kinase (HSVTK) to tumor cells has been accomplished with the aid of both viral and nonviral vectors [161, 162]. Patients with mesothelioma were treated with first-generation Ad vectors expressing HSVTK [163–165].

39.6.2 Gene Therapy for Chronic Lung Disease

Gene therapy is the novel tool for the treatment of chronic respiratory disease such as cystic fibrosis, emphysema, and asthma. Therefore, gene transfer has become the foremost priority of these diseases. Pulmonary emphysema might be produced due to lack of α1-antitrypsin (AAT), produced abundantly by the liver which acts as an anti-protease and counteracts the effect of neutrophil and other pro-inflammatory molecules which are excreted at sites of inflammation [166, 167]. Current protein therapy needs high capital expenditure (requiring weekly intravenous administration) and may increase the risk of viral transmission in recipients from human serum [167]; to overcome the limitation of protein therapy, gene therapy is one of the alternative techniques. In previous studies first-generation Ad vectors helped to detect AAT in bronchioalveolar fluid for only 1 week post-administration [168]. The problem faced in transient gene expression was vector-induced inflammatory responses that are now improved with the use of helper-dependent Ad and suitable promoter sequences, but still the desired vector is needed for the use of Ad to cure chronic diseases. Cationic liposomes have been employed in non-blinded study in which human AAT gene complexed with cationic lipid (DOTMA/DOPE) was delivered to the nasal epithelium of patients with AAT deficiency [169–173]. Cystic fibrosis (CF) is a chronic lung disease which has gained most attention in the area of gene therapy. This disease is mainly caused due to alteration in the gene called cystic fibrosis transmembrane conductance regulator (CFTR), resulting in abnormal release, inflammation, recurring infections, and ultimately premature death [174]. As a treatment CFTR replacement lung gene therapy has been extensively and intensively investigated, and it is hypothesized that CFTR gene transfers to the epithelial cells of the small airways, where the disease first originates likely to offer therapeutic benefit [175, 176]. Apart from usual barriers during gene transfer to the lung, CF lung suffers abnormal secretions of mucus, from abnormally thick mucus secretions, bacterial colonization, and inflammatory milieu. CF sputum and bronchioalveolar lavage fluid have been shown to block the transport of small particles [177] and reduce viral and nonviral gene transfer [177–179].

39.7 Cutting-Edge Technologies in Treatment of Lung Diseases

Promising technologies, such as scRNAseq, induced pluripotent stem cells (iPSCs), and 3D bioprinting, may help to develop winning cell therapies for diseases of the respiratory system [180–185]. These strategies help to stimulate the endogenous cellular pathways for restoration of the diseased lung. The combination of PSC technology and power of 3D bioprinting can greatly precede the target of in vitro lung development of the diseased lung. Understanding the elementary biological procedures and purpose of this information will help to develop safe and effective lung regenerative medicine in the future [181, 182]. Delivery of drug to the pulmonary system is a challenge, and therefore attention has been shifted for drug delivery systems based on nanocarrier [186]. These drug delivery systems provide sustained delivery drugs with reduced toxicity, reduced therapeutic dose, and enhanced compliance of patients. One type of delivery system is the nanostructured lipid carrier (NLC) which is formed by combining liquid and solid lipids with different ratios [187]. NLCs have displayed therapeutic effects in various diseases. A study demonstrated that NLC administration formulated by a technique called double emulsion showed significant decrease in neutrophil numbers offering effects of anti-inflammation in mice model of acute lung injury [188]. The major concern in patients related to increased healthcare costs is poor adherence [189]. Strategies till date used for combating poor adherence include electronic inhalers and self-management tools (such as web-based and mobile applications to record symptoms and monitor lung function). In patients with cystic fibrosis (CF), chipped nebulizers can provide objective date- and time-stamped adherence data. Wireless smart inhalers have been developed that directly send data to a website related to health [190]. There are various techniques to control physiological parameters such as mobile applications that determine "peak flow," "exhaled nitric oxide fraction (F_{eNO})," and physical activity (Table 39.1) [209].

39.8 Future Prospective

Understanding the mechanism of lung disease provides research evidences that help to develop novel drug that may provide clinical symptomatic relief in various lung diseases. Even though numerous drugs targeting different molecule are under investigation, a better understanding of pathophysiological targets and their mechanism in disease is needed. The genetic variability in animal models has helped to understand the function of particular genes or clusters of gene in disease pathway. The alteration of molecular pathway in the pathogenesis of various lung diseases may be investigated on genetic basis, whereas response to medication is still a challenging area for research. Currently in vitro approaches require suitable animal models;

Table 39.1 Novel pharmacological therapies for lung diseases

Disease	Target	Drug	Mechanism	Reference
Lung adenocarcinoma	TMEM16A	TMEM16A inhibitor, CaCCinh-A01	"Suppresses cancer cell growth by inhibiting TMEM16A"	[191]
Lung cancer	Nitric oxide	Truncated deguelin	"Arrested the cell cycle at G2/M phase and suppressed Hsp90 function"	[192]
Lung cancer	H460 and LTEP-A-2	Diplatin	"Inhibited H460 and LTEP-A-2 xenograft tumors via augmentation of Fas-mediated apoptosis"	[193]
COPD	Pulmonary inflammation	3,4,5-Trihydroxycinnamic acid	"Exert anti-inflammatory and antioxidant effects by upregulation of Nr2"	[194]
Pulmonary disease (COPD, asthma)	mRNA	Small interfering RNA (siRNA)	"siRNA can induce posttranscriptional gene silencing by inhibition of respective mRNA"	[195]
COPD	Phosphodiesterase (PDE) 4	PDE4 inhibitor roflumilast with CHF6001	"Significantly reduced TNF-α secretion of the chemokines CCL2 and CCL4 in alveolar macrophages (AM)"	[196]
COPD	Inflammation	Ginsenoside	"Ginsenoside Rg3 ameliorated acute exacerbation of COPD by suppressing neutrophil migration"	[197]
Lung fibrosis	Phosphodiesterase (PDE) 4	AA6216, PDE 4 inhibitor	"Inhibited lung fibrosis by significantly inhibiting TGF-β1 production by THP-1 cells"	[198]
Allergic asthma	Receptor activator of nuclear factor kappa-B ligand (RANKL)	RANKL antagonist	"Blockade of RANKL/RANK and NF-κB signaling pathways"	[199]
Asthma	Interleukin-36 receptor	IL-36RN (interleukin-36 receptor antagonist)	"Significantly suppressed the expression of pro-inflammatory factors"	[200]
Asthma	Inflammation	Glabridin (a flavonoid, especially found in root of *Glycyrrhiza glabra*)	"Attenuated airway inflammation and hyperresponsiveness in a mice model of ovalbumin-induced asthma"	[201]

Acute respiratory distress syndrome	Inflammation and pulmonary fibrosis	Stem cell-derived exosomes	"Anti-inflammation, alveolar remodeling and fibrosis prevention"	[202]
Acute respiratory distress (COVID-19)	Inflammation and oxidative stress	Pirfenidone	"Pirfenidone could inhibit apoptosis, downregulate ACE receptors expression, decrease inflammation by several mechanisms, and ameliorate oxidative stress"	[203]
Cystic fibrosis	Myeloperoxidase (MPO)	Myeloperoxidase (MPO) inhibitor, AZM198	"Decreases oxidative stress in epithelial lining fluid (ELF) of transgenic β-epithelial sodium channel (βENaC)-overexpressing mice"	[204]
Idiopathic pulmonary fibrosis	Janus kinase 3 (JAK3) signaling	Thieno[3, 2-*d*]pyrimidines (JAK3 inhibitor)	"Anti-inflammatory"	[205]
Idiopathic pulmonary fibrosis	Cathepsin	4-Pyridyl derivative of asperphenamate 3 (cathepsin inhibitor)	"Anti-fibrotic"	[206]
Pulmonary fibrosis	ZEB1/E-cadherin pathway	AAV9-shPIM1	"Alleviated bleomycin-induced pulmonary fibrosis by inhibiting PIM1 (serine/threonine protein kinase)"	[207]
Pulmonary fibrosis	CX3C receptor 1-mediated autophagy via Akt signaling pathway	Akt1 inhibitor (A-674563)	"Significantly decreased macrophage autophagy and fibrosis in hyperoxia mice models"	[208]

therefore, the challenge is to develop suitable models that may precisely reiterate diseases related to the human respiratory system, counting the effect of lifestyle and environment on these conditions.

39.9 Conclusion

Respiratory diseases still remain a desired unmet area of clinical need, and it is a big challenge whether present treatment strategy is refined enough to meet industry and patients need. In the current chapter, we explored numerous novel opportunities and challenges for the near future related to genomics and molecular characterization of lung disease, lung injury and repair, translational lung research, and the role of microbiome in lung disease along with cutting-edge technologies in the management of lung disease.

References

1. Khaltaev N, Axelrod S (2019) Chronic respiratory diseases global mortality trends, treatment guidelines, life style modifications, and air pollution: preliminary analysis. J Thorac Dis 11 (6):2643
2. Boubour A, Paradkar M, Thakur KT (2020) Global and regional burden of tuberculosis and tuberculous meningitis. In: Tuberculous meningitis. Academic Press, pp 1–15
3. Maarman GJ, Shaw J, Allwood B (2020) Pulmonary hypertension in majority countries: opportunities amidst challenges. Curr Opin Pulm Med 26(5):373–383
4. Viegi G, Maio S, Fasola S, Baldacci S (2020) Global burden of chronic respiratory diseases. J Aerosol Med Pulm Drug Deliv 33(4):171–177
5. Duan RR, Hao K, Yang T (2020) Air pollution and chronic obstructive pulmonary disease. Chron Dis Transl Med 17(3):395–401
6. McAllister DA, Liu L, Shi T, Chu Y, Reed C, Burrows J, Adeloye D, Rudan I, Black RE, Campbell H, Nair H (2019) Global, regional, and national estimates of pneumonia morbidity and mortality in children younger than 5 years between 2000 and 2015: a systematic analysis. Lancet Glob Health 7(1):e47–e57
7. Serebrisky D, Wiznia A (2019) Pediatric asthma: a global epidemic. Ann Glob Health 85(1):6
8. Mattiuzzi C, Lippi G (2019) Current cancer epidemiology. J Epidemiol Glob Health 9 (4):217–222
9. Perry R, Braileanu G, Palmer T, Stevens P (2019) The economic burden of pediatric asthma in the United States: literature review of current evidence. PharmacoEconomics 37(2):155–167
10. Artigas MS, Wain LV, Tobin MD (2012) Genome-wide association studies in lung disease. Thorax 67(3):271–273
11. Morrell NW, Aldred MA, Chung WK, Elliott CG, Nichols WC, Soubrier F, Trembath RC, Loyd JE (2019) Genetics and genomics of pulmonary arterial hypertension. Eur Respir J 53 (1):1801899
12. Guignabert C, Humbert M (2020) Targeting transforming growth factor beta receptors in pulmonary hypertension. Eur Respir J 57(2):2002341
13. Williamson JD, Sadofsky LR, Hart SP (2015) The pathogenesis of bleomycin-induced lung injury in animals and its applicability to human idiopathic pulmonary fibrosis. Exp Lung Res 41(2):57–73

14. Molet S, Belleguic C, Lena H, Germain N, Bertrand CP, Shapiro SD, Planquois JM, Delaval P, Lagente V (2005) Increase in macrophage elastase (MMP-12) in lungs from patients with chronic obstructive pulmonary disease. Inflamm Res 54(1):31–36
15. Shapiro SD, Goldstein NM, Houghton AM, Kobayashi DK, Kelley D, Belaaouaj A (2003) Neutrophil elastase contributes to cigarette smoke-induced emphysema in mice. Am J Pathol 163(6):2329–2335
16. Gooptu B, Ekeowa UI, Lomas DA (2009) Mechanisms of emphysema in α1-antitrypsin deficiency: molecular and cellular insights. Eur Respir J 34(2):475–488
17. Shapiro SD (2007) Transgenic and gene-targeted mice as models for chronic obstructive pulmonary disease. Eur Respir J 29(2):375–378
18. Mallakin A, Kutcher LW, McDowell SA, Kong S, Schuster R, Lentsch AB, Aronow BJ, Leikauf GD, Waltz SE (2006) Gene expression profiles of Mst1r-deficient mice during nickel-induced acute lung injury. Am J Respir Cell Mol Biol 34(1):15–27
19. Maeda Y, Davé V, Whitsett JA (2007) Transcriptional control of lung morphogenesis. Physiol Rev 87(1):219–244
20. Raj JU, Aliferis C, Caprioli RM, Cowley AW Jr, Davies PF, Duncan MW, Erle DJ, Erzurum SC, Finn PW, Ischiropoulos H, Kaminski N (2007) Genomics and proteomics of lung disease: conference summary. Am J Physiol Lung Cell Mol Physiol 293(1):L45–L51
21. Lumb AB (2016) Nunn's applied respiratory physiology eBook. Elsevier Health Sciences, Amsterdam
22. Tripathi P, Aggarwal A (2006) NF-kB transcription factor: a key player in the generation of immune response. Curr Sci Bangalore 90(4):519
23. Grommes J, Soehnlein O (2011) Contribution of neutrophils to acute lung injury. Mol Med 17 (3–4):293
24. Lee I-T, Yang C-M (2013) Inflammatory signalings involved in airway and pulmonary diseases. Mediat Inflamm 2013:791231
25. Nelson RJ (2004) Seasonal immune function and sickness responses. Trends Immunol 25 (4):187–192
26. Nelson RJ, Demas GE (1996) Seasonal changes in immune function. Q Rev Biol 71 (4):511–548
27. Colgan SP, Curtis VF, Campbell EL (2013) The inflammatory tissue microenvironment in IBD. Inflamm Bowel Dis 19(10):2238
28. Dantzer R (2001) Cytokine-induced sickness behavior: where do we stand? Brain Behav Immun 15(1):7–24
29. Hotamisligil GS (2006) Inflammation and metabolic disorders. Nature 444(7121):860–867
30. Chung K, Adcock I (2008) Multifaceted mechanisms in COPD: inflammation, immunity, and tissue repair and destruction. Eur Respir J 31(6):1334–1356
31. Hartupee J, Mann DL (2016) Role of inflammatory cells in fibroblast activation. J Mol Cell Cardiol 93:143–148
32. Baylis D, Bartlett DB, Patel HP, Roberts HC (2013) Understanding how we age: insights into inflammaging. Longev Healthspan 2(1):8
33. Gharibi V, Ebrahimi MH, Soleimani E, Khanjani N, Fakherpour A, Abadi MBH (2020) The role of oxidative stress in pulmonary functions in bakers exposed to flour dust. Int J Occup Saf Ergon:1–21
34. Coleman JW (2001) Nitric oxide in immunity and inflammation. Int Immunopharmacol 1 (8):1397–1406
35. Van der Vliet A, Eiserich JP, Cross CE (2000) Nitric oxide: a pro-inflammatory mediator in lung disease? Respir Res 1(2):1
36. Gaston B, Drazen JM, Loscalzo J, Stamler JS (1994) The biology of nitrogen oxides in the airways. Am J Respir Crit Care Med 149:538–551
37. Beckman JS, Koppenol WH (1996) Nitric oxide, superoxide, and peroxynitrite: the good, the bad, and the ugly. Am J Phys 271:C1424–C1437

38. Grisham MB, Jourd'Heuil D, Wink DA (1999) Nitric oxide. I. Physiological chemistry of nitric oxide and its metabolites: implications in inflammation. Am J Phys 276:G315–G321
39. Ischiropoulos H, Al-Mehdi AB, Fisher AB (1995) Reactive species in ischemic rat lung injury: contribution of peroxynitrite. Am J Phys 269:L158–L164
40. Akaike T, Noguchi Y, Ijiri S, Setoguchi K, Suga M, Zheng YM, Dietzschold B, Maeda H (1996) Pathogenesis of influenza virus-induced pneumonia: involvement of both nitric oxide and oxygen radicals. Proc Natl Acad Sci U S A 93:2448–2453
41. Tsuji C, Shioya Y, Hirota Y, Fukuyama N, Kurita D, Tanigaki T, Ohta Y, Nakazawa H (2000) Increased production of nitrotyrosine in lung tissue of rats with radiation-induced acute lung injury. Am J Phys 278:L710–L725
42. Savill J, Hogg N, Ren Y, Haslett C (1992) Thrombospondin cooperates with CD36 and the vitronectin receptor in macrophage recognition of neutrophils undergoing apoptosis. J Clin Invest 90:1513–1522
43. Matute-Bello G, Liles WC, Radella F 2nd, Steinberg KP, Ruzinski JT, Jonas M, Chi EY, Hudson LD, Martin TR (1997) Neutrophil apoptosis in the acute respiratory distress syndrome. Am J Respir Crit Care Med 156:1969–1977
44. Albertine KH, Soulier MF, Wang Z, Ishizaka A, Hashimoto S, Zimmerman GA, Matthay MA, Ware LB (2002) Fas and fas ligand are upregulated in pulmonary edema fluid and lung tissue of patients with acute lung injury and the acute respiratory distress syndrome. Am J Pathol 161:1783–1796
45. Matute-Bello G, Liles WC, Steinberg KP, Kiener PA, Mongovin S, Chi EY, Jonas M, Martin TR (1999) Soluble Fas ligand induces epithelial. cell apoptosis in humans with acute lung injury (ARDS). J Immunol 163:2217–2225
46. Neff TA, Guo RF, Neff SB, Sarma JV, Speyer CL, Gao H, Bernacki KD, Huber-Lang M, McGuire S, Hoesel LM, Riedemann NC, BeckSchimmer B, Zetoune FS, Ward PA (2005) Relationship of acute lung inflammatory injury to Fas/FasL system. Am J Pathol 166:685–694
47. Matute-Bello G, Liles WC, Radella F, Steinberg KP, Ruzinski JT, Hudson LD, Martin TR (2000) Modulation of neutrophil apoptosis by granulocyte colony-stimulating factor and granulocyte/macrophage colony-stimulating factor during the course of acute respiratory distress syndrome. Crit Care Med 28:1–7
48. Aggarwal A, Baker CS, Evans TW, Haslam PL (2000) G-CSF and IL-8 but not GM-CSF correlate with severity of pulmonary neutrophilia in acute respiratory distress syndrome. Eur Respir J 15:895–901
49. Asano M, Nishizawa M, Nagata S (1991) Three individual regulatory elements of the promoter positively activate the transcription of the murine gene encoding granulocyte colony-stimulating factor. Gene 107:241–246
50. Severgnini M, Takahashi S, Tu P, Perides G, Homer RJ, Jhung JW, Bhavsar D, Cochran BH, Simon AR (2005) Inhibition of the Src and Jak kinases protects against lipopolysaccharide-induced acute lung injury. Am J Respir Crit Care Med 171:858–867
51. Orlicek SL, Hanke JH, English BK (1999) The src family-selective tyrosine kinase inhibitor PP1 blocks LPS and IFN-gamma-mediated TNF and iNOS production in murine macrophages. Shock 12:350–354
52. Lowell CA, Berton G (1998) Resistance to endotoxic shock and reduced neutrophil migration in mice deficient for the Src-family kinases Hck and Fgr. Proc Natl Acad Sci U S A 95:7580–7584
53. Alonzi T, Maritano D, Gorgoni B, Rizzuto G, Libert C, Poli V (2001) Essential role of STAT3 in the control of the acute-phase response as revealed by inducible gene inactivation (correction of activation) in the liver. Mol Cell Biol 21:1621–1632
54. Akira S (2000) Roles of STAT3 defined by tissue-specific gene targeting. Oncogene 19:2607–2611
55. Schumann RR, Kirschning CJ, Unbehaun A, Aberle HP, Knope HP, Lamping N, Ulevitch RJ, Herrmann F (1996) The lipopolysaccharide-binding protein is a secretory class 1 acute-phase

protein whose gene is transcriptionally activated by APRF/STAT/3 and other cytokineinducible nuclear proteins. Mol Cell Biol 16:3490–3503

56. Takeda K, Clausen BE, Kaisho T, Tsujimura T, Terada N, Forster I, Akira S (1999) Enhanced Th1 activity and development of chronic 370 CHOPRA ET AL enterocolitis in mice devoid of STAT3 in macrophages and neutrophils. Immunity 10:39–49
57. Simon A, Rai U, Fanburg B, Cochran B (1998) Activation of the JAK-STAT pathway by reactive oxygen species. Am J Phys 44:C1640–C1652
58. Haddad JJ, Land SC (2002) Redox/ROS regulation of lipopolysaccharide-induced mitogen-activated protein kinase (MAPK) activation and MAPK-mediated TNF-alpha biosynthesis. Br J Pharmacol 135:520–536
59. Nakamura K, Hori T, Sato N, Sugie K, Kawakami T, Yodoi J (1993) Redox regulation of a src family protein tyrosine kinase p56lck in T cells. Oncogene 8:3133–3139
60. Rahman I, Biswas SK, Kode A (2006) Oxidant and antioxidant balance in the airways and airway diseases. Eur J Pharmacol 533(1–3):222–239
61. Comhair SA, Erzurum SC (2002) Antioxidant responses to oxidant-mediated lung diseases. Am J Phys Lung Cell Mol Phys 283(2):L246–L255
62. Kinnula VL, Everitt JI, Mangum JB, Chang L-Y, Crapo JD (1992) Antioxidant defense mechanisms in cultured pleural mesothelial cells. Am J Respir Cell Mol Biol 7:95–103
63. St Clair DK, Oberley T, Ho YS (1991) Overproduction of Mn-superoxide dismutase modulates paraquat-mediated toxicity in mammalian cells. FEBS Lett 293:199–203
64. Surinrut P, Shaffer J, Marsh J, Heintz NH, Mossman BT (1993) Transfection of a human catalase gene ameliorates asbestos-induced cytotoxicity in hamster tracheal epithelial cells. Am Rev Respir Dis 147:A205. (Abstract)
65. Janssen YM, Marsh JP, Driscoll KE, Borm PJ, Oberdörster G, Mossman BT (1994) Increased expression of manganese-containing superoxide dismutase in rat lungs after inhalation of inflammatory and fibrogenic minerals. Free Radic Biol Med 16(3):315–322
66. Quinlan T, Spivack S, Mossman BT (1994) Regulation of antioxidant enzymes in lung after oxidant injury. Environ Health Perspect 102(suppl 2):79–87
67. Kinnula VL, Crapo JD (2003) Superoxide dismutases in the lung and human lung diseases. Am J Respir Crit Care Med 167(12):1600–1619
68. Ventrice EA, Marti-Sistac O, Gonzalvo R, Villagra A, Lopez-Aguilar J, Blanch L (2007) Molecular and biophysical mechanisms and modulation of ventilator-induced lung injury. Med Intensiva 31:73–82. https://doi.org/10.1016/S0210-5691(07)74779-4
69. Bianchi ME (2007) DAMPs, PAMPs and alarmins: all we need to know about danger. J Leukoc Biol 81:1–5
70. Bhatia M, Moochhala S (2004) Role of inflammatory mediators in the pathophysiology of acute respiratory distress syndrome. J Pathol 202:145–156. https://doi.org/10.1002/path.1491
71. Kubo H (2011) Molecular basis of lung tissue regeneration. Gen Thorac Cardiovasc Surg 59:231–244. https://doi.org/10.1007/s11748-010-0757-x
72. González-López A, Albaiceta GM (2012) Repair after acute lung injury: molecular mechanisms and therapeutic opportunities. Crit Care 16(2):209
73. Lindsay CD (2011) Novel therapeutic strategies for acute lung injury induced by lung damaging agents: the potential role of growth factors as treatment options. Hum Exp Toxicol 30:701–724. https://doi.org/10.1177/0960327110376982
74. Arrowsmith J, Miller P (2013) Trial watch: phase II and phase III attrition rates 2011-2012. Nat Rev Drug Disc 12:569
75. Wehling M (2009) Assessing the translatability of drug projects: what needs to be scored to predict success? Nat Rev Drug Discov 8:541–546
76. Paul SM, Mytelka DS, Dunwiddie CT et al (2010) How to improve R&D productivity: the pharmaceutical industry's grand challenge. Nat Rev Drug Discov 9:203–214
77. Mestre-Ferrandiz J, Sussex J, Towse A (2012) The R&D cost of a new medicine. Office of Health Economics, London

78. Southan C, Williams AJ, Ekins S (2013) Challenges and recommendations for obtaining chemical structures of industry-provided repurposing candidates. Drug Discov Today 18:58–70
79. Tobinick EL (2009) The value of drug repositioning in the current pharmaceutical market. Drug News Perspect 22:119–125
80. Barnes PJ (2013) Theophylline. Am J Respir Crit Care Med 188:901–906
81. To Y, Ito K, Kizawa Y et al (2010) Targeting phosphoinositide-3-kinase-δ with theophylline reverses corticosteroid insensitivity in COPD. Am J Resp Crit Care Med 182:897–904
82. Allarakhia M (2013) Open-source approaches for the repurposing of existing or failed candidate drugs: learning from and applying the lessons across diseases. Drug Design Devel Ther 7:753–766
83. Kobayashi Y, Wada H, Rossios C et al (2013) A novel macrolide solithromycin exerts superior anti-inflammatory effect via NF-κB inhibition. J Pharmacol Exp Ther 345:76–84
84. Kobayashi Y, Wada H, Rossios C et al (2013) A novel macrolide/fluoroketolide, solithromycin (CEM-101), reverses corticosteroid insensitivity via phosphoinositide 3-kinase pathway inhibition. Br J Pharmacol 169:1024–1034
85. Holmes AM, Solari R, Holgate ST (2011) Animal models of asthma: value, limitations and opportunities for alternative approaches. Drug Discov Today 16:659–670
86. Beigel JH, Nam HH, Adams PL, Krafft A, Ince WL, El-Kamary SS, Sims AC (2019) Advances in respiratory virus therapeutics–A meeting report from the 6th isirv Antiviral Group conference. Antivir Res 167:45–67
87. Neher RA, Dyrdak R, Druelle V, Hodcroft EB, Albert J (2020) Potential impact of seasonal forcing on a SARS-CoV-2 pandemic. Swiss Med Wkly 150(1112):w20224
88. Bi J, Min Z, Yuan H, Jiang Z, Mao R, Zhu T, Liu C, Zeng Y, Song J, Du C, Chen Z (2020) PI3K inhibitor treatment ameliorates the glucocorticoid insensitivity of PBMCs in severe asthma. Clin Transl Med 9(1):1–13
89. van der Velden JL, Alcorn JF, Chapman DG, Lundblad LK, Irvin CG, Davis RJ, Butnor K, Janssen-Heininger YM (2020) Airway epithelial specific deletion of Jun-N-terminal kinase 1 attenuates pulmonary fibrosis in two independent mouse models. PLoS One 15(1):e0226904
90. Ren HY, Grove DE, De La Rosa O, Houck SA, Sopha P, Van Goor F, Hoffman BJ, Cyr DM (2013) VX-809 corrects folding defects in cystic fibrosis transmembrane conductance regulator protein through action on membrane-spanning domain 1. Mol Biol Cell 24(19):3016–3024
91. Accurso FJ, Rowe SM, Clancy JP, Boyle MP, Dunitz JM, Durie PR, Sagel SD, Hornick DB, Konstan MW, Donaldson SH, Moss RB (2010) Effect of VX-770 in persons with cystic fibrosis and the G551D-CFTR mutation. N Engl J Med 363(21):1991–2003
92. Liu Q, Gao Y, Ci X (2019) Role of Nrf2 and its activators in respiratory diseases. Oxidative Med Cell Longev 2019:7090534
93. O'Dwyer DN, Dickson RP, Moore BB (2016) The lung microbiome, immunity, and the pathogenesis of chronic lung disease. J Immunol 196(12):4839–4847
94. Dickson RP, Erb-Downward JR, Huffnagle GB (2013) The role of the bacterial microbiome in lung disease. Expert Rev Respir Med 7(3):245–257
95. Hilty M, Burke C, Pedro H, Cardenas P, Bush A, Bossley C, Davies J, Ervine A, Poulter L, Pachter L, Moffatt MF (2010) Disordered microbial communities in asthmatic airways. PLoS One 5(1):e8578
96. Millares L, Ferrari R, Gallego M, Garcia-Nuñez M, Pérez-Brocal V, Espasa M, Pomares X, Monton C, Moya A, Monsó E (2014) Bronchial microbiome of severe COPD patients colonised by Pseudomonas aeruginosa. Eur J Clin Microbiol Infect Dis 33:1101–1111
97. Hewitt R, Farne H, Ritchie A, Luke E, Johnston SL, Mallia P (2016) The role of viral infections in exacerbations of chronic obstructive pulmonary disease and asthma. Ther Adv Respir Dis 10:158–174
98. Papi A, Bellettato CM, Braccioni F, Romagnoli M, Casolari P, Caramori G, Fabbri LM, Johnston SL (2006) Infections and airway inflammation in chronic obstructive pulmonary disease severe exacerbations. Am J Respir Crit Care Med 173:1114–1121

99. Rohde G, Wiethege A, Borg I, Kauth M, Bauer TT, Gillissen A, Bufe A, Schultze-Werninghaus G (2003) Respiratory viruses in exacerbations of chronic obstructive pulmonary disease requiring hospitalisation: a case-control study. Thorax 58:37–42
100. Mallia P, Message SD, Gielen V, Contoli M, Gray K, Kebadze T, Aniscenko J, Laza-Stanca V, Edwards MR, Slater L et al (2011) Experimental rhinovirus infection as a human model of chronic obstructive pulmonary disease exacerbation. Am J Respir Crit Care Med 183:734–742
101. Sze MA, Dimitriu PA, Hayashi S, Elliott WM, McDonough JE, Gosselink JV, Cooper J, Sin DD, Mohn WW, Hogg JC (2012) The lung tissue microbiome in chronic obstructive pulmonary disease. Am J Respir Crit Care Med 185:1073–1080
102. Mortaz E, Adcock IM, Ricciardolo FL, Varahram M, Jamaati H, Velayati AA, Folkerts G, Garssen J (2015) Anti-Inflammatory Effects of Lactobacillus rahmnosus and Bifidobacterium breve on cigarette smoke activated human macrophages. PLoS One 10:e0136455
103. Tomosada Y, Chiba E, Zelaya H, Takahashi T, Tsukida K, Kitazawa H, Alvarez S, Villena J (2013) Nasally administered Lactobacillus rhamnosus strains differentially modulate respiratory antiviral immune responses and induce protection against respiratory syncytial virus infection. BMC Immunol 14:40
104. Marri PR, Stern DA, Wright AL, Billheimer D, Martinez FD (2013) Asthma-associated differences in microbial composition of induced sputum. J Allergy Clin Immunol 131 (2):346–352
105. Huang YJ, Nelson CE, Brodie EL, DeSantis TZ, Baek MS, Liu J, Woyke T, Allgaier M, Bristow J, Wiener-Kronish JP, Sutherland ER (2011) Airway microbiota and bronchial hyperresponsiveness in patients with suboptimally controlled asthma. J Allergy Clin Immunol 127(2):372–381
106. Raghu G, Rochwerg B, Zhang Y, Garcia CA, Azuma A, Behr J, Brozek JL, Collard HR, Cunningham W, Homma S et al (2015) An official ATS/ERS/JRS/ALAT clinical practice guideline: treatment of idiopathic pulmonary fibrosis. An update of the 2011 clinical practice guideline. Am J Respir Crit Care Med 192:e3–e19
107. Knippenberg S, Ueberberg B, Maus R, Bohling J, Ding N, Tort Tarres M, Hoymann HG, Jonigk D, Izykowski N, Paton JC et al (2015) Streptococcus pneumoniae triggers progression of pulmonary fibrosis through pneumolysin. Thorax 70:636–646
108. Lasithiotaki I, Antoniou KM, Vlahava VM, Karagiannis K, Spandidos DA, Siafakas NM, Sourvinos G (2011) Detection of herpes simplex virus type-1 in patients with fibrotic lung diseases. PLoS One 6:e27800
109. Pulkkinen V, Bruce S, Rintahaka J, Hodgson U, Laitinen T, Alenius H, Kinnula VL, Myllärniemi M, Matikainen S, Kere J (2010) ELMOD2, a candidate gene for idiopathic pulmonary fibrosis, regulates antiviral responses. FASEB J 24:1167–1177
110. Molyneaux PL, Cox MJ, Willis-Owen SA, Mallia P, Russell KE, Russell AM, Murphy E, Johnston SL, Schwartz DA, Wells AU et al (2014) The role of bacteria in the pathogenesis and progression of idiopathic pulmonary fibrosis. Am J Respir Crit Care Med 190:906–913
111. Folcik VA, Garofalo M, Coleman J, Donegan JJ, Rabbani E, Suster S, Nuovo A, Magro CM, Di Leva G, Nuovo GJ (2014) Idiopathic pulmonary fibrosis is strongly associated with productive infection by herpesvirus saimiri. Mod Pathol 27:851–862
112. Collard HR, Moore BB, Flaherty KR, Brown KK, Kaner RJ, King TE Jr, Lasky JA, Loyd JE, Noth I, Olman MA et al (2007) Acute exacerbations of idiopathic pulmonary fibrosis. Am J Respir Crit Care Med 176:636–643
113. Han MK, Zhou Y, Murray S, Tayob N, Noth I, Lama VN, Moore BB, White ES, Flaherty KR, Huffnagle GB, Martinez FJ, Investigators COMET (2014) Lung microbiome and disease progression in idiopathic pulmonary fibrosis: an analysis of the COMET study. Lancet Respir Med 2:548–556
114. Varney VA, Parnell HM, Salisbury DT, Ratnatheepan S, Tayar RB (2008) A double blind randomised placebo controlled pilot study of oral cotrimoxazole in advanced fibrotic lung disease. Pulm Pharmacol Ther 21:178–187

115. O'Dwyer DN, Armstrong ME, Trujillo G, Cooke G, Keane MP, Fallon PG, Simpson AJ, Millar AB, McGrath EE, Whyte MK et al (2013) The Toll-like receptor 3 L412F polymorphism and disease progression in idiopathic pulmonary fibrosis. Am J Respir Crit Care Med 188:1442–1450
116. Trujillo G, Meneghin A, Flaherty KR, Sholl LM, Myers JL, Kazerooni EA, Gross BH, Oak SR, Coelho AL, Evanoff H et al (2010) TLR9 differentiates rapidly from slowly progressing forms of idiopathic pulmonary fibrosis. Sci Transl Med 2:57ra82
117. Lafyatis R, Farina A (2012) New insights into the mechanisms of innate immune receptor signalling in fibrosis. Open Rheumatol J 6:72–79
118. Margaritopoulos GA, Antoniou KM, Karagiannis K, Samara KD, Lasithiotaki I, Vassalou E, Lymbouridou R, Koutala H, Siafakas NM (2010) Investigation of Toll-like receptors in the pathogenesis of fibrotic and granulomatous disorders: a bronchoalveolar lavage study. Fibrogenesis Tissue Repair 3:20
119. Papanikolaou IC, Boki KA, Giamarellos-Bourboulis EJ, Kotsaki A, Kagouridis K, Karagiannidis N, Polychronopoulos VS (2015) Innate immunity alterations in idiopathic interstitial pneumonias and rheumatoid arthritis-associated interstitial lung diseases. Immunol Lett 163:179–186
120. Samara KD, Antoniou KM, Karagiannis K, Margaritopoulos G, Lasithiotaki I, Koutala E, Siafakas NM (2012) Expression profiles of Toll-like receptors in non-small cell lung cancer and idiopathic pulmonary fibrosis. Int J Oncol 40:1397–1404
121. Stenbit AE, Flume PA (2011) Pulmonary exacerbations in cystic fibrosis. Curr Opin Pulm Med 17:442–447
122. Ramsey BW (1996) Management of pulmonary disease in patients with cystic fibrosis. N Engl J Med 335:179–188
123. Gold R, Carpenter S, Heurter H, Corey M, Levison H (1987) Randomized trial of ceftazidime versus placebo in the management of acute respiratory exacerbations in patients with cystic fibrosis. J Pediatr 111:907–913
124. Hurley MN, Ariff AH, Bertenshaw C, Bhatt J, Smyth AR (2012) Results of antibiotic susceptibility testing do not influence clinical outcome in children with cystic fibrosis. J Cyst Fibros 11:288–292
125. Smith AL, Fiel SB, Mayer-Hamblett N, Ramsey B, Burns JL (2003) Susceptibility testing of Pseudomonas aeruginosa isolates and clinical response to parenteral antibiotic administration: lack of association in cystic fibrosis. Chest 123:1495–1502
126. Price KE, Hampton TH, Gifford AH, Dolben EL, Hogan DA, Morrison HG, Sogin ML, O'Toole GA (2013) Unique microbial communities persist in individual cystic fibrosis patients throughout a clinical exacerbation. Microbiome 1:27
127. Carmody LA, Zhao J, Schloss PD, Petrosino JF, Murray S, Young VB, Li JZ, LiPuma JJ (2013) Changes in cystic fibrosis airway microbiota at pulmonary exacerbation. Ann Am Thorac Soc 10:179–187
128. Stressmann FA, Rogers GB, Marsh P, Lilley AK, Daniels TW, Carroll MP, Hoffman LR, Jones G, Allen CE, Patel N et al (2011) Does bacterial density in cystic fibrosis sputum increase prior to pulmonary exacerbation? J Cyst Fibros 10:357–365
129. Cox MJ, Allgaier M, Taylor B, Baek MS, Huang YJ, Daly RA, Karaoz U, Andersen GL, Brown R, Fujimura KE et al (2010) Airway microbiota and pathogen abundance in age-stratified cystic fibrosis patients. PLoS One 5:e11044
130. Zhao J, Schloss PD, Kalikin LM, Carmody LA, Foster BK, Petrosino JF, Cavalcoli JD, VanDevanter DR, Murray S, Li JZ et al (2012) Decade-long bacterial community dynamics in cystic fibrosis airways. Proc Natl Acad Sci U S A 109:5809–5814
131. Aitken ML, Limaye A, Pottinger P, Whimbey E, Goss CH, Tonelli MR, Cangelosi GA, Dirac MA, Olivier KN, Brown-Elliott BA et al (2012) Respiratory outbreak of Mycobacterium abscessus subspecies massiliense in a lung transplant and cystic fibrosis center. Am J Respir Crit Care Med 185:231–232

132. Esther CR Jr, Esserman DA, Gilligan P, Kerr A, Noone PG (2010) Chronic Mycobacterium abscessus infection and lung function decline in cystic fibrosis. J Cyst Fibros 9:117–123
133. Sanguinetti M, Ardito F, Fiscarelli E, La Sorda M, D'Argenio P, Ricciotti G, Fadda G (2001) Fatal pulmonary infection due to multidrug-resistant Mycobacterium abscessus in a patient with cystic fibrosis. J Clin Microbiol 39:816–819
134. Boehler A, Estenne M (2003) Post-transplant bronchiolitis obliterans. Eur Respir J 22 (6):1007–1018
135. Gill DR, Davies LA, Pringle IA, Hyde SC (2004) The development of gene therapy for diseases of the lung. Cell Mol Life Sci 61(3):355–368
136. Hazinski TA, Ladd PA, DeMatteo CA (1991) Localization and induced expression of fusion genes in the rat lung. Am J Respir Cell Mol Biol 4:206–209
137. Nagahiro I, Mora BN, Boasquevisque CH, Scheule RK, Patterson GA (2000) Toxicity of cationic liposome-DNA complex in lung isografts. Transplantation 69(9):1802–1805
138. Shoji F, Yonemitsu Y, Okano S, Yoshino I, Nakagawa K, Nakashima Y et al (2003) Airway-directed gene transfer of interleukin-10 using recombinant Sendai virus effectively prevents post-transplant fibrous airway obliteration in mice. Gene Ther 10:213–218
139. Jeppsson A, Pellegrini C, O'Brien T, Miller VM, Tazelaar HD, McGregor CG (1998) Transbronchial gene transfer of endothelial nitric oxide synthase to transplanted lungs. Ann Thorac Surg 66:318–324
140. Mora BN, Boasquevisque CH, Boglione M, Ritter JM, Scheule RK, Yew NS et al (2000) Transforming growth factor-beta1 gene transfer ameliorates acute lung allograft rejection. J Thorac Cardiovasc Surg 913–920(75):119
141. Schmid RA, Stammberger U, Hillinger S, Gaspert A, Boasquevisque CH, Malipiero U et al (2000) Fas ligand gene transfer combined with low dose cyclosporine A reduces acute lung allograft rejection. Transpl Int 13(suppl 1):S324–S328
142. Kita Y, Suzuki K, Nogimura H, Takahashi T, Kazui T (2003) CTLA4Ig-gene transfection inhibits obliterative airway disease in rats. Ann Thorac Surg 75:1123–1127
143. Greenberger JS, Epperly MW, Gretton J, Jefferson M, Nie S, Bernarding M et al (2003) Radioprotective gene therapy. Curr Gene Ther 3:183–195
144. Epperly MW, Guo HL, Jefferson M, Nie S, Gretton J, Bernarding M et al (2003) Cell phenotype specific kinetics of expression of intratracheally injected manganese superoxide dismutase-plasmid/liposomes (MnSOD-PL) during lung radioprotective gene therapy. Gene Ther 10:163–171
145. Carbone DP (1997) The biology of lung cancer. Semin Oncol 24:388–401
146. Salgia R, Skarin AT (1998) Molecular abnormalities in lung cancer. J Clin Oncol 16:1207–1217
147. Mukhopadhyay T, Roth JA (1996) Antisense regulation of oncogenes in human cancer. Crit Rev Oncog 7:151–190
148. Caplen NJ (2003) RNAi as a gene therapy approach. Expert Opin Biol Ther 3:575–586
149. Roth JA, Nguyen D, Lawrence DD, Kemp BL, Carrasco CH, Ferson DZ et al (1996) Retrovirus-mediated wild-type p53 gene transfer to tumors of patients with lung cancer. Nat Med 2:985–991
150. Nemunaitis J, Swisher SG, Timmons T, Connors D, Mack M, Doerksen L et al (2000) Adenovirus-mediated p53 gene transfer in sequence with cisplatin to tumors of patients with nonsmall-cell lung cancer. J Clin Oncol 18:609–622
151. Schuler M, Herrmann R, De Greve JL, Stewart AK, Gatzemeier U, Stewart DJ et al (2001) Adenovirus-mediated wild-type p53 gene transfer in patients receiving chemotherapy for advanced nonsmall-cell lung cancer: results of a multicenter phase II study. J Clin Oncol 19:1750–1758
152. Nemunaitis J, Cunningham C, Buchanan A, Blackburn A, Edelman G, Maples P et al (2001) Intravenous infusion of a replication-selective adenovirus (ONYX-015) in cancer patients: safety, feasibility and biological activity. Gene Ther 8:746–759

153. Swisher SG, Roth JA, Nemunaitis J, Lawrence DD, Kemp BL, Carrasco CH et al (1999) Adenovirus-mediated p53 gene transfer in advanced nonsmall-cell lung cancer. J Natl Cancer Inst 91:763–771
154. Gautam A, Densmore CL, Waldrep JC (2000) Inhibition of experimental lung metastasis by aerosol delivery of PEI-p53 complexes. Mol Ther 2:318–323
155. Densmore CL, Kleinerman ES, Gautam A, Jia SF, Xu B, Worth LL et al (2001) Growth suppression of established human osteosarcoma lung metastases in mice by aerosol gene therapy with PEI-p53 complexes. Cancer Gene Ther 8:619–627
156. Gautam A, Densmore CL, Melton S, Golunski E, Waldrep JC (2002) Aerosol delivery of PEI-p53 complexes inhibits B16-F10 lung metastases through regulation of angiogenesis. Cancer Gene Ther 9:28–36
157. Pluda JM (1997) Tumor-associated angiogenesis: mechanisms, clinical implications, and therapeutic strategies. Semin Oncol 24:203–218
158. Sauter BV, Martinet O, Zhang WJ, Mandeli J, Woo SL (2000) Adenovirus-mediated gene transfer of endostatin in vivo results in high level of transgene expression and inhibition of tumor growth and metastases. Proc Natl Acad Sci U S A 97:4802–4807
159. Denny WA (2003) Prodrugs for gene-directed enzyme-prodrug therapy (suicide gene therapy). J Biomed Biotechnol 2003:48–70
160. Vile RG, Hart IR (1993) Use of tissue-specific expression of the herpes simplex virus thymidine kinase gene to inhibit growth of established murine melanomas following direct intratumoral injection of DNA. Cancer Res 53:3860–3864
161. Fukunaga M, Takamori S, Hayashi A, Shirouzu K, Kosai K (2002) Adenoviral herpes simplex virus thymidine kinase gene therapy in an orthotopic lung cancer model. Ann Thorac Surg 73:1740–1746
162. Molnar-Kimber KL, Sterman DH, Chang M, Kang EH, ElBash M, Lanuti M et al (1998) Impact of preexisting and induced humoral and cellular immune responses in an adenovirus-based gene therapy phase I clinical trial for localized mesothelioma. Hum Gene Ther 9:2121–2133
163. Sterman DH, Kaiser LR, Albelda SM (1998) Gene therapy for malignant pleural mesothelioma. Hematol Oncol Clin North Am 12:553–568
164. Sterman DH, Molnar-Kimber K, Iyengar T, Chang M, Lanuti M, Amin KM et al (2000) A pilot study of systemic corticosteroid administration in conjunction with intrapleural adenoviral vector administration in patients with malignant pleural mesothelioma. Cancer Gene Ther 7:1511–1518
165. Pierce JA (1997) Alpha1-antitrypsin augmentation therapy. Chest 112:872–874
166. Rosenfeld MA, Siegfried W, Yoshimura K, Yoneyama K, Fukayama M, Stier LE et al (1991) Adenovirus-mediated transfer of a recombinant alpha 1-antitrypsin gene to the lung epithelium in vivo. Science 252:431–434
167. Canonico AE, Conary JT, Meyrick BO, Brigham KL (1994) Aerosol and intravenous transfection of human alpha-1-antitrypsin gene to lungs of rabbits. Am J Respir Cell Mol Biol 10:24–29
168. Brigham KL, Lane KB, Meyrick B, Stecenko AA, Strack S, Cannon DR et al (2000) Transfection of nasal mucosa with a normal alpha1-antitrypsin gene in alpha-1-antitrypsindeficient subjects: comparison with protein therapy. Hum. Gene Ther 11:1023–1032
169. Stecenko AA, Brigham KL (2003) Gene therapy progress and prospects: alpha-1-antitrypsin. Gene Ther 10:95–99
170. Song S, Morgan M, Ellis T, Poirier A, Chesnut K, Wang J et al (1998) Sustained secretion of human alpha-1-antitrypsin from murine muscle transduced with adeno-associated virus vectors. Proc Natl Acad Sci U S A 95:14384–14388
171. Song S, Scott-Jorgensen M, Wang J, Poirier A, Crawford J, Campbell-Thompson M et al (2002) Intramuscular administration of recombinant adeno-associated virus 2 alpha1

antitrypsin (rAAV-SERPINA1) vectors in a nonhuman primate model: safety and immunologic aspects. Mol Ther 6:329–335
172. Pilewski JM, Frizzell RA (1999) Role of CFTR in airway disease. Physiol Rev 79:S215–S255
173. Trapnell BC, Chu CS, Paakko PK, Banks TC, Yoshimura K, Ferrans VJ et al (1991) Expression of the cystic fibrosis transmembrane conductance regulator gene in the respiratory tract of normal individuals and individuals with cystic fibrosis. Proc Natl Acad Sci U S A 88:6565–6569
174. Engelhardt JF, Yankaskas JR, Ernst SA, Yang Y, Marino CR, Boucher RC et al (1992) Submucosal glands are the predominant site of CFTR expression in the human bronchus. Nat Genet 2:240–248
175. Jacquot J, Puchelle E, Hinnrasky J, Fuchey C, Bettinger C, Spilmont C et al (1993) Localization of the cystic fibrosis transmembrane conductance regulator in airway secretory glands. Eur Respir J 6:169–176
176. Sanders NN, De Smedt SC, Van Rompaey E, Simoens P, De Baets F, Demeester J (2000) Cystic fibrosis sputum: a barrier to the transport of nanospheres. Am J Respir Crit Care Med 162:1905–1911
177. Stern M, Caplen NJ, Browning JE, Griesenbach U, Sorgi F, Huang L et al (1998) The effect of mucolytic agents on gene transfer across a CF sputum barrier in vitro. Gene Ther 5:91–98
178. Perricone MA, Rees DD, Sacks CR, Smith KA, Kaplan JM, St George JA (2000) Inhibitory effect of cystic fibrosis sputum on adenovirus-mediated gene transfer in cultured epithelial cells. Hum Gene Ther 11:1997–2008
179. Virella-Lowell I, Poirier A, Chesnut KA, Brantly M, Flotte TR (2000) Inhibition of recombinant adeno-associated virus (rAAV) transduction by bronchial secretions from cystic fibrosis patients. Gene Ther 7:1783–1789
180. Hawkins FJ, Suzuki S, Beermann ML, Barillà C, Wang R, Villacorta-Martin C (2020) Derivation of airway basal stem cells from human pluripotent stem cells. Cell Stem Cell 28 (1):79–95.e8. https://doi.org/10.1101/2020.02.21.959395
181. Jacob A, Morley M, Hawkins F, McCauley KB, Jean JC, Heins H (2017) Differentiation of human pluripotent stem cells into functional lung alveolar epithelial cells. Cell Stem Cell 21 (4):472–488
182. Firth AL, Menon T, Parker GS, Qualls SK, Lewis BM, Ke E (2015) Functional gene correction for cystic fibrosis in lung epithelial cells generated from patient iPSCs. Cell Rep 12 (9):1385–1390
183. Ikonomou L, Herriges MJ, Lewandowski SL, Marsland R 3rd, Villacorta-Martin C, Caballero IS (2020) The in vivo genetic program of murine primordial lung epithelial progenitors. Nat Commun 11(1):635
184. Montoro DT, Haber AL, Biton M, Vinarsky V, Lin B, Birket SE (2018) A revised airway epithelial hierarchy includes CFTR-expressing ionocytes. Nature 560(7718):319–324
185. Plasschaert LW, Zilionis R, Choo-Wing R, Savova V, Knehr J, Roma G (2018) A single-cell atlas of the airway epithelium reveals the CFTR-rich pulmonary ionocyte. Nature 560:377–381
186. Patil TS, Deshpande AS (2018) Nanostructured lipid carriers-based drug delivery for treating various lung diseases: a state-of-the-art review. Int J Pharm 547(1–2):209–225
187. Gaba B, Fazil M, Ali A, Baboota S, Sahni JK, Ali J (2015) Nanostructured lipid (NLCs) carriers as a bioavailability enhancement tool for oral administration. Drug Deliv 22 (6):691–700
188. Kardara M, Hatziantoniou S, Sfika A, Vassiliou AG, Mourelatou E, Magkou C, Armaganidis A, Roussos C, Orfanos SE, Kotanidou A, Maniatis NA (2013) Caveolar uptake and endothelial-protective effects of nanostructured lipid carriers in acid aspiration murine acute lung injury. Pharm Res 30(7):1836–1847
189. Helvaci A, Izgu N, Ozdemir L (2020) Relationship between symptom burden, medication adherence and spiritual well-being in patients with chronic obstructive pulmonary disease. J Clin Nurs 29(13–14):2388–2396

190. Blakey JD, Bender BG, Dima AL, Weinman J, Safioti G, Costello RW (2018) Digital technologies and adherence in respiratory diseases: the road ahead. Eur Respir J 52 (5):1801147
191. Guo S, Qiu L, Chen Y, Wang X, Ma B, Qu C, Cui J, Zhang H, Xing C, Zhan Y, An H (2020) TMEM16A-inhibitor loaded pH-responsive nanoparticles: a novel dual-targeting antitumor therapy for lung adenocarcinoma. Biochem Pharmacol 178:114062
192. Zhou H, Dong Y, Ma X, Xu J, Xu S (2020) Development of a novel truncated deguelin derivative possessing nitric oxide donor as a potential anti-lung cancer agent. Fitoterapia 146:104670
193. Wang L, Zeng X, Ye J, Chen X, Lin X (2020) Diplatin, a novel water-soluble platinum complex, inhibits lung cancer growth via augmentation of Fas-mediated apoptosis. Eur J Pharmacol 879:173128
194. Min JH, Kim MG, Kim SM, Park JW, Chun W, Lee HJ, Oh SR, Ahn KS, Lee JW (2020) 3,4,5-Trihydroxycinnamic acid exerts a protective effect on pulmonary inflammation in an experimental animal model of COPD. Int Immunopharmacol 85:106656
195. Kandil R, Merkel OM (2019) Pulmonary delivery of siRNA as a novel treatment for lung diseases. Ther Deliv 10(4):203–206
196. Lea S, Metryka A, Li J, Higham A, Bridgewood C, Villetti G, Civelli M, Facchinetti F, Singh D (2019) The modulatory effects of the PDE4 inhibitors CHF6001 and roflumilast in alveolar macrophages and lung tissue from COPD patients. Cytokine 123:154739
197. Guan X, Yuan Y, Wang G, Zheng R, Zhang J, Dong B, Ran N, Hsu ACY, Wang C, Wang F (2020) Ginsenoside Rg3 ameliorates acute exacerbation of COPD1 by suppressing neutrophil migration. Int Immunopharmacol 83:106449
198. Matsuhira T, Nishiyama O, Tabata Y, Kaji C, Kubota-Ishida N, Chiba Y, Sano H, Iwanaga T, Tohda Y (2020) A novel phosphodiesterase 4 inhibitor, AA6216, reduces macrophage activity and fibrosis in the lung. Eur J Pharmacol 885:173508
199. Liu XG, Li J, Zheng LJ, Han B, Huang F (2020) Interleukin-36 receptor antagonist alleviates airway inflammation in asthma via inhibiting the activation of Interleukin-36 pathway. Int Immunopharmacol 81:106200
200. Gregorczyk I, Maślanka T (2020) Blockade of RANKL/RANK and NF-κB signalling pathways as novel therapeutic strategies for allergic asthma: a comparative study in a mouse model of allergic airway inflammation. Eur J Pharmacol 879:173129
201. Dogan MF, Parlar A, Cam SA, Tosun EM, Uysal F, Arslan SO (2020) Glabridin attenuates airway inflammation and hyperresponsiveness in a mice model of ovalbumin-induced asthma. Pulm Pharmacol Ther 63:101936
202. Xu B, Chen SS, Liu MZ, Gan CX, Li JQ, Guo GH (2020) Stem cell derived exosomes-based therapy for acute lung injury and acute respiratory distress syndrome: a novel therapeutic strategy. Life Sci 254:117766
203. Seifirad S (2020) Pirfenidone: a novel hypothetical treatment for COVID-19. Med Hypotheses 144:110005
204. Dickerhof N, Huang J, Min E, Michaëlsson E, Lindstedt EL, Pearson JF, Kettle AJ, Day BJ (2020) Myeloperoxidase inhibition decreases morbidity and oxidative stress in mice with cystic fibrosis-like lung inflammation. Free Radic Biol Med 152:91–99
205. Zhu Y, Zheng X, Wang C, Sun X, Sun H, Ma T, Li Y, Liu K, Chen L, Ma X (2020) Synthesis and biological activity of thieno [3, 2-d] pyrimidines as potent JAK3 inhibitors for the treatment of idiopathic pulmonary fibrosis. Bioorg Med Chem 28(2):115254
206. Yuan L, Zou C, Ge W, Liu Y, Hu B, Wang J, Lin B, Li Y, Ma E (2020) A novel cathepsin L inhibitor prevents the progression of idiopathic pulmonary fibrosis. Bioorg Chem 94:103417
207. Zhang X, Zou Y, Liu Y, Cao Y, Zhu J, Zhang J, Chen X, Zhang R, Li J (2020) Inhibition of PIM1 kinase attenuates bleomycin-induced pulmonary fibrosis in mice by modulating the ZEB1/E-cadherin pathway in alveolar epithelial cells. Mol Immunol 125:15–22
208. Chen Y, Zhang H, Li F, Wang X (2020) Inhibition of CX3C receptor 1-mediated autophagy in macrophages alleviates pulmonary fibrosis in hyperoxic lung injury. Life Sci 259:118286
209. Maniscalco M, Vitale C, Vatrella A, Molino A, Bianco A, Mazzarella G (2016) Fractional exhaled nitric oxide-measuring devices: technology update. Med Devices (Auckl) 9:151

SPRINGER NATURE

GPSR Compliance

The European Union's (EU) General Product Safety Regulation (GPSR) is a set of rules that requires consumer products to be safe and our obligations to ensure this.

If you have any concerns about our products, you can contact us on ProductSafety@springernature.com

In case Publisher is established outside the EU, the EU authorized representative is:

Springer Nature Customer Service Center GmbH
Europaplatz 3
69115 Heidelberg, Germany

The manufacturer's authorised representative in the EU is Springer Nature Customer Service Centre GmbH, Europaplatz 3, 69115 Heidelberg, Germany. If you have any concerns regarding our products, please contact ProductSafety@springernature.com

Printed and bound by CPI Group (UK) Ltd, Croydon, CR0 4YY
30/09/2026
02225294-0020